The Red Sea

Springer Earth System Sciences

For further volumes:
http://www.springer.com/series/10178

Najeeb M.A. Rasul · Ian C.F. Stewart
Editors

The Red Sea

The Formation, Morphology, Oceanography and Environment of a Young Ocean Basin

هيئة المساحة الجيولوجية السعودية
SAUDI GEOLOGICAL SURVEY
www.sgs.org.sa

Springer

Editors
Najeeb M.A. Rasul
Saudi Geological Survey
Center for Marine Geology
Jeddah
Saudi Arabia

Ian C.F. Stewart
Stewart Geophysical Consultants Pty. Ltd.
Adelaide, SA
Australia

ISSN 2197-9596 ISSN 2197-960X (electronic)
Springer Earth System Sciences
ISBN 978-3-662-45200-4 ISBN 978-3-662-45201-1 (eBook)
DOI 10.1007/978-3-662-45201-1

Library of Congress Control Number: 2014956496

Springer Heidelberg New York Dordrecht London

Printed on acid-free paper

Springer-Verlag GmbH Berlin Heidelberg is part of Springer Science+Business Media (www.springer.com)

Preface

The Red Sea is unique in all respects, including its tectonic history, environment and biology. It is a young ocean basin that along its length has undergone or is undergoing the transition from a continental rift to true oceanic seafloor spreading, the nature of which is still open to vigorous debate. In addition, due to its semi-enclosed nature and location within an arid region, the environment is affected by high evaporation rates that together with limited contact with the Indian Ocean results in high temperatures and salinities. Lower sea levels in the past have also led to extensive evaporite deposition within its basin, while brines and metallic deposits in the axial deeps have been the subject of considerable research; the metalliferous muds may be exploited at some stage in the future. The conditions in the Red Sea have in turn governed the flora and fauna in the sea and along the coast, notable among which are the extensive coral reefs that fringe the sea. The adjacent areas are undergoing rapid development that together with the associated changes are placing some stress on the environment in many areas.

Various topics, from the geology to the past and present environment and the effects of human activities are examined in this volume, which aims to present the current thinking and summaries of research in each field of study. Each chapter aims to give a reasonably comprehensive overview of its subject matter, including useful reference lists for further study. The chapters in the volume were presented at a workshop held in Jeddah, Saudi Arabia, from February 3 to February 5, 2013, under the auspices of the Saudi Geological Survey (SGS). We wish to thank Dr. Said J. Alqahtani of Jubran Holding for assistance in funding the workshop. The support of the Survey in the preparation of this volume is greatly appreciated, and we would like to thank all those who have been involved in its production. We would especially like to thank Dr. Zohair A. Nawab, President, Dr. Abdullah M. Alattas, Assistant President, Mr. Abdullah F. Al-Khattabi, Chief Geologist and Mr. Nasir S. Aljahdli, Director, Survey Department of the Saudi Geological Survey for their encouragement in planning this volume. Colleagues at the SGS Center for Marine Geology, and Mr. Louiesito Abalos are also thanked for making the workshop a success. The contributions of the many technical referees to improving the contents of the chapters as well as the assistance of Ms. Radhika Sree of Springer and Dr. Geoff Bailey in preparing this volume for publication are also greatly appreciated.

Najeeb M.A. Rasul
Ian C.F. Stewart

Contents

Introduction to the Red Sea: Its Origin, Structure, and Environment

Najeeb M.A. Rasul, Ian C.F. Stewart, and Zohair A. Nawab

Location, Bathymetry and Statistics

The Red Sea, also known as Bahr al Ahmar in Arabic, is a semi-enclosed, elongated warm body of water about 2,000 km long with a maximum width of 355 km, a surface area of roughly 458,620 km^2, and a volume of $\sim$250,000 km^3 (Head 1987). It includes 3.8 % of the world's coral reefs (Sea Around Us 2007). The Red Sea is bounded by nine countries and incised by numerous coastal lagoons and a large number of islands and extensive groups of shoals, and at its northern end is bifurcated by the Sinai Peninsula into the Gulf of Aqaba and the Gulf of Suez (Fig. 1). The International Hydrographic Organization (IHO 1953) defines the limits of the Red Sea as follows: In the north, the southern limits of the Gulf of Suez, along a line running from Ras Muhammad (27° 43′N) to the southern point of Shadwan Island (34° 02′E) and thence westward on a parallel (27° 27′N) to the coast of Africa, and also along a line running from Ras al Fasma south-westerly to Requin Island (27° 57′N, 34° 36′E) through Tiran Island to the south-west point thereof and thence westward on a parallel (27° 54′N) to the coast of the Sinai Peninsula. In the south, the boundary of the Red Sea is taken to be along a line joining Husn Murad (12° 40′N, 43° 30′E) and Ras Siyyan (12° 29′N, 43° 20′E).

A bathymetric and topographic map prepared from various sources is presented in Fig. 2. The sea is connected to the Arabian Sea and Indian Ocean via the Gulf of Aden in the south through the narrow Strait of Bab-al-Mandab, which has a minimum with of only 30 km, where the main channel is about 310 m deep and 25 km wide at Perim Island (Morcos 1970). Although the Hanish Sill at 13° 44′N has a maximum depth of only 137 m and forms the shallowest part of the channel (Werner and Lange 1975), it is likely that the Red Sea has remained connected to the Gulf of Aden for at least the past 400,000 years (Lambeck et al. 2011). However, during the Last Glacial Maximum (LGM), the water depth over the Hanish Sill is estimated to have fallen to only 25–33 m (Biton et al. 2008; Lambeck et al. 2011), with considerable effects on the Red Sea circulation and ecology (Trommer et al. 2011). Lowering of the sea level may also have facilitated human migration across the southern Red Sea (Rohling et al. 2013; Bailey, this volume).

The Red Sea has three distinct depth zones; shallow shelves less than 50 m in depth, deep shelves depth ranging between 500 and 1,000 m, and the central axis with depths between 1,000 and 2,900 m. Approximately 40 % of the Red Sea is quite shallow, under 100 m in depth, and about 25 % is under 50 m deep. The shallow water comprises fringing reefs a few km wide and barrier reefs, whereas the deeper water slopes are irregularly marked by coral reefs. The continental slope has an irregular profile, with a series of steps down to about 500 m depth. The 15 % of the Red Sea that forms the narrow axial trough is over 1,000 m in depth and contains the bathymetric depressions known as "Deeps" that were formed by the spreading of the sea. In this elongated narrow basin because of the tectonic down-faulting, the depth abruptly increases and Ahmad et al. (1995) indicate that the maximum recorded depth in the central axis is 2,920 m, while a figure of 3,040 m in the central trough located off the Port of Sudan has also been reported (Touliabah et al. 2010; Eltyab 2004 and several uncited reports). However, from recent data along the axis, Augustin et al. (2014) suggest that the western Suakin Deep, with a depth of 2,860 m at 19.6°N, is the deepest part of the Red Sea Rift. The deeps are filled with hot saline brines (Manheim 1974; Hartmann et al. 1998a, b) and some contain metalliferous sediments (Blanc and Anschutz 1995; Scholten et al. 2000) such as in the most studied one, the Atlantis II Deep (Coulibay et al. 2006).

In the north, the Red Sea bifurcates into the Gulf of Aqaba and Gulf of Suez, where it connects to the Mediterranean Sea via the Suez Canal. The Gulf of Aqaba is 160–180 km long and 19–25 km wide, narrow in the north

N.M.A. Rasul (✉) · Z.A. Nawab
Saudi Geological Survey, Jeddah, Saudi Arabia
e-mail: najeeb_rasul@hotmail.com; rasul.nm@sgs.org.sa

I.C.F. Stewart
Stewart Geophysical Consultants Pty. Ltd, Adelaide, South Australia, Australia
e-mail: stewgeop@senet.com.au

N.M.A. Rasul and I.C.F. Stewart (eds.), *The Red Sea*,
Springer Earth System Sciences, DOI 10.1007/978-3-662-45201-1_1,

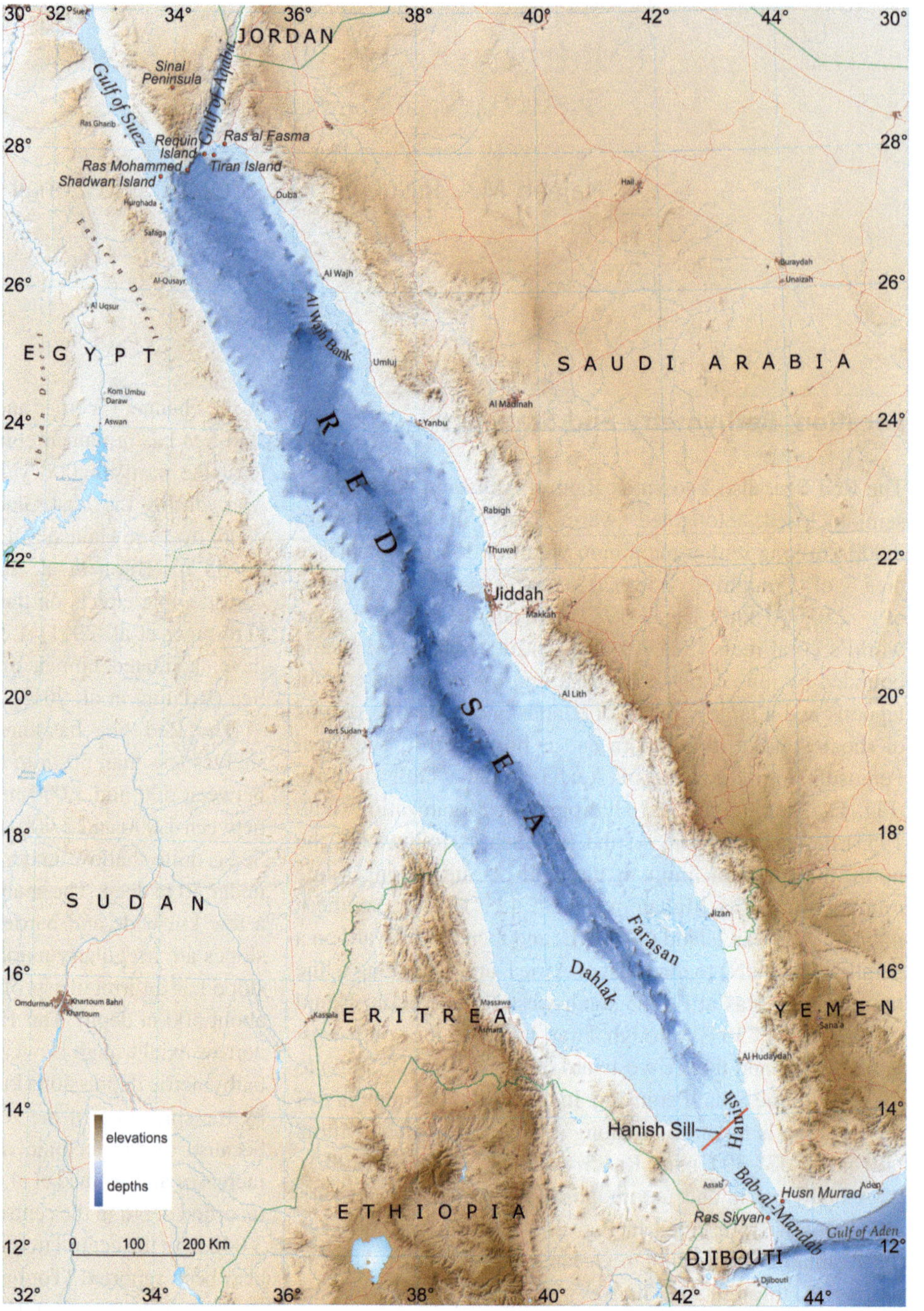

Fig. 1 Geographic map of the Red Sea area, where *darker colours* indicate greater depths or higher elevations

and widening to the south with maximum depths of 1,850 m towards the east. Shelves and coastal plains are absent, with a sharp increase in depth. The Gulf of Suez is a 300 km long, 25–60 km wide, shallow flat-bedded basin with depths ranging between 50 and 75 m. Depths increase towards the south but remain under the 100 m mark at the confluence of the Red Sea and do not exceed 200 m.

The Red Sea has over a thousand islands with the two largest groups lying in the south, the Farasan Islands in the east and the Dahlak Archipelago with over 350 islands in the western Red Sea. The Hanish and Zugar Islands are located in the Yemeni waters, with three recently formed volcanic islands (Jebel at Tair, 2007–2008; Jónsson and Xu, this volume) as well as one formed in the December 2011 to January 2012 eruption (named "Sholan" Island) near Rugged Island in the southern Red Sea, and one resulting from the September to November 2013 eruption (named "Jadid" Island). Several are well known, such as Tiran Island near the mouth of the Gulf of Aqaba and Shadwan Island at the entrance to the Gulf of Suez. Recent eruptions are discussed by Jónsson and Xu in this volume. Numerous other islands and shoals of varying dimensions are present in the Red Sea,

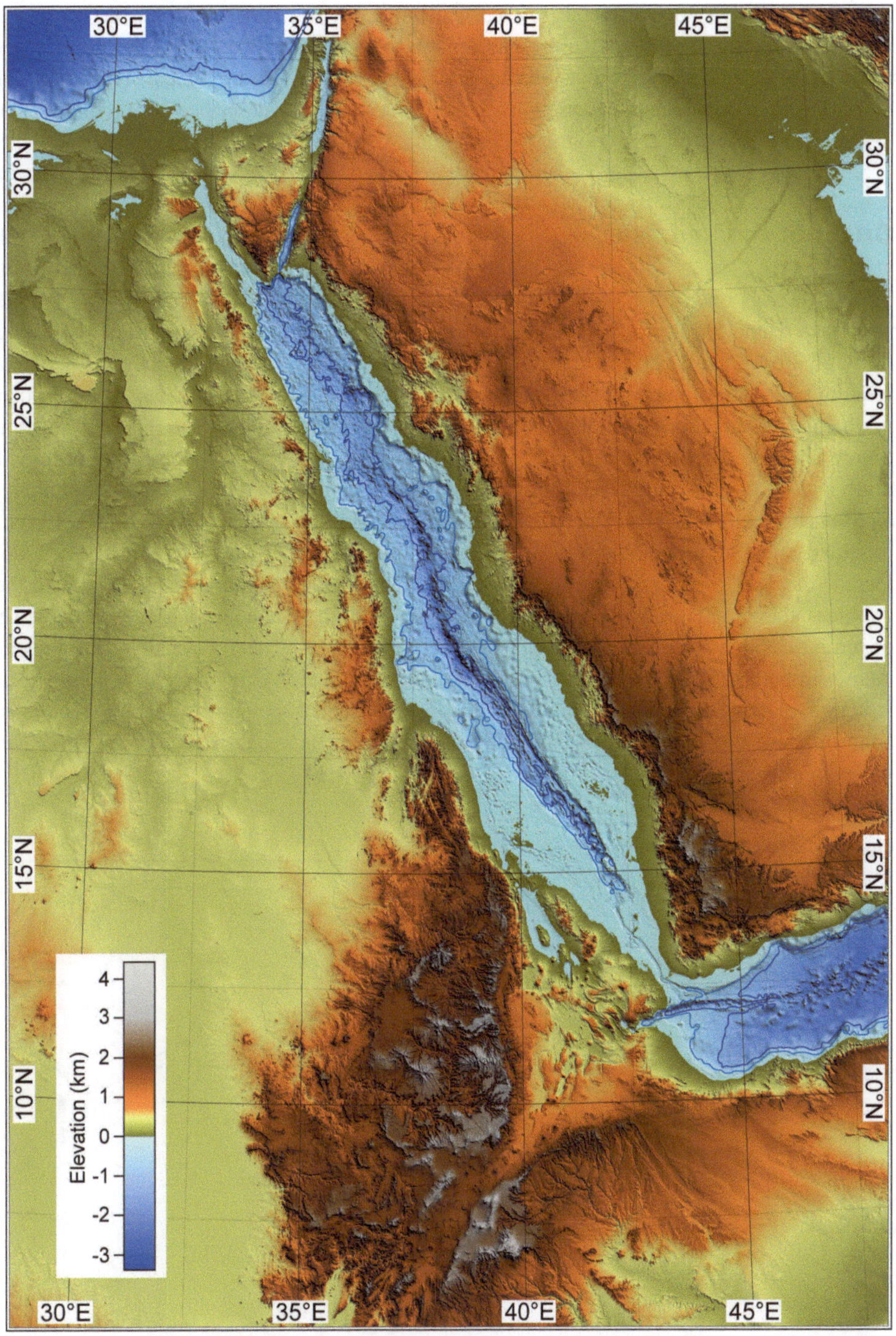

Fig. 2 Topography and bathymetric map. Bathymetric contours are at 500 m interval (*Courtesy* Marco Ligi)

such as Shiab Al Kabir off the coast of Jeddah which is well known but not studied in detail. The Al Wajh Bank in the north with numerous shoals and islands is also known for its unique biodiversity (Bruckner et al. 2011), and the Dhalak archipelago and the Farasan Islands in the southern Red Sea are also well known for the various biological assemblages.

The origin of the name of the Red Sea is of interest as there is no definitive explanation. Possibly, early sailors gave the name to the sea because of the remarkable colour of the water generated by the interaction of the shadows surrounding the sea, due to the stretches of hills and reddish brown mountains rich in certain minerals along the coast, the exotic corals and shoals in the water and the extensive stretches of golden yellowish glowing sand under the sun in the adjacent desert. A more likely explanation could be the changing colour of its water as it does attain a red hue seasonally. Normally the Red Sea has a rich blue-green colour, but when the microscopic red algae (red tide) that bloom seasonally and live in the surface water die off the water colour turns into a reddish brown pigment, hence the name. This phenomenon of cyanobacteria *Trichodesmium erythraceum* bloom is quite common in the Red Sea, as described by Ismael in this volume.

Origin of the Red Sea

The Red Sea is one of the youngest oceanic zones on earth and was created by slow seafloor spreading. Together with the Gulf of Aqaba-Dead Sea transform fault, it forms the western boundary of the Arabian plate, which is moving in a north-easterly direction. The plate is bounded by the Bitlis Suture and the Zagros fold belt and subduction zone to the north and north-east, and the Gulf of Aden spreading centre and Owen Fracture Zone to the south and southeast (Fig. 3 modified after Stern and Johnson 2010). Along most of its length, the Red Sea forms a rift through the Precambrian Arabian-Nubian shield, which was accreted from a number of terranes with very diverse lithologies in the Neoproterozoic from about 870–560 Ma. Reviews of the formation and structure of the shield are given by Nehlig et al. (2002), Johnson and Woldehaimanot (2003), Stern and Johnson (2010), Johnson et al. (2011), and Johnson and Kattan (2012), and the general geology of the shield including the more recent formation and structure of the Red Sea is described by Grainger (2007). Sultan et al. (1993) indicate that the Arabian and Nubian sections of the shield have remained as rigid plates during the Red Sea rifting, and the site of the present Red Sea was probably already a zone of structural weakness in the late Precambrian, with the breakup and rifting controlled by pre-existing fault systems (Makris and Rihm 1991; Bosworth et al. 2005).

The trough of the Red Sea apparently formed in at least two complex phases of plate motion. Spreading may have been initiated in the Eocene, with continental rifting in the Oligocene (about 30 Ma) associated with massive and rapid (1.5 My) eruption of flood basalts in Ethiopia and Yemen at around 30 Ma (Cochran 2005), followed by rifting and seafloor spreading between the African and Arabian tectonic plates (Wright et al. 2006). The Gulf of Suez opened up about 30 Ma, and the northern part of the Red Sea opened up about 20 Ma (Miocene). The splitting of the plates may have been caused by the injection of magma into the rift rather than movement along tectonic faults. Well-developed and mature spreading centres are present in the central Red Sea, but it is not known whether the mantle ascent is active

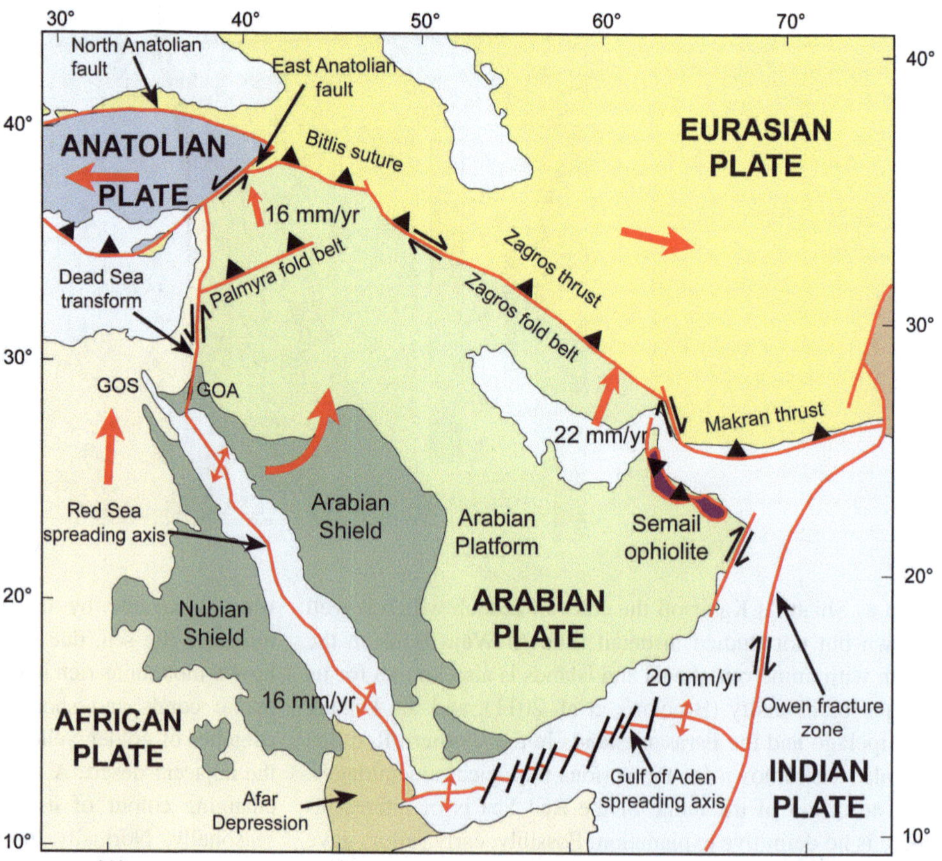

Fig. 3 Main tectonic features of the Arabian Peninsula and surrounding areas (modified after Stern and Johnson 2010)

(Camp and Roobol 1992) or rather passive (Bonatti 1985). Late stage rifting may be observed to the north of 23° 30′N (Cochran 1983; Martinez and Cochran 1988), whereas a transition zone exists in the southern Red Sea between 20° and 23° 30′N (Botz et al. 2011). The greater opening in the south has resulted in a trough that reaches depths of more than 2,900 m in the Atlantis II Deep. Hydrothermal activity on the seabed (hot brines), formation of new basalt, and the recent formation of a volcanic island in the southern Red Sea as well as seismic activity are evidence of the on-going seafloor spreading. In the area between 15° and 20°, active seafloor spreading is prominent and is characterised by the formation of oceanic crust of Mid-Ocean Ridge Basalt (MORB) composition during the last 3 Ma (Altherr et al. 1988; Eissen et al. 1989). The topography of the seabed in the Red Sea also suggests that in the central to southern area (axial trough), the spreading has been more prominent compared to the northern trough.

In the Red Sea, the central spreading axis is opening at an average rate of about 1.6 cm yr^{-1}, with various estimates of the pole of rotation of the Arabian Plate such as 26°N, 21°E (Le Pichon 1968; cf. Fournier et al. 2010) and 34.6°N, 18.1°E (Sultan et al. 1993). Based on GPS data, a recent estimate of the geodetic Euler vector of Arabia relative to Nubia is 31.7°N, 24.6°E, with anticlockwise rotation of 0.37° Myr^{-1} (Reilinger et al. 2006; ArRajehi et al. 2010). This results in the rate of opening of the Sea being greater in the south than in the north, as indicated by the width of the Red Sea between the shoulders of the Precambrian basement. These are about 200 km apart in the north at latitude 27°N and 350 km apart in the south at latitude 17°N.

It is generally accepted that true seafloor spreading now occurs in the Red Sea south of about 24°N and may occur in the north as well (e.g. Girdler 1991). However, there is still some debate as to whether widening of the northern half involves true seafloor spreading or mainly just stretching and thinning of continental crust, with ductile extension or shear on low-angle faults rather than the generation of true oceanic crust (e.g. Cochran and Martinez 1988; Martinez and Cochran 1988; Makris and Rihm 1991; Cochran 2005), and the transition to spreading is also discussed by Augustin et al. (2014) and Mitchell and Park (2014). However from geological reconstructions across the Red Sea, Sultan et al. (1993) suggest that the amount of continental crust underlying the Red Sea is small. Recent correlations of the geology in eastern Egypt and north-western Saudi Arabia are also consistent with "near coast-to-coast" (15–30 km gap between present coastlines) reconstructions of the Red Sea opening in which most of the northern Red Sea is underlain by oceanic crustal material (Kozdroj et al. 2011).

The opening of the Red Sea is considered to have taken place in two stages (Mutwalli 2009) that correspond to two distinct volcanic episodes along the western margin of the Arabian plate (Coleman et al. 1983; Camp and Roobol 1992). The earlier stage of doming and rifting of the Shield occurred along a NNW-trending rift system from about 30–14 Ma (Oligocene to Middle Miocene) that eventually led to the formation of the Red Sea and corresponded to the eruption of the older Tertiary harrats or lava fields and included the emplacement of the coastal dike system that parallels the Red Sea. The Cenozoic to recent lava fields (harrats) of western Saudi Arabia and the adjacent part of Africa that are associated with opening of the Red Sea are shown in Fig. 4 (after Roobol and Al-Rehaili 1998). The southern end of the Red Sea joins the Gulf of Aden spreading centre and the northern end of the East African Rift Zone at a triple junction in the Afar region, which may be underlain by an upwelling deep mantle convection plume (e.g. Daradich et al. 2003), resulting in extensive volcanism in the region, although the existence of mantle plumes is still open to debate (e.g. Anderson 2013). From the geology in south-west Saudi Arabia, it has also been suggested that little uplift in fact occurred at the time of the initial stage of continental rifting (Schmidt et al. 1982). This was followed by the second stage of rifting during at least the past 5 million years, and possibly the past 10 million years, that resulted in further uplift and the formation of the Red Sea escarpment inland from the coast as well as the 200 km long Gulf of Aqaba and its 1,000 km extension as the Dead Sea Rift Zone or transform fault (Girdler and Southren 1987; Bayer et al. 1988; McGuire and Bohannon 1989; Bohannon et al. 1989; Johnson 1998). Within Saudi Arabia, the coastal fault system that parallels the Red Sea can be subdivided into two distinct stages corresponding to pre- and syn-rift stages of the opening of the Red Sea (Roobol and Kadi 2008). The shield was attenuated in pre-rift times to form steeply dipping faults that created a horst and graben topography. Later, syn-rift rotational normal faults formed during the first stage of opening of the Red Sea due to gravity collapse of the Precambrian shield along the margins of the newly opened Red Sea. This resulted in a prominent fault scarp separating the coast plain from the Precambrian shield and the formation of sedimentary basins adjacent to the shield with in excess of 5 km of sediments in some places. In the early stages of rifting, prior to a permanent connection to the Gulf of Aden, thick evaporite deposits accumulated in the Red Sea, and on the shelf and marginal areas, these deposits are overlain by Recent sediments. The sea only turned into an open marine environment when the Gulf of Suez in the north and Indian Ocean in the south became connected in the Pliocene.

The Red Sea has a complex tectonic history of uplift and subsidence and has made the transition from rifting to true seafloor spreading (Bruckner et al. 2011). The sea is widening at a slow rate and it is considered that the sea is an ocean in the making, the development of which resembles

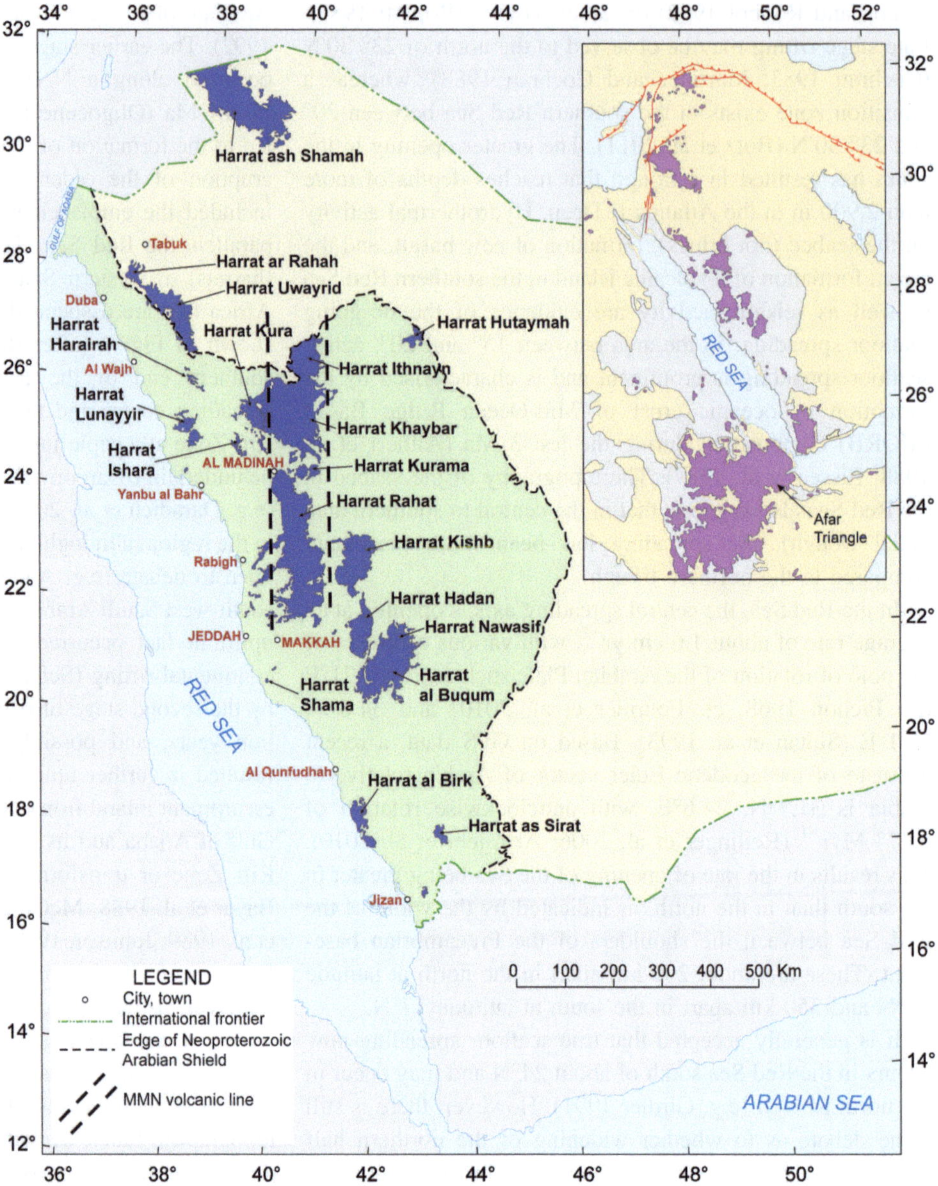

Fig. 4 Cenozoic lava fields (harrats) of western Saudi Arabia and volcanic areas (after Roobol and Al-Rehaili 1998)

that of the Atlantic Ocean in its early stage. The mechanism and rate of opening is still subject to debate and the formation and nature of the Red Sea crust is discussed in this volume by Bosworth, Ligi et al., Mitchell, Reilinger et al., Ehrhardt and Hubscher, and Bonatti et al.

Onshore Topography

The Precambrian shield on both sides of the Red Sea, including either side of the Gulfs of Suez and Aqaba, exhibits topographic shoulders. The uplift has been dated to between 13.8 and 5 Ma from fission track ages of apatite (Bohannon et al. 1989), and these dates suggest that 2.5–4 km of uplift occurred along the Arabian escarpment during the past 13.8 million years. The escarpment crest on the eastern side of the Red Sea is higher than that on the west, possibly due to upwelling asthenosphere and Tertiary volcanism during an early stage of rifting (Dixon et al. 1989). The topographic asymmetry may also be a global phenomenon associated with ocean ridges and rifts, due to the westward drift of the lithosphere relative to the underlying mantle. Doglioni et al. (2003) proposed a mechanism to account for this that requires depleted and lighter asthenosphere generated below the ocean ridge to be shifted eastwards relative to the lithosphere resulting in a density deficit below the eastern flank so that the lithosphere rises or is buoyed up. Within the Arabian Peninsula, the surface has a gentle regional slope to the north-east away from the crest of the escarpment, with similar tilting in Ethiopia, but limited or no tilting in Sudan and Egypt (Bohannon et al. 1989).

The width of the relatively shallow (less than 100 m deep) shelf region in the southern Red Sea is much greater and with more subdued relief compared to that in the northern half of the Red Sea. Although the Red Sea decreases in width at its southern end to only 29 km at the Strait of Bab-al-Mandab, the distance between the escarpments of the shoulder uplifts continues to increase in the south. Much of the area between the escarpments there is dry land and forms the Afar Depression (Fig. 3). The stratigraphy, volcanism, and rifting of the Afar and surrounding area including the Gulf of Aden are described in detail by Beyene and Abdelsalam (2005) and Bosworth et al. (2005). The area has active volcanism and includes the North-Afar or Danakil Depression, which is an active rift zone where the crust is only 15–20 km thick (Prodehl and Mechie 1991) and which lies up to 155 m below sea level. The Depression is probably the present location of the spreading centre and rifting in the south rather than along the axis of the actual Red Sea (Bosworth et al. 2005). Although there is still some debate about the nature of the underlying crust in the Danakil Depression, it is probable that the axial volcanic regions are at least floored by oceanic crust, as may also be the case over wider areas (Bosworth et al. 2005). The depression has significant seismicity and is separated from the main Red Sea basin in the south by the 500 km long Danakil horst that has a mean elevation of 400 m and includes Precambrian basement as well as later overlying sediments (Tesfaye et al. 2003).

Geophysical Data

Summaries of the early geophysical surveys, the onshore and offshore stratigraphy, and their tectonic implications were given by Drake and Girdler (1964) and Skipwith (1973), although some of the concepts about the spreading of the Red Sea have been modified with more recent data and analyses. An early compilation of marine magnetic and gravity data was given by Allan (1970), although the detail was very limited, especially away from the axial trough. Geophysical surveys and analyses of the whole Red Sea are summarised in the compilation by Makris et al. (1991a), and a more recent overview of the surface geophysical data, volcanism, and structure of the northern Red Sea is given by Cochran (2005).

While numerous geophysical surveys, including magnetic, gravity, and seismic work have been undertaken over the past 50 years, there are as yet no really detailed magnetic and surface or airborne gravity data sets covering the whole region that would aid in resolving the nature of the crust offshore, especially that underlying the northern half of the Red Sea. Detailed surface or airborne gravity and magnetic data is not generally available over much of the Red Sea, and there is little geophysical data for the transition zone offshore. Although airborne magnetic and gravity surveys have been flown over the Saudi Arabian side of the Red Sea in recent years for exploration purposes, these data are not available for general use.

The first comprehensive compilation of shipborne and airborne magnetic surveys and other geophysical data over the Red Sea was that by Hall (1979), which also included an interpretation of the main features in the magnetic trends and profiles as well as a summary of the many previous reconstructions of the opening of the Red Sea. Aeromagnetic surveys that cover most of the Saudi Arabian shield were carried out from the early 1960s through 1982, and a summary of their compilation and interpretation is given by Blank and Andreasen (1991). A revised and more complete regional compilation of the aeromagnetic data is available for western Saudi Arabia and the central Red Sea (Zahran et al. 2003), as shown in the reduced-to-the-pole image in Fig. 5. This includes regional aeromagnetic data between Sudan and Saudi Arabia carried out by the Saudi-Sudanese Red Sea Commission (RSC) in the mid 1970s (Izzeldin 1987) that covers the central one-third of the Red Sea and shows the magnetic lineations and north-east-trending transform faults expected in seafloor spreading up to about 23° 30′N latitude.

Although there are numerous faults and boundaries evident within the shield in the magnetic maps, such as the north-west-trending Najd fault system, these have probably been inactive during most of the Phanerozoic. Within about 100 km of the Red Sea coast, there are prominent dikes in the magnetic data (Blank 1977) that parallel virtually the entire length of the coast in Saudi Arabia (cf. Fig. 5) and continue into Yemen (Davison et al. 1994). They are rarely exposed at the surface, where they are usually 200–300 m thick and nearly vertical (Roobol and Stewart 2009). K–Ar dates show the dikes to mainly have ages from 24 to 20 Ma (Camp and Roobol 1992), and they were emplaced during the initial phase of rifting (Pallister 1987), but are largely inactive now. The north-north-west Red Sea coastal fault system can be traced along the entire length of the Red Sea and locally shows Tertiary base metal mineralisation (Hayes et al. 2002). A similar fault system exists on the African side of the Red Sea and can be seen on Landsat and Google Earth images. Further inland from the coastal dike system (about 400 km from the coast) the north-west-trending pre-rift Ja'adah gabbro dike that underlies the Tabuk graben in north-west Saudi Arabia (Phoenix Corporation 1985) is also prominent in the magnetic images.

In the central Red Sea, several north-east-trending transform faults can be observed, and these trends appear to continue up to 200 km into the shield. It is possible that these trends existed prior to rifting and may have controlled the location of the later transforms and offsets in the spreading

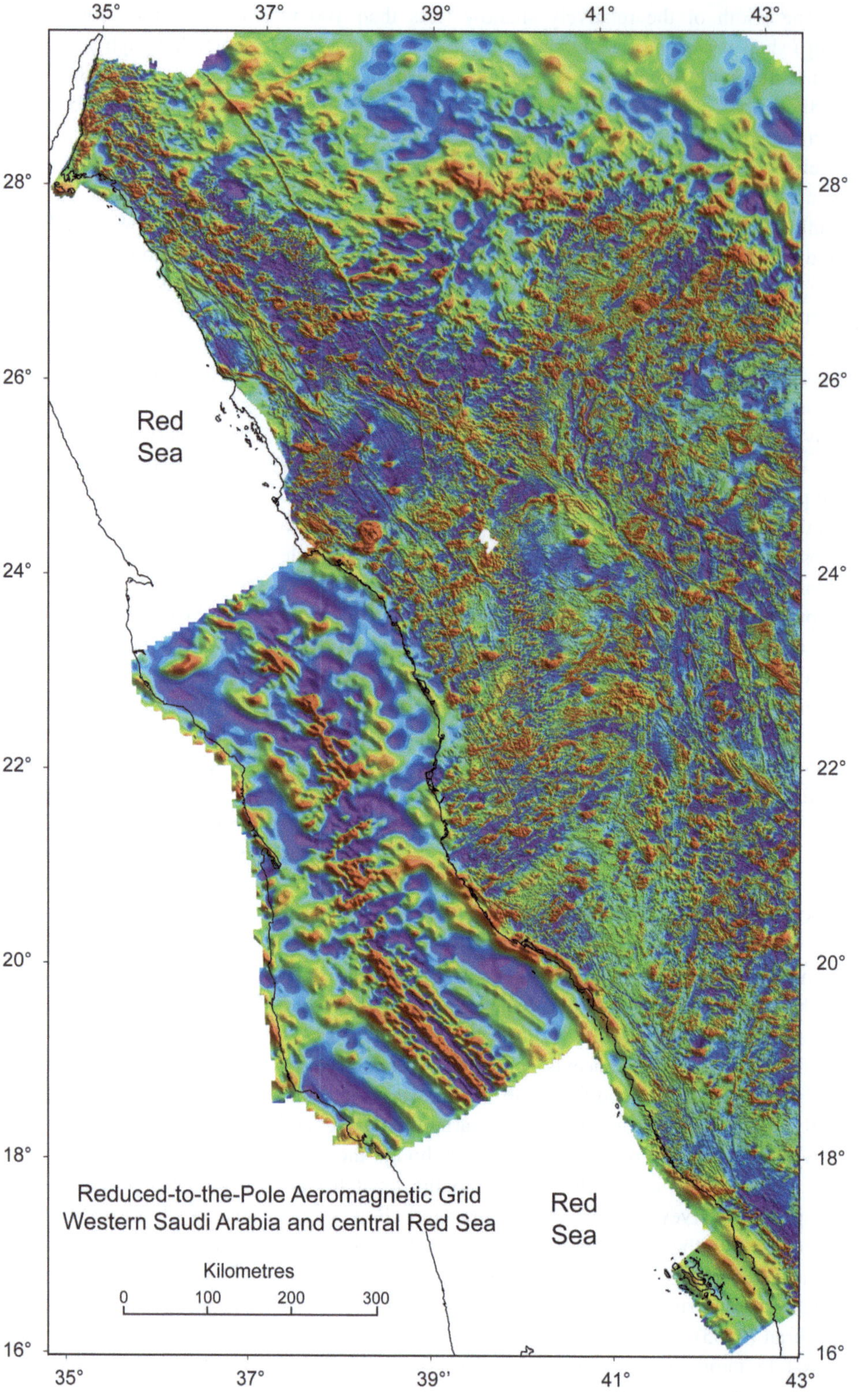

Fig. 5 Reduced-to-the-pole image of aeromagnetic compilation for western Saudi Arabia and the central Red Sea; high values are *red* and low values are *blue* or *purple*. The hole in the data at 24.4°N, 39.6°E coincides with the city of Al Madinah

centre. In the magnetic images, there are the types of lineations expected in normal oceanic seafloor spreading to the south of these transform faults (south of about 20°N), as also observed in the southern Red Sea (Hall 1989). Further north the striping is less evident in the magnetic data, and this may be due to evaporites that allowed the seafloor to cool more slowly resulting in a reduced magnetisation (Hall 1979) or could be caused by hydrothermal alteration (Augustin et al. 2014). Airborne magnetic profiles from several surveys across the continent-ocean transition zone along parts of the

Saudi Arabian coast are discussed by Hall (1979), who suggests that subtle lineations observed in the magnetic data parallel to the axis of the northern Red Sea (not shown here) are in fact consistent with seafloor spreading and oceanic crust over a width of 170 km at 27°N. That is, Hall (1979) indicates that the continent-ocean boundary is not far from the present coastline, including in north-west Saudi Arabia, while in the south (south of about 20°N), the transition zone may even lie inshore. From the magnetic data, Meshref (1990) also considers that the seafloor spreading is occurring in the northern Red Sea. Saleh et al. (2006) correlate long-wavelength gravity and magnetic trends with structure and also suggest that oceanic crust occurs over most of the width of the northern Red Sea.

The gravity map of the Red Sea obtained from satellite data has been discussed by Stewart and Johnson (1994), and a more recent version of the satellite free-air gravity from revised TOPEX data (Sandwell and Smith 2009) is shown in Fig. 6, where the colour interval is 4 milligals. The simple Bouguer gravity obtained by applying corrections for water depth based on a Bouguer density of 2.2 gm cm^{-3} (cf. Izzeldin 1987) is given in Fig. 7, with a colour interval of 6 milligals. Although this is no real substitute for actual surface or airborne gravity measurements as far as the details are concerned, some of the main structural trends can be seen. The axial trough is a conspicuous gravity maximum, as expected from a zone of thin hot crust, but this is less well defined to the north of 24°N, which may indicate that the generation of oceanic crust is less well developed in the north. The transform faults can also be inferred in the gravity in the zone of offsets in the spreading axis from about 20°N to 23°N. The gravity minima on either side of the Red Sea to the south of 20°N may be due to zones of Middle Miocene evaporites, which can exceed 3 km in thickness (Girdler and Underwood 1985).

Numerous crustal heat flow measurements have been carried out in the Red Sea, mainly near the axial trough and the deep brine pools, with considerable variations within the deeps (Hall 1979; Makris et al. 1991c). Girdler and Evans (1977) indicate that high heat flow is observed over the whole Red Sea, with a mean of 467 mW m^{-2} in the axial trough, while on the shelf areas away from the Deeps the mean is 111 mW m^{-2}, which is much higher than the global average of 59 mW m^{-2} (Chapman and Pollack 1975). In the most northerly part of the Red Sea, the heat flow increases from about 125 mW m^{-2} near the coast to more than 250 mW m^{-2} in the axial depression (Martinez and Cochran 1989). Within the deep brine pools, the heat flow values are about 250–600 mW m^{-2} (Anschutz et al. 1999). A compilation of the heat flow data for the region by the International Heat Flow Commission may be found at http://www.heatflow.und.edu/ (accessed on November 2013).

The transition from the continental to the oceanic regime is of considerable interest. The lithosphere beneath the shield in Saudi Arabia is about 140 km thick (Hansen et al. 2006) and may thin to around 40–60 km near the Red Sea (Camp and Roobol 1992). Perhaps the most detailed study of the crust in western Arabia was the seismic refraction line carried out across the southeastern shield from Riyadh to the Farasan Islands (Gettings et al. 1986), in which it was shown that the crustal thickness decreases fairly abruptly from about 40 km over virtually the whole width of the shield to only around 15–20 km at the coastal margin (cf. Makris et al. 1983, 1991b; Davison et al. 1994). This is consistent with modelling of many gravity profiles across the coastal plain along much of the Red Sea coast in Saudi Arabia (Stewart and Johnson 1994, and unpublished data), which also used unpublished seismic reflection sections for additional control on the analyses.

Considerable seismicity is associated with the opening of the Red Sea, and the regional seismicity is discussed in two chapters by El Hadidy and Ogubazghi and Goitom in this volume. The seismicity of Egypt is described by Badawy (2005) and that of the Sinai Peninsula is given by Abdel-Rahman et al. (2009). The earthquake activity is quite variable along the length of the Red Sea, with much of the seismicity taking place in the Gulfs of Aqaba and Suez and the northernmost Red Sea, with localised zones of seismicity elsewhere. A good idea of the spatial variation in activity is given by the rate of release of seismic moment, which for an earthquake is a product of fault plane area, the fault slip or displacement, and the shear modulus of the rock and is probably the most physically realistic measure of the size of an earthquake. Using 26,250 earthquakes greater than magnitude 2 since 1950, the logarithm of the cumulative seismic moment in 10 km cells (in units of Nm) is shown in Fig. 8, in which it can be seen that there are appreciable variations even along the axis of the Red Sea. At both the southern and northern ends of the Gulf of Aqaba, there is relatively low activity and significant gaps in seismicity also appear to exist along the spreading axis. In the axial regions with apparently low seismicity, most of the spreading occurs with aseismic slip on the transform faults at the spreading centre, presumably due to local variations in temperature (thin hot crust in the aseismic areas?) and hydrothermal conditions in the uppermost 10 km or so (cf. Roland et al. 2010).

Onshore there tends to be fairly diffuse low-level seismicity over most of western Arabia and Egypt, especially within a few hundred km of the Red Sea, with more intense activity in the south, including the Afar region and Yemen. In Egypt, a localised zone of higher activity centred on 23.7°N, 32.9°E is associated with the Aswan Dam. Note that the zone of apparent intense activity to the north-east of the

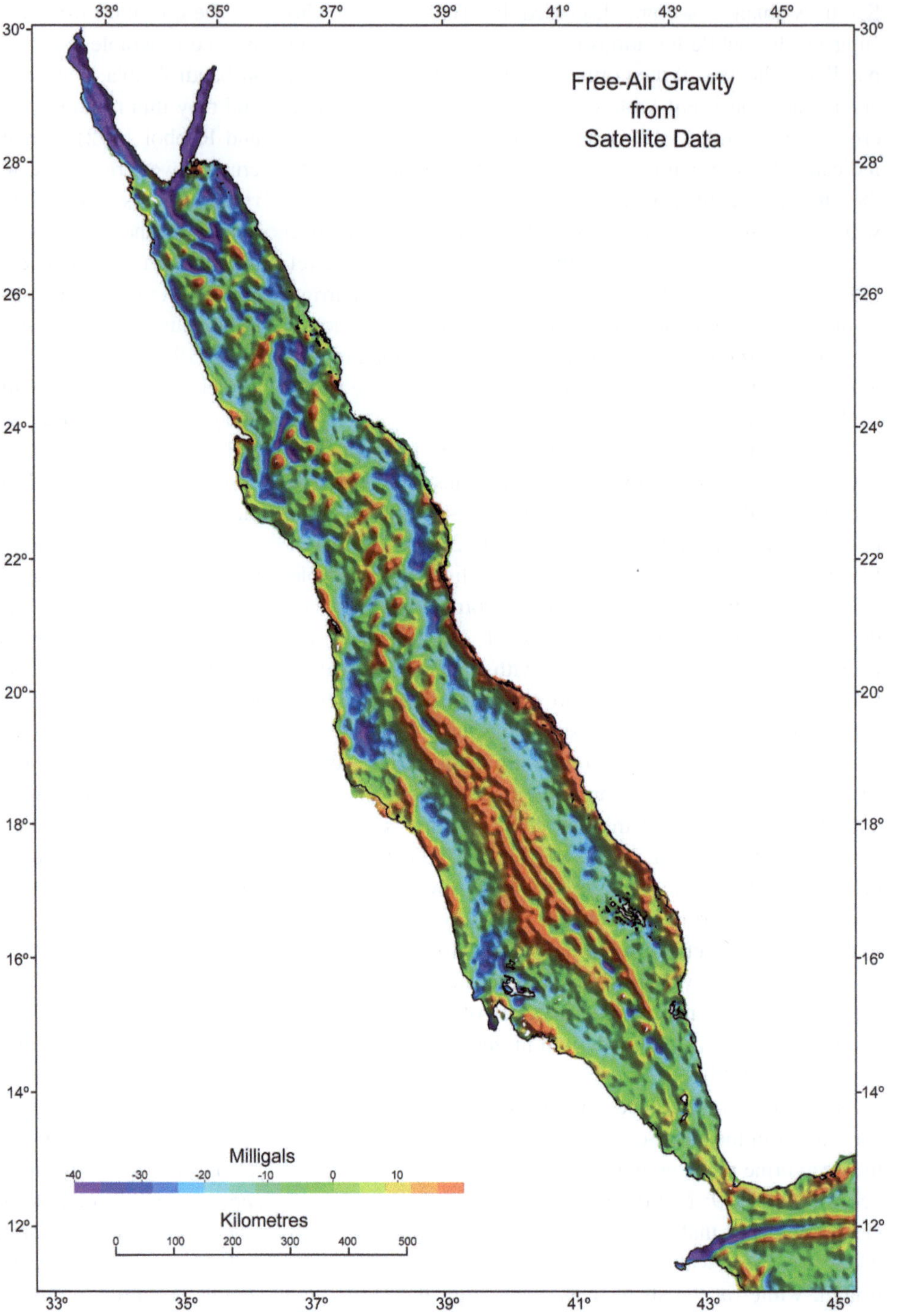

Fig. 6 Free-air satellite gravity over the Red Sea (TOPEX data; Sandwell and Smith 2009)

Gulf of Aqaba near the northern margin of Fig. 8 towards 37°E is probably an artefact largely due to blasting associated with phosphate mining in Jordan. A quite localised area of activity in the Harrat Lunayyir region (cf. Fig. 4) at about 25.3°N, 37.8°E, to the north of Yanbu in Saudi Arabia, is due to an episode of very recent dike intrusion that commenced in 2009 (Pallister et al. 2010; Baer and Hamiel 2010).

Gulfs of Aqaba and Suez

At the northern end of the Red Sea, the Gulf of Suez (or Clysmic Rift) is of economic importance as it hosts numerous oil and gas fields, while the Gulf of Aqaba is a region undergoing active tectonic movement with considerable associated seismic hazard. The Gulf of Aqaba is part

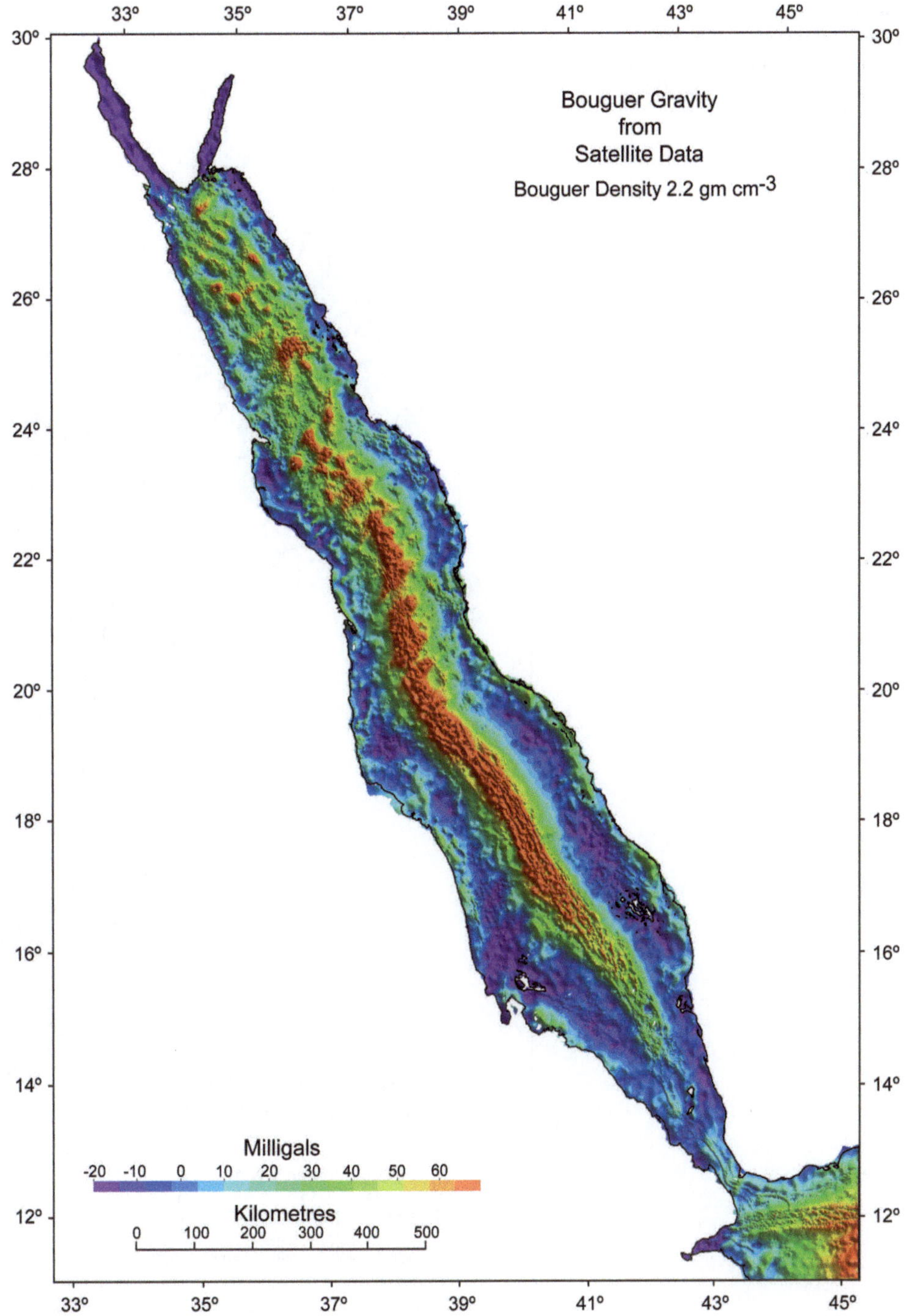

Fig. 7 Bouguer gravity derived from free-air data via a simple reduction using a rock density of 2.2 gm cm^{-3} (cf. Izzeldin 1987)

of a left lateral transform fault system that forms the northwest boundary of the Arabian Plate and connects the Red Sea, where seafloor spreading occurs, with the Zagros-Taurus zone of continental collision. The Gulf of Aqaba is essentially a number of subparallel sinistral transform faults with associated pull-apart basins or rhomb-shaped grabens that are in places more than 10 km deep (Ben-Avraham et al. 1979, 2008) with a maximum water depth of about 1,850 m resulting from the oblique motion along the fault system over the past 14 million years (Makris and Rihm 1991). Ben-Avraham (1985) notes that it "is of particular interest because it is one of the two places in the world (the Gulf of California-Imperial Valley is the other), where a mid-ocean ridge system changes into a transform system and runs into a continent". Ben-Avraham et al. (2008) estimate the sinistral motion on the Gulf of Aqaba and Dead Sea Rift to be about 5–7 mm yr^{-1}. GPS measurements yield a current slip rate on the Wadi Araba Fault section of the Rift immediately to the

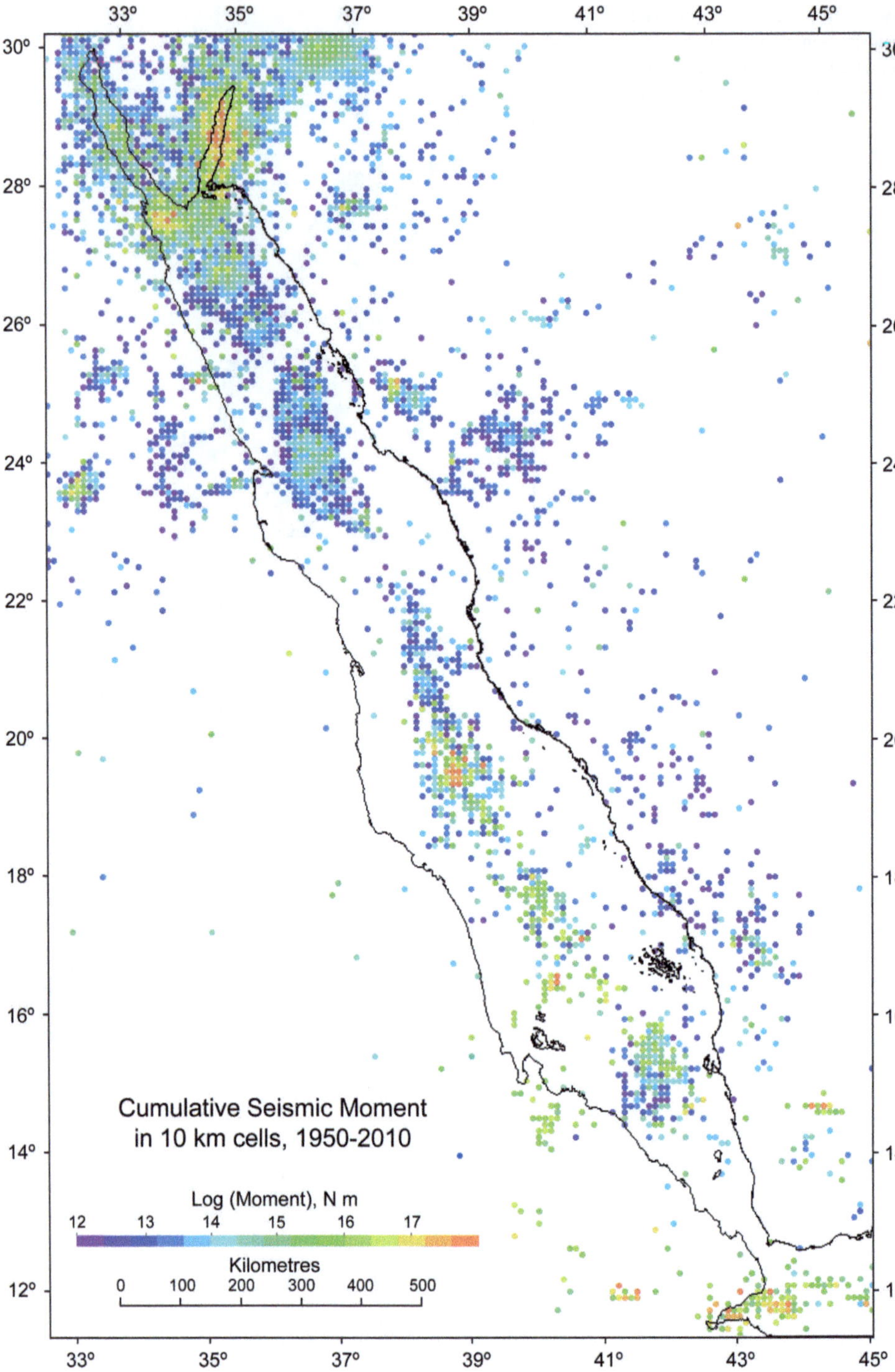

Fig. 8 Earthquake seismic moment (a measure of energy release) summed over 10 km cells using seismicity data from 1950 to 2010

north of the Gulf of Aqaba of 4.9 ± 0.4 mm yr^{-1} (Al Tarazi et al. 2011). The stages in which the pull-apart basins in the Gulf of Aqaba may have formed and become wider with time due to rotation of the Arabian Plate are discussed by Roobol and Stewart (2009). From alignments of the Cenozoic sediments across the southern end of the Gulf, Roobol and Stewart (2009) suggest that a total of about 115 km of sinistral displacement has occurred by movement along the Aqaba-Levant fault system. The mean heat flow in the Gulf averages about 80 mW m^{-2} and increases from north to south, which is consistent with propagation of the Red Sea rifting into the Dead Sea Rift (Ben-Avraham and Von Herzen 1987).

The Gulf of Aqaba is one of the most seismically active areas along the western margin of the Arabian plate. Its 1,000 km extension as the Aqaba-Levant fault system, which includes the Dead Sea Rift Zone, has a long history of seismic activity (Al Amri et al. 1991; Ben-Avraham et al. 1979, 2008; Hamiel et al. 2009), and a paleoseismic study of the Lebanese section by Daëron et al. (2007) indicated that there have been 10–12 major earthquakes in the past 12,000 years. A magnitude 7.3 earthquake in 1995 occurred just off the Saudi Arabian coast in the Gulf (Roobol et al. 1999), and a 20 km long section of coastline to the south of Haql showed the effects of rupture (Williams et al. 2001).

The late Cenozoic Gulf of Suez Basin is less spectacular than the Gulf of Aqaba in the effects due to tectonic movements, with relatively shallow water depths (locally depths reach 50–70 m in the troughs). The Gulf was originally formed during the early Paleozoic as a narrow embayment of the Tethys that was rejuvenated during formation of the East African rift system and Red Sea in the Paleogene (Farhoud 2009). About 35 km of extension occurred in the southern Gulf of Suez (Bosworth et al. 2005), and it is basically a failed continental rift that remains floored by continental crust, with a complex extensional system of blocks that have rotated along low-angle or listric fault planes (Bosworth 1995), with three distinct depocentres for sedimentation. Rifting in the central Gulf commenced about 23–25 Ma and terminated at the northern end of the Gulf at a major east-west Eocene structural boundary, with more diffuse extension further north (Bosworth and McClay 2001). Since the middle Miocene, the widening of the Red Sea at its northern end has largely been accommodated by the Gulf of Aqaba-Dead Sea transform, but continuing seismicity in the southern Gulf indicates that there is still some ongoing extensional stress and associated normal faulting. A review of the pre-rift, syn-rift, and post-rift stratigraphy of the Gulf is given by Farhoud (2009), including an interpretation of aeromagnetic data in terms of faulting, basins, and accommodation zones. The syn- and post-rift sediments range in age from Miocene through Pleistocene, and the maximum thickness of sediments in the Morgan basin in the southern Gulf is around 8,000 m (Farhoud 2009).

The Red Sea Deeps and Brine Pools

There are twenty-five bathymetric depressions or "Deeps" in the axial trough of the Red Sea. The deeps are located between 1,400 and 2,400 m water depth where transform faults intersect the spreading axes, creating a restricted environment where brines and sediments can accumulate. They contain mineralised hot brine pools derived from the leaching of the underlying Miocene salt deposits (Bertram et al. 2011) and hydrothermal activity giving rise to a large quantity of metalliferous deposits (e.g. Au, Ag, Cu, Zn, and Pb) and sediments in the deeps. Some of the well-known ones are the Atlantis II Deep, the Discovery Deep, and the Oceanographer Deep (Fig. 9). However, the geology of the deeps varies significantly from basalts to evaporite deposits of varying thickness, from high-temperature hyper-saline brines to metalliferous sediments and even chimneys while others have no significant ore mineralisation (Anschutz et al. 2000; Schmidt et al. 2003; Gurvich 2006; Dekov et al. 2007; Pierret et al. 2010). Some brine pools (e.g. Atlantis II Deep) are stratified with several layers of uniform temperature and salinity with stepwise changes in conditions beginning at the sea water–brine interface (Antunes et al. 2011). The hydrothermal activity, brines, and salt flows are described in this volume by Anschutz, Schmidt et al., Feldens and Mitchell, and Hovland et al.

The base and precious metal mineralisation contained within the sediments and hot brines in the deeps are currently being investigated for their commercial potential. There has been considerable interest in exploration for and exploitation of the metalliferous deposits from the Atlantis II Deep, which is the deepest and contains the richest deposits. The Saudi-Sudanese Red Sea Commission (RSC) was established in the mid-1970s to assess the economic potential of the metalliferous deposits in the Atlantis II basin from a depth of over 2,000 m. A German company, Preussag A.G., was commissioned to conduct a five-year exploration program and technical feasibility study and the encouraging data led to a joint venture between Diamond Fields International Ltd., from Canada, and Manafa International, Saudi Arabia, to undertake investigative activities to explore the potential metalliferous deposits. The environmental aspects of mining these deposits are discussed by Thiel et al. in this volume.

Coastal Geomorphology

Plateaux and mountains rise steeply to more than 1,000 m above sea level to the north of Jeddah and up to 3,660 m in Yemen. The coastal plain is about 2–50 km wide and slopes up gently to the east until it meets the mountains, which are deeply cut by valleys, although streams flowing in the uplands generally fail to cross the coastal plain to reach the sea. The Red Sea has a barrier reef that is located 10–40 km off the coast of Saudi Arabia and is about 400 km long and several km wide. The seabed platform lies at a depth of 30–60 m, on which sit many steep-sided patch reefs. Almost continuous marginal coral reefs occur along much of the coast from the Gulf of Aqaba to the Strait of Bab-al-Mandab. A similar description of the reef morphology also applies to the African side of the Red Sea (Chiffings 2003). Sand dunes of various dimensions are present along the entire stretch of the Red Sea.

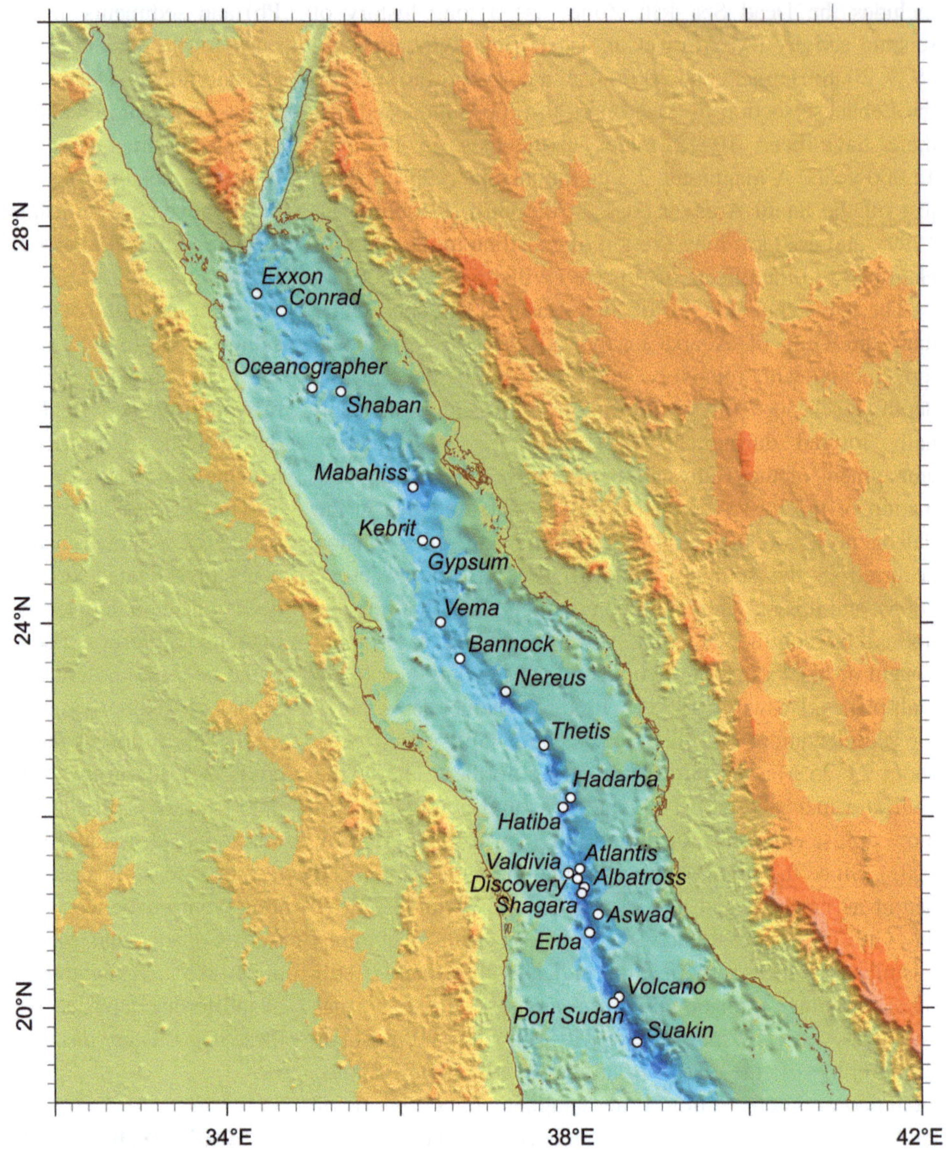

Fig. 9 Location of Deeps along the Red Sea rift axis (*Courtesy* Marco Ligi and Nico Augustin)

Most sandy shorelines in the northern Red Sea are narrow beaches adjoining narrow coral reef flats. Broader beaches occur where the reef edge swings further out from shore, leaving a lagoon. The diversity of rocky shores is significantly higher than that of sandy beaches or mud, although the biomass may be less. Off the Egyptian Red Sea coast, 20 % is formed by rocky, erosional, and wave-cut cliffs. Much of the rocky intertidal zone of the north occurs in erosion notches of fossil cliffs. These provide a more moist and sheltered habitat than do the horizontal expanses of intertidal rock that are common in the south and support a greater range of fauna (Chiffings 2003). In the southern Red Sea, the beaches are wider and sandy because of the sediment input from the seasonal wadis. The continental shelf is therefore wider in the southern Red Sea and may stretch to over 50 km offshore.

Wadis (Seasonal Streams)

Wadis are seasonal streams formed by water draining from the mountains. Since they are only occasionally active, most rarely reach the sea or drain into the sharms. The beds of the wadis are dried up from the extremely high temperatures so that a short rainfall fails to generate a flow and most of the water is soaked up by the bed. However, during the monsoonal rains, the numerous wadis that remain dried up throughout most of the year are flooded, passing through the numerous channels that drain into the Red Sea or into the sharms (Höltz and Zötl 1978). Wadis are one of the most important plant habitats as their floors are often filled with well-drained silts, sands and gravels that may hold considerable amounts of moisture to feed the vegetation, and farming is common in some wadis. River input is absent in the Red Sea because of the small watershed (Siddall et al. 2004); however, one of the larger wadis, Baraka (Tokar), in Sudan is active for 40–70 d yr^{-1} mostly during autumn and drains into the Red Sea with a water discharge of 970×10^6 m^3 at 18.5° (Trommer et al. 2011).

Coastal Lagoons and Sabkhas

he coastal areas along the Red Sea are incised by lagoons of different sizes and shapes and a few are drained by wadis (seasonal streams), while sabkhas (supra-tidal deposits) are common, as described by Rasul in this volume. Coastal sabkhas are supra-tidal flats that are formed in low-energy environments near high tide level and are seasonally flooded. They are about a metre above sea level and common along the Red Sea coast and are composed of evaporite minerals, namely salt (NaCl), gypsum, and anhydrite. Algal mats which are complex associations of cyanophytes, bacteria, and diatoms (Chiffings 2003) are associated with the sabkhas; they are a few centimetre thick, beneath which a black reducing layer is common. The sediment texture of the sabkhas ranges between sand and clay and has abundant salt provided by sea water during tidal fluctuations. The surfaces of the sabkhas are soft and crumbling because of the quick loss of moisture and are almost always devoid of higher plant life due to the very saline waters below the salt crust and the poor drainage. The surfaces of sabkhas are flat and smooth, with gradients of less than 0.5 m km^{-1}. During dry periods, capillary water evaporates and concentrates to a saline brine, with eventual precipitation of salts near the surface. Water rising through pore spaces of the sand increases lubrication and produces a soft quicksand of low load bearing strength.

Sharms

Sharms are drowned river valley estuaries often known as *marsas* or *khawrs*, also referred to as lagoons and coastal inlets, are described by Rasul in this volume. The coastline of the Red Sea is indented with a large number of lagoons and coastal inlets all along its length. Most lagoons have narrow entrances typically 20–25 m in depth where they cut through coralline limestone ridges and range in depth between 3 and 25 m, and they stretch up to 25 km in length and 5 km in width. Some lagoons are drained directly by wadis, while some now have no input from the wadis due to obstruction in recent years. The peripheries of some lagoons are flat and form sabkhas that are covered by a thin sheet of water stretching almost 3 km inland during high tide, and some have raised terraces and during high tide undercutting of the rock-exposure or raised coral reefs takes place. At present, no active erosion in the sharms is observed and flood waters from the escarpment only rarely reach them. Beyond the coral reef, the sharm floors, on average, abruptly fall to 50 m below sea level. Most lagoons have coral reefs and shoals that are often populated by mangrove stands and seabed covered with algae and seagrass.

Early ideas on the formation of sharms suggested that they were contemporaneous with the coral growth now exposed in the flanks of the sharm entrances, the idea being that silty water from wadi runoff inhibited the growth of corals in the central channel. However, other evidence suggests a later breaching of the fossil reef (Vincent 2008). For example, it has been argued that the bifurcation of the sharms behind the beach ridge represents lagoons that existed before the reefs were breached. Isotopic dating of the raised reefs breached by the sharms indicates that they are older than 40,000 years and may be as old as 146,000 years BP. The origin of the sharms requires a coincidence of low sea level and pluvial conditions such that wadis flowed to the Red Sea (Brown et al. 1989). Wadi flow to a sea level lower than that at present would allow erosional breaching of the elevated reef and at the same time allow scouring of some of the soft sediment built up in the lateral lagoons. The sharms were formed 12,000–8,000 years ago during the last pluvial episode when sea level rose from −60 to −20 m. At the maximum low sea level of −120 m, about 18,000 years ago, the climate was arid and after about 8,000 years BP the sea level was too high for sharms to form. Sharms might well have formed during other interglacial periods but they have been filled in by coral growth (Vincent 2008). Some of the lagoons have been declared to be environmentally sensitive (maritime protected areas).

Salt-Pans and Salt Marshes

Salt-pans are common only in the northern and southern areas of the Red Sea. In order to extract salt, sea water is trapped in an artificial pond and the water is naturally evaporated so that salt crystals are formed. The salt harvesting sites are mostly in the coastal region where the tidal range is usually high, and once the water is trapped, levee-like structures are made to hold the sea water. These ponds are also feeding grounds for various species of water fowl.

Salt marshes are found in parts of the central and southern Red Sea where sewage is discharged. Many new marsh communities occur as a result of sewage outfalls along the coast of Saudi Arabia and near Port Sudan including areas where pollution is common. Enrichment not only stimulates marsh development, but in the case of the Red Sea, also adds significant nutrient loads. Almost 3 % of the Egyptian coast comprises salt marsh, and salt marshes in the north-western Red Sea are present around Ras Mohammed, Abu-Monqar, Wadi el-Gemal islands, and Gebel Elba (Chiffings 2003), while they are sparse elsewhere along the Red Sea.

Offshore Islands

There are over a thousand islands in the Red Sea, most of which have been named but not explored due to inaccessibility. The Farasan Islands off Jizan, Saudi Arabia, is a group of 126 islands, while the Dahlak group, off Massawa, Eritrea, has over 354 islands. Only a few of the islands are inhabited; 3 in the Farasan Islands, 10 in the Dhalak Archipelago, and the Kamaran Islands in Yemen. The vast majority of these islands are small and are composed of coralline limestone, but most islands in the central Red Sea are of volcanic origin. Although these islands do not have any economic importance, some of them have gained a strategic importance because of their locations near sea lanes and straits and some are used as anchorages for passing vessels. The number of islands near the eastern coast of the Red Sea is higher than on the western side. Saudi Arabia has the largest number of islands followed by Eritrea.

There are four types of islands in the Red Sea:

Coral reef islands are formed from corals that grow in tropical areas in shallow clear waters, high water temperature, and salinity in the absence of fresh water input. Some of these coral reef islands are the Farasan Islands, the Dahlak Islands, and Kamaran Island, in addition to most small ones located along the coasts.

Continental islands are formed after the separation of the landmass from the neighbouring outcrop resulting from faulting or erosion. They are longitudinal and stretch parallel to the coast. Tiran, Sanafir, Hatiba, Shadwan, Jubal, and Mukawwar Islands are typical examples.

Sandy islands are made-up of sand-size sediment and originate from sandbars resulting from waves in low tidal areas. They grow on coral reef barriers and are longitudinal. They are widespread along the coast and are mostly located in the central part of the eastern Red Sea coast. Examples are Shebara, Kashran, Al Taweelah, and Al Thara Islands.

Volcanic islands are formed by eruptions of volcanoes on the ocean floor. They are located in the centre of the Red Sea due to opening along the axial trough, mainly in the southern part. Al Zabarjad, Al Tayr, Al Zubeir, Haneesh, and Breem are well known, in addition to other islands along the coasts of Eritrea and Saudi Arabia in the region of Hurrat Al Barak and Al Kahma, and Jabal Dahban and Jabal Kudumbal Islands occur to the south of the Al Qunfudah area.

Sediments in the Red Sea

The Red Sea is unique among the seas of the world since no river flows into it. Due to the absence of rivers and permanent streams, terrigenous material is only supplied to the Red Sea during periods of heavy rainfall and flash floods. Frequent dust storms over the Red Sea contribute a significant quantity of silt to fine quartz sand. Aeolian (wind-transported) and calcareous biogenic sediments form a significant contribution to the marine realm. The continental shelf in general is narrow, especially in the north, but is somewhat broader in the south and is partly formed by the occasional sediment input from the wadis that drain into the sea. The Red Sea is veneered with only limited amounts of clay to sand-size detrital material supplied by the wadis to the sea because of lack of rain and controlled erosion of igneous and metamorphic source rocks. Coarse-grained (gravel to sand) material is rich in carbonate material of biogenic origin abundant in the form of forams, gastropods, pelecypods, cephalopods, coralline algae, and coral debris generated mostly by human activity and to some extent by fish biting on the corals for food.

However, the southern part of the Red Sea is relatively richer in detrital material because the diagnostic carbonate materials are diluted by the terrigenous material brought in by the numerous wadis that drain into the sea intermittently compared to those located in the north where the wadis are inactive for years because of extended dry spells and arid desert conditions. Pebble to gravel to sand-size materials are common in the thalweg of the wadis and are similar in composition to the source rocks. The sediments in the lagoons along the length of the Red Sea show variations in the composition and mineralogy depending on the connectivity of the lagoon with the wadis and the sub-environment. Sand storms, water currents, a relatively small tidal range, and wind action affect the transport of sediment, mostly from north to south in line with the prevailing wind and current directions.

Mineral Resources

A lack of rainfall and absence of permanent rivers along the Red Sea limit the erosion of igneous and metamorphic source rocks and control the distribution of terrigenous material. The sand and gravel tend to be rich in carbonate material of biogenic origin, while limited amounts of quartz, feldspars, mica (biotite and muscovite), and heavy minerals are supplied to the sea. The common clay minerals in the Red Sea are chlorite, montmorillonite, smectite, illite, and kaolinite. Heavy minerals rich in magnetite are found in limited amounts on the beaches in the northern and southern parts of the Saudi Red Sea. Metalliferous deposits of commercial value, the product of the opening of the Red Sea, are present in the deeps. For example, the Atlantis II deep holds one of the biggest offshore deposits ready to be exploited (Anschutz et al. 2000). NaCl from the Red Sea water and evaporite minerals, for example gypsum, from the sabkhas, are common exploitable products.

Environment of the Red Sea

Since historical times, the Red Sea has been used for shipping, mainly via the Gulf of Suez and Suez Canal, with vessels of all sizes from dhows to very large crude carriers (VLCC). The coasts are subjected to multiple and often conflicting uses, ranging from recreational beaches and resorts, to fishing, for consumption (desalination plants), to use as a receptacle for municipal and industrial wastes. The Red Sea marine environment is affected by human activities along the coastal areas due to urbanisation and coastal development (for example, dredge and landfill operations in the vicinity of most major cities), industries including thermal power and desalination plants and oil refineries, cement factories, recreation and tourism, waste water treatment facilities (sewage and solid waste), dumping of unwanted wastes such as plastic and household items and industrial refuse, chemical, petro-chemical and fertilizer industries, coastal mining and quarrying activities, oil bunkering and ballast water discharge, and habitat modification such as the filling and conversion of wetlands (UNEP 1997). The presence of algal mats in the coastal area and sabkhas engulfing the coral reefs is common especially in the areas where human interference and anthropogenic input to the marine environment is severe.

Considerable funding from both international and domestic agencies and organizations has been made available to study the Red Sea in order to understand its formation, associated features, and resources. Numerous research vessels with multiple research objectives have visited the Red Sea. In this volume of the Red Sea book, current research has been addressed by contributors who are actively studying the Red Sea in a variety of disciplines. The tectonics, coastal geomorphology, sediment types and dispersal patterns, mineralogical make-up, coral reefs and their preservation, water exchange with the Indian Ocean via the Gulf of Aden and the Strait of Bab-al-Mandab, the formation of recent volcanic islands in the southern Red Sea, environmental aspects and the role of coastal lagoons for shrimp farming including aqua and mariculture have drawn particular attention recently.

Desalination Plants along the Red Sea

There is a huge demand for desalinated water to meet domestic consumption requirements. The water from the Red Sea is also used by industries for cooling purposes, and hot brine and treatment chemicals (chlorine and anti-scalants) are discharged back into the sea and both the thermal and saline inputs affect the environment. For example, the discharged water may cause mortality of the coral reefs and environmental pollution in the coastal waters. There are over 18 desalination plants along the Saudi coast of the Red Sea and the Kingdom is now the largest producer of desalinated water (RSTSAP 2006) with a present capacity of more than 3 million m^3/d.

Pollution

The long flushing time of 200 years makes the Red Sea vulnerable to pollution. The major sources of pollution in the Red Sea are related to land-based activities, sewage discharge, oil spills, and shipping activities and transportation including ballast water discharge. Rapid population growth and inadequate treatment and disposal facilities, poorly or untreated sewage, and sewage dumped by ships cause contamination to the Red Sea and coastal areas (PERSGA 1998). The input of nutrient-rich sewage results in eutrophication of the coastal waters around some major cities, ports, and tourist facilities (Gerges 2002). Pollution from solid waste is a major problem around urban centres and shipping lanes (Gladstone et al. 1999). Chemical pollution, although limited to the vicinity of industrial zones with effluents from mines, desalination plants, and chemical industries, is a severe problem. In the sediments of the coastal environment of the Ash Shuqayq area 6 % of the chlorite is from the local desalination plant and can harm humans and the ecosystem (Alharbi et al. 2012). Povelsen et al. (2003) reported sediment contamination with heavy metals, while Risk et al. (2009) reported the effects of sewage discharge on black coral in the vicinity of Jeddah. Contamination of beaches in the Gulf of Suez by tar balls and oil slicks is common (PERSGA 1998; Sheppard 2000) and maritime pollution caused by shipping is common in the Gulf of Aqaba.

Red Sea Biology

The warm tropical water of the Red Sea hosts some of the most spectacular coastal and marine environments of the world with a rich biodiversity. The sea is known for its biological characteristics including its fauna and flora, particularly coral reefs and numerous fish species, marine turtles, and a number of unique marine habitats, such as mangroves, seagrass beds, algae, salt-pans, and salt marshes. The shallow shelves and coastal areas along the Red Sea are known to have over two hundred species of hard and soft corals inhabited by over two thousand invertebrate species and more than 1,350 species of fish (Goren and Dor 1994), as well as crabs, turtles, and dolphins, most of which are indigenous to the Red Sea. Seagrass on the seabed and mangrove stands (mostly *Avicennia marina* plus 3 other species) are sporadically distributed along the shoreline and on numerous islands.

Coral Reefs

The Red Sea coral reefs are amongst the most interesting and productive reefs in the world with more than 50 genera of hermatypic corals, comprising 180–200 species (Sheppard and Sheppard 1991; Sheppard et al. 1992). An extensive study along the central-northern Red Sea coastline of Saudi Arabia increased this to a probable 260, based on recently identified species and range extensions (De Vantier et al. 2000). They are one of the important resources of tropical sea-shores and the most diverse ecosystem in the Red Sea. They perform valuable functions as protective barriers since they retain sand on the beaches, serve as habitats for a variety of marine communities, and serve as spawning and nursery grounds for numerous species of reef fish.

Coral reefs are common in the central and northern half of the Red Sea, where they are well developed and drop steeply into deep water, although thinner reefs occur in the Gulf of Aqaba and some other northern shores. The greatest development occurs in offshore barrier reefs and in fringing reefs. In the south, reefs are found in shallow and relatively turbid water and are less well developed. However, well-developed reefs are found around the Farasan Islands, the Dahlak Islands, and Kamaran Island. Algal reefs occur in the southern Red Sea in low-energy environments; they support a dense brown algal cover and provide important hard substrate sandy areas. A longitudinal series of coral reefs lie along the axis of the Red Sea on ridges resulting from normal faulting and upward movement of underlying salt deposits. Numerous atolls are also found mostly on the ridges (UNEP 1997).

There are 13 principal coral communities in the Red Sea. Most show considerable localisation that is correlated with latitude but linked with gross changes in coastal bathymetry and morphology. The reefs are common at depths between 5 and 20 m. Coral cover is usually less than 50 %, but in sheltered areas, one or two species, especially *Porites,* may cover 80 % of the substrate (Chiffings 2003). There are 194 species of corals recorded along the Red Sea coast of Saudi Arabia, while about 30 coral species have been recognised in the Gulf of Aqaba and about 80 near Jeddah, with less than 50 near Al Birk, and less than 10 on the southernmost Saudi inshore reefs. Five areas along the Saudi coast are noted for their extensive coral reefs: the Tiran islands, Al Wajh Bank, the area north of Yanbu, the coastline between Obhur and Thuwal, north of Jeddah, and the outer Farasan bank. The Gulf of Aden also supports surprisingly rich and complex reefs despite upwelling water and sandy shorelines (Behairy et al. 1992). The total number of coral reef species in the southern Red Sea and the Gulf of Aden is estimated to be around 130 (UNEP 1997).

The main reasons for the good development of the reef systems along the Red Sea are because of its great depth and an efficient water circulation pattern. The over 2,000 km stretch of the coastline is populated with coral reefs where the fringing reefs are 5,000–7,000 years old and are largely formed of stony acropora and porites corals. The reefs form platforms and sometimes lagoons along the coast and occasionally other features such as cylinders (such as the Blue Hole at Dahab). There are many offshore reefs including several true atolls. Many of the unusual offshore reef formations defy classic (i.e. Darwinian) coral reef classification schemes and are generally attributed to the high levels of tectonic activity that characterise the area.

Increasing population and the associated social activity along the shores of major cities have caused damage to coral formations and their environment including the flora and fauna. Industrial cities along the Red Sea coast are now densely populated, and development and maritime transport, oil spills, land filling, pollutant discharges and effluent from sewage treatment plants and treated water from desalination plants are all threats to the coral reefs of the Red Sea. In general, the fringing reefs are in good condition with the exception of those adjacent to major cities. Coral reefs and their environment are discussed in this volume by Bruckner and Dempsey, Nasr, Klaus, Mansour and Madkour, Rowlands and Purkis, and Sawall and Al-Sofyani.

Fish

About 1,350 fish species have been recorded from the Red Sea, among them about 30 species of sharks and 25 species of rays. Within the Red Sea there are major differences in assemblage composition between areas north and south of latitude 20°N. There is a general increase in productivity from the north, in the Gulf of Aqaba, to the south towards the Saudi-Yemen border. Over 74 % of the annual Red Sea landings come from the southern section between Al Lith and the Yemeni border. However, the highest fish species diversity and density is found on coral reefs. The difference in assemblage is due to the differences in water quality and nutrient loading and the north–south gradients in temperature, salinity, and turbidity levels. The Gulfs of Aqaba and Suez support distinctive fish assemblages. Those of the Gulf of Suez share greater affinities with southern Red Sea assemblages than with the Gulf of Aqaba, probably due to its shallow and turbid water. New fish species continue to be discovered every year; most are shared with the Indian Ocean, but because of the Red Sea's semi-enclosed nature, the degree of endemism, species that occur only here and nowhere else, is exceptionally high. It is estimated to be about 11 % for continental shelf species (species occurring above 200 m depth) and 17 % for deep-water and deep-sea species. Because of constantly high temperatures of about 22 °C down to below 2,000 m, and the shallow sill at the Red Sea's southern entrance, there are no primary deep-sea fishes. The deep sea is inhabited by fish species originating from shallow water that became adapted to the deep-sea environment.

Dugongs

Dugong dugon, also known as sea-cows, inhabits the Red Sea and Gulf of Aden. They are quiet, harmless animals that are mostly found in sheltered, shallow and isolated areas, and lagoons with seagrass beds. The dugong distribution in the Red Sea is not continuous, since populations are found in isolated channels and bays. They are rare but occasionally reported in the Gulf of Aqaba, scarce in the Gulf of Suez, reported regularly but are not common in the Sudanese Red Sea, and are very rare along the Eritrean coast. In Saudi Arabia, dugongs are concentrated in three locations, the Al Wajh Bank, the Al Lith, and Jizan areas.

Deep-Sea Mollusks

Mollusks are one of the most conspicuous macrobenthic components of the deep-sea basin fauna in terms of abundance and diversity. Revised data from historical and recent cruises in the Red Sea show the diversity of the various species. To date, the number of species inhabiting the Red Sea below 400 m exceeds 200 taxa with a dominance of gastropoda, followed by bivalvia. Although the distribution pattern and the environment they inhabit has never been investigated in detail, the deep-sea benthos faces unique hydrological conditions in the Red Sea, as discussed in this volume by Janssen and Taviani.

Marine Turtles

The Red Sea hosts a great variety of marine turtles, described in this volume by Mancini et al. Five of the seven world's species can be found here, including the green turtle (*Chelonia mydas*), the hawksbill turtle (*Eretmochelys imbricata*), the loggerhead turtle (*Caretta caretta*), the leatherback turtle (*Dermochelys coriacea*), and the olive ridley turtle (*Lepidochelys olivacea*). However, only the green and hawksbill turtles are considered common throughout the region and are known to both feed and nest here. Green turtles can be found usually within shallow, coastal bays with extensive seagrass meadows that constitute their main food item; on the other hand, hawksbill turtles are usually found in association with coral reefs where they feed on corals and sponges. Both species have a fundamental role in keeping their habitats healthy and actual threats to marine turtle populations might represent a threat to the entire ecosystem. Important nesting grounds are located in the Tiran Islands, Al Wajh Bank and Farasan Islands, and the south Sinai of Egypt (PERSGA 2006), Dahlak Islands, Ras Sharma and Dhobah (Miller 1989; PERSGA 1998). The region between Jabal Aziz Island and Perim in the southern Red Sea is the most important nesting ground for hawksbill turtles in the Arabian region (PERSGA 2006).

Sea Slugs

Sea slugs are often flamboyant creatures living in nearly all marine habitats of the Red Sea—they can be found on coral reefs, in coral rubble and debris and on rocky flats, in seagrass beds, and ploughing through soft substrates, in search of prey or mates. As they are predators of most marine creatures and plants and need to defend themselves since the majority have lost their shells, they have developed a vast arsenal of chemical weapons, which are of great interest in biomedical research. Sea slugs were described from the Red Sea as early as 1775, and since then the number of species recorded has increased steadily (Yonow 2008, see this volume).

Phytoplankton

Phytoplankton play a fundamental role in the marine food web, biogeochemical cycling, and climatic processes. In the Red Sea, phytoplankton are an important food supply for coral reef ecosystems as they transfer energy to higher levels of the marine food web. Although the Red Sea is considered an oligotrophic sea, sporadic blooms in the open oceanic parts and/or the coastal reef areas that reach mesotrophic levels occur, and there is evidence of seasonal blooms taking place during spring months in the north-western region. There are 88 species of phytoplankton dinoflagellates in the Red Sea. The distribution increases eastward and the winter trend seems to be one to two orders of magnitude greater than in summer (Chiffings 2003). Water temperature and wind play an important role in the productivity of the Red Sea. They act synergistically to control the nutrient level at the euphotic zone (through convection) and facilitate the horizontal transportation of nutrients (through advection).

Mangrove Stands

Four species of mangrove have been recorded from the Red Sea, including *A. marina*, *Rhizophora mucronata, Bruguiera gymnorhiza,* and *Ceriops tagal.* However, only two species *A. marina* and *R. mucronata* are common and have been reported by recent surveys in the region, where *A. marina* is the most abundant and occurs in all mangrove areas in the region. Most mangrove stands are mono-specific stands of *A. marina*. Typically, mangroves grow as narrow or, rarely, broad forests along the coastal shoreline, on the coasts of nearshore and offshore islands, and fringing tidal creeks and channels (locally known as khors, sharms, or mersas). They flourish in brackish water, extreme temperatures, and high salinities. Stands are generally thin, 50–100 m wide, but they may extend from less than 100 m to more than 20 km and range in height between 5 and 7 m. Mangrove stands are distributed more or less along the entire Red Sea coast of Saudi Arabia where their areal extent has been estimated to be 200 km^2 (Vincent 2008). The most dense and largest stands are found in the south (Mandura et al. 1987; Price et al. 1987). Although mangrove stands of the Red Sea are generally too small for commercial harvesting of forest products, they play significant ecological roles such as a nursery function for several fish and marine invertebrate species, as coral reef and coastal protection, as well as supporting marine and terrestrial biodiversity. Mangroves in the region are discussed by Khalil in this volume.

Seagrass

Eleven species of seagrass have been reported in the Red Sea, ranging from mid-water to depths of about 70 m (El Shaffai 2011), due to the high transparency of the Red Sea water. The most commonly recorded species are *Halodule stipulacea, Halodule uninervis, Thalassodendron cliatum, Syringidium isetifolium,* and *Halophila ovalis*. Of the 11 seagrass species in the entire Red Sea, ten have been recorded along the Saudi coast, while two species are found in the Gulf of Aqaba, where temperatures are cooler and the substrate is not suitable for seagrass growth. Nine species of seagrass have been reported from the Yemeni Red Sea coast, only three of which were recorded from the Gulf of Aden coast. Seagrass beds generally occur in protected shallow areas in lagoons and bays and are inhabited by a diverse fauna and are more common in the shallow southern Red Sea. The distribution of seagrass is primarily controlled by water transparency, seabed type, water movement, salinity, and temperature. The major groups inhabiting seagrass beds include mollusks, polychaetes, crustaceans, echinoderms, and fishes, with perhaps about 10 % of the species in seagrass beds occurring nowhere else. The standing crop and productivity of Red Sea seagrass beds are comparable to that reported from other tropical regions of the world. As elsewhere in the Red Sea region, seagrass beds stabilize nearshore sediments, provide a juvenile habitat for a range of commercially important crustaceans and fishes, are a source of food for significant species (e.g. dugong, turtle), and probably export nutrients and energy to adjacent subtidal systems.

Algae (Seaweeds)

More than 500 taxa of benthic algae have been recorded from the Red Sea. Shallow coral reef areas of the northern and central Red Sea are often seasonally dominated by filamentous green, small brown and tuft-forming (calcareous-coralline) red algae. Perennial brown algae, such as *Sargassum*, *Cystoseira*, and *Hormophysa*, are dominant over shallow, hard substrates in the southern Red Sea. Macroalgae often form the major cover of hard substrates in areas too turbid for coral growth. For many species, vigorous water movement is essential, although dense algal growth also occurs on unconsolidated substrate, helping to stabilize it (Chiffings 2003). Generally, there is no specific information on the status of algal communities in the Red Sea. However, in the Farasan Islands, algal reefs may be built by calcareous algae under a low-energy environment on coarse sediment at depths between 2 and 4 m (Bruckner et al. 2011).

Benthic and Planktonic Foraminifera

The Red Sea benthic and planktonic foraminifera have been studied by many authors (Said 1950; Reiss et al. 1977; Bahafzallah 1979; Abou-Ouf et al. 1988; Abou-Ouf and El-Shater 1991; Hottinger et al. 1993; Haunold et al. 1997; Abu-Zied et al. 2011, 2012; Abu-Zied 2012). The intertidal and lagoonal benthic foraminifera of the Red Sea are dominated by symbiont-bearing species such as *Peneroplis planatus*, *Sorites orbiculus*, *Neorotalia calcar*, *Coscinospira hemprichii* and *Monalysidium acicularis,* and others (e.g. *Quinqueloculina,* cf. *Q. Limbata, Q. costata, Spiroloculina communis,* and *Elphidium striatopunctatum*) (Reiss and Hottinger 1984; Abu-Zied et al. 2011; Abu-Zied and Bantan 2013) and are controlled by environmental factors such as dissolution and bioerosion in calm waters, abrasion and breakage in the nearshore sediments, pH, temperature, salinity, substrate types, and algae (Abu-Zied et al. 2011). These species show a good correlation with tidal elevation, predicting sea level changes in the Shoaiba lagoon with a precision of ±0.16 m (Abu-Zied and Bantan 2013).

The deep Red Sea benthic foraminifera are dominated by species such as a *Buccella granulata-Gyroidinoidessoldanii-Bolivina persiensis* assemblage, indicating the predominance of oligotrophic, highly oxygenated bottom waters between depths of 650–1,300 m, while a *Melonis novozealandicum-Spirophthalmidium acutimargo* assemblage was recorded in the deep and bathyal-slope sediments indicating its tolerance for wider ranges of environmental conditions (Edelman-Fürstenberg et al. 2001; Abu-Zied and Bantan 2013). The deepest parts of the Red Sea are dominated by *Brizalina spathulata* and *Melonis novozealandicum* assemblages indicating the existence of temporary oxic conditions.

Modern Red Sea planktonic foraminifera live in oligotrophic surface waters and are dominated by species such as *Globigerinoides sacculifer, G. ruber,* and *Globigerinella siphonifera. G. sacculifer* dominates both living and fossil (sediment) assemblages of the more saline northern Red Sea, and then decreases southwards to be overcome by *G. ruber* (Auras-Schudnagies et al. 1988, 1989; Siccha et al. 2009; Abu-Zied and Bantan 2013). Both the benthic and planktonic foraminifera in the Red Sea amplify climatic signals and have been used by many authors for paleoceanographic and paleoclimatologic purposes (e.g. Schoell and Risch 1976; Halicz and Reiss 1981; Reiss and Hottinger 1984; Rohling 1994; Almogi-Labin et al. 1996; Rohling et al. 1998; Geiselhart 1998; Fenton et al. 2000; Siddall et al. 2004; Badawi et al. 2005; Edelman-Furstenberg et al. 2009).

Oceanography (Climatology)

The Red Sea is located in the subtropical zone within an area described as desert and is an extraordinary marine body of water partially isolated from the open ocean. The Hanish Sill, about 100 km to the north of the Strait of Bab-al-Mandab, has a minimum depth of only 137 m and hence the Red Sea is cut off from the currents of the Indian Ocean, resulting in the lack of a well-developed tidal system. The Red Sea is replenished by the ocean, but it is a slow process and can take up to 6 years to renew the surface waters and up to 200 years to renew the entire Red Sea (Maillard and Soliman 1986; Sheppard et al. 1992).

Warm to very warm surface water temperatures in all parts of the Red Sea basin are recorded and vary seasonally between 22 and 32 °C. During the summer, the temperature averages 26 °C in the north and 30 °C in the south, with only about 2–4 °C variation during the winter, averaging 23 °C. However, locally temperatures as high as 34 °C in the Shoaiba lagoon have been recorded. The warm water temperatures and very high salinities typical in the Red Sea make it one of the hottest and saltiest bodies of sea water in the world (Belkin et al. 2009), at least 4 ‰ greater than the world ocean average. A very small tidal range, almost a tideless sea at the nodal point in the central Red Sea, and the lack of any permanent rivers flowing into the sea make the Red Sea a unique environment. The characteristics of the Red Sea waters are influenced by its geographic location in an arid zone, being narrow and shallow at Bab-al-Mandab, as well as the wind dynamics and seasonal reversal of monsoon winds over the southern Red Sea and the intrusion of upwelled water from the Gulf of Aden during the south-west monsoon.

The climate of the Red Sea is the result of two distinct monsoon seasons; a north-easterly monsoon and a south-westerly monsoon. Monsoon winds occur because of the differential heating between the land surface and sea. The Red Sea water mass is exchanged with the Arabian Sea and Indian Ocean via the Gulf of Aden. This physical factor reduces the effect of high salinity caused by evaporation that exceeds precipitation and cold water in the north and relatively hot water in the south. Although the meteorological conditions are relatively uniform, the extremely high evaporation rate leads to the formation of salinity fronts, in which temperature fronts tend to develop caused by wind-induced upwelling whose effect is amplified by steep bathymetry and morphological features (orographic). Three groups of fronts have been identified from north to south: (1) the Egypt-Saudi Arabia front, (2) the Sudan-Saudi Arabia front, and (3) the Eritrea-Yemen front (Belkin et al. 2009). The Al Lith area forms an approximate climatic northern border to the area in south-western Saudi Arabia which is influenced by the south-west and north-west monsoon seasons.

Precipitation and Evaporation

The Red Sea is characterised by tropical to subtropical humid conditions. Onshore desert conditions are characterised by a very hot arid climate and very low precipitation averaging 6 cm y^{-1} (Behairy et al. 1991) and evaporation rate around 2 m y^{-1} (Pedgley 1974; Maillard and Soliman 1986; Sofianos et al. 2002; Siddall et al. 2004). The rainfall is mostly in the form of showers of short duration, often associated with thunderstorms and occasionally with dust storms. The isolated rains occur mainly during the winter months, although several years may pass by without any precipitation. Desert depressions occasionally invade the northern Red Sea in spring and are particularly characterised by dust in the atmosphere and poor visibility (Morcos 1970). The varying rainfall amounts to as little as 10–15 mm y^{-1} over the sea, a few mm per year along the coastline, and up to 180 mm y^{-1} at Suakin (19°N) has been reported (UNEP 1997). Over the broad coastal Tihama plains of the Red Sea rainfall averages 16 mm y^{-1} at Al Wajh, 63 mm yr^{-1} in Jeddah, and 63 mm y^{-1} in the Jizan region (Bruckner et al. 2011).

Evaporation greatly exceeds precipitation and is between 1.4 and 2 m yr^{-1} (Hastenrath and Lamb 1979). High water temperatures and wind stress increase evaporation and also result in increased salinity, especially in the northern part of the Red Sea. The average relative humidity varies from a maximum of 68 % during summer to a minimum of 50 % during winter and is significantly higher during the day time than at night. Evaporation over the Red Sea is seasonally affected by the atmospheric circulation pattern that is related to the Indian Ocean monsoon (Rohling et al. 2013).

Water Transparency

Since major fluvial input is lacking in the Red Sea, the turbidity level is almost non-existent, making the Red Sea water more transparent and increasing the light penetration. Growth of microalgae is limited because of the high salt content that also makes the water of the Red Sea very clear. The lack of sediment input from various anthropogenic or natural sources also makes the water transparency higher, recorded at over 28 m near Shiab Al Kabir, 24 km off Saudi Arabia near Jeddah, while the transparency is over 45 m in the area close to Port Sudan.

Salinity

The high salinity arises because of the enclosed nature of the sea, higher water and air temperatures, low precipitation and high evaporation, the absence of major rivers draining into the sea, and the limited exchange of water with the Indian Ocean which has a lower salinity. The Red Sea is the hottest and saltiest body of sea water on earth and a unique environment that thrives with coral reefs supporting its habitats. Due to the high evaporation and low precipitation, surface water salinities range from about 36.5 near Bab-al-Mandab to about 40–41 ‰ at the northern end of the main Red Sea basin (Alraddadi 2013). However, locally a salinity as high as 52 ‰ in the Shoaiba lagoon has been recorded. The northward increase of salinity is attributed to the high rate of evaporation and mixing of low saline Gulf of Aden surface inflow with the more saline deep waters through turbulence. The salinity at every latitude is generally higher in summer than in winter, and the annual variation decreases from north to south, being more than 1 ‰ in the north to about 0.5 ‰ in the south (Abdullah 1985; Morcos and Varely 1990).

Tidal Range

The tides in the Red Sea are diurnal (twice a day) and the tidal range varies between 60 cm in the north at the confluence of the Gulf of Suez and the Red Sea and 90 cm in the south at the Gulf of Aden, but fluctuates between 20 and 30 cm at the nodal points near Jeddah in the eastern and Port Sudan in the western Red Sea as discussed here by Pugh and Abualnaja. The central Red Sea (Jeddah area) is therefore almost tideless, and as such the annual water level changes are more significant. Because of the small tidal range, the water during high tide inundates the coastal sabkhas as a thin sheet of water extending up to a few hundred metres rather than inundating the sabkhas through a network of channels. However, south of Jeddah in the Shoiaba area, the water from the lagoon may cover the adjoining sabkhas as far as 3 km inland, whereas north of Jeddah in the Al Kharrar area the sabkhas are covered by a thin sheet of water for up to 2 km. The prevailing north and north-eastern winds influence the movement of water in the coastal inlets to the adjacent sabkhas, especially during sand storms and occasionally during spring tides. Winter mean sea level is 50 cm higher than in summer (Behairy et al. 1992).

Water Currents

The Red Sea is known for its unpredictable currents and strong winds. In the Red Sea, detailed current data are lacking, in part because they are weak and variable both spatially and temporally and are governed by the wind. During the summer, north-westerly winds drive surface water south for about four months at a velocity of 15–20 cm s^{-1}, whereas in winter the flow is reversed resulting in the inflow of water from the Gulf of Aden into the Red Sea; the latter predominates, resulting in an overall drift to the northern end of the Red Sea. Generally, the velocity of the tidal current is between 50–60 cm s^{-1} with a maximum of 1 m s^{-1} at the entrance of the Al Kharrar lagoon, for example, where the thalweg is scoured resulting in a well-sorted sediment veneer (Rasul et al. 2010). However, the range of the north-north-east current along the Saudi coast is 8–29 cm s^{-1} but the entrances of most lagoons have high current velocities. The main surface drifts are slow moving and are easily modified and even reversed by local effects and by small tides (UNEP 1997). Tidal velocities passing through constrictions caused by reefs, sand bars, and low islands and shoals commonly exceed 1–2 m s^{-1}, as observed in the Al Masturah area 200 km north of Jeddah.

Currents in the Red Sea largely result from density gradients in the water column. Surface water density in the Red Sea rises with a fall in water temperature to the north and increased evaporation causing increased salinity. Decreasing temperatures and evaporation in the Gulf of Suez result in the formation of dense water that turns under and is returned southward in the deep Red Sea. In winter (September–June), surface water is driven into the Red Sea by prevailing winds from the south-east, beneath which there is a deep outward flow of denser saline water. During summer, prevailing winds change to the north-west, driving the upper water layer out of the Red Sea over a deeper inward flow from the Gulf of Aden. Net outflows are 10 % higher in salinity and balance the effects of evaporation in the Red Sea (Chiffings 2003).

Wind Regime

The wind regime is characterised by both seasonal and regional variations in speed and direction with the average speed generally increasing northward (Patzert 1974). With the exception of the northern part of the Red Sea, which is dominated by persistent north-westerly winds with speeds ranging between 7 and 12 km h^{-1} (Rao and Behairy 1986), the rest of the Red Sea and the Gulf of Aden is subjected to the influence of regular and seasonally reversible winds. However, occasional winds from this direction are slightly more frequent in summer than in winter. Southerly winds occasionally blow during winter months only (UNEP 1997). The Red Sea circulation pattern is governed by eddies and sub-gyres generated by wind forcing that vary both spatially and temporally (Sofianos and Johns 2007). The Red Sea response to wind forcing depends strongly on wind direction; along the axis of the Red Sea winds do not interact with the surrounding topography, whereas cross-axis winds interact with high, steep mountains along the Red Sea, resulting in a highly structured wind field favourable to oceanic eddy formation (Clifford et al. 1997). Wind speeds from 6.7 to 9.3 m s^{-1} blowing from the south or south-south-east in the southern half of the Red Sea have been recorded (Morcos 1970; Patzert 1974). Waves are mostly generated by wind and tidal action. In shallow waters, wind is the driving force in the Red Sea for transporting sediments either as suspension or as bed-load. Wind-induced currents play an important role in the Red Sea in initiating the process of re-suspension of bottom sediments and transfer of materials from sites of dumping to sites of burial in a quiescent environment of deposition. Wind-generated currents are important in determining the sediment dispersal pattern and its role in the erosion and accretion of the coastal rock exposures and submerged coral reefs. During summer, the prevailing winds flow in a southerly direction along the entire length of the Red Sea, influencing clockwise airflow in the Arabian Sea. This generates strong south-westerly winds, leading to cool, nutrient-rich upwelling resulting in the development of Ecklonia kelp beds in places that inhibit the development of coral reefs (Chiffings 2003). In winter, the general air current is from north-east, whereas in summer it is from the north or north-west in the northern region. However, in the south, the wind is mostly from west or south-west. In general, the air current is more or less parallel to the coastal belt of the Red Sea.

Dust Storms

The area along the Red Sea is characterised by tropical to subtropical humid conditions. Desert depressions occasionally invade the northern Red Sea in spring and are particularly characterised by dust in the air which is very dry and the intense convection lifts vast amounts of dust and very fine sand-to-silt-size material into the atmosphere and gives rise to poor visibility and hazy conditions (Morcos 1970). Frequent dust storms pass through the Red Sea causing low visibility and generally hazy conditions. During its transit, the sediment load is dropped off over the sea and the Arabian Peninsula, with some on the African continent. The Red Sea receives aeolian material in variable proportions from the easternmost Sahara, Sudan, and Saudi Arabia (Hickey and Goudie 2007; Jiang et al. 2009). In summer, the eastward blowing jet stream is from the Sudan area (mainly through the Tokar area), whereas in winter the westward blowing wind from Saudi Arabia also contributes dust to the Red Sea (Rohling et al. 2013). Strong easterly zonal winds from Egypt also contribute aeolian material to the Red Sea (Jiang et al. 2009).

References

Abdel-Rahman K, Al-Amri AMS, Abdel-Moneim E (2009) Seismicity of the Sinai Peninsula, Egypt. Arab J Geosci 2:103–118

Abdullah AM (1985) Hydrological structure and circulation of waters of the Red Sea. University of Moscow, Moscow (in Russian)

Abou-Ouf M, El-Shater A (1991) The relationship between the environmental conditions of the Jeddah coast, Red Sea, and benthic foraminifera. J King Abdulaziz Univ Mar Sci 2:49–64

Abou-Ouf M, Rao NVND, Tag RJ (1988) Benthic foraminifera from littoral sediments of Al-Lith-Al Qunfidhah coast, southern Red Sea. Indian J Mar Sci 17:217–221

Abu-Zied RH, Bantan RA (2013) Hypersaline benthic foraminifera from the Shuaiba Lagoon, eastern Red Sea, Saudi Arabia: their environmental controls and usefulness in sea level reconstruction. Mar Micropaleontol 103:51–67

Abu-Zied RH (2012) Effect of the Red Sea brine-filled deeps (Shaban and Kebrit) on the composition and abundance of benthic and planktonic foraminifera. Arab J Geosci. doi:10.1007/s12517-012-0641-3

Abu-Zied RH, Basaham AS, El-Sayed MA (2012) Effect of municipal wastewaters on bottom sediment geochemistry and benthic foraminifera of two Red Sea coastal inlets, Jeddah, Saudi Arabia. Environ Earth Sci. doi:10.1007/s12665-012-1751-7

Abu-Zied RH, Bantan RA, Basaham AS, El Mamoney MH, Al-Washmi HA (2011) Composition, distribution, and taphonomy of nearshore benthic foraminifera of the Farasan Islands, southern Red Sea Saudi Arabia. J Foramin Res 41:349–362

Ahmad F, Sultan SAR, Moammar MO (1995) Residual transport velocities during winter, within the Atlantis II deep area of the central Red Sea. Oceanol Acta 18(3):385–388

Al Amri AM, Schult FR, Bufe CG (1991) Seismicity and aeromagnetic features of the Gulf of Aqaba (Elat) region. J Geophys Res 96 (B12):179–185

Alharbi OA, Phillips MR, Williams AT, Gheith A, Bantan RA, Rasul NM (2012) Desalination impacts on the coastal environment: Ash Shuqayq, Saudi Arabia. Sci Total Environ 421–422:163–172

Allan TD (1970) Magnetic and gravity fields over the Red Sea. Philos Trans Roy Soc Lond A 267:153–180

Almogi-Labin A, Hemleben C, Meischner D, Erlenkeuser H (1996) Response of Red Sea deep-water agglutinated foraminifera to water mass changes during the late quaternary. Mar Micropaleontol 28:283–297

Alraddadi TM (2013) Temporal changes in the Red Sea circulation and associated water masses. University of Southampton, Ocean and Earth Science, PhD Thesis, 198 pp

Al Tarazi E, Abu Rajab J, Gomez F, Cochran W, Jaafar R, Ferry M (2011) GPS measurements of nearfield deformation along the southern Dead Sea fault system. Geophys Geochem Geosyst 12 (12):17. doi:10.1029/2011GC003736

Altherr R, Henjes-Kunst F, Puchelt H, Baumann A (1988) Volcanic activity in the Red Sea axial trough—evidence for a large mantle diapir? Tectonophysics 150:121–133

Anderson DL (2013) The persistent mantle plume myth. Aust J Earth Sci 60:657–673

Anschutz P, Blanc G, Chatin F, Geiller M, Pierret M-C (1999) Hydrographic changes during 20 years in the brine-filled basins of the Red Sea. Deep Sea Res Part 1 Oceanogr Res Pap 46:1779–1792

Anschutz P, Blanc G, Monnin C, Boulègue J (2000) Geochemical dynamics of the Atlantis II Deep (Red Sea): II. Composition of metalliferous sediment pore water. Geochim Cosmochim Acta 64:3995–4006

Antunes A, Ngugi DK, Stingl U (2011) Microbiology of the Red Sea (and other) deep-sea anoxic brine lakes. Environ Microbiol Rep 3(4):416–433

ArRajehi A, McClusky S, Reilinger R, Daoud M, Alchalbi A, Ergintav S, Gomez F, Sholan J, Bou-Rabee F, Ogubazghi G, Haileab B, Fisseha S, Asfaw L, Mahmoud S, Rayan A, Bendik R, Kogan L (2010) Geodetic constraints on present-day motion of the Arabian Plate: implications for Red Sea and Gulf of Aden rifting. Tectonics 29, TC3011 10. doi:10.1029/2009TC002482

Augustin N, Devey CW, van der Zwan FM, Feldens P, Tominaga M, Bantan RA, Kwasnitschka T (2014) The rifting to spreading transition in the Red Sea. Earth Planet Sci Lett 395:217–230. doi:10.1016/j.epsl.2014.03.047

Auras-Schudnagies A, Kroon D, Ganssen G, Hemleben C, Van Hinte JE (1988) Biogeographic evidence from planktonic foraminifers and pteropods for Red Sea anti-monsoonal surface currents. In: Kroon D, Brummer GJA (eds) Planktonic foraminifers as tracers of ocean-climate history. Free University Press, Amsterdam, pp 203–227

Auras-Schudnagies A, Kroon D, Ganssen G, Hemleben C, Van Hinte JE (1989) Distributional pattern of planktonic foraminifers and pteropods in surface waters and top core sediments of the Red Sea, and adjacent areas controlled by the monsoonal regime and other ecological factors. Deep Sea Res 36:1515–1533

Badawi A, Schmiedl G, Hemleben C (2005) Impact of late Quaternary environmental changes on deep-sea benthic foraminiferal faunas of the Red Sea. Mar Micropaleontol 58:13–30

Badawy A (2005) Seismicity of Egypt. Seismol Res Lett 76:149–160

Baer G, Hamiel Y (2010) Form and growth of an embryonic continental rift: InSAR observations and modelling of the 2009 western Arabia rifting episode. Geophys J Int 13. doi:10.1111/j.1365-246X.2010.04627.x

Bahafzallah AAK (1979) Recent benthic foraminifera from Jiddah Bay, Red Sea (Saudi Arabia). Neues Jahrbuch für Geologie und Palaeontologie (Monatshefte und Abhandlungen) 7:385–398

Bayer HJ, Hotzl H, Jado AR, Roscher B, Voggenreiter W (1988) Sedimentary and structural evolution of the northwest Arabian Red Sea margin. Tectonophysics 153:137–151

Behairy AKA, Sheppard CRC, El-Sayed MK (1992) A review of the geology of coral reefs in the Red Sea. UNEP Reg Seas Rep Stud 152:39

Behairy AKA, Rao DNVN, El-Shater A (1991) A siliciclastic coastal sabkha, Red Sea coast, Saudi Arabia. J Fac Mar Sc King Abdulaziz Univ 2:65–77

Belkin IM, Cornillon PC, Sherman K (2009) Fronts in large marine ecosystems. Prog Oceanogr 81:223–236

Ben-Avraham Z, Von Herzen RP (1987) Heat flow and continental breakup: the Gulf of Elat (Aqaba). J Geophys Res 92(B2):1407–1416

Ben-Avraham Z (1985) Structural framework of the Gulf of Elat (Aqaba), northern Red Sea. J Geophys Res 90(B1):703–716

Ben-Avraham Z, Garfunkel Z, Lazar M (2008) Geology and evolution of the southern Dead Sea Fault with emphasis on subsurface structure. Ann Rev Earth Planet Sci 36:357–387

Ben-Avraham Z, Garfunkel Z, Almagor G, Hall JH (1979) Continental breakup by a leaky transform: the Gulf of Elat (Aqaba). Science 206:214–216

Bertram C, Krätschell A, O'Brien K, Brückmann W, Proelss A et al (2011) Metalliferous sediments in the Atlantis II Deep—assessing the geological and economic resource potential and legal constraints. Resour Policy 36:315–329

Beyene A, Abdelsalam MG (2005) Tectonics of the Afar Depression: a review and synthesis. J Afr Earth Sc 41:41–59

Biton E, Gildor H, Peltier WR (2008) Red Sea during the Last Glacial Maximum: implications for sea level reconstruction. Paleoceanography 23, PA1214. doi:10.1029/2007PA001431

Blanc G, Anschutz P (1995) New stratification in the hydrothermal brine system of the Atlantis II Deep, Red Sea. Geology 23:543–546

Blank HR, Andreasen GE (1991) Compilation and interpretation of aeromagnetic data for the Precambrian Arabian shield, Kingdom of Saudi Arabia. Saudi Arabian Directorate General of Mineral Resources Open-File Report USGS-OF-10-8, 54 pp

Blank HR (1977) Aeromagnetic and geological study of Tertiary dikes and related structures on the Arabian margin of the Red Sea. In: Saudi Arabian Directorate General of Mineral Resources; Red Sea Research 1970–1975, Bulletin no. 22:G1–18

Bohannon RG, Naeser CW, Schmidt DL, Zimmermann RA (1989) The timing of uplift, volcanism, and rifting peripheral to the Red Sea: a case for passive rifting? J Geophys Res 94:1683–1701

Bonatti E (1985) Punctiform initiation of seafloor spreading in the Red Sea during transition from a continental to an oceanic rift. Nature 316:33–37

Bosworth W, McClay K (2001) Structural and stratigraphic evolution of the Gulf of Suez Rift, Egypt: a synthesis. In: Ziegler PA, Cavazza W, Robertson AHF, Crasquin-Soleau S (eds) Peri-Tethys Memoir 6: Peri-Tethyan Rift/Wrench Basins and Passive Margins. Memoires du Museum National d'Histoire Naturelle, Paris 186:567–606

Bosworth W (1995) A high-strain rift model for the southern Gulf of Suez (Egypt). Geol Soc London Spec Publ 80:75–102
Bosworth W, Huchon P, McClay K (2005) The Red Sea and Gulf of Aden basins. J Afr Earth Sci 43:334–378
Botz R, Schmidt M, Kus J, Ostertag-Henning C, Ehrhardt A, Olgun N, Garbe-Schonberg D, Scholten J (2011) Carbonate recrystallisation and organic matter maturation in heat-affected sediments from the Shaban Deep, Red Sea. Chem Geol 280(1–2):126–143
Brown GF, Schmidt DL, Huffman AC (1989) Geology of the Arabian Peninsula, shield area of western Saudi Arabia. U.S. Geological Survey Professional paper 560A, 188 pp
Bruckner A, Rowlands G, Riegl B, Purkis S, Williams A, Renaud P (2011) Khaled bin Sultan Living Oceans Foundation Atlas of Saudi Arabian Red Sea Marine Habitats. Panoramic Press, Phoenix, 274 pp
Camp VE, Roobol MJ (1992) Upwelling asthenosphere beneath western Arabia and its regional implications. J Geophys Res 97 (B11):15255–15271
Chapman DS, Pollack HN (1975) Global heat flow: a new look. Earth Planet Sci Lett 28:23–32
Chiffings T (2003) Marine region 11: Arabia Seas. A global representative system of marine protected areas. http://www.deh.gov.au/coasts/mpa/nrsmpa/global/volume3/chapter11.html
Clifford M, Horton C, Schmitz J, Kantha LH (1997) An oceanographic nowcast/forecast system for the Red Sea. J Geophys Res 102 (C11):25101–25122
Cochran JR (1983) A model for development of Red Sea. Am Assoc Pet Geol Bull 67(1):41–69
Cochran JR (2005) Northern Red Sea: nucleation of an oceanic spreading center within a continental rift. Geochem Geophys Geosyst (G3) 6(3):34, Q03006. doi:10.1029/2004GC000826
Cochran JR, Martinez F (1988) Evidence from the northern Red Sea on the transition from continental to oceanic rifting. Tectonophysics 153:25–53
Coleman RG, Gregory RT, Brown GF (1983) Cenozoic volcanic rocks of Saudi Arabia. Saudi Arabian Deputy Ministry for Mineral Resources, Open-File Report USGS-OF-03-93, 82 pp
Coulibay AS, Anschutz P, Blanc G, Malaizé B, Pujol C, Fontanier C (2006) The effect of paleo-oceanographic changes on the sedimentary recording of hydrothermal activity in the Red Sea during the last 30,000 years. Mar Geol 226:51–64
Daëron M, Klinger Y, Tapponier P, Elias A, Jacques E, Sursock A (2007) 12,000-year-long record of 10 to 13 paleoearthquakes on the Yammouneh Fault, Levant Fault System, Lebanon. Bull Seismol Soc Am 97(3):749–771
Daradich A, Mitrovica JX, Pysklywec RN, Willett SD, Forte AM (2003) Mantle flow, dynamic topography, and rift-flank uplift of Arabia. Geology 31:901–904
Davison I, Al-Kadasi M, Al-Khirbash S, Al-Subbary AK, Baker J, Blakey S, Bosence D, Dart C, Heaton R, McClay K, Menzies M, Nichols G, Owen L, Yelland A (1994) Geological evolution of the southeastern Red Sea Rift margin, Republic of Yemen. Am Assoc Pet Geol Bull 106:1474–1493
De Vantier L, Turak E, Al-Shaikh K, De'ath G (2000) Coral communities of the central-northern Saudi Arabian Red Sea. Fauna Arab 18:23–66
Dekov VM, Scholten JC, Botz R, Garbe-Schönberg C-D, Stoffers P (2007) Fe-Mn-(hydr)oxide carbonate crusts from the Kebrit Deep, Red Sea: precipitation at the seawater/brine redoxcline. Mar Geol 236:95–119
Dixon TH, Ivins ER, Franklin BJ (1989) Topographic and volcanic asymmetry around the Red Sea: constraints on rift models. Tectonics 8:1193–1216
Doglioni C, Carminati E, Bonatti E (2003) Rift asymmetry and continental uplift: Tectonics 22(3):1024. doi:10.1029/2002TC001459
Drake CL, Girdler RW (1964) A geophysical study of the Red Sea. Geophys J Roy Astron Soc 8:473–495
Edelman-Fürstenberg Y, Almogi-Labin A, Hemleben C (2009) Palaeoceanographic evolution of the central Red Sea during the late Holocene. Holocene 19:117–127
Edelman-Fürstenberg Y, Scherbacher M, Hemleben C, Almogi-Labin A (2001) Deep-sea benthic foraminifera from the central Red Sea. J Foramin Res 31:48–59
Eissen J-P, Juleau T, Joron J-L, Duprè B, Humler E, Al Mukhamedov A (1989) Petrology and geochemistry of basalts from the Red Sea axial rift at 18° north. J Petrol 30:791–839
El Shaffai A (2011) Studies on the seagrass ecosystem in Wadi El Gemal National Park, Red Sea. M.Sc. thesis, Suez Canal University, Ismailia
Eltyab MM (2004) Review of the trochus fishery in Sudan SPC. Trochus Inf Bull 11:5–7
Farhoud K (2009) Accommodation zones and tectono-stratigraphy of the Gulf of Suez, Egypt: a contribution from aeromagnetic analysis. GeoArabia 14:139–162
Fenton M, Geiselhart S, Rohling EJ, Hemleben C (2000) Aplanktonic zones in the Red Sea. Mar Micropaleontol 40:277–294
Fournier M, Chamot-Rooke N, Petit C, Huchon P, Al-Kathiri A, Audin L, Beslier M-O, d'Acrement E, Fabbri O, Fleury J-M, Khanbari K, Lepvrier C, Leroy S, Maillot B, Merkouriev S (2010) Arabia-Somalia plate kinematics, evolution of the Adel-Owen-Carlsberg triple junction, and opening of the Gulf of Aden. J Geophys Res 115:B04102. doi:10.1029/2008JB006257
Geiselhart S (1998) Late quaternary paleoceanographic and paleoclimatologic history of the Red Sea during the last 380,000 years: evidence from stable isotopes and faunal assemblages. Ph.D. thesis, Institut und Museum für Geologie und Paläontologie der Universität Tübingen, 87 pp
Gerges MA (2002) The Red Sea and Gulf of Aden action plan—facing the challenges of an ocean gateway. Ocean Coast Manag 45:885–903
Gettings ME, Blank HR, Mooney WD, Healy JH (1986) Crustal structure of southwestern Saudi Arabia. J Geophys Res 91:6491–6512
Girdler RW (1991) The Afro-Arabian rift system—an overview. Tectonophysics 197:139–153
Girdler RW, Underwood M (1985) The evolution of early oceanic lithosphere in the southern Red Sea. Tectonophysics 116:95–108
Girdler RW, Evans TR (1977) Red Sea heat flow. Geophys J Royal Astron Soc 51:245–251
Girdler RW, Southren TC (1987) Structure and evolution of the northern Red Sea. Nature 330:716–721
Gladstone W, Tawfiq N, Nasr D, Andersen I, Cheung C, Drammeh H, Krupp F, Lintner S (1999) Sustainable use of renewable resources and conservation in the Red Sea and Gulf of Aden: issues, needs and strategic actions. Ocean Coast Manag 42:671–697
Goren M, Dor M (1994) Updated check list of the fishes of the Red Sea. Israel Academy of Science and Humanities, Jerusalem, 120 pp
Grainger DJ (2007) The geologic evolution of Saudi Arabia: a voyage through space and time. Saudi Geological Survey, Jeddah, 246 pp (with 9 booklets)
Gurvich EG (2006) Metalliferous sediments of the world ocean. Fundamental theory of deep-sea hydrothermal sedimentation. Springer, Berlin, 416 pp
Halicz E, Reiss Z (1981) Paleoecological relations of foraminifera in a desert-enclosed sea—the Gulf of Aqaba (Elat), Red Sea. Mar Ecol 2:15–34
Hall SA (1979) A total intensity magnetic anomaly map of the Red Sea and its interpretation. Saudi Arabian Directorate General of Mineral Resources, USGS Project Report 275, 260 pp

Hall SA (1989) Magnetic evidence for the nature of the crust beneath the southern Red Sea. J Geophys Res 94:12267–12279

Hamiel Y, Amit R, Begin ZB, Marco S, Katz O, Salamon A, Zilberman E, Porat N (2009) The seismicity along the Dead Sea fault during the last 60,000 years. Bull Seismol Soc Am 99:2020–2026

Hansen S, Schwartz S, Al-Amri A, Rodgers A (2006) Combined plate motion and density-driven flow in the asthenosphere beneath Saudi Arabia: evidence from shear-wave splitting and seismic anisotropy. Geology 34:869–872

Hartmann M, Scholten JC, Stoffers P, Wehner F (1998a) Hydrographic structure of brine-filled deeps in the Red Sea—new results from the Shaban, Kebrit, Atlantis II, and Discovery Deep. Mar Geol 144:311–330

Hartmann M, Scholten JC, Stoffers P (1998b) Hydrographic structure of brine-filled deeps in the Red Sea: correction of Atlantis II Deep temperatures. Mar Geol 144:331–332

Hastenrath S, Lamb PJ (1979) Climatic Atlas of the Indian Ocean, part 2. The ocean heat budget. University of Wisconsin Press, Madison

Haunold TG, Baal C, Piller WE (1997) Benthic foraminiferal associations in the northern Bay of Safaga, Red Sea. Egypt Marine Micropaleontology 29:185–210

Hayes TS, Sutley SJ, Kadi KA, Balkhiyour MB, Siddiqui AA, Beshir Z, Hashem HI (2002) Jabal Dhaylan zinc-lead deposits, geological setting, genesis and 1996–2000 exploration progress. Saudi Geological Survey Open-File Report SGS-OF-2001-5, vol 1: Text, Appendix 1 and Plates; vol 2: Appendices 2, 3, 4 and 5; 71 pp

Head SM (1987) Red Sea fisheries. In: Edwards AJ, Head SM (eds) Red Sea: key environments. Pergamon Press, Oxford, pp 363–382

Hickey B, Goudie AS (2007) The use of TOMS and MODIS to identify dust storm source areas: the Tokar delta (Sudan) and the Seistan basin (south west Asia). In: Goudie AS, Kalvoda J (eds) Geomorphological variations. P3 K, Prague, pp 37–57

Höltz H, Zötl JG (1978) Climatic changes during the Quaternary Period. In: Al-Sayari SS, Zötl JG (eds) Quaternary Period in Saudi Arabia. Springer, Vienna, pp 310–311

Hottinger L, Halicz E, Reiss Z (1993) Recent Foraminiferida, Gulf of Aqaba, Red Sea. Opera Academia Scientiarum et Artium Slovenica, Classis IV: Historia Naturalis 33, Paleontological Institute "Ivan Rakovec"

International Hydrographic Organization (1953) Limits of Oceans and Seas, 3rd edn. International Hydrographic Organization, PANGAEA, Bremerhaven

Izzeldin YA (1987) Seismic, gravity and magnetic surveys in the central part of the Red Sea; their interpretation and implications for the structure and evolution of the Red Sea. Tectonophysics 143:269–306

Jiang H, Farrar JT, Beardsley RC, Chen R, Chen C (2009) Zonal surface wind jets across the Red Sea due to mountain gap forcing along both sides of the Red Sea. Geophys Res Lett 36:L19605. doi:10.1029/2009GL040008

Johnson PR (1998) Tectonic map of Saudi Arabia and adjacent areas: Saudi Arabian Deputy Ministry for Mineral Resources, Technical Report USGS-TR-98-3 (IR 948)

Johnson PR, Woldehaimanot B (2003) Development of the Arabian-Nubian Shield: perspectives on accretion and deformation in the northern East African Orogen and the assembly of Gondwana. In: Yoshida M, Windley BF, Dasgupta S (eds) Proterozoic East Gondwana: Supercontinent assembly and breakup, Special Publications. Geological Society, London, 206:289–325

Johnson PR, Andresen A, Collins AS, Fowler AR, Fritz H, Ghebreab W, Kusky T, Stern RJ (2011) Late Cryogenian-Ediacaran history of the Arabian-Nubian Shield: a review of depositional, plutonic, structural, and tectonic events in the closing stages of the northern East African Orogen. J Afr Earth Sc 61:167–232

Johnson PR, Kattan FH (2012) The geology of the Saudi Arabian shield. Saudi Geology Survey, Jeddah, 479 pp

Kozdroj W, Kattan FH, Kadi KA, Al Alfy ZSA, Oweiss KA, Mansour MM, Johnson PR (2011) SGS-EMRA trans-Red Sea project for geological and structural correlation between the central Eastern Desert Terrane (Egypt) and Midyan Terrane (Saudi Arabia). Saudi Geological Survey Open-File Report SGS-TR-2011-5

Lambeck K, Purcell A, Flemming NC, Vita-Finzi C, Alsharekh AM, Bailey GN (2011) Sea level and shoreline reconstructions for the Red Sea: isostatic and tectonic considerations and implications for hominin migration out of Africa. Quatern Sci Rev 30:3542–3574

Le Pichon X (1968) Sea-floor spreading and continental drift. J Geophys Res 73:3661–3697

Maillard C, Soliman G (1986) Hydrography of the Red Sea and exchanges with the Indian Ocean in summer. Oceanol Acta 9:249–269

Makris J, Allam A, Mokhtar T, Basahel A, Dehghani GA, Bazari M (1983) Crustal structure in the northwestern region of the Arabian shield and its transition to the Red Sea. King Abdulaziz Univ Bull Fac Earth Sci 6:435–447

Makris J, Mohr P, Rihm R (eds) (1991a) Red Sea: birth and early history of a new oceanic basin. Tectonophysics Special Issue 198:468

Makris J, Henke CH, Egloff F, Akamaluk T (1991b) The gravity field of the Red Sea and East Africa. Tectonophysics 198:369–381

Makris J, Tsironidis J, Richter H (1991c) Heatflow density distribution in the Red Sea. Tectonophysics 198:383–393

Makris J, Rihm R (1991) Shear-controlled evolution of the Red Sea: pull apart model. Tectonophysics 198:441–466

Mandura AS, Saifullah SM, Khafaji AK (1987) Mangrove ecosystem of southern Red Sea coast of Saudi Arabia. Proc Saudi Biol Soc 10:165–193

Manheim FT (1974) Red Sea geochemistry. Init Rep Deep Sea Drilling Proj 23:975–998

Martinez F, Cochran JR (1988) Structure and tectonics of the northern Red Sea: catching a continental margin between rifting and drifting. Tectonophysics 150:1–32

Martinez F, Cochran JR (1989) Geothermal measurements in the northern Red Sea: implications for lithospheric thermal structure and mode of extension during continental rifting. J Geophys Res 94:12239–12265

McGuire AV, Bohannon RG (1989) Timing of mantle upwelling: evidence for a passive origin for the Red Sea rift. J Geophys Res 94:1677–1682

Meshref WM (1990) Tectonic framework. In: Said R (ed) The geology of Egypt. A.A. Balkema Publishers, Rotterdam, pp 113–155

Miller JD (1989) An assessment of the conservation status of marine turtles in Saudi Arabia. Volume 1, MEPA coastal and marine management series, Report no. 9, Jeddah, 209 pp

Mitchell NC, Park Y (2014) Nature of crust in the central Red Sea. Tectonophysics. doi:10.1016/j.tecto.2014.04.029

Morcos SA, Varley A (eds) (1990) Red Sea, Gulf of Aden and Suez Canal. A bibliography on oceanographic and marine environmental research. Alesco-Persga and UNESCO, 198 pp

Morcos SA (1970) Physical and chemical oceanography of the Red Sea. Oceanogr Mar Biol Ann Rev 8:73–202

Mutwalli SA (2009). Direct relationship between Red Sea rifting and formation of the Atlantis II deep. In: Proceedings 5th Task Force Workshop, Abu Dhabi, 1–6 Dec 2009

Nehlig P, Genna A, Asfirane F (2002) A review of the Pan-African evolution of the Arabian Shield. GeoArabia 7:103–124

Pallister JS, McCausland WA, Jónsson S, Lu Z, Zahran HM, El Hadidy S, Aburukbah A, Stewart ICF, Lundgren PR, White RA, Moufti MRH (2010) Broad accommodation of rift-related extension recorded by dike intrusion in Saudi Arabia. Nature Geosci 3:705–712. doi:10.1038/NGEO966 (with supplementary information, 5 pp)

Pallister JS (1987) Magmatic history of Red Sea rifting: perspective from the central Saudi Arabian coastal plain. Geol Soc Am Bull 98:400–417

Patzert WC (1974) Wind-induced reversal in Red Sea circulation. Deep Sea Res 21:109–121

Pedgley DE (1974) An outline of the weather and climate of the Red Sea. Publ Centre National Pour L'Exploitation des Oceans, Actes du Colloques 2:9–21

PERSGA (1998). Strategic action programme for the Red Sea and Gulf of Aden. Regional Organisation for the Conservation of the Environment of the Red Sea and Gulf of Aden. www.persga.org/about/projects/sap/overview.asp

PERSGA (2006) The State of the Marine Environment, report for the Red Sea and Gulf of Aden. In: Wilkinson G, Facey R, Hariri K (eds) PERSGA, Jeddah, 241 pp

Phoenix Corporation (1985) The interpretation of an airborne geophysical survey of the cover rocks region. Saudi Arabian Deputy Ministry for Mineral Resources, Kingdom of Saudi Arabia, 76 pp

Pierret MC, Clauer N, Bosch D, Blanc G (2010) Formation of thetis metal-rich sediments in the absence of brines, Red Sea. J Geochem Explor 104:12–26

Povelsen E, Alshabrawy MM, Mohammed AS, El-Seoud AA (2003) Heavy metals and hazardous organic pollutants in sediment and mussels in the Gulf of Suez, 1999 and 2001. Egypt Environ Aff Agency. http://www.eeaa.gov.eg/English/main/Env2003/Day2/Solidwaste/presentationpovlesen.pdf

Price ARG, Medley PAH, McDowall RJ, Dawson-Sheppard AR, Hogarth PJ, Ormond RFG (1987) Aspects of mangal ecology along the Red Sea coast of Saudi Arabia. J Nat Hist 21:449–464

Prodehl C, Mechie J (1991) Crustal thinning in relationship to the evolution of the Afro-Arabian rift system: a review of seismic refraction data. Tectonophysics 198:311–327

Rao DNVN, Behairy AKA (1986) Nature and composition of shore-zone sediments between Jeddah and Yanbu, Eastern Red Sea. Mar Geol 70:287–305

Rasul MA, Abushosha M, Al-Hazmi O, Al-Nomani S, Bahareth F, Widinly N, Akbar A, Qutub AS, Farawati R (2010) Bathymetric survey and ground-truthing of the Al-Kharrar lagoon, Red Sea coast of Saudi Arabia. Saudi Geological Survey Technical Report SGS-TR-2008-8, 77 pp

Reilinger R, McClusky S, Vernant P, Lawrence S, Ergintav S, Cakmak R, Ozener H, Kadirov F, Guliev I, Stepanyan R, Nadariya M, Hahubia G, Mahmoud S, Sakr K, ArRajehi A, Paradissis D, Al-Aydrus A, Prilepin M, Guseva T, Evren E, Dmitrotsa A, Filikov SV, Gomez F, Al-Ghazzi R, Karam G (2006) GPS constraints on continental deformation in the Africa-Arabia-Eurasia continental collision zone and implications for the dynamics of plate interactions. J Geophys Res 111:29. doi:10.1029/2005JB004051 B05411

Reiss Z, Leutenegger S, Hottinger L, Fermont WJJ, Meulenkamp JE, Thomas E, Hansen HJ, Buchardt B, Larsen AR, Drooger CW (1977) Depth-relations of recent larger foraminifera in the Gulf of Aqaba-Elat. Utrecht Micropaleontol Bull 15:1–244

Reiss Z, Hottinger L (1984) The Gulf of Aqaba, ecological micropaleontology: ecological studies 50. Springer, Berlin, pp 1–354

Risk MJ, Sherwood OA, Nairn R, Gibbons C (2009) Tracking the record of sewage discharge off Jeddah, Saudi Arabia, since 1950, using stable isotope records from antipatharians. Mar Ecol Prog Ser 397:219–226

Rohling EJ, Grant KM, Price AP, Larrasoana JC (2013) Paleoclimate variability in the Mediterranean and Red Sea regions during the last 5,000,000 years. Curr Anthropol 54. http://www.jstor.org/stable/10.1086/673882

Rohling EJ (1994) Review and new aspects concerning the formation of eastern Mediterranean sapropels. Mar Geol 122:1–28

Rohling EJ, Fenton M, Jorissen FJ, Bertrand P, Ganssen G, Caulet JP (1998) Magnitudes of sea-level lowstands of the past 500,000 years. Nature 394:162–165

Roland E, Behn MD, Hirth G (2010) Thermal-mechanical behavior of oceanic transform faults: implications for the spatial distribution of seismicity. Geochem Geophys Geosyst 11:15. doi:10.1029/2010GC003034

Roobol MJ, Al-Rehaili M (1998) Geohazards along the Makkah-Madinah-Nafud (MMN) volcanic line. Saudi Arabian Deputy Ministry for Mineral Resources Technical Report BRGM-TR-98-9, 20 pp (11 figs., 1 table)

Roobol MJ, Kadi K (2008) Cenozoic faulting in the Rabigh area, central-west Saudi Arabia (including the sites of King Abdullah Economic City and King Abdullah University of Science and Technology). Saudi Geological Survey Technical Report SGS-TR-2008-6, 12 pp (1 fig., 2 plates)

Roobol MJ, Stewart ICF (2009) Cenozoic faults and recent seismicity in northwest Saudi Arabia and the Gulf of Aqaba region. Saudi Geological Survey Technical Report SGS-TR-2008-7, 35 pp (35 figs., 2 apps., 10 plates)

Roobol MJ, Al-Rehaili M, Arab N, Celebi M, Halawani MA, Janjou D, Kazi A, Martin C, Sahl M, Showail A (1999) The Gulf of Aqaba earthquake of 22 November, 1995: Its effects in Saudi Arabia. Saudi Arabian Deputy Ministry for Mineral Resources Technical Report BRGM-TR-99-16, 67 pp (49 figs., 4 tables, 1 appendix)

RSTSAP (2006). Red Sea tourism strategy and action plan; 2006, 948 pp. http://redsea.sct.gov.sa/reports/Final%20Report%20Draft%2009.05.06.pdf. Accessed 12 Jan 2012

Said R (1950) The distribution of foraminifera in the northern Red Sea. J Foramin Res 1:9–29

Saleh S, Jahr T, Jentzsch G, Saleh A, Abou Ashour NM (2006) Crustal evaluation of the northern Red Sea rift and Gulf of Suez, Egypt from geophysical data: 3-dimensional modelling. J Afr Earth Sc 45:257–278

Sandwell DT, Smith WHF (2009) Global marine gravity from retracked Geosat and ERS-1 altimetry: ridge segmentation versus spreading rate. J Geophys Res 114:B01411. doi:10.1029/2008JB006008

Schmidt M, Botz R, Faber E, Schmitt M, Poggenburg J, Garbe-Schönberg D, Stoffers P (2003) High-resolution methane profiles across anoxic brine-seawater boundaries in the Atlantis-II, Discovery, and Kebrit deeps (Red Sea). Chem Geol 200:359–376

Schmidt DL, Hadley DG, Brown GF (1982) Middle Tertiary continental rift and evolution of the Red Sea in southwestern Saudi Arabia. Saudi Arabian Deputy Ministry for Mineral Resources, Open-File Report, USGS-OF-03-6, 56 pp

Schoell M, Risch H (1976) Oxygen and carbon isotope analyses on planktonic foraminifera of core VA 01-188 P (Southern Red Sea). Geol Jahrb D17:15–32

Scholten JC, Stoffers P, Garbe-Schönberg D, Moammar M (2000) Hydrothermal mineralization in the Red Sea. In: Cronan DS (ed) Handbook of Marine Mineral Deposits. CRC Press, LLC, pp 369–395

Sea Around Us (2007) A global database on marine fisheries and ecosystems. Fisheries Centre, University of British Columbia, Vancouver, Canada. www.seaaroundus.org/lme/SummaryInfo.aspx?LME=33

Sheppard CRC (ed) (2000) The Red Sea: Seas at the millennium—an environmental evaluation, vol 2. Elsevier Science, The Netherlands

Sheppard C, Price A, Roberts C (1992) Marine ecology of the Arabian region: patterns and processes in the extreme tropical environments. Academic Press, The Netherlands

Sheppard CRC, Sheppard ALS (1991) Corals and coral communities of Arabia. Fauna of Saudi Arabia 12:3–170

Siccha M, Trommer G, Schulz H, Hemleben C, Kucera M (2009) Factors controlling the distribution of planktonic foraminifera in the Red Sea and implications for the development of transfer functions. Mar Micropaleontol 72:146–156

Siddall M, Smeed DA, Hemleben C, Rohling EJ, Schmelzer I, Peltier WR (2004) Understanding the Red Sea response to sea level. Earth Planet Sci Lett 225:421–434

Skipwith P (1973) The Red Sea and coastal plain of the Kingdom of Saudi Arabia, a review. Saudi Arabian Directorate General of Mineral Resources, Technical Record TR-1973-1, 149 pp

Sofianos SS, Johns WE (2007) Observations of the summer Red Sea circulation. J Geophys Res 112(C6):C06025. doi:10.1029/2006JC003886

Sofianos SS, Johns W, Murray SP (2002) Heat and freshwater budgets in the Red Sea from direct observations at Bab el Mandab. Deep-Sea Res II 49(7–8):1323–1340

Stern RJ, Johnson PR (2010) Continental lithosphere of the Arabian Plate: a geologic, petrologic, and geophysical synthesis. Earth Sci Rev 101:29–67

Stewart ICF, Johnson PR (1994) Satellite gravity and Red Sea tectonics. Saudi Arabian Deputy Ministry for Mineral Resources, Open-File Report USGS-OF-94-10 (IR-877), 22 pp (8 figs)

Sultan M, Becker R, Arvidson RE, Shore P, Stern RJ, El Alfy Z, Attia RI (1993) New constraints on Red Sea rifting from correlations of Arabian and Nubian Neoproterozoic outcrops. Tectonics 12:1303–1319

Tesfaye S, Harding DJ, Kusky TM (2003) Early continental breakup boundary and migration of the Afar triple junction, Ethiopia. Geol Soc Am Bull 115:1053–1067

Touliabah EH, Wafaa S, Mohammed AE, Kuchari G, Abdulwassi NIH (2010) Phytoplankton composition at Jeddah coast—Red Sea, Saudi Arabia in relation to some ecological factors. J Mar Sci King Abdulaziz Univ 22(1):115–131

Trommer G, Siccha M, Rohling EJ, Grant K, van der Meer MTJ, Schouten S, Baranowski U, Kucera M (2011) Sensitivity of Red Sea circulation to sea level and insolation forcing during the last interglacial. Clim Past 7:941–955. doi:10.5194/cp-7-941-2011

UNEP (1997). Assessment of land-based sources and activities affecting the marine environment in the Red Sea and Gulf of Aden. UNEP Regional Seas Reports and Studies no. 166

Vincent P (2008) Saudi Arabia: an environmental overview. Taylor and Francis, London, 309 pp

Werner F, Lange K (1975) A bathymetric survey of the sill area between the Red Sea and the Gulf of Aden. Geol Jahrb D13:125–130

Williams VS, Al-Rehaili MH, Al-Sulaimani G, Showail AA, Bahloul S, Zahrani M (2001) Preliminary report on investigations of neotectonics and paleoseismicity on the east coast of the Gulf of Aqaba. Saudi Geological Survey Open-File Report SGS-OF-2001-2, 11 pp

Wright TJ, Ebinger C, Biggs J, Ayele A, Yirgu G, Keir D, Stork A (2006) Magma-maintained rift segmentation at continental rupture in the 2005 Afar dyking episode. Nature 442:291–294. doi:10.1038/nature04978

Yonow N (2008) Sea Slugs of the Red Sea. Pensoft Publishers, Sofia, 304 pp

Zahran HM, Stewart ICF, Johnson PR, Basahel MH (2003) Aeromagnetic-anomaly maps of central and western Saudi Arabia. Saudi Geological Survey Open-File Report SGS-OF-2002-8, 6 pp (1 fig., 1 table, 4 sheets) (scale 1:2,000,000)

The Red Sea: Birth of an Ocean

Enrico Bonatti, Anna Cipriani, and Luca Lupi

Abstract

Nowhere on the present-day Earth can the transition from a continental to an oceanic rift be observed and studied better than in the Red Sea region, where three rifts in different stages of evolution meet in a triple point located in the Afar region. A thermal and/or compositional mantle plume may have risen from the upper mantle below Afar already at ~30 Ma, and may have triggered, at least in part, the rifting process. The axial area of the rifts is marked by intense seismicity. While the East African is a fully continental rift, the Gulf of Aden rift experienced oceanic crust accretion between Arabia and Somalia starting already at 17 Ma with a progressive westward propagation that impacted against Africa in the Afar Triangle starting at <1 Ma. The axial zone of oceanic crustal accretion in the Gulf of Aden is segmented by several small (<30 km) offsets, including two major transform-fracture zones, the Socotra (offset ~50 km) and the Alula-Fartak (offset 180 km). Spreading is asymmetric, faster in the northern (Arabia) side (11–13 mm/a) than in the southern (Somalia) side (8 mm/a). The Afar Triangle is a topographically depressed region, located between the continental blocks of Nubia, Somalia, and the Danakil Alps, that separate it from the southern Red Sea. It is an area of thin crust, seismicity related to extension, and intense intrusive and extrusive, mostly basaltic, magmatism. Intrusive basaltic magmatism appears to be important in triggering the rifting process in Afar. Northern Afar displays basaltic ranges oriented parallel to the axis of the Red Sea, such as the Erta Ale, with a crestal permanent lava lake. These ranges represent an incipient oceanic accretionary plate boundary separating Africa from Arabia. At the northern tip of Afar, the plate boundary is displaced to the axial zone of the southern Red Sea, an elongated basin oriented ~N30°W. Its southern part is characterized by an axial rift valley floored by oceanic basalt and accompanied by parallel Vine-Matthews magnetic anomalies,

E. Bonatti (✉) · A. Cipriani
Lamont Doherty Earth Observatory, Columbia University, Palisades, New York, USA
e-mail: enrico.bonatti@bo.ismar.cnr.it

E. Bonatti
Istituto di Scienze Marine, CNR via Gobetti 101, 40129 Bologna, Italy

A. Cipriani
Dipartimento di Scienze Chimiche e Geologiche, Università di Modena e Reggio Emilia, L.go S. Eufemia 19, 41100 Modena, Italy

L. Lupi
Centro di Documentazione e Studi della Dancalia, Pisa, Italy

N.M.A. Rasul and I.C.F. Stewart (eds.), *The Red Sea*,
Springer Earth System Sciences, DOI 10.1007/978-3-662-45201-1_2,

suggesting initial oceanic crust accretion at ~5 Ma, although alternative interpretations suggest that the entire width of the southern Red Sea is underlain by oceanic crust. Moving still farther north, the axial valley becomes discontinuous and the initial accretion of oceanic crust appears to take place in discrete cells that become younger northward. Propagation from these initial nuclei will result in a continuous axial zone of oceanic accretion. Some of these axial "deeps" are the locus of intense hydrothermal activity and metallogenesis. Moving north, the oceanic rift impacts against the Zabargad fracture zone, a major topographic-structural feature that crosses the Red Sea in a NNE direction, offsetting its axis by nearly 100 km. Zabargad island, located at the SSW end of the fracture zone, exposes a sliver of sub-Red Sea lithosphere, including mantle peridotite bodies, Pan-African granitic gneisses criss-crossed by basaltic dykes, gabbro intrusions, and a sedimentary sequence starting with pre-rift Cretaceous deposits. North of the Zabargad Fracture zone, the Red Sea lacks an axial rift valley; it probably consists of extended thinned and faulted continental crust injected by gabbros and basaltic dykes. The activation of the NNE-trending Aqaba-Dead Sea fault at about 14 Ma has deactivated rifting in the Gulf of Suez. Basalt chemistry suggests that the degree of melting of the Red Sea subaxial mantle decreases from south to north, in parallel with a decreasing spreading rate and a lesser influence of the Afar plume.

Introduction

For over two billion years, the Earth's internal engine has fueled a cyclical process, whereby continents assemble in a single "Supercontinent", and then gradually fragment and disperse again. An even cursory look at a globe shows that at present we are in a phase when a number of continental blocks are scattered seemingly at random on the surface of our Planet. Reconstructions of the geological past suggest that continental masses were last assembled in a single Supercontinent, called Pangea, roughly from 250 to 150 Ma. An entire cycle of continental breakup, dispersal, and reassembly lasts roughly 500 million years. A fundamental step in this "Wilson Cycle" (named after the great Canadian earth scientist Tuzo Wilson, a key figure in the Plate Tectonics scientific revolution) is the splitting of a continent and the birth and growth of a new ocean.

Understanding the processes that occur during the transition from a continental to an oceanic rift, the earliest stages of seafloor spreading, and the formation of passive margins, has been for years a major challenge in the Earth sciences. Nowhere in today's Earth are these events better displayed that in the Red Sea region where Arabia is in the process of splitting form Africa (Figs. 1, 2 and 3). Here, within one and the same region, we find a network of rift zones in different stages of evolution, from fully continental (East African), to "pre-oceanic" (Afar and northern Red Sea), to oceanic (southern Red Sea and Gulf of Aden). The three rifts meet in the Afar region in a classical "triple point". A thermal and/or compositional mantle plume may have risen below this region as early as at ~30 Ma, causing massive volcanism. Plate tectonic reconstructions suggest that, while emplacement of oceanic crust initiated as early as at 17 Ma in the eastern Gulf of Aden, it started more recently (at ~5 Ma) in the Southern Red Sea, but has not yet started in the northern Red Sea. Although different parts of this system have been the object of many studies, a number of key aspects remain poorly known. Adjacent to the southern part of the Red Sea on the African side lies the northern Afar Triangle, a piece of dry land partly below sea level, located between the Eritrean-Ethiopian plateaus and a coastal range (Danakil Alps) that reaches up to over 2,000 m above sea level.

The Red Sea has been the object of some very early geological reasoning. For instance, Herodotus, discussing the Red Sea in his "Histories" (~400 BC), provides us with some of the earliest speculations concerning the importance of "geological time" (Rawlinson 1858):

> …Suppose now that the Nile should change its course and flow into this gulf, the Red Sea: what is to prevent it from being silted up by the river within, say, twenty thousand years? Personally I think even ten thousand would be enough. That being so, surely in the vast stretch of time which has passed before I was born, a much bigger gulf than this could have been turned into dry land by the silt brought down by the Nile, for the Nile is a great river and does, in fact, work great changes…

Jumping over 2,000 years, we reach Alfred Wegener, who early in the last century pioneered the concept of Continental Drift. Wegener realized that the Red Sea is a young oceanic rift formed by the separation of Arabia from Africa. He also understood that the Afar Triangle obstructs any attempt to fit the African and Arabian coastlines; he therefore suggested that the Afar region must have developed by volcanism that took place during and after the separation of the two continental blocks: A remarkable intuition, particularly since at that time information on the geology of Afar was very scant. It is worth citing a few

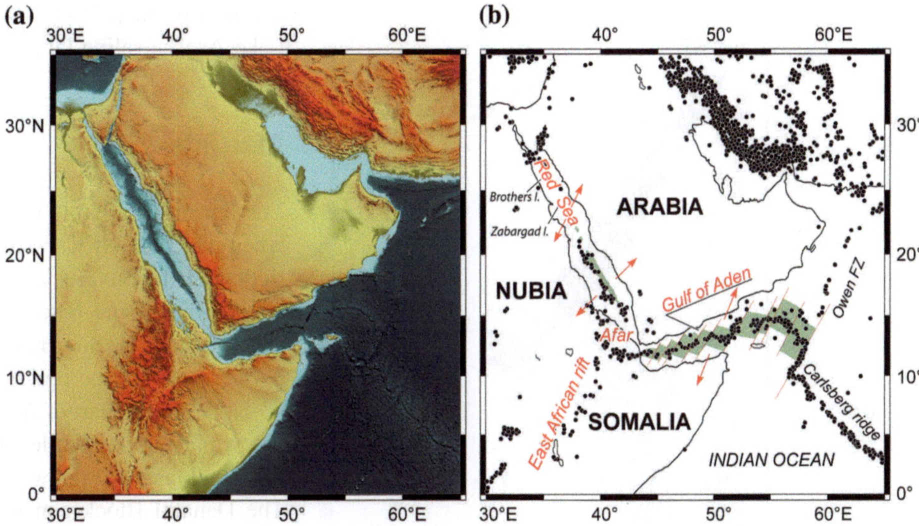

Fig. 1 **a** Red Sea–Afar–Gulf of Aden–East African Rift region. **b** Distribution of earthquakes. Areas underlain by oceanic crust are *dark shaded*. Spatial analysis and mapping were performed using the GMT (Wessel and Smith 1995) and PLOTMAP (Ligi and Bortoluzzi 1989) packages

paragraphs and reproducing an illustration (our Fig. 2) from the 4th edition of Wegener's classic "The Origin of Continents and Oceans" (Wegener 1929).

> "The finest examples of such faults are provided by the East African grabens [rift valleys]. They belong to a large fault system which can be traced northwards through the Red Sea, the Gulf of Aqaba and the Jordan Valley to the edge of the Taurus fold range (Fig. 50). According to recent investigations, these faults are continuing south-wards also, as far as the Cape Province, but the finest examples of them are to be found in eastern Africa. Neumayr-Uhlig (183) describes them approximately as follows:
>
> From the mouth of the Zambesi, a 50–80 km wide graben of this type, runs north, including the Shiré River and Lake Nyasa, then turns northwest and disappears. In its place, close by and parallel to it, begins the graben of Lake Tanganyka, whose grandeur is indicated by the fact the depth of the lake is 1700–2700 m, but the height of the wall-like precipice 2000–2400, even 3000 m. In its northerly continuation, this rift valley includes the Russisi River, Lake Kivu, Lake Edward and Lake Albert." He goes on "The margins of the valley appear ridged as if the earth had burst here with a certain upward movement of the fault margins as they were suddenly released. This peculiar protrusive shape of the edges of the plateau may well be connected with the fact that immediately east of Lake Tanganyka the sources of the Blue Nile are located, while the lake itself drains into the Congo River. A third prominent rift valley begins east of Lake Victoria, including Lake Rudolf farther north and turning near Abyssinia towards the northeast, where it continues on to the Red Sea on one side and the Gulf of Aden on the other. In the coastal region and in the interior of what was formerly German east Africa, these faults mostly take the form of step faults whose eastern side is downthrown.
>
> Of special interest is the large triangle, marked in Fig. 50 by dots in the same way as the valley floors, which lies in the angle formed by Abyssinia and the Somali peninsula (between Ankober, Berbera and Massawa). This relatively flat, low-lying area is composed entirely of recent volcanic lavas. Most authors regard it as a vast broadening of the rift floor. This idea is suggested particularly by the shape of the coastlines on either side of the Red Sea, whose otherwise accurate parallelism is spoilt by this salient; if one cuts this triangle out, the opposite corner of Arabia fits perfectly into the gap."

The intuitions of Wegener have been confirmed by field work carried out in the Gulf of Aden, Afar, and the Red Sea particularly in the second half of last century, in parallel with the establishment of the theory of Plate Tectonics. Some of this early work includes Cochran (1981) and Girdler and Styles (1978) for the Gulf of Aden, the C.N.R.-C.N.R.S. Afar Team (1971) for the Afar Rift, and Girdler and Styles (1974), Coleman (1974) and Cochran (1983) for the Red Sea.

In this chapter, we will provide a brief introduction to our knowledge of the Gulf of Aden, Afar, and the Red Sea Rifts.

Gulf of Aden Rift

Of the three rifts that converge toward the Afar region, the Gulf of Aden rift involved the gradual development of a new accretionary plate boundary between Arabia and Somalia. The axial zone of the Gulf of Aden displays oceanic spreading segments (Fig. 3) offset frequently by transform faults (Cochran 1981) and floored by MORB-type basalts, with a mild Afar Plume signature (Cann 1970; Schilling et al. 1992). Vine-Matthews magnetic stripes accompany the axial oceanic zone, indicating a spreading direction (N37°E) oblique to the general trend of the spreading center (N90°E). In addition to several small (<30 km) transform offsets, at least two major transform-fracture zones offset the Gulf of Aden axial rift zone, that is, the Socotra FZ at 55°E (offset ~50 km) and the Alula-Fartak FZ at 52°E (offset 180 km) (Tamsett and Searle 1990). Both these transforms have a strong topographic signature, including "nodal deeps"; their extension merges with offsets of the Somali and Arabian continental margins, suggesting that they were active already

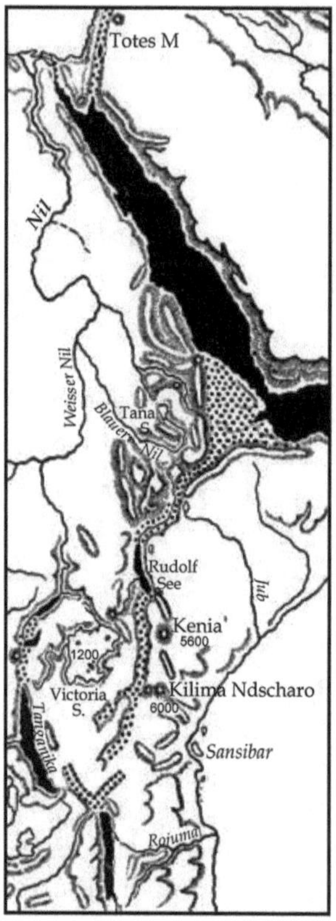

Fig. 2 Map of the Red Sea region, taken from Fig. 50 of Alfred Wegener "Origin of Continents and Oceans", 4° edition (1929)

during the initial opening. The oldest conjugate magnetic anomaly of the eastern part of the Gulf of Aden (east of the Alula-Fartak F.Z.) has been identified as anomaly 5d, suggesting that seafloor spreading in this part of the Gulf of Aden started already at 17.6 Ma (Leroy et al. 2004). Spreading in the eastern Gulf of Aden has been strongly asymmetric (Leroy et al. 2004), faster on the northern (Arabian) side (11–13 mm yr^{-1}) than on the southern (Somalia) side (8 mm yr^{-1}) (d'Acremont et al. 2006). The oldest anomalies become progressively younger moving from east to west, indicating that the initial emplacement of oceanic crust propagated westward. Moving west from the Alula-Fartak FZ, the oldest magnetic anomalies are ~10 Ma, while west of about 45°E the zone of oceanic crust accretion becomes limited to a narrow axial strip. West of ~44°E regular magnetic stripes have not been clearly identified; axial emplacement of oceanic crust may have just started (Hébert et al. 2001).

The Gulf of Aden rift propagates then into the Gulf of Tadjoura and penetrates into the Afar region in the Goubet-Asal rift (Fig. 4), a ~40 km long, highly seismic segment, with intense axial basaltic volcanism and a system of extensional fissures and faults (Fig. 5) flanking the axial neovolcanic zone (Harrison et al. 1975; Manighetti et al. 1997, 1998; Pinzuti et al. 2013). The Goubet-Asal rift opens at a rate of 1.6 cm yr^{-1} in a N40°E direction, and terminates in Lake Asal, a saline lake whose surface is below sea level.

Afar Rift

The Afar region is the locus of a "textbook" triple junction, where the East African Rift, the Gulf of Aden Rift, and the Red Sea Rift converge and meet (Fig. 1). The entire region is covered by fissure basalts, scattered central volcanic cones, and major volcanic ranges, located particularly in the northern part of the region.

The northern part of Afar is a depression (with wide areas below present-day sea level) located between the Ethiopian–Eritrean plateau to the west and the Danakil Alps to the east. The Danakil Block consists of units of Precambrian continental lithosphere overlain by a pre-rift sedimentary sequence including Mesozoic sandstones and limestones. The Danakil Block rotated anticlockwise by about 23° since its separation from the Ethiopian Plateau, being pinned to Africa near the Gulf of Zula and to Arabia in the Bab-al-Mandab area. En échélon elongated volcanic ranges oriented parallel to the Red Sea rift occupy the axial zone of northern Afar. The northernmost and largest is the Erta Ale range, first described by Barberi and Varet (1977), made of recent basalts, and displaying a permanent lava lake on its summit (Fig. 6). An unusually shallow (~1 km) magma chamber has been identified below the summit of the Erta Ale (Pagli et al. 2012). South of the Erta Ale we find two other parallel axial basaltic ranges, that is, Tat Ali and Alayta (Figs. 4 and 6). The basaltic floor of the depression is dissected by extensional faults and fissures oriented parallel to the Red Sea axis; it is affected by widespread fissural basaltic volcanism (Fig. 7a, b) and intense hydrothermal activity. A relatively large lake, lake Afrera (Fig. 6b), lies south of Erta Ale; it is fed by hydrothermal springs and its salinity reaches 158 g/l (Martini 1969). The surface of the lake lies about 100 m below sea level. Unpublished bathymetric lines revealed a >60 m deep trench oriented NNE in the lake's southern basin; thus, the lake's floor may include the lowest point of the entire Afar region. NW of Erta Ale, close to the scarp of the Ethiopian Plateau, iron-manganese deposits have been described; they were probably formed by submarine hydrothermal activity (Bonatti et al. 1972). North of the Erta Ale range the axial part of the rift is covered by a flat salt plain (Fig. 8a); the thickness of the evaporites is >1 km. These evaporites may be related to various episodes of marine invasion and desiccation of northern Afar. The evaporites are uplifted and metamorphosed by hydrothermal activity in a limited area (called Dallol) located near the center of the salt plain (Fig. 8b, c).

Dozens of small (few hundred meters high, up to 1 km in diameter) circular volcanoes are scattered in the northern/central Afar region. They are characterized by a low (<0.5)

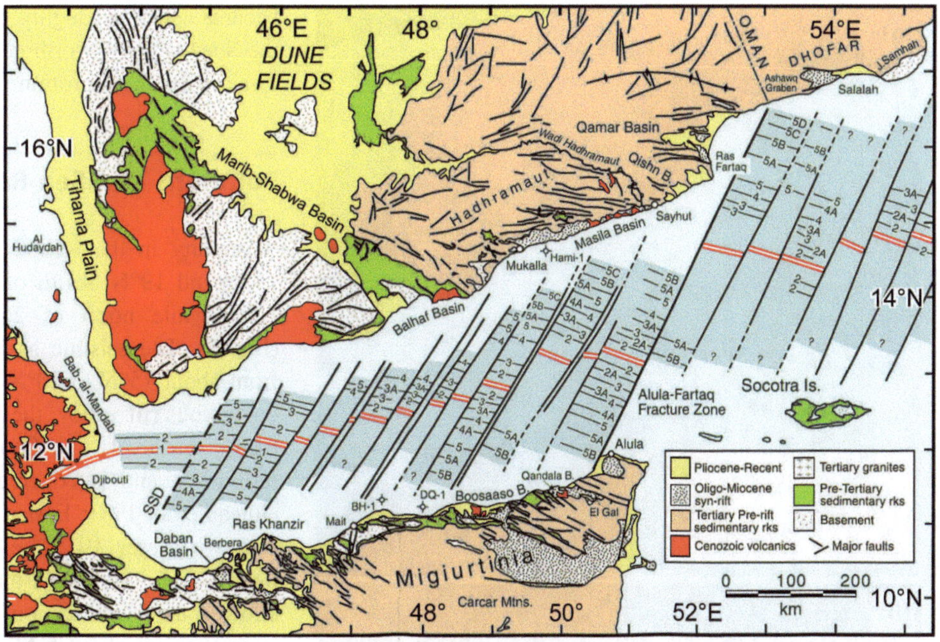

Fig. 3 Gulf of Aden spreading centers, transform faults and magnetic anomalies (modified from Bosworth et al. 2005)

height/diameter ratio, lower than that of subaerial pyroclastic cones. They are made prevalently of shattered basaltic glass fragments (hyaloclastites), frequently well stratified (Figs. 9 and 10), containing also dispersed crystallized basalt blocks. These volcanoes were probably formed by submarine basaltic eruptions. One of them, located close to the northern shore of Lake Afrera, is capped by shallow water marine carbonates (Fig. 10b). Facies analysis as well ^{14}C and U-Th dating of these carbonates suggest they were deposited roughly 30 ka, before the last desiccation of northern Afar (Bonatti et al. 1971). At least one of these small volcanoes has a flat top (Fig. 9b–d) and resembles a "mini-guyot". The flat top is not due to erosion and truncation at sea level, as in classical Pacific guyots, but rather to horizontal deposition of basaltic pyroclastics injected into sea water by explosive submarine eruptions (Bonatti and Tazieff 1970). Another volcano has flanks consisting of stratified hyaloclastites due to submarine eruptions, capped by dark massive basalts that were probably erupted subaerially (Fig. 10a).

The axial basaltic ranges of northern Afar are oriented NNW, parallel to the Red Sea axis; however, they are offset in a few places along NNE lineaments (Barberi et al. 1974) that may correspond to offsets in the scarps of the Ethiopian Plateau. Small volcanic cones have been observed aligned NNE, at an angle to the direction of the axial rift. In contrast to the axial ranges, they are composed of alkali basalts, occasionally containing ultramafic xenoliths.

The axial zone of northern Afar marks probably an incipient plate boundary between the Nubian and Arabian plates. Thus, although geographically part of Africa, the eastern portion of northern Afar, including the Danakil Alps, belongs to the Arabian plate. The distribution of seismic events suggests that the accretionary plate boundary is displaced at around 14°N from the northern tip of Afar to the Red Sea axis, along what is probably an embryonic transform system (Fig. 1).

A series of seismic-magmatic events starting in September 2005 might clarify how extension and rupturing of the crust may occur in Afar (Wright et al. 2006; Ebinger et al. 2008; Keir et al. 2009). A large number of seismic events clustered in a limited area between the Gabbo and Dabbahu volcanic systems of Central Afar (Fig. 1). Seismicity reached a maximum from September 24–26, with magnitudes up to 5.6. At the same time, small basaltic eruptions took place along fissures close to the Dabbahu volcano, and a number of fractures opened along a 60 km segment between the two volcanoes (Fig. 7c, d). Magma intruded along this 60 km long segment formed a megadyke located between 2 and 9 km below the surface, that weakened the lithosphere leading to extension and breaking of the crust (Keir et al. 2009). Additional intrusive-volcanic events followed (Ferguson et al. 2010, 2013a, b), suggesting that opening of a rift can be "active", that is, determined initially by magmatic events at depth, that can result in breakup of the crust and extensional faulting close to the surface.

Red Sea

The Red Sea fills a long (about 2,250 km) and narrow (up to 350 km wide) depression that separates Africa from Arabia (Fig. 2). Much of the Red Sea is only a few hundred meters

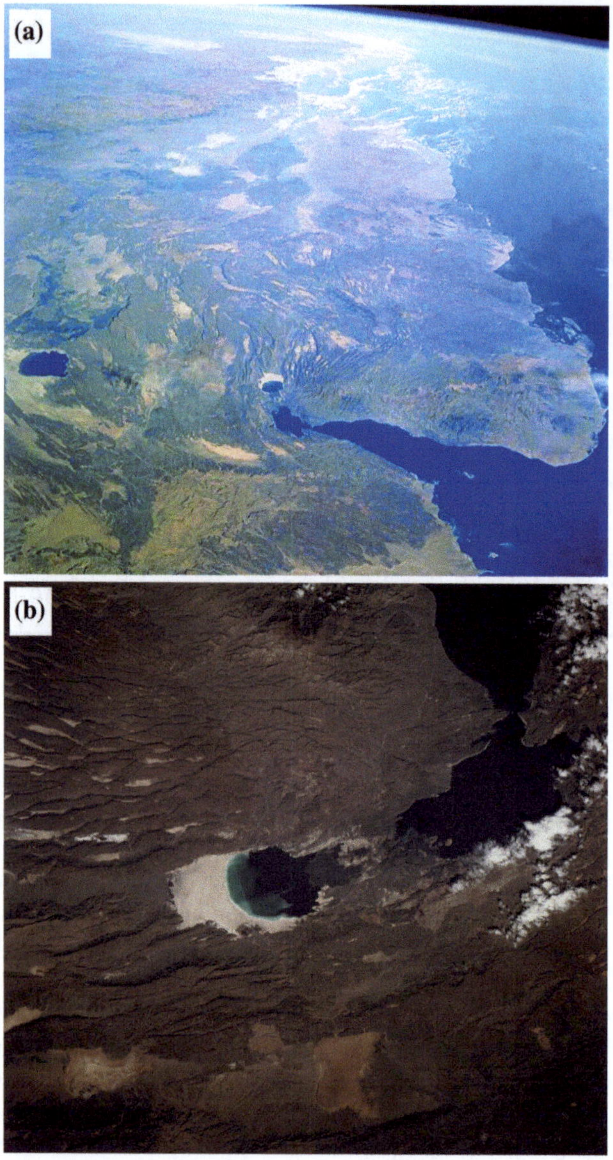

Fig. 4 **a** Satellite image of the Afar region (view toward north), showing the Gulf of Tadjoura and the Ghoubbet–Asal rift, where the Gulf of Aden rift impacts the African continent. **b** Image of the Ghoubbet–Asal rift

deep, except for the floor of a narrow valley that runs almost continuously along its axis, reaching depths of over 2,000 m. The Red Sea connects in the south with the Gulf of Aden–Indian Ocean through the Strait of Bab-al-Mandab, where the seafloor reaches a maximum depth of 120 m. The average strike of the Red Sea from Suez to the Strait of Bab-al-Mandab is N30°W; however, the orientation of the Red Sea axial zone shows a few discontinuities (Crane and Bonatti 1987), with stretches oriented ~5°–10°N, that is, parallel to the Aqaba–Dead Sea transform (Fig. 1), that became active at roughly 14 Ma, causing the deactivation of the Suez rift (Bosworth et al. 2005). The Zabargad Fracture Zone is the most prominent of these discontinuities.

Southern Red Sea Region

Seismicity in the Red Sea axial zone is intense between 17°N and 19°N; south of 15°N it is displaced in northern Afar, while north of 20°N seismicity is nearly absent (Fig. 1). The seismic axial zone of the southern Red Sea from about 15°N to 20°N displays a ~300 km long nearly continuous rift valley, the floor of which reaches ~2,200 m below sea level. The floor of the valley is nearly free of sediments, and is covered by basalt with MORB affinity (Juteau et al. 1983; Essien et al. 1989). A detailed magnetic survey (Roeser 1975) revealed a pattern of well-formed, parallel Vine-Matthews magnetic anomalies (Fig. 11a) that allowed the identification of a set of isochrons. The oldest anomalies are about 5 Ma at around 18°N but are younger toward the tips of the axial trough stretch, hinting at the possibility of axial propagation of the zone of oceanic crust accretion. Spreading rates range between 0.6 and 1.0 cm yr^{-1}. Spreading appears to be asymmetric at around 17°N, with rates about 70 % higher on the western, African Side. According to Girdler and Styles (1974) and to LaBrecque and Zitellini (1985), accretion of oceanic crust and sea floor spreading occurred in the southern Red Sea in two stages, the first taking place already at about 30 Ma. This view, based in part on the presence of magnetic anomalies along the margins of the Red Sea, implies that the entire width of the Red Sea is underlain by oceanic crust. Other authors interpret "marginal" magnetic anomalies as caused by basaltic dykes emplaced in the stretched continental crust, similar to those outcropping in the coastal areas of Saudi Arabia. This alternative view implies a southern Red Sea oceanic crust accretion limited to <5 Ma (Cochran 1983; Bonatti 1985).

Multideeps Red Sea Region

Moving north from about 20°N, the axial zone of the Red Sea is characterized by a number of discontinuous deeps with a complex morphology and depths reaching over 2,000 m. The two largest are Atlantis II Deep and Discovery Deep. They became the focus of much interest in the 1960s because it was found that they contain dense, hot brines as well as hydrothermal metalliferous deposits (Degens and Ross 1969). Some of the brines have temperatures >50 °C. This discovery occurred years before hydrothermal activity

Fig. 5 Open fissures in basalts moving progressively away from the axial neo-volcanic zone in the Ghoubbet–Asal rift. Basalts in **a**, **b** display a "polygonal" surface typical of submarine eruptions. **c** Open fissure; **d** fissure with vertical offset

and metallogenesis were observed along mid-ocean ridges. The presence of dense, hot brines in topographic depressions within the deeps may be related to sea water having circulated within evaporite deposits and having acquired a high salinity. In this, Red Sea hydrothermal fluids differ from those observed along mid-ocean ridges that generally are not highly saline. Deposits from the brines include (Fig. 12) abundant Fe-sulfides, often enriched in Zn and Cu, Fe-hydroxides (goethite), Fe-montorillonite, Mn-oxides and hydroxides, manganosiderite and anhydrite (Bischoff 1969).

Fig. 6 **a** Satellite image of Afar. The Erta Ale, Tat Ali, and Alayata volcanic systems are clearly visible in the inset. **b** Erta Ale ridge. Zone in the vicinity of the Erta Ale, with areas below sea level in cyan. **c** Lava lake on the summit of Erta Ale

Transitional Red Sea Region

Moving north from about 21°N, we enter a region where a number of discrete axial troughs with high amplitude magnetic anomalies are separated by inter-trough zones that lack an axial rift valley and magnetic anomalies and have a thick continuous sediment cover across the axis. The width and length of the trough segments decrease moving north. From south to north, the axial deeps are known as Hadarba, Thetis, Nereus, and Bannock (Fig. 11b). Ligi et al. (2012, this volume) have described the Thetis and Nereus segments. The Thetis deep originated from the coalescence of three sub-basins that become smaller and shallower moving from south to north. Accretion of oceanic crust started at about 2.2 Ma in the southernmost sub-basin where a ~4 km deep magma chamber has been detected, 1.5 Ma in the central sub-basin, and <0.78 Ma in the northern sub-basin, in a pattern suggesting northward propagation of the axial zone of oceanic crust accretion. Initial oceanic crust accretion occurred at ~2 Ma in the center of the Nereus segment (Ligi et al. 2011, 2012). The Bannock Deep lacks a steep axial rift valley but displays a gentle graben-like morphology with a continuous sediment cover, except for a 500 m high relief protruding from the sediment surface that gives rise to a magnetic anomaly. Basaltic dolerite was sampled from this high, probably a basaltic seamount (Bonatti et al. 1984). The Thetis–Nereus intertrough zone lacks a rift valley and magnetic anomalies and shows a thick sediment cover, while the Thetis–Hadarba intertrough zone is narrower but may

Fig. 7 **a**, **b** Fissure basalt flow in central Afar. **c**, **d** Open fissure resulting from dyke intrusion in 2005, southern Afar (from Keir et al. 2009)

expose fragments of continental basement (Ligi et al. 2012, this volume).

The age and degree of evolution of the axial oceanic cells decrease northward, from Hadarba to Tethis to Nereus to Bannock (Fig. 11b). Note that the axial zone deeper than 1,000 m narrows where the axial "oceanic" segments are present (Fig. 11b), suggesting that the rifting process becomes focused in a narrow axial zone as soon as oceanic crust starts to accrete. Basalt chemistry, distribution of magnetic anomalies and data on crustal thickness suggest an initial "active" burst in each cell (Ligi et al. 2011), with a high degree of mantle melting and rapid seafloor spreading, followed by a steady-state "passive" pattern of spreading (Fig. 13).

The pattern of discrete axial segments of oceanic crust accretion has led to the hypothesis that breakup of the continental crust starts with the evolution of initial cells of mantle upwelling and melting, followed by axial propagation that leads to a nearly continuous axial spreading zone, as observed along mid-oceanic ridges (Bonatti 1985). A possible explanation for the development of initial axial cells of mantle upwelling calls for a subaxial hot low-viscosity mantle zone below a denser, colder, high-viscosity layer, a situation that can be treated as a Raleigh–Taylor instability leading to discrete, equidistant upward injections of the hot, low-density low-viscosity mantle. This mantle-driven pattern may be disrupted by pre-existing lithospheric-tectonic anomalies, such as the Zabargad Fracture Zone.

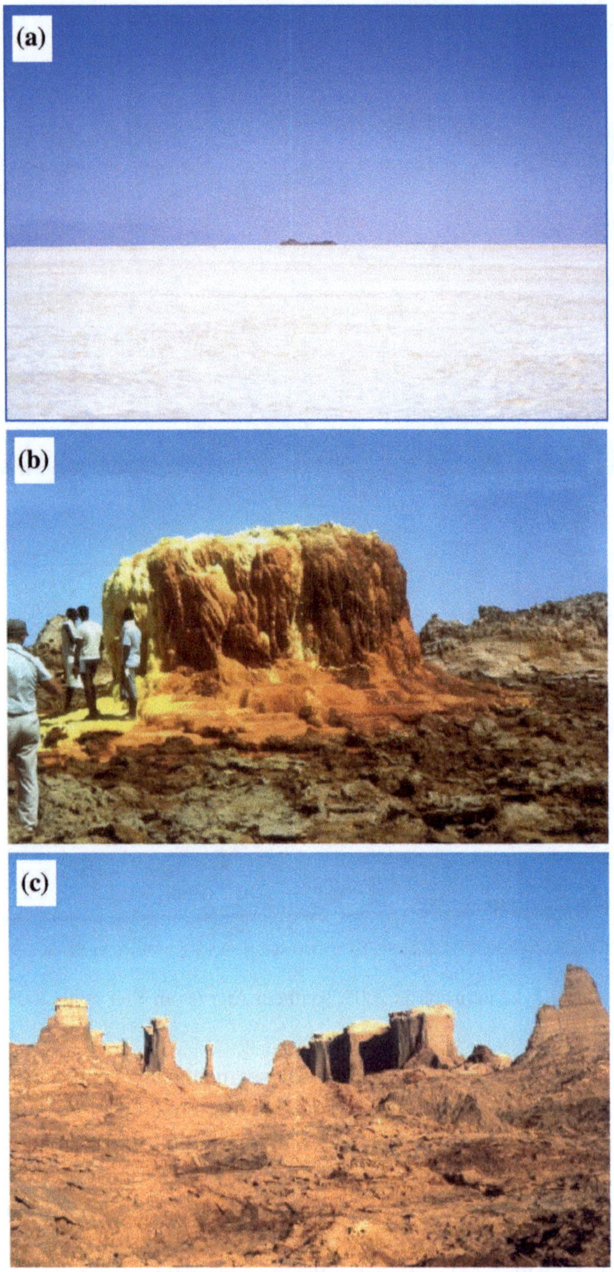

Fig. 8 **a** Salt plain in northern Afar. **b**, **c** Evaporites and hydrothermal mineralizations from the Dallol area, northern Afar

Zabargad Fracture Zone and Northern Red Sea

The orientation of the Red Sea axis, as determined by the axial zone within the 1,000 m isobaths, changes moving N at roughly 24°N from a NNW- to a NNE-direction. It then picks up again its NNW orientation at roughly 26°N (Fig. 1). The NNE-oriented stretch crosses the Red Sea from Zabargad Island at the SSW end to the Mabahiss basin at the NNE end of the fracture zone. NNE-oriented bathymetric and structural elements characterize this stretch, which we call the Zabargad Fracture Zone. The Red Sea axis is displaced laterally by about 100 km along the fracture zone, that can be considered the precursor to a major oceanic transform. The Zabargad F.Z. may act as a "locked zone", as defined by Courtillot (1982), against which the N-propagating axial oceanic rift is impacting (Fig. 11b).

Zabargad Island, located at the SSW end of the Fracture Zone, exposes an uplifted block of sub-Red Sea lithosphere, including three distinct upper mantle peridotite bodies, a gabbro intrusion, and a section of Pan-African continental granitoid gneisses criss-crossed by basaltic dykes, in addition to a probably Lower Cretaceous well stratified sandstone-limestone unit (Fig. 14), an evaporite unit and uplifted Pleistocene reefs (Bonatti et al. 1981, 1986; Bosworth et al. 1996). The dyke-injected continental granitic gneisses probably provide an example of what the Red Sea crust may look like outside the axial oceanic segments (Fig. 14b). The Zabargad mantle peridotites provide an example of "pre-oceanic", rather undepleted mantle bodies (Fig. 14a). The uplift of the island may be due in part to transpression occurring along the SSW side of the fracture zone, while transtension at the NNE-segment may have produced the Mabahiss pull-apart basin, floored by young basalts.

North of the Zabargad FZ we have no clear evidence of an "organized" emplacement of oceanic crust, although a few scattered deeps with basaltic and hydrothermal activity have been identified, such as Shaban Deep, Kebrit Deep, and Oceanographer Deep. Seismic reflection profiles as well as drilling suggest that the northern Red Sea is probably floored by attenuated continental crust dissected by basaltic dykes and intrusions (Cochran 2005; Bosworth et al. 2005), overlain by a several km thick sedimentary sequence ranging from early Tertiary to the present.

Two small islets ("The Brothers") located in the northern Red Sea at around 26°N (Fig. 1) provide interesting hints as to the processes of pre-oceanic rift magmatism. They expose gabbros with elemental and isotopic compositions similar to MORB that crystallized at relatively low pressure (Fig. 13a), suggesting that basaltic melts with oceanic affinity intrude the base of a thinned continental crust before the breakup and the injection at the seafloor of new oceanic crust (Bonatti and Seyler 1987).

Fig. 9 **a** Submarine volcano made by basaltic hyaloclastites, central Afar. **b–d** Flat-top guyot made of stratified basaltic hyaloclastites, central Afar

Magmatism Along the Red Sea

^{8}Na and ^{8}Fe concentrations in the Red Sea axial rift basalts (i.e., Na and Fe contents corrected for the effects of differentiation according to Klein and Langmuir, 1987), imply: (1) south to north decreasing degree of partial melting, from about 18 % at 18°N to less than 10 % at 26°N (Haase et al. 2000), and (2) crossing of the solidus at pressures decreasing from about 2.5 to less than 1.5 GPa going from south to north. The northward decreasing degree of partial melting correlates with a decreasing spreading rate from 17 to 10 mm y^{-1} (Reilinger et al. this volume). Thus, the ranges of magma composition are comparable to those of slow spreading oceanic ridges. Shallowing of the solidus toward the north suggests a cooler mantle beneath the zone of late-stage continental rifting than beneath the zone of active spreading.

Basalt $^{87}Sr/^{86}Sr$ and $^3He/^4He$ ratios decrease and the $^{143}Nd/^{144}Nd$ ratio increases moving along the axis from south to north (Altherr et al. 1988; Moreira et al. 1996), possibly due to a decreasing influence of the Afar plume.

Fig. 10 **a** Volcano made of submarine basaltic hyaloclastites (*yellowish*) capped and mantled by subaerial basalt (*black*). **b** Submarine basaltic volcano capped by marine carbonates (northern Afar)

Moreover, the elemental and isotopic composition even of those basaltic melts emplaced in the attenuated crust of the northern Red Sea through, for instance in the Shaban and Mabahiss Deeps (Haase et al. 2000), do not show contamination by continental lithosphere components. The cooler mantle temperatures in the north are probably due to the fact that relatively shallow mantle from beneath the extending continental lithosphere is melting while relatively hot material from greater depths in the upper mantle is ascending beneath the spreading axis in the south.

Red Sea Stratigraphy

The Red Sea basin contains a thick sedimentary sequence that has been identified by seismic reflection profiling, by drilling for hydrocarbons, by scientific DSDP-ODP drilling and conventional coring. The sequence includes from top-down the following units: (a) Plio–Pleistocene biogenous–terrigenous deposits, that lie above a seismic reflector S, representing the top of an ubiquitous evaporite unit (Ross and Schlee 1973). The thickness of the post-evaporitic deposits reaches more than 100 m; their composition changes with climatic variations that affected sea level, intermittent communication with the Mediterranean and/or the Indian Ocean, biological productivity, and terrigenous input. (b) Evaporite deposition took place throughout the entire Red Sea basin starting at around 14 Ma in the Serravallian and reaching a thickness of thousands of meters (Mitchell et al. 2010; Bosworth this volume). Halite and anhydrite-gypsum were the main mineral phases; they are locally interbedded with sandstones and shales as well as with carbonate platform deposits, indicating intervals of normal marine deposition. The late Miocene was again dominated by evaporite deposition, and in the late Messinian, the Red Sea may have experienced stages of desiccation, similar to basins of the Mediterranean. (c) Pre-evaporitic synrift deposits, widespread throughout the Red Sea basin, include limestones, sandstones, and shales of various thickness, predominantly of shallow-marine origin ranging in age from Neogene to early Miocene. An example of pre-rift deposits is the Lower Cretaceous sandstone/limestone unit exposed on Zabargad Island (Bosworth et al. 1996).

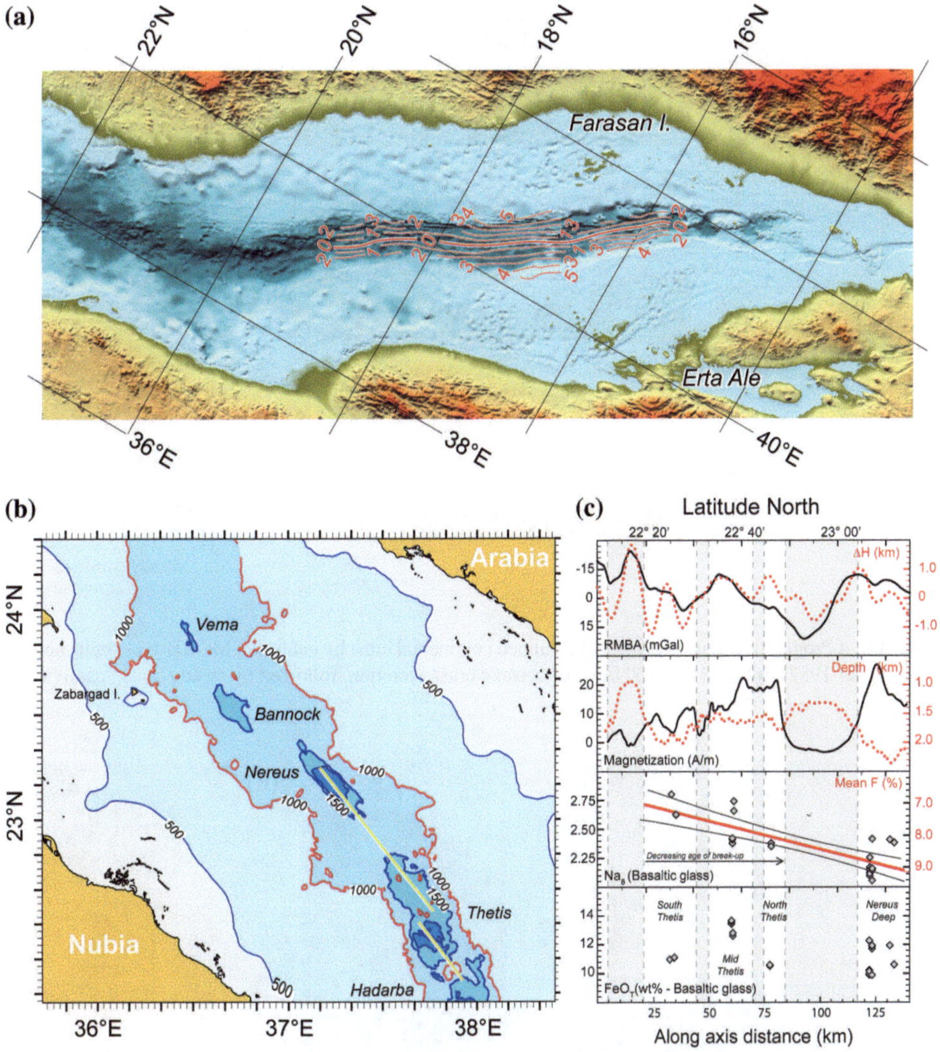

Fig. 11 **a** Axial zone of the southern Red Sea with lineated magnetic anomalies (modified from Cochran 1983). **b** Transitional zone of the Red Sea, with discrete axial rift valley segments separated by intertrough zones. *Note* how the zone limited by the 1,000 m isobath narrows where the axial segments of oceanic crust accretion are present. Contour interval 500 m. *Yellow solid line* indicates along-axis profiles shown in (**c**). **c** Thetis and Nereus along-axis variations of residual mantle Bouguer anomaly (RMBA) and inferred crustal thickness (Δh), rock magnetization and bathymetry, basaltic glasses Na_8 and mean degree of melting, and iron content (modified from Ligi et al. 2012)

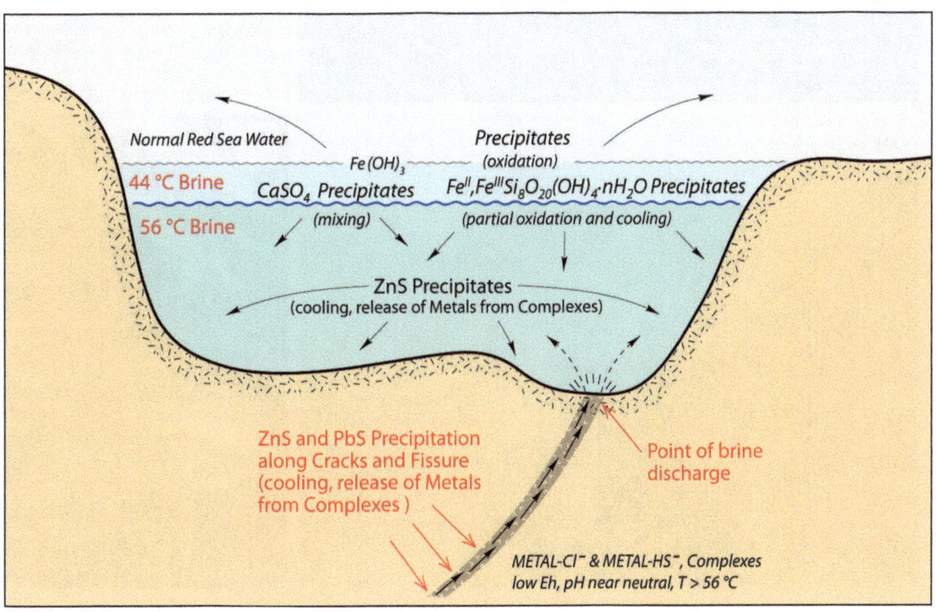

Fig. 12 Hydrothermal activity and metallogenesis at the Atlantis II Deep, central Red Sea (modified from Bischoff 1969)

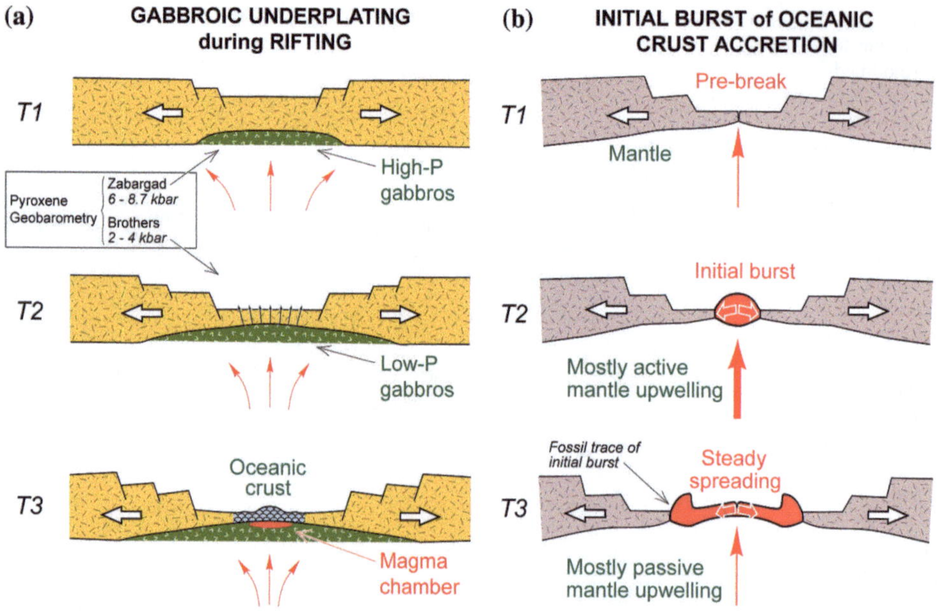

Fig. 13 **a** Progressive underplating of the thinned continental crust by gabbroic, MORB-type melts before initial accretion of oceanic crust (Bonatti and Seyler 1987). **b** Initial "active" burst of oceanic crust accretion, followed by steady-state "passive" sea floor spreading (Ligi et al. 2011)

Fig. 14 Zabargad Island (location in Fig. 11b). **a** A portion of the southern peridotite hill; uplifted reef limestones are also visible. **b** Pan-African granitic gneiss criss-crossed by basaltic dykes. **c** Cretaceous sandstone-limestone stratified deposits. **d** Pan-African granitic gneiss

Conclusions

The Gulf of Aden-Afar-Red Sea region displays a spectacular range of geological situations that illustrate different modes and different stages in the processes of splitting of a continent and of birth of a new ocean. This article provides an introduction to the variety of important problems that can be tackled in this unique region of our Planet. Some of these problems have been addressed more in depth in other chapters of this volume.

Acknowledgments This research was sponsored by the PRIN2012 Programme (Project 20125JKANY_002) and was supported by the Saudi Geological Survey and the Italian Consiglio Nazionale Ricerche. We are grateful to M. Ligi and Najeeb M.A. Rasul for valuable help in putting together this chapter, and to D. Bernoulli and G. Ottonello for helpful and constructive reviews.

References

Altherr R, Henjes-Kunst F, Puchelt H, Baumann A (1988) Volcanic activity in the Red Sea axial trough: evidence for a large mantle diapir? Tectonophysics 150:121–133

Barberi F, Bonatti E, Marinelli G, Varet J (1974) Transverse tectonics during the split of a continent: data from the Afar rift. Tectonophysics 23:17–29

Barberi F, Varet J (1977) Volcanism of Afar—small scale plate tectonics implications. Geol Soc Am Bull 88:1251–1266

Bischoff JL (1969) Red sea geothermal brine deposits. In: Degens ET, Ross DA (eds) Hot Brines and recent heavy metal deposits of the Red Sea. Springer, New York, pp 338–401

Bonatti E (1985) Punctiform initiation of seafloor spreading in the Red Sea during transition from continental to an oceanic rift. Nature 316:33–37

Bonatti E, Colantoni P, Della Vedova B, Taviani M (1984) Geology of the Red Sea transitional region (22°–25°N). Acta Oceanol 7:385–398

Bonatti E, Emiliani C, Ostlund G, Rydell HS (1971) Final desiccation of the Afar rift, Ethiopia. Science 172:468–470

Bonatti E, Fisher DE, Joensuu O, Rydell HS, Beyth M (1972) Iron-manganese-barium deposit from the Northern Afar rift (Ethiopia). Econ Geol 67:717–730

Bonatti E, Hamlyn P, Ottonello G (1981) Upper mantle beneath a young oceanic rift—peridotites from the island of Zabargad (Red-Sea). Geology 9:474–479

Bonatti E, Ottonello G, Hamlyn PR (1986) Peridotites from the island of Zabargad (Red Sea). J Geophys Res 91:599–631

Bonatti E, Seyler M (1987) Crustal underplating and evolution in the Red Sea rift. J Geophys Res 92:12083–12821

Bonatti E, Tazieff H (1970) Exposed guyot from the Afar rift, Ethiopia. Science 168:1087–1089

Bosworth W, Darwish M, Crevello P, Taviani M, Marshak S (1996) Stratigraphic and structural evolution of Zabargad Island (Red Sea, Egypt) since the early cretaceous. In: El Youssef SA (ed) Proceedings of the 3rd international conference on geology of the Arab World, vol 1, pp 161–190

Bosworth W, Huchon P, McClay K (2005) The Red Sea and Gulf of Aden basins. J Afr Earth Sc 43:334–378

Cann JR (1970) New model for the structure of the ocean crust. Nature 226:928–930

C.N.R.-C.N.R.S. (1971) Géologie de l'Afar septentrional (Dankalie). Editions C.N.R.S., 46 pp

Cochran JR (1981) The Gulf of Aden: structure and evolution of a young ocean basin and continental margin. J Geophys Res 86:263–287

Cochran JR (1983) A model for development of the Red Sea. Bull Am Assoc Pet Geol 67:41–69

Cochran JR (2005) Northern Red Sea: nucleation of an oceanic spreading center within a continental rift. Geochem Geophys Geosyst 6:Q03006. doi:10.1029/2004GC000826

Coleman RG (1974) Geologic background of the Red Sea. In: Whitmarsh RB, Weser OE, Ross DA et al (eds) Initial reports of the Deep Sea drilling project. vol 23, U.S. Government Printing Office, Washington, DC, pp 813–819

Crane K, Bonatti E (1987) The role of fracture zones during early Red Sea rifting: structural analysis using space shuttle radar and LANDSAT imagery. J Geol Soc London 144:407–420

Courtillot V (1982) Propagating rifts and continental breakup. Tectonics 1:239–250

d'Acremont E, Leroy S, Maia M, Patriat P, Beslier MO, Bellahsen N, Fournier M, Gente P (2006) Structure and evolution of the eastern Gulf of Aden: insights from magnetic and gravity data. Geophys J Int 165:786–803

Degens ET, Ross DA (eds) (1969) Hot Brines and recent heavy metal deposits in the Red Sea. Springer, New York, 600 pp

Ebinger CJ, Keir D, Ayele A, Calais E, Wright TJ, Belachew M, Hammond JO, Campbell E, Buck WR (2008) Capturing magma intrusion and faulting processes during continental rupture: seismicity of the Dabbahu (Afar) rift. Geophys J Int 174:1138–1152

Essien J-P, Juteau T, Joron J-J, Dupre B, Humler E, Al'Mukhamedov A (1989) Petrology and geochemistry of basalts from the Red Sea axial rift at 18°N. J Petrol 30:791–839

Ferguson DJ, Barnie TD, Pyle DM, Oppenheimer C, Yirgu G, Lewi E, Kidane T, Carn S, Hamling IG (2010) Recent rift-related volcanism in Afar, Ethiopia. Earth Planet Sci Lett 292:409–418

Ferguson DJ, Calvert AT, Pyle DM, Blundy JD, Yirgu G, Wright TJ (2013a) Constraining timescales of focused magmatic accretion and extension in the Afar crust using lava geochronology. Nature Commun 4:1416. http://dx.doi.org/10.1038/ncomms2410

Ferguson DJ, Maclennan J, Bastow ID, Pyle DM, Jones SM, Keir D, Blundy J, Plank T, Yirgu G (2013b) Melting during late-stage rifting in Afar is hot and deep. Nature 499:70–73. http://dx.doi.org/10.1038/nature12292

Girdler RW, Styles P (1974) Two stage seafloor spreading. Nature 247:7–11

Girdler RW, Styles P (1978) Seafloor spreading in the western Gulf of Aden. Nature 271:615–617

Haase KM, Muhe R, Stoffers P (2000) Magmatism during extension of the lithosphere: geochemical constraints from lavas of the Shaban Deep, northern Red Sea. Chem Geol 166:225–239

Harrison GGA, Bonatti E, Stieltjes L (1975) Tectonism of axial valleys in spreading centers: data from the Afar rift. In: Pilger A, Rosler A (eds) Afar depression in Ethiopia. Stuttgart, Germany, pp 178–198

Hébert H, Deplus C, Huchon P, Khanbari K, Audin L (2001) Lithospheric structure of a nascent spreading ridge inferred from gravity data: the western Gulf of Aden. J Geophys Res 106:26345–26363

Juteau T, Essien J-P, Monin AS, Zonenshain LP, Sorokhtin OG, Matveenkov VV, Almukhamedov AI (1983) Structure et petrologie du rift axial de la mer Rouge vers 18° Nord: resultats de la campagne sovietique de plongees avec submersible (1980). Bull des Centres de Rech Explor-Prod Elf-Aquitaine 7:217–231

Keir D, Hamling I, Ayele A, Calais E, Ebinger C, Wright TJ, Jacques E, Mohammed K, Hammond JOS, Belachew M, Baker E, Rowland JV, Lewi E, Bennati L (2009) Evidence for focused magmatic accretion at segment centers from lateral dike injections captured beneath the Red Sea rift in Afar. Geology 37:59–62

Klein EM, Langmuir CH (1987) Global correlations of ocean ridge basalt chemistry with axial depth and crustal thickness. J Geophys Res 92:8089–8115

LaBrecque JL, Zitellini N (1985) Continuous sea-floor spreading in the Red Sea: an alternative interpretation of magnetic anomaly patterns. Am Assoc Pet Geol Bull 69:513–524

Leroy S, Gente P, Fournier M, d'Acremont E, Patriat P, Beslier M-O, Bellahsen N, Maia M, Blais A, Perrot J, Al-Kathiri A, Merkouriev S, Fleury J-M, Ruellan P-Y, Lepvrier C, Huchon P (2004) From rifting to spreading in the eastern Gulf of Aden: a geophysical survey of a young oceanic basin from margin to margin. Terra Nova 16:185–192

Ligi M, Bortoluzzi G (1989) PLOTMAP: Geophysical and geological applications of good standard quality cartographic software. Comput Geosci 15:519–585

Ligi M, Bonatti E, Bortoluzzi G, Cipriani A, Cocchi L, Caratori Tontini F, Carminati E, Ottolini L, Schettino A (2012) Birth of an ocean in the Red Sea: initial pangs. Geochem Geophys Geosyst 13:Q08009. doi:10.1029/2012GC004155

Ligi M, Bonatti E, Caratori Tontini F, Cipriani A, Cocchi L, Schettino A, Bortoluzzi G, Ferrante V, Khalil SM, Mitchell NC, Rasul N (2011) Initial Burst of Oceanic Crust accretion in the Red Sea due to edge-driven mantle convection. Geology 39:1019–1022. doi:10.1130/G32243.1

Manighetti I, Tapponnier P, Courtillot V, Gruszow S, Gillot PY (1997) Propagation of rifting along the Arabia-Somalia plate boundary: the gulfs of Aden and Tadjoura. J Geophys Res 102:2681–2710

Manighetti I, Tapponnier P, Gillot PY, Jacques E, Courtillot V, Armijo R, Ruegg JC, King G (1998) Propagation of rifting along the Arabia-Somalia plate boundary: into Afar. J Geophys Res 103:4947–4974

Martini M (1969) La geochimica del lago Giulietti (Ethiopia). Rendiconti della Societa Italiana di Mineralogia e Petrologia XXV:1–16

Mitchell NC, Ligi M, Ferrante V, Bonatti E, Rutter E (2010) Submarine salt flows in the central Red Sea. Geol Soc Am Bull 122:701–713

Moreira M, Valbracht PJ, Staudacher T, Allegre CJ (1996) Rare gas systematics in Red Sea ridge basalts. Geophys Res Lett 23:2453–2456

Pagli C, Wright TJ, Ebinger CJ, Yun S, Cann JR, Barnie T, Ayele A (2012) Shallow axial magma chamber at the slow-spreading Erta Ale Ridge. Nat Geosci 5:284–288. doi:10.1038/ngeo1414

Pinzuti P, Humler E, Manighetti I, Gaudemer Y (2013) Petrological constraints on melt generation beneath the Asal Rift (Djibouti) using quaternary basalts. Geochem Geophys Geosyst 14:2932–2953. doi:10.1002/ggge.20187

Rawlinson GMA (1858) The history of Herodotus, book II. John Murray, London 616 pp

Roeser HA (1975) A detailed magnetic survey of the southern Red Sea. Geol Jahrb D13:131–153

Ross DA, Schlee J (1973) Shallow structure and geologic development of the southern Red Sea. Geol Soc Am Bull 84:3827–3848

Schilling J-G, Kingsley RH, Hanan BB, McCully BL (1992) Nd–Sr–Pb isotopic variations along the Gulf of Aden: evidence for Afar mantle plume-continental lithosphere interaction. J Geophys Res 97:10927–10966

Tamsett D, Searle R (1990) Structure of the Alula-Fartak Fracture Zone, Gulf of Aden. J Geophys Res 95:1239–1254

Wegener A (1929) Die Entstehung der Kontinente und Ozeane, 4th edn. Friedrich vieweg and sohn, Braunschweig

Wessel P, Smith WHF (1995) New version of the Generic Mapping Tools released: EOS, Trans Am Geophys Union 76:329

Wright TJ, Ebinger CJ, Biggs J, Ayele A, Yirgo G, Keir D, Stork A (2006) Magma-maintained rift segmentation at continental rupture in the 2005 Afar diking episode. Nature 442:291–294

Geological Evolution of the Red Sea: Historical Background, Review, and Synthesis

William Bosworth

Abstract

The Red Sea is part of an extensive rift system that includes from south to north the oceanic Sheba Ridge, the Gulf of Aden, the Afar region, the Red Sea, the Gulf of Aqaba, the Gulf of Suez, and the Cairo basalt province. Historical interest in this area has stemmed from many causes with diverse objectives, but it is best known as a potential model for how continental lithosphere first ruptures and then evolves to oceanic spreading, a key segment of the Wilson cycle and plate tectonics. Abundant and complementary datasets, from outcrop geology, geochronologic studies, refraction and reflection seismic surveys, gravity and magnetic surveys, to geodesy, have facilitated these studies. Magnetically striped oceanic crust is present in the Gulf of Aden and southern Red Sea, active magma systems are observed onshore in the Afar, highly extended continental or mixed crust submerged beneath several kilometers of seawater is present in the northern Red Sea, and a continental rift is undergoing uplift and exposure in the Gulf of Suez. The greater Red Sea rift system therefore provides insights into all phases of rift-to-drift histories. Many questions remain about the subsurface structure of the Red Sea and the forces that led to its creation. However, the timing of events—both in an absolute sense and relative to each other—is becoming increasingly well constrained. Six main steps may be recognized: (1) plume-related basaltic trap volcanism began in Ethiopia, NE Sudan (Derudeb), and SW Yemen at ~31 Ma, followed by rhyolitic volcanism at ~30 Ma. Volcanism thereafter spread northward to Harrats Sirat, Hadan, Ishara-Khirsat, and Ar Rahat in western Saudi Arabia. This early magmatism occurred without significant extension or at least none that has yet been demonstrated. It is often suggested that this "Afar" plume triggered the onset of Aden–Red Sea rifting, or in some models, it was the main driving force. (2) Starting between ~29.9 and 28.7 Ma, marine syn-tectonic sediments were deposited on continental crust in the central Gulf of Aden. Therefore, Early Oligocene rifting is established to the east of Afar. Whether rifting propagated from the vicinity of the Sheba Ridge toward Afar, or the opposite, or essentially appeared synchronously throughout the Gulf of Aden is not yet known. (3) By ~27.5–23.8 Ma, a small rift basin was forming in the Eritrean Red Sea. At approximately the same time (~25 Ma), extension and rifting commenced within Afar itself. The birth of the Red Sea as a rift basin is therefore a Late Oligocene event. (4) At ~24–23 Ma, a new phase of volcanism, principally basaltic dikes but also layered gabbro and granophyre bodies, appeared nearly synchronously throughout the entire Red Sea, from Afar and Yemen to northern Egypt. The result was that the Red Sea rift briefly linked two very active volcanic centers covering 15,000–25,000 km^2 in the north and >600,000 km^2 in the south. The presence of the "mini-plume" in

W. Bosworth (✉)
Apache Egypt Companies, 11 Street 281, New Maadi
Cairo, Egypt
e-mail: bill.bosworth@apachecorp.com

N.M.A. Rasul and I.C.F. Stewart (eds.), *The Red Sea*,
Springer Earth System Sciences, DOI 10.1007/978-3-662-45201-1_3,

northern Egypt may have played a role somewhat analogous to Afar vis-à-vis the triggering of the dike event. The 24–23 Ma magmatism was accompanied by strong rift-normal extension and deposition of syn-tectonic sediments, mostly of marine and marginal marine affinity. The area of extension in the north was very broad, on the order of 1,000 km, and much narrower in the south, about 200 km or less. Throughout the Red Sea, the principal phase of rift shoulder uplift and rapid syn-rift subsidence followed shortly thereafter. Synchronous with the appearance of extension throughout the entire Red Sea, relative convergence between Africa and Eurasia slowed by about 50 %. (5) At ~14–12 Ma, a transform boundary cut through Sinai and the Levant continental margin, linking the northern Red Sea with the Bitlis–Zagros convergence zone. This corresponded with collision of Arabia and Eurasia, which resulted in a new plate geometry with different boundary forces. Red Sea extension changed from rift normal (N60°E) to highly oblique and parallel to the Aqaba–Levant transform (N15°E). Extension across the Gulf of Suez decreased by about a factor of 10, and convergence between Africa and Eurasia again dropped by about 50 %. In the Afar region, Red Sea extension shifted from offshore Eritrea to west of the Danakil horst, and activity began in the northern Ethiopian rift. (6) These early events or phases all took place within continental lithosphere and formed a continental rift system 4,000 km in length. When the lithosphere was sufficiently thinned, an organized oceanic spreading center was established and the rift-to-drift transition started. Oceanic spreading initiated first on the Sheba Ridge east of the Alula-Fartaq fracture zone at ~19–18 Ma. After stalling at this fracture zone, the ridge probably propagated west into the central Gulf of Aden by ~16 Ma. This matches the observed termination of syn-tectonic deposition along the onshore Aden margins at approximately the same time. At ~10 Ma, the Sheba Ridge rapidly propagated west over 400 km from the central Gulf of Aden to the Shukra al Sheik discontinuity. Oceanic spreading followed in the south-central Red Sea at ~5 Ma. This spreading center was initially not connected to the spreading center of the Gulf of Aden. By ~3 to 2 Ma, oceanic spreading moved west of the Shukra al Sheik discontinuity, and the entire Gulf of Aden was an oceanic rift. During the last ~1 My, the southern Red Sea plate boundary linked to the Aden spreading center through the Gulf of Zula, Danakil Depression, and Gulf of Tadjoura. Presently, the Red Sea spreading center may be propagating toward the northern Red Sea to link with the Aqaba–Levant transform. However, important differences appear to exist between the southern and northern Red Sea basins, both in terms of the nature of the pre- to syn-rift lithospheric properties and the response to plate separation. If as favored here no oceanic spreading is present in the northern Red Sea, then it is a magma-poor hyperextended basin with β factor >4 that is evolving in many ways like the west Iberia margin. It is probable that the ultimate geometries of the northern and southern Red Sea passive margins will be very different. The Red Sea provides an outstanding area in which to study the rift-to-drift transition of continental disruption, but it is unlikely to be a precise analogue for all passive continental margin histories.

Introduction

The Red Sea has long been recognized as one component of a continent scale rift system that reaches from the Dead Sea to Mozambique. This lead to the popularization of the term "Afro-Arabian rift system" by geologists mapping its different segments (Baker 1970; Khan 1975; Kazmin 1977). It was suggested that the Red Sea, and similarly the Gulf of Aden, were oceanic rifts at divergent plate boundaries, with a triple junction located at Afar (Fig. 1; Gass 1970; McKenzie et al. 1970; Girdler and Darracott 1972; Burke and Dewey 1973; Le Pichon and Francheteau 1978). Application of plate tectonics has been a central theme in studies of the Red Sea since then.

In 1972, the R/V *Glomar Challenger* conducted Leg 23B in the Red Sea and drilled six wells (sites 225–230) in and nearby the axial trough (Whitmarsh et al. 1974). These wells confirmed that basalts and fluids with mantle-derived lead isotopes have been emplaced into the sediment column along the Red Sea axis and supported the hypothesis that the Red Sea was an evolving oceanic rift. As part of this project, Coleman (1974) compiled a geologic map of the entire Red Sea basin, volcanism of the rift margins, and axial hydrothermal deeps. Based on geological observations, Coleman

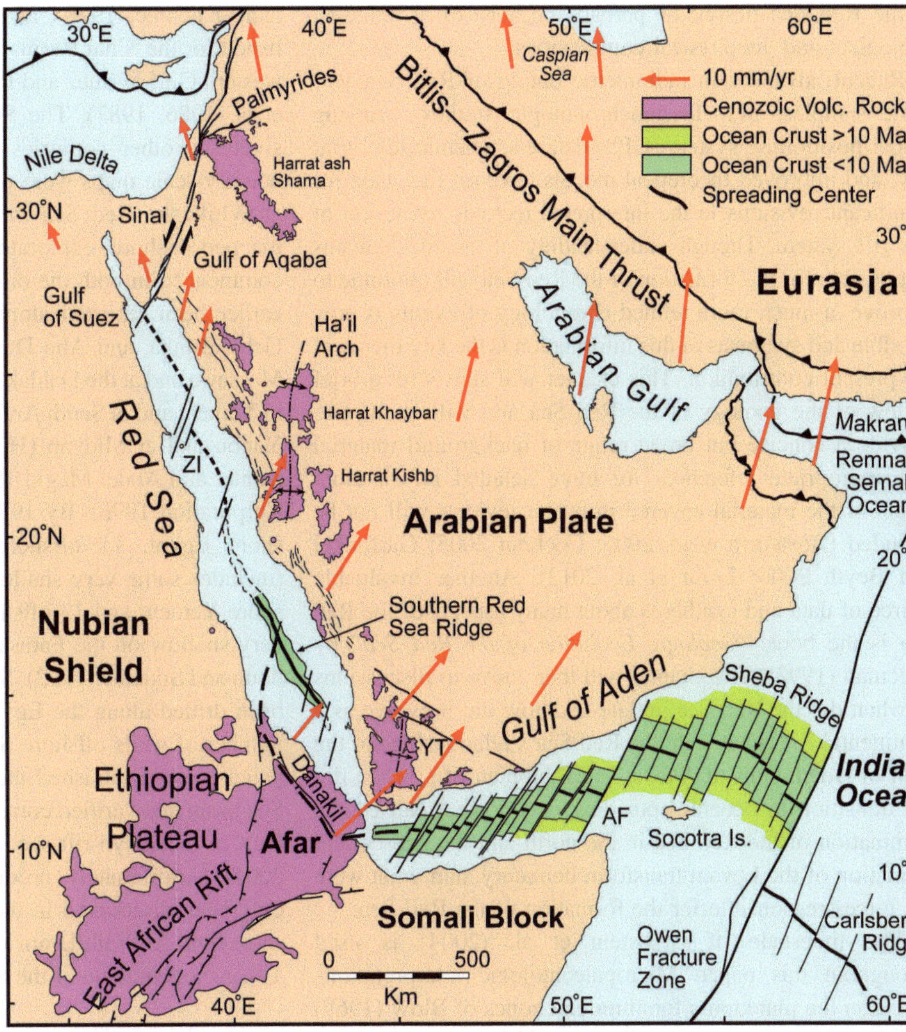

Fig. 1 Tectonic features of the greater Red Sea rift system, including the northern Ethiopian (East African) rift, Afar, and the Gulfs of Aden, Aqaba and Suez. After Bosworth et al. (2005). *AF* Alula-Fartak fracture zone, *YT* Yemen traps, *ZI* Zabargad Island. *Red arrows* are GPS velocities in a Eurasia-fixed reference frame from ArRajehi et al. (2010). Albers conical equal area projection

argued that oceanic crust was restricted to the axial trough in the southern Red Sea and that the rest of the margin was continental crust intruded by tholeiitic gabbros and basalt dike swarms. Similar interpretations had been developed from geophysical datasets (Girdler 1958; Drake and Girdler 1964; Lowell and Genik 1972). This contrasts with the McKenzie et al. (1970) model that envisioned a pre-rift coast-to-coast restoration with the entire Red Sea underlain by oceanic crust. Numerous intermediate models had also been proposed (Girdler 1966, 1970; Girdler and Darracott 1972).

The most convincing evidence of oceanic rifting was the recognition of striped magnetic anomalies along the southern Red Sea axis (Phillips 1970; Girdler and Styles 1974; Röser 1975; Searle and Ross 1975; Hall et al. 1977; Cochran 1983). These authors did not agree on the age of the anomalies when compared to world-wide magnetostratigraphy, but they did concur that the striping represents the signature of true oceanic spreading. Many other types of geophysical and geochemical data supported the oceanic rift scenario, and more details of these will be presented later in this chapter.

Just as it is firmly thought that the southern Red Sea is presently an oceanic rift, there is clear consensus that the northern end of the rift—the Gulf of Suez—is purely continental in character (Steckler 1985; Jarrige et al. 1986; Courtillot et al. 1987; Girdler and Southren 1987; Joffe and Garfunkel 1987; Moretti and Chénet 1987). The intervening areas—the central and northern Red Sea—have not met with similar agreement. In outcrop and in offshore exploratory wells, there are clear similarities with the Gulf of Suez, both in terms of stratigraphy and underlying basement lithologies (Colemen 1974; Tewfik and Ayyad 1984; Barakat and Miller 1984; Beydoun 1989; Beydoun and Sikander 1992; Bosworth 1993). Along the axial trough, however, there are some similarities with the southern Red Sea axis and some important differences (discussed below). Suffice it to say that the northern and central Red Sea are transitional between aborted continental rifting in the north and well-defined oceanic rifting in the south, though these areas may not represent a simple chronological progression now captured at different points in its evolution. Hence, the broad interest in the Red Sea despite-or perhaps because of-its inherent geological and geophysical complexities.

Recent advances in radiometric dating of Red Sea volcanic samples, new thermochronologic studies, growing global positioning system (GPS) datasets, seismic tomography, and improved theoretical models have set the stage for significant revisions in the interpreted tectonic evolution of this rift system. Though understanding of the mechanisms responsible for the formation of the Red Sea will continue to improve, a much more refined chronology of events is now possible and synthesis of this information is the key theme of the present contribution. This chapter will start with a brief review of the geology of the Red Sea and will attempt to provide a concise but broad range of background material and appropriate references for more detailed information. Much of the material covered in recent reviews will not be included (Bosworth et al. 2005; Cochran 2005; Garfunkel and Beyth 2006; Lazar et al. 2012). Another invaluable source of data and syntheses about many aspects of the Red Sea is the book "*Geologic Evolution of the Red Sea*" by Coleman (1993). The chapter will then move to discussions of when did the Red Sea initiate and how did it evolve as a continental rift, what were the Red Sea's relationships to the Gulf of Aden, Gulf of Aqaba, and Ethiopian rifts, when did the transition to oceanic spreading occur, what caused the termination of the Red Sea in the north and the subsequent formation of the Levant transform boundary, and what were the forces responsible for the formation of the Red Sea.

The timescale of Gradstein et al. (2004) is used throughout this paper. Micropaleontologic interpretations based on the planktonic foraminiferal zones of Blow (1969) and the calcareous nannofossil zones of Martini (1971) have been adjusted to this timescale.

Geologic and Tectonic Setting

Background

Modern broad-scale geologic mapping of the margins of the Red Sea began in earnest as a cooperative program between the US Geological Survey and the Kingdom of Saudi Arabia that resulted in coverage and integration of data from all of the Arabian Peninsula (USGS-Arabian American Oil Company 1963; Geukens 1966; Greenwood and Bleackley 1967; Bender 1975; Brown et al. 1989). Brown (1972) produced a tectonic interpretation of these data that included the Red Sea coastal plain and bathymetry of the axial trough. Ethiopia and Djibouti also received extensive attention due to interest in the Afar flood basalts/plume, the Danakil Horst, and the Strait of Bab-al-Mandab (Brinckmann and Kursten 1969; Clin and Pouchan 1970; Barberi et al. 1971; Kazmin 1973). Further mapping was completed in Yemen (Grolier and Overstreet 1978), and similar compilations were published for the Sinai Peninsula (Eyal et al. 1980) and later the western Gulf of Suez and Red Sea margin of Egypt (Klitzsch et al. 1986, 1987). The Sudan was not mapped as extensively as other countries bordering the Red Sea, though country-scale maps were produced (Vail 1975, 1978).

While the Red Sea margins were being systematically mapped onshore, exploratory drilling for hydrocarbons had commenced in both the onshore and offshore. Oil seeps had earlier been reported along the coastline at Gebel el Zeit, Gebel Tanka, and Abu Durba in the Gulf of Suez, north of Massawa and at the Dahlak Islands in the Sudan, near Zeidiye in Yemen, and in Saudi Arabia at the Farasan Islands, north of Yanbu and at Midyan (Hume et al. 1920; Beydoun 1989; Bunter and Abdel Magid 1989; Egyptian General Petroleum Corporation 1996). By 1987, 10 wells had been drilled offshore Egypt, 11 offshore Sudan, 18 offshore Ethiopia (includes some very shallow on the Dahlak Islands), 5 offshore Yemen, and 13 offshore Saudi Arabia (includes some very shallow on the Farasan Islands) (Beydoun 1989; Beydoun and Sikander 1992). Since then, 4 deepwater wells have been drilled along the Egyptian margin and an undisclosed number of wells offshore Saudi Arabia. These drilling campaigns have established the general stratigraphy of the Red Sea basin and verified correlations with the Gulf of Suez and Gulf of Aden syn-rift sections (reviewed in Bosworth et al. 2005). Unfortunately, productive hydrocarbon systems have only been discovered in the Midyan (both onshore and offshore) and Al Wajh/Umm Luj basins of Saudi Arabia and the Tokar–Suakin delta of the Sudan.

Africa's Other Rifts and the Red Sea

The evolution of the Red Sea should be considered in the context of the entire African plate. This can be discussed from several perspectives. Africa is a remnant of Gondwana, which began to break up during the Late Carboniferous starting with "Karoo" rifts in southern and east Africa (Groenewald et al. 1991; Bumby and Guiraud 2005). Diachronous Early Permian rifting occurred along the northern African margin from Morocco to Egypt, resulting in the initiation of the Neotethyan seaway (Stampfli and Borel 2002). By the Middle Triassic, seafloor spreading was probably occurring in the eastern Mediterranean basin (Robertson et al. 1996; Stampfli et al. 2001). Rifting began along the Central Atlantic margins in the Late Triassic (Davison 2005). During the Jurassic, extension spreads to the inboard basins of the Neotethyan margin (Guiraud and Bosworth 1999), the Blue Nile rift in Sudan (Wycisk et al. 1990; Bosworth 1992), the Marib-Shabwa basin in Yemen (Bott et al. 1992), the Nogal rift of Somalia (Granath 2001), the Lamu embayment of Kenya (Reeves et al. 1987) and continued in the Karoo Lugh-Mandera basin of

Kenya–Somalia–Ethiopia (Ali Kassim et al. 2002). Seafloor spreading initiated along the Central Atlantic margins in the Early Jurassic (~180 Ma; Klitgord and Schouten 1986; Davison 2005) and in the Somali basin in the Middle to Late Jurassic (pre-157 Ma; Rabinowitz et al. 1983). By the Late Jurassic, faulting was active in the Benue trough in Nigeria (Guiraud 1993) and was present throughout the South Atlantic basin by the Early Cretaceous (Rabinowitz and LaBreque 1979). Seafloor spreading initiated in the South Atlantic in the Neocomian to Aptian, progressing from south to north (Uchupi 1989) and in the equatorial Atlantic in the Late Aptian (~115 Ma; Basile et al. 2005). By this time, the present geometry of the African plate was essentially established. Through the Cretaceous and locally into the Cenozoic, however, continental extension continued in many African basins, particularly within the Benue trough, the Termit basin of Niger, the Dobaa basin of Chad, the south Sudan rifts, the Anza trough, the Sirte basin of Libya, the Western Desert of Egypt, and the basins of Yemen and Somalia (Fairhead 1988; Genik 1992; Guiraud and Maurin 1992; Bosworth 1992, 1994; Janssen et al. 1995; Bumby and Guiraud 2005; Guiraud et al. 2005). In summary, the breakup/dispersal of the Gondwana segment of Pangaea lasted several hundred million years. The Red Sea–Gulf of Aden rift system, with the East African rifts, is part of this ongoing process (Bumby and Guiraud 2005; Bosworth et al. 2005).

The driving mechanism for the breakup of a supercontinent such as Pangaea has been attributed to thermal blanketing of the underlying mantle and related plume or hotspot activity impinging on the base of the lithosphere (Anderson 1982). Slab retreat and slab pull on surrounding plate boundaries have also been suggested as associated principal driving forces (Davies and Richards 1992; Lithgow-Bertelloni and Richards 1998; Collins 2003; Reilinger and McClusky 2011). Both plumes and nearby subduction zone processes probably played key roles in the evolution of the Red Sea (reviewed in Bosworth et al. 2005). Plumes weaken continental lithosphere and therefore may strongly influence the location of lithospheric failure, while plate boundary forces can drive extension.

This brief overview illustrates the dominant tectonic activity in Africa over the past 300 My—rifting and formation of passive continental margins. However, it has been suggested that this general theme has been modified during the most recent 30 My of its progression (Burke 1996). Early in the past century, geologists recognized that Africa possesses a distinctive basin and swell geomorphology (Krenkel 1922, 1957; Argand 1924; Holmes 1965). Following Krenkel, Burke and Wilson (1972) suggested that this topography was generated by mantle processes, while the African plate was essentially at rest with respect to the underlying pattern of mantle circulation. "Stationary Africa" arose ~30 Ma (Burke 1996) and has been attributed to intensified collision between Africa and Eurasia (Bailey 1992, 1993) and/or the impingement and eruption of the Afar plume, which possibly in conjunction with the Principe and other plumes acted to pin the base of the lithosphere to the upper mantle (Burke 1996). Burke suggested that the arresting of Africa plate motion led to a surge in intraplate volcanism and to renewed rifting, that is, formation of the Red Sea–Gulf of Aden–East Africa rift system (Afro-Arabia rift system). From about the end of the Cretaceous up until the development of stationary Africa, intraplate extension was absent or very localized across Africa, a gap in the rifting history of ~35 Ma. This gap was a period of major peneplanation (Burke 1996; Burke and Gunnell 2008) including the development of extensive laterites and paleosols across the Arabian–Nubian shield (Coleman 1993). This African surface was the backdrop on which the Red Sea was formed.

Geomorphology, Axial Deeps, and Rift Shoulder Uplift

The most prominent geomorphologic features of the Red Sea are the extreme deeps along the basin axis and the high elevations along most of its rifted shoulder (Fig. 2). Mapping compiled by Laughton (1970) revealed a continuous main axial trough that extends from just south of Ras Mohammed in southernmost Sinai to the vicinity of the Zubayr Islands offshore Yemen. The main trough is generally 1000 m or greater in depth and is accentuated by isolated deeps that exceed 2000 m (Degens and Ross 1969; Monin et al. 1981, 1982). The axial deeps contain hot brine pools and basaltic cones, some of which have been studied by detailed side beam sonar and direct sampling (Pautot 1983; Pautot et al. 1984; Bicknell et al. 1986) and are the subject of other chapters in this book. The deeps are intimately related to the evolution of oceanic spreading centers in the Red Sea (Cochran 1983; Bonatti 1985; Martinez and Cochran 1988; Cochran and Martinez 1988, reviewed in Cochran 2005; Cochran and Karner 2007).

Cochran (2005) compiled all the available bathymetric data for the Northern Red Sea (Fig. 2b) and integrated this with gravity and magnetic data to produce a new model for the nucleation of an oceanic spreading center at the axial depression. An interesting detail of his bathymetric map (present in older maps but not as refined) is the fact that the axial trough strikes 5°–10° more to the northwest than the overall trend of the marine basin. This may reflect an adjustment to extension parallel to the Gulf of Aqaba transform boundary (Bosworth et al. 2005; Lazar et al. 2012, discussed below) and is an expected consequence of the two Eulerian pole Red Sea opening models proposed by many authors (e.g., Joffe and Garfunkel 1987 and references therein).

The marine shelf and coastal plains of the Red Sea are variable in width and were formed by a complex interplay of tectonic, sedimentary, and biotic activity. The present climate

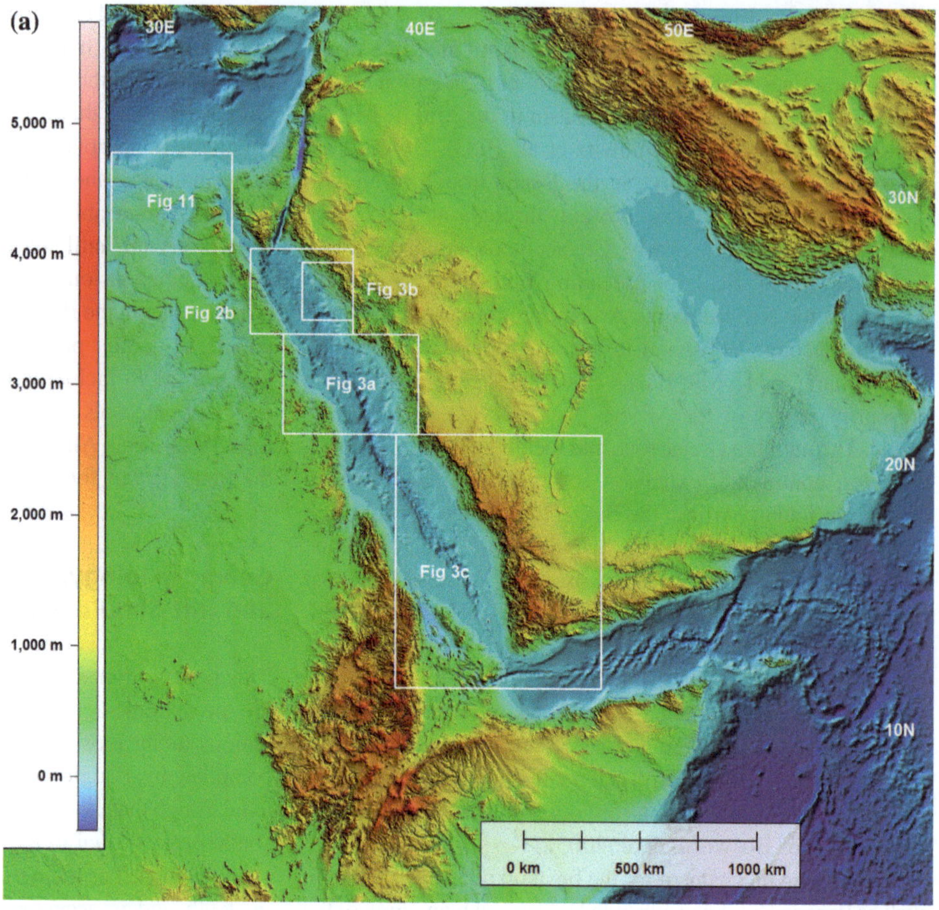

Fig. 2 Topography and bathymetry of the Red Sea area (geographic projection). **a** From GLOBE Task Team et al. (1999) (onshore) and Smith and Sandwell (1997) (offshore; Seasat radar altimetry derived bathymetry). Compared with Fig. 1 for positions of plate boundaries, Red Sea volcanic terranes and place names. Locations of other figures are given. **b** Detailed bathymetry of northern Red Sea from Cochran (2005). Onshore is same dataset as in (**a**). Topographic–bathymetric profile runs from 26°15′N, 33°30′E to 28°N, 37°E through the Conrad deep

of the entire Red Sea region is arid, with the exception of localized mountainous areas where increased rainfall due to orographic forcing results in increased vegetation (Zahran and Willis 2009). These areas are sometimes referred to as "mist oases" such as at Mt. Elba, Egypt (1,435 m; Kassas and Zahran 1971). No permanent rivers enter into the basin, though intermittent rivers such as the Barka (or Baraka) in the Red Sea Hills of Sudan (Fig. 3a) exist (Beydoun 1989). The Barka forms the Tokar delta at the coastline, and during the wetter pluvial stages of the Pleistocene, this probably was a constantly flowing river. Similarly, many dry wadis that now reach the Red Sea and deliver little sedimentation would have been more significant input points in the past. This is an important concept in the hydrocarbon exploration strategies employed in both the Gulf of Suez and the Red Sea (Richardson and Arthur 1988; Lambiase and Bosworth 1995). Offshore from Tokar several exploration wells have been drilled, and significant (though presently noncommercial) quantities of gas were discovered in a very thick sedimentary package that has been referred to as the Suakin delta.

The southern coastlines of both the Arabian and African Red Sea margins are curvilinear and paired and reflect the initial rift geometry of the basin. North of about 24° latitude however on both sides of the basin, the shorelines are linear (Figs. 2b, 3b), and as discussed below, this does not reflect the original shape of the rift. It has been suggested that the straight coasts may be the result of young basement-involved faulting (Bosworth 1994; Bosworth and Burke 2005). The youngest faults actually observed in reflection seismic data offshore from these areas generally detach within the Middle and Late Miocene halite units (Mougenot and Al-Shakhis 1999; Bosworth and Burke 2005). Gravity-driven detached faults would probably produce a scalloped coastline, and there is not any active seismicity near the coast to support the basement-fault model. The straight northern coasts remain somewhat enigmatic.

Inland from coastal plains, or in many areas immediately adjacent to the coastline, high elevations are encountered along most of the Red Sea. From Taif (Saudi Arabia) to Taizz (Yemen) along the Arabian margin, the high terrain is bounded by an unbroken erosional escarpment with the highest peaks exceeding 3,000 m (Fig. 3a, c; Spohner and Oleman 1986; Bohannon 1986; Coleman 1993). In Ethiopia, the escarpment marks the eastern edge of the Ethiopian plateau, with elevations in excess of 2,000 m (Fig. 2a). The geomorphology in Afar is made more complex by the presence of the Danakil horst near the coast and the Afar depression itself (Figs. 1, 2a, 3c; Barberi et al. 1972). Along

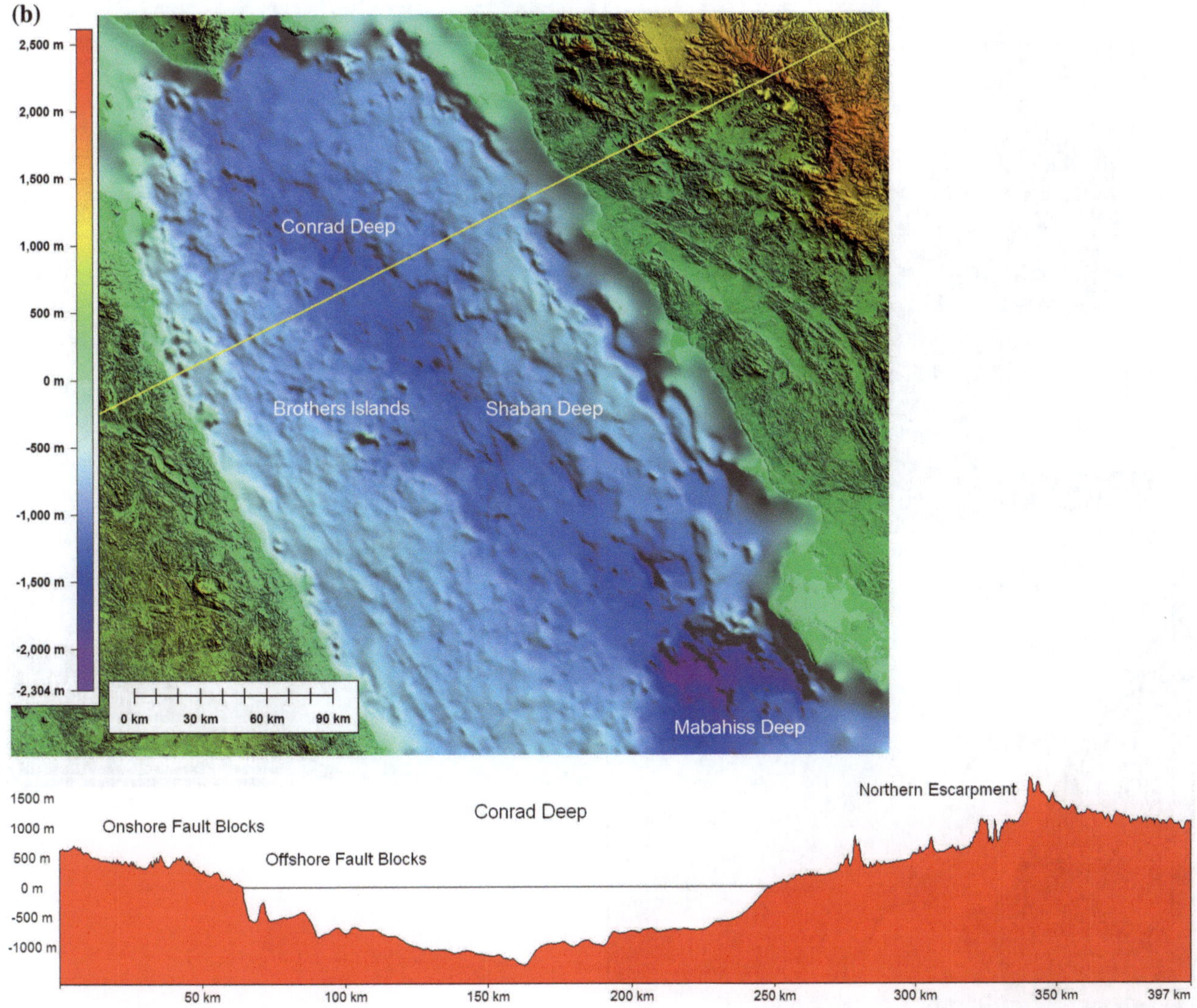

Fig. 2 (continued)

the Sudan margin, the elevation in the Red Sea Hills is more subdued but still typically greater than 1,000 m. This magnitude of relief continues north to Egypt and along the western margin of the Gulf of Suez, with isolated basement peaks reaching 1,400 to more than 2,000 m (Gebel Gharib 1,757 m; Gebel Shaayib Al-Banat 2,187 m). The elevation on Sinai is greater, with Mt. Sinai itself 2,285 m and nearby Mt. St. Catherine 2,629 m (Figs. 2a, 4).

The timing and causes of the uplift of the margins of the Red Sea and Gulf of Suez have long been debated. Gass (1970) and later other workers suggested that the region of Afar was uplifted prior to rifting during the Oligocene or even earlier. Apatite fission track cooling dates were first obtained for the basement complex of Sinai (Kohn and Eyal 1981) and indicated that an important period of denudation occurred between ~27 and 20 Ma. However, most of the dates clustered at 22–20 Ma. Data from the western margin of the Gulf were similarly centered at 22 ± 1 Ma (Omar et al. 1989). Along the Saudi Arabian southern Red Sea margin, Bohannon (1986) suggested that uplift was coeval with faulting at about 25–23 Ma, with a total of ~3.5 km of unroofing. Later interpretations by Bohannon et al. (1989) envisioned initial erosion and uplift at 20 Ma, but with at least 2.5 km of the total occurring after 13.8 Ma. Apatite (U-Th)/He thermochronometry from exposed footwall blocks in the central Saudi Arabian margin indicated rapid exhumation at ~22 Ma and a second distinct pulse about 8 Ma later (Szymanski 2013).

For the Yemeni margin, exhumation was identified at 17–16 Ma (Menzies et al. 1992, 1997), but on the conjugate margin in Eritrea, Abbate et al. (2002) found a broad range of fission track ages (400–10 Ma). Modeling did suggest, however, a major crustal cooling event driven by denudation at 20 Ma. Ghebreab et al. (2002) also found cooling ages along the Eritrean margin north of Danakil clustering between 23 and 17 Ma.

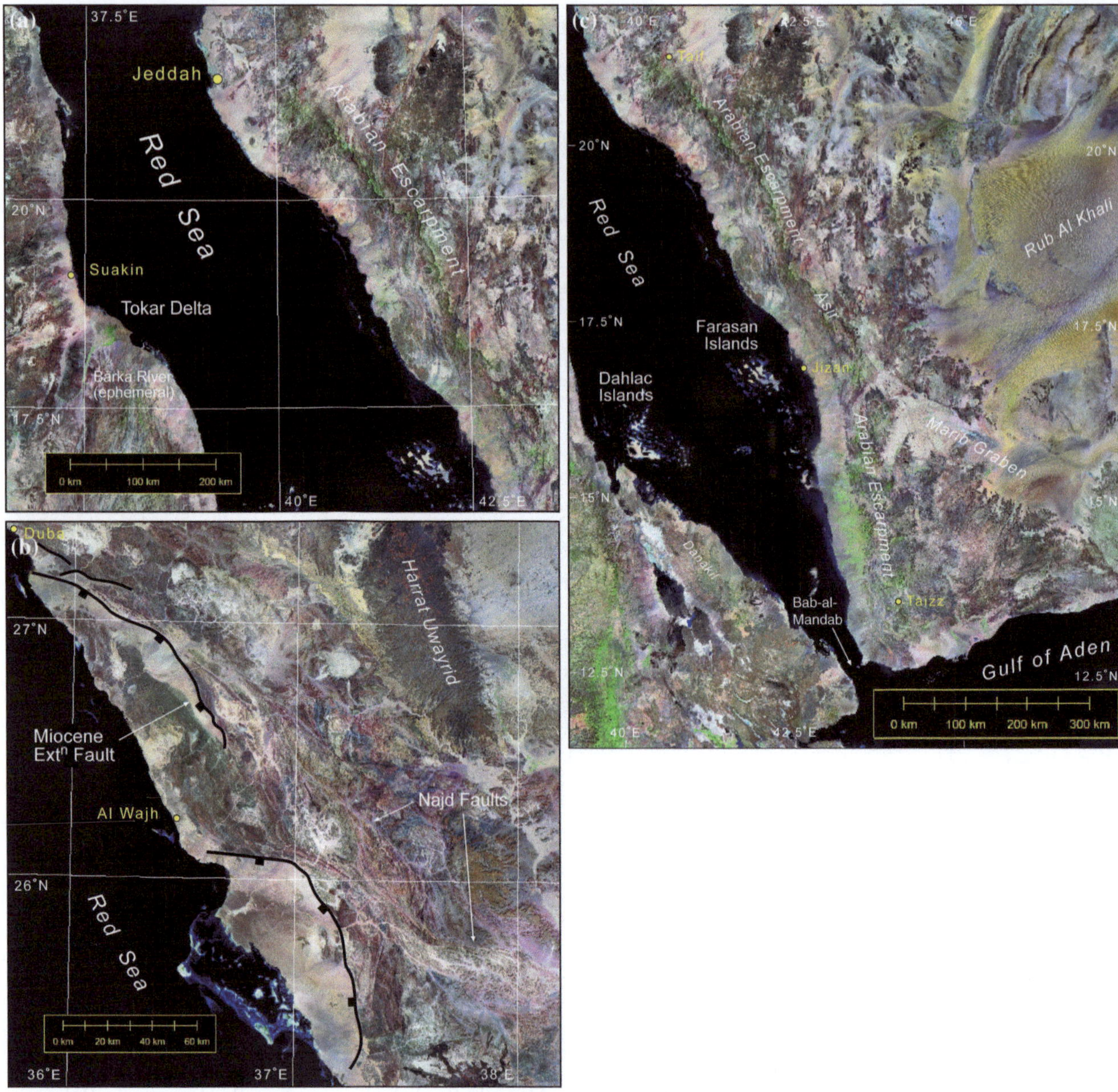

Fig. 3 Landsat 7 mosaics of the Red Sea margins (transverse Mercator projections): **a** The Arabian escarpment south of Jeddah, Saudi Arabia, and the Tokar delta at the mouth of the ephemeral Barka River south of Suakin, Sudan. **b** Miocene extensional faults near Duba and Al Wajh, Saudi Arabia that reactivated parts of the Neoproterozoic Najd fault system. **c** The Arabian escarpment from Taif, Saudi Arabia to Taizz, Yemen. Imagery is from the NASA Stennis Space Center GeoCover project (MDA Federal 2004)

Taken in their entirety, the fission track and (U-Th)/He data suggest that the rift flanks of the Red Sea began denudation at about 24–23 Ma, at least locally, and by about 22–20 Ma, fairly continuous rift shoulders were present. Some fission track data have been produced that suggest an even earlier unroofing event, at ~34 Ma—in the Late Eocene (Steckler and Omar 1994; Omar and Steckler 1995). These data come from the western Gulf of Suez and Egyptian Red Sea margins. Bosworth and McClay (2001) suggested that this phase of uplift, if significant, was more likely related to the Late Eocene Syrian arc compressional phase that is well documented in these same areas (Guiraud and Bosworth 1999).

Large-Scale Basin Geometry

Utilizing wells drilled in the offshore Egyptian Red Sea margin, Tewfik and Ayyad (1984) and Barakat and Miller (1984) established that the Miocene stratigraphy of the

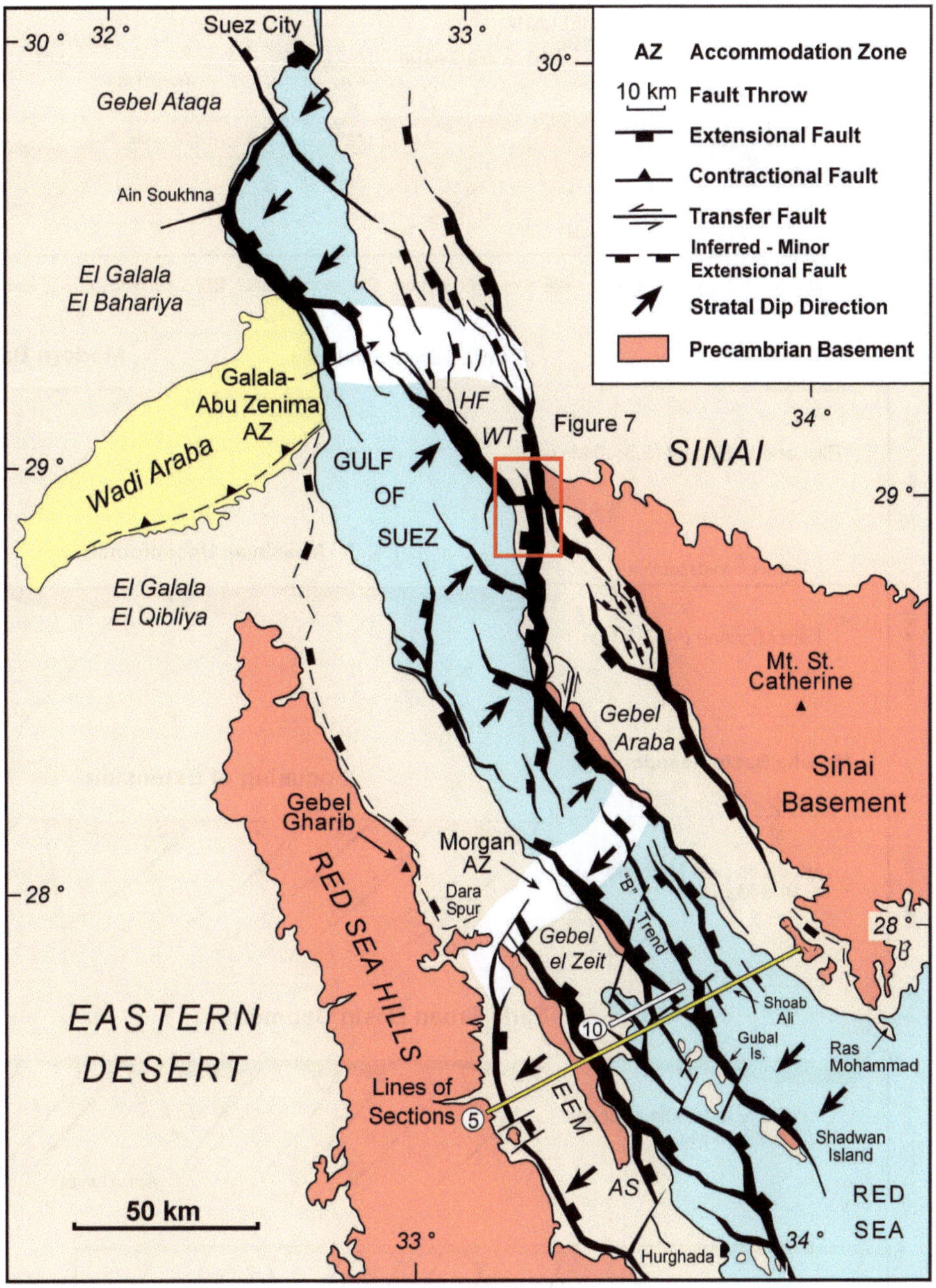

Fig. 4 Structure of the Gulf of Suez continental rift basin thought to be representative of the early structure of the Red Sea (after Khalil 1998; Bosworth and McClay 2001). *AS* Abu Shaar el Qibli, *EEM* Esh el Mellaha basement range and SW-dipping basin, *HF* Hammam Faraun fault block

northern Red Sea was remarkably similar to that of the Gulf of Suez. This confirmed the idea that the heavily explored—there are now nearly 4,000 wells drilled—and well-exposed Gulf of Suez could be utilized as an analogue for the early structural and sedimentologic history of the northern Red Sea.

On a regional scale, the Gulf of Suez consists of three sub-basins, each ~100 km in length and ~50–90 km in width (Fig. 4). Internally, each of these sub-basins consists of nested, rotated fault blocks (Fig. 5). Structural complexity and degree of stratal rotation increase systematically toward the south (Colleta et al. 1988; Patton et al. 1994). Moustafa (1976) analyzed available dipmeter and other subsurface data and recognized that within each sub-basin, the regional dip of beds is consistently in one direction, and hence, the overall geometry is that of a large-scale half graben broken up into a series of smaller half grabens. In the southern sub-basin, dip is generally to the southwest; in the central sub-basin, to the northeast; and in the northern sub-basin, again to the southwest. Hence, the "polarity" of the basins flips back and forth along the rift axis in what Moustafa referred to as dip domains. Similar large-scale reversals in structural dip, or major offsets of sub-basins with similar dip, have been identified in most continental rifts. The boundaries between these sub-basins have been named accommodation

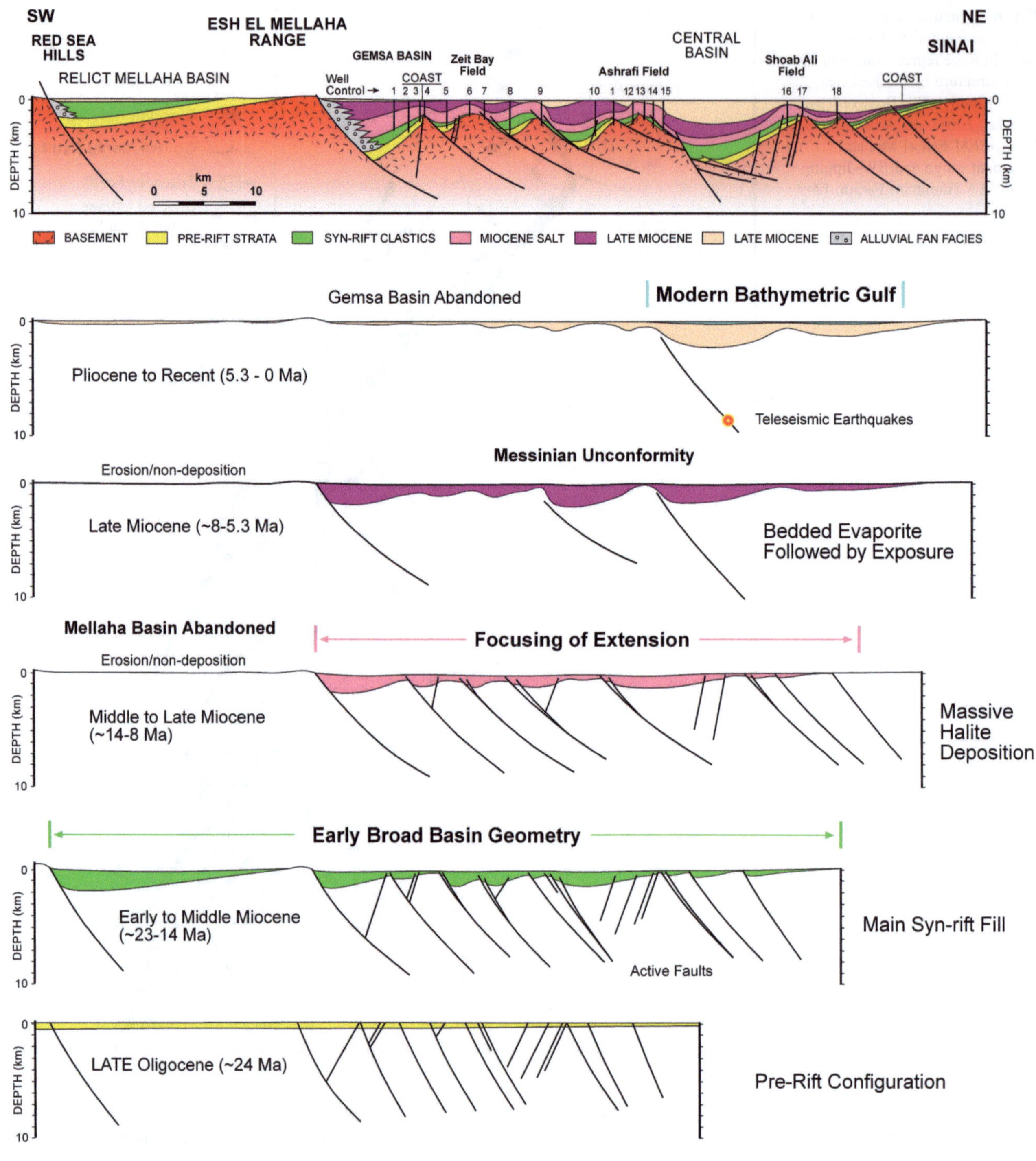

Fig. 5 Structural cross section of the southern Gulf of Suez and sequential views of sediment accumulation and sites of active faulting (after Bosworth 1994, 1995). Location is shown in Fig. 4

zones (Bosworth 1985; Rosendahl et al. 1986) or transfer zones (Morley et al. 1990; Moustafa 1997). In my original discussion of these common features, it was specifically noted that I was referring to crustal-scale sub-basin boundaries; there are smaller structures in most rifts that are better described as transfer faults (Gibbs 1984) or relay ramps (Larsen 1988; Peacock and Sanderson 1991).

The alternating half-graben geometry theme is not restricted to the Gulf of Suez. Along the Red Sea margin of Egypt, another polarity reversal has been recognized in outcrop in the vicinity of Quseir (Fig. 6; Jarrige et al. 1990; Younes and McClay 2002; Khalil and McClay 2009). North of the Quseir accommodation zone, structural dip is to the southwest and represents a regional continuation of the

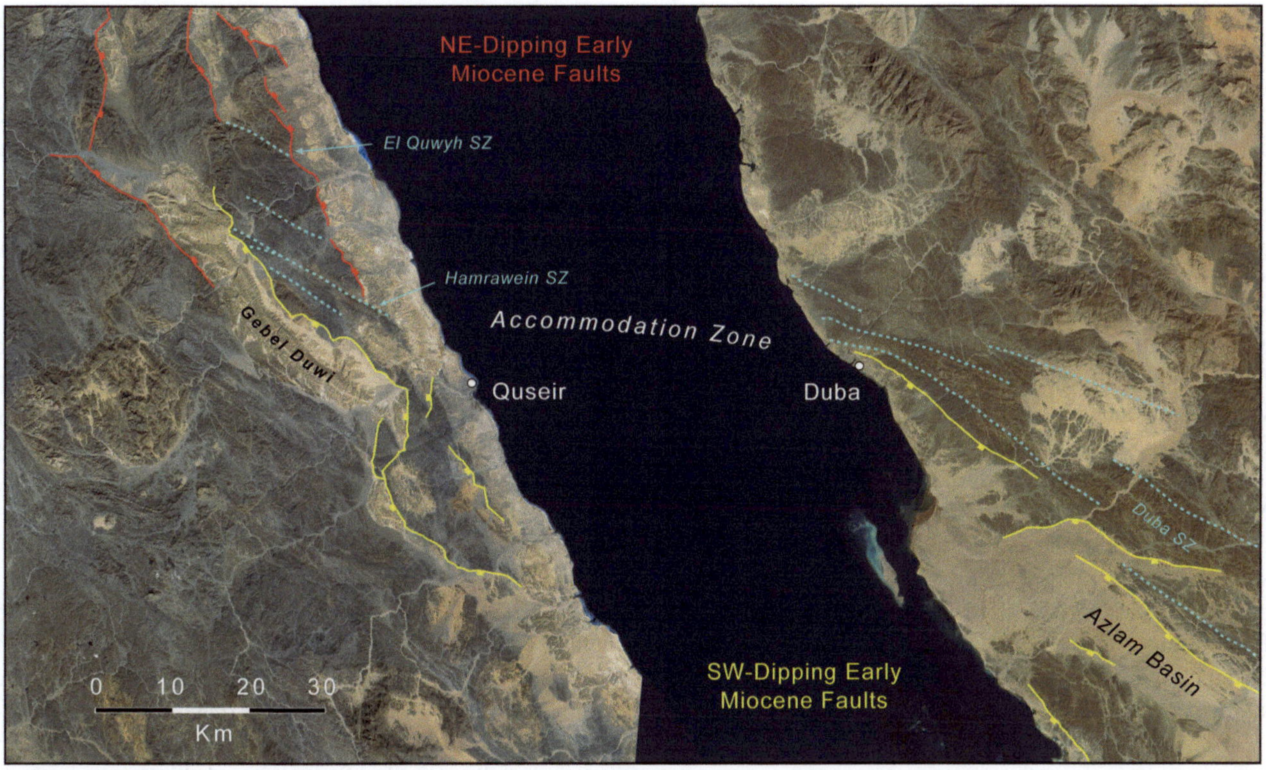

Fig. 6 Principal structural features of the Quseir–Duba accommodation zone after restoring the Egyptian and Saudi Arabian margins to a pre-rift configuration at circa 23 Ma [restoration from Bosworth and Burke (2005); Africa is held stationary in this view with present-day north to the top]. Late Neoproterozoic Najd shear zones are shown as *blue dashed lines* [Egyptian terminology is from Khalil and McClay (2009)]. Miocene extensional faults are shown in *red* or *yellow* depending on direction of dip. Faults are plotted on SPOT imagery on the Egyptian margin and QuickBird imagery on the Saudi margin courtesy of Google Earth

southern Gulf of Suez sub-basin. At Gebel Duwi, the structural dip changes to the northeast. On the Saudi margin, the same structural change occurs south of Duba. After restoring the Red Sea to an Early Miocene configuration (see below), the Gebel Duwi and Duba accommodation zones are linked and pass through the area of the Brothers Islands (Bosworth 1994; Bosworth and Burke 2005). Fantozzi and Sgavetti (1998) recognized the presence of accommodation zones in outcrops along the paired margins of the Gulf of Aden, so this can reasonably be interpreted to have been a general attribute of the entire Red Sea–Gulf of Aden continental rift system.

At the scale of individual fault blocks, the Gulf of Suez and northern Red Sea margins display a great variety of sizes and styles of structuring (Angelier 1985; Jarrige et al. 1986; Colletta et al. 1988; Moretti and Colletta 1988; Perry and Schamel 1990; Bosworth 1995). The largest fault blocks are generally positioned on the outboard margins of the basin, as at Esh el Mellaha in the Gulf of Suez which has a length of 80 km and a width of 25 km (Fig. 4). The syn-rift stratigraphy present on these large blocks is typically only that produced during the earliest phase of extension, and then, their bounding faults ceased moving and the sub-basins were abandoned (Fig. 5). These “relict basins” (Bosworth 1994) are present along both sides of the Gulf of Suez and the northern Red Sea (Szymanski 2013). Stratal rotation in the relict basins is typically only 10–20°. More interior to the rift, fault block dimensions are smaller, the amount of rotation typically systematically increases toward the rift axis, and the movement on faults continued longer into the syn-rift history. Stratal dip along the axis of the southern Gulf of Suez locally exceeds 50° (Bosworth 1995).

In outcrop, vertical fault geometry is generally difficult to constrain—exposures of fault planes are simply too limited, except in rare cases of extreme topography. Both interpretations of basement-involved listric, detached faulting and bookshelf or domino-style faulting have been presented for many parts of the Red Sea and Gulf of Suez (Davison et al. 1994; Geoffroy et al. 1998; Perry and Schamel 1990; McClay et al. 1998). In some oil fields in the southern Gulf of Suez, basement faults have been penetrated by multiple wellbores. Some of these fault planes display abrupt changes in dip with depth, and others are most simply interpreted as listric in profile (Bosworth 1995; Bosworth et al. 2012). Correctly interpreting fault geometry is critical to estimates of horizontal extension and also for proper placement of exploration and development wells.

Along-strike fault geometry is much better documented in most exposed rift settings, and this is certainly the case in the Gulf of Suez and Red Sea. Surface traces of many of the larger extensional faults show a characteristic zigzag pattern, particularly where the footwall block is crystalline basement (Fig. 6; Jarrige et al. 1986; McClay et al. 1998). In some cases, subsurface data (wells, seismic) suggest that this is partly erosion of fault line scarps along pre-existing basement fractures and faults. However, many of the angular changes in strike are real and represent intersection of the predominantly NW–SE to NNW–SSE striking extensional faults with hard-linkage transfer faults/cross-faults of a variety of orientations. In the Gulf of Suez, the largest cross-faults are ~NNE–SSW and are important both on Sinai and in the Eastern Desert (Bosworth 1995; Bosworth et al. 1998; McClay et al. 1998). NE–SW striking cross-faults, perpendicular to the extensional faults, are often present in subsurface interpretations but are generally less important in outcrop. In the northern Red Sea, Neoproterozoic WNW–ESE striking Najd shear zones play the most significant role in modifying Miocene fault geometry (Figs. 3b, 6; Younes and McClay 2002; Khalil and McClay 2009).

Stratigraphy

The age and distribution of the pre-Red Sea stratigraphy of the Arabian Peninsula and northeast Africa have been extensively documented from outcrop studies and in the subsurface of the Gulf of Suez (Beydoun 1978; Hadley and Schimdt 1980; Klitzsch 1990; Schandelmeier and Reynolds 1997; Issawi et al. 1999; Ziegler 2001, reviewed in Bosworth et al. 2005; Guiraud et al. 2005). However, none of the offshore wells drilled in the Red Sea has penetrated any definitive Paleozoic or Mesozoic section (except immediately at the Sudanese coastline; Bunter and Abdel Magid 1989). Only at Zabargad Island (Fig. 1) is there exposed ~200 m of marine strata, dated as Early Cretaceous in age (Bosworth et al. 1996). As the richest source rocks in the Gulf of Suez are pre-rift limestones, the lack of pre-rift Red Sea sedimentary rocks has always been considered a major exploration risk in this basin (Beydoun 1989; Beydoun and Sikander 1992).

The thickness of the pre-Red Sea stratigraphic section increases both toward the far north and south (see Bosworth et al. 2005 their Fig.9). In Egypt, this is due to the large wedge of sedimentary rock developed south of Paleo- and Neotethys; in Eritrea, it is related especially to a thick Jurassic section associated with the opening of the Indian Ocean. In both areas, the pre-rift stratigraphy reaches about 2,500 m in total.

The literature on the syn-rift stratigraphy of the Red Sea is also voluminous. Unlike the pre-rift strata, the offshore Red Sea has provided a wealth of information from the few wellbores that have been drilled, particularly with regard to the age of the stratigraphic units. This has been reviewed by Tewfik and Ayyad (1984), Barakat and Miller (1984), Miller and Barakat (1988), Bunter and Abdel Magid (1989), Hughes and Beydoun (1992), Coleman (1993), Hughes and Filatoff (1995), Hughes et al. (1999), and Hughes and Johnson (2005). The closely related stratigraphy of the Gulf of Suez is summarized in Evans (1988), Richardson and Arthur (1988), Hughes et al. (1992), Patton et al. (1994), Wescott et al. (1997), McClay et al. (1998), Plaziat et al. (1998), and Bosworth and McClay (2001).

The base of the syn-rift section includes heterogeneous and laterally discontinuous beds of sandstone and conglomerate in most parts of the rift system (Fig. 7). The basal beds locally contain clasts of basalt or interdigitations of basalt flows or pyroclastics (e.g., Hadley et al. 1982; Schmidt et al. 1983; Sellwood and Netherwood 1984). The depositional environments include lacustrine and fluvial to marginal marine. In offshore Eritrea, this section has been dated as late Chattian, ~27.5–23.0 Ma (Hughes et al. 1991). Further north along the Saudi and Sudanese margins and into the Gulf of Suez, the oldest strata that are definitively syn-rift are Aquitanian in age, ~23.0–20.4 Ma (data reviewed in Bosworth et al. 2005). The top of this rift initiation package often contains shallow marine limestone and thin evaporite beds (Saoudi and Khalil 1986; Evans 1988; Bosworth et al. 1998).

The transition into the main phase of syn-rift sedimentation is diachronous between some individual fault blocks but regionally occurs at the base of the Burdigalian at ~20.4 Ma. This corresponds to when the fission track data indicate the presence of a well-established and rapidly rising rift shoulder and supports a thermomechanical linkage between extension-driven subsidence and flank uplift (Steckler 1985; Joffe and Garfunkel 1987; Steckler et al. 1988). Subsidence rates increased dramatically, with up to 1,500 m of open marine Globigerina-bearing shale and marl and turbiditic sandstone deposited in axial sub-basins (Fig. 9a). In the Gulf of Suez, the Burdigalian marls often contain total organic carbon (TOC) of 1.5–2.2 % and constitute an important hydrocarbon source rock (Alsharhan 2003). This may also be the case in the Red Sea, but the distribution of appropriate facies is not well constrained (Beydoun 1989). TOC measured in wells along the Egyptian Red Sea margin is 0.67–1.14 % (Barakat and Miller 1984) and would not be considered a significant source rock by most petroleum geologists.

The Burdigalian Red Sea and Gulf of Suez sections also contain thick sandstone facies that constitute important reservoir objectives. This is the case for the two largest known fields, Morgan and Belayim in the central Gulf (Egyptian General Petroleum Corporation 1996). The depositional environments of the sandstones are submarine fan and channel complexes in the basin axes that are thought to have been sourced from structurally controlled point sources

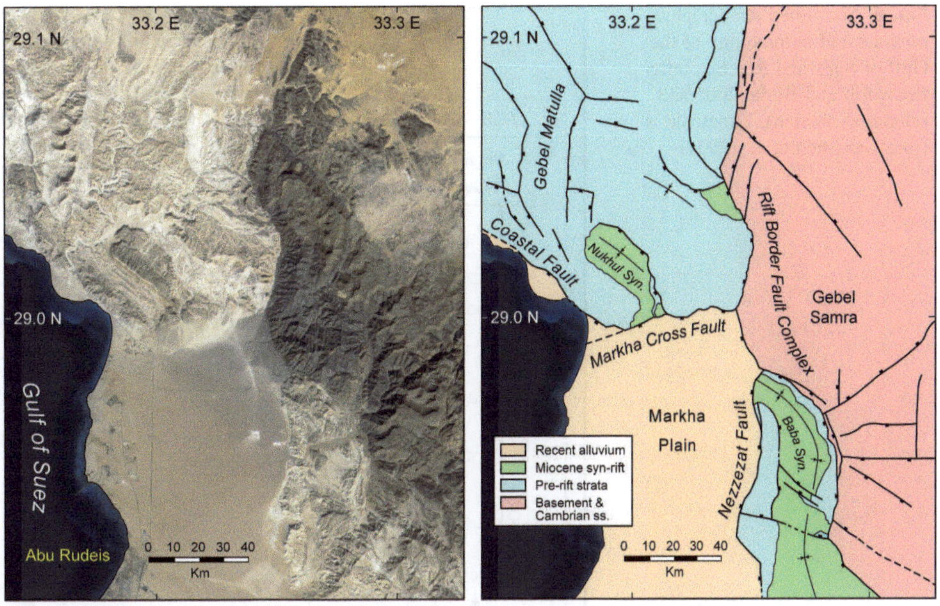

Fig. 7 High-resolution QuickBird satellite image of part of the border fault complex of the central Gulf of Suez (*image acquired and processed for Apache Corporation by Spatial Energy*). Outcrop geology is simplified from Khalil (1998) and Bosworth et al. (2012). Location is shown in Fig. 4. *Syn.* syncline. Transverse Mercator projection

along the basin margins (Lambiase and Bosworth 1995; Wescott et al. 1997; Khalil and McClay 2009). In more proximal fault blocks, the Burdigalian section often contains conglomerates and sandstones deposited as alluvial fans and fan deltas (Fig. 8b; Sharp et al. 2000; Young et al. 2000).

The main syn-rift fill continued into the early Middle Miocene (Langhian) and a second brief period of evaporite deposition occurred throughout most of the rift system (Fig. 8). Deep marine conditions then reappeared but were short lived as nearly the entire basin began depositing evaporites at about 14 Ma (early Serravallian). The principal evaporite minerals are halite and anhydrite/gypsum, and these are interbedded with conglomerate, sandstone, and shale depending on the position within the various sub-basins. Another brief period of normal marine deposition returned in the late Serravallian during which extensive carbonate platforms were developed (Fig. 9c; Bosworth et al. 1998; Cross et al. 1998). The Late Miocene was typified throughout by restricted marine evaporite deposition and interbedded conglomerate, sandstone, and shale similar to the early Serravalian.

Within salt walls and domes, the massive halite sections sometimes exceed 3 km thickness. Age control within the various evaporite-dominated formations is rare, but the uppermost shale interbeds have yielded broad ranging "Late" Miocene ostracods and calcareous nannoplankton (El-Shafy 1992). These beds are capped by an unconformity that extends throughout the Gulf of Suez and Red Sea which is in turn overlain by Pliocene strata. This unconformity—the "S" reflector of Ross et al. (1973) and the "top Zeit" of industry parlance—is therefore essentially equivalent to the Messinian unconformity of the Mediterranean (Coleman 1993). Like the Mediterranean, the Red Sea must have been mostly a dry basin during the Late Messinian.

Salt Flowage

The formation of salt domes and walls has been studied in both the subsurface and along the margins of the Red Sea and Gulf of Suez (Hassan and Dashlouty 1970; Ross and Schlee 1973; Mulder et al. 1975; Khedr 1984; Miller and Barakat 1988; Patton et al. 1994; Bosworth 1995; Heaton et al. 1995; Bosence et al. 1998, reviewed in Orszag-Sperber et al. 1998). Both local salt overhangs and larger salt canopies have been described (Fig. 10). Consensus is that flowage of the Middle to Late Miocene massive halite began soon after its deposition and subsequently impacted the distribution of younger sediments. In general, the configuration of the salt walls in the Gulf of Suez follows that of the underlying extensional fault blocks and hard-linkage transfer faults (Bosworth 1995).

In many oil fields in the Gulf of Suez, the ultimate top seal is the massive salt, and hence, mapping this surface is a key aspect of exploration. As in other salt basins, the high seismic velocities of the evaporites and their complex geometry make imaging the sub-salt structure and stratigraphy exceptionally difficult. Some progress in addressing this issue has been reported in recent years (Mougenot and Al-Shakhis 1999; Musser et al. 2012).

Volcanicity

The volcanic rocks of the Red Sea basin provide critical information about the timing of rifting, the nature of the subcrustal mantle, the possible connections between the Afar plume and rifting, the transition from continental to oceanic rifting, and the onset of seafloor spreading. Coleman (1993)

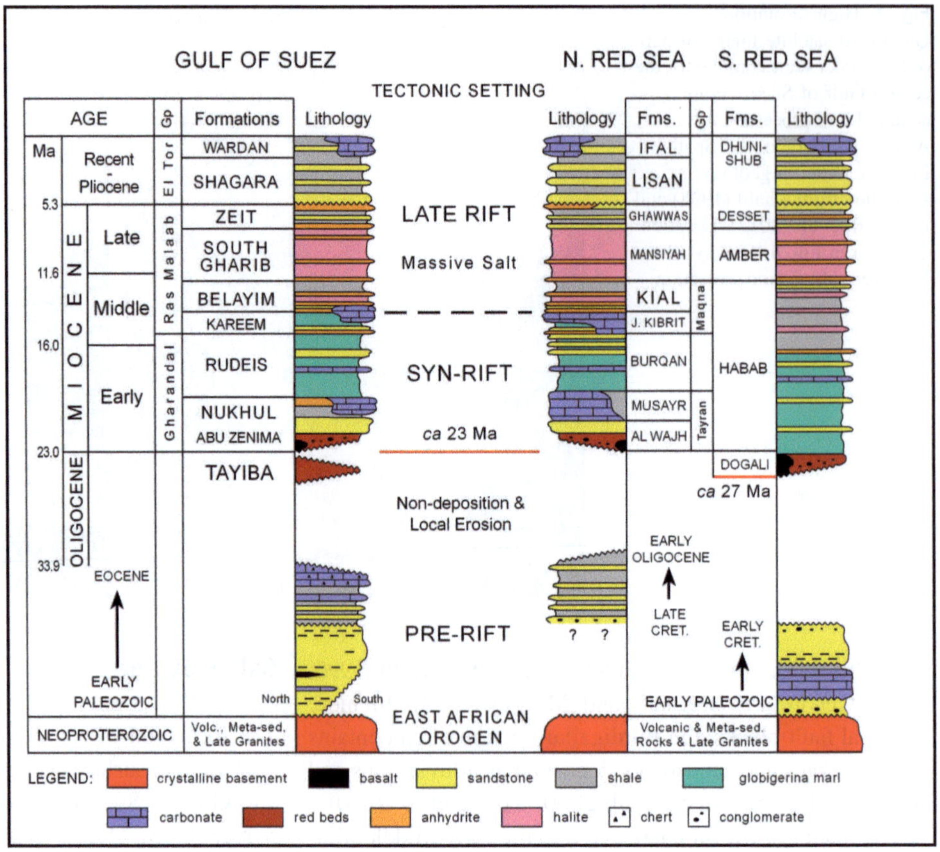

Fig. 8 Simplified stratigraphic sections and terminology of the Gulf of Suez and Red Sea (after Bosworth and Burke 2005 and references therein). Timescale is from Gradstein et al. (2004)

has provided a thorough introduction to this material, and a recent review is provided by Szymanski (2013). In terms of chronology, volcanism can be discussed in four phases: (1) the Afar and Yemen trap basalts and Older Harrats; (2) sheeted dike complexes of the Saudi Arabian margin and time-equivalent isolated dikes and flows of Sinai and northern Egypt; (3) the Younger Harrats; and (4) post-Miocene volcanism of the Red Sea axial trough.

$^{40}Ar/^{39}Ar$ age dating has shown that eruption of the Afar basalts/trachytes began at 31 Ma, followed by rhyolitic eruptions ~1 Ma later and then a period of ~5 Ma of much more subdued outflowings of basalts and ignimbrites (Zumbo et al. 1995; Rochette et al. 1997; Chernet et al. 1998; George et al. 1998; Ukstins et al. 2002; and Coulié et al. 2003). Smaller occurrences of rhyolites on the southern Sudanese margin were similarly dated at 30 Ma (Kenea et al. 2001) and basalts at the Older Harrat Hadan 28–26 Ma (Sebai et al. 1991; Féraud et al. 1991). The Yemeni traps show the same sequence as Afar, with basalts beginning at 31 Ma followed by massive ignimbrites from 30 to 26 Ma (Baker et al. 1994, 1996; Ukstins et al. 2002; Coulié et al. 2003).

The second phase of Red Sea volcanism principally involved intrusion of NW–SE striking basaltic dikes along the Yemen and Saudi Red Sea margins, Sinai, and northern Egypt. In southern Saudi Arabia, the dikes are exposed with coeval layered gabbro and granophyre (McGuire and Coleman 1986; Coleman 1993). The dikes in Saudi Arabia were intruded over a brief time span from ~24 to 22 Ma (Table 1; Sebai et al. 1991). Basaltic dikes in Sinai and the vicinity of Cairo correspond to the same ages and fed an area of flows that covered ~15,000 km^2 (preserved) to perhaps 25,000 km^2 (original) (Figs. 11 and 12). Though authors differ on the interpretation of the Cairo basalts, field relationships suggest that they were erupted more or less as a single geologic event similar to the initial basaltic volcanism at Afar (Bosworth et al. 2015) and can be thought of as a short-lived "mini-plume." Recent analyses of trace elements and Sr–Nd–Pb–Hf isotopes suggest that these sub-alkaline basalts were derived from mixing of an Afar plume-like source with metasomatized continental lithosphere (Endress et al. 2011). In the Gulf of Suez, the circa 23 Ma basalts are found associated with the very basal syn-rift sedimentary units (Fig. 12a; Sellwood and Netherwood 1984; Bosworth and McClay 2001; Jackson 2008). The weighted mean average plateau and integrated total fusion ages for the Red Sea dikes and flows of Egypt are 23.0 ± 0.1 and 22.8 ± 0.1 Ma, respectively (Table 1), essentially at the Oligocene–Miocene transition (Gradstein et al. 2004). The weighted mean average of Sebai et al.'s dates for Saudi Arabia is slightly younger at 21.8 ± 0.03 Ma, but these

Fig. 9 Field photographs of syn-rift strata from the northern Red Sea rift system (Egypt). **a** Early Miocene Rudeis Formation shale and thin sandstone, the main syn-rift fill facies of the basin. From the dip slope of the Hammam Faraun fault block (Fig. 4); **b** proximal fan-delta sandstone and conglomerate near the rift border fault complex and deformed by a syn-rift fault propagation fold at Wadi Baba (Fig. 7); **c** Middle Miocene reef platform and carbonate talus overlying crystalline basement at Abu Shaar el Qibli northwest of Hurghada (Fig. 4). These carbonate rocks are time stratigraphic equivalents of the Belayim Formation of Fig. 8

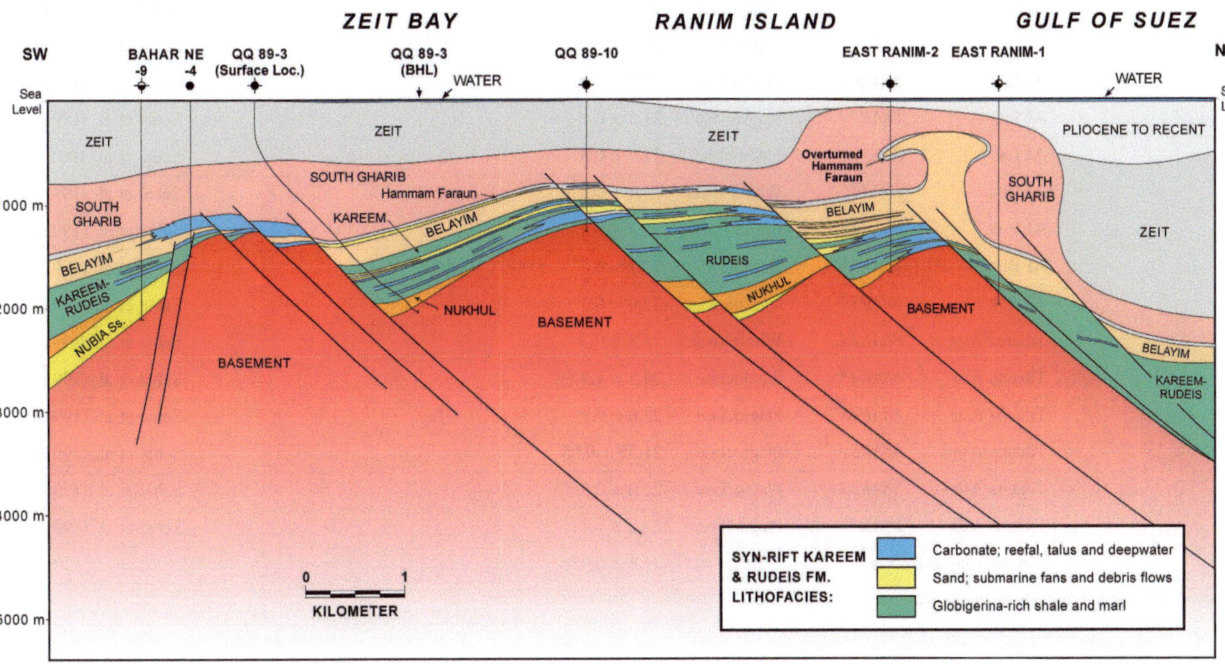

Fig. 10 Structural cross section of the salt ridge at East Ranim. The lithofacies of the main syn-rift Rudeis and Kareem Formations are shown by *color coding*. The South Gharib and Belayim Formations contain massive halite that has flowed to produce a classic *mushroom-*shaped structure with a salt overhang. East Ranim-2 penetrated the Hammam Faraun Member three times, once in an overturned orientation. Location is shown in Fig. 4

Table 1 Compilation of Ar^{40}/Ar^{39} data for the Red Sea dike event at the Oligocene–Miocene boundary

Location	Country	Outcrop	Sample Name	Mineral	Weighted mean plateau age	Integrated total fusion age	Source
North	Egypt	Cairo		Whole rock	22.4[a]		Lotfy et al. (1995)
		Cairo		Whole rock	22.6[a]		Lotfy et al. (1995)
		Gebel Qatrani		Whole rock	23.67 ± 0.15		Kappelman et al. (1992)
		Gebel Qetrani		Whole rock	23.68 ± 0.14		Kappelman et al. (1992)
		Gebel Qatrani		Whole rock	23.62 ± 0.16		Kappelman et al. (1992)
		Gebel Qetreni	*08-GQ-01*	Whole rock	21.7 ± 0.3	22.0 ± 0.8	Bosworth et al. (2015)
		Gebel Gatrani	*08-GQ-02*	Whole rock	21.4 ± 0.3	20.0 ± 0.3	Bosworth et al. (2015)
		Gebal Qatrani	*08-GQ-03*	Whole rock	23.2 ± 0.4	24.0 ± 0.8	Bosworth et al. (2015)
		West Faiyum	*WFAY-1*	Whole rock	23.5 ± 0.7	24.0 ± 0.5	Bosworth et al. (2015)
		West Faiyum	*WFAY-2*	Whole rock	22.1 ± 0.4	21.7 ± 0.6	Bosworth et al. (2015)
		Bahariya Oasis	*09-WD-01*	Whole rock	23.1 ± 0.2	22.9 ± 0.2	Bosworth et al. (2015)
		Bahariya Oasis	*09-WD-02*	Whole rock	23.6 ± 0.2	23.2 ± 0.2	Bosworth et al. (2015)
		Bahariya Oasis	*09-WD-03*	Whole rock		20.8 ± 0.2	Bosworth et al. (2015)
		Bahariya Oasis	*09-WD-04a*	Whole rock		24.2 ± 0.2	Bosworth et al. (2015)
		Bahariya Oasis	*09-WD-04b*	Whole rock		25.0 ± 0.3	Bosworth et al. (2015)
		Qaret Dab'a	*09-WD-05*	Whole rock	23.4 ± 0.4	23.1 ± 0.4	Bosworth et al. (2015)
		Wadi Tayiba	*06-GZ-02*	Whole rock		23.1 ± 0.9	Bosworth et al. (2015)
		Wadi Nukhul	*06-GZ-03*	Whole rock	24.4 ± 0.3	23.0 ± 3.0	Bosworth et al. (2015)
		Weighted mean average			23 0 ± 0 1	22.8 ± 0.1	
South	Saudi Arabia	Al Wajh	*89A53*	Plagioclase	22.50 ± 0.06		Sebai et al. (1991)
		AI Lith	*88AR1*	Plagioclase	22.3 ± 0.6		Sebai et al. (1991)
		Al Lith	*AS82*	Plagioclase	23.3 ± 0.5		Sebai et al. (1991)
		Al Lith	*88A4*	Amphibole	24.0 ± 0.2		Sebai et al. (1991)
		AI Lith	*88AR67*	Plagioclase	21.1 ± 0.2		Sebai et al. (1991)
		AI Lith	*88AR69*	Plagioclase	21.5 ± 0.1		Sebai et al. (1991)
		AI Lith	*AS85*	Plagioclase	21.7 ± 0.3		Sebai et al. 1991
		AI Birk	*89A44*	Whole rock	23.9 ± 0.3		Sebai et al. (1991)
		AI Birk	*89A44*	Biotite	22.3 ± 0.5		Sebai et al. (1991)
		AI Birk	*88AR55*	Hornblende	21.1 ± 0.5		Sebai et al. (1991)
		AI Birk	*88AR53*	Hornblende	22.0 ± 0.3		Sebai et al. (1991)
		Tihama Asir	*88AR14*	Whole rock	21.8 ± 0 3		Sebai et al. (1991)
		Tihama Asir	*88AR48*	Plagioclase	24.5 ± 1.3		Sebai et al. (1991)
		Tihama Asir	*88AR43*	Plagioclase	22.6 ± 0.9		Sebai et al. (1991)
		Tihama Asir	*88AR8*	Plagioclase	21.25 ± 0.05		Sebai et al. (1991)
		Tihama Asir	*88AR37*	Plagioclase	22.0 ± 0.5		Sebai et al. (1991)
		Tihama Asir	*88AR35*	Plagioclase	22.2 ± 1.1		Sebai et al. (1991)
		Weighted mean average			21.8 ± 0.03		

[a] not used in weighted average as no standard deviation provided

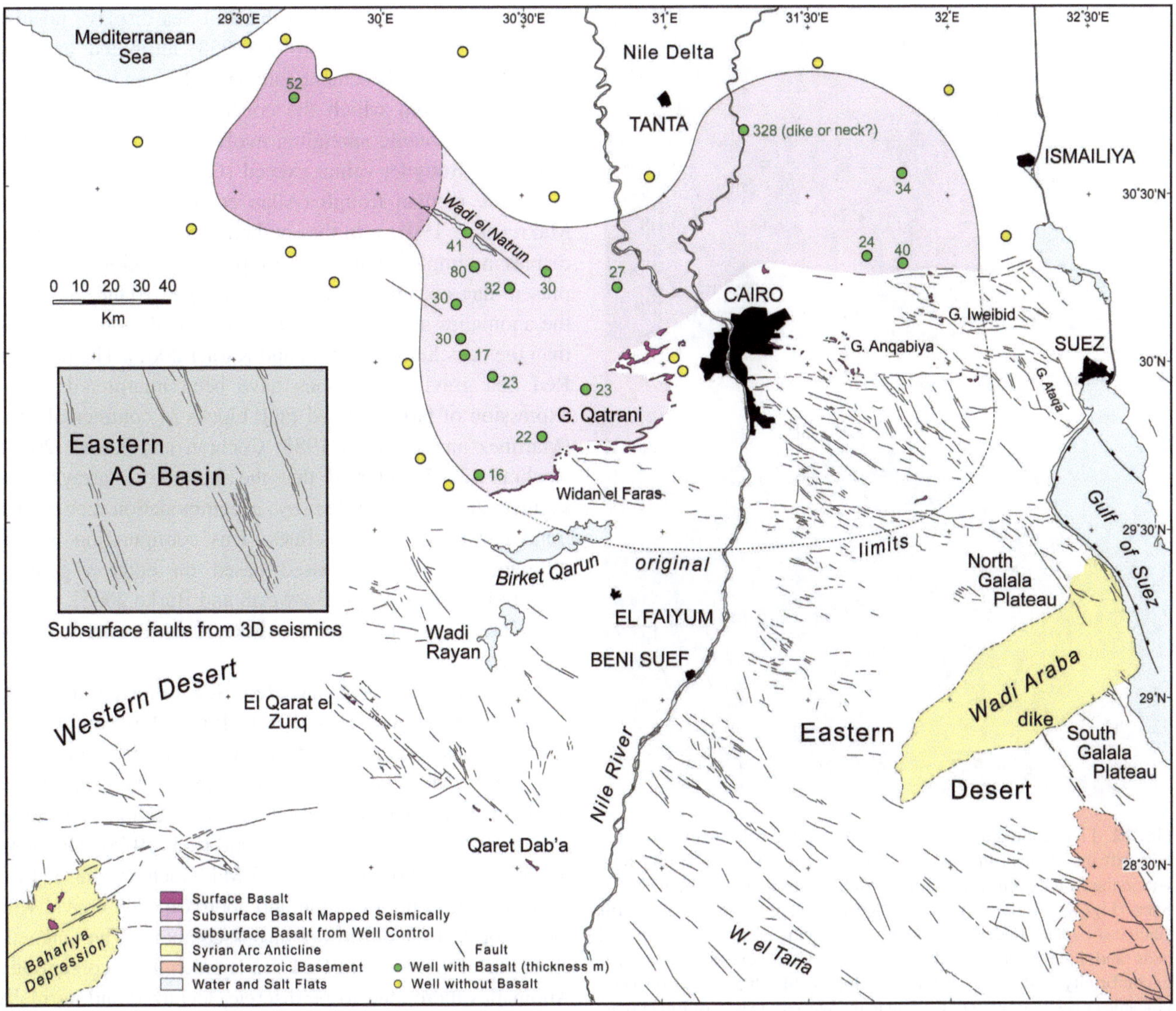

Fig. 11 Cairo basalt province and faulting associated with the northern limits of the greater Red Sea rift system. The *purple shaded* region shows the extent of 23–22 Ma basalt flows in the subsurface, with an area of approximately 15,000 km^2 (would have covered 20,000–25,000 km^2 prior to erosion). Darker shading is the area of flows mapped by Williams and Small (1984) in 2D reflection seismic data. Lighter shading is the area of flows based solely on well penetrations (thickness of flow is annotated). *Gray shaded box* shows basal Miocene-age subsurface faults in a merge of several large-scale 3D seismic surveys. After Bosworth et al. (2015). Transverse Mercator projection

basalts were emplaced over a much greater area, and this average may include several closely related eruptive events. The age of the onset of dike intrusion was statistically the same in both the Gulf of Suez and Saudi margin.

After the 24–22 Ma dike event, volcanism ceased in most parts of the Red Sea basin. Exceptions are Afar, where a complex igneous history continued to the present day (Barberi et al. 1972; Varet 1978; Zanettin et al. 1978; Berhe 1986; Vellutini 1990; Tefera et al. 1996), and Harrat Ishara (north of Madinah) which experienced basaltic volcanism from ~17 to 14.5 Ma (Szymanski 2013). In the Middle Miocene, volcanism returned to far-reaching areas of the Arabian Peninsula: Harrat ash Shaam, Jordan at ~13 Ma (Ilani et al. 2001), and the Younger Harrats of Saudi Arabia—Uwayrid ~12 Ma, Khaybar ~11 Ma, and Rahat ~10 Ma (Coleman et al. 1983; Coleman 1993 and references therein). Volcanism continues to the Recent in these and other Younger Harrats, often with north–south aligned vents and dikes. These orientations support the interpretation that in the Arabian shield, the maximum horizontal stress has been north–south during post-Miocene times (Bosworth and Strecker 1997). This contrasts with the central platform areas of Arabia (Ghawar province) where the maximum horizontal stress is presently ENE–WSW (Ameen 2014).

Fig. 12 Field photographs of circa 23 Ma basaltic volcanism associated with rift initiation in the northern Red Sea rift system. **a** Basalt flow within the Abu Zenima Formation at Wadi Tayiba in the northern Gulf of Suez (Fig. 4) overlain by Early Miocene Nukhul Formation syn-rift siliciclastic rocks; underlain by Oligocene Tayiba Formation pre-rift red beds and Late Eocene Tanka Formation white limestone. Base-rift unconformity is shown by white line; **b** NW–SE aligned monogenetic volcanic cones southwest of Faiyum in the Western Desert of Egypt (part of Qarat el Zurq–Qaret Dab'a trend in Fig. 11)

Typical mid-ocean ridge basalts (MORBs) of the Red Sea rift axis have been observed and sampled in situ by submersible dives (Monin et al. 1982; Juteau et al. 1983), and closely related tholeiites are exposed subaerially at the active axial shield volcano of Jabal at Tair Island in the very southern Yemeni Red Sea (Mattash 2008). The ongoing volcanism of the Red Sea axial trough is discussed in detail in other chapters of this book.

Lithospheric and Crustal Structure

Geophysical studies of the Red Sea have been reviewed by Cochran (2005), Cochran and Karner (2007), and Lazar et al. (2012). Gravity observations initiated with the coastal survey of von Triulzi (1898) and then the shipboard expedition of Vening Meinesz (1934). These data led to the fundamental observation that the Red Sea displays positive gravity anomalies that Girdler (1958) attributed to mafic intrusions into the basement complex. This can be thought of as the seed from which the concept that the Red Sea is undergoing oceanic spreading evolved (Lazar et al. 2012). Maximum Bouguer values exceed 100 mGal and are located along the median trough (Allan et al. 1964; Allan 1970; Makris et al. 1991a). In the southern Red Sea, the anomalies display a simple, approximately parallel relationship to the present-day coastlines. North of Zabargad Island, however, the anomalies are less continuous and trend more northwest than the coastlines (Cochran and Karner 2007). The northern Red Sea gravity anomalies have been interpreted as the expression of large, rotated fault blocks of continental crust (Martinez and Cochran 1988; Cochran and Karner 2007). Cochran (2005) inferred that the fault block arrays were segmented and separated by accommodation zones that trend normal to the coastlines. This configuration is very different than that proposed based on outcrop geology (Fig. 6; Bosworth 1994; Bosworth and Burke 2005), though the general concept of along-strike rift segmentation is in agreement.

Magnetic studies of the Red Sea followed the early gravity investigations (Allan et al. 1964; Drake and Girdler 1964). It was soon accepted that linear magnetic anomalies that are present on both sides of the axial deep in the southern Red Sea were produced by seafloor spreading (Allan 1970; Röser 1975; Searle and Ross 1975). The onset of spreading is thought to be $\sim$5 Ma (Cochran 1983) at the Miocene–Pliocene transition and approximately coeval with the major Zeit unconformity discussed above. The magnetic character of the northern Red Sea is much more contentious. Anomalies there tend to be discrete, localized and normally magnetized (Cochran et al. 1986; Martinez and Cochran 1988; Guennoc et al. 1988), and therefore must have been produced within the past 780 ka (Brunhes chronozone) based on the geomagnetic polarity timescale (Shackleton et al. 1990; Gee and Kent 2007). These young features are interpreted to be individual volcanoes erupted within the northern axial deeps (e.g., Bannock and Shaban Deeps, Fig. 2b; Bonatti et al. 1984; Pautot et al. 1984) or localized intrusions. Martinez and Cochran (1988) similarly interpreted a sub-sea sediment piercing structure near the Brothers Islands to be a volcano built on the abyssal plain west of the axial trough. It has also been suggested that high-frequency, discontinuous but linear magnetic anomalies are present in the northern Red Sea dataset and that these were formed by seafloor spreading (Saleh et al. 2006).

Seismic refraction profiling has provided important constraints on the configuration of the Moho along the Red Sea margins of Egypt (Makris et al. 1981; Gaulier et al. 1988) and Saudi Arabia (Mooney et al. 1985; Prodehl 1985; Milkereit and Fluh 1985; Blank et al. 1986; Gettings et al.

1986). In the north, the un-thinned continental crustal thickness is ~40 km in both Arabia and Africa. At the coastlines, the crust has been thinned to ~20 km, though the limited data that are available suggest that the change is more abrupt on the Egyptian margin (see compilation in Voggenreiter et al. 1988). In southern Saudi Arabia, the 1,070 km US Geological Survey refraction line extends from the Farasan Islands to near Riyadh and has received a variety of interpretations. There is general agreement that the continental crust is ~38–45 km thick beneath the Arabian shield and that the Moho there is approximately horizontal. At the coastal plain, Mooney et al. (1985) interpreted an abrupt thinning of the crust to ~20 km and then further gradual tapering to less than 10 km thickness beneath the Farasan Islands. Prodehl (1985) modeled a much more gradual rise in the depth to the Moho starting west of the Asir Range and reaching ~14 km at Farasan. Milkeriet and Fluh (1985) interpreted abrupt thinning similar to Mooney et al. (1985) but with the presence of an intra-crustal high-velocity zone beneath the coastal plain. These various studies suggest that the crust along the shelf of the northern Red Sea is about 50 % of its original thickness and in the southern Red Sea (Saudi margin) about 25–30 %.

Early compilations of heat flow data demonstrated that the entire Red Sea basin is hotter than the worldwide average of ~60 mW/m^2 and that the magnitude of this anomaly increases toward the axial trough (Girdler 1970; Scheuch 1976; Girdler and Evans 1977). Heat flow along the margins of the northern Red Sea is typically about 125 mW/m^2 (Martinez and Cochran 1988), whereas in the Eastern Desert of Egypt, the average is slightly more than 70 mW/m^2 (Morgan et al. 1980). Along the Red Sea axis, values of 250–350 mW/m^2 are common with a maximum of ~400 mW/m^2 (Martinez and Cochran 1988; Makris et al. 1991b). The observed heat flow profiles have been modeled with respect to a variety of geometric and kinematic rift histories (simple shear, pure shear), and the general conclusions are that the zone of crustal thinning must have widened as extension initially progressed and then starting at ~5 Ma dramatically narrowed to a presently active width of ~20 km (Buck et al. 1988; Martinez and Cochran 1989). A simple shear lithospheric-scale detachment geometry (Wernicke 1985; Voggenreiter et al. 1988) could not reproduce the value or distribution of heat flow, and neither simple shear nor pure shear models resulted in shallow mantle melting. This suggested that an additional source of heat such as convection was required, in agreement with studies by McGuire and Bohannon (1989) of mantle zenoliths. These authors found that the temperature beneath western Arabia at a depth of 40 km is ~900 °C, much too high to be explained by the present measured shallow geothermal gradient. Hence, the postulated mantle upwelling must be very young, synchronous with Neogene rifting rather than prior, or there would have been time for equilibration of the shallow heat flow.

Plate Kinematics and Red Sea Restorations

Early models of the Red Sea basin as a pair of rifted continental margins focused on how much oceanic crust had been generated and resulted in two end members: (1) Young oceanic crust is restricted to the axial trough, and opening is therefore restricted to ~85 km (Girdler 1958; Drake and Girdler 1964; Hutchinson and Engels 1970, 1972; Lowell and Genik 1972; Ross and Schlee 1973) versus (2) the entire width of the Red Sea was formed by seafloor spreading and a pre-rift restoration should be coastline to coastline (McKenzie et al. 1970). Numerous intermediate models were subsequently developed (reviewed in Girdler and Whitmarsh 1974).

It was later generally accepted that oceanic spreading only started in the Red Sea about 5 Ma (discussed below), though continental rifting began in the Oligocene or Early Miocene. The early separation of Arabia from Africa therefore involved stretching of continental lithosphere or some other mechanism for generation of surface area, which needed to be incorporated in plate restorations (reviewed in Le Pichon and Francheteau 1978; Coleman 1993). Le Pichon and Francheteau (1978) interpreted the instantaneous pole of opening for the Red Sea from the magnetic lineaments of the southern axial trough and compared this with the total opening (Eulerian pole) of McKenzie et al. (1970) and key geologic constraints. They concluded that the movement of Arabia relative to Africa has been essentially stable since the Early Miocene, but that the total opening can only be on the order of 3°–4° rather than the 6° of McKenzie et al. Hence, they estimated total opening to be 150–200 km at 19°N latitude or about 65–115 km of nonspreading center-generated new surface. Numerous refinements and adjustments have been made to the LePichon and Francheteau and earlier reconstructions (Cochran 1983; Girdler and Underwood 1985; Bohannon 1986; Joffe and Garfunkel 1987).

GPS datasets for Arabia and adjacent plates are now very robust and indicate that Arabia is moving ~20.6 mm/yr north relative to a Eurasia-fixed reference frame (Fig. 1; ArRajehi et al. 2010). This agrees with larger scale plate circuit estimates for Arabia–Eurasia convergence for the past ~22 Ma (McQuarrie et al. 2003). Similarly, convergence between Africa (Nubia) and Eurasia of ~6.6 mm/yr has occurred for the past ~11 Ma. The agreement between geodetically derived, present-day instantaneous velocities, and plate tectonic, geologic-term velocities suggests that restoration of the Red Sea–Gulf of Aden rift system can be reasonably produced if an independent key basin parameter is known: either (a) time of rift initiation or (b) the starting separation of some unchanged feature on both margins of the basin. ArRajehi et al. (2010) assumed coastline-to-coastline restoration (similar geometrically to McKenzie et al. but not necessarily all oceanic crust) and derived an age of initiation for Red Sea rifting of 24 ± 2.2 Ma. Detailed discussion of the significance of GPS results to Red Sea restorations is presented by Reilinger et al. in this volume.

Red Sea Rift Chronology

Continental Rifting

The initiation of continental rifting in the Red Sea is directly tied to the geologic history of Afar and the Gulf of Aden (reviewed in Bosworth et al. 2005 and references therein). Extension began in the eastern Gulf of Aden by ~30 Ma (late Early Oligocene), and the entire gulf was active by the mid-Oligocene though precise data are lacking (Fig. 13a; Roger et al. 1989; Hughes et al. 1991; Watchorn et al. 1998). In the southernmost Red Sea, offshore Eritrea, the base of the syn-rift section is dated as ~27.5 to ~23.0 Ma (Late Oligocene; Fig. 13b; Hughes et al. 1991). Similarly in the Afar region, major extension is interpreted to have begun by ~25 Ma (Fig. 13c; Barberi et al. 1972, 1975; Zanettin et al. 1978).

For the rest of the Red Sea, north as far as the Gulf of Suez and Cairo district of Egypt, definitive basal syn-rift strata are paleontologically dated to be ≤23 Ma (Early Miocene, Hughes and Beydoun 1992; Bosworth et al. 2005). As discussed above, these sedimentary rocks are interbedded with or contain detritus of basalts that are radiometrically dated 24–22 Ma (Table 1). Oligocene strata are locally present, the Tayiba Formation and lateral equivalents (Fig. 8). But these Oligocene red beds do not display a syn-tectonic relationship to Gulf of Suez faulting (e.g., Khalil 1998; Jackson et al. 2006; Jackson 2008). Recent paleontologic studies in western Sinai suggest that the base of the oldest syn-rift units is locally latest Oligocene (Hewaidy et al. 2012 reviewed in El Atfy et al. 2013). If correct, this is compatible with the uncertainty in the age of the basal basalts and the assertion that these basalts mark the onset of rifting. More paleontologic data for the southern Red Sea are needed, and better resolution within the "Late Oligocene" would be helpful, but it appears that continental extension invaded the entire Gulf of Aden and Afar region during the Early to Late Oligocene and then stalled briefly (perhaps 2–3 My). At the Oligocene–Miocene transition, dike intrusion and faulting shot from Eritrea to Egypt with no recognizable time difference—without discernible propagation (Fig. 13c; Richardson and Arthur 1988; Omar and Steckler 1995).

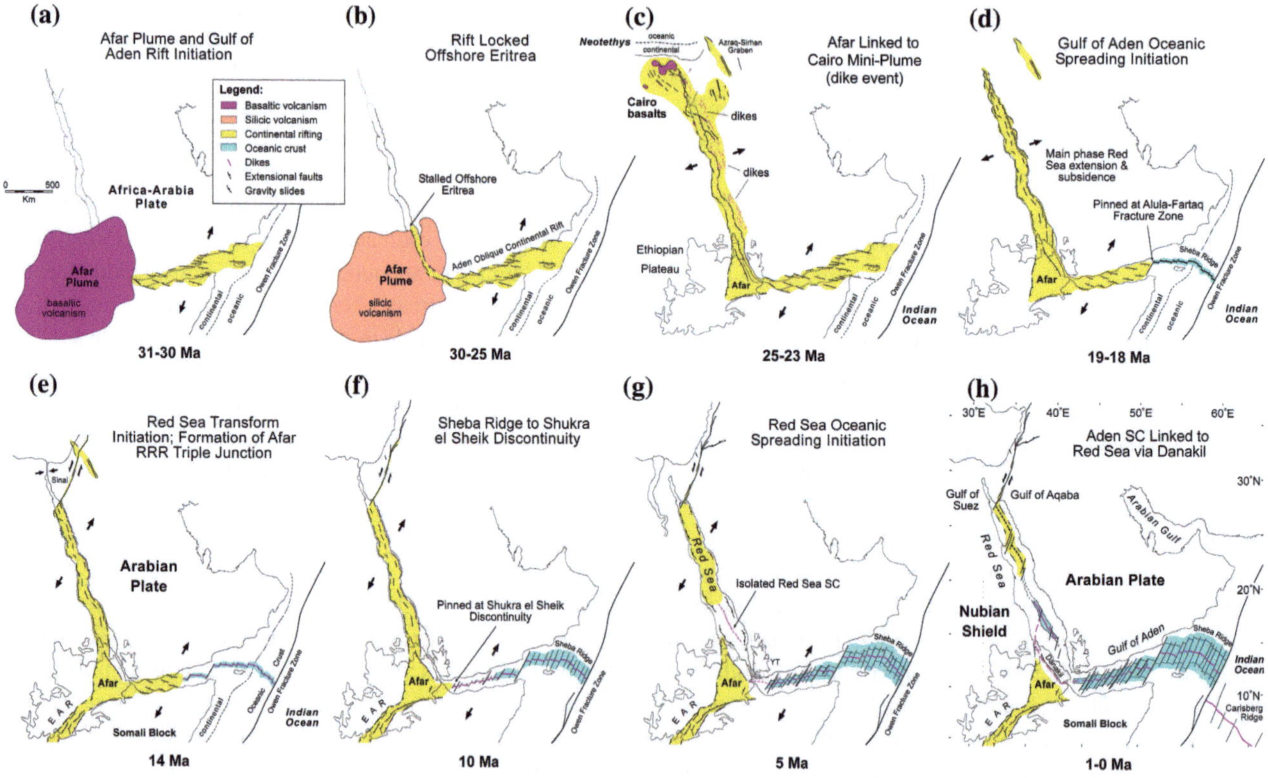

Fig. 13 Synthesis of continental and oceanic rift initiation and propagation in the greater Gulf of Aden–Red Sea rift system. Simplified from Bosworth et al. (2005). EAR = East African Rift; SC = spreading center

Where documented, the earliest syn-rift strata in the Gulf of Aden overlie marine pre-rift rocks of Oligocene or older age (Hughes et al. 1991; Robertson and Bamakhalif 1998; Bosworth et al. 2005). This suggests that this rift initiated near sea level. The evidence is not as definitive for the Afar region. A regional peneplain separates the pre-plume and 31 Ma trap basalts throughout Ethiopia and Yemen, and in northern Eritrea and parts of Yemen, this is marked by extensive laterites (Canuti et al. 1972; Zanettin et al. 1978; Davison et al. 1994; Sagri et al. 1998). Bohannon (1986) and Coleman (1993) considered this as evidence against significant pre-plume or pre-rift doming. However, pre-plume laterites are not known from central Afar itself, so the area of doming could have been fairly localized, or it may have been synchronous with the eruption of the earliest trap basalts (Burke 1996, reviewed in Şengör 2001).

The oldest syn-rift section in the southern Red Sea (offshore Eritrea) is marine and demonstrates a Late Oligocene seaway connection with the Gulf of Aden (Hughes et al. 1991; Hughes and Beydoun 1992). This paleogeographic detail limits how much Afar doming could have occurred along the focus of rifting. At the northern end of the Red Sea in the Gulf of Suez and vicinity of Cairo, stratigraphic relationships are similar though the rocks are somewhat younger: The Oligocene pre-rift strata are a mixture of shallow marine and low-relief fluvial facies, and with the exception of a few localized, very thin nonmarine beds, the syn-rift fill is essential all marine from the start of extension (Sellwood and Netherwood 1984; Richardson and Arthur 1988; Jackson 2008). The Cairo basalts are very consistent in thickness over a very large area, also suggesting a low-relief, flat pre-rift geometry (Bosworth et al. 2015). Garfunkel (1988) estimated that the maximum pre-rift erosion in western Sinai was a few hundred meters and that there was no evidence of uplift specific to the future rift axis.

Most geologic evidence favors superposition of Gulf of Aden and Red Sea rifting on a generally peneplaned surface that was near or in some areas below sea level. At the onset of extension, fault block rotation resulted in only very localized uplift, best documented in the Gulf of Suez (Garfunkel and Bartov 1977; Sellwood and Netherwood 1984; McClay et al. 1998; Carr et al. 2003). A total of 2–3 Myr after rift initiation, this situation changed dramatically. At about 20 Ma in the Gulf of Suez, depositional environments shifted from marginal to open marine with water depths of 200 m or more in sub-basin axes (Fig. 13d). Total compaction corrected subsidence rates increased by a factor of two or more (Steckler 1985; Moretti and Colletta 1987; Evans 1988; Richardson and Arthur 1988; Steckler et al. 1988). As discussed above, apatite fission track data indicate that this main phase of syn-rift subsidence was accompanied by unroofing and uplift of the rift shoulders along the entire length of the Red Sea rift system. These authors have proposed thermomechanical models linking the two processes.

The next major phase in the Red Sea continental rift history was marked by the onset of the Aqaba–Levant transform boundary (Fig. 13e). The arguments concerning the timing of this event are diverse (reviewed in Bosworth and McClay 2001; Bosworth et al. 2005) and generally suggest onset of left lateral shear in the Middle Miocene at ~14–12 Ma. Some authors have inferred a somewhat earlier initiation for the transform; for example, Garfunkel and Beyth (2006) proposed ~18–17 Ma in response to the beginning of oceanic spreading in the Gulf of Aden (discussed below). If the Gulf of Aqaba movement started in the Middle Miocene, then it corresponds to the time of collision of Arabia with Eurasia (Şengör and Yilmaz 1981; Hempton 1987; Woodruff and Savin 1989; Burke 1996). Offset of geologic features limits the total slip on this transform to be ~107 km of which 45 km is Pliocene to Recent (Quennell 1951, 1958). These data provide important kinematic constraints for Red Sea palinspastic restorations. With the onset of Aqaba–Levant motion, extension across the northern Red Sea changed from rift normal (NE–SW) to highly oblique and parallel to the transform (NNE–SSW, Fig. 13e).

Movement on the Aqaba–Levant transform is interpreted to have caused minor counterclockwise rotation of the new Sinai micro-plate and local compression and uplift in the northern Gulf of Suez (Fig. 13e; Patton et al. 1994). Middle Miocene sea level was also somewhat lowered (Haq et al. 1987), and the combined effect was that for the first time since the earliest Miocene the Gulf of Suez–Neotethyan seaway connection was largely severed. Open marine deposition was replaced by evaporitic conditions throughout most of the Gulf of Suez and northern Red Sea at ~12 Ma (intra-Serravallian; see stratigraphic discussion above). In the southern Red Sea, open marine shale was deposited offshore from Eritrea until the early part of the Late Miocene (Savoyat et al. 1989; Hughes and Beydoun 1992). The connection to the Gulf of Aden through Bab-al-Mandab therefore persisted somewhat longer, but by ~10 Ma (lower Tortonian), massive halite and anhydrite were being deposited throughout the Red Sea.

Oceanic Rifting

The easternmost and oldest segment of the oceanic spreading system of the Gulf of Aden–Red Sea rift is called the Sheba Ridge (Fig. 13d; Mathews et al. 1967). Spreading at the Sheba Ridge is thought to have started at ~19–18 Ma (Sahota 1990; Leroy et al. 2004), about 12 My after the onset of continental rifting in this same area. Spreading was pinned at the Alula-Fartaq fracture zone for a few million years and then moved a few hundred kilometers west by ~16 Ma (Fig. 13e). By ~10 Ma, the spreading center had reached the Shukra el Sheik discontinuity in the western Gulf of Aden and was pinned again (Fig. 13f; Manighetti et al. 1997). Only after ~2 Ma did spreading propagate into Afar (Audin 1999; Hébert et al. 2001; Audin et al. 2004).

Oceanic spreading did not propagate from the western Gulf of Aden or Afar into the Red Sea. Rather, as discussed above based on magnetic striping, the oldest Red Sea spreading center appeared within the southern Red Sea continental rift at about 17°N latitude (Fig. 13g; Allan 1970; Röser 1975; Searle and Ross 1975). The age of initiation of spreading in this region was ~5 Ma (Cochran 1983). Since that time, complex volcanic activity and extensional faulting has continued within Afar west of the Danakil Alps (reviewed in Tefera et al. 1996; Redfield et al. 2003; Garfunkel and Beyth 2006). The Red Sea is linking to the Afar depression through the Gulf of Zulu and similarly to the Gulf of Aden through the Gulf of Tadjoura (Fig. 13h).

The length of the Gulf of Aden–Afar (subaerial)–Red Sea oceanic spreading center from the Carlsberg Ridge to the end of magnetic striping at about 19°N latitude in the southern Red Sea is ~2,800 km (Fig. 13h). As briefly outlined here and noted by many workers, the ridge propagated in pulses, but the average rate since inception at ~19 Ma is ~150 km/My. By comparison, the earlier continental rift system reaches from just east of Socotra Island–Ras Sharbithat (Fig. 13c; Stein and Cochran 1985) to NW of Cairo, a distance of ~3,900 km. The continental rift also developed in segments, and within each segment, age dating cannot resolve any discernible propagation. However, the entire rift was completed from 30 to 23 Ma or within 7 My. The average rate of advancement was ~550 km/My or nearly four times as rapid as growth of the later oceanic rift.

Discussion and Synthesis

The Red Sea offers great insight into how continental rifts form and subsequently can evolve into oceanic basins. But is this system a suitable model for many rift settings, or is it perhaps very unique or an end member type? As discussed at the introduction to this chapter, the Red Sea is both another phase in the long history of breakup of Gondwana, and the product of a tectonic environment shaped by impingement of large mantle plumes (particularly Afar) at the base of the African lithosphere. Adding complexity to this background is the ongoing collision of Africa–Arabia with Eurasia, and the modifications to plate boundaries produced by movement on the Aqaba–Levant transform margin.

Rift Driving Forces

Theoretical models and analyses of continental rifts often focus on two fundamental questions: (1) what are the driving forces for rifting? and (2) how do rifts grow (propagate) laterally? Milanovsky (1972) divided continental rifts into those associated with continental "platforms" and others found in young fold belts. The platform rifts were either typified by broad doming and abundant alkaline volcanism, or no doming and little or no volcanism. Sengör and Burke (1978) elaborated on these fundamental differences and proposed the terms "active" and "passive" rifting based on their interpretation of the underlying dynamics of these two rift types. In active rifts, extension is driven by upwelling mantle convection currents, while passive rifts are the result of far-field extensional stresses arising from lithospheric plate movement and plate boundary interactions. Utilizing finite element modeling, Dunbar and Sawyer (1988) proposed that both volcanic-rich domed rifts and non-volcanic crevice style rifts could be formed by regional extensional stresses without involvement of mantle currents. In their models, the surficial manifestations of rifting were controlled by the positions of weakness in the pre-rift lithosphere, whether in the crust or upper mantle. In a lithospheric plate subjected to extensional forces, uplift occurred above a mantle weakness, whereas fault-bounded basins formed above crustal weakness. Various combinations of these pre-existing weaknesses can be envisioned, and they might be laterally offset from each other or vertically superimposed. Buck (2006) however has argued that normal continental lithosphere may be too strong to rift without magmatic dike intrusion, and therefore, there must be an appropriate combination of regional extension and a source of sufficient magma. He suggested that both the Red Sea and Ethiopian Rift developed as magma-assisted rifts, whereas the model is difficult to apply to the Gulf of Aden where syn-rift dikes are absent.

Based on a synthesis of the timing of geologic events and the evolving geometry of the Arabian region, Bosworth et al. (2005) favored the interpretation that the principal driving forces for rifting in both the Gulf of Aden and Red Sea were far-field stresses, principally slab pull beneath the approaching Urumieh–Dokhtar arc (McQuarrie et al. 2003). Bosworth et al. considered the trigger for rifting to be the impingement of the Afar plume at ~31 Ma, and the actual onset of full-length Red Sea rifting the ~24–22 Ma dike event. The Bosworth et al. interpretation for the Red Sea is therefore compatible with Buck's theoretical considerations. The geometric association of the Gulf of Aden, Red Sea, and Ethiopian Rift with Afar—the classic Afar rift–rift–rift triple junction (Mohr 1970; McKenzie et al. 1970; Burke and Dewey 1973)—proves a role for the plume in rifting here.

It is critical, however, to recognize that during eruption of the Afar flood basalts, a triple junction did not exist (Fig. 13a) and that the various arms of the rift system evolved thereafter at discernibly different geologic times (Bosworth et al. 2005).

Reilinger and McClusky (2011) have synthesized geodetic (GPS) and plate tectonic observations for the Nubia–Arabia–Eurasia plate system and the Mediterranean basin. Their analysis further supports the interpretation that the primary driving force in this system derives from subduction of Neotethyan lithosphere beneath Eurasia (Jolivet and Faccenna 2000; McQuarrie et al. 2003; Faccenna et al. 2013). They correlate the initiation of extensional tectonics in the Mediterranean (Alboran, Belearic, and Aegean basins) with the ~50 % decrease in the rate of convergence between Africa and Eurasia at 24 ± 4 Ma, the onset of rifting in the Red Sea (they include continental rifting in the Gulf of Aden, which is incorrect as it is older; see discussion above). A second ~50 % decrease in convergence occurred when oceanic rifting developed along the entire Gulf of Aden and movement shifted from the Gulf of Suez to the Aqaba–Levant transform margin at 11 ± 2 Ma. This was also coeval with intensified extension in numerous Mediterranean basins.

Rift Propagation

Early models of "rift" propagation often really focused on how oceanic rifts (i.e., spreading centers/mid-ocean ridges) progress laterally through a precursor zone of continental extension (e.g., Vink 1982; Courtillot 1982; Martin 1984). Important questions included how much continental extension predates breakthrough of the oceanic rift, how the extension was physically accommodated (faulting versus igneous intrusion), the resulting diachronous nature of the continent–ocean boundary, and the relative roles of rigid plates and micro-plates (Burke and Whiteman 1973; Burke and Dewey 1974; Hey 1977; LePichon and Sibuet 1981). In the first step of these models, the onset of continental rifting is instantaneous across the ruptured plate. In later steps, the oceanic rift propagates over time through the extended lithosphere (generally toward a pole of rotation). An additional observation by Martin (1984) was that an oceanic spreading center could appear within a segment of stretched continental lithosphere without connection to a propagating ridge which is of relevance to the spreading center of the southern Red Sea.

Instantaneous rupture of the lithospheric plate is a requirement if plate tectonics is applied rigorously at all scales of observation. Alternatively if some degree of distributed deformation is allowed, then the rift zone can propagate by a process of "unzipping" (Omar and Steckler 1995). It is important to emphasize that Omar and Steckler were specifically discussing the onset of continental rifting; regardless of how this early separation occurred, the later development of an oceanic rift could proceed by a similar or different geometrical/kinematic history. Based on fission track data discussed above, Omar and Steckler concluded that continental rifting of the Red Sea occurred in two phases at ~34 Ma and 25–21 Ma and that in both instances the rifting was simultaneous along its entire length. No propagation could be discerned, and the tenets of rigid plate tectonics were upheld.

The end members of rift propagation models are illustrated in Fig. 14a, b. A minor modification of the propagating rift is the inclusion of a finite, approximately constant width rift valley that does the actual propagation. For the Red Sea and Gulf of Aden, Bosworth et al. (2005) suggested this would have been on the order of 60–80 km. Courtillot (1982) emphasized the importance of "locked zones" that impede the lateral growth of rifts. This process can occur in either continental or oceanic rift systems and is particularly well displayed in the oceanic rift history of the Gulf of Aden (Manighetti et al. 1997) as discussed above. Wijk and Blackman (2005) referred to this as a stalled rift mode. The stalling or locking of a rift would result in pulses of extension along a rift system (Fig. 14c). Between successive locked zones, the rift could either propagate laterally or instantaneously rupture the lithosphere (hybrid of either Fig. 14a or b). The locked zones could correspond to the positions of pre-rift lithospheric structures that might be favorably oriented to evolve into transform faults in the ultimate oceanic rift system. If unfavorably oriented, they might result in distributed deformation and not lead directly to a simple transform geometry (Fig. 14c).

Timing of rifting for the Gulf of Aden and Red Sea summarized above suggests that these basins experienced a stalled rift mode not only during oceanic rift propagation but also during the precursor continental rifting. By the Late Oligocene (~27.5 Ma), extension and syn-rift sedimentation were occurring throughout the Gulf of Aden and into the southernmost Red Sea offshore present-day Eritrea. Rifting did not move north, based on presently available data, until about 3.5 My later. When the next rift segment developed, it jumped immediately to northern Egypt based on the age of the associated dike event (Table 1), as proposed by Omar and Steckler (1995). Lyakhovsky et al. (2012) have emphasized that an important parameter during rift propagation is the long-term memory effect of fractured rocks. It is possible that during the 3.5 My stall, the pre-Red Sea lithosphere was being pre-conditioned/weakened for rifting by a period of focused brittle deformation without significant surface extension.

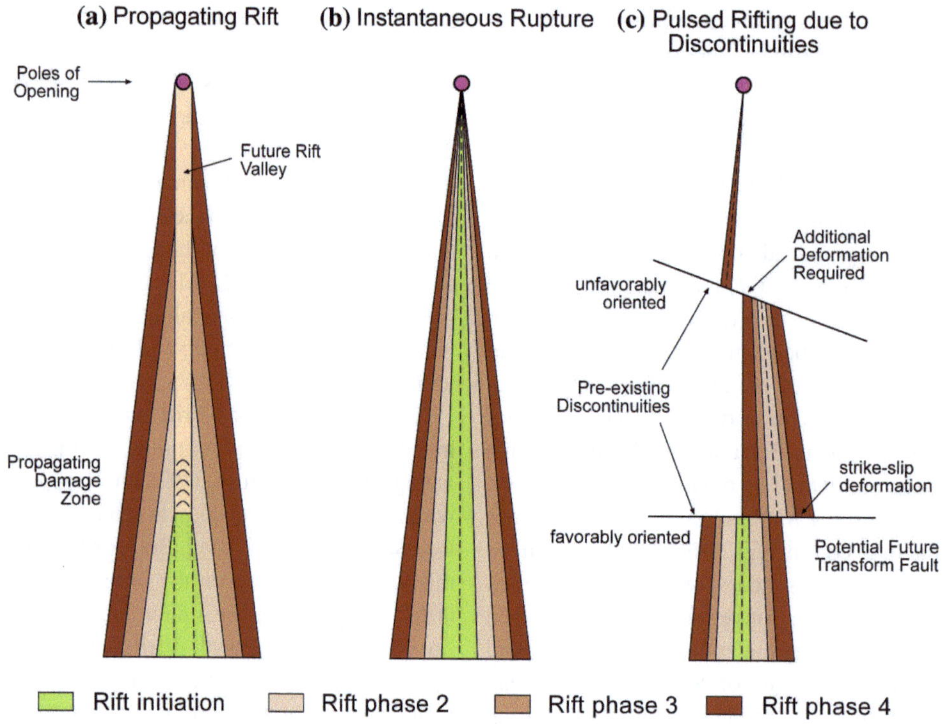

Fig. 14 Modes of continental rift propagation. **a** and **b** are traditional models of propagating (unzipping) and instantaneously rupturing rifts; **c** is a pulsed rift history in which the basin undergoes phases of stalling or locking at pre-existing discontinuities. After passing a locked zone, the rift can switch to the style of either (**a**) or (**b**). The stalled mode is common in oceanic rifts (Courtillot 1982; Van Wijk and Blackman 2005). The colors are meant to show the overall width of the rift growing through time (stretching of the lithosphere) and not age of accreted material which would be the case in oceanic rifts. The Gulf of Aden–Red Sea rift system developed as in (**c**), with individual segments rupturing instantaneously as far as geologic data can presently discern

Northern and Southern Red Sea Comparison

A plausible Red Sea regional model can be summarized as follows: (1) The early phase of purely continental rifting is represented by the Gulf of Suez; (2) a continental rift that is beginning to show some expression of oceanic rifting is seen in the northern Red Sea; and (3) a young oceanic basin that has entered the early phase of drift is found in the southern Red Sea (see discussions in Cochran 1983, 2005; Bonatti 1985; Martinez and Cochran 1988; Girdler 1991; Coleman 1993). In line with this often accepted progression of rifting styles, Cochran (2005) interpreted the presence of large, rotated fault blocks in the northern Red Sea with faults that sole into a zone of plastic creep in the lower crust. This keeps the Moho flat but allows the observed high upper crustal relief. Through time extension is focused at the rift axis, the lithosphere thins rapidly and melt is generated. One product of this is the small axial volcanoes present in some of the axial deeps, which with further extension and magmatism would coalesce into cells of seafloor spreading.

It is now recognized that some of the Earth's magma-poor continental margins did not develop an oceanic spreading center until they were truly hyperextended with continental mantle rocks exposed at the seafloor (Whitmarsh et al. 2001; Henning et al. 2004; Sutra and Manatschal 2012 and references therein). Developed originally with data and interpretations from the west Iberia margin and Alpine Tethyan exposures, the models proposed for these margins invoke development of a wide zone of upper crustal rift basins and rotated fault blocks above ductile, distributed deformation in a weak lower or middle crust. Deformation history and geometry are complex and varied, but a common theme is that eventually concave downward exhumation faults bring subcontinental mantle rocks to the seafloor. An oceanic spreading center is delayed until later and ultimately forms in a position controlled by weaknesses in the subcontinental lithospheric mantle and the thermal structure of the rising asthenospheric mantle. The juncture between the hyperextended crust (≤10 km) and normal crust (~30 km) is commonly abrupt and is referred to as the "necking zone" (Mohn et al. 2012). The northern Red Sea at the Brothers Islands is ~190 km wide today (Fig. 2b). In a pre-rift configuration, this area was ~45 km wide (Fig. 6). Ignoring the role of the low-strain onshore fault blocks and assuming purely mechanical extension, this gives a β factor of greater than four—this is certainly a candidate for a hyperextended terrane. As discussed above, geophysical data suggest that there is also a necking zone along both coastlines though most pronounced on the Egyptian margin (Voggenreiter et al. 1988).

Much less is known about the structure of the southern Red Sea, though there is no question that the amount of opening is considerably greater than in the north (Le Pichon and Francheteau 1978; Joffe and Garfunkel 1987). A free-air gravity anomaly map derived from satellite altimetry (Smith and Sandwell 1997) presented by Cochran and Karner (2007) lacks evidence of prominent linear rift-parallel gravity highs and lows like those of the north, and this is interpreted to reflect a lack of large, rotated fault blocks. Cochran and Karner (2007) suggested that the southern Red Sea resembles the West African and Brazilian passive continental margins in this respect, where regional syn-rift sag basins dominate (Karner et al. 2003). They proposed that proximity to the Afar plume weakened the lithosphere beneath the southern Red Sea (Burke 1996; Courtillot et al. 1999) and that this is resulting in widely differing responses to extension in the north and south.

In addition to the effects of distance from the Afar plume, other authors have described the role that the Aqaba–Levant transform boundary (Dead Sea fault) has played in differentiating the behavior of the northern Red Sea from that of the south (Ben Avraham 1985, 1987; Ben-Avraham and Von Herzen 1987; Ben-Avraham et al. 2008; Lazar et al. 2012). It seems very plausible that the presence of this plate boundary should have a significant impact on mantle convection beneath the northern Red Sea and perhaps the mode of crustal deformation.

Figure 13c shows a compilation of where continental extension was occurring at ~24 Ma, when the main Red Sea dike event took place and deformation reached the margin of Neotethys in the north. In the southern and central Red Sea, crustal effects of rifting (faulting, volcanism) never exceeded about 200 km in width. In the north, however, deformation was distributed over a much broader area, from the Western Desert of Egypt to the Azraq–Sirhan graben in Jordan and Israel or on the order of 1,000 km (Segev and Rybakov 2011; Lyakhovsky et al. 2012; Szymanski 2013; Bosworth et al. 2015). Significant Red Sea extension or volcanism is not observed north of the Egyptian Mediterranean coastline. If the Mediterranean basin overlies oceanic lithosphere, it would generally be too strong to rupture (e.g., Vink et al. 1984). Steckler and ten Brink (1986) suggested that this is why the Red Sea plate boundary subsequently shifted to the Aqaba–Levant transform and abandoned the Gulf of Suez (see further discussion in Lyakhovsky et al. 2012). The very broad zone of early northern Red Sea extension might reflect initial interaction of the Red Sea rift with this impenetrable oceanic barrier.

Two Plume Model

The 24–22 Ma basaltic volcanism of northern Egypt is not sufficiently voluminous to be considered the product of a mantle plume in a traditional sense. However, the minimum preserved subsurface area of 15,000 km^2 and an average thickness of ~30 m (Fig. 11) give an erupted volume of 450 km^3 to which should be added the smaller flows of the northern Gulf of Suez and Bahariya Oasis. As discussed above, trace element and Sr–Nd–Pb–Hf isotope analyses suggest that these sub-alkaline basalts were derived from mixing of an Afar plume-like source with metasomatized spinel-facies continental lithosphere (Endress et al. 2011). Endress et al. relate this and other North African mid-Cenozoic magmatic activity to upwelling mantle material from the northern margins of the South African Superplume. Bosworth et al. (2015) refer the northern Egypt volcanism to the effects of a Cairo "mini-plume" (Fig. 15).

Field and subsurface observations indicate that the Cairo and Gulf of Suez volcanism were coeval with the onset of NE–SW extension and the deposition of syn-rift sediments (see

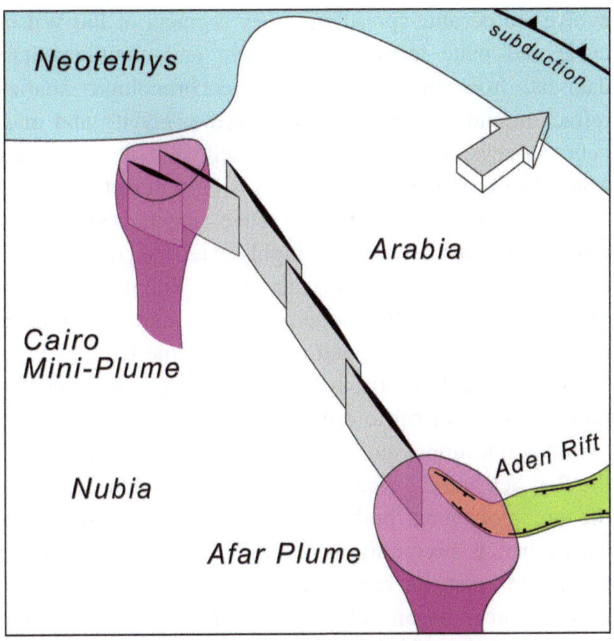

Fig. 15 Two-plume model for the origin of the Red Sea continental rift. Basaltic dikes (shown in *gray*) that link the Afar plume to the Cairo mini-plume at 24–22 Ma are drawn schematically. At the surface, they are much smaller, narrower, and numerous; their geometry at depth is unknown. The feathering into a broader zone of intrusion in the north however is real

above). For most of the area of the Cairo basalts, the amount of extension is very small, and structuring did not significantly impact the distribution of the main flows. But initial rift faults did occupy the area of just east of Bahariya to the east side of the Gulf of Suez, with the basaltic volcanism focused at the middle of this broad zone. Only minor basaltic dikes and flows are actually associated with the ultimate focus of extension along the Gulf of Suez and Northern Red Sea. For these reasons, it seems that rifting did not generate the basalts, but rather the basalts may have helped control the direction of the rift as it shot north from Eritrea, perhaps in conjunction with a stress concentration at the bend in the Levant–Mediterranean continental margin (Burke 1996). The Cairo mini-plume may have acted as a trigger for the last phase of Red Sea rift propagation, similar to the role proposed for the Afar plume for Gulf of Aden continental rifting (Bosworth et al. 2005).

Summary

The Red Sea is part of an extensive rift system that includes from south to north the oceanic Sheba Ridge, the Gulf of Aden, the Afar region, the Red Sea, the Gulf of Aqaba, the Gulf of Suez, and the Cairo basalt province. Historical interest in this area has stemmed from many causes with diverse objectives, but it is best known as a potential model for how continental lithosphere first ruptures and then evolves to oceanic spreading, a key segment of the Wilson cycle and plate tectonics. Abundant and complementary datasets, from outcrop geology, geochronologic studies, refraction and reflection seismic surveys, gravity and magnetic surveys, to geodesy, have facilitated these studies. Magnetically striped oceanic crust is present in the Gulf of Aden and southern Red Sea, active magma systems are observed onshore in the Afar, highly extended continental or mixed crust submerged beneath several kilometers of seawater is present in the northern Red Sea, and a continental rift is undergoing uplift and exposure in the Gulf of Suez. The greater Red Sea rift system therefore provides insights into all phases of rift-to-drift histories.

Many questions remain about the subsurface structure of the Red Sea and the forces that led to its creation. However, the timing of events, both in an absolute sense and relative to each other, is becoming increasingly well constrained. Six main steps may be recognized: (1) Plume-related basaltic trap volcanism began in Ethiopia, NE Sudan (Derudeb), and SW Yemen at ∼31 Ma, followed by rhyolitic volcanism at ∼30 Ma. Volcanism thereafter spread northward to Harrats Sirat, Hadan, Ishara-Khirsat, and Ar Rahat in western Saudi Arabia. This early magmatism occurred without significant extension or at least none that has yet been demonstrated. It is often suggested that this "Afar" plume triggered the onset of Aden–Red Sea rifting, or in some models, it was the main driving force. (2) Starting between ∼29.9 and 28.7 Ma, marine syn-tectonic sediments were deposited on continental crust in the central Gulf of Aden. Therefore, Early Oligocene rifting is established to the east of Afar. Whether rifting propagated from the vicinity of the Sheba Ridge toward Afar, or the opposite, or essentially appeared synchronously throughout the Gulf of Aden is not yet known. 3) By ∼27.5–23.8 Ma, a small rift basin was forming in the Eritrean Red Sea. At approximately the same time (∼25 Ma), extension and rifting commenced within Afar itself. The birth of the Red Sea as a rift basin is therefore a Late Oligocene event. (4) At ∼24–23 Ma, a new phase of volcanism, principally basaltic dikes but also layered gabbro and granophyre bodies, appeared nearly synchronously throughout the entire Red Sea, from Afar and Yemen to northern Egypt. The result was that the Red Sea rift briefly linked two very active volcanic centers covering 15,000–25,000 km^2 in the north and >600,000 km^2 in the south. The presence of the "mini-plume" in northern Egypt may have played a role somewhat analogous to Afar vis-à-vis the triggering of the dike event. The 24–23 Ma magmatism was accompanied by strong rift-normal extension and deposition of syn-tectonic sediments, mostly of marine and marginal marine affinity. The area of extension in the north was very broad, on the order of 1,000 km, and much narrower in the south, about 200 km or less. Throughout the Red Sea, the principal phase of rift shoulder uplift and rapid syn-rift subsidence followed shortly thereafter. Synchronous with the appearance of extension throughout the entire Red Sea, relative convergence between Africa and Eurasia slowed by about 50 %. (5) At ∼14–12 Ma, a transform boundary cut through Sinai and the Levant continental margin, linking the northern Red Sea with the Bitlis–Zagros convergence zone. This corresponded with collision of Arabia and Eurasia, which resulted in a new plate geometry with different boundary forces. Red Sea extension changed from rift normal (N60°E) to highly oblique and parallel to the Aqaba–Levant transform (N15°E). Extension across the Gulf of Suez decreased by about a factor of 10, and convergence between Africa and Eurasia again dropped by about 50 %. In the Afar region, Red Sea extension shifted from offshore Eritrea to west of the Danakil horst, and activity began in the northern Ethiopian rift. (6) These early events or phases all took place within continental lithosphere and formed a continental rift system 4,000 km in length. When the lithosphere was sufficiently thinned, an organized oceanic spreading center was established and the rift-to-drift transition started. Oceanic spreading initiated first on the Sheba Ridge east of the Alula-Fartaq fracture zone at ∼19–18 Ma. After stalling at this fracture zone, the ridge probably propagated west into the central Gulf of Aden by ∼16 Ma. This matches the observed termination of syn-tectonic deposition along the onshore Aden margins at approximately the same time. At ~10 Ma, the Sheba Ridge rapidly propagated west over 400 km from the central Gulf of Aden to the Shukra al Sheik discontinuity. Oceanic spreading followed in the south-central Red Sea at ~5 Ma. This spreading center was initially not connected to the spreading center of the Gulf of Aden. By ~3–2 Ma, oceanic spreading moved west of the Shukra al Sheik discontinuity, and the entire Gulf of Aden was an oceanic rift. During the last ~1 My, the southern Red Sea plate boundary linked to the Aden spreading center through the Gulf of Zula, Danakil Depression, and Gulf of Tadjoura.

Presently, the Red Sea spreading center may be propagating toward the northern Red Sea to link with the Aqaba–Levant transform. However, important differences appear to exist between the southern and northern Red Sea basins, both in terms of the nature of the pre- to syn-rift lithospheric properties and the response to plate separation. If as favored here no oceanic spreading is present in the northern Red Sea, then it is a magma-poor hyperextended basin with β factor >4 that is evolving in many ways like the west Iberia margin. Testing, this hypothesis will require the acquisition of deep reflection seismic profiles from margin to margin and the drilling of more offshore wells (Steckler et al. 2001). It is probable that the ultimate geometries of the northern and southern Red Sea passive margins will be very different. The Red Sea provides an outstanding area in which to study the rift-to-drift transition of continental disruption, but it is unlikely to be a precise analogue for all passive continental margin histories.

Acknowledgments Philippe Huchon and Ken McClay collaborated with me on an earlier synthesis of the Red Sea and Gulf of Aden, and their past research and interpretations have heavily influenced what is presented in this chapter. Daniel Stockli, Daniel Helgeson, and Michael Cosca are co-researchers on the Cairo plume and circa 23 Ma dike event, and their efforts and perseverance are very much appreciated. James Cochran very kindly provided me with his digital compilation of northern Red Sea bathymetric data. I have had the great pleasure to work with many scientists in the field in the Gulf of Suez and Red Sea and I thank them all. Nickolas Raterman and William Wescott kindly reviewed the manuscript and suggested many important improvements. Stockli and Raterman impressed upon me the similarities of magma-poor hyperextended margins and the northern Red Sea. Kevin Burke has been my patient mentor regarding all things rifting, plumes, and the Red Sea in particular. I sincerely thank him for this.

References

Abbate E, Balestrieri ML, Bigazzi G (2002) Morphostructural development of the Eritrean rift flank (southern Red Sea) inferred from apatite fission track analysis. J Geophys Res 107(B11):11

Ali Kassim M, Carmignani L, Conti P, Fantozzi PL (2002) Geology of the Mesozoic-Tertiary sedimentary basins in southwestern Somalia. J Afr Earth Sci 34:3–20

Allan TD (1970) Magnetic and gravity fields over the Red Sea. Philos Trans R Soc Lond 267:153–180

Allan TD, Charnock H, Morelli C (1964) Magnetic, gravity and depth surveys in the Mediterranean and Red Sea. Nature 204:1245–1248

Alsharhan AS (2003) Petroleum geology and potential hydrocarbon plays in the Gulf of Suez rift basin, Egypt. Am Assoc Pet Geol Bull 87:143–180

Ameen MS (2014) Fracture and in-situ stress patterns and impact on performance in the Khuff structural prospects, eastern offshore Saudi Arabia. Mar Pet Geol 50:166–184

Anderson DL (1982) Hotspots, polar wander, Mesozoic convection, and the geoid. Nature 297:391–393

Angelier J (1985) Extension and rifting: the Zeit region, Gulf of Suez. J Struct Geol 7:605–612

Argand E (1924) La Tectonique de l'Asie. In: Proceedings of the 13th international geological congress, vol 1, pp 171–372

ArRajehi A, McClusky S, Reilinger R, Daoud M, Alchalbi A, Ergintav S, Gomez F, Sholan J, Bou-Rabee F, Ogubazghi G, Haileab B, Fisseha S, Asfaw L, Mahmoud S, Rayan A, Bendik R, Kogan L (2010) Geodetic constraints on present-day motion of the Arabian Plate: implications for Red Sea and Gulf of Aden rifting. Tectonics 29

Audin L (1999) Pénétration de la dorsale d'Aden dans la depression Afar entre 20 et 4 Ma. PhD thesis, Université de Paris 7 et Institut de Physique du Globe de Paris, Paris, 278 pp

Audin L, Quidelleur X, Coulié E, Courtillot V, Gilder S, Manighetti I, Gillot P-Y, Tapponnier P, Kidane T (2004) Paleomagnetism and K-Ar and 40Ar/39Ar ages in the Ali Sabieh area (Republic of Djibouti and Ethiopia): constraints on the mechanism of Aden ridge propagation into southeastern Afar during the last 10 Myr. Geophys J Int 158:327–345

Bailey DK (1992) Episodic alkaline igneous activity across Africa; implications for the causes of continental break-up. Geol Soc London Spec Publ 68:91–98

Bailey DK (1993) Carbonate magmas. J Geol Soc London 150:637–651

Baker BH (1970) The structural pattern of the Afro-Arabian rift system in relation to plate tectonics. Philos Trans Royal Soc London Series A Math Phys Sci 267:383–391

Baker J, Menzies M, Snee L (1994) Stratigraphy, $^{40}Ar/^{39}Ar$ geochronology and geochemistry of flood volcanism in Yemen. Mineral Mag 58A:42–43

Baker J, Snee L, Menzies M (1996) A brief Oligocene period of flood volcanism in Yemen: implications for the duration and rate of continental flood volcanism at the Afro-Arabian triple junction. Earth Planet Sci Lett 138:39–55

Barakat H, Miller P (1984) Geology and petroleum exploration, Safaga Concession, northern Red Sea, Egypt. In: Proceedings of the 7th exploration seminar, Egyptian General Petroleum Corporation, Cairo, pp 191–214

Barberi F, Giglia G, Marinelli G, Santacroce R, Tazieff H, Varet J (1971) Carte geologique du Nord de l'Afar, scale 1:500,000. CNR-CNRS, La Celle-St Cloud

Barberi F, Borsi S, Ferrara G, Marinelli G, Santacroce R, Tazieff H, Varet J (1972) Evolution of the Danakil Depression (Afar, Ethiopia) in light of radiometric age determinations. J Geol 80:720–729

Barberi F, Santacroce R, Varet J (1975) Structural evolution of the Afar triple junction. In: Pilger A, Rösler A (eds) Afar depression of Ethiopia, Proceedings of an international symposium on the afar region and related rift problems, vol 1. Bad Bergzabern FR, Germany, 1–6 April 1974. E. Schweizerbartsche Verlagsbuchhandlung, Stuttgart, pp 38–54

Basile C, Mascle J, Guiraud R (2005) Phanerozoic geological evolution of the equatorial atlantic domain. In: Catuneanu O, Guiraud R, Eriksson P, Thomas B, Shone R, Key R (eds) Phanerozoic evolution of Africa. J Afr Earth Sci 43:275–282

Ben-Avraham Z (1985) Structural framework of the Gulf of Elat (Aqaba)—northern Red Sea. J Geophys Res 90:703–726

Ben-Avraham Z (1987) Rift propagation along the southern Dead Sea rift (Gulf of Elat). Tectonophysics 143:193–200

Ben-Avraham Z, Von Herzen RP (1987) Heat flow and continental breakup: the Gulf of Elat (Aqaba). J Geophys Res 92:1407–1416

Ben-Avraham Z, Garfunkel Z, Lazar M (2008) Geology and evolution of the southern Dead Sea fault with emphasis on subsurface structure. Annu Rev Earth Planet Sci 36:357–387

Bender F (1975) Geology of the arabian peninsula, Jordan. U.S. Geological Survey Professional Paper 560-D

Berhe SM (1986) Geologic and geochronologic constraints on the evolution of the Red Sea-Gulf of Aden and Afar depression. J Afr Earth Sci 5:101–117

Beydoun ZR (1978) Southern Arabia and northern Somalia: comparative geology. Philosophical Transactions of the Royal Society (London) Series A, pp 267–292

Beydoun ZR (1989) Hydrocarbon prospects of the Red Sea-Gulf of Aden; a review. J Pet Geol 12:125–144

Beydoun ZR, Sikander AH (1992) The Red Sea-Gulf of Aden: re-assessment of hydrocarbon potential. Mar Pet Geol 9:474–485

Bicknell JD, MacDonald KC, Miller SP, Lonsdale PF, Becker K (1986) Tectonics of the Nereus Deep, Red Sea: a deep tow investigation of a site of initial rifting. Mar Geophys Res 8:131–148

Blank HR, Mooney WD, Healy JH, Gettings MA, Lamson RJ (1986) A seismic refraction interpretation of the eastern margin of the Red Sea depression, Southwest Saudi Arabia. U.S. Geological Survey Open-File Report 86–257, 20 pp

Blow WH (1969) Late middle Eocene to recent planktonic foraminiferal biostratigraphy. In: Bronniman R (ed) Proceedings of the 1st international conference on planktonic microfossils, vol 1. Geneva, 1967, Brill, Leiden, pp 199–421

Bohannon RG (1986) Tectonic configuration of the western Arabian continental margin, southern Red Sea. Tectonics 5:477–499

Bohannon RG, Naeser CW, Schmidt DL, Zimmerman RA (1989) The timing of uplift, volcanism, and rifting peripheral to the Red Sea: a case for passive rifting? J Geophys Res 94:1683–1701

Bonatti E (1985) Punctiform initiation of seafloor spreading in the Red Sea during transition from a continental to an oceanic rift. Nature 316:33–37

Bonatti E, Colantoni P, Della Vedova B, Taviani M (1984) Geology of the Red Sea transitional zone (22°N–25°N). Oceanol Acta 7:385–398

Bosence DWJ, Al-Awah MH, Davison I, Rosen BR, Vita-Finzi C, Whittaker E (1998) Salt domes and their control on basin margin sedimentation: a case study from the Tihama plain, Yemen. In: Purser BH, Bosence DWJ (eds) Sedimentation and tectonics in Rift Basins—Red Sea—Gulf of Aden. Chapman and Hall, London, pp 448–466

Bosworth W (1985) Geometry of propagating continental rifts. Nature 316:625–627

Bosworth W (1992) Mesozoic and early Tertiary rift tectonics in East Africa. Tectonophysics 209:115–137

Bosworth W (1993) Nature of the Red Sea crust: a controversy revisited: comment. Geology 21:574–575

Bosworth W (1994) A model for the three-dimensional evolution of continental rift basins, north-east Africa. Geol Rundsch 83:671–688

Bosworth W (1995) A high-strain rift model for the southern Gulf of Suez (Egypt). In: Lambiase JJ (ed) Hydrocarbon habitat in Rift basins. Geological Society, London, Special Paper 80, pp 75–102

Bosworth W, Burke K (2005) Evolution of the Red Sea—Gulf of Aden rift system. In: Post PJ, Rosen NC, Olson DL, Palmes SL, Lyons KT, Newton GB (eds) Petroleum systems of divergent continental margin basins. 2005 Gulf Coast Section SEPM Foundation 25th Bob F. Perkins Annual Research Conference, Houston, 4–7 Dec, 2005, CD-ROM, pp 342–372

Bosworth W, McClay K (2001) Structural and stratigraphic evolution of the Gulf of Suez rift, Egypt: A synthesis. In: Ziegler PA, Cavazza W, Robertson AHF, Crasquin-Soleau S (eds) Peri-Tethys Memoir 6: Peri-Tethyan Rift/Wrench basins and passive margins, vol 186. Mémoires du Muséum National d'Histoire Naturelle de Paris, pp 567–606

Bosworth W, Strecker MR (1997) Stress field changes in the Afro-Arabian rift system during the Miocene to recent period. Tectonophyscis 278:47–62

Bosworth W, Crevello P, Winn RD Jr, Steinmetz J (1998) Structure, sedimentation, and basin dynamics during rifting of the Gulf of Suez and northwestern Red Sea. In: Purser BH, Bosence DWJ (eds) Sedimentation and Tectonics of Rift basins: Red Sea-Gulf of Aden. Chapman and Hall, London, pp 77–96

Bosworth W, Darwish M, Crevello P, Taviani M, Marshak S (1996) Stratigraphic and structural evolution of Zabargad Island (Red Sea, Egypt) since the Early Cretaceous. In: Youssef El SA (ed) Proceedings of the 3rd international conference on geology of the Arab World, vol 1, pp 161–190

Bosworth W, Huchon P, McClay K(2005) The Red Sea and Gulf of Aden Basins. In: Catuneanu O, Guiraud R, ErikssonP, Thomas B, Shone R, Key R(eds) Phanerozoic evolution of Africa. J Afr Earth Sci 43:334–378

Bosworth W, Khalil S, Clare A, Comisky J, Abdelal H, Reed T, Kokkoros G (2012) Integration of outcrop and subsurface data during the development of a naturally fractured Eocene carbonate reservoir at the East Ras Budran concession, Gulf of Suez, Egypt. In: Spence GH, Redfern J, Aguilera R, Bevan TG, Cosgrove JW, Couples GD, Daniel J-M (eds) Advances in the study of fractured reservoirs, vol 374. Geological Society, London, Special Publications, 27 pp

Bosworth W, Stockli DF, Helgeson DE, Cosca M (2015) Integrated outcrop, 3D seismic, and geochronologic interpretation of Red Sea dike-related deformation (ca. 23 Ma) in the Western Desert, Egypt—the role of the Cairo "mini-plume" (in press)

Bott WF, Smith BA, Oakes G, Sikander AH, Ibrahim AI (1992) The tectonic framework and regional hydrocarbon prospectivity of the Gulf of Aden. J Pet Geol 15:211–243

Brinckmann J, Kursten M (1969) Geological sketchmap of the Danakil depression, 1: 250,000 scale. Bundesanstaldt für Bodenforschung, Hanover

Brown GF (1972) Tectonic map of the Arabian Peninsula: Map AP-2, scale 1: 4,000,000. Saudi Arabian Directorate General of Mineral Resources, Jeddah

Brown GF, Schmidt DL, Huffman ACJ (1989) Geology of the Arabian Peninsula, shield area of western Saudi Arabia. U.S. Geological Survey Professional Paper 560-A, 188 pp

Buck WR (2006) The role of magma in the development of the Afro-Arabian rift system. In: Yirgu G, Ebinger CJ, Maguire PKH (eds) The Afar volcanic province within the East African Rift System. Geological Society Special Publication, London, vol 259, pp 43–54

Buck WR, Martinez F, Steckler MS, Cochran JR (1988) Thermal consequences of lithospheric extension: pure and simple. Tectonics 7:213–234

Bumby AJ, Guiraud R (2005) The geodynamic setting of the Phanerozoic basins of Africa. J Afr Earth Sc 43:1–12

Bunter MAG, Abdel Magid AEM (1989) The Sudanese Red Sea; 1, New developments in stratigraphy and petroleum-geological evolution. J Pet Geol 12:145–166

Burke K (1996) The African plate. S Afr J Geol 99:341–409

Burke K, Dewey JF (1973) Plume-generated triple junctions: key indicators in applying plate tectonics to old rocks. J Geol 81:406–433

Burke K, Dewey JF (1974) Two plates in Africa in the cretaceous? Nature 249:313–316

Burke K, Gunnell Y (2008) The African erosion surface; a continental-scale synthesis of geomorphology, tectonics, and environmental change over the past 180 million years. Geological Society of America Memoir, vol 201, 66 pp

Burke K, Whiteman AJ (1973) Uplift, rifting and the break-up of Africa. In: Tarling DH (ed) Proceedings of the NATO conference on continental drift, Newcastle, vol 2. Academic Press, London, pp 735–745

Burke K, Wilson JT (1972) Is the African plate stationary? Nature 239:387–390

Canuti P, Gregnanin A, Piccirillo EM, Sagri M, Tacconi P (1972) Volcanic intercalation in the Mesozoic sediments of the Kulubi area (Harrar, Ethiopia). Bollettino della Societa Geologica Italiana 91:603–614

Carr ID, Gawthorpe RL, Jackson CAL, Sharp IR, Sadek A (2003) Sedimentology and sequence stratigraphy of early syn-rift tidal sediments: the Nukhul Formation, Suez Rift. Egypt. J Sediment Res 73:407–420

Chernet T, Hart WK, Aronson JL, Walter RC (1998) New age constraints on the timing of volcanism and tectonism in the northern Main Ethiopian Rift-southern Afar transition zone (Ethiopia). J Volcanol Geoth Res 80:267–280

Clin M, Pouchan P (1970) Carte Geologique du Territore Française des Afars et des Issas, 1:200,000 scale. Centre d'Etudes Geologique et de Development du T.F.A.I, Djibouti

Cochran JR (1983) A model for development of the Red Sea. Bull Am Assoc Petrol Geol 67:41–69

Cochran JR (2005) Northern Red Sea: nucleation of an oceanic spreading center within a continental rift. Geochem Geophys Geosyst 6:Q03006

Cochran JR, Karner GD (2007) Constraints on the deformation and rupturing of continental lithosphere of the Red Sea: the transition from rifting to drifting. In: Karner GD, Manatschal G, Pinheiro LM (eds) Imaging, mapping and modelling continental lithosphere extension and breakup, vol 282. Geological Society, London, Special Publications, pp 265–289

Cochran JR, Martinez F, Steckler MS, Hobart MA (1986) Conrad Deep: a new northern Red Sea deep: origin and implications for continental rifting. Earth Planet Sci Lett 78:18–32

Cochran JR, Martinez F (1988) Evidence from the northern Red Sea on the transition from continental to oceanic rifting. Tectonophysics 153:25–53

Coleman RG (1974) Geologic background of the Red Sea. In: Whitmarsh RB, Weser OE, Ross DA et al (eds) Initial reports of the deep sea drilling project, vol 23. Government Printing Office, Washington, pp 813–819

Coleman RG (1993) Geologic evolution of the Red Sea. Oxford monographs on geology and geophysics, vol 24. Oxford University Press, Oxford,186 pp

Coleman RG, Gregory RT, Brown GF (1983) Cenozoic volcanic rocks of Saudi Arabia. USGS Open-File Report 83-788, 82 pp

Colletta B, Le Quellec P, Letouzey J, Moretti I (1988) Longitudinal evolution of the Suez rift structure (Egypt). Tectonophysics 153:221–233

Collins WJ (2003) Slab pull, mantle convection, and Pangaean assembly and dispersal. Earth Planet Sci Lett 205:225–237

Coulié E, Quidelleur X, Gillot P-Y, Courtillot V, Lefèvre J-C, Chiesa S (2003) Comparative K-Ar and Ar/Ar dating of Ethiopian and Yemenite Oligocene volcanism: implications for timing and duration of the Ethiopian traps. Earth Planet Sci Lett 206:477–492

Courtillot V (1982) Propagating rifts and continental breakup. Tectonics 1:239–250

Courtillot V, Armijo R, Tapponnier P (1987) Kinematics of the Sinai triple junction and a two phase model of Arabia-Africa rifting. In: Coward MP, Dewey JF, Hancock PL (eds) Continental extensional tectonics, vol 28. Geological Society, London, Special Publication, pp 559–573

Courtillot V, Jaupart C, Manighetti I, Tapponnier P, Besse J (1999) On causal links between flood basalts and continental breakup. Earth Planet Sci Lett 166:177–195

Cross NE, Purser BH, Bosence DWJ (1998) The tectono-sedimentary evolution of a rift margin carbonate platform: Abu Shaar, Gulf of Suez, Egypt. In: Purser BH, Bosence DWJ (eds) Sedimentation and tectonics of rift basins: Red Sea-Gulf of Aden. Chapman and Hall, London, pp 271–295

Davies GF, Richards MA (1992) Mantle convection. J Geol 100:151–206

Davison I (2005) Central Atlantic margin basins of North West Africa: Geology and hydrocarbon potential (Morocco to Guinea). J Afr Earth Sci 43:254–274

Davison I, Al-Kadasi M, Al-Khirbash S, Al-Subbary AK, Baker J, Blakey S, Bosence D, Dart C, Heaton R, McClay K, Menzies M, Nichols G, Owen L, Yelland A (1994) Geological evolution of the southeastern Red Sea Rift margin, Republic of Yemen. Geol Soc Am Bull 106:1474–1493

Degens ET, Ross DA (eds) (1969) Hot brines and recent heavy metal deposits in the Red Sea a geochemical and geophysical account. Springer, New York, 571 pp

Drake CL, Girdler RW (1964) A geophysical study of the Red Sea. Geophys J Roy Astron Soc 8:473–495

Dunbar JA, Sawyer DS (1988) Continental rifting at pre-existing lithospheric weaknesses. Nature 333:450–452

Egyptian General Petroleum Corporation (1996) Gulf of Suez Oil Fields (a comprehensive overview). The Egyptian General Petroleum Corporation, New Maadi, 736 pp

El Atfy H, Brocke R, Uhl D (2013) A fungal proliferation near the probably Oligocene/Miocene boundary, Nukhul Formation, Gulf of Suez, Egypt. J Micropalaeontology 32:183–195

El-Shafy AA (1992) Miocene-Pliocene boundary in the Gulf of Suez-Region, Egypt. In: Proceedings of the 10th petroleum exploration and production conference vol 1. Cairo, November, 1990, Egyptian General Petroleum Corporation, Cairo, pp 213–232

Endress C, Furman T, Abu El-Rus MA, Hanan BB (2011) Geochemistry of 24 Ma basalts from NE Egypt: source components and fractionation history. In Van Hinsbergen DJJ, Buiter SJH, Torsvik TH, Gaina C, Webb SJ (eds) The formation and evolution of Africa: a synopsis of 3.8 Ga of Earth history vol 357. Geological Society Special Publications, London, pp 265–283

Evans AL (1988) Neogene tectonic and stratigraphic events in the Gulf of Suez rift area, Egypt. Tectonophysics 153:235–247

Eyal M, Bartov Y, Shimron AE, Bentor YK (1980) Geological map of the Sinai, scale 1:500,000. Geological Survey of Israel, Jerusalem

Faccenna C, Becker TW, Jolivet L, Keskin M (2013) Mantle convection in the Middle East: reconciling afar upwelling, Arabia indentation and Aegean trench rollback. Earth Planet Sci Lett 375:254–269

Fairhead JD (1988) Mesozoic plate tectonic reconstructions of the central South Atlantic Ocean: the role of the West and Central African rift system. Tectonophysics 155:181–191

Fantozzi PL, Sgavetti M (1998) Tectonic and sedimentary evolution of the eastern Gulf of Aden continental margins: new structural and stratigraphic data from Somalia and Yemen. In: Purser BH, Bosence DWJ (eds) Sedimentation and tectonics of Rift Basins: Red Sea—Gulf of Aden. Chapman and Hall, London, pp 56–76

Féraud G, Zumbo V, Sebai A, Bertrand H (1991) $^{40}Ar/^{39}Ar$ age and duration of tholeiitic magmatism related to the early opening of the Red Sea rift. Geophys Res Lett 18:195–198

Garfunkel Z (1988) Relation between continental rifting and uplifting: evidence from the Suez rift and northern Red Sea. Tectonophysics 150:33–49

Garfunkel Z, Bartov Y (1977) The tectonics of the Suez rift. Geol Surv Isr Bull 71:44–50

Garfunkel Z, Beyth M (2006) Constraints on the structural development of Afar imposed by the kinematics of the major surrounding plates. In: Yirgu G, Ebinger C, Maguire PKH (eds) The structure and evolution of the East African Rift System in the Afar volcanic province, vol 259. Geological Society Special Publication, London, pp 23–42

Gass IG (1970) The evolution of volcanism in the junction area of the Red Sea, Gulf of Aden and Ethiopian rifts. Philos Trans Royal Soc London A 267:369–382

Gaulier JM, Le Pichon X, Lyberis N, Avedik F, Geli L, Moretti I, Deschamps A, Salah H (1988) Seismic study of the crust of the northern Red Sea and Gulf of Suez. Tectonophysics 153:55–88

Gee JS, Kent DV (2007) Source of oceanic magnetic anomalies and the geomagnetic polarity timescale. Treatise Geophys 5:455–507

Genik GJ (1992) Regional framework, structural and petroleum aspects of rift basins in Niger, Chad and the Central African Republic (C.A.R.). Tectonophysics 213:169–185

Geoffroy L, Huchon P, Khanbari K (1998) Did Yemeni tertiary granites intrude neck zones of a stretched continental upper crust? Terra Nova 10:196–200

George R, Rogers N, Kelley S (1998) Earliest magmatism in Ethiopia: evidence for two mantle plumes in one flood basalt province. Geology 26:923–926

Gettings ME, Blank HR, Mooney WD, Healy JH (1986) Crustal structure of southwestern Saudi Arabia. J Geophys Res 91:6491–6512

Geukens F (1966) Geology of the Arabian Peninsula, Yemen. U.S. Geological Survey Professional Paper 560-B

Ghebreab W, Carter A, Hurford AJ, Talbot CJ (2002) Constraints for timing of extensional tectonics in the western margin of the Red Sea in Eritrea. Earth Planet Sci Lett 200:107–119

Gibbs AD (1984) Structural evolution of extensional basin margins. J Geol Soc London 142:609–620

Girdler RW (1958) The relationship of the Red Sea to the East African Rift system. Quar J Geol Soc London 114:79–105

Girdler RW (1966) The role of translational and rotational movements in the formation of the Red Sea and Gulf of Aden. In: The world rift system. Geological Survey of Canada, Canada paper 66-14, pp 65–77

Girdler RW (1970) A review of Red Sea heat flow. Philos Trans Roy Soc London Series A Math Phys Sci 267:191–203

Girdler RW (1991) The Afro-Arabian rift system; an overview. Tectonophysics 197:139–153

Girdler RW, Darracott BW (1972) African poles of rotation. Comments Earth Sci Geophys 2:131–138

Girdler RW, Evans TR (1977) Red sea heat flow. Geophys J Roy Astron Soc 51:245–251

Girdler RW, Southren TC (1987) Structure and evolution of the northern Red Sea. Nature 330:716–721

Girdler RW, Styles P (1974) Two-stage Red Sea floor spreading. Nature 247:7–11

Girdler RW, Underwood M (1985) The evolution of early oceanic lithosphere in the southern Red Sea. Tectonophysics 116:95–108

Girdler RW, Whitmarsh RB (1974) Miocene evaporites in Red Sea cores, their relevance to the problem of the width and age of oceanic crust beneath the Red Sea. In: Whitmarsh RB, Weser OE, Ross DA et al (eds) Initial reports of the deep sea drilling project, vol 23. Government Printing Office, Washington, pp 913–921

GLOBE Task Team and others (Hastings, David A., Paula K. Dunbar, Gerald M. Elphingstone, Mark Bootz, Hiroshi Murakami, Hiroshi Maruyama, Hiroshi Masaharu, Peter Holland, John Payne, Nevin A. Bryant, Thomas L. Logan, J.-P. Muller, Gunter Schreier, and John S. MacDonald) (eds) (1999) The global land one-kilometer base elevation (GLOBE) digital elevation model, Version 1.0. National Oceanic and Atmospheric Administration, National Geophysical Data Center, 325 Broadway, Boulder, Colorado 80303, USA

Gradstein FM, Ogg JG, Smith AG, Bleeker W, Lourens LJ (2004) A new geologic time scale, with special reference to Precambrian and Neogene. Episodes 27:83–100

Granath JW (2001) The Nogal rift of Northern Somalia: Gulf of Aden. Reactivation of a Mesozoic rift. In: Ziegler PA, Cavazza W, Robertson AHF, Crasquin-Soleau S (eds) Peri-Tethys Memoir 6: Peri-Tethyan Rift/Wrench Basins and Passive Margins vol 186, Mémoires du Muséum national d'Histoire naturelle de Paris, pp 511–527

Greenwood JEGW, Bleackley D (1967) Geology of the Arabian Peninsula, Aden Protectorate. U.S. Geological Survey Professional Paper 560-C, pp 1–96

Groenewald PB, Grantham GH, Watkeys MK (1991) Geological evidence for a Proterozoic to Mesozoic link between southeastern Africa and Dronning Maud Land, Antarctica. J Geol Soc London 148:1115–1123

Grolier MJ, Overstreet WC (1978) Geologic Map of Yemen Arab Republic (San'a). U.S. Geological Survey, Miscellaneous Inf. Series, Map 1-1143-B, scale 1:500,000

Guennoc P, Pautot G, Coutelle A (1988) Surficial structures of the northern Red Sea axial valley from 23°N to 28°N: time and space evolution of neo-oceanic structures. Tectonophysics 153:1–23

Guiraud M (1993) Late Jurassic rifting—early cretaceous rifting and late cretaceous transpressional inversion in the upper Benue basin (NE Nigeria. Bull Centres Rech Expor Prod Elf Aquitaine 17:371–383

Guiraud R, Bosworth W (1999) Phanerozoic geodynamic evolution of northeastern Africa and the northwestern Arabian platform. Tectonophysics 315:73–108

Guiraud R, Maurin J-C (1992) Early Cretaceous rifts of western and Central Africa: an overview. Tectonophysics 213:153–168

Guiraud R, Bosworth W, Thierry J, Delplanque A (2005) Phanerozoic geological evolution of Northern and Central Africa: an overview. In: Catuneanu O, Guiraud R, Eriksson P, Thomas B, Shone R, Key R (eds) Phanerozoic evolution of Africa. J Afr Earth Sci 43: 83–143

Hadley DG, Schmidt DL (1980) Sedimentary rocks and basins of the Arabian shield and their evolution. Inst Appl Geol Bull King Abdulaziz University 4:26–50

Hadley DG, Schmidt DL, Coleman (1982) Summary of tertiary investigations in western Saudi Arabia, current work by the U.S. Geological Survey, and recommended future studies. U.S. Geological Survey Open File Report USGS-OF-03-5

Hall SA, Andreason GE, Girdler RW (1977) Total intensity magnetic anomaly map of the Red Sea and adjacent coastal areas, a description and preliminary interpretation, vol 22. Saudi Arabia Directorate General Mineral Resources Bulletin, Red Sea Research 1970–1975: F1–F15

Haq BU, Hardenbol J, Vail PR (1987) Chronology of fluctuating sea-levels since the Triassic. Science 235:1156–1167

Hassan F, El-Dashlouty S (1970) Miocene evaporites of Gulf of Suez region and their significance. Am Assoc Pet Geol Bull 54:1686–1696

Heaton RC, Jackson MPA, Bamahmoud M, Nani ASO (1995) Superposed Neogene extension, contraction and salt canopy emplacement in the Yemeni Red Sea. In: Jackson MPA, Roberts DG, Snelson S (eds) Salt Tectonics: a global perspective, vol 65. American Association of Petroleum Geologists Memoir, pp 333–351

Hébert H, Deplus C, Huchon P, Khanbari K, Audin L (2001) Lithospheric structure of a nascent spreading ridge inferred from gravity data: the western Gulf of Aden. J Geophys Res 106:26345–26363

Hempton MR (1987) Constraints on Arabian plate motion and extensional history of the Red Sea. Tectonics 6:687–705

Henning AT, Sawyer DS, Templeton DS (2004) Exhumed upper mantle within the ocean-continent transition of the northern West Iberia margin: evidence from prestack depth migration and total tectonic subsidence analyses. J Geophys Res 109:B05103

Hewaidy AA, Farouk S, Ayyad HM (2012) Nukhul formation in Wadi Baba, southwest Sinai Peninsula, Egypt. GeoArabia 17:103–120

Hey R (1977) A new class of "pseudofaults" and their bearing on plate tectonics; a propagating rift model. Earth Planet Sci Lett 37:321–325

Holmes A (1965) Principles of physical geology, 2nd edn. Thomas Nelson Ltd, Edinburgh, 1288 pp

Hutchinson RW, Engels GG (1970) Tectonic significance of regional geology and evaporite lithofacies in northeastern Ethiopia. Philos Trans Roy Soc London Series A Math Phys Sci 267:313–329

Hutchinson RW, Engels GG (1972) Tectonic evolution in the southern Red Sea and its possible significance to older rifted continental margins. Geol Soc Am Bull 83:2989–3002

Hughes GW, Beydoun ZR (1992) The Red Sea-Gulf of Aden: biostratigraphy, lithostratigraphy and palaeoenvironments. J Pet Geol 15:135–156

Hughes GW, Filatoff J (1995) New biostratigraphic constraints on Saudi Arabian Red Sea pre- and syn-rift sequences. In: Al-Husseini MI (ed) Middle east petroleum geosciences, Geo'94, vol 2. Gulf PetroLink, Bahrain, pp 517–528

Hughes GW, Johnson RS (2005) Lithostratigraphy of the Red Sea region. GeoArabia 10:49–126

Hughes GW, Abdine S, Girgis MH (1992) Miocene biofacies development and geological history of the Gulf of Suez, Egypt. Mar Pet Geol 9:2–28

Hughes GW, Perincek D, Grainger DJ, Abu-Bshait A-J, Jarad A-RM (1999) Lithostratigraphy and depositional history of part of the Midyan region, northwestern Saudi Arabia. GeoArabia 4:503–541

Hughes GW, Varol O, Beydoun ZR (1991) Evidence for Middle Oligocene rifting of the Gulf of Aden and for Late Oligocene rifting of the southern Red Sea. Mar Pet Geol 8:354–358

Hume WF, Madgwick TG, Moon FW, Sadek H (1920) Preliminary geological report on the Gebel Tanka area, vol 4. Petroleum Research Bulletin, 16 pp

Ilani S, Harlavan Y, Tarawneh K, Rabba I, Weinberger R, Ibrahim K, Peltz S, Steinitz G (2001) New K-Ar ages of basalts from the Harrat Ash Shaam volcanic field in Jordan: implications for the span and duration of the upper-mantle upwelling beneath the western Arabian plate. Geology 29:171–174

Issawi B, El Hinnawi M, Francis M, Mazhar A (1999) The Phanerozoic geology of Egypt: a geodynamic approach. Egypt Geological Survey Paper 76, 462 pp

Jackson CAL (2008) Sedimentology and significance of an early syn-rift paleovalley, Wadi Tayiba, Suez Rift, Egypt. J Afr Earth Sci 52:62–68

Jackson CAL, Gawthorpe RL, Leppard CW, Sharp IR (2006) Rift-initiation development of normal fault blocks: insights from the Hammam Faraun fault block, Suez rift, Egypt. J Geol Soc London 163:165–183

Janssen ME, Stephenson RA, Cloetingh S (1995) Temporal and spatial correlations between changes in plate motions and the evolution of rifted basins in Africa. Geol Soc Am Bull 107:1317–1332

Jarrige JJ, Ott d'Estevou P, Burollet PF, Montenat C, Prat P, Richert JP, Thiriet JP (1990) The multistage tectonic evolution of the Gulf of Suez and northern Red Sea continental rift from field observations. Tectonics 9:441–465

Jarrige JJ, Ott d'Estevou P, Burollet PF, Thiriet JP, Icart JC, Richert JP, Sehans P, Montenat C, Prat P (1986) Inherited discontinuities and Neogene structure: the Gulf of Suez and northwestern edge of the Red Sea. Philos Trans R Soc Lond A317:129–139

Joffe S, Garfunkel Z (1987) Plate kinematics of the circum Red Sea—a re-evaluation. Tectonophysics 141:5–22

Jolivet L, Faccenna C (2000) Mediterranean extension and the Africa–Eurasia collision. Tectonics 19:1095–1106

Juteau T, Eissen J, Monin AS, Zonenshain (Zonenshayn) LP, Sorokhtin OG, Matveenkov (Matveyenko) VV, Almukhamedov A (1983) Structure et petrologie du rift axial de la Mer Rouge vers 18 degrees Nord: Resultats de la campagne sovietique de plongees avec submersible (1980). Bulletin Des Centres De Recherches Exploration-Production Elf-Aquitaine 7, pp 217–231

Kappelman J, Simons L, Swisher CC III (1992) New age determinations for the Eocene-Oligocene boundary sediments in the Fayum depression, northern Egypt. J Geol 100:647–668

Karner G, Driscoll NW, Barker DHN (2003) Syn-rift regional subsidence across the West African continental margin: the role of lower plate ductile extension. In: Arthur TJ, MacGregor DS, Cameron NR (eds) Petroleum geology of Africa: new themes and developing technologies, vol 207. Geological Society Special Publications, London, pp 105–129

Kassas M, Zahran MA (1971) Plant life on the coastal mountains of the Red Sea, Egypt. J Indian Bot Soc Golden Jubilee 50A:571–589

Kazmin VG (1973) Geological map of Ethiopia, scale 1:2,000,000. Ministry of Mines, Geological Survey of Ethiopia, Addis Ababa

Kazmin VG (1977) Characteristics of geodynamic evolution of the Afro-Arabian rift system. Izd. Nauka, Sib. Otd, Novosibirsk (in Russian)

Kenea NH, Ebinger CJ, Rex DC (2001) Late Oligocene volcanism and extension in the southern Red Sea Hills, Sudan. J Geol Soc London 158:285–294

Khalil SM (1998) Tectonic evolution of the eastern margin of the Gulf of Suez, Egypt. PhD thesis, Royal Holloway, University of London, 349 pp

Khalil SM, McClay KR (2009) Structural control on syn-rift sedimentation, northwestern Red Sea margin, Egypt. Mar Pet Geol 26:1018–1034

Khan MA (1975) The Afro-Arabian rift system. Science Progress (1916) 62: 207–236

Khedr E (1984) Sedimentological evolution of the Red Sea continental margin of Egypt and its relationship to sea-level changes. Sed Geol 39:71–86

Klitgord KD, Schouten H (1986) Plate kinematics of the Central Atlantic. In: Vogt PR, Tucholke BE (eds) The Western North Atlantic Region (The Geology of North America, v. M). Geological Society of America, pp 351–378

Klitzsch E (1990) Chapter 21. Paleozoic. In: Said R (ed) The geology of Egypt. Balkema, Rotterdam, pp 393–406

Klitzsch E, List FK, Pohlmann G, Handley R, Hermina M, Meissner G (eds) (1986, 1987) Geological map of Egypt: 1:50,000 scale, 20 sheets. Conoco and Egyptian General Petroleum Corporation, Cairo

Kohn BP, Eyal M (1981) History of uplift of the crystalline basement of Sinai and its relation to opening of the Red Sea as revealed by fission track dating of apatites. Earth Planet Sci Lett 52:129–141

Krenkel E (1922) Die Bruchzonen Ostafrikas: Tektonik, Vulkanismus, Erdbeben und Schwereanomalien. Verlag von Gebrüder Borntraeger, Berlin, 184 pp

Krenkel E (1957) Geologie und Bodenschätze Afrikas, 2nd edn. Geest and Portig, Leipzig, 597 pp

Larsen P (1988) Relay structures in a Lower Permian basement-involved extension system, East Greenland. J Struct Geol 10:3–8

Lambiase JJ, Bosworth W (1995) Structural controls on sedimentation in continental rifts. In: Lambiase JJ (ed) Hydrocarbon habitat in rift basins, vol 80. Geological Society Special Publication, London, pp 117–144

Laughton AS (1970) A new bathymetric chart of the Red Sea. Philos Trans Roy Soc London Series A Math Phys Sci 267:21–22

Lazar M, Ben-Avraham Z, Garfunkel Z (2012) The Red Sea—new insights from recent geophysical studies and the connection to the Dead Sea fault. J Afr Earth Sc 68:96–110

Le Pichon X, Francheteau J (1978) A plate tectonic analysis of the Red Sea—Gulf of Aden area. Tectonophysics 46:369–406

Le Pichon X, Sibuet J (1981) Passive margins: a model of formation. J Geophys Res 86:3708–3720

Leroy S, Gente P, Fournier M, d'Acremont E, Patriat P, Beslier M-O, Bellahsen N, Maia M, Blais A, Perrot J, Al-Kathiri A, Merkouriev S, Fleury J-M, Ruellan P-Y, Lepvrier C, Huchon P (2004) From rifting to spreading in the eastern Gulf of Aden: a geophysical survey of a young oceanic basin from margin to margin. Terra Nova 16:185–192

Lithgow-Bertelloni C, Richards MA (1998) The dynamics of Cenozoic and Mesozoic plate motions. Rev Geophys 36:27–78

Lotfy HI, Van der Voo R, Hall CM, Kamel OA, Abdel Aal AY (1995) Palaeomagnetism of Early Miocene basaltic eruptions in the areas east and west of Cairo. J Afr Earth Sci 21:407–419
Lowell JD, Genik GJ (1972) Sea-floor spreading and structural evolution of southern Red Sea. Am Assoc Pet Geol Bull 56:247–259
Lyakhovsky V, Segev A, Schattner U, Weinberger R (2012) Deformation and seismicity associated with continental rift zones propagating toward continental margins. Geochem Geophys Geosyst 13:Q01012
Makris J, Allam A, Moller L (1981) Deep seismic studies in Egypt and their interpretation. EOS Trans Am Geophys Union 62:230
Makris J, Henke CH, Egloff F, Akamaluk T (1991a) The gravity field of the Red Sea and East Africa. Tectonophysics 198:269–381
Makris J, Tsironidis J, Richter H (1991b) Heatflow density distribution in the Red Sea. Tectonophysics 198:383–393
Manighetti I, Tapponnier P, Courtillot V, Gruszow S, Gillot PY (1997) Propagation of rifting along the Arabia-Somalia plate boundary: the Gulfs of Aden and Tadjoura. J Geophys Res 102:2681–2710
Martin AK (1984) Propagating rifts: crustal extension during continental rifting. Tectonics 3:611–617
Martinez F, Cochran JR (1988) Structure and tectonics of the northern Red Sea: catching a continental margin between rifting and drifting. Tectonophysics 150:1–32
Martinez F, Cochran JR (1989) Geothermal measurements in the northern Red Sea: implications for lithospheric thermal structure and mode of extension during continental rifting. J Geophys Res 94:12239–12266
Martini E (1971) Standard tertiary and quaternary calcareous nannoplankton zonation. In: Farinacci A (ed) Proceedings of the 2nd planktonic conference, Roma, 1970, Tecnoscienza, Roma, pp 739–785
Mattash MA (2008) Younger volcanic fields of Yemen with focus on Jabal at-Tair active volcano in the Red Sea. In: International Geological Congress Abstracts, vol 33, Abstract 1287199
Mathews DH, Williams CA, Laughton AS (1967) Mid-ocean ridge in the mouth of the Gulf of Aden. Nature 215:1052–1053
McClay KR, Nicols GJ, Khalil SM, Darwish M, Bosworth W (1998) Extensional tectonics and sedimentation, eastern Gulf of Suez, Egypt. In: Purser BH, Bosence DWJ (eds) Sedimentation and tectonics of rift basins: Red Sea-Gulf of Aden. Chapman and Hall, London, pp 223–238
McGuire AV, Bohannon RG (1989) Timing of mantle upwelling; evidence for a passive origin for the Red Sea Rift. J Geophys Res 94:1677–1682
McGuire AV, Coleman RG (1986) The Jabal Tirf layered gabbro and associated rocks of the Tihama Asir Complex, SW Saudi Arabia. J Geol 94:651–665
McKenzie DP, Davies D, Molnar P (1970) Plate tectonics of the Red Sea and East Africa. Nature 226:243–248
McQuarrie N, Stock JM, Verdel C, Wernicke BP (2003) Cenozoic evolution of Neotethys and implications for the causes of plate motions. Geophys Res Lett 30:2036
Federal MDA (2004) Landsat GeoCover ETM + 2000 Edition Mosaics. USGS, Sioux Falls, South Dakota 2000
Menzies MA, Baker J, Bosence D, Dart C, Davidson I, Hurford A, Al Kadasi M, McClay K, Nichols G, Al Subbary A, Yelland A (1992) The timing of magmatism, uplift and crustal extension: preliminary observations from Yemen. In: Storey BC, Alabaster T, Pankhurst RJ (eds) Magmatism and the causes of continental break-up, vol 68. Geological Society Special Publication, London, pp 293–304
Menzies M, Gallagher K, Yelland A, Hurford AJ (1997) Volcanic and non-volcanic rifted margins of the Red Sea and Gulf of Aden: crustal cooling and margin evolution in Yemen. Geochimica Cosmochimica Acta 61:2511–2527
Milanovsky EE (1972) Continental rift zones: their arrangement and development. Tectonophysics 15:65–70
Milkereit B, Fluh ER (1985) Saudi Arabian refraction profile; crustal structure of the Red Sea-Arabian shield transition. Tectonophysics 111:283–298
Miller PM, Barakat H (1988) Geology of the Safaga concession, northern Red Sea, Egypt. Tectonophysics 153:123–136
Mohn G, Manatschal G, Beltrando M, Masini E, Kusznir N (2012) Necking of continental crust in magma-poor rifted margins: evidence from the fossil Alpine Tethys margins. Tectonics 31:TC1012
Mohr PA (1970) The Afar triple junction and sea-floor spreading. J Geophys Res 75:7340–7352
Monin AS, Litvin VM, Podreazhansky AM, Sagalevich AM, Sorokhtin OG, Voitov VI, Yastrebov VS, Zonenshain LP (1982) Red Sea submersible research expedition. Deep Sea Res 29:361–373
Monin AS, Plakhin EA, Podrazhansky AM, Sagalevich AM, Sorokhtin OG (1981) Visual observations of the Red Sea hot brines. Nature 291:222–225
Mooney WD, Gettings ME, Blank HR, Healy JH (1985) Saudi Arabian seismic-refraction profile; a traveltime interpretation of crustal and upper mantle structure. Tectonophysics 111:173–246
Moretti I, Chénet PY (1987) The evolution of the Suez rift: a combination of stretching and secondary convection. Tectonophysics 133:229–234
Moretti I, Colletta B (1988) Fault block tilting: the Gebel Zeit example, Gulf of Suez. J Struct Geol 10:9–19
Morgan P, Swanberg CA, Boulos FK, Hennin SF, El Sayed AA, Basta NZ (1980) Geothermal studies in Northeast Africa. Ann Geol Surv Egypt 10:971–987
Morley CK, Nelson RA, Patton TL, Munn SG (1990) Transfer zones in the east African rift system and their relevance to hydrocarbon exploration in rifts. Am Assoc Pet Geol Bull 74:1234–1253
Mougenot D, Al-Shakhis AA (1999) Depth imaging sub-salt structures: a case study in the Midyan Peninsula (Red Sea). GeoArabia 4:335–463
Moustafa AM (1976) Block faulting of the Gulf of Suez. In: 5th Egyptian General Petroleum Organization Exploration Seminar, Cairo, 19 pp (unpublished report distributed as paper copies at the conference)
Moustafa AR (1997) Controls on the development and evolution of transfer zones: the influence of basement structure and sedimentary thickness in the Suez rift and Red Sea. J Struct Geol 19:755–768
Mulder CJ, Lehner P, Allen DCK (1975) Structural evolution of the Neogene salt basins in the eastern Mediterranean and the Red Sea. Geol Mijnbouw 54:208–221
Musser JA, Al-Amri M, Al-Omar W, Lafon SK, Arifin M (2012) Wide azimuth towed streamer 3-D seismic in the Red Sea. Middle East geosciences conference and exhibition 10, 4–7 March, Bahrain, abstracts
Omar GI, Steckler MS (1995) Fission track evidence on the initial rifting of the Red Sea: two pulses, no propagation. Science 270:1341–1344
Omar GI, Steckler MS, Buck WR, Kohn BP (1989) Fission-track analysis of basement apatites at the western margin of the Gulf of Suez rift, Egypt: evidence for synchronicity of uplift and subsidence. Earth Planet Sci Lett 94:316–328
Orszag-Sperber F, Harwood G, Kendall A, Purser BH (1998) A review of the evaporites of the Red Sea-Gulf of Suez rift. In: Purser BH, Bosence DWJ (eds) Sedimentation and tectonics of rift basins: Red Sea-Gulf of Aden. Chapman and Hall, London, pp 409–426
Patton TL, Moustafa AR, Nelson RA, Abdine SA (1994) Tectonic evolution and structural setting of the Suez Rift. In: Landon SM (ed) Interior rift basins. American Association of Petroleum Geologists Memoir 59, pp 7–55
Pautot G (1983) Les fosses de la Mer Rouge: Approche geomorphologique d'un stade initial d'ouvertur oceanique realisee a l'aide du Seabeam. Oceanol Acta 6:235–244

Pautot G, Guennoc P, Coutelle A, Lyberis N (1984) Discovery of a large brine deep in the northern Red Sea. Nature 310:133–136
Peacock DCP, Sanderson DJ (1991) Displacements, segment linkage and relay ramps in normal fault zones. J Struct Geol 13:721–733
Perry SK, Schamel S (1990) The role of low-angle normal faulting and isostatic response in the evolution of the Suez rift, Egypt. Tectonophysics 174:159–173
Phillips JD (1970) Magnetic anomalies in the Red Sea. Philos Trans Roy Soc London Series A Math Phys Sci 267:205–217
Plaziat JC, Montenat C, Barrier P, Janin MC, Orszag-Sperber F, Philobbos E (1998) Stratigraphy of the Egyptian syn-rift deposits: correlations between axial and peripheral sequences of the north-western Red Sea and Gulf of Suez and their relations with tectonics and eustacy. In: Purser BH, Bosence DWJ (eds) Sedimentation and Tectonics in Rift Basins—Red Sea—Gulf of Aden. Chapman and Hall, London, pp 211–222
Prodehl C (1985) Interpretation of a seismic-refraction survey across the Arabian Shield in western Saudi Arabia. Tectonophysics 111:247–282
Quennell AM (1951) Geology and mineral resource of (former) Transjordan. Colonial Geol Miner Res 2:85–115
Quennell AM (1958) The structural and geomorphic evolution of the Dead Sea Rift. Quar J Geol Soc London 114:1–24
Rabinowitz PD, Coffin MF, Falvey D (1983) The separation of Madagascar and Africa. Science 220:67–69
Rabinowitz PD, LaBreque J (1979) The Mesozoic South Atlantic Ocean and evolution of its continental margins. J Geophys Res 84:5973–6003
Redfield TF, Wheeler WH, Often M (2003) A kinematic model for the development of the Afar depression and its paleogeographic implications. Earth Planet Sci Lett 216:383–398
Reeves CV, Karanja FM, Macleod IN (1987) Geophysical evidence for a failed Jurassic Rift and triple junction in Kenya. Earth Planet Sci Lett 81:231–299
Reilinger R, McClusky S (2011) Nubia-Arabia-Eurasia plate motions and the dynamics of Mediterranean and Middle East tectonics. Geophys J Int 186:971–979
Richardson M, Arthur MA (1988) The Gulf of Suez–northern Red Sea Neogene rift: a quantitive basin analysis. Mar Pet Geol 5:247–270
Robertson AHF, Bamakhalif KAS (1998) Late Oligocene–early Miocene rifting of the northeastern Gulf of Aden: basin evolution in Dhofar (southern Oman). In: Ziegler PA, Cavazza W, Robertson AHF, Crasquin-Soleau S (eds) Peri-Tethys Memoir 6: Peri-Tethyan Rift/Wrench Basins and Passive Margins, Mémoires du Muséum National d'Histoire Naturelle de Paris 186, pp 641–670
Robertson AHF, Dixon JE, Brown S, Collins A, Morris A, Pickett E, Sharp I, Ustaömer T (1996) Alternative tectonic models for the Late Palaeozoic–Early Tertiary development of Tethys in the Eastern Mediterranean region. In: Morris A, Tarling DH (eds) Palaeomagnetism and Tectonics of the Mediterranean Region. Geological Society Special Publication, London, 105, pp 239–263
Rochette P, Tamrat E, Féraud G, Pik R, Courtillot V, Kefeto E, Coulon C, Hoffmann C, Vandamme D, Yirgu E (1997) Magnetostratigraphy and timing of the Oligocene Ethiopian traps. Earth Planet Sci Lett 164:497–510
Roger J, Platel JP, Cavelier C, Bourdillon-de-Grisac C (1989) Données nouvelles sur la stratigraphie et l'histoire géologique du Dhofar (Sultanat d'Oman). Bulletin de la Société géologique de France 8:265–277
Rosendahl BR, Reynolds DJ, Lorber PM, Burgess CF, McGill J, Scott D, Lambiase JJ, Derksen SJ (1986) Structural expressions of rifting: Lessons from Lake Tanganyika, Africa. In: Frostick LE, Renaut RW, Reid I, Tiercelin JJ (eds) Sedimentation in the African Rifts, vol 25. Geological Society Special Publications London, pp 29–43
Röser HA (1975) A detailed magnetic survey of the southern Red Sea. Geologie Jahrbuch 13:131–153
Ross DA, Schlee J (1973) Shallow structure and geologic development of the southern Red Sea. Geol Soc Am Bull 84:3827–3848
Ross DA, Whitmarsh RB, Ali SA, Boudreaux JE, Coleman RG, Fleisher RL, Girdler RW, Manheim F, Matter A, Ngrini C, Stoffers P, Supko PR (1973) Red Sea drillings. Science 179:377–380
Sagri M, Abbate E, Azzaroli A, Balestrieri ML, Benvenuti M, Bruni P, Fazzuoli M, Ficcarelli G, Marcucci M, Papini M, Pavia G, Reale V, Rook L, Tecle TM (1998) New data on the Jurassic and Neogene sedimentation in the Danakil Horst and northern Afar Depression, Eritrea. In: Crasquin-Soleau S, Barrier É (eds) Peri-Tethys Memoir 3: stratigraphy and evolution of Peri-Tethyan platforms, vol 177. Mémoires du Muséum National d'Histoire Naturelle de Paris, pp 193–214
Sahota G (1990) Geophysical investigations of the Gulf of Aden continental margins: geodynamic implications for the development of the Afro-Arabian Rift system. PhD thesis, University College, Swansea, UK
Saleh S, Jahr T, Jentzsch G, Saleh A, Abou Ashour NM (2006) Crustal evaluation of the northern Red Sea rift and Gulf of Suez, Egypt from geophysical data: 3-dimensional modeling. J Afr Earth Sci 45:257–278
Saoudi A, Khalil B (1986) Distribution and hydrocarbon potential of Nukhul sediments in the Gulf of Suez. In: Proceedings of the 7th exploration seminar, vol 1, Egyptian General Petroleum Corporation, Cairo, March 1984, pp 75–96
Savoyat E, Shiferaw A, Balcha T (1989) Petroleum exploration in the Ethiopian Red Sea. J Pet Geol 12:187–204
Schandelmeier H, Reynolds PO, (1997) Palaeogeographic–Palaeotectonic Atlas of North-Eastern Africa, Arabia and Adjacent Areas. Balkema, Rotterdam, 160 pp, 17 plates
Scheuch J (1976) Preliminary heat flow map of the Red Sea and an attempt to provide a geological-geophysical interpretation. In: Pilger A, Roesler A (eds) Afar between Continental and Oceanic Rifting, vol 2, E. Schweizerbart Verlagsbuchhandl (Naegele u. Obermiller), Stuttgart: pp 171−174
Schmidt DL, Hadley DG, Brown GF (1983) Middle Tertiary continental rift and evolution of the Red Sea in southwestern Saudi Arabia. U.S. Geological Survey Open File Report 83-0641
Searle RC, Ross DA (1975) A geophysical study of the Red Sea axial trough between 20.5° and 22°N. Geophys J Roy Astron Soc 43:555–572
Sebai A, Zumbo V, Féraud G, Bertrand H, Hussain AG, Giannerini G, Campredon R (1991) 40Ar/39Ar dating of alkaline and tholeiitic magmatism of Saudi Arabia related to the early Red Sea rifting. Earth Planet Sci Lett 104:473–487
Segev A, Rybakov M (2011) History of faulting and magmatism in the Galilee (Israel) and across the Levant continental margin inferred from potential field data. J Geodyn 51:264–284
Sellwood BW, Netherwood RE (1984) Facies evolution in the Gulf of Suez area: sedimentation history as an indicator of rift initiation and development. Modern Geol 9:43–69
Şengör AMC (2001) Elevation as indicator of mantle-plume activity. In: Ernst RE, Buchan KL (eds) Mantle plumes: their identification through time. Geological Society of America Special Paper 352, pp 183–225
Şengör AMC, Burke K (1978) Relative timing of rifting on earth and its tectonic implications. Geophys Res Lett 5:419–421
Şengör AMC, Yilmaz Y (1981) Tethyan evolution of Turkey: a plate tectonic approach. Tectonophysics 75:181–241
Shackleton NJ, Berger A, Peltier WR (1990) An alternative astronomical calibration of the lower Pleistocene timescale based on ODP Site 677. Trans Roy Soc Edinburgh Earth Sci 81:251–261
Sharp IR, Gawthorpe RL, Underhill JR, Gupta S (2000) Fault propagation folding in extensional settings: examples of structural

style and synrift sedimentary response from the Suez rift, Sinai, Egypt. Geol Soc Am Bull 112:1877–1899

Smith WHF, Sandwell DT (1997) Global sea floor topography from satellite altimetry and ship depth soundings. Sci Mag 277:1956–1962

Spohner R, Oleman P (1986) Topography, Red Sea Region (1:8,000,000). German Research Foundation, Karlsruhe

Stampfli GM, Borel GD (2002) A plate tectonic model for the Paleozoic and Mesozoic constrained by dynamic plate boundaries and restored synthetic oceanic isochrons. Earth Planet Sci Lett 196:17–33

Stampfli GM, Mosar J, Favre P, Pillevuit A, Vannay J-C (2001) Permo-Mesozoic evolution of the western Tethys realm: the Neo- Tethys East Mediterranean Basin connection. In: Ziegler PA, Cavazza W, Robertson AHF, Crasquin-Soleau S (eds) Peri-Tethys Memoir 6: Peri-Tethyan Rift/Wrench Basins and Passive Margins, vol 186, Mémoires du Muséum national d'Histoire naturelle de Paris, pp 51–108

Steckler MS (1985) Uplift and extension at the Gulf of Suez: indications of induced mantle convection. Nature 317:135–139

Steckler MS, Omar I (1994) Controls on erosional retreat of the uplifted rift flanks at the Gulf of Suez and northern Red Sea. J Geophys Res 99:1–2

Steckler MS, ten Brink US (1986) Lithospheric strength variations as a control on new plate boundaries: examples from the northern Red Sea region. Earth Planet Sci Lett 79:120–132

Steckler MS, Berthelot F, Lyberis N, Le Pichon X (1988) Subsidence in the Gulf of Suez: implications for rifting and plate kinematics. Tectonophysics 153:249–270

Steckler M, Cochran J, Bosworth W (2001) Rupturing of continental lithosphere in the northern and central Red Sea/Gulf of Suez region: workshop report. Margins Newsl 6:5–9

Stein CA, Cochran JR (1985) The transition between the Sheba Ridge and Owen Basin: rifting of old oceanic Lithosphere. Geophys J Roy Astron Soc 81:47–74

Sutra E, Manatschal G (2012) How does the continental crust thin in a hyperextended rifted margin? insights from the Iberia margin. Geology 40:139–142

Szymanski E (2013) Timing, kinematics and spatial distribution of Miocene extension in the central Arabian margin of the Red Sea rift system. PhD thesis, University of Kansas, Lawrence, 430 pp

Tefera M, Chernet T, Haro W (1996) Explanation of the geological map of Ethiopia. Ethiopian Institute of Geological Surveys, Addis Ababa 3, 79 pp

Tewfik N, Ayyad M (1984) Petroleum exploration in the Red Sea shelf of Egypt. Proc. 6th Exploration Seminar, Egyptian General Petroleum Corporation and Egypt Petroleum Exploration Society, Cairo, March 1982, 1, pp 159–180

Uchupi E (1989) The tectonic style of the Atlantic Mesozoic rift system. J Afr Earth Sci 8:143–164

Ukstins IA, Renne PR, Wolfenden E, Baker J, Ayalew D, Menzies M (2002) Matching conjugate volcanic rifted margins: $^{40}Ar/^{39}Ar$ chrono-stratigraphy of pre- and syn-rift bimodal flood volcanism in Ethiopia and Yemen. Earth Planet Sci Lett 198:289–306

U.S. Geological Survey-Arabian American Oil Company (1963) Geologic map of the Arabian Peninsula. U.S.G.S. Miscellaneous Geological Investigations Map I-270-A, 1:2,000,000 scale. U.S. Geological Survey, Washington

Vail JR (1975) Geologic map; the Democratic Republic of Sudan and adjacent areas, scale 1:2,000,000. Survey Review, Tolworth

Vail JR (1978) Outline of the geology and mineral deposits of the Democratic Republic of the Sudan and adjacent areas. Overseas geology and mineral resources 49 (includes the Geologic Map of Sudan, scale 1:2:000,000, north and south sheets 1203A and 1203B), 68 pp

Van Wijk JW, Blackman DK (2005) Dynamics of continental rift propagation: the end-member modes. Earth Planet Sci Lett 229:247–258

Varet J (1978) In: Gasse F (ed) Geology of central and southern Afar (Ethiopia and Djibouti Republic). CNRS, Paris, 118 pp

Vellutini P (1990) The Manda-Inakir Rift, Republic of Djibouti: a comparison with the Asal Rift and its geodynamic interpretation. Tectonophysics 172:141–153

Vening Meinesz FA (1934) Gravity Expeditions at Sea. vol 2 (1923–1932). Publications Netherlands Geodetic Commission, Delft

Vink GE (1982) Continental rifting and the implications for plate tectonic reconstructions. J Geophys Res 87:10677–10688

Vink GE, Morgan WJ, Zhao W-L (1984) Preferential rifting of continents: a source of displaced terranes. J Geophys Res 89:10072–10076

Voggenreiter W, Hötzl H, Mechie J (1988) Low-angle detachment origin for the Red Sea Rift system? Tectonophysics 150:51–75

Von Triulzi AE (1898) Expedition S.M. Schiff Pola in das Rothe Meer, nördliche Hälfte, Wissenschaftliche Ergebnisse, XII, Relative Schwerebestimmung. Denkschriften der Kaiserlichen Akademie der Wissenschaften, Mathematisch – Naturwissenschaftliche Klasse 65, pp 131–206

Watchorn F, Nichols GJ, Bosence DWJ (1998) Rift-related sedimentation and stratigraphy, southern Yemen (Gulf of Aden). In: Purser BH, Bosence DWJ (eds) Sedimentation and tectonics of rift basins: Red Sea-Gulf of Aden. Chapman and Hall, London, pp 165–189

Wernicke B (1985) Uniform-sense normal simple shear of the continental lithosphere. Can J Earth Sci 22:108–125

Wescott WA, Krebs WN, Dolson JC, Ramzy M, Karamat SA, Moustafa T (1997) Chronostratigraphy, sedimentary facies, and architecture of tectono-stratigraphic sequences: an integrated approach to rift basin exploration, Gulf of Suez, Egypt. In: Shanley KW, Perkins BF (eds) Gulf coast section SEPM foundation 18th annual research conference, Shallow Marine and Nonmarine Reservoirs, 7–10 Dec 1997, pp 377–399

Whitmarsh RB, Manatschal G, Minshull TA (2001) Evolution of magma-poor continental margins from rifting to seafloor spreading. Nature 413:150–154

Whitmarsh RB, Weser OE, Ross DA et al. (1974) Initial reports of the Deep Sea Drilling Project, Red Sea, vol 23, U.S. Government Printing Office, Washington, 1180 pp

Williams G, Small J (1984) A study of the Oligo-Miocene basalts in the Western Desert. In: Proceedings of the 7th exploration seminar, Egyptian General Petroleum Corporation, March 1984, pp 252–268

Woodruff F, Savin SM (1989) Miocene deepwater oceanography. Paleoceanography 4:87–140

Wycisk P, Klitzsch E, Jos C, Reynolds O (1990) Intracratonal sequence development and structural control of Phanerozoic strata in Sudan. Berl Geowiss Abh A 120:45–86

Younes AI, McClay KR (2002) Development of accommodation zones in the Gulf of Suez-Red Sea rift, Egypt. Am Assoc Pet Geol Bull 86:1003–1026

Young MJ, Gawthorpe RL, Sharp IR (2000) Sedimentology and sequence stratigraphy of a transfer zone coarse-grained delta, Miocene Suez Rift, Egypt. Sedimentology 47:1081–1104

Zahran MA, Willis AJ (2009) The vegetation of Egypt, 2nd edn. Springer, Netherlands, 437 pp

Zanettin B, Justin-Visentin E, Piccirillo EM (1978) Volcanic succession, tectonics and magmatology in central Ethiopia. Atti MemAccad Patavina Sci Lett Arti 90:5–19

Ziegler MA (2001) Late Permian to Holocene paleofacies evolution of the Arabian plate and its hydrocarbon occurrences. GeoArabia 6:504–665

Zumbo V, Féraud G, Bertrand H, Chazot G (1995) $^{40}Ar/^{39}Ar$ chronology of Tertiary magmatic activity in southern Yemen during the early Red Sea-Aden rifting. J Volcanol Geoth Res 65:265–279

Seafloor Spreading Initiation: Geophysical and Geochemical Constraints from the Thetis and Nereus Deeps, Central Red Sea

Marco Ligi, Enrico Bonatti, and Najeeb M.A. Rasul

Abstract

The rifting apart of continents involves interaction of tectonic and magmatic events that reflect the strain-rate and temperature-dependent processes of solid state deformation and decompression melting within the Earth. The spatial and temporal scales over which these mechanisms localize extensional strain, allowing continental rifts to evolve towards seafloor spreading, remain controversial. Here we show the role played by magmatism during the transition from a continental to an oceanic rift based on geophysical and geochemical data from the Thetis and Nereus Deeps, the two northernmost oceanic cells in the central Red Sea. The Thetis segment is made by coalescence of three sub-cells that become shallower, narrower and younger from south to north. Magnetic data reveal that the initial emplacement of oceanic crust is occurring today in the Thetis northern basin and in the southern tip of Nereus. The intertrough zones that separate the Thetis "oceanic" cell from the Nereus cell to the north, and the Hadarba cell to the south, contain thick sedimentary sequences and relicts of continental crust. A seismic reflection profile running across the central part of the southern Thetis basin shows a ~5 km wide reflector about 3.2 km below the axial neovolcanic zone, interpreted as marking the roof of a magma chamber or melt lens and as a last step in a sequence of basaltic melt intrusion from pre-oceanic continental rifting to oceanic spreading. The spatial evolution of mantle melting processes across Thetis and Nereus is evaluated from the chemical composition of 22 basaltic glasses sampled along 100 km of the rift axis. Trace and major element compositions corrected for crystallization show relationships with age of initial emplacement of the oceanic crust and preserve a clear signal of mantle melting depth variations. While Zr/Y and $(Sm/Yb)_n$ decrease, Na*/Ti* increases slightly from south to north. Na_8 correlates positively with Fe_8, and Zr/Y and $(Sm/Yb)_n$ with both Fe_8 and Na_8. This indicates that an increase in the degree of melting corresponds to a decrease in the mean pressure of melting, suggesting active mantle upwelling beneath Thetis and Nereus. The inferred sharp rift-to-drift transition marked by magmatic activity with typical MORB signature and a relatively high degree of mantle melting, with no contamination by continental lithosphere, suggests that lower crust and mantle lithosphere had already been replaced by active upwelling asthenosphere before separation of the Nubian and Arabian plates.

M. Ligi (✉) · E. Bonatti
Istituto di Scienze Marine, CNR, via Gobetti 101, 40129 Bologna, Italy
e-mail: marco.ligi@bo.ismar.cnr.it

E. Bonatti
Lamont Doherty Earth Observatory, Columbia University, Palisades, NY 10964, USA

N.M.A. Rasul
Saudi Geological Survey, Jeddah 21514, Saudi Arabia

N.M.A. Rasul and I.C.F. Stewart (eds.), *The Red Sea*,
Springer Earth System Sciences, DOI 10.1007/978-3-662-45201-1_4,

Introduction

The transition from a continental to an oceanic rift has occurred in various phases of the history of the Earth; it is a major step in the cycle whereby continents assemble in a single "supercontinent" and then fragment and disperse again (Wilson 1966). As a continental rift zone evolves, the tectonic and volcanic processes associated with crustal extension become confined to axial magmatic rift zones leading to oceanic spreading and the birth of a new ocean. Rift systems range from "narrow" to "wide" (Hopper and Buck 1998), and from "volcanic" to "non-volcanic" (Boillot et al. 1980; Eldholm and Grue 1994; Whitmarsh et al. 2001; Reston and Morgan 2004; Huismans and Beaumont 2011) resulting from the interaction of structural and magmatic processes during continental rifting and rupturing. However, the relationships between rifting and magmatism, the initial generation of oceanic crust and the transition from rifting to spreading are still poorly constrained (Lizarralde et al. 2007; Armitage et al. 2009).

The Red Sea/Gulf of Aden system is an ideal "natural laboratory" where these questions can be uniquely addressed. The East African-Afar-Gulf of Aden-Red Sea Region contains a network of rifts in different stages of evolution, from fully continental (East African) to fully oceanic (Gulf of Aden), with Afar and the Red Sea in intermediate stages (Girdler 1985; Bonatti 1985; Sultan et al. 1992; Voggenreiter and Hotzl 1989; Cochran and Karner 2007; Bonatti et al. this volume). Since late Oligocene-Miocene this rift system has been fragmenting the Arabian-Nubian shield, which consists of ancient island arc and back arc basins, continental microplates and interlaying ocean basins, welded together when Gondwana was assembled during the Proterozoic Pan African orogenic cycle (Kroner et al. 1987; Sultan et al. 1993; Stern 1994; Ghebreab 1998). The impingement from below of the Afar mantle plume, starting roughly 34 Ma (Hofmann et al. 1997; Storey 1995; Ebinger and Sleep 1998), may have influenced the evolution of the rift system. Plate tectonic reconstructions suggest that, while emplacement of oceanic crust initiated as early as $\sim$17 Ma in the Gulf of Aden (Fournier et al. 2010), it started more recently ($\sim$5 Ma) in the southern Red Sea (Chu and Gordon 1998), and has not yet started in the northern Red Sea (Bonatti 1985; Cochran et al. 1991; Cochran 2005).

Although the different parts of this system have been the object of many studies (Drake and Girdler 1964; Ross and Schlee 1973; Girdler and Styles 1974; Searle and Ross 1975; Barberi and Varet 1977; Cochran 1981; Pautot 1983; Bonatti 1985; Izzeldin 1987, 1989; Coleman and McGuire 1988; Guennoc et al. 1988, 1990; Altherr et al. 1990; Makris and Rhim 1991; Ghebreab 1998; Bosworth et al. 2005; Cochran 2005; D'Acremont et al. 2006; Cochran and Karner 2007; Mohriak and Leroy 2013), several key aspects of the transition from a continental to an oceanic rift remain poorly known: (a) pre-rift reconstructions of the continental margins (e.g. Bosworth et al. 2005) and Euler poles of opening of the Red Sea (Nubia-Arabia) are still debated; (b) to what extent is the break up of continental lithosphere in the Red Sea—Gulf of Aden regions related to the Afar mantle plume and driving flows in the asthenosphere (Alvarez 2010; Höink et al. 2011; Chang et al. 2011) versus far-field extensional stresses due to slab-pull (e.g. ArRajehi et al. 2010); (c) do the degree of melting and opening rate increase in an "active" burst of accretion as soon as the continental lithosphere breaks, as suggested by Ligi et al. (2011)?; (d) what role do lower crustal intrusions play in accommodating extension and in the evolution of melts supplied from the underlying asthenosphere (Bonatti and Seyler 1987; White et al. 2008; Thybo and Nielsen 2009)?; (e) how do crust-forming melts evolve during the transition and the first stages of sea floor spreading?; (f) how do melting processes in a young oceanic rift (Red Sea) differ from melting below a mature Mid Ocean Ridge?; (g) is there a systematic change in melt composition along the Red Sea axis that parallels the presumed south to north propagation of the oceanic rift?; (h) which viable geodynamic mechanisms could result in the exhumation of mantle rocks at the Zabargad Island offshore Egypt?. Addressing these questions should lead ultimately to a better understanding of changes in thermal structure and composition of the upper mantle as it upwells beneath rifts that evolve from continental to oceanic.

The southern Red Sea has a more or less continuous axial zone with basaltic (oceanic) crust and lineated Vine-Matthews magnetic anomalies (Fig. 1b; Roeser 1975; Stern and Johnson 2010; Ravat et al. 2011). The emplacement of oceanic crust started in this area roughly 5 Ma (Bosworth et al. 2005). In some cases the emplacement of oceanic crust is associated with emissions of hydrothermal fluids and formation of metalliferous deposits (Bäcker and Schoell 1972). Moving northward the structural set up of the axial region becomes more complicated, consisting of a complex pattern of axial deeps carpeted by oceanic crust separated by non-magnetic sediment (including evaporites) covered relay zones. It is still matter of controversy whether the crust in the inter-trough zones and outside the axial zone of the central and southern Red Sea is oceanic (e.g. Girdler 1985; Hall 1989; Sultan et al. 1992; AlMalki et al. 2014; Mitchell and Park 2014; Augustin et al. 2014; Mitchell this volume) or exhumed mantle (Voggenreiter and Hotzl 1989; Rowan 2014) or continental dissected and "impregnated" by numerous magma intrusions (i.e., transitional crust; Bonatti 1985; Bosworth et al. 2005; Cochran and Karner 2007; Ligi et al. 2012). Moving further north the axial deeps become smaller and regularly spaced; they then disappear in the northernmost region of the Red Sea (Fig. 1), although scattered basaltic intrusions have been observed there (Cochran 2005). This pattern is interrupted by a few major

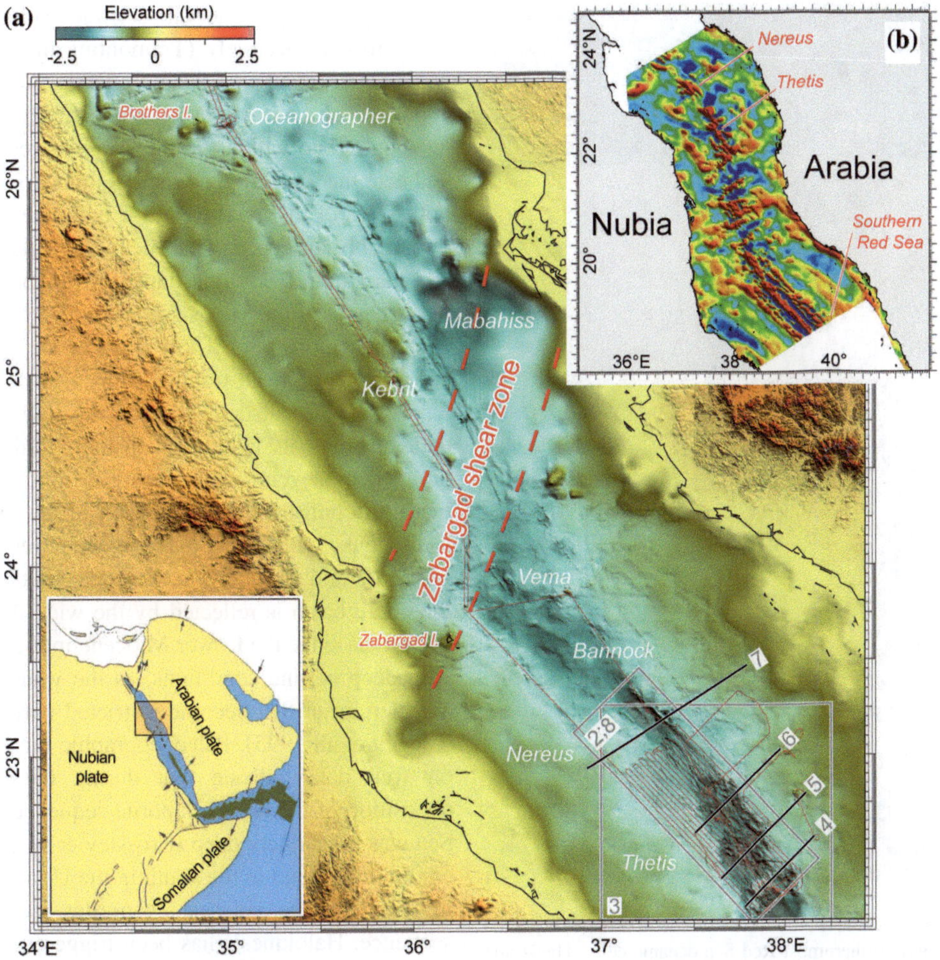

Fig. 1 a Topography of the northern-central Red Sea. Location of major deeps is indicated (*white text*). *Thick red dashed lines* limit the Zabargad shear zone. *Thin red lines* indicate ship tracks during RS05 geophysical data acquisition. The *black solid lines* show the shot locations of the multichannel seismic profiles TL-25M, TL-23M, TL-21M and CGG-3, as shown in Figs. 4, 5, 6 and 7, respectively. The numbered *gray boxes* refer to other figures of Thetis and Nereus topography given in this paper. *Bottom left inset* simplified plate tectonic framework of the Red Sea/Gulf of Aden area: dark areas, oceanic crust. **b** Reduced-to-the-pole magnetic anomaly map over the southern central Red Sea (modified after Stern and Johnson 2010). Linear magnetic stripes flanking the rift axis and indicating oceanic seafloor spreading are clearly visible up to 19°N. Spatial analysis and mapping were carried out using the PLOTMAP package (Ligi and Bortoluzzi 1989)

structures that cross the Red Sea in a near NNE–SSW orientation (roughly parallel to the Dead Sea fault; Saleh et al. 2006). The northernmost of these transverse features (the Zabargad shear zone) constitutes the northern limit of the region with axial oceanic deeps (Fig. 1).

We report here on the transition from a continental to an oceanic rift as it occurs today in the central Red Sea, where in the winter of 2005 we carried out multibeam, magnetic and multichannel seismic reflection surveys as well as bottom rock sampling with the *R/V Urania* (cruise RS05) (Ligi et al. 2011, 2012), focusing on the two northernmost Red Sea axial oceanic segments, that is, the Thetis and Nereus Deeps (Figs. 1 and 2). Details on data and methods can be found in Ligi et al. (2012) (Fig. 3).

Geophysical Constraints

The Thetis and Nereus rift valleys, emplaced within the thick sedimentary-evaporitic sequence ubiquitous in the Red Sea, are separated by a shallower inter-trough zone. Our single and multichannel seismic profiles have identified three different seismic facies enclosing the Miocene evaporites and the Plio-Quaternary deposits (Figs. 4, 5, 6 and 7), similar to those observed elsewhere in the Red Sea (Phillips and Ross 1970; Izzeldin 1987; Guennoc et al. 1990). The upper unit is composed of a rather thin Plio-Quaternary sedimentary sequence overlying, almost everywhere unconformably, a Miocene unit composed mostly of evaporites. The two units

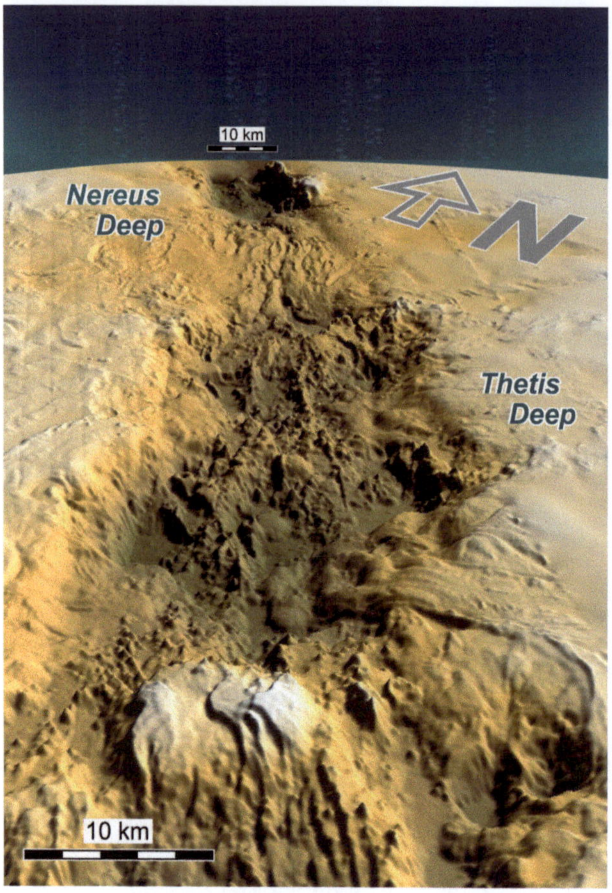

Fig. 2 Birth of an ocean in the central Red Sea. The 3D image shows the morphology of the two northernmost Red Sea oceanic cells, Thetis and Nereus Deeps. *Dark and light colours* indicate depths >1,500 and <1,000 m, respectively. Shading light from N30°E. The Thetis and Nereus rift valleys, emplaced within a thick sedimentary-evaporitic sequence, are separated by a "non-oceanic" inter-trough zone. Each basin is characterized by a highly magnetic axial volcanic ridge and by numerous scattered small volcanoes. Salt flows down-slope from the Arabian margin (*right side*), flooding the Thetis basaltic floor (*dark colours*)

are separated by a strong reflector named "S" (Phillips and Ross 1970). The Miocene evaporites below reflector "S" show two different seismic facies. One is rich in reflectors corresponding to layered evaporites representing bedding of anhydrite, gypsum, and shales, with an average seismic velocity ranging between 3.0 and 3.5 km/s (Izzeldin 1982). The other includes regions almost devoid of reflectors corresponding to rock salt with an average velocity of 4.2–4.6 km/s. The layered evaporites are always surrounded and underlain by rock salt with salt increasing toward the axial trough (Figs. 5, 6 and 7). The thickness of hemipelagic cover and evaporitic layers has been estimated from refraction data of Tramontini and Davis (1969), our own multi-channel (RS05) and single-channel (cruises MR79 and MR83) seismic reflection data and those of Izzeldin (1982, 1987, 1989), adopting average interval seismic velocities of 2 and 3.75 km/s, respectively (Tramontini and Davis 1969). The post-evaporitic sequence has a nearly uniform thickness of about 0.2 km, except in the centre of the Nereus-Thetis intertrough zone where the thickness increases up to 0.6 km and near the axial trough where it thins down to <0.1 km. On both sides of the axial trough it is thinner than average and remarkably disturbed. The thickness of the Miocene section is a maximum below the flanks of the main depression and in the centre of the Thetis-Nereus inter-trough zone, where it can reach more than 5 km (Ligi et al. 2012), thinning as approaches the axial troughs (Figs. 4, 5, 6 and 7). The tectonic stage of salt deposition in the Red Sea is critical for understanding its geological history. Red Sea stratigraphy, outlined by several wells during drilling exploration of the margins, confirms the wide extent of the Miocene evaporite formations, with thick massive evaporites intercalated by anhydrite and shale layers. Tectonic subsidence throughout the entire basin is reflected by the wide deposition of thick evaporites since the Lower Miocene in the northern Red Sea with deep marine conditions, as the water supply from the Mediterranean Sea become restricted (Crossley et al. 1992; Rouchy et al. 1995). Paleogeographic reconstructions based on well data indicate that during the Middle Miocene deposition of the main evaporite sequence, most of the Red Sea was moderately deep (Crossley et al. 1992). These deep conditions extend to the south in the Upper Miocene and are in line with the potassic evaporites from the top of the sequence. Halokinesis has been triggered by clastic progradation and extensional tectonics after only a few million years since salt deposition in the Red Sea, forming autochthonous and allochthonous structures (Davison et al. 1994; Mohriak and Leroy 2013). The combination of the original distribution of the salt and the bulk movement of salt basinward due to gravitational failure results in a concentration of salt in distal portions of the margin (Rowan 2014; Feldens and Mitchell this volume; Ehrhardt and Huebscher this volume). In the Thetis and Nereus region salt diapirs and salt walls are separated by the incipient spreading centre where gravity-driven salt, covered by a thin layer of Plio-Quaternary sediments (Mitchell et al. 2010), flows down-slope from the Arabian margin flooding the basaltic floor (Figs. 2 and 3). Some contribution of salt generated from axial hydrothermal activity, as suggested by Hovland et al. (this volume), cannot be excluded.

Basin-wide gridded gravity data, bathymetry and sediment thickness were used to calculate the mantle Bouguer anomaly (MBA). The calculated MBA, reflecting mostly crustal thickness variations and thermal structure of the mantle beneath the rift axis, displays axial minima centered on the Thetis and Nereus basins, separated by a positive

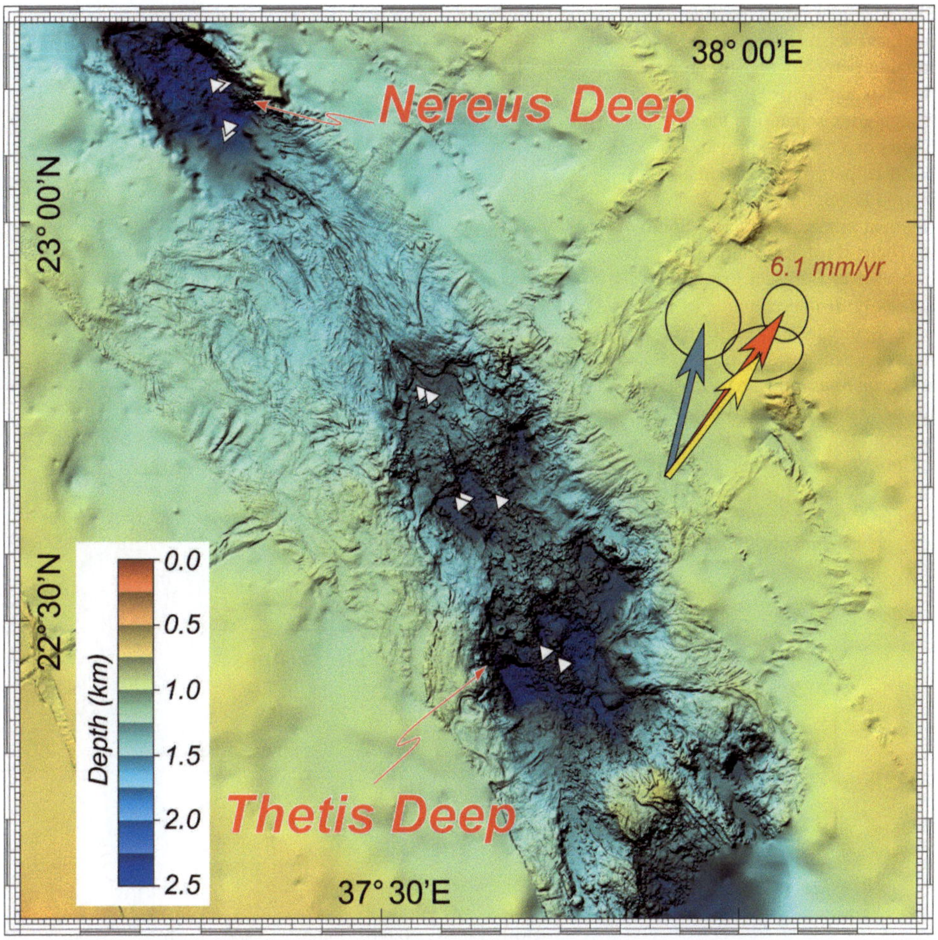

Fig. 3 *Shaded relief* image from swath bathymetry (source of light from NE) showing that Thetis segment consists of three right-stepped sub-basins. No evidence of left-lateral shear deformation is present in the inter-trough zones suggesting that dike intrusions from mid-segment melt injections maintain along axis segmentation in this young sea-floor spreading centre, as also observed in Afar (Wright et al. 2006). Grid resolution 25 m. *Color bar* in km below sea level. *Dark gray, orange* and *yellow arrows* indicate velocity vectors of Arabia relative to Nubia plate from NUVEL1 (DeMets et al. 1994; McClusky et al. 2003; Chu and Gordon 1998), with full spreading rates of 9.7, 9.4 and 12.1 mm/a, respectively. *White filled triangles* location of samples shown in Table 1

anomaly at the inter-trough zone. Inversion of the MBA (Fig. 8a) after removal of the predicted thermal gravity field (Ligi et al. 2008), and magnetic anomaly data (Fig. 8c) resulted in the areal distribution of crustal thickness and of the intensity of rock magnetization, as well as in crustal ages and spreading rates (Fig. 8b, d). A constant thickness magnetized layer is commonly assumed to undergo magnetic inversions. A variable thickness of the magnetized layer will not change the position of the transition between normal and reverse magnetic field; it will affect only the magnitude of magnetization (a greater thickness implies lower magnetization for a given magnetic anomaly). Assuming a magnetized layer of constant thickness (1 km) in the 3-D inversion (Fig. 8d), we observe that magnetization at zero-age is much higher in the northern tip of Thetis and in the southern tip of Nereus than elsewhere. Given that the iron content of basalts from Thetis and Nereus does not correlate with the intensity of magnetization (Ligi et al. 2012), we infer that the thickness of the magnetized layer is greater in these regions. A greater thickness of extrusives may not imply a greater crustal thickness, but crustal thickness inferred from gravity confirms this view (Fig. 8b).

Thetis Deep

A nearly complete multibeam coverage of the Thetis Deep revealed a complex structure formed probably by the presence of three discrete axial cells of initial emplacement of oceanic crust. The three axial rift valley systems are displaced en-echelon with mean width and length, and maximum depth decreasing from south to north (Fig. 3). Each of the three sub-basins is generally sediment-free and is characterized by linear axial volcanic ridges associated with strong magnetic anomalies as well as by scattered small (few hundred meters high) central volcanoes (Figs. 3 and 8). Each of the three sub-basins is separated by a complex transfer zone, including structures normal to the Red Sea orientation (Fig. 3), probably precursors to small (short-offset) transform faults. The flanks of the three sub-basins (i.e. of the entire Thetis Deep) consist of several hundred meters high scarps (Figs. 2 and 3), with evaporite tectonics (i.e. down-slope slumping), superimposed on normal faulting and playing a major role in determining the morphology (Mitchell et al. 2010). The asymmetry of the Thetis valley, with the steepest

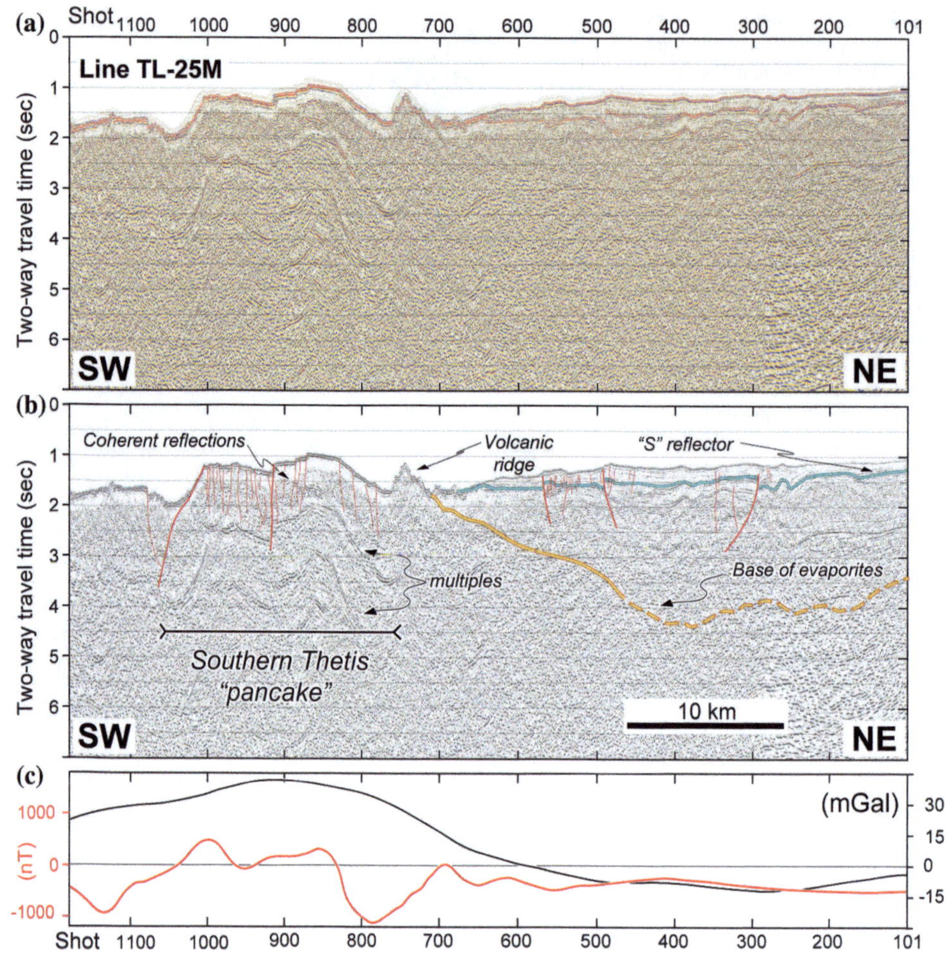

Fig. 4 Finite-difference time migrated seismic line TL-25M running perpendicular to the Red Sea axis and across the inter-trough zone separating the Southern Thetis Basin with the northern part of the Hadarba-Hatiba Deep, the "oceanic" segment immediately to the South of Thetis. Profile location is indicated in Fig. 1. **a** Processed data only; **b** interpretation of the processed data. *Red lines* mark faults and fractures. The *yellow* and *cyan lines* indicate the base of miocene evaporites and the horizon "S", respectively. **c** Gravity and magnetic profiles: reduced-to-the-pole magnetic anomaly, *red solid line*, and free air gravity anomaly, *black solid line*

flanks on the Nubian margin, is mostly due to giant salt flows from the Arabian margin, flooding the deepest parts of the basin from the NE (Figs. 2, 3, 5 and 6).

Thetis Southern Sub-basin

The Thetis southern sub-basin, a ~30 km-long, ~25 km-wide and ~2,200 m-deep axial valley, is floored by a sharp-crested axial neovolcanic ridge and by numerous small central volcanoes (~200 m high, a few hundred m in diameter), seemingly distributed at random. The southern tip of the axial volcanic ridge impacts against a large (~10 km diameter, ~1 km high) pancake-like sub-circular relief, affected by extensional faults oriented in a Red Sea direction (Figs. 2 and 3). This feature dominates the inter-trough zone separating the southern Thetis basin from the northern part of the Hadarba-Hatiba Deep, the "oceanic" segment immediately to the south of Thetis. A 5 km-long 300 m-high volcanic ridge running parallel to the Red Sea axis separates the sub-circular structure from the eastern flank of the rift valley. Several small cone-shaped central volcanoes affect the flanks of the sub-circular relief, while a few of them are located along normal faults on its almost flat summit. Seismic reflection profile TL-25M, running perpendicular to the rift axis, crosses the Thetis-Hadarba inter-trough zone and extends over the entire sub-circular relief. Line TL-25M shows that a number of normal faults offsetting the seafloor disrupt this structure at depth, suggesting recent tectonic activity. The major faults, dipping SW, have large throws and accommodate the vertical displacement (Fig. 4). Below a highly reflective seafloor, interpreted as an erosional surface, some coherent reflections at a depth of ~0.5 s reveal a seismic unit with well stratified lithology. This, a relatively thick crust, and a low intensity of magnetization (Fig. 8), suggest it is not a young "oceanic" basaltic construction, but the remnant of a Pan African crustal fragment similar to those described from the West Iberian margin (Manatschal et al. 2001). This remnant of continental crust may act as a "locked zone" (Courtillot 1982) separating Thetis from the next axial oceanic segment to the south, that is, the Hadarba-Hatiba Deep (Fig. 1). The intense volcanism affecting its flanks in contrast with the few isolated volcanic cones on its summit suggests that the sub-circular bounding faults are the preferred pathways for magma injections rather than the Red Sea trending normal faults (Fig. 3).

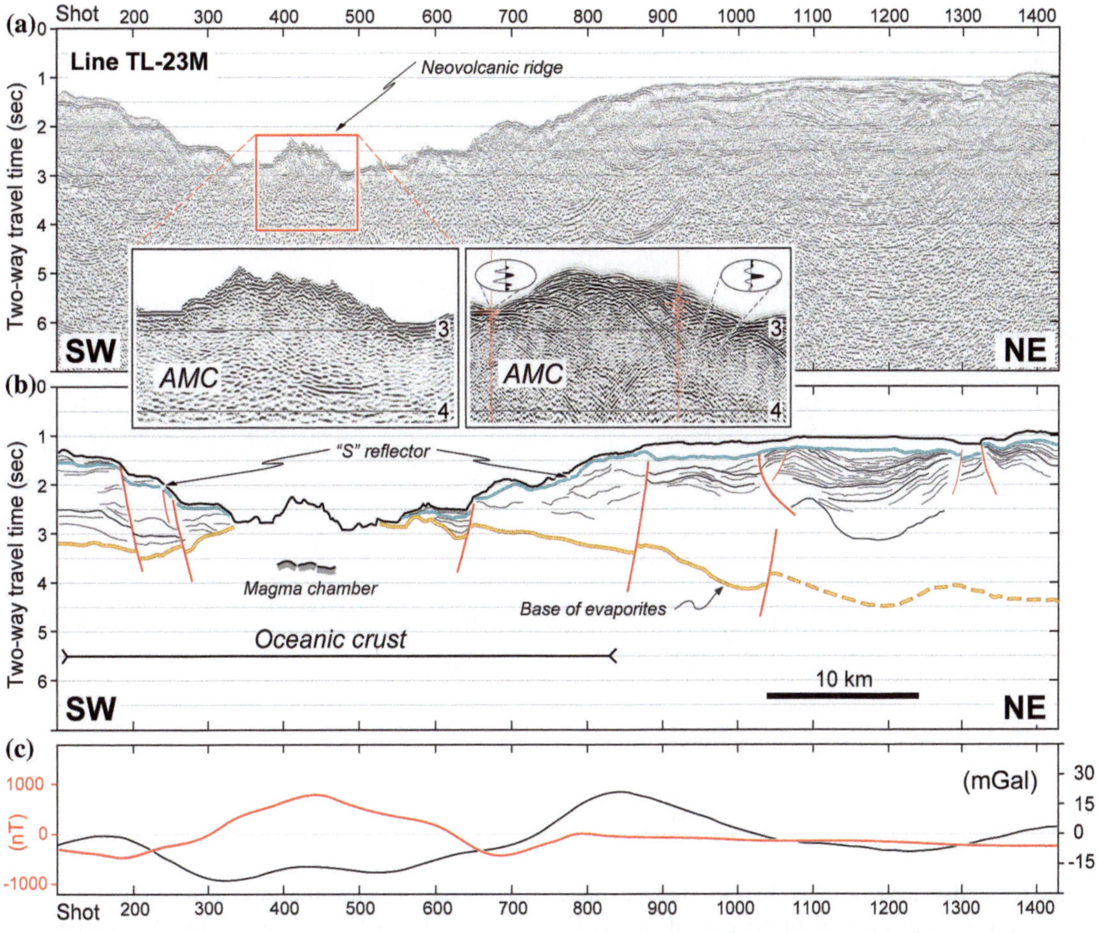

Fig. 5 Multichannel seismic line TL-23M running perpendicular to the axis of the southern Thetis basin. Profile location is indicated in Fig. 1. **a** Finite-difference time migrated section. *Left inset* detail focusing on the interpreted magma chamber. *Right inset* stack section. The reflector from the top of the axial magma chamber (*AMC*) has polarity opposite to that of the seafloor reflector due to seismic velocity reduction caused by presence of melt below. Wiggle traces show the polarity of the different reflections. **b** Interpreted line drawing (symbols as in Fig. 4). The extension of the oceanic crust beneath evaporites is inferred from magnetic data. Gravity-driven deformation of the miocene evaporites due to a combination of overall basin-ward tilt of the extending salt layer and differential sedimentary loading causes the development of an allochthonous salt nappe that moves toward the rift axis overlying the Quaternary oceanic crust. The non-total flooring of the axis by evaporites implies a flowage velocity lower than spreading rate. **c** Gravity and magnetic profiles: reduced-to-the-pole magnetic anomaly, *red solid line*, and free air gravity anomaly, *black solid line*

Seismic profile TL-23M, running across the central part of the southern Thetis basin (Fig. 5), shows a ~5 km-wide reflector about 1.3–1.5 s below the axial neo-volcanic zone. The polarity of this reflection is the reverse of that of the seafloor (Fig. 5), suggesting it is related to a negative velocity contrast. We interpret it as marking the roof of a magma chamber or melt lens, similar to those identified below several mid-ocean ridges (Detrick et al. 1987; Collier and Sinha 1990; Calvert 1995; Singh et al. 2006a, b). Assuming a 4.5 km/s acoustic velocity for the upper oceanic crust at Thetis (Tramontini and Davis 1969), this reflector is ~3.2 km below the seafloor, implying that the top of the inferred magma chamber lies deeper than those identified below fast spreading ridges such as the East Pacific Rise (Detrick et al. 1987). Magma supply is one of the major controlling factors in determining the presence and the depth of magma chambers beneath ridges (Morgan and Chen 1993). At a mid-ocean ridge segment far from a hot-spot, crustal thickness and in turn magma supply, depend mostly on spreading rate (Chen 1992). This supports the idea (Purdy et al. 1992) that the depth of sub-ridge magma chambers is inversely proportional to spreading rate (Fig. 9).

Based on two-dimensional magnetic forward modelling, Ligi et al. (2012) suggested an initial emplacement of oceanic crust as early as 2.2 Ma, with an average full spreading rate of ~16 mm/a, ranging from 23 mm/a during Matuyama (2.59–0.78 Ma) down to 8 mm/a during Brunhes (<0.78 Ma). The intervals of higher spreading rates correspond to areas of thicker crust (Fig. 8b). Moving north within the southern sub-basin the negative anomalies flanking the axial anomaly disappear, implying a younger initial emplacement of oceanic crust (Fig. 8c). The valley floor displays a second en-echelon

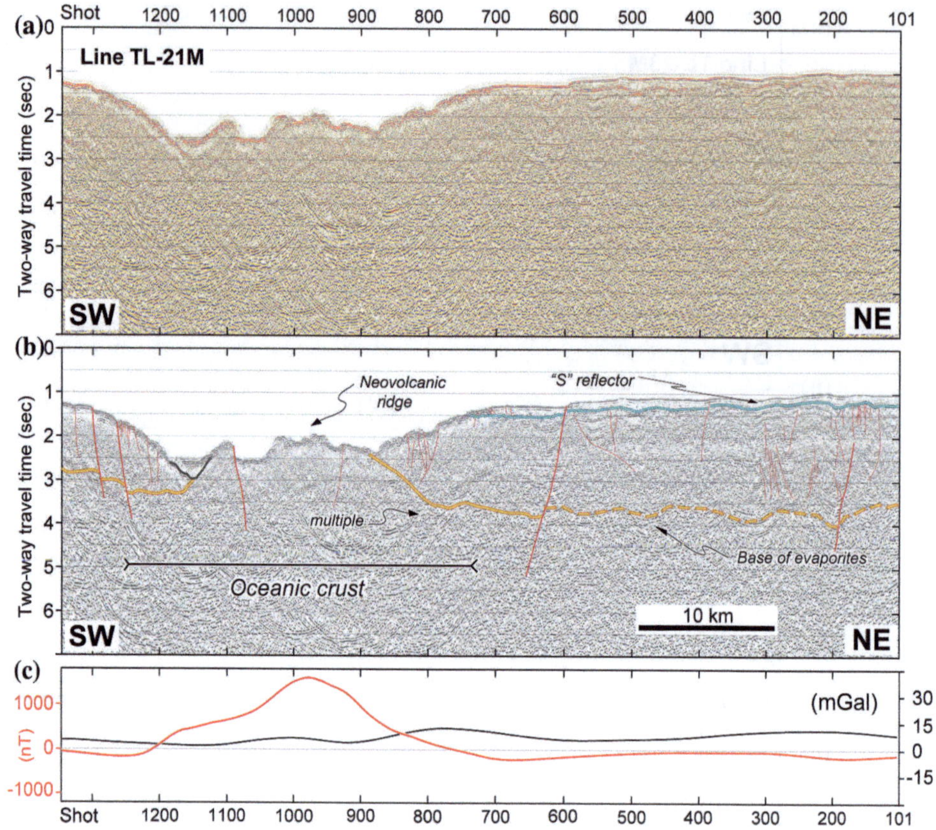

Fig. 6 Finite-difference time migrated seismic line TL-21M running perpendicular to the axis of the Thetis "central sub-basin. Profile location is indicated in Fig. 1. **a** Processed data only; **b** interpretation of the processed data. Symbols are given in Fig. 4. The extension of the oceanic crust beneath evaporites is inferred from magnetic data. **c** Gravity and magnetic profiles: reduced-to-the-pole magnetic anomaly, *red solid line*, and free air gravity anomaly, *black solid line*

axial volcanic ridge displaced eastward by ~10 km and lined up with the axial neo-volcanic zone of the central sub-basin. This pattern suggests that the oceanic crust in the overlapping zone formed by over-shooting of multiple dike intrusions from mid-segment melt injections, maintaining an along-axis segmentation during early stages of ocean basin formation (Wright et al. 2006).

Thetis Central and Northern Sub-basins

The Thetis "central sub-basin", a ~15 km long, ~20 km wide and ~1,900 m deep axial valley, shows a well developed strongly magnetic axial neo-volcanic zone in addition to scattered small volcanoes (Figs. 3 and 8). Seismic line TL-21M, running across the central part of the basin, shows that the neo-volcanic zone grows within a rift valley flanked by a pair of master faults, 10 km apart and with large throws accommodating a vertical displacement of ~300–500 m (Fig. 6). In contrast, the Thetis "northern sub-basin" lacks a clearly defined linear neovolcanic zone. An embryonic volcanic axial ridge is growing in the southern portion of the sub-basin from coalescence of central volcanoes aligned along axial normal faults, while the northern portion of the axial region is characterized by a single 150 m-throw NE-dipping normal fault (Fig. 3).

Based on two-dimensional magnetic forward modelling, Ligi et al. (2012) suggested for the northern sub-basin a recent (<0.78 Ma) initial emplacement of oceanic crust, and for the central basin as early as ~1.6 Ma with an average spreading rate of 15 mm/a.

Nereus Deep

Nereus is a ~50 km long, ~2,500 m deep axial rift valley flanked by >1 km-high walls with step-like morphology due to normal faults partly masked by "salt tectonics" at segment ends (Fig. 3). The ~12 km wide valley floor is dissected by a ~250 m high neovolcanic axial ridge composed of fresh oceanic-type basalts (Bonatti et al. 1984), with a strong central magnetic anomaly. At the mid-segment point, it is flanked by symmetrical reversed anomalies (Matuyama chron), that vanish toward the segment ends (Fig. 8c). Crustal thickness is at a maximum corresponding to reversed magnetization (Fig. 8d); moving south-east-ward the maximum thickness shifts toward the basin axis. Based on magnetic forward modelling, Ligi et al. (2012) suggested that the Nereus "oceanic" cell started about ~2 Ma, roughly at the same time of the initial activation of the southern Thetis cell. The average full spreading rate is 20 mm/a, with a maximum of 26 mm/a during Matuyama, slowing down to 10 mm/a during Brunhes.

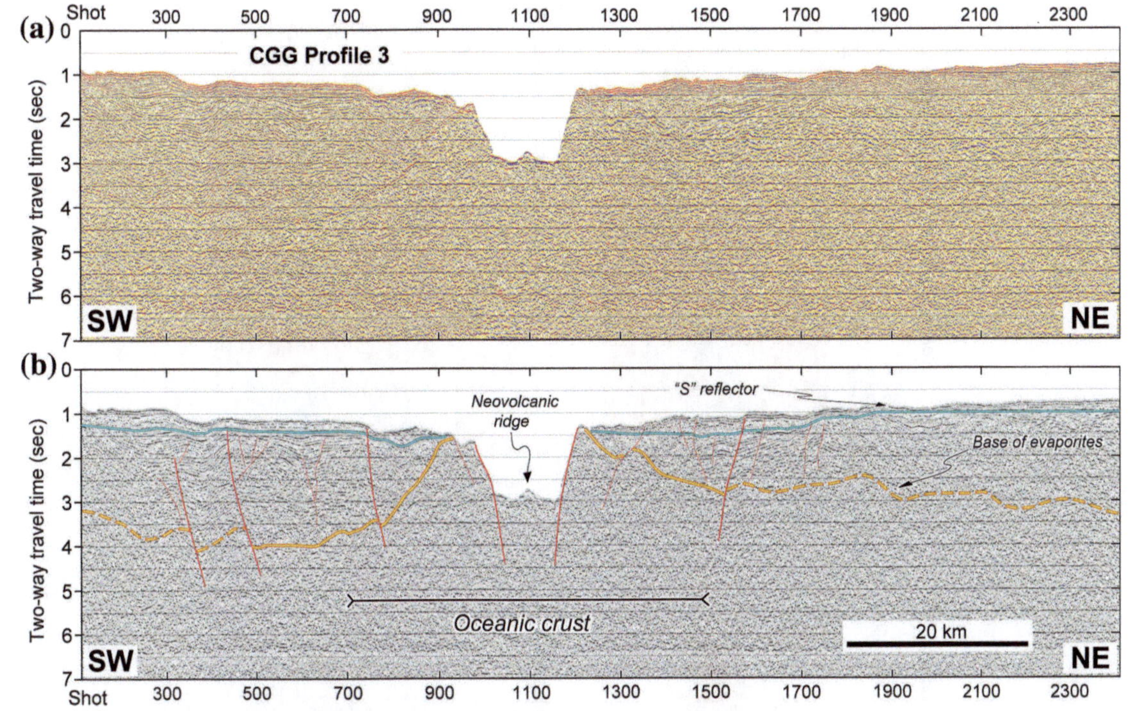

Fig. 7 Time migrated section of Profile 3 from Izzeldin (1989) running perpendicular to the axis of the Nereus Deep. Profile location is indicated in Fig. 1. A scanned copy of the stack section shown in Fig. 3 of Izzeldin (1987) was converted to seismic traces, band pass filtered and time migrated adopting a constant *P*-wave water velocity of 1,500 m/s. Profile location is indicated in Fig. 1. **a** Processed data only; **b** interpretation of the processed data (symbols as in Fig. 4). The extension of the oceanic crust beneath evaporites is inferred from magnetic data. Evaporite flowage toward the rift axis is fully blocked by igneous fault rotated blocks flanking the rift valley

Axial propagation is suggested by younger seafloor toward the segment tips. Reprocessed seismic line CGG 3 of Izzeldin (1987), running across the central part of the Nereus axial trough, indicates that the thick sediment pile is interrupted at the trough walls. The trough is slightly asymmetric with the NE wall being more elevated. Magnetic data indicate an igneous basement for the two tilted blocks bounding the axial valley. Basaltic walls of the axial trough are also suggested by volcanoclastic-pelagic carbonate breccias dredged from the walls and by carbonate-volcanoclastic turbidites observed in gravity cores from the valley floor and forming centimetre to decimetre thick layers, never reported elsewhere in the Red Sea (Taviani 1998). These gravity flows that include basalt fragments from the valley walls, probably due to brecciation along the fault scarps, and remobilized unstable sediment from the rift shoulders, may have been triggered by earthquakes; in fact events with magnitude >5 are not infrequent in the Red Sea.

The Nereus and Thetis Deeps are separated by a $\sim$35 km long gentle axial depression, carpeted by sediments and devoid of volcanism and of linear magnetic anomalies (Fig. 8c). The inter-trough seafloor is affected by curvilinear ridge-trough topography aligned in a general Red Sea direction (Fig. 3). Seismic reflection profiles suggest they are due to folding of the sedimentary sequence, caused by evaporite flowing toward the rift axis, that carried passively the overlying hemipelagic layer. RTP magnetic anomalies range from 0 to 100 nT over the entire inter-trough region and are unrelated to the relief (Fig. 8c). Magnetic modeling, assuming a magnetized layer with thickness and susceptibility similar to those observed at the Nereus and Thetis deeps below a 4–5 km-thick pile of amagnetic deposits, predicts a strong axial magnetic positive anomaly with amplitudes up to 1,000 nT. Thus, the observed low-amplitude positive magnetic anomalies may be due either to a lack of a magnetized layer underlying the Miocene and Plio-Quaternary deposits or to the blanketing of the magnetic signal caused by opposite magnetization of the overlying thick evaporitic sequence. We tested these hypotheses through two dimensional forward magnetic modeling. Assuming a negative susceptibility value of −0.02 SI for the salt layer, the magnetic signal from the basaltic layer will be nullified if it lies below a >8 km-thick evaporitic sequence, that is, twice the thickness of the evaporites estimated by several authors for this region [4–5 km by Ligi et al. (2012) and Mitchell et al. (2010); 3–3.5 km by Izzeldin (1987); 4.3–5 km by Tramontini and Davis (1969)]. In addition, assuming a 4–5 km thick salt layer, the salt susceptibility should change up to −0.35 SI to fit the observed magnetic anomalies, an unrealistic value for salt rocks (given the

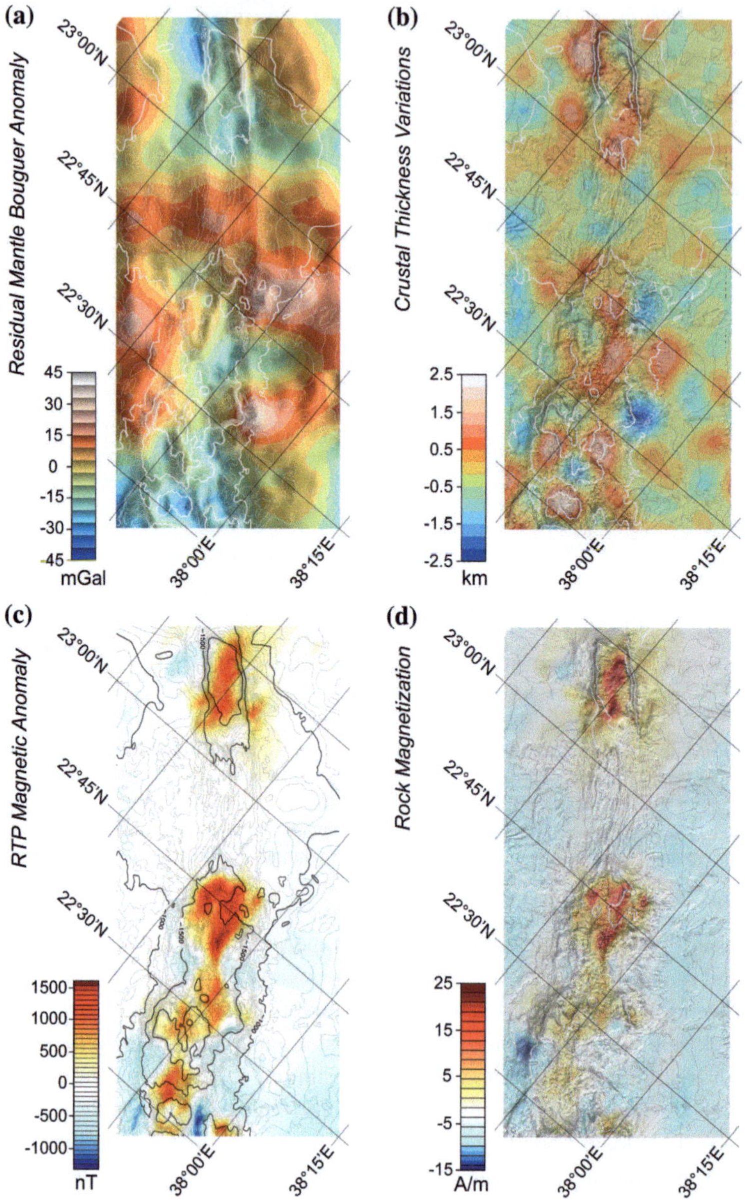

Fig. 8 Inversion of gravity and magnetic data superimposed on bathymetry (data processing details in Ligi et al. 2012). **a** Residual mantle Bouguer anomaly (RMBA) map generated by subtracting from the computed free-air anomalies the mantle thermal contribution to the gravity field and the attraction of sea floor topography, sediments and evaporites, and of the crust-mantle interface, assuming 5 km-thick crust and densities of 1,040, 1,900, 2,200, 2,690 and 3,330 kg m^{-3} for seawater, sediments, evaporites, crust and mantle, respectively. **b** Inferred variations of crustal thickness over a 5 km-thick crust obtained after downward continuation of a low-pass filtered set (features with wavelength <12 km were filtered out) of the computed residual anomalies to a depth of 8 km below sea level, assuming a crust-mantle density contrast of 640 kg/m^3. **c** Reduced to the pole magnetic anomaly map. Observed magnetic data, corrected for IGRF and levelled based on cross-over errors, are reduced to the North Pole by phase shifting them through an angle determined from the inclination of the present and the remanent magnetic field directions. Declinations and inclinations of the present field are estimated from IGRF. Remanent inclination is assumed to be that of an axial geocentric dipole, with the pole coinciding with the present spin axis. **d** Intensity of rock magnetization obtained by inversion of reduced-to-the-pole magnetic anomalies using the Fourier method of Caratori Tontini et al. (2009), assuming 1 km-thick magnetic layer parallel to the bottom of the evaporite layer

average susceptibility for halite is −0.01 SI, Telford et al. 1990). These results suggest that stretched and thinned continental crust, probably injected by diffuse basaltic intrusions, lies below Miocene evaporitic and Plio-Quaternary biogenic sediments in the Thetis-Nereus inter-trough zone.

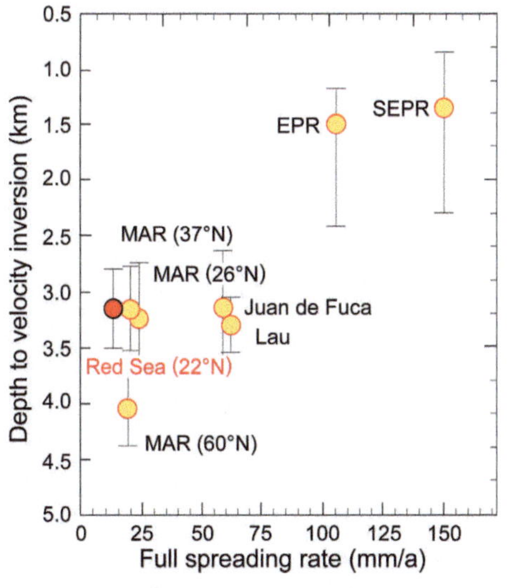

Fig. 9 Depth of mid-ocean-ridge axial magma chamber versus spreading rate. *Yellow filled circles* indicate magma chamber depths from Detrick et al. (1987), Collier and Sinha (1990), Calvert (1995) and Singh et al. (2006a, b). *Red circle* shows the depth of the southern Thetis magma chamber

Geochemical Constraints

The chemical variations of major and trace elements for the basaltic glasses we collected from the neo-volcanic zone of Thetis and Nereus are reported in Ligi et al. (2012). To discuss variations in melt generation across the Thetis and Nereus Deeps we have to take into account magmatic differentiation and the geochemical characteristics of the parental magmas.

Along axis major elements variability shows different patterns for Thetis and Nereus. Major element contents of Nereus samples do not show clear trends proceeding from south to north, probably due to the short distance separating the sample sites. Moving from the Thetis southern sub-basin to the northern sub-basin, while SiO_2 slightly increases, a maximum for FeO, TiO_2 Na_2O and K_2O and a minimum for MgO, Al_2O_3 and CaO is recorded in samples from the central sub-basin. The data indicate that the decrease in MgO observed in the central sub-basin is accompanied by increases in FeO, TiO_2 and Na_2O, and decreases in CaO and Al_2O_3 consistent with gabbroic fractionation involving initial removal of olivine, followed by extraction of three-phase assemblage olivine-plagioclase-clinopyroxene. In addition to the changes in melt composition produced during crystallization, the difference in major elements contents at a given MgO value may reflect changes in magma generation processes, such as variation of pressure and extent of melting, mantle source heterogeneities or contamination and assimilation effects.

In order to correct for the effects of differentiation, we calculated Na_8, Fe_8, Ti_8 and $(H_2O)_8$, that is, Na_2O, FeO, TiO_2 and H_2O normalized to 8 wt% MgO (Klein and Langmuir 1987; Plank and Langmuir 1992; Taylor and Martinez 2003), assuming olivine-plagioclase-clinopyroxene fractionation. In order to limit possible errors introduced by this correction, we used only analyses with 5.5–8.5 % MgO (Table 1). Samples from Thetis are characterized by mean and standard deviation values of $Na_8 = 2.54 \pm 0.17$ %, $Fe_8 = 10.61 \pm 0.53$ %, $(H_2O)_8 = 0.20 \pm 0.04$ %, $SiO_2 = 50.04 \pm 0.41$ %, and $Na^*/Ti^* = 2.30 \pm 0.33$ (*Na and Ti from normalized Na_8 and Ti_8). Samples from Nereus are less homogeneous and are characterized by lower Na_8 (2.21 ± 0.12 %), $(H_2O)_8$ (0.12 ± 0.03 %) and SiO_2 (49.86 ± 0.46 %), and higher Fe_8 (10.90 ± 10.81 %) and Na^*/Ti^* (2.47 ± 0.37 %) mean values (Table 1). Figure 10 shows along axis variations of Fe_8 and Na_8 across the Thetis and Nereus Deeps. While Fe_8 values do not show a clear trend along axis, Na_8 decreases from the Thetis southern sub-basin to the northern sub-basin and from the centre toward the southern tip of Nereus. Basalt Na_8 is inversely related to the mantle degree of melting by which the basalt was generated (Klein and Langmuir 1987). Thus, oceanic basalts in the younger Thetis northern sub-basin were generated by a mantle degree of melting higher than that of the basalts produced today in the more evolved Thetis central and southern sub-basins. The Nereus basalts degree of melting decreases toward the axial mid-segment point. Since Fe_8 largely varies in proportion to mean pressure of melting, the positive correlation of basalt Fe_8 with Na_8 both at Thetis and Nereus, shown in Ligi et al. (2012), implies that high degrees of melting are associated with low average pressures of melting.

Rare earth elements (REE) variability across the Thetis and Nereus Deeps is shown in Fig. 11. Basalt REE, high field strength elements (HFSE), Na and Ti are commonly used to constrain the mantle melting depths (Fram et al. 1998; Wang et al. 2002; Ferguson et al. 2013). Heavy REE and HFSE favour garnet structure, because their distribution coefficients change during melting of spinel versus garnet peridotites (Ellam 1992). For a given melting column, melts in equilibrium with garnet will produce high Zr/Y and $(Sm/Yb)_n$. These ratios will be gradually reversed as the melting column enters into the spinel stability field. Given that the partition coefficient K_{Na} of cpx/melt decreases with decreasing pressure, while K_{Ti} of cpx/melt and of garnet/melt remains constant or increases, the Na/Ti increases with decreasing mean pressure of melting. Figure 10 show the variations of Zr/Y, $(Sm/Yb)_n$ and Na^*/Ti^* across Thetis and Nereus Deeps. While Zr/Y and $(Sm/Yb)_n$ decrease, Na^*/Ti^* increases slightly from south to north, which is consistent with the depth variations of pressure of melting. Na_8

Table 1 Compositional parameters for basaltic glasses sampled along the axis of Thetis and Nereus

Sample ID	Latitude	Longitude	Depth	Mg#	Na_8	Fe_8	Ti_8	$(H_2O)_8$	Na*/Ti*	Zr/Y	(Sm/Yb)n	La/Nb	Ba/Nb
RS05-07-01b	22.41500	37.76950	1,732	58.69	2.80	10.76	1.34	0.24	2.58	2.67	1.05	1.09	5.96
RS05-08-01	22.43150	37.74400	1,842	58.96	2.62	11.18	1.24	0.23	2.61	2.57	1.07	1.17	6.50
RS05-08-02a	22.43150	37.74400	1,842	58.72	2.63	11.07	1.29	0.23	2.52	2.77	1.04	1.00	6.41
RS05-03-02	22.63100	37.67900	1,537	46.90	2.38	10.09	1.30	0.20	2.26	2.65	1.02	1.02	5.69
RS05-03-03	22.63100	37.67900	1,537	46.77	2.41	10.31	1.34	0.19	2.22	2.66	1.03	1.03	5.89
RS05-03-12a	22.63100	37.67900	1,537	46.26	2.42	10.27	1.55	0.19	1.94	2.68	1.07	1.05	6.02
RS05-02-01a	22.63550	37.62400	1,690	52.49	2.74	11.36	1.89	0.22	1.80	2.66	1.04	1.06	5.58
RS05-02-01b	22.63550	37.62400	1,690	52.58	2.66	11.04	1.79	0.22	1.84	2.64	1.04	1.05	5.49
RS05-10-02a	22.76600	37.57450	1,570	59.07	2.38	10.01	1.11	0.13	2.65	2.11	0.97	1.09	7.14
RS05-10-02b	22.76600	37.57450	1,570	58.84	2.35	10.01	1.15	0.13	2.53	2.12	0.97	0.98	6.55
MR83-D3-1	23.11120	37.27890	1,975	61.95	2.12	10.22	0.92	0.11	2.85	1.33	0.76	0.91	8.22
MR83-D3-B1	23.11120	37.27890	1,975	61.54	2.15	10.48	0.92	0.09	2.90	1.29	0.84	0.92	8.87
MR83-D3-3A	23.11120	37.27890	1,975	62.15	2.18	10.19	0.89	0.09	3.02	1.27	0.78	0.92	8.36
MR83-D3-3B	23.11120	37.27890	1,975	61.33	2.17	10.55	0.93	0.09	2.90	1.26	0.78	0.95	8.84
MR83-D3-4B	23.11120	37.27890	1,975	55.36	2.26	11.69	1.31	0.16	2.13	1.82	0.83	0.79	7.78
MR83-D3-4C	23.11120	37.27890	1,975	53.60	2.12	10.84	1.19	0.13	2.21	1.93	0.84	0.75	7.60
MR79-34-11A	23.19730	37.28250	1,749	59.14	2.38	10.22	1.00	0.14	2.96	1.59	0.84	0.94	7.22
*MR83-003	23.11120	37.27890	1,975	58.60	2.41	12.30	1.59		1.87				
*MR83-003-1	23.11120	37.27890	1,975	61.69	2.06	9.96	0.91		2.81				
*MR83-003-4A	23.11120	37.27890	1,975	55.38	2.15	10.97	1.28		2.07				
*MR83-003-6A	23.11120	37.27890	1,975	55.36	2.16	11.05	1.26		2.12				
*MR79-034-A	23.19730	37.28250	1,749	57.79	2.41	12.32	1.60		1.87				

correlates positively with Fe_8 (Ligi et al. 2012). In addition, Zr/Y and $(Sm/Yb)_n$ correlate positively with both Fe_8 and Na_8 (Fig. 12), while Na*/Ti* values show a slightly negative correlation. This indicates that an increase in the degree of melting corresponds to a decrease in the mean pressure of melting, suggesting an active mantle upwelling beneath Thetis and Nereus rather than a passive mantle flow. In fact, if the higher degree of mantle melting were due solely to a hotter passively upwelling mantle, peridotite solidus and initiation of melting would occur deeper in the mantle, thus increasing the average pressure of melting. Two Zr/Y and $(Sm/Yb)_n$ different positive trends can be distinguished looking separately at the samples from Thetis and from Nereus (Fig. 12). The two slopes imply that Nereus basalts originated from a slightly higher mantle degree of melting and a wider range of melting pressures than Thetis basalts. This suggests that each oceanic cell may tap mantle with slightly different fertility and/or temperature.

During continental rifting and rupturing contamination and assimilation processes may be an important factor in determining variations in major and trace element composition. Sr and Nd isotopic ratios and incompatible trace elements are used to assess the role of crustal contamination along the axis of Thetis and Nereus (Fig. 13). Rocks from the Thetis northern basin and the southern tip of Nereus, where the initial emplacement of oceanic crust is occurring today (Ligi et al. 2012), as well as the more differentiated rocks with longer residence time in magma chambers, should show the best evidence for crustal contamination. Figure 13 shows the mean values of $^{143}Nd/^{144}Nd$ and $^{87}Sr/^{86}Sr$ of basaltic glasses from Thetis and Nereus (data from Altherr et al. 1990; Volker and McCulloch 1993; Antonini et al. 1998). The Nd–Sr isotopic diagram indicates an absence of continental lithospheric components in the genesis of Thetis and Nereus basalts and suggests a source having slightly enriched "Depleted Mid-ocean-ridge Mantle" (DMM) composition (according to Workman and Hart 2005). The mantle source affinity is also suggested by highly incompatible trace element ratios such as Ba/Nb and La/Nb shown in Fig. 13b, and by REE variability that can be explained by varying the mean degree and mean pressure of melting within the spinel stability field from a source having DMM composition.

Discussion

Thetis and Nereus are the two northernmost "oceanic" segments (Fig. 3), with axial troughs carpeted by MORB and with Vine-Matthews magnetic anomalies. Magnetic profiles

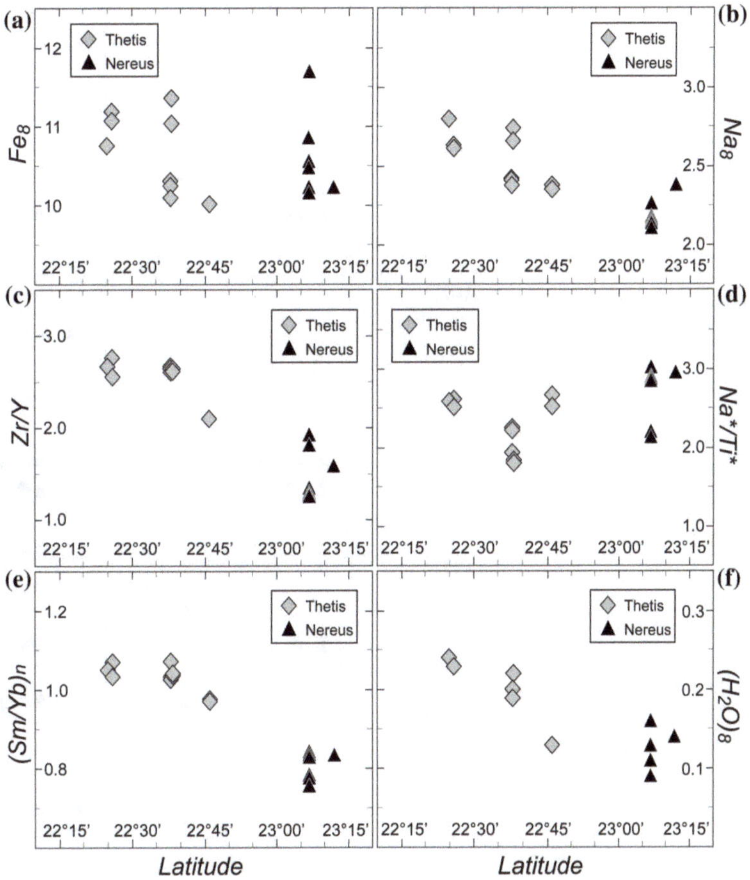

Fig. 10 Along axis variations of geochemical parameters suggested as proxy of degree and pressure of mantle melting. NaO and FeO compositions at a common 8% MgO were calculated using the equations of Klein and Langmuir (1987), and for TiO_2 and H_2O that of Taylor and Martinez (2003). Chondrite normalized $(Sm/Yb)_n$ ratios were obtained using chondritic values from Anders and Grevesse (1989). **a** Fe_8 of Thetis and Nereus basaltic glasses do not show a clear trend along the rift axis. **b** Na_8 decreases from the southern Thetis basin to the southern tip of Nereus, suggesting an increase of mean degree of mantle melting. **c–f** Variations of Zr/Y, Na*/Ti*, $(Sm/Yb)_n$, $(H2O)_8$ along Thetis and Nereus rift axis. These variations are qualitatively consistent with a shallow melting region beneath the rift axis, mostly within the spinel-peridotite stability field

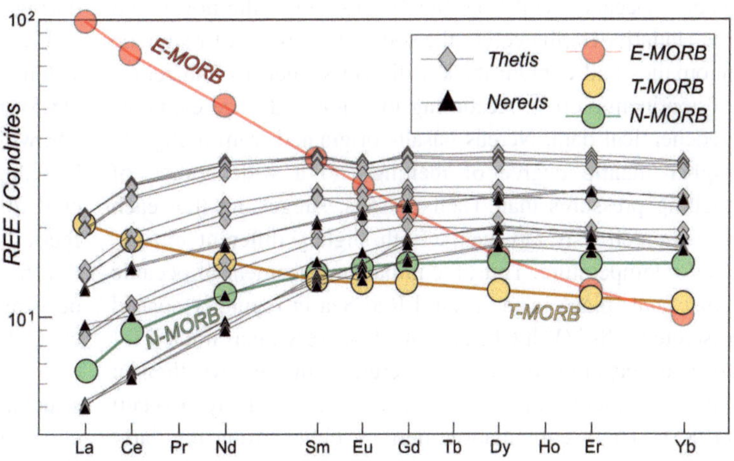

Fig. 11 REE pattern for Thetis and Nereus basaltic glasses. Normalization to chondritic values is from Anders and Grevesse (1989). Enriched E-MORB, transitional T-MORB and normal N-MORB patterns are from Langmuir et al. (1992)

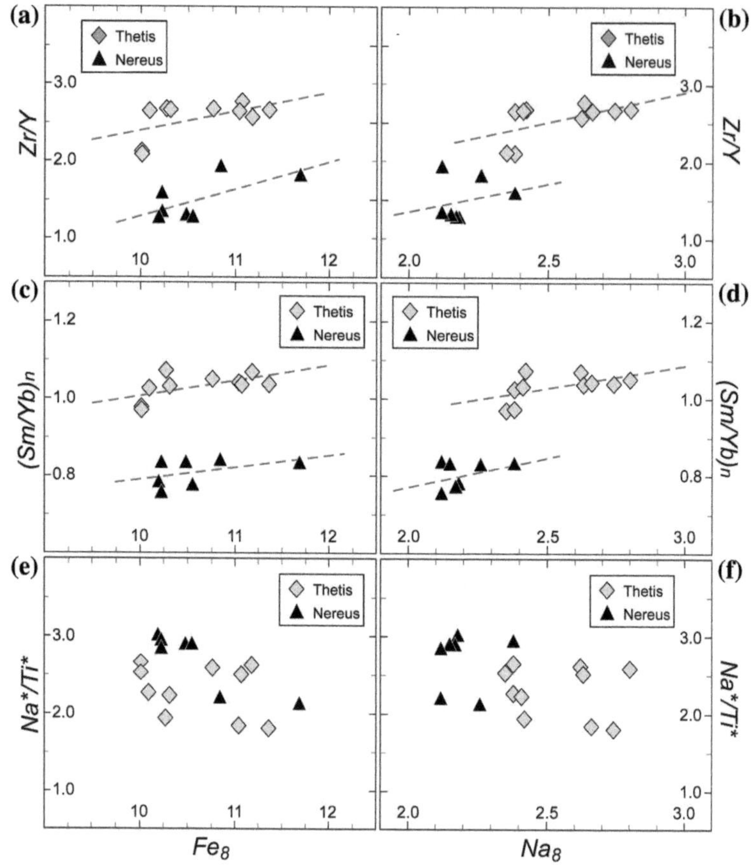

Fig. 12 Distribution of Zr/Y, $(Sm/Yb)_n$ and Na*/Ti* ratios versus Fe_8 (**a**, **c**, **e**) and Na_8 (**b**, **d**, **f**) in MORB glasses along the axis of Thetis and Nereus. Nereus and Thetis basaltic glasses display similar range of mean melting pressure. Nereus basalts originated from degrees of mantle melting higher than Thetis basalts, suggesting sources with different compositions. Zr/Y and $(Sm/Yb)_n$ show weak positive correlations (*red dashed lines*) with both Fe_8 and Na_8

from the northern part of Hadarba-Hatiba Deep, the oceanic cell immediately south of Thetis, indicate an initial oceanic accretion at ~3 Ma (Izzeldin 1987), while those from Thetis and Nereus are at ~2.2 and ~2 Ma, respectively (Ligi et al. 2011). The Thetis segment is made by coalescence of three sub-cells that become shallower, narrower and younger from south to north. The "intertrough zones" that separate the Thetis "oceanic" cell from the Nereus cell to the north, and the Hadarba-Hatiba cell to the south, are devoid of magnetic anomalies and contain thick sediment sequences and relicts of continental crust. According to major and trace elements geochemical data, Nereus basalts originated from a slightly higher mantle degree of melting and a wider range of melting pressures than Thetis basalts, suggesting that each oceanic cell may tap mantle with slightly different fertility and/or temperature. Thus, the initial emplacement of oceanic crust takes place in the central Red Sea in regularly spaced discrete "cells" (Whitehead et al. 1984) serving as nucleii for axial propagation of oceanic accretion, that evolve then in linear segments of spreading, as suggested by Bonatti (1985). This segmentation derives either from regularly spaced nuclei of upwelling asthenosphere (Choblet and Parmentier 2001; Wang et al. 2009), or from an initial structural segmentation inherited from pre-existing structural "accommodation zones" (Dixon et al. 1987; Corti et al. 2003), as observed in the East African Rift (Bosworth 1989; Makris and Rhim 1991; Ghebreab 1998), or most likely from a combination of the two processes.

Models predict that cool continental lithosphere and slow continental rifting (8–16 mm/a) lead to negligible melt production (White and McKenzie 1989; Bown and White 1994; Schmeling 2010). The low plate separation velocity during rifting, on average close to the modern value of ~12.2 mm/yr (Chu and Gordon 1998; Reilinger et al. this volume), and the lack of synrift volcanism (Bosworth et al. 2005) in the central Red Sea would suggest a magma poor rift, with a transition from continental to oceanic rift marked by exhumed mantle and initial alkaline magmatism (Jagoutz et al. 2007). Exposed peridotites in magma poor rifts such as Iberia-Newfoundland or the Alps show a complex interaction between melt and lithospheric mantle with widespread occurrence of refertilized plagioclase peridotites (Müntener et al. 2004, 2010; Le Roux et al. 2007). Direct observation of exhumed deep lithosphere in the Red Sea, along with the structures and the petrology associated with the formation of conjugate passive margins, is possible only on the island of Zabargad and on the Brothers Islets (Fig. 1). Zabargad represents an emerged sliver of sub-Red Sea lithosphere, probably uplifted by transpressive/transtensive tectonics along the Zabargad Fracture Zone, and provides a sample of

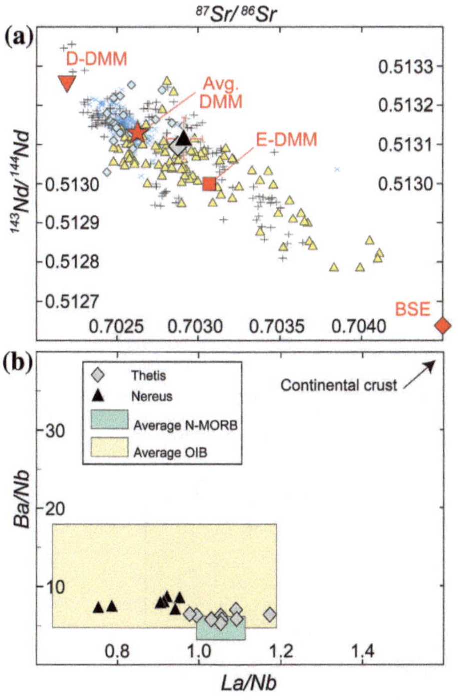

Fig. 13 Crustal contamination and source heterogeneity. **a** Average Sr–Nd isotopic compositions of Thetis (*gray filled diamond*. Data from Altherr et al. (1990)) and Nereus (*black filled triangle*. Data from Volker and McCulloch (1993) and Antonini et al. (1998)). Basaltic glasses are superimposed on Sr–Nd isotopic global MORB data (*yellow triangles* Indian Ocean mid-ocean ridges; *cyan diamonds* Juan De Fuca Ridge; *plus symbol* Mid Atlantic Ridge; *cross symbol* East Pacific Rise) of Workman and Hart (2005). *Red filled diamond* and *star* indicate isotopic composition of the bulk silicate Earth (BSE) and of the average depleted mantle (DMM), respectively. D-DMM: 2σ depleted DMM, and E-DMM: 2σ enriched DMM. The error bars (2σ) for Thetis and Nereus samples are also indicated (*red lines*). **b** Ba/Nb versus La/Nb for Thetis and Nereus basaltic glasses (*gray diamonds* and *black triangles*, respectively). OIB and N-MORB fields are from Weaver (1991)

continent-ocean transitional lithosphere (Bonatti et al. 1981, 1983, 1986; Nicolas et al. 1987). Zabargad exposes mantle-derived peridotites of sub-continental affinity in faulted contact with mafic-felsic granulitic gneisses of lower crustal origin, intruded by basaltic dykes. The gabbroic rocks were suggested to be originally part of a basic layered complex which crystallized at relatively high pressure (Bonatti and Seyler 1987). The Brothers are two islets located off-axis in the northern Red Sea. Previous work has shown that below a thin carbonate cap, the islets consist of MORB-like gabbroic rocks that crystallized at relatively low-P, cut by doleritic dykes (Bonatti and Seyler 1987). The Brothers Islets may represent tectonically uplifted fragments of gabbroic intrusions underplating thinned and stretched continental crust before initiation of seafloor spreading.

Geochemical data from Thetis and Nereus show a rift-to-drift transition marked by magmatic activity with typical MORB signature and a relatively high degree of mantle melting, with no contamination by continental lithosphere (Fig. 13). Most of the observed geochemical variability of Thetis and Nereus basaltic glasses can be explained by varying the mean degree and mean pressure of melting from a source having DMM composition. Basalts geochemistry suggests also "active" mantle upwelling probably enhanced by secondary sub-rift mantle convection related to the sharp horizontal thermal gradient between the thick/cold continental lithosphere and the hot melt-rich low-viscosity ascending asthenosphere (Buck 1986; Buck and Su 1989; Scott and Stevenson 1989; Sotin and Parmentier 1989; Keen and Boutilier 1995; Boutilier and Keen 1999; Nielsen and Hopper 2004). The presence of a magma chamber ~3.2 km below the neo-volcanic zone of the Thetis southern sub-basin (Fig. 5) implies the crystallization of lower oceanic crust intrusives as a last step in a sequence of basaltic melt intrusion from pre-oceanic continental rifting to oceanic spreading.

Although several questions remain open, the results above allow us to outline the evolution of the central Red Sea rift and the transition from continental rift to oceanic seafloor spreading resulting from the interactions between structural and magmatic processes (Fig. 14). Starting in Late Oligocene (~30 Ma), far field tectonic forces related to the inception of the Zagros subduction (Bosworth this volume) induced in the Red Sea region stretching and thinning of the lithosphere, with formation of extensional shear zones along the margins and passive upwelling of the asthenosphere beneath the rift axis. In addition, during continental rifting and proceeding for 20–15 Ma, a thermal/compositional anomaly probably related to the Afar plume (Ebinger and Sleep 1998) ponded beneath the lithosphere. This enabled deep partial melting of the rising asthenosphere and melt migration toward the rift axis with melt underplating the continental lithosphere and probably feeding swarm dikes (with an age of 24–22 Ma) that run along the entire Red Sea coast of Arabia (Bosworth et al. 2005). Melt solidification generated high-pressure gabbroic rocks (Fig. 14a) similar to those exposed on Zabargad Island (Bonatti and Seyler 1987). Increased temperatures deepened the solidus and enhanced short-lived small-scale convection by locally decreasing viscosity and increasing buoyancy of the upwelling asthenosphere (Ligi et al. 2011). Melt percolation and storage generated low-P gabbroic intrusions and basaltic dykes, exemplified by gabbros and dolerites from the Brothers Islets (Bonatti and Seyler 1987), that formed a wedge-shaped zone of softened and weakened lithospheric mantle along the axial zone beneath the thinned upper continental lithosphere (Fig. 14b). Active upwelling of the deeper asthenosphere, favoured by the softened axial zone, forced the final rupturing of the upper continental lithosphere. Mantle partial melting, occurring mostly in the spinel-peridotite stability field, generated aggregated N-MORB

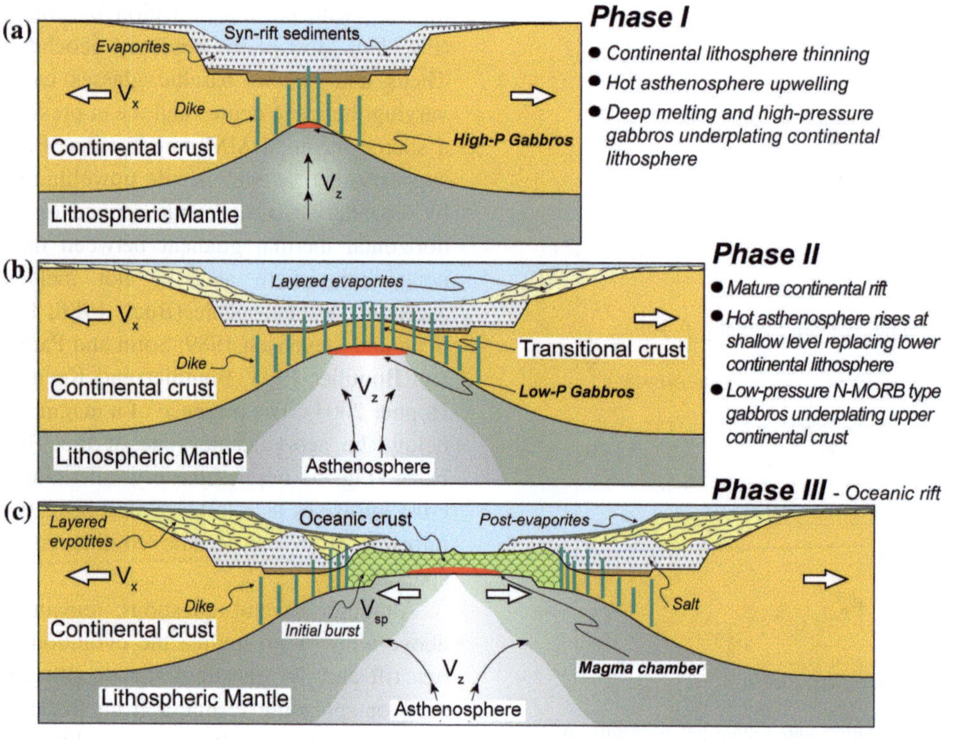

Fig. 14 Cartoon showing the continental to oceanic rift transition in the central Red Sea. **a** Continental rift induced in the Red Sea region stretching and thinning of the lithosphere with formation of extensional shear zones along the margins and passive upwelling of the asthenosphere beneath the rift axis. This enabled deep partial melting of the rising asthenosphere and melt migration toward the rift axis with melt underplating the continental lithosphere. Melt solidification generated high-pressure gabbroic rocks similar to those exposed on Zabargad Island (Bonatti and Seyler 1987). **b** Increased temperatures deepen the solidus and enhance short-lived small-scale convection by locally decreasing viscosity and increasing buoyancy of the upwelling asthenosphere (Ligi et al. 2011). Melt percolation and storage generated low-P gabbroic intrusions and basaltic dykes exemplified by gabbros and dolerites from the Brothers Islets (Bonatti and Seyler 1987). **c** Continental rupturing and seafloor spreading initiation. Active upwelling of the deeper asthenosphere forced the final rupture of the upper continental lithosphere. Mantle partial melting generated aggregated N-MORB melts that migrated and were extracted at the rift axis, forming shallow olivine-gabbro intrusions and basaltic lava flows

melts that migrated and were extracted at the rift axis, forming shallow olivine-gabbro intrusions and basaltic lava flows (Fig. 14c).

Emplacement of MORB with no contamination by lithospheric material in the axial part of the Central Red Sea during the very early stages of seafloor spreading implies replacement of lower crust and mantle lithosphere by active upwelling asthenosphere before breakup. In the Afar Depression, extension is dominated by diking with the density of dike-induced faulting increasing toward the rift axis. Most of the plate opening in this zone of incipient rupture is achieved by magma intrusions (Ebinger et al. 2010; Keir et al. 2011). When the volume of dyke intrusions and extension rates are relatively high, repeated intrusions in the same axial zone can potentially thermally weaken the crust and promote ductile stretching and thinning (Bastow and Keir 2011; Daniels et al. 2014). On the contrary, when magma volumes and rates of extension are low and the melt does not repeatedly intrude the same axial zone, dyke intrusions may instead increase the strength of the crust compositionally (Beutel et al. 2010; Daniels et al. 2014). Thus, during the late stage of rifting and during the initial phase of melt intrusion, episodic dike injections accommodate the extension in the upper crust, maintain its thickness and eventually increase its strength, delaying upper crust rupturing (Daniels et al. 2014). In the ductile melt-rich lower crust, the extension is accommodated continuously by a combination of magmatic addition and viscous flow (Wright et al. 2012), favouring the rise at shallow levels of the asthenosphere (Rychert et al. 2012). This is consistent with a model where depth-dependent extension, due to decoupling between the upper and lower lithosphere, results in mantle-lithosphere necking breakup before crustal-necking breakup (Kuznir and Karner 2007; Huismans and Beaumont 2011). A similar conceptual model for break-up in the Red Sea has been recently proposed by Mohriak and Leroy (2013) with very rapid thinning of the continental crust from onshore to offshore and incipient spreading in the axial trough, separating two Late Miocene salt basins.

Conclusions

Magnetic anomalies show a south to north time progression of the initial emplacement of oceanic crust within the Red Sea system. Mantle upwelling and melting may be affected by the south-north decreasing opening rate of the Red Sea and by the influence of the Afar plume, also decreasing from south to north. The northward propagation of oceanic accretion is today impacting against the Zabargad Fracture Zone, a morphotectonic feature that intersects the Red Sea in a NNE–SSW direction offsetting its axis.

Geochemistry of Thetis and Nereus basaltic glasses suggests a rather sharp rift-to-drift transition marked by magmatic activity with a typical MORB signature and a relatively high degree of mantle melting, with no contamination by continental lithosphere, suggesting that the lower crust and mantle lithosphere had already been replaced by active upwelling asthenosphere before separation of the Nubian and Arabian plates. Gabbros from the Brothers and Zabargad islands suggest that continental break up in the northern Red Sea, a relatively non-volcanic rift, is preceded by intrusion of oceanic-type basaltic melts that crystallize at progressively shallower crustal depths as rifting progresses towards continental break-up. A magma chamber or melt lens was detected $\sim$3.5 km below the axial neovolcanic zone in the southern Thetis basin. The presence of a few kilometers deep subrift magma chamber soon after the initiation of oceanic spreading implies the crystallization of lower oceanic crust intrusives as a last step in a sequence of basaltic melt intrusion from pre-oceanic continental rifting to oceanic spreading. Thus, oceanic crust accretion in the Red Sea rift starts at depth before continental break up, emplacement of oceanic basalt at the sea floor, and development of Vine-Matthews magnetic anomalies, pointing to a rift model where the lower continental lithosphere has been replaced by upwelling asthenosphere before continental rupturing. This model would imply depth-dependent extension due to decoupling between the upper and lower lithosphere with mantle-lithosphere-necking breakup before crustal-necking breakup. This mode of initial oceanic crust accretion may have been common in Mesozoic Atlantic-type rifts, in addition to wider, amagmatic, Iberian-type continent-ocean zones of transition, which may be considered an end-member example of how continents breakup (Péron-Pinvidic et al. 2013), but cannot be directly applicable to the many basins in the South and North Atlantic, and in the Red Sea (Ligi et al. 2012; Mohriak and Leroy 2013).

Acknowledgments This research was sponsored by the PRIN2012 Programme (Project 20125JKANY_002). The work was supported by the Saudi Geological Survey and the Italian Consiglio Nazionale Ricerche. We are grateful to D. Keir and W.U. Mohriak for their helpful and constructive reviews.

References

Altherr R, Henjes-Kunst F, Puchelt H, Baumann A (1990) Volcanic activity in the Red Sea axial trough: evidence for a large mantle diapir. Tectonophysics 150:121–133

AlMalki KA, Betts PG, Ailleres L (2014) Episodic sea-floor spreading in the Southern Red Sea. Tectonophysics 617:140–149

Alvarez W (2010) Protracted continental collisions argue for continental plates driven by basal traction. Earth Planet Sci Lett 296:434–442

Anders E, Grevesse N (1989) Abundances of the elements: Meteoritic and solar. Geochim Cosmochim Acta 53:197–214

Antonini P, Petrini R, Contin G (1998) A segment of sea-floor spreading in the central Red Sea: basalts from the nereus deep (23° 00′–23° 20′ N). J Afr Earth Sci 27:107–114

Armitage JJ, Henstock TJ, Minshull TA, Hopper JR (2009) Lithospheric controls on melt production during continental breakup at slow rates of extension: application to the North Atlantic. Geochem Geophys Geosyst 10:Q06018. doi:10.1029/2009GC002404

ArRajehi A, McClusky S, Reilinger R, Daoud M, Alchalbi A, Ergintav S, Gomez F, Sholan J, Bou-Rabee F, Ogubazghi G, Haileab B, Fisseha S, Asfaw L, Mahmoud S, Rayan A, Bendik R, Kogan L (2010) Geodetic constraints on present-day motion of the Arabian plate: implications for Red Sea and Gulf of Aden rifting. Tectonics 29:TC3011. doi:10.1029/2009TC002482

Augustin N, Devey CW, van der Zwan FM, Feldens P, Tominaga M, Bantan RA, Kwasnitschka T (2014) The rifting to spreading transition in the Red Sea. Earth Planet Sci Lett 395:217–230

Bäcker H, Schoell M (1972) New deeps with brines and metalliferous sediments in the Red Sea. Nature 240:153–158

Barberi F, Varet J (1977) Volcanism of Afar—small scale plate tectonics implications. Geol Soc Am Bull 88:1251–1266

Bastow ID, Keir D (2011) The protracted development of the continent-ocean transition in Afar. Nat Geosci 4:248–250

Beutel E, van Wijk J, Ebinger C, Keir D, Agostini A (2010) Formation and stability of magmatic segments in the main Ethiopian and Afar rifts. Earth Planet Sci Lett 293:225–235

Boillot G, Grimaud S, Mauffret A, Mougenot D, Kornprobst J, Mergoil-Daniel J, Torrent G (1980) Ocean–continent boundary off the Iberian margin: a serpentinite diapir west of the Galicia bank. Earth Planet Sci Lett 48:23–34

Bonatti E (1985) Punctiform initiation of seafloor spreading in the Red Sea during transition from continental to an oceanic rift. Nature 316:33–37

Bonatti E, Clocchiatti R, Colantoni P, Gelmini R, Marinelli G, Ottonello G, Santacroce R, Taviani M, Abdel-Meguid AA, Assaf HS, El Tahir MA (1983) Zabargad (St. John) Island: an uplifted fragment of Sub-Red Sea lithosphere. J Geol Soc London 14D:667–690

Bonatti E, Colantoni P, Della Vedova B, Taviani M (1984) Geology of the Red Sea transitional region (22°–25°N). Oceanol Acta 7:385–398

Bonatti E, Hamlyn P, Ottonello G (1981) Upper mantle beneath a young oceanic rift—peridotites from the island of Zabargad (Red-Sea). Geology 9:474–479

Bonatti E, Ottonello G, Hamlyn PR (1986) Peridotites from the island of Zabargad (Red Sea). J Geophys Res 91:599–631

Bonatti E, Seyler M (1987) Crustal underplating and evolution in the Red Sea rift. J Geophys Res 92:12083–12821

Bosworth W (1989) Basin and range style tectonics in East Africa. J Afr Earth Sc 8(2–4):191–201

Bosworth W, Huchon P, McClay K (2005) The Red Sea and Gulf of Aden basins. J Afr Earth Sci 43:334–378

Boutilier RR, Keen CE (1999) Small scale convection and divergent plate boundaries. J Geophys Res 104:7389–7403

Bown JW, White RS (1994) Variation with spreading rate of oceanic crustal thickness and geochemistry. Earth Planet Sci Lett 121:435–449

Buck WR (1986) Small-scale convection induced by passive rifting: the cause for uplift of rift shoulders. Earth Planet Sci Lett 77:362–372

Buck WR, Su W (1989) Focused mantle upwelling below mid-ocean ridges due to feedback between viscosity and melting. Geophys Res Lett 16(7):641–644

Calvert AJ (1995) Seismic evidence for a magma chamber beneath the slow-spreading Mid-Atlantic Ridge. Nature 377:410–413

Caratori Tontini F, Cocchi L, Carmisciano C (2009) Rapid 3-D forward model of potential fields with application to Palinuro Seamount magnetic anomaly (southern Tyrrhenian Sea, Italy). J Geophys Res 114:B02103. doi:10.1029/2008JB005907

Chang SJ, Merino M, Van der Lee S, Stein S, Stein CA (2011) Mantle flow beneath Arabia offset from the opening Red Sea. Geophys Res Lett 38:L04301. doi:10.1029/2010GL045852

Chen YJ (1992) Oceanic crustal thickness versus spreading rate. Geophys Res Lett 19:753–756

Choblet G, Parmentier EM (2001) Mantle upwelling and melting beneath slow spreading centers: effects of variable rheology and melt productivity. Earth Planet Sci Lett 184:589–604

Chu D, Gordon RG (1998) Current plate motions across the Red Sea. Geophys J Int 135:313–328

Cochran JR (1981) The Gulf of Aden: structure and evolution of a young ocean basin and continental margin. J Geophys Res 86:263–287

Cochran JR (2005) Northern Red Sea: nucleation of an oceanic spreading center within a continental rift. Geochem Geophys Geosyst 6:Q03006. doi:10.1029/2004GC000826

Cochran J, Gaulier J-M, LePichon X (1991) Crustal structure and the mechanism of extension in the Northern Red Sea: constraints from gravity anomalies. Tectonics 10:1018–1037

Cochran JR, Karner GD (2007) Constraints on the deformation and rupturing of continental lithosphere of the Red Sea: the transition from rifting to drifting. In: Karner GD, Manatschal G, Pinheiro LM (eds) Imaging, mapping and modeling continental lithosphere extension and breakup, Special Publication 282. Geological Society of London, London, pp 265–289

Coleman RG, McGuire AV (1988) Magma systems related to the Red Sea opening. Tectonophysics 150:77–100

Collier J, Sinha M (1990) Seismic images of a magma chamber beneath the Lau Basin back-arc spreading centre. Nature 346:646–648

Corti G, Van Wijk J, Bonini M, Sokoutis D, Cloethingh S, Innocenti F, Manetti P (2003) Transition from continental break-up to punctiform seafloor spreading: how fast, symmetric and magmatic. Geophys Res Lett 30:6–9

Courtillot V (1982) Propagating rifts and continental breakup. Tectonics 1:239–250

Crossley R, Watkins C, Raven M, Cripps D, Carnell A, Williams D (1992) The sedimentary evolution of the Red Sea and Gulf of Aden. J Pet Geol 15:157–172

D'Acremont E, Leroy S, Maia M, Patriat P, Berslier MO, Bellahsen N, Fournier M, Gente P (2006) Structure and evolution of the eastern Gulf of Aden: insights from magnetic and gravity data. Geophys J Int 165:786–803

Daniels KA, Bastow ID, Keir D, Sparks RSJ, Menard T (2014) Thermal models of dyke intrusion during the development of continent-ocean transition. Earth Planet Sci Lett 285:145–153

Davison I, Al-Kadasi M, Al-Kihrbash A, Baker J, Blakey S, Bosence D, Dart C, Heaton R, McClay K, Menzies M, Nichols G, Owen L, Yelland A (1994) Geological evolution of the southeastern Red Sea margin, Republic of Yemen. Geol Soc Am Bull 106:1474–1493

DeMets C, Gordon RG, Argus DF, Stein S (1994) Effect of recent revisions to the geomagnetic reversal time scale on estimates of current plate motions. Geophys Res Lett 21:2191–2194

Detrick RS, Buhl P, Vera E, Mutter J, Orcutt J, Madsen J, Brocher T (1987) Multi-channel seismic imaging of a crustal magma chamber along the East Pacific Rise. Nature 326:35–41

Dixon TH, Stern RJ, Hussein IM (1987) Control of Red Sea rift geometry by pre-Cambrian structure. Tectonics 6:551–571

Drake CL, Girdler RW (1964) A geophysical study of the Red Sea. Geophys J Roy Astron Soc 8:473–495

Ebinger C, Ayele A, Keir D, Rowland J, Yirgu G, Wright T, Belachew M, Hamling I (2010) Length and timescales of rift faulting and magma intrusion: the Afar rifting cycle from 2005 to present. Annu Rev Earth Planet Sci 38:439–466

Ebinger CJ, Sleep NH (1998) Cenozoic magmatism throughout East Africa resulting from impact of a single plume. Nature 395:788–791

Eldholm O, Grue K (1994) North Atlantic volcanic margins: dimensions and production rates. J Geophys Res 99:2955–2968

Ellam RM (1992) Lithospheric thickness as a control on basalt geochemistry. Geology 20:153–156

Ferguson DJ, Maclennan J, Bastow ID, Pyle DM, Jones SM, Keir D, Blundy JD, Plank T, Yirgu G (2013) Melting during late-stage rifting in Afar is hot and deep. Nature 499:70–73

Fournier M, Chamot-Rooke N, Petit C, Huchon P, Al-Kathiri A, Audin L, Beslier M-O, d'Acremont E, Fabbri O, Fleury J-M, Khanbari K, Lepvrier C, Leroy S, Maillot B, Merkouriev S (2010) Arabia-Somalia plate kinematics, evolution of the Aden-Owen-Carlsberg triple junction, and opening of the Gulf of Aden. J Geophys Res 115:B04102. doi:10.1029/2008JB006257

Fram MS, Lesher CE, Volpe AM (1998) Mantle melting systematics: transition from continental to oceanic volcanism on the southeast greenland margin. Proc ODP Sci Results 152:373–386

Ghebreab W (1998) Tectonics of the Red Sea region reassessed. Earth-Sci Rev 45:1–44

Girdler RW (1985) Problems concerning the evolution of oceanic lithosphere in the Northern Red Sea. Tectonophysics 116:109–122

Girdler RW, Styles P (1974) Two stage seafloor spreading. Nature 247:7–11

Guennoc P, Pautot G, Coutelle A (1988) Surficial structures of the northern Red Sea axial vally from 23°N to 28°N: time and space evolution of neo-oceanic structures. Tectonophysics 153:1–23

Guennoc P, Pautot G, Leqentrec MF, Coutelle A (1990) Structure of an early oceanic rift in the northern Red-Sea. Oceanol Acta 13:145–157

Hall SA (1989) Magnetic evidence for the nature of the crust beneath the Southern Red Sea. J Geophys Res 94:12267–12279

Hofmann C, Courtillot V, Feraud G, Rochette P, Yirgu G, Ketefo E, Pik R (1997) Timing of the Ethiopian flood basalt event and implications for plume birth and global change. Nature 389:838–841

Höink T, Jellinek AM, Lenardic A (2011) Viscous coupling at the lithosphere-asthenosphere boundary. Geochem Geophys Geosyst 12:Q0AK02. doi:10.1029/2011GC003698

Hopper JR, Buck WR (1998) Styles of extensional decoupling. Geology 26:699–702

Huismans R, Beaumont C (2011) Depth-dependent extension, two-stage breakup and cratonic underplating at rifted margins. Nature 473(7345):74–78

Izzeldin AY (1982) On the structure and evolution of the Red Sea. PhD thesis, IPG, Strasbourg, France, 164 pp

Izzeldin AY (1987) Seismic, gravity and magnetic surveys in the central part of the Red Sea: their interpretation and implications for the structure and evolution of the Red Sea. Tectonophysics 143:269–306

Izzeldin AY (1989) Transverse structures in the central part of the Red Sea and implications on early stages of oceanic accretion. Geophys J Int 96:117–129

Jagoutz O, Müntener O, Manatschal G, Rubatto D, Peron-Pinvidic G, Turrin BD, Villa IM (2007) The rift-to-drift transition in the North Atlantic: a stuttering start of the MORB machine? Geology 35:1087–1090

Keen CE, Boutilier RR (1995) Lithosphere-asthenosphere interactions below rifts. In: Banda E, Torne M, Talwani M (eds) Rifted ocean-continent boundaries. Kluwer Academic, Norwell, pp 17–30

Keir D, Belachew M, Ebinger CJ, Kendall J, Hammond JOS, Stuart GW, Ayele A (2011) Mapping the evolving strain field in the Afar triple junction using crustal anisotropy. Nat Commun 2(285). doi:10.1038/ncomms1287

Klein EM, Langmuir CH (1987) Global correlations of ocean ridge basalt chemistry with axial depth and crustal thickness. J Geophys Res 92:8089–8115

Kroner A, Stern RJ, Dawoud AS, Compston W, Reischmann T (1987) The Pan-African continental-margin in northeastern Africa—evidence from a geochronological study of granulites at Sabaloka, Sudan. Earth Planet Sci Lett 85:91–104

Kuznir NJ, Karner GD (2007) Continental lithospheric thinning and breakup in response to upwelling divergent mantle flow: application to the woodlark, newfoundland and Iberia margin. In: Karner GD, Manatschal G, Pinheiro LM (eds) Imaging, mapping and modelling continental lithosphere extension and breakup, Special Publication 282. Geological Society of London, London, pp 389–419

Langmuir CH, Klein EM, Plank T (1992) Petrological systematics of mid-ocean ridge basalts: constraints on melt generation beneath ocean ridges. In: Morgan JP, Blackman DK, Sinton JK (eds) Mantle flow and melt generation at Mid-Ocean ridges. geophysical monograph 71. American Geophysical Union, Washington DC, pp 183–280

Le Roux V, Bodinier JL, Tommasi A, Alard O, Dautria JM, Vauchez A, Riches AJV (2007) The Lherz spinel lherzolite: refertilized rather than pristine mantle. Earth Planet Sci Lett 259:599–612

Ligi M, Bortoluzzi G (1989) PLOTMAP: geophysical and geological applications of good standard quality cartographic software. Comput Geosci 15:519–585

Ligi M, Cuffaro M, Chierici F, Calafato A (2008) Three-dimensional passive mantle flow beneath mid-ocean ridges: an analytical approach. Geophys J Int 175:783–805

Ligi M, Bonatti E, Caratori Tontini F, Cipriani A, Cocchi L, Schettino A, Bortoluzzi G, Ferrante V, Khalil SM, Mitchell NC, Rasul N (2011) Initial burst of oceanic crust accretion in the Red Sea due to edge-driven mantle convection. Geology 39:1019–1022. doi:10.1130/G32243.1

Ligi M, Bonatti E, Bortoluzzi G, Cipriani A, Cocchi L, Caratori Tontini F, Carminati E, Ottolini L, Schettino A (2012) Birth of an ocean in the Red Sea: initial pangs. Geochem Geophys Geosyst 13:Q08009. doi:10.1029/2012GC004155

Lizarralde D, Axen GJ, Brown HE, Fletcher JM, González-Fernández A, Harding AJ, Holbrook WS, Kent GM, Paramo P, Sutherland F, Umhoefer PJ (2007) Variation in styles of rifting in the Gulf of California. Nature 448:466–469

Makris MJ, Rhim R (1991) Shear controlled evolution of the Red Sea: pull-apart model. Tectonophysics 198:441–466

Manatschal G, Froitzheim N, Rubenach M, Turrin BD (2001) The role of detachment faulting in the formation of an ocean-continent transition: insights from the Iberia Abyssal Plain. In: Wilson RCL, Whitmarsh RB, Taylor B, Froitzheim N (eds) Non-Volcanic rifting of continental margins: a comparison of evidence from land and sea, Special Publication 187. Geological Society of London, London, pp 405–428

McClusky S, Reilinger R, Mahmoud S, Ben Sari D, Tealeb A (2003) GPS constraints on Africa (Nubia) and Arabia plate motions. Geophys J Int 155:126–138

Mitchell NC, Ligi M, Ferrante V, Bonatti E, Rutter E (2010) Submarine salt flows in the central Red Sea. Geol Soc Am Bull 122:701–713

Mitchell NC, Park Y (2014) Nature of crust in the central Red Sea. Tectonophysics 628:123–139. http://dx.doi.org/10.1016/j.tecto.2014.04.029

Mohriak WU, Leroy S (2013) Architecture of rifted continental margins and break-up evolution: insights from the South Atlantic, North Atlantic and Red Sea-Gulf of Aden conjugate margins. In: Mohriak WU, Danforth A, Post PJ, Brown DE, Tari GC, Nemcok M, Sinha ST (eds) Conjugate divergent margins, Special Publication 369. Geological Society of London, London, pp 497–535. doi:10.1144/SP369.17

Morgan PJ, Chen YJ (1993) Dependence of ridge-axis morphology on magma supply and spreading rate. Nature 364:706–708

Müntener O, Manatschal G, Desmurs L, Pettke T (2010) Plagioclase peridotites in ocean–continent transitions: refertilized mantle domains generated by melt stagnation in the shallow mantle lithosphere. J Petrol 51:255–294. doi:10.1093/petrology/egp087

Müntener O, Pettke T, Desmurs L, Meier M, Schaltegger U (2004) Refertilization of mantle peridotite in embryonic ocean basins: trace element and Nd-isotope evidence and implications for crust-mantle relationships. Earth Planet Sci Lett 221:293–308

Nicolas A, Boudier F, Montigny R (1987) Structure of Zabargad Island and early rifting of the Red sea. J Geophys Res 92:461–474

Nielsen TK, Hopper JR (2004) From rift to drift: mantle melting during continental breakup. Geochem Geophys Geosyst 5:Q07003. doi:10.1029/2003GC000662

Pautot G (1983) Red Sea deeps—a geomorphological study by Seabeam. Oceanol Acta 6:235–244

Péron-Pinvidic G, Manatschal G, Osmundsen PT (2013) Structural comparison of archetypal Atlantic rifted margins: a review of observations and concepts. Mar Pet Geol 43:21–47

Phillips JD, Ross DA (1970) Continuous seismic reflection profiles in the Red Sea. Philos Trans R Soc Lond 267:143–152

Plank T, Langmuir CH (1992) Effects of the melting regime on the composition of oceanic crust. J Geophys Res 97:19749–19770

Purdy GM, Kong LSL, Christeson GL, Solomon SC (1992) Relationship between spreading rate and the seismic structure of mid-ocean ridges. Nature 355:815–817

Ravat D, Salem A, Abdelaziz AMS, Elawadi E, Morgan P (2011) Probing magnetic bottom and crustal temperature variations along the Red Sea margin of Egypt. Tectonophysics 510:337–344

Reston TJ, Morgan JP (2004) Continental geotherm and the evolution of rifted margins. Geology 32:133–136

Ross DA, Schlee J (1973) Shallow structure and geologic development of the southern Red Sea. Geol Soc Am Bull 84:3287–3848

Roeser HA (1975) A detailed magnetic survey of the southern Red Sea. Geol Jahr D13:131–153

Rowan MG (2014) Passive-margin salt basins: hyperextension, evaporite deposition, and salt tectonics. Basin Res 26:154–182

Rouchy JM, Noel D, Wali AMA, Aref MAM (1995) Evaporitic and biosiliceous cyclic sedimentation in the miocene of the Gulf of Suez—depositional and diagenetic aspects. Sed Geol 94:277–297

Rychert CA, Hammond JOS, Harmon N, Kendall JM, Keir D, Ebinger C, Bastow ID, Ayele A, Belachew M, Stuart G (2012) Volcanism in the Afar Rift sustained by decompression melting with minimal plume influence. Nat Geosci 5:406–409. doi:10.1038/ngeo1455

Saleh S, Jahr T, Jentzsch G, Saleh A, Ashour NMA (2006) Crustal evaluation of the northern Red Sea rift and Gulf of Suez, Egypt from geophysical data: 3-dimensional modeling. J Afr Earth Sc 45:257–278

Scott DR, Stevenson DJ (1989) A self-consistent model of melting, magma migration and buoyancy-driven circulation beneath mid-ocean ridges. J Geophys Res 94:2973–2988

Searle RC, Ross DA (1975) A geophysical study of the Red Sea axial trough between 20.5° and 22°N. Geophys J Roy Astron Soc 43:555–572

Schmeling H (2010) Dynamic models of continental rifting with melt generation. Tectonophysics 480:33–47

Singh SC, Crawford WC, Carton H, Seher T, Combier V, Cannat M, Canales JP, Dusunur D, Escartin J, Miranda JM (2006a) Discovery of a magma chamber and faults beneath a Mid-Atlantic Ridge hydrothermal field. Nature 442:1029–1032

Singh SC, Harding AJ, Kent GM, Sinha MC, Combier V, Bazil S, Tong CH, Pye JW, Barton PJ, Hobbs RW, White RS, Orcutt JA (2006b) Seismic reflection images of Moho underlying melt sills at the East Pacific Rise. Nature 442:287–290

Sotin CJ, Parmentier EM (1989) Dynamical consequences of compositional and thermal density stratification beneath spreading centers. Geophys Res Lett 16:835–838

Stern RJ (1994) Arc assembly and continental collision in the Neoproterozoic East African Orogen: implications for the consolidation of Gondwanaland. Annu Rev Earth Planet Sci 22:319–351

Stern RJ, Johnson PR (2010) Continental lithosphere of the Arabian plate: a geologic, petrologic, and geophysical synthesis. Earth-Sci Rev 101:29–67

Storey BC (1995) The role of mantle plumes in continental break-up: case-histories from Gondwanaland. Nature 377:301–308

Sultan M, Becker R, Arvidson RE, Shore P, Stern RJ, El Alfy Z, Guinness EA (1992) Nature of the Red Sea crust: a controversy revisited. Geology 20:593–596

Sultan M, Becker R, Arvidson RE, Shore P, Stern RJ, El Alfy Z, Attia RI (1993) New constraints on Red-Sea rifting from correlations of Arabian and Nubian Neoproterozoic outcrops. Tectonics 12:1303–1319

Taylor B, Martinez F (2003) Back-arc basin basalt systematics. Earth Planet Sci Lett 210:481–497

Taviani M (1998) Axial sedimentation of the Red Sea transitional region (22°–25° N): pelagic, gravity flow and sapropel deposition during the late quaternary. In: Purser BH, Bosence DWJ (eds) Sedimentation and tectonics of rift basins: Red Sea—Gulf of Aden. Chapman and Hall, London, pp 467–478

Telford WN, Geldart LP, Sheriff RE (1990) Applied geophysics, 2nd edn. Cambridge University Press, Cambridge, 770 pp

Thybo H, Nielsen CA (2009) Magma compensated crustal thinning in continental rift zones. Nature 457:873–876

Tramontini C, Davis D (1969) A seismic refraction survey in the Red Sea. Geophys J Roy Astron Soc 17:225–241

Voggenreiter W, Hotzl H (1989) Kinematic evolution of the southwestern Arabian continental margin: implications for the origin of the Red Sea. J Afr Earth Sc 8:541–564

Volker F, McCulloch MT (1993) Submarine basalts from the Red Sea: New Pb, Sr, and Nd isotopic data. Geophys Res Lett 20:927–930

Wang K, Plank T, Walker JD, Smith EI (2002) A mantle melting profile across the Basin and Range, SW USA. J Geophys Res 107: ECV5. doi:10.1029/2001JB000209

Wang Y, Forsyth DW, Savage B (2009) Convective upwelling in the mantle beneath the Gulf of California. Nature 462:499–502

Weaver BL (1991) The origin of ocean island end-member compositions: trace element and isotopic constraints. Earth Planet Sci Lett 104:381–397

White RS, McKenzie D (1989) Magmatism at rift zones: the generation of volcanic continental margins and flood basalts. J Geophys Res 94:7685–7729

White RS, Smith LK, Roberts AW, Christie PAF, Kusznir NJ, The Rest of the iSIMM Team (2008) Lower-crustal intrusion on the North Atlantic continental margin. Nature 452:460–465

Whitehead JA, Dick HJB, Schouten H (1984) A mechanism for magmatic accretion under spreading centres. Nature 312:146–148

Whitmarsh RB, Manatschal G, Minshull TA (2001) Evolution of magma-poor continental margins from rifting to seafloor spreading. Nature 413:150–154

Wilson JT (1966) Did the Atlantic close and then re-open? Nature 211:676–681

Workman RK, Hart SR (2005) Major and trace element composition of the depleted MORB mantle (DMM). Earth Planet Sci Lett 231:53–72

Wright TJ, Ebinger CJ, Biggs J, Ayele A, Yirgo G, Keir D, Stork A (2006) Magma-maintained rift segmentation at continental rupture in the 2005 Afar diking episode. Nature 442:291–294

Wright T, Sigmundsson F, Ayele A, Belachew M, Brandsdottir B, Calais E, Ebinger C, Einarsson P, Hamling I, Keir D, Lewi E, Pagli C, Pedersen R (2012) Geophysical constraints on the dynamics of spreading centres from rifting episodes on land. Nat Geosci 5:242–249

The Northern Red Sea in Transition from Rifting to Drifting-Lessons Learned from Ocean Deeps

Axel Ehrhardt and Christian Hübscher

Abstract

The transition from continental rifting to seafloor spreading can be observed along the 2,000 km length of the Red Sea Rift system. Whereas the southern Red Sea shows seafloor spreading since 5 Ma and the central part gives evidence of a transitional stage, the northern Red Sea is thought to represent the latest stage of continental rifting. Ocean deeps along the rift axis are considered to be first seafloor spreading cells that will accrete sometime in the future to a continuous spreading axis. The northern Red Sea deeps are isolated structures often associated with single volcanic edifices in comparison with the further developed larger central Red Sea deeps where small spreading ridges are active. Our analysis of the northern Red Sea deeps showed that not all deeps can be related to initial seafloor spreading cells. Two types of ocean deeps were identified: (a) volcanic and tectonically impacted deeps that opened by a lateral tear of the Miocene evaporites (salt) and Plio-Quaternary overburden; (b) non-volcanic deeps built by subsidence of Plio-Quaternary sediments due to evaporite subrosion processes. These deeps develop as collapse structures. The volcanic deeps can be correlated with their positions in NW–SE-oriented segments of the Red Sea which are consequently termed volcanic segments. The N–S segment, linking the volcanically active NW–SE segments, is termed as "non-volcanic segment" as no volcanic activity is known, in agreement with the magnetic data that show no major anomalies. Accordingly, the deep that was analyzed in this segment is interpreted as a collapse-related structure. However, collapse-type ocean deeps are not limited to the non-volcanic segments as subrosion processes due to hydrothermal circulation are possible at any part of the axial depression. The combined interpretation of bathymetry and seismic reflection profiles gives further insight into lateral salt gliding. Salt rises are present where the salt flows above basement faults. The internal reflection characteristic of the salt changes laterally from reflection-free to stratified, which suggests significant salt deformation during the salt deposition. Acoustically transparent halite accumulated locally and evolving rim synclines were filled by stratified evaporite facies.

A. Ehrhardt (✉)
Bundesanstalt für Geowissenschaften und Rohstoffe (BGR), Federal Institute for Geosciences and Natural Resources, Hannover, Germany
e-mail: axel.ehrhardt@bgr.de

C. Hübscher
Center for Earth System Research and Sustainability Institute of Geophysics, University of Hamburg, Hamburg, Germany

N.M.A. Rasul and I.C.F. Stewart (eds.), *The Red Sea*,
Springer Earth System Sciences, DOI 10.1007/978-3-662-45201-1_5,

Table 1 Main parameters for northern Red Sea deeps derived mainly from high-resolution multi-beam bathymetric, magnetic, and MCS data sets

	Conrad Deep	Shaban Deep	Kebrit Deep
Coordinates	27°03′N/34°43′E	26°15′N/35°24′E	24°43′N/36°17′E
Size	10 × 2 km	6 × 5 km	4 × 3 km
Volume	~1.5 km^3	~6.2 km^3	~2.5 km^3
	(1,340 m)	(1,250 m)	(1,240 m)
Brine volume	~0.4 km^3	~4.2 km^3	~0.2 km^3
Magnetic anomaly	Yes (2)	Yes (1)	No

The volume of each deep was calculated from the bottom of the deep to the quoted depth. The top of the brine body was calculated from the seismic lines using constant velocity $v_w = 1.5$ km s^{-1}

Introduction

The Red Sea Rift is a unique spot on Earth to observe the transition from continental rifting to seafloor spreading. Along the 2,000 km long axis of the Red Sea, all stages from rifting to drifting can be examined (see Fig. 1). Because direct basement observations are prevented by area-wide evaporite sequences (e.g., Bosworth et al. 2005; Whitmarsh et al. 1974) that overly the crystalline basement, near surface structures above the evaporite layers are investigated in order to get information about the inherent basement dynamics. Ocean deeps, that is, isolated pools with a size of some tens of square kilometers and depths of few hundreds of meters that penetrate the seafloor within the Red Sea are considered to be an accompanying phenomenon of the process from continental rifting to seafloor spreading (Bonatti 1985; Cochran and Martinez 1988; Martinez and Cochran 1988) which makes the meaning of the ocean deeps important in order to understand the plate tectonic setting of this region. Many Red Sea ocean deeps are associated with magmatism, either directly observed as volcanoes or indirectly as magnetic dipole anomalies. These deeps were suggested to represent initial seafloor spreading cells (Bonatti 1985). Taking into account the thick sedimentary cover that generally characterizes the entire Red Sea basin and especially the massive Miocene evaporite layers in the main trough (Cochran et al. 1991), the analysis of surficial structures such as the Red Sea deeps represents one way to extract information about the status of the rifting and the nature of the buried and masked northern Red Sea crust.

In this paper, we compare area-wide geophysical data such as magnetic data and satellite altimetry data with regionally acquired multi-channel seismic and hydroacoustic data from the northern Red Sea oceanic deeps (Fig. 1). First, we review what we know about the Conrad Deep (CD) published by Ehrhardt et al. (2005). The CD as well as the Shaban Deep (SD) is located in the NW–SE segment of the northern Red Sea. The "magnetic quiet" N–S segment hosts the Kebrit Deep (KD). From these individual studies, we infer the genesis of the deeps, associated with salt tectonics and fluid flow.

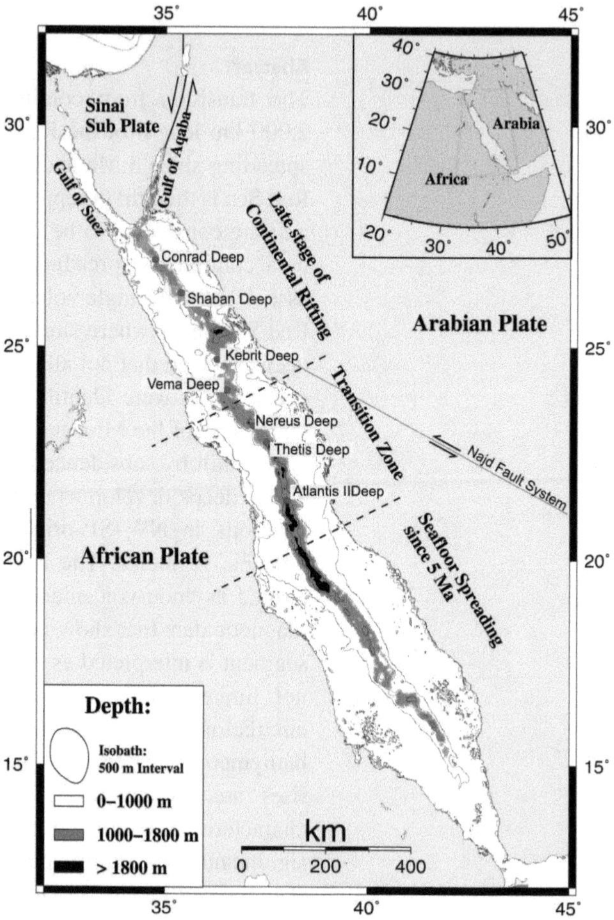

Fig. 1 The Red Sea can be separated along its strike into three provinces. The southern province is characterized by organized seafloor spreading, the central part is in a transition stage, and the northern province of the Red Sea is most probably in the late stage of continental rifting (Cochran et al. 1986). *Red dots* mark the position of prominent Red Sea deeps. Old tectonic lineaments such as the Neoproterozoic Pan-African Najd fault system terminate against the Red Sea rift (Bosworth et al. 2005)

Geological Setting

For the geological setting of the Red Sea, we refer to Bosworth (this volume). In this chapter, we focus our attention on the analysis of the Red Sea deeps. Since the 1960s, several ocean deeps were discovered along the axis of the Red Sea (Bäcker and Schoell 1972; Degens and Ross 1969; Miller et al. 1966). Because of size, age, and formation history, we distinguish between central and northern Red Sea deeps. The larger central Red Sea deeps are located in the axial trough and are separated by inter-trough zones (Cochran and Martinez 1988; Searle and Ross 1975). They are floored by young basaltic crust and exhibit magnetic anomalies not older than 1.7 Ma. The northern Red Sea deeps are smaller and form only isolated deeps within the axial depression. Some of them are accompanied by volcanic activity. Many Red Sea deeps contain bottom-water brines and metalliferous sediments pointing to hydrothermal circulation of seawater (Botz et al. 2011; Hartmann 1980; Scholten et al. 1991; Winckler et al. 2000). The largest and most prominent deep is the Atlantis II Deep, located in the central part of the Red Sea in the vicinity of other large deeps such as the Chain Deep and Discovery Deep. Other prominent deeps are the Tethys and Nereus Deeps further north, but still in the central part of the Red Sea.

Transition from Rifting to Seafloor Spreading

According to its development, the Red Sea can be separated into three parts from south to north (Cochran et al. 1986). In the southern part, seafloor spreading since ca. 5 Ma forms a continuous mid-ocean ridge system (MOR) (Röser 1975). The inception of spreading becomes progressively younger to the north. In the central Red Sea, the MOR becomes discontinuous forming single deeps (oceanized cells) that are separated by inter-trough zones (Searle and Ross 1975) where the spreading commenced not earlier than 1.7 Ma (Cochran et al. 1986). In the northern Red Sea, only isolated smaller deeps are distributed along an axial depression (Table 1 and Fig. 1). The axial depression is the northward prolongation of the axial trough of the central Red Sea. It is a characteristic feature of the northern Red Sea and bisects the main trough with water depths of about 1,200 m. The main trough is covered in the most areas of the Red Sea with huge amounts of Miocene evaporites with an average thickness of about 3,000 m (Girdler and Styles 1974; McKenzie et al. 1970; Mitchell et al. 1992; Searle and Ross 1975) that mask the underlying basement on seismic data. The top of the evaporites is seismically imaged as a prominent reflector that was named S-Reflector (Searle and Ross 1975; Whitmarsh et al. 1974) which can be identified easily in most areas of the Red Sea. The alignment of the coastal shelves, the main trough, and the axial depression is shown in Fig. 2.

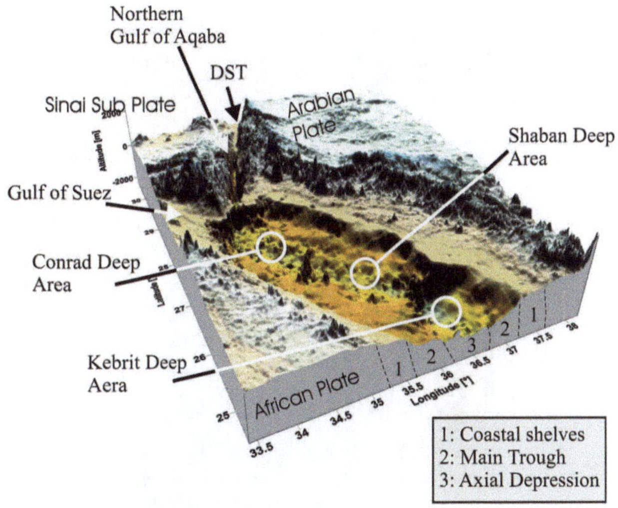

Fig. 2 3D block diagram showing northern Red Sea bathymetry (data from GEBCO) and characterization of seafloor morphology along an east–west cross section. The positions of the surveyed deeps are marked with white circles

Opening of the Red Sea became increasingly focused to the axial depression that represents the tectonically most active part of the rift (Cochran et al. 1991).

Segmentation of the Northern Red Sea

Although the Red Sea is still in the transition from continental rifting to seafloor spreading, it is already possible to observe segmentation along the rift. The bathymetric trend of the axial depression separates the northern Red Sea into segments; one NW–SE segment from 25°N to 28°N and one N–S-trending segment from 24°N to 25°N (Fig. 3). The N–S segment is also known as the Zabargad Fracture Zone (Bonatti et al. 1984; Marshak et al. 1992). The N–S segment links the NW–SE segment in the north with another NW–SE segment in the south so that the axial depression forms a kind of en-echelon pattern (Figs. 1 and 3). Segmentation along passive continental margins is also observed in Baffin Bay (Skaarup et al. 2006) where a sequence of volcanic and non-volcanic segments was identified. Similar segmentation is also known from the South Atlantic Ocean (Franke et al. 2007).

Two of the examined northern Red Sea deeps (Conrad and Shaban Deeps, CD and SD) are associated with magnetic anomalies. This indicates the presence of highly magnetic rocks, which are interpreted as magmatic intrusions and extrusions (Fig. 3 and Table 1). The Kebrit Deep (KD) is located in the N–S-oriented segment and is not correlated with

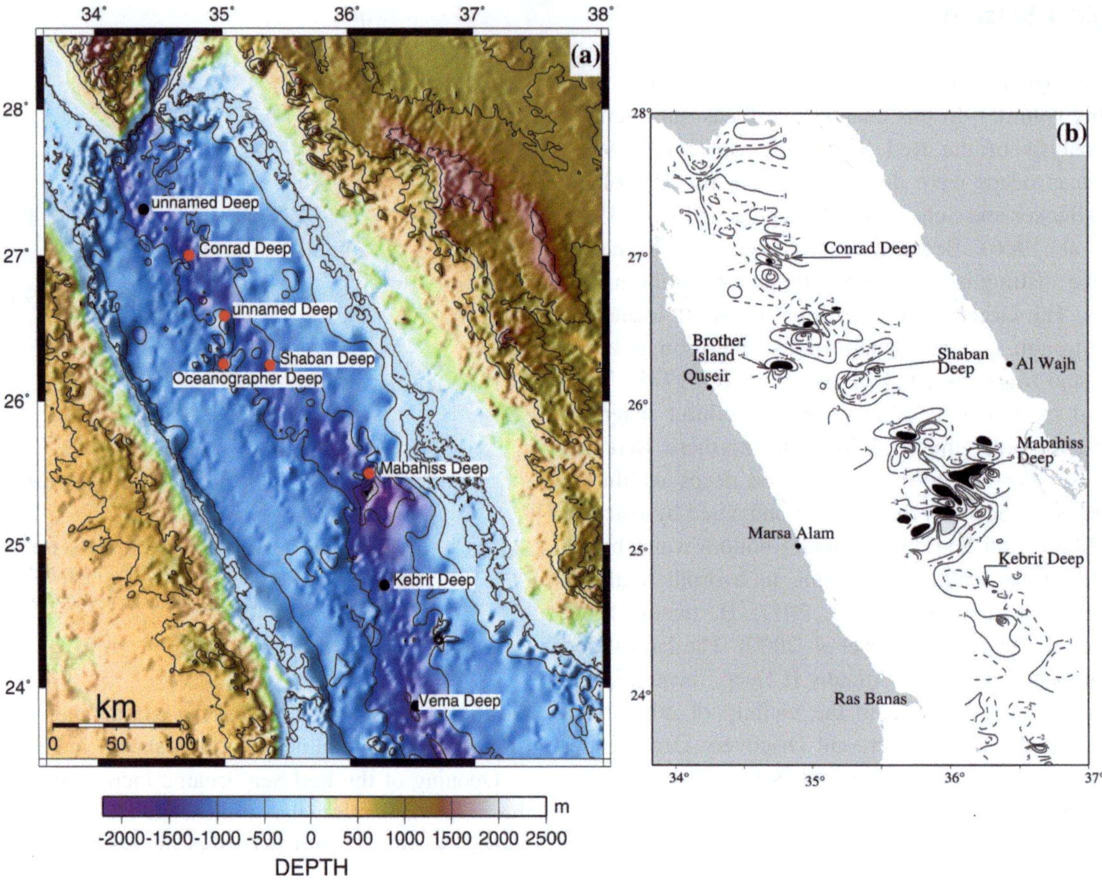

Fig. 3 **a** Bathymetric map (from satellite altimetry; Sandwell and Smith 1997). **b** magnetic map of the northern Red Sea (after Coutelle et al. 1991). Isolated magmatic dipole anomalies are interpreted as igneous intrusions into the basement. Strong magnetic anomalies are found at the Shaban and Conrad Deeps as well as Oceanographer, Mabahiss and unnamed deeps (*red dots*). Other deeps are not associated with strong magnetic anomalies such as Kebrit and Vema Deeps (*black dots*)

distinct magnetic anomalies (Table 1, Figs. 1 and 3). This segment represents a magnetic quiet zone in contrast to the more magnetic segment to the north. The southern adjoining segment comprises deeps with pronounced rift structures and initial spreading activity, for example the Nereus Deep (Antonini et al. 1998). As the northern part of the Red Sea rift is close to the stage of seafloor spreading and the extension already is focused to the axial depression, the heat flow is anomalously high. Heat flow between 250 and 350 mW/m^2 was measured in the axial depression (Martinez and Cochran 1988), which is about 6–10 times higher than the world mean (Makris and Rihm 1991). This high heat flow is further evidence of recent igneous activity which probably was also responsible for hydrothermal activity. These hydrothermal systems cause the deposition of the abundant metalliferous sediments in and near the deeps and the hyper-saline brine bodies within the deeps (Cocherie et al. 1994).

The Northernmost Red Sea Deeps

The CD and SD are located in the NW–SE-trending segment of the northern Red Sea (Figs. 1 and 3) and were discovered in 1986 and 1984 (Cochran et al. 1986; Pautot et al. 1984), respectively. The CD is one of the northernmost Red Sea deeps. Only one deep is reported north of the CD (at 27°19′ N, 35°23′E), but without any significant magnetic anomaly (Guennoc et al. 1988).

The CD was discovered during a multi-component survey, including seismic, hydro-acoustic, magnetic, gravity, and heat flow measurements, by the R/V Conrad in 1984 (Cochran et al. 1986). The CD is conspicuous because of its elongated N20°E-trending shape and the fact that it is located between two magnetic dipole-type anomalies NE and SW of it. The strike of the deep is almost parallel to the strike of the Gulf of Aqaba and the Dead Sea Transform, although the location of deep is a southward prolongation of the Gulf of Aqaba (Figs. 1, 2, and 3). Cochran et al. (1986) modeled two intrusion bodies in order to explain the anomalies and dated the emplacement of the intrusions to be only 40,000 years ago. Based on the high-resolution multi-channel seismic and multi-beam bathymetric data, Ehrhardt et al. (2005) linked the trend and shape of the CD to deep-rooted transtensional tectonics. They further suggested that the enhanced heat transfer between the two magmatic intrusions lowered the viscosity of the Miocene evaporites, thus supporting the evolution of the deep. Since heat flow values change rapidly and maximum values of 605 mWm^{-2} were reported from CD area (Cochran et al. 1986). Ehrhardt et al. (2005) suggested hydrothermal circulation and salt dissolution as the source of the brine.

The southeastern part of the SD was discovered in the early 1980s by the Preussag cruises (see e.g., Bäcker et al. 1975), but the SD in its entire extent was mapped in 1984. It is named after the Arab month of Shaban and is sometimes also termed the Jean Charcot Deep. The SD is more or less rhombic in shape and associated with a dipole-type magnetic anomaly that is caused by a magmatic extrusion (Pautot et al. 1984). This volcanic edifice forms an elongated NW–SE-trending ridge in the center of the deep. The SD is split up into four sub-basins that are filled with hyper-saline brine. Geochemical analyses of the brine water and the surrounding sediments result in contradictory interpretations of its formation. While Blum and Puchelt (1991) suggest that hydrothermal circulation was responsible for brine formation, Cocherie et al. (1994) analyzed REE signatures of SD metalliferous sediments and concluded that the SD brine water is seawater dominated, established by mainly lateral dissolution of Miocene evaporites.

The KD, located in the N–S segment of the northern Red Sea (Figs. 1, 2 and 3), was discovered in 1974 (Bäcker et al. 1975) and was revisited several times in order to provide long-term datasets of the brine body (Blum and Puchelt 1991; Hartmann et al. 1998; Winckler et al. 2001). Its name (Kebrit is Arabic for sulfur) points to hydrothermal activity. Blum and Puchelt (1991) discovered black smokers at the flanks of the deep and sampled them. Some KD chimneys consisted of pure sulfur. The KD is oval shaped and includes a hyper-saline brine body (e.g., Hartmann et al. 1998). Smooth slopes of the deeps are intersected by graben- or half-graben-like structures. The origin of the KD hyper-saline brine body is also controversial, as Blum and Puchelt (1991) propose hydrothermal circulation, but Winckler et al. (2001) prefer that the brines formed by the lateral solution of Miocene evaporites. At the link between the NW-SE segment and the N–S segment the largest northern Red Sea deep, the Mabahiss Deep developed. It is associated with a high amount of magmatic activity (Guennoc et al. 1988).

Methods and Observations

Our study is based on high-resolution multi-channel seismic, multi-beam and parametric echosounder data collected during R/V Meteor Cruises M44/3 and M52/3 in 1999 and 2002. Three northern Red Sea deeps (CD, SD, and KD) were investigated (Ehrhardt and Hübscher 2003; Hübscher et al. 2000). The close spacing of the seismic lines led to complete bathymetric mapping of the seafloor by the swath echosounder and enabled interpolation between the 2D-seismic lines (Fig. 4). These data are discussed together with previously published magnetic data (Coutelle et al. 1991) and gravity inferred from satellite altimetry data (Sandwell and Smith 1997).

The three northern Red Sea deeps show major differences in size and shape and in the accompanying geophysical anomalies. The bathymetry of the three northern Red Sea deeps is displayed in Figs. 4, 5, and 6. Selected 2D lines will illustrate the subsurface structure of the deeps and their surroundings. The main parameters regarding the deeps are listed in Table 1.

Conrad Deep

The CD is the northernmost Red Sea deep and is located within the axial depression of the northern Red Sea (Figs. 3 and 5). The elongated deep is 10.5 km long and 2 km wide and has an overall strike of N20°E, which is sub-parallel to the trend of the Gulf of Aqaba and thus the Dead Sea Transform (cf. Figs. 3 and 5). It comprises two sub-basins showing a small left stepping offset. Its maximum depth is 1,520 m which is approx. 200 m deeper than the surrounding seafloor. Its slopes are steepest on the longitudinal sides and distributed in an en-echelon pattern. The Klauke Deep to the west is approximately 150 m deeper than the inter-deep area. Circular and elongated volcanic edifices SW of the CD are grouped along the western flank of the axial depression. The appearance of the seafloor in the vicinity of the CD is different from NW to SE. That to the NW shows a gentle stepwise decrease of water depth from south to north and is obviously affected by the presence of the Klauke Deep. That to the SE shows two parallel to sub-parallel ridge-like structures, called the salt rise and salt wall (Fig. 5).

The seismic data resolve the uppermost Miocene evaporites and the Pliocene–Quaternary (PQ) overburden (Figs. 6, 7, 8 and 9). PQ sediments are as thick as 150–250 m which is

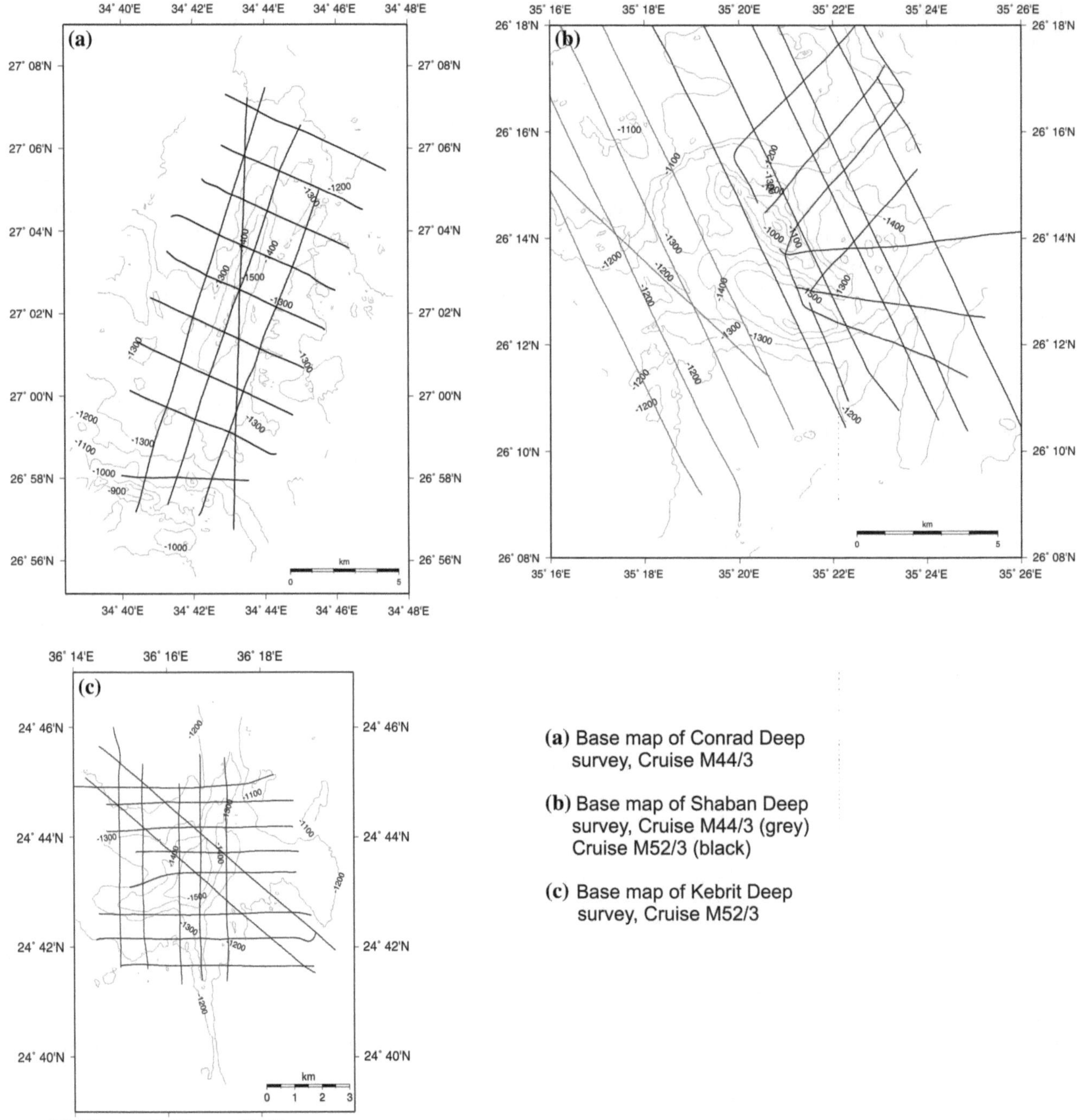

Fig. 4 Base maps of the multi-channel seismic surveys in the areas of the Conrad Deep (**a**), Shaban Deep (**b**) and Kebrit Deep (**c**). The contour maps show where deeps have been imaged by multi-channel seismic reflection data (MCS). Note the dense MCS line spacing with an average distance of less than 2 km

typical for the Red Sea (Guennoc et al. 1988). The S-reflection corresponding to the top of the evaporites is clearly visible in all seismic lines. The S-reflector is not continuous across the CD. It terminates against the CD and the salt wall, respectively. On its southeastern side, the S-reflector is about 200 ms "deeper" compared to the northwestern side. The longitudinal margins of the CD are bounded by normal faults with inclination angles of 16° and 11°, respectively.

We identified the following characteristics of the CD:

- The Pliocene–Quaternary supra-salt succession is subdivided into parallel layered lower and an upper succession which shows wavy and divergent internal reflections.

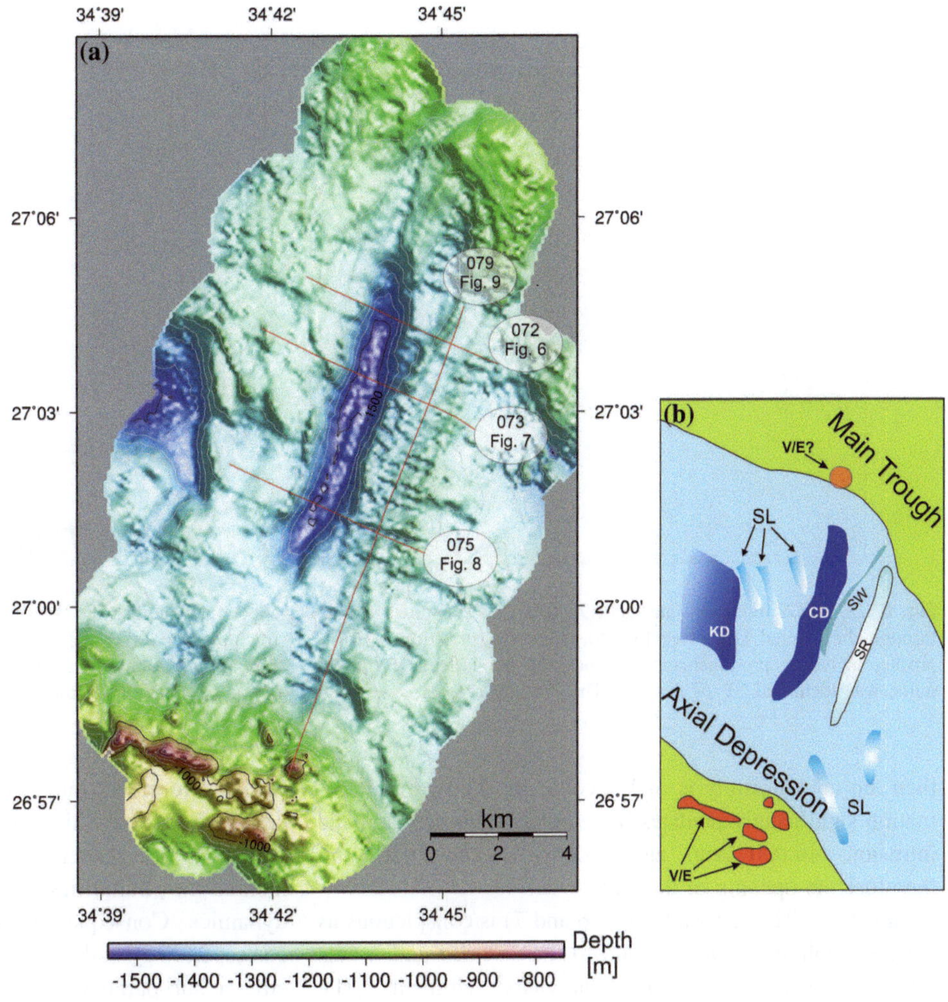

Fig. 5 Bathymetry map of the Conrad Deep (CD) area. The CD is located between the slopes of the axial depression. The southwestern slope is characterized by volcanic extrusions (see V/E). Volcanic extrusions on the northeastern slope are not imaged but the magnetic data give strong evidence for the presence of a volcanic/magmatic body (see Fig. 3). Note the second deep (Klauke Deep) west of the CD. *Red lines* and corresponding annotations show the location of the following seismic lines. The *right-hand figure* shows a sketch of the bathymetry with all structures quoted before (*CD* Conrad Deep; *KD* Klauke Deep; *SW* Salt wall; *SR* Salt rise; *V/E* Volcanic extrusion; *SL* Streamlines)

- Elongated salt rises—both straight and arcuate—create topographic ridges.
- The salt wall emerges from the evaporite sequence. It pierces the seafloor and is not covered by PQ sediments.
- High-amplitude reflector is located close to the CD (Fig. 8).
- At the SW tip of the CD, close to the slope of the axial depression, volcanic extrusions are observed in both seismic records and the bathymetry (Figs. 5 and 9).
- The reflection characteristics from the evaporites beneath the S-reflection change laterally from well-stratified to reflection-free (Fig. 8).

Pliocene–Quaternary succession: The appearance of Plio-Quaternary sediments varies from well-layered (Fig. 8, southeastern side) to chaotic (Fig. 6, northwestern side). Based on this seismic character, a subdivision into two seismic units is possible, pre-kinematic, and syn-kinematic. On the SE side of the deep, over the elongated salt rise, the change from a divergent to a concordant reflection pattern marks the onset of the salt tectonic deformation that led to the formation of the salt rise. The distinction between both units corresponds probably to the Units I and II sampled during DSDP Leg XXIII sites 225, 227, and 228 (Stoffers and Ross 1974).

Salt Rise: The salt rise on the SE side of the CD (Figs. 5, 6, 7, 8 and 9) strikes SW–NE and parallel to the strike of the CD. This rise structure could be the result of lateral salt flowage over a step in the underlying basement. Although the seismic data do not show evidence of the basement, we interpret these salt rises as the consequence of salt flow above basement faults in analogy to numerous instances from the Permian Central European Basin System where Zechstein salt ridges are frequently underlain by basement faults (see Warren 2006 and references therein).

The onset of the formation of the salt rise is marked by the dividing unconformity in the overlying Plio-Quaternary deposits. It is important to note that the age of the unconformity only refers to the development of the rise and not to the opening of the CD, other than previously explained by Ehrhardt et al. (2005). If the seismic units divided by the unconformity can be correlated to Units I and II of DSDP Leg XXIII sites 225, 227, and 228 (Stoffers and Ross 1974),

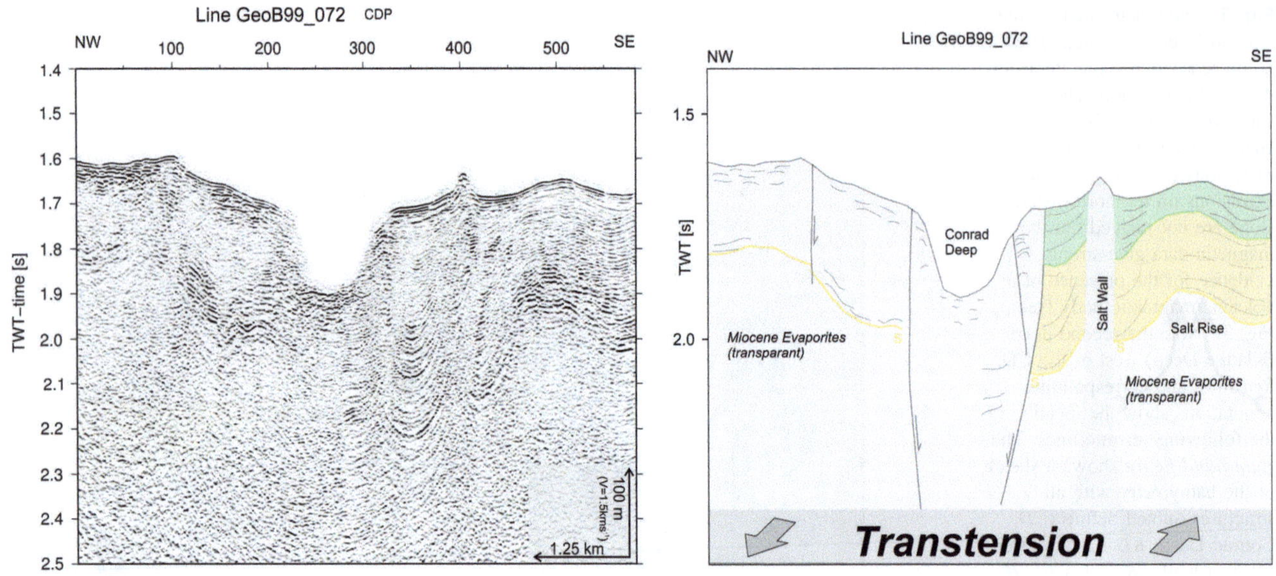

Fig. 6 *Left* MCS reflection line GeoB99_072, NW–SE cross section through the Conrad Deep (see Fig. 5). *Right* interpretation of MCS profile, including pre-kinematic (*yellow*) and syn-kinematic (*green*) units, was identified SE of the CD. The pre-unit reveals concordant reflection patterns; the syn-unit (*green*) is characterized by onlap structures and divergent reflection patterns. Both units are cut by a vertical transparent unit (salt wall). Basement-induced salt tectonics are a good explanation for the elongated salt rise

then the unconformity is of Pliocene age. Thus, the deformation significantly predates the emplacement of magmatic intrusions 40,000 years ago (Cochran et al. 1986) and therefore the opening of the CD.

Salt Wall: The salt wall (Figs. 6 and 7) is conspicuous as it trends oblique to the CD and to the salt rise and it cuts through the entire Pliocene–Quaternary sediments. The seismic lines and the bathymetry (Fig. 5) reveal that the salt wall strikes sub-parallel to the strike of the CD; thus, it intersects the deep, building the SE slope of the deep in its southern sub-basin. The salt wall acts obviously as backstop for the Plio-Quaternary sediments. Where this backstop is absent, the slope is not stabilized and mass wasting into the deep occurred along listric fault planes (Fig. 8).

An evolutionary model which could explain such a feature by hydrothermally driven ascent of hot brines is presented by Hovland et al. (this volume). Hydrothermal circulation above basement faults started and the salt wall was created due to salt precipitation out of the expelling fluids. The general possibility of hydrothermal circulation of salty fluids had been suggested before by Hovland et al. (2006) and was confirmed by laboratory- and molecular-scale dynamic simulations (Hovland et al. 2006). The occurrence of hydrothermal liquids is supported by the report of unusual high heat flow values (605 mW m^{-2}) at the northern end of the deep (Cochran et al. 1986).

Arcuate salt rises: The arcuate shape of salt rises and topographic ridges that terminate obliquely against the CD slopes are not related to deep-rooted faults. Mitchell et al. (2010) presented multi-beam bathymetric data from the central Red Sea which showed that salt flows downslope similar to ice-glaciers. Talbot and Pohjola (2009) documented the similarities between ice streams and salt glacier dynamics. Consequently, we see these salt rises and the corresponding bathymetric ridges as streamlines which emerge sub-parallel to the flowing direction.

High-amplitude Reflector: Consistent to this model is the observation of a high-amplitude reflector (HA, see Fig. 8b).

The high-amplitude reflector is not phase-reversed implying a strong positive acoustic impedance contrast caused by precipitated salt. Similar processes were observed by Netzeband et al. (2006) in the Mediterranean. Hydrothermal circulation caused salt precipitation near the seafloor (because of cooling) and the development of salt bodies or salt filled pore volumes that increased the bulk density and interval velocity.

Volcanic Extrusions: Southwest of the CD at the slope of the axial depression several volcanic extrusions were mapped by seismic (Fig. 9), bathymetric (Fig. 5) and magnetic methods (Fig. 3). Probably the volcanic extrusions ascended along the normal faulted slope of the axial depression. Similar volcanic bodies are likely at the northeastern end of the CD, also at the slope of the axial depression. However, they are not imaged by seismic data and only magnetic data are available as evidence (Fig. 3).

Lessons learned from the CD: The CD is located within the axial depression of the northern Red Sea. The deep disrupts the top of the Miocene evaporites, but it does not

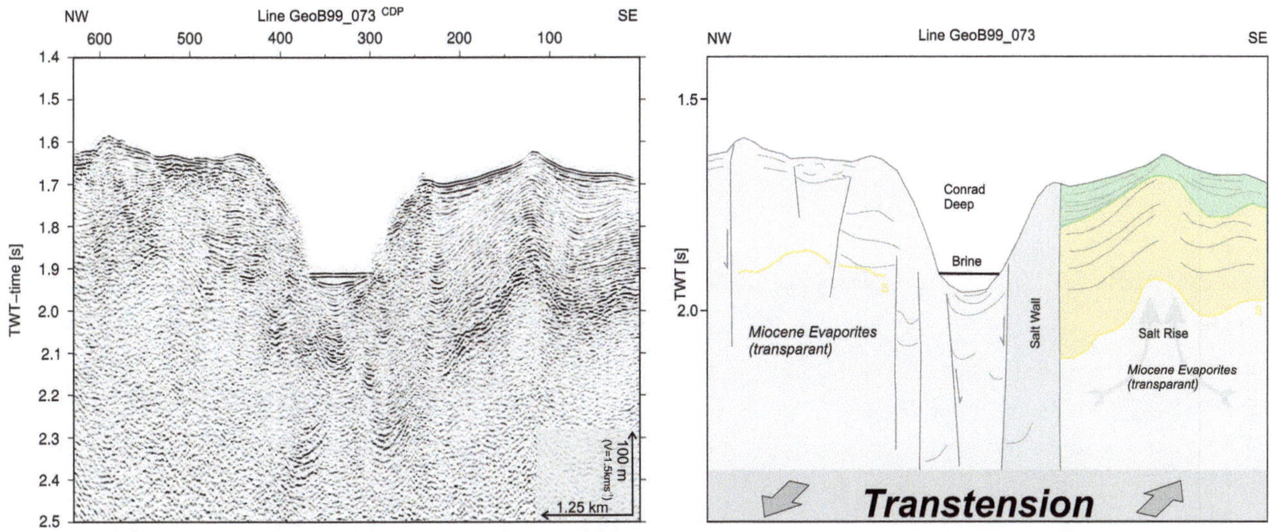

Fig. 7 *Left* MCS reflection line GeoB99_073, NW–SE MCS profile across the Conrad Deep (see Fig. 5). *Right* Interpretation of pre-kinematic (*yellow*) and syn-kinematic (*green*) units occurs SE of the deep. The salt wall builds the SE slope of the deep and forms a backstop for Plio-Quaternary sediments

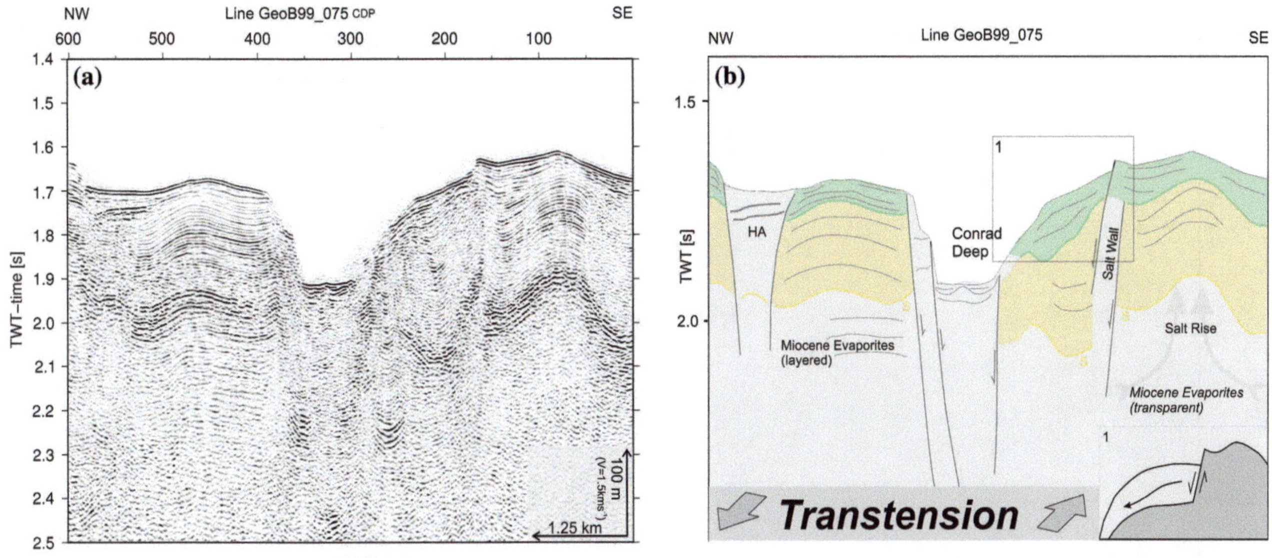

Fig. 8 *Left* MCS reflection line GeoB99_075, NW–SE MCS reflection profile across the Conrad Deep (see Fig. 5). *Right* Interpretation of MCS profile. Pre-kinematic (*yellow*) and syn-kinematic (*green*) units were identified SE and NW of the deep. SE slope collapsed without the salt/mud wall as backstop. On the NW side, a high-amplitude reflector (HA) is located next to the layered Plio-Quaternary sediments

reach crystalline basement. In comparison with the central Red Sea deeps, it is in a very juvenile stage. However, the position of the CD is not arbitrary; it developed above the junction of two tectonic lineaments, the NNW–SSE Red Sea Rift and an oblique, almost perpendicular (parallel to the Aqaba trend) transform fault. Magmas used this zone of weakness to ascend. In Late Pleistocene time, magma ascended at the margins of the axial depression to form both intrusions and volcanoes. Temperature radiation decreased the salt viscosity between the intrusions. As a result of the ongoing rifting of the Red Sea, a transtensional regime was established between both magmatic intrusions/extrusions causing the opening of the CD (Ehrhardt et al. 2005). Salt flow above the basement faults formed the salt rises. Salt flow toward the axial depressions formed streamlines analogous to glaciers.

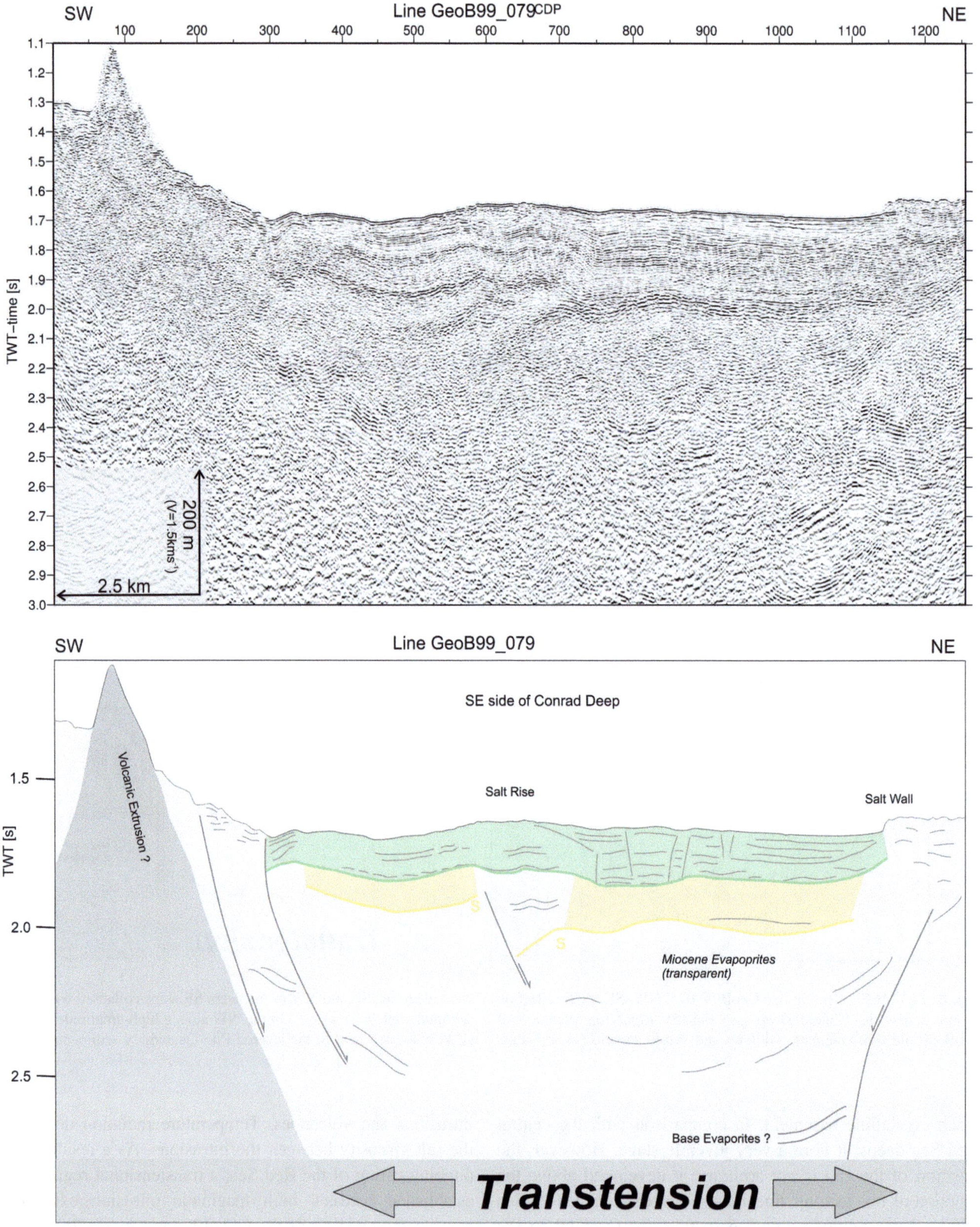

Fig. 9 *Top* MCS reflection line GeoB99_079, SW–NE cross section (see Fig. 5). *Bottom* interpretation of MCS profile, including pre-kinematic (*yellow*) and syn-kinematic (*green*) units on the SE area of the deep. The seismic line is parallel to the Red Sea extension

Shaban Deep

The 6 by 5 km wide SD is located in the central axial depression (Fig. 10). In the center of the SD, a volcano built an elongated NW–SE ridge, rising to a level of about 900 m bsl. The SD is subdivided into four sub-basins. The two northern sub-basins are shallower, with maximum water depths of about 1,400 m, and the southern sub-basins are deeper with maximum water depths of more than 1,600 m. The northwestern escarpment consists of a prominent, about 100-m-high rim anticline. The seafloor north of the SD is smoother than that to the south. The eastern side is also affected by two major grabens, striking NE–SW. The transition from the smooth northern domain of the SD to the southern deeper situated domain can be traced across the SD where the difference in depth is reflected by the depth of the sub-basins.

The bathymetry reflects two main tectonic trends. The NW–SE trend, parallel to the axial depression of the Red Sea and the NE–SW trend that crosscuts the SD. This trend builds a smaller ridge that is perpendicular to the volcanic ridge and several elongated NE–SW ridges and valleys in the east of the SD.

The sub-surface structure of the SD area is best seen on the NNW–SSE MCS reflection lines (Figs. 11, 12, 13 and 14). The typical thickness of Plio-Quaternary sediments, around 150–250 m, is equivalent to 200–300 ms TWT and covers the evaporites. Similar to the CD region, the suprasalt succession reveals two sequences with different reflection characteristics. The S-reflection was identified on all lines as marking the top of the Miocene evaporites. It is significantly deeper in the south compared to the northern margin of the SD. In the north, it terminates against the slope approximately at 1.7 s TWT. A brine reflector is detected within all four SD sub-basins and in similar water depths. East of the SD, a low frequency reflection probably images the base of the Miocene evaporites (Figs. 13 and 14). The two-way travel time is comparable to base reflections known from old industry data (Preussag data from the 1980s; Wiedicke-Hombach, pers. comm.). Similar to the CD area further north, two units of the Pliocene–Quaternary suprasalt succession and the lateral variation of the evaporites, well-stratified or transparent, are observed. The rim anticline is a peculiar structure that looks similar to the CD salt wall.

The following features characterize the SD:

- A central volcanic ridge that strikes along the axial rift.
- A perpendicular SSW–NNE striking ridge across the SD.
- The northern sub-basins reveal a rim anticline.
- Straight and arcuate grabens south and southeast of the SD.
- Topographic ridges along the rift axis.
- The reflection characteristics from evaporites beneath the S-reflection change laterally from well-stratified to transparent (Fig. 11).

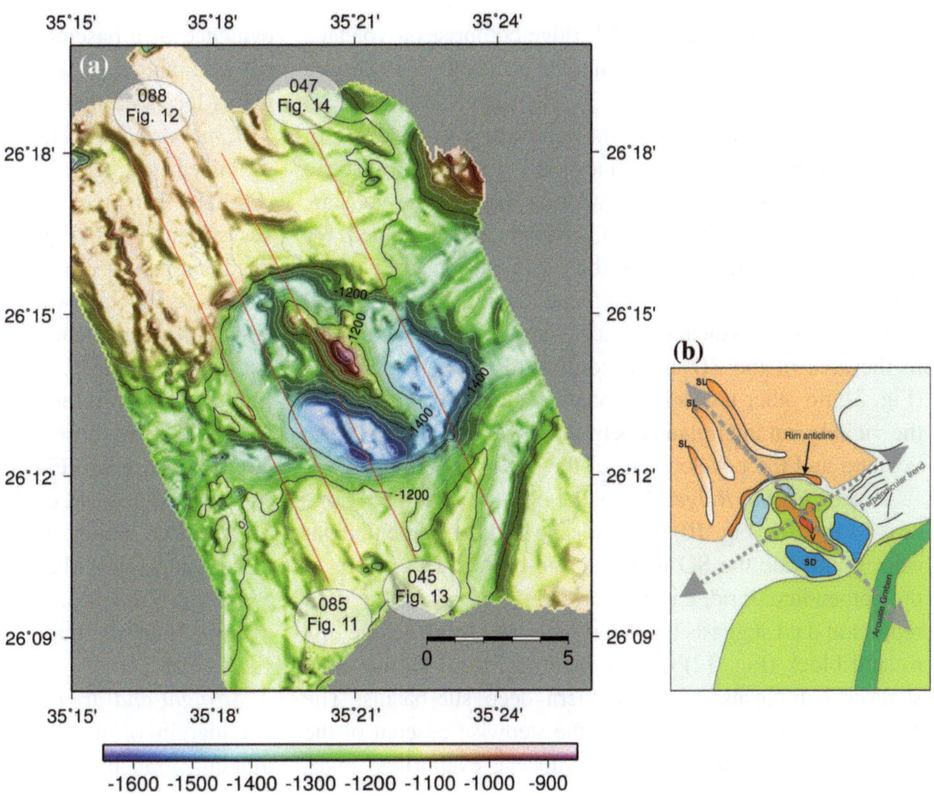

Fig. 10 **a** High-resolution bathymetric map of the Shaban Deep (SD). *Red lines* indicate seismic reflection profiles in Figs. 11, 12, 13 and 14. The SD is characterized by the volcano in the center of the deep. Note the difference in depth between the northern and southern SD. A WSW–ENE-trending graben seems to cross the SD. The northern part of the SD is bounded by an anticline. The south is influenced by SW–NE-trending graben- or valley-like structures that coalesce with the WSW–ENE graben. **b** The sketch illustrates the mentioned structures. Note the *dashed lines* showing the two trends normal to each other. *SL* Streamline

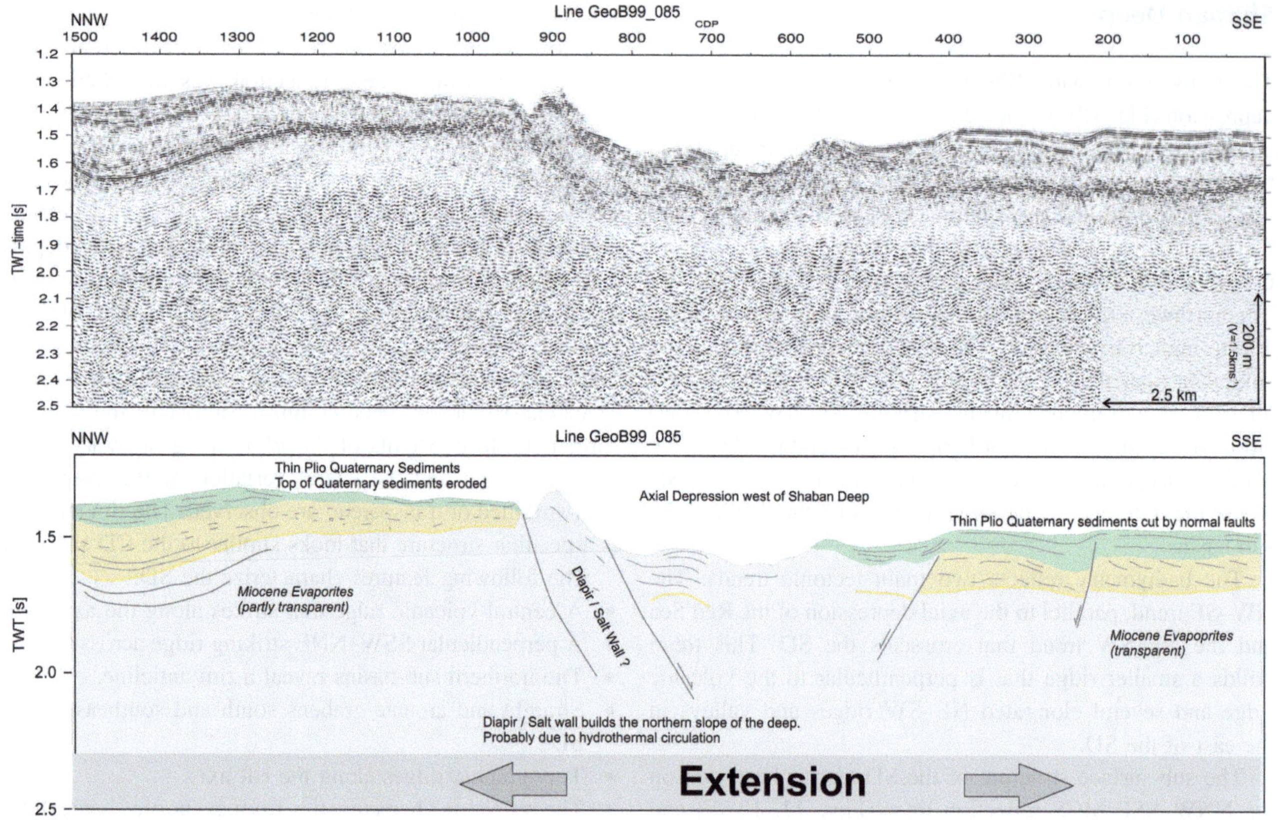

Fig. 11 *Top* MCS reflection line GeoB99_085, NNW–SSE section crossing the western extension of the SD. The area experiences N–S extension. *Bottom* Interpretation of MCS profile. Extensional faulting is identified in the Plio-Quaternary sediments. The anticline structure at the northern rim of the SD is probably caused by hydrothermal circulation similar to that seen in the CD

Volcanic Ridge: The NW–SE ridge comprises a volcanic extrusion (Haase et al. 2000; Pautot et al. 1984) that causes a magnetic dipole anomaly (Fig. 3). The ridge crosses the SD from NW to SE and rises in the center to a water depth of 900 m (Fig. 10). Seismic lines only touch the volcano (Fig. 13) but show clearly the western slope of the volcanic cone. The shape of the volcano is narrow and elongated. No lava flows could be observed; thus, it is unclear how long it was active. Basaltic rock fragments found within the SD point to a young volcanic edifice. As the volcano is limited to the area of the SD and according to the magnetic map (Fig. 3), no other volcanic or magmatic activity is present in the vicinity, an interrelation between the emplacement of the volcano and the development of the SD is obvious.

Perpendicular Ridge: A smaller SW–NE-trending ridge crosses the SD (Fig. 10). The volcano and the perpendicular ridge subdivide the SD into four sub-basins. The nature of the perpendicular ridge is unclear. Interpretation of the MCS reflection data suggests that the ridge is an elevated edge of a rotated block (Fig. 12) or may be a step from the northern shallow sub-basins to the southern deep sub-basins. The extensional normal faulting and the stepwise descent of the S-reflection (i.e., the top of the Miocene evaporites) gives evidence of a basement fault perpendicular to the Red Sea rift axis, and it is likely that this junction of tectonic lineaments localized the volcanic activity.

Rim anticline: The northern margin of the SD is a prominent elongated structure forming a rim anticline (Fig. 10). This diapir-like or wall-like structure is sometimes very pronounced (Figs. 11 and 12) and is sometimes subdued and part of the slope. In any case, it serves as backstop similar to the salt wall of the CD and prevents Plio-Quaternary sediment and the Miocene S-reflector from terminating into the deep. Two scenarios explain the conspicuous rim: (a) Similar to the discussion of salt flow in the CD area, we take these observations as evidence for southward-directed salt creep (Fig. 15). The elongated ridges represent streamlines, and salt thickening due to shortening creates the rim anticline. South of the deep, the salt flows away from the SD causing salt withdrawal and therefore a decreased slope angle; (b) A diapir or salt wall similar to the CD is formed by hydrothermal circulation and salt re-precipitation (Figs. 11, 12, 13 and 14).

Straight and arcuate graben: The eastern and southeastern margin of the SD is transected by several SW–NE-trending graben structures (Fig. 10). Seismic lines across

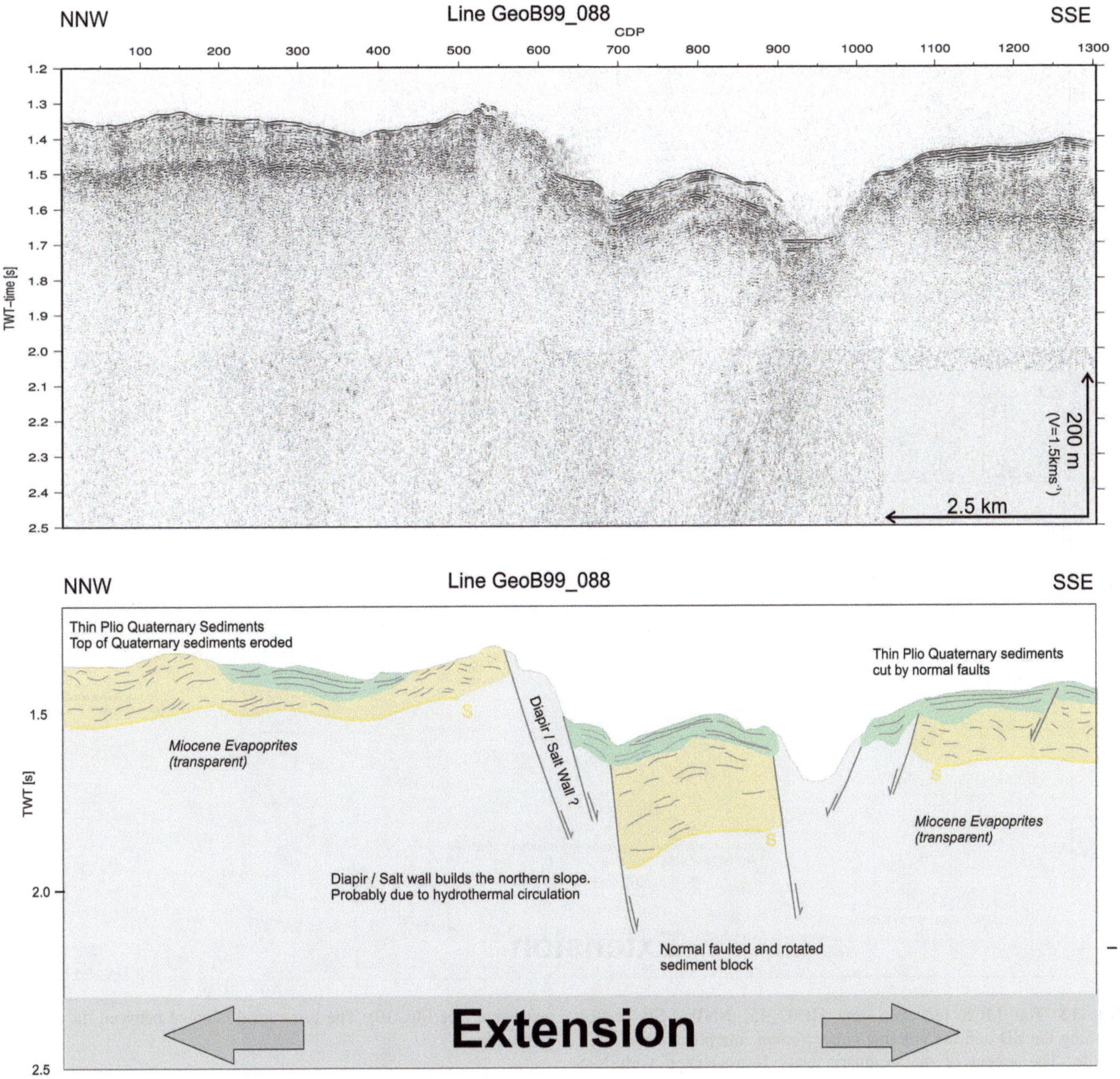

Fig. 12 *Top* MCS reflection line GeoB99_088, NNW–SSE section crossing the western part of the SD (see Fig. 10). *Bottom* Interpretation of MCS profile. N–S extension caused block rotation along normal faults in the area of the SD. The Plio-Quaternary sediments are rather thin with two identified units

these grabens (Fig. 14) show thick and faulted Plio-Quaternary sediments. The arcuate graben (Figs. 10 and 14) is probably the result of an underlying strike-slip fault that compensates SW–NE transform motion.

Lessons learned from Shaban Deep: Similar to the CD, the SD is located at a junction of the rift axis and a perpendicular fault. While the CD emerged between two volcanoes, the SD developed around a central volcanic ridge. We suggest that hydrothermal circulation associated with the ascent of magmas dissolved the salt and opened the SD. Additionally, the preexisting SW–NE-trending fault plays a vital part which is depicted in a sketch (Fig. 16). The left side shows the situation before the onset of magmatic activity. The SW–NE-orientated fault zone was most probably a strike-slip fault with an additional normal component of faulting (in order to explain the different depths between the northern and southern parts). The right side of Fig. 16 illustrates the development of the SD in the successive stages of the ascent of the magmatic body. The intrusion used the preexisting fault for its ascent (Fig. 16a). Because of the lateral replacement of the sediments by the magmatic cone, an extensional regime perpendicular to the strike of the fault,

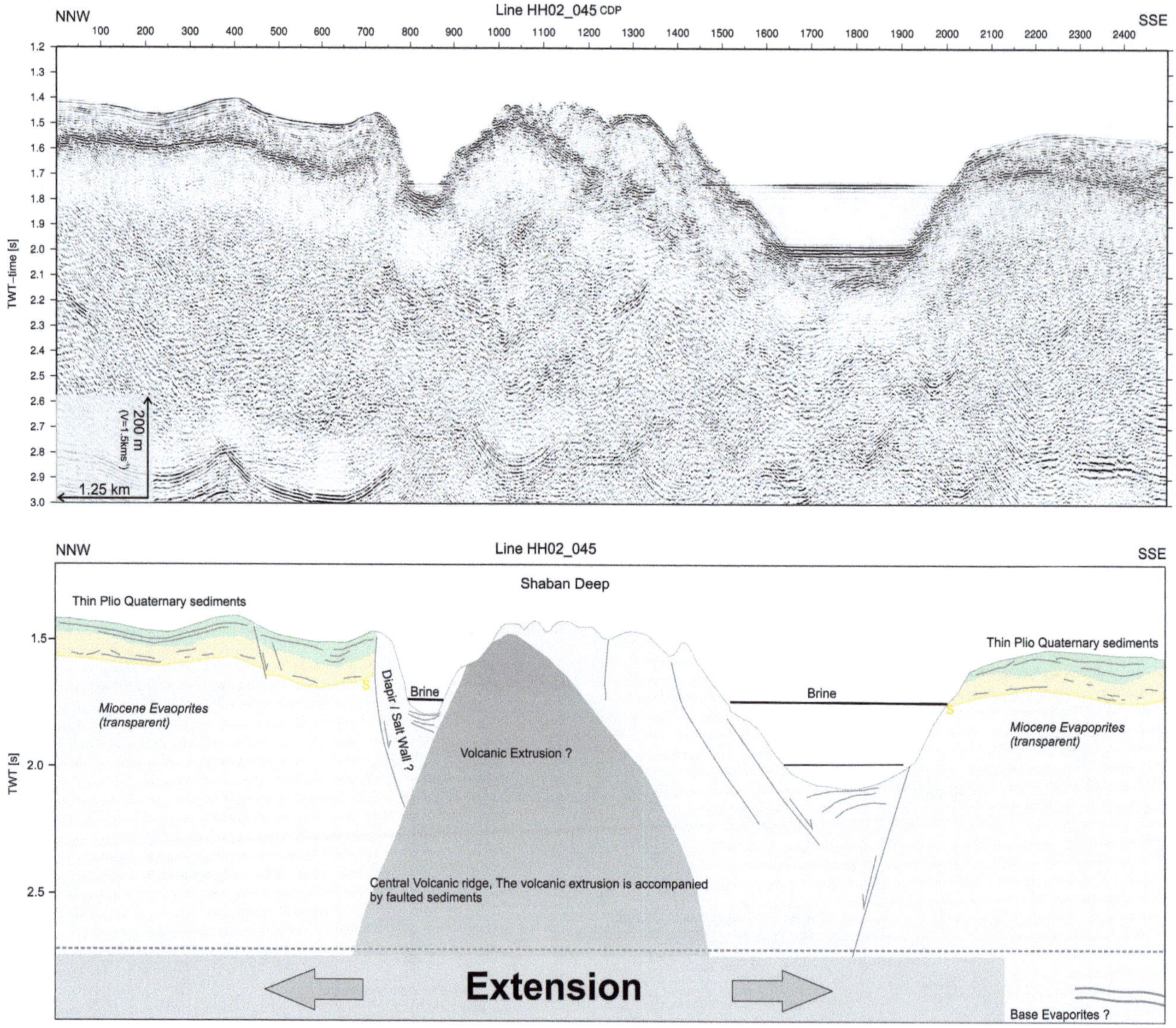

Fig. 13 *Top* MCS reflection line HH02_45, NNW–SSE section crossing the SD and the volcanic cone. *Bottom* Interpretation of MCS profile. The ascent of the magmatic body is associated with N–S extension (see Fig. 10). The brine level is equal between the different sub-basins

that is, in NW–SE direction was initiated (Fig. 16b). Heat from the magma and associated hydrothermal circulation lowered the viscosity of the surrounding evaporites and re-mobilized these. The strike-slip displacement along the preexisting fault was blocked by the volcano; thus, it side-stepped to the south around the barrier (Fig. 16d). This scenario could explain the SW–NE striking fault that is separated from the volcano, the circular shape of the deep, and the faulted SE-part of the survey area. Similar to the CD, rift-related extension, intersecting fault systems, and the emplacement of magmatic intrusion/extrusion are the main causes forming the deep.

Kebrit Deep

Compared to the CD and SD, the KD (KD; Fig. 17) has a more oval and asymmetric shape. It is located in the Red Sea axial depression but in contrast to the CD and SD, the KD lies within the N–S-trending segment (see Fig. 3). It has smooth slopes that are intersected by two W–E-trending grabens and half grabens, and a N–S-trending graben that crosses the deep but is slightly offset to the east. All these grabens coalesce in the center of the KD comparable to a star-shaped pattern. It is reasonable that this is the expression

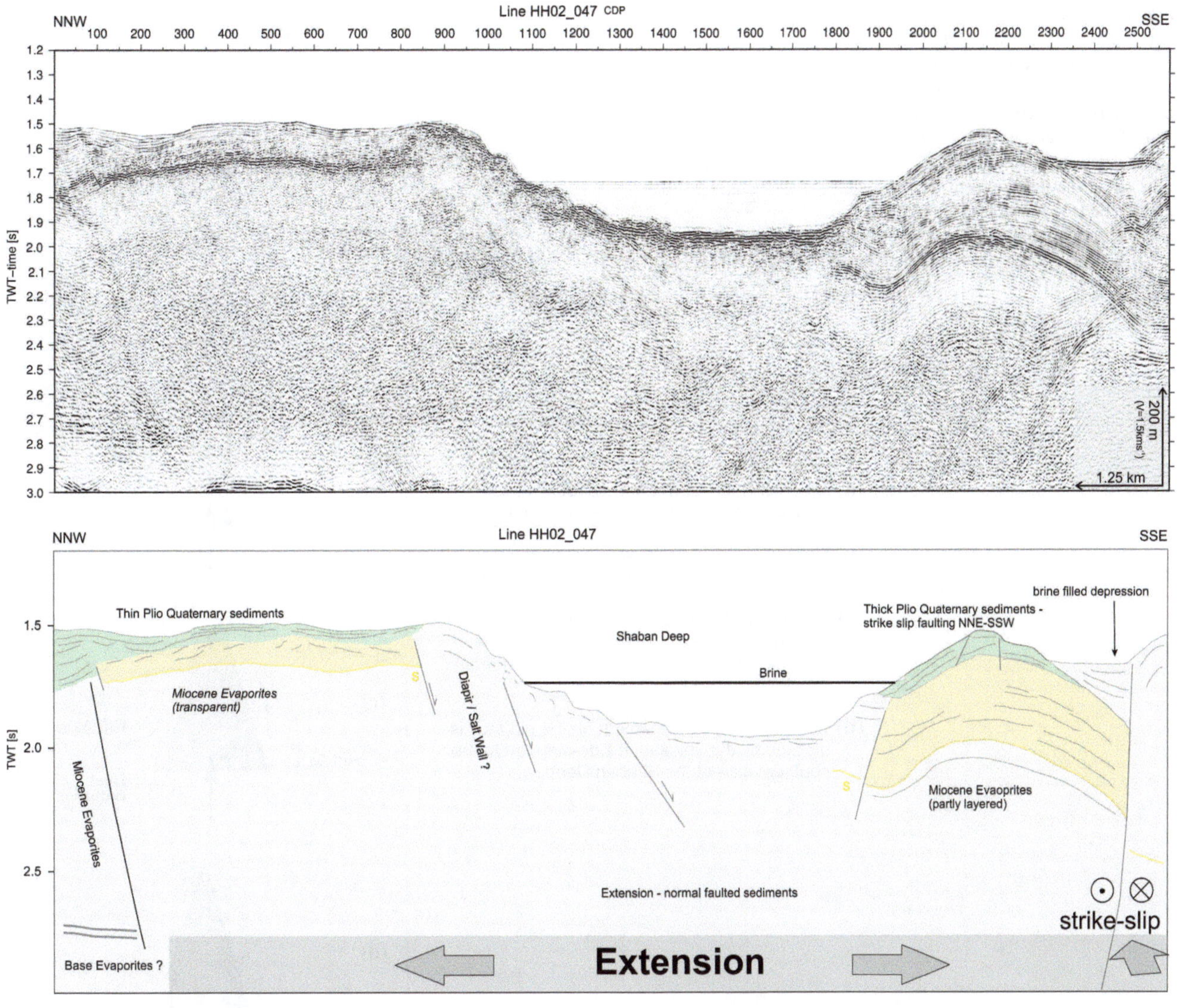

Fig. 14 *Top* MCS reflection line HH02_47, NNW–SSE section crossing the eastern part of the SD (see Fig. 10). *Bottom* Interpretation of MCS profile. Extensional normal faulting is complemented by strike-slip faulting in the southern part

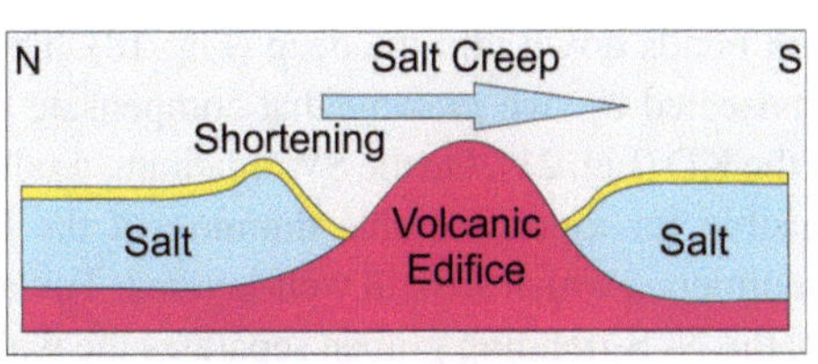

Fig. 15 Sketch of a N–S section across the SD explaining the bulge-like salt wall and the decreased slope angle in the south by lateral salt flow from north to south along the axial depression

of a through-going N–S-trending fault as the surface expression of the rift axis. The morphology of the surrounding seafloor west of the N–S-trending graben is regular and smooth but irregular and rough east of it.

The KD differs in many details from the CD and SD. Apart from size, shape, and location, the internal structure also shows major differences (see Table 1).

The following features characterize the KD:

- Star-shaped pattern.
- Smooth slopes, no escarpment-like margin.
- Prominent differences between western and eastern provinces as well as seafloor to the north and south.
- The reflection characteristics of evaporites beneath the S-reflection changes laterally from well-stratified to transparent.

Star-shaped pattern: There are four main grabens that radiate from the center of the KD. The two W–E-trending grabens developed because of subsidence due to subrosion (salt dissolution) (Fig. 21). The bathymetric map (Fig. 17)

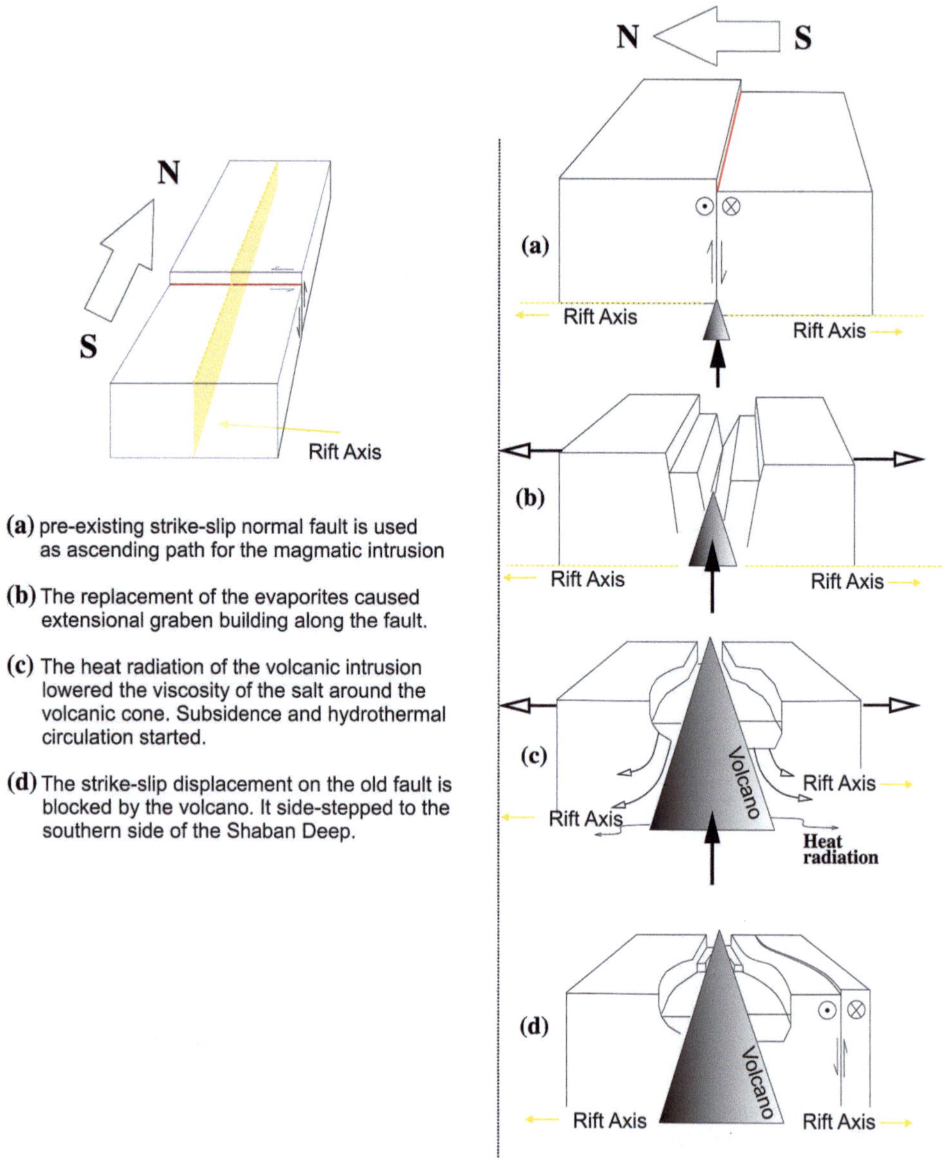

Fig. 16 Cartoon of possible development of the SD. *Left* Bird's-eye view from the south on a preexisting strike-slip normal fault (*red*) that is crossing the Red Sea rift axis (*yellow*). *Right side* Cross section sequence illustrating the development of the SD. See the location of the *red line* for orientation

shows that graben depth decreases toward the western end of the survey area and the grabens terminate just to the west. The N–S-trending grabens are through going and it is possible that these grabens are the surface expression of a strike-slip fault.

Smooth slopes: The slopes into the KD are smooth and have no escarpment-like margins (Figs. 18 and 19). The entire sedimentary succession bends down into the deep. The prominent brine layer could only be a result of hydrothermal circulation (Fig. 20).

Prominent differences between western and eastern provinces as well as north and south: The seismic lines clearly show a different appearance of the NW, SW, SE, and NE quadrants of the KD. The NW quadrant shows regular layered sediments on top of the Miocene evaporites. This thick succession bends down into the deep (Fig. 18). The western side is transected by two grabens that compensate the subsidence of the KD (Fig. 21). On the SW quadrant, a salt anticline is responsible for a considerable thinning of the Plio-Quaternary sediments, which are still well-layered. Toward the SE quadrant, the N–S-trending graben separates the well-layered sediments from underlying chaotic structures above a high-amplitude reflector (see Fig. 19) (Bosworth and Strecker 1997). Below this reflector, well-stratified sediments cover the S-reflector. The NW quadrant is dominated by an arcuate anticline with a diameter of 2.5 km caused by vertical salt mobilization. High heat flow and hydrothermal circulation may have remobilized the salt. To the west, separated by the N–S graben, there is a sharp transition to the well-layered, concordant sediments of the NW quadrant.

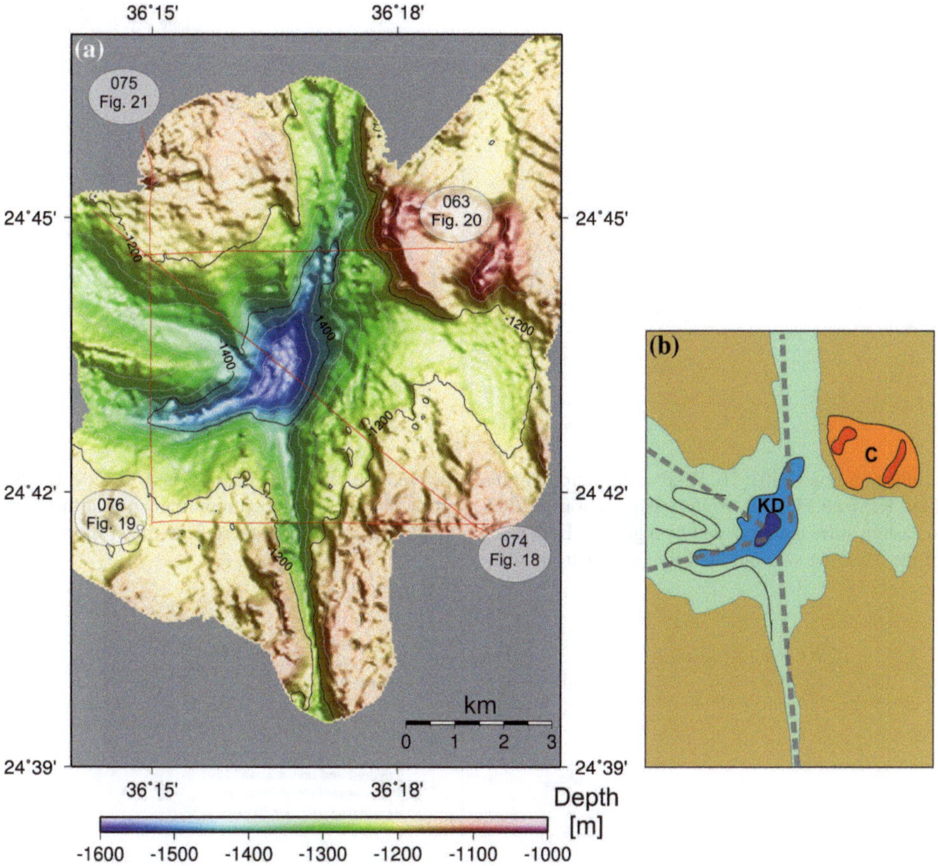

Fig. 17 **a** High-resolution bathymetric map of the Kebrit Deep (KD). *Red lines* mark the position of the seismic lines shown in Figs. 18, 19, 20, and 21. The KD has an oval shape with grabens emanating from the deep in a star-shaped pattern. **b** Sketch of the main structures of the Kebrit Deep. *Black dashed lines* star-shaped pattern. The N–S lines are slightly offset to the east. A circular structure (C) in the northeast overtops the surrounding of the KD

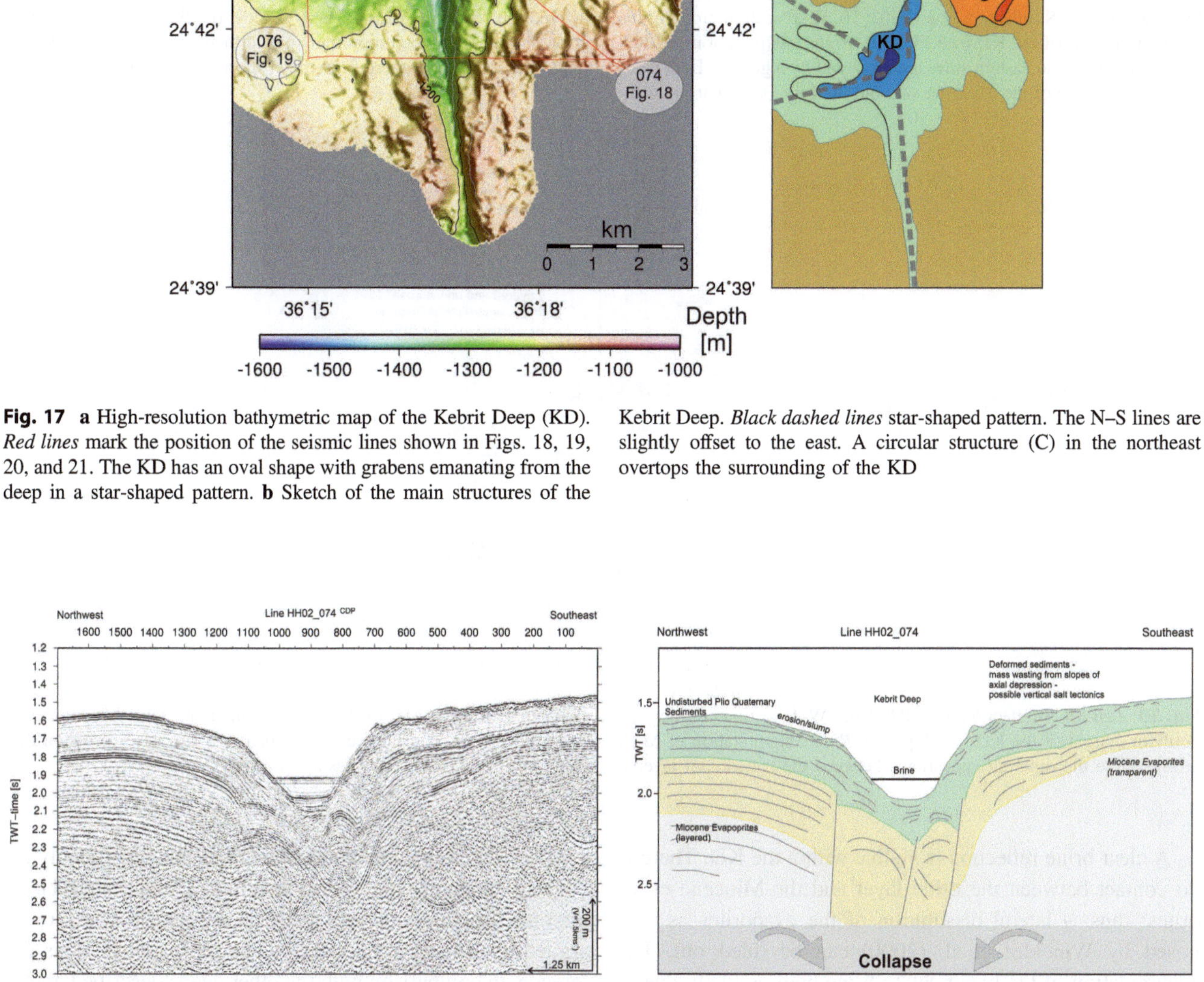

Fig. 18 *Left* MCS reflection line HH02_74, NW–SE section crossing the KD (see Fig. 17). *Right* Interpretation of MCS profile. Note the thick Plio-Quaternary sediment sequences that bend down into the deep. The Miocene S-reflector does not crop out in the deep. The NW corner of the KD comprises an abundant sequence of concordant Plio-Quaternary sediments (approx. 600 m). The deep represents a collapse structure, most likely due to dissolution of the Miocene salt beneath. In the NW, the evaporites are well-stratified but transparent to the SE

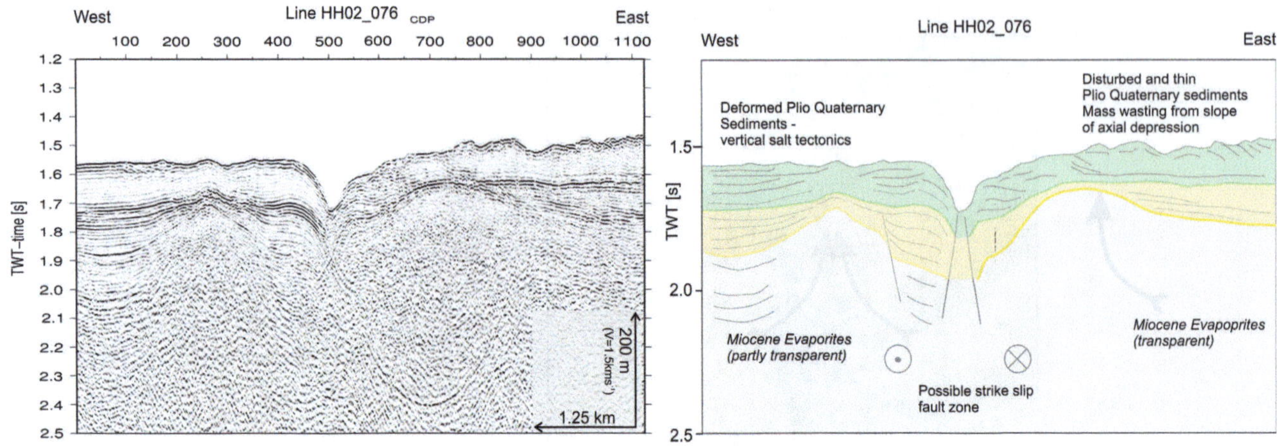

Fig. 19 *Left* MCS reflection line HH02_76, W–E section crossing the southern margin of the KD (see Fig. 17). *Right* Interpretation of MCS profile. Western and eastern sides are separated by a graben. Both sides show an anticline of the Miocene salt, but the Plio-Quaternary is well-layered on the western side and chaotic on the eastern side. It is possible that the graben represents the surface structure of a strike-slip fault. The reflection pattern from the evaporites varies from west to east

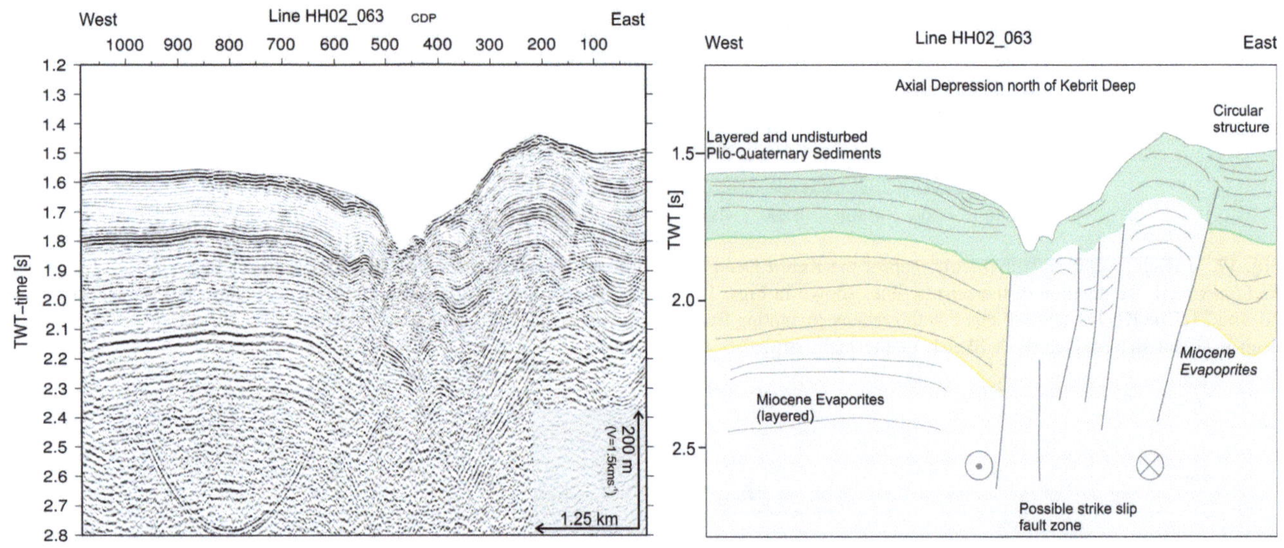

Fig. 20 *Left* MCS reflection line HH02_63: W–E section crossing the northern margin of the KD (see Fig. 17). *Right* Interpretation of MCS profile. Note the different appearance between the concordant layered sediments on the western side and the faulted and folded sediments on the eastern side. The intervening graben is a surface expression of a strike-slip fault. The evaporites are layered

A clear brine reflection is visible within the KD. There is no contact between the brine layer and the Miocene evaporites; thus, a lateral dissolution of the evaporites as proposed by Winckler et al. (2000) can be ruled out. The accumulation of the hyper-saline brine body is likely due to hydrothermal circulation. No evidence for volcanic activity is seen in the seismic data. The magnetic map of Coutelle et al. (1991) shows only slight magnetic anomalies (Fig. 3).

Lessons learned from the KD: Although the three mentioned deeps (CD, SD, and KD) are all northern Red Sea brine deeps, they have little in common. The KD is not associated with magnetic anomalies and thus is probably not affected by magmatic intrusions (Fig. 2 and Table 1). Its internal structure suggests that it is a collapse structure, as argued by Bäcker et al. (1975). The most conspicuous difference, in comparison with the other deeps described above, is the smooth and continuously dipping slopes without scarps. The entire Plio-Quaternary sequence and the top of the evaporites are affected by the subsidence. The cause of the subsidence is linked with dissolution of the Miocene evaporites, because of hydrothermal circulation. A basement fault system such as the N–S graben could have initiated the

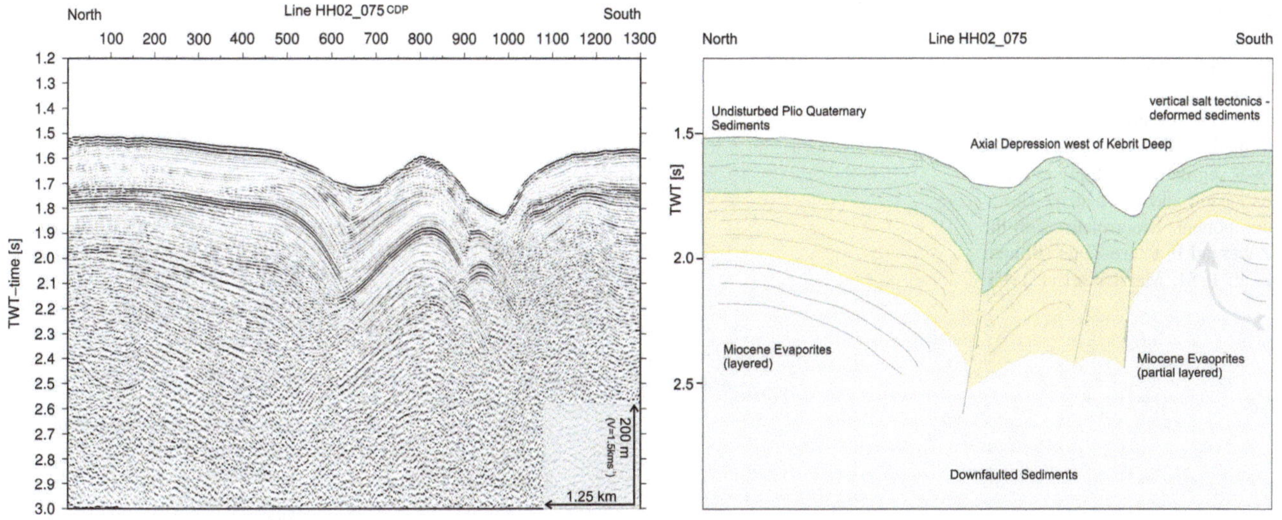

Fig. 21 *Left* MCS reflection line HH02_75, N–S section crossing the western margin of the KD (see Fig. 17). *Right* Interpretation of MCS profile. The northern side shows well-layered, slightly divergent sediments. Two graben structures separate the thick Plio-Quaternary sediments from thinner sediments in the south. Vertical salt tectonics influenced the southern side. The evaporites show a southward divergent reflection pattern in the northern segment

thermal activity. It is conceivable that this position had an impact on the development of the hydrothermal circulation cell and thus supported the formation of the KD. The circulation of seawater dissolved the evaporites beneath the deep and caused continuous collapse of the overlying sedimentary cover. Hydrothermal circulation and the absence of a magmatic body is also assumed by Blum and Puchelt (1991) who analyzed the deposits of the KD as being of low temperature origin. These facts support the approach that subrosion caused the development of the KD without any significant alteration of the ductile properties of the evaporites.

The KD developed most likely as a collapse structure due to hydrothermal circulation and subrosion (salt dissolution) of the evaporites below the deep. Thus, it developed independently of volcanic extrusions/intrusions and the tectonic activity in the axial depression. Because of the higher heat flow in the center of the Red Sea (axial depression) (Cochran et al. 1986; Martinez and Cochran 1988), it is likely that the necessary hydrothermal circulation cell was established in the axial depression.

Discussion

Using the morphology, internal structure, associated magnetic anomalies and the location of the northern Red Sea deeps, we are able to determine to types of deeps with regard to their development. The first type is based on the CD and SD, representing deeps in the NW–SE segment of the northern Red Sea, and the second is based on the KD, located in the N–S segment of the northern Red Sea (Table 1, Figs. 1, 2 and 3).

Segmentation-Controlled Development

The investigated Red Sea deeps were grouped into intrusion-related deeps (CD and SD) and collapse-type deeps (KD). This classification correlates with the location of the deeps. The intrusion-related deeps are in the NW–SE segment of the northern Red Sea and the collapse-type KD is found in the N–S segment. The magnetic anomaly map (Fig. 3) shows more anomalies in the NW–SE segment. In addition to the already discussed SD and CD, Martinez and Cochran (1988) mapped an unnamed deep at 26° 31′N/35° 00′E associated with a magnetic anomaly which probably developed similarly to CD and SD. The N–S segment does not reveal any significant magnetic anomaly that points to magmatic activity. This is in agreement with the collapse-type development of the KD. Other deeps, such as the Vema Deep at the southern end of the N–S segment (Figs. 3 and 22), exhibit similar patterns to the KD, for example the shallow and smooth slopes and continuously dipping sediments (Fig. 4 in Guennoc et al. 1988). The N–S segment was already mapped as the Zabargad Fracture Zone (Bonatti et al. 1984; Marshak et al. 1992) and may act as a transform zone. Smaller transform zones may exist south of the CD (Ehrhardt et al. 2005) and north of the SD at the Brothers Islands and correlate probably with the accommodation zones identified by Cochran (2005), Cochran et al. (1991), and Cochran and Martinez (1988).

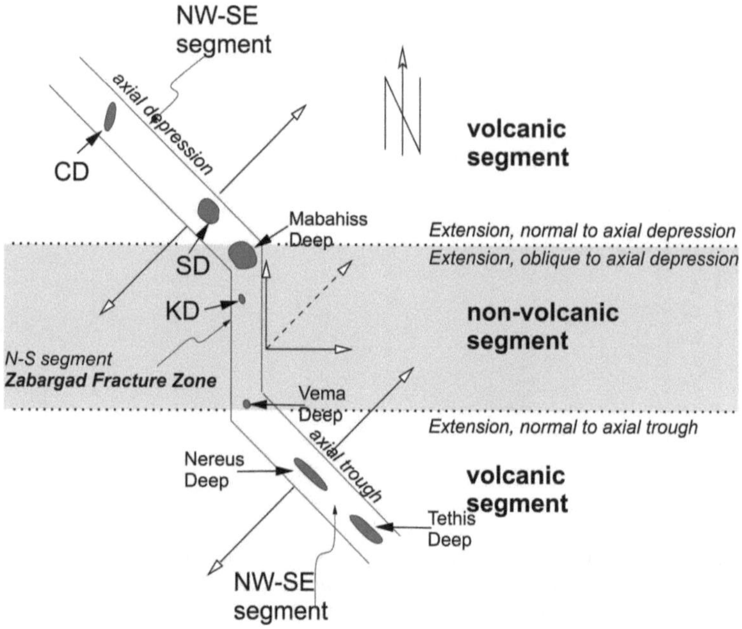

Fig. 22 Sketch of the segmentation of the northern Red Sea. The NW–SE segments are oriented normal to the Red Sea extension direction and are termed "volcanic" segments. The N–S segment is oblique to the extension and has a strike-slip component. It correlates with the Zabargad Fracture Zone (Bonatti et al. 1984; Marshak et al. 1992)

Further to the south, another NW–SE segment of the central Red Sea adjoins the studied N–S segment (Zabargad Fracture Zone) (Fig. 22). It exhibits well-developed ocean deeps such as the Nereus Deep that forms a 40-km-long trough with a central ridge from which tholeiitic basalts were recovered (Antonini et al. 1998). This segment comprises deeps that are associated with magmatic activity, like in the northern NW–SE segment, but the deeps are further developed, in agreement with the model of a rift that propagated from north to south (Bonatti 1985; Martinez and Cochran 1988). Thus, a correlation between types of the deeps with their location exists and magma-related deeps are limited to the NW–SE segments of the Red Sea.

Following this approach, the development of the northern Red Sea deeps is classified into intrusion-related deeps and collapse-type deeps. This classification correlates also with the geographical allocation of the deeps (see Table 1). Magmatic activity and the following development of intrusion-related type of deeps are restricted to the NW–SE segment, thus to the segment where the Red Sea extension is normal to the strike of the axial depression. The KD and the Vema Deep which are located in the N–S segment are collapse-type deeps. The lack of magmatic intrusions in the N–S segment can be explained by the reduced extension due to the oblique strike of this segment in comparison with the main extension direction. The reduced extension could be insufficient to promote magmatic intrusions. The significant strike-slip component is probably compensated along the apparent strike-slip fault that crosses the KD. The narrow appearance of the N–S fault (Fig. 17) shows no major extension at the seafloor. It is reasonable that the extension is masked by lateral salt flow similar to the model of Augustin et al. (2014) for the central Red Sea. Thus, no intrusion-related deep such as the CD and SD is mapped in the N–S segment. The deeps in the N–S segment such as KD and Vema Deep developed as collapse structures. However, this type of deep is not necessarily limited to the N–S segment, as it is only dependent on the existence of a hydrothermal circulation cell, and this could be anywhere along the rift. Furthermore, the described types of deeps represent end-member models. Deeps of "mixed origin" are possible.

After classifying the deeps as direct and indirect rift-related structures, it is next possible to extract information about the rift. The NW–SE segment is oriented normal to Red Sea extension. With continued extension along the axial depression (Cochran et al. 1991), the NW–SE segment was pierced by magmatic intrusions. Red Sea deeps developed between or around them. This is in agreement with the model of Bonatti (1985) who correlated the conspicuous constant spacing of the magnetic anomalies with Raleigh-Taylor instabilities according to density/viscosity inversions in the upper mantle. Thus, the NW–SE segments are termed "volcanic segments."

In contrast, the N–S segment of the northern Red Sea is not associated with major magnetic anomalies and seems to be unaffected by magmatic intrusions. However, it is possible that intrusive bodies are present with temperatures above the magnetic blocking temperature, but in general the temperature of the top of the crystalline basement should be below the Curie temperature. Although high heat flow values were reported from the northern Red Sea (Martinez and Cochran 1989), the temperature rises gently because of the good thermal conductivity of the evaporites (~ 5.5 W K^{-1} m^{-1}) (Cochran and Martinez 1988). The temperature at the

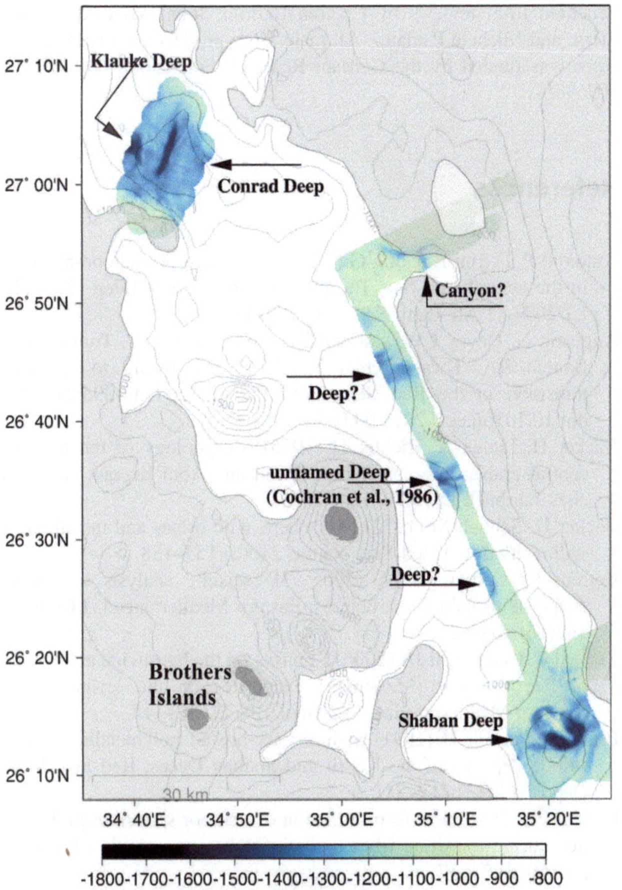

Fig. 23 Bathymetric track of Meteor Cruise M44/3. This line shows the bathymetric link between the two survey areas of the M44/3 cruise. Although the track is slightly offset from the axial depression, several deep-like areas were detected. *Grey colors* indicate water depths above 1,000 m and *white colors* depths of more than 1,000 m

sediment-basement interface is calculated to be about 300 °C, using a three-layer model and representative parameters for heat flow, thermal coefficient, and layer thickness (see also Cochran and Martinez 1988). The deeps that were discovered so far in the N–S segment have the topographic characteristics of collapse structures caused by subrosion (salt dissolution) of Miocene evaporites from below. Thus, the N–S segment is termed a "non-volcanic" segment that links the "volcanic" NW–SE segments (Fig. 22).

This sequence of volcanic and non-volcanic passive continental margin segments can be observed also at other oceans such as Baffin Bay between Baffin Island (North America) and Greenland (Skaarup et al. 2006) or the South Atlantic Ocean (Franke et al. 2007).

As the collapse-type Red Sea deeps are only dependent on a hydrothermal circulation spot, it is conceivable that they also developed in the NW–SE "volcanic" segment, independently from magmatic activity and thus not correlated with known magnetic anomalies. This is supported by the existence of the unnamed deep north of the CD (Guennoc et al. 1988) that reveals evidence for a collapse-type deep. In addition, the processing of the bathymetric data of the transits of R/V Meteor Cruise M44/3 points to some more deeps in the NW–SE segment (Fig. 23). This map illustrates the combined swath echosounder data from the cruises M44/3, M52/3 and transits. At 26° 35′N/35° 11′E, the unnamed deep mentioned above could be identified (Cochran et al. 1986), besides other depressions that point to deeps with water depths of more than 1,350–1,400 m.

Seismostratigraphy of Evaporites

The influence of buoyancy on driving salt tectonics has long been overestimated (Hudec and Jackson 2007). Instead, in many settings, evaporite deformation is primarily related to gravitational forces as a consequence of basement tilt (*gliding*) (Brun and Fort 2011; Hudec and Jackson 2007). Netzeband et al. (2006) stated that salt deformation due to basin tilt may start during deposition. The divergent reflection pattern beneath a flat salt top as revealed in Fig. 21 (KD) corroborates the assumption of syn-depositional salt tectonics.

The lateral variation between layered and transparent evaporites is also intriguing (Figs. 8, 11,18, 19, 20 and 21). Seismic data from the Atlantis II Deep show the same phenomenon (Fig. 5 of Mitchell et al. 2010). Generally, a vertical succession of evaporites of alternating transparent and stratified sequences is well known from the Messinian evaporites of the Levant Basin (Bertoni and Cartwright 2005, 2006; Dümmong and Hübscher 2011; Netzeband et al. 2006). In the Levant Basin, the vertically varying reflection characteristics imply a vertical change of evaporite facies, in which the transparent sequences are typically considered as halite and the stratified sequences are carbonates or sulfates (gypsum or anhydrite). Such alternating successions are well known and reflect the transition from low-to-high salinity or vice versa (e.g., Warren 2006). The lateral presence of transparent as well as stratified evaporites as observed in the Red Sea is uncommon but not unique, for example, it is also known from the Santos Basin off Brazil (Mohriak et al. 2012). In the Santos Basin, the evaporites are generally tabular, but internally seismically transparent diapirs alternate horizontally with folded and stratified evaporites. Similar to the Red Sea, the Santos evaporites developed along a continental rift system. Since the top of the evaporites is just a little deformed compared to the underlying salt, halokinesis must have started during the precipitation phase. In contrast to post-depositional salt diapirism, for example in the Zechstein Basins, the rim synclines were not filled with clastic sediments; instead, they were filled with evaporites other than the transparent halite.

We conclude that the lateral change of evaporite facies is the consequence of rift- or salt load-induced basin tilt, resulting halite gliding and deformation and the infill of evolving accommodation space between halite folds (e.g., diapirs) by evaporite facies other than halite, for example, gypsum or carbonates.

Conclusions

As a result of this work, we conclude that rift-related extension and faulting induced magmatic intrusions and extrusions, respectively. Where rifting and magmatic activity coalesce, the development of the northern Red Sea deeps is preferred. As magmatic-related deeps such as the CD, SD, or Nereus Deep are only mapped in the NW–SE segments of the northern Red Sea, the NW–SE segments were termed "volcanic" segments. The "volcanic" segments are linked by a "non-volcanic" N–S segment that is oblique to the NW–SE segments and acts probably as transform zone. Collapse-type deeps such as the KD are not linked to magmatic activity and are not necessarily limited to the "non-volcanic" segments of the Red Sea. The present multi-beam data suggest that deeps are more abundant than previously known.

The observation of the widespread salt sediments showed, as opposed to most of the worldwide known salt deposits where vertical variations in seismic reflection characteristics reflect deposition of varying salt facies that the northern Red Sea evaporites reveal a lateral variation from stratified to transparent. Accommodation space for stratified evaporites evolved already during salt deposition due to syn-depositional salt tectonics, most likely due to basin floor tilt caused by rifting or salt load. The bathymetry is not only controlled by plate tectonics and volcanism. The sediment covered salt flows as an analogue to ice-glaciers and builds streamlines that are formed sub-parallel to the flow direction. Salt rises are present where the salt flows above basement faults.

In order to broaden the knowledge of the Red Sea deeps and their relation to the rifting, extended detailed bathymetric data sets must be acquired and added to seismic data acquisition at crucial locations, such as, for example, the different deeps. This would be helpful for the understanding of the Red Sea rifting process and to understand the complex behavior of evaporites in tectonically active regions such as the Red Sea.

Acknowledgments We would like to thank Jürgen Pätzold (University of Bremen) for providing the bathymetric data. Thanks also go to Volkard Spiess (University of Bremen) for providing the seismic acquisition equipment during M44/3 cruise and Mohammed Salem (University of Bremen) for editing of the bathymetric data of the CD. We are grateful to the crew of the RV Meteor for the technical assistance provided during cruise M44/3 and M52/3. The manuscript benefited from reviews by Claudia Bertoni, John Woodside, Robert Stern, and Michele Paulatto. All maps were produced with GMT. This work was funded by the German Research Foundation (DFG) Grant HU698/4.

References

Antonini P, Petrini R, Cotin G (1998) A segment of seafloor-spreading in the central Red Sea: Basalts from the Nereus Deep (23°00′N–23°20′N). J Afr Earth Sci 27(8):107–114

Augustin N, Devey CW, van der Zwan FM, Feldens P, Tominaga M, Bantan RA, Kwasnitschka T (2014) The rifting to spreading transition in the Red Sea. Earth Planet Sci Lett 395:217–230. doi:10.1016/j.epsl.2014.03.047

Bäcker H, Lange K, Richter H (1975) Morphology of the Red Sea central graben between Subair Island and Abul Kizaan. Geologisches Jahrbuch D 13:79–123

Bäcker H, Schoell M (1972) New deeps with brines and metalliferous sediments in the Red Sea. Nature 240(6):153–158

Bertoni C, Cartwright JA (2005) 3D seismic analysis of circular evaporite dissolution structures, Eastern Mediterranean. J Geol Soc 162(16):909–926

Bertoni C, Cartwright JA (2006) Controls on the basinwide architecture of late miocene (Messinian) evaporites on the levant margin (Eastern Mediterranean). Sed Geol 188(21):93–114

Blum N, Puchelt H (1991) Sedimentary-hosted polymetallic massive sulfide deposits of the Kebrit and Shaban Deeps, Red Sea. Miner Deposita 26(3):217–227

Bonatti E (1985) Punctiform initiation of seafloor spreading in the Red Sea during transition from a continental to an oceanic rift. Nature 316(6023):33–37

Bonatti E, Colantoni P, Della Vedova B, Taviani M (1984) Geology of the Red Sea transitional region (22°N–25°N). Oceanol Acta 7 (4):385–398

Bosworth W, Huchon P, McClay K (2005) The Red Sea and Gulf of Aden Basins. J Afr Earth Sci 43(1–3):334–378

Bosworth W, Strecker MR (1997) Stress field changes in the Afro-Arabian rift system during the miocene to Recent period. Tectonophysics 278(1–4):47–62

Botz R, Schmidt M, Kus J, Ostertag-Henning C, Ehrhardt A, Olgun N, Garbe-Schonberg D, Scholten J (2011) Carbonate recrystallisation and organic matter maturation in heat-affected sediments from the Shaban Deep, Red Sea. Chem Geol 280(1–2):126–143

Brun J-P, Fort X (2011) Salt tectonics at passive margins: geology versus models. Mar Pet Geol 28(6):1123–1145

Cocherie A, Calvez JY, Oudin-Dunlop E (1994) Hydrothermal activity as recorded by Red Sea sediments: Sr–Nd isotopes and REE signatures. Mar Geol 118(3–4):291–302

Cochran JR (2005) Northern Red Sea: nucleation of an oceanic spreading center within a continental rift. Geochem Geophys Geosyst G3 6(3):34

Cochran JR, Gaulier J-M, Le Pichon X (1991) Crustal structure and the mechanism of extension in the northern Red Sea: constraints from gravity anomalies. Tectonics 10:1018–1037

Cochran JR, Martinez F (1988) Evidence from the northern Red Sea on the transition from continental to oceanic rifting. Tectonophysics 153(1–4):25–53

Cochran JR, Martinez F, Steckler MS, Hobart MA (1986) Conrad Deep: a new northern Red Sea deep: origin and implications for continental rifting. Earth Planet Sci Lett 78(1):18–32

Coutelle A, Pautot G, Guennoc P (1991) The structural setting of the Red Sea axial valley and deeps: implications for crustal thinning processes. Tectonophysics 198 (2–4): 395–396, 398, 401–409

Degens ET, Ross DA (1969) Hot brines and recent heavy metal deposits in the Red Sea. Springer, New York, 600 pp

Dümmong S, Hübscher C (2011) Levant Basin—Central Basin. In: Lofi J, Déverchère J, Gaullier V, Gillet H, Gorini C, Guennoc P, Loncke L, Maillard A, Sage F, Thinon I (eds) Seismic atlas of the Messinian Salinity Crisis markers in the Mediterranean and Black Seas. Mémoires de la Société Géologique de France, no. 179, and World Geological Map Commission, Paris, 72 pp

Ehrhardt A, Hübscher C (2003) Preliminary results—geophysics. Meteor Berichte. In: Pätzold J, Bohrmann G, Hübscher C (eds) Meteor Berichte M52 Black Sea, Mediterranean—Red Sea, vol 3 (2), pp 3–21

Ehrhardt A, Hübscher C, Gajewski D (2005) Conrad Deep, northern Red Sea: development of an early stage ocean deep within the axial depression. Tectonophysics 411(21):19–40

Franke D, Neben S, Ladage S, Schreckenberger B, Hinz K (2007) Margin segmentation and volcano-tectonic architecture along the volcanic margin off Argentina/Uruguay, South Atlantic. Marine Geology 244(1–4):46–67

Girdler RW, Styles P (1974) Two sage Red Sea floor spreading. Nature 247:5

Guennoc P, Pautot G, Coutelle A (1988) Surficial structures of the northern Red Sea axial valley from 23° N to 28°N: time and space evolution of neo-oceanic structures. Tectonophysics 153(1–4):1–23

Haase KM, Mühe R, Stoffers P (2000) Magmatism during extension of the lithosphere: geochemical constraints from lavas of the Shaban Deep, northern Red Sea. Chem Geol 166(3–4):225–239

Hartmann M (1980) Atlantis II Deep geothermal brine system. Hydrographic situation in 1977 and changes since 1965. Deep Sea Res Part A Oceanogr Res Pap 27(2):161–164, IN3–IN4, pp 165–171

Hartmann M, Scholten JC, Stoffers P, Wehner F (1998) Hydrographic structure of brine-filled deeps in the Red Sea—new results from the Shaban, Kebrit, Atlantis II, and discovery deep. Mar Geol 144 (4):311–330

Hovland M, Rueslåtten HG, Johnsen HK, Kvamme B, Kuznetsova T (2006) Salt formation associated with sub-surface boiling and supercritical water. Mar Pet Geol 23(8):855–869

Hübscher C, Böke W, Gutowski M, Kästner R, Salem M (2000) Preliminary results of Leg M44/3—very high resolution multichannel reflection seismics. Meteor Berichte. In: Pätzold J, Halbach PE, Hempel G, Weikert H (eds) Meteor Berichte—Eastern Mediterranean—Northern Red Sea, pp 85–108

Hudec MR, Jackson MPA (2007) Terra infirma: understanding salt tectonics. Earth Sci Rev 82(1–2):1–28

Makris J, Rihm R (1991) Shear-controlled evolution of the Red Sea: pull apart model. Tectonophysics 198(2–4):441–466

Marshak S, Bonatti E, Brueckner H, Paulsen T (1992) Fracture zone tectonics at Zabargad Island, Red Sea (Egypt). Tectonophysics 216 (6):379–385

Martinez F, Cochran J (1989) Geothermal measurements in the northern Red Sea: Implications for lithospheric thermal structure and mode of extension during continental rifting. J Geophys Res 94 (B9):12239–12266

Martinez F, Cochran JR (1988) Structure and tectonics of the northern Red Sea: catching a continental margin between rifting and drifting. Tectonophysics 150(1–2):1–31

McKenzie DP, Davies D, Molnar P (1970) Plate tectonics of the Red Sea and East Africa. Nature 226(5242):243–248

Miller AR et al (1966) Hot brines and recent iron deposits in deeps of the Red Sea. Geochim Cosmochim Acta 26(24):1029

Mitchell DJW, Allen RB, Salama W, Abouzakm A (1992) Tectono-stratigraphic framework and hydrocarbon potential of the Red Sea. J Pet Geol 15(2):23

Mitchell NC, Ligi M, Ferrante V, Bonatti E, Rutter E (2010) Submarine salt flows in the central Red Sea. Geol Soc Am Bull 122(5/6):13

Mohriak WU, Szatmari P, Anjos S (2012) Salt: geology and tectonics of selected Brazilian basins in the global context. Geol Soc Lond Spec Publ 363(28):131–158

Netzeband GL, Hübscher CP, Gajewski D (2006) The structural evolution of the Messinian evaporites in the Levantine Basin. Mar Geol 230(3–4):249–273

Pautot G, Guennoc P, Coutelle A, Lyberis N (1984) Discovery of a large brine deep in the northern Red Sea. Nature 310(5973):133–136

Röser HA (1975) A detailed geomagnetic survey of the southern Red Sea. Geol Jahrbuch 13:23

Sandwell DT, Smith WHF (1997) Marine gravity anomaly from geosat and ERS-1 altimetry. J Geophys Res 102(B5):10039–10054

Scholten JC, Stoffers P, Walter P, Plüger W (1991) Evidence for episodic hydrothermal activity in the Red Sea from the composition and formation of hydrothermal sediments, Thetis Deep. Tectonophysics 190(1):109–117

Searle RC, Ross DA (1975) A geophysical study of the Red Sea axial trough between 20.5° and 22°N. Geophys J Roy Astron Soc 43 (2):555–572

Skaarup N, Jackson HR, Oakey G (2006) Margin segmentation of Baffin Bay/Davis Strait, eastern Canada based on seismic reflection and potential field data. Mar Pet Geol 23(1):127–144

Stoffers P, Ross DA (1974) Sedimentary history of the Red Sea. In: Whitmarch RB, Weser OE, Ross DA, et al (eds) Initial report DSDP 23, pp 849–865

Talbot CJ, Pohjola V (2009) Subaerial salt extrusions in Iran as analogues of ice sheets, streams and glaciers. Earth Sci Rev 97:167–195

Warren JK (2006) Evaporites: sediments, resources and hydrocarbons. Springer, Berlin

Whitmarsh RB, Weser OE, Ross DA et al (1974) Initial reports of the deep sea drilling projects. U.S. Gov Printing Off Washington DC 23B:35–56

Winckler G, Aeschbach-Hertig W, Kipfer R, Botz R, Rübel AP, Bayer R, Stoffers P (2001) Constraints on origin and evolution of Red Sea brines from helium and argon isotopes. Earth Planet Sci Lett 184:671–683

Winckler G, Kipfer R, Aeschbach-Hertig W, Botz R, Schmidt M, Schuler S, Bayer R (2000) Sub sea floor boiling of Red Sea brines: new indication from noble gas data. Geochim Cosmochim Acta 64 (9):1567–1575

Lineaments in Gravity Data of the Red Sea

Neil C. Mitchell

Abstract

The structure of the crust beneath the Red Sea is obscured by thick evaporites, impeding progress in understanding the processes associated with continental rifting and leaving the nature of the crust (whether oceanic or continental) still controversial. Here, version 18.1 of the marine gravity field derived from satellite altimetry measurements by Sandwell and Smith (1997) is examined because the gravity field has the potential to reveal structure associated with the topography of basement beneath the evaporites and with crustal density variations. Comparing the satellite-derived data with gravity data from expeditions of RRS *Shackleton* in 1979 and RV *Conrad* in 1984, discrepancies are found to have standard deviations of 6.1 and 4.9 mGal, respectively, somewhat higher than parts of the gravity data from the open oceans. Coherent features in maps of these discrepancies suggest that some systematic errors in version 18.1 of the gravity field still remain. Nevertheless, they appear not to affect the short-wavelength structure of the data because simple image processing reveals some striking structural features in plan-view. The satellite-derived gravity data are enhanced by showing them with artificial shading and as directional second derivatives. The maps reveal lineaments that cross the central Red Sea. Many of them die out towards the coastlines and have convex-north-west shapes. These features are interpreted as evidence for migrating volcanic segments of oceanic crust, here suggesting a 1.5 mm year^{-1} along-axis movement of the sub-axial asthenosphere away from the Afar plume. This migration velocity is modest compared with the plate-opening rate ($\sim$12 mm year^{-1} here) and compared with velocities deduced from V-shaped ridges near Iceland and the Azores.

Introduction

Resolving the structure beneath the Red Sea is important for addressing questions such as what type of crust lies beneath the sea and how oceanic fracture zones relate to structures on land. Of the former issue, Cochran (2005) and Cochran and Karner (2007) have argued convincingly that the northern Red Sea is floored by stretched continental crust, while other authors have shown that the crust in the southern Red Sea is probably mostly oceanic (Allan 1970; Cochran 1983; Roeser 1975; Vine 1966) except near the coasts (Egloff et al. 1991). In the intervening central Red Sea, high crustal velocities from seismic refraction experiments suggest that oceanic crust exists within about 65 km of the deeps (Davies and Tramontini 1970; Drake and Girdler 1964; Egloff et al. 1991; Tramontini and Davies 1969). The smooth character of a basement reflector in seismic reflection data also suggests oceanic crust (Izzeldin 1987). However, the lower amplitudes of magnetic anomalies away from the deeps may indicate instead that these are areas of continental crust (Bonatti 1985; Cochran 1983; Ligi et al. 2011, 2012, this volume).

In Mitchell and Park (2014), the depth of a crustal seismic refractor recorded in the central Red Sea east of Thetis Deep (Davies and Tramontini 1970; Tramontini and Davies 1969)

N.C. Mitchell (✉)
School of Earth, Atmospheric and Planetary Sciences, University of Manchester, Williamson Building, Oxford Road, Manchester M13 9PL, UK
e-mail: neil.mitchell@manchester.ac.uk

N.M.A. Rasul and I.C.F. Stewart (eds.), *The Red Sea*,
Springer Earth System Sciences, DOI 10.1007/978-3-662-45201-1_6,

was found to co-vary with free-air gravity anomalies derived from satellite altimetry data (Sandwell and Smith 2009). The rate of co-variation is compatible with the effect of topography of basement with a density contrast of crustal rocks with evaporites. This offers the possibility that marine gravity data can allow us to "see" through the evaporites to resolve some of the underlying structure.

Subsequent to the original data release of Sandwell and Smith (1997), their estimates of the free-air marine gravity field have improved with re-tracking of radar waveforms, better filtering methods and increasing coverage of the oceans by successive altimeter missions (Sandwell et al. 2013; Sandwell and Smith 2009). Although tracking error can be expected to be a problem in areas near coasts of narrow seas such as in the Red Sea, sea surface waves produce noise in altimeter data (Goff 2009), so the limited fetch and therefore smaller wave heights in the Red Sea than in broader oceans may partly compensate for the effect of tracking error. In the present study, version 18.1 of the data is studied, which is referred to as "SSv18".

The available terrestrial gravity field around the Red Sea has only modest resolution and is not useful for comparing with fine structure in the marine gravity field, so instead elevation data (Becker et al. 2009) derived from the Shuttle Radar Topography Mission (SRTM) were used in the present study to fill in land areas of the following maps. Strictly speaking, elevation data cannot be compared simply with gravity data because the topography of the Arabian-Nubian shield is the result of a long history of tectonics, erosion and other geological processes, whereas the gravity anomalies represent effects of topography of the crust under the Red Sea and variations in density and thickness of the crust. Nevertheless, the elevation data can provide some context for the offshore gravity data.

Datasets

Gravity data were collected on RRS *Shackleton* in 1987 with a LaCoste and Romberg marine gravity meter (Girdler and Southren 1987) and during RV *Robert Conrad* cruise 2507 in 1984 with a Bell BGM-3 (Cochran and Martinez 1988). Little information exists for the *Shackleton* data but, for the *Conrad* data, Cochran and Martinez (1988) found internal crossing errors of only 1 mGal in areas navigated by Loran-C and <3 mGal in the worst cases. (Given the "spikey" character of the *Conrad* data shown here, these values were presumably computed after data filtering.) These gravity data were obtained from the National Geophysical Data Center (www.ngdc.noaa.gov).

The SSv18 gravity field was sampled along the ship tracks, and the shipboard gravity measurements were then subtracted from those sampled values. The resulting differences are shown as profiles plotted towards N030°W in Figs. 1 and 2 for the *Shackleton* and *Conrad* cruises, respectively. Some spikes in Fig. 1 at the ends of lines appear to be mostly due to centripetal accelerations occurring with course changes. Given that other spikes occur in both datasets, the shipboard values were first filtered with a 4-km running median average filter before computing the statistics and other graphs outlined below. This 4-km filter width was necessary to ensure that a few samples populate the filter, whereas the theoretical spatial resolution limit of the marine gravity field is about equal to the depth to the density variations causing the anomalies (Turcotte and Schubert 1982), here 1–6 km depth below sea level to the evaporite/basement interface in the central Red Sea (Egloff et al. 1991; Izzeldin 1987, 1989; Tramontini and Davies 1969), so ~1–6 km

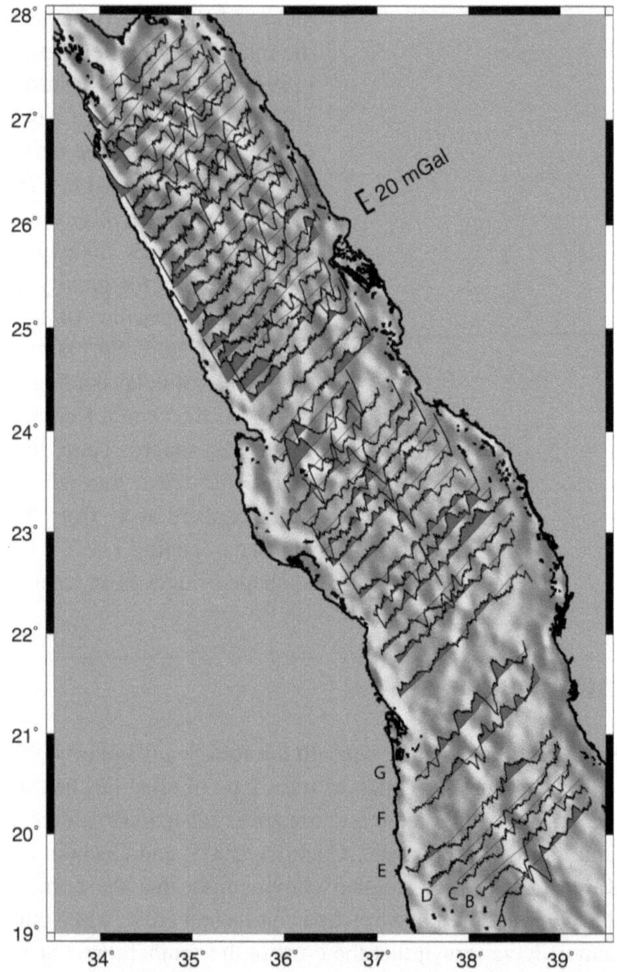

Fig. 1 Map of the Red Sea showing differences in free-air gravity anomaly between the Sandwell and Smith (2009) anomalies derived from satellite altimetry (SSv18) and the shipboard anomalies collected on RRS *Shackleton* (i.e., SSv18 minus the *Shackleton* data). The differences are plotted as profiles towards N030°W with the scale shown upper-right and with positive differences shaded. Background is a *shaded*-relief image of SSv18 illuminated from N060°E

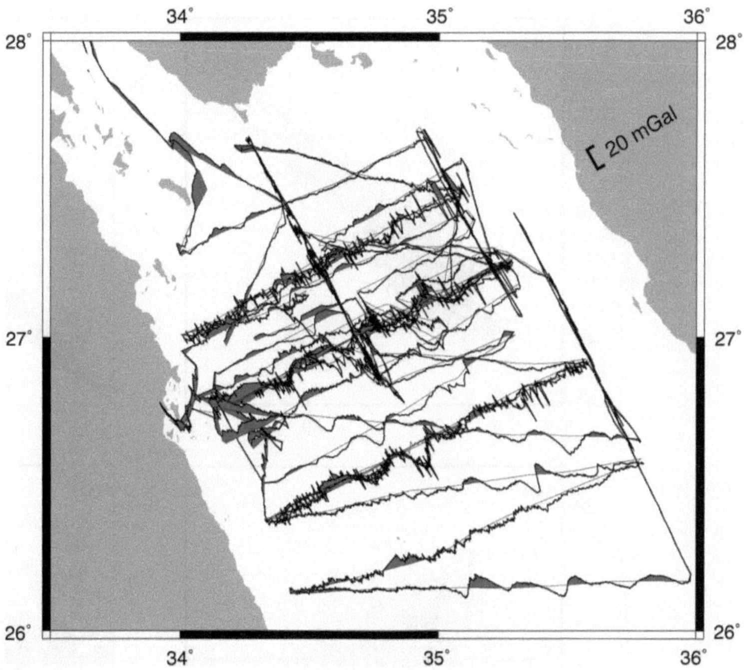

Fig. 2 Difference in free-air gravity anomaly between the SSv18 and the shipboard data collected on RV *Conrad* during cruise 2507, as Fig. 1

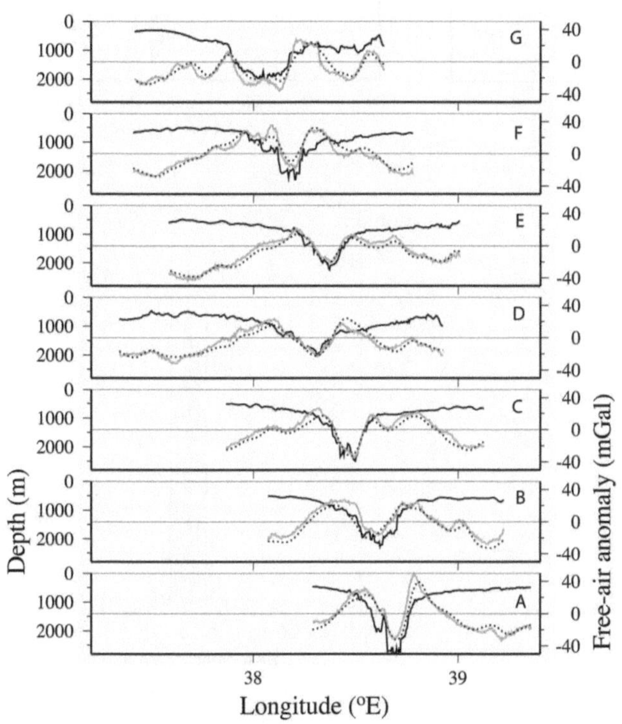

Fig. 3 Profiles of bathymetry (*bold lines*), free-air gravity shipboard measurements (*grey lines*) and free-air gravity sampled from SSv18 (*dotted lines*) for RRS *Shackleton* profiles (**A–G**) marked in Fig. 1

represent the range of the expected theoretical resolution. Figure 3 shows the shipboard and sampled gravity data along seven of the *Shackleton* profiles located in Fig. 1, revealing discrepancies over short wavelengths, including a tendency to flatten depressions or highs owing to the poorer resolution of the satellite-derived data.

In Fig. 4a, b, the variation of SSv18 with the filtered shipboard data is shown. The discrepancies between the two datasets are enlarged in Fig. 4c, d. Although the co-variation appears good, some large positive free-air anomaly (FAA) values have a negative bias and some large negative FAA values have a positive bias in the SSv18 data. This is an effect of the poorer resolution of the altimetry-derived data.

Figure 5 suggests that the differences are roughly normally distributed. The mean differences are −1.8 and −1.2 mGal in the *Shackleton* and *Conrad* data, respectively; note that these datasets have been reduced using different reference ellipsoids [International Association of Geodesy 1967 formula for the *Shackleton* data, Potsdam for the *Conrad* data and Earth Gravitational Model 2008 (Pavlis et al. 2012) for the SSv18 data (Sandwell and Smith 2009)]. The corresponding standard deviations are 6.1 and 4.9 mGal, the lower *Conrad* value possibly representing superior performance of the Bell BGM-3 gravity meter on the *Conrad* compared with the older LaCoste and Romberg meter that was installed on RRS *Shackleton*. These values are similar to the 5.6 mGal standard deviation found by Ligi et al. (2012) for a more limited area in the central Red Sea around Thetis Deep, but larger than differences between SSv18 and values from shipboard gravity in the open oceans, which have been shown locally to reach as small as 2.03 mGal (Sandwell and Smith 2009). Comparisons of the newer (version 21) gravity

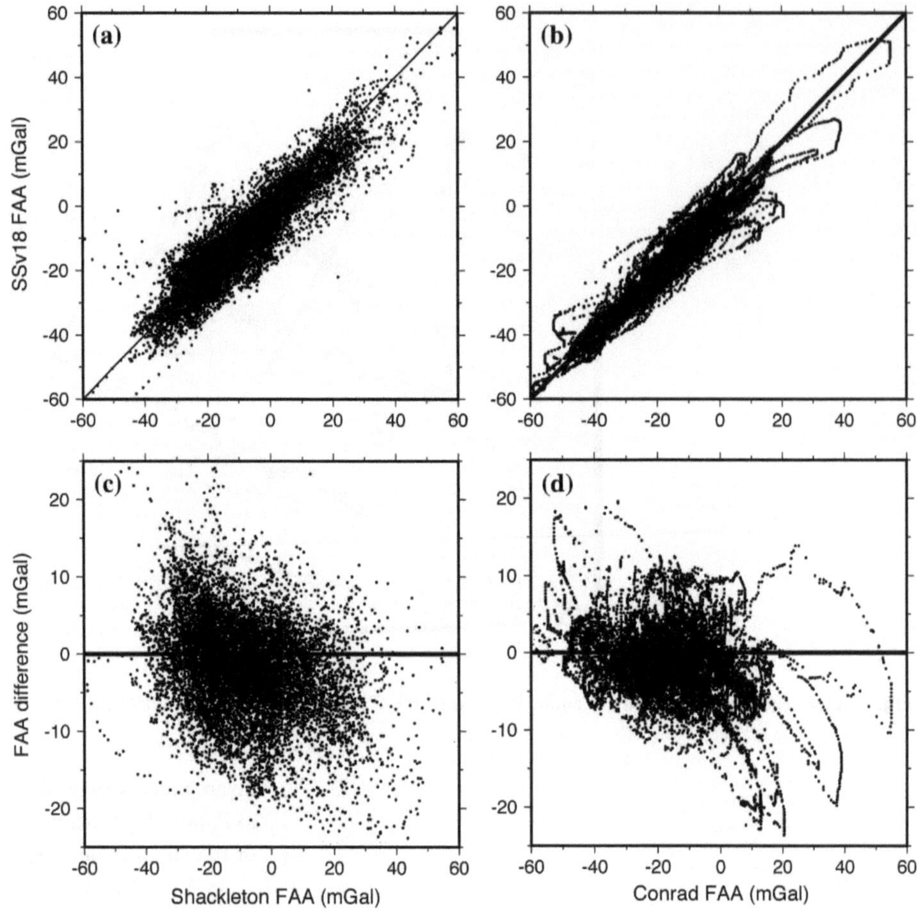

Fig. 4 Scatterplots of SSv18 free-air anomaly (FAA) versus shipboard values collected on **a** RRS *Shackleton* and **b** RV *Conrad*. Graphs **c** and **d** show differences between SSv18 and the shipboard values of (**a**) and (**b**), respectively (i.e., SSv18 minus shipboard value)

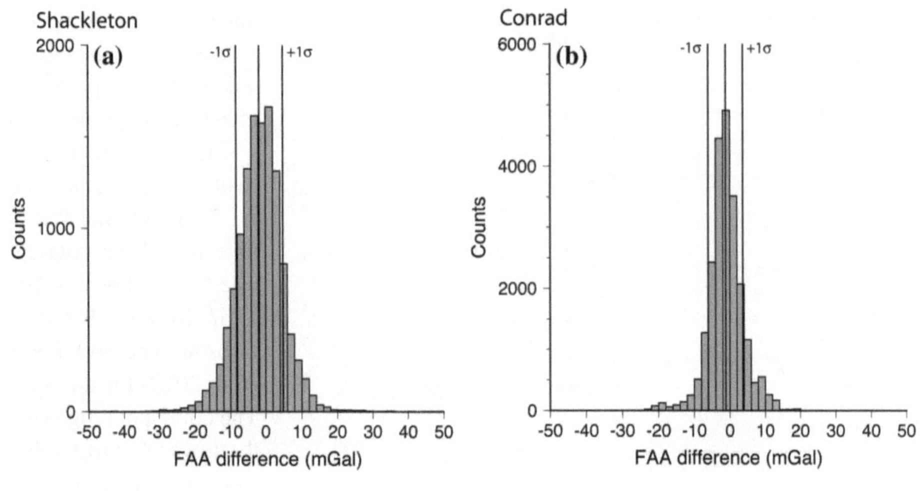

Fig. 5 Histograms of differences between the SSv18 and the shipboard free-air anomaly data from **a** RRS *Shackleton* and **b** RV *Conrad*. *Vertical lines* show the means and standard deviations of the differences

model with ship gravity measurements collected globally (National Geophysical Data Center data) by Sandwell et al. (2013) suggested a most likely median absolute difference of 2.75 mGal. For comparison with a dataset from another inland sea, a modest 2.35 mGal standard deviation was found between SSv18 and shipboard gravity data from Hudson Bay (Keating and Pinet 2013).

Somewhat more of a concern for structural studies, the difference maps in Figs. 1 and 2 show some non-random components (differences correlating between lines). Figure 6 is a colour-coded map of the differences. Whereas some narrow anomalies correlating between lines are probably caused by the satellite data resolution mentioned earlier (e.g. at 23.5°N mid-way between the coasts), large areas of blue

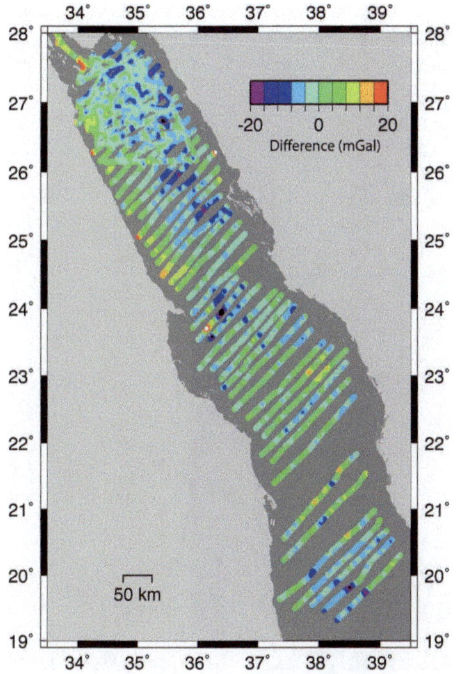

Fig. 6 Differences of the SSv18 gravity field and the *Shackleton* and *Conrad* shipboard gravity data (SSv18 minus shipboard). Shipboard data were first de-spiked with a 4-km along-track median filter

(−10 mGal difference) occur on the east side at 25°–26°N, centre at 24°N and south of the dataset south of 20°N. The 24°N central anomaly lies in an area of strong gravity field gradients (Styles and Gerdes 1983). Some coherent positive anomalies of +10 mGal occur close to the Egyptian coast. The magnitudes of these systematic biases are larger than those in the Hudson Bay study (Keating and Pinet 2013), where they typically reach only 4 mGal. Their origins are unclear. Sandwell (pers. comm. 2013) has suggested that these could be edge effects associated with converting altimetry deflections of the vertical offshore to gravity anomalies. Nevertheless, these anomalies are small compared with the >100 mGal full range of the SSv18 dataset and the appearance of structural features in the images presented below suggests that the gravity anomalies do not result in false structures appearing in plan-view.

Image Processing

To show elevations and the SSv18 data together, elevation values from the SRTM (Becker et al. 2009) were first rescaled so that the rescaled relief had a similar local variation to the gravity (a factor of 0.06 was found to achieve

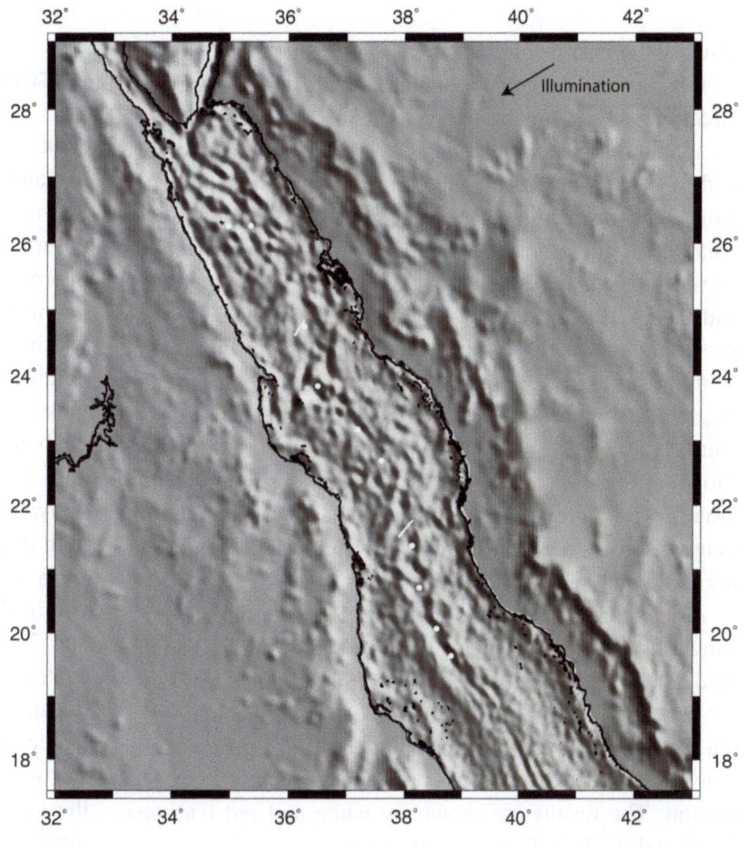

Fig. 7 The marine gravity field (SSv18) and land elevation data both shown in shaded relief with the artificial illumination from N060°E. Two *white bars* within the sea represent 3 m y of plate-tectonic movement predicted using the Chu and Gordon (1998) plate rotation pole. *Open circles* show a selection of the deeps of Pautot et al. (1984) to help readers locate features. From south to north, these are the Suakin, Port Sudan, Erba, Atlantis II, Thetis, Nereus, Vema, Kebrit, Jean Charcot and Oceanographer deeps

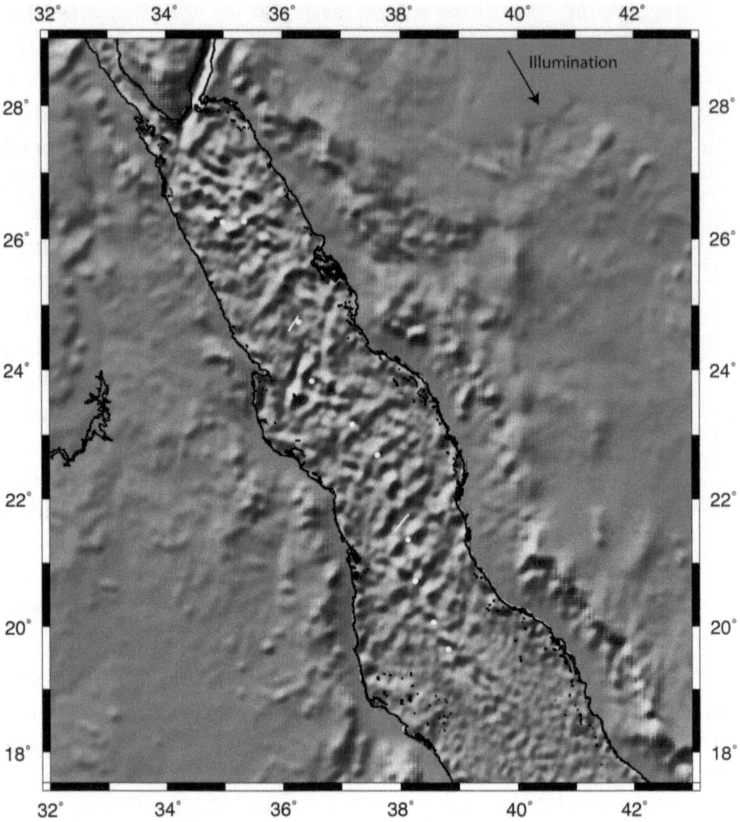

Fig. 8 As Fig. 7, with illumination from N030°W

visually the best result). Figures 7 and 8 show shaded-relief images of that combined dataset with the illumination directions shown in the upper-right in each figure. Before completing this operation, the elevation data were smoothed with a 0.25°-wide cosine-tapered two-dimensional filter to reduce the distraction of fine erosional geomorphology of escarpment gullies, etc. As the marine free-air gravity field is correlated with the relief of a basement refractor (Mitchell and Park 2014), basement structures beneath the evaporites should be correctly displayed with artificial illumination from the northerly compass direction (i.e., an apparent ridge in gravity should represent a ridge in the underlying basement, not a valley). The directional second derivatives shown in Figs. 9 and 10 were computed in a similar fashion, though with a grey-scale range that is narrower than the full range in order to enhance small curvature values representing more subtle data ridges or valleys.

A selection of the deeps given by Pautot et al. (1984) is shown by white and red dots on Figs. 7, 8, 9 and 10 so that readers can locate features on the maps (see caption of Fig. 7 for listing). To represent the trends expected of Nubia-Arabia plate motion, the Chu and Gordon (1998) plate rotation pole was used to rotate two points representing 3 m y of motion. The results are shown by white and red bars near 25°N and 22°N in Figs. 7, 8, 9 and 10.

Observations

Structures lying sub-parallel to the overall trends of the Red Sea and coasts are revealed in Figs. 7 and 9. Within the Red Sea north of 26°N, a series of NW–SE lineaments can be observed in Fig. 7. These correspond with the structures interpreted as continental rift blocks by Cochran (2005) [Compare also with the orientations of the extensional faults within Saudi Arabia in Figs. 3b and 6 of Bosworth (this volume)]. Elsewhere, Fig. 7 reveals longer lineaments mostly along the axis, although structures are visible near the Arabian coast at 22°N and on the African coast 22°–24°N and around Zabargad Island (near 24°N). The second derivative in Fig. 9 tends to enhance smaller inflections in both datasets to reveal further lineaments.

In Fig. 8, prominent transverse structures are visible within the Red Sea at 24°N (the Zabargad Fracture Zone) and 25°N. Further transverse structures are visible in the central Red Sea from 24°N to 20.5°N. At 20.5°N, a major lineament can be seen crossing the entire Red Sea and this feature is aligned with lineaments in elevation data on the surrounding land. A further lineament running N–S occurs in the marine gravity data on the Arabian side at 20°N and other lineaments on the corresponding African side, but

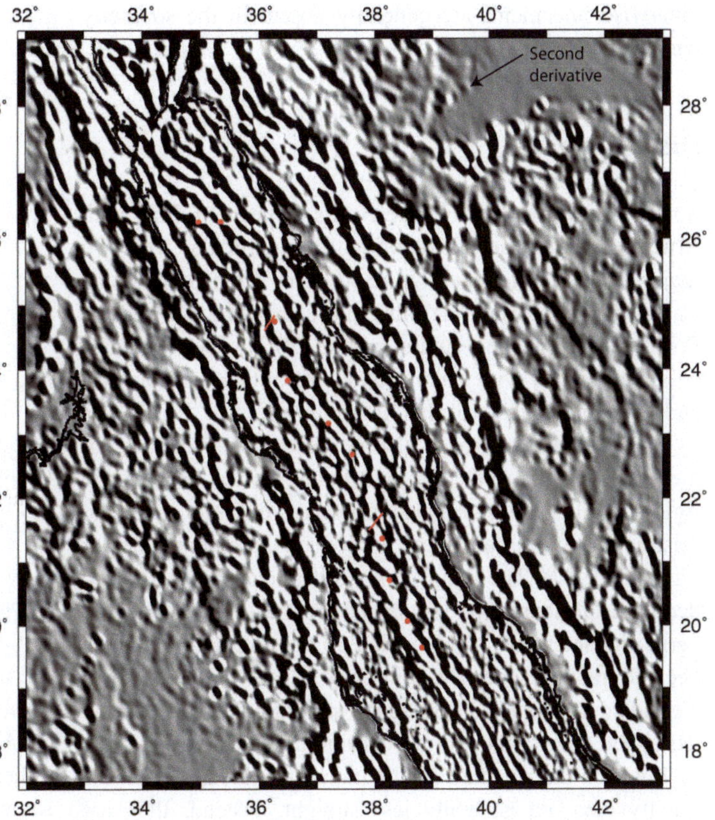

Fig. 9 Second derivative of SSv18 offshore and elevation data (Becker et al. 2009) onshore computed from N060°E (*black* represents downward-curved areas such as ridge crests and *white* represents upward-curved areas such as valleys or breaks in slope). *Red* annotation: 3 m y plate motion and deeps as Fig. 7

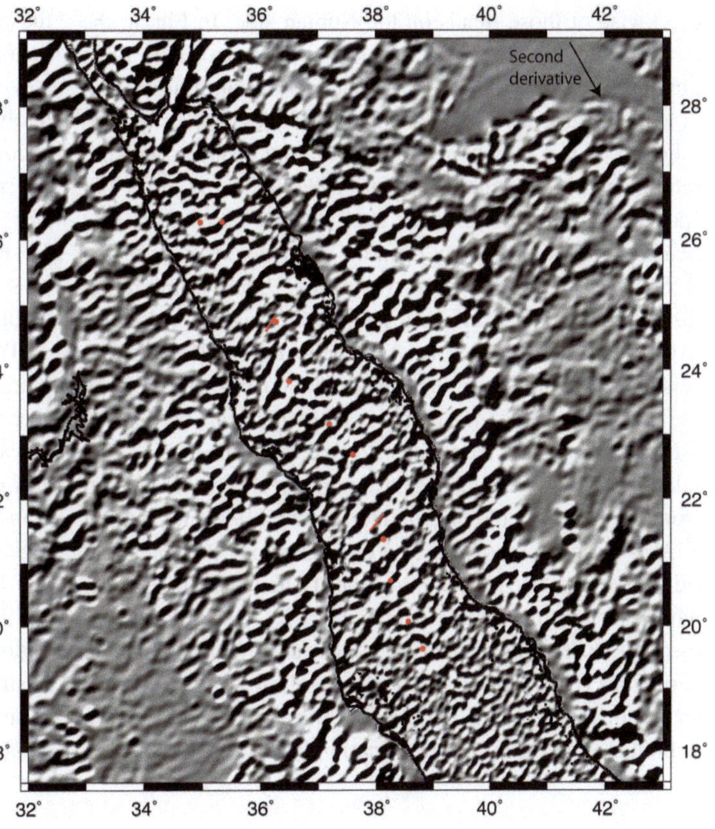

Fig. 10 Second derivative of SSv18 offshore and elevation data (Becker et al. 2009) onshore computed from N030°W. *Red* annotation: 3 m y plate motion and deeps as Fig. 7

otherwise lineaments are generally absent in the southerly part of the gravity dataset.

Discussion

Some transverse gravity lineaments appear to link with structures within the shields on land, in particular with two major ancient plate suture zones (Camp 1984; Stoeser and Camp 1985): the Halaib-Yanbu suture around 24°N and the Port-Sudan-Bi'r Umq suture around 20°N. These sutures coincide roughly with major protrusions of the Arabian coast and with indentations of the African coast.

The region south of 20°N has fewer transverse lineaments. A seismic refraction experiment carried out here by Egloff et al. (1991) revealed primarily oceanic crust and seismic reflection data show a subdued basement reflector (Izzeldin 1987). The Egloff et al. (1991) data show a 2-km-deep crustal depression in oceanic crust at the major transverse lineament at 20°N, which is therefore probably an oceanic fracture zone. The continuity of this feature makes it a good candidate for a continental structure that has directly led to the development of an oceanic transform fault.

Between there and 23°N, the transverse lineaments in the gravity data are generally less straight. Instead, they are slightly convex-NW, lying clockwise from the trend expected of plate motion on the Arabian side and anticlockwise of those trends on the Nubian side. In Fig. 8, the trends generally do not continue to the coast. This morphology is expected of non-transform discontinuities and volcanic segments on slow-spreading ridges, which tend to migrate in the direction expected of the mantle hot spot reference frame (Schouten et al. 1987) or away from hot spots (Briais and Rabinowicz 2002). In the case of the Red Sea, the migration is away from the Afar region, as expected if SAA is migrating north-westwards. Such a migration direction has been inferred from low seismic velocities extending north-west from Afar beneath the Red Sea in seismological studies of the upper mantle (Park et al. 2007).

The procedure of Schouten et al. (1987) can be used to estimate the rate of along-axis flow of SAA. In Fig. 11, the relative velocity of Nubia and Arabia is shown, computed from the rotation pole of Chu and Gordon (1998) for a point at 21.5°N on the axis (Fig. 7). Velocities along the spreading ridge are expected to lie along the dashed line in Fig. 11, which bisects the Nubia-Arabia velocity [assuming symmetrical spreading, which is suggested by the magnetic anomalies (Chu and Gordon 1998; Izzeldin 1987)] and with an orientation passing through the deeps. In contrast with Schouten et al. (1987), no distinction is made here between the mantle and SAA velocities. To estimate the direction of vectors between Nubia and SAA and between Arabia and SAA, the orientations of three segments were interpreted on either side of the axis from the near-axis trends in Fig. 8 (one segment immediately south of the white bar at 21.5°N and two segments north of there). Those orientations are shown by the grey lines in Fig. 11. The lines do not intersect along the dashed line, a result probably of the limited extents of data to make these measurements and data variability. Nevertheless, the velocity of the SAA can be estimated most likely to lie in the centre of the velocity space region encompassed by these lines along the dashed line, that is, at the solid circle marked "sub-axial asthenosphere" in Fig. 11. The velocity of SAA relative to the axis is thus around 1.5 mm year^{-1}, a slow rate compared with the plate velocities (around 12 mm year^{-1} full separation rate of Arabia and Nubia here (Chu and Gordon 1998).

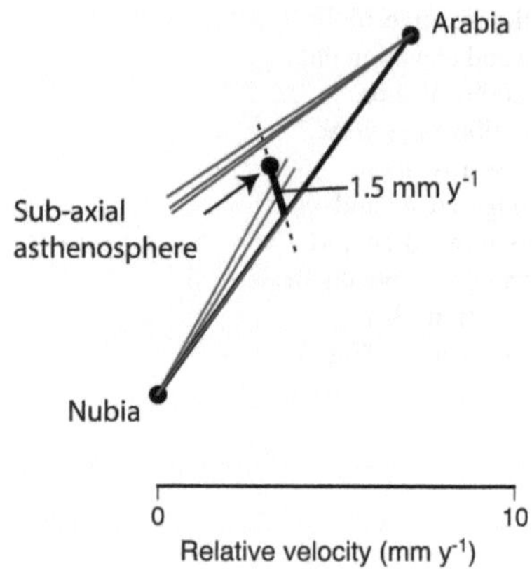

Fig. 11 Velocity-space diagram for the migrating segments of the central Red Sea. *Bold circles* marked Nubia and Arabia represent the relative separation of the two plates computed from the rotation pole of Chu and Gordon (1998) for a point at 21.5°N on the axis (*white bar* in Fig. 7). *Dashed line* represents velocities along the spreading ridge. *Grey lines* represent measured orientations of spreading segments. Remaining *bold solid circle* is the estimated SAA velocity

To put this SAA movement in context relative to the Afar plume, Fig. 12 locates the central Red Sea area of migrating spreading segments on a three-dimensional view of the elevation data (Becker et al. 2009). The region of strongly elevated topography in Nubia and southern Arabia (higher than 1.5 km or red in Fig. 12) may reflect the northern-most region of magmatism and/or most important dynamic uplift and lithospheric heating by the Afar plume (Ebinger et al. 1989). The migrating segments lie to the north-west of this region. Comparison with evidence of the mantle plume from seismology is problematic because the present models of seismic velocity are derived from data with poor ray coverage under the Red Sea. Nevertheless, a study of shear-

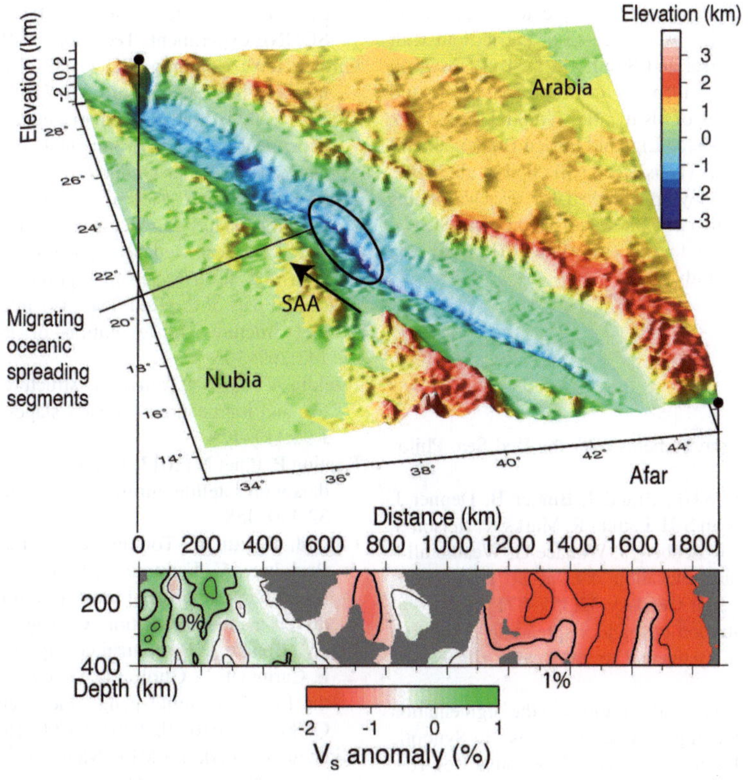

Fig. 12 Context of the proposed northward migrating oceanic spreading segments. Three-dimensional projection of the elevation/bathymetry model of Becker et al. (2009) is shown with the area of migrating segments marked (*ellipse*). These migrations imply a modest north-north-west velocity of the upper mantle (*arrow*). The central Red Sea corresponds with a region of the upper mantle with intermediate shear-wave velocities shown by the lower cross section, which is from a model of Park et al. (2007) based on teleseismic arrivals in Arabia. The velocity section is shown contoured every 0.5 % and represents structure between the two *solid circles* marked on the 3D view. *Grey* areas of section have poor teleseismic arrival constraints

wave velocity (V_s) by Park et al. (2007) was based on teleseismic arrivals recorded at stations distributed across Saudi Arabia and provides a view that is probably mimicked under the Red Sea. A vertical section of their model (Fig. 12) reveals a ~1.5 % increase in upper mantle V_s going north-west from southern Arabia beneath the coast. The area of migrating segments lies adjacent to intermediate V_s in the Park et al. (2007) model. Estimates of upper mantle temperature from sodium oxide contents of spreading centre lavas ("Na8.0") suggest a commensurate 60 °C decrease from 18°N to 26°N (Haase et al. 2000).

The area of modest 1.5 mm year^{-1} migration therefore appears to lie at the periphery of plume-influenced mantle. It also contrasts with regions of more rapid mantle velocity implied by V-shaped ridges around spreading centres (presumably located more proximal to their respective plumes). For example, such ridges around the Reykjanes Ridge suggest velocities away from the Icelandic plume of 87–282 mm year^{-1} (Poore et al. 2009) and those around the Mid-Atlantic Ridge south of the Azores suggest 60 mm year^{-1} (Cannat et al. 1999).

Conclusions

Simple image processing (shaded-relief and directional second derivatives or curvature) applied to version 18.1 of the marine gravity field of Sandwell and Smith (2009) reveals lineaments within the Red Sea, likely representing crustal structures beneath the evaporites. Lineaments crossing the central Red Sea are not straight and are more like migrating oceanic spreading segments observed elsewhere. Their orientations here imply a modest 1.5 mm year^{-1} migration of the SAA away from the Afar plume. Combined with other evidence of the region of influence of the plume (elevations, axial lava $Na_{8.0}$ and V_s from teleseismic arrivals in Arabia), the area of migrating lineaments lies at the periphery of the Afar plume.

Acknowledgments This project is largely based on data derived from satellite altimetry and generously provided online by David Sandwell and Walter Smith. Those researchers and the agencies providing the *Shackleton* and *Conrad* gravity data are thanked for making these data freely available. The figures and many of the plate-tectonic and image processing calculations in this article were created or carried out with the "GMT" software system (Wessel and Smith 1991). Bill Bosworth kindly shared copies of geological maps of the Red Sea coasts and Yongcheol Park kindly gave a copy of his mantle V_s model. This work was initiated by discussions with Marco Ligi and Enrico Bonatti aboard RV *Urania* during a project on the Thetis Deep (Mitchell et al. 2010a, b). Jim Cochran and Walter Mooney are thanked for reviews of this paper. Najeeb M.A. Rasul and others of the Saudi Geological Survey are also thanked for an invitation to the enjoyable 2013 Red Sea workshop, which motivated this study.

References

Allan TP (1970) Magnetic and gravity fields over the Red Sea. Philos Trans R Soc A267:153–180

Becker JJ, Sandwell DT, Smith WHF, Braud J, Binder B, Depner J, Fabre D, Factor J, Ingalls S, Kim S-H, Ladner R, Marks K, Nelson S, Pharaoh A, Trimmer R, Von Rosenberg J, Wallace G, Weatherall P (2009) Global bathymetry and elevation data at 30 arc seconds resolution: SRTM30_PLUS. Mar Geodesy 32:355–371

Bonatti E (1985) Punctiform initiation of seafloor spreading in the Red Sea during transition from a continental to an oceanic rift. Nature 316:33–37

Briais A, Rabinowicz M (2002) Temporal variations in the segmentation of slow to intermediate spreading mid-ocean ridges. 1. Synoptic observations based on satellite altimetry data. J Geophys Res 107, Paper 2098. doi:10.1029/2001JB000533

Camp VE (1984) Island arcs and their role in the evolution of the western Arabian shield. Geol Soc Am Bull 95:913–921

Cannat M, Briais A, Deplus C, Escartín J, Georgen J, Lin J, Mercouriev S, Meyzen C, Muller M, Pouliquen G, Rabain A, da Silva P (1999) Mid-Atlantic Ridge-Azores hotspot interactions: along-axis migration of a hotspot-derived event of enhanced magmatism 10 to 4 Ma ago. Earth Planet Sci Lett 173:257–269

Chu D, Gordon RG (1998) Current plate motions across the Red Sea. Geophys J Int 135:313–328

Cochran JR (1983) A model for the development of the Red Sea. Am Assoc Petrol Geolo Bull 67:41–69

Cochran JR (2005) Northern Red Sea: nucleation of an oceanic spreading center within a continental rift. Geochem Geophysics Geosyst 6, Paper Q03006. doi:10.01029/02004GC000826

Cochran JR, Karner GD (2007) Constraints on the deformation and rupturing of continental lithosphere of the Red Sea: the transition from rifting to drifting. In: Karner GD, Manatschal G, Pinheiro LM (eds) Imaging, mapping and modelling continental lithosphere extension and breakup, Special Publication 282. Geological Society, London, pp 265–289

Cochran JR, Martinez F (1988) Evidence from the northern Red Sea on the transition from continental to oceanic rifting. Tectonophysics 153:25–53

Davies D, Tramontini C (1970) The deep structure of the Red Sea. Philos Trans R Soc A267:181–189

Drake CL, Girdler RW (1964) A geophysical study of the Red Sea. Geophys J Roy Astron Soc 8:473–495

Ebinger CJ, Bechtel TD, Forsyth DW, Bowin CO (1989) Effective elastic plate thickness beneath the East African and Afar plateaux and dynamic compensation of uplifts. J Geophys Res 94:2883–2901

Egloff F, Rihm R, Makris J, Izzeldin YA, Bobsien M, Meier K, Junge P, Noman T, Warsi W (1991) Contrasting structural styles of the eastern and western margins of the southern Red Sea: the 1988 SONNE experiment. Tectonophys 198:329–353

Girdler RW, Southren TC (1987) Structure and evolution of the northern Red Sea. Nature 330:716–721

Goff JA (2009) Statistical characterization of Geosat altimetry noise: dependence on environmental parameters. Geochem Geophys Geosyst 10, Paper Q08007. doi:10.1029/2009GC002569

Haase KM, Mühe R, Stoffers P (2000) Magmatism during extension of the lithosphere: geochemical constraints from lavas of the Shaban Deep, northern Red Sea. Chem Geol 166:225–239

Izzeldin AY (1987) Seismic, gravity and magnetic surveys in the central part of the Red Sea: their interpretation and implications for the structure and evolution of the Red Sea. Tectonophysics 143:269–306

Izzeldin AY (1989) Transverse structures in the central part of the Red Sea and implications on early stages of oceanic accretion. Geophys J 96:117–129

Keating P, Pinet N (2013) Comparison of surface and shipborne gravity data with satellite-altimeter gravity data in Hudson Bay. Lead Edge 32:450–458

Ligi M, Bonatti E, Tontini FC, Cipriani A, Cocchi L, Schettino A, Bortoluzzi G, Ferrante V, Khalil S, Mitchell NC, Rasul N (2011) Initial burst of oceanic crust accretion in the Red Sea due to edge-driven mantle convection. Geology 39:1019–1022

Ligi M, Bonatti E, Bortoluzzi G, Cipriani A, Cocchi L, Caratori Tontini F, Carminati E, Ottolini L, Schettino A (2012) Birth of an ocean in the Red Sea: initial pangs. Geochem Geophys Geosys 13, Paper Q08009. doi:10.1029/2012GC004155

Mitchell NC, Park Y (2014) Nature of crust in the central Red Sea and topographic control of evaporite flowage. Tectonophys. 628:123–139. doi:10.1016/j.tecto.2014.04.029

Mitchell NC, Ligi M, Farrante V, Bonatti E, Rutter E (2010a) Submarine salt flows in the central Red Sea. Geol Soc Am Bull 122:701–713.

Mitchell NC, Schmidt M, Ligi M (2010b) Comment on "Formation of thetis deep metal-rich sediments in the absence of brines, Red Sea" by Pierret. J Geochem Expl 108:112–113

Park Y, Nyblade AA, Rodgers AJ, Al-Amri A (2007) Upper mantle structure beneath the Arabian Peninsula and northern Red Sea from teleseismic body wave tomography: implications for the origin of Cenozoic uplift and volcanism in the Arabian Shield. Geochem Geophys Geosys 8, Paper Q06021. doi:10.1029/2006GC001566

Pautot G, Guennoc P, Coutelle A, Lyberis N (1984) Discovery of a large brine deep in the northern Red Sea. Nature 310:133–136

Pavlis NK, Holmes SA, Kenyon SC, Factor JK (2012) The development and evaluation of the Earth Gravitational Model 2008 (EGM2008). J Geophys Res 117, Paper B04406. doi:10.1029/2011JB008916

Poore H, White N, Jones S (2009) A Neogene chronology of Iceland plume activity from V-shaped ridges. Earth Planet Sci Lett 283:1–13

Roeser HA (1975) A detailed magnetic survey of the southern Red Sea. Geol Jahrb 13:131–153

Sandwell DT, Smith WHF (1997) Marine gravity anomaly from Geosat and ERS-1 satellite altimetry. J Geophys Res 102:10039–10054

Sandwell DT, Smith WHF (2009) Global marine gravity from retracked Geosat and ERS-1 altimetry: ridge segmentation versus spreading rate. J Geophys Res 114, Paper B01411. doi:10.1029/2008JB006008

Sandwell D, Garcia E, Soofi K, Wessel P, Chandler M, Smith WHF (2013) Toward 1-mGal accuracy in global marine gravity from CryoSat-2, Envisat, and Jason-1. The Leading Edge, pp 892–899

Schouten H, Dick HJB, Klitgord KD (1987) Migration of mid-ocean ridge volcanic segments. Nature 326:835–839

Stoeser DB, Camp VE (1985) Pan-African microplate accretion of the Arabian Shield. Geol Soc Am Bull 96:817–826

Styles P, Gerdes KD (1983) St. John's Island (Red Sea): a new geophysical model and its implications for the emplacement of ultramafic rocks in fracture zones and at continental margins. Earth Planet Sci Lett 65:353–368

Tramontini C, Davies D (1969) A seismic refraction survey in the Red Sea. Geophys J Roy Astron Soc 17:225–241

Turcotte DL, Schubert G (1982) Geodynamics: applications of continuum physics to geological problems. Wiley, New York 450 pp

Vine FJ (1966) Spreading of the ocean floor—new evidence. Science 154:1405–1415

Wessel P, Smith WHF (1991) Free software helps map and display data. Eos Trans Am Geophys Union 72:441

Geodetic Constraints on the Geodynamic Evolution of the Red Sea

Robert Reilinger, Simon McClusky, and Abdullah ArRajehi

Abstract

We use geodetic, plate tectonic, and geologic observations to quantitatively reconstruct the geologic evolution of the Red Sea and Gulf of Aden since separation of Arabia from Africa in the Late Oligocene. Rifting initiated at 22 ± 3 Ma roughly simultaneously along the full strike of the proto-Red Sea and Gulf of Aden. Rifting began along pre-existing zones of weakness associated with a Pan-African Precambrian collisional suture shortly after the Afro-Arabia Plate was weakened by impingement of the African hot spot (~30 Ma). The initial phase of continental rifting followed a roughly linear trend from the Gulf of Suez in the north, to the Bab-al-Mandab in the south where the Afar Triple Junction (junction of Red Sea, Gulf of Aden, and East African rifts) was located at that time. The initial rate of extension across the rift was roughly half the present-day rate. At 11 ± 2 Ma, the rate of rifting doubled to the present-day rate (24 ± 1 mm/year in the south [~12°N] and 7 ± 1 mm/year in the north [~27°N]) and the configuration of rifting changed in both the northern and southern Red Sea. This time corresponds to the initiation of ocean spreading (i.e., complete severing of the continental lithosphere and intrusion of rift basalts) along the full extent of the Gulf of Aden. The changes in the S Red Sea involved the propagation of the Afar Triple Junction westward to its present location (~11.5°N, 42°E), the transfer of rifting from the S Red Sea (Bab-al-Mandab) to the more N–S-oriented Danakil Depression, and accompanying CCW rotation of the Danakil Block with respect to Africa. In the northern Red Sea, rifting transferred from the Gulf of Suez to the more N–S-oriented Gulf of Aqaba/Dead Sea fault system. The rate of rifting has not changed significantly since that time (i.e., 11 ± 2 Ma). The initiation of rifting at 22 ± 3 Ma corresponds temporally with slowing of Africa–Eurasia convergence by a factor of ~2 and the changes at 11 ± 2 Ma with a second phase of slowing of Africa–Eurasia convergence, while Arabia–Eurasia convergence has remained roughly unchanged since >30 Ma. These observations are consistent with simple models where changes in Africa–Arabia–Eurasia relative plate motions are the fundamental cause of post-Oligocene Middle East and Mediterranean tectonics. Based on the simultaneity between full ocean spreading along the Gulf of Aden and a doubling of the extension rate across the Red Sea, and the

R. Reilinger (✉)
Department of Earth Atmospheric and Planetary Sciences, Massachusetts Institute of Technology, Cambridge, MA, USA
e-mail: reilinge@erl.mit.edu

S. McClusky
Research School of Earth Sciences, Australia National University, Canberra, Australia

A. ArRajehi
King Abdulaziz City for Science and Technology, Riyadh, Saudi Arabia

N.M.A. Rasul and I.C.F. Stewart (eds.), *The Red Sea*,
Springer Earth System Sciences, DOI 10.1007/978-3-662-45201-1_7,

change to more N–S-oriented rifting in both the northern and southern Red Sea, we hypothesize that slowing of Africa–Eurasia convergence resulted from a decrease in slab pull on the African Plate across the evolving AR-AF plate boundary.

Introduction

Since the recognition and initial quantification of the Plate Tectonics hypothesis in the 1960s (e.g., Cox 1973), an outstanding question has concerned the forces driving plate motions (e.g., Elsasser 1971). Progress has been made in understanding both the balance of forces acting on the plates (e.g., Forsyth and Uyeda 1975; Hager and O'Connell 1981; Conrad and Lithgow-Bertelloni 2002) and the response of the ocean lithosphere to these forces (e.g., Parsons and Richter 1980; Gordon et al. 2008). However, the relative importance of the forces acting on the plates and the response of the continental lithosphere to those forces remain under discussion due to limited observational constraints (e.g., England and McKenzie 1983; Spence 1987; Turcotte and Shubert 2002; Thatcher 2003; Le Pichon and Kreemer 2010).

New observations, analysis strategies, and modeling capabilities provide opportunities to advance understanding of global geodynamics (i.e., the study of the forces acting on the Earth and the deformational response to those forces). Great strides have been made in seismic instrumentation and analysis strategies allowing mapping of seismic structures (tomography) and the alignment of seismic velocities (anisotropy) deep in the Earth (e.g., Rawlinson et al. 2010; Long and Becker 2010, and references therein). Likewise, major advances have been made in remote sensing and image processing (e.g., Forte et al. 2012), as well as global satellite mapping of land and sea topography (e.g., Farr et al. 2007; Becker et al. 2009), gravity and gravity changes (GRACE and GOCE; e.g., Tapley et al. 2004; Pail et al. 2010), surface deformation (InSAR; e.g., Ryder and Burgmann 2008; Walters et al. 2011; Hetland et al. 2012; GPS; see references in this chapter), areal mapping of landforms to cm resolution and precision (LIDAR; http://lidar.cr.usgs.gov/), rock dating techniques allowing precise estimates of rates of structural offsets (e.g., Yeats et al. 1996; Kozaci et al. 2009, and references therein), and modeling/computational strategies that allow investigation of a wide range of hypotheses (http://www.geodynamics.org/cig/community/documents/reference).

A major focus of our research has been to use the Global Positioning System (GPS) to map directly the relative motions and deformations of the African, Arabian, and Eurasian plate system with the expectation that this completely new, quantitative information would provide an improved basis to constrain geodynamic processes. This plate system provides unique opportunities to address these problems because it encompasses the complete range of tectonic processes including subduction (Hellenic and Cyprus arcs), ocean rifting (Gulf of Aden, central Red Sea), continental rifting (e.g., Gulf of Aqaba, S Red Sea/Afar, East African rift), a young continental collision (Arabia–Eurasia), and major continental transform faults (e.g., Dead Sea and North Anatolian faults), among others (Fig. 1). In addition to their importance for geodynamic studies, these investigations provide a quantitative basis for determining fault behavior and the earthquake deformation cycle (Thatcher 1993; Wang et al. 2012), an important aspect of seismic and tsunami hazard evaluation. It is within this broader context that we examine the geodynamic evolution of the Red Sea.

Since the mid-1990s, GPS has been used to observe directly present-day crustal motions associated with the opening of the Red Sea (McClusky et al. 2000; ArRajehi et al. 2010; McClusky et al. 2010, and references therein). In this chapter, we present the most recent results of these ongoing studies focusing on new observations that more precisely constrain present-day deformation along the full extent of the Red Sea. We investigate the relationship of present Nubia–Arabia separation with plate tectonic and structural studies of rifting since the Late Oligocene (~20–30 Ma). We use these relationships to suggest a geodynamic scenario for the initiation and evolution of the Red Sea.

Tectonic Setting of the Red Sea and Gulf of Aden

A detailed description of the tectonic evolution of the Red Sea is provided by Bosworth in this volume. Briefly, prior to the Oligocene, the Nubian, Somalian, and Arabian plates formed a single plate (Afro-Arabian Plate) that was subducting northward along the southern margin of Eurasia (Dercourt et al. 1986). During the Late Oligocene (~35–20 Ma), the Arabian plate separated from Africa along the Red Sea and Gulf of Aden rifts (Cochran 1981; Chu and Gordon 1998; Redfield et al. 2003; Bosworth et al. 2005; DeMets et al. 2010; d'Acremont et al. 2010). The East African rift system that separates the African Plate into the Nubian and Somalian plates, together with the Red Sea and Gulf of Aden, form the Afar rift–rift–rift Triple Junction (Fig. 1) (e.g., McKenzie et al. 1970; Hempton 1987; Joffe and Garfunkel 1987; Le Pichon and Gaulier 1988; Ebinger and Casey 2001; McQuarrie et al. 2003; Garfunkel and Beyth 2006). Formation of the Triple Junction followed the massive eruption of flood basalts. This massive volcanism is believed to be due to the African Plume that is also thought to be responsible for the high elevation of

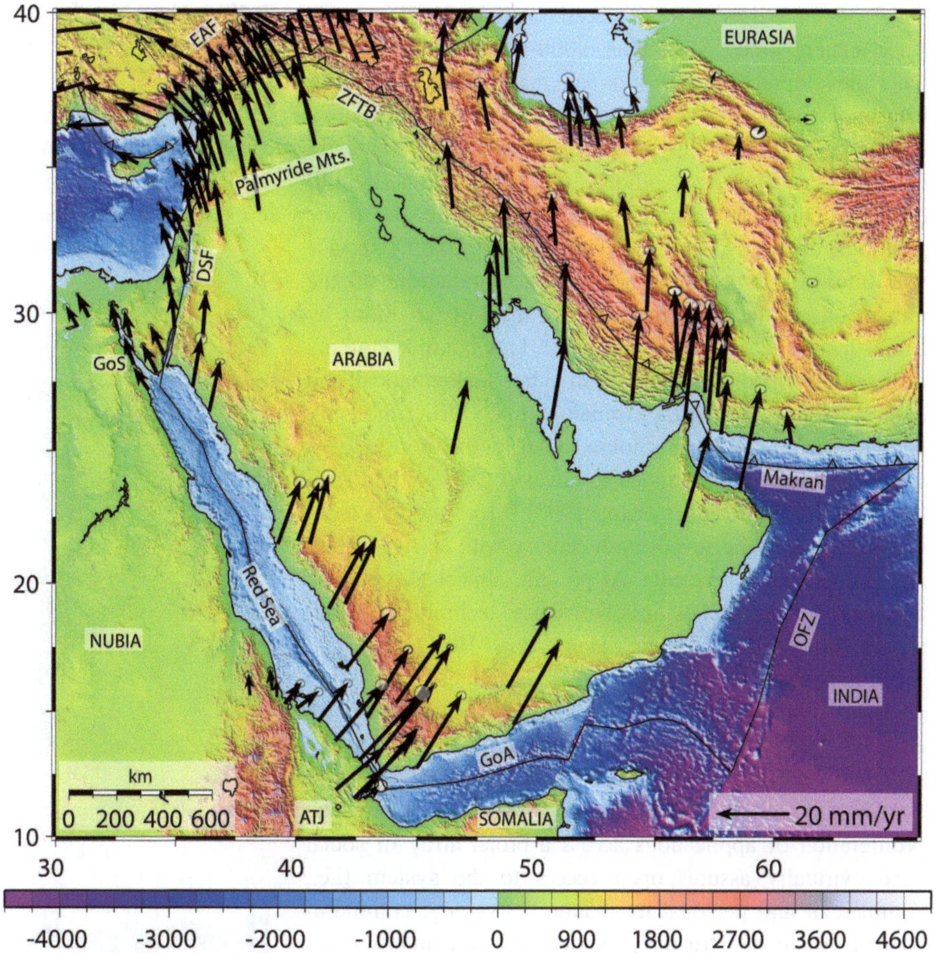

Fig. 1 GPS velocities and 1-sigma uncertainties with respect to Eurasia in the Arabia–Africa–Eurasia zone of plate interaction. Not all velocities are shown for clarity

the Ethiopian Plateau. The role of the African Plume in driving or initiating rifting remains an area of active investigation (e.g., Bialas et al. 2010; Faccenna et al. 2013).

Subduction of Nubian ocean lithosphere (Neotethys) continues at present in the eastern Mediterranean along the Hellenic and Cyprus arcs, and on the eastern side of the Arabian Plate, along the Makran trench (Fig. 1). Collision of the Arabian continent with Eurasia may have begun as early as 34 Ma in the Zagros, but ocean closure is estimated to have occurred by 20 Ma, prior to the separation of Arabia from Africa (McQuarrie et al. 2003, McQuarrie and van Hinsbergen 2013; Mouthereau et al. 2012). This collision is estimated to have resulted in 200–500 km of convergence distributed across a broad continental collision zone in eastern Turkey, the Zagros/Iran, and the Caucasus region (e.g., Jolivet and Faccenna 2000; McQuarrie and van Hinsbergen 2013). It is a major driver of the active tectonics of the eastern Mediterranean region (e.g., Sengor et al. 1985; Allen et al. 2004; Barazangi et al. 2006) and the devastating earthquakes that have affected this area throughout recorded history (e.g., Ambraseys and Jackson 1998). Furthermore, the kinematics of the separation of Arabia from Africa, and the continuing continental collision with Eurasia, offer opportunities to evaluate the role of different forces in driving/resisting Arabia plate motion (Jolivet and Faccenna 2000; McQuarrie et al. 2003; Bellahsen et al. 2003; Reilinger and McClusky 2011). Precise constraints on Red Sea rifting and Arabia Plate motion are therefore important for evaluating earthquake, tsunami, and volcanic hazards (e.g., Pallister et al. 2010) along Arabia's plate boundaries (Red Sea/Gulf of Aden rifts, Dead Sea fault, East Anatolian fault, Zagros fold/thrust belt, Makran subduction, and Owen Fracture zone, as well as throughout the Arabia–Eurasia collision zone) and for constraining the dynamics of plate motions and deformations.

Geodesy and the Global Positioning System

Geodesy is the study of the Earth's shape and gravity field and changes to them. Geodetic and survey techniques are essential for navigation, mapping, and infrastructural development and as such have been continuously developed since ancient times. More than 4,000 years ago, the Egyptians developed predecessors of many later surveying instruments and may have developed more extended positioning capabilities to align major sites in the known world (Tompkins 1997). Greek scholars (e.g., Pythagoras, Euclid) developed the mathematical basis to determine and relate positions, distances, angles, and areas on the Earth's surface that allowed surprisingly accurate determination of the dimensions of the Earth from astronomical observations by Eratosthene circa 240 BC (Aujac 2001). Claudius Ptolemy further developed geographic and other mathematical and natural sciences during the Roman period. During the early Islamic period, Islamic scholars held to the notion of a spherical Earth and continued to develop geodesy and related fields of mathematics (King 1993). More modern techniques began in the seventeenth century with spirit leveling and distance measurements. At present, almost all geodetic and survey approaches involve space techniques. Chief among these is the GPS (e.g., Hager et al. 1991).

GPS consists of a system of 32 satellites 20,000 km above the Earth's surface that complete 2 orbits of the Earth each 24 h (http://tycho.usno.navy.mil/gpscurr.html). The satellites are operated by the US Department of Defense in cooperation with the Interagency GPS Executive Board. The proliferation of applications across a broad array of social needs virtually assures open access to the system [i.e., commercial and personal navigation, surveying, communications, weather, structure monitoring, and many other engineering and scientific applications; National Research Council (1995)]. Other Global Navigation Satellite Systems (GNSS) have been developed by Russia (GLONASS), a European consortium (Galileo), Japan (QZSS), and China (Beidou), but these systems are not yet contributing significantly to measuring solid Earth processes.

The basic principles involved in GPS precise positioning are provided in most modern geodesy texts, including King et al. (1985) and Misra and Enge (2011). The satellites transmit two microwave frequencies that allow observers on the ground to determine their position with an accuracy of 1–2 mm in horizontal coordinates and 3–10 mm in height by recording data over a 24-h period. The keys to obtaining these precisions from regional measurements are highly accurate timing provided by atomic clocks on the GPS satellites, precise orbital positions for the satellites provided by the International GNSS Service (IGS) (determined from a global network of observing stations) and processing software that uses elaborate mathematical models to account for the Earth's rotation, solid Earth and ocean tides, and the ionospheric and atmospheric delays of the GPS signal, among other factors that influence position estimates.

Estimating Surface Motions from GPS Observations

GPS measurements can be done in survey mode (sGPS) where the GPS antenna is precisely positioned with respect to a survey marker and observed temporarily (i.e., one to a few days or more; Fig. 2a). By repeating such measurements

Fig. 2 **a** GPS survey system using the Tech2000™ mast setup. **b** GPS continuous station located in the KACST Observatory, Ain Namas, KSA

periodically over the course of a few years (annually or less frequently depending on motion rates), we are able to estimate how the position has changed during the observation period. Alternately, GPS continuously recording stations (cGPS) allow estimation of position on a daily basis, or more frequently (Fig. 2b). Even in this case, reliable estimates of long-term (i.e., secular) velocities require a minimum of 2.5 years of observations because annual and semiannual systematic errors can bias estimates of steady-state motion (Blewitt and Lavellee 2002). We use both types of GPS observations for the studies presented here.

While the precision of our site velocities varies with observation period, the GPS horizontal velocities we determine using the GAMIT-GLOBK processing software (Herring et al. 2013) have 1-sigma uncertainties in the range of 0.4–1.0 mm/year. Because extension rates across the Red Sea vary from 7 to 24 mm/year from north to south, these precisions allow us to investigate details of rifting processes. We do not consider vertical motions in this study because they are generally smaller than horizontal motions in rift settings, and the GPS constraints are less precise.

Velocity estimates are determined in a global reference frame, that is, with respect to other stations on the Earth's surface. The reference frame is determined and maintained (updated) by the International Terrestrial Reference Frame (ITRF) Service (Altamimi et al. 2007) using well-positioned stations (i.e., long history of well-behaved observations) located around the globe and accounting for motions of the Earth's tectonic plates. We determine site velocities within the ITRF (2008), but we present them in different reference frames to better illustrate the nature of plate motions and strains (i.e., deformation). In particular, we use reference frames fixed to the Eurasian, Arabian, Nubian, and Somalian plates to investigate motions within these plates and across their common boundaries.

Motion of the Arabian Plate

Figure 1 shows GPS velocities throughout a broad area encompassing the Arabian Plate and bordering regions of adjacent plates including Nubia, Somalia, and the complexly deforming continental collision zone with Eurasia north of the Bitlis–Zagros suture zone. The velocities are shown with respect to the highly stable Eurasian Plate (determined from 12 stations on EU with a WRMS residual velocity of 0.6 mm/year). We present the velocities in a Eurasian reference frame because of the large size and internal stability of Eurasia. In addition, Eurasia has a unique position within the Plate Tectonic framework of the hemisphere serving as the "backdrop" (i.e., overriding plate) for subduction of the Nubian, Arabian, Indian, and Australian plates along its southern boundary, as well as plate systems in the western Pacific (Reilinger and McClusky 2011; their Fig. 1). As such, it provides the best perspective (i.e., reference frame) from which to investigate active geodynamic processes. Subduction along the Makran in SE Iran, right lateral strike slip on N–S striking structures in eastern Iran, and broadly distributed strain in the Zagros Mountains (e.g., Vernant et al. 2004) are well illustrated and quantified in this representation. In regard to dynamic processes driving plate motions and deformations, a revealing feature of this perspective is the rapid motions of the Arabian Plate and the neighboring Anatolian and Aegean regions, rapid motions that occur within the confines of the slowly moving, large Nubian and Eurasian plates. This observation alone requires large and apparently dominant contributions from sub-lithospheric processes in driving observed plate motions and interactions (e.g., sinking of the lithosphere along subduction zones, delamination of the overriding lithosphere, mantle upwelling).

Deformation Along Arabia's Plate Boundaries

To examine more closely deformation along the Red Sea and other boundaries of the Arabian Plate, in Fig. 3 we plot the velocity field in a reference frame fixed to the Arabian Plate. We determine this reference frame by minimizing in a least squares sense the GPS velocities indicated by the purple error ellipses in the figure. Predominantly left-lateral, strike-slip motion at ~4–5 mm/year characterizes the full extent of the Dead Sea fault (DSF) from the Gulf of Aqaba to SE Turkey (e.g., Gomez et al. 2007; Le Beon et al. 2008; Al Taraza et al. 2011; Sadeh et al. 2012). The DSF joins the East Anatolian fault (EAF) and offshore Cyprus arc near its northern end forming the complex Karaman–Maras Triple Junction. The EAF is also predominantly a left-lateral, strike-slip fault with little if any fault-normal compression (Reilinger et al. 2006). The absence of fault-normal compression is interesting given the clear morphological evidence (i.e., the topography) for indentation of the Arabian Plate into southern Eurasia (Fig. 1). The absence of fault-normal compression at present appears to require a change from initial continental collision and associated indentation and uplift to lateral escape of the Anatolian Plate.

Continuing east, the northern Zagros are characterized by right-lateral, strike-slip faulting with increasing compression to the SE. Southeast of the Strait of Hormuz, the last remnants of the Neotethys Ocean lithosphere are subducting along the Makran trench at rates that increase to the east reaching >20 mm/year. The eastern part of the Makran trench experienced an M_w = 8.1 earthquake in 1945 attesting to its capability to generate great earthquakes (Byrne et al. 1992).

The Owen Fracture zone separates the SE part of Arabia from the adjacent Indian Plate. At this longitude, the Arabian Plate is converging faster with Eurasia than the Indian Plate,

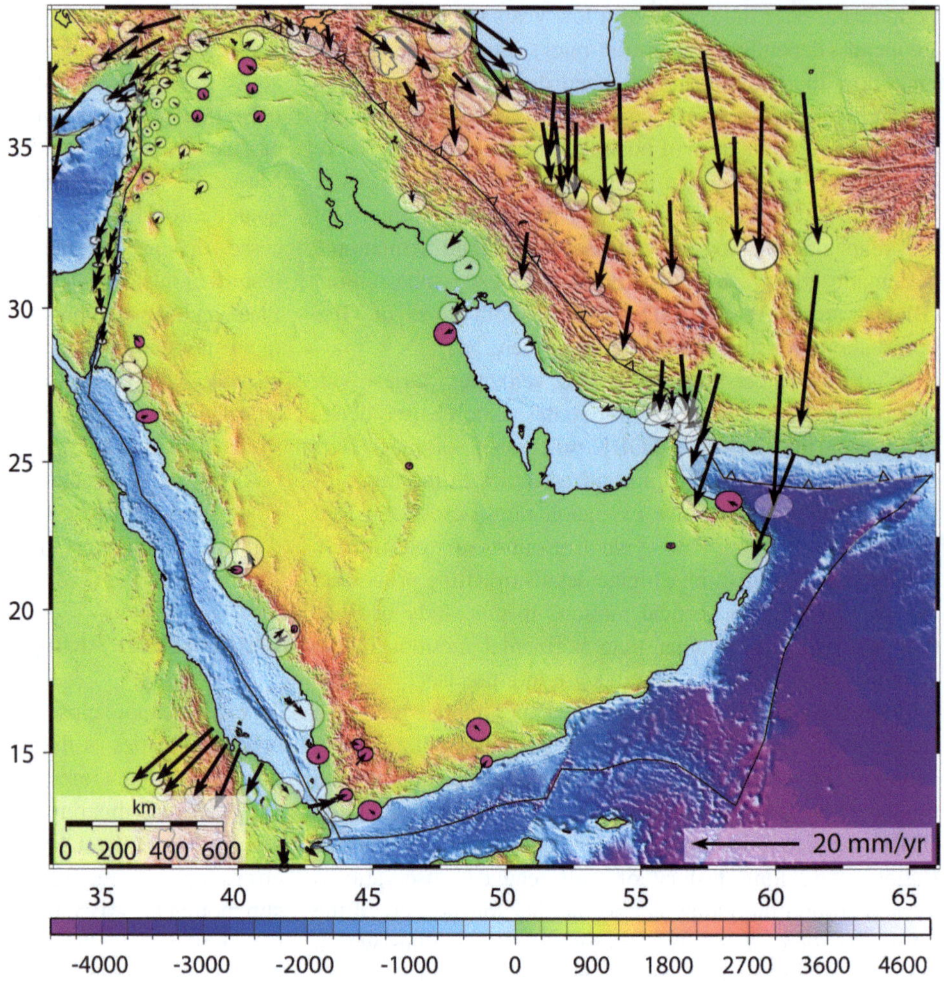

Fig. 3 GPS velocities and 95 % confidence ellipses shown in an Arabia Plate reference frame. The reference frame is determined by minimizing the velocities for sites with purple error ellipses. Base map as in Fig. 1

resulting in right lateral strike slip and compression on the Fracture zone (Reilinger et al. 2006; Fournier et al. 2008).

Ocean rifting along the Gulf of Aden separates the Arabian Plate from Somalia. Figure 4a shows GPS-derived motions in a Somalia-fixed reference frame (3 stations and WRMS 0.9 mm/year). Extension rates increase from ~15 to 25 mm/year from the western end of the Gulf to the intersection with the Owen Fracture zone. The direction of extension closely parallels the orientation of fracture zones in the Gulf, suggesting that it has not changed substantially (<20°) since ocean spreading began along the full extent of the Gulf at ~11 ± 2 Ma.

Separation of Arabia from Nubia and Opening of the Red Sea Basin

Figure 4b shows GPS velocities with respect to Nubia. The Nubian reference frame is determined from 12 GPS sites widely spread on the Nubian Plate. The WRMS residual velocity from this reference frame for stations on Nubia is ~0.6 mm/year (McClusky et al. 2003; Saria et al. 2013). Considerable spatial geodetic coverage is available along the Red Sea in Saudi Arabia, the Sinai, and Gulf of Aqaba, and across the southern Red Sea in Eritrea, Djibouti, and Yemen. The rate of extension varies from 7 mm/year across the northernmost Red Sea (27°N) to about 24 mm/year in the south near the Bab-al-Mandab (13°N). Extension is roughly perpendicular to the strike of the Red Sea at the latitude of the Bab-al-Mandab and becomes increasingly more oblique (left lateral) toward the north. At the southern end of the Gulf of Aqaba, motion of Arabia with respect to the adjacent Sinai is 4.5 mm/year left lateral and 2.2 mm/year extension; the fault becomes almost pure left lateral strike slip at the northern end of the Gulf (southernmost end of the Dead Sea fault).

Relationship Between Present-Day GPS and Geologic Plate Tectonic Motions

The geodetic observations we present constrain plate motions and deformations over time scales of 3–20 years. How present-day geodetic motions are related to long-term plate tectonic motions and cumulative geologic offset rates on plate

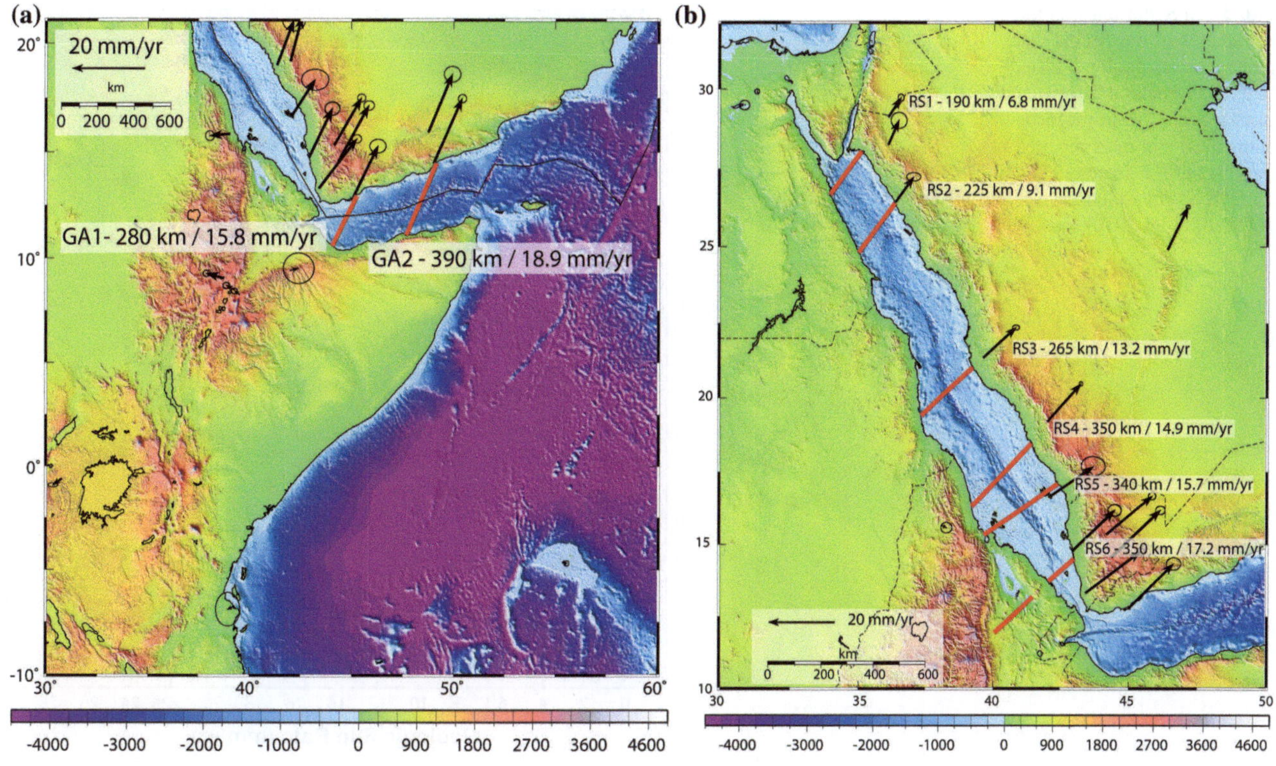

Fig. 4 **a** GPS velocities and 95 % confidence ellipses with respect to Somalia along the southern coast of the Arabian Plate on a bathymetric and topographic map of the southern Red Sea and Gulf of Aden. *Bold lines* across the Gulf of Aden rift are estimates of basin width at the location and direction of the GPS velocities on the Arabian coast. Basin widths ($W(t)$) and observed Arabia Plate velocities (V(gps)) used to estimate the age and timing of extension in Fig. 7 are annotated. **b** As Fig. 4a with GPS velocities with respect to Nubia. Base maps as in Fig. 1. From ArRajehi et al. (2010)

boundaries are not necessarily obvious. Fortunately, plate tectonic estimates of relative motions of the Afro-Arabian Plate (i.e., prior to breakup in the Late Oligocene) with respect to Eurasia and motion of Arabia with respect to Africa have been reported from plate reconstructions (McQuarrie et al. 2003; DeMets et al. 2010). Dated offset features along the DSF, Owen Fracture zone, and Main Recent Fault provide additional constraints on geologic deformation rates. Offsets along the Red Sea and Gulf of Aden, as well as changes in the geometry of extension in both the northern and southern Red Sea, provide further opportunities to investigate the relationship between geodetic and geologic plate motions.

Figure 5 shows a plot of reported slip rates on Arabia's plate boundaries versus rates deduced from a block model constrained by GPS only (modified from Reilinger et al. 2006). Uncertainties on geologic and plate tectonic rates include both uncertainties in individual estimates and variations between estimates reported for the same structures from separate studies. As indicated in the figure, rates agree at the 10–15 % level and are within measurement uncertainties.

Figure 6 shows a plot of the motion of a point on the Afro-Arabian Plate with respect to Eurasia as a function of time since 56 Ma as reported by McQuarrie et al. (2003) from plate reconstructions. From 56 to 25 Ma, the single plate converged with Eurasia at ~32 mm/year. At roughly 25 Ma, the plate began to split along the proto-Red Sea as a result of slowing of the African Plate's northward motion with respect to Eurasia (to ~14 mm/year). Arabia's rate of convergence with Eurasia also slowed at this time from 32 to 21 mm/year, but remained more rapid than Africa resulting in opening of the Red Sea. Although poorly constrained due to the wide temporal separation in the magnetic anomalies used to constrain plate motions between 20 and 5 Ma, McQuarrie et al. (2003) identified a second episode of slowing of Nubia–Eurasia convergence (to ~7 mm/year) and an associated increase in the rate of extension across the Red Sea at roughly 11 Ma. This overall scenario is generally consistent with that reported by Le Pichon and Gaulier (1988) from earlier plate reconstructions.

Superimposed on the plate tectonic motions shown in Fig. 6 are present-day geodetic rates extrapolated to geologic time. Comparison of these rates indicates that the rate of convergence of Arabia with Eurasia has not changed significantly since the initial breakup of the Afro-Arabian Plate at ~25 Ma. Similarly, Nubia's rate of convergence with Eurasia has remained approximately constant since ~11 Ma. Based on the agreement between the GPS plate motions and geologic estimates from fault offsets over a few

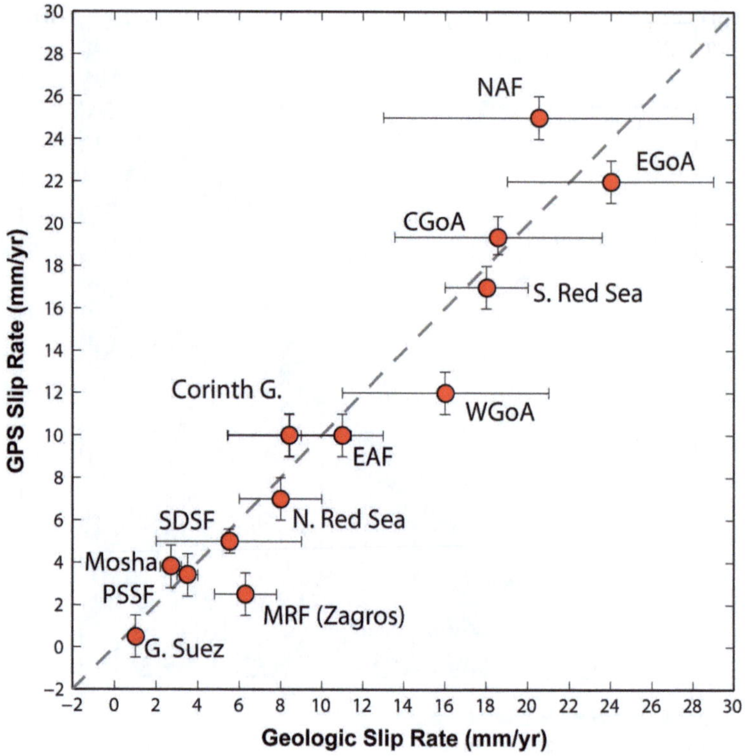

Fig. 5 GPS slip rates on the boundaries of the Arabia Plate plotted versus geologic (dated fault and structural offsets) and plate tectonic (from magnetic anomalies) slip rates reported in the literature. Agreement is <10–15 % level (modified from ArRajehi et al. 2010)

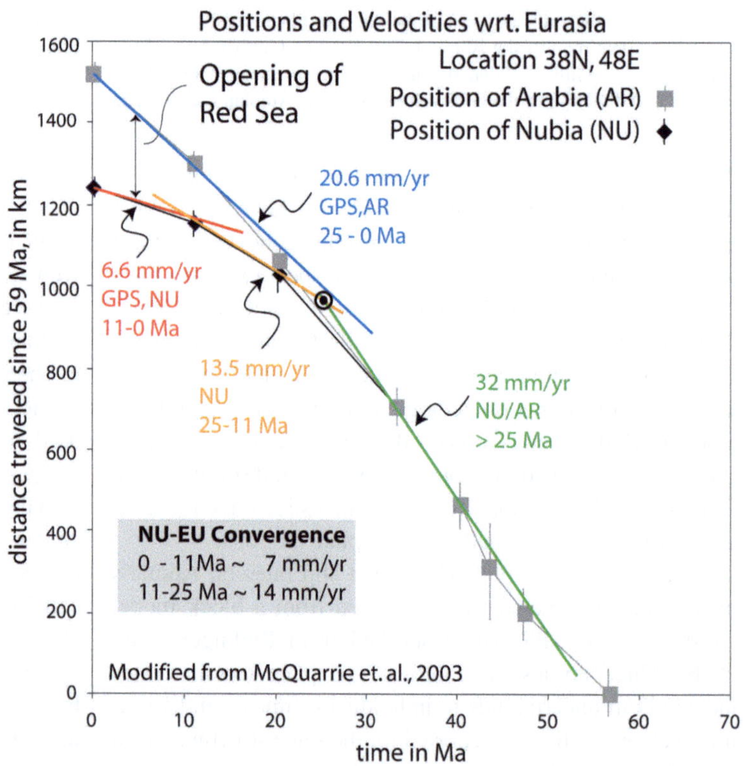

Fig. 6 Motion of a point on the Arabia–Africa Plate with respect to Eurasia as a function of time since 56 Ma compared to the geodetic rates extrapolated to this time (modified from McQuarrie et al. 2003; ArRajehi et al. 2010). Geodetic and plate tectonic rates are equal within uncertainties for Arabia (AR) to ≥21 Ma and Nubia (NU) to ≥11 Ma. The rate of Nubia-Eurasia convergence for 25–11 Ma and 11–0 Ma are indicated in the *inset*

thousand to a few hundred thousand year timescales, as well as with plate tectonic motions averaged over a few to 10 s of millions of years (Fig. 6) (a remarkable temporal extrapolation by a factor of $\sim 10^6$!), we use the precise GPS rates and structural offsets to investigate the geodynamic evolution of the Red Sea basin.

Age of Red Sea and Gulf of Aden Rifts

In Fig. 4a, b, we estimate the width of the Gulf of Aden and Red Sea basins at the locations of our best-determined GPS velocities on the western and southern coasts of the Arabian Plate. We use the shoreline as a proxy for the width to eliminate bias and assure uniformity of the estimated width along strike. Figure 7 shows a plot of the width of the Red Sea and Gulf of Aden basins versus the observed GPS velocity with respect to Nubia (Red Sea) and Somalia (Gulf of Aden). The linear relationship between velocity and basin width is well established. We estimate that the increase in Nubia motion with respect to Arabia initiated at ~11 Ma. From Fig. 6, we estimate a pre-11 Ma rate of opening of the Red Sea (20.6–13.5 mm/year) that was half the present-day GPS opening rate (20.6–6.6 mm/year). Assuming that the change in Nubia–Arabia motion was a change in rate only as suggested by McQuarrie et al. (2003), total extension of the Red Sea will then be

$$W_{(t)} = W_{(\text{pre11})} + W_{(\text{post11})} \quad (1)$$

$$W_{(t)} = [0.5V_{(\text{gps})} \cdot (T - 11)] + [V_{(\text{gps})} \cdot 11] \quad (2)$$

where $W_{(\text{pre11})}$ and $W_{(\text{post11})}$ are the amount of opening of the Red Sea pre- and post-11 Ma, $W_{(t)}$ is the total measured width of the Red Sea and Gulf, $V_{(\text{gps})}$ is the present-day opening velocity (rate) of Arabia with respect to Nubia or Somalia at the location of $W_{(t)}$, and T is the time of initial rifting. Solving for T,

$$T = \left(\frac{W_{(t)}}{V_{(\text{gps})}} - 5.5\right) \Big/ 0.5 \quad (3)$$

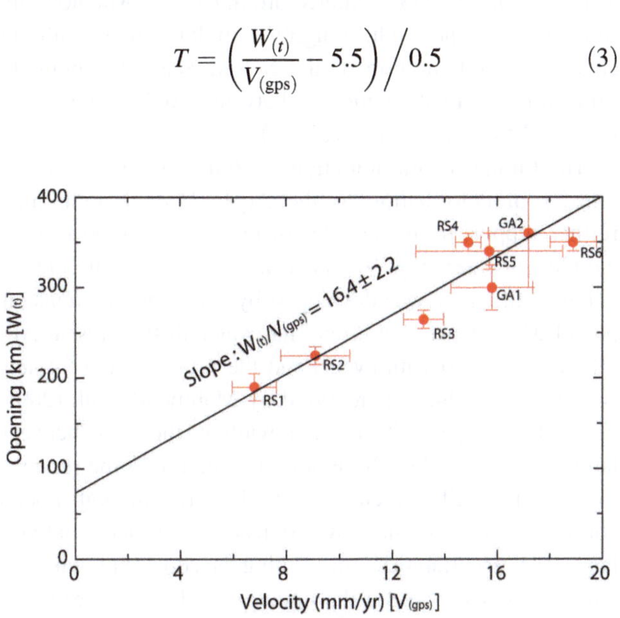

Fig. 7 Plot of the GPS Arabia Plate motion rate with respect to Nubia (V(gps)) versus the width of the Red Sea and Gulf of Aden rifts (W(t)) at each GPS station location (Fig. 4a, b). The well-defined linear relationship between rift width and rate of extension is consistent with rifting initiating roughly simultaneously (±3 M year) along the Red Sea and the West Gulf of Aden at 22 ± 3 Ma (see text for discussion)

which is independent of any uniform bias introduced by using the shoreline to estimate the width of the basins. From the slope of $W_{(t)}$ versus $V_{(\text{gps})}$ in Fig. 7 (~16.4 ± 2.2 M year), we estimate a time of initial opening of the Red Sea of ~22 ± 3 Ma. This age is on the young side, but is consistent within uncertainties with many geologic estimates for the initiation of the main phase of rifting (e.g., 20–30 Ma; Omar and Steckler 1995; McQuarrie et al. 2003; Redfield et al. 2003; Garfunkel and Beyth 2006). In addition, the uncertainty on the slope of $W_{(t)}$ versus $V_{(\text{gps})}$ suggests that rifting initiated over a relatively short time (±3 M year, or ±~10–15 % of the total time of rifting) along the full length of the Red Sea rift basin, a result that is supported by fission track analyses dating the early rift flank uplifts (Omar and Steckler 1995).

Changes in the Style of Rifting in the Southern and Northern Red Sea

The change in Nubia–Arabia relative motion at 11 ± 2 Ma also corresponds to a change in the character of rifting in both the southern and northern Red Sea. In the southern Red Sea, rifting bifurcates at about 17°N latitude with one branch of extension continuing along the main rift axis toward the Bab-al-Mandab and the other extending southward into the Danakil Depression/Afar volcanic region (Fig. 1). Extension rates decrease approximately linearly toward the SE along the initial rift axis and increase correspondingly southward along the Danakil/Afar volcanic depression resulting in counterclockwise rotation of the intervening Danakil Block. Together the two branches of the rift accommodate total Arabia–Nubia extension.

We use the GPS velocities along the Danakil Block to quantify deformation in the S Red Sea in terms of a rotation pole and rate of the Danakil Block with respect to Nubia (i.e., Danakil Block–Nubia Euler Vector). Figure 8 shows a reconstruction of the southern Red Sea that involves an initial 10° clockwise back-rotation of the Danakil Block about the geodetic Euler pole (black star) that results in contact between unextended terrains in the northernmost Danakil Block and the adjacent Nubia crust at about 5 Ma (i.e., 10° at 1.9°/M year). We hypothesize that prior to 5 Ma, the Danakil Block–Nubia rotation pole was located near the northern end of the block, south of its present location (red star in Fig. 8). The northward translation of the Euler pole at 5 Ma resulted from the northernmost part of the Danakil Block separating from Nubia at that time. We estimate the pre-5 Ma rotation rate by setting the Danakil Block–Nubia rate to zero at the northern end of the Block (i.e., at location of the contact of unextended terrains), while the rate near the southern end continued at the Arabian Plate rate. This back-rotation results in closure of the Danakil–Afar Depression without substantial overlap of unextended terrains at

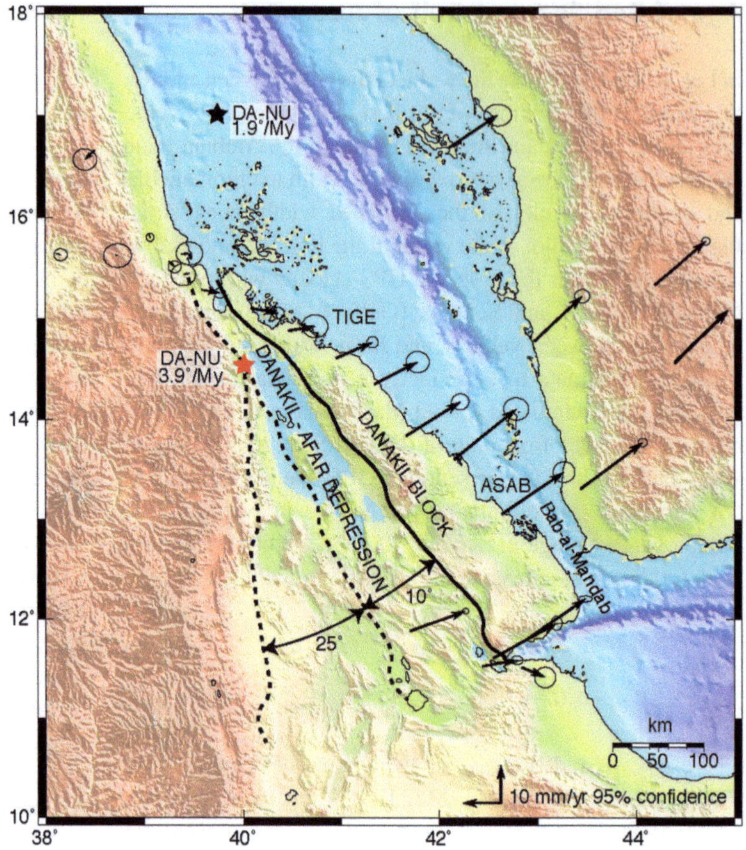

Fig. 8 Back-rotation of the western side of the Danakil Block (represented by the *solid black line*) around the GPS Danakil–Nubia (DA–NU) rotation pole (*black star* at ~17°N) to initial overlap of unextended terrains (at ~15°N) after 10° rotation (~5 Ma; easternmost *dashed line*). A second back-rotation around a proposed pole near the location of initial overlap (*red star*) of an additional 25° (westernmost *dashed line*) closes the Danakil Depression. Full closure is achieved at 11 ± 2 Ma (see text for discussion)

11 ± 2 Ma (i.e., 5 Ma + 25° at 3.9°/M year and estimating a 15 % uncertainty based on rotation rate uncertainties and variations in total closure along strike). Within the resolution of our observations, the change in the configuration of extension in the southernmost Red Sea occurred simultaneously with the 11 ± 2 Ma increase in the overall rate of Arabia–Nubia extension (Fig. 6).

We use a similar approach to estimate the timing of initiation of the southern DSF, the Gulf of Aqaba, and the Gulf of Suez rift. Initial Red Sea opening extended northwest along the Gulf of Suez before transferring to its present location along the Gulf of Aqaba and DSF (e.g., Le Pichon and Gaulier 1988). Total post-Late Miocene offset on the southern DSF is estimated to be 45 ± 5 km (e.g., Hempton 1987). The geodetic slip rate along the southern DSF is 4.4 ± 0.3 mm/year (e.g., Reilinger et al. 2006). This indicates that the fault initiated at 10 ± 2 Ma, assuming a constant average slip rate based on the steady motion of Arabia with respect to Nubia during the past 11 Ma (Fig. 6), and the consistency of geodetic and geomorphic slip rates for the southern DSF (Fig. 5; Klinger et al. 2000; Niemi et al. 2001). In Fig. 9, we estimate similar timing for the initiation of extension and strike-slip motion on the Gulf of Aqaba from the width of the Gulf (11 ± 3 Ma) and the structural offset of the Sinai Peninsula from the adjacent section of the Arabian Plate (11 ± 2 Ma). We conclude that the most recent phase of activity along the DSF initiated around 11 ± 2 Ma along the full strike of the fault, roughly simultaneously with the change in configuration of the S Red Sea, and about the same time as relative motion between Arabia and Nubia increased by a factor of 2 (Fig. 6).

The timing for the initiation of Red Sea rifting, and the transfer of deformation to the Aqaba-DSF system, finds further support from estimates of the time required to open the Gulf of Suez (Fig. 9). As with the Red Sea, the present width of the Suez was developed by 2 phases of extension, pre-11 Ma extension at a rate that was half the present AR-NU rate (2.5 ± 0.4 mm/year) and the present rate of extension across the Suez as reported by Mahmoud et al. (2005) of 1 ± 0.5 mm/year. The average width of the Suez (between latitude 27.8 and 29.7°N) is 48 km. The total time required to open the Gulf is then 26 ± 7 M year. Although poorly constrained due to the low extension rates and relatively large rate uncertainties, this timing is consistent with our independent estimate for the initiation of Red Sea extension at 22 ± 3 Ma.

To summarize, geodetic and plate tectonic constraints on the rates of relative plate motion and the structure of the Nubia–Arabia plate boundary (Red Sea) are consistent with the initiation of rifting at 22 ± 3 Ma and a change to a more

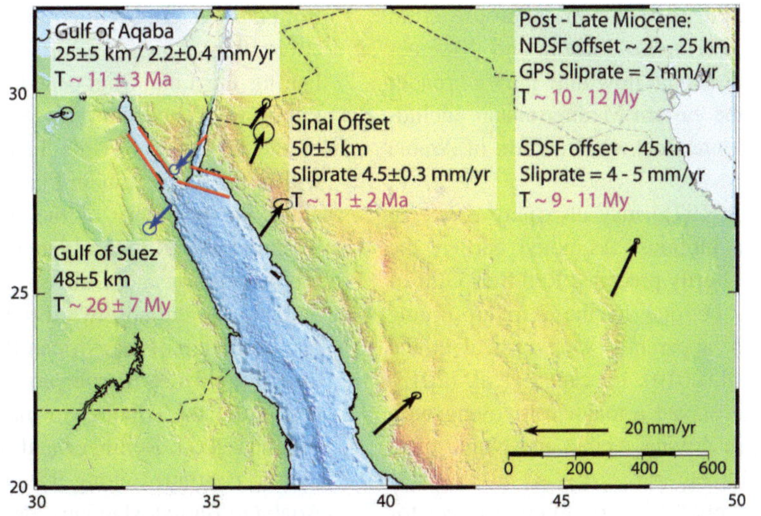

Fig. 9 Estimates of the age of the Dead Sea fault system from fault slip rates deduced from geodetic measurements [from the block model of Reilinger et al. (2006)] and structural offsets. We also show that 26 ± 7 Ma is the estimated time for the initiation of extension across the Gulf of Suez assuming that the pre-11 Ma Arabia–Nubia rate was focused across the Gulf from the time of initiation to 11 Ma (i.e., half the present Arabia–Nubia geodetic rate) and using the present extension rate reported by Mahmoud et al. (2003) for 11 Ma to present (see text for discussion). Velocity vectors show motions with respect to Nubia and their 95 % confidence ellipses (modified from Reilinger and McClusky 2011)

N–S orientation in both the northern (i.e., from the Suez to the Aqaba/DSF) and southern Red Sea (from the main rift axis to the Danakil/Afar Depression) at 11 ± 2 Ma. The apparent simultaneity and similar spatial orientation of these changes strongly suggest they result from changes in Arabia–Nubia plate motions.

Geodynamic Evolution of the Africa–Arabia–Eurasia Plate System

The relationships between geodetic and plate tectonic motions support a kinematic scenario for the opening of the Red Sea basin involving a first stage of continental extension that initiated in the Late Oligocene–Early Miocene along the full extent of the present-day Red Sea. Opening followed the impact of the African Plume at the proto-Afar Triple Junction then located near the Bab-al-Mandab. Initial extension was concentrated along the central axis of the Red Sea extending along the Gulf of Suez to the Bab-al-Mandab. Extension across the Gulf of Aden started at approximately this same time. The initiation of extension across the East African rift that forms the third rift of the Afar rift–rift–rift Triple Junction is uncertain. McKenzie et al. (1970) and Tesfaye et al. (2003) suggest that rifting started roughly simultaneously with the Red Sea and Gulf of Aden. Wolfenden et al. (2004) suggest a later initiation of ~ 11 Ma that we note here coincides with the second phase of accelerated Red Sea rifting. If extension began at the same time across the East African rift as the Red Sea, it was very slow compared with the Red Sea even during this slow period of Nubia–Arabia extension, as is clear from the relatively narrow width of the East African rift in N Ethiopia. Overall, this scenario is consistent with reconstructions based primarily on geologic observations (e.g., Bosworth et al. 2005; Garfunkel and Beyth 2006).

Early studies of global plate dynamics identified forces associated with subduction and mantle upwelling along mid-ocean ridges as likely the most important contributors to the balance of forces acting on the plates, with mantle flow contributing to or retarding plate motion (Forsyth and Uyeda 1975; Parsons and Richter 1980). Ridge "push" that drives the plates apart is due to the thermally driven higher elevation of the ridge system and the injection of magma due to the upwelling of basalt forming new ocean crust. Subduction of ocean lithosphere can drive plate motions through "pulling" the trailing plate as the subducted plate sinks into the mantle under its own weight (because it cools and becomes more dense with age). In addition, the subducting ocean lithosphere may "suck" the overriding plate toward the retreating trench [i.e., by inducing flow in the mantle wedge (Turcotte and Shubert 2002)] opening a back-arc basin if the subducting plate is converging more slowly than it sinks (e.g., Royden 1993; Conrad and Lithgow-Bertelloni 2002), a process thought to be responsible for the opening of the Aegean Sea and other Mediterranean basins (Jolivet and Faccenna 2000; Reilinger and McClusky 2011).

Toroidal flow (i.e., flow around an axis vertical to the Earth's surface) around the edges of the subducting Nubian lithosphere along the Hellenic trench may also contribute to crustal deformation at the edges of the Arc (Perouse et al. 2012; Tiryakioglu et al. 2013). Le Pichon and Kreemer (2010) suggest that toroidal mantle flow may play a broader role in the kinematics of the eastern Mediterranean, including driving present-day counterclockwise rotation of Arabia and Anatolia (Fig. 1).

Mantle plumes (Morgan 1971) may also play a role in driving or retarding plate motions. As noted earlier, the massive flood basalts that shortly preceded Red Sea–Gulf of Aden rifting, thought to be due to the African Plume (Ebinger and Sleep 1998), apparently were critical in the rifting process (Bialas et al. 2010). Faccenna et al. (2013) suggest the African plume played a major role, along with Hellenic trench rollback, in driving Arabia and Nubia plate motion and interaction with Eurasia.

We investigate qualitatively the role of different forces for the dynamics of the AR-AF-EU plate system based on the simple notion that changes in plate motions (i.e., accelerations) require changes in the force balance on the plates. While the African Plume was important for initiating rifting of Arabia from Africa (e.g., Bosworth et al. 2005; Garfunkel and Beyth 2006), whether rifting was driven by the plume (e.g., Faccenna et al. 2013) or the plume only weakened the Afro-Arabian Plate allowing other forces to drive plate separation remains uncertain. The observation that NU-EU convergence slowed while AR-EU convergence changed much less seems contrary to the notion of the Plume driving separation since this mechanism would tend to drive the plates apart more or less symmetrically. That is, while slowing of AF-EU convergence could conceivably result from a southward push from the Plume, this mechanism would be expected to accelerate the northward motion of Arabia. In fact, Arabia's northward motion with respect to Eurasia also slowed at the time of rifting (Fig. 6) although less than Nubia. The reduction in the rate of Arabia–Eurasia convergence at the time of rifting appears not to be due to resistance from continental collision since collision occurred following the initiation of rifting (e.g., McQuarrie et al. 2003; Bosworth et al. 2005; Okay et al. 2010). Furthermore, the plume-driven rifting hypothesis cannot easily account for a second episode of slowing of NU-EU convergence and the systematic changes around the boundaries of the Arabian Plate at $\sim$11 Ma.

We suggest the following dynamic scenario that attributes AR-AF-EU plate motions and inter-plate deformation to forces associated directly with subduction of ocean lithosphere.

1. Prior to the initiation of extension across the Red Sea and Gulf of Aden, convergence of the AF/AR Plate with Eurasia was driven by subduction pull by the Neotethys slab, the same forces that we suggest drive convergence of the Indian and Australian plates with Eurasia (Reilinger and McClusky 2011). At that time, subduction was occurring along the entire Nubia–Arabia plate boundary from the western Mediterranean to the Makran (Dercourt et al. 1986).
2. During the Late Oligocene/Early Miocene, the African Plume weakened the Afro-Arabian Plate at the location of the proto-Afar Triple Junction. Extensional stresses within the Afro-Arabian Plate caused by subduction pull along the boundary of the Arabian Plate with Eurasia fractured the Afro-Arabian continental plate along the proto-Red Sea and Gulf of Aden that in turn caused a reduction in the northward pull and hence the northward rate of motion of the African Plate. At approximately this time, roughly N–S extension initiated in the Mediterranean from the Alboran to the Aegean seas (Jolivet and Facenna 2000; Reilinger and McClusky 2011).
3. Rifting continued along the initial Red Sea rift (Gulf of Aqaba to Bab-al-Mandab) for about 11 M year with little (i.e., <±10 to 15 %) change in rate or orientation.
4. At about 11 ± 2 Ma, ocean spreading (i.e., complete severing of the continental lithosphere and injection of mid-ocean ridge basalts) extended along the full length of the Gulf of Aden rift separating the Arabian Plate from the Somalian part of the African Plate. The Red Sea continued as a continental rift. Complete severing of the AR-SOM plate boundary further reduced the slab pull force transmitted to the Somalian part of the African Plate.
5. The reduced pull transmitted to Somalia caused a second episode of slowing of Africa's rate of convergence with Eurasia and accordingly an increase in the rate of rifting along both the Red Sea and Gulf of Aden. Because the Gulf of Aden is oriented approximately E–W, the reduced northward pull on Africa caused an increase in the northward component of AR-AF motion. This change in turn caused a change to more N–S rifting in both the northern and southern Red Sea (Suez to Gulf of Aqaba and Bab-al-Mandab to the Danakil/Afar, respectively). Approximately simultaneously (Tortonian), there was a change in the configuration of extension in the Mediterranean basins (Jolivet and Faccenna 2000; Reilinger and McClusky 2011, and references therein).
6. The small amount of extension across the EAR and its present slow rate of rifting are consistent with the absence of subduction boundaries around the periphery of the Somalia Plate.

This dynamic scenario that attributes plate motions to slab pull and slab suction (opening of the Mediterranean basins) provides an internally consistent, intuitively simple explanation for the post-Late Oligocene kinematic evolution of the Arabia–Africa–Eurasia plate system.

Acknowledgments We thank those individuals and organizations that established and maintain the global GPS tracking network. We are grateful to UNAVCO for assistance with the installation of the CGPS stations in Halyat Amar and Namas, KSA, and for logistical support for GPS survey observations. The chapter benefited substantially from careful and constructive reviews by Charles DeMets, Rebecca Bendick, and Nicolas Bellahsen. This research was supported in part by KACST and NSF grants EAR-0337497, EAR-0305480, EAR-0635702, and EAR-0947969 to MIT.

References

Allen M, Jackson J, Walker R (2004) Late cenozoic reorganization of the Arabia-Eurasia collision and the comparison of short-term and long-term deformation rates. Tectonics 23, TC2008. doi:10.1029/2003TC001530

Altamimi Z, Collilieux X, Legrand J, Garayt B, Boucher C (2007) ITRF 2005: a new release of the international terrestrial reference frame based on time series of station positions and earth orientation parameters. J Geophys Res 112:B09401. doi:10.1029/2007JB004949

Al Taraza E, Rajab JA, Gomez F, Cochran W, Jaafar R, Ferry M (2011) GPS measurements of near-field deformation along the southern Dead Sea fault system. Geochem Geophys Geosyst 12. doi:10.1029/2011GC003736

Ambraseys NN, Jackson J (1998) Faulting associated with historical and recent earthquakes in the eastern Mediterranean region. Geophys J Int 133:390–406

ArRajehi A, McClusky S, Reilinger R, Daoud M, Alchalbi A, Ergintav S, Gomez F, Sholan J, et al (2010) Geodetic constraints on present-day motion of the Arabian plate: implications for Red Sea and Gulf of Aden rifting. Tectonics 29, TC3011. doi:10.1029/2009TC002482

Aujac G (2001) Eratosthène de Cyrène, le pionier de la geographie. Édition du CTHS, Paris 224 pp

Barazangi M, Sandvol E, Seber D (2006) Structure and tectonic evolution of the Anatolian plateau in eastern Turkey. In: Dilek Y, Pavlides S (eds) Postcollisional tectonics and magmatism in the Mediterranean region and Asia. Geological Society of America Special Paper 409, pp 463–474, doi:10.1130/2006.2409(22)

Becker JJ, Sandwell DT, Smith WHF et al (2009) Global bathymetry and elevation data at 30 arc seconds resolution: SRTM30_PLUS. Mar Geodesy 32(4):355–371

Bellahsen N, Faccenna C, Funiciello F, Daniel JM, Jolivet L (2003) Why did Arabia separate from Africa? Insights from 3D laboratory experiments. Earth Planet Sci Lett 216:365–381

Bialas R, Buck WR, Qin R (2010) How much magma is required to rift a continent? Earth Planet Sci Lett 292:68–78

Blewitt G, Lavallee D (2002) Effect of annual signals on geodetic velocity. J Geophys Res 107(B7):2145. doi:10.1029/2001JB000570

Bosworth W, Huchon P, McClay K (2005) The Red Sea and Gulf of Aden basins. J Afr Earth Sc 43:334–378

Byrne D, Sykes LR, Davis DM (1992) Great thrust earthquakes and aseismic slip along the plate boundary of the Makran subduction zone. J Geophys Res 97(B1):449–478

Chu D, Gordon G (1998) Current plate motions across the Red Sea. Geophys J Int 135:313–328

Cochran JR (1981) The Gulf of Aden: structure and evolution of a young ocean basin and continental margin. J Geophys Res 86:263–288

Conrad CP, Lithgow-Bertelloni C (2002) How mantle slabs drive plate tectonics. Science 298(5591):207–209

Cox A (1973) Plate tectonics and geomagnetic reversals. W. H. Freeman and Company, San Francisco, 702 pp ISBN 0-7167-0258-4

d'Acremont E, Leroy S, Maia M, Gente P, Autin J (2010) Volcanism, jump and propagation on the Sheba ridge, eastern Gulf of Aden: segmentation evolution and implications for oceanic accretion processes. Geophys J Int 180:535–551

DeMets C, Gordon RG, Argus D (2010) Geologically current plate motions. Geophys J Int 181:1–80

Dercourt J, Zonenshain LP, Ricou L-E, Kazmin VG, Le Pichon X, Knipper AL, Grandjacquet C, Sbortshikov IM et al (1986) Geological evolution of the Tethys Belt from the Atlantic to the Pamirs since the Lias. Tectonophysics 123:241–315

Ebinger C, Sleep N (1998) Cenozoic magmatism throughout east Africa resulting from impact of a single plume. Nature 395:788–791

Ebinger C, Casey M (2001) Continental break-up in magmatic provinces: an Ethiopian example. Geology 29:527–530

Elsasser WM (1971) Sea-floor spreading as thermal convection. J Geophys Res 76(5):1101–1112

England PC, McKenzie DP (1983) A thin viscous sheet model for continental deformation. Geophys J Roy Astron Soc 73:523–532

Faccenna C, Becker TW, Jolivet L, Keskin M (2013) Mantle convection in the Middle East: reconciling Afar upwelling, Arabia indentation and Aegean trench rollback. Earth Planet Sci Lett 375:254–269

Farr TG et al (2007) The shuttle radar topography mission. Rev Geophys 45, RG2004. doi:10.1029/2005RG000183

Forsyth DW, Uyeda S (1975) On the relative importance of the driving forces of plate motion. Geophys J R Astron Soc 43:163–200

Forte AM, Cowgill E, Bernardin T, Kreylos O, Hamann B (2012) Late Cenozoic deformation of the Kura fold-thrust belt, southern Greater Caucasus. Geol Soc Am Bull 122:465–486

Fournier M, Camot-Rooke N, Petit C, Fabri O, Huchon P, Maillot B, Lepvrier C (2008) In situ evidence for dextral active motion at the Arabia-India plate boundary. Nat Geosci 1:54–58

Garfunkel Z, Beyth M (2006) Constraints on the structural development of Afar imposed by the kinematics of the major surrounding plates. Geol Soc London, Spec Publ 259. doi:10.1144/GSL.SP.2006.259.01.04

Gomez F, Karam G, Khawlie M, McClusky S, Vernant P, Reilinger R, Jaafar R, Tabet C, Khair K, Barazangi M (2007) Global positioning system measurements of strain accumulation and slip transfer through the restraining bend along the Dead Sea fault system in Lebanon. Geophys J Int 168:1021–1028

Gordon RG, Argus DF, Royer J-Y (2008) Space geodetic test of kinematic models for the Indo-Australian composite plate. Geology 36:827–830

Hager BH, O'Connell RJ (1981) A simple global model of plate dynamics and mantle convection. J Geophys Res 86(B6):4843–4867

Hager BH, King RW, Murray MH (1991) Measurements of crustal deformation using the global positioning system. Annu Rev Earth Planet Sci 19:351–382

Hempton MR (1987) Constraints on Arabia plate motion and extensional history of the Red Sea. Tectonics 6:687–705

Herring TA, King RW, McClusky SC (2013) Introduction to GAMIT/GLOBK, Release 10.5. MIT, Cambridge 48 pp

Hetland EA, Muse P, Simons M, Lin YN, Agram PS, DiCaprio CJ (2012) Multiscale InSAR time series (MInTS) analysis of surface deformation. J Geophys Res 117:B02404. doi:10.1029/2011JB008731

Joffe S, Garfunkel Z (1987) Plate kinematics of the circum Red Sea: a re-evaluation. Tectonophysics 141:5–12

Jolivet L, Faccenna C (2000) Mediterranean extension and the Africa-Eurasia collision. Tectonics 19:1095–1106

King DA (1993) Astronomy in the service of Islam. Variorum, Aldershot, UK

King RW, Masters EG, Rizos C, Stolz A, Collins J (1985) Surveying with GPS, monograph 9. School of Surveying, The University of New South Wales, Kensington 128 pp

Klinger Y, Avouac JP, Abou Karaki N, Dorbath L, Bourles D, Reyss JL (2000) Slip rate of the Dead Sea transform fault in the northern Araba valley (Jordan). Geophys J Int 142:755–768

Kozacı Ö, Dolan JF, Finkel RC (2009) A late Holocene slip rate for the central North Anatolian fault at Tahtaköprü, Turkey, from cosmogenic 10Be geochronology: implications for fault loading and strain release rates. J Geophys Res 114:B01405. doi:10.1029/2008JB005760

Le Beon M, Klinger Y, Amrat AQ, Agnon A, Dorbath L, Baer G, Ruegg J-C, Charade O, Mayyas JO (2008) Slip rate and locking depth from GPS profiles across the southern Dead Sea Transform. J Geophys Res 113:B11403. doi:10.11029/2007JB005280

Le Pichon X, Gaulier JM (1988) The rotation of Arabia and the levant fault system. Tectonophysics 153:271–294

Le Pichon X, Kreemer C (2010) The Miocene to present kinematic evolution of the Eastern Mediterranean and Middle East and its implications for dynamics. Annu Rev Earth Planet Sci 38:323–351. doi:10.1146/annurev-earth-040809-152419

Long MD, Becker TW (2010) Mantle dynamics and seismic anisotropy. Earth Planet Sci Lett 297:341–354. doi:10.1016/j.epsl.2010.06.036

Mahmoud S, Reilinger R, McClusky S, Vernant P, Tealeb A (2005) GPS evidence for northward motion of the Sinai block: implications for E. Mediterranean tectonics. Earth Planet Sci Lett 238:217–227

McClusky S, Balassanian S, Barka A, Demir C, Ergintav S, Georgiev I, Gurkan O, Hamburger M (2000) GPS constraints on plate kinematics and dynamics in the eastern Mediterranean and Caucasus. J Geophys Res 105:5695–5719

McClusky S, Reilinger R, Mahmoud S, Ben Sari D, Tealeb A (2003) GPS constraints on Africa (Nubia) and Arabia plate motion. Geophys J Int 155:126–138

McClusky S, Reilinger R, Ogubazghi G, Amleson A, Haleab B, Vernant P, Sholan J, (2010) Kinematics of the southern Red Sea–afar triple junction and implications for plate dynamics. Geophys Res Lett 37. doi:10.1029/2009GL041127

McKenzie DP, Davies D, Molnar P (1970) Plate tectonics of the Red Sea and East Africa. Nature 226:243–248

McQuarrie N, Stock JM, Verdel C, Wernicke BP (2003) Cenozoic evolution of Neotethys and implications for the causes of plate motions. Geophysi Res Lett 30. doi:10.1029/2003GL017992

McQuarrie N, van Hinsbergen DJJ (2013) Retrodeforming the Arabia-Eurasia collision zone: age of collision versus magnitude of continental subduction. Geology 41:315–318

Misra P, Enge P (2011) Global positioning system: signals, measurements, and performance. Ganga-Jamuna Press, Lincoln. ISBN 0-909544-1-7

Morgan WJ (1971) Convection plumes in the lower mantle. Nature 230:42–43

Mouthereau F, Lacombe O, Vergés J (2012) Building the Zagros collisional orogen: timing, strain distribution and the dynamics of Arabia/Eurasia plate convergence. Tectonophysics 532–535:27–60

National Research Council (U.S.) (1995) Committee on the future of the global positioning system. National Academy of Public Administration, Washington DC

Niemi TM, Zhang H, Atallah M, Harrison JBJ (2001) Late pleistocene and holocene slip rate of the Northern Wadi Araba fault, Dead Sea transform, Jordan. J Seismol 5:449–474

Okay AI, Zattin M, Cavazza W (2010) Apatite fission-track data for the Miocene Arabia-Eurasia collision. Geology 38:35–38

Omar GI, Steckler MS (1995) Fission track evidence on the initial rifting of the Red Sea: two pulses, no propagation. Science 270:1341–1344

Pail R, Goiginger H, Schuh W, Höck E, Brockmann J, Fecher T, Gruber T, Mayer-Gürr T, Kusche J, Jäggi A, Rieser D (2010) Combined satellite gravity field model GOCO01S derived from GOCE and GRACE. Geophys Res Lett 37:5–10

Pallister JS, McCausland WA, Jónsson S, Lu Z, Zahran H, El-Hadidy S, Aburukbah A, Stewart ICF, Lundgren PR, White RA, Moufti MRH (2010) Broad accommodation of rift-related extension recorded by dyke intrusion in Saudi Arabia. Nat Geosci 3:705–712

Parsons B, Richter FM (1980) A relation between the driving force and geoid anomaly associated with mid-ocean ridges. Earth Planet Sci Lett 51:445–450

Pérouse E, Chamot-Rooke N, Rabaute A, Briole P, Jouanne F, Georgiev I, Dimitrov D (2012) Bridging onshore and offshore present-day kinematics of central and eastern Mediterranean: Implications for crustal dynamics and mantle flow. Geochem Geophys Geosyst 13. doi:10.1029/2012GC004289

Rawlinson N, Pozgay S, Fishwick S (2010) Seismic tomography: a window into deep earth. Phys Earth Planet Inter 178:101–135

Redfield TF, Wheeler WH, Often M (2003) A kinematic model for the development of the Afar depression and its paleogeographic implications. Earth Planet Sci Lett 216:383–398

Reilinger R, McClusky S, Vernant P, Lawrence S, Ergintav S, Cakmak R, Ozener H, Kadirov F (2006) GPS constraints on continental deformation in the Africa-Arabia-Eurasia continental collision zone and implications for the dynamics of plate interactions. J Geophys Res 111:BO5411. doi:10.1029/2005JB004051

Reilinger R, McClusky S (2011) Nubia-Arabia-Eurasia plate motions and the dynamics of Mediterranean and Middle East tectonics. Geophys J Int 186:971–979

Royden L (1993) The tectonic expression of slab pull at continental convergent boundaries. Tectonics 12:303–325

Ryder I, Burgmann R (2008) Spatial variations in slip deficit on the central San Andreas fault from InSAR. Geophys J Int 175(3):837–852

Sadeh M, Hamiel Y, Ziv A, Bock Y, Fang P, Wdowinski S (2012) Crustal deformation along the Dead Sea transform and the Carmel Fault inferred from 12 years of GPS measurements. J Geophys Res 117. doi:10.1029/2012JB009241

Saria E, Calais E, Altamimi Z, Willis P, Farah H (2013) A new velocity field for Africa from combined GPS and DORIS space geodetic Solutions: Contribution to the definition of the African reference frame (AFREF). JGeophys Res 118. doi:10.1002/jgrb.50137

Sengor AMC, Gorur N, Saroglu F (1985) Strike-slip faulting and related basin formation in zones of tectonic escape: Turkey as a case study. In: Biddle KT, Christie-Blick N (eds) Strike-slip deformation, basin formation, and sedimentation, Special Publication Society of Economic Paleontologists and Mineralogists, vol 37, pp 227–264

Spence W (1987) Slab pull and the seismotectonics of subducting lithosphere. Rev Geophys 25. doi:10.1029/RG025i001p00055

Tapley BD, Bettadpur S, Ries JC, Thompson PF, Watkins MM (2004) GRACE measurements of mass variability in the Earth system. Science 305:503–505

Tesfaye S, Kusky TT, Harding D (2003) Early continental breakup boundary and migration of the Afar triple junction, Ethiopia. Geol Soc Am Bull 115:1053–1067

Thatcher W (1993) The earthquake cycle and its role in the long-term deformation of the continental lithosphere. Ann Geofis 36:13–24

Thatcher W (2003) GPS constraints on the kinematics of continental deformation. Inter Geol Rev 45:191–212

Tiryakioglu I, Floyd M, Erdogan S, Gulal E, Ergintav S, McClusky S, Reilinger R (2013) GPS constraints on active deformation in the Isparta Angle region of SW Turkey. Geophys J Int 195:1455–1463

Tompkins P (1997) Secrets of the Great Pyramid: two thousand years of adventure and discoveries surrounding the great pyramid of cheops. BBS Publishers, New York 432 pp

Turcotte DL, Schubert G (2002) Geodynamics, 2nd edn. Cambridge University Press, Cambridge

Vernant P, Nilforoushan F, Hatzfeld D, Abbassi M, Vigny C, Masson F, Nankali H, Martinod J, Ashtiani A, Bayer R, Tavakoli F, Chéry J (2004) Contemporary crustal deformation and plate kinematics in Middle East constrained by GPS measurements in Iran and northern Oman. Geophys J Int 157:381–398

Walters RJ, Holley RJ, Parsons B, Wright T (2011) Interseismic strain accumulation across the North Anatolian Fault from Envisat InSAR measurements. Geophys Res Lett 38. doi:10.1029/2010GL046443

Wang K, Hu Y, He J (2012) Deformation cycles of subduction earthquakes in a viscoelastic Earth. Nature 484. doi:10.1038/nature11032

Wolfenden E, Ebinger C, Yirgub G, Deino A, Ayalew D (2004) Evolution of the northern Main Ethiopian rift: birth of a triple junction. Earth Planet Sci Lett 224:213–228

Yeats RS, Sieh KE, Allen CR (1996) Geology of earthquakes. Oxford University Press, UK, 576 pp, ISBN 13: 9780195078275

Seismicity and Seismotectonic Setting of the Red Sea and Adjacent Areas

Salah El-Hadidy Youssef

Abstract

It was thought that the Kingdom of Saudi Arabia was characterized by low seismic activity and that the Arabian shield was relatively stable. After establishing the Saudi National Seismic Network and occurrence of some felt earthquakes in the Arabian Shield, this assumption changed, especially for the western part of the shield. A study of the most recent felt earthquakes in the Red Sea coastal plain area that were recorded by the Saudi National Seismic Network can help in better understanding the Red Sea rifting. Moment tensor inversion of the waveform data offers a unique insight into the active tectonics of the region as well as providing information for seismic hazard analyses of the western part of Saudi Arabia and the Gulf of Aqaba. Moment tensor analyses of recent earthquakes in the Red Sea coastal plain show that some Cenozoic faults are seismically active, with normal faulting under NE–SW tension and NW–SE compression. This is seen in the northern Red Sea, Gulf of Aqaba, and the Badr and Harrat Lunayyir earthquakes, whereas published moment tensor solutions of the largest earthquakes in and around the Arabian plate indicate that most of the earthquakes in eastern Saudi Arabia reflect reverse faulting with NE forces due to collision with the Eurasian plate. There is also an E–W tension axis acting on the Makkah–Madinah–Nafud volcanic line, in agreement with the fault plane solution of the Qunfudah earthquake. It is recommended that two seismically active regions on either side of the Red Sea (Abu Dabbab and Harrat Lunayyir) should be studied further, including their relation to the rifting of the Red Sea. The Saudi National Seismic Network plays an important role in a better understanding of the seismicity and seismotectonic setting of Saudi Arabia in particular and the Arabian plate in general, including providing data for hazard evaluations.

Introduction

Four major tectonic plates (Arabia, Eurasia, India, and Africa) and one smaller tectonic block (Anatolia) are responsible for seismicity in the Middle East. Geologic development of the region is the result of tectonic processes that include subduction, large-scale transform faulting, compressional mountain building, and crustal extension. Along the eastern margin of the Mediterranean region, the complex interaction between the African, Arabian, and Eurasian plates caused the Red Sea Rift, a spreading center between the African and Arabian plates, with a spreading rate of approximately 10 mm/yr in the north and 16 mm/yr in the south (Chu and Gordon 1998). The Arabian plate came into existence 25–30 Ma, when what is now the Arabian Peninsula, Syria, Jordan, Iraq, and westernmost Iran began to separate from the African continent because of rifting along the Red Sea and Gulf of Aden.

The Arabian plate has moved northeast from Egypt and Sudan, north from Somalia, and has rotated counterclockwise about a point in the vicinity of the Gulf of Suez. Such movement is accommodated by compression and strike-slip faulting along the Bitlis and Zagros zones, where the

S.E.-H. Youssef (✉)
National Center for Earthquakes and Volcanoes, Saudi Geological Survey, Jeddah, Saudi Arabia
e-mail: el_hadidy.sy@sgs.org.sa

N.M.A. Rasul and I.C.F. Stewart (eds.), *The Red Sea*,
Springer Earth System Sciences, DOI 10.1007/978-3-662-45201-1_8,

Arabian plate collides with and subducts beneath the Eurasian plate, and by strike-slip displacement along the Dead Sea transform. At the present time, the northern part of the Arabian plate moves northwest, with respect to the Eurasian plate, at a rate of 20 + 3 mm/yr. Because they are regions of extension, the southern, southwestern, and southeastern margins of the Arabian plate have weak to moderate earthquake activity; the compressive northerly and northeasterly margins, conversely, are regions of strong earthquake activity. Overall, the plate has moved as much as 350 km away from Africa, depending on where the margins of the initial rift are placed with respect to present-day exposure of Precambrian basement, and on how much stretched continental crust is believed to remain in the Red Sea basin, and depending on how much hybrid continental–oceanic and true oceanic crust is inferred beneath the Red Sea shelves and axial trough. Displacement along the Dead Sea transform separating the northeasternmost part of the African plate from the Arabian plate and linking extension in the Red Sea with compression in the Bitlis zone is believed to be about 110 km.

Tectonic Setting of the Red Sea

The Red Sea Rift System is one of the world's largest active rift systems, which comprises a variety of rifting stages starting from initial faulting and advancing through several stages of continental rifting. It began about 30 million years ago, separating the western edge of the Arabian Plate from Africa (Camp and Roobol 1992). The rifting of the Red Sea was initiated in the Oligocene, with thermally driven uplift and domal arching of the Arabian–Nubian Shield. Some studies (Wernicke 1985; Voggenreiter et al. 1988; McGuire and Bohannon 1989) suggest that the Red Sea developed as a passive rift, where extensional stresses due to the far-field body forces are accommodated on low-angle detachment planes extending through the entire lithosphere below the rift. This results in passive upwelling of asthenospheric material below the rift and asymmetric thinning, where the thinnest lithosphere is laterally offset from the rift axis. Other studies (McKenzie 1978; Bellahsen et al. 2003) argue that the Red Sea is an active rift, where the lithosphere is thermally eroded by flow in the underlying asthenosphere, requiring the presence of hot, ascending material. In this case, the rift flanks are thermally uplifted, and the area of greatest lithospheric thinning is coincident with the rift axis. It has also been suggested that these two end member models are not mutually exclusive; rifting in the Red Sea may have been initiated by passive processes, followed by more recent active processes associated with mantle upwelling (Camp and Roobol 1992; Ebinger and Sleep 1998; Daradich et al. 2003). In the active rifting model, the rift event is preceded by a period of uplift; however, in the passive rifting model, associated uplift postdates the rifting event (McGuire and Bohannon 1989). Geologic and fission track data show that western Arabia was at or below sea level prior to 30 Ma and that the uplift of the Red Sea margin postdates the initiation of rifting by about 5–10 million years (McGuire and Bohannon 1989). These results demonstrate that the Red Sea began as a passive rift (Wernicke 1985; Voggenreiter et al. 1988; McGuire and Bohannon 1989).

The Saudi Arabian Red Sea coastal margin has a complex geomorphologic history reflecting the interactions among structure, uplift, erosion, and volcanism. The major landscape features comprise a coastal plain up to 50 km in width, an erosional escarpment, a region of hills and mountains between the coastal plain and escarpment that are the remnants of erosional retreat of the escarpment, and one or more erosional surfaces east of the escarpment (broadly referred to as the Najd Pediplain) that include in places exhumed remnants of the surface that developed on the Arabian Shield at the end of the Precambrian (Huchon et al. 1991; Davison et al. 1994). Structurally, the margin includes the Cenozoic Red Sea basin, which underlies the Red Sea and most of the coastal plain, continental crust composed of Precambrian rocks of the Arabian Shield, small basins inland from the Red Sea that contain Cenozoic sedimentary rocks and reflect local extensions, fractures inland from and subparallel to the coastline that are intruded by Cenozoic dikes, and large fields of Cenozoic to Recent basalt (harrats). These were extruded unconformably onto the shield and on sedimentary rocks of the Red Sea basin (Huchon et al. 1991; Davison et al. 1994). The contact between the Red Sea basin and Arabian shield is partly faulted and partly depositional (Bohannon 1986).

The Arabian Peninsula forms a single tectonic plate, the Arabian Plate, with active boundaries as shown by the earthquake seismicity. Figure 1 shows the locations of historical and instrumental earthquakes in and around the Arabian Peninsula up to 2013 for earthquakes with magnitude above 3 on the Richter scale (Saudi Geological Survey database). It is obvious that more than 90 % of earthquake epicenters are concentrated along the plate boundary, in addition to scattered seismic activity within the Arabian Shield. A few of these earthquakes are related to the Cenozoic faults as well as reactivation of the existing Precambrian faults, and others are associated with the volcanic activity in the Cenozoic lava fields (harrats).

Earthquakes in the vicinity of Saudi Arabia are now well monitored since the establishment of the Saudi National Seismic Network that currently (as of July 2014) has 164 broadband seismic stations using satellite data transmission distributed across the Kingdom (Fig. 2), having started in 2003 with only 13 stations. It includes local seismic networks installed in the volcanic regions to monitor the earthquakes

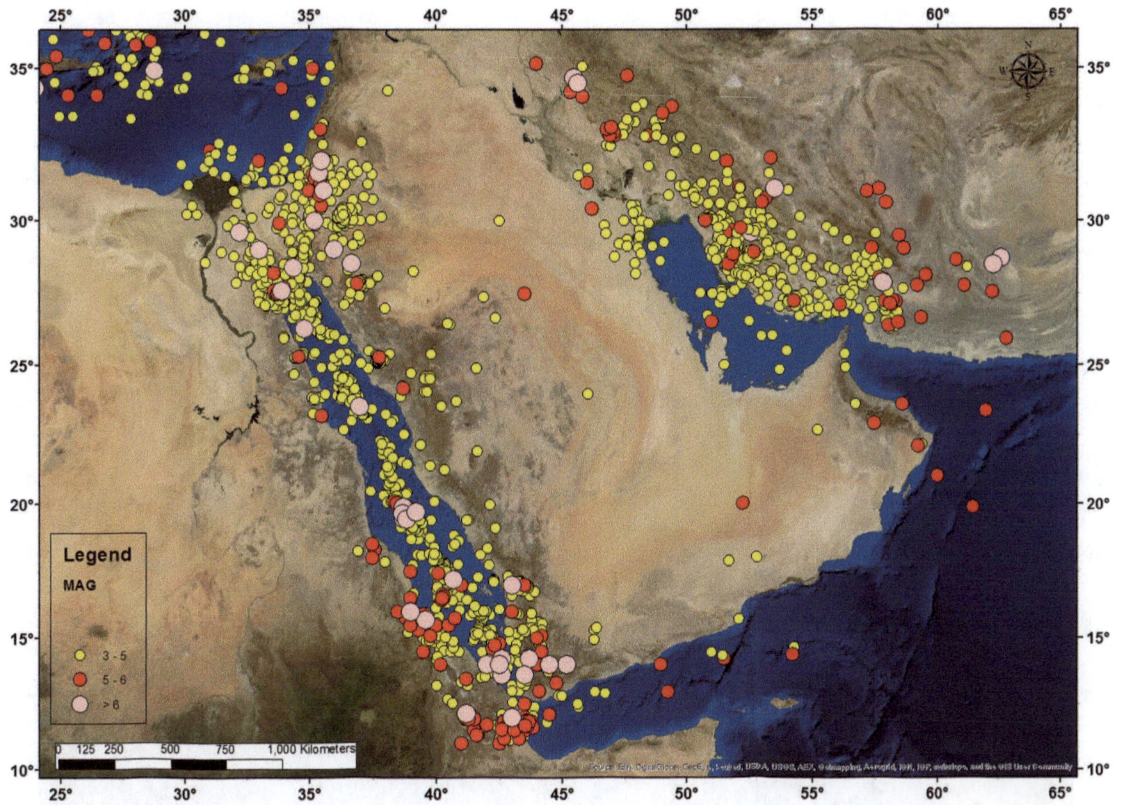

Fig. 1 Seismicity up to 2013 including historical and instrumental earthquakes above Ml 3 (Saudi geological survey database)

due to magmatic activity, in Harrats Lunayyir (Al-Shaqah) north of Yanbu, Rahat (near Madinah), and Khaybar (north of Madinah). The network also includes strong motion, geothermal, and GPS instruments. The network is intended to provide information for public safety, emergency management, seismic hazard quantification for both natural and man-made earthquakes, and engineering applications. The network now enables the detection and location of any earthquake greater than magnitude 2 within much of the area of Saudi Arabia, and smaller events can be detected within the areas of better coverage, especially in the western shield.

Some Recent Felt Earthquakes

The northern Red Sea is characterized by the frequent occurrence of felt earthquakes, and magnitude 5 or 6 events are not uncommon along the axis of the Red Sea, although usually they do not pose a significant hazard onshore. Probably the most notable instrumentally recorded earthquake to affect Saudi Arabia was the magnitude 7.2 event in the Gulf of Aqaba on November 22, 1995, that had a maximum intensity of VII, and caused some damage in the towns of Haql and Al Bad and was felt up to 500 km away in Al Madinah (Roobol et al. 1999). El Hadidy et al. (1998) studied the earthquake activity along the Gulf of Aqaba after this earthquake. About 12,000 events were recorded during this seismic sequence, where the earthquake epicenters and the fault plane solution of the main shock correlate well with the elongation of the Gulf of Aqaba.

The most recent earthquake sequences occurred in 1969, 1972, and 2001. The 1969 Shadwan sequence, characterized by a maximum magnitude mb 6.0 and Ms 6.8, occurred in the north near Shadwan Island and caused major damage in many localities of Egypt where 3 people were killed, 15 injured, 2 mosques collapsed, and 30 buildings were destroyed. The main shock was widely felt up to 300 km from the epicenter (Egypt, Dead Sea, Jordan Valley, Saudi Arabia). The maximum intensity ranged from IX to X, especially at Shadwan Island, Tawaila, and Jubabl (Maamoun et al. 1978). Offshore, about 10 km west of the rupture area, a coral reef emerged a few meters above sea level after the earthquake, which indicates a predominantly dip-slip movement in good agreement with the fault plane solution of the main shock of this sequence (Tables 1 and 2, Fig. 6). The second sequence in the same area occurred in 1972, when about 19 earthquakes were recorded with a maximum magnitude of Ml 5.5. The focal mechanism solution of the largest event shows pure normal faulting with tension axes having a NE–SW direction which is in good agreement with

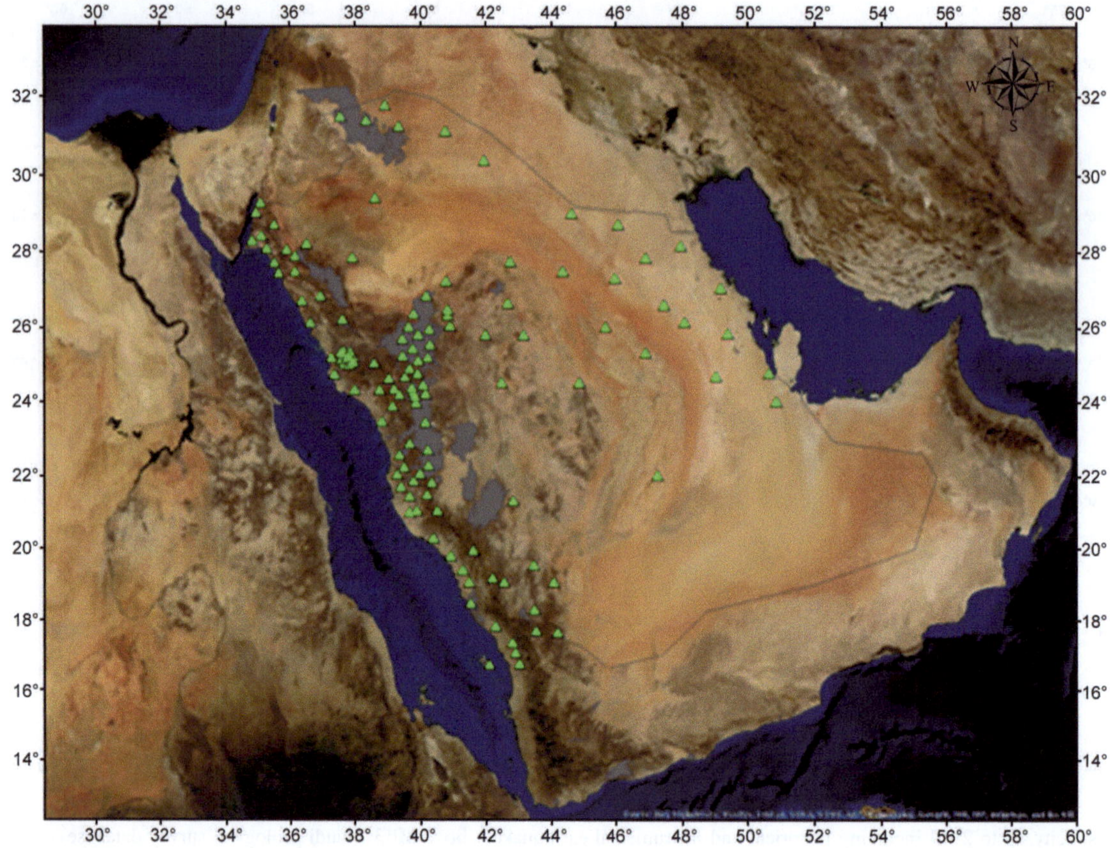

Fig. 2 Seismograph stations of the Saudi National Seismic Network

Table 1 Parameters of the earthquakes used for focal mechanisms and plotted in Fig. 6

No.	Date			Location		Depth (km)	Magnitude
	D	M	Y	Lat.	Long.		Ml
1	31	03	1969	27.610	33.91	6.2	6.8
2	28	06	1972	27.700	33.8	6.1	5.5
3	22	11	1995	28.826	34.8	7	6.9
4	22	06	2004	27.822	36.93	11	5.2
5	19	05	2009	25.282	37.76	7.8	5.4
6	27	08	2009	24.126	38.75	5.6	5.1
7	29	08	2011	19.395	41.26	15	4.4
8	23	01	2014	17.33	42.66	9.6	5.1
9	25	01	2014	17.345	42.71	15	3.7

the tension stress acting on the Red Sea (Tables 1 and 2, Fig. 6). From August 14 to September 21, 2001, a swarm of 600 earthquakes northeast of Shadwan Island was recorded by the Egyptian National Seismic Network (ENSN), with magnitudes ranging from 0.6 to 4.6. The fault plane solution of the largest event in this sequence shows normal faulting with a strike-slip component (Tables 1 and 2, Fig. 6) and tension axes trending NW–SE, perpendicular to the tension axes of the 1969 and 1972 events.

These three sequences occurred at the intersection of three major structural trends, the Gulf of Aqaba–Dead Sea Rift, the Gulf of Suez, and the main Red Sea rift. The Red Sea is divided into three distinct zones; each zone represents different stages of development (Cochran and Martinez 1988). The present active tectonics is characterized by the concentration of extension and deformation in the axial depression and by the emplacement of a number of large intrusions. The intrusions occur at distinct sites located at the

Table 2 Focal mechanism parameters of the events listed in Table 1

No.	Date			Fault planes						Stress axes				Reference
	D	M	Y	NP1			NP2			*P*-axis		*T*-axis		
				St.	Dip	Rake	St.	Dip	Rake	St.	Dip	St.	Dip	
1	31	03	1969	294	37	–	113	53	–	19	82	203	08	Hu: blue beach ball
2	28	06	1972	288	40	–	121	51	–	75	82	205	05	Hu: blue beach ball
3	22	11	1995	196	59	–	294	77	–	159	31	62	12	H: blue beach ball
4	22	06	2004	304.3	40.9	−87.3	–	–	–	09.9	85.5	212.7	04.2	Damegh: light blue beach ball
5	19	05	2009	140	40	−90	320	50	−90	230	83	50	05	Present work: blue beach ball
6	27	08	2009	335	70	−50	87	44	−150	49	188	37	15	Present work: red ball
7	29	08	2011	15	50	−65	159	46	−117	352	71	88	2	Present work: black beach ball
8	23	01	2014	165	90	5	75	85	180	300	04	165	85	Present work: green beach ball
9	25	01	2014	256	81	155	350	65	10	305	11	210	24	Present work: green beach ball

Hu Huang and Soloman (1987); *H* Haasib (1990); Aldamegh et al. (2009); the beach balls representing the focal mechanisms are shown in Fig. 6

intersection of transfer zones with the faults bounding the rotated blocks in the marginal areas, and at a depth located in the axial depression halfway between transfer zones.

On August 27 and 28, 2009, two earthquakes occurred with magnitudes Ml 5.1 and 3.8, respectively, about 45 km north of Badr and 65 km east of Yanbu in western Saudi Arabia. Moment tensor inversion of the largest event using software of Herrmann and Ammon (2004) showed pure normal faulting and a preferred fault plane solution with a NW–SE trend (Tables 1 and 2, Fig. 6), in good agreement with a known NW–SE fault. The epicenter is located within the fault system that separates the Red Sea coastal plain from the Red Sea foothills and resulted from early extension of the crust. The event suggests that this Cenozoic fault has been reactivated.

A cluster of earthquake epicenters (maximum magnitude 3.1) occurred in May 18, 2008, in a populated area about 100 km north of Abha city in SW Saudi Arabia, within the Precambrian Shield above the Red Sea escarpment. Although they were minor earthquakes, some were felt by the local population. The seismic activity was located on a tightly folded Precambrian volcano-sedimentary belt running NNE to SSW, where there are many Precambrian faults. It is likely that some form of crustal extension is operating there, causing a reactivation of some of the existing Precambrian faults under EW tensional stresses within the belt of volcano-sedimentary rocks (El Hadidy et al. 2012).

In May 29, 2008, a small earthquake with Ml 3.1 occurred about 80 km to the east of Jizan city in SW Saudi Arabia, near the Yemen border, where a belt about 1,000 km long and 200 km wide is characterized by many Cenozoic faults that cut the Tertiary Yemen Volcanic Group as well as the Pliocene–Quaternary volcanic rocks. This belt of Cenozoic faults extends over the border with Saudi Arabia to the north of Jizan to about 17°N and is highly active with many historic earthquakes that have caused much damage and many casualties.

Between August 29 and August 31, 2011, four earthquakes occurred with a maximum magnitude of Ml 4.4 northeast of Al-Qunfudah (19.40°N, 41.26°E), in Makkah Municipality. From moment tensor inversion (Herrmann and Ammon 2004) the largest event showed normal faulting (Tables 1 and 2, Fig. 6), and the preferred fault plane solution has a NS trend with an EW tension axis, which is similar to the Makkah–Madinah–Nafud) volcanic trend as well as the graben system in the region. The epicenter is on the Red Sea coastal plain a short distance west or seaward of the Master Listric Fault that separates the Cenozoic coastal plain fault system from the escarpment of the Precambrian shield outcrop.

From April 14, 2009, and continuing for more than three months, an earthquake swarm occurred in Harrat Lunayyir (Al-Shaqah), in the Saudi Arabian Shield, about 50 km inland from the Red Sea coast. This swarm is related to magmatic activity, a dike intrusion episode that failed to reach the surface (Pallister et al. 2010). Thousands of earthquakes have been recorded with a maximum magnitude Ml of 5.4 that was felt in Madinah. This seismic activity is aligned in a NW–SE trend along the existing NW-trending dike as shown in Fig. 3, with focal depths from 3 km to 15 km. The fault plane solutions obtained from moment tensor inversion of the 30 largest events show a NE–SW tension axis and NW–SE compression axis (Tables 1 and 2, Fig. 6). The regional stress field obtained from the focal mechanism solutions shows three types of motions: pure

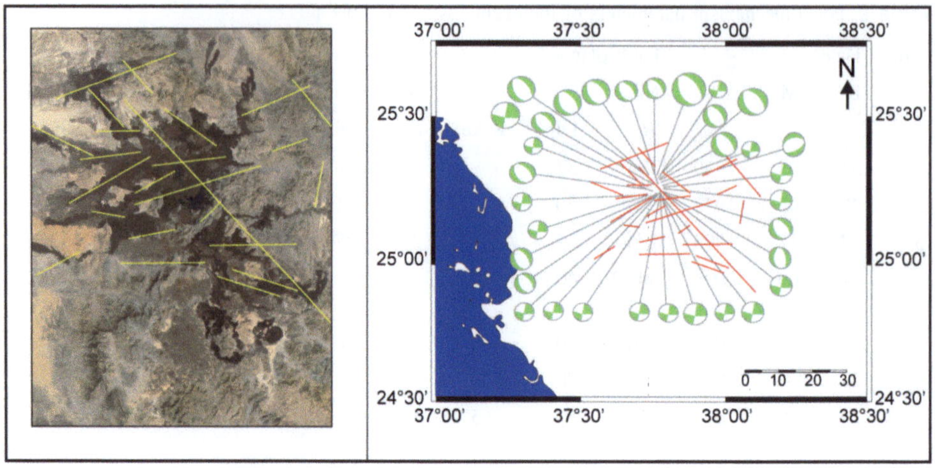

Fig. 3 Seismicity of Harrat Lunayyir (Al-Shaqah) (*left*) and fault plane solutions of the largest 30 earthquakes (*right*). The *red lines* indicate the faults and NW–SE dike

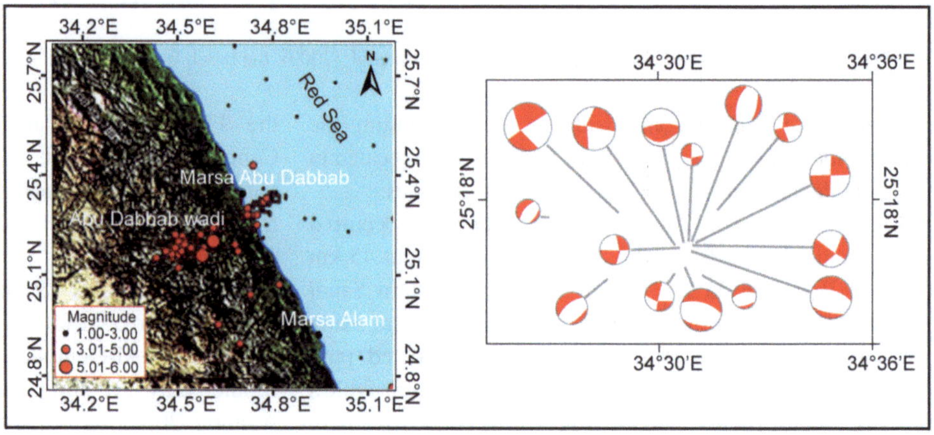

Fig. 4 Seismicity of the Abu Dabbab area (*left*) and fault plane solutions of 15 earthquakes (*right*) using moment tensor techniques (after Hosny et al. 2011)

normal faults, strike-slip faults, and reverse faults (Fig. 3). The seismic and magmatic activity caused crustal deformation and fractures that reached up to 12 km from the epicentral area.

In the western Red Sea coastal plain (Egyptian Eastern Desert), there is a seismically active area called "Abu Dabab," at about the same latitude as Harrat Lunayyir on the eastern side of the Red Sea. This area is characterized by frequent earthquake swarms, and the seismicity at Abu Dabbab is accompanied by distinct rumbling sounds similar to distant quarry blasts, as noted in traditions passed down over the centuries by Bedouin in the area. The name itself, "Abu Dabbab," means "earthquake sounds" (from Abu: father and Dabbab: knocking). This region has been studied by many authors. Fairhead and Girdler (1970) analyzed the seismicity of the Red Sea for the period 1953–1968 using the data of the Worldwide Standardized Seismograph Network (WWSSN), and observed an event of magnitude mb 6 on November 12, 1955, for which the epicenter was located in the Precambrian rocks of the Egyptian Red Sea Hills. Hosny et al. (2011) studied an earthquake swarm in the Abu Dabbab region, located in the Egyptian Nubian Shield in the western Red Sea; the earthquake epicenters show a NE–SW trend perpendicular to the Red Sea as shown in Fig. 4, which includes the moment tensor inversion solutions of the largest events. The tension axis has a NW–SE trend, and the compression axis has a NE–SW trend. The focal depths are located at the brittle–ductile transition at depths of about 9–10 km, where the shallow transition zone may be due to the repeated intrusion of magmatic bodies in the lower crust beneath the Abu Dabbab area (Hosny et al. 2011). Many studies provide evidence for a volcano-sedimentary belt close to the Abu Dabbab region (Fig. 5). The eastern desert in Egypt is made up of low-grade metamorphosed ophiolitic and island arc-related rocks tectonically emplaced over poly-metamorphosed and poly-deformed basement (El Gaby et al. 1990; Greiling et al. 1994; Fritz et al. 1996; Neumayr et al. 1996). The core complexes and the major tectonic trends on both sides of the Red Sea strike NW–SE and are related to a crustal-scale sinistral shear zone, the Najd Fault System. As elsewhere in the Central Eastern Desert, the Wadi Mubarak belt comprises a thick succession of low-grade volcano-sedimentary rocks of arc

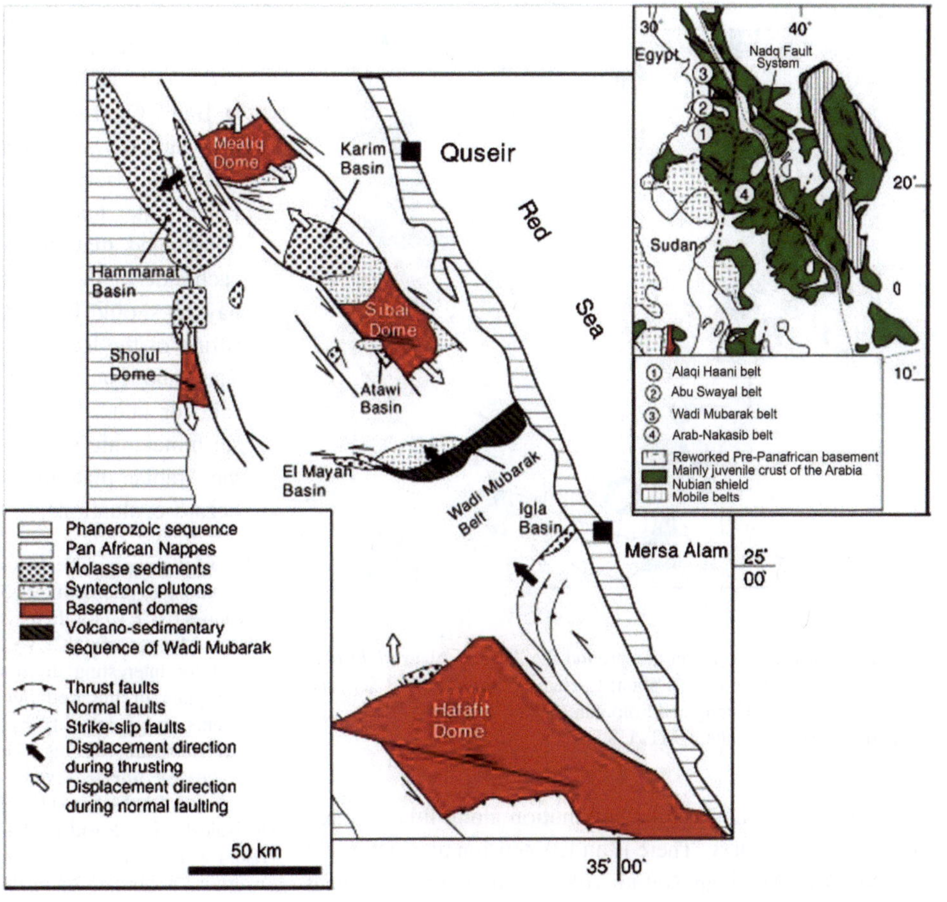

Fig. 5 Tectonic map of the Central Eastern Desert showing the core complexes and the location of the Wadi Mubarak belt (modified after Fritz et al. 1996). The *inset* shows the NW–SE striking Najd Fault System and other deformation belts in the Arabian–Nubian Shield

and back-arc affinity (Akaad et al. 1995) that are discordant to the general NW–SE trend of the Najd Fault System. The Wadi Mubarak belt contains two types of major structure: thrusts with intermediate dip and steep shear zones with a dominant strike-slip motion composed of phyllonites and sheared metavolcanics that define both structural elements and which indicate that the area has been exposed to magmatic activity.

On January 23, 2014, seismic activity started in southwest Saudi Arabia about 50 km NE of Jizan and 13 km east of Baysh. Fifty-three earthquakes were recorded, the largest one with magnitude Ml 5.1, and the two largest aftershocks with Ml 3.7 and 3.5. The focal mechanisms of the main shock and the largest aftershock obtained from the inversion of *P*-wave first arrival polarities show normal faulting with a strike-slip component (Tables 1 and 2, Fig. 6), and the preferred fault plane solution has a NE–SW trend, which is in good agreement with the distribution of the aftershocks as well as the known NE–SW faulting in the region. This event was widely felt in many villages and cities around the epicentral area. The maximum intensity observed ranged between V and VI in the epicentral area on the Modified Mercalli Intensity Scale, with a peak ground acceleration calculated at 297 cm/s^2.

Discussion

It was thought that the Kingdom of Saudi Arabia was characterized by low seismic activity and that the Arabian shield was relatively stable. After establishing the Saudi National Seismic Network and the occurrence of some felt earthquakes in the Arabian Shield, this assumption changed, especially for the western part of the shield.

This recent seismic activity indicates that some Cenozoic faults in the Red Sea coastal plain are seismically active. From the focal mechanisms of the largest earthquakes in and around the Arabian plate (Fig. 6), it appears that most of the earthquakes in western Saudi Arabia are on normal faults under NE–SW tension and NW–SE compression, as seen in the northern Red Sea, Gulf of Aqaba, and the Badr and Harrat Lunayyir earthquakes. The fault plane solutions in eastern Saudi Arabia reflect reverse faulting with NE forces due to collision with the Eurasian plate. In the northwest, the Red Sea Rift terminates at the southern boundary of the Dead Sea Transform Fault, a strike-slip fault that accommodates differential motion between the African and Arabian plates. Both the African and Arabian plates are moving in a NNE direction; the Arabian plate is moving slightly faster,

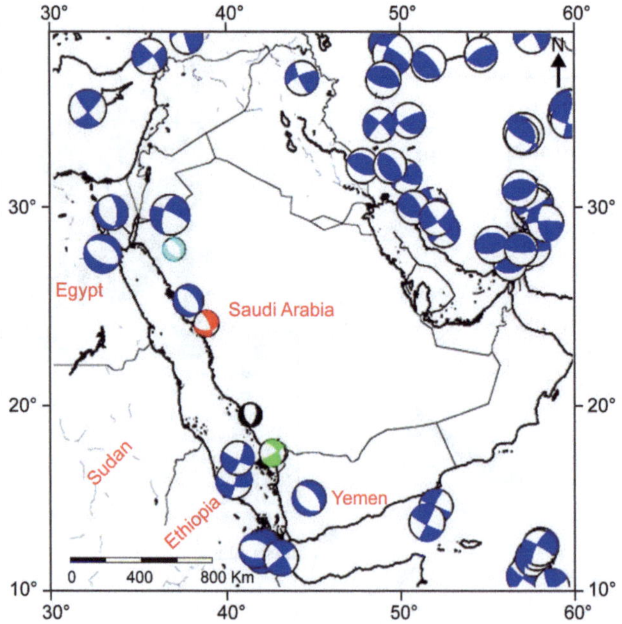

Fig. 6 Focal mechanisms in and around the Arabian plate (*light blue* Aldamegh et al. 2009; coastal plain: El-Hadidy, northern Red Sea, and Gulf of Aqaba: *Hu* Huang and Soloman (1987); *H* Haasib (1990); and around the Arabian plate, CMT, USGS)

resulting in left-lateral strike-slip motion along this segment of the plate boundary. There is an EW tension axis acting on the Makkah–Madinah–Nafud volcanic line, in agreement with the EW tension axis from the fault plane solution of the Qunfudah earthquake (black beach ball in Fig. 6).

Conclusion

1. Felt earthquakes occur in the Red Sea coastal plain, and especially in the Gulf of Aqaba, and small events are now well monitored with new seismograph networks.
2. The seismicity shows that some Cenozoic faults within the Red Sea coastal plain are seismically active and must be considered in seismic hazard assessments.
3. The fault plane solutions of the western Saudi Arabia earthquakes show normal faults and normal faults with strike-slip components in some cases.
4. From the fault plane solutions, tensional forces with a NE–SW direction and NW–SE compressional forces are in good agreement with the extensional forces causing opening of the Red Sea as well as the forces influencing motion of the Arabian Plate to the NE.
5. There is a belt of potentially active Cenozoic faults 1000 km long and 200 km wide along the SW margin of the Arabian Peninsula. Although this belt is largely within Yemen, some of the faults extend into Saudi Arabia.
6. It was believed that Saudi Arabia was stable with low levels of seismic activity, but from recent felt earthquakes and swarm activity at Harrat Lunayyir, and earthquakes at Badr, Qunfudah, Abha, and Jizan, we can conclude that parts of western Saudi Arabia can be considered as seismically active.
7. For a better understanding of Red Sea rifting, it is recommended that the two seismically active regions on each side of the Red Sea (Abu Dabbab and Harrat Lunayyir) should be studied, including their relation to the rifting of the Red Sea.
8. The Saudi National Seismic Network plays an important role for better understanding of the seismicity and the seismotectonic setting of Saudi Arabia in particular and the Arabian plate in general, including providing data for hazard evaluations.

Acknowledgments The author acknowledges and appreciates the cooperation, of the staff of the National Center of Earthquakes and Volcanoes (NCEV) for their kind help, with special thanks to Eng. Hani Zahran, Manger of (NCEV). The author wishes to thank Dr. John Roobol for interesting discussions about the Cenozoic faults in the coastal plain of the Red Sea, and also Dr. Ian Stewart for useful comments and help with the manuscript. Our thanks extend to Dr. Robert Herrmann for his kind help in carrying out moment tensor inversions for selected events using his software. Deepest thanks are extended to Dr. Mahmoud El-Hadidy for producing the figures and to Dr. Najeeb M.A. Rasul for his patience while we finished the chapter. Also, I would like to express my deep thanks to Dr. Walter Mooney for his critical review of the manuscript.

References

Akaad MK, Noweir AM, Abu El Ela AM (1995) The volcano sedimentary association and ophiolites of Wadi Mubarak, eastern Desert, Egypt. In: Proceedings international conference 30 years Cooper. Geological Survey Egypt, Special Publication 69, pp 231–248

Aldamegh KS, Abou Elenean KM, Hussein HM, Rodgers AJ (2009) Source mechanisms of the June 2004 Tabuk earthquake sequence, Eastern Red Sea margin, Kingdom of Saudi Arabia. J Seismolog 13:561–576. doi:10.1007/s10950-008-9148-5

Bellahsen N, Faccenna C, Funiciello F, Daniel JM, Jolivet L (2003) Why did Arabia separate from Africa? Insights from 3-D laboratory experiments. Earth Planet Sci Lett 216:365–381

Bohannon RG (1986) Tectonic configuration of the western Arabian continental margin, Southern Red Sea. Tectonics 5(4):477–499

Camp VE, Roobol MJ (1992) Upwelling asthenosphere beneath western Arabia and its regional implications. J Geophys Res 97:15255–15271

Chu D, Gordon RG (1998) Current plate motions across the Red Sea. Geophys J Int 135:313–328

Cochran JR, Martinez F (1988) Evidence from the northern Red Sea on the transition from continental to oceanic rifting. Tectonophysics 153:25–53

Daradich A, Mitrovica JX, Pysklywec RN, Willett SD, Forte AM (2003) Mantle flow, dynamic topography, and rift-flank uplift of Arabia. Geology 31:901–904

Davison I, Al-Kadasi M, Al-Khirbash A, Baker J, Blakey S, Bosence D, Dart C, Heaton R, McClay K, Menzies M, Nichols G, Owen L, Yelland A (1994) Structural evolution of the southeastern Red Sea margin, Republic of Yemen. Geol Soc Am Bull 106:1474–1493

Ebinger CJ, Sleep NH (1998) Cenozoic magmatism throughout east Africa resulting from impact of a single plume. Nature 395:788–791

El Gaby S, List FK, Tehrani R (1990) The basement complex of the Eastern Desert and Sinai. In: Said R (ed) The geology of Egypt. Balkema, Rotterdam, pp 175–184

El Hadidy S, Mohamed AA, Albert RNH (1998) Characteristics of earthquake activity along the Gulf of Aqaba. Bull Ain Shams Sci 36:111–150

El Hadidy S, Zahran H, Roobol MJ, Stewart ICF, Farouk M (2012) Recent earthquake activity in western Saudi Arabia. In: 7th Gulf seismic forum, 2012, Saudi Geological Survey, Jeddah (abstract)

Fairhead JD, Girdler RW (1970) The seismicity of the Red Sea, Gulf of Aden and Afar triangle. Philos Trans R Soc A 267:49–74

Fritz H, Wallbrecher E, Khudier AA, Abu El Ela F, Dallmeyer RD (1996) Formation of Neoproterozoic metamorphic core complexes during oblique convergence, Eastern Desert, Egypt. J Afr Earth Sci 23:311–329

Greiling RO, Abdeen MM, Dardir AA, El Akhal H, El Ramly MF, Kamal El Din GM, Osman AF, Rashwan AA, Rice AH, Sadek MF (1994) A structural synthesis of the Proteorozoic Arabian-Nubian Shield in Egypt. Geol Rundsch 83:484–501

Haasib G (1990) Study of the focal mechanisms for recent earthquakes in Egypt and their tectonic implications. MSc thesis, Faculty of Science, Sohag University, Egypt

Herrmann R, Ammon C (2004) Computer programs in seismology (CPS), source inversion. http://www.eas.slu.edu/pub/rbh

Hosny AA, El Hady S, Guidarelli M, Panza M (2011) Source moment tensors of the earthquake swarm in Abu-Dabbab area, south-east Egypt. Rend Fis Acc Lincei. doi:10.1007/s12210-0110158-9

Huang P, Soloman S (1987) Centroid depth and mechanisms of mid-ocean ridge. J Geophys Res 92:1361–1383

Huchon P, Jestin F, Cantagrel JM, Gaulier JM, Al Khirbash S, Gafaneh A (1991) Extensional deformations in Yemen since Oligocene and the Afar triple junction. Annual Tectonicae 5:141–163

Maamoun ME, El-Kashab HM (1978) Seismic study of the Shadwan (Red Sea) earthquake. Academy of Scientific Research and Technology, Helwan Institute of Astronomy and Geophysics, Bulletin no 171

McGuire A, Bohannon R (1989) Timing of mantle upwelling: evidence for a passive origin for the Red Sea Rift. J Geophys Res 94:1677–1682

McKenzie D (1978) Some remarks on the development of sedimentary basins. Earth Planet Sci Lett 40:25–32

Neumayr P, Mogessie A, Hoinkes G, Puhl J (1996) Geological setting of the Meatiq metamorphic core complex in the Eastern Desert of Egypt based on amphibolite geochemistry. J Afr Earth Sc 23 (3):331–345

Pallister JS, McCausland WA, Jónsson S, Lu Z, Zahran HM, El Hadidy S, Aburukbah A, Stewart ICF, Lundgren PR, White RA, Moufti MRH (2010) Broad accommodation of rift-related extension recorded by dike intrusion in Saudi Arabia. Nature Geosci, 8 pp, with supplementary information, 5 pp. doi:10.1038/NGEO966

Roobol MJ, Al-Rehaili M, Arab N, Celebi M, Halawani MA, Janjou D, Kazi A, Martin C, Sahl M, Showail A (1999) The Gulf of Aqaba earthquake of 22 November 1995: its effects in Saudi Arabia. Saudi Arabian Deputy Ministry for Mineral Resources Technical Report BRGM-TR-99-16, 67 pp, 49 figures, 4 tables, 1 appendix

Voggenreiter W, Hötzl H, Jado AR (1988) Red Sea related history of extension and magmatism in the Jizan area (southwest Saudi Arabia): indication for simple-shear during Red Sea rifting. Geol Rundsch 77:257–274

Wernicke B (1985) Uniform-sense normal simple shear of the continental lithosphere. Can J Earth Sci 22:108–125

Seismicity of the Western Side of the Southern Red Sea

Ghebrebrhan Ogubazghi and Berhe Goitom

Abstract

The seismicity of the southern Red Sea is driven by the geological and geophysical processes taking place at and around the Afar Triple Junction, where the Red Sea, East African, and Gulf of Aden Rifts meet. From this plate tectonic node, the Nubian, Arabian, and Somali Blocks are moving away from each other. The consequent structures being predominantly extensional, almost all focal mechanisms of earthquakes in the region express normal faulting, although local tectonic complications are revealed in subordinate strike-slip mechanisms. The b-value, which indicates regional stress conditions and corresponding modalities of energy release, ranges between 0.5 and about 1.5. Seismic episodes are mostly of the swarm type, comprising small to moderate magnitude earthquakes at shallow focal depths. These episodes are occasionally accompanied by volcanism, either basaltic or silicic. Four recent major seismic events in the region, and their significant economic and social consequences, are briefly considered. As the economic development of the region proceeds, the impacts of these natural hazards will inevitably become more severe. This invites the preparation of appropriate mitigation procedures.

Introduction

Despite a relative paucity of seismological recording stations around the Afar region itself, its continuing seismic and volcanic activity has begun to attract attention. The southern Red Sea has been studied by Fairhead and Girdler (1970), El-Isa and Al Shanti (1989), and Sakr et al. (2005); the Afar region by Hofstetter and Beyth (2003), Ayele et al. (2005, 2007a), Ebinger et al. (2008, 2010), and Keir et al. (2009); the wider Horn of Africa by Gouin (1979) and Ayele and Kulhanek (1997); and the East African Rift System as a whole by Fairhead and Girdler (1972), Kebede and Kulhanek (1991, 1994), and Keir et al. (2006). The present study concentrates on the region comprising the coastal zone (offshore and onshore) of the southwestern Red Sea and the northeastern sector of the Afar Depression (AD). As is now well known, Afar is a tectonic triple junction where the Red Sea, the Gulf of Aden, and the East African Rift System meet (Barberi et al. 1972; Tesfaye et al. 2003). It is a focus of lithospheric extension where the Arabian, Nubian, and Somali plates are moving away from each other (Chu and Gordon 1998; McClusky et al. 2010).

The AD and its surroundings show abundant evidence of historical and ongoing seismic and volcanic activity. Prominent during the last half-century have been the 1961 earthquake sequence at Karakore in western Afar (Gouin 1979), the 1969 sequence at Sardo in central Afar (Dakin et al. 1971), the 1989 sequence in the Dobi graben east of Sardo (Sigmundsson 1992; Jacques et al. 1999), the 1993 Bada swarm in southern Eritrea (Ghebreab and Solomon 1994; Ogubazghi et al. 2004), the 2002 sequence in northern Afar (Ayele et al. 2007a) and the 2005 volcanic and seismic activities in the same region (Wright et al. 2006; Ayele et al. 2007b; Ebinger et al. 2010), and other events (e.g., Hagos

G. Ogubazghi (✉) · B. Goitom
Department of Earth Sciences, Eritrea Institute of Technology, P.O. Box 12676, Asmara, Eritrea
e-mail: ogubazghi_ghebrebrhan@yahoo.com

G. Ogubazghi
Department of Geology, University of Botswana, Private Bag, 00704 Gaborone, Botswana

N.M.A. Rasul and I.C.F. Stewart (eds.), *The Red Sea*,
Springer Earth System Sciences, DOI 10.1007/978-3-662-45201-1_9,

et al. 2006a). The Erta'Ale volcano in northern Afar is continuously active with a permanent lava lake (Barberi and Varet 1970) and continues to show changes in lake height, recently overflowing (Field et al. 2012). The most recent large volcanic eruption came in 2011 from Nabro volcano in northeastern Afar, associated with numerous small-to moderate-sized earthquakes. It is a spectacular reminder that the region is prone to a high level of seismicity and occasional related volcanism, either basaltic or silicic.

In this article, emphasis is placed on the coastal sector of the region and some important events within Afar (events in the Gulf of Tadjoura are not included as the intense seismicity of that region is separately related to Gulf of Aden seafloor spreading and rifting).

Tectonic Setting

Detailed description of the tectonics of the Red Sea is given by Bosworth (this volume). The tectonic structure of the area under study is controlled by the complex dynamics of the three plate motions diverging from the Afar triple junction. The separation of the Arabian plate from the Nubian plate across the Red Sea Rift (RSR) occurs at an axial spreading zone in the Rift. However, at 17.5°N 40.0°E (Chu and Gordon 1998; McClusky et al. 2010), there is a southward bifurcation of the spreading, accompanied by a transform offset (Barberi and Varet 1977). One spreading line continues southeastward along the Red Sea axis, but progressively weakens toward Bab-al-Mandab and fails to reach the Gulf of Aden Rift (GAR). The other branch projects SSW-ward into northern and then central Afar where it curves to merge with the western extension of the Gulf of Tadjura Rift (GTR). The seismicity pattern confirms this tectonic geometry (Fig. 2). The block/plate model introduced by McClusky et al. (2010), and derived from GPS measurements, is also consistent with this general setup (see Fig. 1). The Danakil Block (DB) is a microplate rotated between the two bifurcated spreading zones (Fig. 1). The geology and kinematics of this block have been discussed by Chu and Gordon (1998), Eagles et al. (2002), Lahitte et al. (2003), and McClusky et al. (2010).

Swarms of normal faults dominate the landscape in the AD (Barberi et al. 1972; Mohr 1972), although subordinate strike-slip faults have been identified (e.g., Abbate et al. 1995).

The main driving force for plate divergence in the Afar region is generally considered to come from the upwelling of a deep mantle plume (the Afar mantle plume) impacting the base of the lithosphere (White and McKenzie 1989; Ebinger and Sleep 1998; George et al. 1998). However, recent high-quality seismic data from the region have failed to detect classic conduit-like plumes (Hammond et al. 2013) and this is an ongoing debate (Rooney et al. 2012; Rychert et al. 2012; Hammond et al. 2013; Ferguson et al. 2013).

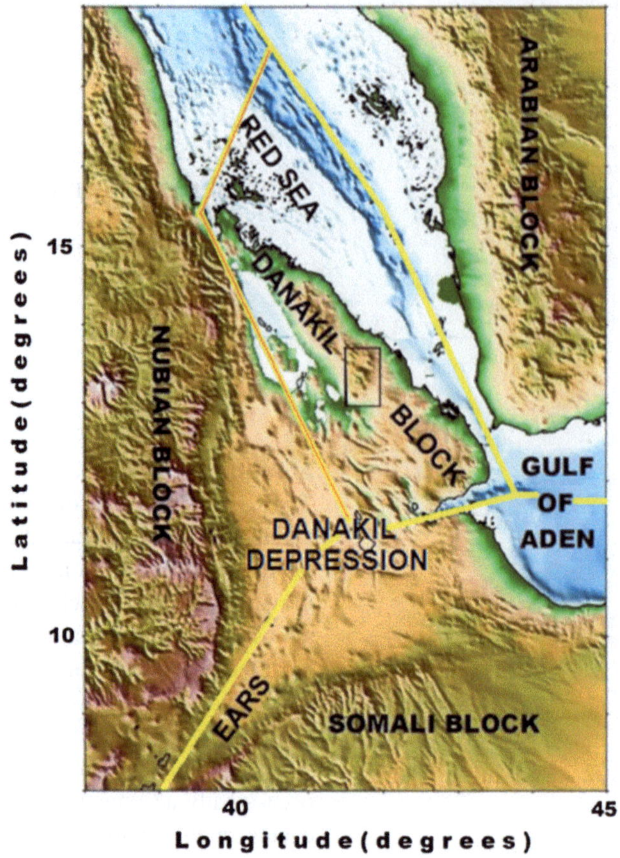

Fig. 1 Tectonic map of the study region. The boundary between blocks/plates, derived from GPS measurements, is indicated by *yellow line* segments (adopted from McClusky et al. 2010). The western part of the Danakil block, shown in *yellow* and *red line* segments, has higher uncertainty in its location. The *rectangle* in the Danakil block encloses the Nabro volcanic range, which was the scene of the 2011 volcanic eruption

The Data Set

The data set for the region may be classified as historic or preinstrumental (for the period before 1900), early instrumental (1900–1963), and late instrumental and contemporary periods (1964 up to present) (Basham and Giardini 1993). Compared to other parts of the globe (see for example Tan et al. 2008), the historical catalog for the Red Sea-Horn of Africa region is a scanty one. Most of the historic volcanic and seismic events of the region were reported to newspapers in Europe by sailors passing through the Red Sea (Gouin 1979). Records of the Orthodox Churches, missionaries, and other foreign settlers in the region, who understood the importance of recording the information, also contributed to this effort. From such sources, Gouin (1979) has carefully analyzed the data and compiled a comprehensive document for the historical seismicity of the region. Nevertheless, his work reveals that most historical earthquakes in the region are poorly constrained in epicentral location and magnitude.

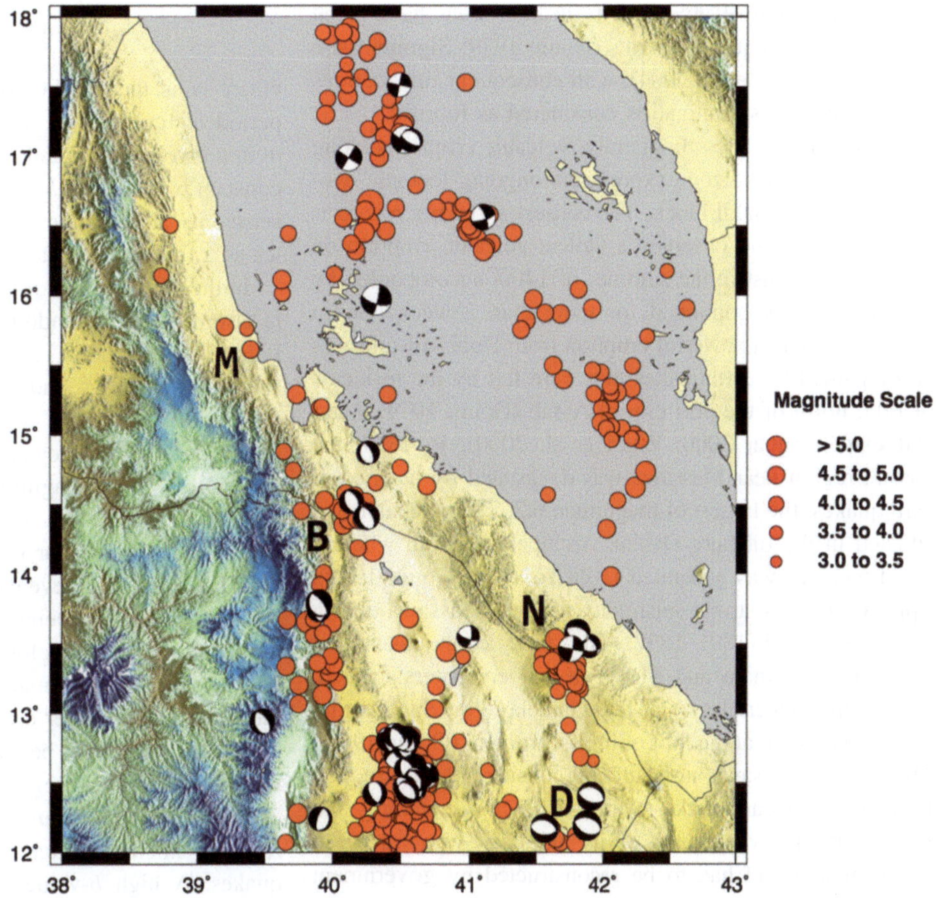

Fig. 2 Epicenters and focal mechanisms of the largest events for the period 1973–2011. The locations of the four earthquake sequences discussed later in the text are indicated by M, B, D, and N, representing Massawa, Bada, Dobi, and Nabro, respectively. Courtesy ISC catalog and Harvard CMT fault plane solutions

During the early instrumental period, detection and location of earthquakes in the southern Red Sea region was possible only for large events recorded at a few remote stations around the globe. Due to the low sensitivities of early seismographs, records were restricted to events with body wave magnitudes (m_B) equal to or greater than 5.5. A potentially important exception to this came from the installation in 1913 of a seismograph in Asmara, in immediate response by the Italian administration to an ongoing strong seismic swarm event. This also served as Italy's contribution to global seismology (Palazzo 1913). The recordings, regrettably since lost, detected 141 local events (Palazzo 1913; Gouin 1979). Due to reliance on relatively few global seismic stations, the accuracy of epicentral location for the southern Red Sea during the early instrumental period was poor, with errors of the order of 200 km or even 500 km (Fairhead and Girdler 1970).

Substantial improvement in epicentral and magnitude determination came in 1963 with the establishment of the Worldwide Standard Seismograph Network (WWSSN) and the introduction of better processing techniques. In the Red Sea region and environs, WWSSN stations commenced operation at Addis Ababa (AAE), Nairobi (NAI), Helwan, near Cairo (HLW), and Jerusalem (JER). During the last decade of the twentieth century, several local stations have been installed in Djibouti, Eritrea, and Ethiopia. And with the establishment in 1993 of the Eastern and Southern Africa Regional Seismological Working Group (ESARSWG) has come a further improvement in lowering the threshold of detectability of seismic events (Dindi et al. 1995). The ESARSWG has brought together the countries along the East African Rift System to merge their respective seismic data and produce a common regional seismological bulletin. This has greatly improved the accuracy of epicentral determination, although the Red Sea region remains sparsely covered by seismological stations compared with other ESARSWG member countries (Dindi et al. 1995).

Seismicity of the Area

Brief Seismic History of the Region

The seismic history of the western side of the southern Red Sea has been presented up to the mid-1970s with comprehensive and thorough analysis by Gouin (1979). The southern Red Sea and northern Afar regions are considered in his section C, and the Eritrean escarpment in section A.

Almost all recent seismic activity in both areas have taken the form of earthquake swarms (Gouin 1979; Sigmundsson 1992; Ogubazghi et al. 2004), with consequent implications for earthquake risk mitigation considered as follows.

Seismicity and, less frequently, volcanic eruptions along the Eritrean coastal zone express the ongoing tectonic evolution of the Danakil block. The earliest surviving record is that of sailors who observed a violent volcanic eruption on the Red Sea coast in the summer of 1400, accompanied by earthquakes that continued for more than a week (Gouin 1979). In 1861, a powerful eruption from Dubi volcano was accompanied by earthquakes that were felt by the residents of Edd, north of the port city of Assab (Gouin 1979; Wiart and Oppenheimer 2000; Wiart et al. 2000). In 1884, the coastal region near Massawa was devastated by a series of earthquakes, the largest of magnitude 6.2. The epicenter was situated in the offshore Dahlak Archipelago, from where a small tsunami was generated (Palazzo 1915; Gouin 1979). Episodes of less strong seismicity affected Massawa during 1886–1887 and 1891–1897. The 1913 earthquakes that shook both Asmara and Massawa had epicenters located, with some uncertainty, in the Gahtelay (or Sabarguma) marginal graben, at the foot of the escarpment (Gouin 1979). The port of Massawa next suffered considerable destruction from a strong earthquake swarm in 1921 (Gutenburg and Richter 1954), when most of its buildings were rendered uninhabitable and had to be reconstructed by government order (Gouin 1979).

Epicentral Distributions

While there is a clear alignment of epicenters along the RSR axial zone north of latitude 17.5°N (Fig. 2), to the south, this zone divides. One branch continues along the RSR axis via volcanic islets, but peters out before reaching Bab-al-Mandab. The other branches SSW via the Dahlak Archipelago to the Gulf of Zula where it then turns SSE-ward along the active volcanic axis of northern Afar. A separate line of epicenters follows the northern Afar plateau margin and its young marginal graben, thus complicating interpretation of historical earthquakes in the Asmara–Massawa area (e.g. Gouin 1979). These epicentral alignments express the active tectonic and geological development of the southern Red Sea and northern Afar regions (e.g., Hofstetter and Beyth 2003).

The *b*-Value

Gutenberg and Richter (1954) introduced an earthquake magnitude-to-recurrence period relationship, given by the equation

$$\mathrm{Log}_{10}N = a - bm$$

where N is the number of earthquakes within a specified period of time having magnitudes equal to or greater than m in a given region, and a and b are positive constants. The constant b, called the b-value, is widely used for the study of seismicity and seismic hazard assessment. The b-value varies as a function of space, time, and also focal depth.

In the computation of the b-value, earthquake occurrence is assumed to be a random process that can be modeled as a Poisson distribution. Thus, dependent events, such as foreshocks, aftershocks, and triggered events, are removed from the data set before the computation of the b-value.

Further, the completeness of the catalog is assessed. If m' is assigned as the magnitude of completeness of a catalog, then it is understood that all earthquake events having magnitudes equal to or greater than m' are included in the set. However, some events with magnitudes below m' may be absent, possibly being below the detection threshold of the local, regional, or global monitoring systems. The value of m' is therefore the smallest magnitude that m can attain in the Gutenberg–Richter earthquake recurrence relationship.

The b-value may be viewed as an indicator of energy release in a given region. A low b-value (less than 1) expresses a region under high stress, where energy is released chiefly through moderately seized or large earthquakes. A high b-value (greater than 1) expresses a low stress regime where energy is released in numerous smaller earthquakes. For this reason, the b-value is treated as a stress meter (Schorlemmer and Wiemer 2004, 2005).

The b-value may be interpreted statistically in the following manner (Lay and Wallace 1995): Assume that the b-value of a given region is equal to 1.0. Further, assume that the number of events per year that occur in the region with magnitudes between 6.0 and 7.0 is 10. Then, the corresponding numbers for magnitudes between 5.0 and 6.0, and between 7.0 and 8.0, are equal on the average to 100 and 1, respectively. Similarly, for b-values more than 1.0 and for an annual occurrence of 10 events with magnitude between 6.0 and 7.0, the corresponding number of events with magnitudes between 5.0 and 6.0, and between 7.0 and 8.0, is equal on the average to more than 100 and less than 1, respectively, and vice versa.

Among the researchers investigating the b-value for the southern Red Sea and neighboring regions, Kebede and Kulhanek (1994) have examined both its spatial and temporal variations. To obtain a reasonably homogenized magnitude, they took data from the National Earthquake Information Services (NEIS) catalog for the period from 1963 to 1990, and obtained a completeness for the region of $m_B \geq 4.6$. The results for the southern Red Sea (SRS) show b-values ranging from a low of 0.5 to a high of 1.5 (Fig. 3).

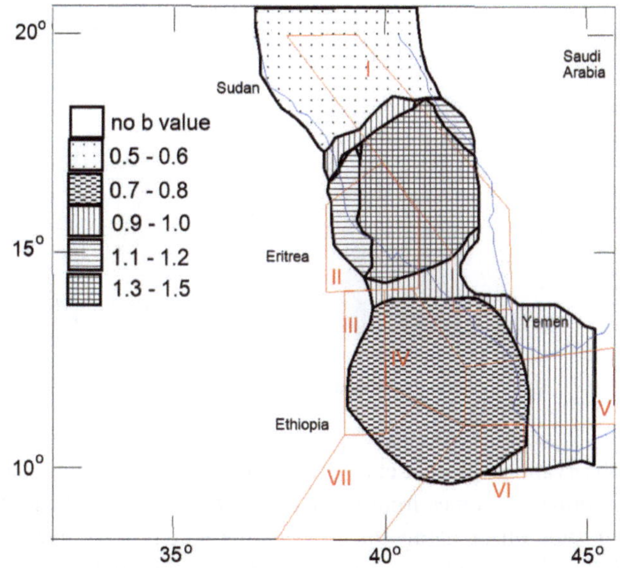

Fig. 3 Spatial distribution of *b*-values (Kebede and Kulhanek 1994), and seismo-tectonic zones outlined in *red* (Hofstetter and Beyth 2003). *Shore lines* are shown in *blue*

Lower values apply mainly to the northern and southern sectors of the SRS, higher values to the central sector.

Ayele and Kulhanek (1997) have repeated this study (for the period 1960–1993), but using an improved database which benefited from updated regional catalogs and the ISC catalog. They obtained a completeness value of $m_B \geq 4.5$ (Ayele 1995). The results for the SRS region show *b*-values ranging from a low of 0.7 to a high of 1.3 (Fig. 4). Lower values again apply, but now with improved resolution, mainly to the northern and southern sectors of the SRS, higher values mainly to the central sector and the plateau–Afar escarpment zone.

Hofstetter and Beyth (2003) have used data from NEIS, Ayele (1995), and Jacques et al. (1999), for the period 1960–2000 to calculate mean *b*-values for the seismically active regions enclosed within the area 4°–20°N; 36°–46°E. Based on seismicity characteristics, they partitioned this area into seven regions. These are shown overlain in Figs. 3 and 4 for comparison purposes. The SRS is incorporated almost wholly within regions I and II, for which Hofstetter and Beyth (2003) obtained *b*-values of 0.95 and 1.07, respectively. Although comparison of the results of Kebede and Kulhanek and Ayele and Kulhanek with those of Hofstetter and Beyth must be treated with caution, since the latter have lumped large parts of the study area together, yet globally the results are in good agreement (Fig. 5).

Focal Mechanisms

Focal mechanisms of events in the southern Red Sea and Afar regions have been comprehensively studied by Hofstetter and Beyth (2003). They provide the source parameters of 125 events, which occurred during 1960–2000. For some of these events, mostly earlier ones, source parameters were obtained from Harvard Centroid Moment Tensor (CMT) fault plane solutions, which are based on global distributions of seismometers. For those events with good waveform data, mostly recent events, source parameters were obtained through waveform inversion. The compilation of the source parameters also includes results obtained from other workers (e.g., Kebede et al. 1989).

The results of Hofstetter and Beyth are generally consistent with the predominant normal fault tectonics of the region. However, the overall kinematics and dynamics of the Afar triple junction include local complications, and there are instances of strike-slip focal mechanisms (McKenzie

Fig. 4 Spatial distribution of *b*-values from Ayele and Kulhanek (1997). *Red* and *blue lines* serve as in Fig. 3, and *GA* stands for Gulf of Aden

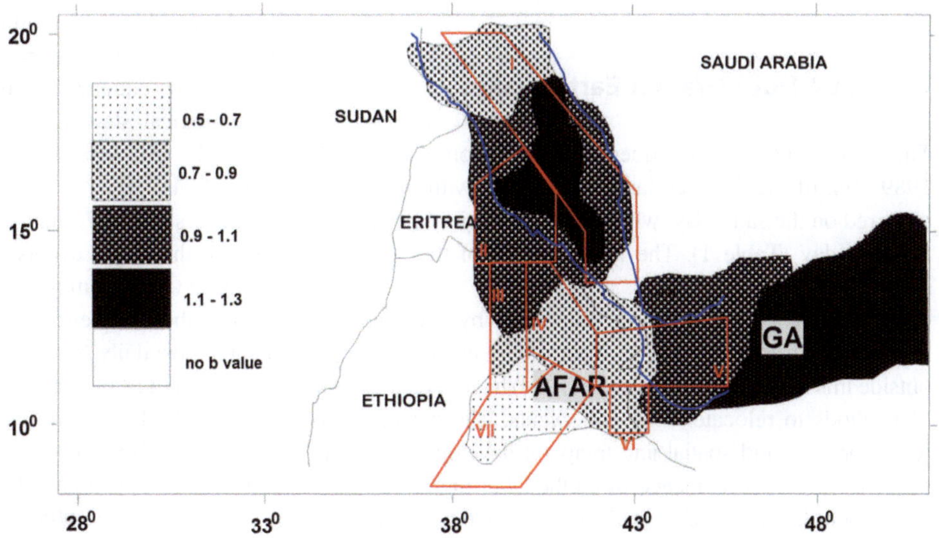

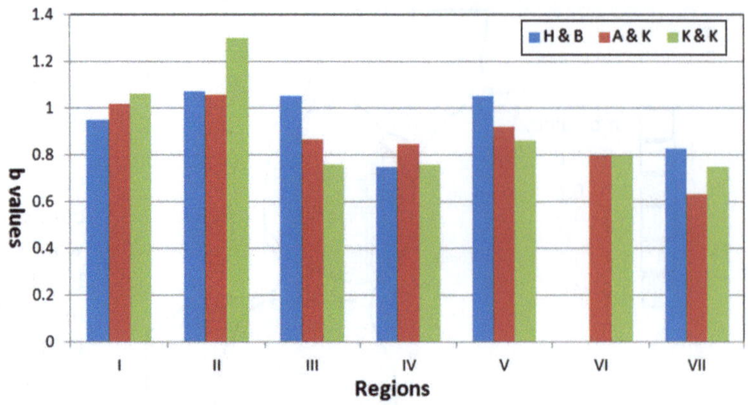

Fig. 5 Comparison of mean *b*-values obtained by Kebede and Kulhanek, Ayele and Kulhanek, and Hofstetter and Beyth for the seven regions defined by Hofstetter and Beyth (2003)

et al. 1970; Abbate et al. 1995; Hofstetter and Beyth 2003; Hagos et al. 2006a). For example, focal mechanisms suggest a nascent transform fault situated immediately south of where the RSR bifurcates at latitude 17.5°N (see Fig. 2).

Some Features and Social and Economic Impacts of Selected Earthquakes

Four recent strong earthquake sequences are considered that impacted on the social and economic life of the region: the 1989 Dobi Graben earthquakes, the 1993 Bada earthquakes, the 2002 Massawa earthquakes, and the 2011 Nabro seismo-volcanic sequence. All four are of the seismic swarm type, lacking a main shock characterized by a magnitude greater by around two than the largest aftershock or foreshock of the sequence (Lay and Wallace 1995). These swarm sequences pose particular hazards to man-made structures, and as discussed below, appropriate mitigation efforts are required to minimize future damaging effects.

The details of the September 18, 2013 (magnitude 5.2), Massawa earthquake are not well scrutinized at this point in time and therefore are not included in this section.

The 1989 Dobi Graben Earthquakes

The first event of this sequence occurred on August 20, 1989. Ten of the fourteen larger events (all with $m_B > 5.0$) occurred on the same day, while the remaining four occurred the next day (Table 1). The largest shock of the sequence had magnitude $m_B = 6.2$.

The sequence was closely monitored by the Djibouti seismological network, but as most of the events occurred outside this network, Jacques et al. (1999) have used a variety of methods to relocate some 300 events. The derived rupturing process and spatial and temporal development of the sequence provided evidence of earthquake migration along the graben. Jacques et al.'s analysis was made before InSAR observations (Wright et al. 2006) were available. Additional information comes from geodetic evidence for 10–20 cm of sinistral offset along the graben, measured some months following the 1989 sequence (Mohr, unpublished data). This is consistent with the bookshelf tectonic model for east-central Afar proposed by Tapponnier et al. (1990).

The 1989 Dobi graben sequence destroyed several bridges and a segment of the road linking the Ethiopian capital Addis Ababa with the Eritrean port city of Assab. The road remained closed to traffic for about one month before temporary repairs were performed. Although there was no reported loss of life, the cost of disruption of services and damages to infrastructure is considered to be high, running into millions of dollars.

The 1993 Bada Earthquakes

The Bada earthquake sequence occurred around the small town of Bada, some 130 km SSE of Massawa and 60 km inland from the Red Sea coast. The epicenters were clustered in the Comayli marginal graben at the foot of the plateau escarpment (Mohr 1975). Following a few foreshocks during late April 1993, the earthquake sequence proper commenced on May 1 with a $m_B = 4.3$ event and numerous smaller events. The sequence was monitored by the ASME (Asmara) seismological station, located 150 km to the north–northwest. The ISC located 18 of the larger events (Ogubazghi et al. 2004).

Figures 6 and 7 show the spatial and temporal distributions of the larger events of the sequence, the former with reference to the Comayli marginal graben. Differing from the Dobi graben sequence, a relatively big event occurred at Bada almost daily for about two weeks from May 1 (Fig. 7). The magnitude of the largest event was $m_B = 5.2$. The *b*-value of the Bada earthquake sequence has been computed from the ASME data to have a value of 1.67 (Ogubazghi et al. 2004). This high value is typical of earthquake swarms (Lay and Wallace 1995).

Table 1 List of the largest events of the 1989 Dobi graben earthquake sequence

No.	Date	Origin time	Lat. (°)	Lon. (°)	m_B	Agency
1	20/08/1989	11:16	11.76	41.85	5.7	IPGP
2	20/08/1989	11:17	11.79	41.78	5.7	IPGP
3	20/08/1989	11:46	11.87	41.71	6.0	IPGP
4	20/08/1989	11:56	11.81	41.83	5.2	IPGP
5	20/08/1989	13:25	11.81	41.76	5.2	IPGP
6	20/08/1989	13:26	11.87	41.72	5.3	IPGP
7	20/08/1989	18:27	11.76	41.63	5.1	IPGP
8	20/08/1989	18:39	11.86	41.73	5.4	IPGP
9	20/08/1989	18:54	11.87	41.59	5.3	IPGP
10	20/08/1989	19:25	11.90	41.69	6.1	IPGP
11	21/08/1989	01:09	11.89	41.76	6.2	IPGP
12	21/08/1989	05:03	11.93	41.57	5.7	IPGP
13	21/08/1989	05:05	11.88	41.65	5.2	IPGP
14	21/08/1989	07:07	11.91	41.65	5.1	IPGP

Data from l'Institut de Physique du Globe de Paris (IPGP) of the University of Paris (Jacques et al. 1999)

A team of geologists that visited the area a few days after the commencement of the sequence witnessed landslides and widespread faulting (Ghebreab and Solomon 1994). There were no reported casualties or damages, although the inhabitants, terrified by the ground motion, fled the area.

The 2002 Massawa Earthquakes

Massawa is the second largest city and the main port of Eritrea with a population of about 50,000. As previously remarked, this important economic center has sustained significant and severe destruction from earthquakes during the last one and half centuries.

A preliminary study of the 2002 earthquake sequence conducted by Ogubazghi (2002) has been based on macroseismic and instrumental data from stations ASME and ABSE, located, respectively, 70 km SSW and 90 km due south from Massawa (Fig. 8). About 150 events of the swarm were located, and the largest of them had a local magnitude M_L = 4.4. Almost all events occurred in the neighborhood of the city and some even within the city (Fig. 8), an uncomfortable portent for future hazards. Table 2 lists the largest events of the sequence.

On this occasion, several city buildings suffered only light damages, such as cracking of walls and falling of interior plaster. While the inhabitants of Massawa were alarmed, no casualties were reported.

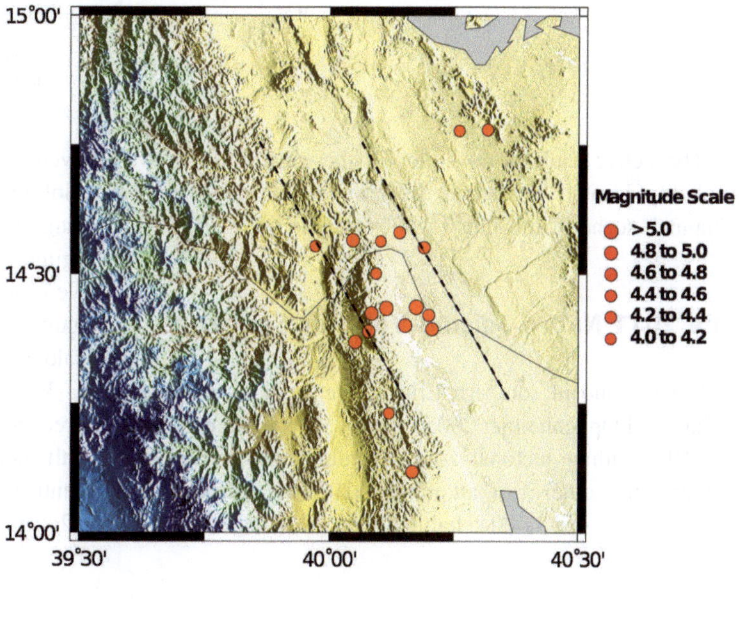

Fig. 6 Spatial distribution of the eighteen largest events of the Bada earthquake sequence. The Comayli marginal graben is outlined between *dotted lines* (adopted from Ogubazghi et al. 2004)

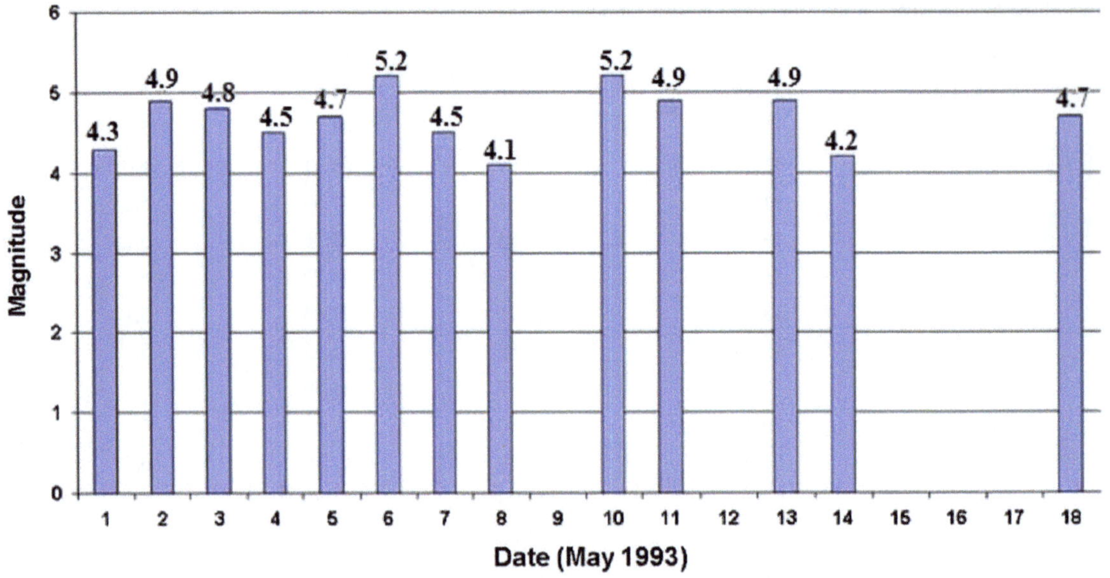

Fig. 7 Time distribution of some of the largest events of the Bada earthquake sequence, with indicated magnitudes (Ogubazghi et al. 2004)

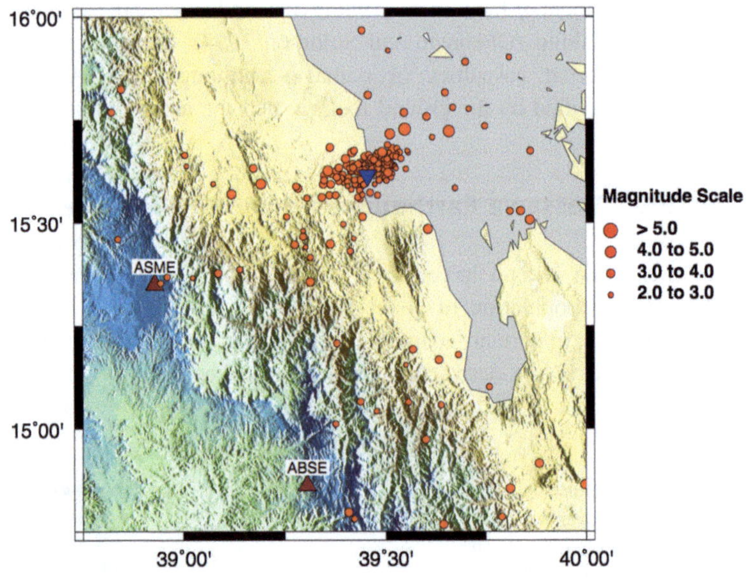

Fig. 8 Epicenters of the Massawa 2002 earthquake sequence for the period January through August 2002, the majority occurring in July. Massawa indicated in *blue*

The active fault responsible for these earthquakes was probably located immediate offshore along the Massawa Channel, scene of the July 1884 event (Gouin 1979).

The 2011 Nabro Seismo-Volcanic Sequence

Nabro volcano in southern Eritrea is a large stratovolcano (Wiart and Oppenheimer 2005), one of several aligned along a NNE-trending tectonic zone. It culminates in a nested caldera, the outer rim attaining 2,200 m altitude. Until Spring 2011, the 3,000 inhabitants of the village of Sireru lived securely on the western floor of the outer caldera (Solomon 2012). Then, on March 31, a strong earthquake of magnitude 4.5 hit the area (Ogubazghi et al. 2014), damaging a number of houses in Sireru (Fig. 9). Although there were no casualties from the shock, according to the inhabitants, a flock of goats were killed by a rockslide down the caldera wall.

We speculate that the earthquake might be considered a precursor to the volcanic eruption which followed two and a half months later. For Nabro, volcano awoke abruptly from centuries-long dormancy to violent eruptive activity on June 12, 2011. About 4 h before the start of the eruption, a swarm

Table 2 List of the largest events of the 2002 Masswa earthquake sequence

No.	Date	Origin time	Lat. (°)	Lon. (°)	M_L	Agency
1	05/07/2002	11:35:31.3	15.653	39.485	4.3	ASM
2	07/07/2002	14:44:36.8	15.620	39.369	3.1	ASM
3	07/07/2002	19:09:24.2	15.626	39.358	3.6	ASM
4	07/07/2002	20:29:13.0	15.670	39.418	3.0	ASM
5	10/07/2002	01:00:11.0	15.635	39.425	3.1	ASM
6	13/07/2002	00:14:21.3	15.726	39.549	4.4	ASM
7	13/07/2002	04:16:14.1	15.654	39.465	3.4	ASM
8	17/07/2002	18:56:43.1	15.722	39.659	4.0	ASM
9	17/07/2002	19:41:54.9	15.715	39.512	3.5	ASM
10	18/07/2002	17:22:16.9	15.644	39.505	3.7	ASM
11	25/07/2002	01:51:40.7	15.655	39.402	3.1	ASM
12	01/08/2002	01:56:10.3	15.448	39.363	3.0	ASM

Data from station ASME at the University of Asmara (Ogubazghi 2002)

Fig. 9 Most houses in Sireru village were built on sedimentary infill on the caldera floor. Poorly built houses like the one shown collapsed. Behind, a better built one sustained only minor cracking

of strong earthquakes commenced. Table 3 lists the larger events (all with $m_B > 4.5$) that were reported by the PDE of the USGS. The largest event in the sequence had a magnitude $m_B = 5.6$.

Sireru eyewitnesses stated that the eruption commenced at about 23:45 h local time (20:45 UTC). Unfortunately, not all the villagers left the area. Sireru village was completely destroyed, buried partly by lava and partly by volcanic ash (Figs. 10 and 11). Seven lives were lost, and about 12,000 inhabitants were displaced from local villages (Solomon 2012). The eruption continued for about one month, ejecting tephra reaching up to a maximum altitude of 22 km. Air traffic in the region was disrupted, and the monetary value of animal life and property destroyed has been estimated in millions of dollars (Solomon 2012).

Discussion and Conclusions

The seismicity of the SRS region is dominated by the extensional processes of the Red Sea, the Gulf of Aden, and the East African Rift Systems. In fact, the AD is virtually the only place in the world where the transition between continental and oceanic rifting is subaerially exposed. The magnitudes of the earthquakes resulting from the tectonic activity in this region range from small to moderate and are shallow in depth. The largest earthquakes that occurred in the SRS region during the last 100 years had magnitudes less than 7.0, confirming this point; they also had shallow depths (Gouin 1979; Hofstetter and Beyth 2003). This unique geologic environment has attracted important scientific expeditions involving at times close to 200 broadband seismographs (e.g., Maguire et al. 2006; Hammond et al. 2013). As a result, at times, high-precision earthquake locations can be achieved (Keir et al. 2006), but these projects typically last a few years at most, and the region is lacking permanent systems for monitoring seismic and volcanic activities. Consequently, the long-term patterns of seismicity are poorly constrained.

It is interesting to note that for the four earthquake sequences considered, the Dobi and Nabro events had higher magnitude events and the distributions of the largest events were crowded to the first few days of the start of the sequences. In the case of the Massawa and Bada sequences, the magnitudes were relatively smaller and the distributions in time of the largest events were relatively wider.

The region has a high level of seismicity and volcanism inviting appropriate mitigation efforts. Seismic hazard studies for the region have been performed by Gouin (1976), Kebede and Asfaw (1996), and by the ESARSWG (Midzi et al. 1999). The latter two assessments were done following

Table 3 List of the events of the 2011 Nabro seismo-volcanic earthquake sequence

No.	Date	Origin time	Lat. (°)	Lon. (°)	m_B	Agency
1	31/03/2011	18:33:39.0	13.11	41.98	4.6	PDE
2	12/06/2011	16:33:11.9	13.51	41.72	4.3	PDE
3	12/06/2011	17:18:09.0	13.26	41.78	4.8	PDE
4	12/06/2011	17:47:21.0	13.43	41.70	4.6	PDE
5	12/06/2011	18:01:20.0	13.09	41.91	4.4	PDE
6	12/06/2011	18:48:53.0	13.54	41.88	4.6	PDE
7	12/06/2011	19:03:39.0	13.42	41.92	4.5	PDE
8	12/06/2011	19:21:50.0	13.21	41.83	5.0	PDE
9	12/06/2011	19:37:41.0	13.30	41.69	4.7	PDE
10	12/06/2011	19:44:16.0	13.42	41.64	4.7	PDE
11	12/06/2011	20:32:39.0	13.40	41.67	5.6	PDE
12	12/06/2011	21:03:25.0	13.50	41.68	5.6	PDE
13	12/06/2011	21:37:13.0	13.13	41.68	4.5	PDE
14	12/06/2011	21:41:57.0	13.42	41.67	4.4	PDE
15	17/06/2011	09:16:12.0	13.23	41.73	5.6	PDE

Preliminary determination of epicenters of the USGS

Fig. 10 After the Nabro eruption, trees around Sireru village were half-buried in more than a meter of volcanic ash. Most of the agricultural land around the volcano was thus destroyed

the guidelines of the Global Seismic Hazard Assessment Program (GSHAP) (Giardini and Basham 1993; Basham and Giardini 1993). These assessments used models and data that had various types of defects, which include ill-defined seismo-tectonic source zones, and poor velocity and attenuation models. Efforts are being made to address those defects, and lately, Midzi et al. (2010) have produced velocity models for the region. On the other hand, to avoid these defects, Hagos et al. (2006b) used the methods of spatially smoothed seismicity and Monte Carlo simulation to estimate the hazard of the region.

With regard to hazard and mitigation efforts, the following points apply to the study area.

Due to the inhomogeneous nature of the upper crust, the attenuation of seismic waves in this region is high, resulting in high seismic intensities to be limited to small areas around the epicenter. Consequently, compared to deeper earthquakes, damages due to shallow events affect smaller areas (around the epicenter) than similar deeper events.

Most of the destructive events in the region have been earthquake swarms. If appropriate precautions are not taken, the resulting risks to man-made structures and to human life

Fig. 11 Satellite photograph of the Nabro trachyte lava flow. The flow reached a length of 20 km and a maximum width of 2 km, with ponding at the lava terminus (courtesy of NASA)

could be high. This may be explained in the following manner. Estimates of seismic hazard are given in the form of peak ground acceleration as a percentage of g (the acceleration due to gravity of the Earth) of the horizontal component with certain confidence intervals (Giardini and Basham 1993). Engineers use such calculated parameters in the design of structures. The imbedded assumption is that the earthquakes are normal events with most of the energy being released by the main shock. Under normal conditions, although aftershocks and foreshocks may cause damages, the main threat for a structure designed to withstand the peak load is coming from the main shock. However, in the case of swarms, a structure designed to withstand the peak load is successively confronted by loads coming from the strong events of the sequence. A structure weakened by the first event of the sequence is weakened further by the next strong event of the sequence and may eventually fail. Thus, in the case of swarms, the main threat is coming from all the larger events. Consequently, adjustment must be made to the computed peak ground acceleration to take this into account by adding an appropriate swarm factor.

With economic development in the offing in the region, vulnerability to seismic hazard is likely to increase with corresponding increase in risk. This is particularly true with the commencement of mining efforts in the AD.

The combination of economic development and considerable seismic and volcanic hazard means that the potential risks in the SRS may be increasing. Yet, the lack of monitoring means that the hazards and associated risks remain poorly constrained. An increase in seismic monitoring would be a first step toward a better understanding and risk mitigation in the SRS region.

Acknowledgments Many thanks to Paul Mohr, Vunganai Midzi, and James Hammond for thoroughly reviewing the manuscript and contributing highly valuable comments. Their inputs have indeed improved the chapter significantly.

References

Abbate E, Passerini P, Zan L (1995) Strike-slip faults in a rift area: a transect in the Afar Triangle, East Africa. Tectonophysics 241:67–97

Ayele A (1995) Earthquake catalogue of the Horn of Africa for the period 1960–1993. Seismological Department, Uppsala University Report 3–95, 32 pp

Ayele A, Kulhanek O (1997) Spatial and temporal variations of seismicity in the Horn of Africa from 1960 to 1993. Geophys J Int 130:805–810

Ayele A, Nyblade AN, Langston CA, Cara M, Leveque J-J (2005) New evidence for Afro-Arabian plate separation in southern Afar. Geological Society, London, Special Publications 259:133–141. doi:10.1144/GSL.SP.2006.259.01.12

Ayele A, Stuart G, Bastow I, Keir D (2007a) The August 2002 earthquake sequence in north Afar: insights into the neotectonics of the Danakil microplate. J Afr Earth Sci 48:70–79

Ayele A, Jacques E, Kassim M, Kidane T, Omar A, Tait S, Nercessian A, Chabalier J-B, King G (2007b) The volcano-seismic crisis in Afar, Ethiopia, starting September 2005. Earth Planet Sci Lett 255:177–187. doi:10.1016/j.epsl.2006.12.014

Barberi F, Tazieff H, Varet J (1972) Volcanism in the Afar depression: its tectonic and magmatic significance. In: Girdler RW (ed) East African Rifts. Tectonophysics, vol 15, pp 19–29

Barberi F, Varet J (1970) The Erta'Ale volcanic range (Danakil depression, northern Afar, Ethiopia). Bull Volc 34:848–917

Barberi F, Varet J (1977) Volcanism of Afar: small scale plate tectonics implications. Bull Geol Soc Am 88:1251–1266

Basham P, Giardini D (1993) Technical guidelines for global seismic hazard assessment (GSHAP). Ann Geophys 36(3–4):15–24

Chu D, Gordon G (1998) Current plate motions across the Red Sea. Geophys J Int 135:313–328. doi:10.1046/j.1365-246X.1998.00658.x

Dakin FM, Gouin P, Searle RC (1971) The 1969 earthquakes in Serdo, central Afar. Bull Geophys Obs 13:19–56

Dindi E, Havskov J, Iranga M, Jonathan E, Lombe DK, Mamo A, Turyomurugyendo G (1995) Potential capability of the East African Seismic stations. Bull Seismol Soc Am 85:354–360

Ebinger CJ, Sleep NH (1998) Cenozoic magmatism throughout Africa. Nature 395:788–791

Ebinger CJ, Keir D, Ayele A, Calais E, Wright TJ, Belachew M, Hammond JOS, Campbell E, Buck WR (2008) Capturing magma intrusion and faulting processes during continental rupture: seismicity of the Dabbahu (Afar) rift. Geophys J Int 174:1138–1152. doi:10.1111/j.1365-246X.2008.03877.x

Ebinger C, Ayele A, Keir D, Rowland J, Yirgu G, Wright T, Belachew M, Hamling I (2010) Length and time scales of rift faulting and magma intrusion: the Afar rifting cycle from 2005 to present. Ann Rev Earth Planet Sci 38:437–464

Eagles G, Gloaguen R, Ebinger CJ (2002) Kinematics of the Danakil microplate. Earth Planet Sci Lett 203(2):607–620

El-Isa ZH, Shanti A (1989) The seismicity and tectonics of the Red Sea and western Arabia. Geophys J Int 97:445–457

Fairhead JD, Girdler R (1970) The seismicity of the Red Sea, Gulf of Aden and Afar triangle. Philos Trans R Soc Lond 267:49–74
Fairhead JD, Girdler R (1972) The seismicity of East African rift system. Tectonophysics 15:115–122
Ferguson DJ, Maclennan J, Bastow ID, Pyle DM, Jones SM, Keir D, Blundy JD, Plank T, Yirgu G (2013) Melting during late-stage rifting in Afar is hot and deep. Nature 499:70–77. doi:10.1038/nature12292
Field L, Barnie T, Blundy J, Brooker RA, Keir D, Lewi E, Saunders K (2012) Integrated field, satellite and petrological observations of the November 2010 eruption of Erta Ale. Bull Volc 74:2251–2271. doi:10.1007/s00445-012-0660-7
George R, Rogers N, Kelley S (1998) Earliest magmatism in Ethiopia: evidence for two mantle plumes in one basalt province. Geology 26:923–926
Ghebreab W, Solomon S (1994) Formation of a recent asymmetric graben-in-graben structure at Bada, the Danakil Depression of Eritrea. Geo-Sci Dev 1:16–18
Giardini D, Basham P (1993) The global seismic hazard assessment program (GSHAP). Ann Geophys 36(3–4):3–13
Gouin P (1976) Seismic zoning in Ethiopia, Bull Geophys Obs, No 14
Gouin P (1979) Earthquake History of Ethiopia and the Horn of Africa. International Development Research Center (IDRC)-118e, Canada, 259 pp
Gutenberg B, Richter CF (1954) Seismicity of the earth and associated phenomena, 2nd ed, Princeton University Press, Princeton, 310 pp
Hagos L, Shomali H, Roberts R (2006a) Re-evaluation of focal depths and source mechanisms of selected earthquakes in the Afar Depression. Geophys J Int 167:297–308
Hagos L, Arvidsson R, Roberts R (2006b) Application of the spatially smoothed seismicity and Monte Carlo methods to estimate the seismic hazard of Eritrea and the surrounding region. Nat Hazards 39:395–418
Hammond JOS, Kendal J-M, Stuart GW, Ebinger CJ, Bastow I, Keir D, Ayele A, Belachew M, Goitom B, Ogubazghi G, Wright TJ (2013) Mantle upwelling and initiation of rift segmentation beneath the Afar Depression. Geology. doi:10.1130/G33925.1
Hofstetter R, Beyth M (2003) The Afar depression: interpretation of the 1960–2000 Earthquakes. Geophys J Int 155:715–732
Jacques E, Ruegg JC, Lepine JC, Tapponnier P, King GCP, Omar A (1999) Relocation of M $\geq$ 2 events of the 1989 Dobi seismic sequence in Afar: evidence for earthquake migration. Geophys J Int 138:447–469
Kebede F, Kim W-Y, Kulhanek O (1989) Dynamic source parameters of the March–May 1969 Serdo earthquakes in Central Afar, Ethiopia, deduced from teleseismic body waves. J Geophys Res 94:5603–5614
Kebede F, Kulhanek O (1991) Recent seismicity of the East African rift system and its implications. Phys Earth Planet Inter 68:259–273
Kebede F, Kulhanek O (1994) Spatial and temporal variations of *b*-values along the East African rift system and southern Red Sea. Phys Earth Planet Inter 83:249–264
Kebede F, Asfaw LM (1996) Seismic hazard assessment for Ethiopia and the neighboring countries. SINET, Ethiop J Sci 19:15–50
Keir D, Stuart GW, Jackson A, Ayele A (2006) Local earthquake magnitude scale and seismicity rate. Bull Seismol Soc Am 96 (6):2221–2230. doi:10.1785/0120060051
Keir D, Hamling IJ, Ayele A, Calais E, Ebinger C, Wright TJ, Jacques E, Mohamed K, Hammond JOS, Belachew M, Baker E, Rowland JV, Lewi E, Bennati L (2009) Evidence for focused magmatic accretion at segment centers from lateral dike injections captured beneath the Red Sea rift in Afar. Geology 37:59–62. doi:10.1130/G25147A.1
Lahitte P, Gillot PY, Courtillot V (2003) Silicic central volcanoes as precursors to rift propagation: the Afar case. Earth Planet Sci Lett 207:103–116
Lay T, Wallace TC (1995) Modern Global Seismology. Academic Press, Waltham, 521 pp
Maguire PKH, Keller GR, Klemperer SL, Mackenzie GD, Keranen K, Harder S, O'Reilly B, Thybo H, Asfaw L, Khan MA, Amha M (2006) Crustal structure of the northern main Ethiopian Rift from the EAGLE controlled source survey; a snapshot of incipient lithospheric break-up. In: Yirgu G, Ebinger CJ, Maguire PKH (ed) The Afar volcanic province within the East African rift system, vol 259. Geological Society Special Publication, London, pp 269–292
McClusky S, Reilinger R, Ogubazghi G, Amleson A, Haileb B, Vernant P, Sholan J, Fisseha S, Asfaw L, Bendick R, Kogan L (2010) Kinematics of the southern Red Sea-Afar Triple Junction and implications for plate dynamics. Geophys Res Lett 37:L05301. doi:10.1029/2009GL041127
McKenzie DP, Davies D, Molnar P (1970) Plate tectonics of the Red Sea and East Africa. Nature 226:248–253
Midzi V, Hilatywayo DJ, Chapola LS, Kebede K, Atakan K, Lombe DK, Turyomurugyendo G, Tugume FA (1999) Seismic hazard assessment in eastern and southern Africa. Ann Geophys 42 (6):1067–1083
Midzi V, Ayele A, ESARSWG (2010) Determination of velocity models for the east and southern Africa region. Afr Geosci Rev 17 (1):21–34
Mohr P (1972) Surface structure and plate tectonics of Afar. In: Girdler RW (ed) East African Rifts. Tectonophysics 15:3–18
Mohr P (1975) Structural elements of the Afar margins. Data from ERTS-1 imagery. Bull Geophys Obs Addis Ababa 15:83–89
Ogubazghi G (2002) Recent seismic activity in the Massawa area: interim report. Internal Publication of the Earth Sciences Department of the University of Asmara, 20 pp
Ogubazghi G, Ghebreab W, Havskov J (2004) Some features of the 1993 Bada earthquake swarm of south eastern Eritrea. J Afr Earth Sci 38:135–143
Ogubazghi G, Goitom B, Kibreab A, King JG (2014) Relocation of the 31 March 2011 earthquake in the Nabro volcanic range of southern Eritrea. Eritrean J Sci Eng 1:93–102
Pallazo L (1913) La statione sismica d'Asmara. Bolletino della Società Sismologica Italiana 17:110–127
Pallazo L (1915) Cronistoria dei terremoti etiopici anteriori all'anno 1913. Bolletino della Società Sismologica Italiana 19:293–345
Rooney TO, Herzberg C, Bastow ID (2012) Elevated mantle temperature beneath East Africa. Geology 40:27–30. doi:10.1130/G32382.1
Rychert CA, Hammond JOS, Harmon N, Kendall JM, Keir D, Ebinger C, Bastow ID, Ayele A, Belachew M, Stuart G (2012) Volcanism in the Afar Rift sustained by decompression melting with minimal plume influence. Nature. doi:10.1038/NGEO1455
Sakr K, Abdel-Monem SM, Khalil H, Mahmoud S, Hamimi Z, Al-Aydrus A, Al Subbary A, Al-Kottbah A, El Ganad I, Al-Gabary A, Al-Wosabi MA (2005) A study of crustal deformation along the Red Sea region using geodetic and seismic data from Egypt and Yemen. Acta Geophysica Polonica 53(1):75–86
Schorlemmer D, Wiemer S (2004) Earthquake statistics at Parkfield: 1. Stationarity of b values. J Geophys Res 109:B12307. doi:10.1029/2004JB003234
Schorlemmer D, Wiemer S (2005) Microseismicity data forecast rupture area. Nature 434:1086
Sigmundsson F (1992) Tectonic implications of the 1989 Afar earthquake sequence. Geophys Res Lett 19:877–880

Solomon A (2012) Summary of activities during the Nabro volcanic crisis. Internal Publication of the Southern Red Sea Region Administration, 20 pp
Tan O, Tapirdamaz MC, Yoruk A (2008) The Earthquake Catalogues for Turkey. Turk J Earth Sci 17:405–418
Tapponnier P, Armijo R, Manighetti I, Courtillot V (1990) Bookshelf faulting and horizontal block rotations between overlapping rifts in southern Afar. Geophys Res Lett 17:1–4
Tesfaye S, Harding DJ, Kusky TM (2003) Early continental breakup boundary and migration of the Afar triple junction. Geol Soc Am 115:1053–1067
White R, McKenzie D (1989) Magmatism at rift zones: the generation of volcanic continental margins and flood basalts. J Geophys Res 94:7685–7729
Wiart P, Oppenheimer C (2000) Largest known historical eruption in Africa: Dubbi volcano, Eritrea, 1861. Geology 28:291–294
Wiart P, Oppenheimer C (2005) Large magnitude silicic volcanism in north Afar: the Nabro volcanic range and Ma'alalta volcano. Bull Volc 67:99–115
Wiart P, Oppenheimer C, Francis P (2000) Eruptive history of Dubbi volcano, northeast Afar (Eritrea), revealed by optical and SAR image interpretation. Int J Remote Sens 21:911–936
Wright TJ, Ebinger C, Biggs J, Ayele A, Yirgu G, Keir D, Stork A (2006) Magma maintained rift segmentation at continental rupture in the 2005 Afar dyking episode. Nature 442:291–294. doi:10.1038/nature04978

Volcanic Eruptions in the Southern Red Sea During 2007–2013

Sigurjón Jónsson and Wenbin Xu

Abstract

The first volcanic eruption known to occur in the southern Red Sea in over a century started on Jebel at Tair Island in September 2007. The early phase of the eruption was energetic, with lava reaching the shore of the small island within hours, destroying a Yemeni military outpost and causing a few casualties. The eruption lasted several months, producing a new summit cone and lava covering an area of 5.9 km^2, which is about half the area of the island. The Jebel at Tair activity was followed by two more eruptions within the Zubair archipelago, about 50 km to the southeast, in 2011–2012 and 2013, both of which started on the seafloor and resulted in the formation of new islands. The first of these eruptions started in December 2011 in the northern part of the archipelago and lasted for about one month, generating a small (0.25 km^2) oval-shaped island. Coastal erosion during the first two years following the end of the eruption has reduced the size of the island to 0.19 km^2. The second event occurred in the central part of the Zubair Islands and lasted roughly two months (September–November, 2013), forming a larger (0.68 km^2) island. The recent volcanic eruptions in the southern Red Sea are a part of increased activity seen in the entire southern Red Sea region following the onset of a rifting episode in Afar (Ethiopia) in 2005.

Introduction

Several volcanic islands are located in the central part of the southern Red Sea between Yemen and Eritrea. The islands belong to Yemen and have been divided into three groups that from north to south are the Jebel at Tair Island, the Zubair group of islands, and the Hanish–Zukur Islands (Fig. 1). In addition, an older small and eroded volcanic island, called Perim, is located in the Strait of Mandeb (Bab-al-Mandab), between Yemen and Djibouti. Other islands in the southern Red Sea are located near the coastlines of Yemen, Saudi Arabia, and Eritrea, but not along the central axis of the Red Sea. These islands include the Farasan and Dahlak reef archipelagos (Fig. 1), located on Miocene sedimentary shelves (Almalki et al. 2014).

S. Jónsson (✉) · W. Xu
King Abdullah University of Science and Technology (KAUST), Thuwal 23955, Saudi Arabia
e-mail: sigurjon.jonsson@kaust.edu.sa

Jebel at Tair is a single, oval-shaped island, that is, ~ 11 km^2 in size and is for the most part covered by basaltic lava. The Zubair group contains just over 10 islands, most of which are small (<2 km in diameter), except the largest island (13 km^2), simply named Zubair Island. The Hanish–Zukur Islands are considerably larger and consist of three main islands, Zukur, and the great and little Hanish Islands, as well as several smaller islets and sea rocks. These islands are also primarily volcanic with evidence of both effusive and explosive activity. The three recent volcanic eruptions on Jebel at Tair and within the Zubair Islands during 2007–2008, 2011–2012, and 2013 were the first known eruptions to occur in the southern Red Sea for more than a century. Several eruptions were documented on Jebel at Tair and Zubair Islands during the eighteenth and nineteenth centuries (Siebert et al. 2010), but the most recent eruption within the Hanish–Zukur Islands is unknown.

The Tair and Zubair eruptions occurred in a remote and mostly uninhabited area where no seismometers or conventional geodetic instruments were installed to monitor volcanic and earthquake activity. Earthquakes located by

N.M.A. Rasul and I.C.F. Stewart (eds.), *The Red Sea*,
Springer Earth System Sciences, DOI 10.1007/978-3-662-45201-1_10,

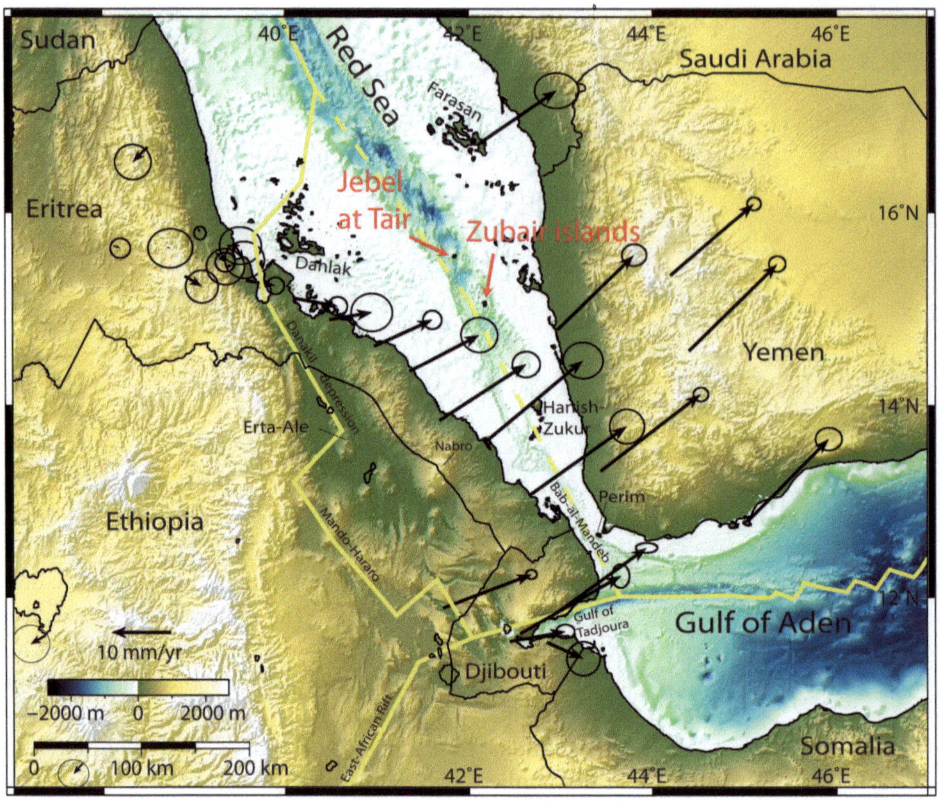

Fig. 1 Map of the southern Red Sea showing the location of Jebel at Tair and Zubair islands. Also shown are Farasan, Dahlak, Hanish–Zukur, and Perim islands and Erta Ale and Nabro volcanoes, GPS velocity vectors with error ellipses with respect to fixed Eurasia (ArRajehi et al. 2010), and plate boundaries (*yellow solid* and *dashed lines*)

regional networks provide limited information, although they show increased seismicity associated with the eruptions (International Seismological Centre 2013). Given the lack of local instrumental recordings and direct observations of the two eruptions, remotely sensed data from satellites provide key information for understanding and explaining the course of events during the eruptions and in the subsequent months. We therefore base our analysis of the volcanic activity on available medium- to high-resolution optical images as well as on images acquired by Synthetic Aperture Radar (SAR) instruments onboard several radar satellites.

Tectonic Background

The Red Sea has been opening since the Arabian plate broke away from the African plate about 24 million years ago (Bosworth et al. 2005). The rifting began with continental stretching and thinning and later progressed to sea-floor spreading. The rate of opening increases from about 7 mm/year in the northern Red Sea to roughly 16 mm/year in the south (ArRajehi et al. 2010; Reilinger et al. this volume). The rifting in the southern Red Sea is not focused on a single boundary, but is separated into two roughly parallel branches (Fig. 1). One branch follows the center of the Red Sea with a decaying opening rate toward the southeast until it dies out somewhere northwest of Mandeb Strait. The other branch follows the Danakil and Afar depressions in Eritrea and Ethiopia and meets up with the East African Rift and the Aden ridge in Djibouti to form a triple junction between the three rifts. GPS measurements show that the opening rate across the Danakil branch gradually increases to the southeast to the full rate between the Arabian and African plates (McClusky et al. 2010; Reilinger et al. this volume). This means that the area east of the Danakil depression and north of the Gulf of Tadjoura forms a separate tectonic block with the southern part of the block near the Mandeb Strait moving along with the Arabian Plate (Fig. 1). The central Red Sea trough becomes narrower and shallower from Jebel at Tair toward the Hanish–Zukur Islands, which is another manifestation of the decreasing opening rate of the Red Sea toward the southeast.

The region near the triple junction has been exceptionally active starting in 2005, with multiple earthquake swarms, volcanic eruptions, and magmatic intrusions. The Dabbahu rifting episode in the Manda-Hararo rift segment in Afar started in 2005 with a massive magma intrusion (Wright et al. 2006). It continued for several years with multiple earthquake swarms and intrusions (Grandin et al. 2010; Hamling et al. 2010). The rifting episode was preceded by an intrusion into the Dallol rift segment, north of Erta Ale volcano in the Danakil depression (Nobile et al. 2012). The Alu-Dalafilla volcano, also part of the Erta Ale volcanic range, erupted in 2008 and Nabro volcano in Eritrea erupted in 2011 (Smithsonian Institution 2008, 2011; Ogubazghi and

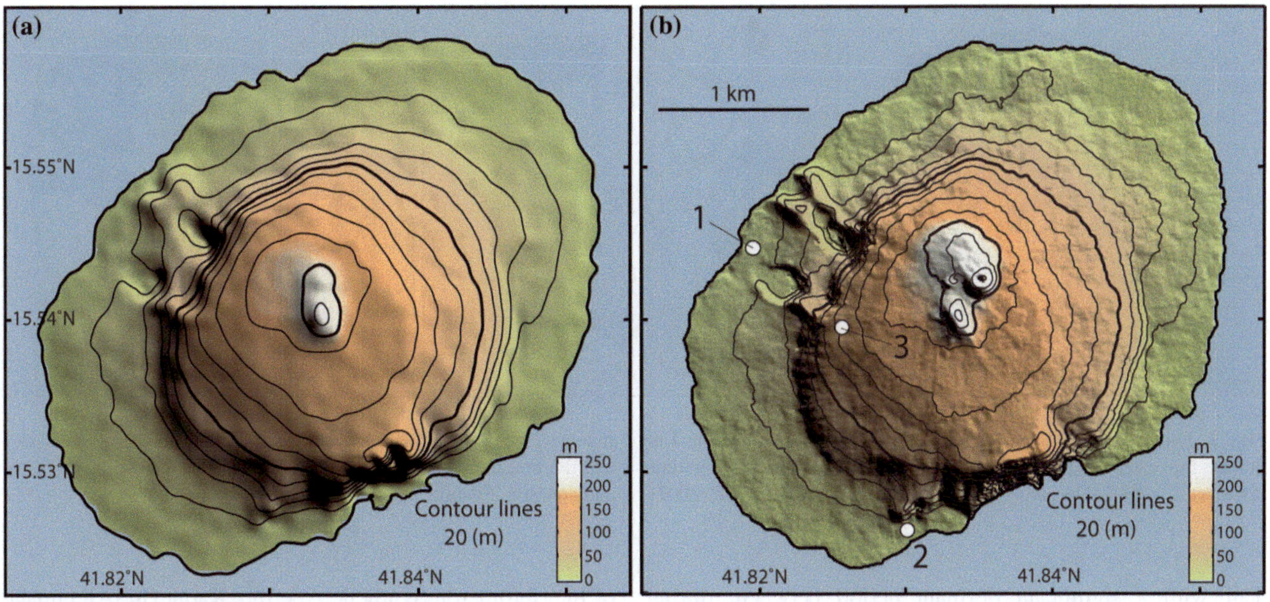

Fig. 2 Topographic maps of Jebel at Tair Island with 20-m contour lines from **a** interpolated SRTM and **b** TanDEM-X data. Numbers *1–3* in **b** indicate locations of buildings shown in Fig. 4

Goitom, this volume). In addition, an intensive earthquake swarm rocked the Gulf of Aden in November 2010 (Ahmed et al. 2012), and further north, an intrusion occurred in Harrat Lunayyir in western Saudi Arabia (Pallister et al. 2010). The three eruptions in the central Red Sea at Jebel at Tair and the Zubair Islands were a part of this general increase in volcanism across the region.

Jebel at Tair Island

The Jebel at Tair Island is located about 90 km from the Yemeni coastline, 120 km from Eritrea and 50 km northwest of the Zubair archipelago (Fig. 1). The island has an elliptical shape, about 4 km long and 3 km wide, covering an area of 11.4 km^2. It consists of a single stratovolcano and its highest point is 258 m above sea level. The volcanic edifice is significantly larger than that, as the average sea depth around the volcano is around 1,000 m, so the total height of the edifice exceeds 1,200 m.

We mapped the topography of the Jebel at Tair Island with satellite radar data from the TanDEM-X mission (Krieger et al. 2013). The TanDEM-X mission consists of two SAR satellites launched in 2007 (TerraSAR-X) and 2010 (TanDEM-X) that fly in a close formation, separated by only a few hundred meters, with one satellite transmitting X-band (3.1 cm) radar signals to the ground and both satellites receiving the reflected signals. With this bi-static configuration, a global digital elevation model (DEM) called WorldDEM is being produced that will be of unprecedented accuracy and coverage of the Earth's land surface (Rossi et al. 2012). We used TanDEM-X image pairs acquired on November 15, 2011 and December 26, 2012 with an incidence angle of ~34° from ascending orbits. Each image pair was precisely coregistered, and interferograms were formed by computing the phase difference between the two images. Then, we unwrapped the interferograms and generated height models, which we geo-referenced and averaged to produce a new DEM of the island (Fig. 2).

Our new TanDEM-X DEM provides much better information about the topography of the island than preexisting height models. The DEM pixel spacing is ~10 m, and the relative vertical accuracy of TanDEM-X DEMs is about 2 m (Krieger et al. 2013). In comparison, the posting of the Shuttle Radar Topography Mission (SRTM) DEM is only ~90 m (Farr et al. 2007) and the global ASTER GDEM around 30 m, but the ASTER GDEM includes only a part of Jebel at Tair Island. The improvement of the new TanDEM-X DEM can be clearly observed when displayed beside the SRTM DEM (here interpolated to 10 m posting) in Fig. 2. Note that the SRTM radar data were acquired in 2000, while the TanDEM-X was in 2011–2012, so the latter DEM includes the topographic changes that occurred as a result of the 2007–2008 eruption (see further analysis below).

The shape of Jebel at Tair Island is fairly symmetrical, with summit scoria cones and gentle slopes out to a radius of about 1 km, which are bounded by steeper slopes that again are surrounded by a gently sloping apron toward the coast. Gass et al. (1973) pointed out that this topographic profile indicates two separate periods of eruptive activity on Jebel at

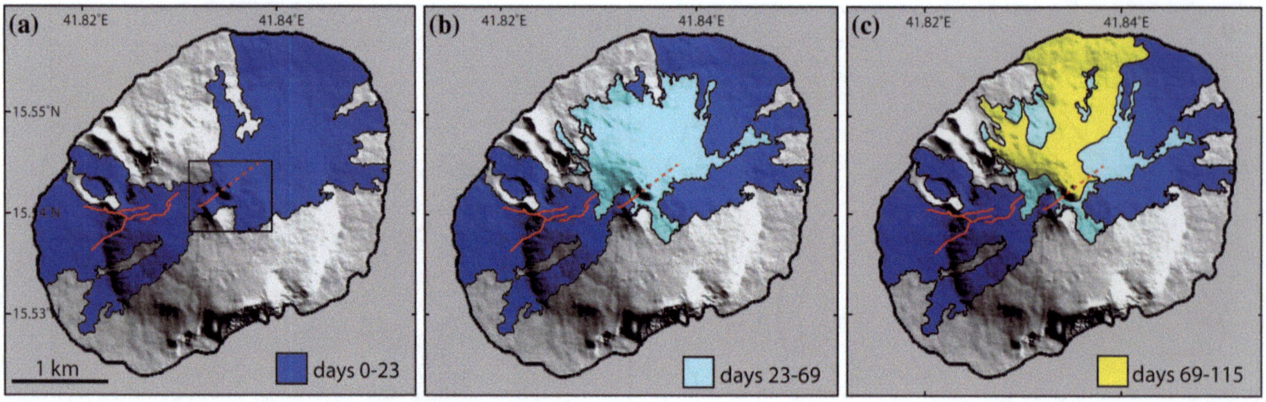

Fig. 3 Evolution of the 2007–2008 lava flow field on Jebel at Tair during the early (September 30, 2007–October 22, 2007), middle (October 23, 2007–December 8, 2007), and late (December 8, 2007–January 23, 2008) parts of the eruption. *Red lines* show eruptive fissures and the box in **a** marks the area covered in Fig. 6 (modified from Xu and Jónsson 2014a)

Tair with a significant dormant period between, during which sea cliffs formed. The sea cliffs are today mostly covered by lava from the second eruptive period, although they are still exposed on the southeastern side of the island.

Most of the surface of Jebel at Tair Island is covered by tholeiitic basaltic lava flows and is cut by numerous open radial fissures, possibly showing a slight orientation preference parallel to the trend of the Red Sea (Gass et al. 1973). Several historical eruptions are known to have occurred on Jebel at Tair before the 2007 activity, although the historical accounts for this region are neither long nor complete. There are reports of three eruptions during the nineteenth century, in 1833, 1863, and in 1883, and one in the eighteenth century, but no eruptions were recorded during the twentieth century (Siebert et al. 2010).

The 2007–2008 Jebel at Tair Eruption

The Jebel at Tair eruption started on September 30, 2007 and lasted at least until mid-January 2008 (Xu and Jónsson 2014a). The early phase of the eruption was energetic with lava extruding from fissures extending from the summit of the island toward both the west and the northeast. During this early phase, lava appears to have reached the shoreline within hours, as fresh lava extending to the sea could be seen during aerial inspection by Yemeni scientists on October 1, 2007 (Jamal M. Sholan, pers. comm. 2012). Lava was seen fountaining from a short fissure extending from the summit area and down to the northeastern flank of the volcano, while only steam was rising from fissures on the west flank. Lava fountaining during the later parts of the eruption was mostly confined within a new summit cone, with lava flowing to the north and to the northeast. The progress and coverage of the lava flow was analyzed by mapping interferometric decorrelation in a series of Interferometric Synthetic Aperture Radar (InSAR) images acquired by the Japanese ALOS satellite prior to, during and after the eruption (Xu and Jónsson 2014a). The total area of the new lava is 5.9 km^2, covering about 50 % of the island (Fig. 3).

Jebel at Tair was the site of a small Yemeni military outpost that was destroyed by the September 2007 eruption, resulting in several casualties (Smithsonian Institution 2007). Comparison between high-resolution (50–60 cm) optical images taken by the Quickbird and Worldview-2 satellites before and after the eruption shows damage and destruction of the military buildings (Fig. 4), with some surrounded or buried in lava, while others were probably damaged by tephra or earthquakes. The first images (Figs. 4a–b and 5) show a building near the northwest coast that was surrounded by the lava flow and with a collapsed roof. Another building, located on the south coast (Fig. 4c–d), was not threatened by the lava flows, but the collapsed roof suggests damage due to earthquake activity. Finally, a lighthouse (Fig. 4e–f), located halfway between the summit and the west coast, appears to have been completely destroyed by lava.

The high-resolution optical images from before and after the eruption reveal a number of other changes on the island (Xu and Jónsson 2014a). The pre-eruption image exhibits several radial fissures and a network of car tracks, which were partially buried by the 2007–2008 lava. The post-eruption images show that the coastline advanced by several tens of meters in a few places where lava flowed into the sea, particularly at the north coast, enlarging the area of the island by 0.13 km^2 (Xu and Jónsson 2014a). Changes in the summit area are especially pronounced, with a new prominent scoria cone and an underlying N60°E trending fissure, extending to the southwest and northeast of the cone (Figs. 5 and 6). This fissure clearly indicates the orientation of the dike feeding the eruption at the summit. More fresh fissures,

Fig. 4 Satellite optical image close-ups of buildings on Jebel at Tair before (*upper row*) and after (*lower row*) the 2007 eruption showing infrastructure damage caused by the volcanic activity. The locations of these three buildings are shown in Fig. 2, numbered *1–3*

Fig. 5 Photographs taken on Jebel at Tair on October 21, 2007, showing a building (*left*) surrounded by fresh lava (see also Fig. 4b) and the erupting summit cone (*right*). (*Photos* Jamal M. Sholan)

with more easterly strikes, can be seen in the post-eruption image further to the west. These fissures were active during the first hours of the eruption. The eruptive fissures also extended further to the northeast, as seen during the helicopter flight on October 1, 2007, but lava from the summit cone later buried them. Therefore, the total length of the eruptive fissures soon after the start of the eruption exceeded 1 km and may have been as long as 2 km, but quickly contracted to a short fissure during the first 24 h of activity. The eruption probably condensed to a single vent within days and continued there for over three months, building up the new summit scoria cone.

As we have DEMs of Jabel at Tair Island from before and after the 2007–2008 eruption (Fig. 2), we can estimate the thickness and volume of the eruptive products by differencing the two DEMs. However, the SRTM DEM is of much lower resolution and quality than the new TanDEM-X DEM, which makes this estimation somewhat inaccurate.

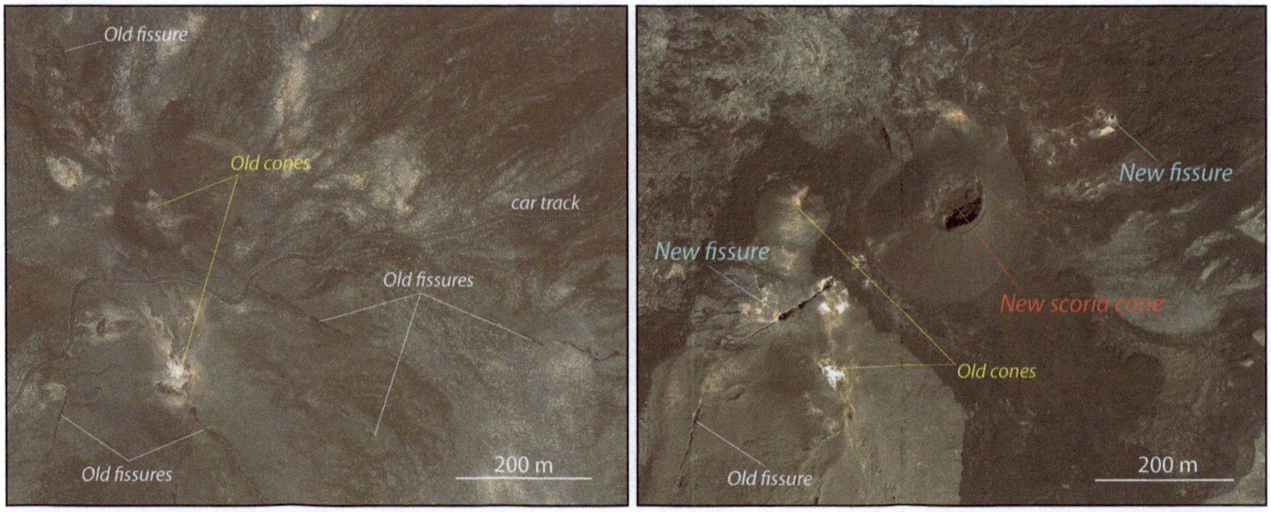

Fig. 6 Satellite optical image close-ups of the summit area of Jebel at Tair before (*left*) and after (*right*) the eruption showing eruptive fissures and a new summit scoria cone (the area covered in this figure is shown in Fig. 3a). The images were taken by the Quickbird (July 6, 2007) and Worldview-2 (October 22, 2011) satellites

We therefore used a series of pre-eruption InSAR data from the ALOS satellite to correct and improve the SRTM DEM (Xu and Jónsson 2014a). We then subtracted the ALOS-corrected SRTM from the new TanDEM-X DEM and produced a map of the height differences (Fig. 7). The height differences indicate that the average thickness of the lava is about 3.8 m, and the subaerial volume is 0.022 ± 0.011 km^3, or a dense-rock-equivalent (DRE) volume of ~ 0.017 km^3, assuming 25 % vesicularity (Xu and Jónsson 2014a). The uncertainty was estimated by comparing the height differences in areas that were not covered by the new lava; the standard deviation of the height differences in these areas is only 1.9 m. The volume estimate does not include the unknown amount of lava that flowed into the sea.

Comparison of satellite radar data from before and after the start of an eruption can be used to derive maps of the co-eruptive ground deformation in areas not covered by new lava flows. The main method is to form radar interferograms (InSAR) from pre- and post-eruption image pairs, where the interferometric phase provides information about the relative line-of-sight displacements between the ground and the radar satellites (Massonnet and Feigl 1998). In addition, offset tracking of features in radar amplitude images or optical images can be used to measure deformation if the displacements are large enough in comparison with the image resolution (Michel et al. 1999). We used InSAR data from both the Japanese ALOS and the European Envisat satellites to map co-eruptive ground displacements on Jebel at Tair along two different line-of-sight directions, as well as horizontal amplitude-image offsets (Xu and Jónsson 2014a). These different one-component displacement measurements,

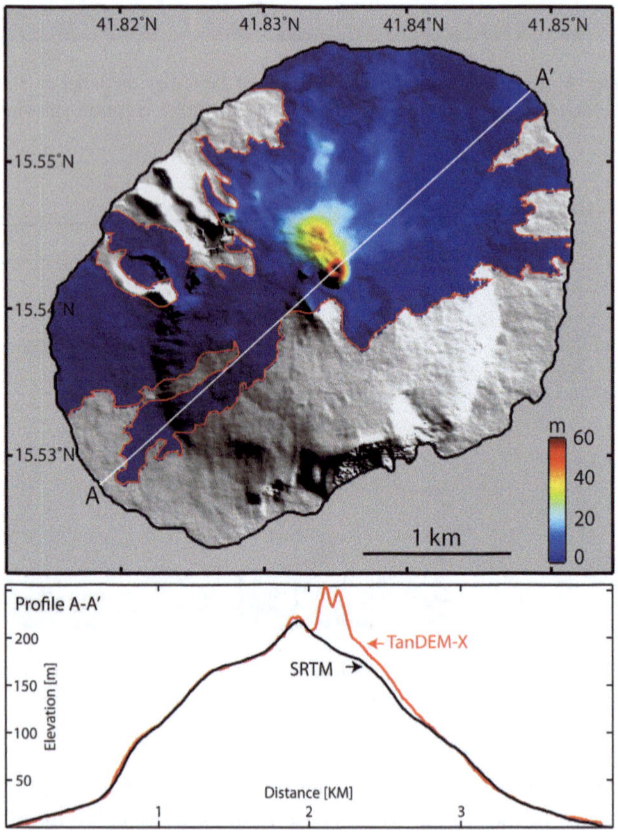

Fig. 7 The thickness of the 2007–2008 lava and other erupted deposits (*top*), estimated from the height differences between pre- and post-eruption DEMs (Fig. 2), see for example the DEM comparison along profile A–A′ (*bottom*) through the new summit cone (partly from Xu and Jónsson 2014a)

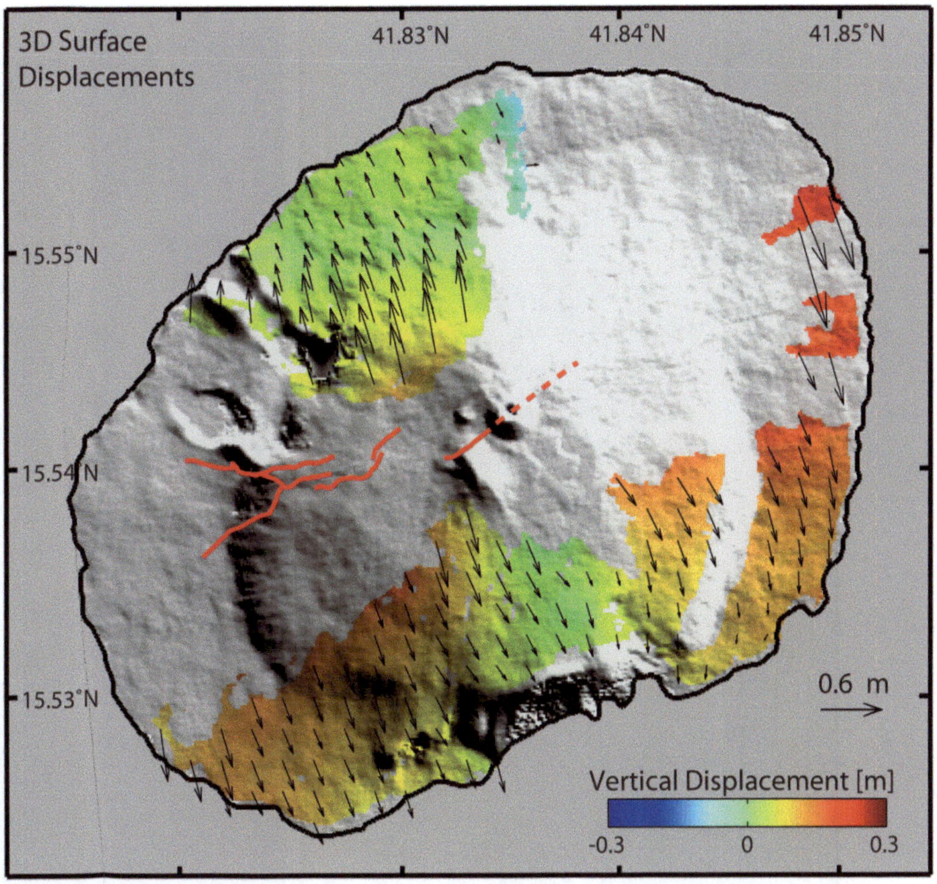

Fig. 8 Map of three-dimensional co-eruption surface displacements derived from satellite radar data, with arrows showing horizontal extension perpendicular to eruptive fissures (*red lines*) and colors indicating vertical movement

that is, the line-of-sight InSAR and the image-offset data, were then combined to invert three-dimensional (3D) displacements on the ground surface (Fig. 8). The map of the 3D deformation shows horizontal displacement of the northwest and southeast flanks of the volcano away from the eruptive fissures, with a maximum extension of over one meter. The vertical displacements are smaller and harder to interpret, although positive displacements (uplift) are seen near the intruding dike in most places. Modeling of this displacement field shows that a near-vertical dike, intersecting the surface at the mapped eruptive fissures, can roughly explain the observed deformation (Xu and Jónsson 2014a).

The northeast orientation of the feeder dike is surprising as it is roughly perpendicular to the axis of the Red Sea rift. Older surface fissures are easily identified in satellite images from before the 2007–2008 eruption, and they exhibit several different radial directions (Gass et al. 1973; Xu and Jónsson 2014a). This indicates that the stress field within the Jebel at Tair volcanic edifice is both isolated from the regional Red Sea stress field and that it has varied with time. Each new dike intrusion seems to modify the local stress field enough such that the next intrusion ascends to the surface with a different orientation. Comparable temporal variations of stress orientations within other volcanoes have recently been found, for example on Fernandina volcano, Galápagos (Chadwick et al. 2011).

Zubair Islands

The Zubair archipelago is located about 50 km southeast of Jebel at Tair, about 50 km from the Yemeni coast and just over 100 km from Eritrea (Fig. 1). These islands are sitting on a shallow (<100 m) north-northwest-oriented platform, that is, ~25 km long and ~10 km wide. The platform is situated within the Red Sea median trough and is separated from the trough flanks by narrow trenches to the west and to the east, which are up to 900 m deep (Gass et al. 1973). Before the 2011–2013 activity, the archipelago consisted of 10 small islands and several sea rocks. The largest island, simply called Zubair Island, is about 5 km long, 3 km wide and covers an area of ~13 km^2. The other islands are significantly smaller, mostly less than 2 km in diameter (Fig. 9). The dominant structural trend found on some of the islands, for example on Zubair Island, is ~N10°W (Gass et al. 1973), or similar to the overall north-northwest orientation of the entire archipelago and the Red Sea rift.

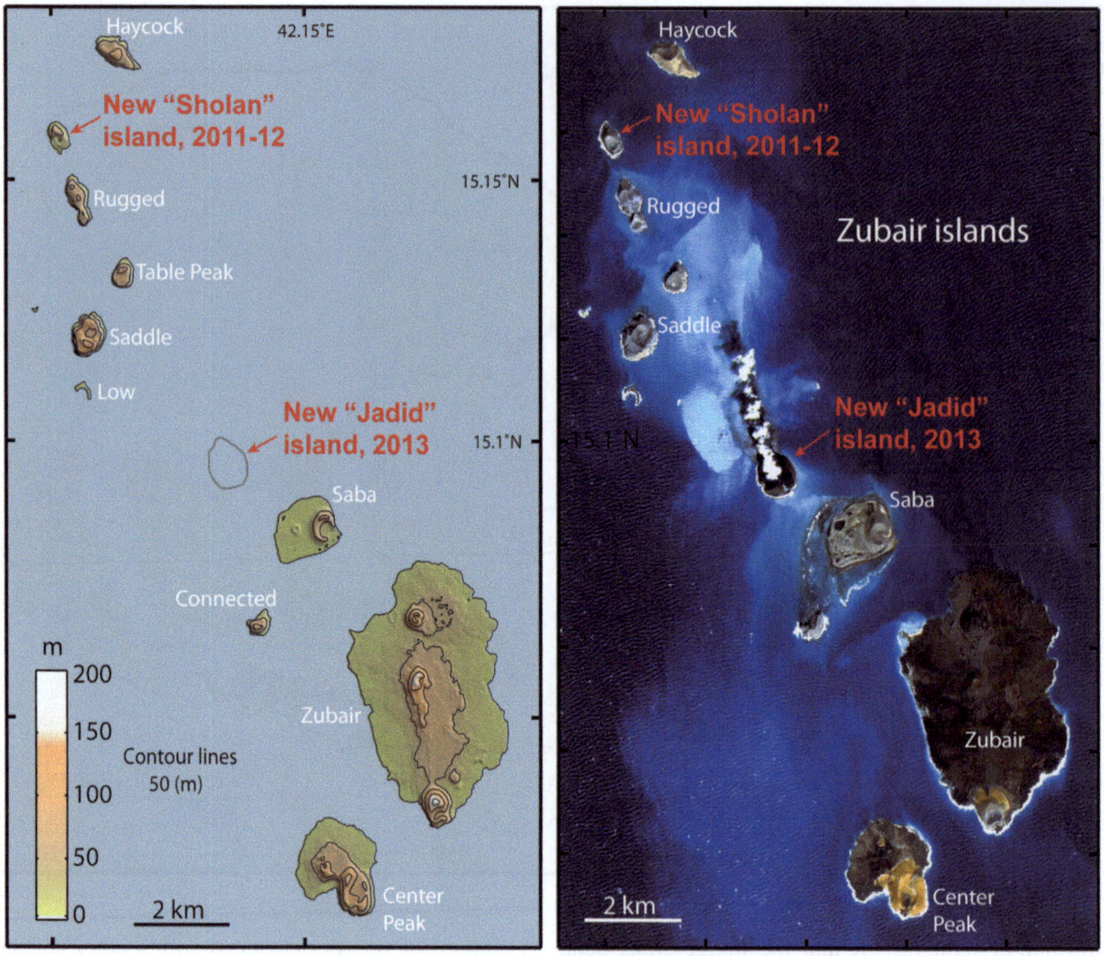

Fig. 9 Topographic map (*left*) of the Zubair archipelago (generated from TanDEM-X data) showing the location of the new islands (Sholan and Jadid) formed during the 2011–2012 and 2013 eruptions. The Landsat-8 image (*right*) is from Nov 8, 2013 and shows steam and ash rising from the new Jadid Island in the central part of the archipelago. The map does not include Quoin Island, which is also considered a part of the archipelago but is located off the map to the northwest

The Zubair Islands are all of volcanic origin and primarily consist of basaltic tuff, agglomerate, and lava (Gass et al. 1973). They were formed in surtseyan eruptions that started as submarine activity and then progressed to explosive phreatomagmatic eruptions, building up a ring of agglomerates and ash. In some cases, the activity developed to a point at which the tuff ring protected the eruptive vent from the seawater, allowing for transition to an effusive lava eruption, similar to that observed during the 1963–1967 Surtsey eruption south of Iceland (Moore 1985). This is evident on Zubair and Centre Peak Islands where lava covers a significant portion of their exposed surfaces (Fig. 9). In other cases, the activity did not progress beyond the explosive phase, such as on Rugged, Saddle, and Connected Islands (Gass et al. 1973), as well as on some other small islands and sea rocks.

There is also evidence for a period of quiescence on some of the islands, separating two phases of eruptive activity (Gass et al. 1973). In these cases, on Centre Peak, Saba, and Zubair Islands, the earlier phase eruptions all progressed to effusive activity, producing lava near the original vent. The activity in the later phase produced scoria cones and more lava. The quiescence is marked by erosional sea cliffs that are in many places covered by volcanic products from the second phase.

Similar to Jebel at Tair, the historical accounts of eruptions are limited and many eruptions may have occurred unnoticed, especially before the opening of the Suez Canal in 1869. Two eruptions are known to have occurred within the Zubair Islands before the recent activity began. Both of them are reported to have taken place on Saddle Island during the nineteenth century, in 1824 and in 1846 (Siebert et al. 2010), although the latter one is classified as uncertain.

The 2011–2012 Zubair Eruption

Yemeni fishermen reported an offshore eruption near the Zubair Islands on December 18, 2011. The surtseyan eruption occurred in shallow seawater between Haycock and Rugged Islands (Fig. 9), and by December 23, 2011, a new island (Sholan Island) had formed at the eruption site. The exact timing of the onset of the eruption is not known. Two earthquakes of magnitude 3.7 and 3.9 occurred on 13 December, and the eruption may have started on the seafloor following these events. However, there are no signs of the activity in satellite images until on 18 December, when a clear plume is seen rising from the eruption site (Smithsonian Institution 2013).

The eruption appears to have started along a short north-northwest-oriented fissure, parallel to the dominant structural trends of the archipelago, as indicated by the elongated activity seen on the satellite image from December 23, 2011 (Fig. 10a). Later, the eruption was restricted to a single vent and continued well into January (Xu and Jónsson 2014b). Strong phreatomagmatic activity can be seen in a video, taken on January 2, 2012 from a Yemeni military helicopter, sending a significant amount of tephra into the sky. The end of the eruption was reported as sometime between January 7, 2012 and January 15, 2012 (Smithsonian Institution 2013). However, high-resolution images taken by the Worldview-2 satellite can further constrain the end of the activity, as an image from January 9, 2012 shows that the eruption was still ongoing but had ended by January 12, 2012 (Fig. 10). Given this information, as well as a MODIS thermal anomaly that was detected on January 11, 2012 at 22:35 UTC (Smithsonian Institution 2013), we time the end of the eruption to the early morning of January 12, 2012. The duration of the subaerial activity was thus 25 days, from December 18, 2011 to January 12, 2012.

The eruption did not progress beyond an explosive phase. The satellite image from January 9, 2012 shows that seawater is in contact with the eruptive vent from the southwest and that explosive activity is still ongoing (Fig. 10b). In addition, no effusive eruption products can be seen in the January 12, 2012 image, immediately after the end of the eruption (Fig. 10c). The new Sholan Island is therefore similar to many of the smaller Zubair Islands that were likely also formed in single explosive eruptions, which probably lasted days to weeks, rather than months to years.

The January 12, 2012 image shows that Sholan Island was about 0.77 km long and 0.52 km wide at the end of the eruption, covering an area of 0.25 km^2 (Xu et al. 2015).

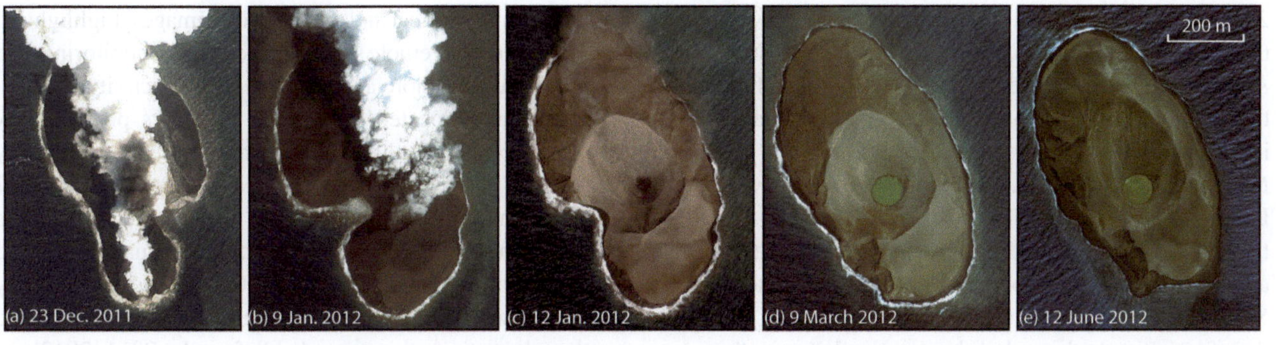

Fig. 10 Series of high-resolution images from the Quickbird and Worldview-2 satellites of the new Sholan Island in the northern Zubair archipelago, showing the 2011–2012 eruption and post-eruptive coastal changes

Fig. 11 Photograph of Sholan Island in the northern Zubair archipelago, taken from the southeast on January 17, 2012 (*photo* Jamal M. Sholan)

Images from subsequent months show how the shape and size of the island changed due to coastal erosion and landslides. This erosion and evidence for landslide activity can be seen in a photograph of the island taken shortly after the end of the eruption (Fig. 11). By March 2012, notable changes to the island's shape had already taken place and a lake had formed in the main crater (Fig. 10d). Further coastal changes can be seen in the June 2012 image (Fig. 10e) with the area of Sholan Island reduced to $\sim 0.24\ \text{km}^2$, and then to $\sim 0.19\ \text{km}^2$ by the end of 2013, or to about 75 % of its size at the end of the eruption.

The 2013 Zubair Eruption

A new volcanic eruption started in the Zubair archipelago on September 28, 2013, when a steam plume and a clear SO_2 anomaly were detected from satellite data (NASA-OMI 2013). The exact start time of the eruption is unknown, but no subaerial activity is visible in satellite images collected on and before September 27, 2013. No earthquakes were detected before or during this activity by stations in Yemen or Eritrea (International Seismological Centre 2013).

This eruption took place further south than the 2011–2012 eruption, between Saddle and Saba Islands (Fig. 9). Within days, a new island (Jadid Island) had formed about 2 km northwest of Saba Island, and the island grew larger as the eruption continued through October and into November. As with the onset of the eruption, the timing of the eruption's end can only be constrained by satellite data. A plume is seen rising from the island in a Landsat-7 image on November 16, 2013 (Fig. 12d), but no activity is visible in a Landsat-8 image from November 24, 2013 (Fig. 12e). Furthermore, MODIS data seem to indicate a small thermal anomaly on 20 November, but no anomaly is seen in the MODIS data from 21 November (NASA-MODIS 2013). We therefore estimate the end of the eruption between these two days and its total duration is ~ 54 days, which is more than twice as long as the 2011–2012 eruption.

The new Jadid Island is also significantly larger than Sholan Island formed during the 2011–2012 eruption; it is about 0.9 km in diameter, with an area of $\sim 0.68\ \text{km}^2$ (Xu et al. 2015). However, similar to the earlier activity, the eruption does not seem to have progressed beyond an explosive phase to an effusive phase, as satellite images late in the eruption, for example on November 8, 2013 and November 16, 2013 (Fig. 12c–d), indicate only explosive activity. In addition, medium-resolution satellite images from after the eruption do not show signs of lava on the new island, although this has yet to be confirmed by field inspection or higher resolution imagery.

Discussion and Conclusions

The Jebel at Tair and Zubair volcanic eruptions occurred in a remote offshore and mostly uninhabited area without seismometers or conventional geodetic instruments. Only three brief visits were paid to the islands by Yemeni scientists during the eruptions, and these were the only direct observations, along with occasional photographs and videos (posted on the Internet) from ships passing by or from Yemeni military personnel. Almost all the information we have about this activity is from optical and radar satellite images, highlighting the importance of remote sensing in detecting, monitoring, and understanding offshore and remote volcanic activity.

Earthquakes located by regional networks provide only limited information, although they show increased seismicity near Jebel at Tair Island as early as September 21, 2007 (International Seismological Centre 2013), nine days before the eruption started. At least four $M_L > 4$ earthquakes occurred during the afternoon of September 30, 2007, which probably coincided with the dike propagating to the surface. Limited earthquake activity was recorded before the 2011–2012 and

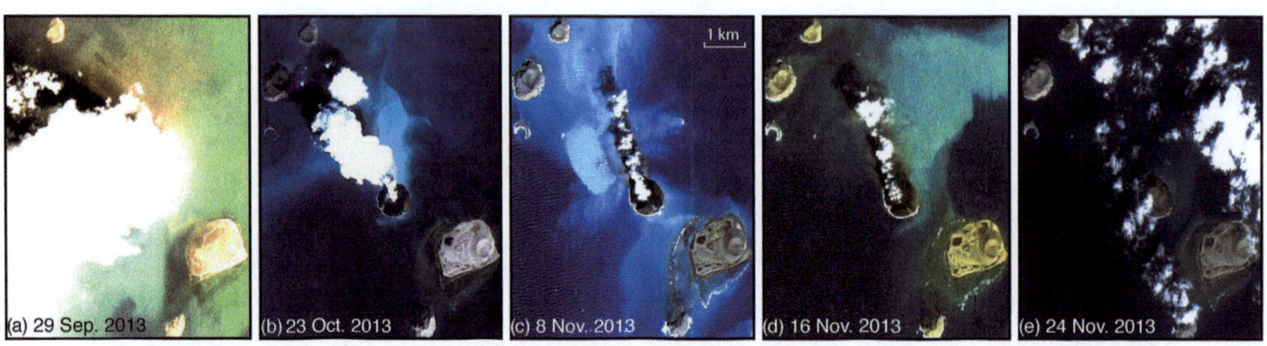

Fig. 12 A series of satellite images of the new Jadid Island in the central part of the Zubair archipelago, showing the 2013 eruption and post-eruptive changes. The images are from the Landsat-7 and Landsat-8 satellites

2013 Zubair eruptions. Only two earthquakes were detected in the area several days before the 2011–2012 eruption started, and no recorded earthquakes coincided with the start of the 2013 eruption. The lack of pre-eruption earthquakes is due to the limited monitoring capability of the current seismic stations, located more than 100 km away on the Yemeni mainland. Therefore, improved seismic monitoring with seismic stations on the islands in the southern Red Sea is needed to forecast and monitor future eruptions.

The eruptions on Jebel at Tair in 2007–2008 and within the Zubair archipelago in 2011–2012 and 2013 were the first eruptions in the southern Red Sea for more than a century. The Zubair Islands had only two known eruptions prior to the 2011–2012 and 2013 activity, according to historical accounts, in 1824 and in 1846, while four previous eruptions are reported on Jebel at Tair, in $\sim$1750, 1833, 1863, and 1883 (Siebert et al. 2010). The reawakening of eruptive activity in the Red Sea coincides with increased volcanic activity across the entire region that started with the large Dabbahu diking event in Afar in 2005 (Wright et al. 2006). Since then, multiple dike intrusions and several eruptions have occurred in Afar, the Red Sea, and the Gulf of Aden (e.g., Hamling et al. 2010; Smithsonian Institution 2008, 2011; Ahmed et al. 2012). While it is not clear how these events are related to one another, the burst of activity since 2005 strongly suggests that there is a link between them. Mapping of transient crustal strain caused by the initial diking event and subsequent dike intrusions in the Dabbahu rifting episode shows that significant strain extends hundreds of kilometers away from the Manda-Hararo rift segments and affects the southern Red Sea and Gulf of Aden (Pagli et al. 2012). While the exact nature of the stress changes caused by the Dabbahu rifting episode in the southern Red Sea islands has not yet been studied in detail, strong stress changes with influence on earthquake occurrence have been reported far away from other rifting episodes (Maccaferri et al. 2013), thus demonstrating the potential for a connection between the Red Sea eruptions and the distant Dabbahu activity.

Acknowledgments We thank Jamal M. Sholan (Seismological and Volcanological Observation Center, Dhamar, Yemen) for providing field photos and information about the Jebel at Tair and Zubair activity. We have named the two new islands in the Zubair archipelago Jadid ("new" in Arabic) and Sholan, as Jamal Sholan was the first Yemeni scientist to visit that island. The Envisat and ERS satellite data were provided by the European Space Agency (ESA) and ALOS data by the Japan Exploration Agency (JAXA) through Category-1 Project #6703. The German Space Agency provided the TanDEM-X data through project XTI-GEOL3441. The Landsat-7 and 8 images were distributed by the Land Processes Distributed Active Archive Center (LP DAAC), located at USGS/EROS, Sioux Falls, SD (http://lpdaac.usgs.gov). The research reported in this publication was supported by King Abdullah University of Science and Technology (KAUST).

References

Ahmed A, Doubre C, Leroy S, Perrot J, Audin L, Rolandone F, Keir D, Al-Ganad I, Khanbari K, Mohamed K, Vergne J, Jacques E, Nercessian A (2012) November 2010 earthquake swarm—western Gulf of Aden (abstract). In: Proceedings of the Afar Rift Consortium conference, Addis Ababa, Ethiopia, 11–13 Jan 2012

Almalki KA, Betts PG, Aillers L (2014) Episodic sea-floor spreading in the Southern Red Sea. Tectonophysics 617:140–149

ArRajehi A, McClusky S, Reilinger R, Daoud M, Alchalbi A, Ergintav S, Gomez F, Sholan J, Bou-Rabee F, Ogubazghi G, Haileab B, Fisseha S, Asfaw L, Mahmoud S, Rayan A, Bendik R, Kogan L (2010) Geodetic constraints on present-day motion of the Arabian Plate: Implications for Red Sea and Gulf of Aden rifting. Tectonics 29: TC3011, doi:10.1029/2009TC002482

Bosworth W, Huchon P, McClay K (2005) The Red Sea and Gulf of Aden Basins. J Afr Earth Sci 43:334–378

Chadwick WW, Jónsson S, Geist DJ, Poland M, Johnson DJ, Batt S, Harpp KS, Ruiz A (2011) The May 2005 eruption of Fernandina volcano, Galapagos: GPS and InSAR observations of a circumferential dike intrusion. Bull Volc 73:679–697

Farr TG, Rosen PA, Caro E, Crippen R, Duren R, Hensley S, Kobrick M, Paller M, Rodriguez E, Roth L, Seal D, Shaffer S, Shimada J, Umland J, Werner M, Oskin M, Burbank D, Alsdorf D (2007) The Shuttle Radar Topography Mission. Reviews of Geophysics 45: RG2004, doi:10.1029/2005RG000183

Gass IG, Mallick DIJ, Cox KG (1973) Volcanic islands of the Red Sea. J Geol Soc Lond 129(3):275–309

Grandin R, Socquet A, Jacques E, Mazzoni N, de Chabalier J-B, King GCP (2010) Sequence of rifting in Afar, Manda-Hararo rift, Ethiopia, 2005–2009: Time-space evolution and interaction between dikes from Interferometric synthetic aperture radar and static stress change modeling. J Geophys Res 115:B10413. doi:10.1029/2009JB000815

Hamling IJ, Wright TJ, Calais E, Bennati L, Lewi E (2010) Stress transfer between thirteen successive dyke intrusions in Ethiopia. Nat Geosci 3:713–717

International Seismological Centre (2013) On-line Bullet, International Seismological Centre, Thatcham. http://www.isc.ac.uk

Krieger G, Zink M, Bachmann M, Bräutigam B, Schulze D, Martone M, Rizzoli P, Steinbrecher U, Antony JW, De Zan F, Hajnsek I, Papathanassiou K, Kugler F, Cassola MR, Younis M, Baumgartner S, López-Dekker P, Prats P, Moreira A (2013) TanDEM-X: a radar interferometer with two formation-flying satellites. Acta Astronaut 89:83–98

Maccaferri F, Rivalta E, Passarelli L, Jónsson S (2013) The stress shadow induced by the 1975–1984 Krafla rifting episode. J Geophys Res 118:1109–1121

Massonnet D, Feigl KL (1998) Radar interferometry and its application to changes in the Earth's surface. Rev Geophys 36:441–500

McClusky S, Reilinger R, Ogubazghi G, Amleson A, Healeb B, Vernant P, Sholan J, Fisseha S, Asfaw L, Bendick R, Kogan L (2010) Kinematics of the southern Red Sea-Afar Triple Junction and implications for plate dynamics. Geophys Res Lett 37:L05301. doi:10.1029/2009GL041127

Michel R, Avouac JP, Taboury J (1999) Measuring ground displacements from SAR amplitude images: application to the Landers earthquake. Geophys Res Lett 26:875–878

Moore JG (1985) Structure and eruptive mechanisms at Surtsey Volcano, Iceland. Geol Mag 122:649–661

NASA-MODIS 2013, The MODIS website. http://modis.gsfc.nasa.gov

NASA-OMI 2013, Global sulfur monitoring home page. http://so2.gsfc.nasa.gov

Nobile A, Pagli C, Keir D, Wright TJ, Ayele A, Ruch J, Acocella V (2012) Dike-fault interaction during the 2004 Dallol intrusion at the northern edge of the Erta Ale Ridge (Afar, Ethiopia). Geophys Res Lett 39:L19305. doi:10.1029/2012GL053152

Pagli C, Wright TJ, Wang H, Calais E, Bennati L, Lewi E (2012) Three-dimensional time-varying crustal velocity and strains in the Afar triangle. In: Proceedings of the Afar Rift Consortium conference, Addis Ababa, Ethiopia, 11–13 Jan 2012

Pallister JS, McCausland WA, Jónsson S, Lu Z, Zahran HM, El-Hadidy S, Aburukbah A, Stewart ICF, Lundgren PR, White RA, Moufti MRH (2010) Broad accommodation of rift-related extension recorded by dyke intrusion in Saudi Arabia. Nat Geosci 3:705–712

Rossi C, Gonzalez FR, Fritz T, Yague-Martinez N, Eineder M (2012) TanDEM-X calibrated raw DEM generation. ISPRS J Photogram Remote Sens 73:12–20

Siebert L, Simkin T, Kimberly P (2010) Volcanoes of the world (3rd ed). Smithsonian Institution, University California Press, Oakland, 551 pp

Smithsonian Institution (2007) Jebel at Tair. In: Bulletin of the global volcanism network, 32(10):2−5

Smithsonian Institution (2008) Jebel at Tair. In: Bulletin of the global volcanism network, 33(4):11−12

Smithsonian Institution (2011) Zubair. Bulletin of the global volcanism network, 36(11):2−4

Smithsonian Institution (2013) Zubair. In: Bulletin of the global volcanism network, 38(4):17−18

Wright TJ, Ebinger C, Biggs J, Ayele A, Yirgu G, Keir D, Stork A (2006) Magma-maintained rift segmentation at continental rupture in the 2005 Afar dyking episode. Nature 442:291–294

Xu W, Jónsson S (2014a) The 2007–2008 volcanic eruption on Jebel at Tair island (Red Sea), observed by satellite radar and optical images. Bull Volc 76:1−14. doi: 10.1007/s00445-014-0795-9

Xu W, Jónsson S (2014b) Two recent surtseyan eruptions in the Zubair archipelago (Red Sea) studied using high-resolution radar and optical satellite remote sensing. Geophysical Research Abstracts, vol 16, EGU2014-1224-1

Xu W, Ruch J, Jónsson S (2015) Birth of two volcanic islands in the southern Red Sea (in revision)

Red Sea Salt Formations—A Result of Hydrothermal Processes

Martin Hovland, Håkon Rueslåtten, and Hans Konrad Johnsen

Abstract

A new conceptual model, called 'the hydrothermal salt model', predicts that salt may accumulate in the marine sub-surface from the hydrothermal circulation of sea water. The hypothesis is based on the physicochemical behaviour of supercritical sea water; when sea water is driven into its supercritical high-temperature and high-pressure domain (407 °C, 298 bars), it loses its solubility for the common sea salts (chlorides and sulphates). Consequently, a spontaneous precipitation of salts takes place in the water-filled pore spaces. The same process may occur when porous rocks containing saline pore water are exposed to sufficiently high temperature and pressure, for example in the subduction of oceanic crust. Salts will also precipitate sub-surface or sub-marine during boiling, in contact with high heat flow sources, such as magma intrusions. Large accumulations of salt are found within the central trough and along both sides of the Red Sea rift. Many of the associated accumulation features are difficult to explain in terms of the conventional 'evaporite' model for salt deposits. The features are as follows: (1) several kilometre thick salt deposits on both flanks of the Red Sea, (2) thick (~ 3 km) salt deposits inside the central graben, (3) dense, hot brines inside some of the central graben deeps (Atlantis II Deep, Conrad Deep, etc.), (4) up to 40-km-long walls and ridges of exposed salt on the northern Red Sea seafloor, (5) tall walls of salt adjacent to the Conrad Deep central graben, and (6) large flows of salt in the Thetis Deep. Furthermore, these huge accumulations of salt took place in a relatively short geological period of time. On the basis of these pertinent structures and features, it is concluded that accumulation, deformation, and transportation of salts may have several drivers and origins, including a process closely associated with hydrothermal activity. Whereas hydrothermally produced salts are lost directly to sea water in mid-ocean spreading hydrothermal systems, they are protected by sediments and ponded high-density brines on the seafloor in deep-spreading centres like that of the Red Sea. Because the Red Sea is the closest active analogue to the rifting and rupturing of 'Atlantic-type' continental lithosphere, it may also be the key to understanding the accumulations of huge underground salt deposits and salt-related structures in some of the world's ancient deep-water rifted margins.

Introduction

There are several kilometre thick sub-sea accumulations of salts in the Red Sea (Savoyat et al. 1989; Ligi et al. this volume; Ehrhardt and Hübscher this volume). Ever since the discovery of these enormous accumulations of sea salts, their mode of formation has represented a challenging puzzle to geologists. The original theory of evaporation was

M. Hovland (✉)
Tech Team Solutions, 4035 Stavanger, Norway
e-mail: Martin.Hovland@techteamsolutions.no

H. Rueslåtten
Lithicon, 7041 Trondheim, Norway

H.K. Johnsen
Det norske oljeselskap ASA, 7011 Trondheim, Norway

N.M.A. Rasul and I.C.F. Stewart (eds.), *The Red Sea*,
Springer Earth System Sciences, DOI 10.1007/978-3-662-45201-1_11,

developed on the basis of observed salt lakes during the early 1900s and later modified during the 1970s and 1980s, into the '*top-down*' 'evaporite' theory (Schmalz 1969; Stoffers and Kühn 1973; Schreiber and Hsü 1980). Briefly, the solar evaporite theory suggests that deep salt water basins with restricted communication to the open ocean became so confined and concentrated that saline deposits precipitate, settle to the bottom, and remain preserved, or that the basin dries out and fills up several times, causing thick accumulation of salts.

For salt thicknesses of some tens of metres, the 'evaporite' explanation may be realistic. However, the accumulation of salt thicknesses of several thousand metres, as found in the Mediterranean, the Red Sea, and other places worldwide, is difficult to explain with the evaporite theory (Schreiber et al. 2007; Warren 2010; Ligi et al. this volume, Bosworth this volume, Feldens and Mitchell this volume). It is especially difficult to explain how large volumes of salt may have accumulated in sedimentary basins such as in south-western Canada, which contain volumes of salts approaching those of the entire world oceans (Hewitt 1962). In addition, the large underground salt bodies, occurring offshore, apparently also defy one of the fundamental laws of geology that '*the present is the key to the past*', as there does not appear to be a location where massive accumulations of salt are currently forming by large-scale solar evaporation.

Another conundrum is that no one has ever managed to numerically model the physicochemical processes responsible for the accumulation of the 'evaporite' salt masses. These and also some additional problems lead sedimentologist Richard Selley (2005) to state: '*This required improbably large volumes of sea water, to produce the resultant evaporites. It is now realized that many evaporites actually form .... by circulating brines'*. Thus, Selley points at some fundamental aspects of salt accumulations—there is 'too much' of it, it forms by circulating brines, and the brines are founded in sub-surface.

Kendall (2005) also adds the 'potash problem': '*The problem with this marine origin is that the chemical and mineralogical characters of most potash deposits depart significantly from those that would be expected from simple seawater evaporation. If the marine origin is correct, then other processes must have been involved to cause the differences*'. Although Kendall (2005) did not realize that the ocean's salt composition changed significantly over time (cf. Lowenstein et al. 2014, Babel and Schreiber 2014), we here point out the possibility that some other processes, in addition to solar evaporation of sea water, may have been active.

Lowell and Rona (2005) have already suggested that the magma chamber in a hydrothermal cell may be insulated from the near surface area by a high-density brine: '*The formation of a brine layer at the base of a hydrothermal system may act as a thermal conductive barrier between the overlying hydrothermal circulation and the magma body and be a salinity source for saline vent fluids'*.

Based on the fact that halite and other salts are barely soluble in supercritical water (ScriW) and a great deal of water must be processed in order to make substantial thicknesses of salt, we suggest that even though solar evaporation is important, there probably is much salt formed by hydrothermal processes. Furthermore, we suggest that the Red Sea basin represents a modern active analogue for the large-scale formation of hydrothermal salt in a rifting tectonic setting.

Warren (2010) defines *Platform evaporites* as '*ancient bedded marine evaporites formed in subsea-level seepage lagoons*"; and *Basin-wide evaporites* (like those in the Red Sea) he defines as "*marine evaporite successions that have filled large-scale tectonically and hydrologically isolated subsea-level sedimentary basins to their hydrological equilibrium level*'. Furthermore, he states that '*...water cannot flow uphill and so such marine-derived evaporite systems must by definition be accumulating in subsea-level depressions*'. But hydrothermal water actually does flow 'uphill' and it is commonly observed that aqueous fluids driven by hydrothermal processes vent up through the seafloor (Lowell and Rona 2005; Coumou et al. 2008).

The hydrothermal salt model first proposed by Hovland et al. (2006a) demonstrates how large masses of halite (and other sea salts) can accumulate by precipitating from supercritical and boiling sea water. This process takes place where sea water is exposed to high heat flow in the subsurface. In the present paper, we will focus on some of the particular features found in the northern and central Red Sea and use them to substantiate and improve the hydrothermal salt model.

The Hydrothermal Salt Model

An important aspect of the hydrothermal salt model is that it allows for the formation of vertical 'piercement' salt structures. These salt structures may also be conduits of hot fluids, including brines, and can therefore act as giant hydrothermal chimneys. A key to understanding the subsurface accumulation of large salt bodies is linked to how brines behave in the sub-surface when pressurized and heated beyond their boiling and critical point. At depths greater than $\sim$3 km ($P > 298$ bars), sea water will no longer boil. When heated to >407 °C at these pressures (Koschinsky et al. 2013), the sea water becomes supercritical and a phase separation occurs, with the formation of a high-density brine, precipitation of solid salt particles, and a supercritical water vapour phase (ScriW) which has a specific gravity of 0.3 g/cm^3 (at the critical point). Above 570 °C, a nearly pure vapour phase coexists with solid salt, and convection is driven exclusively by the temperature gradient (Geiger et al. 2005). The inability of ScriW to dissolve common sea salts results in the accumulation of salt particles in the conduits whenever brines move into the supercritical domain.

Experimental observations of salt precipitating from supercritical saline waters were first published by Tester et al. (1993). During experiments with the phase behaviour of synthetic brines above their critical points, they visually observed 'shock crystallization' of NaCl and Na_2SO_4 through a sapphire window in their pressure cell. At temperatures above 405 °C, a 'cloud' formed inside the pressure chamber as the solubility of the salts dropped dramatically over a temperature range of only a few °C. The resulting solids were found to consist of 'highly amorphous kernel-shaped particles' of sizes between 10 and 100 μ (Tester et al. 1993).

These observations demonstrate the potential for precipitation of salt from sea water circulating in faults and fractures close to a heat source in tectonically active basins such as the Red Sea. The natural domain for ScriW on Earth is deep in the sub-surface and inside deep-sea hot vents (Von Damm et al. 2003; Simakin and Ghassemi 2003), well hidden from direct observation. Salt may also precipitate in subduction zones, when porous rocks containing saline pore water are exposed to sufficiently high temperature and pressure. By combining numerical molecular modelling with basin modelling, it was possible to simulate how large amounts of solid salts could be produced and accumulated hydrothermally by supercritical 'out-salting' (Hovland et al. 2006a, b). This model also includes the formation of thick deposits of layered salts that may form either sub-aerially or sub-marine in the vicinity of volcanism (high heat flow), with an analogy to the salt pan of Lake Asale, Eritrea, adjacent to the Red Sea and the active volcano Erta Ale, and to the 'salt volcanoes' of Dallol (Talbot 2007).

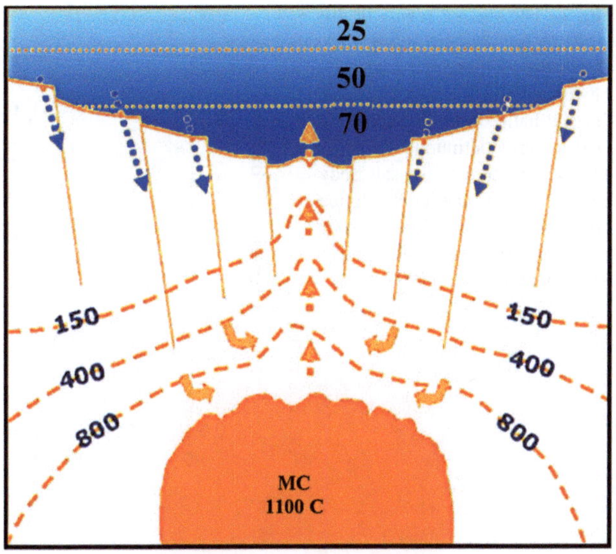

Fig. 1 A conceptual sketch (not to scale) illustrating the hydrothermal salt model in the sub-surface. It is based on the inferred sub-surface conditions below the Atlantis II Deep in the Red Sea (Ramboz et al. 1988; Zierenberg and Holland 2004). *Arrows* indicate the forced convection of water, vapour, and particles, driven by the strong heat flow gradient (Hovland et al. 2006a, b). *MC* = magma chamber. All numbers are temperatures in °C. The depth to the seafloor in the Atlantis II deep is 2,100 m, and the sub-surface depth to the top of the magma chamber is inferred to be 1.5–2 km (based on Hovland et al. 2006a)

Numerical Modelling of Sub-surface Salt Accumulation

A conceptual digital model of a hydrothermal system in the Atlantis II Deep of the Red Sea was constructed. This model includes a shallow magma chamber that causes a sufficiently high heat flow to drive a convection cell of sea water through a package of sedimentary rocks (Fig. 1). Based on the physical mechanisms that govern the supercritical behaviour of sea water, a numerical simulation was performed for the hydrothermal flow. The simulation model results in the accumulation of salt that precipitates along the flow lines within the supercritical region (Fig. 1) (Hovland et al. 2006a, b).

In the zone closest to the magma chamber, temperatures will be beyond 800 °C which is close to the melting point of halite (Fig. 1). According to these simulation results, the supercritical water vapour will rise vertically in the central reflux zone, and the flow will have highest velocity immediately above the magma chamber due to the buoyancy forces (Hovland et al. 2006a).

The zones with supercritical out-salting require accommodation space for large masses of silt- /sand-sized solid salt particles (Tester et al. 1993), as modelled in the Red Sea analogy. However, as the accommodation space for the solid salt fills up, it is assumed that the hydraulic force exerted by the buoyant ScriW will force its way upwards and also displace the salt particles and brine to form large diapirs, ridges, and domes, as observed in the Red Sea. Furthermore, when the system cools down to sub-critical conditions, solid salt will start to dissolve in the circulating brine and be transported upwards to re-precipitate as solid salt upon further cooling. This is also a refining process, because the various salts have very different solubilities, both in hot and cold water. Thus, the more soluble salt species (e.g. $MgCl_2$) are transported further, and may even reach up to the sea bottom without precipitating.

Piercing salt bodies are frequently reported to be covered by anhydrite or gypsum. Calcium sulphate (anhydrite) in the sea water has relatively low solubility compared to that of sodium chloride (halite). In evaporation settings, calcium sulphate will precipitate next to the precipitation of calcium

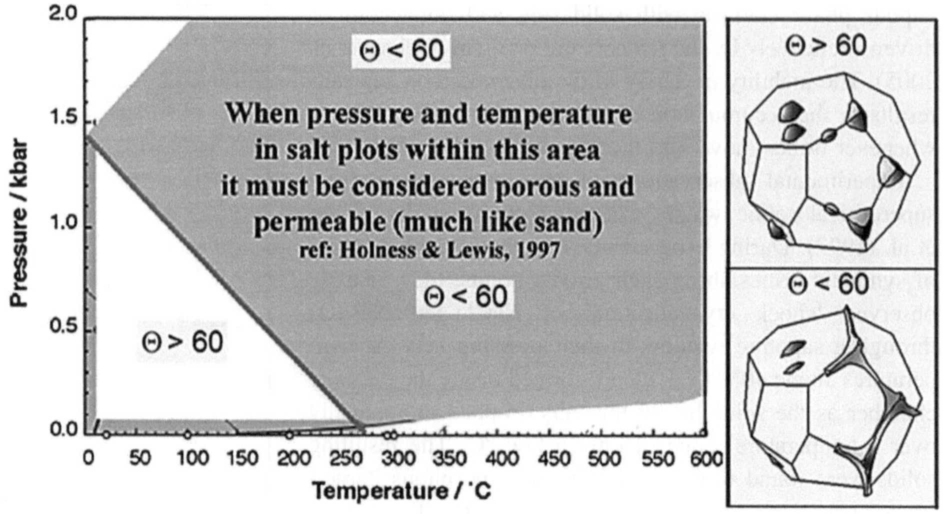

Fig. 2 The *shaded region* in this pressure–temperature space shows the PT conditions where Lewis and Holness (1996) found that the halite-H_2O dihedral angles were continuous strings and the structure of the salt body becomes permeable (crystal figure *lower right*). Illustration based on Lewis and Holness (1996) and Holness and Lewis (1997)

carbonate. Furthermore, the surfaces of exposed salt bodies in contact with sea water (boundary layer) will attain salt concentrations at saturation levels, and this may also cause precipitation of calcium sulphates from the sea water (Katz et al. 1981). Such a coating of precipitated sulphate minerals (anhydrite/gypsum) will protect the salt from further dissolution by sea water.

Anhydrite layers are also observed under salt bodies. This may result from the dissolution of a halite formation that contained some anhydrite (B.C. Schreiber pers. com. 2013). Both the lower layers of the Zechstein evaporites and the Cretaceous halite of Thailand (Khorat Basin) have about 1–1.5 m of anhydrite residue below them, usually called the 'basal' anhydrite. The trace elements and stable isotopes of these layers are out of keeping with normal marine salts (B. C. Schreiber pers. com. 2013).

Tectonically active environments with high sedimentation rates such as in the Red Sea may offer the necessary protection to submarine precipitated salts. The Red Sea is characterized by a relatively high sedimentation rate of about 11 cm per 10^3 years (Wall and Warren 1969) and has near tropical climatic conditions with high surface evaporation, which also helps to keep the general salinity high.

The 'Holness Theory'

In general, salt formations are treated as being nearly impermeable for brines and hydrocarbons. However, according to laboratory work first carried out by Lewis and Holness (1996), halite buried at depths greater than ~3 km and elevated temperatures becomes permeable due to altered dihedral angles between the salt crystals: *In sedimentary basins with normal geothermal gradients, halite bodies at depths exceeding 3 km will contain a stable interconnected brine-filled porosity, resulting in permeabilities comparable to those of sandstones'* (Lewis and Holness 1996). Thus, halite occurring at depths of more than ~3 km and temperatures above ~200 °C has a continuous pore system filled with liquid water (sub-critical) and therefore relatively high permeabilities. This is caused by the altered dihedral angles between the salt crystals (Fig. 2). Consequently, supersaturated brines can easily be transported through salt formations at these depths (Schoenherr et al. 2007a, b). Fluids are also observed to flow through shallower, but deep-rooted buried salt, for example salt domes or 'diapirs' (Aharon et al. 1992; Sassen et al. 2004; Gay et al. 2011). This must be due to flow through near-vertical conduits inside the salt body, kept open by a continuous flow of hot brines and condensation (fresh) water.

This water-saturated inter-connected pore system is also reducing the effective stresses in the salt body, which now is influenced by the liquid pressure, and not only the overburden pressure, that is the salt crystals are partly supported by the pore liquids, and the system is attaining a more ductile behaviour. It is well known that the seismic imaging (*p*-waves) of large salt bodies is difficult. This may well be caused by the change in seismic impedance when going from the upper 'brittle salt' with higher sonic velocities to a more ductile and permeable salt deeper down with lower sonic velocities. Thus, the so-called *decollement* seen on many interpreted seismic images of deep-lying salt may actually represent this transition zone, which causes a seismic artefact rather than another geological entity. Instead of representing a low-angle surface on which the salt has been sliding and deforming, it may rather represent the transition zone between ductile (below) and brittle (above) salt (Lewis and Holness 1996; Holness and Lewis 1997).

Pertinent Red Sea Features

There are several salt-associated features in the Red Sea which are difficult to explain in terms of the conventional 'evaporite' salt model (Stoffers and Kühn 1973; Schreiber et al. 2007; Warren 2010; Feldens and Mitchell this volume; Ehrhardt and Hübscher this volume). Some of these features are as follows:

1. Several kilometre thick salt deposits on both flanks of the Red Sea (Searle and Ross 1975; Gaulier et al. 1988; Savoyat et al. 1989).
2. Thick salt deposits inside the central graben (Cochran 2005; Ligi et al. this volume; Feldens and Mitchell this volume; Ehrhardt and Hübscher this volume).
3. Dense, hot brines in some of the central grabens (Red Sea 'Deeps') (Zierenberg and Holland 2004; Pirajno 2009).
4. Long walls and ridges of salt exposed to sea water in the northern Red Sea (Mart and Ross 1987), including the Conrad Deep, Thetis Deep, and Nereus Deep central grabens (Ehrhardt et al. 2005; Ligi et al. this volume).
5. Large flows of salt in the Thetis Deep (Mitchell et al. 2010a; Feldens and Mitchell this volume).

Hot Brine Pools

Along the centre of the 2,000 km long, narrow Red Sea, there are about 25 local 'deeps'. These active spreading centres (Mitchell et al. 2010b) are hosting hydrothermal circulation, and the deepest troughs contain hot brines (Ehrhardt and Hübscher this volume). The discovery of the dense, high-temperature brines came as a surprise to the Swedish deep-sea Albatross expedition in 1948. According to Lowell et al. (1995), the first evidence of hydrothermal activity at a submerged seafloor spreading centre appeared as an anomaly in temperature and salinity in hydrographic data. Subsequent sampling of the water column and seafloor also revealed that mid-water acoustic reflections were produced by the impedance contrast caused by hydrothermal brines containing suspended metal sulphides (Lowell and Germanovich 1997).

Three pools of hot brines are found within the Atlantis II Deep (Degens and Ross 1969), and sediments underlying these brines are enriched in heavy metals. The Atlantis II Deep is defined by the 2,000 m isobath. It is 14 km long, about 5 km wide, and contains approximately 5 km^3 of density-stratified brines, which are actively accumulating through the discharge of hydrothermal solutions (Pirajno 2009). The temperatures of the hydrothermal fluids that enter the Deep are variably estimated to range from approximately 210 °C to as high as ~450 °C. These hot brines have high concentrations of Fe, Mn, Zn, Cu, Pb, Co, Ba, Li, and Si which precipitate as layers of sulphides and silicates (Pirajno 2009). This indicates sea water contact with the mantle (peridotite), also suggesting that some of the circulating sea water may participate in serpentinization reactions.

Although the Atlantis II Deep has been intensely studied, it has not been possible to locate the hydrothermal vents. Pirajno (2009) explains this fact by the presence of soft muddy sediments trapped in the brine pools and by abundant precipitation of sulphides, sulphates, and silicates. In addition, we suggest there may be a mixture of saturated brines that are precipitating salts upon cooling. This will effectively prevent the imaging and detection of focussed discharge plumes into the water column. However, black smokers associated with pure sulphur chimneys were documented at the flanks of the Kebrit Deep (Blum and Puchelt 1991).

The actual heat flow values of the Red Sea deeps were investigated by Anschutz et al. (1999) and were calculated to be somewhere between 250 and 600 mW/m^2. Furthermore, they noted that the range of heat flow values corresponds to bottom heat flow values that have been reported earlier for the axial zone of the Red Sea. Whereas other studied brine pools indicate stable (steady-state) temperature and salinity conditions, it was found that the Valdivia Deep brine increased by 4.1 °C and 10 ‰, respectively, between 1972 and 1992. This suggests that the deeps in the Red Sea undergo stages of heating [e.g. with hydrothermal brine discharge, Anschutz et al. (1999)] and stages of cooling. These processes of heating and cooling in the deeps have important implications for underground fluid flow in the hydrothermal salt model, which we discuss later (Simakin and Ghassemi 2003).

The northernmost active deeps in the Red Sea are the Conrad, Kebrit and Shaban Deeps, located up to 500 km north of the Atlantis II Deep (Blum and Puchelt 1991, Ehrhardt and Hübscher this volume). According to Bonatti (1985), this part of the Red Sea consists predominantly of stretched and thinned continental crust with a few tectonically formed isolated deeps that mark initial stages of seafloor spreading with local emplacement of oceanic crust. Recent bathymetry and shallow seismic surveys of these three deeps have revealed some very interesting features (e.g. Ehrhardt and Hübscher this volume). These data provide new evidence about how thick layers of evaporites may dissolve, deform, flow, re-precipitate, and collapse, mainly due to hydrothermal circulation, causing subrosion (salt dissolution) and re-precipitation of salts, processes that are still active with subrosion in these deeps.

Salt-Cored Ridges, Northern Red Sea

In the northern Red Sea, Mart and Ross (1987) described numerous 'diapirs' occurring in water depths ranging between 700 and 1,300 m. They are typically elongated, attaining lengths exceeding 40 km and widths of 4 km, and

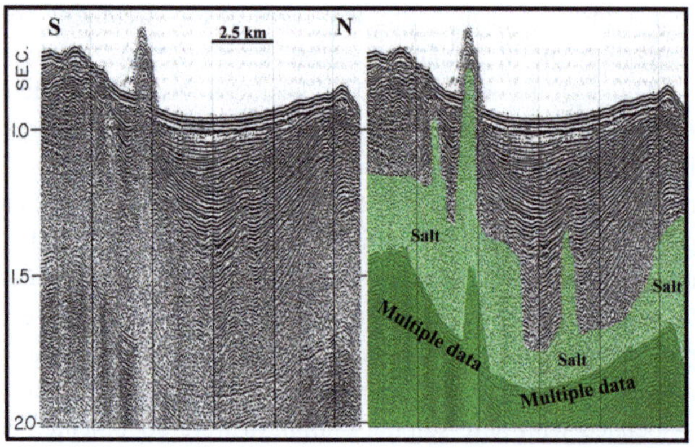

Fig. 3 A 2D-seismic image from the northern Red Sea (based on Mart and Ross 1987). The image crosses over one of the many ridges, consisting of salt (probably halite with an outer layer of sediments). The ridge rises about 250 m above the surrounding seafloor, and its base is about 500 m wide. Such ridges can be up to 40 km long (cf. Fig. 4). The interpreted image on the right shows inferred sub-surface salt, in *green colour*. Multiple seismic reflections are shown in *dark green*

they are covered by up to 1 km of overburden. These 'diapirs' commonly ascend either along the boundary faults of extentional rift half-grabens or into the rift floors. One of the most pronounced positive features is about 500 m wide at the base and rises at least 250 m above the sediment-covered seabed (Fig. 3, for the general location see Fig. 4). Some 'diapirs' are also topped by negative 'collapse' structures, one of which measures about 2 km across and 75 m in depth. Mart and Ross (1987) interpreted these to represent salt dissolution pits (Fig. 5).

Because of their anomalous appearance, Mart and Ross (1987) question whether these structures represent conventional salt diapirs. They are very difficult to understand using the conventional halokinetic explanations and interpretations, particularly because they seem to have been 'squeezed' up into the water column without being severely leached. Another possibility is that they represent viscous pressure ridges ('salt injectites') with surface liquid outflow and corresponding crater formation, but these normally only occur in compressional settings, involving sedimentation onto salt layers (Ings and Beaumont 2010).

However, as a consequence of the hydrothermal salt theory, we suggest they represent ductile and overpressured salt originating underground, which has been pushed upwards from the sub-surface formation zones along active deep faults associated with tectonic extension. Thus, we interpret these structures as salt injectites (extrusives) caused by the lack of sub-surface accommodation space and high fluid pressure. An apparent analogy to these features would be sand injectites, described from, among other locations, the North Sea (Løseth et al. 2012). The exposed portions of the salt bodies are probably protected against re-dissolution in sea water by a cap of anhydrite and/or sediments. The locally observed depressions in their summit either represent dissolution craters, where the capping sediments or anhydrites have been breached (Mart and Ross 1987), or

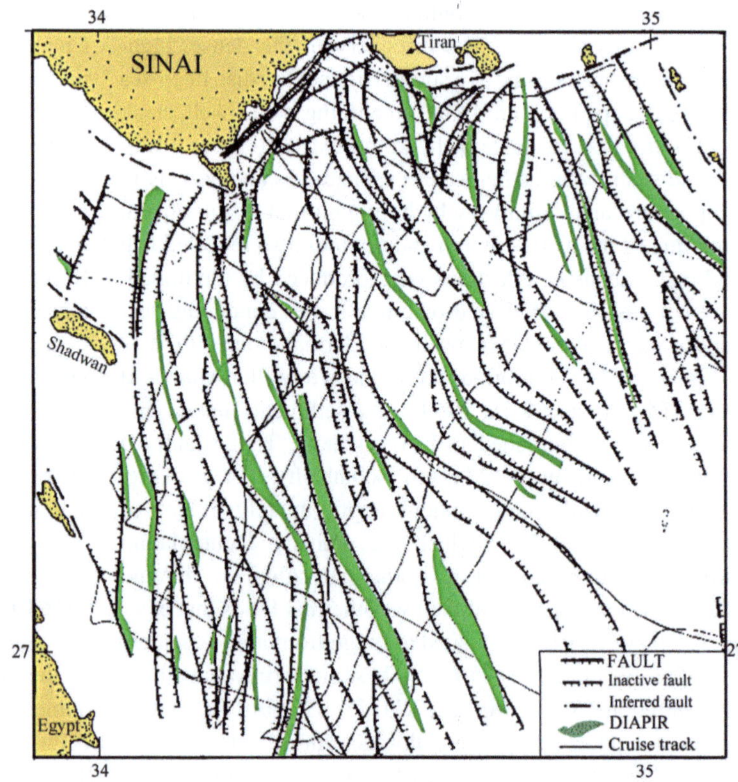

Fig. 4 Map of up to 40-km-long salt-cored ridges (in *green*) in the northern Red Sea (from Mart and Ross 1987). Note that the ridges are generally parallel to the spreading axis of the northern Red Sea. This suggests that they form along deep faults associated with the tectonic extension process

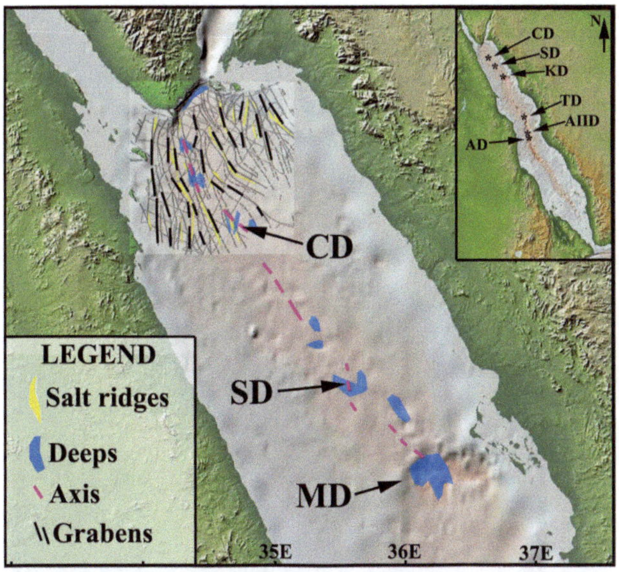

Fig. 5 *Shaded* relief topographic map of the northern Red Sea. The area of the northern Red Sea mapped by Mart and Ross (1987) is *inset*, with *salt-cored ridges* shown in *yellow* (cf. Figs. 3 and 4). Some of the 'deeps' in the northern Red Sea are also indicated in *dark blue* (Cochran 2005; Ehrhardt et al. 2005). *CD* Conrad Deep, *SD* Shaban Deep, *MD* Mabahiss Deep, *KD* Kebrit Deep, *TD* Thetis Deep, *AIID* Atlantis II Deep, and *AD* Albatross Deep

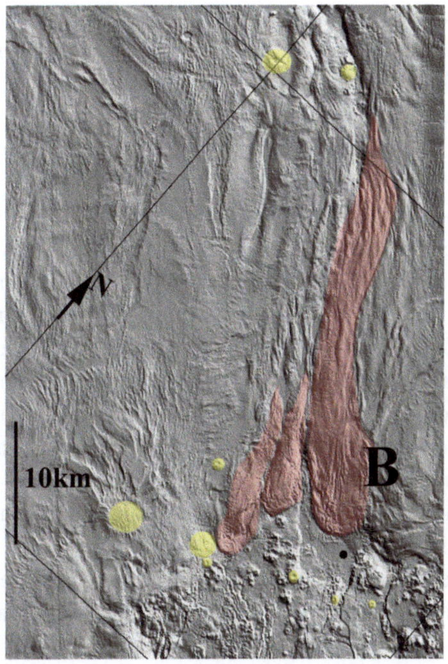

Fig. 6 *Shaded* relief presentation of bathymetric data of the central graben of the Thetis Deep (modified from Mitchell et al. 2010a). There is evidence of massive salt flows. In this image, the Thetis Deep, proper, is seen as the flat-floored depression with rugged topography to the south-east. Although only three of the large salt flow lobes (according to Mitchell et al. 2010a) have been highlighted here (in *pink*), there are many more that can be identified on this image. Note the salt flow (named 'B' by Mitchell et al. 2010a) flowing down the slope on the *right*, foreground

expulsion craters, where fluids emanate from a conduit and locally erode the outlet areas. These structures are assumed to represent archetypal hydrothermally associated salt ridges, some of which, or parts of which, may be acting as fluid conduits in a high heat flow tectonic setting. As these features have, to our knowledge, previously not been explained in this manner, we suggest that their formation is highly relevant for understanding the underground accumulation of hydrothermal solid salts in the Red Sea and elsewhere, as for example in the Guaymas Basin (Gulf of California spreading zone), where sub-surface salt masses have been documented by 2D-seismic and gravity investigations (Miller et al. 2009).

Massive Salt Flows

The hydrothermal salt model predicts that large quantities of silt-/sand-sized salt particles will precipitate underground and possibly make their way to the surface, near or within the central grabens. As previously mentioned, the reason why they have to surface is either the lack of underground accommodation space or high fluid pressures. As soon as the underground pressure attains a value greater than the local lithostatic pressure, hydraulic fracturing will occur and the fractures will act as conduits towards the surface. The flow will most probably follow the nearest available extensional fault system.

Surprisingly large flow-like features were identified on topographic data from the Thetis Deep of the central Red Sea (Fig. 6). These features are interpreted by Mitchell et al. (2010a) and Feldens and Mitchell (this volume) being the result of evaporite flowage, based on their estimated 200–500 m reliefs. Because the regional superficial hemipelagic sediment layer is much thinner than 200 m, the only possibility is fluidized evaporites, that is 'salt flows' (Mitchell et al. 2010a, b). The flow features were discovered using swath (multi-beam echo sounder) mapping.

The flowing material is suspected to consist of a water slurry with particulate solid salt or a concentrated precipitating brine and moving solid salt at a velocity of at least 1 m per year, according to Mitchell et al. (2010a) and Feldens and Mitchell (this volume). One of these flows, flow 'B', is apparently issuing out of the eastern wall of the Thetis Deep and flowing downhill, into the central graben (Fig. 6). The flows are recognized by their smooth surface morphology which contrasts with the rugged morphology typical of volcanic ridges and cones on the floor of the Deep. Mitchell et al. (2010a) and Feldens and Mitchell (this volume) interpret these salt flows to occur as a consequence of Miocene evaporite dissolution by hydrothermal activity. We suggest that the salt was formed in the sub-surface supercritical out-salting zone (Hovland et al. 2006a) and emanates at the flanks of the deep as salt injectites, as previously discussed.

Perhaps the most unexpected aspect of the results of Feldens and Mitchell (this volume) is the extensiveness of the flows and that they seem to originate from the flanks of the graben, rather than from point-sources within the graben. However, by carefully analysing the detailed bathymetry, it can be seen that there are also numerous apparent dome-shaped features both inside the central graben and along some of the slopes leading down to it (Fig. 6). Some of these domes are interpreted by us to represent either normal volcanic hot vents (hydrothermal sulphide mounds) or 'salt volcanoes', much like mud volcanoes (Hovland et al. 1997), where mixtures of liquefied salt particles and sediment (volcanic) grains well up to the surface from the underground supercritical outsalting zones.

Some of these salt flows occur as apparently over 20-km-long drifts along the central valley of the Red Sea (graben trough) and certainly do not appear as typical landslide deposits. From their morphology, they can be described as flow deposits, blankets, or drifts.

Salt Ridges in the Conrad Deep

Although there are numerous salt-cored ridges along the flanks of the northern Red Sea (Mart and Ross 1987), it came as a surprise to discover similar ridges of apparently exposed massive salt crossing over one of the hot spreading centres (deeps) of the Red Sea. According to Ehrhardt et al. (2005) and Ehrhardt and Hübscher (this volume), the Conrad Deep in the northern Red Sea is a poorly understood basin, associated with rifting and shear accommodation (Fig. 7). It has a series of anomalous features, especially associated with salt formations, which Ehrhardt et al. (2005) and Ehrhardt and Hübscher (this volume) suspected to be salt diapirism, by applying the conventional evaporite model. On 2D-seismic data over this area, there are 3,000-m-thick evaporites (salt deposits) (Cochran et al. 1986). These are characterized by having a high-amplitude low-frequency reflection, located 200–300 ms two-way time (TWT) below the seafloor. Depending on the facies at the upper part of the Miocene evaporites (halite, anhydrite, or gypsum), the internal reflectivity of this sediment body alternates from seismically transparent to a layered pattern (Ehrhardt et al. 2005 and Ehrhardt and Hübscher this volume, Ligi et al. this volume).

Ehrhardt and Hübscher (this volume) suggest that the lateral change of evaporite facies (e.g. transparent to stratified) is due to halite folding (diapirism) and deposition of other salts, for example gypsum, carbonate, and anhydrite (stratified). We agree that the halite bodies are acoustically transparent, but suggest that the seismic transparency of

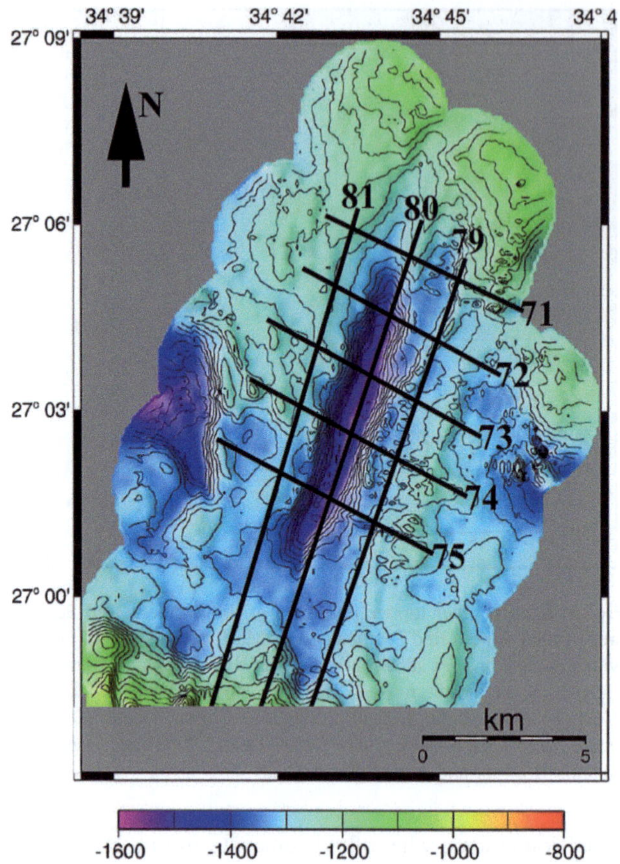

Fig. 7 *Shaded* relief topographic map of the Conrad Deep (modified from Ehrhardt et al. 2005), showing the location of the high-resolution seismic survey lines, some of which have been interpreted in Fig. 8

halite may be caused by both depositional (e.g. hydrothermal processes) and tectonic processes, as explained herein (see discussion).

The Conrad Deep is characterized by steep slopes on both flanks (9°–18°) and is composed of at least two sub-basins, of which the south-western basin is shallower and separated from the deeper north-eastern basin (Fig. 7). A brine layer was detected by Ehrhardt et al. (2005) in both these basins. Salt diapirs or volcanic extrusions are located at the southern flank of the axial depression. They are 150–190 m higher than the surrounding seafloor of the main trough (Figs. 7 and 8). The elongated rise that strikes parallel to the deep was called the '*SE-Rise*' by Ehrhardt et al. (2005) (Fig. 8) and was concluded to be an uplift related to the emplacement of a magmatic body at the flanks of the Conrad Deep axial depression.

Another feature of the Conrad Deep is suspected to represent the development of an elongated diapiric wall, called the '*Salt Wall*' (Ehrhardt et al. 2005). It runs obliquely to the strike of the Conrad Deep's SE-shoulder (Fig. 8) and strikes N34ºE along its SW-extension into the deep and acts partly as the SE slope of the deep (Fig. 8). This rise must have occurred

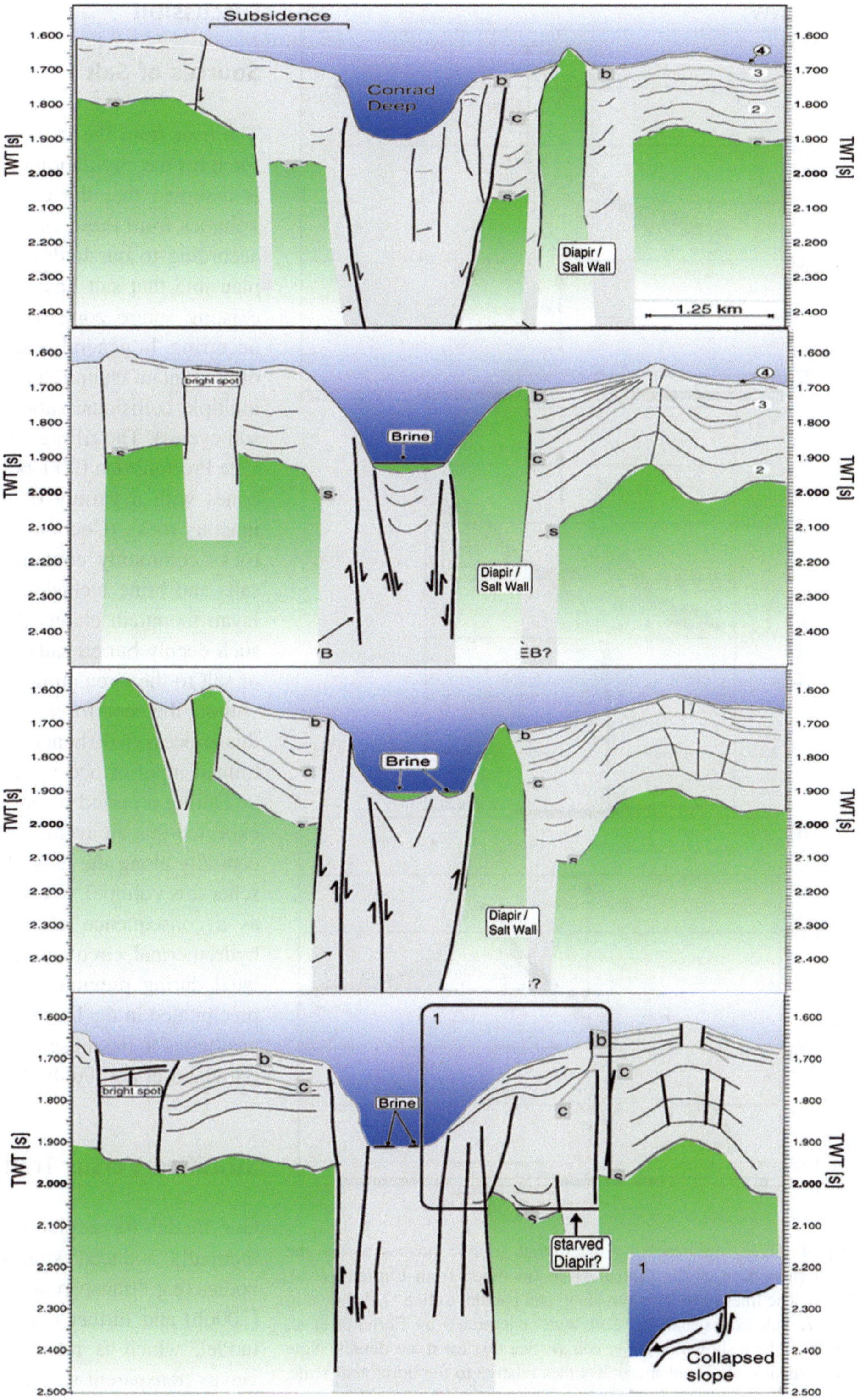

Fig. 8 A series interpreted parallel 2D seismic sections crossing the Conrad Deep in the Red Sea. They have been interpreted by Ehrhardt et al. (2005) and have been reproduced here with the inferred salt bodies (interpreted by Ehrhardt et al. 2005) enhanced in *green* colour, by us. The four seismic lines are, from top to bottom (i.e. from north to south), as follows: line G_99_072, −073, −074, and −075 (see Fig. 7 for location). The interpretation of seismic line —075 (bottom) has been slightly modified in accordance with Ehrhardt and Hübscher (this volume) with a salt wall at the head of the indicated slump feature

most recently and is still active. Thus, it is inferred by Ehrhardt et al. (2005) and Ehrhardt and Hübscher (this volume) that 'salt diapirism' is still active on the flanks of the Conrad Deep (Figs. 8 and 9). However, in our interpretation, this is evidence of ongoing underground subrosion combined with hydrothermal salt production (re-precipitation) and the surface expression of underground salt accommodation, that is rock salt extrusion or salt injection. In our model, the lack of underground accommodation space leads to upward migration of the salt, as will be discussed later.

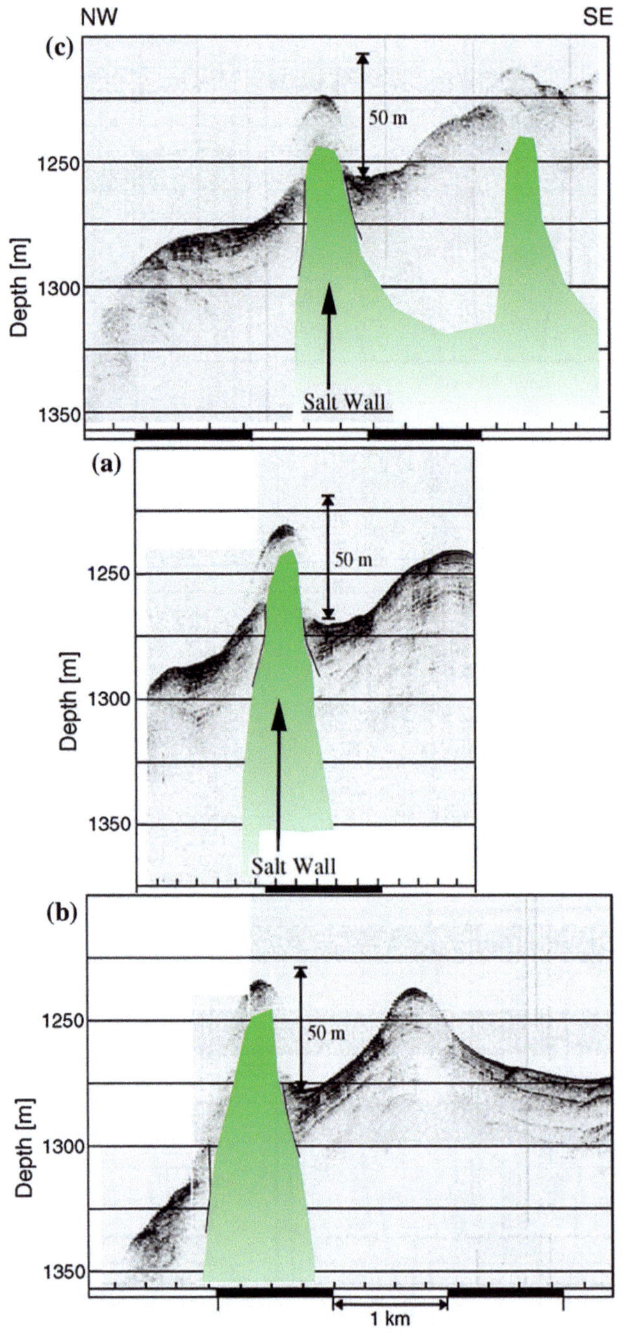

Fig. 9 Three high-resolution parasound seismic images across the northern portion of the Conrad Deep (modified from Ehrhardt et al. 2005). The lines have been run along and parallel to line '72' shown in Fig. 7. Salt ridges and the '*Salt Wall*' interpreted by Ehrhardt et al. (2005) is here shown in *green colour*. See text for more details. Note the vertical scale is enhanced 20 times relative to the horizontal scale

We thus infer that the anomalies documented by Ehrhardt et al. (2005) and Ehrhardt and Hübscher (this volume) are the results of locally produced hydrothermal salt, underground. The '*SE-Rise*' and '*Salt Wall*' structures could, therefore, probably more exactly be termed as salt extrusions/intrusions, or even rock salt injectites (Figs. 7, 8 and 9).

Discussion

Sources of Salt

The heat from the magma chambers represents the driving force for the circulation of sea water and brines. However, it is assumed that the energy needed to generate large salt volumes from these circulating brines may not be sufficient according to our hydrothermal model. Therefore, it is also plausible that salt has accumulated sub-surface in the pre-existing suture zone where the current Red Sea rifting is occurring. In general, a suture zone represents the root of an old mountain chain (Burke 1996) and is the consequence of multiple collisions, subductions, and rifting episodes (Wilson cycles). The rifting zone of the Red Sea began with two Late Precambrian (900–600 Ma) north–south trending suture zones with a variety of granitic, metamorphic, and mafic igneous rocks (Cochran and Karner 2007). These types of rocks commonly contain abundant accumulations of solid salts and brine inclusions, as found in the Alps and Himalayan mountain chains (Warren 2010). It is suggested that such deeply buried salts may contribute significant volumes of salt to the circulating sea water during the rifting process without the need for vast amount of heat energy input. Also, this process will benefit from a semi-enclosed basin with limited input of ocean water.

During a period of only ~0.5 Ma (in the Miocene), it is expected that an average of 3,000 m of salts were deposited centrally along the axis of the Red Sea (Ehrhardt and Hübscher this volume). We suggest that this was partly achieved as a consequence of increased heat flow and associated hydrothermal circulation, where subterranean salt accumulated during previous Wilson cycles was remobilized and precipitated in the basin, together with sea water salts, partly analogous to the present conditions in Lake Asale and Dallol (Hovland et al. 2006b; Talbot 2007).

Stratified Versus Transparent Salt Bodies

Our model for explaining the difference between hydrothermally induced stratified salt versus amorphous salt bodies (e.g. 'transparent salt') was provided in Hovland et al. (2006b) and further commented on by Talbot (2007). This model, which is relevant for the discussion of stratified versus transparent salt in the Red Sea proper (Ehrhardt and Hübscher this volume), is based on re-interpretations of seismic data across the Dallol 'salt volcanoes' and parts of Lake Asale, Eritrea: '*In our interpretative model of the Lake Asale system, it is inferred that the region close to the individual heat-sources (at some 3.2 km depth) are characterized by intersecting faults and fracture systems. This rock volume may also contain hot injection sills. Realistic temperatures in this region may be in the order 400-600 °C, which means that the temperatures in the intervening volume between the surface volcanoes and the heat sources span from supercritical conditions at depth to boiling at depths shallower than 2.8 km. Hot brines are venting in the 'salt-volcanoes' and deposit their brines and solid materials on the crater rims, on their slopes and on the surrounding salt flats where they form thick, near-horizontal evaporite layers*' (Hovland et al. 2006b). When these bodies of salt are fully submerged as in the Red Sea, their appearance on seismic data will be transparent for the 'salt volcanoes', which are basically constituted mostly of halite, and stratified, and may be composed of many different types of salt and sediments as also suggested by Ehrhardt and Hübscher (this volume).

Accumulation in the ScriW Zone

When the brines flowing in the recharge zones are entering into the supercritical domain of sea water, a spontaneous precipitation of salt occurs in the fracture zones. According to Lewis and Holness (1996), these salt deposits are permeable to the supercritical water vapour. Obviously, the accommodation space in the fractures is limited, and the continuous deposition of salt will fill up the entire fracture volume in the supercritical zone. The hydraulic pressure will increase, and eventually the salt and brines will be transported upwards by the refluxing water vapour in a process analogue to the formation of mud volcanoes. The water vapour will drive the process by exerting 'a gas lift' on the slurry of brines and salt particles.

The amount of salts that will accumulate in the zones of supercritical water conditions will depend on the following: (1) the volumes of circulating sea water or brines, (2) the volume of the underground fracture system and/or (3) the possibility for the produced salts to escape upwards or deform their surroundings as injectites, and, finally, (4) the longevity of the energy source, that is the size of the magma chambers.

Cooling, Dissolution, and Transport of Salt

Over time, the recharge zones where salt and brine accumulated will gradually be cooled down to sub-critical conditions. Also, the reflux zone may be partly clogged by precipitating minerals. This leads to changes in the flow patterns, which may imply that more of the reflux occurs laterally along the fault zones. At shallower depths than about 3 km (298 bars), or at temperatures lower than 407 °C, the brines become sub-critical, and boiling will occur. Boiling, combined with upward flow, induces cooling and condensation. Condensation of supercritical vapour produces pure water which has the potential to dissolve salts in the reflux conduits or emerge at the seafloor as fresh water flows. This assumption is supported by field observations of reflux zones, where the venting waters have salinities about half of that for sea water (Jarrard 2003).

During the cooling stage, salt will start to dissolve in the circulating water and be transported upwards as hot, dense brines. In the Red Sea, it is observed that the build-up of the salt ridges to a large extent occurs along the borders of the deeps. Also, the observed salt flows occur from the flanks of the deeps, suggesting an upward flow and injection of salt through faults flanking the central graben of the Red Sea (Fig. 8).

This upwards migration of water and brines with dynamically crystallizing and dissolution of salts keeps vertical conduits open within the salt bodies (Lewis and Holness 1996; Schoenherr et al. 2007a, b). Salt flows documented by Mitchell et al. (2010a) streaming out on the central graben floor in many different configurations further support this model.

That salt domes (diapirs) also in the Gulf of Mexico act as conduits for anomalously heated water, combined with methane gas and oil, has been demonstrated by Ruppel et al. (2005) and Gay et al. (2011). These observations are from a mud volcano located on top of a large sub-surface salt dome in the Mississippi Canyon, near the Ursa and Mars hydrocarbon fields. Also, the Chapopote mud volcano in the southern Gulf of Mexico (MacDonald et al. 2004; Hovland et al. 2005) strongly suggests that some salt domes act as deeply sourced hydrothermal chimneys.

Already in 1991, Hardie (1991) suggested the existence of three classes of evaporites, based on their geochemical composition: (i) from non-marine waters, with key minerals Na_2CO_3, (ii) from mixtures of marine and non-marine waters, with key minerals $MgSO_4$ and Na_2SO_4, and, not least, (iii) from hydrothermal and basinal brines, with key minerals $KCl \pm CaCl_2$.

The dissolution of accumulated salts occurs in sub-critical conditions in hot water. This will lead to faster dissolution of the more soluble sea salts, for example the magnesium chlorides. A further refinement takes place during the re-precipitation of salts in the upward flow, where the precipitation occurs according to the specific solubilities of the different salts at the various temperatures.

Finally, solid salt may also form within the gradually cooling brine pools located on the seafloor, fed by the brine-venting geysers (Ramboz et al. 1988) and by flowing solid salt from the graben flanks (Mitchell et al. 2010a). If warm, dense brines form in these pools, they will eventually cool, and the common sea salts (e.g. halite) will also precipitate and accumulate on the seafloor. The 'mother salt' below the Atlantis II Deep contains remnants of aligned halite chevron textures which are primary sub-aqueous textures (Whitmarsh et al. 1974). This also points in the direction of salt formation in submarine brine pools. According to J.K. Warren (pers. com. 2005), these halites will re-crystallize into bi-axially compressed halokinetic textures; a set of textures clearly different from hydrothermal salt precipitates found in fractures and layers beneath the brine-filled pools of the Atlantis II Deep. This interpretation is also in agreement with that of Orszag-Sperber et al. (1998), who state that evaporites deposited within an active rift system may be characterized by sub-aquatic facies because the irregular, structurally controlled, coastal high relief areas preclude the development of sabkha-type evaporites.

Accommodation Space and the 'Holness Zone'

According to Lewis and Holness (1996) and Holness and Lewis (1997), the precipitating salts will still be permeable for brines depending on the PT conditions up to a certain depth. When the brine cools, salt will precipitate and crystallize and exert a pressure on the overburden salts and sediments. This is suspected to be an important displacement mechanism for deep salt and for the formation of salt domes. '*Our results explain recent observations of major salt-fluid interactions at depth, and suggest that deep-rooted salt diapirs may act as conduits for basinal formation water*' (Lewis and Holness 1996). We term this PT-dependent transition zone for halite 'The Holness Zone'. In the hydrothermal model for salt formation, one of the consequences of the Holness Zone is that any upward-migrating water from below this zone will increase the pressure and may cause hydraulic fracturing of the above lying impermeable rock salt, as also pointed out by Lewis and Holness (1996), Holness and Lewis (1997) and Schoenherr et al. (2007b).

Do We Understand Salt Tectonics (Deformation)?

Numerous geologists and geophysicists are currently battling with the understanding of salt deposition and tectonism (deformation), especially in deep-water sediment basins. A very interesting paper by Davison et al. (2012) analyses and discusses the salt deposition, loading, and 'gravity draining' in the Santos and Campos rift basins offshore Brazil, where huge amounts of sub-salt oil have been found. They reach the rather surprising overall conclusion: '*Early salt movement occurred in the Santos Basin where layered evaporite reflectors indicate that diapirs started to grow during salt deposition, and salt was deposited in the surrounding rim synclines. Salt glaciers were also developed which were subsequently buried by later salt deposition, indicating that salt deformation (downslope salt drainage) was occurring during salt deposition*'. Contrary to our model, they probably assume that the 'downslope salt drainage' took place aerially, that is not under perhaps 1 km of sea water, as in the Red Sea. The reason why these workers do not cite the conventional models of Warren (2010) and others is that very few, if any, viable models exist with which to interpret the development of deep-water salt basins (Figs. 10 and 11).

Even when it comes to the relatively simple model for salt diapirism (Warren 2010), there is little understanding of how emplacement occurs. Figure 12 is a modern 2D-seismic image of a salt diapir (for illustration purpose only). It shows a salt body that looks 'classical'. However, the salt body cannot have behaved according to the generally understood diapiric models for salt deformation, where the body rises through the enclosing sediments due to density differences. A closer look at the seismic section suggests that the salt body has risen upwards, somehow fuelled locally by a deeper-lying salt reservoir. The exact mechanism is still not fully understood, but we can clearly see that the root of the salt piercement structure is attached to the basement (above a fault?) and that it is only the local sediments surrounding the diapir that have subsided. The sediments further away do not seem to have been affected by the movement. This is contradictory to the conventional model of salt diapirism, where an extensive salt layer is supposed to be squeezed towards

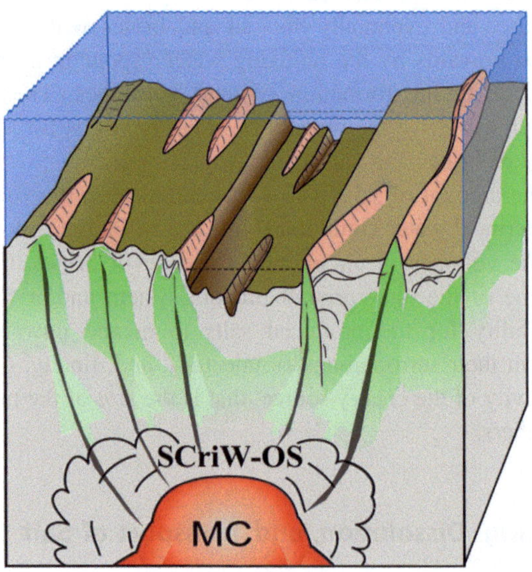

Fig. 10 Conceptual sketch (not to scale) of the Conrad Deep (based on Ehrhardt et al. 2005; Cochran 2005). The Conrad Deep is anomalous in that it has salt ridges adjacent to it and crossing it (indicated as *pink* and *green ridges*). The ridges are schematically illustrated in the block diagram. In our model, the elongated ridges represent salt injectites originating from the underground supercritical out-salting zone ('SCriW-OS') surrounding the magma chamber (MC). Note also that this Deep has a warm brine pool indicated by the *thin, broken horizontal lines* in the central graben. Note also the so-called salt wall (Ehrhardt et al. 2005) shown in the foreground on the right side of the central graben. Suspected major extensional normal faults are indicated as *grey zones* underlying the salt domes and ridges

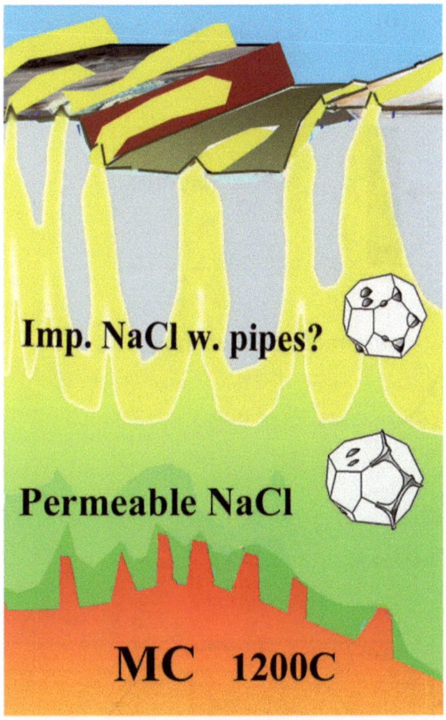

Fig. 11 The accommodation space for the precipitating salts will gradually be filled up and the internal pressure will increase. Above the Holness permeability zone, a solid salt body will form, which will move upwards due to tectonic and hydraulic pressures as well as re-deposition of dissolved salts from deeper layers. This salt body will contain open conduits caused by high fluid pressures through which fluids may flow

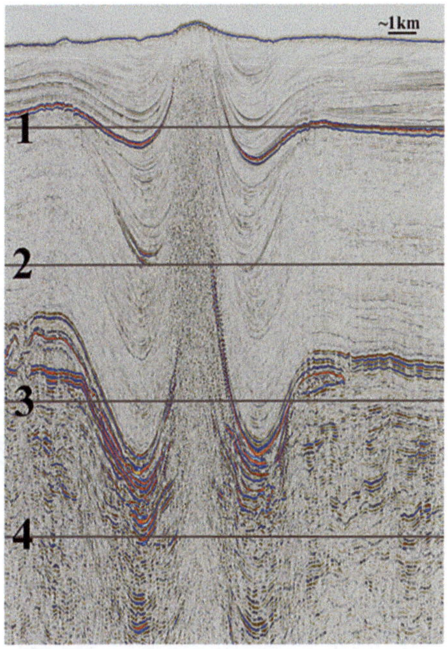

Fig. 12 A salt piercement structure ('diapir') seen on a modern 2D-seismic image (from the Barents Sea). The vertical scale is in seconds, two-way travel time. The obvious rim syncline surrounding the salt feature suggests that mass stored in the vicinity at depth has been transported away and upwards, and perhaps even deposited in the rim syncline as another mineral (magnesium salts?)

the salt diapir by the weight of the overburden sediments (Vendeville 2005).

Observations from Oman: Salt and Bitumen

Our hydrothermal model for salt deposition, movement, and refining is based on the basic physical properties of ScriW. But there is one more aspect of ScriW not mentioned yet, - that of hydrothermally associated oil (Simoneit 1993). In the onshore infra-Cambrian South Oman Salt Basin, there are oil deposits found within so-called intra-salt carbonate stringers (Schoenherr et al. 2007a, b). According to these workers, the oil has been dissolved in high-temperature/high-pressure fluids (not specified, but presumed by us to be ScriW) and transported as 'migrabitumens' through the salt structures into the intra-salt carbonate stringers (Fig. 13). This is a hydrothermally driven process that may take place in the supercritical domain of water, because ScriW is fully miscible with oil. Figure 13 is a modified version from Schoenherr et al. (2007a), which is presented here, for discussion.

In their model for how the oil was emplaced inside the carbonate stringers occurring inside the salt domes, they conclude the following (see Fig. 13): (i) the influx of focused high-temperature fluids must have caused significant local thermal anomalies in the Ara Group, (ii) this process inevitably increased the fluid volume within the carbonate stringers leading to a strong pore pressure pulse which most likely contributed to their overpressures, (iii) the overpressures were partly released by the migration of oil from the carbonate stringers into the surrounding Ara Salt as shown by Schoenherr et al. (2007b), and (iv) during this process, the oil must have moved preferentially into the grain boundary triple junction tubes of the salt by exceeding the capillary entry pressure, resulting in diffusively dilated zones in the Ara Salt (Fig. 13a, b).

This last argument (iv) has no importance in a supercritical environment, because *the oil is fully dissolved in supercritical water vapour, and there is no capillary entry pressure to overcome,* that is, it is ScriW (mixed with oil) displacing ScriW in the salt. This high-temperature/high-pressure hydrofracturing model of Schoenherr et al. (2007a) is remarkably similar and fully compatible with our hydrothermal salt model.

Future Studies

In this communication, we have suggested that much of the salt in the Red Sea has been produced underground as a consequence of hydrothermal processes. Our theory is

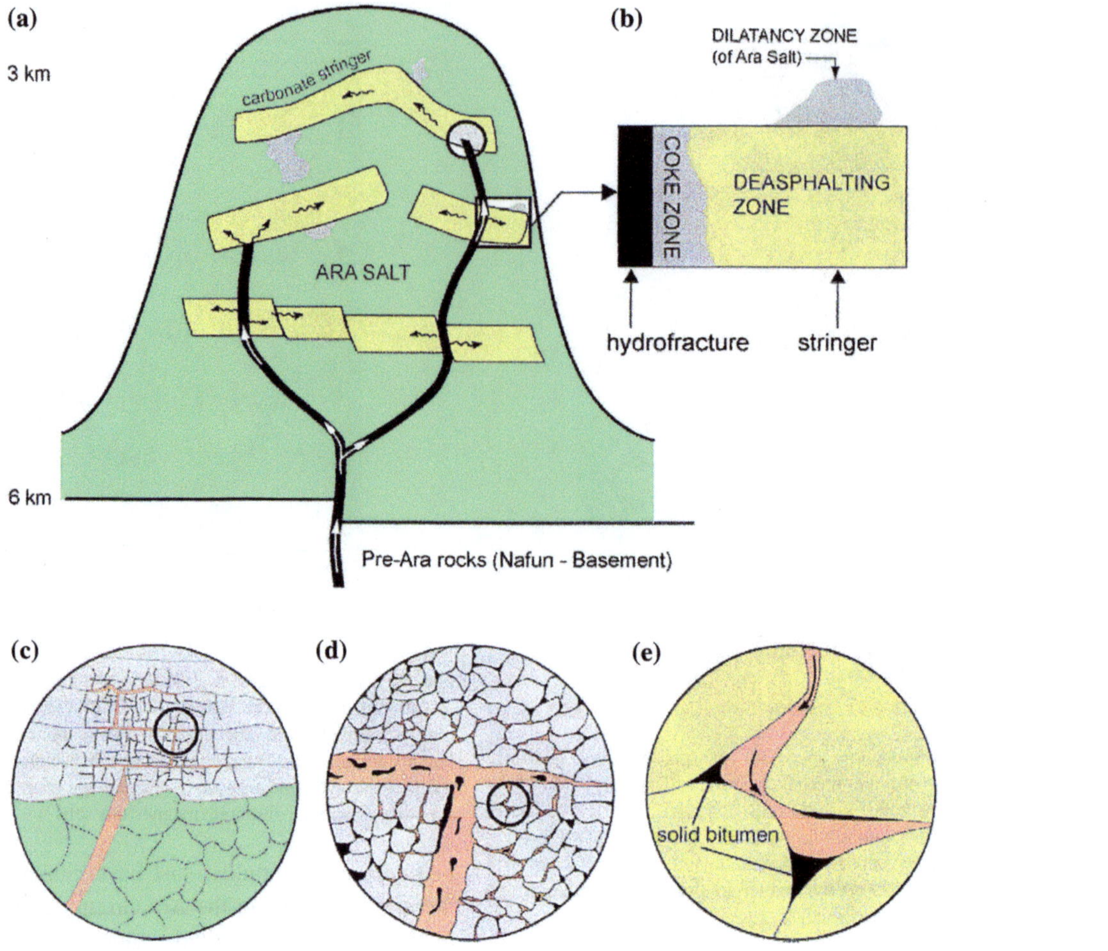

Fig. 13 Sketch based on Schoenherr et al. (2007a) that efficiently describes how salt domes may form and how hydrocarbons charge into carbonate stringers inside the salt dome, by migration of ScriW. **a** The hydrothermal fluids, including ScriW and dissolved hydrocarbons, enter the salt dome at depth (*black*). **b** At the carbonate stringer, the incoming fluids increase the pore pressure in the stringer, resulting in a pressure release and dilatancy zone in the surrounding salt, in accordance with the Holness theory. **c** A schematic illustration showing the salt/carbonate interface (see *circle* in **a**). When the fracture hits a carbonate stringer the fluids form fractures in the carbonate. **d** Detailed view of such a carbonate fracture system at the grain scale (see *black circle* in **c**). Here, liquid hydrocarbons probably condense as the ScriW cools down to subcritical phase. **e** Detailed view at the micron scale. The remaining hot hydrothermal fluids (condensed water?) flow through the carbonate pore network, leaving behind coke-like solid bitumen by thermal cracking, as observed in the South Oman Salt Basin

mainly based on two very important laboratory observations: (i) the nature of supercritical water (Tester et al. 1993) and (ii) the Holness Theory, which proves that halite at elevated pressures and temperatures has a permeability equivalent to that of sand (Holness and Lewis 1997).

Even though there are many observations from the Red Sea and elsewhere that support our hydrothermal salt theory, it has still not been tested in all ways possible. Therefore, we hope that future drilling campaigns, sampling, and mapping campaigns in the Red Sea will target some of the features we have interpreted herein. Carefully sampled salt from the 'salt flows' and from the salt domes and diapirs need to be scrutinized with chemical and stable isotope analyses in order to confirm or debunk some of our suggestions and interpretations.

It is our hope that this communication, therefore, will eventually lead to a clearer narration of the salt/evaporite conundrum. We think the model offers a new way of interpreting ancient deep-water rift-associated salt bodies, for example those found in the deep-water Gulf of Mexico and in the Atlantic Santos Basin, off Brazil. And we fully agree with the following comments made in the introductory chapter of the book 'Evaporites Through Space and Time': '*In addition, a suggestion, presented since the Florence Conference, proposes that large volumes of 'evaporites' may actually be due to and form from rift-sourced and hydrothermal fluids* (Hovland et al. 2006a, b). *This concept certainly must be addressed in the future. Evaporites still represent an uncertain narrative, with further clarity coming slowly as we gain additional information as we learn more about old and new deposits*' (Schreiber et al. 2007).

Conclusions

The basic components of a modified hydrothermal salt model are outlined based mainly on combining old and modern observations from the depths of the Red Sea and elsewhere. Perhaps the most promising aspect of the model is that it not only manages to explain how large volumes of massive salt can form underground during submerged rifting, but also how both cumuliform rock salt (mainly halite) may form adjacent to and simultaneously with stratiform salts, interlayered with muddy and sandy sediments transported into the basin by rivers. The following prerequisites are necessary for our model to work:

1. A submerged zone of high heat flow, for example above a hot magma chamber (1,200 °C) at a depth of 500–2,000 m below the seafloor.
2. A zone with ScriW conditions above the heat source and more than 200 m of sediment cover.
3. Water depth of 1–3 km.
4. Forced sea water convection through faults and fractures.

These conditions need to be in place at least for a period of 10 k year to obtain significant volumes of salt.

All of these prerequisites are easily obtained in young rift basins. It is suggested that the Red Sea provides a number of salt features that can be explained by the hydrothermal salt model. The modified model can be summarized as follows: the rifting of continental crust will generate the formation of:

(a) Grabens and half-grabens.
(b) Alkalic lakes and lacustrine (hydrothermal) salts may form (e.g. east African rifts, Afar, Lake Asale, etc.).
(c) If sea water fills the deep penetrating faults and fractures, solid particulate salts will form in the deep supercritical out-salting zones.
(d) This salt will exert pressure on its surroundings and will be squeezed upwards and form salt ridges and salt domes (injectites) on the seafloor or on the surface (provided there is a dry climate).
(e) Brines and particulate salt will flow upwards and exit through the seafloor inside grabens, or along graben flanks where it will crystallize under sub-critical conditions.
(f) Hot, dense brine lakes will form inside some of the grabens.
(g) Sediments and dense highly concentrated brines will protect the solid salt from re-dissolution by sea water.
(h) Low-solubility salts (e.g. anhydrite and gypsum precipitated from the sea water) will also protect the exposed salts from sea water dissolution.
(i) Sediments will eventually cover the seafloor salts and protect them from dissolution by normal sea water.

Many of the pertinent observations of salt features in the Red Sea can be explained by the modified hydrothermal out-salting model, thus offering an alternative to the conventional evaporite model for salt accumulations. The main strength of this model is that it allows the formation of sub-surface salt bodies, without any climatic restrictions. In this way, it opens up the possibility to explain the deposition of fluvial transported silicate sediments in the same period as the evaporites are formed. The salt bodies may be protected from dissolution by a cap of sediments, anhydrite, carbonates, or just a hyper-saline brine pond on the seafloor. Furthermore, the new salt model can be applied for interpreting salt formations observed on extraterrestrial bodies, such as planet Mars (Montgomery et al. 2009; Hovland et al. 2009).

Acknowledgments We would like to thank Najeeb M.A. Rasul for inviting us to participate in the Red Sea book project, Christian Hübscher for very interesting and engaged discussions, and Bjørn Jamtveit for inviting us to present the associated salt dome model at the 26th Kongsberg deep-earth seminar, 2013. We also wish to thank our two reviewers, Charlotte B. Schreiber and Christopher Talbot for their thorough, constructive and encouraging comments.

References

Aharon P, Roberts HH, Snelling R (1992) Submarine venting of brines in the deep Gulf of Mexico: observations and geochemistry. Geology 20:483–486

Anschutz P, Blanc G, Chatin F, Geiller M, Pierret M-C (1999) Hydrographic changes during 20 years in the brine-filled basins of the Red Sea. Deep Sea Res Part I 46(10):1779–1792

Babel M, Schreiber BC (2014) Geochemistry of evaporites and evolution of seawater. In: Treatise on Geochemistry Elsevier Ltd, New York, pp 483–560. http://dx.doi.org/10.1016/B978-0-08-095975-7.00621-5

Blum N, Puchelt H (1991) Sedimentary-hosted polymetallic massive sulfide deposits of the Kebrit and Shaban Deeps, Red Sea. Miner Deposita 26:217–227

Bonatti E (1985) Punctiform initiation of seafloor spreading in the Red Sea during transition from a continent to an oceanic rift. Nature 316:33–37

Burke K (1996) The African plate. S Afr J Geol 99:339–409

Cochran JR (2005) Northern Red Sea: nucleation of an oceanic spreading center within a continental rift. Geochem Geophys Geosyst, AGU, Q03006. doi:10.1029/2004GC000826

Cochran JR, Karner GD (2007) Constraints on the deformation and rupturing of continental lithosphere of the Red Sea: the transition from rifting to drifting. Geol Soc London Spec Publ 282:265–289

Cochran JR, Martinez F, Steckler MS, Hobart MA (1986) Conrad deep: a new Northern Red Sea deep. Origin and implications for continental rifting. Earth Planet Sci Lett 78:18–32

Coumou D, Driesner T, Heinrich CA (2008) The structure and dynamics of mid-ocean ridge hydrothermal systems. Science 321:1825–1828

Davison I, Anderson L, Nuttall P (2012) Salt deposition, loading and gravity drainage in the Campos and Santos salt basins. In: Alsop GI et al. (eds) Salt tectonics, sediments and prospectivity, special publication 363. Geological Society of London, pp 159–173

Degens ET, Ross DA (1969) Hot brines and recent heavy metal deposits in the Red Sea. Springer, New York, pp 535–541

Ehrhardt A, Hübscher C, Gajewski D (2005) Conrad deep, northern Red Sea: development of an early stage ocean deep within the axial depression. Tectonophysics 411:19–40

Gaulier JM, LePichon X, Lyberis N, Avedik F, Geli L, Moretti I, Deschamps A, Salah H (1988) Seismic study of the crust of the northern Red Sea and Gulf of Suez. Tectonophysics 116:55–88

Gay A, Takano Y, Gilhooly WP III, Berndt C, Heeschen K, Suzuki N, Saegusa S, Nakagawa F, Tsunogai U, Jiang SY, Lopez M (2011) Geophysical and geochemical evidence of large scale fluid flow within shallow sediments in the eastern Gulf of Mexico, offshore Louisiana. Geofluids 11:34–47

Geiger S, Driesner T, Heinrich CA, Matthäi SK (2005) On the dynamics of NaCl-H_2O fluid convection in the Earth's crust. J Geophys Res 110:B07101

Hardie LA (1991) On the significance of evaporates. Ann Rev Earth Planet Sci 19:131–168

Hewitt DF (1962) Salt in Ontario. Industrial mineral report no 6, Ontario Department of Mines, 40 pp

Holness MB, Lewis S (1997) The structure of halite-brine interface inferred from pressure and temperature variations of equilibrium dihedral angles in the halite–H_2O–CO_2 system. Geochim Cosmochim Acta 61(4):795–804

Hovland M, Hill A, Stokes D (1997) The structure and geomorphology of the Dashgil mud volcano, Azerbaijan. Geomorphology 21:1–15

Hovland M, MacDonald I, Rueslåtten H, Johnsen HK, Naehr T, Bohrmann G (2005) Chapopote asphalt volcano may have been generated by supercritical water. EOS 86(42):397–402

Hovland M, Kutznetsova T, Rueslåtten H, Kvamme B, Johnsen HK, Fladmark GE, Hebach A (2006a) Sub-surface precipitation of salts in supercritical seawater. Basin Res 18(2):221–230

Hovland M, Rueslåtten H, Johnsen HK, Kvamme B, Kutznetsova T (2006b) Salt formation associated with sub-surface boiling and supercritical water. Mar Pet Geol 23:855–869

Hovland M, Rueslåtten H, Johnsen HK, Fichler C (2009) Hydrothermal evaporites—from the Conrad Deep, via Dallol, to Elysium Planitia (Abstract). International association of sedimentologists (IAS) annual meeting, Alghero, Sardinia, Book of Abstracts

Ings SJ, Beaumont C (2010) Shortening viscous pressure ridges, a solution to the enigma of initiating salt 'withdrawal' minibasins. Geology 38(4):339–342

Jarrard RD (2003) Subduction fluxes of water, carbon dioxide, and potassium. Geochem Geophys Geosyst 4(5):8905

Katz A, Starinsky A, Taitel-Goldman N, Beyth M (1981) Solubilities of gypsum and halite in the Dead Sea and in its mixtures with seawater. Limnolo Oceanogr 26(4):709–716

Kendall AC (2005) Evaporites. In: Selley RC, Cocks LRM, Plimer IR (eds) Encyclopedia of geology, vol 5. Elsevier, Amsterdam, pp 94–97

Koschinsky A, Garbe-Schönberg D, Sander S, Schmidt K, Gennerich H-H, Strauss H (2013) Hydrothermal venting at pressure-temperature conditions above the critical point of seawater, 5°S on the Mid-Atlantic Ridge. Geology 36(8):615–618

Lewis S, Holness M (1996) Equilibrium halite-H_2O dihedral angles: High rock-salt permeability in the shallow crust? Geology 24:432–434

Lowell RP, Germanovich LN (1997) Evolution of a brine-saturated layer at the base of a ridge-crest hydrothermal system. J Geophys Res 102:10245–10255

Lowell RP, Rona P (2005) Hydrothermal activity. In: Selley RC, Cocks LRM, Plimer IR (eds) Encyclopedia of geology, vol 5. Elsevier, Amsterdam, pp 362–372

Lowell RP, Rona PA, Von Herzen RP (1995) Seafloor hydrothermal systems. J Geophys Res 100:327–352

Lowenstein TK, Kendall B, Anbar AD (2014) The geologic history of seawater. In: Treatise on geochemistry Elsevier Ltd, New York, pp 569–621. http://dx.doi.org/10.1016/B978-0-08-095975-7.00621-5

Løseth H, Rodrigues N, Cobbold PR (2012) World's largest extrusive body of sand? Geology. doi:10.1130/G33117.1

MacDonald IR et al (2004) Asphalt volcanism and chemosynthetic life in the Campeche Knolls, Gulf of Mexico. Science 304:999–1002

Mart Y, Ross DA (1987) Post-miocene rifting and diapirism in the northern Red Sea. Mar Geol 74:173–190

Miller NC, Lizarralde D, Harding A, Kent G (2009) Constraints on early Gulf of California rifting from seismic images across the eastern margin of Guaymas Basin. AGU fall meeting, Abstract, Poster T31A-1783

Mitchell NC, Ligi M, Ferrante V, Bonatti E, Rutter E (2010a) Submarine salt flows in the central Red Sea. Geol Soc Am Bull 122 (5/6):701–713

Mitchell NC, Schmidt M, Ligi M (2010b) Comment on formation of Thetis deep metal-rich sediments in the absence of brines, Red Sea by Pierret et al (2010). J Geochem Explor 108: 112–113

Montgomery DR, Som SM, Jackson MPA, Schreiber BC, Gillespie AR, Adams JB (2009) Continental-scale salt tectonics on mars and the origin of Valles Marineris and associated outflow channels. Geol Soc Am Bull 121(1/2):117–133. doi:10.1130/B26307.1

Orszag-Sperber F, Harwood G, Kendall A, Purser BH (1998) A review of the evaporites of the Red Sea-Gulf of Suez rift. In: Purser BH, Bosence DWJ (eds) Sedimentation and Tectonics of rift basins: Red Sea-Gulf of Aden. Chapman and Hall, London, pp 409–426

Pirajno F (2009) Hydrothermal processes and mineral systems. Springer, Berlin, 1250 pp

Ramboz C, Oudin E, Thisse Y (1988) Geyser-type discharge in Atlantis II Deep, Red Sea: evidence of boiling from fluid inclusions in epigenetic anhydrite. Can Mineral 26:765–786

Ruppel C, Dickens GR, Castellini DG et al (2005) Heat and salt inhibition of gas hydrate formation in the northern GoM. Geophys Res Lett 32:L04605

Sassen R, Roberts HH, Carney R, Milkov A, DeFreitas DA, Lanoil B, Zhang C (2004) Free hydrocarbon gas, gas hydrate, and authigenic minerals in chemosynthetic communities of the northern Gulf of Mexico continental slope: relation to microbial processes. Chemical Geology 205(3–4):195–217

Savoyat E, Shiferaw A, Balcha T (1989) Petroleum exploration in the Ethiopian Red Sea. J Pet Geol 12:187–204

Schmalz RF (1969) Deep-water evaporite deposition: a genetic model. Am Assoc Petrol Geol 53:758

Schoenherr J, Littke R, Urai JL, Kukla PA, Rawahi Z (2007a) Polyphase thermal evolution in the infra-Cambrian Ara Group (South Oman Salt Basin) as deduced by maturity of solid reservoir bitumen. Org Geochem 38:1293–1318

Schoenherr J, Urai JL, Kukla PA, Littke R, Schléder Z, Larroque J-M, Newall MJ, Al-Abry N, Al-Siyabi H, Rawahi Z (2007b) Limits to the sealing capacity of rock salt: a case study of the infra-Cambrian Ara salt from the south Oman salt basin. Am Assoc Petrol Geol 91 (11):1541–1557

Schreiber BC, Hsü KJ (1980) Evaporites. In: Hobson GD (ed) Developments in petroleum geology, vol 2. Applied Science Ltd, London, pp 87–138

Schreiber BC, Lugli S, Babel M (2007) Evaporites through space and time. Geol Soc London Spec Publ 285:1–13

Searle RC, Ross DA (1975) A geophysical study of the Red Sea axial trough between 2.5° and 22°N. Geophys J Roy Astron Soc 43:555–572

Selley RC (2005) Mineralogy and classification. In: Selley RC, Cocks LRM, Plimer IR (eds) Encyclopedia of geology, vol 5. Elsevier, Amsterdam, pp 27–37

Simakin A, Ghassemi A (2003) Salt loaded heat pipes: steady-state operation and related heat and mass transport. Earth Planet Sci Lett 215:411–424

Simoneit BR (1993) Aqueous high-temperature and high-pressure organic geochemistry of hydrothermal vent systems. Geochima et Cosmochima Acta 57:3231–3243

Stoffers P, Kühn R (1973) Red Sea evaporites: a petrographic and geochemical study. In: Whitmarsh RB et al Leg 23, Deep Sea drilling project final report, Texas A and M University, pp 821–847

Talbot CJ (2007) Hydrothermal salt—but how much? Discussion. Marine and petroleum geology. doi:10.1016/J.Marpetgeo.2007.05.005

Tester J, Holgate HR, Armellini FJ, Webley PA, Killilea WR, Hong GT, Berner HE (1993) Supercritical water oxidation technology. In: Tedder DW, Pohland FG (eds) Emerging technologies in hazardous waste management III. American Chemical Society, Washington, DC, pp 35–76

Vendeville BC (2005) Salt tectonics driven by sediment progradation: part I mechanics and kinematics. Am Assoc Pet Geol Bull 89:1071–1079

Von Damm KL, Lilley MD, Shanks III WC, Bockington M, Bray AM, O'Grady KM, Olson E, Graham A, Proskurowski G, the SouEPR science party (2003) Extraordinary phase separation and segregation in vent fluids from the southern East Pacific Rise. Earth Planet Sci Lett 206(3–4):365–378

Wall D, Warren JS (1969) Dinoflagellates in Red Sea piston cores. In: Degens ET, Ross DA (eds) Hot brines and recent heavy metal deposits in the Red Sea. Springer, New York Inc, pp 317–347

Warren JK (2010) Evaporites: sediments, resources and hydrocarbons. Springer, Berlin, 998 pp

Whitmarsh RB, Weser PE, Ross DA (1974) Initial reports of the Deep Sea drilling project, vol 23. US Government Printing Office, Washington, pp 821–847

Zierenberg RA, Holland ME (2004) Sedimented ridges as a laboratory for exploring the subsurface biosphere. In: Wilcock WSD, DeLong EF, Kelley DS, Baross JA, Cary SC (eds) The subseafloor biosphere at mid-ocean ridges. American Geophysical Union, Washington, DC, pp 305–323

Salt Flows in the Central Red Sea

Peter Feldens and Neil C. Mitchell

Abstract

The central Red Sea is a nascent oceanic basin. Miocene evaporites, kilometers in thickness, were deposited during its continental rifting phase and early seafloor spreading. With further seafloor spreading, increasing dissolution due to increasing hydrothermal circulation as well as normal fault movements removed lateral constraint of the evaporites at the walls of the axial rift valley. Because halite is a ductile material that forms a large part of the evaporite sequence, the evaporites started to move downslope, passively carrying their hemipelagic sediment cover. Today, flowlike features comprising Miocene evaporites are situated on the top of younger magnetic seafloor spreading anomalies. Six salt flows, most showing rounded fronts in plan view, with heights of several hundred meters and widths between 3 and 10 km, are identified by high-resolution bathymetry and DSDP core material around Thetis Deep and Atlantis II Deep, and between Atlantis II Deep and Port Sudan Deep. The relief of the underlying volcanic basement likely controls the positions of individual salt flow lobes. On the flow surfaces, along-slope and downslope ridge and trough morphologies parallel to the local seafloor gradient have developed, presumably due to extension of the hemiplegic sediment cover or strike-slip movement within the evaporites. A few places of irregular seafloor topography are observed close to the flow fronts, interpreted to be the result of dissolution of Miocene evaporites, which contributes to the formation of brines in several of the deeps. Based on the vertical relief of the flow lobes, deformation is taking place in the upper part of the evaporite sequence. Considering a salt flow at Atlantis II Deep in more detail, strain rates due to dislocation creep and pressure solution creep were estimated to be 10^{-14} 1/s and 10^{-10} 1/s, respectively, using given assumptions of grain size and deforming layer thickness. The latter strain rate, comparable to strain rates observed for onshore salt flows in Iran, results in flow speeds of several mm/year for the offshore salt flows in certain locations. Thus, salt flow movements can potentially keep up with Arabia–Nubia tectonic half-spreading rates reported for large parts of the Red Sea.

Introduction

Halite is a ductile material that already deforms under low differential stresses. In response to gravitational stresses, it can form allochthonous salt sheets (Hudec and Jackson 2007). (We use the term "salt" here in the loose sense commonly used in petroleum geology, to signify originally evaporitic sequences whose rheology is strongly dependent on their halite components.) Recent as well as past allochthonous salt sheets from the onshore and offshore environment are known from more than 35 basins worldwide (Hudec and Jackson 2006), including in Iran, Tunisia,

P. Feldens (✉)
Institute of Geosciences, Kiel University, Otto-Hahn Platz 1, 24118 Kiel, Germany
e-mail: pfeldens@geophysik.uni-kiel.de

N.C. Mitchell
School of Earth, Atmospheric and Environmental Sciences, University of Manchester, Williamson Building, Oxford Road, Manchester M13 9PL, UK

N.M.A. Rasul and I.C.F. Stewart (eds.), *The Red Sea*,
Springer Earth System Sciences, DOI 10.1007/978-3-662-45201-1_12,

Algeria, Germany, Morocco, Canada, South America, Khazakstan, Ukraine, Yemen, Australia, and the Gulf of Mexico (Mohr et al. 2007). For the most part, the sources of allochthonous salt sheets are salt diapirs extruding toward the surface (Hudec and Jackson 2007). Allochthonous salt sheets may be found beneath the surface, either completely or partly buried by clastic sediments, or have spread directly at the surface without a cover of siliciclastic sediments (Hudec and Jackson 2006; Talbot and Pohjola 2009). The latter is the case in Iran, where the best preserved onshore extrusive salt sheets, originally called "salt glaciers" (Lees 1927) are located. However, to avoid redundancy (Talbot and Pohjola 2009), we use the term salt flow (Mitchell et al. 2010). The viscous fluidlike behavior of the Iranian salt flows is dominated by the weak rheology of their halite component (comparable to the behavior of ice glaciers) (Talbot and Pohjola 2009) and results in flow speeds of cm/year to dm/day during surges (Jackson and Talbot 1986).

However, not all salt flows are related to extruding diapirs. Instead, salt flows may develop where evaporites have lost their lateral constraint, observed for evaporites deposited in basins during the continental rifting phase predating the onset of seafloor spreading, for example in the early Atlantic Ocean (Pautot et al. 1966). These evaporites, now situated off the coasts of Africa and South America (Talbot 1993), form important barriers for migrating hydrocarbons. Thus, the early behavior of these evaporites may be of economical interest but is poorly known, as the salt has since been covered by thick siliciclastic sequences, lowering the resolution of acoustic imaging techniques. Further, the salt has been subsequently remobilized and deformed.

The Red Sea serves as an analog of the situation of the early Atlantic Ocean: Here, evaporites were deposited during the Miocene in a continental rift valley (Stoffers and Kühn 1974; Girdler and Whitmarsh 1974). With the onset of seafloor spreading, evaporites overlying the spreading center were dissolved due to hydrothermal circulation, volcanism, and direct contact with seawater during the continuing spreading. In combination with normal fault movements, the evaporites adjacent to the spreading center lost their lateral constraint, and moved downslope due to differential stress exerted by their own weight and the weight of the sedimentary roof. The existence of mobile salt flows situated along the Red Sea spreading axis was already suggested several decades ago (Girdler and Whitmarsh 1974). This suggestion was based on Miocene evaporites that were encountered overlying younger oceanic crust which was identified by its magnetic properties. Corresponding morphologic evidence for salt flows was recently described (Mitchell et al. 2010) at the walls of Thetis Deep. Thus, the Red Sea offers the opportunity to study recent submarine salt flows close to the seafloor surface, especially in the walls of the deeps along the central Red Sea.

High-resolution bathymetric data along the central Red Sea trough became recently available (Schmidt et al. 2011; Ligi et al. 2012). This chapter will explore the available bathymetric data along the Central Red Sea spreading center (Fig. 1) and highlight the typical morphology associated with salt flows in the Red Sea. Further, some basic constraints on flow speed and speculations on deformation mechanisms acting within the salt are developed.

Regional Setting

The Red Sea is currently transitioning from continental to oceanic rifting. Continental rifting is still observed in its northern part, while recently formed oceanic seafloor is observed in the south and within several deeps along the central Red Sea, presumably developed since the Pliocene (Coleman 1993). Detailed information on the opening of the Red Sea and its current structure is given in the articles by Bosworth (this volume), Ligi et al. (this volume), and Ehrhardt (this volume) and by Coleman (1993). In the central Red Sea, evaporites were deposited prior to seafloor spreading, during the Miocene, in the continental rifting phase (Girdler and Whitmarsh 1974). Several seawater incursions left sequences of clay and carbonate deposits. Subsequent evaporation produced evaporites that mainly comprise halite and anhydrite (Stoffers and Kühn 1974; Coleman 1993). Throughout the sequence, interbedded volcanic material exists (Stoffers and Kühn 1974). Evaporite deposition ended approximately with the onset of seafloor spreading. The upper surface of the evaporites is recognized basinwide as a distinct reflector in reflection seismic data ("S-Layer"; Coleman 1993). The thickness of the evaporites reaches 3–4 km in places (Mitchell et al. 1992; Coleman 1993). Based on seismic refraction experiments, the thickness of the evaporites adjacent to the axial trough at the latitude of the Thetis deeps still exceeds 1 km (Tramontini and Davies 1969; Mitchell et al. 2010; Mitchell, this volume), but the nature of the crust beneath those near-trough evaporites is still under discussion. The evaporites are covered by few hundred meters of hemipelagic sediment (Ross and Schlee 1973).

Salt Flows Along the Red Sea Spreading Axis

Observations

Within Thetis Deep (Fig. 2), located at 22°30′ N 37°46′ E, Mitchell et al. (2010) described several key features of submarine salt flow morphology. Generally, the center of Thetis Deep shows a rough morphology including several cone-shaped and ridgelike volcanic structures. In contrast, the walls of the Thetis Deep have a less rugose appearance.

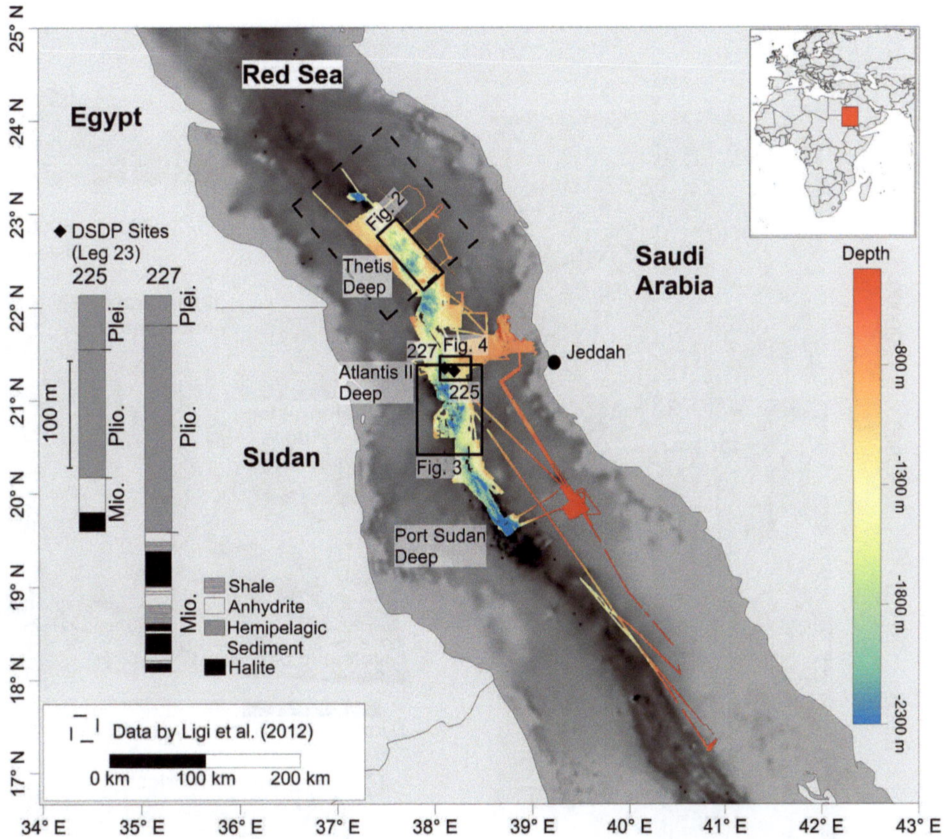

Fig. 1 Overview of high-resolution bathymetric data available in the central Red Sea between Port Sudan Deep, Atlantis II Deep, and Thetis Deep, overlying low-resolution bathymetry. The simplified stratigraphy of the Deep Sea Drilling (DSDP) Sites 225 and 227, including the upper part of the evaporite sequence, is displayed

Here, three rounded flow fronts ("A," "B," and "C" in Fig. 2) are recognized. Features B and C protrude from the eastern wall into Thetis Deep, while feature A is located at the northeast boundary to the northern intertrough zone. At features B and C, separate flow lobes can be observed. The widths of the fronts are approximately 3 km at A and 10 km at B and C. Relief heights of the flow front are on the order of 200–500 m. The flow lobes exhibit several steps corresponding to the steps in the adjacent volcanic basement (Fig. 2). Downslope ridges and troughs exist with wavelengths of approximately 1 km and are oriented parallel to the direction of maximum seafloor gradient (Fig. 2). Partly curved fabrics are observed (Fig. 2). Along slope, the ridges and troughs have wavelengths of approximately 200–500 m and heights of less than 20 m (Fig. 2). Along- and downslope ridges are partly superimposed. Mitchell et al. (2010) observed an area of increased seafloor roughness within an embayment of one of the flow lobes.

Morphology similar to the salt flows in Thetis Deep exists further south along the Red Sea spreading axis. An example is displayed in Fig. 3 for the area between 21.3° N and 20.5° N, showing the axial trough and its eastern wall. The central axial trough is characterized by rough seafloor, exhibiting a multitude of cone-shaped volcanic structures and faults that are mainly oriented SE–NW. Compared to most of the central trough, the sedimentary seafloor surface at its eastern side appears smooth. Three features with rounded fronts (white "D," "E," and "F" in Fig. 3) are recognized. Due to incomplete bathymetric data, their upslope extents (outside the central axial trough) are not known. The widths of the fronts, parallel to the axial trough, are approximately 15 km for feature D, 9 km for feature E, and 5 km for feature F. In the south, volcanic structures occur on the along-slope sides of flowlike features D and E and rise to elevations above the flows (for example, where profiles B–B′ and C–C′ cross). Downslope ridge and trough morphology, with amplitudes in the range of 10–30 m and wavelengths of 400–500 m are notably observed in the narrow section of feature D, which lies between two volcanic features (see profile D–D′). To the south of feature D, the amplitude increases to a maximum of 100 m. Near the rounded front, surface lineaments appear to rotate toward the northwest. Although difficult to determine due to the rough topography, the elevation of the distal part of feature D against the surrounding volcanic basement is approximately 500 m at its southern end and 600 m at its northern end (profile A–A′). A prominent cone-shaped volcano, rising approximately 600 m, is situated between flowlike features D and E (where profiles C–C′ and B–B′ cross). Notably, several curved ridges, convex toward the east, with heights of approximately 15 m, are observed directly toward the east of the volcano (best observed south of "E" in Fig. 3). Feature E itself exhibits a stepwise

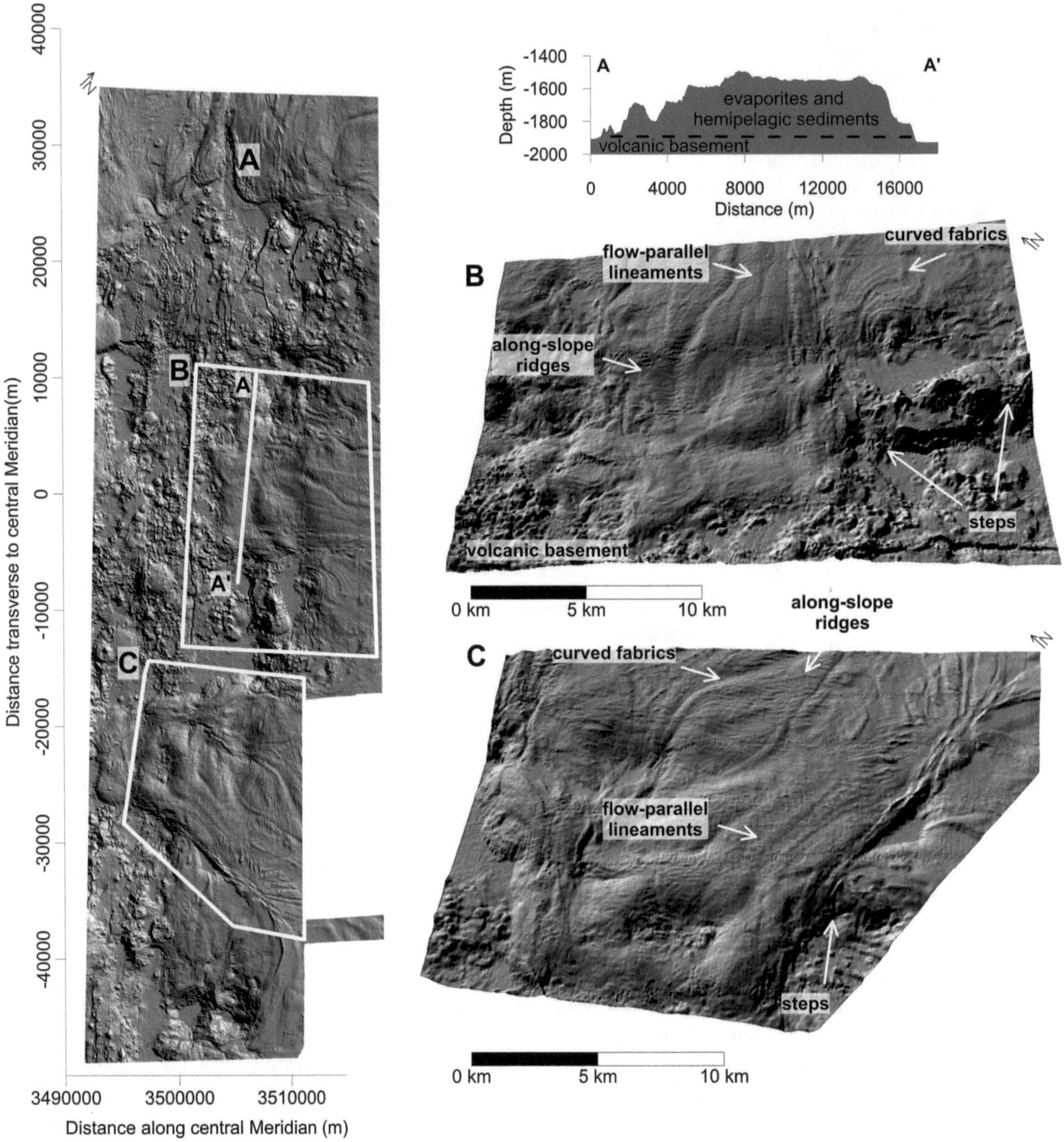

Fig. 2 Shaded relief image of multibeam data in the Thetis Deep (multibeam data of Ligi et al. 2012). Image resolution is 25 m, *vertical exaggeration* of the insets *B*, and *C* is 2:1. Illumination is from 216°/86°. Projected in oblique Mercator projection with the *central line* defined by the coordinates 38.33° N/23.14° E and 37.06° N/22.05° E. Three flowlike features *A*, *B*, and *C* are identified. Refer to text for further discussion of morphological features

morphology; the contact with the volcanic basement is situated at approximately 2,250 m, the first step rises to 1,800 m, and the second rises to a depth of 1,200 m (profile E–E′, Fig. 3).

Feature F is situated directly at the southeastern wall of the famous Atlantis II Deep (Fig. 3, detailed view in Fig. 4), located approximately 125 km south of the Thetis Deep. A major difference between Thetis Deep and Atlantis II Deep is the existence of highly saline brine within the latter (Schmidt et al., this volume). The formation of the brine is connected to local hydrothermal circulation cells that cause a partial dissolution of Miocene evaporites (Anschutz and Blanc 1995).

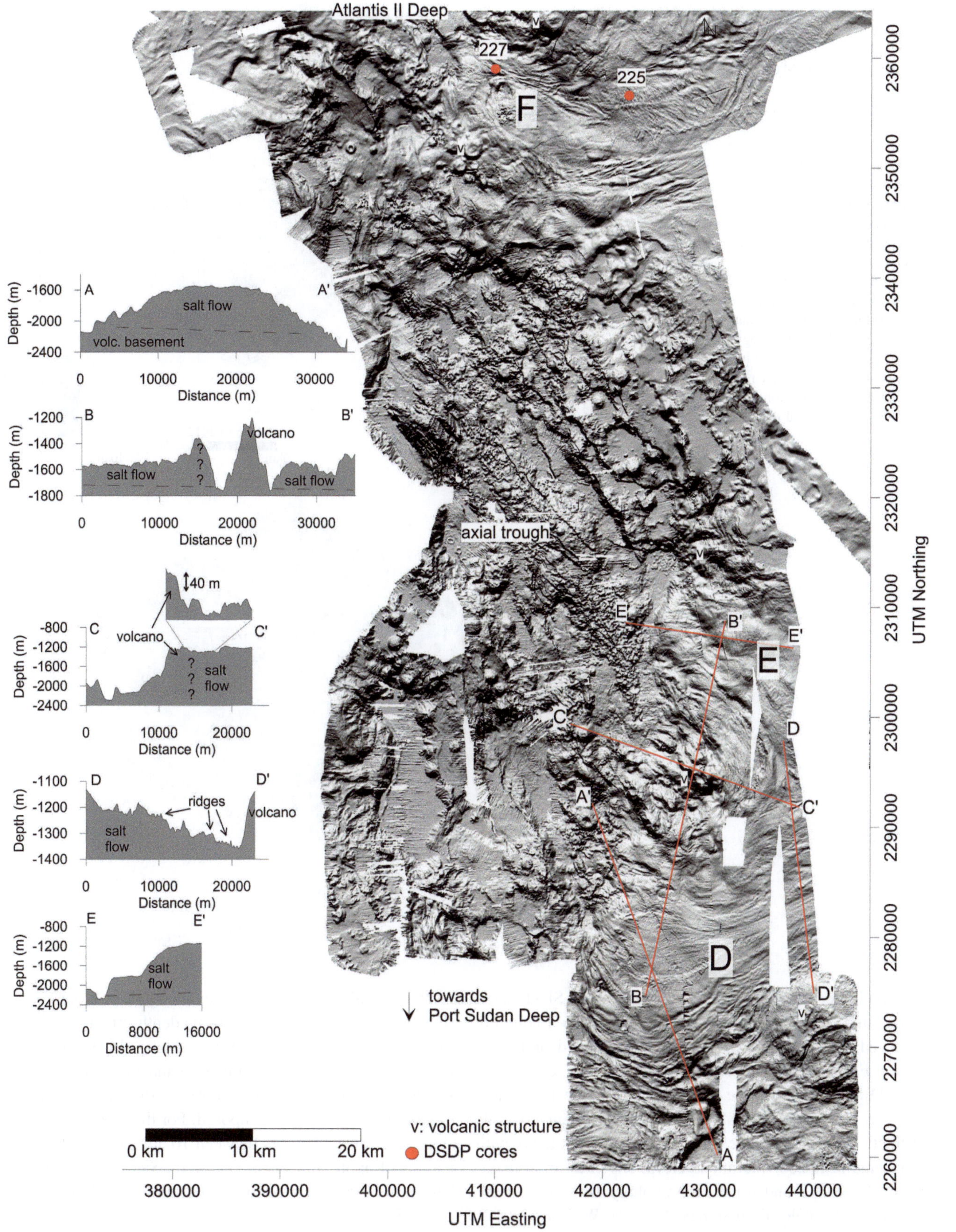

Fig. 3 Shaded relief image of multibeam data south of Atlantis II Deep. Three flowlike features *D*, *E*, and *F* are recognized

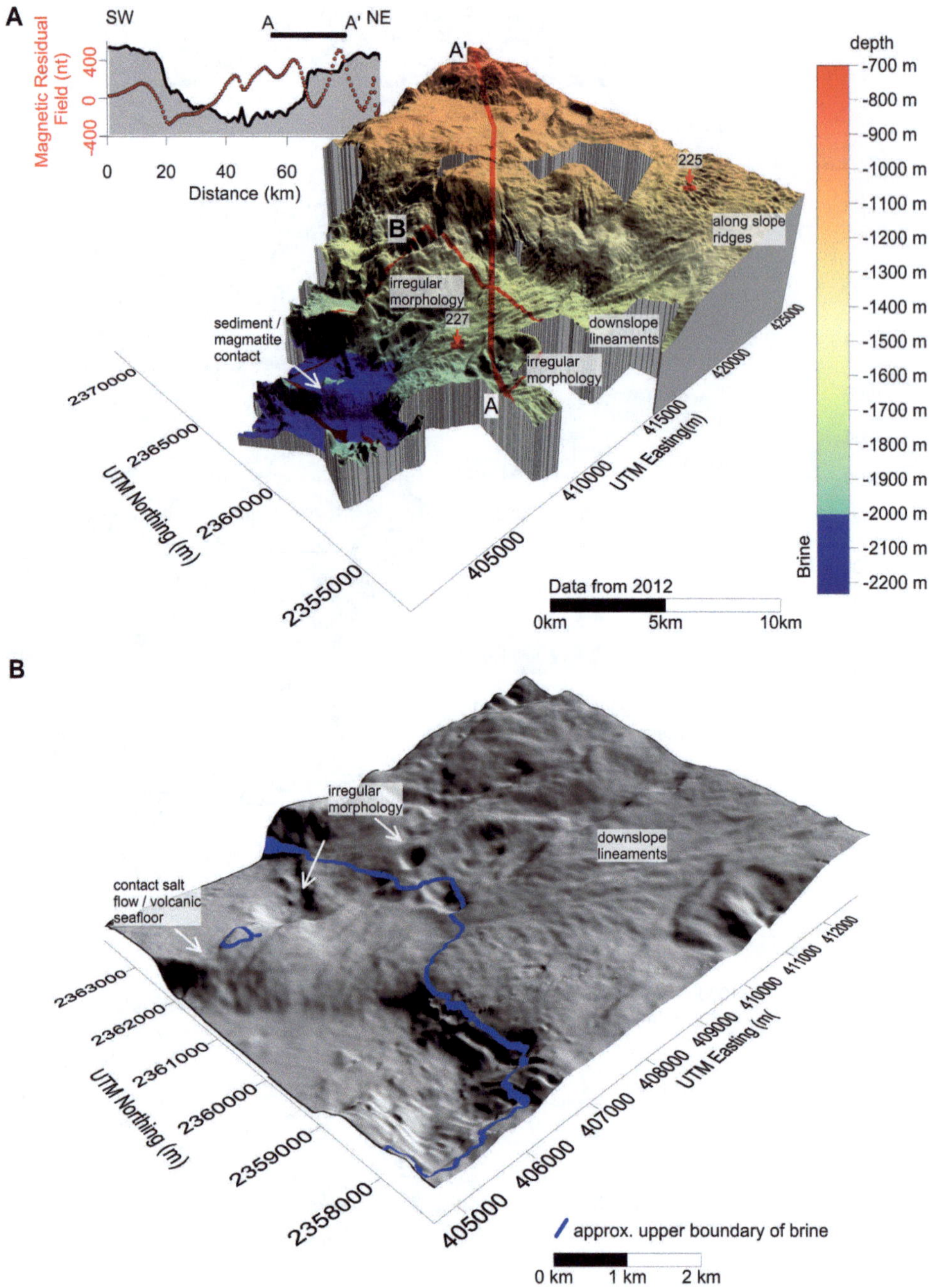

Fig. 4 *Top* Three-dimensional image of the southeastern wall of Atlantis II Deep. Vertical exaggeration is 6:1. Magnetic data display anomalies beneath the evaporites recovered from DSDP Site 227. The magnetic data and associated databases are available from the National Geophysical Data Center, National Oceanic and Atmospheric Administration, US Department of Commerce, http://www.ngdc.noaa.gov/. *Bottom* Shaded relief image of the proximal part of the salt flow. Close to the brine filling Atlantis II Deep, irregular morphology is observed. The depth of the brine level is taken from Hartmann et al. (1998). Illumination is from 346°/39°. The vertical exaggeration is 2:1

Atlantis II Deep is characterized by a set of NW–SE-directed faults, and cone-shaped volcanic features are frequently observed (Figs. 3 and 4). Arguably, the rounded outline of feature F is more diffuse and difficult to recognize than features D and E. Again, the distal part of feature F, situated between two areas of elevated volcanic basement, exhibits distinct downslope ridge and trough morphology with wavelengths of 200–400 m and elevations of 20 m at maximum. The ridge and trough morphology fades out at the contact to the volcanic basement (Fig. 3). With a depth of approximately 2008 m below sea level (Hartmann et al. 1998; Schmidt et al., this volume), the brine–seawater interface is located slightly above the sediment/volcanic seafloor contact (Fig. 4). At approximately the same depth, irregular seafloor topography (Fig. 4) is observed. Along-slope ridges and troughs are observed at the more proximal part of feature F (at DSDP Site 225 in Figs. 3 and 4). Along-slope and downslope ridges appear partly superimposed. For the most part, along-slope ridges have heights between 10 and 20 m, with wavelengths of approximately 300–500 m. Steps in the volcanic basement morphology continue beneath feature F (to the east of "F" in Fig. 3).

During the DSDP program in the early 1970s (DSDP Leg 23), several sites were drilled (Figs. 1 and 3) at the location of feature F. The recovered cores show that the top of the Miocene evaporite sequence lies at a depth below seafloor of approximately 150–200 m beneath a cover of partially deformed hemipelagic sediments. Magnetic data (Fig. 3, Izzeldin 1987) indicate that magnetic anomalies are present beneath the area covered by evaporites today (Fig. 4).

Identification of Evaporite Flows

The interpretation of the lobate-shaped sedimentary features at the walls of Thetis Deep as salt flows by Mitchell et al. (2010) was based on the combined analysis of seismic and bathymetric data. Headwalls, which are generally indicative of abrupt slope failures (e.g., Völker et al. 2011; Krastel et al. 2012), are absent in the walls of the Red Sea deeps. A volcanic origin of the flowlike features is rejected based on observed folded layering in seismic data. Additionally, the relief of the flowlike features A to C is larger than the hemipelagic sediment thickness, indicating that evaporites are part of the observed flow lobes. Thus, features A to C are interpreted as being caused by moving evaporites, passively carrying the hemipelagic sediments above them. The hypothesis may be made that the features D, E, and F are related to salt movement as well. By themselves, the morphological similarity to the salt flows observed at Thetis Deep might be only a coincidence, but at Atlantis II Deep, the DSDP cores clearly demonstrate an evaporite origin. The occurrence of only minor deformation of hemipelagic sediment samples with DSDP cores 225 and 227 (Girdler and Whitmarsh 1974) and the absence of headwalls show that no large-scale slope failures exist at the site of feature F. While older low-amplitude magnetic anomalies in the northern Red Sea may relate to mafic intrusions (Cochran 1983), the magnetic anomalies in the Atlantis II Deep were previously interpreted to result from seafloor spreading. At 21° and 22° N, magnetic anomalies include at least anomalies 2A and 2, respectively (Izzeldin 1987; Chu and Gordon 1998), corresponding to the Pliocene/Pleistocene epoch (Mankinen and Dalrymple 1979). The magnetic anomalies are situated beneath the evaporites sampled at the base of DSDP Sites 225 and 227 that are of Miocene age (Stoffers and Kühn 1974; see also Fig. 4). Therefore, it appears reasonable that evaporite flowage occurs at Atlantis II Deep, and generally these flows exhibit a characteristic morphology, including rounded fronts, downslope and along-slope ridge and trough features, escarpments in the volcanic basement that continue beneath the salt flows, as well as irregular topography that resembles dissolution structures. Unfortunately, no geophysical data are available over features D and E (Fig. 3), but the structures are also interpreted as salt flows based on their morphological similarities to the Thetis Deep salt flows.

Morphology Related to Salt Flows

The morphology of the hemipelagic cover on top of the moving evaporites includes downslope ridges and troughs, along-slope ridges, areas of rough topography, as well as steplike morphology at the flow fronts.

Downslope ridge and trough morphologies ("flowstripes") are observed in many regimes of fast ice flow worldwide (Campbell et al. 2008), but the physical explanation for their origin is unclear (Gudmundsson et al. 1998; Glasser and Gudmundsson 2012). Within ice glaciers, such flowstripes form as a response to either subglacial morphology within fast-flowing glaciers, shear margins—for example due to converging ice flows—or to lateral compression due to narrowing cross sections (Bindschadler 1998; Glasser and Gudmundsson 2012). No converging salt flows are observed based on the available bathymetric data. The transfer of subglacial undulations through ice glaciers involves fast glacier flow and basal slippage (Glasser and Gudmundsson 2012), while in salt flows frictional boundaries dominate (Talbot and Pohjola 2009). Furthermore, flowstripes transferring from subglacial basins commonly have wavelengths of at least the glacier's thickness. For the downslope ridges at feature F, this is not the case. However, both lateral compression and increasing flow velocity downflow cause longitudinal extension (Glasser and Gudmundsson 2012). Such behavior may be observed at sites D and F. At D, the flow narrows between rising volcanic basement; at F, the number of downslope ridges increases directly beneath a step of the flow surface. The downslope ridges may also be related to strike-slip movements in the evaporite caused by the rough topography expected at the base of the flows, similar to nunataks observed in glaciology. At least some of the lineaments may be explained by strike-slip movements, as strike-slip-like features have been observed in seismic data of the Thetis Deep and separate flow lobes are observed at the flow fronts at B and C (Mitchell et al. 2010, see also Fig. 2).

In this regard, the partly observed along-slope ridges approximately parallel to the central trough may be created due to extension of the hemipelagic cover, for example, related to changing morphology of the subflow basement. Similar features were observed during submarine spreading following landslides (Micaleff et al. 2007), where sediment extends due to underlying, deforming material, and an along-slope ridge and trough morphology is created where shear stresses exceed the strength of strata, which then separate into individual blocks. However, such morphology is not always observed in the vicinity of steps in the flow surface. A reason may be that extension is not always accommodated by formation of individual blocks, but by thinning of the hemipelagic cover (Mitchell et al. 2010).

While the origin of the downslope and along-slope ridges observed on the proposed salt flows cannot be definitely resolved, their striking morphological similarity, where present, with features observed on ice glaciers or related to submarine mass movement indicates that sediment flow actually takes place in the vicinity of the Red Sea central trough. Hence, these observations reinforce the interpretation of features D and E as salt flows. The salt flow at D and E seems to be steered by the local volcanic basement. For example, the prominent volcano at the cross section of C–C′ and B–B′ (Fig. 3) is forcing the flow to diverge, causing the formation of the two separate lobes at D and E. Ridges with a convex shape toward the east—of sedimentary origin based on their texture—lying immediately east of the volcano-shaped feature (profile C–C′) are interpreted as compressional ridges. In contrast, steep depressions are situated to the north and south of the volcano (profile B–B′), eventually forming a no-flow zone with regard to salt movement. Further indications for the importance of the local basement controlling the salt flows are ridges rotating toward the NW at the distal (NW) end of feature D. As cross section A–A′ demonstrates, the volcanic basement in the central trough is deeper toward the northwest of feature D than toward the south and the curved fabrics indicate a preferred downslope movement toward the NW. Generally, both along-slope and downslope ridges and troughs attributed to the salt flows are located either perpendicular or parallel to the seafloor gradient. This agrees with a formation induced by gravitational stresses, as would be expected from downslope-flowing salt (Mitchell et al. 2010).

Finally, it is interesting to note that the salt flow F continues beneath the top brine level (Fig. 4). This irregular topography observed close to the flow front (Fig. 4) may be related to the dissolution of evaporites (Pilcher and Blumstein 2007), contributing to the formation of the brine in the enclosed Atlantis II Deep. Similar small areas of irregular seafloor are observed on feature C (Mitchell et al. 2010).

Mechanical Behavior of Salt Flows

Introduction

The deformation mechanisms of halite (rock salt) occurring as a response to the local stress field range from cataclastic deformation (breakage) to crystal–plastic movement (recrystallization and creep), and pressure solution creep (Urai et al. 2008). Under most natural conditions, halite shows a ductile behavior when its elastic limit is exceeded. The related plastic deformation can include movement within a crystal ("dislocation creep") as well as mass transfer by diffusion ("pressure solution creep," Green 1984). Dislocation creep involves movements within the crystal lattice along glide planes that cause the formation of subgrains. It is the dominant deformation process under higher temperatures, differential stresses, and dry conditions (Urai et al. 2008). However, under low differential stresses and available interstitial water, pressure solution creep is important (De Meer et al. 2002). This type of creep involves dissolution at grain boundaries exposed to high stress, the transport of the dissolved material along the grain boundaries, and the precipitation at boundaries under low stress. Pressure solution creep can cause considerably higher strain rates than dislocation creep (Jackson and Talbot 1986). Therefore, the amount of intergrain water (Talbot and Pohjola 2009) and grain size (Urai et al. 2008) significantly influences mechanical behavior. The effectiveness of pressure solution creep increases with the amount of intergranular water and decreasing grain size (i.e., with increasing grain total surface area and decreasing diffusion distance).

The mechanical behavior of extrusive salt flows (locally called "namakiers") has been studied in southern Iran. Namakiers show varied advancement rates (Talbot and Pohjola 2009) under low differential stresses of 0.03–0.25 MPa (Wenkert 1979; Talbot and Rogers 1980). Namakiers move at several cm to m per year, but can surge with speeds of centimeters to decimeters per day during heavy rain (Jackson and Talbot 1986; Talbot et al. 2000). Correspondingly, measured strain rates vary widely between 10^{-9} and 10^{-11}. Pressure solution creep active in fine-grained mylonitic shear zones is considered the dominant deformation process involved in onshore salt flow.

Generally, strain rate due to dislocation creep for extrusive onshore salt flows follows a power law with differential stress (Schléder and Urai 2007):

$$\dot{\varepsilon}_{DC} = 8.1 \times 10^{-5} e^{(Q_{DC}/RT)} (\sigma_1 - \sigma_3)^{3.4} \quad (1)$$

where Q_{DC} = the activation energy energy for dislocation creep ($-56,000$ J mol^{-1}), T = temperature in Kelvin, R = universal gas constant, and $\sigma_1 - \sigma_3$ = differential stress (Pa).

The strain rate for pressure solution creep (based on experiments on porous and dense polycrystalline aggregates of synthetic rock salt) may be estimated from (Schléder and Urai 2007; Urai et al. 2008):

$$\dot{\varepsilon}_{ps} = 4.7 \times 10^{-4} e^{(Q_{ps}/RT)} \frac{(\sigma_1 - \sigma_3)}{TD^3} \quad (2)$$

where D = grain diameter (mm) and Q_{ps} = activation energy for pressure solution creep ($-24,530$ J mol^{-1}).

Equations 1 and 2 were developed for halite, but large parts of the evaporite sequence in the Red Sea are comprised of anhydrite (see DSDP core stratigraphy in Fig. 1). The mechanical behavior of anhydrite is less well understood, especially under wet conditions and low stresses and strain rates that are expected in salt flows located at or near the surface (Dell'Angelo and Olgaard 1995; Urai et al. 2008). Available flow laws were derived at high pressures and strain rates (Dell'Angelo and Olgaard 1995; Heidelbach et al. 2001) and are likely not applicable for salt flows within the Red Sea. However, generally anhydrite is stronger than halite (Ross and Bauer 1992; Zulauf et al. 2009), so strain should be dominantly accommodated within the halite layer (Ross et al. 1987; Ross and Bauer 1992) under shear stresses parallel to the layering. Therefore, for further estimations, we consider the halite component of the evaporite sequence only.

Estimation of Stresses

The movement of submarine salt flows is a response to the acting differential stress that is determined by the density and geometry of the salt and the hemipelagic cover. The morphological features on the salt flows are either parallel or perpendicular to the direction of maximum seafloor gradient, indicating that the movement—at least of the upper part of the evaporite sequence—is driven by gravitation (tectonic forcing can be neglected). According to descriptions of the cores from DSDP Sites 225 and 227, the upper approximately 100-m evaporite sequence comprises interbedded halite and anhydrite, with minor shale (Whitmarsh et al. 1974, Fig. 1). The following estimates of the stresses within the upper evaporites are based on DSDP Site 227 in the SE wall of Atlantis II Deep. Based on the available morphological and geophysical information, this site is drilled in a salt flow snout near feature "F" in Fig. 3.

The vertical stress component σ_1 is estimated with

$$\sigma_1 = \sum_{i=1}^{\text{layer}} (\rho_i - \rho_w) g h_i \tag{3}$$

with ρ_w: density of water (g/cm^3), ρ_{layer}: density of sediment/evaporite (g/cm^3), g: gravitational acceleration (m/s^2), and h: thickness of layers (m).

If the evaporites are assumed to behave linearly elastically and isotropically, the horizontal stress σ_3 may be roughly estimated with (Bjørlykke 2010, p. 285):

$$\sigma_3 = \frac{v}{1 - v} \sigma_1 \tag{4}$$

with v being the Poisson ratio (the ratio between extension/contraction along σ_1 and contraction/extension along σ_3). v is 0.253 for halite and 0.273 for anhydrite (Gercek 2007). Thus, the horizontal stress is approximated with 0.75 * σ_1 within the evaporite layer.

For the evaporites situated in the upper part of the stratigraphic sequence, we assume that tectonic influence is negligible and that no gliding takes place at the evaporite base (Talbot and Pohjola 2009). Further, as the pore water pressures are unknown, we cannot calculate correct effective stresses. Compacted layers of low permeability may significantly influence vertical stresses. Assuming a sediment layer depth of 150 m with density ρ of 1.6 g/cm^2, a halite thickness of 100 m with a density of 2.2 g/cm^2, a water density of 1 g/cm^3 and g = 10 m/s^2, σ_1 and σ_3 in the upper region of the evaporite sequence are 2.1, and 1.6 MPa, respectively. The differential stress $\sigma_1 - \sigma_3$ is then 0.5 MPa.

Discussion

Location and Thickness of the Deformation Zone

Several possible depths of the main deformation zone within the stratigraphy may occur. As the thickness of the complete evaporite sequence can exceed several kilometers and immediately outside these deeps is at least 1 km (Tramontini and Davies 1969), strain could be focused in the lower parts of the evaporite sequence. Here, higher temperature and eventual hydrothermal water supports deformation. However, this cannot be the deformation mechanism responsible for the formation of the salt flows A–F observed in the Thetis and around Atlantis II Deep (Figs. 2 and 3); the flow fronts are only several hundred meters high, and thus, movement must be located in the upper part of the evaporite sequence. Here, flow could be homogenous, with the evaporites moving en masse, as reported for some allochthonous salt sheets (Fletcher et al. 1995). For the discussion of dislocation creep, it will be assumed that the complete evaporite sequence is deforming (which likely results in an overestimation of the resulting strain rate); that is, the deforming thickness equals the vertical relief of approximately 500 m minus a hemipelagic sediment thickness of approximately 200 m.

However, movement along thin shear zones is also possible; the formation of smaller grains within shear zone mylonites would enhance the effectiveness of pressure solution creep, allowing for higher deformation rates. Indeed, an earlier analysis of DSDP Leg 23 halite samples found only minor deformation (Stoffers and Kühn 1974) in the upper 100 m of the evaporite sequence, indicating the main deformation zone may be located beneath the depth drilled by the DSDP. Shear zones may form due to synsedimentary variations of mechanical properties, e.g., due to different mineralogical compositions within the evaporite sequence (Schléder and Urai 2007). The shear zones may further be connected to shale layers (Mitchell et al. 2010) observed, e.g., in DSDP Sites 225 and 227 (Fig. 1) within the evaporites (Whitmarsh et al. 1974) because of (a) the slow release of water from the shale due to its ongoing compaction, causing a weakening of the surrounding halite or (b) shear within the shale itself caused by high pore pressures and resulting low shear strength, existing due to the very slow drainage of water that results from the extremely low permeability of the surrounding evaporites. During the DSDP drilling campaign, shale layers were mostly reported adjacent to anhydrite, though shale adjacent to halite also exists (Stoffers and Kühn 1974). Their occurrence may also be expected in the evaporite sequence below the depth drilled by the DSDP.

Without adequate samples, it cannot be determined with certainty whether slippage within the shale layer took place. However, exceedingly high pore water pressures within the shale could cause a reduced effective stress and reduced shear strength, favoring such slippage. Generally, the rate of shale compaction is rapid at shallow burial depth and then decreases with increasing depth (Magara 1980). Commonly, the Athy equation was used to calculate the porosity–depth relationship of shale (Athy 1930):

$$\Phi = \Phi_0 e^{-cZ} \quad (5)$$

with Φ = porosity, Φ_0 = initial porosity, c = constant [1/m], approx. 10^{-3} for marine sediments (Brückmann 1989), and Z = burial depth in m.

While the Athy equation was found to be unsatisfactory for larger burial depths (Goulty 1998; Burrus 1998), it gives reliable results for shallow burial depths of up to 1 km (Wetzel 1986; Burrus 1998).

According to Eq. 5, the porosity of shale after 280 m of burial is reduced to approx. 75 % of its initial value. The initial porosity of clayey silt to pelagic clay may be assumed to be 0.7–0.8 (Meade 1966; Einsele 1989). Assuming an original porosity of 0.75, the expected porosity at 280 m burial is 0.56, with empirical data suggesting a porosity of 0.58 (Einsele 1989).

This indicates that the shale within Site 227, with a measured porosity of 0.49 (Table 1), is not undercompacted, and water was efficiently drained. Pathways for water removal within the shale could be formed by an observed intense brecciation of the shale that was attributed to salt flowage (Stoffers and Kühn 1974), while an increase in permeability of rock salt was observed due to microcracking or small faults during deformation under low temperature and mean effective stresses (Urai et al. 1986; Schléder et al. 2008). Thus, slippage in the shale itself appears less likely. Nevertheless, higher porosity and water contents in anhydrite directly adjacent to the shale indicate a potential weakening of the surrounding anhydrite (Table 1)—the same may take place where halite is situated adjacent to shale. Here, the high water content would increase the efficiency of deformation by pressure solution creep.

Table 1 Selection of porosity-depth values reported for DSDP Site 227 (Manheim et al. 1974)

Depth *m*	Porosity	H_2O %	Composition
252	0.006	0.2	Anhydrite
282	0.066	2.43	Anhydrite
283	0.493	22.4	Shale

Estimation of Strain Rates

Dislocation Creep

Strain rates due to dislocation creep were estimated using Eq. 1. Although the differential stress and the temperature both increase with depth, we use constant values to simplify the analysis. The temperature at the evaporite surface is 311 K at Site 225 and 320 K at Site 227, with a temperature gradient of 92° and 117° per kilometer depth (Girdler et al. 1974; Girdler and Whitmarsh 1974). Thus, we assume a temperature of 330 K for the upper 300 m of the evaporite sequence. With a differential stress of 0.5 MPa, dislocation creep would thus occur at $\sim 10^{-14}$ 1/s.

Pressure Solution Creep

The strain rates due to pressure solution creep were estimated using Eq. 2. The critical parameter here is the grain size *D*, which is poorly known and strongly dominates the pressure solution creep equation. The majority of halite sampled during the DSDP campaign was coarse grained (Stoffers and Kühn 1974), suggesting that pressure solution creep would be ineffective. However, as suggested earlier, the main deformation zone may be located beneath the evaporite sequence sampled by the DSDP. Furthermore, few areas comprising halite with 1 mm grain size exist even in the DSDP cores. Interestingly, they are located adjacent to shale layers, which include veins filled with halite precipitated from pore waters (Stoffers and Kühn 1974), an indication of solution precipitation. For comparison, halite grain sizes in onshore mylontic zones are ~ 0.6 mm (Schléder and Urai 2007). Using the differential stress of 0.5 MPa and T = 330 K, pressure solution should occur at strain rates of approx. 5×10^{-10} 1/s for 0.6 mm grains, 10^{-10} for 1 mm grains, and 10^{-13} for 1 cm grains.

Strain Rate Constraints

Rotation of Anhydrite Layers

Strain rates for the salt flows at the southeastern wall of Atlantis II Deep can be estimated from anhydrite layers encountered throughout the evaporites in the DSDP cores (Stoffers and Kühn 1974). If these beds were originally deposited horizontally as seems reasonable, they have subsequently been rotated between 40° to 60° at Site 227 and 10° to 25° at Site 225 (Whitmarsh et al. 1974). The rotational strain would thus be given by (Twiss and Moores 1992):

$$E_s = \tan \varphi \qquad (6)$$

Assuming that deformation started with seafloor spreading at 2.4 Ma (Stoffers and Kühn 1974), the strain rate has thus been approximately 1.5×10^{-14} 1/s at Site 227 and 4×10^{-14} 1/s at Site 225. Strain rates of these orders of magnitude agree with the estimated strain rate from dislocation creep (Fig. 5).

Seafloor Spreading Rates

Seafloor spreading at Atlantis II and Thetis Deep occurs at 12 mm/year (Chu and Gordon 1998). With a half-spreading rate of 6 mm/year, approximately 15 km of oceanic crust has been generated since the onset of seafloor spreading at 2.4 Ma as indicated by magnetic anomalies (Stoffers and Kühn 1974). Recent bathymetric data show that salt fronts advance well into the central trough at the latitude of DSDP Site 227 (feature F), in Thetis deep (feature C) and south of Atlantis II Deep (Feature D), and possibly the central axis in the intertrough areas (Augustin et al. 2014). Some of this evaporite would have dissolved [demonstrated by brine including dissolved Miocene evaporite components in Atlantis II Deep (Anschutz and Blanc 1995)], so the rate from seafloor spreading may be a minimum. On the other hand, in many areas, the flow fronts do not reach the centers of the deeps, demonstrating spatially different flow speeds. For flows advancing into the central trough the half-spreading rate are used to constrain flow speed.

Generally, the approximate evaporite flow speed may be calculated with (Twiss and Moores 1992)

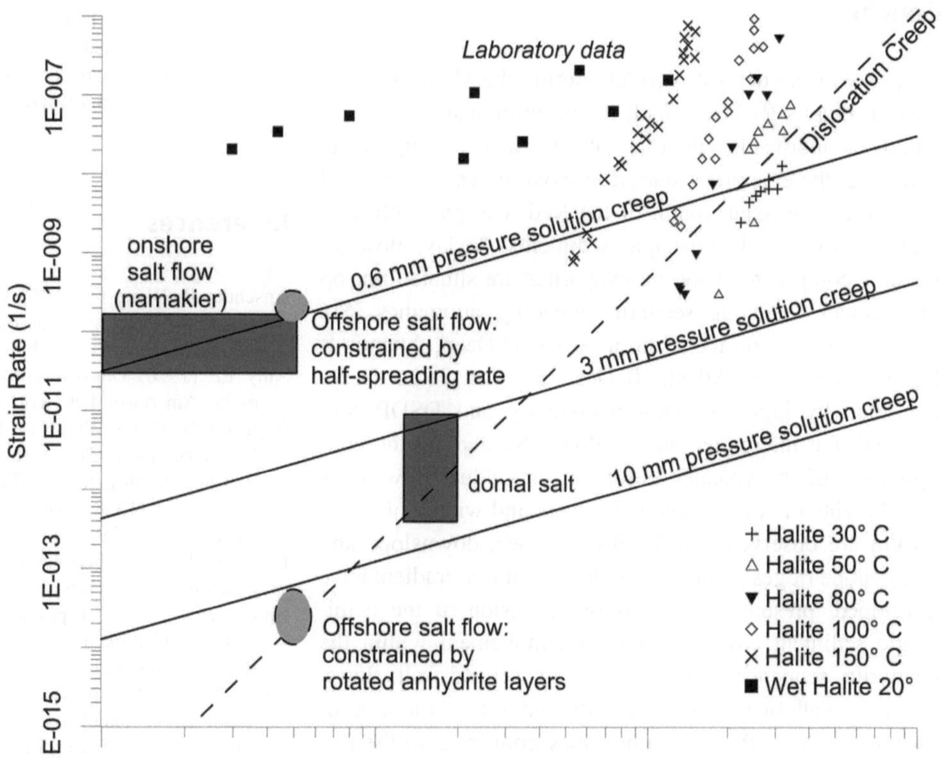

Fig. 5 Comparison of pressure solution creep and dislocation creep strain rate/differential stress relationship estimated for offshore salt flows (*gray ellipsoids*), compared with data derived for onshore namakiers and domal salt (*gray rectangles*) and laboratory data. Constraints for offshore salt flows are made based on the rotation of anhydrite layers, and the local half-spreading rate. Refer to the text for further discussion. Modified from Schléder and Urai (2007)

$$\text{Flow Speed [m/s]} = \text{Shear Strain Rate [1/s]} * \text{Shear Zone Thickness [m]} \quad (7)$$

For dislocation creep, the shear zone thickness was (probably) overestimated at 300 m, resulting in a flow speed of 0.1 mm/year. This is less than the seafloor spreading half-rate by two orders of magnitude. For pressure solution creep, the water-influenced halite thickness adjacent to shales is less than 4 m based on measurements of water content at DSDP site 277 (Manheim et al. 1974). Assuming a 1 mm grain size [the finest halite grain size reported for the DSDP cores (Girdler and Whitmarsh 1974)] over a cumulative depth extent of 50 cm throughout the evaporite/shale sequence, the strain rate required to keep up with the half-spreading rate is 2×10^{-10} 1/s. For comparison, individual mylonitic shear zones observed onshore reach approximately 5 cm in thickness, but several exist at each namakier (Schléder and Urai 2007). While these are only rough estimates, it shows that salt flow movement speed due to pressure solution creep, with estimated strain rates between 1×10^{-10} 1/s and 5×10^{-10} 1/s (Fig. 5) for fine-grained halite, is potentially able to explain how the flows keep up with seafloor spreading at certain locations.

Conclusion

Evaporites were deposited widely during the Miocene in the area of today's Red Sea. With the onset of seafloor spreading, the evaporites lost their lateral constraint. Because halite is ductile, the evaporites started to move downslope toward the developing axial trough of the Red Sea, passively carrying a cover of hemipelagic sediments. Today, flowlike features comprising Miocene evaporites are situated on top of younger magnetic seafloor spreading anomalies. Six flowlike features on the seafloor of Thetis Deep, Atlantis II Deep, and between Atlantis II Deep and Port Sudan Deep, identified by high-resolution bathymetry and DSDP core material, are interpreted as salt flows. Steered by the configuration of the volcanic basement, individual flow lobes with heights of several hundred meters and widths of 3 and 10 km are observed. On the flow surface, downslope and along-slope ridges parallel to the local seafloor gradient have developed, presumably caused by extension of the hemiplegic sediment cover or strike-slip movement within the evaporite sequence. At the front of several salt flows, irregular seafloor points to local dissolution. In the case of Atlantis II Deep, this dissolution may contribute to the formation of the local brine. Based on the conditions known for Atlantis II Deep, differential stresses within the upper evaporites are $\sim$0.5 MPa. The corresponding deformation was assumed to take place within the upper part of the evaporite sequence, as the vertical reliefs of the salt flows do not exceed a few hundred meters. Based on physical parameters measured during the DSDP drilling campaign, strain rates due to dislocation creep were estimated to be 10^{-14} 1/s. However, this rate is low relative to rates expected of seafloor spreading, so advance of several flow lobes well into the Red Sea central trough cannot be explained by dislocation creep, but indicates the work of solution and precipitation processes ("pressure solution creep"). Depending on halite grain size, water content, and the formation of thin mylonitic shear zones (all uncertain parameters), strain rates of 10^{-10} 1/s caused by pressure solution creep appear possible, similar to deformation measured in extrusive onshore salt flows in Iran. Such strain rates result in flow speeds of several mm/year for offshore salt flows in certain locations. Thus, salt movements can potentially keep up with half-spreading rates reported for large parts of the Red Sea.

Acknowledgement We thank the reviewers L.M. Pinheiro and J. Gardner for constructive and helpful reviews. This chapter is the result of the Jeddah Transect Project, a collaboration between King Abdulaziz University and GEOMAR Helmholtz-Center for Ocean Research that was funded by King Abdulaziz University (KAU) Jeddah, Saudi Arabia, under Grant No. T-065/430. The authors, therefore, acknowledge with thanks KAU technical and financial support. We wish to thank M. Ligi and E. Bonatti who gave us access to multibeam data of Thetis Deep. Nico Augustin and Rashad Bantan collected and processed large parts of the remaining multibeam data. Mark Schmidt made ship time available for additional salt flow-related surveys and contributed with valuable comments. Further, we wish to thank masters and crews of FS Poseidon and FS Pelagia for their invaluable assistance during our surveys.

References

Anschutz P, Blanc G (1995) Origin of fluids and the evolution of Atlantis II Deep hydrothermal system, Red Sea: strontium isotope study. Geochim Cosmochim Acta 59(23):4799–4808

Athy LF (1930) Density, porosity, and compaction of sedimentary rocks. Am Assoc Pet Geol Bull 14(1):1–24

Augustin N, Colin WD, van der Zwanm F, Feldens P, Bantan R, Kwasnitschka T (2014) The rifting to spreading transition in the Red Sea. Earth Planet Sci Lett 395:217–230

Bindschadler R (1998) Monitoring ice sheet behavior from space. Rev Geophys 36(1):79–104

Bjørlykke K (2010) Petroleum geoscience: from sedimentary environments to rock physics. Springer, Berlin

Brückmann W (1989) Typische Kompaktionsabläufe mariner Sedimente und ihre Modifikation in einem rezenten Akkretionskeil (Barbados Ridge). Tübinger Geowissenschaftliche Arbeiten, Reihe A Nr. 5, 135 pp

Burrus J (1998) Overpressure models for clastic rocks, their relation to hydrocarbon expulsion: a critical reevaluation. In: Law BE, Ulmishek GF, Slavin VI (eds) Abnormal pressures in hydrocarbon environments. American Association of Petroleum Geologists Memoir, pp 35–63

Campbell I, Jacobel R, Welch B, Petterson R (2008) The evolution of surface flow stripes and stratigraphic folds within Kamb ice stream: why don't they match? J Glaciol 54(186):421–427

Chu D, Gordon RG (1998) Current plate motions across the Red Sea. Geophys J Int 135(2):313–328

Cochran JR (1983) A model for the development of the Red Sea. Am Assoc Pet Geol Bull 67:41–69

Coleman RG (1993) Geologic evolution of the Red Sea. Oxford Monogr Geol Geophys 24, 186 pp

De Meer S, Drury MR, De Bresser JHP, Pennock GM (2002) Current issues and new developments in deformation mechanisms, rheology and tectonics, vol 200. Geological Society, London, pp 1–27 (Special Publications)

Dell'Angelo L, Olgaard DL (1995) Experimental deformation of fine-grained anhydrite: evidence for dislocation and diffusion creep. J Geophys Res 100(B8):15425–15440

Einsele G (1989) In-situ water contents, liquid limits, and submarine mass flows due to a high liquefaction potential of slope sediment (results from DSDP and subaerial counterparts). Geol Rundsch 78:821–840

Fletcher CF, Hudec MR, Watson IA (1995) Salt glacier and composite sediment-salt glacier models for the emplacement and early burial of allochthonous salt sheets. In: Jackson MPA, Roberts DG, Snelson S (eds) Salt tectonics a global perspective. American Association of Petroleum Geologists Memoir, pp 77–108

Gercek H (2007) Poisson's ratio values for rocks. Int J Rock Mech Min Sci 44(1):1–13

Girdler RW, Whitmarsh RB (1974) Miocene evaporites in Red Sea cores, their relevance to the problem of the width and age of oceanic crust beneath the Red Sea. In: Whitmarsh RB, Weser OE, Ross DA (eds) Initial reports of the Deep Sea Drilling Project, vol 23. US Government Printing Office, Washington, pp 913–921

Girdler RW, Erickson AJ, Von Herzen R (1974) Downhole temperature and shipboard thermal conductivity measurements aboard d/v Glomar Challenger in the Red Sea. In: Whitmarsh RB, Weser OE, Ross DA (eds) Initial reports of the Deep Sea Drilling Project, vol 23. US Government Printing Office, Washington, pp 879–886

Glasser NF, Gudmundsson GH (2012) Longitudinal surface structures (flowstripes) on Antarctic glaciers. Cryosphere 6:383–391

Goulty NR (1998) Relationships between porosity and effective stress in shales. First Break 16(12):413–419

Green HW (1984) Pressure solution creep: some causes and mechanisms. J Geophys Res 89(B6):4313–4318

Gudmundsson GH, Raymond CF, Bindschadler R (1998) The origin and longevity of flow stripes on Antarctic ice streams. Ann Glaciol 27:145–152

Hartmann M, Scholten JC, Stoffers P, Wehner F (1998) Hydrographic structures of brine-filled deeps in the Red Sea—results from the Shaban, Kebrit, Atlantis II, and Discovery Deep. Mar Geol 144:311–330

Heidelbach F, Stretton IC, Kunze K (2001) Texture development of polycrystalline anhydrite experimentally deformed in torsion. Int J Earth Sci (Geol Rundsch) 90:118–126

Hudec MR, Jackson MPA (2006) Advance of allochthonous salt sheets in passive margins and orogens. Am Assoc Pet Geol Bull 90:1535–1564

Hudec MR, Jackson MPA (2007) Terra infirma: understanding salt tectonics. Earth-Sci Rev 82:1–28

Izzeldin AY (1987) Seismic, gravity and magnetic surveys in the central part of the Red Sea: their interpretation and implications for the structure and evolution of the Red Sea. Tectonophysics 143:269–306

Jackson MPA, Talbot CJ (1986) External shapes, strain rates, and dynamics of salt structures. Geol Soc Am Bull 97(3):305–323

Krastel S, Wynn RB, Georgiopoulou A, Geersen J, Heinrich R, Meyer M, Schwenk T (2012) Large-scale mass wasting on the northwest African continental margin: some general implications for mass wasting on passive continental margins. Adv Nat Technol Hazards Res 31:189–199

Lees GM (1927) Salzgletscher in Persien. Mitteilungen der deutschen Gesellschaft (Wien) 20:29–34

Ligi M, Bonatti E, Bortoluzzi G, Cipriani A, Cocchi L, Tontini FC, Carminati E, Ottolini L, Schettino A (2012) Birth of an ocean in the Red Sea: initial pangs. Geochem Geophys Geosyst 13(8):1–29

Magara K (1980) Comparison of porosity-depth relationships of shale and sandstone. J Pet Geol 3:175–185

Manheim FT, Dwight L, Belastock RA (1974) Porosity, density, grain density, and related physical properties of sediments from the Red Sea drill cores. In: Whitmarsh RB, Weser OE, Ross DA (eds) Initial reports of the Deep Sea Drilling Project, vol 23. US Government Printing Office, Washington, pp 887–907

Mankinen EA, Dalrymple GB (1979) Revised geomagnetic polarity time scale for the interval 0-5 m.y. B.P. J Geophys Res 84(B2):615–626

Meade RH (1966) Factors influencing the early stages of the compaction of clays and sands – review. SEPM J Sediment Res 36(4):1085–1101

Micaleff A, Masson DG, Berndt C, Stow DAV (2007) Morphology and mechanics of submarine spreading: a case study from the Storegga Slide. J Geophys Res 112(F03023):1–12

Mitchell DJW, Allen RB, Salama W, Abouzakm A (1992) Tectono-stratigraphic framework and hydrocarbon potential of the Red Sea. J Pet Geol 15(2):187–210

Mitchell NC, Ligi M, Ferrante V, Bonatti E, Rutter E (2010) Submarine salt flows in the central Red Sea. Geol Soc Am Bull 122(5–6):701–713

Mohr M, Warren JK, Kukla PA, Urai JL, Irmen A (2007) Subsurface seismic record of salt glaciers in an extensional intracontinental setting (Late Triassic of northwestern Germany). Geology 35 (11):963–966

Pautot G, Auzende JM, LePichon X (1966) Continuous deep salt layer along North Atlantic margins related to early phase of rifting. Nature 227:351–354

Pilcher RS, Blumstein RD (2007) Brine volume and salt dissolution rates in Orca Basin, northeast Gulf of Mexico. Am Assoc Pet Geol Bull 91:823–833

Ross JV, Bauer SJ (1992) Semi-brittle deformation of anhydrite-halite shear zones simulating mylonite formation. Tectonophysics 213:303–320

Ross DA, Schlee J (1973) Shallow structure and geologic development of the southern Red Sea. Geol Soc Am Bull 84(12):3827–3848

Ross JV, Bauer SJ, Hansen FD (1987) Textural evolution of synthetic anhydrite-halite mylonites. Tectonophysics 140:307–326

Schléder Z, Urai JL (2007) Deformation and recrystallization mechanisms in mylonitic shear zones in naturally deformed extrusive Eocene-Oligocene rocksalt from Eyvanekey plateau and Garmsar hills (central Iran). J Struct Geol 29(2):241–255

Schléder Z, Urai JL, Nollet S, Hilgers C (2008) Solution-precipitation creep and fluid flow in halite: a case study of Zechstein (Z1) rocksalt from Neuhof salt mine (Germany). Int J Earth Sci (Geol Rundsch) 97:1045–1056

Schmidt M, Devey C, Eisenhauer A (eds) (2011) FS Poseidon Fahrtbericht/Cruise Report P408—The Jeddah Transect; Jeddah – Jeddah, Saudi Arabia, 13.01.-02.03.2011 IFM-GEOMAR Report 46

Stoffers P, Kühn R (1974) Red sea evaporites: a petrographic and geochemical study. In: Whitmarsh RB, Weser OE, Ross DA (eds) Initial reports of the Deep Sea Drilling Project, vol 23. US Government Printing Office, Washington, pp 821–847

Talbot CJ (1993) Spreading of salt structures in the Gulf of Mexico. Tectonophysics 228:151–166

Talbot CJ, Pohjola V (2009) Subaerial salt extrusions in Iran as analogues of ice sheets, streams and glaciers. Earth Sci Rev 97:155–183

Talbot CJ, Rogers EA (1980) Seasonal movements in a salt glacier in Iran. Science 208:395–397

Talbot CJ, Medvedev S, Alavi M, Shahrivar H, Heidari E (2000) Salt extrusion rates at Kuh-e-Jahani, Iran: June 1994 to November 1997, vol 174. Geological Society, London (Special Publications), pp 93–11

Tramontini C, Davies D (1969) A seismic refraction survey in the Red Sea. Geophys J Int 17(2):225–241

Twiss RJ, Moores ME (1992) Structural geology. Palgrave Macmillan, Basingstoke

Urai JL, Spiers CJ, Lister GS (1986) Weakening of rock salt by water during long-term creep. Nature 324:554–557

Urai J, Schléder Z, Spiers C (2008) Flow and transport properties of salt rocks. In: Littke R, Bayer U, Gajewski D, Nelskamp S (eds) Dynamics of complex intracontinental basins: the Central European Basin system. Springer, Berlin, pp 277–290

Völker D, Scholz F, Geersen J (2011) Analysis of submarine landsliding in the rupture area of the 27 February 2010 Maule earthquake, Central Chile. Mar Geol 288:79–89

Wenkert DD (1979) The flow of salt glaciers. Geophys Res Lett 6 (6):523

Wetzel A (1986) Sedimentphysikalische Eigenschaften als Indikatoren für Ablagerung, Diagenese und Verwitterung von Peliten. Geowiss Fakultät Univ Tübingen, 135 pp

Whitmarsh RB, Weser OE, Ross DA (eds) (1974) Initial reports of the Deep Sea Drilling Project, vol 23. US Government Printing Office, Washington

Zulauf G, Zulauf J, Bornemann O, Kihm N, Peinl M, Zanella F (2009) Experimental deformation of a single-layer anhydrite in halite matrix under bulk constriction. J Struct Geol 31:460–474

Geochemical Classification of Brine-Filled Red Sea Deeps

Mark Schmidt, Radwan Al-Farawati, and Reiner Botz

Abstract

The major geochemical characteristics of Red Sea brine are summarized for 11 brine-filled deeps located along the central graben axis between 19°N and 27°N. The major element composition of the different brine pools is mainly controlled by variable mixing situations of halite-saturated solution (evaporite dissolution) with Red Sea deep water. The brine chemistry is also influenced by hydrothermal water/rock interaction, whereas magmatic and sedimentary rock reactions can be distinguished by boron, lithium, and magnesium/calcium chemistry. Moreover, hydrocarbon chemistry (concentrations and $\delta^{13}C$ data) of brine indicates variable injection of light hydrocarbons from organic source rocks and strong secondary (bacterial or thermogenic) degradation processes. A simple statistical cluster analysis approach was selected to look for similarities in brine chemistry and to classify the various brine pools, as the measured chemical brine compositions show remarkably strong concentration variations for some elements. The cluster analysis indicates two main classes of brine. Type I brine chemistry (Oceanographer and Kebrit Deeps) is controlled by evaporite dissolution and contributions from sediment alteration. The Type II brine (Suakin, Port Sudan, Erba, Albatross, Discovery, Atlantis II, Nereus, Shaban, and Conrad Deeps) is influenced by variable contributions from volcanic/magmatic rock alteration. The chemical brine classification can be correlated with the sedimentary and tectonic setting of the related depressions. Type I brine-filled deeps are located slightly off-axis from the central Red Sea graben. A typical "collapse structure formation" which has been defined for the Kebrit Deep by evaluating seismic and geomorphological data probably corresponds to our Type I brine. Type II brine located in depressions in the northern Red Sea (i.e., Conrad and Shaban Deeps) could be correlated to "volcanic intrusion-/extrusion-related" deep formation. The chemical indications for hydrothermal influence on Conrad and Shaban Deep brine can be related to brines from the multi-deeps region in the central Red Sea, where volcanic/magmatic fluid/rock interaction is most obvious. The strongest hydrothermal influence is observed in Atlantis II brine (central multi-deeps region), which is also the hottest Red Sea brine body in 2011 (~68.2 °C).

Introduction

Brine from the Red Sea deeps is slightly acidic (pH ~ 5–7), nearly saturated with respect to halite, and enriched relative to normal Red Sea deep water in, for example, Ca, K, Si, Ba, and trace metals such as Fe, Mn, and Zn (e.g., Pierret et al. 2001; Gurvich 2006). The salty brine bodies are mainly formed by seawater having dissolved the Miocene evaporites which build up to several-kilometer-thick deposits under the

M. Schmidt (✉)
GEOMAR Helmholtz Centre for Ocean Research Kiel, Wischhofstr. 1-3, 24148 Kiel, Germany
e-mail: mschmidt@geomar.de

R. Al-Farawati
Faculty of Marine Sciences, King Abdulaziz University, P.O. Box 80207, Jeddah 21589, Saudi Arabia

R. Botz
Institute of Geosciences, University Kiel, Ludewig-Meyn-Str. 10, 24118 Kiel, Germany

N.M.A. Rasul and I.C.F. Stewart (eds.), *The Red Sea*,
Springer Earth System Sciences, DOI 10.1007/978-3-662-45201-1_13,

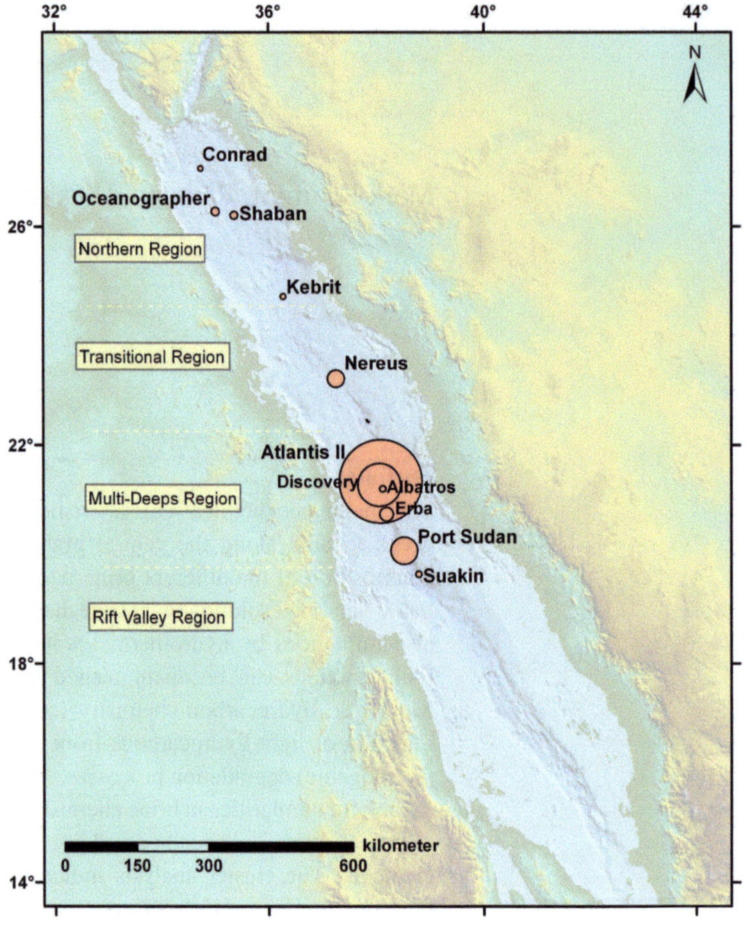

Fig. 1 Red Sea area including the studied brine-filled deeps (*orange circles*). Geotectonic classification of Red Sea segments (*yellow dotted lines*) is adapted from Bonatti (1985). Symbol size of brine-filled deeps reflects the difference between warm brine temperature and Red Sea deepwater temperature (~21.7 °C)

normal pelagic/terrigenous sediment of the Red Sea (e.g., Craig 1969). The temperatures of most of the brine pools are slightly higher (23–26 °C) than the Red Sea deep water (21.7 °C). However, brines exist which are strongly influenced by subsurface heating forming brines with temperatures up to 68.2 °C (Atlantis II Deep, ~21°N in Fig. 1).

The highly saline (chlorinity of up to ~155 ‰) and dense (up to ~1.2 g/cm^3) brine bodies are normally separated from Red Sea deep water (S ~ 40.7 ‰; σ ~ 1.04 g/cm^3) by sharp interfaces, and transport across the brine–seawater interface is mainly controlled by diffusion (e.g., Anschutz and Blanc 1996; Winckler et al. 2001; Schmidt et al. 2003; Swift et al. 2012). The deep brine pools are strictly anaerobic, and iron and manganese oxides are precipitated above the brine–seawater interface of metal-rich brine (e.g., Scholten et al. 2000; Schmidt et al. 2003; Anschutz, this issue). High gas concentrations (CO_2, hydrocarbons) can be determined in the brine reaching, for example, methane saturation (at 1 atm.) in the Kebrit Deep brine (Faber et al. 1998). Stable carbon and hydrogen isotope analyses of dissolved hydrocarbons in Red Sea brine clearly indicate a thermogenic origin of natural hydrocarbon gas which was probably released by the degradation of organic matter in the deep subsurface (e.g., Schenk et al. 2010), at generally high geothermal gradients (e.g., Girdler and Evans 1977). However, hydrothermal formation of higher hydrocarbons and of aromatic compounds from recent organic debris is also possible in the Kebrit or the Atlantis II Deeps as discussed in Botz et al. (2007), and Simoneit et al. (1987), and Wang et al. (2011), respectively.

The chemistry of brine in the Red Sea is not only controlled by the original Miocene salt composition (Schoell and Faber 1978), but also by various physical/chemical processes occurring at the sediment/brine and brine/seawater boundaries. A simplified sketch of relevant processes at the related interfaces originally presented by Schmidt et al. (2006) is given in Fig. 2. Detrital (turbidites) sediment input, planktonic debris sedimentation, volcanic rock alteration, and (bio)geochemical redox cycles (i.e., Mn and Fe precipitation and dissolution, bacterial sulfate reduction, organic matter oxidation, hydrocarbon oxidation) are the dominating processes that can change the original brine chemistry. Brine formation and evolutionary changes were discussed by several authors in the context of hydrographic, (isotope) geochemical, or microbial data evaluation (e.g., Craig 1969; Schoell and Faber 1978; Anschutz et al. 1999; Pierret et al.

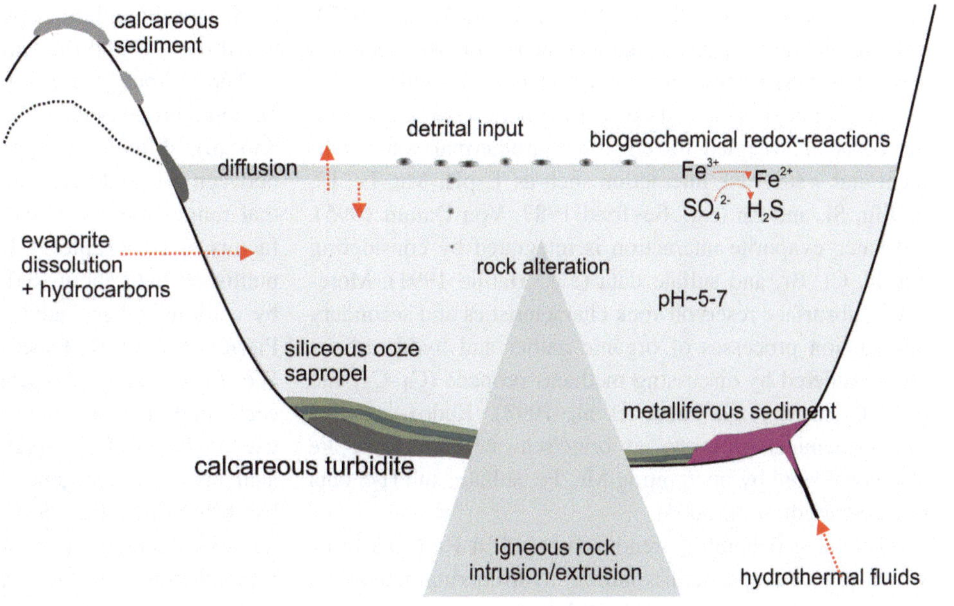

Fig. 2 Model of brine-filled Red Sea deeps [modified after Schmidt et al. (2006)]. *Red arrows* mark transport/reaction processes that mainly determine the type of brine

2001; Winckler et al. 2001; Schmidt et al. 2006; Antunes et al. 2011). Anschutz et al. (1999), for example, discussed Red Sea brine formation and evolution based on salinity and temperature profiles measured across brine–seawater interfaces. Hydrothermally active and non-hydrothermally active brine formation could be classified for the period between the early 1970s and 1992, and the authors concluded that both types of brine are "extremely" stable once the brine pool was formed within a depression at the seafloor. Pierret et al. (2001) discussed several chemical parameters (i.e., major and trace elements; oxygen, hydrogen, and strontium isotope data) of different Red Sea brine bodies and concluded that "A striking feature is the absence of a relationship between the position of the deeps along the axis and their evolutionary maturity." However, based on noble gas composition of Red Sea brine, Winckler et al. (2001) identified two principally different "geochemical fingerprints" that reflect the geological setting of the deeps along the axis of the Red Sea. These fingerprints point to (1) subsurface MORB-like hydrothermal activity with the addition of mantle-derived components to the brine and (2) crustal and sedimentary fluids migrating from depth into the brine-filled depressions. Furthermore, in contrast to most other authors, Hovland et al. (2006 and this issue) postulated a different pathway for evaporite and brine formation in the Red Sea, which is controlled by supercritical phase separation of heated seawater.

This study summarizes recent information about brine chemistry of 11 deep brine bodies located between 19 and 27°N along the central Red Sea graben axis (Fig. 1), including the newly formed brine of the Conrad Deep, which is the northernmost brine pool discovered in the Red Sea (Ehrhardt et al. 2005). Beside a more detailed discussion of selected geochemical parameters, we use multi-element statistical approaches that summarize the main processes determining the brine chemistry, for example, evaporite dissolution, hydrothermal input/alteration, and secondary biogeochemical processes. New hydrographic and geochemical data presented here were derived from joint German–Saudi–Egyptian–Sudanese research projects conducted during the last decade (e.g., www.jeddah-transect.org).

Data Collection and Evaluation of Selected Geochemical Parameters

Geochemical and hydrographic data from the Red Sea deeps were obtained from sampling during research cruises ME44/3 (RV Meteor 1999, Pätzold et al. 2000), ME52/3 (RV Meteor 2002, Pätzold et al. 2003), RS05 (RV Urania 2005, Bonatti et al. 2005), P408-2 (RV Poseidon 2011, Schmidt et al. 2011), and 64PE350-1 (RV Pelagia 2012, Schmidt et al. 2013). Hydrographic data (i.e., pressure, temperature, conductivity, sound velocity) were determined by using the standard Seabird 9 plus-CTD and a Sea and Sun-CTD which was built and calibrated for ambient temperatures of up to 80 °C. Water and brine sampling was conducted by using Niskin rosette water samplers.

Brine samples were taken immediately after recovery of the water sampler. Samples were stored in pre-cleaned PE bottles for inorganic tracer studies. Headspace vials were filled for trace gas measurements. Small Zinsser plastic vials were filled for anion determination. All samples were handled according to our standard geochemical methods (http://www.geomar.de/en/research/fb2/fb2-mg/benthic-biogeochemistry/mg-analytik/). Subsequent inductive coupled plasma–optical emission spectrometry (ICP-OES), ion chromatography (IC), gas chromatography (GC), and isotope ratio mass spectrometry (cf-IRMS) analyses were conducted accordingly.

Geochemical parameters discussed here were selected on the basis of accepted tracers for hydrothermal water/sediment and water/rock interaction such as T, pH, Mn, Fe, Li, B, Mg, Si, and Zn (e.g., Seyfried 1987; Von Damm 1995). Seawater–evaporite interaction is integrated by considering Na, K, Cl, Br, and sulfate data (e.g., Hardie 1991). Moreover, subsurface reservoir rock characteristics and secondary degradation processes of organic matter and hydrocarbons are considered by discussing methane–propane (C_1–C_3) and $\delta^{13}C$–C_1 data (e.g., Faber et al. 1998). Redox-sensitive biogeochemical processes at brine/seawater interfaces are also considered by interpreting Mn, Fe, sulfate, and H_2S data (e.g., Schmidt et al. 2003).

General geochemical trends observed in Red Sea brine which reflect evaporite dissolution, hydrothermal alteration, or hydrocarbon formation/degradation are discussed by using element–element cross-plots. Mixing trends and deviations from trends are discussed here. However, the geochemical data set presented in our study comprises a matrix of 13 × 22, which strongly suggests a statistical data evaluation to identify correlation structures or similarities between the various Red Sea brine bodies. The statistical interpretation of similarity of brine compositions was performed by using cluster analysis (Statistica®). A tree diagram is presented where Euclidean distances (median clustering) of standardized brine data/clusters are expressed by the order of amalgamation. Selected brine parameters are as follows: T, pH, Na, Cl, Br, Ca, Mg, Si, K, B, Li, Mn, Fe, Zn, Sr, DIC, CH_4, SO_4, H_2S, $\delta^{13}C$-DIC, $\delta^{13}C$–CH_4, $C_1/[C_2 + C_3]$.

Brine Characterization

Recent Data Summary of Chemical Characteristics of Red Sea Brine

Major (Na, K, Ca, Mg) and trace elements (Li, B, Si, Mn, Fe, Sr, Zn), anions (Cl, Br, SO_4), hydrocarbon (isotope)composition, inorganic (isotope)carbon composition, H_2S, temperature, and pH data of brine samples measured in this study are presented in Table 1. The units in mg/l are used in both Table 1 and Fig. 3 to make comparison with the majority of older (from the 1960s to 1980s) published Red Sea literature easier. Only selected chemical parameters are plotted in concentration bar charts in Fig. 3. The concentration bars for each parameter are organized from north (left) to south (right) along the *x*-axis of Fig. 3. In order to visualize possible linkages to the geotectonic sections, which are defined in Fig. 1 (after Bonatti 1985), a red dotted line marks the boundary between the "multi-deeps and transitional region" and the "northern region" (cf. Fig. 1).

The various Red Sea brine pools analyzed in this study exhibit a broad range of variable maximum thicknesses (11–336 m), different temperatures (23–68.2 °C), pH values between 5.0 and 7.7, and chemical element concentrations that range from near seawater concentration to enrichment factors of 10^6 (Table 1). In summary, our data resemble the published data about Red Sea brine composition published by various other authors (e.g., Degens and Ross 1969; Pierret et al. 2001; Faber et al. 1998; Antunes et al. 2011). We do not want to discuss the chemical characteristics of each single Red Sea brine pool in detail, as we would rather use a statistical interpretation approach by using cluster analysis to present general geochemical trends for the studied brine-filled Red Sea deeps (Fig. 1). However, a few remarks should be mentioned here based on the summary of brine chemistry in Fig. 3 and Table 1.

Brine, for example, rich in magnesium (e.g., Oceanographer Deep brine), could indicate primary brine contributions from late-stage seawater evaporation that is near the bischofite saturation (Krupp 2005), or secondary brine from equilibration of seawater with bischofite and kainite evaporate deposits (Wallmann et al. 1997). The dissolution of the latter evaporate deposits is usually accompanied by low pH values down to 4.5 and Mg concentrations of about 100 times higher than seawater Mg concentrations (Wallmann et al. 1997). However, the highest Mg concentrations measured in Red Sea brine (6,821 mg/l, Oceanographer Deep, Table 1) is only a factor of <5 higher than Red Sea deep water (Table 1). Therefore, two additional likely scenarios for Mg enrichment in Red Sea brine should be considered: (1) dissolution of small amounts of intercalated Mg-rich evaporite minerals such as polyhalite (Stoffers and Kühn 1974) and/or (2) dissolution of abundant detrital carbonaceous material (Pierret et al. 2001). Both processes would be accompanied by an increase in Ca and Sr concentrations and in dissolved inorganic carbon, whereas the dissolution of polyhalite also increases K concentrations in brine. The enrichment of Ca, Mg, K, Sr, and DIC compared to Red Sea deep water (Table 1) was determined in Oceanographer Deep brine and, for example, to a minor extent in Shaban Deep brine (Table 1). Both deeps are located at nearly the same northern latitude (Fig. 1); however, their brines exhibit distinct differences also in other geochemical parameters such as Fe, Si, and B concentrations or hydrocarbon and sulfur species occurrence (Table 1).

Among others, the variation in SO_4 concentrations of all measured brines is pronounced (maximum − minimum = 3,693 mg/l) and the brines can be divided into two groups based on SO_4 concentrations; brine with up to 1.6 times higher concentrations compared to RSDW, i.e., brines of the Conrad, Shaban, Erba, Port Sudan, and Suakin Deeps;

Table 1 Summary of measured chemical brine compositions sampled from Red Sea deeps (19–27°N)

Brine	T (°C)	pH	Na (mg/l)	Cl (mg/l)	Br (mg/l)	Ca (mg/l)	Mg (mg/l)	Si (mg/l)	K (mg/l)	B (mg/l)	Li (mg/l)	Mn (mg/l)	Fe (mg/l)	Sr (mg/l)	Zn (mg/l)	SO_4 (mg/l)
Conrad[a]	23.0	6.1	98,751	163,781	150	987	2,455	1.2	2,647	18.0	2.79	0.01	0.47	31.2	0.026	4,685
Oceanographer[a]	24.9	5.6	100,083	167,227	372	3,455	6,821	1.0	3,014	49.7	5.57	18.07	0.55	95.8	0.072	991
Shaban-N[b]	22.8	5.7	109,995	173,738	n.m.	791	2,381	5.3	1,927	7.8	1.11	3.26	11.81	22.0	n.m.	4,954
Shaban-S[b]	24.5	6.0	111,385	176,234	n.m.	888	2,065	5.7	1,907	7.8	0.92	2.13	6.44	25.0	n.m.	4,656
Kebrit[b]	23.4	5.5	110,483	182,075	278	2,267	2,944	13.1	1,435	27.4	2.65	8.90	0.36	59.0	0.030	1,929
Nereus[a]	30.0	7.4	81,766	146,473	121	9,250	1,557	0.1	3,027	9.1	1.15	0.05	0.65	59.3	0.025	1,081
Atlantis II-SW[c]	68.2	5.0	107,540	184,000	122	5,960	861	27.9	3,397	10.7	8.40	88.97	99.89	53.2	3.723	1,032
Atlantis II-N[c]	66.0	5.0	106,100	182,571	117	6,010	861	26.8	3,356	10.9	7.75	90.92	84.80	53.4	3.534	1,043
Discovery[c]	45.0	6.6	106,570	178,036	118	5,806	894	3.2	3,318	10.6	6.30	64.53	1.00	53.1	0.362	1,024
RSDW[c]	21.7	8.1	12,583	20,711	71	497	1,598	0.0	368	5.3	0.20	0.07	0.50	9.1	0.015	3,206
Port Sudan[c]	35.9	6.6	87,009	136,679	116	1,428	1,723	17.1	1,818	8.2	0.52	7.16	7.03	25.1	0.081	4,589
Erba[c]	26.1	7.4	60,279	94,948	108	1,185	1,736	4.4	1,209	6.9	0.38	18.71	1.34	22.9	0.050	4,529
Suakin[c]	23.7	7.7	55,561	92,633	119	2,290	1,528	3.4	1,385	7.4	0.36	32.35	0.02	19.6	0.123	3,463
Albatross[c]	23.9	6.8	99,888	163,394	124	4,488	1,187	2.8	2,188	9.7	1.90	26.37	0.26	48.0	0.095	1,430

Brine	DIC (mg/l)	CH_4 (nl/l)	H_2S (ml/l)	$\delta^{13}C$-DIC (‰ VPDB)	C1/(C2 + C3)	$\delta^{13}C$–CH_4 (‰ VPDB)
Conrad[a]	102.0	144,090	n.d.	−10.9	5	−47.6
Oceanographer[a]	96.0	4,921,375	15.1	−8.7	21	−48.3
Shaban-N[b]	165.6	19,456	n.d.	−2.3	9	−10.3
Shaban-S[b]	75.6	7,340	n.d.	−4.3	6	−34.1
Kebrit[b]	115.2	9,576,414	24.0	−11.8	47	−22.4
Nereus[a]	2.4	3,810	n.d.	−9.4	77	n.m.
Atlantis II-SW[c]	31.8	103,451	n.d.	−6.7	48	−16.2
Atlantis II-N[c]	31.8	100,859	n.d.	−6.7	54	−16.1
Discovery[c]	8.9	17,437	n.d.	−12.9	105	−7.0
RSDW[c]	29.6	60	n.d.	0.3	20	−32.0
Port Sudan[c]	19.1	12,697	n.d.	−10.0	762	11.2
Erba[c]	13.0	16,392	n.d.	−7.5	570	−13.9
Suakin[c]	10.8	7,291	n.d.	−6.6	263	−2.6
Albatross[c]	5.2	8,854	n.d.	−10.2	168	17.6

n.m. not measured
n.d. not detectable (below detection limit)
[a] RV Urania cruise RS05 (Bonatti et al. 2005)
[b] RV Meteor cruises M44/3 and M52/3 (Pätzold et al. 2000, 2003)
[c] RV Poseidon P408-2 and RV Pelagia 64PE350-1 cruises (Schmidt et al. 2011, 2013)

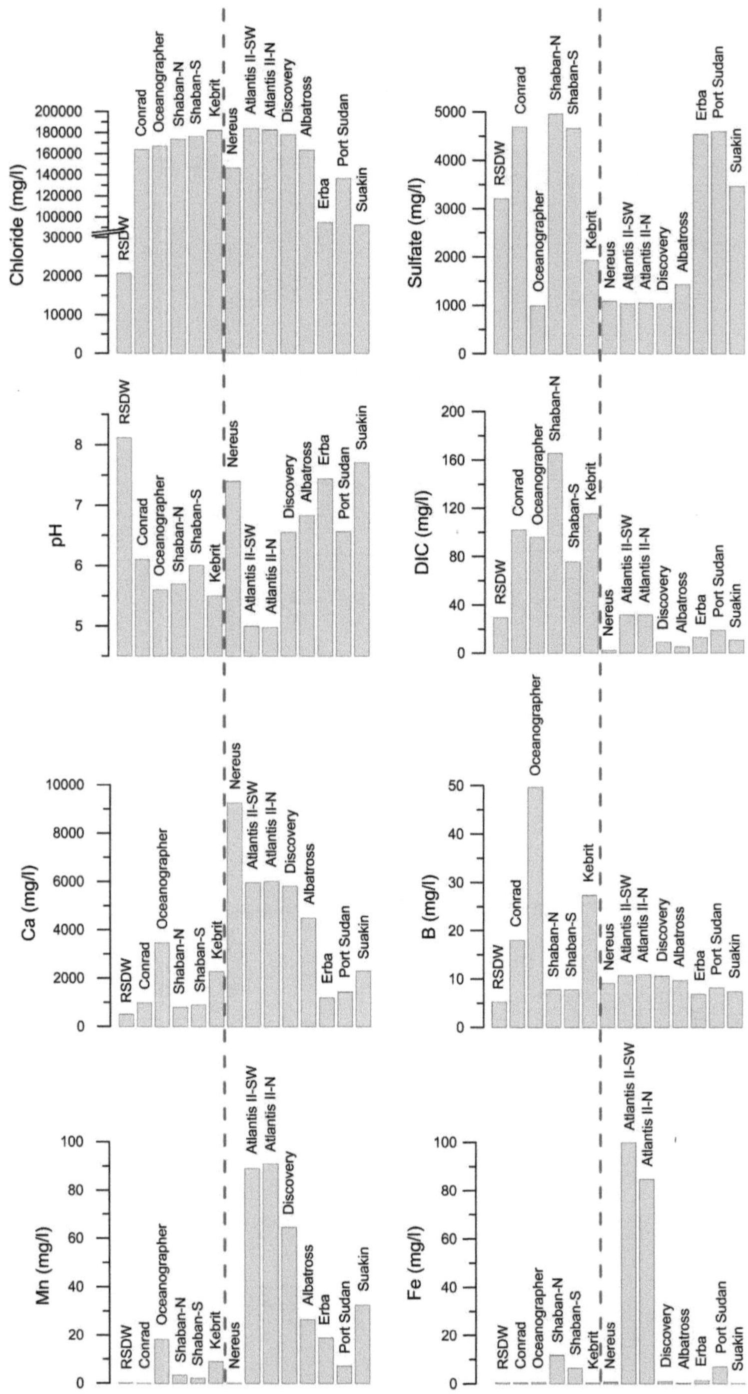

Fig. 3 Chemical composition of Red Sea brine (selected parameters). The deeps are organized along the *x*-axis from North (*left*) to South (*right*). The *red dotted line* marks a tectonic boundary separating the "multi-deeps and transitional region" from the "northern region" according to Bonatti (1985)

and brines with about a factor of 0.3 lower concentrations compared to RSDW, i.e., those found in the Oceanographer, Kebrit, Nereus, Atlantis II, Discovery, and Albatross Deeps (Fig. 3, Table 1). The level of SO_4 is possibly balanced (inversely correlated) by the Ca concentration in brine. This can be caused by gypsum (or anhydrite) precipitation when the solubility product of [Ca × SO4] is increased in acidic brine by, for example, carbonate dissolution. This might be the reason for large gypsum crystals which can be found at the brine sediment interface in the Kebrit Deep, where carbonaceous detritus can be dissolved from sediment (Scholten et al. 2000). However, in general, sulfate in brine can also be controlled by inorganic/microbial sulfate reduction or hydrothermal water/rock interaction in tectonically active regions such as the Red Sea (see discussion below).

The sulfur cycle in Red Sea brine plays also a major role in controlling the iron (II) concentration of brine, as, for example, iron released from water/rock alteration would

Table 2 Element composition of Atlantis II-SW brine (lower convective brine layer "LCL") within the last 4 decades

Year	Temperature (°C)	Fe (mg/kg)	Mn (mg/kg)	Cu (mg/kg)	Zn (mg/kg)	Reference
1965	55.9	58	72	0.830	2.5	Miller et al. (1966)
1965	55.9	81	82	0.260	5.4	Brewer and Spencer (1969)
1966	56.49	90	87	0.375	7.0	Brooks et al. (1969)
1971	59.17	86	82	0.075	3.5	Hartmann (1973)
1976	61.74	81	82	0.021	3.0	Danielsson et al. (1980)
1977	61.5	81	81	0.001	3.0	Hartmann (1985)
1997	67.2	74	84	n.m.	3.0	Garbe-Schönberg et al. (1999)
2011	68.2	74	78	n.m.	n.m.	Schmidt et al. (2011)
2012	68.1	74	79	0.059	3.5	Schmidt et al. (2013)

Concentrations for 1997–2012 are calculated by using a measured density of 1.19 (pycnometric determination)

precipitate immediately as iron sulfide due to the high concentration of hydrogen sulfide in the Kebrit or Oceanographer Deeps (Table 1). Alternatively, the hydrogen sulfide that in Atlantis II brine can be formed by sulfate reduction would precipitate as iron sulfide due to the extremely high iron (II) concentrations (Table 1, Fig. 3). The high temperatures of about 400 °C postulated for the hydrothermal circulation cell in the subsurface of the Atlantis II Deep (Winckler et al. 2000), or even brine temperatures of 68 °C (Table 1), would enhance the hydrothermal alteration processes of magmatic rocks (Seyfried 1987), releasing large amount of iron or manganese into the brine (Fig. 3, Table 1). Moreover, brine temperatures are controlling the solubility products of metal-rich minerals (e.g., pyrite), and amorphous components (e.g., ferrihydrite) of the underlying sediments (Scholten et al. 2000), or of the brine/seawater interfaces (Schmidt et al. 2003). For instance, the Zn concentrations of about 3.5 mg/l which were measured in Atlantis II Deep brine (Tables 1 and 2) fit nicely to the experimentally determined sphalerite solubility at the respective chlorinity and temperature conditions (Barrett and Anderson 1988).

Evaporite Dissolution

Miocene evaporite deposits of the Red Sea trough mainly consist of halite that can build up to 3,000-m-thick deposits. As documented in large coastal outcrops (e.g., El-Anbaawy et al. 1992), and in a few deep-sea drilling cores (i.e., Stoffers and Kühn 1974), the halite is intercalated by, for example, anhydrite, polyhalite, and organic-rich silt and clay stone layers. Assuming that sodium and chloride concentrations of Red Sea brine are mainly controlled by halite rehydration, the measured Na and Cl concentrations of brine should plot along a straight line in a sodium/chloride concentration cross-plot (Fig. 4). Most of our measured data (Table 1) plot along a theoretical line (dashed red line in Fig. 4). This line represents a calculated mixing line between Red Sea deep water and halite-saturated seawater (enrichment factor of about 10). Most of the various brine bodies show a rather uniform Na/Cl ratio of about 0.61 (molar ratios of 0.93–0.98), equal or slightly higher than the Red Sea deep water value of about 0.608 (molar ratio of 0.94). However, the Na/Cl ratio of Atlantis II and Nereus brine is about 0.9 and less (0.86), respectively. The negative deviations from the mixing line in Fig. 4 and molar ratios of Na/Cl between 0.86 and 0.9 most probably indicate minor Cl contributions from various congruent dissolving mineral phases intercalated in the evaporite deposits in the Atlantis II area, such as polyhalite, sylvite, and carnallite (Stoffers and Kühn 1974).

Hydrothermal Influence on Brine Chemistry

Numerous scientific observations have been published on hydrothermalism occurring worldwide in tectonically active regions such as mid-ocean ridges (e.g., Baker et al. 2002), mid-plate volcanoes (e.g., Hekinian et al. 2004), and arc/backarc systems (e.g., Reeves et al. 2011). Experimental and theoretical work on hydrothermal alteration processes was performed in order to study both the compositions of hydrothermal fluids (e.g., Seyfried 1987; von Damm et al. 1985) and to deduce their formation conditions.

The Red Sea is an area of active seafloor spreading where hydrothermal sediments, often covered by anoxic brine layers, can be found within seafloor depressions located along the central Red Sea graben (e.g., Scholten et al. 2000). In particular, thick Fe- and Mn-rich sediments (with associated trace metals) that may even be of economic interest are found on the bottom of the Atlantis II deep (e.g., Bertram et al. 2011). The individual brines within the various Red Sea deeps show hydrothermal influence to various intensities (e.g., Pierret et al. 2001). The present work, however, does not focus on individual brines and their complex formation conditions. We rather use selected element ratios in evaluating the overall hydrothermal influence on Red Sea deeps

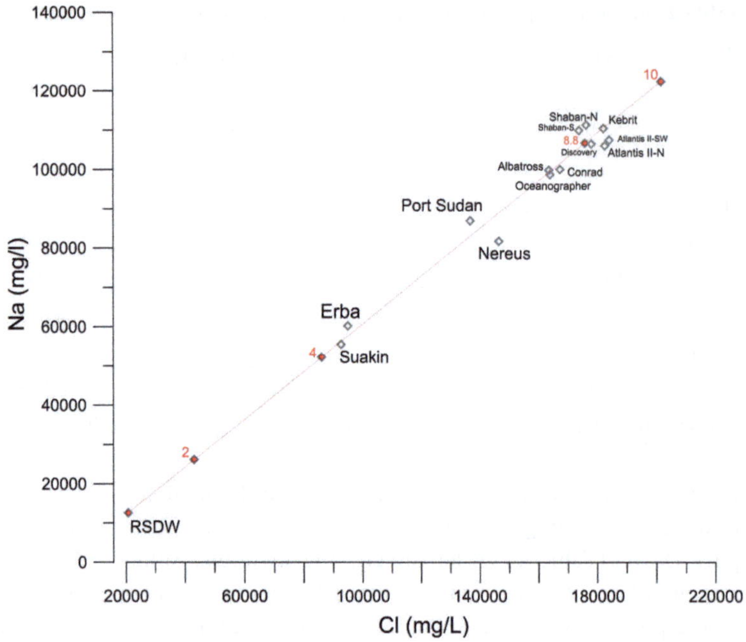

Fig. 4 Na and Cl concentrations of Red Sea brine are plotted along a mixing line between Red Sea deep water and nearly saturated RSDW (saturated with respect to halite at an enrichment factor of about 10). The enrichment factors (labeled in *red*) were calculated by using PHREEQC-Pitzer calculation (Parkhurst and Appelo 2013). The highest measured NaCl concentrations of Red Sea brines (Table 1), plot around the enrichment factor of 8.8, which is close to halite saturation

by summarizing geochemical data of a number of brine bodies in diagnostic plots (Figs. 5 and 6, Table 1). This is an oversimplification as, for instance, we do not consider or quantify variable hydrothermal influx and redox processes on short timescales, which also control hydrothermal trace concentrations (Fig. 2). We interpret the brine as a buffered system reflecting the general geological setting with slow element exchange processes as postulated by various other authors (e.g., Anschutz et al. 1999; Winckler et al. 2001). This rough approach to generalizing the different scenarios of hydrothermal influence on Red Sea brine seems to be even more justified, as the geochemical composition of the Atlantis II brine, which exhibits the strongest temperature change of all brine-filled Red Sea deeps, appears to be rather stable in its hydrothermal element composition within the last 4 decades as, for example, demonstrated by the Fe, Mn, and Zn content of Atlantis II LCL brine (Table 2).

Nevertheless, in order to minimize the influence of brine–seawater redox cycles in our evaluation of hydrothermal influences on brine chemistry, we selected boron/lithium and magnesium/calcium in order to distinguish between hydrothermal fluids from volcanic rock alteration and fluids mainly influenced by sedimentary strata. Both boron concentrations and isotopes are powerful tools in characterizing the origin of deep fluids (e.g., You and Gieskes 2001). The B content of Red Sea deep water is 5.3 mg/l (Table 1), whereas the B concentrations in the brines are significantly enriched up to 49.7 mg/l. This highest B value was measured in the brine of the Oceanographer Deep that is located off-axis in the northern Red Sea (Tooms et al. 1973), where deep subsurface sediment alteration is most likely to occur rather than large-scale element input through hydrothermal circulation. However, we cannot exclude, for example, boracite dissolution from evaporites being an additional boron and magnesium source in the Oceanographer, Kebrit, and Conrad Deeps (Table 1).

All other brine bodies do not show relevant changes of B with Mg (Table 1).

Another geochemical parameter that is used in classifying the various brines is the Mg/Ca ratio (Fig. 5). The Mg content of Red Sea deep water is 1,598 mg/l (Table 1). Pure hydrothermal fluids approach zero Mg concentrations as this element is consumed by subsurface water/rock reactions (e.g., clay formation, von Damm et al. 1985, and numerous others since). In contrast to Mg, the Ca content of hydrothermal brine from the Red Sea is probably of a more complex nature since the sources and sinks of Ca in brine are manifold, such as anhydrite dissolution/precipitation, detrital carbonate dissolution, or feldspar alteration. Monnin and Ramboz (1996) investigated hydrothermal influence on selected Red Sea brines. Their findings are mainly based on anhydrite saturation calculations of brine and interstitial waters, and general sedimentary observations. The authors concluded that anhydrite undersaturation seems to characterize brines with a hydrothermal signature and suggested that the Atlantis II, Discovery, and Suakin Deeps contain hydrothermal components. However, the situation of the latter deep still remains unclear as Pierret et al. (2001) on the base of their geochemical brine data concluded that the Suakin brine has no hydrothermal input. Nevertheless, Fig. 5 shows that the cross-plot of Mg/Ca ratios versus the B concentrations seems to be promising in classifying Red Sea brine. In accordance with the published ideas about Red Sea brine formation in general (e.g., Pierret et al. 2001;

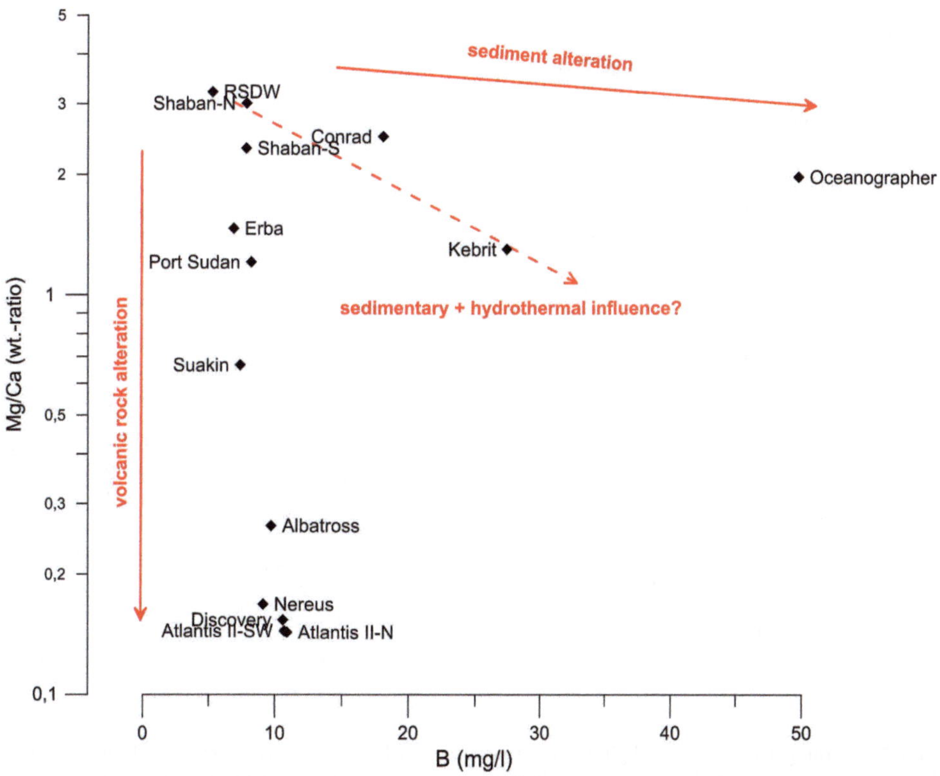

Fig. 5 Boron concentration versus Mg/Ca ratio is plotted for different Red Sea brine bodies. The increasing hydrothermal influence (volcanic rock and sediment alteration) is indicated by the *red arrows*

Winkler et al. 2001), the brines of our study show significant tendencies toward sediment-influenced systems, deep hydrothermal fluid sources (volcanic rock alteration) or mixtures of both. The brine classification scheme used in Fig. 5 also supports the conclusion of Monnin and Ramboz (1996) that the Suakin brine is influenced by hydrothermal processes.

In addition to our brine classification in Fig. 5, the plot of the B/Li molar ratios versus the Li concentrations (Fig. 6), which was established by Haffert et al. (2013) as classification scheme for fluids of different origins, also indicates that Red Sea brines are probably best explained by various mixtures of seawater with fluids that derived from sediment-covered high-temperature vents (HTV) and mid-ocean ridge (MOR) high-temperature vents, respectively.

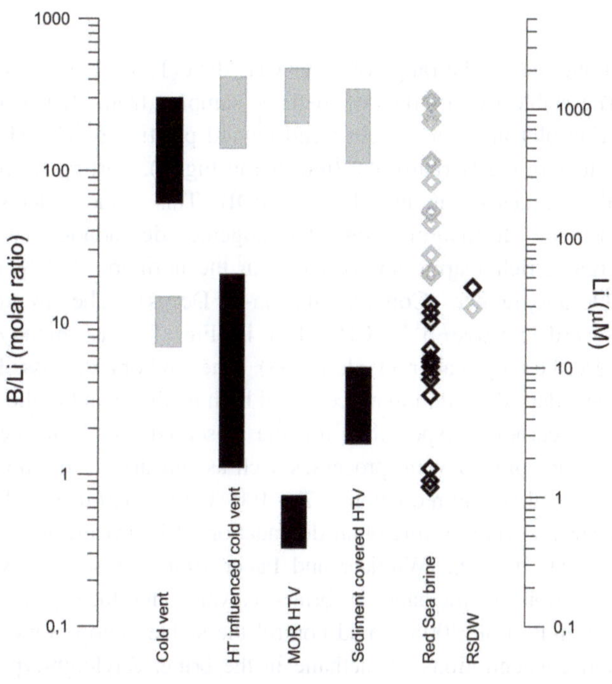

Fig. 6 Classification of vent fluids by using boron and lithium data according to Haffert et al. (2013). *Black symbols* and *bars* indicate B/Li ratios. *Gray-shaded symbols* indicate Li concentration data. Red Sea brine data cover the whole range from "mid-ocean ridge high-temperature vents" (MOR HTV), over "sediment-covered high-temperature vents" to "high-temperature influenced cold vent"

Hydrocarbon Formation and Degradation

A striking feature of Red Sea brines is their enrichment in light hydrocarbons (e.g., Burke et al. 1981; Faber et al. 1998; Schmidt et al. 2003). Methane concentrations between 4 and 9,600 µl/L were determined during this study (Table 1). The ratio of methane to higher hydrocarbons ($C_1/[C_2 + C_3]$) is in the range of 5–570. This ratio probably indicates a thermogenic origin of light hydrocarbons (Fig. 7), which is also

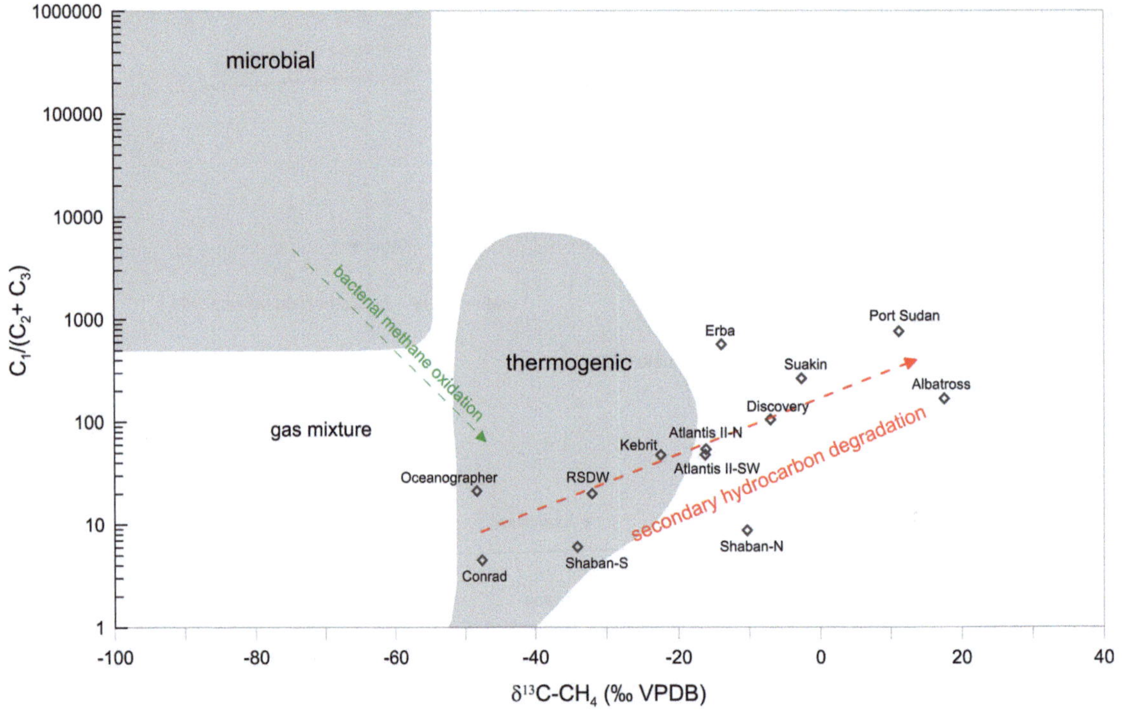

Fig. 7 Thermogenic sources of light hydrocarbon gases in Red Sea brine. The *red arrow* follows a trend indicating secondary C_1–C_3 oxidation processes with efficient ethane and propane degradation. The *green arrow* indicates the trend of bacterial methane oxidation that relatively enriches ethane and propane to methane (Whiticar and Faber 1986)

supported by the range of δ^{13}C–CH_4 data (Table 1, Fig. 7). The stable isotope data of methane sampled from Red Sea brine plot along a straight trend toward positive δ^{13}C–CH_4 values in the Bernard plot (red line in Fig. 7), starting from δ^{13}C-values of about −48 ‰ VPDB. This value reflects methane formation from thermogenic degradation of organic-rich marine source rocks in the northern Red Sea (Oceanographer, Conrad, Shaban-S Deeps). The trend toward positive δ^{13}C–CH_4 data in Fig. 7 was already described by Faber et al. (1998). The authors discussed secondary degradation processes of both methane and higher hydrocarbons responsible for the observed trend in the Bernard plot. Several processes such as subsurface thermochemical sulfate reduction at $T > 100$ °C (e.g., Krouse et al. 1988), secondary microbial degradation of hydrocarbons at $T < 60$ °C (e.g., Whiticar and Faber 1986), or secondary oxidation of methane at brine–seawater interfaces (e.g., Schmidt et al. 2003) could control the stable isotope ratios and concentrations of methane in the brine. Rayleigh-type isotope fractionation describes oxidation of methane in a concentration–isotope cross-plot (Fig. 8), by using a fractionation factor $\alpha = 1.027$ (Faber et al. 1998). One can follow the degree of degradation (100–0 % of residual methane) when starting at different levels of initial concentrations of methane (x-axis), and a uniform δ^{13}C-initial value of −48 ‰.

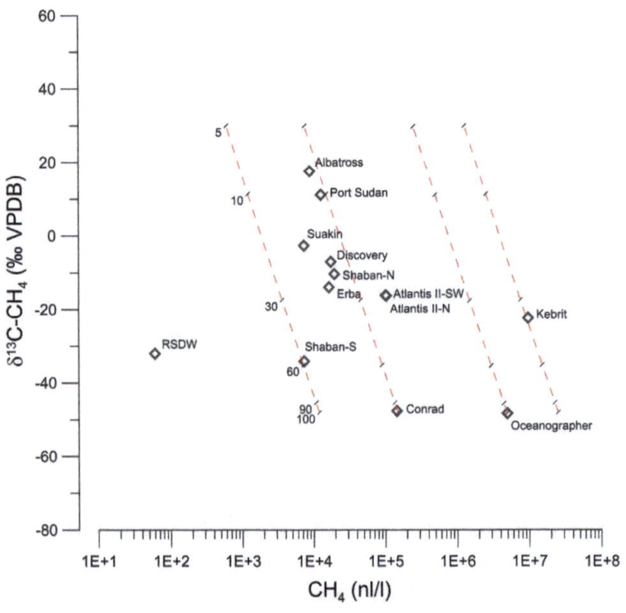

Fig. 8 Decrease in methane dissolved in brine bodies with increasing δ^{13}C–CH_4 values observed in Red Sea brines. Rayleigh fractionation (oxidation) lines (*dotted red lines*) starting with different concentration levels of methane but with the same carbon isotope signature (i.e., −48 ‰ VPDB). The calculated amount of residual methane is given in percent. The isotope fractionation is calculated with the fractionation factor $\alpha = 1.027$ (Faber et al. 1998)

A uniform initial value should be a good first approach; however, the same or similar source rock conditions (type of organic matter, maturity) at depth would be a prerequisite. Due to the lack of information about source rock composition along the Red Sea central axis, we avoid discussing it; however, most of the brine bodies of the Red Sea line up along one oxidation line starting with the Conrad Deep (northern Red Sea), in which a recently formed brine pool was discovered in 1999 (Pätzold et al. 2000). Following the isotope fractionation approach in Fig. 6, more than 70 % of an initial methane loading would have been oxidized to yield the methane concentrations and isotope values measured in brines located in the multi-deep area of the central Red Sea (Fig. 1). The highest initial natural gas loading is indicated for the Kebrit and Oceanographer Deeps (25 and 5 ml/L, respectively) that are located in the northern region (Fig. 1).

Brine Classification

Statistical Classification of Brine-Filled Deeps

The statistical cluster analyses of geochemical brine data, as was explained in Sect. "Data Collection and Evaluation of Selected Geochemical Parameters", indicate grouping of the investigated Red Sea deeps in this study into two neighboring types (Figs. 9 and 10). Type I group brine (i.e., Kebrit and Oceanographer brines, Fig. 9) is mainly determined by a high salinity, low pH, low temperature, and low trace metal concentrations (Table 1). The high amount of H_2S in Type I brine is probably the main reason for particularly low heavy metal concentrations in the brine (e.g., Weber and Gurskii 1982; Schmidt et al. 2003). Type I brine-filled deeps (Kebrit and Oceanographer Deeps) are located slightly off-axis with thick Plio-Quaternary sediment deposits on top of the Miocene evaporite (Ehrhardt 2005). Based on selected parameters, as already discussed above, hydrothermal fluid derived from water/volcanic rock interaction appears to be of minor importance. This is especially true as surface sediments in both deeps do not show indications of former hydrothermal element enrichments (Scholten et al. 2000). However, massive sulfides have been described for the brine/seawater sediment contact in the Kebrit Deep (Blum and Puchelt 1991).

A second group that is represented by the Atlantis II Deep clearly shows indications of hydrothermal influence (Type II, Fig. 10). High element concentrations such as Mn, Fe, and Zn are attributed to subsurface water/volcanic rock interaction which on the other hand also determines low Mg and low sulfate concentrations (Seyfried 1987). The Atlantis II brine that is known to reflect the strongest hydrothermal influence observed in Red Sea deeps exhibits also the highest temperatures measured in the lower convective brine layer (LCL) compared to all other brine-filled Red Sea deeps (Table 1, e.g., Hartmann et al. 1998a, b).

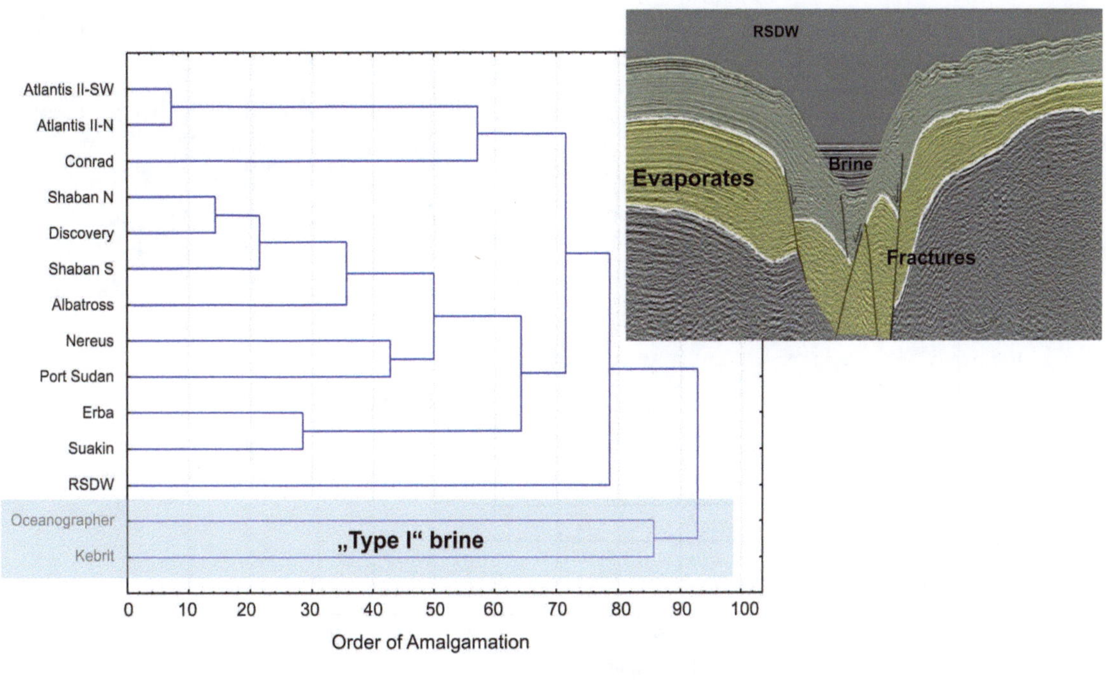

Fig. 9 Tree diagram grouping brine-filled Red Sea deeps by statistical cluster analysis of geochemical data (selected parameters from Table 1). Kebrit and Oceanographer Deeps (Type I brine) are linked to "collapse structure" Red Sea deeps based on the seismic data interpretation of Ehrhardt (2005). Seismic image of the Kebrit Deep is modified after Ehrhardt (2005)

Correlation of Geochemical and Seismic Characterization

Based on seafloor morphology and seismic reflection data acquisition, Ehrhardt (2005) presented a first approach to classifying "ocean deep" formations in the Red Sea. The Kebrit Deep, which is located above deep-reaching strike-slip faults, is characterized by a smooth subsidence of a sequence of Plio-Quaternary sediments on top of evaporate deposits (Ehrhardt 2005). Magmatic anomalies are absent in the Kebrit Deep area, and Ehrhardt (2005) classified the Kebrit depression as a "collapse-type deep," where hot fluid injection at deep-reaching faults could have initiated the subsurface evaporite dissolution (Fig. 9). Our geochemical brine characterization summarized in the previous chapters strengthens the idea about a "collapse-type" deep formation for both "Type I" deeps (i.e., the Kebrit and Oceanographer Deeps).

The Conrad Deep, which is influenced by a magmatic intrusion, and the Shaban Deep, which exhibits a magmatic ridge in its center, are classified as "intrusion-/extrusion-related" deeps (Ehrhardt 2005; Ehrhardt et al. 2005). The two deeps are geochemically described as "Type II" in the statistical tree diagram (Fig. 10). All deeps characterized by "Type II" brine show more or less strong indications for hydrothermal water/rock interaction that is probably linked to magmatic/volcanic activity. We do not want to discuss the different tectonic models of the various segments of the Red Sea rifting/spreading zones; however, our geochemical classification of brine-filled deeps aligned along the Red Sea central graben axis (19–27°N) would support the ideas of Bonatti (1985) that the deep formations could be a punctiform expression of the start of seafloor spreading in the Red Sea.

Conclusion

Various geochemical data evaluation approaches (i.e., concentration cross-plots, isotope/concentration plots, statistical cluster analyses) are applied on 13 investigated Red Sea brine bodies filling 11 depressions located in the central Red Sea graben between 19 and 27°N. Summarizing the measured element concentrations of the investigated brine bodies, the salt content in Red Sea brine varies between about 140 and 255 ‰. The brine chemistry is mainly controlled by halite dissolution documented by the high sodium and chloride concentration of up to 111 and 182 g/l, respectively, and molar Na/Cl ratios >0.94 (RSDW ratio).

However, the different brine bodies also exhibit variable degrees of influence by magmatic rock and sediment alteration. Based on boron, lithium, Ca, and Mg concentration data evaluation, the Kebrit and Oceanographer Deep brines are mainly influenced by its sedimentary setting, which means that warm fluids interacting with buried sediments changed the original evaporite solution chemistry. This is

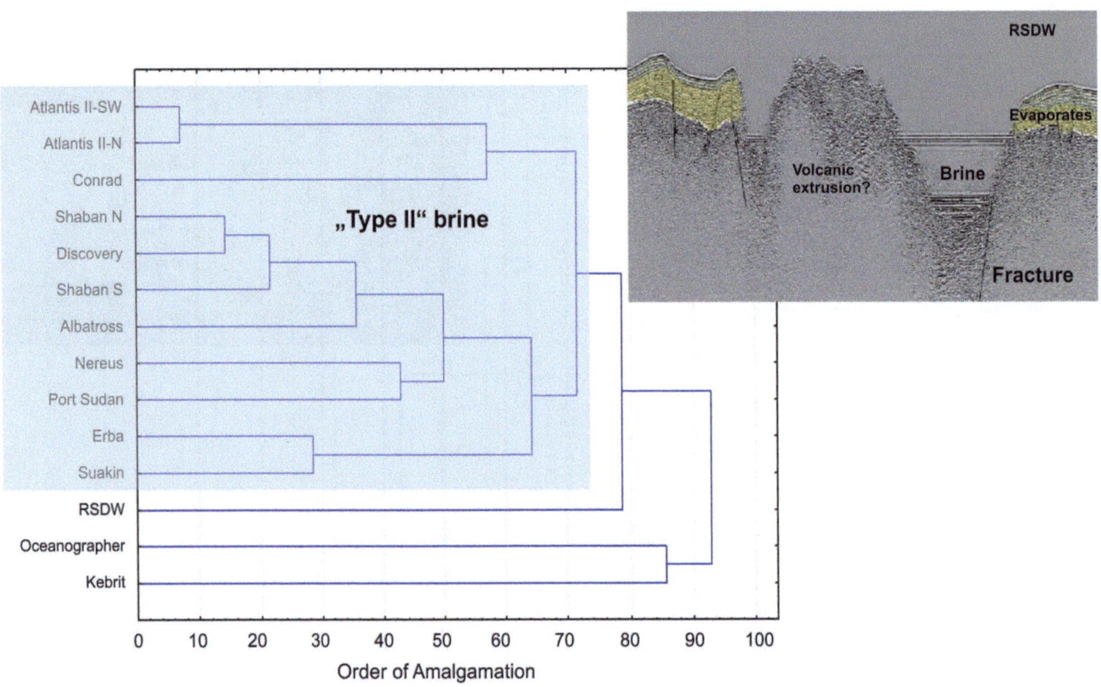

Fig. 10 Tree diagram grouping brine-filled Red Sea deeps by statistical cluster analysis of geochemical data (selected parameters from Table 1). Deeps filled with Type II brine (*blue shaded*) are linked to "intrusion-/extrusion-related" deep formation based on the seismic data interpretation of Ehrhardt (2005). Seismic image of the Shaban Deep is modified after Ehrhardt (2005)

also indicated by the high hydrocarbon content of the brine and its thermogenic origin from probably deeply buried marine source rocks. Moreover, secondary degradation of hydrocarbons by either subsurface thermogenic sulfate reduction or bacterial activity is obvious not only in the Kebrit Deep. Influences by hydrothermal volcanic/magmatic rock alteration on all other investigated brine bodies are demonstrated by a slight increase in boron concentrations with strong decreasing Mg/Ca ratios. Moreover, the high Li content (>1,000 μM) and low B/Li ratio, for example, in the Atlantis II brine, correspond to a published chemical classification for high-temperature hydrothermal fluids.

Mainly, two different brine groups can be inferred by statistical cluster analyses based on the measured 22 geochemical parameters. Two brine-filled deeps (Kebrit and Oceanographer Deeps) are related to a sedimentary low-temperature hosted system (Type I group). Nine brine-filled deeps (Suakin, Port Sudan, Erba, Albatross, Discovery, Atlantis II, Nereus, Shaban, Conrad) can be related to a high-temperature magmatic rock alteration system (Type II group). The Atlantis II Deep brine may define the end member of the Type II system, whereas all the other brine bodies related to the Type II group show minor influences by high-temperature magmatic rock alteration. The Type I and II brine systems based on the statistical geochemical classification can be correlated with a published geophysical/geomorphological classification scheme for the Red Sea depressions, which differentiates "collapse structure-related" and "magmatic/volcanic intrusion-/extrusion-related" Red Sea deeps. The geochemical and geophysical relationship of Type II deeps located in the northern Red Sea (e.g., Conrad and Shaban Deeps) to Type II deeps located in the multi-deep area (e.g., Atlantis II Deep) may support the idea that deep formation in the northern region is a punctiform expression of early seafloor spreading.

Acknowledgment Multilateral scientific, logistic, and political efforts during the last decade made it possible to collect this comprehensive geochemical data set for brine in the Red Sea. However, the main analytical work and data evaluation have been conducted during the last 3 years within the Jeddah Transect Project (www.jeddah-transect.org). The collaboration for the Jeddah Transect Project between King Abdulaziz University (KAU) Jeddah, Saudi Arabia, and the Helmholtz Center for Ocean Research GEOMAR, Kiel, was funded by King Abdulaziz University, under Grant no. T-065/430-DSR. The authors therefore acknowledge with thanks KAU technical and financial support. We also thank reviewers E. Faber and J. Scholten for their valuable suggestions and comments.

References

Anschutz P, Blanc G (1996) Heat and salt fluxes in the Atlantis II Deep (Red Sea). Earth Planet Sci Lett 142:147–159

Anschutz P, Blanc G, Chatin F, Geiller M, Pierret MC (1999) Hydrographic change during 20 years in the brine-filled basins of the Red Sea. Deep Sea Res 46:1779–1792

Antunes A, Ngugi DK, Stingl U (2011) Microbiology of the Red Sea (and other) deep-sea anoxic brine lakes. Environ Microbiol Rep 3 (4):416–433

Baker ET, Hey RN, Lupton JE, Resing JA, Feely RA, Gharib JJ, Massoth GJ, Sansone FJ, Kleinrock MC, Martinez F, Naar DF, Rodrigo C, Bohnenstiehl D, Pardee D (2002) Hydrothermal venting along Earth's fastest spreading center: East Pacific Rise, 27.5°–32.3° S. J Geophys Res 107(B7). http://dx.doi.org/10.1029/2001JB000651

Barrett TJ, Anderson GM (1988) The solubility of sphalerite and galena in 1–5 m NaCl solutions to 300 °C. Geochim Cosmochim Acta 52:813–820

Bertram C, Krätschell A, O'Brien K, Brückmann W, Proelss A, Rehdanz C (2011) Metalliferous sediments in the Atlantis II Deep—assessing the geological and economic resource potential and legal constraints. Resour Policy 36(4):315–329

Blum N, Puchelt H (1991) Sedimentary hosted poly-metallic massive sulfide deposits of the Kebrit and Shaban Deeps, Red Sea. Miner Deposita 26:217–227

Bonatti E (1985) Punctiform initiation of seafloor spreading in the Red Sea during transition from a continental to an oceanic rift. Nature 316:33–37

Bonatti E, Bortoluzzi G, Calafato A, Cipriani A, Ferrante V, Ligi M, Lopez Correa M, Redini F, Barabino G, Carminati E, Mitchell N, Sichler B, Schmidt M, Schmitt M, Rasul N, Al Nomani S, Bahareth F, Khalil S, Farawati R, Gitto D, Raspagliosi M (2005) Geophysical, geological and oceanographic surveys in the Northern Red Sea. Report on the morphobathymetric, magnetometric, oceanographic, coring and dredging investigations during cruise RS05 aboard R/V Urania. ISMAR-CNR Cataloging-In-Publication data: ISMAR Bologna technical report no. 94, 34 pp

Botz R, Schmidt M, Wehner H, Hufnagel H, Stoffers P (2007) Organic-rich sediments in brine-filled Shaban- and Kebrit Deeps, Northern Red Sea. Chem Geol 244:520–553

Brewer PG, Spencer D (1969) A note on the chemical composition of the Red Sea brines. In: Degens ET, Ross DA (eds) Hot brines and recent heavy metal deposits in the Red Sea. Springer, New York, pp 174–179

Brooks RR, Kaplan IR, Peterson MNA (1969) Trace element composition of the Red Sea geothermal brine and interstitial water. In: Degens ET, Ross DA (eds) Hot brines and recent heavy metal deposits in the Red Sea. Springer, New York, pp 180–203

Burke RA, Brooks JM, Sackett WM (1981) Light hydrocarbons in Red Sea brines and sediments. Geochim Cosmochim Acta 45:627–634

Craig H (1969) Geochemistry and origin of the Red Sea brines. In: Degens ET, Ross DA (eds) Hot brines and recent heavy metal deposits in the Red Sea. Springer, New York, pp 208–242

Danielsson LG, Dyrssen D, Graneli A (1980) Chemical investigations of Atlantis-II and Discovery brines in the Red Sea. Geochim Cosmochim Acta 44:2051–2065

Degens ET, Ross DA (1969) Hot brines and recent heavy metal deposits in the Red Sea: a geochemical and geophysical account. Springer, New York, 600 pp

Ehrhardt LA (2005) Seismic and hydroacoustic studies of surficial sediment tectonics along the northern Red Sea Rift and the Dead Sea Transform fault. GEO-LEO e-docs/Geophysik, 126 pp

Ehrhardt A, Hübscher C, Gajewski D (2005) Conrad Deep, Northern Red Sea: development of an early stage ocean deep within the axial depression. Tectonophysics 411:19–40

El-Anbaawy MIH, Al-Aawah MAH, Al-Thour KA, Tucker ME (1992) Miocene evaporites of the Red Sea rift, Yemen Republic: sedimentology of the Salif halite. Sed Geol 81:61–71

Faber E, Botz R, Poggenburg J, Schmidt M, Stoffers P, Hartmann M (1998) Methane in Red Sea brines. Org Geochem 29(1–3):363–379

Garbe-Schönberg CD, Scholten J, Stoffers P, Moammar M (1999) Trace element chemistry of brines and brine-interfaces in the

Atlantis-II, Chain, and Discovery Deep in the Red Sea. In: Second symposium on Red Sea marine environment, Jeddah. King Abdulaziz University, Saudi Arabia

Girdler RW, Evans TR (1977) Red Sea heat flow. Geophys J Int 51 (1):245–251

Gurvich EG (2006) Metalliferous sediments of the World Ocean fundamental theory of Deep-Sea hydrothermal sedimentation. Springer, Heidelberg, 416 pp

Haffert L, Haeckel M, Liebetrau V, Berndt C, Hensen C, Nuzzo M, Reitz A, Scholz F, Schönfeld J, Perez-Garcia C, Weise SM (2013) Fluid evolution and authigenic mineral paragenesis related to salt diapirism—the Mercator mud volcano in the Gulf of Cadiz. Geochim Cosmochim Acta 106:261–286

Hardie LA (1991) On the significance of evaporites. Annu Rev Earth Planet Sci 19:131–168

Hartmann M (1973) Untersuchungen von suspendiertem Material in den Hydrothermallaugen des Atlantis-II-Tiefs. Geol Rundsch 62:742–754

Hartmann M (1985) Atlantis-II Deep geothermal brine system. Chemical processes between hydrothermal brines and Red Sea Deep Water. Mar Geol 64:157–177

Hartmann M, Scholten JC, Stoffers P, Wehner F (1998a) Hydrographic structure of brine-filled deeps in the Red Sea—new results from the Shaban, Kebrit, Atlantis II, and Discovery Deep. Mar Geol 144:311–330

Hartmann M, Scholten JC, Stoffers P (1998b) Hydrographic structure of brine-filled deeps in the Red Sea: correction of Atlantis II Deep temperatures. Mar Geol 144:331–332

Hekinian R, Cheminee JL, Stoffers P (2004) Oceanic hotspots: intraplate submarine magmatism and tectonism. Springer, Berlin, 480 pp

Hovland M, Kuznetsova T, Rueslåtten H, Kvamme B, Johnsen HK, Fladmark GE, Hebach A (2006) Sub-surface precipitation of salts in supercritical seawater. Basin Res 18:221–230

Krouse HR, Viau CA, Eliuk LS, Ueda A, Halas S (1988) Chemical and isotopic evidence of thermochemical sulphate reduction by light hydrocarbon gases in deep carbonate reservoirs. Nature 333:415–419

Krupp RE (2005) Formation and chemical evolution of magnesium chloride brines by evaporite dissolution—implications for evaporite geochemistry. Geochim Cosmochim Acta 69(17):4283–4299

Miller AR, Densmore CD, Degens ET, Hathaway JC, Manheim FT, McFarlin PF, Pocklington R, Jokela A (1966) Hot brines and recent iron deposits in deeps of the Red Sea. Geochim Cosmochim Acta 30:341–359

Monnin C, Ramboz C (1996) The anhydrite saturation index of the ponded brines and sediment pore waters of the Red Sea deeps. Chem Geol 127:141–159

Parkhurst DL, Appelo CAJ (2013) Description of input and examples for PHREEQC version 3—a computer program for speciation, batch-reaction, one-dimensional transport, and inverse geochemical calculations. U.S. Geological Survey Techniques and Methods, Book 6, Chap. A43, 497 pp. Available only at http://pubs.usgs.gov/tm/06/a43/

Pätzold J, Halbach PE, Hempel G, Weikert H (2000) Östliches Mittelmeer—Nördliches Rotes Meer 1999, Cruise No. 44, 22 January–16 May 1999. METEOR-Berichte, Universität Hamburg, 00-3, 240 pp

Pätzold J, Bohrmann G, Hübscher C (2003). Black Sea—Mediterranean—Red Sea, Cruise No. 52, January 2–March 27, 2002. METEOR-Berichte, Universität Hamburg, 03-2, 178 pp

Pierret MC, Bosch D, Clauer N, Blanc G, France-Lanord C (2001) Chemical and isotopic (87Sr/86Sr, δ18O, δD) constraints to the formation processes of Red-Sea brines. Geochim Cosmochim Acta 65:1259–1275

Reeves EP, Seewald JS, Saccocia P, Bach W, Craddock PR, Shanks WC, Sylva SP, Walsh E, Pichler T, Rosner M (2011) Geochemistry of hydrothermal fluids from the PACMANUS, Northeast Pual and Vienna Woods hydrothermal fields, Manus Basin, Papua New Guinea. Geochim Cosmochim Acta 75(4):1088–1123

Schenk CJ, Charpentier RR, Klett TR, Brownfield ME, Kirschbaum MA, Pitman JK, Cook TA, Tennyson ME (2010) Assessment of undiscovered oil and gas resources of the Red Sea Basin Province. U.S. Geological Survey Fact Sheet 2010-3119, 2 pp

Schmidt M, Botz R, Faber E, Schmitt M, Poggenburg J, Garbe-Schönberg D, Stoffers P (2003) High-resolution methane profiles across anoxic brine-seawater boundaries in the Atlantis-II, Discovery, and Kebrit deeps (Red Sea). Chem Geol 200:359–376

Schmidt M, Botz R, Aeschbach-Hertig W, Bayer R, Schmitt M, Boettcher M, Stoffers P, Bonatti E (2006) Biogeochemistry of brines in the Northern Red Sea. EGU General Assembly 2006, Vienna. Geophysical research abstracts, vol 8. EGU06-A-04400

Schmidt M, Devey C, Eisenhauer A (2011) FS Poseidon Fahrtbericht/Cruise Report P408—the Jeddah Transect; Jeddah—Jeddah, Saudi Arabia, 13.01.-02.03.2011 IFM-GEOMAR report 46. IFM-GEOMAR, Kiel, 80 pp

Schmidt M, Al-Farawati R, Al-Aidaroos A, Kürten B (2013) RV PELAGIA Fahrtbericht/Cruise Report 64PE350/64PE351—JEDDAH-TRANSECT; 08.03.-05.04.2012 Jeddah-Jeddah, 06.04-22.04.2012 Jeddah-Duba. GEOMAR Report, N. Ser. 005, GEOMAR Helmholtz Centre for Ocean Research Kiel, 154 pp. doi:10.3289/GEOMAR_REP_NS_5_2013

Schoell M, Faber E (1978) New isotopic evidence for the origin of Red Sea brines. Nature 275:436–438

Scholten JC, Stoffers P, Garbe-Schönberg D, Moammar M (2000) Hydrothermal mineralization in the Red Sea. In: Cronan DS (ed) Handbook of Marine Mineral Deposits. CRC Press, Boca Raton, pp 369–395

Seyfried WE Jr (1987) Experimental and theoretical constraints on hydrothermal alteration processes at mid-ocean ridges. Annu Rev Earth Planet Sci 15:317–335

Simoneit BRT, Grimalt JO, Hayes JM, Hartman H (1987) Low temperature hydrothermal maturation of organic matter in sediments from the Atlantis II deep, Red Sea. Geochim Cosmochim Acta 51 (879):894

Stoffers P, Kühn R (1974) Red Sea evaporites: a petrographic and geochemical study. In: Whitmarsh RB, Weser OE, Ross DA et al (eds) Initial reports of the Deep Sea Drilling Project, vol 23. U.S. Government Printing Office, Washington DC, pp 821–847

Swift SA, Bower AS, Schmitt RW (2012) Vertical, horizontal, and temporal changes in temperature in the Atlantis II and Discovery hot brine pools, Red Sea. Deep-Sea Res I 64:118–128

Tooms JS, Holmes R, Horowitz A (1973) Confirmation of Ostapoff's fourth brine hole, Red Sea. Nat Phys Sci 241:161–162

Von Damm KL (1995) Controls on the chemistry and temporal variability of seafloor hydrothermal fluids. In: Humphris SE, Zierenberg RA, Mullineaux LS, Thomson RE (eds) Seafloor hydrothermal systems: physical, chemical, biological, and geological interactions. AGU monograph series no. 91, American Geophysical Union, Washington DC, pp 222–247

Von Damm KL, Edmond JM, Grant B, Measures CI, Walden B, Weiss RF (1985) Chemistry of submarine hydrothermal solutions at 21°N, East Pacific Rise. Geochim Cosmochim Acta 49(11):2197–2220

Wallmann K, Suess E, Westbrook GH, Winckler G, Cita MB, and the MEDRIFF consortium (1997) Salty brines on the Mediterranean seafloor. Nature 387:31–32

Wang Y, Yang J, Lee OO, Dash S, Lau SCK, Al-Suwailem A, Wong TYH, Danchin A, Qian PY (2011) Hydrothermally generated aromatic compounds are consumed by bacteria colonizing in Atlantis II Deep of the Red Sea. Int Soc Microb Ecol J 5:1652–1659

Weber WW, Gurskii YN (1982) Maltene formation in present sediments of the Kebrit brine depressions of the Red Sea. Geologiya Nefti Gaza 1:29–33

Whiticar MJ, Faber E (1986) Methane oxidation in sediment and water column environments: isotope evidence. Org Geochem 10:759–768

Winckler G, Kipfer R, Aeschbach-Hertig W, Botz R, Schmidt M, Schuler S, Bayer R (2000) Sub sea floor boiling of Red Sea Brines—new indication from noble gas data. Geochim Cosmochim Acta 64:1567–1575

Winckler G, Aeschbach-Hertig W, Kipfer R, Botz R, Rübel AP, Bayer R, Stoffers P (2001) Constraints on origin and evolution of Red Sea brines from helium and argon isotopes. Earth Planet Sci Lett 184:671–683

You CF, Gieskes JM (2001) Hydrothermal alteration of hemi-pelagic sediments: experimental evaluation of geochemical processes in shallow subduction zones. Appl Geochem 16:1055–1066

Hydrothermal Activity and Paleoenvironments of the Atlantis II Deep

Pierre Anschutz

Abstract

The Atlantis II Deep is a 65 km^2 topographic depression located in the axial trough of the Red Sea at 2,000 m depth. The depression traps 17 km^3 of hot and dense brines fed by hydrothermal fluids. This chapter reviews numerous data collected during the last 50 years. Chemical and isotopic data suggest that the processes that lead to the formation of the Atlantis II Deep brines are similar to those that produce open ridge black smoker fluids, but the recharging fluid is sea water in the case of sediment-free ridges, whereas it is sea water that has dissolved evaporites in the case of the Atlantis II Deep. The monitoring of temperature indicates that the heat flux was 0.54×10^9 W between 1965 and 1995. After 1995, the heat flux became 10 times lower. The substratum of the Atlantis II Deep consists of MORB-type basalts, which are covered with 0- to 30-m-thick metalliferous sediments. The solid fraction contains biogenic calcareous and/or siliceous components and silico-clastic detrital particles diluted by metalliferous sediment, which consists of metal oxides, sulphides, carbonates, sulphates, and silicates that precipitated from the hydrothermal fluids. The redox interface between the deepest brine layer and sea water is a major place of mineral precipitation. During glacial periods before the Holocene, the redox boundary was located above the brine–sea water boundary, so that hydrothermal metals spread over a large area of the Red Sea bottom.

Introduction

The Atlantis II Deep is one of the most remarkable seafloor settings of the planet. It is the largest known present-day submarine hydrothermal environment, consisting of a 65-km^2 topographic depression located in the axial trough of the Red Sea at 2,000 m depth (Fig. 1). The depression traps hot and dense brines fed by hydrothermal fluids, the sources of which have not yet been sampled. Base and precious metal mineralization is contained with a 10- to 30-m-thick sequence of sediments. The Atlantis II Deep was discovered in 1963 (Miller 1964), and hydrographic parameters were measured from 1966 onwards (Degens and Ross 1969). Prior to 1970, the Atlantis II Deep was the only known seafloor hydrothermal mineral deposit. High-temperature hydrothermal venting sources on a sediment-free mid-ocean ridge, forming "black smokers", were discovered on the East Pacific Rise in 1979 (Spiess et al. 1980). After 35 years of exploration using increasingly sophisticated oceanographic technology, hydrothermal activity is known to discharge hot fluids at various localities all along the 60,000 km mid-oceanic ridge system (Rona 2008).

Several major scientific findings have resulted from the study of mid-ocean ridge hydrothermal systems that have contributed to our understanding of processes of ore mineral formation, sea water–oceanic crust interaction, and the influence of these processes on global seawater chemistry. Hydrothermal circulation results from the penetration of cold sea water through hot and fractured oceanic crust that is created at the axis of spreading ridges. As sea water circulates close to a heat source (either a magma chamber or hot rock), its chemical composition is strongly modified through reaction with the volcanic rocks, leading to the formation of

P. Anschutz (✉)
Laboratory EPOC, CNRS, University of Bordeaux, 33405 Talence, Bordeaux, France
e-mail: pierre.anschutz@u-bordeaux.fr

N.M.A. Rasul and I.C.F. Stewart (eds.), *The Red Sea*,
Springer Earth System Sciences, DOI 10.1007/978-3-662-45201-1_14,

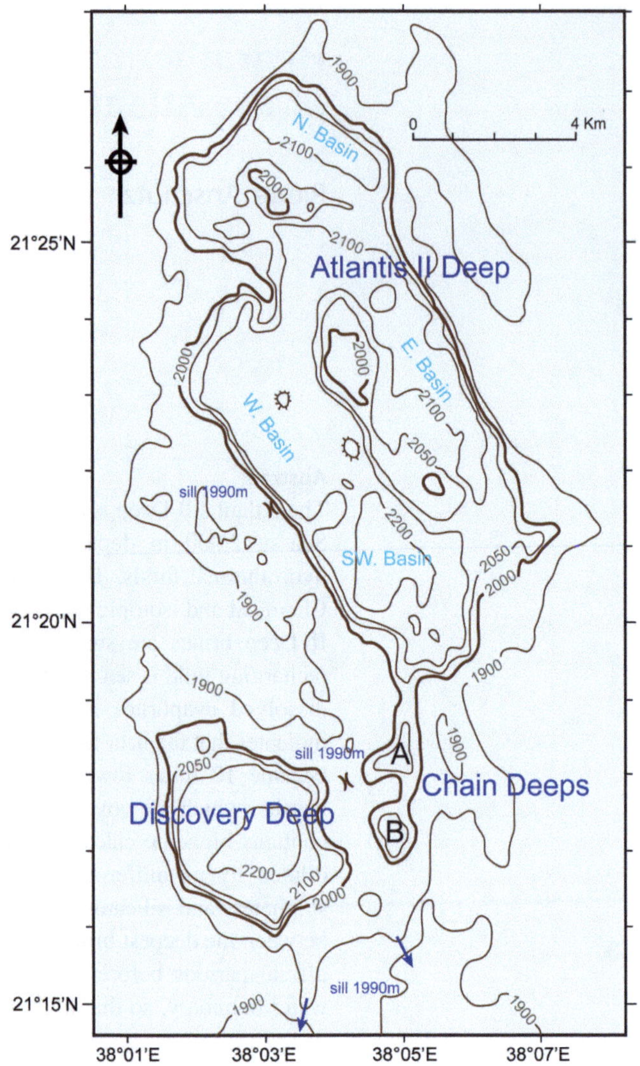

Fig. 1 Bathymetric map of the Atlantis II Deep area. SW, W, E, N: South-western, western, eastern, and northern basins. Contour lines have been traced from Seabeam investigations (Pautot 1983) and corrected with sound velocity calculations for Red Sea water and brines. The 2,000 m contour line corresponds to the top of the present-day layered brine-filled deep. The 2,050 m line shows the extension of the Atlantis II Deep lower brine (LCL). Southward *arrows* indicate the direction of the overspill of the dense brine pool, based on the present-day topography

hydrothermal fluids. The hot, anoxic, and metal-enriched fluids rise rapidly back to the seafloor where they mix with the ambient sea water, forming chimneys by the precipitation of metal sulphides (Shanks 2001; German and Von Damm 2003). A wide range of mineral deposits consisting of massive sulphide mounds and accumulations of Fe–Mn oxyhydroxides and silicates result from this hydrothermal discharge, both along the global ridge crest and at subduction zones (Koski et al. 2003).

The Atlantis II Deep hydrothermal system is unique in that fluids circulate not only through the crust but also through evaporites. This creates hydrothermal fluids that are dense brines and are trapped in a topographic depression. Monitoring the temperature and salinity provides precise estimates of heat and salt fluxes. Secondly, hydrothermal minerals form stratified muddy metalliferous sediments that allow us to study the evolution of hydrothermal activity over several thousand years.

Methods

This chapter reviews numerous data collected during the last 50 years. Hydrography of the Atlantis II Deep has been monitored intermittently with the most recent paper reporting complete T and S profiles (Swift et al. 2012). Prior to that, the last measurements were made by the RV Sonne in 1997. The main hydrographical parameters of concern are temperature, salinity, and the depth of brine layers. Temperature data of the different investigations can be confidently compared. Salinity data are more difficult to compare, because they have been obtained through different methods, such as conductivity (Brewer et al. 1969), chlorinity Mohr–Knudsen titration (e.g. Hartmann et al. 1998; Schmidt et al. 2003), or refractometry (e.g. Blanc and Anschutz 1995), and they appear either in volumetric units or in mass units. Data on chlorinity have been converted into salinity in ‰ using the Danielsson et al.

(1980) formulae, where $S = 1.67\ Cl + 4.02$. Volumetric units (g/L or mol/L) have been converted into mass units (g/kg = ‰, or mol/kg) using density calculations for NaCl aqueous solutions. All these calculations and conversions may induce an error greater than 3 ‰ in salinity data.

Some of the most remarkable features of the Atlantis II brine pool are the sharp transitions between the convective brine layers, where density gradients occur across less than 1 m vertical distance. The absolute depth of these transitions, which allows calculation of the volume of each brine layer, has been determined from acoustic data measurements corrected for the higher sound velocities in the dense brines (Hartmann 1980; Anschutz and Blanc 1996), from the length of the CTD cable (e.g. Bubnov et al. 1977), or directly from the pressure measurement with the CTD (e.g. Schoell and Hartmann 1973). Therefore, the position of the transition layers can be compared from one study to another as long as a margin of error of 1 or 2 m is taken into account.

The metalliferous sediments that underlie the brine pool were studied extensively during the 1970s to evaluate their economic potential. Hundreds of sediment cores were collected with the research vessels Wando River and Valdivia in 1969, 1971, and 1972 and enabled Bäcker and Richter (1973) to establish a reference stratigraphy for the metalliferous sediments. Two cores were collected in 1985 during the hydrotherm cruise on the RV Marion Dufresne (Blanc et al. 1986) and were very representative of the Bäcker and Richter stratigraphy. Core 683 was collected in the southwest basin at 2,174 m depth (21° 20.75′ N, 38° 04.85′ E) and core 684 in the western basin at 2,110 m depth (21° 22.30′ N, 38° 03.55′ E). Both cores reached the basaltic basement and contained the entire sedimentary sequence, without any of the breccia or slump deposits described in some other cores of the Atlantis II Deep (Bäcker and Richter 1973). Data presented here come mostly from the cores 683 and 684, because they were the subject of numerous detailed studies of mineralogy and chemical composition (Badaut et al. 1990; Anschutz and Blanc 1995a, b; Blanc et al. 1998), micropalaeontology (Anschutz and Blanc 1993), pore water composition (Anschutz et al. 2000), and isotopic composition (Dupré et al. 1988; Anschutz et al. 1995; Rouxel et al. 2012).

Formation of Atlantis II Deep Brines

Composition of the deep brine

The Atlantis II Deep hot brine is an anoxic, metal-rich, salty solution isolated from sea water because of its high in situ density of 1,186 ± 5 kg m^{-3} (Anschutz et al. 1998). The chemical composition of the brine reflects the origin of the hydrothermal solution and the chemical processes that occur within the brine or at the brine–sea water interface. The salinity values of the brine measured between 1966 and 2010 were obtained from different techniques. Published values range up to 250 and 270 ‰. The chloride concentration is 158 g kg^{-1} (Hartmann et al. 1998) or 4.47 ± 0.06 mol kg^{-1} (Anschutz et al. 2000; Schmidt et al. 2003). The Na^+ concentration is 4.17 ± 0.04 mol kg^{-1}. This is about 7 times higher than Red Sea bottom water. Compared to the seawater major salt content, the hot brine is also enriched in dissolved calcium, potassium, and strontium, but it is depleted in dissolved magnesium and sulphate. Table 1 shows some properties of sea water, of Atlantis II Deep brines, and of an East Pacific Rise black smoker (Von Damm 1990). Black smoker fluids are hot, and the salinity remains close to that of sea water. There are many examples of hydrothermal vents with salinities much higher and much lower than sea water due to phase separation at depth (German and Von Damm 2003). However, salinities as high as that of the Atlantis II brine never have been recorded for open ridge hydrothermal solutions. The hydrothermal end member is highly depleted in dissolved magnesium and sulphate and enriched in calcium and dissolved metals. The isotopic ratio of dissolved strontium is lower than that of sea water and close to that of oceanic crust basalts. Except for salinity, the brines of the Atlantis II Deep show similar trends to open ridge hydrothermal solutions. This suggests that brines are derived from the leaching of hot basalts by salty solutions. Salt probably comes from interstitial waters of Miocene evaporite, which is a major formation of the Neogene sedimentary sequence of the Red Sea rift. Helium and argon isotopes measured in the brine have a mantle-derived origin (Lupton et al. 1977; Winckler et al. 2001), which confirms that fluids that feed the Atlantis II Deep have interacted with newly injected fresh basaltic material.

Table 1 Chemical composition of Red Sea deep water, Atlantis II Deep lower brine, and a black smoker of the East Pacific Rise in mmol/kg

	T (°C)	Cl^-	Ca^{2+}	Mg^{2+}	SO_4^{2-} (mmol/kg)	Fe^{2+}	Mn^{2+}	Sr^{2+}	$^{87}Sr/^{86}Sr$
Red Sea water[a]	21.9	626	10.7	56.3	33.0	<	<	0.12	0.70916
Atlantis II brine[a]	62	4528	116	24.6	7.59	0.99	1.49	0.57	0.70710
EPR 21°N[b]	350	489	21.5	0	0	0.8–2.4	0.6–1.0	0.09	0.703

[a] 1985 data (Anschutz and Blanc 1995; Anschutz et al. 1995)
[b] Von Damm (1990)

The strontium concentration and the $^{87}Sr/^{86}Sr$ ratio of the hot brine in 1985 were 0.68 mmol L^{-1} and 0.7070, respectively (Blanc et al. 1995). Similar values were obtained from samples collected in 1978 (Zierenberg and Shanks 1986) and in 1992 (Pierret et al. 2001). The strontium isotopic ratio is between that of Red Sea bottom water (0.70916) or interstitial waters of upper Miocene evaporite samples recovered from DSDP Site 227 (0.70894, Zierenberg and Shanks 1986), and basalts from the Red Sea axial rift (0.70292 ± 0.00023, Eissen et al. 1989; Volker et al. 1993; Ligi et al.; this volume). A mass balance based on these data indicates that the hot brine Sr can stem from interstitial waters of the Miocene evaporites that leached hot basalts with a water–rock ratio of 2 or 3, which is similar to some values calculated for oceanic hydrothermal systems on open ridges such as the East Pacific Rise (Albarède et al. 1981). Hence, the processes that lead to the formation of the Atlantis II Deep brines are similar to those that produce black smoker fluids, but the recharging fluid is sea water in the case of sediment-free ridges, whereas it is sea water that has dissolved evaporites in the case of the Atlantis II Deep. While a recent input of fluid into the Atlantis II Deep is indicated by the temperature and salinity evolution of the brines, the exact site of the hydrothermal brine venting has never been sampled. The nature of the metalliferous sediments that cover the bottom of the deep and the distribution of the Mn/Fe ratio dissolved in the lower brine suggest that the hydrothermal vent source is located in the south-west basin (Bäcker and Richter 1973; Zierenberg and Shanks 1983; Oudin et al. 1984). The deepest parts of the eastern and the northern basins have depressions in which the temperature of the brine can be more than 5 °C lower than that of the main brine pool (Schoell and Hartmann 1973; Blanc and Anschutz 1995; Stoffers et al. 1998). These cooler brines are trapped in the deep parts because they are dense. They probably represent relict brines from the recent past. Since the salinity is the same as the overlying hot brine, the salinity most likely remained constant during the modern episode of hydrothermal activity, which suggests that the fluid that fed the system had the salinity of the brine.

The ionic strength of the brine sampled in 1985 was 6.28 (Anschutz et al. 2000). Because of this high value, the calculation of the saturation state of the brine with respect to solid phases needs thermodynamic calculations with a water–rock interaction model based on Pitzer's ion interaction approach (Pitzer 1979; Greenberg and Moller 1989). The saturation indices reveal that the brine is undersaturated with respect to halite, gypsum, and anhydrite at in situ temperature (Monnin and Ramboz 1996; Anschutz et al. 2000), suggesting that the major salt concentrations are not controlled by in situ equilibrium of the brine pool with respect to salt minerals. The brine is close to equilibrium with halite at a temperature close to that of Red Sea bottom waters (22 °C), which supports the hypothesis that the salinity of the incoming brine has been reached through equilibrium with evaporite minerals at seawater temperature before the heating of the solution through fractured Red Sea basaltic crust.

Although the ventilation of Red Sea bottom waters is restricted to deep-water formation in the Gulf of Suez during winter (Maillard 1974; Sofianos and Johns 2002), Red Sea deep waters are oxic with values between 100 and 130 µmol L^{-1} dissolved O_2 at 1,900 m depth, just above the Atlantis II Deep (Weikert 1982; Hartmann et al. 1998; Sofianos and Johns 2007; Swift et al. 2012). The lower brine is anoxic but does not contain dissolved sulphides. The Atlantis II Deep is a unique spot in the ocean that is anoxic, but not euxinic. The absence of dissolved O_2 and S(-II) forms allows dissolved transition metals to accumulate in the brine. Dissolved Fe^{2+} and Mn^{2+} reach millimolar levels, and dissolved Zn reaches several tens of micromoles per litre (Hartmann 1985; Pierret et al. 2001).

Structure of the Brine Pool in the Atlantis II Deep

The brines in the Atlantis II Deep occur in several horizontally uniform, well-mixed layers, with the hottest and saltiest water at the bottom, separated from the successively cooler and fresher layers above by sharp interfaces. The observed layering is due to double-diffusive convection (Turner 1969). The Atlantis II Deep is one of the outstanding locations in the world where this phenomenon can be observed. The brine layers are called lower convective layer (LCL) for the deepest one and upper convective layers 1, 2, 3, and 4 (UCL1, UCL2, UCL3, and UCL4) for the shallower brines. In 2008, the LCL brine was at a temperature of 68.3 °C and at a salinity of 252 ‰ (Swift et al. 2012). The brine extended from the deepest part of the deep (2,200 m depth) to 2,048 ± 1 m depth (Fig. 1). The position of this LCL–UCL1 transition zone has not changed significantly during the last 45 years (Table 2). The surface area of this interface is 51.7 km^2, according to the Seabeam map obtained during the MEROU cruise of the RV Jean Charcot in 1978 (Pautot 1983) after correction of sound velocities along the water column (Anschutz and Blanc 1996), corresponding to a volume of hydrothermal brine of 3.94 km^3. The UCL1 has been monitored since 1966. Its upper limit was at about 2,020 m depth until 1992, and its surface area was 60.2 km^2. The volume of this brine body was 1.4 km^3. The upper interface was at 2,027 m depth in 2008. UCL1 has a salinity and temperature intermediate between LCL and sea water. Thinner well-mixed convective layers occur up to 2,000 m depth. The surface area of the deep at this depth is about 65.3 km^2. It corresponds to the topographic limit of

Table 2 Temperatures, salinities, and depths of the brine interfaces of the Atlantis II Deep lower convective layer (*LCL*) and upper convective layers 1 and 2 (*UCL*1 and *UCL*2) recorded between 1965 and 2008

Reference	Date	T LCL (°C)	S LCL (‰)	T UCL1 (°C)	S UCL1 (‰)	T UCL2 (°C)	S UCL2 (‰)	Top LCL (m)	Top UCL1 (m)	Top UCL2 (m)
Miller et al. (1966)	1965-02	55.92	263		136					
Munns et al. (1967), Brewer et al. (1969)	1966-11	56.49	263	44.26	133			2,049	2,015	
Ross (1972)	1971-03	59.2	257	49.7	143			2,049	2,019	
Schoell and Hartmann (1973)	1971-04	59.2	266.3	50.3	141.1			2,049	2,023	
Schoell and Hartmann (1978), Hartmann (1980)	1972-02	59.9	266.2	51	141.5			2,049	2,019	
Bubnov et al. (1977)	1976-03	61.74	266	48.39	143.2	44.7	119	2,049	2,019	
Bubnov et al. (1977)	1976-06	61.88	272	48.87	144.7	44.7	119	2,049	2,019	
Hartmann (1980), Schoell and Hartmann (1978)	1977-11	61.52	266.5	49.95	143.4	45.82	115.8	2,048	2,020	2,014
Monin and Plakhin (1982)	1980-02	61.73	270	52.28	145.1	48.16	113.6	2,047	2,021	2,013
Blanc and Anschutz (1995)	1992-09	66.05	270	55.1	153	49.5	118	2,047	2,022	2,006
Hartmann et al. (1998)[a]	1995-03	67.2	269	57.1	149	52	116	2,047	2,028	2,016
Schmidt et al. (2003)	1997-07	67.27	262	56.7	148	52.2	114	2,046	2,025	2,013
Swift et al. (2012)[b]	2008-10	68.28	252	57.32	162	51.25	118	2,048	2,027	2,013

[a] Depths uncertain (CTD failure)
[b] *S* from conductivity using equation of state of sea water

the Atlantis II brine *sensus stricto*. Above this depth, the Atlantis II Deep system remains a topographic depression that includes the Discovery and the Chain Deeps. The top of the transition zone is at 1,903 m depth, where the temperature and salinity of Red Sea bottom waters are found. Above this depth, dense brines can no longer accumulate, because they spill over the pass at the southward opening to the deeper and wider region of the Red Sea axial rift zone. Therefore, the 1,900 m depth contour represents the uppermost boundary of the Atlantis II Deep system, which extends over 142 km^2. The volume of the transition zone between 2,000 and 1,900 m depth is 11 km^3. The transition zone between convective brines and sea water shows breaks in the temperature profile which can be ascribed to topography. The sharp gradient of properties at 1,900 m depth is due the presence of the pass at this depth at the southern end of the system (Fig. 1). A jump in the temperature gradient at 1,990 m depth occurs at the maximum depth where the Atlantis II Deep, the Wando Terrace, and Discovery Deep are connected. Above this depth, the area connected to the Atlantis II Deep increases by 20 km^2 (i.e. by 30 %). This surface is not directly above the hot brine, but above the bottom of the Wando Terrace and the cooler brine of the Discovery Deep (45 °C, Swift et al. 2012). Therefore, 1,990 m is the depth where the heat and salt transported from the LCL are not only dispersed vertically, but also to a great extent laterally.

The salinity of the UCL1 in 1992 (153 g/kg) would correspond to a mixture of 49 % of LCL and 51 % of sea water, whereas the heat content of the UCL1 would correspond to a mixing ratio of 86 % of LCL and 14 % of sea water, as deduced from temperature and specific heat of each brine (Anschutz and Blanc 1996). Other chemical components, such as dissolved silica (Anschutz and Blanc 1995b), also show non-conservative mixing, with a contribution of the LCL silica to the UCL1 of 68 %. Therefore, a simple mixing model does not fit with observations and does not explain the composition of the upper brine layers.

The formation and persistence of layers similar to those observed in the Atlantis II Deep are the consequence of the stratification with respect to the two different components, heat and salt, the diffusivities of which are different. In a fluid in which both temperature and salinity increase with depth, buoyant elements use the potential energy of temperature to rise against the stabilizing effect of salt, which diffuses less rapidly than heat. Parcels of buoyant fluid can only raise a finite height before losing their positive buoyancy. A fluid element will lose heat faster than salt, and buoyancy forces will drive it back towards its initial position, so convection does not extend throughout the depth of the fluid (Huppert 1971; Turner 1973). If the heating rate from below is sufficiently large, a convective layer at the bottom grows while transferring some heat to the fluid above. With time, this causes another convective layer to form, and eventually, a sequence of layers develops, separated by interfaces across which heat and salt are transferred by molecular diffusion (Huppert and Linden 1979). This phenomenon of double-diffusive convection was invoked for the Atlantis II Deep on the basis of the physical parameters of the brines (Turner 1969; McDougall 1984; Anschutz et al. 1998). A diffusive interface separating two convective layers having temperature and salinity differences of ΔT and ΔS will only be stable if the density ratio $R_\rho = \beta\Delta S/\alpha\ \Delta T$ is greater than about 2, where α and β are the proportional density changes due to unit changes in temperature and salinity, respectively. It was shown that the convection persists up to $R_\rho = 28$ (Turner 1985). During the last 45 years, values of the density ratio were always greater than 14 at the LCL–UCL1 interface (Anschutz et al. 1998), since the salinity step ΔS was always above 110 ‰, and the temperature step ΔT remained below 13 °C. R_ρ was the highest in 1972 (21.28), because the value of ΔT was the lowest (8.30 °C). Therefore, the thin interfaces of the Atlantis II Deep can certainly be considered to be stable double-diffusive interfaces. However, the density ratios of the Atlantis II Deep interfaces are above the maximum values that have been investigated in laboratory experiments. Therefore, the mode of property transfer in the Atlantis II Deep system is not well known. Some dissolved species of the brines, such as Na, Ca, and K, show concentration versus depth profiles that are equal or close to conservative behaviour in comparison with Cl (Danielsson et al. 1980; Hartmann 1985). This argues more favourably for a transfer of dissolved species from one brine layer to another by mixing rather than by diffusion. But other species, such as SO_4 or Mg, show non-conservative behaviour (Hartmann 1985), indicating that diffusion must also be important. Uncharged species such as H_4SiO_4, the form of dissolved silica in Atlantis II Deep conditions, does not enter in the electroneutrality budget of the brine. Therefore, the molecular diffusion mode can be shown by the silica profile better than by charged species profiles, since H_4SiO_4 can be independently transferred from one brine to another according to its own diffusion coefficient, whereas charged species have to diffuse mutually. The value of the tracer diffusion coefficient of H_4SiO_4 at 25 °C is 2.2×10^{-5} cm^2/s (Applin 1987). The mutual tracer coefficient of Na and Cl is 1.61×10^{-5} cm^2/s (Li and Gregory 1974). The ratios between diffusivities (2.2:1.61) is very close to the ratio between the calculated contributions of LCL to UCL1 for Si and Cl (68 %:49 %), suggesting that molecular diffusion is the main driver of exchange between the brine layers. Heat and salt diffusing above the UCLs (above 2,000 m depth) disperse into Red Sea deep water by vertical and lateral mixing through the transition zone, where the density ratio is low (<2) and turbulent transport becomes more important than diffusion in this less stratified water.

Evolution of Atlantis II Deep Current Hydrothermal Activity

In 1966, the layered structure of the hot brines was observed from high-resolution vertical profiling of T and S (Munns et al. 1967), and it has continued to be present in all later observations. The lower brine body was already present in 1965, but its temperature was 12.3 °C lower than in 2008 (Fig. 2; Table 2). The temperature increased at a mean rate of 0.526 °C yr^{-1} (0.25×10^9 W) between 1965 and 1976. Between 1980 and 1995, the mean rate was lower, with a value of 0.362 °C yr^{-1} (0.17×10^9 W), and the value dropped to 0.09 °C yr^{-1} (0.043×10^9 W) after 1997. Numbers in brackets represent the flux of heat H in the LCL during the period considered, calculated according to $H = \Delta T \times C_P \times \rho \times V)/\Delta t$, where C_P is the specific heat of the brine, ρ is its density, V is its volume, and Δt is the time interval (Anschutz and Blanc 1996). Between 1966 and 1972, only two layers were recognized (LCL and UCL1). The temperature of the UCL1 increased rapidly during this time interval at a rate of 1.34 °C yr^{-1}, corresponding to a heat input of 0.24×10^9 W (Fig. 3). During this time interval, no temperature increase was observed above the UCL1 brine. The top of the temperature anomaly of the transition zone was located at about 1,950 m depth, that is 50 m below the depth reached by the temperature anomaly 20 years later. Between 1972 and 1976, the temperature of the UCL1 decreased by 2.6 °C, but much heat accumulated above. The temperature of the LCL decreased by 0.36 °C between June 1976 (Bubnov et al. 1977) and November 1977 (Schoell and Hartmann 1978). During this period, a new layer UCL2 appeared and most of the heat accumulated in the depth range 2,020–1,950 m (0.37×10^9 W). Between 1977 and 1995, heat continued to accumulate in the upper part of the Atlantis II Deep above the ULC1 at a rate of 0.33×10^9 W. A new layer UCL3 was recorded in 1992 (Blanc and Anschutz 1995). After 1995, no more heat accumulated above the LCL. The comparison between the 1992 data (Blanc and Anschutz 1995) and 2008 data (Swift et al. 2012) shows that the geometry of the brine layers changed during this interval (Fig. 2), while the mean temperature did not vary. The temperature anomaly reached the 1,903 m depth boundary at the beginning of the 1990s. Therefore, almost all additional heat supplied in the Atlantis II Deep was confined to the depression and was not dispersed in the sea, so that a precise heat balance can be calculated. The heat loss at the wall rock was negligible (Anschutz and Blanc 1996). The heat flux in the whole Atlantis II Deep hydrothermal field was almost constant during this period and amounted to 0.54×10^9 W. This is equivalent to the flux of 10 hydrothermal vents as measured at 21°N on the East Pacific Rise (Macdonald et al. 1980), on the Juan de Fuca ridge (Converse et al. 1984), and in the

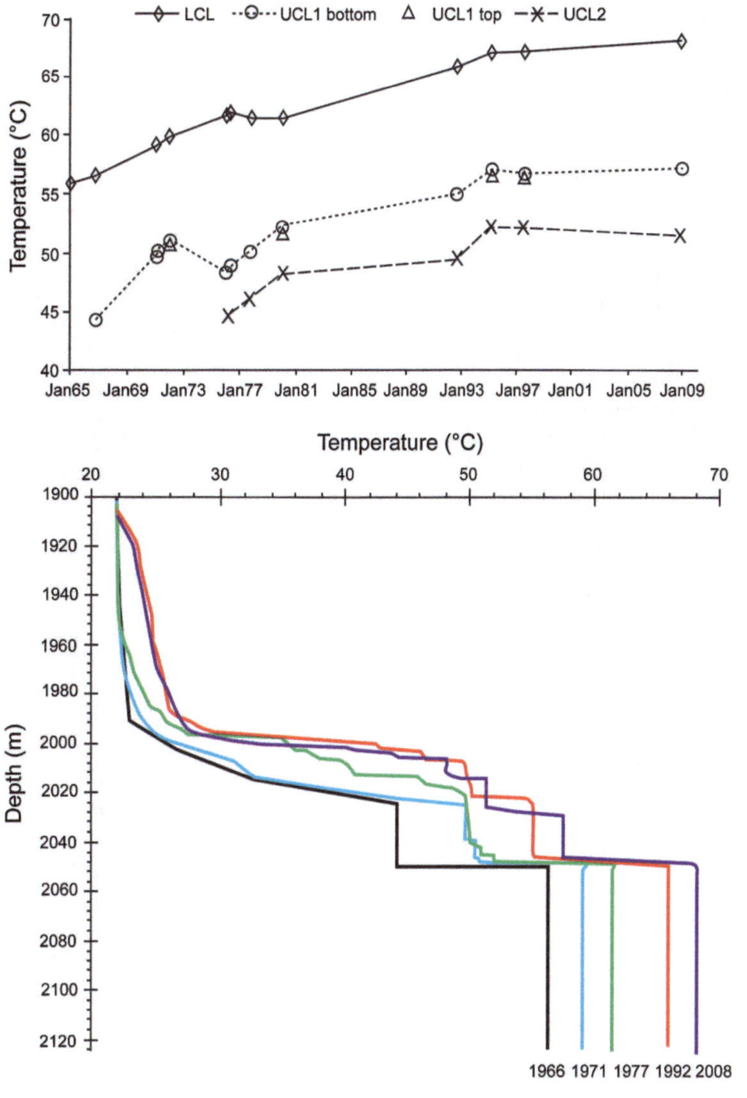

Fig. 2 *Top* evolution of the temperature of LCL, UCL1, and UCL2. *Bottom* vertical profiles of temperature in 1966, 1971, 1977, 1992, and 2008. Data after Miller et al. (1966), Munns et al. (1967), Schoell and Hartmann (1973, 1978), Bubnov et al. (1977), Monin and Plakhin (1982), Blanc and Anschutz (1995), Hartmann et al. (1998), Schmidt et al. (2003), and Swift et al. (2012)

Rainbow field on the mid-Atlantic ridge (German et al. 2010). Assuming that the heat flux of 0.54×10^9 W was constant before 1966, Anschutz and Blanc (1996) calculated that this hydrothermal episode began after 1937 (Fig. 3) and it therefore lasted approximately 60 years. After 1995, the heat flux became 10 times lower than before, and the total heat flux dropped to 0.05×10^9 W. This suggests that the modern hydrothermal episode stopped recently. However, the persistence of the brine over thousands of years, revealed by continuous predominance of metalliferous particles in the Holocene sediment pile, indicates that hydrothermal pulses have occurred several times in the past, suggesting that the thermal and structure of the brine pool evolved continuously in the past.

The salt content of the brine pool also increased with time. The salt mass balance calculation is less precise than the heat balance, because salinity profiles are more difficult to compare from one data set to another. The salinity of the LCL was 260 ± 3 ‰ between 1965 and 1971 and increased to 266 and 270 ‰ between 1971 and 1995. A salinity of 252 ‰ was reported for the 2008 cruise (Swift et al. 2012), but this value was based on uncorrected conductivity measurements. The salinity of the UCL1 was 133 ‰ in 1966, and it remained at 143 ± 2 ‰ between 1971 and 1980. The value reached 153 ‰ in 1992 and 162 ‰ in 2008. The mean salinity of the upper layers of the brine pool increased like the temperature during the 1971–1995 time interval. After 1995, the salt content of the transition zone did not change significantly. A mean salt budget based on the salinity evolution between 1966 and 1992 indicates that the salt flux of the UCL1 and of the 2,020- to 1,900-m-depth interval was by 40 ± 10 kg s^{-1}, and 170 ± 30 kg s^{-1}, respectively (Anschutz and Blanc 1996), corresponding to a total salt input of 210 ± 40 kg s^{-1}. The input of salty solutions into the LCL did not change the salinity significantly. Fluid addition in the LCL would increase the volume of the brine pool. The

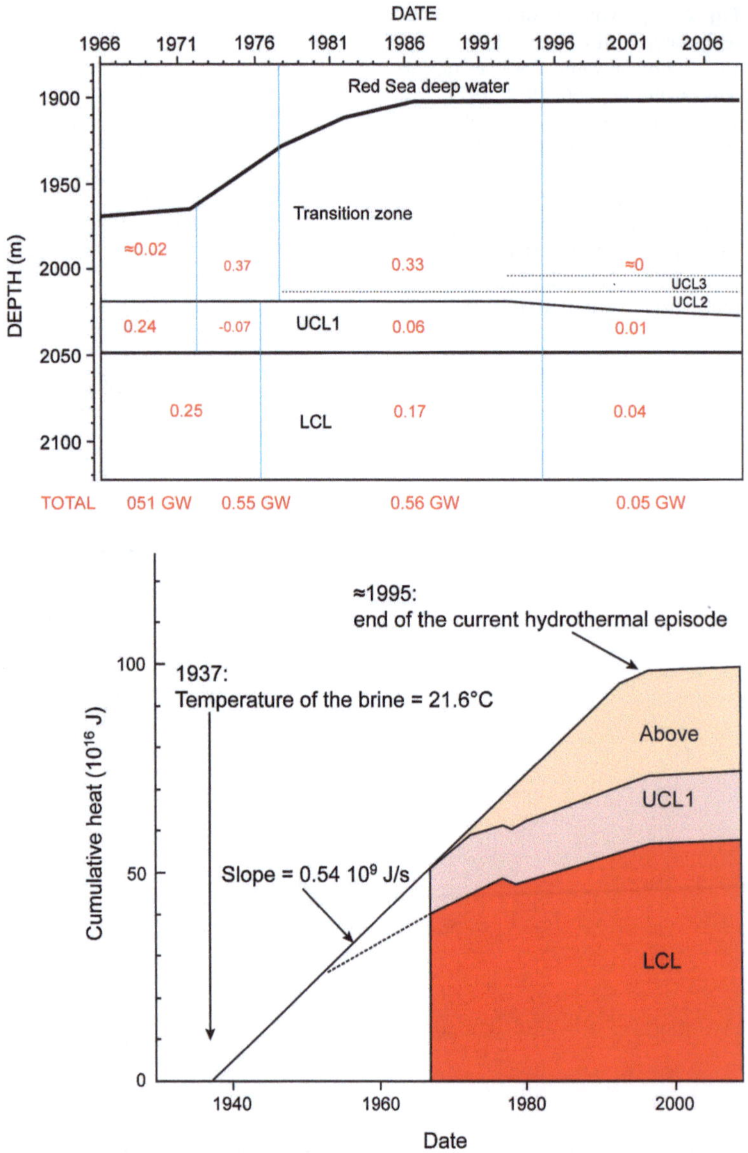

Fig. 3 Evolution of the Atlantis II Deep heat flux. *Top* the heat fluxes in each brine layer are in red ($\times 10^9$ W). They were derived from temperature change between time intervals, the specific heat, the density, and the volume of the brine layers. The numbers at the bottom represent the heat supplies in the whole system. *Black lines* represent sharp interfaces between well-mixed brine layers and the upper limit of the transition zone. Between 1972 and 1976, a new layer and a sharp interface were created above UCL1. This induced an important heat and salt transfer above UCL1 and an increase in the rate of elevation of the transition zone. *Bottom* accumulation of heat in the Atlantis II Deep system compared to the heat contained in a cold (21.6 °C) brine

monitoring of the LCL–UCL1 interface indicates that its position remained at 2,048 m depth to within 2 m. Therefore, the maximum increase in thickness of the brine during the 1966–2008 period was 2 m, which still represents a volume increase of about 0.1 km^3. Considering that the overall heat and salt fluxes entered the Atlantis II Deep from a source of fluids located at the bottom of the LCL during the 1965–1995 hydrothermal episode, Anschutz et al. (1998) showed that simple mixing of LCL brine with upper brines or purely diffusive processes from below or the combination of both could not explain the evolution of the physical properties of the upper layers. The system has been fed by at least two fluid sources: first, a hot and saline brine that has entered the deep directly into the LCL and has supplied more than two-thirds of the overall heat and a part of the salt to the upper brines though a complex combination of diffusion and mixing, and second, a downflowing fluid mixed with sea water that has entered the deep from above and has supplied most of the salt contained in the UCLs and the transition zone.

Hydrothermal Activity and Mineral Precipitation Over the Last 23,000 Years

The substratum of the Atlantis II Deep consists of MORB-type basalts, which are covered with 0- to 30-m-thick metalliferous sediments (Bäcker and Richter 1973). The solid fraction of the sediment is of two different origins. The first is the normal Red Sea pelagic sediment, which consists of biogenic calcareous and/or siliceous components and silico-clastic detrital particles. In the Atlantis II Deep, this fraction

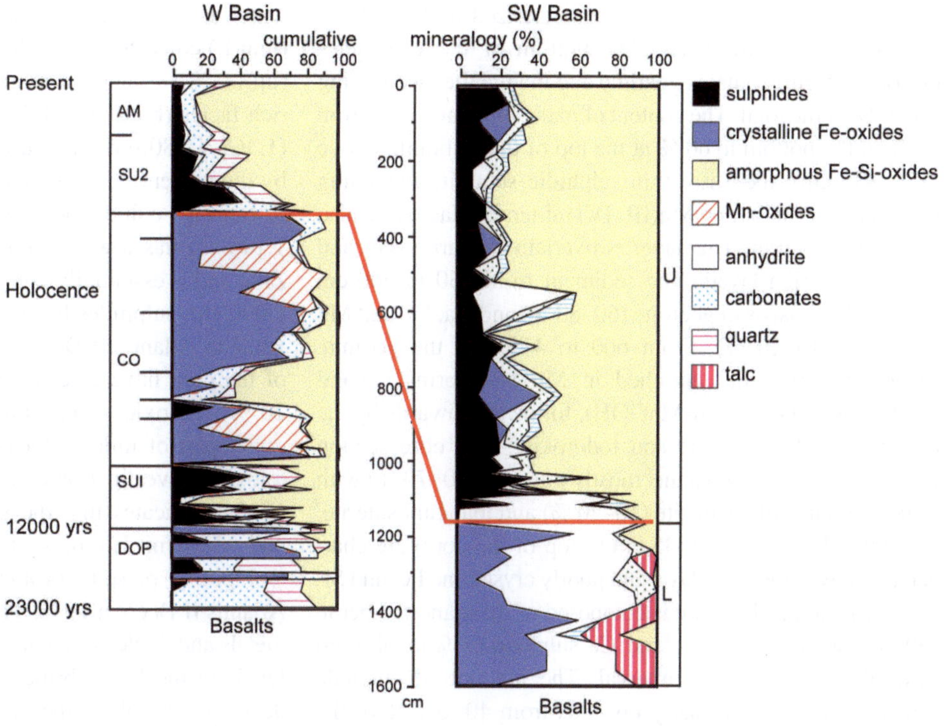

Fig. 4 Distribution of minerals versus depth in cores 683 and 684, expressed as weight percentage of dry and salt-free sediments. Lithological units defined by Bäcker and Richter (1973) are shown for core 684: DOP, lower sulphide (SU1), central oxide (CO), upper sulphide (SU2), and amorphous silicate (AM)

is diluted by metalliferous sediment, which consists of metal oxides, sulphides, carbonates, sulphates, and silicates that precipitated from the hydrothermal fluids. The metalliferous fraction is dominant in the Holocene deposits and minor below. The total estimated reserves of metals are approximately 94×10^6 tons (Mt), including 1.9 Mt Zn, 0.43 Mt Cu, 5,400 t Co, 3,750 t Ag, and 47 t Au (Guney et al. 1984). Holocene sediments are composed of several sedimentary units that were defined in the early 1970s (Bäcker and Richter 1973). Each unit contains fine-grained banded sediment of several cm thick layers, which suggests that hydrothermal activity and physical–chemical properties of the brine changed with time, each layer recording the evolution of these conditions.

The two cores, core 684 from the western basin and core 683 from the southwest basin, were collected during the hydrotherm cruise in 1985 with the RV Marion Dufresne. The basaltic basement in core 684 in the western basin occurs at 13.8 m depth below the seafloor and in core 683 in the southwest basin at 16 m depth below the seafloor (Fig. 4).

A succession of 5 units is observed in cores collected in the W, N, and E basins of the Atlantis II Deep. In core 684, the detrital–oxide–pyrite (DOP) facies is a 2.4-m-thick sequence of sediment with layers that contain mainly carbonates and silico-clastic material. It was deposited during the last glacial maximum period between 12 and 23 ky. Pteropod and foraminifera shells are rare, and the carbonates present mainly consist of a rhombohedral solid solution of Fe–Mn–Ca carbonates. Diatom frustules and radiolarian tests are associated with chalcedony and quartz. Clay minerals are detrital clays similar to those found in sediments collected outside the brine pool (kaolinite, illite, and chlorite). Pyrite may represent 20 wt% of the solid fraction. The isotopic signature of sulphur from the pyrite of the DOP facies indicates that it has been formed by bacterial reduction of seawater sulphate during early diagenetic processes of organic matter mineralization (Kaplan et al. 1969; Shanks and Bischoff 1980). The high level of pyrite content in the DOP facies indicates that the Atlantis II Deep was under the influence of hydrothermal iron inputs during the Pleistocene. Moreover, three interlayers of about 20 cm thickness are composed of more than 75 % of goethite and lepidocrocite, two iron oxyhydroxides, suggesting that a brine pool was already present in the deep before the Holocene.

The boundary between the DOP and SU1 corresponds approximately to the Pleistocene–Holocene limit, as deduced from ^{14}C dates obtained on bulk carbonate fractions from a pteropod-rich layer (Hackett and Bischoff 1973). Zone SU1 contains sulphide minerals, mostly pyrite, sphalerite, and chalcopyrite. These minerals represent in some layers more than 50 % of the dry sediment. Layers with more than 20 % of diatoms and radiolarian test also occur in SU1 and represent well-preserved portions of sapropels that formed in the Red Sea during the Younger Dryas and the early Holocene (Taviani 1998). The presence of the brine pool is probably responsible for the preservation of biogenic silica deposited during these periods (Anschutz and Blanc 1995b; Seeberg-Elverfeldt et al. 2004).

The central oxic unit (CO) is characterized by Fe and Mn oxides and oxyhydroxides. The bottom of the CO facies (1,000–875 cm) contains goethite and manganite, accounting for 80 % of the total. The content of manganite increases from 14 % at the bottom to 60 % at the top of this subunit. Above (875–835 cm), there is a thin sulphidic subunit, containing manganosiderite but no Mn(III, IV) minerals. Haematite also occurs in this unit. This layer is overlain by carbonate- and iron silicate-rich layers. The sediment from 750 to 600 cm essentially consists of goethite (60–80 %) and Ca, Fe, and Mn carbonates (6–20 %). From 600 to 415 cm, the goethite (α-FeOOH) becomes enriched in Mn and forms a solid solution with groutite (α-MnOOH), followed upwards by the occurrence of manganite and todorokite. The core section from 415 to 335 cm contains mostly goethite (50–70 %) with minor amounts of haematite (10–20 %) and manganosiderite.

Units SU2 and AM (335 cm to top of the core) are characterized by sulphides, clays, and poorly crystalline Fe- and Si-bearing material. The base is composed of manganous siderite (50 %), sulphide (16 %), and Ca sulphate (3 %), followed upwards by a ZnS-rich interval. The amount of sulphide minerals decreases regularly upwards from 40 to 7 % of the solid fraction. At the top of the core (140–0 cm), the sulphides are X-ray amorphous and are most likely iron monosulphides (Pottorf and Barnes 1983). Samples also contain up to 5 % anhydrite. The Fe- and Si-bearing poorly crystallized fraction comprises up to 85 % of the solid fraction.

The lithostratigraphy in the SW basin is different from the other basins. The sediment sequence was deposited during the Holocene period, suggesting that the most recent volcanic activity of the Atlantis II Deep occurred in the SW basin (Anschutz and Blanc 1993). Only two main units appear in the SW basin, with a lower unit enriched in anhydrite and crystalline Fe oxides and a thick upper unit, which is similar to the SU2 and AM units. The lower unit (unit L) consists of anhydrite (12–70 %), talc and serpentine (up to 28 %), and Fe oxides (20–60 %), with a magnetite-rich facies (bottom to 1,365 cm) and a haematite-rich facies (1,365–1,180 cm). The upper unit (unit U) is characterized by the presence of sulphides and abundant clay and Fe- and Si-bearing compounds. The lower part from 1,180 to 1,100 cm has low sulphide content (<2 %), predominantly ZnS, and is essentially composed of anhydrite (78–90 %). At 1,090 cm, sulphides become abundant and reach the highest value in Atlantis II Deep sediments (55 %). The upper 10 m of the core have a relatively constant sulphide content (7–20 %). Fe oxides and anhydrite appear sporadically in the upper part of unit U. Facies correlation from micropaleontological investigation and mineralogy (Anschutz and Blanc 1995a) indicates that the unit U corresponds to units SU2 and AM. Unit L represents a part of the facies CO.

Figure 5 presents profiles of dissolved Fe and Mn in the Atlantis II Deep in 1992. The brines are enriched in dissolved metals and depleted in sulphides. Fe and Mn are at millimolar levels in the lower brine. The concentration of Fe and Mn decreases sharply in the transition zone above the brine pool, likely due to oxidation. Redox reactions that are energetically feasible can be determined by thermodynamic equilibrium calculations. Figure 6 shows the common half redox reactions of electron acceptor reduction and associated equilibrium constants for one electron transfer at standard conditions for major species involved in mineral precipitation in the Atlantis II Deep. All these reactions can be theoretically combined to give redox reactions. When the energy yield is favourable, the oxidant species of a redox reaction is that of the half reaction that has the highest log K, and the electron donor species is that of the half reaction with the lowest log K. The standard condition for concentration is 1 M,

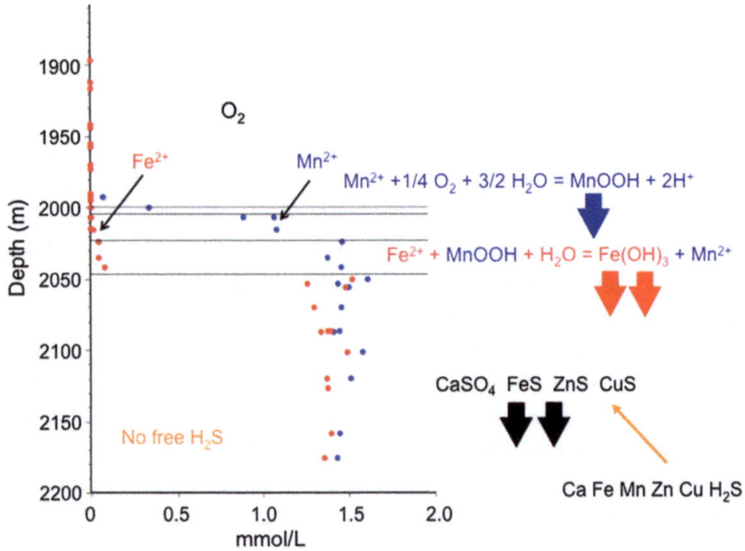

Fig. 5 Vertical profiles of dissolved Mn and Fe in the Atlantis II Deep from samples collected in 1992. Processes of mineral formation are shown. Dissolved Mn is oxidized with dissolved O_2 in the transition zone. Particulate Mn(III,IV) species act as oxidizing agents for Fe(II) in the anoxic UCL1. Fe(III) particles accumulate in the anoxic lower brine, because strong reducing agents, such as free S(-II) forms, are titrated when the hot brine enters the less hot LCL brine. Particulate metal sulphide and sulphate form through quenching in the vicinity of the hydrothermal brine source

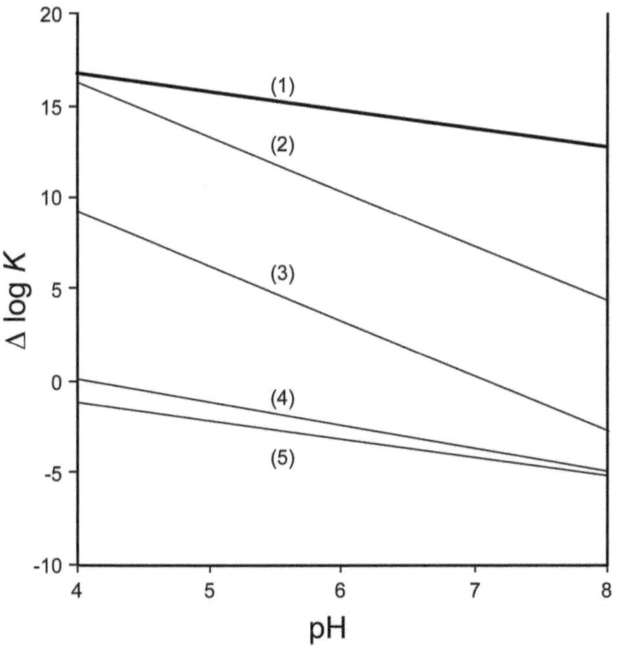

Fig. 6 Variation of $\log K(\Delta \log K)$ with pH and with an activity of 10^{-3} for Fe^{2+} and Mn^{2+} for selected half redox reactions. All these reactions may be associated in pairs. The energy yield is favourable when the oxidant species of a redox reaction is that of the half reaction that has the highest log *K*, and the electron donor species is that of the half reaction with the lowest log *K*. For example, MnO(OH) is an energetically favourable oxidant for Fe^{2+}, H_2S, and CH_4. (1) $1/4O_2 + H^+ + e^- \rightarrow 1/2H_2O$, (2) $MnO(OH) + 3H^+ + e^- \rightarrow Mn^{2+} + 2H_2O$, (3) $Fe(OH)_3 + 3H^+ + e^- \rightarrow Fe^{2+} + 3H_2O$, (4) $1/8SO_4^{2-} + 5/4H^+ + e^- \rightarrow 1/8H_2S + 1/2H_2O$, (5) $1/8CO_2 + H^+ + e^- \rightarrow 1/8CH_4 + 1/4H_2O$

including H^+. The stoichiometry of these half reactions shows that the proton activity has a significant effect on log *K*. That is why we can represent the variation of log *K* (ΔlogK) as a function of pH (e.g. Luther et al. 1997; Bethke et al. 2011) in order to get the equilibrium value at the pH of the Atlantis II Deep brines. The pH of the LCL is between 5 and 6 (Danielsson et al. 1980; Pierret et al. 2001). Oxidation of Fe^{2+} with MnO(OH) is a thermodynamically feasible process. This process has been proposed by several authors to explain the removal of Fe^{2+} and a part of the reductive dissolution of particulate Mn during early diagenesis of marine sediments (Myers and Nealson 1988; Postma 1985; Hyacinthe et al. 2001), or in stratified lake waters (Davison 1993; Crowe et al. 2008), and in the Atlantis II Deep (Hartmann 1985; Danielsson et al. 1980; Blanc et al. 1998).

The high concentration of dissolved Fe and Mn in the brine and the presence of dissolved O_2 in the Red Sea deep water and in the transition zone down to 2,000 m depth (Swift et al. 2012) suggest that metals are oxidized when they diffuse upwards and Fe and Mn oxide particles sink to the bottom. However, kinetically Fe(II) oxidizes faster than Mn(II). That is why only iron oxides, but not Mn oxides, can accumulate in the present-day deep brine (Fig. 5). Iron oxides can be buried in the sediment in the absence of reducing compounds for Fe(III) minerals. Indeed, dissolved sulphide is absent in the LCL, but dissolved methane occurs at a concentration of about 4 μmol/L (Faber et al. 1998). Reduction in iron oxide coupled with anaerobic methane oxidation has been observed recently in marine sediments (Beal et al. 2009). The reaction probably has to be triggered by micro-organisms, which are probably absent in the hot LCL brine. This may explain the large accumulation of iron oxide in the sediment despite micromolar methane concentrations. Fe(II) is a reducing agent for Mn oxides. Therefore, Mn oxides may accumulate on the bottom only when dissolved Fe is depleted in the brine. Such a situation occurs at present in the UCL1 brine (Hartmann 1985), and it occurred during the CO facies deposition period. The precipitation of sulphides occurs by quenching of hydrothermal fluids in the vicinity of discharge points, because of a drastic drop in solubility of metal sulphide minerals when the temperature decreases (Helgeson 1964). Dissolved sulphide or free hydrogen sulphide most likely is present in the incoming brine (Hartmann 1973). However, the fact that hydrogen sulphide has not been detected in the LCL may be due to its rapid removal from the hydrothermal vent by reaction with dissolved transition metals that are in excess relative to sulphide (Fig. 5). Anhydrite ($CaSO_4$) can form when Ca-rich hydrothermal fluid mixes with the sulphate-bearing LCL brine, or it can precipitate when the brine is conductively heated by the incoming fluid.

Transformations of primary minerals through diagenetic processes eventually modify the mineralogy (Blanc et al. 1998; Anschutz et al. 2000). These transformations mostly consist of re-crystallisation of primary Fe and Mn oxides and amorphous phases in more crystallized minerals, the dissolution of primary biogenic carbonates, and the precipitation of Fe–Mn–Ca carbonate solid solutions.

Paleo-Oceanography of the Red Sea and Hydrothermal Sediments

Holocene pelagic sediments from the Red Sea, remote from the deep metalliferous basins, consist of about 50 % detrital silicates such as quartz, clay minerals, and volcanic fragments, and 50 % carbonaceous tests of coccoliths, foraminifera, pteropods, and other organisms (Bäcker 1976). Stratigraphic correlations within Red Sea sediments can be obtained by using radiocarbon dates, micropaleontology, and oxygen isotopes on planktonic foraminifera. The frequency distribution of planktonic and benthic foraminifera in cores from the central Red Sea shows that Holocene sediments consist of a zone characterized by a high total number of

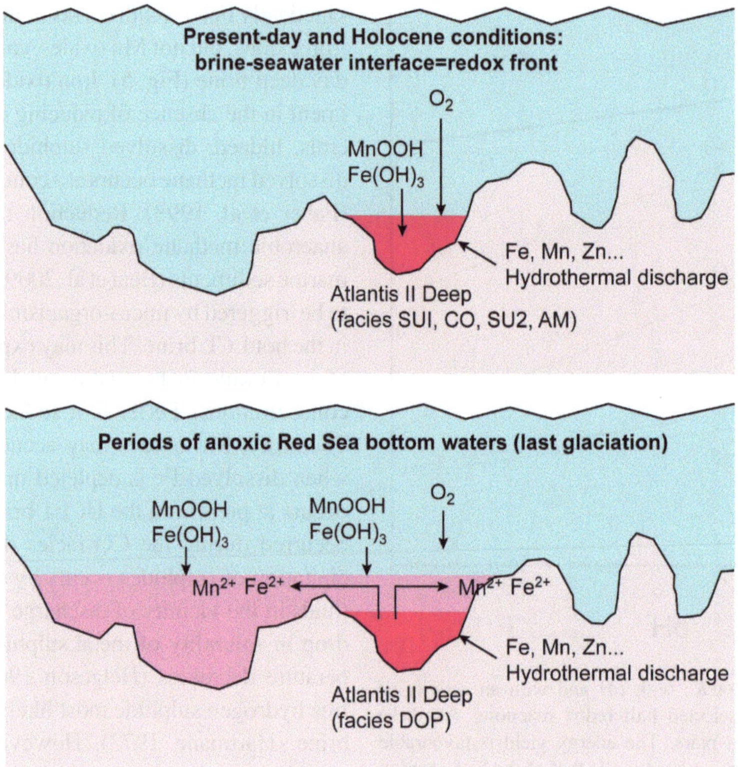

Fig. 7 Schematic cross-section along the axial zone of the Red Sea showing the position of the pycnocline (brine–sea water interface), and the position of the redox front, where metals are oxidized and precipitate. *Top* Holocene, the brine–sea water interface corresponds to the redox front where most of the metals supplied by hydrothermal fluids precipitate. The metalliferous sediments are confined in the hydrothermal deeps. *Bottom* During the last glaciation, bottom waters were anoxic and dissolved metals were able to diffuse outside the brine-filled deeps. They precipitated at the interface between oxic and anoxic sea water located between 1,000 and 1,600 m depth. When dissolved sulphide accumulated in the anoxic Red Sea bottom waters because of bacterial sulphate reduction, only Mn was able to diffuse

planktonic foraminifera, decreasing rapidly downcore to a minimum where the sediment age is close to 11 ky (Locke and Thunell 1988). Below this, Red Sea cores contain an interval characterized by low abundance or absence of planktonic foraminifera and relative high abundance of pteropods (Berggren and Boersma 1969; Reiss et al. 1980). The base of this "aplanktonic zone" occurs around 20 ky B. P. (Reiss et al. 1980). Core 684 contains a similar "aplanktonic" interval located between 1,340 cm in the DOP facies and somewhere in unit SU1 (Anschutz and Blanc 1993). The aplanktonic zone corresponds to the last glacial maximum. (Hemleben et al. 1996; Fenton et al. 2000). During this period, planktonic foraminifers became extinct due to the high salinity levels ($\sim$50 ‰). The anomalously high salinity values are attributed to a drastic reduction in water exchange between the Red Sea and the Indian Ocean through the Strait of Bab-al-Mandab due to lowered sea level, greater evaporation, and to some extent also due to an increased aridity in the area (Siddall et al. 2003).

Sediments of the last glacial period of the Atlantis II Deep (DOP facies), and of other brine-filled deeps, contain less hydrothermal metals than Holocene sediments. Unlike cores collected in brine-filled deeps, the sedimentary flux of metal was higher during the last glacial period than later in "normal" pelagic sediments of the Red Sea deposited below 1,600 m depth (Coulibaly et al. 2006). The distribution of hydrothermal particles suggests that hydrological changes occurred during the late Quaternary in the Red Sea. During the Holocene, metals dissolved in anoxic brines precipitated at the brine–sea water interface, so that metals accumulated only in the deeps. During the last glacial period, hydrothermal metals spread over a large area, suggesting that the redox boundary was not located at the brine–sea water interface, but above. The Red Sea bottom water was probably anoxic (Fig. 7). Red Sea bottom water anoxia probably occurred as a consequence of higher stratification of the water column due to sea-level change. This interpretation suggests that the observed variation in metal concentrations in the sediment recording of the Atlantis II Deep is not only related to fluctuations of the hydrothermal fluid discharge, but also to the varying redox conditions in the Red Sea bottom water. Since both the Atlantis II Deep and the Suakin Deep recorded lower fluxes of metals during the last glacial maximum, we suggest that the body of anoxic bottom water was in connection with both hydrothermal systems. The connection is possible when the redox boundary is located at 1,600 m depth or above.

The Atlantis II Deep as an Analogue of the Archaean Ocean

The Atlantis II basin has been considered as an analogue of massive sulphide deposits. It is the only known modern deposit that is analogous to laminated Fe-rich chemical sediments, such as banded iron formation (BIF) (Laurila et al. 2014). The present-day deep is anoxic but not euxinic; it is a unique marine oxygen-free environment enriched in dissolved iron but not in dissolved sulphide. Such properties are comparable to the oceans that formed the BIF between 3,600 and 1,800 Ma. The Atlantis II Deep may be examined as a modern and unique natural oceanic analogue of the Precambrian ocean. The formation of BIF is believed to be due to oxidative precipitation of dissolved Fe(II) of hydrothermal origin accumulated in an anoxic ocean depleted in dissolved sulphate. The study of Fe isotopes in BIF or contemporary iron sulphides shows that this period is characterized by a large isotopic variability that is not found in samples from Proterozoic and Phanerozoic times (Anbar and Rouxel 2007). This variability is probably due to a reservoir effect and a Rayleigh distillation during the precipitation of iron-bearing minerals. This interpretation suggests that the evolution of iron isotopes is particularly sensitive to iron concentrations in sea water, and in this sense, the iron isotopes are a new tool to track the oxygenation of Precambrian oceans. However, work on current anoxic systems is needed to better explain the magnitude of the fractionation of iron in the Archaean and Paleoproterozoic sediments. New results show that the Atlantis II Deep sediments present a range of Fe isotope signature as large as that in Precambrian formations (Rouxel et al. 2012), suggesting that it is a unique place where Raleigh distillation of iron can be studied in a present-day marine environment.

Acknowledgments I wish to thank Mark Hannington, Tea Laurila, Susan Humphris, and Pat Shanks for their critical comments in order to improve the manuscript. Louiesito Abalos is thanked for helping me out in producing the figures. I thank my colleagues from the laboratory EPOC, particularly Gérard Blanc for all the work we have done together on the Atlantis II Deep.

References

Albarède F, Michard A, Minster JF, Michard G (1981) $^{87}Sr/^{86}Sr$ ratios in hydrothermal waters and deposits from the East Pacific Rise at 21° N. Earth Planet Sci Lett 55:229–236

Anbar AD, Rouxel O (2007) Metal stable isotopes in paleoceanography. Annu Rev Earth Planet Sci 35:717–746

Anschutz P, Blanc G (1993) L'histoire sédimentologique de la fosse Atlantis II (mer Rouge). Les apports de la micropaléontologie. Compte Rendus de l'Académie des Sci Paris série II 317:1303–1308

Anschutz P, Blanc G (1995a) Chemical mass balances in metalliferous deposits from the Atlantis II Deep (Red Sea). Geochim Cosmochim Acta 59:4205–4218

Anschutz P, Blanc G (1995b) Geochemical dynamics of the Atlantis II Deep (Red Sea): silica behavior. Mar Geol 128:25–36

Anschutz P, Blanc G (1996) Heat and salt fluxes in the Atlantis II Deep (Red Sea). Earth Planet Sci Lett 142:147–159

Anschutz P, Blanc G, Monnin C, Boulègue J (2000) Geochemical dynamics of the Atlantis II Deep (Red Sea). II: pore water composition of metalliferous sediments. Geochim Cosmochim Acta 64:3995–4006

Anschutz P, Blanc G, Stille P (1995) Origin of fluids and the evolution of the Atlantis II Deep hydrothermal system (Red Sea): Sr isotope study. Geochim Cosmochim Acta 59:4799–4808

Anschutz P, Turner JS, Blanc G (1998) The development of layering, fluxes through double-diffusive interfaces, and location of hydrothermal sources of brines in the Atlantis II Deep: Red Sea. J Geophys Res Oceans 103:27809–27819

Applin KR (1987) The diffusion of dissolved silica in dilute aqueous solution. Geochim Cosmochim Acta 51:2147–2151

Bäcker H (1976) Fazies und chemische Zusammensetzung rezenter Ausfällungen aus Mineralquellen im Roten Meer. Geol Jahrb D17:151–172

Bäcker H, Richter H (1973) Die rezente hydrothermal-sedimentäre Lagestätte Atlantis II Tief im Roten Meer. Geol Rundsch 62:697–741

Badaut D, Blanc G, Decarreau A (1990) Variation des minéraux argileux ferrifères, en fonction du temps et de l'espace, dans les dépôts métallifères de la fosse Atlantis II en Mer Rouge. Compte Rendus de l'Académie des Sci Paris série II 310:1069–1075

Beal EJ, House CH, Orphan VJ (2009) Manganese- and iron-dependent marine methane oxidation. Science 325:184–187

Berggren WA, Boersma A (1969) Late Pleistocene and Holocene planktonic foraminifera from the Red Sea. In: Degens ET, Ross DA (eds) Hot brines and recent heavy metal deposits in the Red Sea. Springer, Berlin, pp 282–298

Bethke CM, Sanford RA, Kirk MF, Jin Q, Flynn TM (2011) The thermodynamic ladder in geomicrobiology. Am J Sci 311:188–210

Blanc G, Anschutz P (1995) New hydrographic situation in the Atlantis II Deep hydrothermal brine system. Geology 23:543–546

Blanc G, Anschutz P, Pierret MC (1998) Metalliferous sedimentation into the Atlantis II Deep: a geochemical insight. In: Purser BH, Bosence DWJ (eds) Sedimentation and tectonics of rift basins: Red Sea-Gulf of Aden. Chapman and Hall, London, pp 505–520

Blanc G, Boulègue J, Badaut D, Stouff P (1986) Premiers résultats de la campagne océanographique Hydrotherm (Mai 85) du Marion-Dufresne sur la fosse Atlantis II (Mer Rouge). Compte Rendus de l'Académie des Sci Paris série II 302:175–180

Blanc G, Boulègue J, Michard A (1995) Isotope compositions of the Red Sea hydrothermal end-member. Compte Rendus de l'Académie des Sci Paris série II 320:1187–1193

Brewer PG, Desmore CD, Munns JW, Stanley RJ (1969) Hydrography of the Red Sea brines. In: Degens ET, Ross DA (eds) Hot brines and recent heavy metal deposits in the Red Sea. Springer, Berlin, pp 138–147

Bubnov VA, Fedorova VS, Shcherbinin AD (1977) New data on brines in the Red Sea. Oceanology 17:395–400

Converse DR, Holland HD, Edmond JM (1984) Flow rates in the axial hot springs of the East Pacific Rise (21° N): implication of massive sulfide deposits. Earth Planet Sci Lett 69:159–175

Coulibaly A, Anschutz P, Blanc G, Malaizé B, Pujol C (2006) The effect of paleo-oceanographic changes on the sedimentary recording

of hydrothermal activity in the Red Sea during the last 30,000 years. Mar Geol 226:51–64
Crowe SA, O'Neill AH, Katsev S, Hehanussa P, Haffner GD, Sundby B, Mucci A, Fowle DA (2008) The biogeochemistry of tropical lakes: a case study from Lake Matano, Indonesia. Limnol Oceanogr 53:319–331
Danielsson LG, Dyrssen D, Graneli A (1980) Chemical investigations of Atlantis II and Discovery brines in the Red Sea. Geochim Cosmochim Acta 44:2051–2065
Davison W (1993) Iron and manganese in lakes. Earth-Sci Rev 34:119–163
Degens ET, Ross DA (1969) Hot brines and recent heavy metal deposits in the Red Sea. Springer, Berlin, 600 pp
Dupré B, Blanc G, Boulègue J, Allègre CJ (1988) Metal remobilization at a spreading centre studied using lead isotopes. Nature 333:165–167
Eissen JP, Juteau T, Joron JL, Dupré B, Humler E, Al' Mukhamedov A (1989) Petrology and geochemistry of basalts from the Red Sea axial rift at 18° North. J Petrol 30:791–839
Faber E, Botz R, Poggenburg J, Schmidt M, Stoffers P, Hartmann M (1998) Methane in Red Sea brines. Org Geochem 29:363–379
Fenton M, Geiselhart S, Rohling EJ, Hemleben C (2000) Aplanktonic zones in the Red Sea. Mar Micropaleontol 40:277–294
German CR, Thurnherr AM, Knoery J, Charlou J-L, Jean-Baptiste P, Edmonds HN (2010) Heat, volume and chemical fluxes from submarine venting: a synthesis of results from the rainbow hydrothermal field, 36° N MAR. Deep Sea Res I 57:518–527
German CR, Von Damm KL (2003) Hydrothermal processes. In: Heinrich DH, Karl KT (eds) Treatise on geochemistry, vol 6. Elsevier, Amsterdam, pp 181–222
Greenberg JP, Moller N (1989) The prediction of mineral solubilities in natural waters: a chemical equilibrium model for the Na-K-Ca-Cl-SO4-H2O system to high concentration from 0 to 250 °C. Geochim Cosmochim Acta 53:2503–2518
Guney M, Nawab Z, Marhoun MA (1984) Atlantis-II-Deep's metal reserves and their evaluation. In: Proceedings of the offshore technology conference Houston, paper 4780(3):33–44
Hackett J, Bischoff JL (1973) New data on the stratigraphy, extent, and geologic history of the Red Sea geothermal deposits. Econ Geol 68:553–564
Hartmann M (1973) Untersuchung von suspendiertem Material in den Hydrothermal-Laugen des Atlantis II Tiefs. Geol Rundsch 62:742–754
Hartmann M (1980) Atlantis II Deep geothermal brine system. Hydrographic situation in 1977 and changes since 1965. Deep Sea Res 27A:161–171
Hartmann M (1985) Atlantis II deep geothermal brine system. Chemical processes between hydrothermal brines and Red Sea deep water. Mar Geol 64:157–177
Hartmann M, Scholten JC, Stoffers P (1998) Hydrographic structure of brine filled deeps in the Red Sea. Correction of the Atlantis II Deep temperatures. Mar Geol 144:331–332
Helgeson HC (1964) Complexing and hydrothermal ore deposition. Pergamon Press, New York, 128 pp
Hemleben C, Meischner D, Zahn R, Almogi-Labin A, Erlenkeuser H, Hiller B (1996) Three hundred eighty thousand year long stable isotope faunal records from the Red Sea: influence of global sea level change on hydrography. Paleoceanography 11:147–156
Huppert HE (1971) On the stability of a series of double-diffusive layers. Deep-Sea Res 18:1005–1021
Huppert HE, Linden PF (1979) On heating a stable salinity gradient from below. J Fluid Mech 95:431–464
Hyacinthe C, Anschutz P, Carbonel P, Jouanneau JM, Jorissen FJ (2001) Early diagenetic processes in the muddy sediments of the Bay of Biscay. Mar Geol 177:111–128
Kaplan IR, Sweeney RE, Nissenbaum A (1969) Sulfur isotope studies on Red Sea geothermal brines and sediments. In: Degens ET, Ross DA (eds) Hot Brines and recent heavy metal deposits in the Red Sea. Springer, Berlin, pp 474–498
Koski RA, German CR, Hein JR (2003) Fate of hydrothermal products from mid-ocean ridge hydrothermal systems: near-field to global perspectives. In: Halbach P, Tunnicliffe V, Hein J (eds) Energy and mass transfer in marine hydrothermal systems. Dahlem University Press, Berlin, pp 317–335
Laurila TE, Hannington MD, Petersen S, Garbe-Schönberg D (2014) Early depositional history of metalliferous sediments in the Atlantis II Deep of the Red Sea: evidence from rare earth element geochemistry. Geochim Cosmochim Acta 126:146–168
Li YH, Gregory S (1974) Diffusion of ions in sea water and in deep-sea sediments. Geochim Cosmochim Acta 38:703–714
Locke S, Thunell RC (1988) Paleaceanographic record of the last glacial/interglacial cycle in the Red Sea and Gulf of Aden. Palaeogeogr Palaeoclimatol Palaeoecol 64:163–187
Lupton JE, Weiss RF, Craig H (1977) Mantle helium in the Red Sea brines. Nature 266:244–246
Luther GW III, Sundby B, Lewis BL, Brendel PJ, Silverberg N (1997) Interactions of manganese with the nitrogen cycle: alternative pathways for dinitrogen formation. Geochim Cosmochim Acta 61:4043–4052
Macdonald KC, Becker FN, Spiess FN, Ballard RD (1980) Hydrothermal heat flux of the black smoker vents on the East Pacific Rise. Earth Planet Sci Lett 4:1–7
Maillard C (1974) Eaux intermediaires et formation d'eau profonde en Mer Rouge. In: L'oceanographie physique de la Mer Rouge, Centre National pour l'Exploitation des Oceans. Paris, pp 105–133
McDougall TJ (1984) Convective processes caused by a dense, hot saline source flowing into a submarine depression from above. Deep-Sea Res 31:1287–1309
Miller AR (1964) Highest salinity in the World Ocean. Nature 203:590–591
Miller AR, Densmore CD, Degens ET, Hathaway JC, Manheim FT, McFarling PF, Pocklington R, Jokela A (1966) Hot brines and recent iron deposits in deeps of the Red Sea. Geochim Cosmochim Acta 30:341–359
Monin AS, Plakhin EA (1982) Stratification and space-time variability of Red sea hot brines. Deep-Sea Res 29:1271–1291
Monnin C, Ramboz C (1996) The anhydrite saturation index of the Red Sea ponded brines and sediment pore waters of the Red Sea deeps. Chem Geol 127:141–159
Munns RG, Stanley RJ, Densmore CD (1967) Hydrographic observations of the Red Sea brines. Nature 214:1215–1217
Myers CR, Nealson KH (1988) Microbial reduction of manganese oxides: interactions with iron and sulfur. Geochim Cosmochim Acta 52:2727–2732
Oudin E, Thisse Y, Ramboz C (1984) Fluid inclusion and mineralogical evidence for high temperature saline hydrothermal circulation in the Red Sea metalliferous sediments: preliminary results. Mar Min 5:3–31
Pautot G (1983) Les fosses de la Mer Rouge: approche géomorphologique d'un stade initial d'ouverture océanique réalisée à l'aide du Seabeam. Oceanol Acta 6:235–244
Pierret MC, Clauer N, Bosch D, Blanc G, France-Lanord C (2001) Chemical and isotopic ($^{87}Sr/^{86}Sr$, $d^{18}O$, dD) constraints to the formation processes of Red-Sea brines. Geochim Cosmochim Acta 65:1259–1275
Pitzer KS (1979) Theory: ion interaction approach. In: Pytckowicz RM (ed) Activity coefficients in electrolyte solutions. CRC Press, Boca Raton, pp 157–208
Postma D (1985) Concentration of Mn and separation from Fe in sediments—I. Kinetics and stoichiometry of the reaction between

birnessite and dissolved Fe(II) at 10 °C. Geochim Cosmochim Acta 49:1023–1033

Pottorf RJ, Barnes HL (1983) Mineralogy, geochemistry, and ore genesis of hydrothermal sediments from Atlantis II Deep, Red Sea. Econ Geol Monogr 5:198–223

Reiss Z, Luz B, Almogi-Labin A, Halicz E, Winter A, Wolf M, Ross DA (1980) Late quaternary paleoceanography of the Gulf of Aqaba (Elat), Red Sea. Quatern Res 14:294–308

Rona PA (2008) The changing vision of marine minerals. Ore Geol Rev 33:618–666

Ross DA (1972) Red Sea hot brine area-revised. Science 175:1455–1456

Rouxel O, Resnais Y, Callac N, Anschutz P (2012) Deep sea metalliferous deposits as modern analogues for ancient marine environments. Mineral Mag 76:2302

Schmidt M, Botz R, Faber E, Schmitt M, Poggenburg J, Garbe-Schonberg D, Stoffers P (2003) High-resolution methane profiles across anoxic brine-sea-water boundaries in the Atlantis-II, Discovery, and Kebrit Deeps (Red Sea). Chem Geol 200:359–375

Schoell M, Hartmann M (1973) Detailed temperature structure of the hot brines in the Atlantis II Deep area (Red Sea). Mar Geol 14:1–14

Schoell M, Hartmann M (1978) Changing hydrothermal activity in the Atlantis II Deep geothermal system. Nature 274:784–785

Seeberg-Elverfeldt IA, Lange CB, Arz HW, Pätzold J, Pike J (2004) The significance of diatoms in the formation of laminated sediments of the Shaban Deep, Northern Red Sea. Mar Geol 209:279–301

Shanks WC III (2001) Stable isotopes in seafloor hydrothermal systems: vent fluids, hydrothermal deposits, hydrothermal alteration, and microbial processes. Rev Mineral Geochem 43:468–525

Shanks WC III, Bischoff JL (1980) Geochemistry, sulfur isotope composition and accumulation rates of the Red Sea geothermal deposits. Econ Geol 75:445–459

Siddall M, Rohling EJ, Almogi-Labin A, Hemleben C, Meischner D, Schmelzer I, Smeed DA (2003) Sea-level fluctuations during the last glacial cycle. Nature 423:853–858

Sofianos SS, Johns WE (2002) An oceanic general circulation model (OGCM) investigation of the Red Sea circulation, 1, exchange between the Red Sea and the Indian Ocean. J Geophys Res 107: C11. doi:10.1029/2001JC001184

Sofianos SS, Johns WE (2007) Observations of the summer Red Sea circulation. J Geophys Res 112:C6. doi:10.1029/2006JC003886

Spiess FN, Ken CM, Atwater T, Ballard R, Carranza A, Cordoba D, Cox C, Diaz Garcia VM, Francheteau J, Guerrero J, Hawkins J, Haymon R, Hessler R, Juteau T, Kastner M, Larson R, Luyendyk B, Macdongall JD, Miller S, Normark W, Orcutt J, Rangin C (1980) East Pacific Rise: hot springs and geophysical experiments. Science 207:1421–1433

Stoffers P et al (1998) Hydrography, hydrothermalism and paleoceanography in the Red Sea. Berichte-Reports aus dem Geologisch—Paläontologischen Institut der Universität Kiel, Deutschland 88, 107 pp

Swift SA, Bower AS, Schmitt RW (2012) Vertical, horizontal, and temporal changes in temperature in the Atlantis II and discovery hot brine pools, Red Sea. Deep-Sea Res I 64:118–128

Taviani M (1998) Axial sedimentation of the Red Sea transitional region (22°–25°N): pelagic, gravity flow and sapropel deposition during the late quaternary. In: Purser BH, Bosence DWJ (eds) Sedimentation and tectonics of rift basins: Red Sea-Gulf of Aden. Chapman and Hall, London, pp 467–478

Turner JS (1969) A physical interpretation of the observations of the hot brine layers in the Red Sea. In: Degens ET, Ross DA (eds) Hot brines and recent heavy metal deposits in the Red Sea. Springer, Berlin, pp 164–173

Turner JS (1973) Buoyancy Effects in Fluids. Cambridge University Press, Cambridge, 367 pp

Turner JS (1985) Multicomponent convection. Annu Rev Fluid Mech 17:11–44

Volker F, Mc Culloch MT, Altherr R (1993) Submarine basalts from the Red Sea: New Pb, Sr and Nd isotopic data. Geophys Res Lett 20:927–930

Von Damm KL (1990) Seafloor hydrothermal activity: black smoker chemistry and chimneys. Annu Rev Earth Planet Sci 18:173–204

Weikert H (1982) The vertical distribution of zooplankton in relation to habitat zones in the area of the Atlantis II Deep, central Red Sea. Mar Ecol Prog Ser 8:129–143

Winckler G, Aeschbach-Hertig W, Kipfer R, Botz R, Rübel AP, Bayer R, Stoffers P (2001) Constraints on origin and evolution of Red Sea brines from helium and argon isotopes. Earth and Planet Sci Lett 184:671–683

Zierenberg RA, Shanks WC III (1983) Mineralogy and geochemistry of epigenetic features in metalliferous sediments, Atlantis II Deep, Red Sea. Econ Geol 78:57–72

Zierenberg RA, Shanks WC III (1986) Isotopic variations on the origin of the Atlantis II, Suakin and Valdivia brines, Red Sea. Geochim Cosmochim Acta 50:2205–2214

Environmental Risks of Mining Metalliferous Muds in the Atlantis II Deep, Red Sea

Hjalmar Thiel, Ludwig Karbe, and Horst Weikert

Abstract

After two companies were awarded a 30-year license for the exploration and exploitation of metalliferous sediments in the Atlantis II Deep (Red Sea) in 2011, we herewith present conclusions and recommendations derived from an environmental risk assessment, the Metalliferous Sediment Atlantis II Deep (MESEDA) study, conducted in the period 1977–1981. For economic reasons, this program was discontinued before final report delivery and fell dormant for 30 years. The effects of environmental disturbances of the benthic and the near-bottom water layer habitats in and around the mining site deserve further and more modern risk assessments. We examine the relevance of our 1981 recommendations and of subsequent publications to the extended period of resource extraction planned for this century and recommend more up-to-date risk assessment investigations and evaluations.

Deep-Sea Mining and Environmental Risk Studies

Environmental risks introduced into the deep sea by the extraction of metalliferous resources were issues from the very onset of industrial invasion into this remote inner space. A summary of the early developments is presented in Thiel et al. (1991). Due to market conditions and the worldwide economic crisis, interest in deep-sea mining has been minimal and kept at low levels during the two previous decades in order to secure claim rights, but in recent years, some recovery of this industry can be observed. This is indicated by an increasing number of exploration contracts for deep-sea mining submitted to the International Seabed Authority (e.g., International Seabed Authority 2013) in the Area, those regions of the deep sea beyond national jurisdiction. For mining the Atlantis II Deep in the central Red Sea (Fig. 1), positioned in the common economic zone of the Kingdom of Saudi Arabia and the Democratic Republic of the Sudan, the Diamond Fields Ltd. of Canada and Manafa of Saudi Arabia consortium has received a 30-year license for exploration and exploitation (Diamond Fields 2011).

The Atlantis II Deep is a depression in the central graben system of the Red Sea, containing high saline and warm water, covering a thick layer of metalliferous muds (Fig. 2) at 2,200 m depth. It was investigated during the mid-1960s (Bäcker and Richter 1973) and became immediately attractive for resource extraction. Initial exploration studies in the Atlantis II Deep date back to the late 1960s and early 1970s. In 1976, the German company Preussag signed an exploration contract with the Saudi-Sudanese Joint Commission for the Exploitation of the Red Sea Resources (Blissenbach and Nawab 1982). Intensive exploration and a mining test were conducted in the years 1977–1981 during the Metalliferous Sediment Development Atlantis II Deep (MESEDA) Program (Karbe 1987). Sediment coring, shipboard procedures for ore beneficiation, and resource evaluation were accompanied by multidisciplinary environmental studies including physical, chemical, biological, and modeling investigations of the oceanic, but also of the coastal neritic and even reef regions, for potential disturbance effects. The importance of an environmental risk assessment in conjunction with exploration activities was repeatedly expressed verbally by members of the Red Sea Commission and Preussag and was also published by various authors, for example, by Mustafa

H. Thiel (✉) · L. Karbe · H. Weikert
Institut für Hydrobiologie und Fischereiwissenschaft der Universität Hamburg, Hamburg, Germany
e-mail: hjalmar.thiel@hamburg.de

N.M.A. Rasul and I.C.F. Stewart (eds.), *The Red Sea*,
Springer Earth System Sciences, DOI 10.1007/978-3-662-45201-1_15,

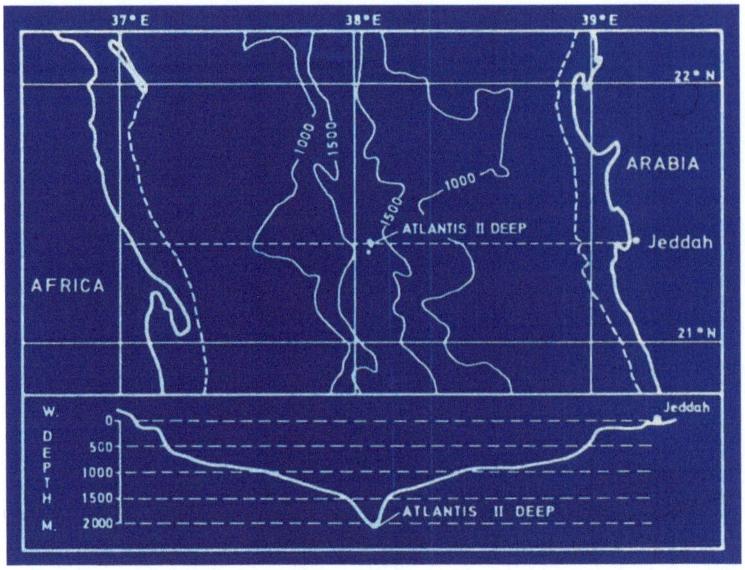

Fig. 1 *Upper part* Map of the central Red Sea showing the location of the Atlantis II Deep in the central graben west of Jeddah. *Lower part* Vertical section across the Atlantis II Deep (*dotted line*)

Fig. 2 Sections of a sediment core from the Atlantis II Deep

and Amann (1978), Mustafa et al. (1980), Nawab (1980), Abu Gideiri (1984); and Nawab (1984). Industry and science also cooperated during the pre-pilot mining test (PPMT) in 1979 (Nawab and Lück 1979; Mustafa and Amann 1980) that included the separation of the metallic components from the sediment by a flotation process on the mining ship (Weber and Ergunalp 1981) and the discharge of tailings back into the sea (Mill 1981; Janke 1981). Financial resources for the environmental risk studies were supplied by the Red Sea Commission and by the German Bundesministerium für Forschung und Technology (Federal Ministry for Research and Technology, today Bundesministerium für Bildung und Wissenschaft, Federal Ministry for Education and Science) in equal amounts. However, one should not overlook the contributions by various universities supplying salaries for scientific and technical personnel, equipment for work at sea, and also for sample and data evaluation in the home laboratories.

Because of declining metal prices, the four-year exploration contract was terminated in the fall of 1981, some five months before the contracted final report delivery was due. While many working reports on cruise results and data evaluations of the various working groups had been independently prepared and published in the Preussag Report Series throughout the investigations, the Red Sea Commission lost interest in the final environmental cooperative risk evaluation and it was proposed to remain an unfinished body of work.

Since all the individual contributions of some 15 scientists were available, summarizing the earlier interim reports and presenting an overall environmental evaluation, the editors of the report decided to complete this task, although with a limited amount of further editing efforts (Karbe et al. 1981/2011). However, this report did not receive a consecutive number in the Preussag Report Series and remained gray literature that was generally unavailable for study and reference. No records exist of how many issues of the report were produced and where they are stored. We have nevertheless referred to the report in our publications (e.g., Karbe 1987; Thiel et al. 1986; Thiel 1987; Weikert 1987).

Together with the developments of mining other resources in the deep sea (polymetallic nodules, massive sulfides, polymetallic/cobalt crusts) in recent years, interest in the extraction of the metalliferous muds from the Atlantis II Deep arose again, and this stimulated renewed consideration of and responsibility for the related environmental problems. Having become aware of the recently revived interest in the exploration and exploitation of the Atlantis II Deep by the companies Diamond Fields and Manafa in 2011, we realized the potential importance of our 1981 report and decided to make it public and more widely available on the Internet after 30 years of dormancy (see www.senckenberg.de/MESEDA). We refer to this report as Karbe et al. (1981/2011).

Two complexes of environmental impacts require particular consideration:

1. The disturbance of the unique multilayered structure of the hot brines in the Atlantis II Deep and the communities living in and around the brine pool and
2. the disturbance of the muddy and rocky seafloor around the Deep, of the deep water in the wider region, and of the biota in these habitats, exerted on them by the discharge, transport, sinking and sedimentation of the fluid, and the particulate phases of the tailings.

The latter are composed of

1. the water masses needed for the transport of the metalliferous mud from its sedimentary source to the ocean surface,
2. unwanted sediment (gange material),
3. upper ocean water added for dilution into the shipboard ore beneficiation process, and
4. organic matter introduced into these processes with the deep and the surface waters.

In Sect. "Conclusions of the 1977–1981 Studies", we reproduce the two complete summarizing chapters (6. Conclusions and 7. Recommendations) of the 1981 report. We also searched the scientific literature for new and relevant results published during the dormancy period (Sects. "Summary of the 1977–1981 Studies" and "MESEDA Results and Recent Publications on Red Sea Oceanography") and ask whether any modification of the conclusions and recommendations detailed in 1981 are essential for an acceptable risk assessment of potential future mining during the second and third decades of this century. Because of tremendous progress in numerical modeling and the development of highly sophisticated instrumentation for oceanographic research, we urge that new risk evaluations on the basis of further risk assessment studies utilizing modern techniques and modeling approaches be conducted.

Conclusions of the 1977–1981 Studies

(Verbatim quotation of Sect. 6 from Karbe et al. (1981/2011, pp. 326–331); information in brackets [ ] is added to the original quotations).

"The presentation of this environmental impact statement marks the conclusion of four years of intense research in the central Red Sea. During this period much has been discovered and many aspects of Red Sea oceanography and biology have been examined in greater detail than was possible during the early pioneering cruises of the middle and late sixties (1960s). However, all the assembled information in this report may only represent a broad description of the gross features of the Red Sea ecosystem as a whole and is the essential foundation to an assessment of the effects of deep-sea mining.

There remain inconsistencies and enigmatic observations in the findings of this study and of necessity the conclusions of this report are tentative and frequently qualified by restrictions. This impact statement may therefore be seen as an interim assessment of the environmental hazards which may arise during the course of mining in the Atlantis II Deep. The findings, conclusions and recommendations of this statement should be subject to revision as further progress is made in the study of the region and as the design of the overall mining system advances to full-scale production.

1. Subsurface, deep-water discharge of tailings is an operational safe-guard of paramount importance to the conservation and protection of marine life in the oceanic Central Red Sea and adjacent coasts and reef areas. There remain unresolved discrepancies between computed characteristics for a long-term continuous discharge of tailings and the observed behavior of slurries pumped intermittently from storage tanks during the Pre-Pilot Mining Test (PPMT). During the proposed Pilot Mining Operation (PMO), every effort should be made to ensure that all aspects of the mining system are operated in a manner representative of continuous full-scale production. At all stages of the PMO and during the initial phase of economic mining, tailings production should be accompanied by monitoring and surveillance programmes.
2. Planktonic organisms living at all depths in the vicinity of the tailings plume(s) are at the most immediate risk from toxic effects of the heavy metals and processing agents present in the discharged waste. Subsequent leaching of the tailings, with the release of zinc, copper, cadmium, mercury and other toxic elements into the water, presents an additional ecological hazard. The monitoring of PPMT operations failed to demonstrate any significant differences between plankton captured with or beyond the tailings plumes. However, these results are inconclusive since monitoring was restricted to a 3-day period preceding the conclusion of the mining test. Under these circumstances it is possible that none of the samples taken outside the detectable plumes was wholly unaffected by previous discharges made at other mine sites in Atlantis II Deep. Thus genuinely uncontaminated 'background' samples suitable for an assessment of short-term effects may not have been obtained during MESEDA II. Leaching tests have also shown that a substantial period may elapse before the onset of

significant dissolution of heavy metal-bearing minerals. The PPMT monitoring results may therefore have failed to reveal the more long-term response of plankton to a residual suspension of tailings.

3. The in situ effect of tailings on oceanic plankton is uncertain, however, confined destruction of the local epipelagic and mesopelagic groups may be of limited significance since many of the organisms that exist in this region are recruited from the Gulf of Aden and waters of the Southern Red Sea. There remains concern as to the fate of plankton which, suffering from clinical effects of sub-lethal toxicity, may transport toxicants to other communities beyond the initial influence of the tailings and thereby establish zones of secondary effects. In addition to the inclusion of substantial inorganic material into the detritus, seston flow through the food web may have a further secondary effect of diminishing the food supply to deep-water benthic life.
4. The replenishment of plankton stock through recruitment from the south would play a major role in the recovery of the Central Red Sea communities. It is therefore essential to establish the population dynamics which regulate the system. Standing stocks and vertical migration have been examined quantitatively during the MESEDA programmes, however, little is known of the background transports which occur and the essential effects with the NW summer monsoon or the mid-SE winter monsoon period from December to January. During the SE monsoon fundamental changes occur in the plankton due to major influx of species from the south. These changes are accompanied by alterations in behavior which may affect vertical migration and zonation throughout the water column. At this time, there is a marked increase in the standing stock below 1,000 m. Since this migration of new immigrants and the replenishment of the local stock is essential to the maintenance of the oceanic community of the region, it is important that in future studies measures are taken to establish that the immigrant population is viable and not a moribund relic of the southern plankton. At present, these seasonal and vertical changes in the plankton may only be described in quantitative terms for parts of the monsoon season; in subsurface waters, observed abundance indicates that there is a natural restraint upon the standing stock, which is low in comparison with that found in other tropical oceanic areas. Below 1,000 m, plankton densities are unusually low and might therefore favor the operation of a deep-water discharge. The effect of toxicants and their ultimate fate during bio-accumulation and transport by the plankton is unknown. The response of plankton to deleterious long-term changes in this oligotrophic environment would require monitoring during both pilot scale and economic mining.
5. The impact of tailings discharge upon benthos would be both physical and toxicological. From the computed spoil area, it may be assumed that tailings are dispersed over 1,500 km^2 of the deep-sea bed. Within this area, there may be identified a region of intense sedimentation which would blanket bottom-living forms creating an azoic zone. Beyond this area of primary tailings deposition and at the periphery of the estimated spoil area, the sediment cover may permit some benthic forms to survive. Excluding toxicological considerations, it may be concluded that if this physical impact of tailings was confined to the specified area and that this area is rendered totally azoic, between 1/25th and 1/50th of the total sea-bed habitat within the Central Graben below 1,500 m [depth] would be destroyed. However, under these circumstances, the portion of the Graben bottom unaffected would be sufficient to sustain the indigenous deep water species and maintain benthic communities of the Red Sea.
6. The residence time of tailings suspension over the seabed is a primary consideration in assessing the toxicological hazard to deep-water benthos. Leaching trials have demonstrated a substantial removal of toxic heavy metals from the solid materials within a period of 2–3 weeks. This process would clearly diminish the toxicological risk of tailings deposited at the edge of the spoil area and may therefore favor survival and promote the recovery of the benthos within the periphery of the deposit. There exists a possible secondary hazard to the benthos due to the continued supply of contaminated seston arising from the bio-accumulation of toxicants within the trophic levels of the epipelagic and mesopelagic zones. The more wide-spread dispersal of such material and the flow of contaminated bottom waters beyond the spoil area pose an incalculable long-term environmental issue, the effect of which may only be assessed by continued monitoring.
7. The release of brines and the associated increase of salinity of bottom waters in the Red Sea may be sufficient to cause localized mortality of biota. The high natural temperature and salinity may account for the oligotrophic conditions which exist in the region such that much of the marine life survives on a marginal existence under stress. Under these circumstances, the biota would be classified as stenohaline [or stenothermal] and would be unable to withstand major changes in salinity or temperature.
8. The total input of heavy metals into the Red Sea from economic mining of the Atlantis II Deep deposit constitutes a significant influx to the basin as [a] whole

irrespective of operational strategies of dispersal. The distribution of such tonnages, as dissolved constituents within the sea, would result in major changes in trace element composition of some water masses. Containment of the major proportion of the heavy metal bearing minerals within an agreed spoil area is therefore the only policy of waste management appropriate to this specific project. The dissolution of minerals and the release of substantial quantities of zinc, copper, cadmium, mercury and other potential toxicants therefore remain a fundamental consideration in this environmental study. The design and operation of tailings discharges must ensure minimum dispersal of the solids, fluids, and leached metals. Deep-water discharge would confine the waste within the Central Graben and, in the absence of significant upwelling, natural bio-geochemical processes of sorption and heavy metal removal may be exploited to limit the dispersal of toxicants beyond an agreed area. The precise definition of this area of tailings influence would require detailed examination of the through-put of heavy metals in the bio-geochemical removal of the contaminants. The PMO would provide the necessary opportunity to monitor the assimilation of heavy metals and subsequent rates of transition to faecal and general detrital material. In this manner an estimate of the effective rates of heavy metal removal may be obtained.

9. Throughout this study there has been an emphasis upon the need for a deep-water discharge of tailings. Examination of biological activity within the epipelagic and mesopelagic zones has led to recommendations that all waste should be restricted to the bottom water below 341 1,000 m. A consideration of the likely effects upon benthos and water chemistry has similarly demonstrated that tailings must be confined to the Central Graben to protect local fisheries and the vulnerable reef and sea bed environment[s] of the coasts and Central Trough. The necessity to minimize the physical area affected by mining in an enclosed oceanic basin has been repeatedly illustrated in this report. However, deep-water discharge would also limit the passage of tailings material through the food web and hence is likely to confine the effects of mining to a limited number of trophic levels within the Red Sea biota. The release of tailings within the zone of plankton migration would subject a larger community of organisms to toxicants and lead to vertical transport of heavy metal[s] to other layers in the water column.
10. Much of the understanding of the Red Sea ecosystem which has arisen out of the MESEDA programme remains only qualitative and restricted to short-term effects of tailings. There remains a clear need for further study of specific aspects of the design of discharge and monitoring the effects of mining. Monitoring of both pilot-scale and economic mining operations will be an essential precaution to avoid unacceptable long-term effects which cannot be anticipated from this study. The major contribution of the environmental programme to date has been to establish background pre-mining conditions in the area which may be used to assess in quantitative terms the effects of bulk tailings discharge in the future. The purpose of further investigation would be to complete studies of seasonal variations in the area, resolve certain inconclusive observations relating to water chemistry and toxicity and to establish more effective techniques for long-term monitoring.
11. An extended programme of research would include synoptic meso-scale and small-scale studies of local currents; measurement of vertical currents; the development of culture techniques for the physiological study of 378 indicator species from the water column; improved methods for the capture of live specimens for tests of salinity and heavy metal tolerance; more detailed monitoring of pelagic fish species; and long-term leaching and bio-accumulation tests."

Summary of the 1977–1981 Studies

(Verbatim Quotation of Sect. 7 from Karbe et al. (1981/2011, pp. 332–334)

"The MESEDA programme, directed to a detailed study of the environmental implications of mining and processing metalliferous muds from the Atlantis II Deep, has provided a wealth of information related to the natural physical, chemical, and biological oceanography of the Central Red Sea. However, investigation of the effects of mining has been restricted to short-term observations of a pre-pilot scale, with the experimental processing plant operating intermittently.

However, mining of the metalliferous muds from the Atlantis II Deep and the disposal of tailings back into the sea presents environmental risks which from careful consideration would seem to be acceptable.

Throughout this report there is a recurrent emphasis upon the need for further direct observation of mining effects. The PMO, planned to be the next step in mining preparation, presents an essential opportunity for sustained long-term monitoring of the environment and the effects of tailings discharge. The following recommendations arise from careful consideration of the Red Sea oceanography and relate to the conduct and surveillance of PMO.

1. The time allowed for responsible preparation of PMO must cover at least a two-year period.
2. For the purpose of establishing relevant monitoring strategies, it is essential that all aspects of PMO, including tailings discharge, are carried out in manner representative of economic mining.

3. Throughout PMO and during economic mining, tailings discharge must be restricted to bathypelagic depth, 1,100 m is proposed as the discharge depth, but subject to any amendment which may arise from further research and monitoring results.
4. During PMO monitoring must include
   - observation of particulate plume development and sedimentation,
   - observation of liquid plume development and its distant extensions,
   - observations on potentially toxic substances in relation to ambient concentrations.
5. Monitoring results from PMO must be used for disposal model verification and correction in order to fully understand the fate of tailings in full scale mining operation and to determine contingencies for emergency events.
6. To avoid later systems' modifications, the schedule of any further environmental programme for PMO-preparation and PMO must include the following tasks:
   - ccontinued analysis of existing hydrographic and current meter records, which at present remains only at a preliminary stage of examination,
   - further study of meso-scale structure in the circulation of the Central Red Sea which is essential to determine the likely fate of tailings dispersed from Atlantis II Deep mining,
   - identification of areas of upwelling and determination of vertical transport rates,
   - continued sampling of sediment surface for iridium trace recovery and disposal test evaluation,
   - more detailed evaluation of plankton samples,
   - concluding studies of seasonal and other long-term temporal and spatial variations in pelagic life and water chemistry,
   - further determinations of heavy metals in water, sediment, and organisms,
   - geochemical analysis of heavy metal speciation, transport, precipitation and long-term dissolution, and leaching of tailings solids,
   - the identification and culture of agreed planktonic and benthic indicator species for physiological examinations of sub-lethal or chronic toxic effects arising from economic mining.
7. Further research must now be directed to the remaining needs of a monitoring programme to be conducted during economic mining. This work must include
   - quantitative examination of replenishment of plankton by species from the Gulf of Aden to establish the likely rates of recovery of populations destroyed by mining,
   - investigations on the exchange of nutrients and organisms between the coastal-neritic and the oceanic regions,
   - studies to discriminate between the effects of mining and the input from other land-derived natural and man-made sources."

MESEDA Results and Recent Publications on Red Sea Oceanography

More than 30 years have passed since data collection, risk evaluation, and report compilation for the MESEDA program were completed. The Red Sea remained in the focus of many research projects, documented in this volume, and results may provoke new views on the evaluation of the early data sets. This does not pertain to the papers resulting from the MESEDA studies (Karbe 1980; Mill 1980; Weikert 1980a, b, 1982; Thiel 1980, 1983; Beckmann 1984; Thiel and Weikert 1984; Theeg 1985), or from summarizing the MESEDA environmental report (Thiel et al. 1986; Karbe 1987; Thiel 1987; Weikert 1987). Whereas a number of publications on the Atlantis II Deep appeared during the period of exploration dormancy, none of these examined further any risk that exploitation of the metalliferous mud might introduce into this unique habitat and its microbial and potential invertebrate life. Only two papers refer to some of the environmental studies and related problems: Whereas the book by Egorov et al. (2012) results from a training exercise of potential and young scientists and ignores almost all pertinent literature, the work of Bertram et al. (2011) essentially represents an economic evaluation and a discussion of the legal situation concerning the Atlantis II Deep resources. They briefly mention, but do not discuss, the "possible adverse environmental impacts" and the "high degree of uncertainty regarding possible environmental risks." They also point to the societal aspects of mining: The potential impacts of mining imply "that the costs faced by society are likely to be larger than the direct costs of resource extraction."

Potential Mining Impacts on the Atlantis II Deep

Because of its unique character as a warm brine pool, the Atlantis II Deep was repeatedly a subject of investigations since the time it was thoroughly examined in the middle of the 1960s. Recent papers were published by, for example, Hartmann et al. (1998), Anschutz et al. (2000), Swift et al. (2012), and Laurila et al. (2014) and in this book by Anschutz (this volume), Hovland et al. (this volume), and Schmidt et al. (this volume).

The Atlantis II Deep constitutes the largest known resource of metalliferous sediment and is therefore under environmental pressure of potential industry penetration and

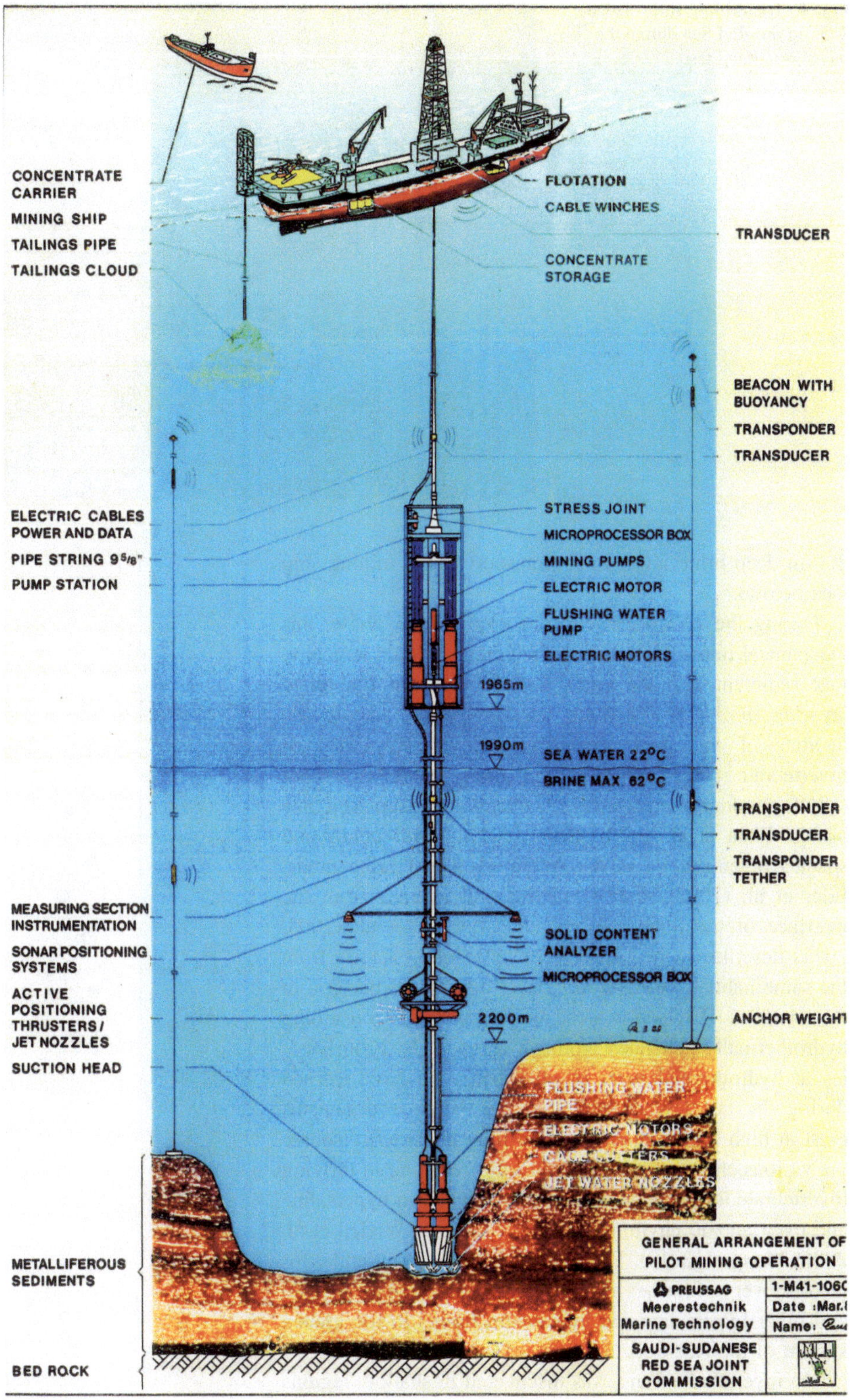

Fig. 3 General arrangement of the pre-pilot mining test (from Preussag marine technology)

commercial exploitation. It is not known which mining technique will finally be applied, but one could assume that this will not deviate much from the methods developed in the 1970s (Fig. 3) and tested in 1979 during the pre-pilot mining test (PPMT; Nawab and Lück 1979). A vibrating suction head was deployed from the drilling ship Sedco 445 (Fig. 4) to suck in the watery surface of the metalliferous muds and its vibration, aided by water jets, loosened the deeper more consolidated sediment. This material was diluted by brine and Red Sea deep water from above the

Fig. 4 The drilling ship "Sedco 445" in the Red Sea during the pre-pilot mining test

200-m deep brine layer for its transport to the mining ship (see below).

During the 15- to 20-year-long exploitation phase, this mechanical impact will disturb the unique layered structure, mix sediment into the brine, and derange the fine-tuned layering of the brines. Such intrusion would disturb the habitats of microbial organisms and probably also of invertebrate communities as yet rarely studied. Trüper (1969) found bacterial sulfate reduction in deep Red Sea brines, Fiala et al. (1990) discovered a novel genus and species of eubacteria in the Atlantis II Deep brine muds, Eder et al. (2002) studied bacteria and archaea from the interface of the Shaban Deep, and Antunes et al. (2008, 2011) described a new lineage of halophilic bacteria from the same habitat. Wang et al. (2011) discovered bacteria in the Atlantis II Deep which are capable of consuming hydrothermally generated aromatic compounds. From studies at hydrothermal vents in all oceans, we have learned during the last few decades that specific microorganisms exist at high temperatures and constitute the energy source via various chemoautotrophic pathways resulting in thriving invertebrate life. Halophiles occur abundantly in hypersaline and even anoxic environments (e.g., Van der Wielen et al. 2005; Ma et al. 2010). There seems to be high potential for the discovery of further microbial life in the Atlantis II Deep and other hypersaline and hyperwarm basins in the graben system of the Red Sea.

No invertebrate fauna was discovered in shallow regions of the upper layer in the brine pool of the Atlantis II Deep. However, Karbe et al. (1981/2011), Thiel et al. (1986), and Karbe (1987) report from the MESEDA studies that there were high concentrations of metals (e.g., Fig. 5) in

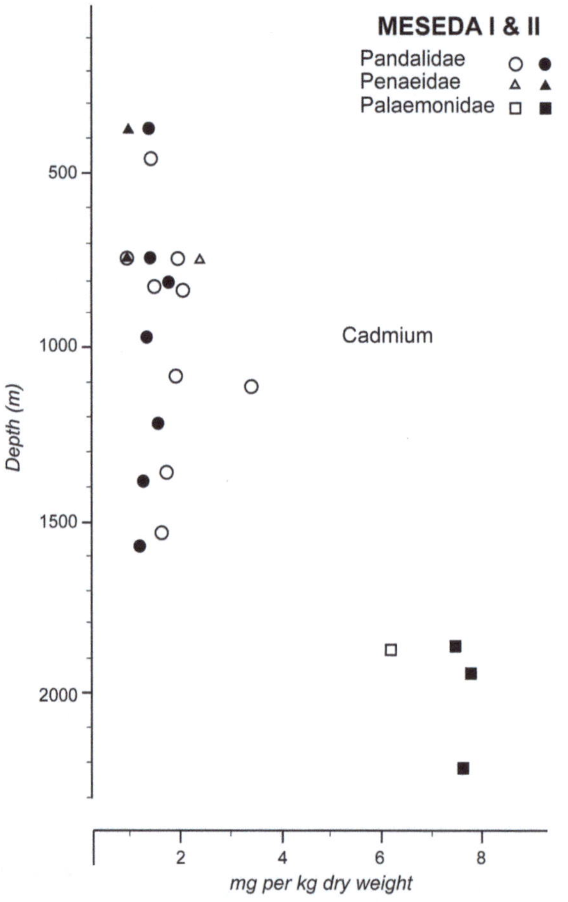

Fig. 5 Concentrations of cadmium in three bottom dwelling shrimp taxa along a depth profile. All sampling stations close to the Atlantis II Deep. Note the high concentrations at great depths. Determinations by atomic absorption spectrometry

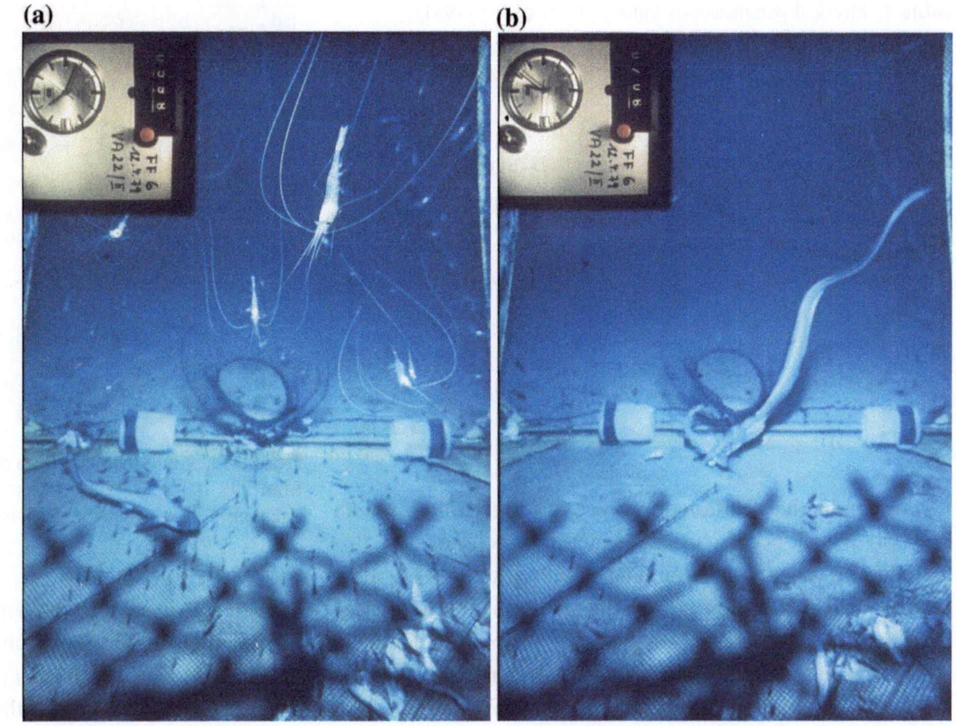

Fig. 6 View into the baited photo-trap at 740 m depth: trap entrance in background center. On the *left* and *right* are baited plastic traps to catch small organisms. *White dots* on the trap floor are the remains of bait (fish). **a** The shark *Iago omanensis* (Norman) and shrimp note the long antennae of the shrimps. **b** *Muraenesox cinereus* (Forsskal) preying on I. *omanensis*—about 2 h later

some species of shrimp and fish (Figs. 6 and 7) living at great depth close to the seafloor, demonstrating faunal adaptation to high levels of metals.

We also wonder whether all the brines in the pool are expelled from below the brine pond or whether hydrothermal vents might additionally exist on the rocky slopes above the upper brine layer with the effluents flowing down to and penetrating the brines from the side and sinking to a depth of equal density. The various brine layers in the pool Swift et al. 2012, Anschutz (this volume), should be able to accommodate such inflow of hydrothermal brines from above the pool surface.

Before any extraction of sediments from the Atlantis II Deep commences, the bacteria and archaea communities in the brine layers and the particulate matter-loaded discontinuity layers between them should be studied thoroughly. The potential occurrence of invertebrate life close to the brines and on the rocky slopes around the brine pool, ultimately associated with hydrothermal vents, should stimulate research activities and discussions on the need for protective measures for the Atlantis II Deep. Even without any hydrothermal effluents as a source for microbial communities or invertebrate species living in association with the Atlantis II Deep, it should be considered whether this unique geochemical setting should be disturbed by the extraction of metal resources or whether it should be preserved for scientific purposes and as a microbial gene pool, also for potential commercial exploitation.

Potential Mining Impacts on the Area Surrounding the Atlantis II Deep

Although the mining process at the seafloor and the associated disturbance may be sealed off to some extent by the overlying brines, metalliferous sediment may penetrate the 200-m brine cover of the ore and contaminate the lower column of the normal Red Sea deep water. However, we assume that this effect would remain relatively limited compared to the massive impact introduced by the discharge of the tailings. For an evaluation of this impact, we apply figures assumed to be vital for economic stability and sustainability some 35 years ago. Developmental trends tend to increase commercial turnover, and therefore, these mining and discharge values are likely to be minimum targets for future exploitation.

The mining process will extract and deliver 100,000 metric tons (mt) of metalliferous mud per day to the mining platform (Fig. 5), including 10,000 mt of solids. This original sediment mass will be diluted with the same amount of seawater, predominantly bottom water from above the 200 m deep brine layer, but also brine water steadily flowing into the extraction pit and replacing the mud removed. On the mining platform, this ore will be subjected to a beneficiation process, thereby separating 300 mt of a metal concentrate daily by flotation. The concentrate will then be shipped to a land-based processing plant for further refining.

Table 1 Physical parameters of tailings, from Lange (1981)

	Range	Average
pH	–	6.0
Solid content (g/l)	18–30	25
Salt content (g/l)	100–200	150
Density of tailing slurry (g/cm^3)	1.06–1.16	1.10
Density of dry solids (g/cm^3)	2.5–4.2	3.2
Temperature at discharge point (°C)	30–40	35
Grain size after dispersion	60 %	<2 μm
	20 %	2–6 μm
	10 %	6–20 μm
	10 %	>20 μm

For conducting the flotation process, further dilution with 200,000 mt of ocean surface water is essential, and this includes 60 kg of plankton and small nekton every day. Flotation is also optimized by the addition of 1,000 g of copper sulfate and 600–1,000 g of potassium amyl xanthate, a thiocarbonate acid, per metric ton of water. Xanthates are poisonous (Block and Part 1986; Block and Wicklund Glynn 1992), but may be recycled to some extent. Mining and flotation will result in the daily production of almost 400,000 mt of wastewater, the tailings, containing 9,700 mt of dry sediment, disrupted plankton and nekton, and flotation chemicals. These tailings would then be discharged back into the sea, day by day for the lifetime of the mine.

Tailings discharge constitutes the most severe complex of environmental problems. It seems rather impossible to imagine the mass extraction, the mass transport, the mass flow through the flotation basins, the mass inflow of floatation chemicals, and finally the discharge of such an enormous amount (400,000 mt of water-sediment slurry) almost every day for 20 or more years. Is this an acceptable process for keeping the Red Sea healthy? To help envision this huge daily mass of original sediment, we suggest visualizing the following comparison: For the transport of 10,000 mt of solids, one would require 460 standard 20-foot (6 m) containers filled to weight loading capacity. When lined up, this corresponds to a container chain that is 2.8 km in length every day for many years. This scenario pertains only to the solids in the tailings and precludes concentration from the slurry, ship transport ashore, and land-based containment.

Lange (1981) compiled data on the range and average of tailings composition (Table 1) and explained the high variability by the vertical and horizontal resource variation in the original deposit. The uptake of varying amounts of brine and deep Red Sea waters also contributes to the wide range in composition and the subsequent effects of tailings distribution after discharge.

For the prediction of tailings distribution after discharge, a gravity flow model (Jancke 1981) and a momentum jet flow model (Mill 1981) were calculated, and during the PPMT in 1979, a test discharge with a release of the tailings in an agreed upon depth of 400 m was achieved. The results of models and tests deviated from each other (Mill 1981; Mill and Jancke 1981; Thiel et al. 1986). Results of the momentum jet flow model were regarded as predicting worst-case conditions. As early as during the 1977 MESEDA survey, the ecologists demanded a discharge depth well below the lower layer of plankton and nekton vertical migration at >800 m depth (Figs. 8 and 9; e.g., Weikert 1980a, b, 1981, 1982; Thiel et al. 1986). Karbe et al. (1981/2011) suggested the discharge into the central graben in at least 1,000 m depth, where only small amounts of plankton and benthos biomasses were found to exist. Discharge at greater depth below the zones of plankton vertical migration had initially provoked severe controversy between scientists and economists, yet to introduce the tailings at a minimum 1,000 m depth became generally accepted (Abu Gideiri 1984; Nawab 1984, 2001; Thiel et al. 1986; Amann 1989). However, the effects of the discharge are not well understood.

Fig. 7 View into the photo-trap back on deck: *M. cinereus* and shrimp from 740 m depth. Not visible: *M. cinereus* carries the swallowed *I. omanensis* in its stomach as seen after dissection

Applying this discharge depth to the momentum jet flow model, Mill (1981) calculated a seafloor blanketing area of 1,500 km^2 over the mining site lifetime, with a sediment cover of up to 4 cm/year at its center. This area is assumed to stretch out to the northwest below 1,500 m depth and would disturb about 2 % of the total graben floor. In the case of additional tailings drift and settlement in a southerly direction, the impacted area would be larger, but the cover less thick. The blanketing of 2–4 % of this deep Red Sea habitat would probably not wipe out any species. Under this limited aspect, the impact was believed to be acceptable (Thiel 1981; Thiel et al. 1986; Karbe 1987) taking into account the morphologically based species concept. A modified evaluation might arise and follow from a modern genetic definition of populations.

However, the particulate phase of the tailings does not constitute the only impact material in the tailings, and the benthic organisms are not the only community components

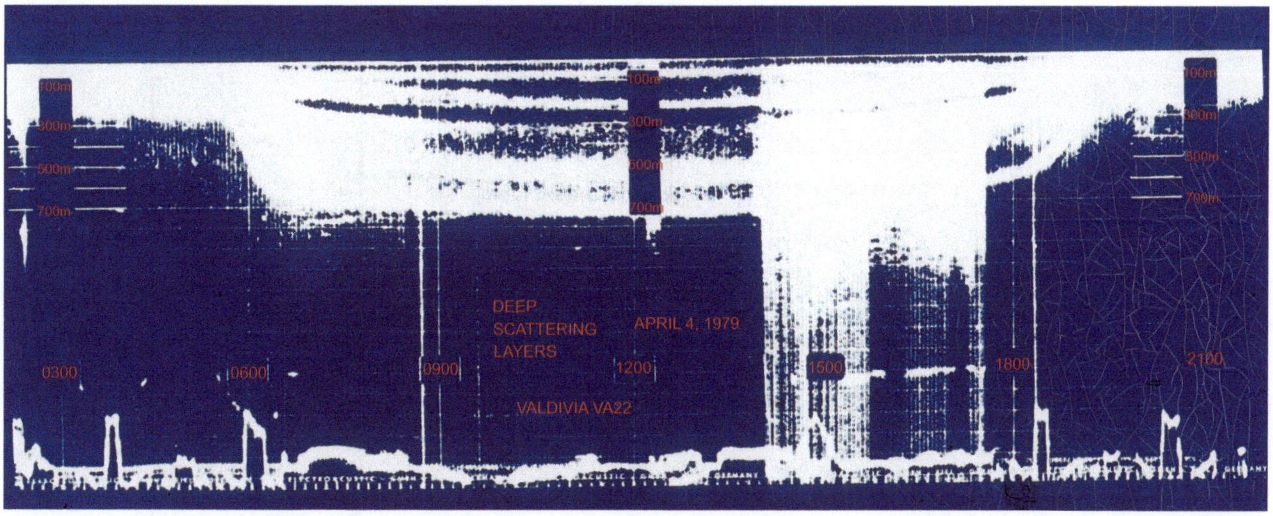

Fig. 8 Echosounder record demonstrating diurnal vertical migration of plankton and nekton

Table 2 Fate of tailings after 40 days of permanent discharge at commercial scale, compiled from data by Mill (1981)

Solid phase	Amounts in suspension	120,000 mt
	Area of plume	260 km^2
	Amounts deposited	260,000 mt
	Area of deposit	1,300 km^2
Fluid phase	Area of plume	13,000 km^2

affected. The discharged water mass, dissolved substances originating from the metalliferous muds and its interstitial water, the flotation chemicals, and probably also the organic matter captured with the added surface water mass need to be considered in any risk evaluation. Table 2 summarizes the predicted effects.

According to Mill's model, the material will flow down to greater depth before a plume spreads out and transport successively merges into the local current. The particulate mass is thought to fall out of its original water body. However, the finer, micron size particulate matter in the tailings (Table 1) that constitutes the overwhelming mass may drift for extended times and distances. Particles in aggregates entering the suction head will presumably separate during ascent to the surface, during the flotation process, or while descending in the discharge tube. About 60 % of the sedimentary matter falls into the range of 1–2 μm, while 20 % measures 2–6 μm. These fine particulates may not settle to the seafloor immediately after release unless some new particle aggregation occurs. A portion of these fine particles may remain in the fluid phase for an extended period of time together with the dissolved substances. The fluid tailings will successively mix with the local water body and will be transported with the currents to adjacent and, probably also, far distant areas.

The model calculations of Mill finally arrive at a steady-state situation for the particulate phase, but it remains unclear what the behavior of the particulate and the fluid phases under permanent tailings inflow at nearly the same position over extended times will be. How will the tailings and the local water mass mix? Will they build up a water body of higher density, and at what distance will the mixed water body lose its individual character? Will the transport be restricted to horizontal directions or drift also in a vertical direction? What is the reaction of planktonic and nektonic organisms to the uptake of dissolved heavy metals and floatation chemicals originating from the tailings?

Karbe and Nasr (1981) and Karbe (1987) describe the results of metal leaching experiments with original sediment and tailings material. They conclude that the tailings are toxic due to dissolved metal components and synergistic effects of these compounds may affect at least non-adapted shallow water organisms. Benthic animals may be subjected to higher metal concentrations originating from leaching of the particulate and settled tailings component. After flotation, a considerable mass of metals remains in the tailings. Daily discharge values of poisonous metals amount to, for example, 30–180 mt of zinc, 20–50 mt of copper, 1,000–3,000 kg of arsenic, 100–500 kg of silver, 50–700 kg of cadmium, and 4–200 kg of mercury (Karbe 1987). Metal concentrations will be diluted after discharge, but not much is known about synergistic effects of various metal components nor of the additional effects of the added copper sulfate and xanthates to enhance metal separation by flotation.

Many questions remain to be answered for an unbiased risk evaluation, and the discharge below 1000 m depth is certainly not sufficient as a precautionary measure. Both reports (Lange et al. 1983; Karbe et al. 1981/2011) point repeatedly to the limitations of the data and of the risk

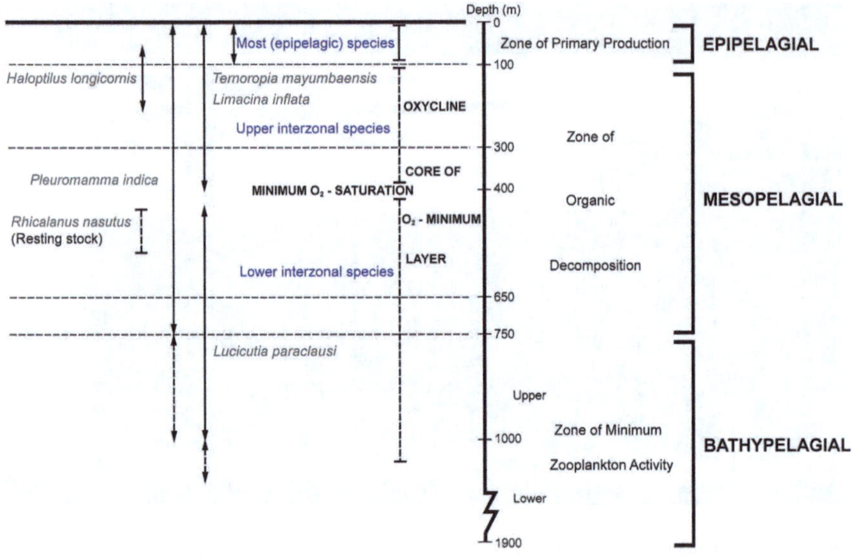

Fig. 9 Structure of the pelagic realm (from Weikert 1982)

evaluation. But the results of research conducted since the finalization of the 1981 environmental risk assessment (Karbe et al. 1981/2011) are of minor relevance to solving the problems.

During the post-MESEDA decades, most of the hydrographic investigations in the Red Sea concentrated on the region of deepwater development in the north or in the south to study the water exchange with the Indian Ocean. Cember (1988) calculated an average water residence time for the total Red Sea of only 36 years, much shorter than anticipated earlier, and probably also affecting the tailings distribution by faster and more distant transport. Only Quadfasel and Baudner (1993) reported on studies in the central Red Sea, describing large gyre structures reaching down to 300 or more meters. These do not have any connection with the gyres discovered in 1,700 m depth in the region of the Atlantis II Deep during the MESEDA campaigns. Furthermore, the physical measurements of Tragon and Garrett (1997) and Meeking and Warner (1999) and the modeling efforts of Sofianos and Johns (this volume) cover only the upper few hundred meters. However, Al-Barakati et al. (2002) describe a three-dimensional model concerning the water circulation in the Red Sea. They also found indications for gyres in shallow and deep regions and strong upwelling in the summer between 17°N and 19°N, that is, south of the Atlantis II Deep. Whether these structures may influence tailings distribution remains an open question. Klevjer et al. (2012) confirm the MESEDA observations (Weikert 1980a, b, 1981, 1982; Thiel et al. 1986; Beckmann 1995) of daily vertical migration down to about 850 m depth for the northern Red Sea, which prompted us as early as in 1977 to demand tailing discharge at a depth of at least 1,000 m into the central graben system (Figs. 8 and 9). Various plankton studies conducted in the central Red Sea (e.g., Weikert and Koppelmann 1993, 1996; Böttger-Schnack et al. 2004) report more details on biomass distribution, life cycles and vertical migration, and confirm the general ecosystem characteristics. Thiel et al. (1987) describe results from respiration studies of the seafloor community and agree with earlier reported limited benthic stock and turnover rates, a characteristic of the warm water and the energy-limited ecosystem.

The results published later than 1981 support earlier findings, but none is directly relevant for tailings disposal. In publications referred to in this section, no reference appeared in relation to Red Sea metalliferous mud extraction and related risk evaluation. This is due to the fact that ore extraction and tailings discharge have not been an issue of concern for more than 30 years. Any additional environmental study was economically not justified after the retreat of the Red Sea Commission. The website of Diamond Fields is scarcely noticed by researchers and does not stimulate environmental considerations. However, all the results presented above would not cause any reduction in further research as demanded in the conclusions and recommendation compiled in the reports on risk evaluation (Karbe et al. 1981/2011; Lange et al. 1983). In contrast, some of the results add new arguments for extended research and precaution.

Other Potential Impacts

During mining activities, environmental impacts may arise from accidents, e.g., transport tube failure, or from waste dumping by ship crews. The breakage of the 2,000-m-long tube for lifting the ore from the seafloor to the ocean surface we regarded as a limited impact. Such failure would be observed almost immediately, and mining would be interrupted. The maximal amount of mud that may spill into the ocean should be restricted to the volume of the lifting tube, about 40 mt. We therefore do not consider such event as a seriously harmful impact.

Severe contamination of the mining sites may occur from all the wastes (refuse and sewage) produced on the mining platform and by any transport or support vessel. This material should be somehow concentrated on board the ships, transported back to the shore, and finally disposed of in land disposal sites. The 1972 London (Dumping) Convention (LC) and the 1996 London Protocol should protect the seas from such contamination. However, for unknown reasons, deep-sea mining is exempted from these international regulations and dumping from ships (cf. Schriever and Thiel 2013) at sea is legally allowable for the mining fleet. Before restrictive paragraphs are issued by the Intermaritime Organization and such regulations have entered into force, we hope that mining and shipping companies will be conscientious enough to protect the oceans and arrange land-based disposal sites.

Environmental Risk Evaluation for the Forthcoming Two or Three Decades

Having been investigators in the MESEDA program and also the editors of the final environmental risk assessment, we feel responsible for the continuation of environmental considerations connecting the environmental studies of the original MESEDA phase with a new phase of environmental research. We frankly express that our 1981/2011 report and its recommendations are insufficient to meet the standards of research and assessments applicable during the second and third decades of this century. The 1981 completed, but never published in full length, environmental report (Karbe et al. 1981/2011) is available on the Internet (see references). It should be critically reviewed and may serve as a basis on which to build future environmental research. Fifteen scientists cooperatively wrote this report at the end of a four-year contract phase expecting that the Red Sea Commission would strive toward Atlantis II Deep exploitation (see Blissenbach and Nawab 1982). The PPMT (including test mining, test ore beneficiation, and test tailings discharge) had been conducted successfully in 1979. While compiling the 1981 report, no indication of policy change was recognizable. All the authors were convinced that the Red Sea Commission and the Preussag company were progressing toward the next developmental phase, the PMO, still a technically scaled-down, but long-term, mining event. This is also expressed in detail in the report on the MESEDA II and III cruises delivered to the German research ministry (Lange et al. 1983, pp. 104 and 110–112). Therefore, many of the 1981 conclusions and recommendations made clear the need for further research and monitoring activities, and related such an advanced risk assessment to a PMO envisaged as expedition MESEDA IV to be conducted only a few years later (see Lange et al. 1980; Mustafa et al. 1980; Mill 1980; Nawab 1980).

The situation today differs from that at the time of the MESEDA report's completion: A new commercial consortium is active in exploration and is presumably preparing for exploitation of the Atlantis II Deep. The respective license has been awarded for a period of 30 years. Mining techniques may be modified, a new PPMT may be necessary, a PMO will follow, but all this will probably happen some years in the future. The intervening time should not only be used for further exploration. Equally important is the undertaking of more sophisticated environmental investigations that take into account new scientific hypotheses and utilize modern techniques. The existence of more diverse biota in the brine pool itself, and probably also in its surroundings above the upper brine surface and in the transition zone toward normal Red Sea deep water, should be intensively investigated. Unknown biota may come to light by submersible or remotely controlled vehicle deployment.

vThe environmental effects of tailings discharge should be re-evaluated on the basis of a much broader knowledge on hydrography. More sophisticated numerical models for the particulate and fluid phase dispersion are now available. Current measurements in and around a potential discharge area should help in a better understanding of long-term water and particle transport, and measurements of the discharge plume would provide better insights into particle sedimentation processes and plume fate. A large-scale tailings discharge experiment with the fluid and the particulate phases labeled for independent fate studies would seem advisable together with more advanced modeling calculations. It would be a serious error to postpone further environmental studies until a new PPMT or even to wait for a new PMO. Effective environmental studies and their evaluation deserve an extended period of time. Discussions should be commenced immediately. A core group of oceanographers of all disciplines should deliver a framework and particularly calculate the timescales for adequate environmental research activities. Industry does not and cannot be expected to have a clear concept of the multiyear timescales required for environmental research and risk evaluation. Therefore, industry does need early interaction with scientists for consultation and advice.

As retired scientists with many years of experience related to deep-sea mining, it is not our aim to formulate a new environmental research program. Yet we feel responsible for making the earlier environmental studies known to industry, science, and the public. The MESEDA studies contributed greatly to our overall knowledge on the Red Sea and considered the environmental problems of metalliferous mud mining under the conditions prevalent more than 30 years ago. The results then obtained are far from being sufficient for a complete and up-to-date impact evaluation, particularly after more than 30 years have passed, and for predictions applicable to two to three decades in the future. It is imperative that a younger generation of marine scientists from all disciplines examines what has been achieved in the past and analyses what should be done in the future for an extensive environmental risk assessment. Also, it is essential that this new group of scientists is endowed with an absolutely independent status. This model of cooperation for deep-sea mining, first verified in MESEDA (see below) and then demanded again by Hoagland et al. (2010), leads to severe and controversial discussions, but also to a common understanding that ultimately serves the protection of the ocean commons. With this in mind, all the results of new environmental investigations and risk assessment should be widely disseminated for discussion in scientific journals and not remain hidden in confidential reports.

We are convinced that the spirit of the Red Sea Commission expressed by Nawab (1984), then its vice president, is still alive and guides the now responsible supervisors: "From the very beginning it was decided that if it were found that mining and/or processing of the metalliferous deposits was going to result either in the destruction of or in severe damage to the environment the whole programme would be shelved."

Acknowledgments We wish to thank Eric J. Foell for streamlining our use of the English language in the manuscript. Louiesito Abalos has kindly adjusted the color images for better printing quality, as all figures were originally prepared more than 30 years ago. Drs. Gerd Schriever, Bettina Martin, and Bernd Christiansen were helpful in the preparation of the photographs. We are most grateful to Dr. Teresa Radziejewska (Poland) and Dr. Rahul Sharma (India) for their helpful reviews of our manuscript. The German contributions to the environmental MESEDA project were supported by the German Ministry of Research and Technology under Grant R 301/R 309 and by the University of Hamburg. We also gratefully acknowledge the positive cooperation with the Red Sea Commission, the Preussag company, particularly with Dr. Harald Bäcker, Chief Scientist of the Preussag team and with all our scientific, technical, and student team members. It has been great fun to work together with scientists from the Republic of the Sudan and the Kingdom of Saudi Arabia. The cooperative spirit of the many individuals from industry, administration, and science remains a positive model for industrial developments concerned about environmental protection, particularly in that no restrictions were placed on the publication of scientific results and our personal views on the potential impacts of mining.

References

Abu Gideiri YB (1984) Impacts of mining on the central Red Sea environment. Deep-Sea Res 31A:823–828

Al-Barakati AMA, James AE, Karakas GM (2002) A three-dimensional hydrodynamic model to predict the distribution of temperature, salinity and water circulation of the Red Sea. J Mar Sci, King Abdulaziz Univ, 13: 3–17

Amann H (1989) The Red Sea pilot project—lessons for future ocean mining. Mar Min 8(1):1–22

Anschutz P, Blanc G, Monin C, Boulegue J (2000) Geochemical dynamics of the Atlantis II Deep (Red Sea.): II. Composition of metalliferous sediment pore waters. Geochim Cosmochim Acta 64 (23):3995–4006

Antunes A, Rainey FA, Wanner G, Taborda M, Pätzold J, Nobre MF, da Costa MS, Huber R (2008) A new lineage of halophilic, wall-less contractile bacteria from a brine-filled deep of the Red Sea. J Bacteriol 190:3580–3587

Antunes A, Ngugi DK, Stingl U (2011) Microbiology of the Red Sea (and other) deep-sea anoxic brine lakes. Environ Microbiol Rep 3 (4):416–433

Bäcker H, Richter H (1973) Die rezente hydrothermal-sedimentäre Lagerstätte Atlantis II Tief im Roten Meer. Geol Rundschau 62 (3):697–741

Beckmann W (1984) Mesozooplankton distribution on a transect from the Gulf of Aden to the Central Red Sea during the winter monsoon. Oceanol Acta 7:87–102

Beckmann W (1995) Der Einfluss der großräumigen Wasseraustauschvorgänge auf den Zooplanktonbestand des Meeres und sein trophisches Gefüge. Dissertation thesis, University of Hamburg, 167 pp

Bertram C, Krätschell A, O'Brien K, Brückmann W, Proelss A, Rehdanz K (2011) Metalliferous sediments in the Atlantis II Deep. Assessing the geological and economic resource potential and legal constraints. Kiel Working Paper No. 1688, Mar 2011 (Kiel Institute for the World Economy), 30 pp

Blissenbach E, Nawab Z (1982) Metalliferous sediments of the seabed: The Atlantis II Deep deposits of the Red Sea. In: Mann-Borgese E, Ginsburg N (eds) Ocean Yearbook 3, nonliving resources. University of Chicago Press, Chicago, pp 77–104

Block M, Part P (1986) Increased availability of cadmium to perfused rainbow trout (*Salmo gairdneri*, Rich.) gills in the presence of complexing agents diethyldithiocarbanate, ethyl xanthate, and isopropyl xanthate. Aquat Toxicol 8:472–476

Block M, Wicklund Glynn A (1992) Influence of xanthates on the uptake of 109Cd by Eurasian dace (*Phoxinus phoxinus*) and rainbow trout (*Oncorhynchus mykiss*). Environ Toxicol Chem 11:873–879

Böttger-Schnack R, Lenz J, Weikert H (2004) Are taxonomic details of relevance to ecologists? An example from oncaeid microcopepods of the Red Sea. Mar Biol 144:1127–1140

Cember RP (1988) On the sources, formation and circulation of Red Sea deep water. J Geophys Res 93(C7):8175–8191

Diamond Fields International Ltd (2011) Atlantis II Red Sea deeps—the world's first deep sea metal mining license. www.diamondfields.com/s/AtlantisII.asp. Accessed Jan 2014

Eder W, Schmidt M, Koch M, Garbe-Schönberg D, Huber R (2002) Procaryotic phylogenetic diversity and geochemical data of the brine—seawater interface of the Shaban Deep, Red Sea. Environ Microbiol 4(11):758–763

Egorov L, Elosta H, Kudla NL, Shan S, Yang KK (2012) Sustainable seabed mining: guidelines and a new concept for Atlantis II Deep. The LRET Collegium 2012 Series 4, XVI + 168 pp. www.southampton.ac.uk/engineering/research/groups/fsi/lrf/lrf_collegium_2012_book_series.page. Accessed Jan 2014

Fiala G, Woese CR, Langworthy TA, Stetter KO (1990) *Flexistipessinus arabici*, a novel genus and species of eubacteria occurring in the Atlantis II deep brines of the Red Sea. Arch Microbiol 154 (2):120–126

Hartmann M, Scholten JC, Stoffers P, Wehner F (1998) Hydrographic structure of brine-filled deeps in the Red Sea—new results from the Shaban, Kebrit, Atlantis II and Discovery Deep. Mar Geol 144: 311–330 and correction 331–332

Hoagland P, Beaulieu S, Tivey MA, Eggert RG, German C, Glowka L, Lin J (2010) Deep-sea mining of seafloor massive sulfides. Mar Policy 34:728–732

International Seabed Authority (2013) International Seabed Authority reaches milestone with the approval of the first exploration plans and institution of overhead charges for administration of contracts. Nineteenth Session (SB/19/18), Kingston, Jamaica, 15–26 July 2013, 11 pp

Jancke K (1981) Computer simulation of a gravity flow. In: Karbe et al. (eds) (1981/2011) Mining of metalliferous sediments from the Atlantis II Deep, Red Sea: pre-mining environmental conditions and evaluation of the risk to the environment. EIS Report to the Saudi-Sudanese Red Sea Joint Commission, Jeddah, pp 276–280

Karbe L (1980) Plankton investigations in an exposed reef of the central Red Sea (Shaab Baraja, Sudan). In: Abu Gideiri YB (ed) Proceedings of the symposium on the coastal and marine environment of the Red Sea, Gulf of Aden and Tropical Western Indian Ocean, 9–14 Jan 1980, vol 2, Khartoum. International Printing House, Khartoum, pp 519–540

Karbe L (1987) Hot brines and the deep sea environment. In: Edwards AJ, Head SM (eds) Key Environment, Red Sea. Pergamon Press, Oxford, pp 70–89

Karbe L, Nasr D (1981) Chemical and toxicological characteristics of the tailings material. In: Karbe et al (eds) (1981/2011) Mining of metalliferous sediments from the Atlantis II Deep, Red Sea: pre-mining environmental conditions and evaluation of the risk to the environment. EIS report to the Saudi-Sudanese Red Sea Joint Commission, Jeddah, pp 245–253

Karbe L, Thiel H, Weikert H, Mill AJB (eds) (1981/2011) Mining of metalliferous sediments from the Atlantis II Deep, Red Sea: pre-mining environmental conditions and evaluation of the risk to the environment. EIS report to the Saudi-Sudanese Red Sea Joint Commission, Jeddah 352 pp. www.senckenberg.de/MESEDA

Klevjer TA, Torres DJ, Kaartvedt S (2012) Distribution and diel vertical movement of mesopelagic scattering layers in the Red Sea. Mar Biol 159:1833–1841

Lange J (1981) Tailings disposal. In: Karbe et al (eds) (1981/2011) Mining of metalliferous sediments from the Atlantis II Deep, Red Sea: pre-mining environmental conditions and evaluation of the risk to the environment. EIS report to the Saudi-Sudanese Red Sea Joint Commission, Jeddah, pp 238–244

Lange J, Bäcker H, Post J, Weber H (1980) Plans and tests for a metal concentration and tailing disposal at sea. In: Abu Gideiri YB (ed) Proceedings of the symposium on the coastal and marine environment of the Red Sea, Gulf of Aden and Tropical Western Indian Ocean, 9–14 Jan 1980, vol 3. International Printing House, Khartoum, pp 65–126

Lange J, Post J, Bäcker H, Karbe L, Thiel H, Weikert H (1983) Abbau von Erzschlämmen des Atlantis II-Tiefs, Rotes Meer: Charakterisierung der aktuellen Umweltbedingungen und Bewertung der Auswirkungen auf das Ökosystem. Forschungsvorhaben MESEDA (Metalliferous Sediments Atlantis II-Deep) R 301 / R 309, Abschlussbericht an das Bundesministerium für Forschung und Technologie, 120 pp

Laurila TE, Hannington MD, Petersen S, Garbe-Schönberg D (2014) Early depositional history of metalliferous sediments in the Atlantis II Deep of the Red Sea: evidence from rare earth element geochemistry. Geochim Cosmochim Acta 126:146–168

Ma Y, Galinski EA, Grant WD, Oren A, Ventosa A (2010) Halophiles 2010: Life in saline environments. Appl Environ Microbiol 76:6971–6981

Meeking S, Warner MJ (1999) Ventilation of Red Sea waters with respect to chlorofluorocarbons. J Geophys Res 104(C5):11087–11097

Mill AJB (1980) Deep ocean mining in the Red Sea. Theory and practice of tailings disposal. In: Abu Gideiri YB (ed) Proceedings of the symposium on the coastal and marine environment of the Red Sea, Gulf of Aden and Tropical Western Indian Ocean, 9–14 Jan 1980, vol 3. International Printing House, Khartoum, pp 33–63

Mill AJB (1981) Computer simulation of discharge from production scale plant. In: Karbe et al (eds) (1981/2011) Mining of metalliferous sediments from the Atlantis II Deep, Red Sea: pre-mining environmental conditions and evaluation of the risk to the environment. EIS Report to the Saudi-Sudanese Red Sea Joint Commission, Jeddah, pp 264–275

Mill AJB, Jancke K (1981) Comparison of model and mining test results. In Karbe et al. (eds) (1981/2011) Mining of metalliferous sediments from the Atlantis II Deep. Red Sea pre-mining environmental conditions and evaluation of the risk to the environment. EIS Report to the Saudi-Sudanese Joint Commission, Jeddah, pp 276–280

Mustafa Z, Amann M (1978) Ocean mining and protection of the marine environment of the Red Sea. In: Ocean Technology Conference, OTC 3188:1199–1266

Mustafa Z, Amann H (1980) Red Sea pre-pilot mining test 1979. In: Ocean Technology Conference, OTC 3874:197–210

Mustafa Z, Nawab Z, Horn R, LeLann F (1980) Role of physical oceanography and environmental studies in the Red Sea. In: Abu Gideiri YB (ed) Proceedings of the symposium on the coastal and marine environment of the Red Sea, Gulf of Aden and tropical Western Indian Ocean, 9–14 Jan 1980, vol 3. International Printing House, Khartoum, pp 8–31

Nawab Z (1980) Introductory speech. In: Abu Gideiri YB (ed) Proceedings of the symposium on the coastal and marine environment of the Red Sea, Gulf of Aden and Tropical Western Indian Ocean, 9–14 Jan 1980, vol 3. International Printing House, Khartoum, pp 1–5

Nawab ZA (1984) Red Sea mining: a new era. Deep-Sea Res 31A:813–822

Nawab Z (2001) Atlantis II Deep: a future deep-sea mining site. In: International Seabed Authority (ed) Proposed technologies for deep seabed mining of polymetallic nodules. Proceedings of the international seabed authority's workshop held in Kingston, Jamaica, 3–6 Aug 1999, pp 295–310

Nawab Z, Lück K (1979) Testförderung von metallhaltigen Schlämmen vom Boden des Roten Meeres. Meerestechnik 10(6):181–187

Quadfasel D, Baudner H (1993) Gyre-scale circulation cells in the Red Sea. Oceanol Acta 16:211–229

Schriever G, Thiel H (2013) Tailings and their disposal in deep-sea mining. In: Proceedings of the 10th (2013) ISOPE ocean mining and gas hydrates symposium. Szczecin, Poland, 22–26 Sept 2013, pp 5–17

Swift SA, Bower AS, Schmitt RW (2012) Vertical, horizontal and temporal changes in temperature in the Atlantis II and Discovery hot brine pools, Red Sea. Deep Sea Res I 64:118–128

Theeg R (1985) Die Aktivität des Elektronen-Transport-Systems Benthischer Lebensgemeinschaften. Dissertation, Universität Hamburg, 195 pp

Thiel H (1980) Community structure and biomass of the benthos in the central deep Red Sea. In: Abu Gideiri YB (ed) Proceedings of the symposium on the coastal and marine environment of the Red Sea, Gulf of Aden and Tropical Western Indian Ocean, vol 3. International Printing House, Khartoum, 9–14 Jan 1980, pp 127–134

Thiel H (1981) Effects on the benthos. In: Karbe et al (eds) (1981/2011) Mining of metalliferous sediments from the Atlantis II Deep, Red Sea: pre-mining environmental conditions and evaluation of the risk to the environment. EIS report to the Saudi-Sudanese Red Sea Joint Commission, Jeddah, pp 313–325

Thiel H (1983) Pteropod shells: another food source for deep sea organisms. Senckenb Marit 15:147–155

Thiel H (1987) Benthos of the deep Red Sea. In: Edwards AJ, Head SM (eds) Key Environment, Red Sea. Pergamon Press, Oxford, pp 112–127

Thiel H, Weikert H (1984) Biological oceanography of the Red Sea oceanic system. Deep Sea Res 31A:829–831

Thiel H, Weikert H, Karbe L (1986) Risk assessment for mining metalliferous muds in the deep Red Sea. Ambio 15:34–41

Thiel H, Pfannkuche O, Theeg R, Schriever G (1987) Benthic metabolism and standing stock in the central and northern deep Red Sea. Publicationi della Stazione Zoologica di Napoli I: Mar Ecol 8:1–20

Thiel H, Foell EJ, Schriever G (1991) Potential environmental effects of deep seabed mining. Bundesminister für Umwelt, Naturschutz und Reaktorsicherheit: Report 102 0 42 46. Berichte aus dem Zentrum für Klima und Meeresforschung der Universität Hamburg, 243 pp

Tragon E, Garrett C (1997) The shallow thermocline circulation of the Red Sea. Deep-Sea Res 44:1355–1376

Trüper HG (1969) Bacterial sulfate reduction in the Red Sea hot brines. In: Degens ET, Ross DA (eds) Hot brines and recent heavy metal deposits in the Red Sea. Springer, Berlin, pp 263–271

Van der Wielen PWJJ, Bolhuis H, Borin S, Daffonchio D, Corselli C, Giuliano L, D'Auria G, de Lange GJ, Huebner A, Varnavas SP, Thomsen J, Tamburrini C, Marty D, McGenity TJ, Timmis KN, BioDeep Scientific Party (2005) The enigma of procariotic life in deep hypersaline anoxic basins. Science 307:121–123

Wang Y, Yang J, Lee OO, Dash S, Lau SC, Al-Suwailem A, Wong TY, Danchin A, Quiam PY (2011) Hydrothermially generated aromatic compounds are consumed by bacteria colonizing in Atlantis II Deep of the Red Sea. Int Soc Microb Ecol J 5:1652–1659

Weber H, Ergunalp D (1981) Processing. In: Karbe et al (eds) (1981/2011) Mining of metalliferous sediments from the Atlantis II Deep, Red Sea: pre-mining environmental conditions and evaluation of the risk to the environment. EIS report to the Saudi-Sudanese Red Sea Joint Commission, Jeddah, pp 235–237

Weikert H (1980a) The oxygen minimum layer in the Red Sea: ecological implications on the occurrence of zooplankton in the area of the Atlantis II Deep. Meeresforschung 28:1–9

Weikert H (1980b) On the plankton of the central Red Sea. A first synopsis of results obtained from cruises MESEDA I and MESEDA II. In: Abu Gideiri YB (ed) Proceedings of the symposium on the coastal and marine environment of the Red Sea, Gulf of Aden and Tropical Western Indian Ocean, 9–14 Jan 1980, vol 3. International Printing House, Khartoum, pp 135–167

Weikert H (1981) Effects on the pelagic organisms. In: Karbe et al (eds) (1981/2011) Mining of metalliferous sediments from the Atlantis II Deep, Red Sea: pre-mining environmental conditions and evaluation of the risk to the environment. EIS report to the Saudi-Sudanese Red Sea Joint Commission, Jeddah, pp 298–312

Weikert H (1982) The vertical distribution of zooplankton in relation to habitat zones in the area of the Atlantis-II-Deep, central Red Sea. Mar Ecol Prog Ser 8:129–143

Weikert H (1987) Plankton and the pelagic environment. In: Edwards AJ, Head SM (eds) Key Environment, Red Sea. Pergamon Press, Oxford, pp 90–111

Weikert H, Koppelmann R (1993) Vertical structural patterns of deep-living zooplankton in the NE Atlantic, the Levantine Sea and the Red Sea. Oceanol Acta 16:163–177

Weikert H, Koppelmann R (1996) Midwater zooplankton profiles from the temperate ocean and particularly landlocked seas. A re-evaluation of interoceanic differences. Oceanol Acta 19:657–664

Calcite and Aragonite Saturation States of the Red Sea and Biogeochemical Impacts of Excess Carbon Dioxide

Ahmed I. Rushdi

Abstract

This chapter discusses the saturation states of the Red Sea with respect to both calcite and aragonite and their possible biogeochemical impacts as a result of ocean carbonate chemistry changes. The saturation levels of the Red Sea surface waters are several-fold supersaturated with respect to calcite and aragonite; they range from 634 to 721 % and from 446 to 488 %, respectively. The saturation levels of the deep waters range from 256 to 341 % with respect to calcite and from 177 to 230 % with respect to aragonite. They generally increase from south to the north. The lowest values of seawater supersaturation with respect to both calcite and aragonite were found at water depths >1,400 m. Changes in the seawater acid–base chemistry due to excess CO_2 emission and oceanic acidification affect the saturation states of calcium carbonate. Based on reported results of the excess CO_2 sink in the northern part of the Red Sea (Krumgalz et al. 1990), the estimated degree of saturation with respect to calcite and aragonite was higher by 1.9 ± 0.4 % at >200 m, 4.9 ± 0.7 % at 200–600 m, and 2.5 ± 0.1 % at >600 m in preindustrial times than in 1982. A projected drop in pH by a 0.1 unit decreases the saturation level by a factor of 1.2, whereas a drop by 0.4 pH unit decreases the saturation level by a factor of 2.1. These changes in saturation levels will have major impacts on the calcifying pelagic and benthic organisms as well as the distribution and depth of coral reefs. Low magnesian calcite and pure calcite are expected to be the dominant carbonate minerals at these low supersaturation levels.

Introduction

The degree of saturation of seawater with respect to calcium carbonate minerals is one of the main factors controlling the mineralogy and morphology of calcium carbonates (Mucci and Morse 1984; Rushdi 1992, 1993, 1995) and their occurrence in seawater (Lowenstam and Epstein 1957; Takahashi 1975; Mackenzie and Pigott 1981; Walter 1984; Burton and Walter 1987; Rushdi et al. 1992). It also affects the precipitation and dissolution rates of the calcium carbonate solid phase at different depths in the oceans (Heath and Culberson 1970; Morse and Berner 1972; Silter et al. 1975; Berner 1976; Feely et al. 1984; Hales and Emerson 1997; Chen et al. 2006). The surface water of the open ocean is usually 200–400 % supersaturated with respect to calcite and aragonite minerals (Wettenberg and Timmerman 1936; Cloud 1962; Pytkowicz and Flower 1967; Broecker et al. 1979; Feely et al. 1984, 1988; Chen et al. 2006). Calcium carbonate minerals in the marine environment are mainly formed by organisms (Chave 1954a). These biogenic carbonate minerals include pure calcite, calcite with various amount of magnesium concentration known as magnesian calcite and aragonite (Chave 1954b; Lowenstam 1955; Stockman et al. 1967; Andersson et al. 2008). Inorganic

A.I. Rushdi (✉)
Chair of Green Energy Research, Food and Agriculture Sciences, King Saud University, Riyadh, Saudi Arabia
e-mail: arushdi@ksu.edu.sa

A.I. Rushdi
Department of Earth and Environmental Sciences, Faculty of Sciences, Sana'a University, Sana'a, Yemen

A.I. Rushdi
College of Earth, Oceanic and Atmospheric Sciences, Oregon State University, Corvallis, OR 97331, USA

N.M.A. Rasul and I.C.F. Stewart (eds.), *The Red Sea*,
Springer Earth System Sciences, DOI 10.1007/978-3-662-45201-1_16,

precipitation of calcium carbonate deposits in seawater is inhibited by the presence of dissolved organic and inorganic chemicals (Simkiss 1964; Pytkowicz 1965, 1973; Suess 1973; Rushdi 1992; Rushdi et al. 1992) although it may form in some special places such as the Bahamas Banks and the hot Arabian coasts (Cloud 1962; Broecker and Takahashi 1966; Pytkowicz 1983; Budd 1988). The two most common forms of calcium carbonate minerals in the marine environment are calcite and aragonite. Aragonite is kinetically favored to precipitate in a solution with a magnesium-to-calcium concentration ratio more than four such as in seawater (Kitano 1964; Rushdi et al. 1992; Rushdi 1993).

There is evidence that biogenic calcium carbonate formation is controlled by the regional and temporal saturation states of surface seawaters (Broecker and Takahashi 1966; Smith and Pesret 1974; Opdyke and Wilkinson 1993; Suzuki et al. 1995; Kleypas et al. 1999; Broecker et al. 2001). Different studies show that there is a positive correlation between precipitation and overgrowth of calcium carbonate by organisms and the saturation states of solutions (Smith and Roth 1979; Gatusso et al. 1999; Takagi 2002; Green et al. 2009, 2013; Waldbusser et al. 2011; Barton et al. 2012). Laboratory investigations have shown that low levels of calcium carbonate saturation decrease the exoskeleton formation of coral reefs and calcifying phytoplankton (Gattuso et al. 1998, 1999; Riebesell et al. 2000; Langdon and Atkinson 2005).

Obviously, the saturation state of seawater with respect to calcium carbonate is important for the formation of biogenic and non-biogenic carbonate minerals; however, less is known about the saturation levels of the Red Sea seawater with respect to calcium carbonate. Therefore, the objectives of this chapter are to describe the saturation levels of the Red Sea with respect to calcite and aragonite, to estimate the possible effects of the projected decrease in seawater pH values as a result of excess fossil fuel emission on the saturation states with respect to both minerals, and to describe the possible impacts of the decrease of saturation levels on the biological and biogeochemical conditions of the Red Sea.

Geological Setting, Formation, and Ecology of the Red Sea

Geological Setting

The Red Sea, an enclosed body of water that lies between 30°N and 12° 30′N, is about 1,932 km long and 280 km in width (Morcos 1970). The narrow southern Strait of Bab-al-Mandab (29 km in width) is the boundary between the Red Sea and the Gulf of Aden. The total surface area of the Red Sea is estimated to range between 438 × 10^3 and 450 × 10^3 km^2 and the volume between 215 × 10^3 and 251 × 10^3 km^3 with an average depth of about 491 m (Morcos 1970; Head 1987; Anderson and Dyrssen 1994). The Red Sea is a shallow sea with the shallowest shelves extending in the southern part. The maximum depth is recorded to be 2,850 m (Edwards and Head 1987).

Formation of the Red Sea

The Quaternary sea level changes have affected the vertical and horizontal locations of contemporary biogenesis of the Red Sea several times, because most of the immediate substrate in all coral seas is biogenic. Therefore, Pleistocene, Holocene, and older processes are very important in the present distribution of some types of habitats (Sheppard et al. 1992; Behairy et al. 1992; Al-Anbaawy 1993). Most importantly are the exact features of the region such as rift formation, upward movement of salt domes, and volcanic foundations (Sheppard et al. 1992). Understanding some of geological processes in the region will clarify some of the ecological features and their geographical scale. The soft substrate in the region is mainly biogenic carbonate.

The Arabian tectonic plate is considered to be a fragment of the very large African plate. Its boundary is marked by the location of the sea floor spreading center which extends through the Gulf of Aqaba and the Red Sea all the way down to the Gulf of Aden and to the Carlsberg Ridge in the Arabian Sea (Friedman 1985; Sheppard et al. 1992; Ligi et al. this volume, Ogubazghi and Goitom; this volume). The Arabian plate is known to move slowly away from Africa with an anticlockwise rotation of about 7° (Tramontini and Davies 1969; Makris et al. 1983; Gaulier et al. 1988; Martinez and Cochran 1988). Hence, because of these geological processes and features, geologists still define the Red Sea as an ocean (Freund 1970; Mckenzie et al. 1970; Girdler and Darrocot 1972; Le Pichon and Francheleau 1978), but this definition is rejected in terms of size. The Red Sea with its oceanic character started to form about 41–34 Ma (Girdler and Styles 1974). Bartov et al. (1980) and Reilinger et al. (this volume) suggested that it started to form about 22–19 Ma. About 2–5 Ma (million years ago), rifting recommenced at rates of up to 2 cm $year^{-1}$ (Braithwaite 1987). The fit of the two sides of the Red Sea is almost perfect from Sinai south to about 15°N. From this point, the fit is ignored. Remarkably, the Afar Triangle in Ethiopia occurs where the rift system diverges into three systems (Yirgu et al. 2006). They are along the length of the Red Sea, into the Arabian Sea and down the African mainland, which shows no spreading center. The Afar Triangle, which is still below the sea level, is a volcanic and basaltic feature and is known as a portion of oceanic floor elevated to become dry land (Sullivan 1974). Wegener (1929) pointed out the fact that if the Afar Triangle was really elevated, then the fit between Arabian and African sides of the Red Sea is reasonably perfect.

Ecology of the Red Sea

The present ecology of the region was driven partly from frequent connection with the Indian Ocean, which brought contemporary fauna and flora to the region. Heavy erosion during the Pliocene and Pleistocene, which brought alluvial materials for benthic biota, also contributed to the ecology of the Red Sea (Sheppard et al. 1992). During the Pleistocene rainy periods, the sea level was about 40–60 m below the present sea level. Thus, very large amounts of alluvial materials reached the areas of fringing reefs. Also, a considerable amount of terrestrial material reached offshore and formed important new substrates and areas of sheltered habitat. The influence of heavy erosion is seen by the presence of major wadi features, which normally developed during the Pleistocene (Behairy et al. 1992). The sea level was similar to the present level about 110–140 Ka. From about 110 Ka to about 30 Ka, the sea level fluctuated from 30 to 60 m below the present sea level (Sheppard et al. 1992). The sea level dropped rapidly to about 120–150 m below the present sea level between the Pleistocene and Holocene. Probably, this was maintained for 1–2 Ka. Then, about 14–15 Ka, the rise of sea level commenced and proceeded to reach the present sea level about 7 Ka. Evidence showed that the Red Sea was hypersaline in the period of about 17 Ka (Gvirtzman et al. 1977; Braithwaite 1987; Sheppard and Sheppard 1991; Sheppard et al. 1992; El-Anbaawy et al. 1992; Taviani 1998a, b; Jassen and Taviani; this volume). During the last 140,000 years, the growth of reefs was irregular, due to changing sea level and alluvial flooding (Sheppard et al. 1992; Behairy et al. 1992). When the sea level was stable at about 20–60 m below the present, it took about eight periods of two thousand years (Potts 1983). Therefore, it was enough time for fringing reefs to develop on the contemporary shoreline and for patch reefs to develop on any older limestone platforms. Erosion of the previous reefs would occur during every drop of sea level, and then alluvial material sheets from coastal mountains would reach the sea and fill the eroded channels. The hot climate in the region enhances the formation of evaporites as well as dolomite and carbonate sediments. The sediments are derived from microfauna, especially foraminifera, corals, and coralline algae (Behairy et al. 1992).

Large islands and archipelagos in the Red Sea are both geologically and biologically very important. The volcanic groups along the central axis are found in the south and central parts of the Red Sea (Sheppard et al. 1992; Behairy et al. 1992; Jonsson and Xu, this volume). The other islands that are formed by limestone support rich soft substrate habitats.

Water Circulation of the Red Sea

The water currents of the Red Sea can be divided into two types: (1) wind-driven currents, which develop surface water currents, and (2) density-gradient currents, which mainly develop deep water currents (Neumann and McGill 1953; Morcos 1970; Edwards 1987). During the summer (from June to September), the prevailing wind through Bab-al-Mandab is northwesterly, which opposes the inward surface flow. During this time of the year, the salinity and temperature increase and reach about 38–39 and over 30 °C, respectively (Maillard and Soliman 1986). This develops an upper layer of low density that travels southward out of the Red Sea over the main section of water mass that flows into the Red Sea. The denser deep water mass continues to leave the Red Sea. In wintertime, the surface water is driven into the Red Sea from the Gulf of Aden by the prevailing winds. This surface water has a salinity of about 36.5 and a temperature value of 25 °C (Sheppard et al. 1992). Beneath this layer is the deep water mass, which is developed by the density-gradient action and outflow through the Strait of Bab-al-Mandab. The northerly drifting Red Sea seawater becomes heavier, because of the salinity increase and the decrease of temperature. This heavy water eventually sinks near the mouth of the Gulf of Suez and returns southward below the thermocline (below about 200–300 m). The returning water mass has a constant temperature of about 21.5 °C and constant salinity of about 40.5 (Couper 1983; Jones et al. 1987).

Chemistry of Carbon Dioxide in Seawater

Acid–Base Chemistry of CO_2

The chemistry of carbon dioxide in seawater solution is a relatively complex system because it occurs in the form of gas, liquid, and solid phases (Stumm and Morgan 1996). The acid–base chemistry is briefly described here, and for further comprehensive information, the reader can consult, for example, Pytkowicz (1983); Stumm and Morgan (1996); DOE (1994); Millero (1995), and Zeebe and Wolf-Gladrow (2001).

$$CO_{2(g)} \overset{H_2O}{\leftrightarrow} CO_{2(l)} + H_2O \leftrightarrow H_2CO_3; \quad K_0 = \frac{[H_2CO_3]}{fCO_2} \tag{1}$$

$$H_2CO_3^* \leftrightarrow H^+ + HCO_3^-; \quad K_1 = \frac{[H^+][HCO_3^-]}{[H_2CO_3^*]} \tag{2}$$

$$HCO_3^- \leftrightarrow H^+ + CO_3^{2-}; \quad K_2 = \frac{[H^+][CO_3^{2-}]}{[HCO_3^-]} \qquad (3)$$

where the subscripts *g* and *l* refer to a gas and a liquid, respectively. Analytically, it is difficult to differentiate between the $CO_{2(l)}$ and H_2CO_3 species; therefore, a hypothetical H_2CO_3*notation is used to express the sum of the $CO_{2(l)}$ and H_2CO_3 concentrations. The bracket represents total stoichiometric concentrations of the exact ionic chemical species, and $f(CO_2)$ is the fugacity of carbon dioxide in the gas phase. The values of the equilibrium constants (i.e., K_0, K_1, and K_2) depend on the temperature, pressure, and salinity (ionic strength) of the seawater, which have been measured by different investigators (e.g., Lyman 1956; Hansson 1973; Mehrbach et al. 1973; Weiss 1974; Dickson and Riley 1979; Dickson 1990).

Direct measurement of the individual species concentrations of the carbon dioxide system in solution is impossible. Therefore, two parameters can be measured and used together with other additional information to describe the carbon dioxide system in seawater. The two parameters include any two of total alkalinity (TA), total carbon dioxide, pH, and fugacity of CO_2 (Pytkowicz 1983; DOE 1994; Millero 1995; Zeebe and Wolf-Gladrow 2001).

Calcite and Aragonite Degree of Saturation Calculation

The degree of saturation of calcite, Ω_c, and of aragonite, Ω_a, is defined as the ratios of the ionic products of the concentrations of calcium and carbonate at in situ temperature, salinity, and pressure to the solubility products of calcite, $K_{sp,c}$, and of aragonite, $K_{sp,a}$, under the in situ conditions. Thus, the Ω_c for calcite is

$$\Omega_c = \frac{(Ca^{2+})(CO_3^{2-})}{K_{sp,c}} \qquad (4)$$

and the Ω_a for aragonite is

$$\Omega_a = \frac{[Ca^{2+}][CO_3^{2-}]}{K_{sp,a}} \qquad (5)$$

where $[Ca^{2+}]$ and $[CO_3^{2-}]$ are, respectively, the concentrations of calcium and carbonate ions in solution. The dissociation solubility products of calcite and aragonite can be calculated based on the work of Mucci (1983). Finally, the degree of saturation is calculated by Eqs. (4) and (5).

When $\Omega > 1$, seawater is supersaturated with respect to calcite or aragonite, and seawater is undersaturated with respect to calcite and aragonite when $\Omega < 1$. Seawater is considered saturated with calcite and aragonite when $\Omega = 1$. Calcium is relatively a conservative element, and its concentration in seawater varies by less than 1.5 %. Thus, calcium-to-salinity ratio in seawater (Culkin and Cox 1966) can be used to estimate the calcium concentration.

Temperature and Salinity Distribution

The surface seawater temperature ranges from 26.00 °C in the north to 31.72 °C in the south [Fig. 1b, data for summer time from the *Marion Dufresne* Cruise 1982 (Beauverger et al. 1985)]. This general pattern of temperature increase from the north to the south is also illustrated in Fig. 2a. The surface seawater temperatures are high near the shoreline of Yemen and increase northward (Rushdi 2014). The vertical distribution of temperature noticeably decreases to about 23 °C at about 200 m depth, and then it gradually decreases to about 22 °C between 200 and 300 m. The water temperature is almost constant below 300 m depth, ranging from 21.41 to 21.62 °C (Figs. 1b and 2a). The increase of surface temperature from north to south was also observed in early wintertime along the shoreline of Yemen (Rushdi 2014).

The Red Sea is one of the saltiest parts of the world oceans, because it is located in an arid region where evaporation exceeds precipitation (Meshal et al. 1984; Osman 1984) and there is no major runoff into the sea except very minor and small seasonal coastal wadis. The mean surface salinity at Perim Island is much the same as that of the western part of the Gulf of Aden (just below 36.5).

The summer surface water salinity increases from 36.88 in the south to 40.20 in the north (Figs. 1c and 2b). The profile of salinity shows a rapid increase with depth in the south and slight increase in the north (Fig. 1c). The salinity is almost constant below 200 m depth and ranges from 40.57 to 40.59. The distribution patterns of temperature and salinity of the Red Sea (Fig. 2a, b) indicate that they are mainly driven by winds and density-gradient variations (Morcos 1970) and mixing process. The surface water layer at ≤50 m depth is obviously affected by northwestern winds into the Gulf of Aden and the mixing process with the water mass inflow into the Red Sea at ~100 m depth through the Strait of Bab-al-Mandab (Morcos 1970; Sheppard et al. 1992). This is indicated by the elevated temperature of 31.72 °C and low salinity of 36.88 (i.e., low-density seawater) of surface water. The deep water mass is likely developed at the upper layer in the northern part of the Red Sea as a result of a density increase in surface waters due to the increase in salinity and decrease in temperature (Morcos 1970; Sheppard et al. 1992). This is illustrated in Fig. 3a,

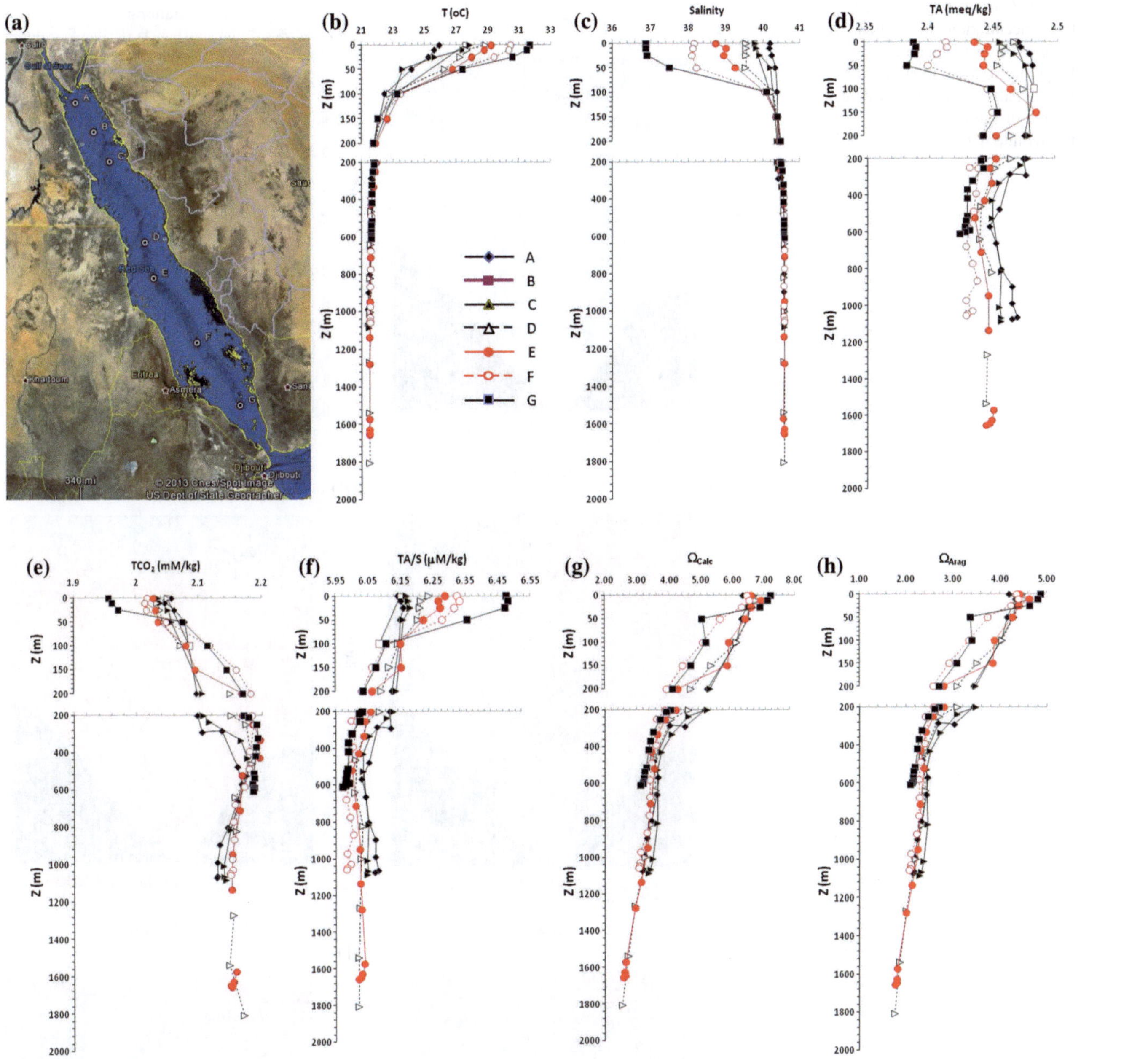

Fig. 1 Plots showing **a** map of station locations; **b** temperature; **c** salinity; **d** total alkalinity; **e** total carbon dioxide; (results of summer 1982; source from Beauverger et al. 1985); **f** total alkalinity-to-salinity ratio; **g** calcite degree of saturation; **h** aragonite degree of saturation; **i** pH and **j** pCO_2 depth profiles of different stations

where the deep water mass, with low temperature (<22 °C) and high salinity (>40.1), is generally found below 300 m depth. A recent study has shown that the salinity of the coastal seawater of Yemen ranges from 36.32 to 37.36 during early winter and increases with depth to a maximum of 40.28–40.35 at depths of more than 100 m (Rushdi 2014). Also, the study has shown that the surface seawater, which is high along the shoreline, increases from 36.49 in the south to >37.10 in the north near Kamaran Island off Yemen.

Both the temperature and salinity distributions are likely affected by the influx of the surface water with high temperature and low salinity from the Gulf of Aden and the deep water mass of low temperature and high salinity similar to the sinking surface water near the Suez Canal in the north (Morcos 1970). The two types of seawater are illustrated in Fig. 3a, where low-salinity and high-temperature surface seawaters and high-salinity and low-temperature deep water masses are shown.

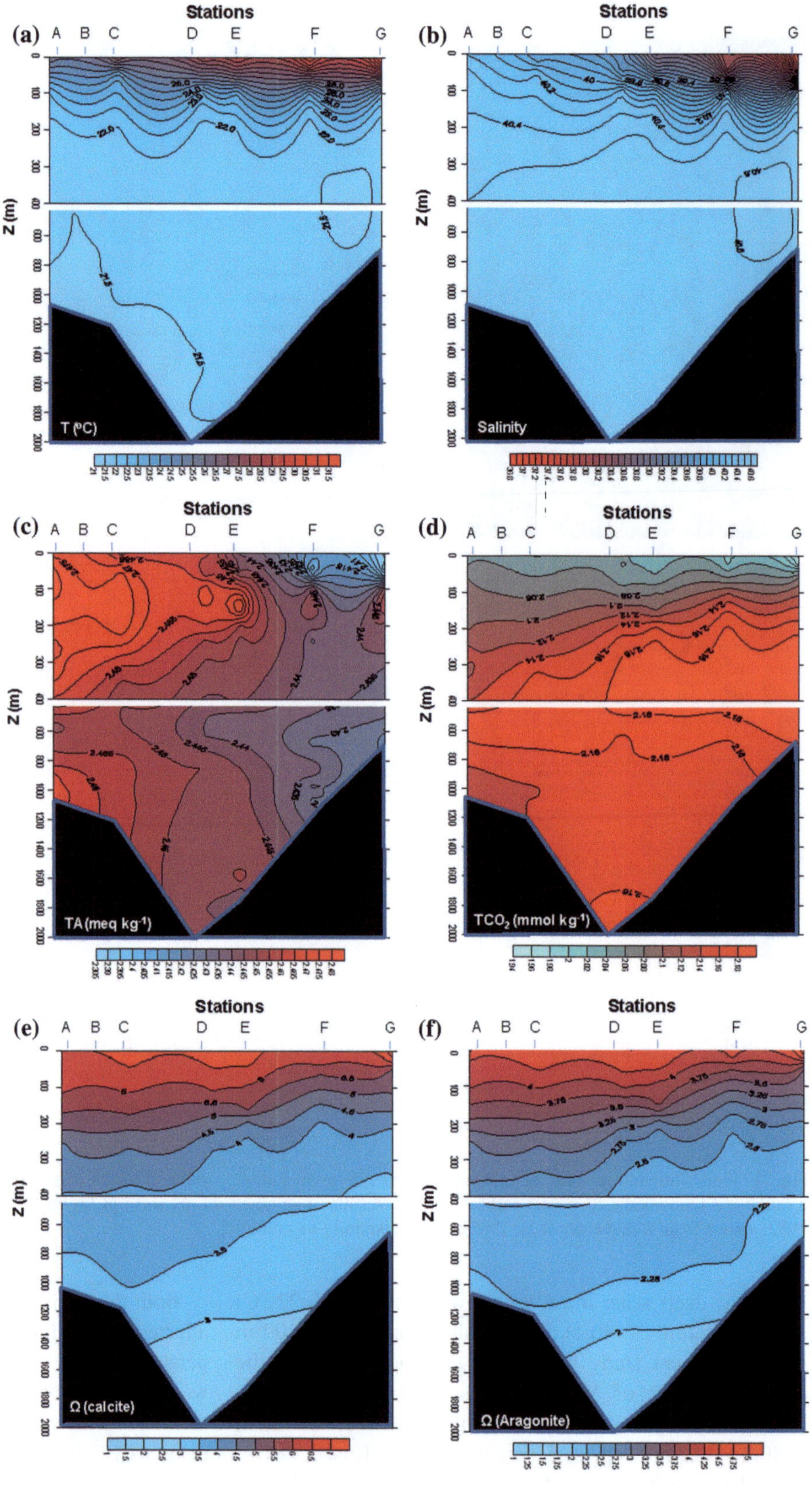

Fig. 2 Cross-sections showing the concentration distribution patterns of: **a** temperatures; **b** salinity; **c** total alkalinity; **d** total carbon dioxide; **e** calcite degree of saturation; and **f** aragonite degree of saturation

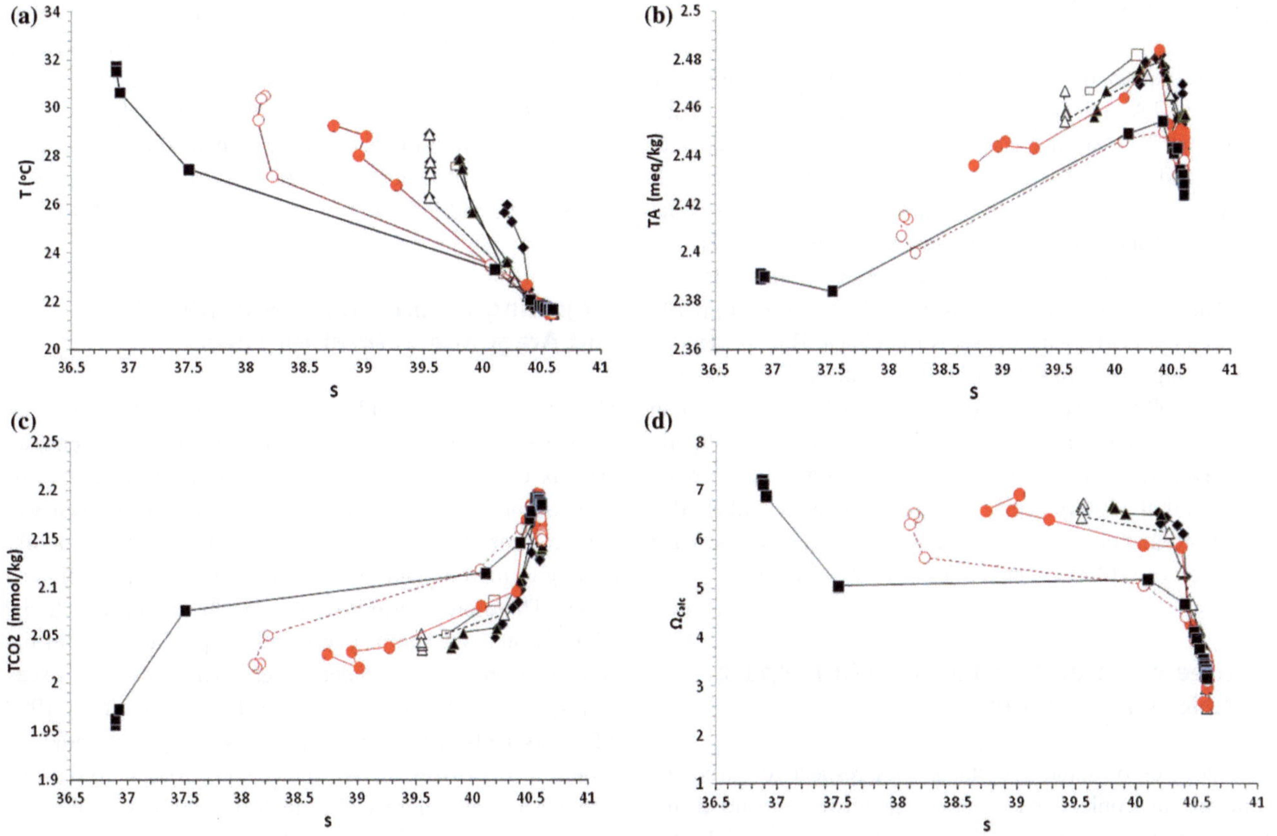

Fig. 3 Plots showing **a** temperature; **b** total alkalinity; **c** total carbon dioxide; and **d** calcite degree of saturation versus salinities

Total Alkalinity and Total Carbon Dioxide Distribution

Generally, the total alkalinity (TA) increases from south to the north as shown in Fig. 1d. The surface seawater TA increases from 2.389 to 2.47 meq kg^{-1} (0–100 m), whereas in deeper waters (>600 m), it increases from 2.424 to 2.466 meq kg^{-1}. In general, TA increases with depth to a maximum value at 100–150 m depth and then slightly decreases (Fig. 1d). At >600 m, it slightly increases with depth and northward (Figs. 1d and 2c). This is presumably due to dissolution of less stable high magnesian calcite in surface waters and re-precipitation of less-soluble calcite and aragonite in deeper water (Rushdi 2014) where the slight decrease in TA between 200 and 600 m is partly due to sinking and mixing of cold water with relatively lower TA values. This variation of TA is also shown in Fig. 3b where the surface layer TA is lower than deep layer TA. The production of calcium carbonate minerals is likely to increase southward in deeper and bottom seawater and northward in surface seawater. This is indicated by the decrease of the ratios of TA to salinity (TA/S) from the south to the north for surface water at <100 m depth and the increase of the ratios from south to north in deeper water (Fig. 1f).

The surface seawater total carbon dioxide (TCO_2) concentration shows a similar distribution trend as TA, with a maximum concentration deeper than TA. This is shown by the high TCO_2 and salinity for deep waters and low TCO_2 and salinity for surface seawater (Fig. 3c). The lowest TCO_2 (1.957 mmol kg^{-1}) is in the south, and the highest value (2.058 mmol kg^{-1}) is in the north. The TCO_2 concentration of the deep seawater layer increases slightly from 2.128 mmol kg^{-1} in the south to 2.186 mmol kg^{-1} in the north (Figs. 1e and 2d). This is presumably due to the organic matter re-mineralization and partly to mixing of surface water with bottom water. The deep seawater in the south is obviously higher in TCO_2 than the surface water layer as shown in Fig. 3c, indicating an older water mass. The cross section (Fig. 2d) shows the water type in the south between 200 and 600 m depth, which likely supports the inflow of water mass from the Gulf of Aden beneath the upper outflow surface water from the Red Sea.

pH and pCO_2 Distributions

The pH values of the Red Sea were estimated from the measured TA and TCO_2 concentrations as explained in the Appendix. The surface seawater pH ranges from 8.289 to 8.236 and generally increases from south to north. It decreases with depth to a minimum (8.020–8.049) at about 400 m and then slightly increases with depth (Fig. 1i). At >100 m depth, the pH generally decreases southward.

vThe pCO_2 of surface seawater (<50 m) ranges from 191 to 344 μatom, where the lowest is found near Bab-al-Mandab and the highest is in the northern part of the Red Sea. The pCO_2 usually increases with depth and reaches the maximum (489–566 μatom) at about 400–500 m and then slightly decreases with depth (Fig. 2j). It is also observed to increase southward at >200 m depth. This pCO_2 increase is likely due to both formation of $CaCO_{3(s)}$ (as will be discussed later) and oxidation of settling organic matter at >200 m depth.

Degree of Saturation Levels with Respect to Calcite and Aragonite

The degrees of saturation states with respect to carbonate minerals in tropical seas control the formation and distribution of biogenic and inorganic calcium carbonate (Kleypas et al. 1999; Broecker et al. 2001). The percent saturation ($\%\Omega = \Omega \times 100$) of surface water of the Red Sea with respect to calcite and aragonite ranges from 634 to 721 % and from 447 to 488 %, respectively. These high levels of saturations with respect to calcite will likely lead to the precipitation of relatively high magnesian (>5 mol% $MgCO_3$) calcite (Rushdi 1989). The saturation levels for both calcite and aragonite decrease with depth (Fig. 1g, h) and generally increase from south to north (Fig. 2e, f). This trend of northward saturation level increase has been observed also for the coastal zone of Yemen (Rushdi 2012, 2014). For deep waters, they range from 256 to 341 % for calcite and from 177 to 230 % for aragonite. Therefore, the entire water body of the Red Sea is supersaturated with respect to both minerals. The data also indicate that physicochemical processes such as mixing and dissolution/precipitation are likely to have dominant effects over biological activities upon the calcite and aragonite saturation levels of the Red Sea seawaters (Fig. 3d).

The levels of the degree of saturation with respect to calcite and aragonite show slight decreases with the increase of salinity (Fig. 3d). This might be due to partial dissolution of calcium carbonate such as high magnesian calcite in the southern part of the Red Sea and re-precipitation of more stable calcite and aragonite toward the north, as mentioned in the previous section.

The results obtained by Krumgalz et al. (1990) during the *Tiran-02* cruise (summer of 1981) in the northern Red Sea were used to estimate the preindustrial calcite and aragonite saturation levels following the method of Feely and Chen (1982). The saturation levels were estimated to be higher by 1.9 ± 0.4 % at the surface layer (<200 m), 4.9 ± 0.7 % at 200–600 m, and 2.5 ± 0.1 % at deep water (>600 m).

Projecting Future Changes in the Calcite and Aragonite Saturation Levels

The abiotic processes affect the carbonate chemistry of the surface seawater. For example, high temperatures enhance the flux of CO_2 from water into the atmosphere due to the decrease of CO_2 solubility, and hence, the TCO_2 decreases. On the contrary, CO_2 is more soluble at lower temperatures, leading to more uptake of CO_2 and higher concentrations of TCO_2. Therefore, warmer waters will have relatively higher $[CO_3^{2-}]$ and saturation levels with respect to calcium carbonate minerals than colder waters. The uptake of excess atmospheric CO_2 by the Red Sea (Krumgalz et al. 1990) affects its acid–base chemistry, because CO_2 is an acidic gas. When it dissolves in seawater, CO_2 reacts with water through a series of reactions to form carbonic acid which dissociates to bicarbonate and hydrogen ions and then bicarbonate ions dissociate to carbonate and hydrogen ions. The dissociated hydrogen ions increase the concentration of $[H^+]$ as well as bicarbonate ion $[HCO_3^{2-}]$ concentration and instantly decrease the pH values and carbonate ion $[CO_3^{2-}]$ concentration. These reactions affect the degree of saturation of seawater with respect to calcium carbonate minerals. In other words, excess atmospheric CO_2 increases surface seawater pCO_2 and decreases the oceanic pH values and carbonate ion concentrations. Time series experiments of the carbon dioxide system such as the Hawaii Ocean time-series (HOT) program, Bermuda Atlantic Time-series Study (BATS), and European Time Series in the Canary Islands (ESTOC) have, respectively, shown long-term pH declines of 0.0019 ± 0.0002 units $year^{-1}$ (Dore et al. 2009), 0.0014 ± 0.0003 units $year^{-1}$ (Bates 2007), and 0.0018 ± 0.0003 units $year^{-1}$ (Santana-Casiano et al. 2007; Gonzalez-Davila et al. 2007). Accordingly, with the lack of serious measures to be taken to reduce the emission of CO_2 into the atmosphere, the present oceanic pH values may drop by 0.1 unit in 50 years and 0.4 unit in 200 years.

In the following, we explore the effect of projected pH decreases of 0.1 and 0.4 upon the Red Sea water saturation levels with respect to calcite and aragonite. A pH drop of 0.1 unit shows that the seawater saturation states of calcium carbonate minerals decrease by a factor of 1.2 (Fig. 4a, c). Therefore, the surface seawater saturation states drop from

635–721 to 528–601 % for calcite and from 447–488 to 373–407 % for aragonite. For deep and bottom waters with a residence time of ~36 years (Cember 1988), the saturation states drop from 256–341 to 213–284 % and from 177–230 to 148–192 % for calcite and aragonite, respectively. Therefore, the Red Sea will still be supersaturated with respect to both calcite and aragonite within the next 50 years. A pH decrease of 0.4 unit can lower the calcite and aragonite saturation states of seawater by a factor of 2.1 (Fig. 4b, d). Hence, the surface seawater of the Red Sea will drop from 635–721 to 302–343 % for calcite and from 447–488 to 213–232 % for aragonite. The bottom water will drop from 256–341 to 122–162 % and from 447–488 to 84–110 % for calcite and aragonite, respectively. The bottom water of the Red Sea may become undersaturated in respect of aragonite in the next century if the same trend of CO_2 emission continues. The 100 % saturation level of the Red Sea with respect to aragonite may move upward and reach the depth of about 1,300 m in the north and about 400 m in the south.

Calcium Carbonate Degree of Saturation Changes and Their Biogeochemical Impacts

Changes in the acid–base chemistry as a result of excess CO_2 dissolution in seawater can affect marine organisms directly and indirectly. The direct effects include organism exoskeleton calcification and photosynthesis. Indirect effects are more related to the changes in the chemical speciation, behavior, and fate of micronutrients, which will affect the availability of these micronutrients for the photosynthesis process (Eisler 2012, and references therein). The geological record indicates that changes in saturation levels of seawater with respect to calcium carbonate minerals have affected calcifying organisms (Royer et al. 2001). Recent works have shown that the decrease of the seawater supersaturation as a result of the increase of atmospheric CO_2 decreased calcium carbonate formation by calcifying organisms such as foraminifera, pteropods, and planktonic larvae of echinoderms (Marubini et al. 2003; Langdon and Atkinson 2005;

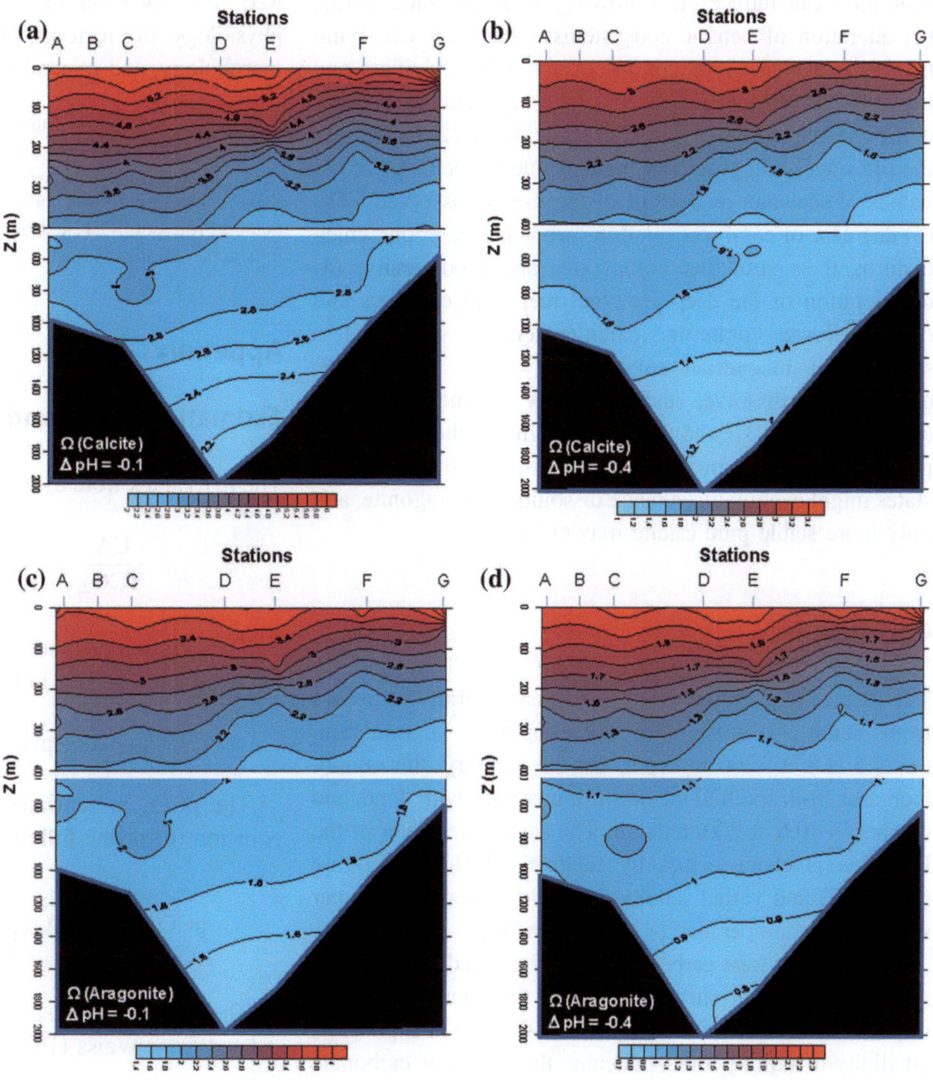

Fig. 4 Cross-sections showing the effects of pH value drops due to excess dissolved CO_2 upon calcite and aragonite degree of saturation levels, **a** effect of −0.1 pH unit drop upon the degree of saturation with respect to calcite, **b** effect of −0.4 pH unit drop upon the degree of saturation with respect to calcite, **c** effect of −0.1 pH unit drop upon the degree of saturation with respect to aragonite, **d** effect of −0.4 pH unit drop upon the degree of saturation with respect to aragonite

Schneider and Erez 2006; Gazeau et al. 2007; Anthony et al. 2008; Jokiel et al. 2008; Comeau et al. 2009; Clark et al. 2009; Lombard et al. 2010; Müller et al. 2010). Clearly, marine benthic environments and their associate organisms are highly affected by the acid–base chemistry changes. Many benthic organisms form skeletal calcium carbonate for protection from predators or for special support. A recent study has demonstrated about a 30 % reduction in growth of the coral *Diploastrea heliopora* in the Red Sea (Cantin et al. 2010). This could be attributed to different processes and environmental changes including possible alteration of seawater saturation levels with respect to calcium carbonate that affected the physiology of the coral. Silverman et al. (2007) have shown that the average calcification rate of coral reefs in the northern part of the Red Sea varies from 60 ± 20 to 30 ± 20 mmol m^{-2} d^{-1} according to the present saturation states of the seawater with respect to aragonite. Therefore, the calcification rates will likely drop by factors of 1.2 and 2.1 in the next 50 and 100 years.

Decreases in the Red Sea calcium carbonate degree of saturation can induce the following biogeochemical shifts: (1) alteration of benthic ecosystems, where non-calcifying benthic organisms may control over benthic calcifiers and possibly the disappearance of calcifying benthos (Widdicombe and Spicer 2008; Kroeker et al. 2011); (2) changes in the life stages and growth of benthos due to decreases of pH values of seawater as a result of the rise of dissolved CO_2; (3) decrease of coral reef calcium carbonate accretion, which disturbs their structural complexity and biodiversity; (4) deterioration of the deep-sea coral reefs and calcifiers and only shallower corals and calcifiers endure; and (5) less stable high magnesian calcite (>12 mol% $MgCO_3$) is expected to dissolve, and only low magnesian calcite (<5 mol% $MgCO_3$) and aragonite might be the main solid phases that can form. Further decreases in the saturation states might eventually enhance dissolution of aragonite, and only more stable pure calcite may exist.

Conclusion

The surface temperature of the Red Sea generally decreases from south to north and with depth to a minimum (~21.5 °C) at >200 m depth. On the contrary, the salinity increases from south to the north. It increases with depth and reaches ~40.5 at 200 m depth. The surface seawater of the Red Sea is characterized by relatively high temperature (>25.0 °C) and varied salinity (<37 to >40) and the deeper water by low temperature (<21.5 °C) and high salinity (>40).

The TA and total carbon dioxide of the Red Sea increase northward for both surface and deep seawaters. They also increase with depth to a maximum at 100–150 m depth. The alkalinity-to-salinity ratio indicates that calcium carbonate production is likely to increase northward in surface waters and southward in deeper waters.

The Red Sea waters are supersaturated with respect to calcite and aragonite. The highest values of supersaturation with respect to both calcite and aragonite are estimated to be in the northern part of the Red Sea, whereas the lowest values of seawater saturation with respect to both minerals are found to decrease with depth and southward. The mineralogy of the carbonate deposits may range from low magnesian calcite, aragonite, to high magnesian calcite.

Increase of dissolved CO_2 in seawater as a result of the present excess CO_2 emission will apparently decrease the degree of saturation by a factor of 1.2 within the next 50 years and by a factor of 2.1 in the next 200 years. This decline in the degree of saturation will apparently affect the biology and biogeochemistry of the Red Sea. These changes will affect mainly calcifying organisms, especially benthos and deeper coral reefs.

Comprehensive studies are needed to investigate the impact of the changes of the calcium carbonate saturation states of the Red Sea water on the growth of calcareous species and physiology of microorganisms as well as the distribution, morphology, and mineralogy of coral reefs of the Red Sea.

Acknowledgment The author thanks Dr. Najeeb M.A. Rasul for the invitation to participate in this book project. This work is partly supported by the National Program for Sciences and Technologies, King Saud University (NPST project no. ENV-842-09). The author also thanks Professors C.T.A. Chen, M. Al-Sayed and B. Hales for their invaluable comments on the first draft of the manuscript.

Appendix

Estimation of pH and pCO_2

The pH values were estimated using the following equation:

$$\frac{CA}{TCO_2} = \frac{[H^+]K_1 + 2K_1K_2}{[H^+]^2 + [H^+]K_1 + K_1K_2} \tag{6}$$

The iteration method is used to estimate the concentration of hydrogen ion $[H^+]$ in the solution. The pH is:

$$pH = -\log[H^+] \tag{7}$$

The pCO_2 in seawater samples can be calculated from equation (6) or the following relationship:

$$pCO_2 = K_0 TCO_2 \left(\frac{[H^+]^2}{[H^+]^2 + [H^+]K_1 + K_1K_2} \right) \tag{8}$$

The solubility constant of CO_2 in seawater can be calculated from Weiss (1974).

References

Al-Anbaawy MIH (1993) Reefal facies and diagenesis of pleistocene sediments in Kamaran Island, Southern Red Sea, Republic of Yemen. Egypt J Geol 37:137–151

Andersson AJ, Mackenzie FT, Bates NT (2008) Life on the margin: implications of ocean acidification on Mg-calcite, high latitude and cold-water marine calcifiers. Mar Ecol Prog Ser 373:265–273

Anderson L, Dyrssen D (1994) Alkalinity and total carbonate in the Arabian Sea. Carbonate depletion in the Red Sea and Persian Gulf. Mar Chem 47:195–202

Anthony KR, Kline DI, Diaz-Pulido G, Dove S, Hoegh-Guldberg O (2008) Ocean acidification causes bleaching and productivity loss in coral reef builders. Proc Natl Acad Sci 105(45):17442–17446

Barton A, Hales B, Waldbusser GG, Langdon C, Feely RA (2012) The pacific oyster, Crassostrea gigas, shows negative correlation to naturally elevated carbon dioxide levels: implications for near-term ocean acidification effects. Limnol Oceanogr 57(3):698–710

Bartov Y, Steinitz G, Eyl M, Eyal Y (1980) Sinistral movement along the Gulf of Aqaba—its age and relation to the opening of the Red Sea. Nature 285:220–222

Bates NR (2007) Interannual variability of oceanic CO_2 and biogeochemical properties in the Western North Atlantic subtropical gyre. Deep Sea Res II 48:1507–1528

Beauverger C, Brunet C, Poisson A (1985) Terres Australes et Antartiques Francaises. Le rapports des Campagnes a la mer. No. 82-04

Behairy AKA, Sheppard CRC, El-Sayed MK (1992) A review of the Geology of coral reefs in the Red Sea. UNEP Regional Seas Reports and Studies No. 152, 41 pp

Berner RA (1976) Equilibrium, kinetics and the precipitation of magnesian calcite from seawater. Am J Sci 278:1435–1477

Braithwaite CJR (1987) Geology and palaeogeography of the Red Sea region. In: Edwards AJ, Head SM (eds) Red Sea. Pergamon Press, Oxford, pp 22–44

Broecker WS, Takahashi T (1966) Calcium carbonate precipitation on the Bahama Banks. J Geophys Res 71:1575–1602

Broecker WS, Takahashi T, Simpson HJ, Peng TH (1979) Fate of fossil fuel carbon dioxide and global carbon cycle. Science 206:409–418

Broecker W, Langdon C, Takahashi T, Peng T-S (2001) Factors controlling the rate of $CaCO_3$ precipitation on Grand Bahama Bank. Glob Biogeochem Cycles 15:589–596

Budd DA (1988) Aragonite-to-calcite transformation during fresh-water diagenesis of carbonates: insights from pore-water chemistry. Geol Soc Am Bull 100:1260–1270

Burton EA, Walter ML (1987) Relative precipitation rates of aragonite and Mg calcite from seawater: temperature or carbonate ion control? Geology 15:111–114

Cantin NE, Cohen AL, Karnauskas KB, Tarrant AM, McCorkle DC (2010) Ocean warming slows coral growth in the central Red Sea. Science 329:322–325

Cember RP (1988) On the sources, formation, and circulation of Red Sea deep water. J Geophys Res 93:8175–8191

Chave KE (1954a) Aspects of biochemistry of magnesium 1. Calcareous and marine organism. J Geol 62:266–283

Chave KE (1954b) Aspects of biochemistry of magnesium 2. Calcareous sediments and rocks. J Geol 62:587–599

Chen CTA, Wang SL, Chou WC, Sheu DD (2006) Carbonate chemistry and projected future changes in pH and $CaCO_3$ saturation state of the South China Sea. Mar Chem 101:277–305

Clark D, Lamare M, Barker M (2009) Response of sea urchin pluteus larvae (Echinodermata: Echinoidea) to reduced seawater pH: a comparison among a tropical, temperate, and a polar species. Mar Biol 156(6):1125–1137

Cloud PE (1962) Behavior of calcium carbonate in seawater. Geochim Cosmochim Acta 26:867–887

Comeau S, Gorsky G, Jeffree R, Teyssié JL, Gattuso J-P (2009) Impact of ocean acidification on a key arctic pelagic mollusc (*Limacina helicina*). Biogeosciences 6(9):1877–1882

Couper A (ed) (1983) The time atlas of the oceans. Time Books, London, 272 pp

Culkin F, Cox RA (1966) Sodium, potassium, magnesium, calcium and strontium in seawater. Deep Sea Res 13:789–804

Dickson AG (1990) Thermodynamics of the dissociation of boric acid in synthetic sea water from 273.15 to 298.15K. Deep-Sea Res 37:755–766

Dickson AG, Riley JP (1979) The estimation of acid dissociation constants in seawater from potentiometric titrations with strong base. 1. The ion product of water-Kw. Mar Chem 7:89–99

DOE (1994) Handbook of methods for the analysis of the various parameters of the carbon dioxide system in sea water; version 2. In: Dickson AG, Goyet C (eds) ORNL/CDIAC-74

Dore JE, Lukas R, Sadler DW, Chruch MJ, Karl DM (2009) Physical and biogeochemical modulation of ocean acidification in the central North Pacific. Proc Nat Acad Sci USA 106:12235–12240

Edwards AJ (1987) Climate and oceanography. In: Edwards AJ, Head SM (eds) Red Sea: key environments. Pergamon Press, Oxford, pp 383–404

Edwards FJ, Head SM (eds) (1987) Red Sea: key environments. Pergamon Press, Oxford, 441 pp

Eisler R (2012) Ocean acidification. A comprehensive overview. CRC Press, Taylor and Francis, Jersey, and Enfield, New Hampshire, 252 pp

El-Anbaawy MIH, Al-aawah MAH, Al-Thour KA, Tucker ME (1992) Miocene evaporates of the Red Sea rift, Yemen Republic: sedimentology of Salif halite. Sed Geol 81:1–71

Feely RA, Chen CTA (1982) The effect of excess CO_2 on the calculated calcite and aragonite saturation horizons in the northeast Pacific. Geophys Res Lett 9(11):1294–1297

Feely RA, Byrne RH, Betzer PR, Gendron JF, Acker JG (1984) Factors influencing the degree of saturation of the surface and intermediate waters of the north pacific ocean with respect to aragonite. J Geophys Res 89:10631–10640

Feely RA, Byrne RH, Acker JG, Betzer PR, Chen CTA, Gendron JF, Lamb MF (1988) Winter-summer variation of calcite and aragonite saturation in northeast pacific. Mar Chem 25:227–241

Freund R (1970) Plate tectonics of the Red Sea and East Africa. Nature 228:453

Friedman GM (1985) Gulf of Elat (Aqaba). Geological and sedimentological framework. In: Friedman GM, Krumbein WE (eds) Hypersaline ecosystems, the Gavish Sabkha. Springer, Berlin, pp 39–71

Gattuso J-P, Frankignoulle M, Bourge I, Romaine S, Buddemeier RW (1998) Effect of calcium carbonate saturation of seawater on coral calcification. Global Planet Change 18(1–2):37–46

Gattuso J-P, Allemand D, Frankignoulle M (1999) Photosynthesis and calcification at cellular, organismal and community levels in coral reefs: a review on interactions and control by carbonate chemistry. Am Zool 39:160–183

Gaulier JM, Le Pichon X, Lyberis N, Avedik F, Geli L, Moretti I, Deschamps A, Salah H (1988) Seismic study of the crust of the northern Red Sea and Gulf of Suez. Tectonophysics 153:55–88

Gazeau F, Quiblier C, Jansen JM, Gattuso J-P, Middelburg JJ, Heip CHR (2007) Impact of elevated CO_2 on shellfish calcification. Geophys Res Lett 34:L07603

Girdler RW, Darracott BW (1972) African poles of rotation. Comments Earth Sci Geophys 2:131–138

Girdler RW, Styles P (1974) Two stage seafloor-spreading. Nature 247:7–11

Gonzalez-Davila M, Santana-Casiano JM, Gonzalez-Davila EF (2007) Interannual variability of the upper ocean carbon cycle in the northeast Atlantic Ocean. Geophys Res Lett 34, L07608. doi:10.1029/2006GL028145

Green MA, Waldbusser GG, Reilly SL, Emerson K, O'Donnel S (2009) Death by dissolution: sediment saturation state as a mortality factor for juvenile bivalves. Limnol Oceanogr 54:1037–1047

Green MA, Waldbusser GG, Hubazc L, Cathcart E, Hall J (2013) Carbonate mineral saturation state as the recruitment cue for settling bivalves in marine muds. Estuaries Coasts 36:18–27

Gvirtzman G, Buchbinder B, Sneh A, Nir Y, Friedman GM (1977) Morphology of the Red Sea fringing reefs: a result of the erosional pattern of the last-glacial low-stand sea level and the following Holocene recolonization. Mem Bur Rech Geol Min 89:480–491

Hales B, Emerson SR (1997) Calcite dissolution in sediments of the Ceara Rise: in situ measurements of porewater O_2, pH, and CO_2(aq). Geochim Cosmochim Acta 61:501–514

Hansson I (1973) A new set of acidity constants for carbonic acid and boric acid in seawater. Deep-Sea Res 20:461–478

Head SM (1987) In: Edwards AJ, Head SM (eds) Key Environment: Red Sea. Pergamon Press, Oxford

Heath GR, Culberson CH (1970) Calcite degree of saturation, rate of dissolution and the compensation depth in the deep oceans. Geol Soc Am Bull 81:3157–3160

Jokiel PL, Rodgers KS, Kuffner IB, Andersson AJ, Cox EF, Mackenzie FT (2008) Ocean acidification and calcifying reef organisms: a mesocosm investigation. Coral Reefs 27(3):473–483

Jones DA et al (1987) Littoral and shallow subtidal environment. In: Edwards AJ, Head SM (eds) Red Sea: key environments. Pergamon Press, Oxford, pp 383–404

Kitano Y (1964) On factors influencing the polymorphic crystallization on calcium carbonates found in marine biological system. In: Miyake Y, Koyama T (eds) Recent researches in the field of hydrosphere, atmosphere and nuclear chemistry. Tokyo

Kleypas JA, Buddemeier RW, Archer D, Gattuso J-P, Langdon C, Opdyke BN (1999) Geochemical consequences of increased atmospheric carbon dioxide on coral reefs. Science 284:118–120

Kroeker KJ, Micheli F, Gambi MC, Martz TR (2011) Divergent ecosystem responses within a benthic marine community to ocean acidification. Proc Natl Acad Sci 108(35):14515–14520

Krumgalz BS, Erez J, Chen CTA (1990) Anthropogenic CO_2 penetration in the northern Red Sea and in the Gulf of Elat (Aqaba). Oceanol Acta 13(3):283–290

Langdon C, Atkinson MJ (2005) Effect of elevated pCO2 on photosynthesis and calcification of corals and interactions with seasonal change in temperature/irradiance and nutrient enrichment. J Geophys Res 110:C09S07

Le Pichon X, Francheteau J (1978) A plate tectonics analysis of the Red Sea-Gulf of Aden area. Tectonophysics 46:369–406

Lombard F, da Rocha RE, Bijma J, Gattuso J-P (2010) Effect of carbonate ion concentration and irradiance on calcification in planktonic foraminifera. Biogeosciences 7(1):247–255

Lowenstam HA, Epstein S (1957) On the origin of sedimentary aragonite needles of the Great Bahamas Bank. J Geol 65:364–375

Lowenstam HA (1955) Aragonite needles secreted by algae and some sedimentary implications. J Sedim Petrol 25:270–272

Lyman J (1956) Buffer mechanism of seawater. PhD thesis, University of California, Los Angeles

Mackenzie FT, Pigott JD (1981) Tectonic controls of Phanerozoic sedimentary rock cycling. J Geol Soc Lond 138:183–196

Maillard C, Soliman G (1986) Hydrography and the Red Sea and exchanges with the Indian Ocean in summer. Oceanol Acta 9:249–469

Makris J, Allam A, Moktar T, Basahel A, Dehghani GA, Bazari M (1983) Crustal structure at the northwestern region of the Arabian shield and its transition to the Red Sea. Bull Fac Earth Sci King Abdulaziz Univ 6:435–447

Martinez F, Cochran JR (1988) Structure and tectonics of the northern Red Sea: catching a continental margin between rifting and drifting. Tectonophysics 150:1–32

Marubini F, Ferrier-Pages C, Cuif J-P (2003) Suppression of skeletal growth in scleractinian corals by decreasing ambient carbonate-ion concentration: a cross-family comparison. Proc Roy Soc Lond B 270(1511):179–184

McKenzie DP, Davies D, Molnar P (1970) Plate tectonics of the Red Sea and East Africa. Nature 226:243–248

Mehrbach C, Culberson CH, Hawley JE, Pytkowicz RM (1973) Measurement of apparent dissociation constants of carbonic acid in seawater at atmospheric pressure. Limnol Oceanogr 18:897–907

Meshal AH, Behairy AKA, Osman MM (1984) Evaporation from coastal and open waters of the central zone of the red sea. Atmos Ocean 22:369–378

Millero FJ (1995) Thermodynamics of the carbon dioxide system in the oceans. Geochim Cosmochim Acta 59:661–677

Morcos SM (1970) Physical and chemical oceanography of the Red Sea. In: Barnes H (ed) Oceanography and marine biology: an annual review 8:73–202

Morse JM, Berner RA (1972) Dissolution kinetics of calcium carbonate in seawater II. A kinetic origin for lysocline. Am J Sci 272:840–861

Mucci A (1983) The solubility of calcite and aragonite in seawater at various salinities, temperatures and 1 atmosphere total pressure. Am J Sci 283:780–799

Mucci A, Morse JW (1984) The solubility of calcite in seawater solutions of various magnesium concentrations, It = 0.697 m at 25 °C and one atmosphere total pressure. Geochim Cosmochim Acta 48:815–822

Müller MN, Schulz KG, Riebesell U (2010) Effects of long-term high CO_2 exposure on two species of coccolithophores. Biogeosciences 7:1109–1116

Neumann AC, McGill DA (1953) Circulation of the Red Sea in early summer. Deep Sea Res 8:223–235

Opdyke BN, Wilkinson BH (1993) Carbonate mineral saturation state and cratonic limestone accumulation. Am J Sci 293:217–234

Osman M (1984) Evaporation from coastal waters off Port Sudan. In: Symposium coral reef environment of the Red Sea, Jeddah, 14–18 Jan 1984, Abstract:7

Potts DC (1983) Evolutionary disequilibrium among Indo-Pacific corals. Bull Mar Sci 33:295–314

Pytkowicz RM (1965) Rates of inorganic calcium carbonate nucleation. J Geol 196–199

Pytkowicz RM (1973) Calcium carbonate retention in supersaturated seawater. Am J Sci 273:515–522

Pytkowicz RM (1983) Equilibria, nonequilibria and natural waters. Wiley, New York, 353 pp

Pytkowicz RM, Flower GA (1967) Solubility of foraminifera in seawater at high pressures. Geochem J 1:169–182

Riebesell U, Zondervan I, Rost B, Tortell PD, Zeebe RE, Morel FMM (2000) Reduced calcification of marine plankton in response to increased atmospheric CO_2. Nature 407:364–376

Royer DL, Berner RA, Beerling DJ (2001) Phanerozoic atmospheric CO2 changes: evaluating geochemical and paleobiological approaches. Earth-Sci Rev 54:349–392

Rushdi AI (1989) Kinetics and equilibrium of magnesian calcite in seawater-like solution. PhD thesis, Oregon State University, 245 pp

Rushdi AI (1992) Mineralogy and morphology of calcium carbonate as a function of magnesium concentration in artificial seawater. J Fac Mar Sci King Abdulaziz Univ 3:13–24

Rushdi AI (1993) Kinetics of calcite overgrowth as a function of magnesium concentrations and supersaturation in artificial seawater. J Fac Mar Sci King Abdulaziz Univ 4:39–54

Rushdi AI (1995) Equilibrium behavior of magnesian calcite mineral: a theoretical approach. J Fac Mar Sci King Abdulaziz Univ 6:41–50
Rushdi AI (2012) Calcite and aragonite saturation states of Kamaran strait surface seawater, Yemen: summer time. Arab J Geosci 5:1093–1102
Rushdi AI (2014) Calcium carbonate saturation levels of the Red Sea coast of Yemen and biogeochemical effects of excess CO_2. In: International conference on the marine environment of the Red Sea, King Abdulaziz University, Jeddah, 10–13 Nov 2014 (in press)
Rushdi AI, Pytkowicz RM, Suess E, Chen CT (1992) The effects of magnesium-to-calcium ratio, in artificial seawater at different ionic products, upon the induction time, and the mineralogy of calcium carbonate: a laboratory study. Geol Rundsch 81:571–578
Santana-Casiano JM, González-Dávila M, Rueda M-J, Llinás O, González-Dávila EF (2007) The interannual variability of oceanic CO_2 parameters in the northeast Atlantic subtropical gyre at the ESTOC site. Glob Biogeochem Cycles 21:GB1015
Schneider K, Erez J (2006) The effect of carbonate chemistry on calcification and photosynthesis in the hermatypic coral Acropora eurystoma. Limnol Oceanogr 51:1284–1293
Sheppard CRC, Sheppard ALS (1991) Corals and coral communities of Arabia. Fauna Saudi Arabia 12:3–191
Sheppard CRC, Price ARG, Roberts C (1992) Marine ecology of the Arabian region. Patterns and processes in extreme tropical environments. Academic Press, London, 359 pp
Silter WV, Be AWH, Berger WH (1975) Dissolution of deep-sea carbonates. Special Publication Cushman Found for Foraminiferal Research no. 13
Silverman J, Lazar B, Erez J (2007) Effect of aragonite saturation, temperature, and nutrients on the community calcification rate of a coral reef. J Geophys Res-Oceans 112, C05004. doi:10.1029/2006JC003770
Simkiss K (1964) The inhibitory effects of some metabolites on the precipitation of $CaCO_3$ from artificial and natural seawater. Journal du Conseil Conseil Permanent International pour l'Exploration de la Mer 29:6–18
Smith SV, Pesret F (1974) Processes of carbon dioxide flux in the Fanning Island Lagoon. Pac Sci 28:225–245
Smith AD, Roth AA (1979) Effect of carbon dioxide concentration on calcification in the Red Coralline Alga Bossiella orbigniana. Mar Biol 52:217–225
Stockman KW, Ginsburg RN, Shinn EA (1967) The production of lime mud by algae in south Florida. J Sedim Petrol 37:348–633
Stumm W, Morgan JJ (1996) Aquatic chemistry: chemical equilibria and rates in natural waters, 3rd edn. Wiley, New York, 1021 pp
Suess E (1973) Interaction of organic compounds with calcium carbonate. Geochim Cosmochim Acta 34:157–168
Sullivan W (1974) Continents in motion: the new earth debate. McGraw-Hill Book Company, New York, 399 pp
Suzuki A, Nakamori T, Kayanne H (1995) The mechanism of production enhancement in coral reef carbonate systems: model and empirical results. Sed Geol 99:259–280
Takagi Y (2002) Otolith formation and endolymph chemistry: a strong correlation between the aragonite saturation state and pH in the endolymph of the trout otolith organ. Mar Ecol Prog Ser 231:237–245
Takahashi T (1975) Carbonate chemistry of seawater and the calcite compensation depth in the ocean. Dissolution of deep-sea carbonates. Spec Publ Cushman Found Foraminiferal Res 13:111–126
Taviani M (1998a) Axial sedimentation of the Red Sea transitional region (22°–25°N): pelagic, gravity flow and sapropel deposition during the late quaternary. In: Purser BH, Bosence DWJ (eds) Sedimentation and tectonics of rift basins: Red Sea-Gulf of Aden. Chapman and Hall, London, pp 467–478
Taviani M (1998b) Post-Miocene reef faunas of the Red Sea: glacio-eustatic controls. In: Purser BH, Bosence DWJ (eds) Sedimentation and tectonics of rift basins: Red Sea-Gulf of Aden. Chapman and Hall, pp 574–582
Tramontini C, Davies D (1969) A seismic refraction survey in the Red Sea. Geophys J Roy Astron Soc 17:2225–2241
Waldbusser GG, Voigt EP, Bergschneider H, Green MA, Newell RIE (2011) Biocalcification in the eastern oyster (*Crassostrea virginica*) in relation to long-term trends in Chesapeake Bay pH. Estuaries Coasts 34:221–231
Walter LM (1984) Magnesian calcite stabilities: a reevaluation. Geochim cosmochim acta 48:1059–1069
Wegener A (1929) The origin of continents and oceans. English translation, 4th rev edn. J Biram, Dover, New York
Weiss R (1974) Carbon dioxide in water and seawater. The solubility of a non-ideal gas. Mar Chem 2:203–215
Wettenberg H, Timmerman E (1936) Uber die sattigung des seewassers an $CaCO_3$. Annalen der Hydrographie und Maritimen Meteorology 64:23–31
Widdicombe S, Spicer JI (2008) Predicting the impact of ocean acidification on benthic biodiversity: what can animal physiology tell us? J Exp Mar Biol Ecol 366(1):187–197
Yirgu G, Ebinger CJ, Maguire PKH (eds) (2006) The Afar Volcanic province within the East African Rift System. Geological Society, London, Special Publications 259:1–6
Zeebe RE, Wolf-Gladrow DA (2001) CO_2 in seawater: equilibrium, kinetics, isotopes, vol 65. Elsevier Oceanography Series, Academic Press, Massachusetts, 341 pp

Lagoon Sediments of the Eastern Red Sea: Distribution Processes, Pathways and Patterns

Najeeb M.A. Rasul

Abstract

The sedimentary characteristics of the lagoons along the Red Sea coast of Saudi Arabia are governed by the potential of the wadis (seasonal streams) to transport sediment, while the sediment distribution is influenced by the water current patterns that are mostly generated by the wind. Human interference and anthropogenic input are the two main issues that change the dynamics and morphology of the lagoons. The absence of permanent fluvial systems keeps the water transparent and helps coral growth, while the exchange of water between the lagoons and the Red Sea is restricted and leads to pollution in the lagoons. The sediment yield varies from north to south depending on the amounts of rainfall. The sediment veneer in the southern half of the Red Sea is mostly terrigenous because of relatively higher input from the wadis, whereas in the northern half where rainfall is low, the numerous wadis contribute terrigenous material in varying amounts. The central part is mainly devoid of terrigenous input where anthropogenic input prevails. The lagoons are mainly blanketed by carbonate of biogenic origin, and therefore the associated Red Sea environment is regarded as a carbonate-rich environment. Some wadis that drain directly into the Red Sea or into the lagoons are either dumping sites or are used as farm land, and lagoons close to major cities are being reclaimed and used as resorts. Although the coastal belt is still somewhat pristine, the health of the marine environment may deteriorate with excessive human interference and irresponsible actions.

Introduction

There are over twenty lagoons along the 1,800 km stretch of the Red Sea coast of Saudi Arabia. These lagoons, often referred to as *sharms, khors or khawrs*, are generally considered to be channels formed by erosion in the Late Pleistocene that were drowned by the post-glacial rise in sea level (Brown et al. 1989). Wadis (seasonal streams) directly drained these lagoons during the last glacial period when the sea level was lower by at least 140 m. An alternative view, proposed by Rabaa (1980), is that the lagoons are remnants of collapse features formed during post-warming (post-glacial?) emergence by selective dissolution of Miocene evaporite beds underlying the younger succession.

The lagoons are elongated, generally shallow and most have gently sloping shorelines and are connected to the Red Sea by a single or multiple channels with steeply dipping flanks. The channels were formed either naturally or are man-made, cutting through a series of barrier islands and reefs (barrier reef system) running parallel to the shoreline (Morley 1975) or raised coral reef terraces of Pleistocene age. However, some lagoons, especially in the northern Red Sea, have almost vertical slopes (reef walls) at the mouth down to 30 m depth composed entirely of coral beds.

The lagoons range in length from a few kilometres to over 20 km with depths ranging from a few metres to over 30 m (Fig. 1). The size and shape of lagoons and associated channels are dependent on the current velocity, wave energy and tidal range. Tides are of primary importance because they provide a periodic exchange of water through channels

N.M.A. Rasul (✉)
Center for Marine Geology, Saudi Geological Survey, Jeddah, Saudi Arabia
e-mail: najeeb_rasul@hotmail.com; rasul.nm@sgs.org.sa

N.M.A. Rasul and I.C.F. Stewart (eds.), *The Red Sea*,
Springer Earth System Sciences, DOI 10.1007/978-3-662-45201-1_17,

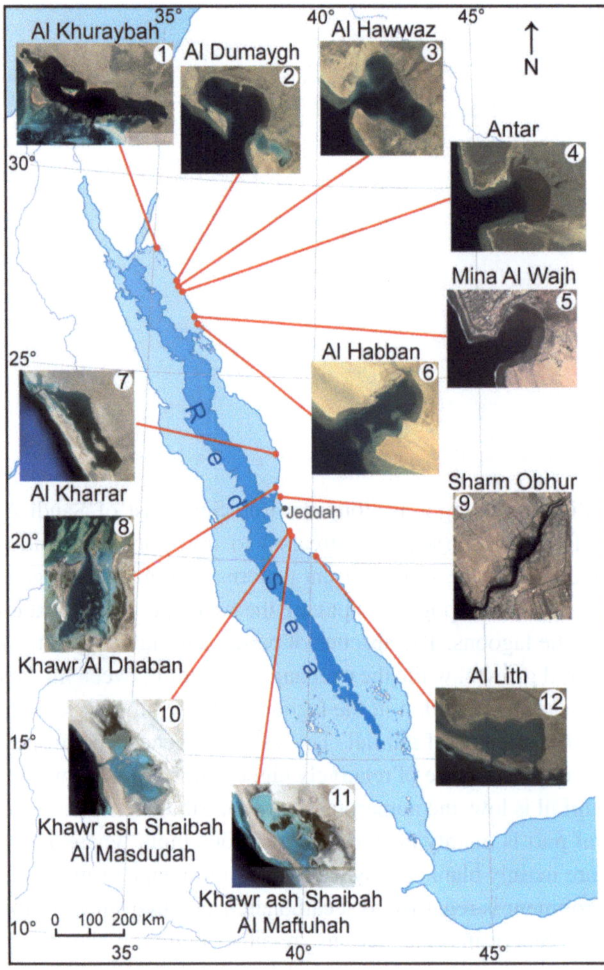

Lagoon		Max. length (km)	Max. width (km)	Area (km²)	Entrance width (m)	Max. depth (m)
1		23.34	4.59	100	16; 36	25
2		1.57	0.71	0.70	25	33
3		1.26	0.65	0.70	2	22
4		0.81	0.40	0.35	24	15
5		0.73	0.43	0.33	28	22
6		1.26	0.50	0.72	42	14
7		18.00	4.86	71.44	150	14
8		6.59	3.58	13.17	37	12
9		11.00	1.20	5.60	250	35
10		4.70	2.46	7.85	14	45
11		8.05	2.70	15.87	9	3
12		6.46	2.83	16.48	58; 44	4

↗ entrance

Fig. 1 Location of the lagoons along the Red Sea coast of Saudi Arabia with their physical parameters

connecting the lagoons to the sea. Currents are usually weak inside the lagoons and are governed by the spring–neap tidal cycles and fluctuation in the water level (Albarakati and Ahmad 2012). Wind stress also plays a dominant but variable role in the process of sediment transport, depending on the strength of the local tides and wind speed (Sultan and Ahmad 1990; Ahmad and Sultan 1992), whereas the strength of the wave energy controls the shape of the lagoons. The margins of the lagoons are bounded by intertidal flats and sabkhas that extend to a few kilometres inland from the shoreline. In some places, elevated sandy berms separate the sabkhas from the lagoons where the low-lying areas are occasionally covered by shallow water during tidal cycles (El-Abd and Awad 1991).

There are over seventy-five wadis along the Red Sea coast of Saudi Arabia; some drain directly into the Red Sea, but some drain into lagoons creating contrasting dynamics in terms of sediment cover and water circulation pattern (Fig. 2). A summary of the drainage basins of selected wadis that are discussed in this chapter is presented in Table 1. Detrital sediments are transported into lagoons by wadis during sporadic rainfall and biogenic materials are either transported by tidal currents from the open sea or are locally produced. The sediment texture in these lagoons ranges from mud to muddy sand to gravel of both terrigenous and biogenic origin, the latter being dominant. Most locally produced skeletal remains of foraminiferal tests and carbonate materials of biogenic origin play a minimal role in understanding the sediment dynamics, whereas the lithogenic fractions of either aeolian or fluvial sources (transported by wadis), although in varying proportions, contribute to understanding sediment dispersal mechanisms, transport patterns and pathways.

Mangrove stands, especially *Avicennia marina*, are scattered on the islands and periphery of some lagoons and are populated by various bird and fish species endemic to the Red Sea. Various species of corals and benthic organisms control the character of sediment and the ecology of islands. Fish, shrimps, dolphins, crabs, dugongs and turtles congregate in the lagoons and use the sheltered environment for spawning,

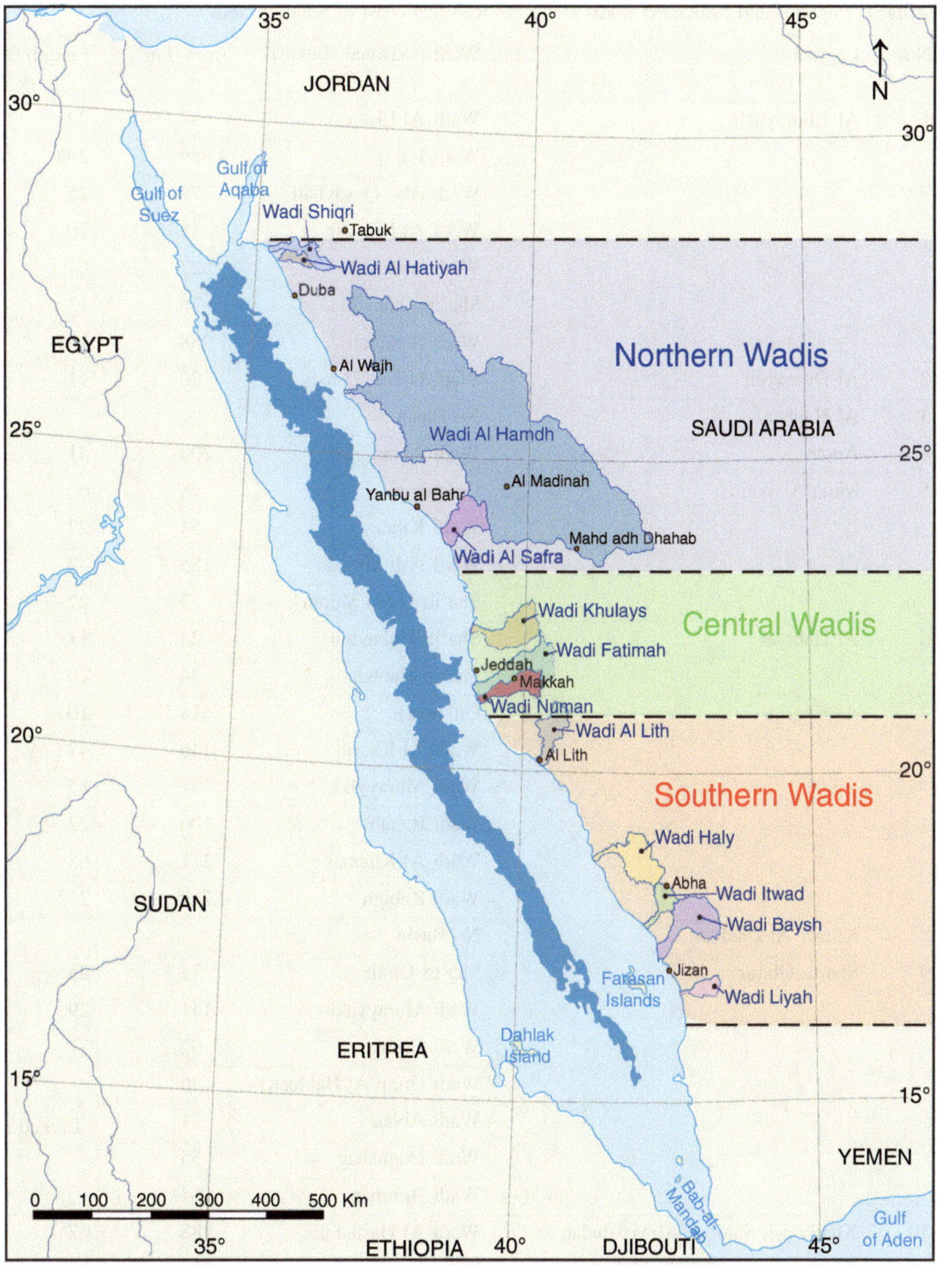

Fig. 2 Selected wadis draining into the Red Sea coast of Saudi Arabia

while some lagoons have been used for fish and shrimp farming (e.g. Khor Al Ghallah). A few are used as recreational amenities especially those located near urban areas (e.g. Sharm Obhur), and some have been used for dumping partially treated sewage through pipelines, for example Al Shabab and Al Arbaeen lagoons (El-Sayed et al. 2011, 2013). Sewage disposal either in the lagoons or their vicinity has affected the corals and ecosystems of the Red Sea (Risk et al. 2009; Abu-Zied et al. 2012). Coastguard stations at the entrances and particularly boat docking facilities inside most lagoons cause the lagoons to undergo environmental stress resulting in loss to the marine community, and also greatly affecting the health of coral species. In addition, desalination plants have also impacted the coastal environments of the Red Sea of Saudi Arabia (Alharbi et al. 2012).

In general, the coastal waters and the transition between open water and land absorb much of the impact of human activities (Coakley and Rasul 2001), including ecotourism, whose links with science and the environment are often negligible (Crockett and Stow 2001). Increasing development and population density along shorelines have impacted marine environments resulting in severe changes in geomorphological and sedimentary character. Some coastal areas along the Red Sea are no exception to the severity of the impact on the sedimentary assemblages. Most lagoons are still not impacted, especially those located in areas where

Table 1 Lagoons and associated wadis along the Red Sea coast of Saudi Arabia

No.	Lagoon	Wadi (seasonal stream)	Area (km^2)	Length (km)	Slope (m/m) (overland slope)	Elevation (m.a.s.I)
1	Al Khuraybah	Wadi Al Lisan	97	22	0.0163	147
		Wadi Ifal	4,889	206	0.1888	2,335
		Wadi Abu Qusaybah	78	25	0.0196	141
		Wadi Al Mihasib	15	10	0.0395	154
		Wadi Aynunah	827	67	0.1993	1,368
		Sha'ip Qunaybi	26	13	0.0340	214
		Wadi Sharmah	1,958	163	0.1370	1,624
2	Al Dumaygh	Wadi Umm 'Ushsh	26	11	0.0800	274
3	Al Hawwaz	No Basin				
4	Antar	Wadi Antar	200	31	0.0915	415
5	Mina Al Wajh	Unknown	8	6	0.0247	36
		Wadi Kibra	54	22	0.0322	199
		Wadi Ash Shijnah	185	42	0.0419	255
		Sha'ib Umm Sidrah	73	22	0.0207	145
6	Al Habban	Sha'ib Hubaybin	24	11	0.0179	84
		Sha'ib Habban	34	20	0.0202	171
7	Al Kharrar	Unknown	14	10	0.0108	22
		Wadi Al Khariq	146	47	0.0361	202
		Wadi Murayyikh	32	15	0.0397	150
		Wadi Rahab	100	27	0.0330	176
		Wadi Al Khamas	371	65	0.0447	314
		Wadi Rabigh	4,739	212	0.0928	1,358
8	Khawr Al Dhaban	No Basin				
9	Sharm Obhur	Sha'ib Uimir	73	25	0.0279	132
		Wadi Murayyakh	141	29	0.0455	320
		Wadi Ghaya	92	Linked to open channel drainage system		
		Wadi Umm Al Hableen	46			
		Wadi Abhar	21			
		Wadi Daghabag	55			
		Wadi Braiman	64			
10	Khawr ash Shaibah Al Masdudah	Wadi Al Hashafat	388	67	0.0294	298
11	Khawr ash Shaibah Al Maftuhah	Unknown	120	37	0.0124	170
		Wadi Al Atwa'	213	45	0.0398	294
		Wadi Al Kharqah	736	92	0.0929	1,194
12	Khor Al Ghallah and Offshore Al Lith	Wadi Sayah	727	85	0.1282	1,551
		Wadi Markub	523	60	0.1198	454
		Wadi Marakh	112	22	0.0946	497
		Wadi Al Ghallah	597	78	0.1586	1,703
		Wadi Khariq Al Bir	169	26	0.0345	177
		Wadi Al Lith	3,157	174	0.2720	2,428

access is minimal. However, the physical and chemical characteristics of sediments have altered with time in areas where uncontrolled human intervention has occurred. Practices impacting the various sub-environments of the Red Sea are addressed by Davenport and Davenport (2006) and Gladstone et al. (2013).

Data Acquisition and Processing

Surficial sediment samples were obtained by either Ponar or Peterson grab samplers from the lagoons. The sediment samplers effectively sample the uppermost 15–30 cm of the bed sediment. On sediment recovery, the sediment colours were recorded and described directly in accordance with the Munsell Soil Chart. The number of samples obtained from each lagoon is presented in Table 2 and in the textural maps of all the lagoons. The procedure for grain size analysis involved pre-treatment and dispersion of the samples as described in British Standard 1377 (BSI 1975). Mechanical sieving and the results of the grain size analysis are presented as a textural classification is in accordance with the method adopted by Folk (1980). GRADISTAT particle size analysis software was used for calculating some particle size statistics (Blott and Pye 2001). Calcium carbonate ($CaCO_3$) was determined by gasometric methods using a 'calcimeter'. Total organic carbon (TOC) was determined by the wet oxidation method of Le Corre (1983).

A limited number of publications have described aspects of the sediments of the lagoons along the Red Sea coast of Saudi Arabia and recently detailed work on selected lagoons has been carried out by the Saudi Geological Survey. Among the 12 lagoons listed (Fig. 1), five have been studied in detail, representing the various sub-environments, either frequently used as resorts, drained by wadis, or with no association to wadis but instead industrially and anthropogenically impacted. A summary of some of these is presented here, and their physical, sedimentological, and hydrological parameters are listed in Table 2.

Sharm Obhur

Sharm Obhur is an elongated erosional lagoon that formed during the Pleistocene by tectonic faulting (Darwin 1962; Berry et al. 1966). The sharm cuts through coralline limestone and conforms to the end of a former fluvial valley inundated by the Red Sea. It forms the lower western part of the Red Sea coastal plain of the Tihama formation (Ali and Hossain 1988). Sharm Obhur has been studied by Dowidar et al. (1978), Behairy et al. (1983), El-Sabrouti (1983), El-Abd and Awad (1992a, b), Basaham and El-Shater (1994), Basaham et al. (2003), Fahmy and Saad (1996), El-Rayis and Eid (1997), Bantan and Rasul (2003), Al-Harbi and Khomayis (2005), Al-Farawati et al. (2008) and Rasul et al. (2009a). These authors present sedimentological data as well as temporal changes in the morphology, geology and sedimentology of the sharm.

Sharm Obhur has seen a remarkable transformation in its environmental setting as well as a series of geomorphological changes since the 1970s because of progressive urbanization of its coastal areas. The head of the sharm in the north was dredged over a kilometre to accommodate recreational facilities, while a bay-like feature at the mouth in the south was constructed that reduced the width of the mouth to about 250 m, leading to changes in water dynamics (Fig. 3). The water exchange between the sharm and the Red Sea has therefore been reduced and rejuvenation is minimized. The middle part of the sharm is the widest, but it has lost over 100 m to land reclamation between 1986 and 2000 and now is $\sim$1,200 m in width. During this time, the area of the Sharm decreased by about 800,000 m^2 representing an average annual loss of about 60,000 m^2 because of infilling (Basaham et al. 2006). This has therefore affected the morphology and composition of the bottom sediments. A concrete open channel system in the wadi leading to the head of the sharm was constructed recently to drain excess rain water to the sharm, which is only likely during heavy rainfall. Several wadis from the east flow towards the sharm and the water collected in the drainage system leads to the head of Sharm Obhur (Table 1).

The sharm is <1 m deep at the head in the north-east, while the thalweg is deep and narrow with a gradual increase in the depth to about 35 m downstream, with an abrupt fall to over 80 m about 100 m offshore (Fig. 3a). The abrupt fall is due to the lack of sediment supplied by Wadi Al Kura. The axis of the sharm is veneered with mixtures of detrital and biogenic fine-grained materials, where the indigenous biogenic marine sediments are locally produced, and detrital components were once transported by the dormant Wadi Al Kura. The sharm contains one of the highest amounts of detrital material compared to other sharms along the eastern Red Sea (Bantan and Rasul 2003). Although minor amounts of sediments are supplied by Wadi Al Kura, the sharm receives detrital sediments by wind, indicated by pitted quartz. The flanks are covered by coral reefs with variations in slope and are covered by varying amounts of very fine sediments. These are fringing-type reefs in which soft corals are abundant compared to branching corals. Among soft corals, *Xenia* and *Heteroxenia* are predominant, whereas in the Scleractinian corals, *Pocillopora* dominates followed by *Stylophora* species. In deeper waters, massive corals such as *Porites* hillocks are present. A considerable part of the reef system is occupied by the sponges of the Callyspongidae family (Jerald Wilson, personal communication). Some coral species are dead, although most survive the environmental stress. The coral reef exposures and the fringing reef pattern of the coastal area continue into the outer part of the sharm (Fahmy and Saad 1996).

The mouth of the sharm is tidally dominated with a current velocity of about 75 cm s^{-1} during spring tide. This is indicative of an inverse estuarine circulation pattern (Pritchard 1952; Nun Vaz et al. 1990). Recent studies show two layers of water flow through the entrance, with surface inflow from the Red Sea into the sharm and bottom outflow

Table 2 Summary of lagoons along the Red Sea coast of Saudi Arabia and their physical and sedimentological parameters

Lagoon	Area (km^2)	Total samples	Depth (m)	Sediment colour (Munsell Chart)	Sediment texture (Folk 1980)	Gravel avg (%)	Sand avg (%)	Mud avg (%)	Mean size (ϕ)	Sorting (ϕ)	Sand composition	$CaCO_3$ % (avg)	TOC % (avg)	Current velocity (cm s^{-1})
Al Khuraybah	24	159	25	2.5Y 6/3 light yellowish brown to GLEY 1 2.5/N black	Mud to muddy sand	2	36	62	2.16 (*fsd*) to 5.91 (*mst*)	0.16 (*vws*) to 2.68 (*vps*)	Dominated by terrigenous	3 to 89 (*71*)	0.17 to 0.98 (*0.37*)	45 (lagoon) to 110 (mouth)
Al Dumaygh	0.71	38	32 (22 at the mouth)	2.5 Y 4/4 olive brown to 5Y 2.5/1 black	Sandy mud to Gravelly sand	1	59	40	3.74 (*vfsd*) to 5.70 (*mst*)	0.16 (*vws*) to 2.17 (*vps*)	Dominated by biogenic in the northwest. Biogenic and terrigenous in the east	29 to 93 (*82*)	0.21 to 0.60 (*0.53*)	25
Al Hawwaz	0.68	37	49 (23 at the mouth)	2.5Y 5/2 greyish brown to GLEY 1 2.5/N black	Sandy mud to gravelly sand	3	67	30	3.71 (*vfsd*) to 5.70 (*mst*)	0.61 (*mws*) to 1.65 (*ps*)	Dominated by biogenic in the east. Biogenic and terrigenous in the north	30 to 92 (*79*)	0.12 to 0.37 (*0.26*)	43
Antar	0.35	12	26 (18 at the mouth	10YR 4/3 brown to 10YR 3/1 very dark grey	Sandy mud to sand	1	56	43	3.74 (*vfsd*) to 5.52 (*mst*)	0.16 (*vws*) to 1.46 (*ps*)	Dominated by terrigenous	13 to 81 (*68*)	0.27 to 1.16 (*0.85*)	No data
Mina Al Wajh	0.33	44	36 (28 at the mouth)	10YR 6/6 brownish yellow to GLEY 1 2.5/N black	Sandy mud to gravelly sand	1	81	18	3.72 (*vfsd*) to 5.73 (*mst*)	0.16 (*vws*) to 1.42 (*ps*)	Dominated by terrigenous and mostly stained	20 to 60 (*49*)	0.37 to 1.97 (0.99)	75
Al Habban	0.72	19	15 (10 from the mouth)	10YR 6/3 pale brown to GLEY 2 2.5/1 greenish black	Sandy mud to gravelly sand	1	54	45	3.70 (*vfsd*) to 5.71 (*mst*)	0.31 (*vws*) to 1.46 (*ps*)	Dominated by biogenic	19 to 86 (*77*)	0.11 to 0.41 (*0.22*)	No data
Al Kharrar	63	119	22	GLEY 1 8/1 (light greenish grey) to Gleyl 3/ (very dark grey)	Mud to gravelly sand	6	64	30	1.36 (*msd*) to 3.21 (*vfsd*)	0.30 (*vws*) to 1.60 (*ps*)	High biogenic in north and high terrigenous in south	2 to 93 (*59*)	0.29 to 2.5 (*0.97*)	18 to 100
Khawr Al Dhaban	16	28	12	10 YR 6/2 light brownish grey to 5 Y 4/1 dark grey	Sandy mud to gravelly muddy sand	3	43	54	1.40 (*msd*) to 2.30 (*fsd*)	1.31 (*ps*) to 0.79 (*ms*)	Dominated by biogenic	90 to 96 (*93*)	0.23 to 0.84 (*0.51*)	29

(continued)

Table 2 (continued)

Lagoon	Area (km^2)	Total samples	Depth (m)	Sediment colour (Munsell Chart)	Sediment texture (Folk 1980)	Gravel avg (%)	Sand avg (%)	Mud avg (%)	Mean size (ϕ)	Sorting (ϕ)	Sand composition	$CaCO_3$ % (avg)	TOC % (avg)	Current velocity (cm s^{-1})
Sharm Obhur (1986)	6	42	35	5 Y 8/1 white to 5 YR 4/3 reddish brown	Mud to gravelly sand	3	46	51	1.30 (*fsd*) to 3.55 (*fsd*)	0.37 (*ws*) to 3.44 (*vps*)	Dominated by terrigenous	40 to 45 (*40*)	0.4 to 0.6 (mud patch 0.9 %)	No data
Sharm Obhur (2003)	6	41	35	2.5Y 6/3 light yellowish brown to 5Y 5/3 olive	Mud to gravelly sand	2	24	74	2.23 (*fsd*) to 4.55 (*fsd*)	0.40 (*ws*) to 1.35 (*ps*)	Dominated by terrigenous	40 to 70 (*50*)	0.2 to 0.4 (mud patch 0.8 % highest 1.4 %)	25 to 80
Sharm Obhur (2009)	6	40	35	GLEY 1 3/1 5GY very dark greenish grey and 5 Y 6/2 light olive grey	Sandy mud to gravelly sand	3	42	55	2.29 (*fsd*) to 4.67 (*cst*)	0.68 (*ms*) to 1.76 (*ps*)	Biogenic and terrigenous	39 to 78 (*54*)	0.19 to 0.37 (*0.29*)	30 to 87
Sharm Obhur (2013)	6	42	35	2.5Y 5/2 greyish brown and 5Y > 2.5/1 black	Mud to gravelly muddy sand	4	44	52	2.90 (fsd) to 4.88 (cst)	0.73 (*ms*) to 2.15 (*vps*)	Terrigenous and biogenic	33 to 61 (*47*)	0.29 to 0.70 (mud patch 0.57 %)	25 to 75
Khawr ash Shaibah Al Masdudah (northern)	8.5	57	7 at the mouth av. 3.5 in the lagoon	10 YR 7/3 (very pale yellow to GLEY 1 3/ (very dark grey)	Gravelly mud to sandy mud	7	64	29	2.87 (fsd) to 0.86 (csd)	0.66 (*mws*) to 1.61 (*ps*)	Mostly biogenic	38 to 92 (*68*)	0.15 to 4.0 (*1.7*)	20 to 50 cm s^{-1} 45 avg at the mouth
Khawr ash Shaibah Al Maftuhah (southern)	7.5	85	15 at the mouth av. 4 in the lagoon	2.5 Y 8/3 (pale yellow to GLEY 1 3/1 (very dark grey)	Gravelly mud to sandy mud	4	60	36	3.42 (vfsd) to 2.87 (fsd)	0.58 (*mws*) to 1.42 (*ps*)	Mostly biogenic	36 to 92 (*75*)	1.08 to 4.95 (*1.92*)	20 to 50 cm s^{-1} 25 avg at the mouth
Khor Al Ghallah and offshore Al Lith	16	121	4 in the lagoon and 59 offshore	GLEY 1 10Y 6/1 greenish grey to GLEY 1 3/1 10Y very dark greenish grey, offshore 10YR 5/2 greyish brown	Mud to gravelly sand in the lagoon, gravelly muddy sand, dominated by sandy mud	1	42	57	0.52 (csd) to 4.24 (est)	0.16 (*vps*) to 1.88 (*ps*)	Mostly terrigenous in the lagoon. Mostly terrigenous in the southeast dominated by biogenic in the west	2 to 40 in the lagoon (3 % in the east). Over 70 % in the southwest offshore up to 40 % in the southeast shallow area.	0.1 to 0.5 in the lagoon. 0.3 to 0.7 offshore	29 at the mouth

* *csd* coarse sand, *msd* medium sand, *fsd* fine sand, *vfsd* very fine sand, *cst* coarse silt, *mst* medium silt
***vws* very well sorted, *mws* moderately well sorted, *ps* poorly sorted, *vps* very poorly sorted

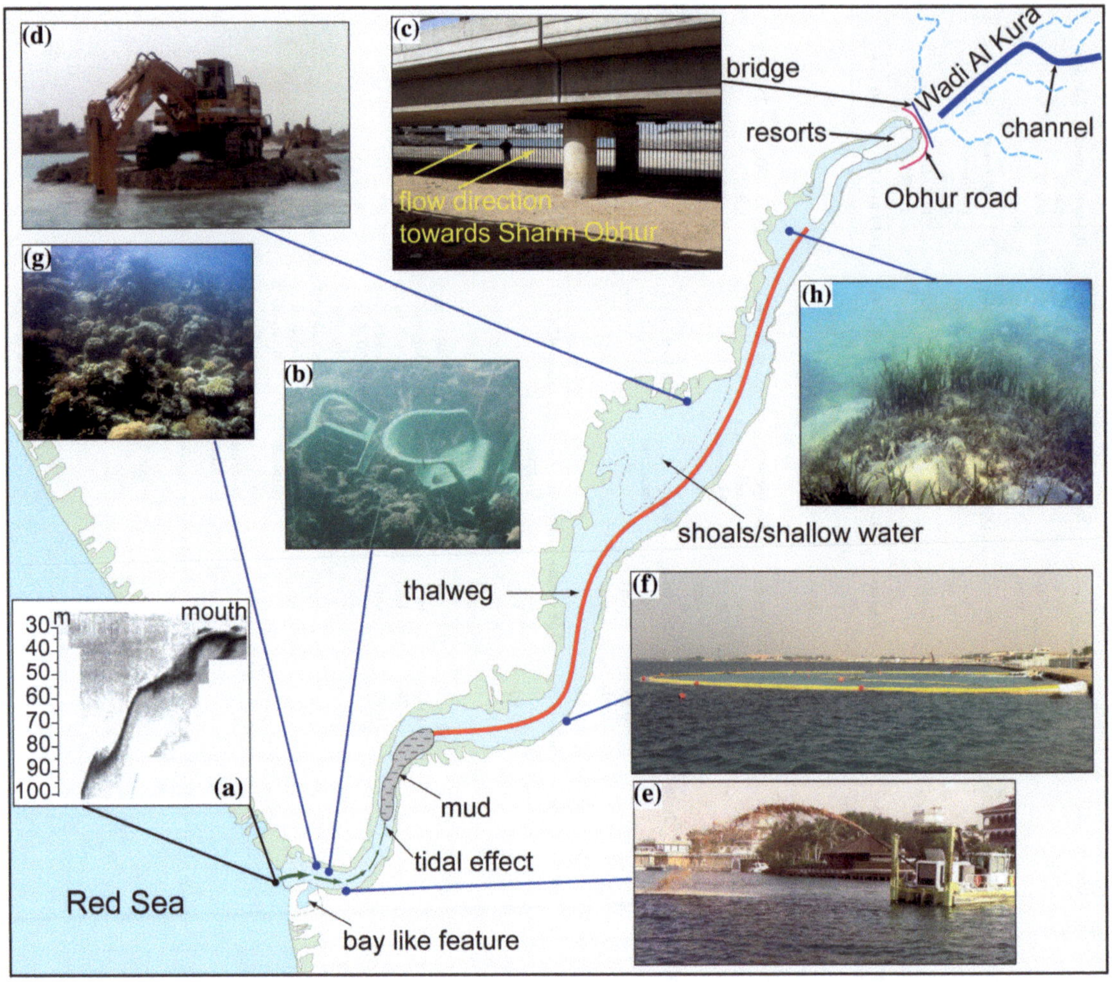

Fig. 3 Sharm Obhur **a** depth increases abruptly at the mouth because of lack of sediment supply; **b** plastic chairs dumped into the sharm along with other anthropogenic material; **c** confluence of the sharm and Wadi Al Kura where barriers have been erected to retain debris; **d** land reclamation in progress; **e** dredging to increase the water depth for boats. The material is dumped into the deep water regardless of the environmental impact; **f** floaters (*in yellow*) deployed to restrict the effect of the sediment plume on the sharm, although the plume eventually disperses due to surface currents; **g** some coral species are not vulnerable to environmental stress and continue to thrive, and **h** seagrass and corals covered with algae show signs of high nutrient input from anthropogenic sources

into the Red Sea. However, in previous studies, three water layers were identified at the entrance: a surface water mass (upper 5 m of the water column) characterized by high temperature and salinity, an intermediate water mass of minimum salinity with a core at 10–20 m water depth, and a bottom (>20 m depth) water mass with maximum salinity. This hydrographic structure develops a two-layer flow, inflow of low-salinity water at both the surface and intermediate depths and outflow of high-salinity water at the bottom (El-Rayis and Eid 1997). Previous studies show that the evaporation rate ranges between 0.6 cm/day in May and 0.8 cm/day in February. The rate of water inflow ranges between 150 $m^3\ s^{-1}$ (November) and 300 $m^3\ s^{-1}$ (February), with the outflow ranging between 150 $m^3\ s^{-1}$ (November) and 400 $m^3\ s^{-1}$ (February) (El-Rayis and Eid 1997), although no recent data are available to substantiate water flow and evaporation rates. The surface water temperature ranges from 24.4 °C in winter to 32.2 °C in summer and increases gradually from the entrance to the head in the north (Basaham et al. 2006). The salinity ranges between 39.1 and 40.2 ‰ and follows a similar pattern to the water temperature (Ahmad and Sultan 1993). However, between 2006 and 2013, the average temperature ranged between 22 °C in winter to over 34 °C in summer, and the salinity averaged 39 and 41 ‰ at the entrance and head, respectively.

Discussion

Coastlines suffer changes over time because of their morphodynamic and sedimentary character, resulting in the spatial distribution of complex sedimentary assemblages

(Holland and Elmore 2008). In the last three decades, urbanization along the sharm has extensively modified the shoreline configuration and sediment characteristics. Recreational activities and domestic pollution have increased resulting in chaotic environmental changes (Fig. 3b). Wadi Al Kura, the old feeder stream or landward extension of the sharm, is presently an infill site and has been cut off by two bridges obstructing the supply of terrigenous material to the sharm. The head has been raised and barriers emplaced to stop debris transfer into the sharm, resulting in the cessation of sediment input (Fig. 3c). During the occasional rainfall, water from several wadis drains in channels to this wadi which may become inundated, but the raised bed at the confluence of the wadi and the sharm holds the water so that only during extreme rainfall does it spill over into the sharm.

Municipal and domestic waste and litter dumping as well as sports activities have caused extreme stress on the environment, especially on corals. The pollutants produced by either human activities or in situ by chemical reactions are likely to be retained by fine sediments. The reduction in width of the sharm and the restricted water exchange with the Red Sea, as well as the relatively weak bottom currents as shown by sediment texture, also lower water quality. Land reclamation is common (Fig. 3d), and dredging is used to increase water depth in front of the residential complexes and marinas to facilitate boat docking (Fig. 3e). Recently storm water has drained into the sharm through pipes, with floaters deployed to restrict sediment dispersal. Unfortunately, the effect of this input could not be controlled because of unsuitable technology (Fig. 3f).

Over 100 species of flora and fauna dwelling in the bottom water were identified between 1974 and 1979 by Bemert and Ormond (1981). Following land reclamation and increased urbanization, several species have been lost (Rifaat et al. 2001). The last decade has witnessed severe deterioration in the bottom environment, and many more species have disappeared (Basaham et al. 2003; Rasul and Qutub 2009). Recent dives in 2014 within and outside the sharm show a sharp decline in exotic species of corals and associated fauna. However, some corals survive under very harsh conditions and are resistant to the various environmental stresses (Fig. 3g). Seagrass beds are common in the northern part and at the mouth, indicating nutrient-rich water supporting their growth (Radwan Farawati, personal communication) possibly related to sewage and litter dumping and the restricted water exchange between the sharm and the Red Sea (Fig. 3h).

Sediment Characteristics

In order to track environmental changes, bottom sediment data from 1986, 2003, 2009, and 2013 are presented (Fig. 4). Significant alteration in the sediment's texture from coarse- to fine-grained through time and mineralogical compositional changes from detrital to biogenic origin have been observed due to cessation of fluvial input from Wadi Al Kura and an increase in human activities. The following sedimentological data show the various processes responsible for the changes over time.

Bottom Sediment Character (1986)

In 1986, the bottom sediment texture ranged between mud and gravelly sand but mostly consisted of muddy sand (Fig. 4a). Mud was present in small amounts close to the head of the sharm in a shallow and sheltered area, and a small elongated patch in the thalweg about 2 km from the sharm mouth covering an area of about 0.022 km^2. Muddy sand followed by sandy mud was supplied by Wadi Al Kura when the wadi directly fed the sharm, and the gravelly sand was mostly of biogenic origin (bioclastic debris). The sediments of the upper part of the sharm were mostly detrital, with relatively small amounts of associated biogenic sand. The sediments were poorly sorted, except in deep water, and the lower half, including the mouth, was moderately sorted. Very fine sands were better sorted at the thalweg, indicating that low-velocity currents were capable of sorting fine sediments.

Bottom Sediment Character (2003)

In 2003, the sediment texture ranged between mud and gravelly sand but was dominated by sandy mud (Fig. 4b). Gravelly muddy sand was present at the mouth, and a patch of mud about 2 km upstream had become much larger since 1986 and covers an area of about 0.127 km^2. The fine sediments at the mouth were re-suspended by strong tidal currents and transported upstream about 2 km during flooding, and during the turn of the tide before the ebb set in the suspended material settled. The sediment veneer had changed from coarse to fine with abundant mud. The sediments were mostly poorly sorted in shallow water due to their coarseness and moderately to well sorted in the deeper water and in the lower half of the sharm.

Bottom Sediment Character (2009)

In 2009, the sediment texture ranged between sandy mud and gravelly sand but was dominated by sandy mud (Fig. 4c). The abundant mud at about 2 km from the mouth in 1986 and 2003 disappeared (Fig. 4a–c). Abundant sand and mud was supplied by the wadi during heavy rainfall in 2009 resulting in major changes in the sediment distribution pattern. Sand of biogenic origin was well sorted at the mouth because of

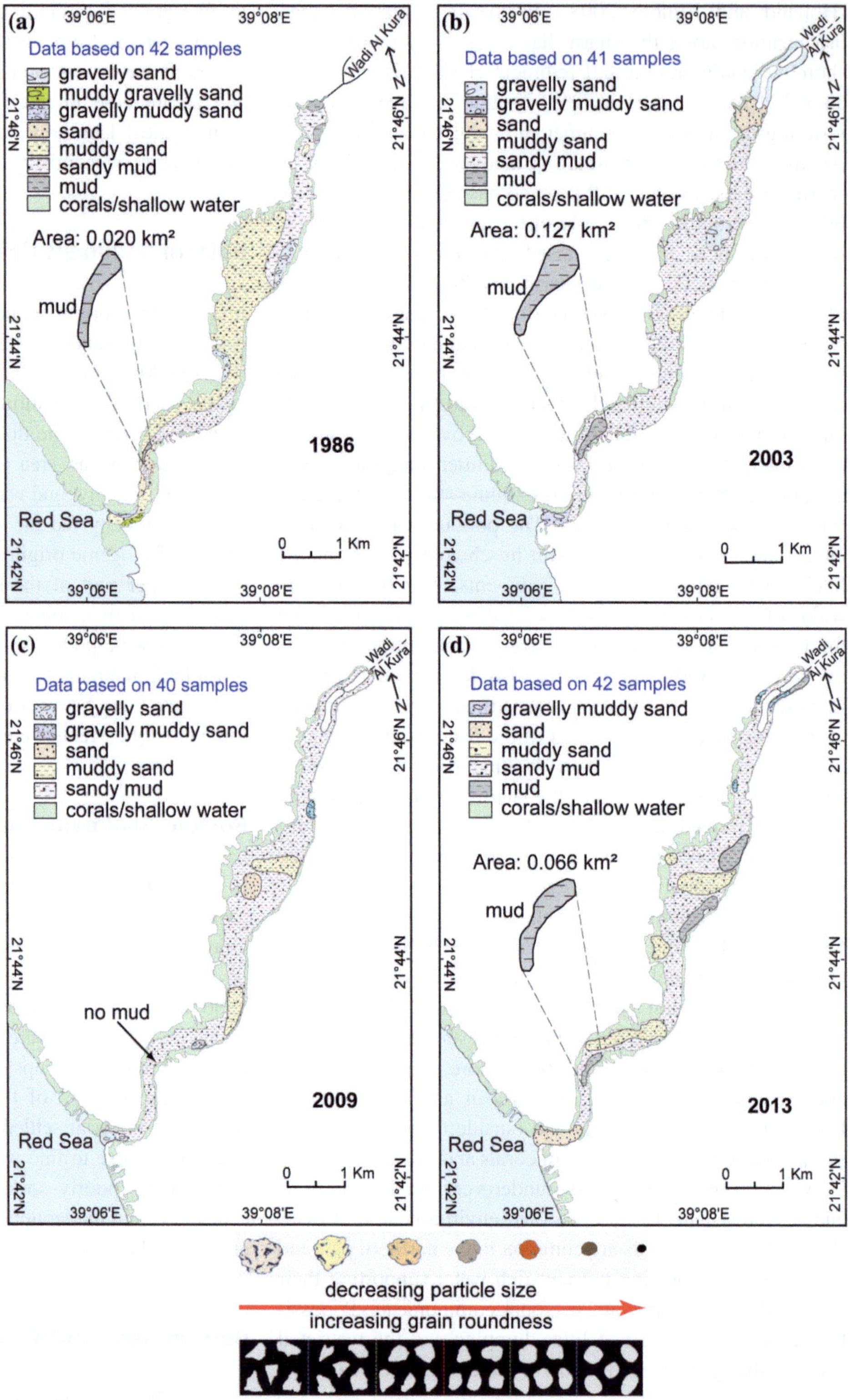

Fig. 4 Sediment distribution and variations in particle size in Sharm Obhur. The sediment veneer shows a fining trend over the last three decades. The textural classification is based on Folk (1980); **a** data are based on 42 sediment samples; **b** data are based on 41 sediment samples; **c** data are based on 40 sediment samples, and **d** data are based on 42 sediment samples

strong tidal currents. The sediments were mostly well sorted in shallow water, while sediments in deeper water in the mid to lower half and at the entrance of the sharm were moderately well sorted. The upper half was poorly to moderately well sorted because of the mixing of fine and coarse sediments of both detrital and biogenic origins.

Bottom Sediment Character (2013)

In 2013, the sediment texture ranged between mud and gravelly muddy sand but the bottom was covered mostly with sandy mud (Fig. 4d). Moderately to well-sorted sand was present at the mouth followed by sandy mud, and abundant mud about 2 km from the mouth covering an area of about 0.066 km^2. In addition, mud was also present at the head and on the central eastern flank. The mud on the eastern side is the recent product of the wadi, transported during rainfall in 2012, while the coarse fraction is of biogenic origin. Sediments are moderately sorted in shallow water, but elsewhere are poorly sorted, due to the flood of 2012 that transported debris and sediments of various sizes, as well as the increasing human influence through land reclamation or routing of storm and groundwater into the sharm via pipelines. The strongly deteriorating water transparency during the flood and groundwater dumping has caused the death of some coral species.

The sharm shows a fining trend in the sediment texture and is now covered mostly with sandy mud of detrital and biogenic origin in deeper waters, especially in the thalweg, while texture is coarse-grained and mostly biogenic in shallow water. The thalweg is 35 m deep and the bottom current is fairly weak facilitating sedimentation of fine-grained sediments under calm conditions with little reworking in the deeper water. The cohesive nature of the terrigenous sediment (wadi source) enhances sediment retention and even strong currents are unable to erode and re-suspend bottom sediment. The influence of reworking is also masked by biological aggregation and the degree of cohesiveness of the detrital sediments that are composed mostly of the clay mineral kaolinite, once supplied by Wadi Al Kura. The interference with natural conditions may lead to pollution and the finer-grained nature of the sediment veneer supports retention of pollutants and deterioration in aquatic life (Basaham et al. 2003). Activities such as jet skiing and movement of boats have greatly increased in recent years and are responsible for re-suspension of near-shore sediments and their transportation to deeper water. The coarse sediment appears to be contributed by coastal erosion and in situ production of biogenic material. The gradual coarsening of near-shore zone sediments is also caused by wave- and wind-generated currents and the energy generated by human activities. In near-shore areas where human influence is limited, the water transparency is higher and the sediments are well sorted. The mouth of the sharm was covered with muddy sand in 1986, gravelly muddy sand in 2003, gravelly sand in 2009, and sand in 2013, and the loss of gravel, but the presence of sand confirms a gradual decrease in the particle size of the sharm sediments. The coarse-grained well-sorted sands are the result of tidal currents that re-suspend fine sediments that are transported either upstream during flooding or to the open sea during ebbing. The limit of influence of the tidal currents is about 2 km from the mouth based on the pattern of mud deposition (Figs. 3 and 4d). However, the sediment at the mouth in 2013 indicates that the terrigenous sand is due to the rainfall of 2012 when sandy mud was transported to the sharm and the mud was lost to the Red Sea by winnowing. Since the sharm is in limited communication with the open sea and because the current velocity is controlled by the barrier (Fig. 3), sand settles out at the mouth and mud is winnowed away. The mouth of the sharm is not as dynamic as that of other sharms such as Al Kharrar lagoon (Rasul et al. 2010a) because of the present configuration of the sharm entrance. Moderately to well-sorted sand prevails at the mouth only because the bottom current of 0.75 m s^{-1} is strong enough to winnow the fine material. A 3D model showing the sediment distribution process and pattern is presented in Fig. 5.

The head, south-eastern part and thalweg show better sorting. At the head, sorting is better because of excess aeolian material, and in the shallow water in the south-eastern half this is due to the effect of wind induced currents. Better sorting of non-cohesive sediments (biogenic origin) under weak current influence is observed. Where finer materials are cohesive, sorting values decrease because strong currents are required to sort the sediments. Where the near-shore sediments show deterioration in sorting, it confirms a multiple source from the fragmentation of corals and anthropogenic input. Although tidal currents tend to be strong at the mouth, the sediments are not very well sorted. The reasons sorting does not improve much are either because the currents are incapable of selective sorting or the current is so strong that fragmentation of the corals and biogenic materials contribute coarser materials to the sediment cover. The latter is probable because of the composition and appearance of the biogenic material which show polished surfaces caused by strong currents but have worn edges, indicating the influence of physical processes on the sedimentary materials. Prolonged exposure of the coral reefs and related organisms to agitated water conditions and mechanical destruction also supply significant sedimentary material to the sharm, but this is mostly confined to the lower end. The lesser degree of sorting is magnified by the addition of coarser material. Poor sorting is indicative of the transport history, hydraulic volatility of transporting agents and multiple sources and the composition of sediments in the sharm.

In general, $CaCO_3$ has increased in the sharm over time, especially at the lower end because of the absence of terrigenous input from Wadi Al Kura, and TOC has increased in the eastern and deep waters. Increased TOC is associated with fine and dark-coloured sediments in a reducing environment that tends to be most severe in stagnant and sheltered areas. $CaCO_3$ ranges between 11 % at the head and 73 % at the mouth, and TOC ranges between 0.2 % at the

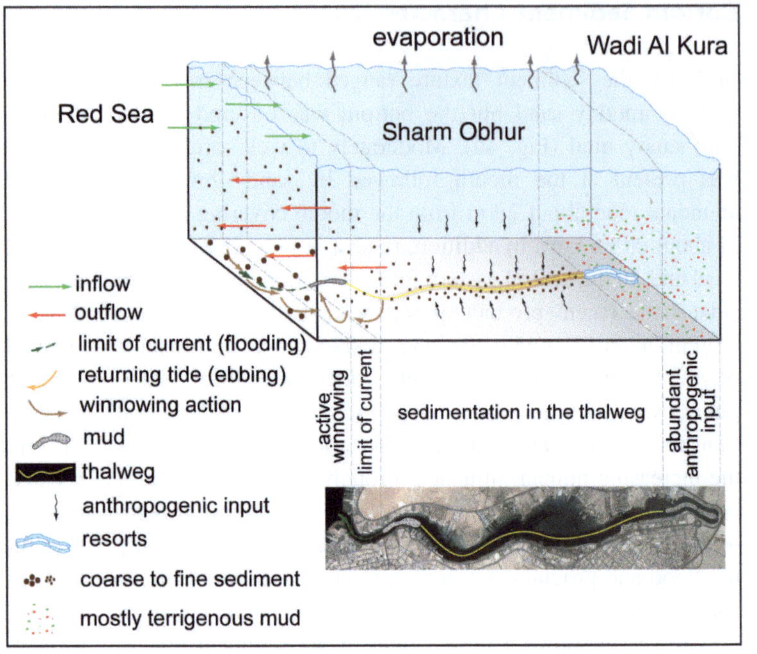

Fig. 5 3D model of the hydrological processes and sediment distribution pattern within Sharm Obhur

mouth to 1.6 % at the head. However, an area ~2 km from the mouth contains 1.2 % TOC due to abundant mud that tends to preserve organic matter well. North-westerly winds also transport debris in the south-easterly direction as a result of which anthropogenic material collects along the eastern shoreline and is further supplemented by human activity resulting in high amounts of TOC.

The chemical properties of dissolved seawater constituents [nutrients, faecal sterols and polyaromatic hydrocarbon (PAHs)] were recently investigated by Al-Farawati et al. (2008). High values of nutrients at the head of Sharm Obhur, in particular nitrite, nitrate and ammonium are attributed to its shallowness and restricted water circulation which enhances accumulation of nutrients, supplemented by additional sources through diffusion of nutrients from interstitial waters. As a result, Al-Farawati et al. (2008) suggest that reducing environments are most likely to develop at the head of the sharm in line with changes in the environment and anthropogenic input.

Present Status and Future Implications

Human interference has distinctly changed the configuration of Sharm Obhur. There is no set pattern of sediment distribution because of human activity at the margins of the sharm and lack of input from Wadi Al Kura. On the contrary, erratic activity has caused variations in the sedimentation pattern as reflected in the texture and sediment composition. The fining trend from abundant muddy sand in 1986 to sandy mud in 2013 may lead to retention of anthropogenic pollutants dumped in the sharm. Land reclamation at the sharm has greatly increased, justifying the need to study the environmental impact on the marine communities, namely the coral reefs and the exotic fish that have declined markedly over the last decade. It is also important to monitor the impact of runoff after heavy rainfall.

Shoaibah Lagoons: Khawr ash Shaibah Al Masdudah (Northern) and Khawr ash Shaibah Al Maftuhah (Southern)

The two hyper-saline lagoons, Khawr ash Shaibah Al Masdudah (northern) and Khawr ash Shaibah Al Maftuhah (southern), commonly referred to as the Shoaibah lagoons, have unique environmental conditions. Limited research has been carried out there by Meshal (1987), Sultan and Ahmad (1990), Ahmad and Sultan (1992), Gheith (1999), Al-Washmi and Gheith (2003), Hariri (2008), Rasul et al. (2010b), Al-Barakati (2010), Abu-Zied et al. (2011), Abu-Zied and Bantan (2013) and Al-Farawati et al. (2014). The lagoons were probably formed by erosion during a sub-pluvial to pluvial Pleistocene phase and were then submerged by the rise in sea level (Braithwaite 1987; Brown et al. 1989). On the western side, the lagoons are bounded by raised coral terraces of Pleistocene age (Marco Taviani, personal communication). No wadis flow into the lagoons, and therefore supply of detrital material is lacking. Most of the sediments are autochthonous carbonates with admixtures of aeolian quartz and evaporite minerals. Remnants of drainage channels of the dormant wadis are present, and small rainwater channels drain into the lagoons but do not play any role in flooding. The wadis that had an influence on the lagoons are listed in Table 1. The northern and southern lagoons are connected to the Red Sea through two separate narrow channels with water depths of 7 and 15 m, respectively (Fig. 6). The depth increases with distance towards the open sea and abruptly drops to over 50 m at the outer reach of the mouth. The northern lagoon has a wide tidally dominated channel, strong currents and unimpaired exchange of water with the open sea, while the southern lagoon has a long narrow channel where the tidal current is strong only until the mouth bends almost 90°, where the tidal current dissipates, reducing its impact on the lagoon and restricting water and sediment movement. The shape of the mouth, narrow entrance, shallow depth and strong currents govern sediment distribution. Shallow depths averaging 3.5 m, wind and tidal stirring are the main forces preventing the lagoons from developing stratification, resulting in a well-mixed body of water with high salinity and temperatures.

The two lagoons are separated by a paved road and connected by drainage pipes that are now buried under sediments (Fig. 6a). The northern lagoon has an elevation of about 6 m at the mouth and low-lying areas surround the lagoons (Fig. 6b, c). The sediment texture in the lagoons ranges between sandy mud to gravelly sand. Very coarse sediment dominates the entrance, and fine to coarse sediments mostly of biogenic origin form the surface sediments of the main lagoons. The biogenic materials are coral debris, coralline algae, molluscan shells, foraminifera, sponges and bryozoans in sand to gravel size fractions, and aeolian quartz in fine sand. The cyclical inundation of low-lying sabkhas by shallow water during flood tides and transfer of evaporite minerals (Fig. 6d) to the lagoon from the adjoining sabkhas during ebbing or rainfall are important in understanding the ecological consequences and sediment transport mechanisms. Carbonate is abundant in the form of calcite and aragonite, and High-Mg calcite indicates the carbonate is recent and formed under shallow water conditions. Halite, gypsum and dolomite are found in different proportions in the sabkhas surrounding the lagoons (Al-Washmi and Gheith 2003). The sediments of the northern lagoon are brown to grey in colour, and in the southern lagoon, the sediment colour ranges from yellow to grey (Table 2). Fine sediments are darker in colour with shades of grey, and the coarse sediments are lighter in colour with shades of yellowish brown to greyish brown. Dark-coloured sediments are found at the periphery of the lagoons and are related to the physicochemical properties and not to external sediment sources. The coarse sediments including shells are stained grey to black because of the reducing environment and formation of authigenic pyrite. Stagnant conditions prevail inside the lagoons because of insufficient water exchange with the open sea, and a lack of rainfall causes hyper-saline conditions.

Mangrove (*Avicennia marina*) stands are scattered around the numerous small islands and margins of the lagoons and act as a source of nutrients to the flora and fauna, also initiating a reducing environment (Fig. 6e, f). Seagrass and macro-algae dominate the shallow parts of the lagoons (Abu-Zied et al. 2011). Gastropods and pelecypods are common and scattered on the tidal flats and in shallow waters. Benthic foraminiferal species that enable sea-level reconstruction have been reported from the lagoon (Abu-Zied and Bantan 2013). The Shoaibah region has suffered severe ecological changes in the past and is characterized by an intensely dry climate where only one wadi drains into the lagoon for a short period each year (Hötzl and Zötl 1978).

Discussion

Sediment colour plays an important role in discriminating various sub-environments in shallow water bodies such as the Shoaibah lagoons. Areas of dark grey sediment are quiescent reducing environments of deposition and stagnation where the currents are almost negligible and incapable of re-suspending and transporting sediments. The reducing environment is further enhanced by restricted water exchange between the Red Sea and the lagoons. Organic debris from mangrove and anthropogenic sources has a pungent smell and adds a dark tone to the sediment, associated with an increase in TOC (Fig. 6f), and is reflected in dark stained shells and coral debris.

The water exchange between the Red Sea and the two lagoons is greatly restricted because of the shape of the mouths where the current is strong, and flushing is restricted especially in the northern lagoon. Coarse sediments at the mouth of the lagoon are lighter in colour and well sorted with minor amounts of finer sediments because strong currents winnow and transport re-suspended sediments either inside the lagoon during flooding or to the open sea during ebbing. A bridge has been constructed recently over the mouth of the northern lagoon reducing the width of the channel to about 5 m, and hence, the exchange of the water between the lagoon and the open sea is minimal despite the strong currents that prevail because of the narrow channel (Fig. 6gi, ii, iii). The shallowness of the lagoons and meteorological forces keep the water agitated causing suspension of fine sediment, resulting in its transport to the lagoon margins by wind coupled with tidal currents. The permanent presence of suspended material in the lagoons affects water clarity. Although the seabed could be seen during Secchi Disk Disappearance Depth (SDDD) data collection, the water is murky due to suspended material. Water clarity increases with depth and increasing sediment size especially at the entrance, where strong tidal currents remove fine sediments and keep the water column free of suspended material, resulting in the

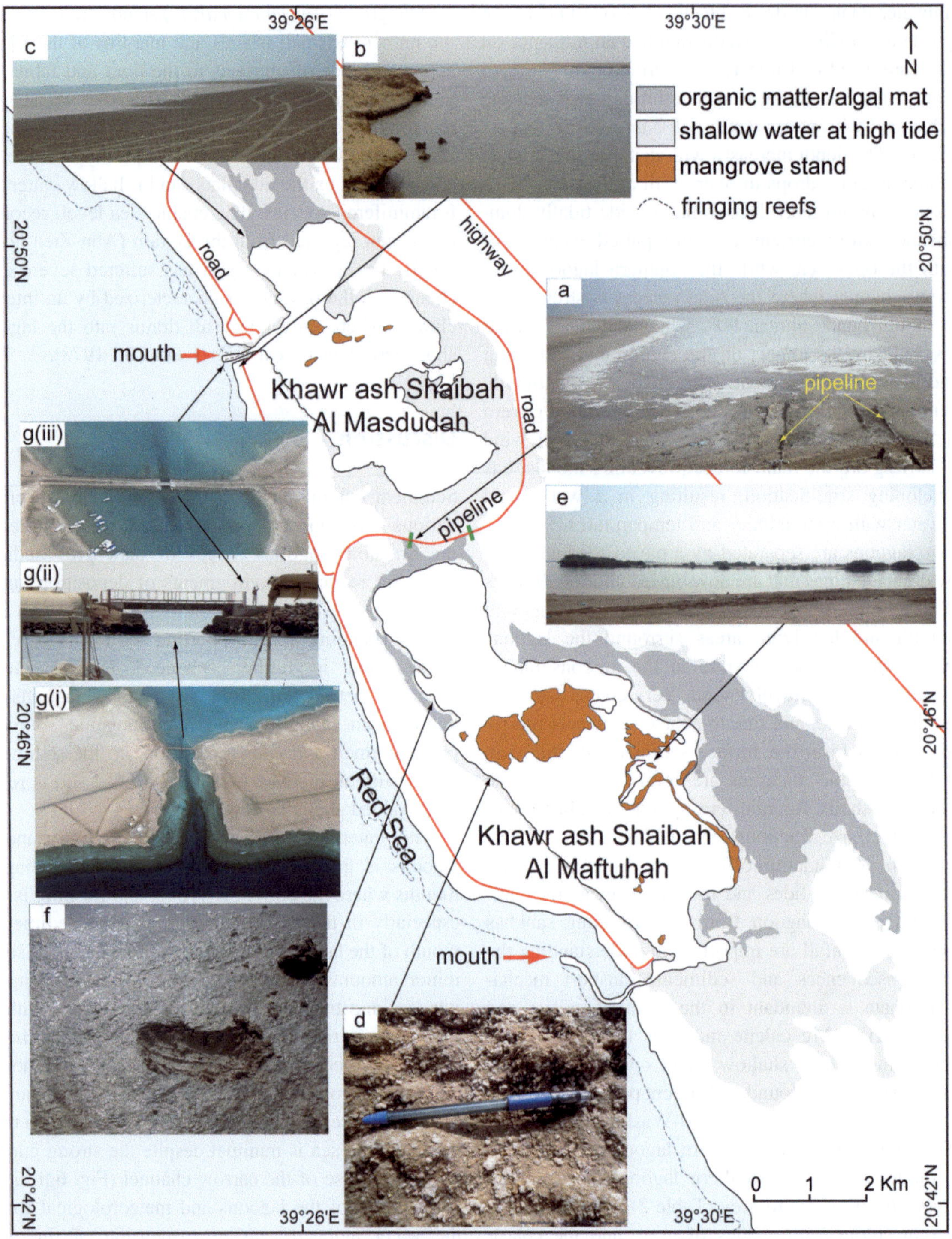

Fig. 6 Two Shoaibah lagoons with subenvironments: **a** rusted pipeline joining the two lagoons; pipelines are clogged with sediments restricting exchange of water between the two lagoons; water covers the shallow area during spring tide and high temperatures increase evaporation resulting in the formation of salt; **b** mouth of the lagoon during high tide. The cliff at the lagoon's entrance is about 6 m high and tapers off at the edge of the entrance; **c** periphery of the lagoon is flat, and so during high tide or strong winds, the sabkha is covered with water; **d** sabkha deposit where salt and gypsum are common; **e** mangrove stand and sabkha are inundated only during very high tide; **f** area inundated with water during tidal cycles. The mud is sticky, and organic debris creates a reducing environment; **g** (*i*) narrow mouth of the lagoon where the sides have coral reefs and the mouth widens; (*ii*) recently constructed bridge has narrowed the entrance to about 5 m and (*iii*) although the entrance has narrowed, the tidal currents show an increased velocity under the bridge where the channel has been deepened to give a smooth flow

highest water clarity. However, the shallow depth in the lagoons increases turbidity and results in a well-mixed body of water with extreme temperatures and salinities.

The flushing time of the lagoon is estimated to be 20 days with current velocities varying between 20 cm s^{-1} in December and 50 cm s^{-1} in July at the entrance (Alaa Barakati, personal communication). The tidal current can be observed in the entire lagoon, but the current velocity weakens as the mouth of the lagoon widens (Fig. 6gi). Strong tidal currents at the entrance play an important role in the redistribution of sediments, but the prevailing northerly and north-westerly winds influence the movement of water from the lagoon to the adjacent sabkhas, particularly during high winds. Previous studies in shallow water have demonstrated that wind-generated waves and their energy flux are important in initiating sediment transport at the entrance of most coastal lagoons (Ahmad et al 2002; Rasul et al. 2010b). Surface waves with periods from 1 to 30 s. are the most energetic waves that aid sediment redistribution in shallow water (Kinsman 1965).

The lagoons have two main sediment sources, biogenic and aeolian, that owe their distribution to transit and reworking processes. A third minor source is the evaporite deposits of the adjoining sabkhas. The sediment veneer ranges between gravelly sand and sandy mud but is dominated by biogenic muddy sand in both lagoons (Fig. 7). The coarse nature of the sediments is the result of constant fragmentation of reefal sediments and high production of molluscs and foraminifera in the lagoon. The northern lagoon is veneered mostly with sand in various size fractions, and gravel dominates the north-eastern side and at the mouth. Fine sediments of biogenic origin dominate the deep and central part of the lagoon. This shows that the current plays an important role in winnowing the fine sediment at the mouth, and the re-suspended mud is transported and deposited in the central part of the lagoon where the current is minimal allowing sediment to settle. The floor of the southern lagoon is also dominated by sand, but the relatively deeper water in the main lagoon is mainly veneered with biogenic mud. The gravel content is higher at the mouth of the lagoon but only at the confluence with the Red Sea, which shows the importance of wind and strong tidal currents that are strong but terminate where the mouth widens (Fig. 6). The coarser texture of the bottom material results from the removal of fine-grained sediment, particularly at the entrance, by the sifting action of strong tidal-, wind- and wave-generated currents (Fig. 7a).

Biogenic carbonate is abundant, followed by detrital and evaporite components. Erosion of reefs contributes most of the carbonate material to the lagoon, and the molluscs are typical products of a marine environment. The distribution of detrital material is mostly controlled by the hydrodynamics and by atmospheric dust brought in by the shamal (wind). The sand dunes in the Middle East are important contributors of aeolian input to the marine environment in the form of pitted quartz that dominates most lagoons. Relatively limited terrigenous material is also supplied to the lagoon by the adjacent coastal plain through the wadis during occasional flash floods (Al-Sayari and Zötl 1978).

Most shell fragments in the lagoons are stained black only on the outer surface (Fig. 7b–e). Similar staining on sediments has been reported in the Al Kharrar lagoon (Rasul et al. 2010a), at sabkhas on the Red Sea coast (Abou-Ouf and El-Shater 1993), on the Indus continental shelf and the Indus Fan sediments (Rasul 1992), the Arabian Gulf (Murray 1966) and the Bahamas (Illing 1954). Ancient and recent calcareous marine sediments stained grey or black with finely divided pyrite have been reported in pellety debris of carbonate sediments from the Middle East, Britain and other areas (Mixon and Pilkey 1976; Sugden 1966). Almost all types of pelitic material may become pyrite-stained. Debris such as ooids and hard shell fragments has staining only on the grain surface. In sedimentology, these blackened shells are regarded as important indicators of environment type (Cramp 1985; Rees 1988). Staining by pyrite is regarded as the end product of anaerobic bacterial activity. When sediment is buried under a pile of terrigenous material, a reducing environment develops due to consumption of oxygen and organic matter by bacteria. According to Sugden (1963), the aerobic bacterial activity, utilizing free oxygen, penetrates the surficial sediment layer. Some organic matter is oxidized and destroyed, and some is buried to depths where free oxygen decreases and then undergoes anaerobic bacterial decay to produce H_2S forming at first ferrous sulphide and subsequently pyrite. A high rate of sedimentation can result in the incomplete conversion to pyrite (Berner et al. 1979). However, a low input of organic matter and its consumption by both aerobic and anaerobic bacteria and low iron input limits pyrite formation (Fisher 1986). Pyrite formation in marine sediments depends upon the amount of iron minerals that are reactive with hydrogen sulphide, the presence of sulphate, and the amount of organic matter available for bacterial decomposition since the conversion of sulphate to hydrogen sulphide is biologically mediated (Berner 1970, 1982). The presence of iron in the sediment will stain grains brown-black during the oxidized state, and during burial and in a reducing environment, grains turn black resulting in iron sulphide in the form of pyrite. It is postulated that the black colour on shell fragments can change to brown when sediment from the reducing environment is exposed to an oxidizing environment by burrowing organisms, erosion or currents (Maiklem 1967). It has been shown experimentally that in a reducing environment, the blackening may occur in a matter of days (Mixon and Pilkey 1976). A pungent smell in sediment also indicates a reducing environment. However, Abou-Ouf and El-Shater (1993) concluded that blackening of shells is due to

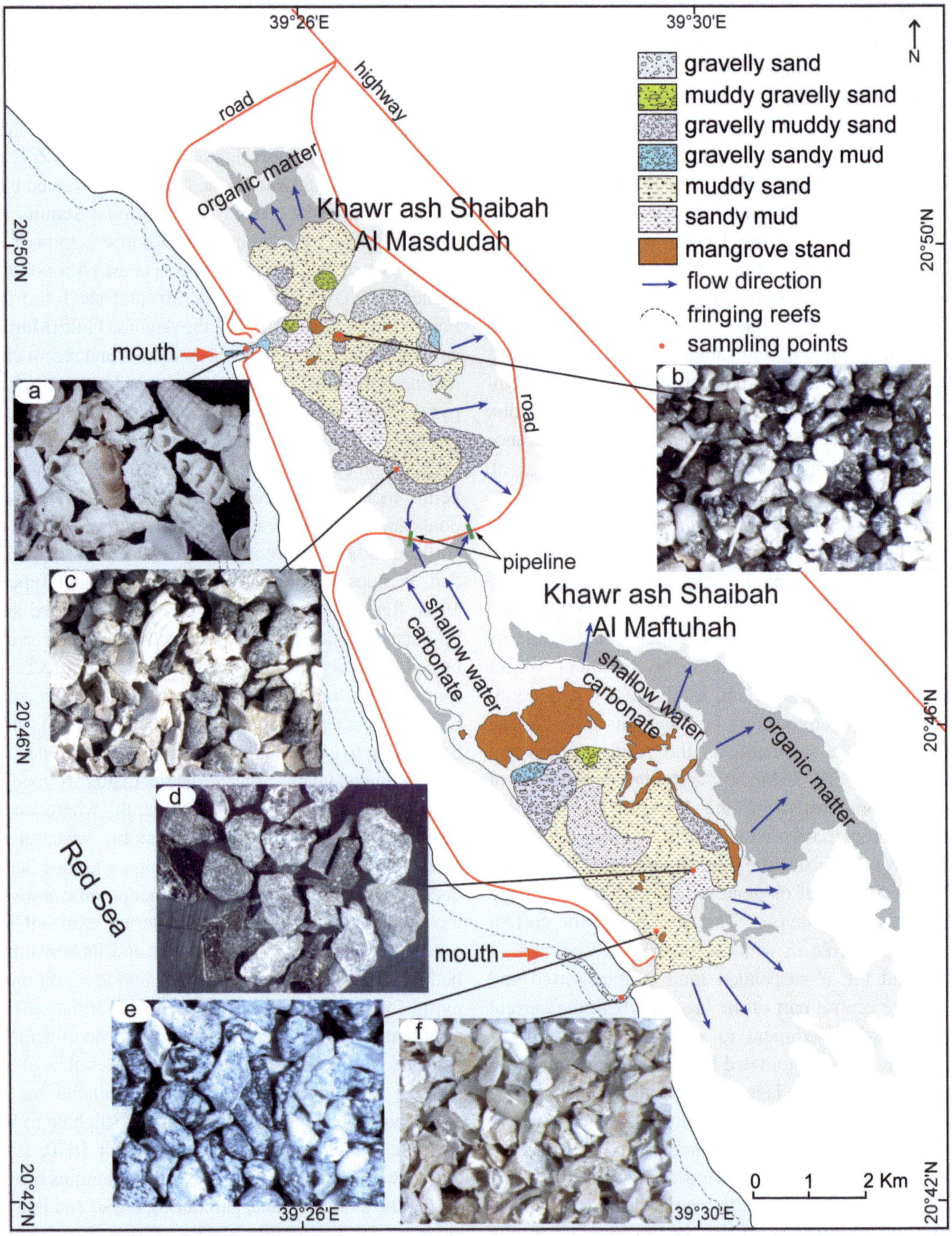

Fig. 7 Sediment distribution in the two Shoaibah lagoons: **a** recent gastropods have a fresh appearance, and the surfaces are polished by water movement; skeletal grain population (shell hash) consists primarily of more or less fragmented gastropod (cerithiids) and bivalve (e.g. *Brachidontes*) shells; **b** *grey to black* stained carbonate material in the vicinity of mangrove stand. Some fresh (*white*) carbonate material (gastropod shells and allochems) is recent and not influenced by a reducing environment; **c** stained and unstained grains indicating a mixed composition of sediment; skeletal component in gravel is dominated by fragmented bivalves, benthic macro-foraminifers and allochems; **d** *dark grey* sand size grains are the product of a reducing environment that gives rise to iron sulphide in the form of pyrite; **e** stained shells and carbonate debris have been severely affected by the reducing environment in the presence of high organic matter from the mangrove stands, and **f** recent biogenic material has a fresh appearance because it is influenced by tidal currents; skeletal component is rich in benthic foraminifers and subordinate ostracods

pyrite in iron-rich clays and sands from the coastal plain (Jado and Hötzl 1984; Jado et al. 1989) and not from the organic matter. In the southern lagoon, biogenic materials are more severely stained black compared to the northern lagoon. This could be due to a lack of exchange of the southern lagoon's water with the Red Sea due to the shape of the mouth and the availability of abundant organic debris from the mangrove stands.

The presence of halite in the sediment is the result of high evaporation and the hyper-saline water has a salinity of 54 %, although salinities of over 60 % have been recorded (Abu-Zied et al. 2011). The lagoons extend into a flat coastal plain that is covered by a few centimetres of saline water during flood tides. The north-north-westerly wind helps in the movement of shallow water over the sabkha. Air temperatures as high as 50 °C in the sabkha areas aid rapid evaporation, causing precipitation of evaporite minerals such as halite and formation of thin crusts and sand-sized halite crystals. Crystals of halite, gypsum and dolomite are abundant in the low-lying sabkhas and are occasionally transported to the lagoon by ebb tides (Fig. 6c).

The $CaCO_3$ content is higher in both the lagoons, especially at the mouths, and decreases gradually inside the lagoons (Table 2). The strong current at the entrances prevents terrigenous sediment from settling out and leaves coarse carbonate as lag deposits (Fig. 7a, f), where the sediment texture becomes coarser with an increase in carbonate content. $CaCO_3$ decreases into the lagoon, where the varying distribution is due to the influx of terrigenous material mostly of aeolian origin and evaporite minerals from the sabkhas diluting the carbonate. The biogenic sediments (shells and carbonates) are comminuted by current activity, thus initiating parturition of fine calcareous mud. Basins in reef-bound coasts without any significant runoff and high aridity are the most probable source of carbonate in the lagoon. Thus, grain size and grain characteristics are primarily controlled by in situ production of carbonate material. In the Red Sea, vigorously growing coral and the presence of fringing and barrier reefs play a very significant role in the high production of reefal and carbonate sediments.

Low values (<0.2 %) of TOC are recorded where the fine sediment is winnowed and lag material in the form of coarser carbonate fails to adsorb the organic material. The distribution of organic matter in general is somewhat higher (<2.5 %) than normal (1–2 %), and higher values are observed where finer sediments dominate and where $CaCO_3$ is low. In isolated areas, particularly close to the mangrove stands, coarse-textured sediment rich in carbonate material tends to show higher amounts since the mangrove supplies organic debris. There is a close relationship between TOC, sediment colour and texture, and a negative relationship between TOC and $CaCO_3$. In general, TOC is high and relates to stagnant conditions in both the lagoons and the high input of decomposing organic debris from the mangrove stands. The coarser sediments, mostly of biogenic origin, are stained grey-black because of the reducing environment confirming that high TOC and staining of shells are inter-related. In general, the fine sediments show high TOC compared to the coarse sediment rich in carbonate debris. The values are compatible with those reported by other workers in the Red Sea (El-Sayed and Hosny 1980; Behairy et al. 1983; Al-Washmi and Rasul 2003; Rasul et al. 2010b).

Dissolved nutrients such as reactive phosphate and nitrite are depleted in the lagoons, similar to the open water of the Red Sea, reflecting poor productivity (Rasul et al. 2010b). The speciation of phosphorus in the surface sediments, especially of the southern lagoon indicates the depositional environment, and apatitic phosphorus is the main form that possibly regulates reactive phosphate in the lagoon (Al-Farawati et al. 2014).

Present Status and Future Implications

Human interference has distinctly changed the configuration of the two Shoaibah lagoons, especially the northern lagoon where the construction of a bridge probably limits the exchange of water between the lagoon and the Red Sea. There is no input of fresh water from the wadis into these lagoons and therefore hyper-saline conditions prevail, and the increase in human activity in terms of boat docking and garbage disposal is on the increase and may lead to retention of anthropogenic pollutants, justifying the need to study the environmental impact on the lagoon.

Khor Al Ghallah and Offshore Al Lith

Khor Al Ghallah is a shallow lagoon fed by Wadi Al Ghallah that currently has limited sediment input because of intense human interference. The maximum depth of Khor Al Ghallah is about 4 m, and at the entrances to the lagoon on the western side between Jazirat Sharifah and Jazirat Sulab and on the eastern side of Jazirat Sulab water depths are 9 and 13 m, respectively (Fig. 8). The seabed off the city of Al Lith has numerous shoals and was at one-time fed by numerous wadis. Several wadis drain into the lagoon, the most important being Wadi Al Ghallah, while Wadi Rahman and Al Lith drain into the open sea and influence offshore sediment transport (Fig. 8; Table 1). The depth offshore increases gradually to 60 m in the south-west because of substantial sediment input, unlike other areas along the Red Sea where the depth increases abruptly because of a lack of sediment influx. During monsoonal rains, the wadis are flooded, passing through channels that drain into the Red Sea (Höltz and Zötl 1978). A number of studies have been carried out in the

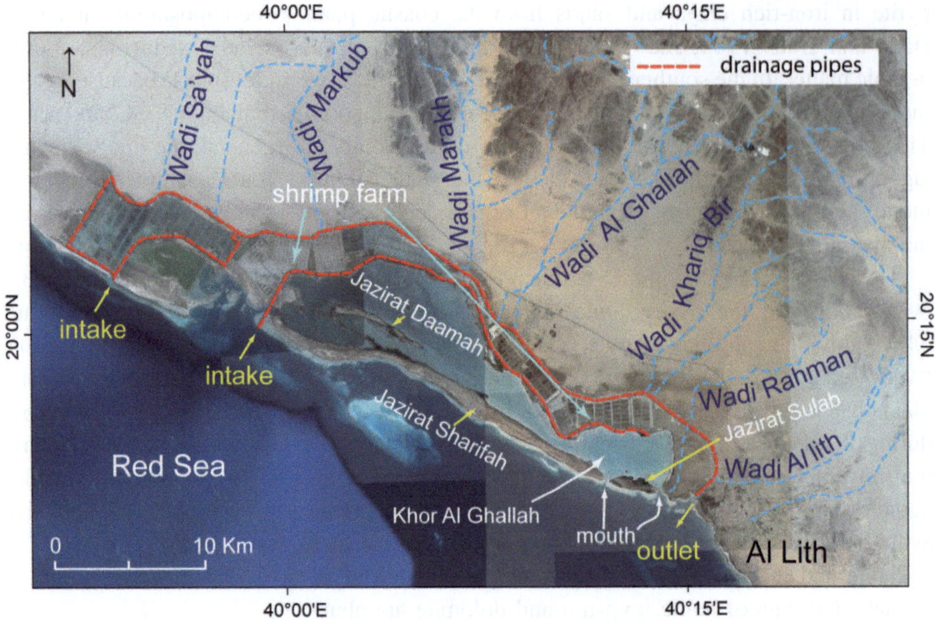

Fig. 8 Khor Al Ghallah and the wadis influencing the coastal and offshore areas

Al Lith region by Hadley and Fleck (1980), Hadley (1975, 1980), Prinz (1983), Pallister (1986), Tag (1986), Abou-Ouf et al. (1988), Heija and Shehata (1989), Tag et al. (1990), Basyoni (1997), Al-Washmi (2002), Al-Washmi et al. (2002), Rasul et al. (2009b).

The coastal zone of Al Lith is a low-lying area that is covered by sediments mostly of aeolian fluviatile marine origin, while shallow marine sediments on the tidal flats are composed of evaporite deposits (Al-Washmi 2002), including sands, clays and clastic carbonates. Geologically the area is part of the Arabian Shield, covered by detrital sediments of the Miocene Baid Formation that is about 30 m thick and composed of conglomerate, sandstone, limestone, marly argillite, chert and basalt, and the Pliocene Bat'Han Formation which is about 35 m thick consisting of conglomerate, sandstone and claystone (Hadley 1975, 1980; Hadley and Fleck 1980; Prinz 1983).

The lagoon is covered with dark-coloured mud to muddy sand, and the offshore area is veneered with sandy mud of terrigenous origin and relatively light-coloured gravelly biogenic sand (Fig. 9, Table 2). The distribution of detrital material is controlled by the local hydrodynamics, and the relative abundance is governed by the particle size and environment of deposition. Very fine sand dominates the lagoon and the offshore areas. The area is influenced by terrigenous input from several wadis draining the Khor Al Ghallah and the onshore area and through several channels connecting the lagoon with the open sea (Fig. 8). Based on the composition of the sand fractions, a model of sediment pathways is presented in Fig. 10.

Discussion

Based on the colour of the sediment in Khor Al Ghallah and offshore Al Lith, two facies have been identified, controlled primarily by provenance: (i) a dark-coloured facies related to terrigenous input, abundant in opaque heavy minerals (Fig. 10a, b) and (ii) a light-coloured facies related to biogenic material rich in carbonate with minor terrigenous influence (Fig. 10a–d). The dark and fine-grained sediments are mostly derived from wadis during occasional rainfall as shown by a relatively higher terrigenous input, where organic debris and iron-rich sediments induce local reducing environments with formation of pyrite, resulting in staining of the carbonate materials (Fig. 10e). Isolated patches of stained foraminifera and shell fragments with pyritic infilling are found in sheltered areas, especially at the foot of the shoals where dark sediments are common. Stained shells and coral debris are related to reducing conditions in quiescent hydrodynamic conditions (Figs. 9 and 10e).

The coarse well-sorted sand and gravel are due to the removal of fines, particularly at the western entrance of the lagoon between Jazirat Sharifah and Jazirat Sulab, by selective sorting generated by strong tidal currents. The fine sediments re-suspended by the current between Jazirat Sharifah and Jazirat Sulab and between Jazirat Sulab and the southern end of the lagoon are transported to the open sea by strong ebb tides that are further enhanced by the narrow channel. At the eastern entrance of the lagoon, the current is relatively weak and loss of sediment takes place when the

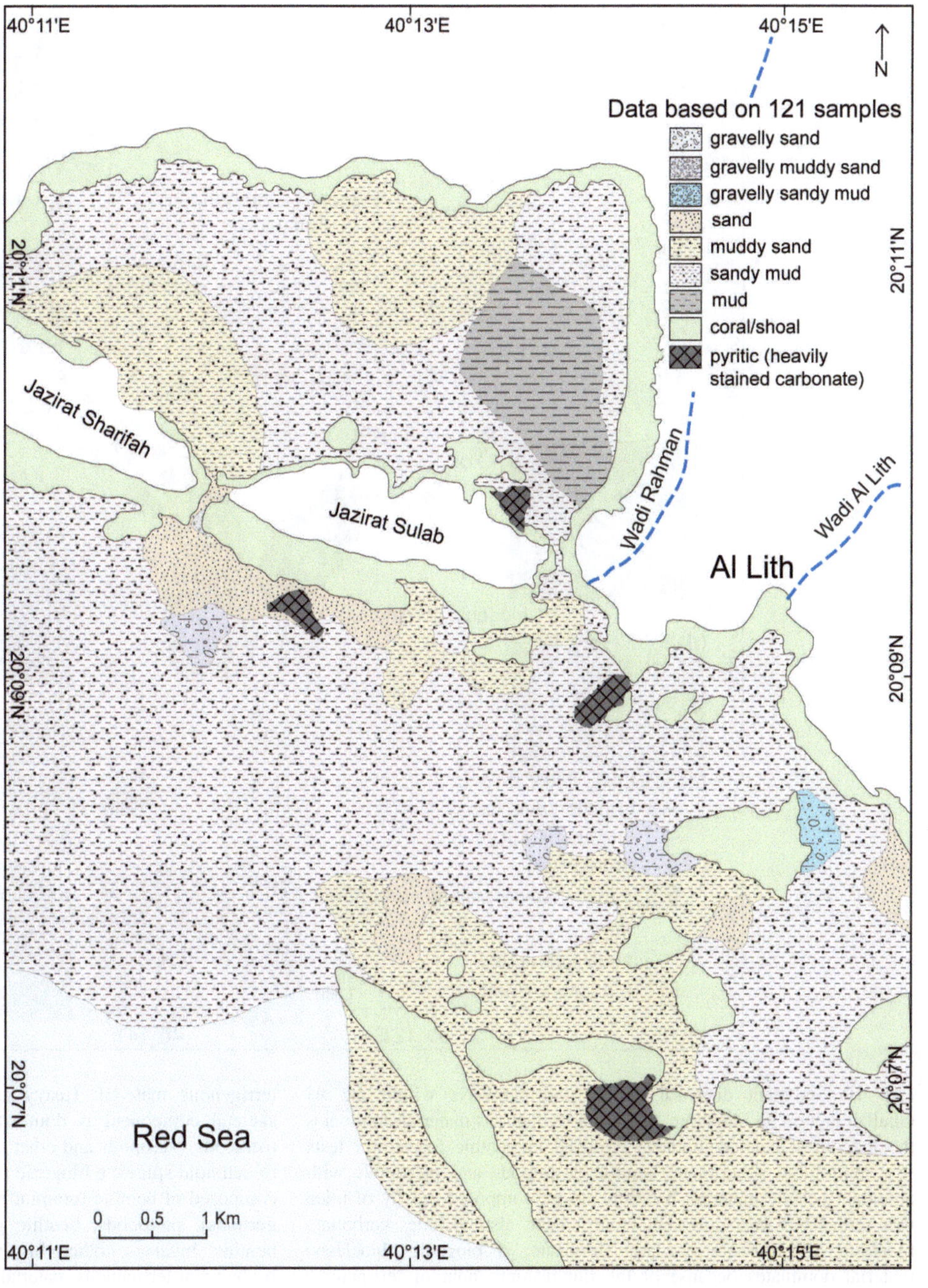

Fig. 9 Sediment distribution in the Al Ghallah lagoon and offshore Al Lith area. The textural classification is based on Folk (1980)

lagoon empties into the open sea. The north-north-westerly winds re-suspend fine sediments that are deposited either on the eastern side and at the margin of the lagoon under quiescent conditions, or are transported to the open water via the narrow channel. Two sedimentary environments are recognized based on the dispersal pattern: (i) a finer terrigenous environment in the south-east, and (ii) a coarser biogenic environment in the west (Fig. 10c, d). In the former, the terrigenous sediments supplied by wadis are dispersed southward (Fig. 10f), whereas the latter environment is veneered with recent biogenic material, and the distribution is governed by strong tidal currents (Fig. 10c, d, g). Sediments in the southern half of the lagoon are a mixture of terrigenous and carbonate materials, while almost pure bioclastic material dominates in the west, especially at the entrance between Jazirat Sharifah and Jazirat Sulab (Fig. 9).

Of the wadis draining into the lagoon, Wadi Al Ghallah is the most important and dominantly influences the sediment distribution. Once the sediment from the wadi enters the lagoon (average depth 2.5 m), the sediments are re-suspended and transported by the strong current generated by the northerly wind in an easterly direction, where sediments

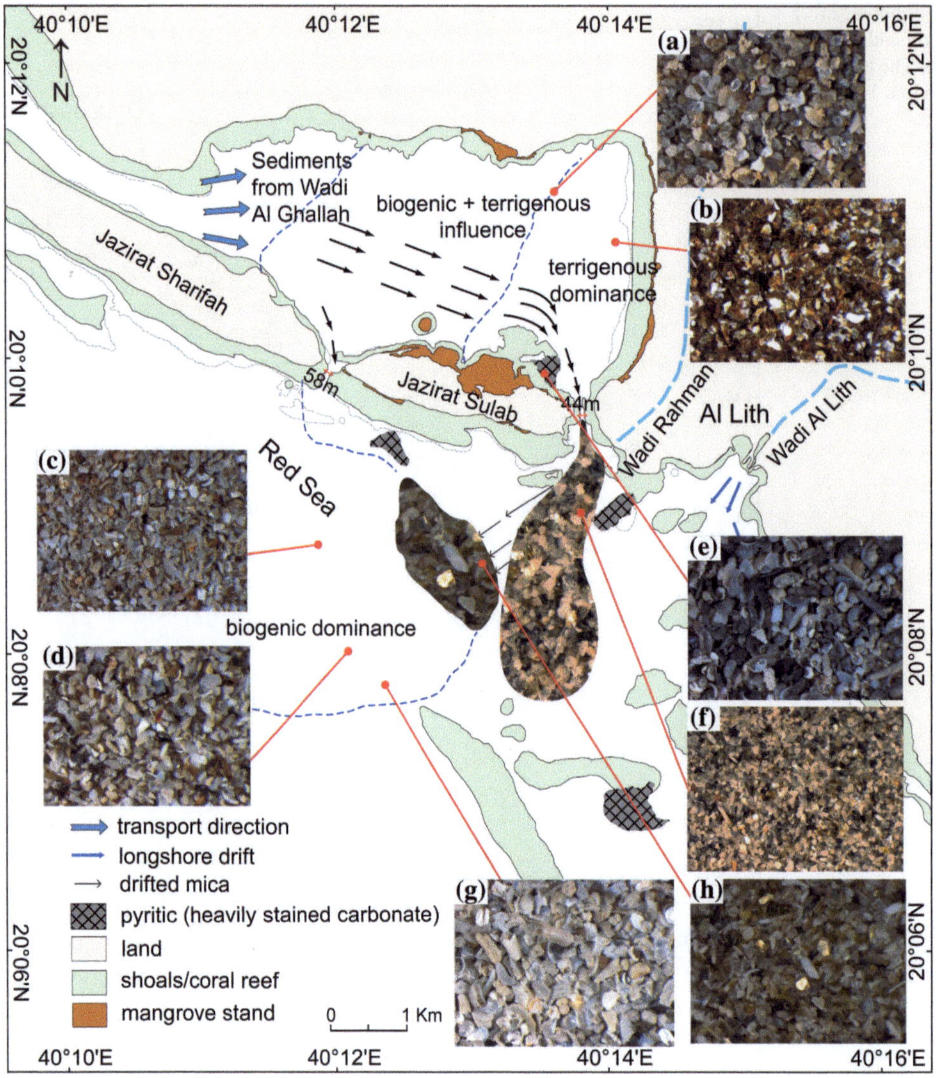

Fig. 10 Sediment dispersal pattern and pathways within the Al Ghallah lagoon and offshore Al Lith: **a** terrigenous material dominates the lagoon; skeletal component enriched in benthic formanifer tests (e.g. miliolids), subordinate benthic gastropods and pteropods with abundant pelagic material; **b** sediments are composed mostly of mica that settles out in a quiescent environment; also includes carbonate bioclasts; **c** carbonate bioclasts dominate; **d** biogenic (bioclasts) material dominates because of the diminishing input of terrigenous material; **e** sediments are stained grey to black because of a reducing environment; skeletal component dominated by benthic molluscs, subordinate ostracods, pteropods and carbonate bioclasts; **f** influence of terrigenous material. Heavy and light minerals are in abundance; skeletal component is dominated by benthic molluscs, subordinate ostracods, pteropods and other carbonate bioclasts including fragments of echinoid spines; **g** biogenic material dominates; skeletal assemblage composed of benthic foraminifers, planktonic foraminifers (e.g. globigerinids), pteropods, benthic gastropods, meroplanktonic stages of benthic bivalves, other bioclasts and higher pelagic input, and **h** abundant terrigenous material; carbonate bioclasts and some benthic foraminifers. However, mica is transported in suspension by surface currents and settles out, and incorporates with the biogenic material under the diminishing effect of current

enter the eastern part via a narrow channel. The re-suspended material also decreases water transparency to about 0.5 m. At the eastern part of the lagoon, the current diminishes appreciably causing deposition of terrigenous material from Wadi Al Ghallah. The terrigenous materials are also transported outside the lagoon in a southerly direction from the eastern opening into the open sea, where the shoals impede sediment transport. Surface currents play a major role in the distribution of mica transported offshore through Wadi Al Ghallah. During sediment transport, mica drifts in a westerly direction either because of the weaker current or with a change in wind direction. Mica mimics clay and stays in suspension for a long time until it settles out, as shown by the presence of mica with biogenic carbonate in the offshore area (Fig. 10h).

Sediments are also transported through secondary wadis present along the periphery of the lagoon. Although Wadi Rahman and Al Lith are important sources of lithogenic material, especially in the eastern part, they do not directly influence the sediment dynamics on the western side. The sediments brought into coastal zones by Wadi Rahman and Al Lith are transported in a south-easterly direction by littoral currents (Al-Washmi et al. 2002), where the dispersal pattern is controlled primarily by the prevailing current. The signature of aeolian quartz is somewhat meagre in the offshore area because the wadis reduce the aeolian influence. The Al Lith sabkhas are bounded on the east by aeolian deposits and are blocked from the Red Sea by coralline limestone (Basyoni 1997). Although the wadis transport most of the quartz, appreciable amounts of aeolian material are transported from the adjoining sabkhas and by severe sand storms that occasionally pass over the Red Sea area.

The lowest carbonate content is in the south-east and in the deeper waters in the south and is mostly derived from biogenic constituents, in places diluted by terrigenous carbonate material mostly derived from Wadi Rahman and Al Lith. The biogenic carbonate content increases with increasing grain size. The carbonate content increases in the western side because of the reduced dominance of terrigenous material (Fig. 9), and a reduction of biogenic material is also caused by climatic factors. The Al Lith area is at the approximate northern climatic limit of the area in south-western Saudi Arabia influenced by two monsoon seasons (NE and SW). The monsoonal wind and rain feed all the wadis substantially in the southern part of Saudi Arabia, and therefore input of terrigenous material plays an important role in diluting the carbonate veneer.

The lack of terrigenous material between the two islands is due to strong tidal currents at the entrance, resulting in fine terrigenous material from Wadi Al Ghallah being transported to the open sea. These currents prevent fine sediment from settling and leave coarse biogenic carbonate as lag deposits, as shown by the sediment texture that becomes coarser with increasing carbonate content. The low or absent carbonate content on the eastern side of the lagoon and in the open sea in the south is due to detrital input from Wadis Al Ghallah, Wadi Rahman and Al Lith. The influx of fluvial material from north to south is restricted to a point where the moving sediments run into the shoals, and the load is deposited.

In general, TOC is higher than normal (>2.0 %), compared to other areas in the Red Sea, and increases with decreasing grain size. In other words, low values are recorded at the northern end where the current winnows the fine sediment and lag material in the form of coarser carbonate fails to adsorb organic material. Most of the area is dominated by values ranging between 0.31 and 0.50 %, and higher values are found in only a few isolated areas, where coarse-textured sediments tend to show higher TOC. This is because the wadis drain through heavily vegetated areas and abundant plant debris supplies organic matter, whereas only scattered patches of vegetation occur on the beach. There is a close relationship between TOC, sediment colour and sediment texture and an inverse relationship between TOC and $CaCO_3$. In general, the finer sediments are richer in TOC compared to coarser sediments rich in carbonate material. However, the status of the lagoon and the influence of the wadis have recently changed due to shrimp farming. The configuration and environment of the lagoon and the influence of Wadi Al Ghallah have greatly altered, with changes in sediment transport pathways and depositional mechanisms in the lagoon and offshore areas (Fig. 8).

Present Status and Future Implications

Human interference has changed the general make-up of the lagoon especially after the construction of a shrimp farm that has retarded the draining of the lagoon by the numerous wadis. It is important that the health of the sediment veneer is studied because of the disposal of the waste water from the shrimp farm in the area. Fresh water input from the wadis into the lagoon is restricted, and therefore, the impact on the environment must be addressed.

Al Khuraybah Lagoon

The Al Khuraibah lagoon is one of the largest and deepest lagoons along the north-east Red Sea coast (Fig. 11, Tables 1 and 2). The geology of the area has been studied by Clark (1987) and the heavy mineral provenance by Nabhan et al. (2010). The Khuraybah region contains Proterozoic stratiform and highly folded volcanic, volcaniclastic and secondary epiclastic rocks of greenschist facies (Sahl 1981). Sedimentary rocks of Mesozoic and Cenozoic age encompass the eastern side of the lagoon. The southern part of the Al Khuraybah Formation consists of reefal limestone that extend along the coastal area, and the upper part of the Khuraibah region consists of limestone and sandstone that overlie the palaeo-relief extending northward and dipping gently under the sand and gravels of the Ifal plain (Nabhan et al. 2010). Part of the coastal Al Khuraybah area is composed of a reefal limestone platform covered by sand and mud. The lagoon is separated from the open water of the Red Sea by many convoluted shoals and coral reefs, and the entrance to the lagoon from the open sea for shallow-draft boats only is restricted to a couple of narrow channels that cut the shoals (coral reefs). The lagoon was once fed by several wadis, although some are now dormant and relict channels can be seen on satellite images (Fig. 12). Wadi Aynunah is one of the major wadis that supply the lagoon with eroded material from the mountains; this wadi is now populated and contains farms (Fig. 11a).

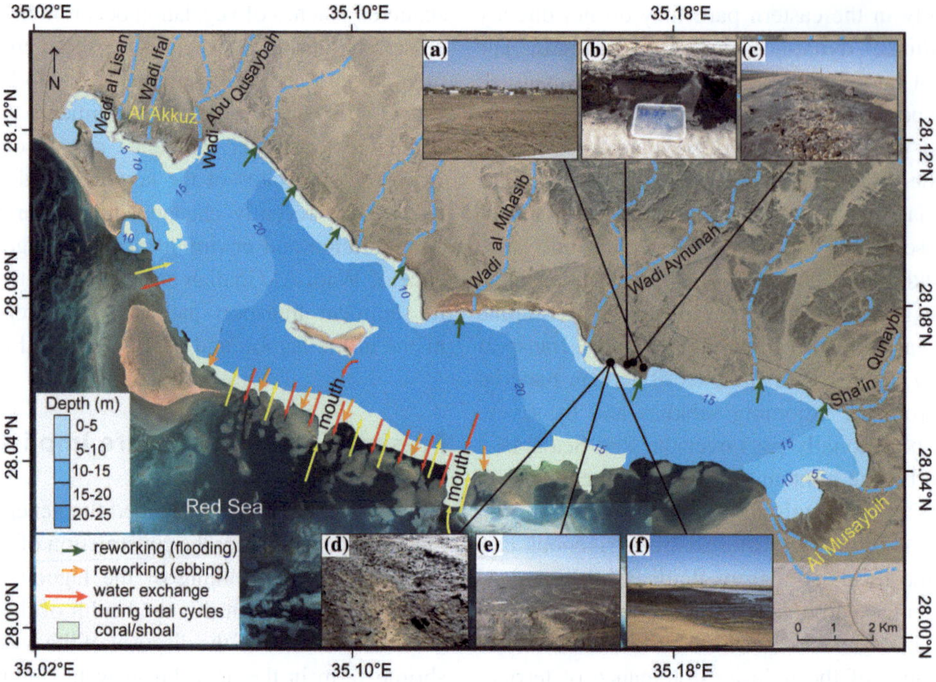

Fig. 11 Bathymetry of Al Khuraybah lagoon and model showing exchange of water between the Red Sea and the lagoon and the process of reworking during flood and ebb tides: **a** Wadi Aynunah, a major wadi supplies the lagoon with eroded material from the mountains, is now populated and used as farm land; **b**, **c** abundant sand size heavy minerals on the beach and intertidal zone consist mainly of biotite, magnetite and ilmenite, giving the sand a black appearance, and **e**, **f** high organic content and decomposed organic debris on the flat low-lying areas are inundated with water during tidal cycles

The lagoon is veneered with mud to muddy sand of both terrigenous and biogenic origin. The abundant mud in the central deeper part of the lagoon is mostly terrigenous, and carbonate is contributed by churning of the seabed sediment by trawling, making the carbonate material even finer (Fig. 12). The re-suspended sediment stays within the lagoon since there is no major outlet to the open sea, and sediment input from the wadis to force sediment out of the lagoon is almost completely absent, whereas in sheltered areas, terrigenous mud and sand are present (Fig. 13). Parts of the coast and shallower areas of the lagoon are veneered with lithogenic sandy mud, while carbonate dominates in other parts of the lagoon where the wadi influence is minimal. Staining of carbonate material is observed in areas where trawling activity is absent. There are well-sorted terrigenous materials, from gravels to cobbles, and wind-blown sand is common along the beach. There are abundant heavy minerals on the beach and offshore, and intertidal zone beach sands consist mainly of biotite, magnetite and ilmenite, giving the sand a black appearance (Fig. 11b, c). Biotite is abundant in the deeper and calmer areas.

In the western part of the lagoon, the sand is mostly of biogenic origin, supplied by fragmentation of reefal limestone and corals. The 0.9 m tidal range results in the transport of mud, and especially mica, in suspension to the open sea during ebbing, while carbonate material of biogenic origin settles at the foot of the coral reef system. During flooding, the water from the Red Sea reaches the lagoon by overspilling; this exchange of water makes the lagoon less saline (39 %) compared to other lagoons where the exchange of water is more limited. However, movement of sediment is restricted because of the shoals. The base of the shoals and Jazirat Umm Shujayrat Island are composed mostly of carbonate materials, and the western side is almost devoid of terrigenous material probably because the flow and transport direction is towards the south-east. Surprisingly, the lagoon lacks gravel-sized material of biogenic origin, indicating that the bulk of the material is from the wadis draining into the lagoon. The beach is covered by algal mats, especially where the influence of Wadi Aynunah was once significant and organic debris together with seagrass was transported to the lagoon and also deposited on the narrow beach in the vicinity of the mouth of Wadi Aynunah (Fig. 11d, e). High organic content and decomposed organic debris were noted on the flat low-lying areas that are inundated with water during spring tides (Fig. 11e, f). At present, pools of water are also stored in the Sharma area where salt is extracted after evaporation. The lagoon and low-lying areas are now

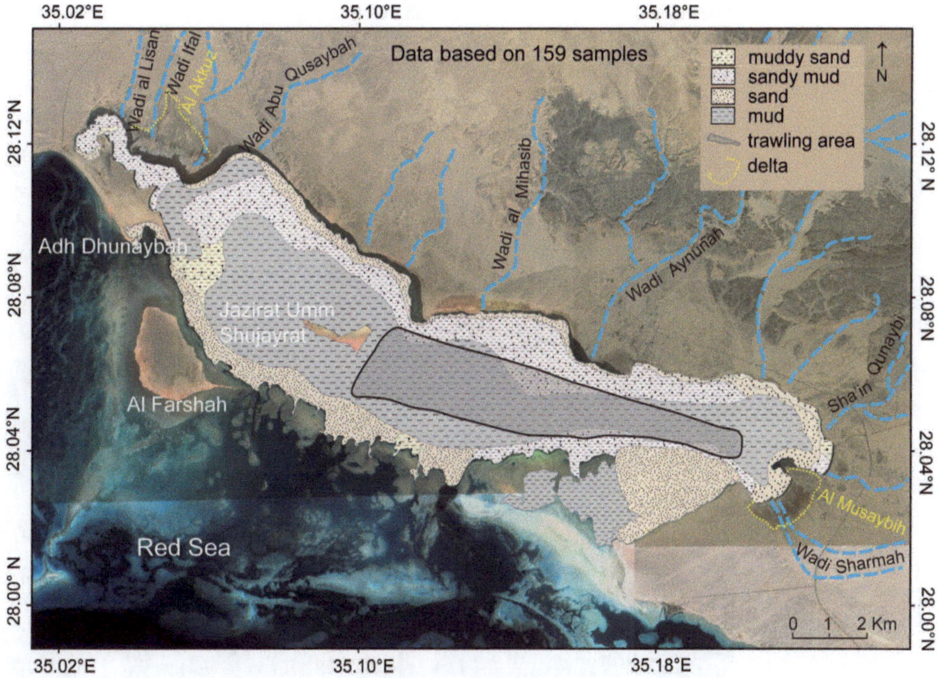

Fig. 12 Sediment distribution map and the wadis feeding the Al Khuraybah lagoon. The textural classification is based on Folk (1980)

fairly polluted by dumping of large amounts of shrimp waste. The composition of the Al Khuraybah lagoon is presented in Fig. 13a–o.

Present Status and Future Implications

The trawling activity has stopped in the lagoon, and therefore, the bed is not as in imbalance as before. Human activity such as salt extraction and boating activity is still common, but the environmental stress has minimized. The input of fresh water to the lagoon is still restricted because of farming practices and construction in the path of the wadi. This might have an impact on the lagoon, but at the present time, the status of the lagoon and the coastal belt is quite healthy.

Al Wajh Lagoons

The lagoons in the northern Red Sea are narrow incisions through coralline ridges extending for a few hundred metres and trending almost perpendicular to the coast. These are sometimes connected to a wadi system of varying dimensions (Table 1). Contemporary coral growth occurs in these lagoons at different depths and the entrance is always over 30 m in depth. At the entrances to these lagoons, the depth often increases abruptly by more than 60 m to a few hundred metres offshore. Five lagoons along the Al Wajh coast were sampled for sedimentological studies, and dives were conducted to study the distribution of corals.

The corals in the Al Wajh area are presently undergoing varying degrees of stress either through natural causes (wadis flushing, winds, currents) or human intervention that control their health and distribution patterns. All the dive sites have steep slopes down to 30 m depth with maximum water clarity, and coral colonies are shaped by the strong northerly wind and water currents. The corals studied were at 5 m water depth. The lagoons house fishing villages and are frequented by visitors, resulting in signs of deterioration in coral growth and health. Some species flourish under severe conditions, but others are vulnerable to slight changes in the environment. The configuration of the lagoons and types of bottom sediments are summarized in Figs. 14, 15, 16, 17, 18 and Table 2.

Al Dumaygh Lagoon

The lagoon is veneered with sandy mud and muddy sand with shoals where the corals are in good health. A coastguard facility is located within the lagoon. The north-western part of the lagoon shows raised coral reef terraces and the south-eastern part has small isolated mangrove stands of *Avicennia marina* and small budding shrubs. A small island separates the lagoon from the open water of the Red Sea, and a narrow 14 m channel allows limited exchange of water only during

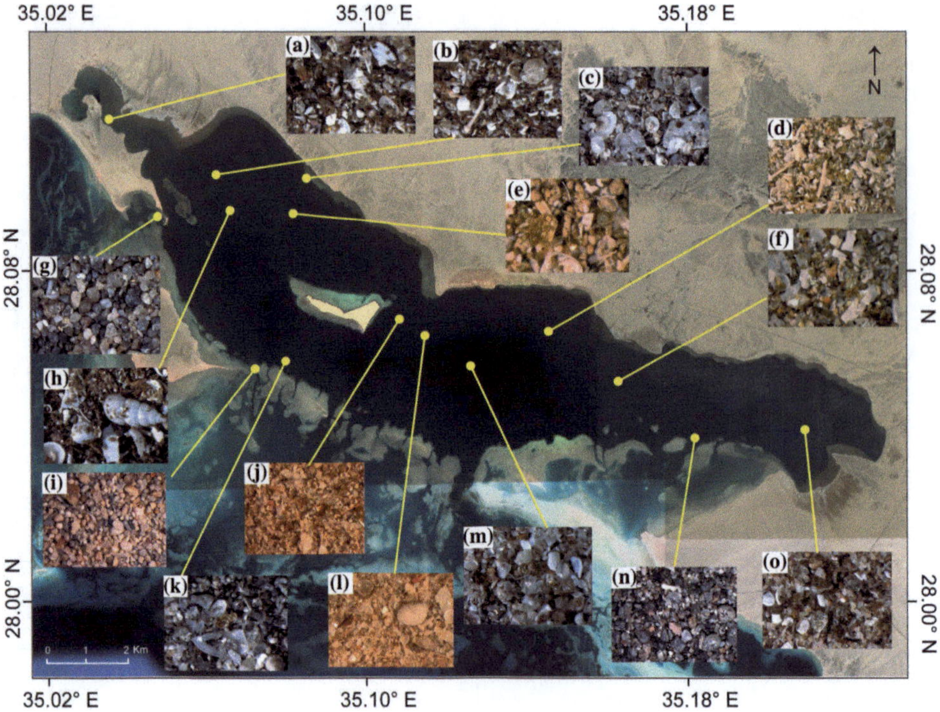

Fig. 13 Sediment composition of the Al Khuraybah lagoon. Terrigenous material composed of abundant mica, quartz, feldspars, magnetite and other opaque heavy minerals is from the numerous wadis draining into the lagoon, while carbonate material is mostly biogenic in nature: **a** mica is common; carbonate skeletal assemblage includes benthic foraminifers, echinoid spines, bivalve fragments and bioclasts; **b** mica is common; carbonate skeletal assemblage dominated by whole and fragmented benthic foraminifers, and bioclasts; **c** devoid of terrigenous material; carbonate skeletal assemblage dominated by whole or fragmented benthic gastropods, bivalves, subordinate echinoid spines and others; **d** mica and haematite and some magnetite; abundant carbonate skeletal assemblage dominated by echinoid spines, bivalves, ostracods, benthic forams and others; **e** mica and haematite present; carbonate skeletal assemblage includes bivalve shells and other bioclasts; **f** abundant mica; carbonate skeletal assemblage includes benthic gastropods, benthic foraminifers, bivalves and others; **g** abundant quartz, feldspars and undetermined stained grains; scarce carbonate skeletal assemblage composed of bioclasts; **h** abundant mica; coarse skeletal assemblage dominated by benthic microgastropods (e.g. *Finella*) and bivalve fragments; **i** some quartz, feldspars and stained material; abundant carbonate skeletal assemblage including fragmented gastropods, bivalves and bioclasts; **j** palimpsest sediments and carbonate skeletal assemblage including benthic foraminifers; **k** abundant mica and quartz; carbonate skeletal assemblage composed of bivalve fragments, benthic gastropods, benthic foraminifers and others; **l** devoid of terrigenous material; abundant carbonate skeletal assemblage including benthic gastropods, bivalves and bioclasts; **m** mica and feldspar present; scarce carbonate skeletal component with benthic foraminifers, echinoid spines and others; **n** opaque heavy minerals with quartz carbonate skeletal assemblage composed of bivalve fragments, benthic gastropods, echinoid spines and others and **o** abundant muscovite and biotite and quartz; rare carbonate skeletal component and bioclasts

high tide when low-lying areas are covered by shallow water. The sediment is mostly of biogenic origin, but aeolian quartz is abundant and eroded material from the raised terraces and fragmented shells are common because of human activity. The sediments are generally poorly sorted except at the entrance where the sediment is moderately well sorted because of strong tidal currents. Terrigenous material in the form of quartz, feldspars and some mica is transported to the lagoon by Wadi Umm Ushsh and its distribution is controlled by the large number of shoals. At 5 m water depth, the corals are healthy and comprise both hard and soft corals. There is a small fishing community, and some local pollution is present because of litter dumping by visitors and fishermen. Recently killed corals are common because of fishing activity, and dead corals make up 20 % of the total. Examples of the types of sediments and corals are presented in Fig. 14a–d.

Al Hawwaz Lagoon

Sandy mud dominates the bottom of this lagoon followed by muddy sand, but a few patches of gravelly sand of biogenic material occur. Shoals are common in the lagoon and control sediment transport. The narrow channel connecting the

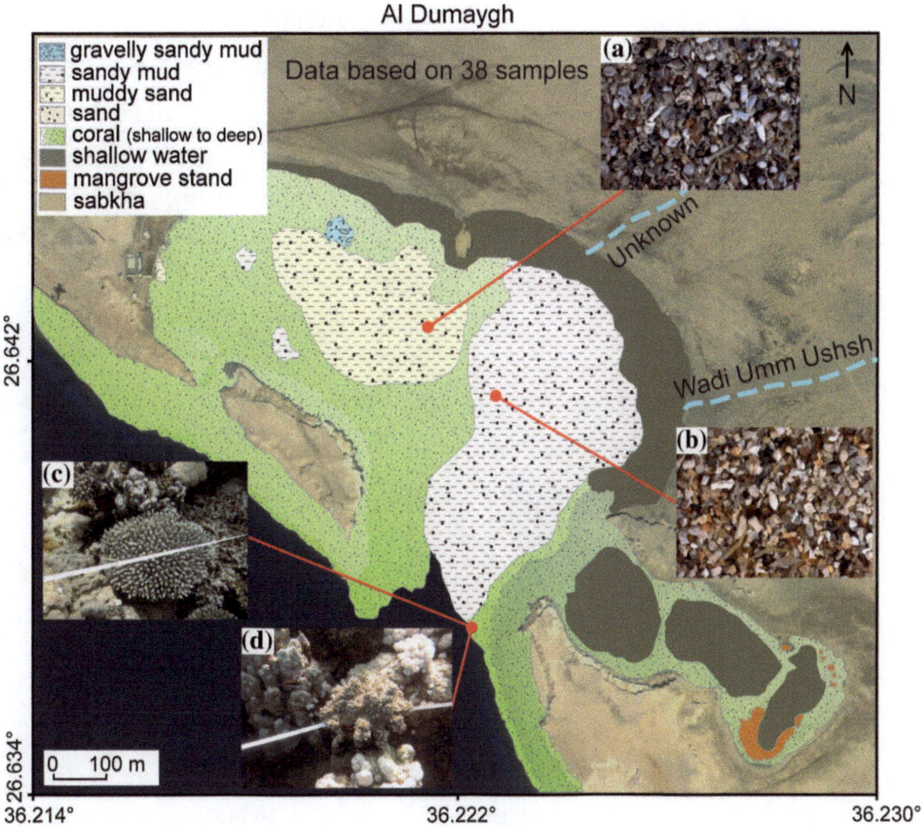

Fig. 14 Al Dumaygh lagoon **a** sediment mostly biogenic; **b** terrigenous material (product of Wadi Umm Ushsh); **c** Scleractinian corals Acropora in the foreground and Porites in the background, respectively, and **d** Alcyoniid and Xeniid soft corals in the foreground, scleractinian coral Porites in the background on the *left side*

lagoon with the Red Sea is not as dynamic as the other lagoons and no wadis flow into the lagoon. At the north-eastern end, there are raised coral terraces in which channels formed by rainfall supply terrigenous mud to the lagoon. The southern part of the lagoon is low-lying and during high tide, part of the surrounding area is inundated with water. A reducing environment stains the biogenic grains because of the restricted exchange of water. At 5 m water depth, corals are healthy and comprise both hard and soft corals. However, some have been impacted by human activities and have died. Examples of the types of sediments and corals are presented in Fig. 15a–d.

Antar Lagoon

This lagoon is fed by Wadi Antar, one of the larger wadis in the northern Red Sea coastal area (Table 1) and the wadi head is connected to numerous distributary channels. The lagoon bottom is veneered with sandy mud and muddy sand. Sand in the eastern part of the lagoon is mostly of terrigenous origin transported by Wadi Antar. Sediment at the entrance is moderately sorted because of current effects. During spring tide, the eastern part is inundated with water as it is flat and low lying and the angle of repose is nominal because of input from the wadi. Abundant mica, quartz and feldspars are contained in the lagoon sediments. At 5 m water depth, the corals are healthy and composed of both hard and soft species but are dominated by hard massive branching corals, some of which have been damaged by either strong waves or fishermen; dead corals make up 20 % of the total. It is an isolated lagoon with little or no influence from tourism. Examples of the types of sediments and corals are presented in Fig. 15a–d.

Mina Al Wajh Lagoon

The head of the lagoon is low-lying and numerous channels drain into the lagoon during rainfall. Wadi Ash Shijnah has a bigger basin, but Sha'ib Umm Sidrah with a smaller basin influences sediment distribution in the lagoon (Table 1). Because the area is flat, sea water inundates the sabkhas and channels during spring tides and transports terrigenous material, mostly in the fine fraction, to the lagoon during ebbing. Aeolian quartz, feldspars, calcite and opaque heavy minerals are abundant in the lagoon. The bottom sediment is mainly muddy sand of terrigenous origin with subordinate biogenic materials. The wide mouth of the lagoon is strongly influenced by wind- and wave-generated currents, supplemented by tidal effects. The sediment at the mouth is well sorted but contains over 60 % carbonate material made up

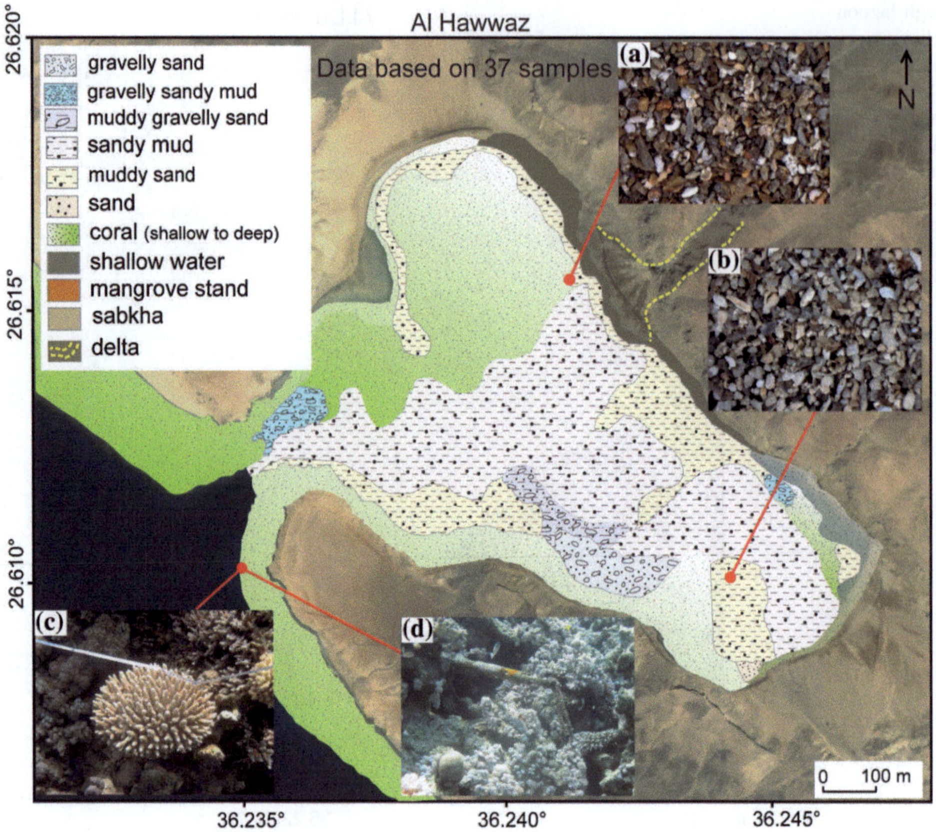

Fig. 15 Al Hawwaz lagoon **a** carbonate skeletal component includes benthic gastropods, benthic foraminifers (e.g. miliolids) and undetermined bioclasts with minor terrigenous influence from the adjoining hills; **b** carbonate skeletal component includes benthic foraminifers (e.g. miliolids) and undetermined bioclasts; sediments are stained black because of a reducing environment; **c** scleractinian coral Acropora in the foreground and Xeniid soft corals in the background on the *left*, and **d** mixed assemblages dominated by Xeniid soft corals Alcyoniid, a massive colony of the scleractinian coral genus *Goniastrea* in the foreground on the *left side*; fishermen in this area are the worst enemy of the corals, and natural factors are also of concern because the impact of sediments from the wadis on the lagoon and strong waves generated by northerly wind cause only the fittest species to survive

mostly of coral debris and shell fragments followed by terrigenous mud supplemented by aeolian quartz. At the head of the lagoon almost 80 % of the bottom sediment is composed of terrigenous material of Sha'ib Umm Sidrah origin. Staining of terrigenous grains and shell fragments is common, due to organic and anthropogenic input. The cohesive terrigenous mud adsorbs pollutants and organic material supporting the formation of a reducing environment. There is a coastguard office at the entrance to the lagoon as well as a fishing village, and the lagoon is frequently visited, resulting in some local pollution. At 5 m water depth, the corals are under stress and dead corals are common because of fishing and other activities. Waste water from the desalination plant located in the north is emptied into the Red Sea, with some impact on the lagoon, contributing to the demise of hard corals. Examples of the types of sediments and corals are presented in Fig. 17a–c.

Al Habban Lagoon

The north-eastern part of this lagoon is connected to a sabkha and residential complex. The two wadis draining into the lagoon do not influence the sediment distribution pattern. The eastern part is shallow, and a channel that drains into the lagoon brings in lithogenic material. The northern periphery is veneered with muddy sand of biogenic or aeolian material. The lagoon is veneered with sandy mud to muddy sand, and staining of biogenic material is locally observed. The walls of shoals/coral beds dip steeply on the north-western and south-western side. At the mouth, the depth increases abruptly, and the bed is mostly covered with muddy biogenic sand that is moderately sorted because of local water currents. At 5 m water depth, the corals are healthy and composed of both hard and soft corals. Some nearly dead

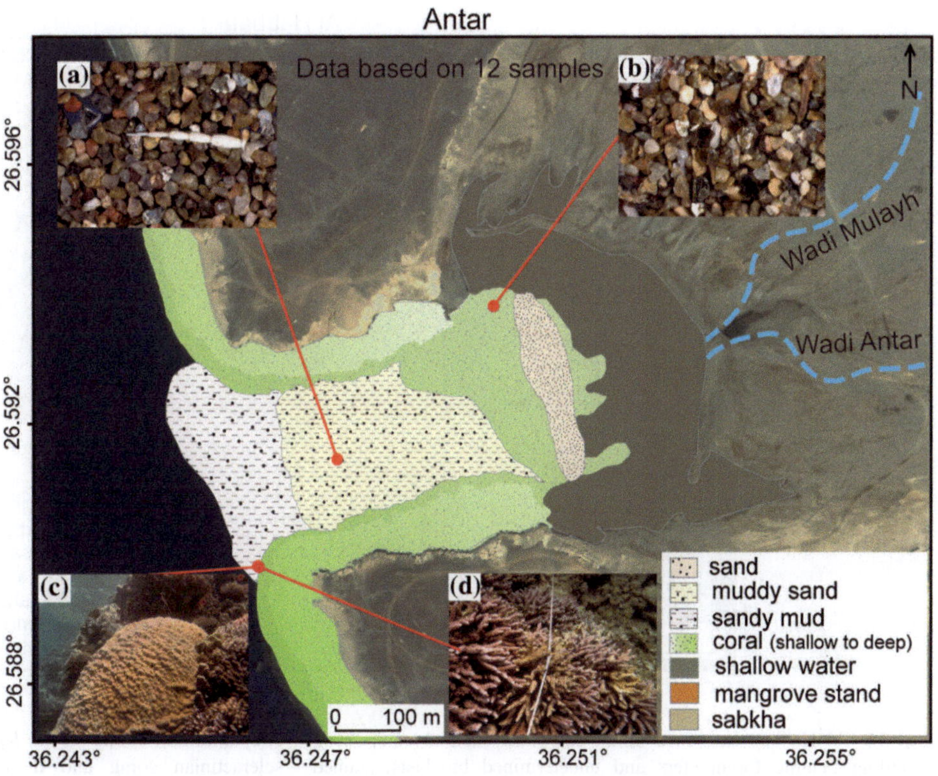

Fig. 16 Antar lagoon **a** carbonate skeletal component with benthic forams and other undetermined bioclasts and biosomes; biogenic material diluted by the sediments from Wadi Mulayh and Wadi Antar; **b** sparse carbonate skeletal component with undetermined bioclasts; biotite and muscovite are supplied by the wadis; **c** massive Porites scleractinian corals, and **d** branching Acropora scleractinian coral

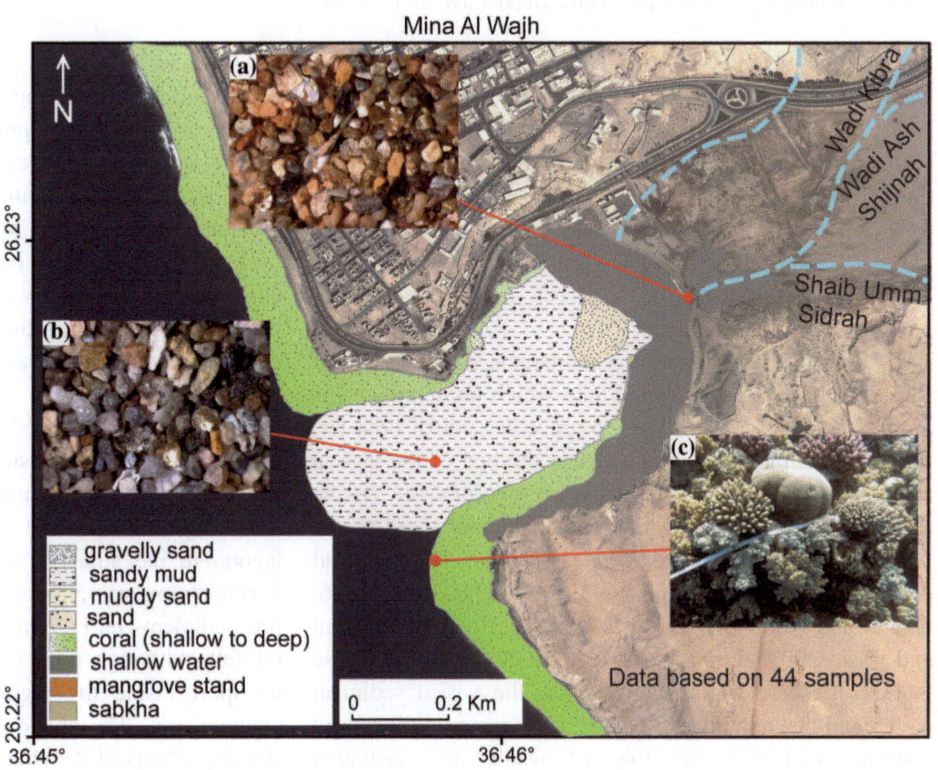

Fig. 17 Mina Al Wajh lagoon **a** sparse carbonate skeletal component with undetermined bioclasts; abundant input from the adjoining wadis; **b** sparse carbonate skeletal component with benthic forams and undetermined bioclasts; carbonate veneer diluted by the terrigenous input from the wadis; **c** *top half* shows digitate scleractinian corals Acropora (pale beige), and a massive Goniastrea, Alcyoniid soft corals at the *bottom half*

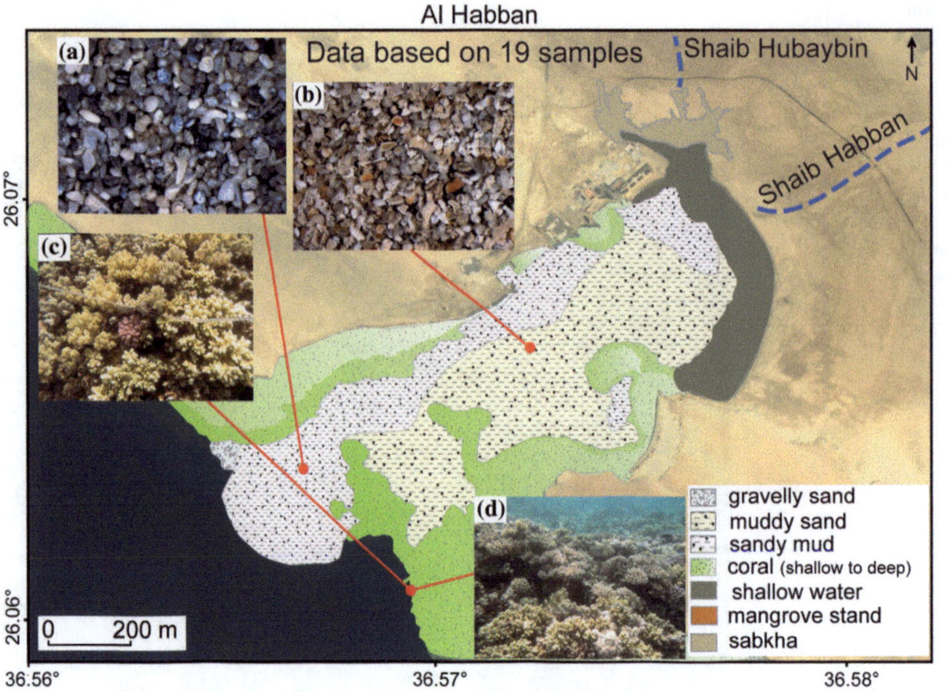

Fig. 18 Al Habban Lagoon **a** minor carbonate skeletal component, including benthic foraminifers and undetermined bioclasts; stained because of reducing environment; **b** carbonate skeletal component includes benthic foraminifers, benthic gastropods and undetermined bioclasts; **c** branching scleractinian coral Acropora (*on the right*) and Alcyoniid soft corals *on the left*) and a purple colony of pocilloporid scleractinian coral, and **d** digitate and branching Acropora and pocilloporid dominated benthic assemblage with some Xeniid soft corals; coral flourishes where human activity is minimal

coral areas have recovered after the removal of the coast-guard station brought about a change in the degree of stress. Examples of the types of sediments and corals are presented in Fig. 14a–d.

Al Kharrar Lagoon

The Al Kharrar lagoon is the second largest lagoon along the Red Sea coast of Saudi Arabia into which numerous wadis drain (Fig. 19, Table 1). Mud to gravelly sand of both terrigenous and biogenic origins veneer the bed of the lagoon (Fig. 20). Numerous studies have been carried out on the sedimentology, mineralogy and faunal assemblage of the Al Kharrar lagoon and its associated sabkhas (El-Abd and Awad 1991; Behairy et al. 1991; Abou-Ouf 1992, 1996; Gheith and Abou-Ouf 1994; Al-Washmi 1999; Al-Washmi and Rasul 2003; Basaham 2008; Rasul et al. 2010a). These include the processes responsible for the varied sediment cover and its relationship to the bathymetry and current patterns, including the role of wadis and sediment provenance in controlling the environment of deposition. A summary of the data is presented in Table 2.

The lagoon is connected to the Red Sea via a channel about 150 m wide, and although artificial connections using pipelines were installed for exchange of water during tidal cycles (Fig. 19a), these have become blocked by sediment and debris. In 2013, new openings were created to flush the nutrient-rich water to the Red Sea because of pollution concerns and excessive algal growth and seagrass in recent years.

The lagoon is bounded by inter-tidal flats and sabkhas in the south that are occasionally covered by shallow water extending about 3 km inland (Fig. 19b). Northerly winds influence the movement of water and sediments from the lagoon to the adjacent sabkhas in the south, particularly during spring tides, and southerly winds and ebb tides bring fine sediments into the lagoon, especially during occasional rainfall. Although tidal forcing is important at the mouth, transport inside the lagoon is essentially a wind-driven phenomenon. The shallow depth combined with the constricting effects of the channel decrease the tidal transport as

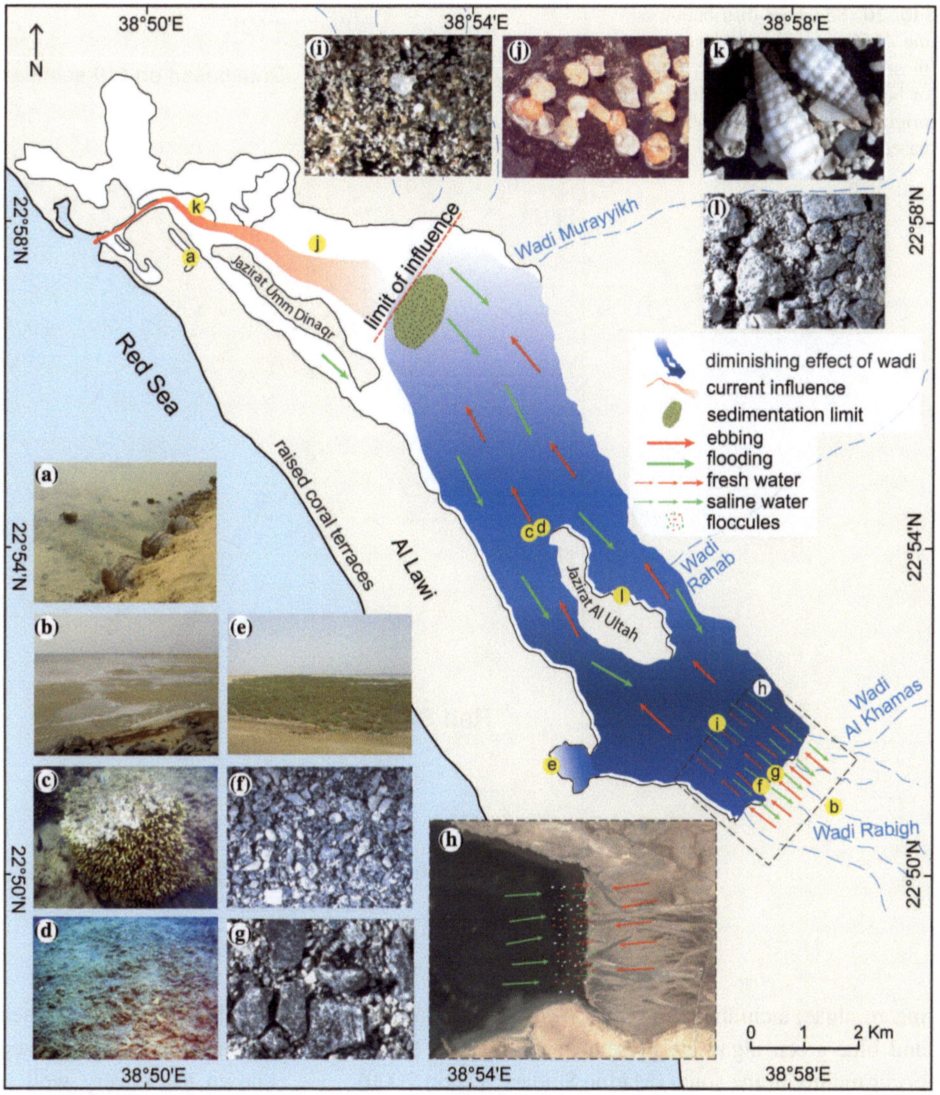

Fig. 19 Al Kharrar lagoon **a** pipeline that once exchanged water between the lagoon and the Red Sea; **b** during spring tides the water travels about 3 km inland, inundating the adjoining sabkhas; **c** coral covered with algae and fine sediments; **d** graveyard of corals resulting from human activities (anchoring, fishing and diving); **e** shrubs in the southern part of the lagoon thrive on nutrient-rich water and stagnant conditions; **f** and **g** shells and coral debris are stained brown to black in a reducing environment; **h** flocculation is initiated by the mixing of fresh and saline water; **i** mixture of aeolian and fluvial quartz; **j** aeolian quartz is common in the northern part of the lagoon where the influence of the wadi is limited; **k** gastropod shell (*Cerithium* sp.) a contemporary input from the Red Sea, and **l** remnant of reefal sediment (calcium carbonate)

distance from the mouth increases, so that wind becomes the primary transport mechanism inside the lagoon. Waves generated by strong local winds winnow fine sediments at the margin of the lagoon and transport sediment to deeper water.

The fine sediments are dark in colour with shades of grey, whereas the coarse sediments are lighter and brown to greyish brown. The sediment veneer mostly comprises coarse-grained sediment in the north and north-west and in shallower water. At the southern end and in deeper water, bottom sediments are dominated by fine-grained sediments. At the mouth, very coarse sand to gravelly sand of biogenic origin dominates because of strong tidal currents that winnow the sediments. Biogenic material is supplied to the lagoon by mechanical breakdown of shells and coral reefs, and by boring marine organisms. Additional biogenic material is supplied to the lagoon by the Red Sea, making the lagoon a poorly sorted environment. Mud to very fine sand of detrital origin is common in the south where the wadi fringes the lagoon.

Corals including *Sarcophyton* and *Poritessolida* and many more are common in the lagoon. Some corals are under stress because of re-suspension of sediments by strong currents, and because of this, coral growth is limited (Fig. 19c). Dead corals are covered with algae and very fine sediments accumulate on the corals following cessation of the currents (Fig. 19d). Human interference has also led to loss of coral reefs, but there are areas where the corals and marine life are in good health.

Mangrove (*Avicennia marina*) stands are scattered around the islands and at the margins, especially in the south, where seagrass also thrives where water from the wadis and lagoon meet. Stagnant conditions prevail at the south-western tip of the lagoon where there is some pollution and the nutrient-rich water supports growth of land shrubs and marine plants including algae and seagrass (Fig. 19e). Several species of

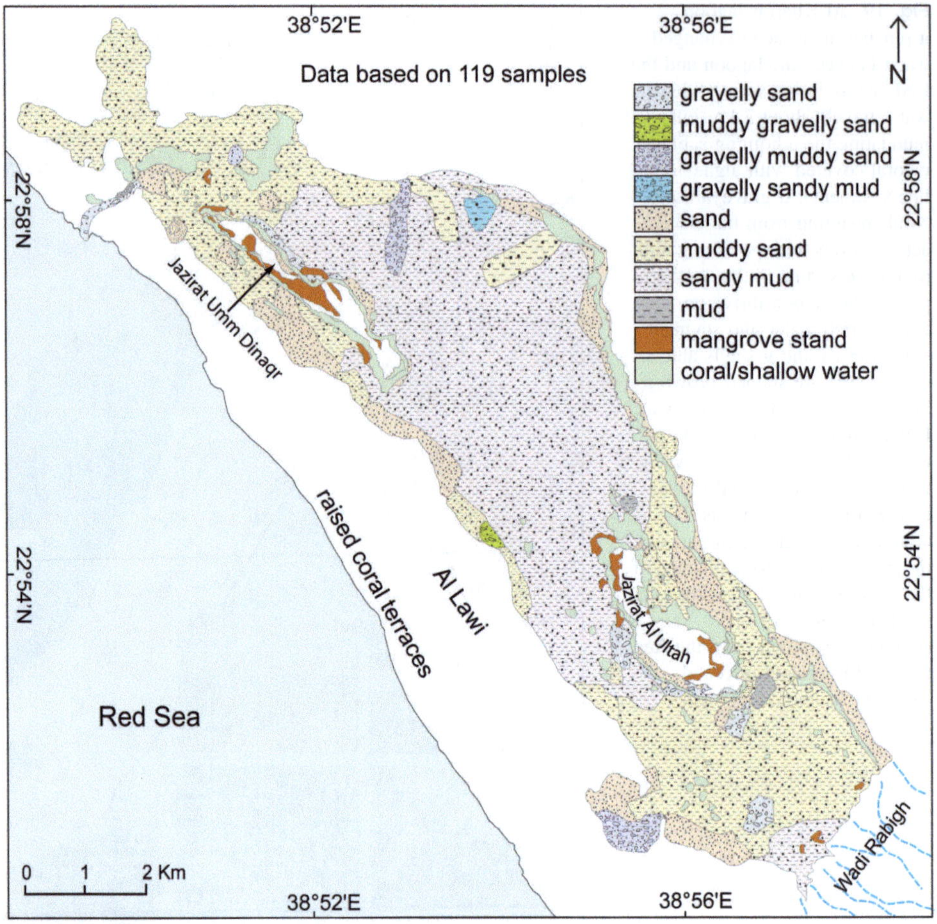

Fig. 20 Sediment distribution in the Al Kharrar lagoon; lagoon is mostly covered with sandy mud of both biogenic and detrital origins. Textural classification is based on Folk (1980)

macro-algae, including *Laurenica*, *Halimeda*, and *Sargassum*, and blue-green algae are present in the lagoon. Halophytes occur mainly in the south (Marine Studies Section 2000).

Discussion

Based on the sediment colour and particle size, two sedimentary environments are identified, controlled primarily by sediment source and physico-chemical properties: (i) a dark-coloured facies in the south, associated with fine terrigenous sediments, and (ii) a light-coloured facies in the north, associated with coarse biogenic sediment. Sandy mud dominates the middle and deeper parts of the lagoon, while muddy sand is common in the southern and northern parts (Fig. 20). The mixing and the coarse nature of the sediment results in poor sorting that is exacerbated by the presence of biogenic materials of varying grain size and grain fragmentation resulting in a bimodal-to-multimodal sediment distribution, especially near the islands where human access is frequent. Although aeolian materials are abundant and known to be very well sorted, the area comprises poorly sorted sediment because of mixing from multiple sources.

The mostly biogenic sediments are stained grey-black because of reducing environments close to the mangrove stands and in areas where stagnation prevails. High sedimentation favours a reducing environment in the presence of high productivity, as observed in the south, and the supply of organic debris by the mangrove stands, seagrass, algae and green foliage around the lagoon results in the dark colour, especially for fine sediments that are more prone to adsorption. The carbonate material is also stained grey-black as a result of the reducing environment that is enhanced by the organic debris. The dark or tarnished shell and coral debris are stained black only on the outer surface (Fig. 19f, g). Similar staining is common in the recent sediments of the Arabian Gulf (Murray 1966), the Bahamas (Illing1954) and the coastal sabkhas of the Red Sea (Abou-Ouf and El-Shater

1993), and in other lagoons along the Red Sea (Rasul et al. 2009a, b, 2010b, 2014).

The coarse light-coloured sediments reflect the removal of fine sediment, particularly at the entrance, by the sifting action of wind- and wave-generated currents and strong tides with velocities of around 1 m s^{-1}. The tidal current becomes stronger as it enters the mouth of the lagoon due to the constricting effect of the narrow channel (about 150 m wide) and diffuses as the entrance widens, resulting in the deposition of fine sediments (Fig. 19). The currents control the bathymetry at the entrance, as well as transporting fine sediments to the margins of the mouth where they settle out to form channel-levee-like structures. The re-suspended sediments at the entrance are also transported into the main lagoon, following the Red Sea saline water trail during flood tide, to be deposited in the deeper water in the lagoon (Fig. 19h). Some sediment is lost to the Red Sea during ebbing and is then transported to the south along the coast that parallels the lagoon (Rasul et al. 2014). During the rainy season when the wadis are most active, mixing of freshwater from the wadis and saline water from the Red Sea occurs, causing flocculation of fine sediments, especially in the south (Fig. 19h); this fine sedimentation is enhanced by the addition of the clay mineral kaolinite that is abundant in sedimentary deposits around the Red Sea and is transported into the lagoon by the wadis, especially through Wadi Rabigh (Rasul et al. 2010a).

Frequent dust storms over the Red Sea contribute a significant quantity of stained quartz of silt to fine sand grain size. The abundant biogenic material is derived from the attrition of corals by strong currents and human interference. Biogenic material dominates except in the south, where the detrital component is supplied to the lagoon via the numerous wadis originating to the south. Sediments brought into the lagoon are supplied by flash floods, wind (aeolian flux) and fragmentation of corals and shells (biogenic input). Since the lagoon is affected by strong tidal currents, the roundness of individual sediment grains, both detrital and biogenic, is important in understanding the effect of currents in governing sediment distribution patterns. Two dispersal pathways evolving from the wadis in the south and running in the north-westerly direction are controlled by ebb tides in general as well by rainfall (Fig. 19).

The degree of roundness of both aeolian and wadi-derived quartz differs significantly (Fig. 19i). The fine-grained aeolian quartz, due to tidal currents, is more rounded especially at the mouth (Fig. 19j), than the coarse-grained fluvial (mostly wadi-derived) quartz in the south that is relatively less rounded in the finer fraction but more rounded in the medium sand size fraction. This is due to movement during the tidal cycles, the shallow water and influence of the wadis. The relatively low roundness of the quartz transported by the wadis indicates that it is perhaps a single-cycle deposit derived from the parent metamorphic and igneous rocks. Both angular and subangular biogenic grains are present, and the low degree of roundness is attributed to constant breaking down by currents and boring by marine organisms. Whole shells and shell fragments are fresher in appearance in the north than in the south (Fig. 19k). The freshness and polished surfaces of shell fragments in the north are caused by constant agitation of water or their recent addition from the Red Sea, and the number of whole shells increases in the south because of reduced water agitation. The weak current does not have much influence on the attrition of biogenic grains nor is it capable of polishing their surfaces; hence, biogenic materials in the southern part of the lagoon are more intact and show a dull lustre.

In the Red Sea, vigorously growing coral, fringing reefs and barriers play a significant role in the very high production of carbonate and reefal sediments. $CaCO_3$ is contributed to the lagoon through the erosion of reefs by wave and current action, destruction of reefs by fishermen and nibbling of corals by fish (Fig. 19l). The source of the carbonate in the Al Kharrar lagoon is undoubtedly the reef system. Basins along reef-bound coasts with significant runoff and high aridity are the most probable source of carbonate material and the grain size, and its characteristics are primarily controlled by the in situ production of $CaCO_3$. A 3D model of the various environments and processes responsible for the sediment distribution in the lagoon is shown in Fig. 21.

The TOC is directly related to particle size and composition of the sediments and the nutrient-rich water in the lagoon. The lowest values of TOC are recorded at the northern end, where the fine sediments are winnowed, and lag material in the form of coarser carbonate fails to adsorb organic material. Higher values are in the south where fine sediments dominate and $CaCO_3$ is relatively low. The area close to the mangrove stands with coarse-textured sediment tends to show higher TOC, as observed in other lagoons, since the organic debris supplied by the mangrove and the fine sediment tends to adsorb TOC that that may be trapped by roots of the mangroves. There is a close relationship between TOC, sediment colour and texture and a negative correlation between TOC and $CaCO_3$.

In general, fine sediments are rich in TOC compared to coarse-grained sediments that contain abundant carbonate debris. Green-red algae in the nutrient-rich water also contribute significant amounts of organic matter to the environment. Dissolved reactive phosphate and nitrite are almost depleted, but there are high levels of dissolved ammonium. High concentrations of nitrite and ammonium in the western part of the lagoon suggest transformations between various forms of nitrogen compounds and may indicate the presence of a suboxic environment. The distribution of heavy metals in the surficial sediments of the lagoons shows the same

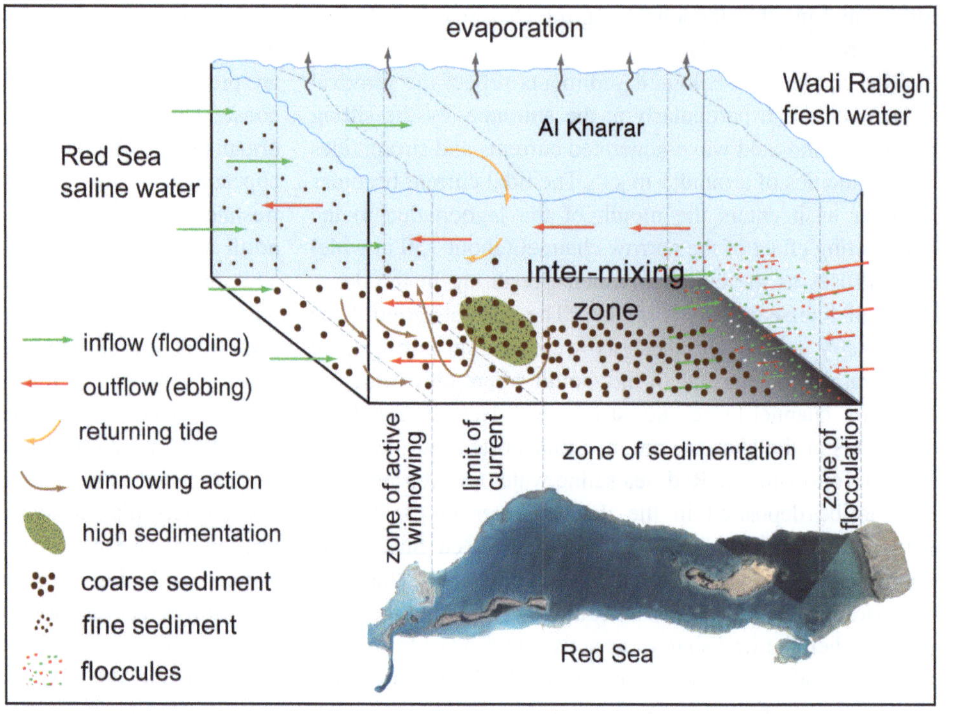

Fig. 21 3D model of the hydrological processes and sediment distribution in the Al Kharrar lagoon

pattern as TOC, with high levels in the southern part of the lagoon, indicating sources from sporadic rainfall (Rasul et al. 2010a).

Present Status and Future Implications

The water of the lagoon was exchanged with the Red Sea through multiple pipes in order to control pollution, but after the choking of the pipes with sediments and artificial materials, the quality of the water deteriorated leading to an increase in the algal bloom and seagrass. Several new openings and modifications made especially at the northern part of the lagoon in order to support an effective exchange of water. An assessment of the water quality and sediment veneer should be taken into consideration to avoid further deterioration.

Khawr Al Dhaban

Khawr Al Dhaban is an isolated shallow lagoon with a maximum depth of 10 m. The channel connecting it with the Red Sea is about 13 m deep with limited access because of numerous shoals and islands separating the lagoon from the Red Sea (Fig. 22). No wadis drain into the lagoon, but water is exchanged with the Red Sea over the shoals during tidal cycles. The sediment texture ranges between sandy mud and gravelly muddy sand, and in general, the veneer is poorly sorted because of weak currents. Sorting improves only marginally at the mouth because of a moderate average current velocity of 0.29 cm s^{-1}. Summer surface water temperatures as high as 34 °C and a salinity of 42 ‰ or even higher have been recorded reflecting the isolation of the lagoon from the Red Sea. The lagoon is veneered with very high carbonate content sediments, averaging 93 % $CaCO_3$ supplied mainly by the fringing reefs and shoals. The sediment fauna includes gastropods, foraminifera and pelecypods. The TOC averages 0.51 % in the bulk sediment but is much higher at 1.79 % in the fine fraction. A reducing environment prevails in the south-western side of the lagoon where the mangrove saplings are affected by water quality, and isolated mangrove stands supply organic debris to the lagoon (Fig. 22a, b). The sediment outside the lagoon in low-lying areas to the south that are inundated by the semi-diurnal tide shows TOC values of 5.6 %, much higher than in the main lagoon, as most of the mangrove debris is either transported or held there where it decomposes and enriches the sediment with organic matter. The southern area is flat, and during high tide, the surrounding area is inundated with a thin sheet of water. The movement of water between the lagoon and the shallow tidal zones transports sediments and organic material, as a result of which a few centimetres of alternating layers of yellowish brown carbonate material and dark rich organic sediments are observed (Fig. 22a, b). The dark-coloured sediment has a pungent smell indicating a reducing environment caused mainly by decomposition of organic debris. Stagnation prevails and organic matter is

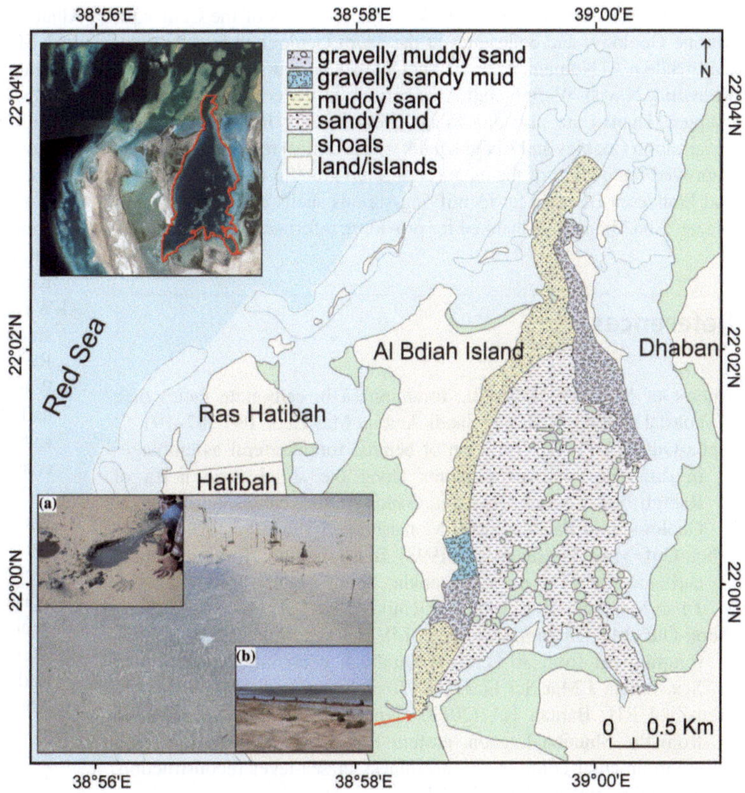

Fig. 22 Khawr Al Dhaban is an isolated body of water with restricted communication with the Red Sea; Shoals are mostly covered with <0.5 m of water, and only during spring tide does the water from the Red Sea move over the shoals to the lagoon. The restricted exchange of water results in a very high salinity of 42 ‰ in the lagoon; **a** sediment is reduced: a scrape shows the sediment is black below the surface. The alternate layering of dark and light-coloured sediments shows its cyclical (spatial and temporal) nature; **b** reducing environment is widespread in the area

incorporated into the sandy mud, augmenting the reduction process and leading to sediment staining.

Conclusion

The sediment transport mechanisms, patterns and pathways to the lagoons are controlled by sediment flux from the wadis, anthropogenic input from sewage treatment plants, aeolian transport and the degree of human intervention, and natural causes including biological productivity in the lagoons. Water currents generated by wind or tidal forcing play important roles in the distribution of sediment. Strong tidal currents affect only the mouth of a lagoon with limited influence on the main lagoon. The sediment composition and texture are also important when it comes to retention of pollutants because fine terrigenous material tends to hold the pollutants better than the biogenic component, and weak currents fail to winnow the floccules and organic-rich materials, resulting in a reduction in quality of the sediment veneer. The lagoon configurations differ significantly from north to south. In the north, the communication of the lagoons with the open sea is mostly through wide openings and the sediment composition is mainly a mixture of biogenic and terrigenous origins, while abundant aeolian material is supplied by dust storms. In the central part, the lagoons are dominated by biogenic material except where wadis drain into them. Here, the lagoons connect to open water through narrow openings, and sedimentation is controlled by strong tidal currents. The lagoons in the south are dominated by terrigenous materials supplied to the lagoons or offshore areas by active wadis. In some areas where the influence of wadis is strong, almost the entire lagoon bottom sediment veneer is lithogenic.

The lagoons in the urban areas have been sites of coastal development for decades, leading to pronounced changes in their morphodynamics, such as at Sharm Obhur. The Al Ghallah lagoon is the recent site of a large shrimp farm that has altered the environmental setting of the surrounding wadis and the lagoon itself. The Al Kharrar lagoon has experienced a bloom in seagrass and macro-algae because of the limited exchange of water with the Red Sea. The two Shoaibah lagoons, especially the northern one, have seen their mouths reduced in width, while Al Khuraybah lagoon has undergone trawling for shrimp for several years, and the wadis, especially Wadi Aynunah, have been used for farms and residences, resulting in a complete change in the environment. The lagoons in the Al Wajh area and Khawr Al Dhaban are less impacted by human activities. All the lagoons with limited human intervention are in near pristine condition. However, the growing trend of reclaiming land along the coastal belt of the Red Sea of Saudi Arabia is a cause for concern if adequate measures for addressing environmental impacts are ignored.

Acknowledgments I wish to thank all the members of the Center for Marine Geology, and colleagues at the Saudi Geological Survey especially Salem Al Nomani, Abdulnasser Qutub, Louiesito Abalos, Mostafa Khurshid, Nawaf Widinly and Ali Tharowi, in the compilation of this chapter. Thanks are also due to Ian Stewart, Guy Rothwell, Geraint Owen, John Coakley and Abdul Khalik whose critical reviews and advice improved the quality of the manuscript. I am grateful to Marco Taviani and Francesca Benzoni for identifying various shells and coral species. Mazen AbuAbdullah is thanked for providing information on the wadis.

References

Abou-Ouf MA (1992) Benthic foraminifera in carbonate facies of a coastal sabkha, Red Sea, Saudi Arabia. Mar Geol 104:187–191

Abou-Ouf MA (1996) Variation of benthic foraminiferal assemblages in different microenvironments along the shore zone north of Rabigh coast, eastern Red Sea, Saudi Arabia. Neues Jahrbuch für Geologie und Paläontologie, Monatshefte 3:29–139

Abou-Ouf MA, El Shater A (1993) Black benthic foraminifera in carbonate facies of a coastal sabkha, Saudi Arabian Red Sea coast. J Fac Mar Sci, King Abdulaziz Univ 4:133–141

Abou-Ouf M, Durgaprasada Rao MVM, Tag RJ (1988) Benthic foraminifera from Al-Leith-Al Qunfidah coast, southeastern Red Sea. Indian J Mar Sci 17:212–217

Abu-Zied RH, Bantan RA (2013) Hypersaline benthic foraminifera from the Shuaiba Lagoon, eastern Red Sea, Saudi Arabia: their environmental controls and usefulness in sea-level reconstruction. Mar Micropaleontol 103:51–67

Abu-Zied RH, Bantan RA, El Mamoney MH (2011) Present environmental status of the Shuaiba Lagoon, Red Sea coast, Saudi Arabia. J Fac Mar Sci, King Abdulaziz Univ 22(2):159–179

Abu-Zied RH, Basaham AS, El-Sayed MA (2012) Effect of municipal wastewaters on bottom sediment geochemistry and benthic foraminifera of two Red Sea coastal inlets, Jeddah, Saudi Arabia. Environ Earth Sci doi:10.1007/s12665-012-1751-7

Ahmad F, Sultan SAR (1992) The effect of meteorological forcing on the flushing of Shuaiba Lagoon on the eastern coast of the Red Sea. J Fac Mar Sci, King Abdulaziz Univ 3(1):3–9

Ahmad F, Sultan SAR (1993) Tidal and sea level changes at Jeddah, Red Sea. Pakistan J Mar Sci 2(2):77–84

Ahmad FC, Khomayis HS, Turki AJ, Rasul NM (2002) A multidisciplinary oceanographic study of three environmentally sensitive sites along the Saudi Arabian coast of the Red Sea. Final Report of the Scientific Research Council, King Abdulaziz University (unpublished), 82 pp

Al Harbi SM, Khomayis HS (2005) Seasonal variations of chlorophyll a and micro-nutrients in Obhur Creek, Red Sea. J Fac Mar Sci King Abdulaziz Univ 16:91–104

Al-Barakati AMA (2010) Application of 2-D tidal model, Shoaiba Lagoon, eastern Red Sea coast. Can J Comput Math Nat Sci Med 1:9–20

Albarakati AMA, Ahmad F (2012) Water column conditions in a coastal lagoon near Jeddah Red Sea. Oceanologia 54(4):675–685

Al-Farawati R, El-Maradny A, Niaz GR (2008) Chemical characteristics of the surface waters for Sharm Obhur, Jeddah, eastern coast of the Red Sea: nutrients, fecal sterols and polyaromatic hydrocarbon. J Fac Mar Sci, King Abdulaziz Univ 19:95–119

Al-Farawati R, El-Sayed MA, Shaban Y, El Maradny AA, Orif M (2014) Phosphorus speciation in the coastal sediments of Khawr ash Shaibah al Masdudah: a coastal lagoon in the eastern Red Sea. Arab Gulf J Sci Res (in press)

Alharbi OA, Phillips MR, Williams AT, Gheith AM, Bantan RA, Rasul NM (2012) Desalination impacts on the coastal environment: Ash Shuqayq, Saudi Arabia. Sci Total Environ 421(422):163–172

Ali KM, Hossain D (1988) Geochemical characteristics of Obhor subsoil. J Fac Mar Sci, King Abdulaziz Univ 1:205–225

Al-Sayari SS, Zötl JG (1978) Quaternary period in Saudi Arabia, vol VI. New York, Springer, 335 pp

Al-Washmi HA (1999) Sedimentological aspects and environmental conditions recognized from the bottom sediments of Al-Kharrar Inlet, eastern Red Sea coastal plain, Saudi Arabia. J Fac Mar Sci, King Abdulaziz Univ 10:71–87

Al-Washmi HA (2002) The grain size characteristics, mineral composition of sediments and geomorphology of Wadi Iyar, Al-Leith area, Red Sea, Saudi Arabia. J Egypt German Soc Zool 39B:11–27

Al-Washmi HA, Gheith AM (2003) Recognition of diagenetic dolomite and chemical surface features of the quartz grains in coastal sabkha sediments of the hypersaline Shuaiba Lagoon, eastern Red Sea coast, Saudi Arabia. J Fac Mar Sci, King Abdulaziz Univ 14:101–112

Al-Washmi HA, Rasul NM (2003) Sedimentary processes of the Quaternary sediments of Al-Kharrar Lagoon, Red Sea, Saudi Arabia. J Environ Sci Univ Mansoura Egypt 26(1):273–296

Al-Washmi HA, Bantan R, Rasul N, Rifaat AE (2002) Study on the geomorphological changes, depositional environments and geochemical properties of sediment veneer in and around Al-Leith area, Red Sea Coast of Saudi Arabia. Scientific Research Council, King Abdulaziz University, unpublished report, 38 pp

Bantan R, Rasul N (2003) Impact of development on the configuration of Sharm Obhur and its bottom sediments. J Environ Sci 26(1):319–337

Basaham AS (2008) Mineralogical and chemical composition of the mud fraction from the surface sediments of Sharm Al-Kharrar, a Red Sea coastal lagoon. Oceanologia 50(4):557–575

Basaham AS, El-Shater A (1994) Textural and mineralogical characteristics of the surficial sediments of Sharm Obhur, Red Sea coast of Saudi Arabia. J Fac Mar Sci King Abdulaziz Univ 5:51–71

Basaham AS, El-Sayed MA, Rasul N, Rifaat AE (2003) Sharm Obhur: sediment characteristics, past and present. Scientific Research Council, King Abdulaziz University, Final report, 108 pp

Basaham AS, Rifaat AE, El-Sayed MA, Rasul N (2006) Sharm Obhur: environmental consequences of 20 years of uncontrolled coastal urbanization. J Fac Mar Sci King Abdulaziz Univ 17:129–152

Basyoni MH (1997) Sedimentological and hydrochemical characteristics of Al-Leith sabkha, Saudi Arabia. J Fac Earth Sci King Abdulaziz Univ 9:75–86

Behairy AKA, El-Rayis OA, Ibrahim AM (1983) Preliminary investigations of some heavy metals in water, sediments and plankton in Obhur Creek (eastern Red Sea). J Fac Mar Sci King Abdulaziz Univ 3:129–139

Behairy AKA, Durgaprasada Rao NVN, El-Shater A (1991) A siliciclastic coastal sabkha Red Sea coast, Saudi Arabia. J Fac Mar Sci King Abdulaziz Univ 2:65–77

Bemert G, Ormond R (1981) Red Sea corals. Kegan Paul International Ltd., London, 193 pp

Berner RA (1970) Sedimentary pyrite formation. Am J Sci 268:1–23

Berner RA (1982) Burial of organic carbon and pyrite sulphur in the modern ocean: its geochemical and environmental significance. Am J Sci 282:451–473

Berner RA, Timothy B, Holden GRJ (1979) Authigenic iron sulphides as paleosalinity indicators. J Sediment Petrol 49:1345–1350

Berry L, Whiteman AJ, Bell SV (1966) Some radiocarbon dates and their geomorphological significance, emerged reef complex of Sudan. Zeitschrift für Geomorphologie 10:119–143

Blott SJ, Pye K (2001) GRADISTAT: a grain size distribution and statistics package for the analysis of unconsolidated sediments. Earth Surf Proc Land 26:1237–1248

Braithwaite CJR (1987) Geology and paleography of the Red Sea region. In: Edwards AJ, Mead SM (eds) Key environments: Red Sea. Pergamon Press, New York, pp 22–24

British Standard Institution (1975) (BSI 1377): Methods of test for soils for civil engineering purposes. British Standards Institution, 143 pp

Brown GF, Schimdt DL, Huffman AC (1989) Shield area of western Saudi Arabia, geology of the Arabian Peninsula. US Geological Survey, professional paper no. 560-A

Clark MD (1987) Geological map of the Al Bad quadrangle, sheet 28A, Kingdom of Saudi Arabia. Ministry of Petroleum and Mineral Resources, Directorate general of mineral resources, Map GM-81C

Coakley JP, Rasul NM (2001) Global contamination issues emerging in coastal regions: implications for the Red Sea ecosystem. J Fac Mar Sci, King Abdulaziz Univ 12:33–46 (Special Issue)

Cramp A (1985) Late pleistocene/holocene sedimentation in the eastern Mediterranean with particular reference to sapropelic units. Ph.D. thesis (unpublished), University of Wales, 160 pp

Crockett D, Stow DAV (2001) Ecotourism, marine science and the environment. J Fac Mar Sci King Abdulaziz Univ 12:13–23 (Special Issue)

Darwin C (1962) The structure and distribution of coral reefs. University of California Press, California

Davenport J, Davenport JL (2006) The coastal alterations due to the artificial and personal leisure transport on coastal environments: a review. Estuar Coast Shelf Sci 67:280–292

Dowidar NM, Rahimudin SA, Aleem AA (1978) Phytoplankton populations in the region of Obhur, Jeddah, Saudi Arabia. Bull Fac Mar Sci, King Abdulaziz Univ 2:271–292

El-Abd YI, Awad MB (1991) Evaporitic sediment distributions in Al-Kharrar sabkhas, west Red Sea of Saudi Arabia, as revealed from the electrical soundings. Mar Geol 97:137–143

El-Abd Y, Awad M (1992a) Sea level changes in Sharm Abhur Red Sea coast of Saudi Arabia as revealed from seismic stratigraphy. Geophys Monogr 11:57–63

El-Abd Y, Awad M (1992b) Effect of sea level changes on the quaternary emergent reef limestone near Sharm Abhur as revealed from geological soundings. Geophys Monogr 11:65–69

El-Rayis OA, Eid FM (1997) Hydrography and water budget of Obhur Creek, Red Sea. J Fac Mar Sci, King Abdulaziz Univ 8:29–45

El-Sabrouti MA (1983) Texture and mineralogy of the surface sediments of Sharm Obhur, West Red Sea coast of Saudi Arabia. Mar Geol 53:103–116

El-Sayed MK, Hosny CF (1980) Sediments of the intertidal zone off Ghardaqa, Red Sea, Egypt. In: Proceedings of the symposium on coastal and marine environments of the Red Sea, Khartoum

El-Sayed MA, El-Maradny AA, Al Farawati R, Yasser AS (2011) Evaluation of the adequacy of a rehabilitation programme, implemented in two Red Sea coastal lagoons, using the hydrological characteristics of surface water. J Fac Mar Sci, King Abdulaziz Univ 22(2):69–108. doi:10.4197/Mar.22-2.6

El-Sayed MA, Al Farawati R, El Maradny AA, Shaban Y, Rifaat AE (2013) Environmental status and nutrients and dissolved organic carbon budget of two Saudi Arabian Red Sea coastal inlets: a snapshot statement. Environ Earth Sci doi:10.1007/s12665-013-2557-y

Fahmy MA, Saad MAH (1996) Temporal and spatial distribution of heavy metals in Obhur creek, a coastal Red Sea water body north of Jeddah. J Fac Mar Sci, King Abdulaziz Univ 7:75–83

Fisher ISJ (1986) Pyrite formation in bio-turbated clays from the Jurassic of Britain. Geochim Cosmochim Acta 50:517–523

Folk RL (1980) Petrology of Sedimentary rocks. Hemphill's, Austin, 182 pp

Gheith AM (1999) Mineralogy and diagenesis of coastal sabkha sediments of the hypersaline lagoons on the eastern coast of the Red Sea, Saudi Arabia. Arab Gulf J Sci Res 17(2):199–219

Gheith AM, Abou-Ouf MA (1994) Textural characteristics, mineralogy and fauna in the shore zone sediments at Rabigh and Sharm Al Kharrar, eastern Red Sea, Saudi Arabia special issue on the Red Sea marine environment. J Fac Mar Sci, King Abdulaziz Univ 7:107–131

Gladstone W, Curley B, Shokri MR (2013) Environmental impacts of tourism in the Gulf and the Red Sea. Mar Pollut Bull 72(2):375–388

Hadley DG (1975) Geology of Al-Qunfidah quadrangle, sheet 19/41c. Kingdom of Saudi Arabia, Directorate General of Mineral Resources, Geological Map GM-19

Hadley DG (1980) Reconnaissance geology of the Musaylim quadrangle sheet 19/40B. Kingdom of Saudi Arabia, Directorate General of Mineral Resources, Map GM-34

Hadley DG, Fleck RJ (1980) Reconnaissance geological map of the Al-Lith quadrangle, sheet 20/40C. Kingdom of Saudi Arabia, Director General of Mineral Resources, Geological Map GM-32

Hariri MS (2008) Effect of hydrographic conditions on the ecology of benthic foraminifera in two different hypersaline lagoons, eastern Red Sea coast, Kingdom of Saudi Arabia. J Fac Mar Sci King Abdulaziz Univ 19:3–13

Heija MA, Shehata WM (1989) Engineering geological aspects of Al Leith sabkha, Saudi Arabia. Abstract, 28th International Geological Congress, Washington, pp 1–6

Holland KT, Elmore PA (2008) A review of heterogeneous sediments in coastal environments. Earth Sci Rev 89:116–134

Hötzl H, Zötl JG (1978) Climate changes during quaternary period. In: Al-Sayari SS, Zötl JG (eds) Quaternary Period in Saudi Arabia. Springer, New York, pp 301–311

Illing LV (1954) Bahaman calcareous sands. Am Assoc Pet Geol Bull 38:1–95

Jado AR, Hötzl H (1984) Quaternary period in Saudi Arabia. Springer, Berlin 361 pp

Jado AR, Hötzl H, Boscher B (1989) Development of sedimentation along the Saudi Arabian Red Sea coast. Earth Sci J King Abdulaziz Univ 3:863–888

Kinsman B (1965) Wind waves, their generation and propagation on the ocean surface. Prentice-Hall, Englewood Cliffs

Le Corre P (1983) Dosage du carbone organique paticulaire. In: Aminot A, Chaussepeid M (eds) Manuel des analyses chimiques en milieu marin. CNEXO, France, pp 203–208

Maiklem WR (1967) Black and brown speckled foraminiferal sand from the southern part of the Great Barrier Reef. J Sediment Petrol 37:1023–1030

Marine Studies Section (2000) Environmental map of Al Kharrar lagoon. Center for Environment and Water Research Institute, King Fahd University of Petroleum and Minerals, Dhahran, Saudi Arabia

Meshal AH (1987) Hydrography of a hypersaline coastal lagoon in the Red Sea. Estuar Coast Shelf Sci 24:167–175

Mixon RB, Pilkey OH (1976) Reconnaissance geology of the submerged and emerged coastal plain province, Cape Lookout area, North Carolina. USGS professional paper 859, U.S. Government Printing Office, Washington, 45 pp

Morley NJF (1975) The coastal waters of the Red Sea. Bull Mar Res Centre 5:1–19

Murray JW (1966) The foraminifera of the Persian Gulf, 4: Khor al Bazam. Paleogeog. Paleoclim. Paleoecology 2:153–169

Nabhan AI, Tayeb OA, Kurdi HA, Sulaimani GS (2010) Exploration for heavy minerals in the beach sands of Al Khuraybah, Northern Red Sea Coast, Saudi Arabia. Saudi Geological Survey Technical Report SGS-TR-2009-5, 27 pp

Nuns Vaz RA, Lennon GW, Bowers DG (1990) Physical behavior of a large, negative or inverse estuary. Cont Shelf Res 10:277–304

Pallister J (1986) Explanatory notes to the geologic map of the Al-Lith quadrangle sheet 20D, Kingdom of Saudi Arabia. Directorate General of Mineral Resources, Map GM-95C

Prinz WC (1983) Geology map of the Al-Qunfudah quadrangle, sheet 19E, Kingdom of Saudi Arabia. Directorate General of Mineral Resources, Map GM-70C

Pritchard DW (1952) Estuarine hydrography. Adv Geophys 1:243–280

Rabaa SMA (1980) Geomorphological characteristics of the Red Sea coast with especial emphasis on the formation of Marsas in the Sudan. In: Proceedings of a symposium on coastal and marine environments of the Red Sea, Gulf of Aden and Tropical Western Indian Ocean. University of Khartoum, Khartoum, Sudan Alesco/RSC 2:53–72

Rasul N (1992) Late Quaternary to present day coarse-grained sedimentation of the Indus fluvial-marine system. Ph.D. thesis (unpublished), University of Wales, 248 pp

Rasul NM, Qutub AS (2009) An investigative survey along the coastal zones of Jeddah for assessing the health of coral reefs. Saudi Geological Survey Technical Report SGS-TR-2008-2, 44 pp

Rasul NM, Al-Nomani S, Al-Hazmi O, Widinly N, Qutub AS, Bantan R (2009a) Bathymetric survey in Sharm Obhur. Saudi Geological Survey Technical Report SGS-TR-2008-11, 14 pp

Rasul NM, Al-Nomani S, Bahareth FZ, Al-Hazmi O, Widinly N, Abushosha M, Qutub AS (2009b) A multi-parameter approach to nearshore seabed mapping of the Al-Lith area. Saudi Geological Survey Technical Report SGS-TR-2008-10, 20 pp

Rasul MA, Abushosha M, Al-Hazmi O, Al-Nomani S, Bahareth F, Widinly N, Akbar A, Qutub AS, Farawati R (2010a) Bathymetric survey and ground-truthing of the Al-Kharrar lagoon, Red Sea coast of Saudi Arabia. Saudi Geological Survey Technical Report SGS-TR-2008-8, 77 pp

Rasul MA, Al-Hazmi O, Abushosha M, Al-Nomani S, Akbar R, Bahareth F, Widinly N, Farawati R, Bantan R, Qutub AS (2010b) Sedimentological and geochemical properties and the application of satellite images in the two Shoaiba lagoons and the associated sabkhas, Red Sea coast of Saudi Arabia. Saudi Geological Survey Technical Report SGS-TR-2008-9, 114 pp

Rasul NM, Qutub AS, Widinly N, Al-Nomani S, Jerais A, Osami A, Abushosha Subahi H (2014) Geological and oceanographical studies in the coastal waters of Rabigh and Masturah regions. Saudi Geological Survey Technical Report (submitted for review)

Rees AWG (1988) Early chemical diagenesis and sedimentology in high energy sublittoral marine environment, Swansea Bay. Ph.D. thesis (unpublished), University of Wales, 339 pp

Rifaat AE, Basaham AS, Al-Washmi HA, Bantan R (2001) Impact of recreational villages, amenity beaches and tourism development on the marine environment off Jeddah. Scientific Research Council, King Abdulaziz University (unpublished report)

Risk MJ, Sherwood OA, Nairn R, Gibbons C (2009) Tracking the record of sewage discharge off Jeddah, Saudi Arabia, since 1950, using stable isotope records from antipatharians. Mar Ecol Ser Prog Ser 397:219–226. doi:10.3354/meps08414

Sahl M (1981) Reconnaissance geology of the Tayyib al Ism quadrangle, Sheet 28/34 B, Kingdom of Saudi Arabia. Saudi Arabian Directorate General of Mineral Resources, Open File Report DGMR-OF-01-18, 24 pp

Sugden W (1963) Some aspects of sedimentation in the Persian Gulf. J Sediment Petrol 38:1–95

Sugden W (1966) Pyrite staining of pellety debris in carbonate sediments from the Middle East and elsewhere. Geol Mag 103:250–256

Sultan SAR, Ahmad F (1990) Flushing of a costal lagoon in the Red Sea. Estuar Coast Shelf Sci 31(3):345–349

Tag RJ (1986) Sedimentology and mineralogy of the Red Sea coast recent sediments between Al-Leith and Al-Qunfudhah, Saudi Arabia. M.Sc. thesis, King Abdulaziz University (unpublished)

Tag RJ, Abu-Ouf M, El-Shater A (1990) Textural characteristics of coastal sediments between Wadi Al-Fagh and Wadi Al-Qunfidah, southeastern Red Sea. Arabian Gulf J Sci Res 8(2):33–47

Sea-Level Changes

David T. Pugh and Yasser Abualnaja

Abstract

Sea levels are always changing, for many reasons. Some changes are rapid, while others take place very slowly. The changes can be local, or extend globally. Sea levels, particularly extremes, are important for coastal flooding and coral reef development, both of which may be impacted by climate change. In this chapter, we review Red Sea sea-level changes, before looking at the various processes involved in more detail and relating them to basin development and dynamics. There is no systematic review of Red Sea levels: most scientific studies have been local and piecemeal; measurements are few and limited to widely spaced harbour facilities. This chapter is a brief overview of sea-level changes and a source of references for further studies. On increasing timescales, we review tidal, weekly, seasonal and long-term changes. Finally, we link to changes of sea level in the recent geological record.

Introduction

Sea levels are always changing, for many reasons. Some changes are rapid, while others take place very slowly. The changes can be local, or extend globally. Sea levels, particularly extremes, are important for coastal flooding and coral reef development, both of which may be impacted by, for example, climate change. In this chapter, we review Red Sea sea-level changes, over timescales from hours to thousands of years, and in doing so make links with several other chapters in this book, from physical oceanography and circulation to geologically recent postglacial reviews. We look at several processes involved in more detail, relating them to basin development and dynamics. This overview makes extensive references to published work and as such will be a useful guide for those planning more specific studies.

There is no published systematic review of Red Sea levels; most scientific studies have been local and piecemeal. Direct sea-level measurements in the region are extremely limited and restricted to a few widely spaced harbour facilities. However, there are other useful sources of information, including shoreline features and, more recently, satellite altimetry measurements.

The time and space scales of sea-level variability are summarized in Fig. 1 (from Pugh and Woodworth 2014). For the Red Sea, the changes due to meteorological effects and those due to tides are roughly of equal magnitude, some tens of centimetres. The longer-term geological timescale changes are many tens of metres and more. In this chapter, on increasing timescales, we review tidal, weekly, seasonal and long-term mean sea-level changes. Finally, we link to changes of sea level in the recent geological record and comment on tsunami generation.

Tides

Red Sea tidal amplitudes are relatively small compared with those of the open ocean and with those generally experienced along coasts near continental shelves. The semidiurnal oscillations predominate with maximum ranges on spring

D.T. Pugh (✉)
National Oceanography Centre, Liverpool, UK
e-mail: d.pugh@me.com

Y. Abualnaja
King Abdullah University of Science and Technology (KAUST), Thuwal, Saudi Arabia

N.M.A. Rasul and I.C.F. Stewart (eds.), *The Red Sea*,
Springer Earth System Sciences, DOI 10.1007/978-3-662-45201-1_18,

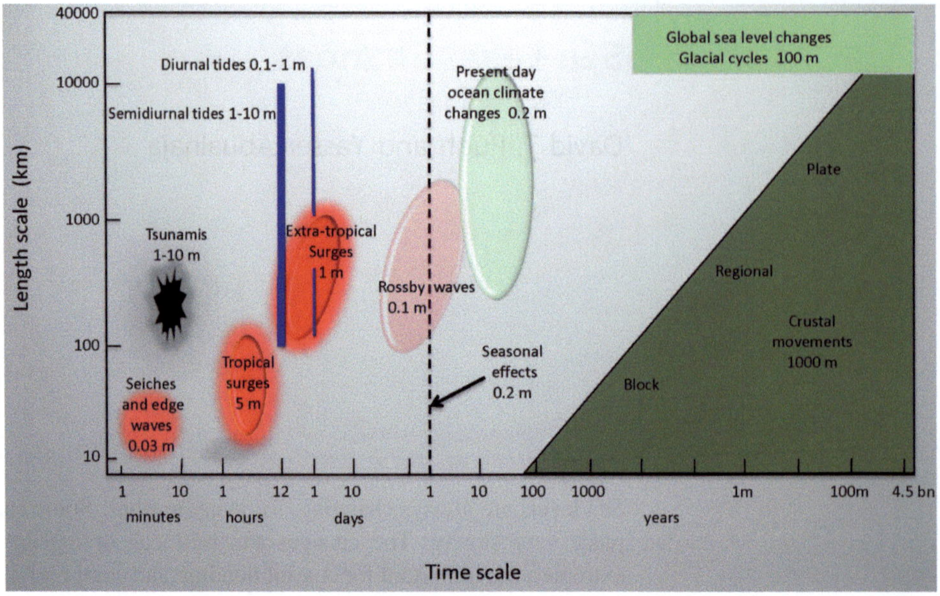

Fig. 1 A conceptual time–space map of reasons for sea-level changes. The effects vary from wind waves over a few seconds, to geological movements over millions of years (from Pugh and Woodworth 2014, with permission)

tides of about one metre in the far north at Suez and in the south near Perim. The length and depth of the Red Sea give half-wavelength dynamics for the semidiurnal tide, which on a rotating Earth results in a wave progression around the amphidrome (a point with zero tidal amplitude) roughly located between Jeddah and Port Sudan, as shown in Fig. 2. The diurnal tides are only a few centimetres in range throughout the Red Sea, being most evident in the records from around Port Sudan, where the semidiurnal ranges are very small due to the local amphidrome. Figure 3a shows a month of sea-level variability (February 1991) at Port Sudan, near the amphidrome, where diurnal tides sometimes dominate, but where the lower-frequency meteorological effects discussed below are equally or more important.

Table 1 shows tidal analyses of a recent 6-month period of sea-level data (actually sub-sea-surface bottom pressures) at three sites, as part of an experiment to investigate Red Sea tides: Port Sudan, Arkylai, some 67 km north of Port Sudan, and Swakin, 55 km to the south. The very low semidiurnal amplitudes show the proximity of an amphidrome, and the north-to-south phase progression shows that the amphidrome is offshore. The diurnal phases are almost identical at all three places. The exact location and movement of this amphidrome are not well known: loss of energy in a reflected south-going Kelvin wave could explain the displacement from the centre towards the coast of Sudan (Pugh 1981), but there are complicating factors, as discussed below.

Despite the small range, the Red Sea has been of interest in the development of scientific ideas about the tides, as discussed by Defant (1961). Because of its long narrow shape and deep sides, the Red Sea was used as an early application of numerical computations of tides, before

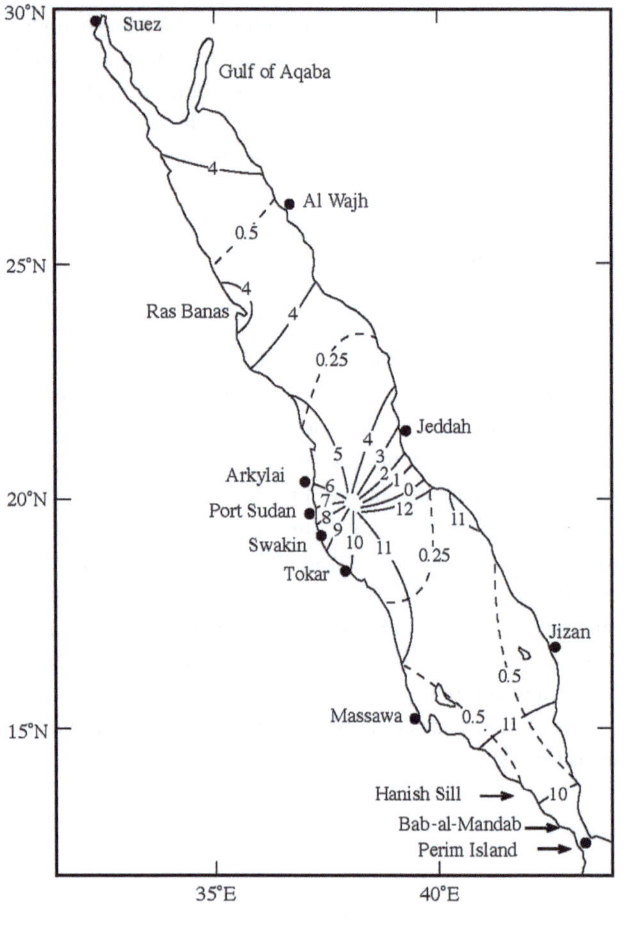

Fig. 2 Map of the principal lunar semidiurnal tide for the Red Sea showing the approximate amphidrome position near Port Sudan. *Solid lines* are Greenwich phase in hours. The *broken lines* are the amplitudes in metres. Maximum amplitudes are at the extreme north and south ends of the Sea

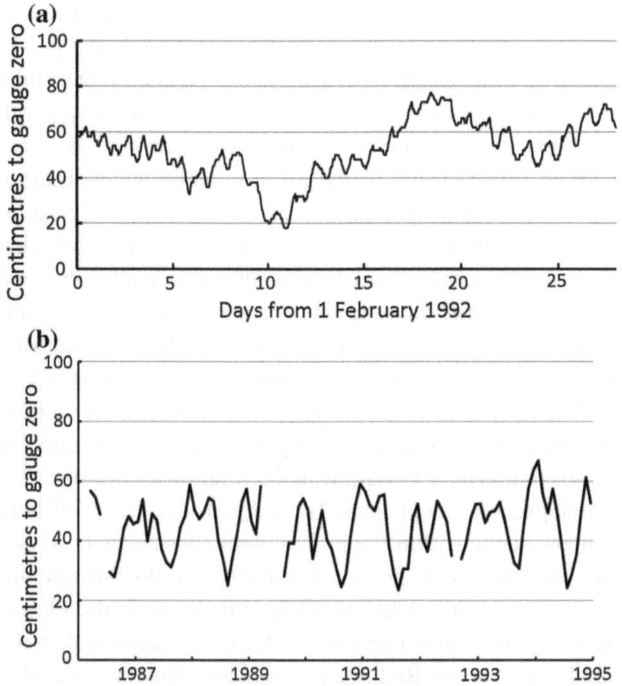

Fig. 3 **a** One month (February 1992) of sea levels at Port Sudan. Note that, because the tides are small, the lower-frequency meteorological effects are very important and **b** monthly mean sea levels at Port Sudan, 1985–1992

modern computer power made it easy and routine (Grace 1930). Proudman (1953) by modelling 39 sections along the axis derived the behaviour of Red Sea tides as seiches in a one-dimensional basin and also studied the influence of direct gravitational forcing within the sea, in both cases without Earth rotation. The relative importance of tidal forcing through the Strait of Bab-al-Mandab, compared with direct gravitational forcing within the Red Sea itself, remained unclear. Defant (1961) called these the *co-oscillating* and *independent* tides respectively; he suggested they are equally important, but also recommended consideration of the frictional energy losses.

Further evidence to understand the Red Sea tides comes from detailed surveys of levels and currents at the southern end of the Red Sea (Jarosz et al. 2005). These show a complicated system of tides with an amphidrome for M_2 between the Perim Narrows and the Hanish Sill. Tides to the south, in the Perim Narrows, are similar to those in the Gulf of Aden. Amplitudes become very small near Assab and then increase northwards to the co-oscillating amplitudes shown in Fig. 2. Using both current and sea-level measurements, Jaroz et al. calculated the tidal energy fluxes. For both K_1 and M_2, there is a flow of tidal energy from the Gulf of Aden into the Strait of Bab-al-Mandab region (300 ± 150 MW and 460 ± 16 MW, respectively). There is probably a small flow of K_1 energy from the Strait into the Red Sea (8 ± 8 MW), but for the semidiurnal tides, the net energy is from the Red Sea into the Gulf (320 ± 200 MW). The Strait is a region of relatively high semidiurnal energy dissipation, with tidal energy flowing in from the Red Sea to the north and from the Gulf of Aden to the south.

The significance of this net outward semidiurnal tidal net energy for the process of overall tide generation in the Red Sea, as discussed by Defant and Proudman, is an interesting scientific question, but it suggests that the tides generated by gravitational forces within the Red Sea are dominant and that the Red Sea is a net exporter of tidal energy. This in turn opens the question of why the central amphidrome is displaced to the Sudanese side and invites examination of the role of energy dissipation in the shallow Gulf of Suez. More detailed analyses are underway to investigate the amphidrome and possible movements, symptomatic, and related to regional tidal dynamics.

To the north, where the Red Sea divides into the Gulf of Aqaba and the Gulf of Suez, the two marginal gulfs have very different tidal regimes. The Gulf of Suez is shallow and interspersed with small islands. It has a mean depth of 36 m and a natural period of 6.7 h; the Gulf of Aqaba has a mean depth of 650 m and a natural period of about 0.9 h (Defant 1961). There is a further amphidrome in the Gulf of Suez, with strong currents and high local tidal energy dissipation.

The Gulf of Aqaba tides oscillate in a similar way to the adjacent Red Sea with a 35-min lag from the entrance at the Strait of Tiran, to the head of the Gulf, a typical standing wave (Monismith and Genin 2004). Tidal currents in the Gulf of Aqaba appear to be associated with internal wave generation, as tidal currents flow through the Strait of Tiran, driving internal tidal waves on the internal density interface; their strength varies considerably throughout the year, associated with varying density stratification.

Table 1 Tidal analyses of sub-sea-surface pressures, 25 February 2013 to 7 September 2013 (noon to noon)

	Latitude (N)	Longitude (E)	stdev	Residual	Ssa		O_1		K_1		M_2		S_2	
					h	g	h	g	h	g	h	g	h	g
Arkylai	20.232	37.204	0.151	0.081	0.173	57	0.018	160	0.031	170	0.027	131	0.013	214
Port Sudan	19.625	37.223	0.160	0.106	0.168	48	0.020	159	0.032	166	0.017	142	0.013	230
Swakin	19.116	37.342	0.156	0.101	0.160	55	0.021	159	0.034	168	0.010	176	0.013	245

Weather Effects

At periods longer than semidiurnal and diurnal tides, the Red Sea levels are influenced by air pressures, winds, evaporation and changes in surface currents. General changes in water circulation are discussed elsewhere in this volume, and sea-level changes are closely linked to these. Here, we discuss only the sea-level changes. The general patterns of winds are different in the north and south of the Sea. The wind direction north of roughly 19°N is persistently from the northwest throughout the year, while the southern region is affected by the seasonally reversing Arabian monsoon system; the prevailing wind in the summer is northerly, whereas in the wintertime, the wind blows from the south to the north (Clifford et al. 1997; Sofianos and Johns 2003).

There are also several local controls acting on the winds on the Red Sea. The surrounding mountain chains of various altitudes greatly influence surface winds; the mountains orographically force the winds to blow approximately parallel to the longitudinal axis of the Red Sea (Patzert 1974). Additionally, the coastal areas are also usually affected by the diurnal sea/land temperature difference, which produces sea breeze in the day and land breeze in the night almost all the year around (Pedgley 1974).

It is useful to look in detail at the variations measured at a mid-located buoy and bottom pressure recorder. The data from a marine buoy, deployed 64 km off the King Abdullah University of Science and Technology (KAUST) shoreline for 26 months (October 2008 to December 2010) along the central Red Sea, show the computed wind stresses. The wind was primarily from the northwest, characteristic of the northern (here 21.5°N) part of the Red Sea (Fig. 4). The wind speed often increased during the day and decreased during the night. Faster wind speeds occur with variations over periods between 2 and 14 days.

A bottom pressure/temperature/conductivity (PTC) instrument was deployed at the same time. In Fig. 5, the power spectra of wind speeds (red) and sea bed pressure (blue) show significant wind energy in the band from 3 to 100 days, with corresponding but generally not obviously matching increases in the energy of sea bed pressure (sea level) variations. There is a possible matching energy peak in both spectra near to 35 days. The bottom pressure and scaled wind speed show that the bottom pressure diurnal variability was smaller than the semidiurnal variability and that the wind speed diurnal variability was larger than the semidiurnal variability. A broad band of pressure and wind power was observed at timescales of 3–14 days. Significant coherence was found between bottom pressure and the alongshore component of the wind stress for periods of less than 14 days with the wind stress leading to the bottom pressure fluctuations. This suggests that the on/offshore component of the Ekman transport (the alongshore wind stress) contributes to bottom pressure variability. The dominant southeastward wind stress had an offshore transport in the Ekman layer (10–30 m) and a set down at the coast. Large-scale wind variability over the Red Sea may generate Kelvin waves that contribute to some of the sea-level variations. Over longer periods, steric effects (salinity changes) and large-scale wind stress will be involved. Limeburner et al. (2012) suggest that the low-frequency variability of 14–43-day fluctuations could be related to the coastal trapped waves propagating along the coast or could be due to inflow to the Red Sea of relatively low salinity water in the south, or variability in the large-scale wind stress. The authors explain that the 14-day bottom pressure variability was due to wind-driven Ekman transport with setup and set down of up to 30 cm due to the local wind stress (Fig. 6).

As another example, further south, sea-level variations at Jizan were strongly affected by the two wind stress components, that is, long-shore (north–south) and cross-shore (east–west) directions considering the orientation of the

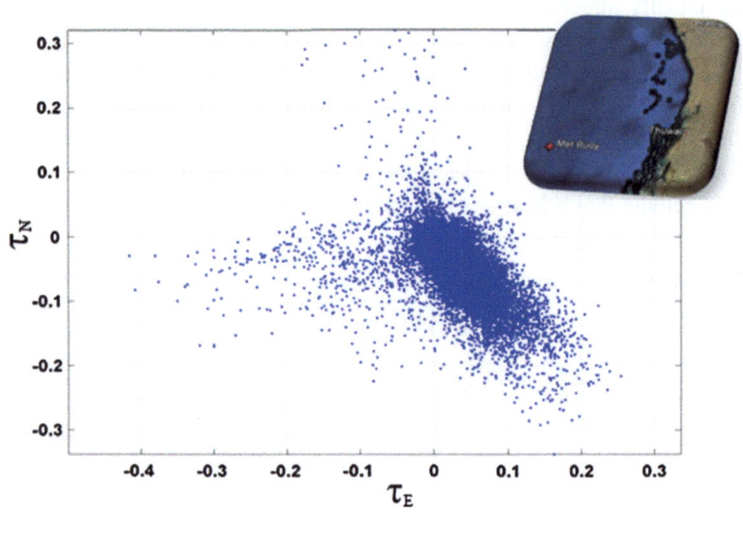

Fig. 4 Scatter plot of hourly wind stress (N/m^2) over two years from a buoy located in the Red Sea 64 km WSW from KAUST (Limeburner et al. 2012)

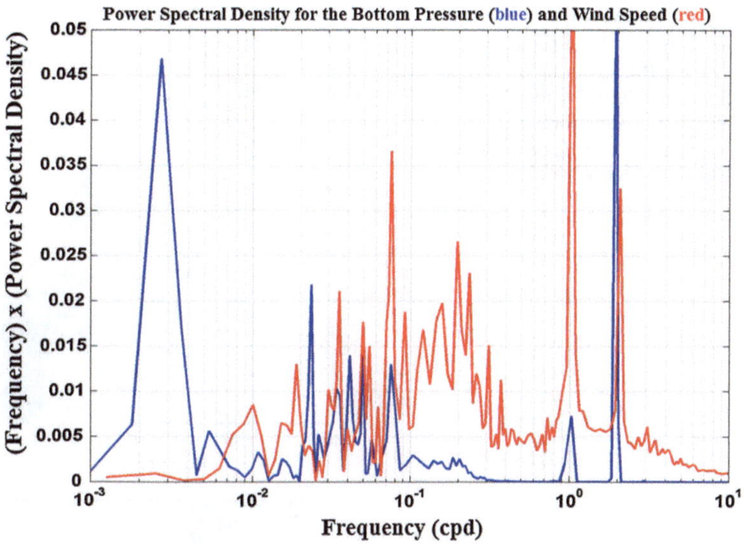

Fig. 5 Bottom pressure and scaled wind speed power spectra. Annual, coastal trapped wave (>14 days), wind (1–14 days) and tidal variability are observed (Limeburner et al. 2012)

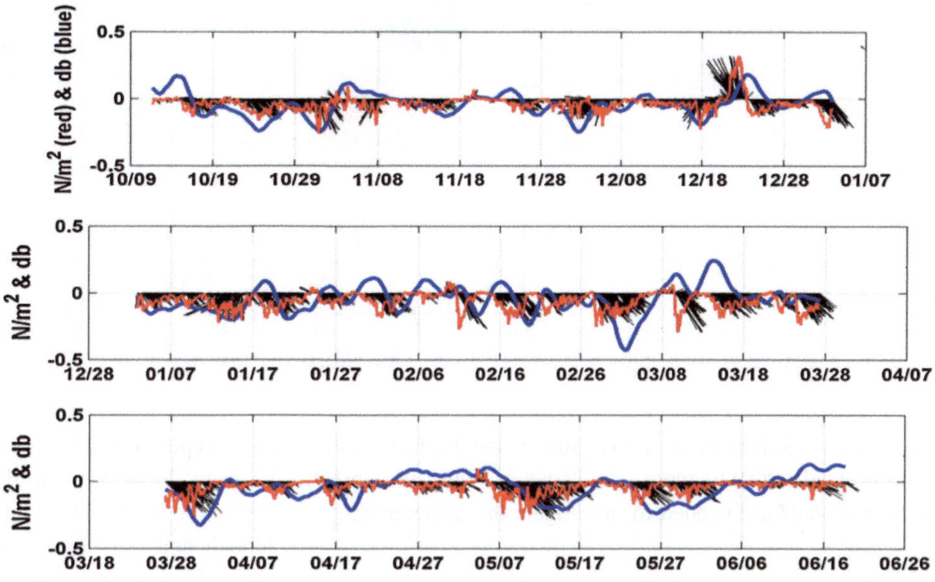

Fig. 6 Buoy wind data stick plot; low-pass-filtered alongshore wind stress leads to "Ekman transport" (*red*), and the *blue line* represents the bottom pressure (Limeburner et al. 2012)

coastline (Abdelrahman 1997; Eltaib 2010). The cross-shore component is directed towards the east causing an increase in sea level all the year around. The minimum value occurs during July when the long-shore component is dominant and directed southward causing a sea-level decrease during summer time. On the other hand, the maximum value occurs during December and January when the long-shore wind stress reverses from the SSE, driving surface currents from the Gulf of Aden into the Red Sea. Patzert (1974) computed the steric effect at Perim in Bab-al-Mandab and showed positive steric effect and high sea level during the winter season.

Sea-level changes associated with eddies can be detected. Satellite-tracked surface drifters, released during the spring of 2010, detected a well-defined anticyclonic eddy around 23°N. Chen et al. (2014) used a finite-volume community ocean model (FVCOM) to investigate the formation of the eddy. The model was driven by the meteorological forcing at the sea surface together with a given tidal elevation at the open boundary and was capable of reproducing the observed anticyclonic eddy with the same location and size. The examination of the anticyclonic eddy shows that wind forcing did not play a critical role in the eddy formation; however, it may significantly impact the structure and intensity of the eddy. South of this eddy, around 19°N, the Tokar Wind Jet, some 150 km south of Port Sudan, during summer season plays a major rule in the formation of a dipolar eddy during the summer season (Figs. 7 and 8). The cyclonic and anticyclonic eddies are highly influenced by the strength of the Jet (Zhai and Bower 2013).

Regionally coherent sea-level changes may be driven by changes outside the Red Sea. These longer-period changes are coherent over a wide area, including Port Sudan on the

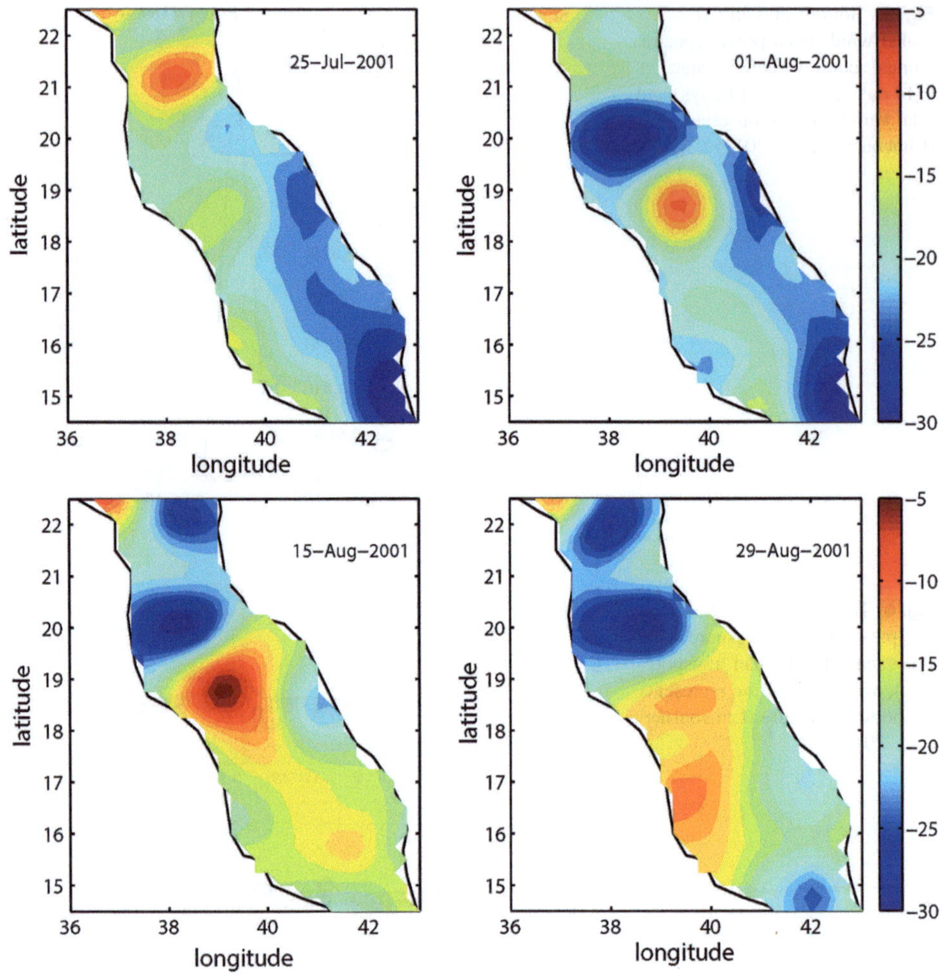

Fig. 7 The sequence development of sea-level anomaly (SLA) during 25 July to 29 August 2001, by the Tokar Wind Jet and the formation of cyclonic and anticyclonic eddies (Zhai and Bower 2013)

west coast (Sultan et al. 1996; Sultan and Elghribi 2003; Pugh et al. 2001); more simultaneous and widespread measurements are needed to investigate this coherence.

Seasonal and Longer Changes

As discussed above, the sea level in the Red Sea has annual and semiannual variations (see Fig. 5 for the dominant annual blue peak and the smaller semiannual peak) due to the seasonal variations in atmospheric parameters, in the evaporation rate and in the seasonal water mass exchange with the Gulf of Aden. These seasonal variations of the sea level account for more than half the total variance in the central and southern parts of the Red Sea (Sultan and Elghribi 2003; Cromwell and Smeed 1998; see also Sultan et al. 1995a).

The annual variability timescale in the central region at KAUST, where the bottom pressure showed approximately 40-cm difference between a high in winter (January–February) and a low in summer (July–August), can be related to wind stress. The semiannual variation is possibly related to the evaporation (Sofianos and Johns 2003; Sultan and Elghribi 2003; Cromwell and Smeed 1998; Sultan et al. 1995a, b, 1996). The amplitudes of the annual variations at Jeddah and Port Sudan are about 20–13 cm, respectively, while the semiannual components are about 10–8 cm. The variations at Jeddah are related to the long-shore wind component only. At Port Sudan, the long-shore and cross-shore components are significantly affecting the sea-level seasonal variations (Sultan et al. 1995b), with the maximum levels occurring in December–February and the lowest levels in June–August.

Normally, in the Northern Hemisphere, the seasonal cycle of sea levels peaks in late summer, as the sea warms, and expands causing steric levels to rise. In the Red Sea, the reverse is true, as is also the case for the Black Sea, where excess of winter freshwater inflow from rivers gives winter highs and dominates over any steric summer expansion. In the case of the Black Sea, there is a long adjustment time for the river inflow to reach a balance with the Mediterranean connection. Similarly, the Red Sea seasonal sea-level changes are dominated by the evaporation/precipitation balance and regional winds, complicated by the narrow

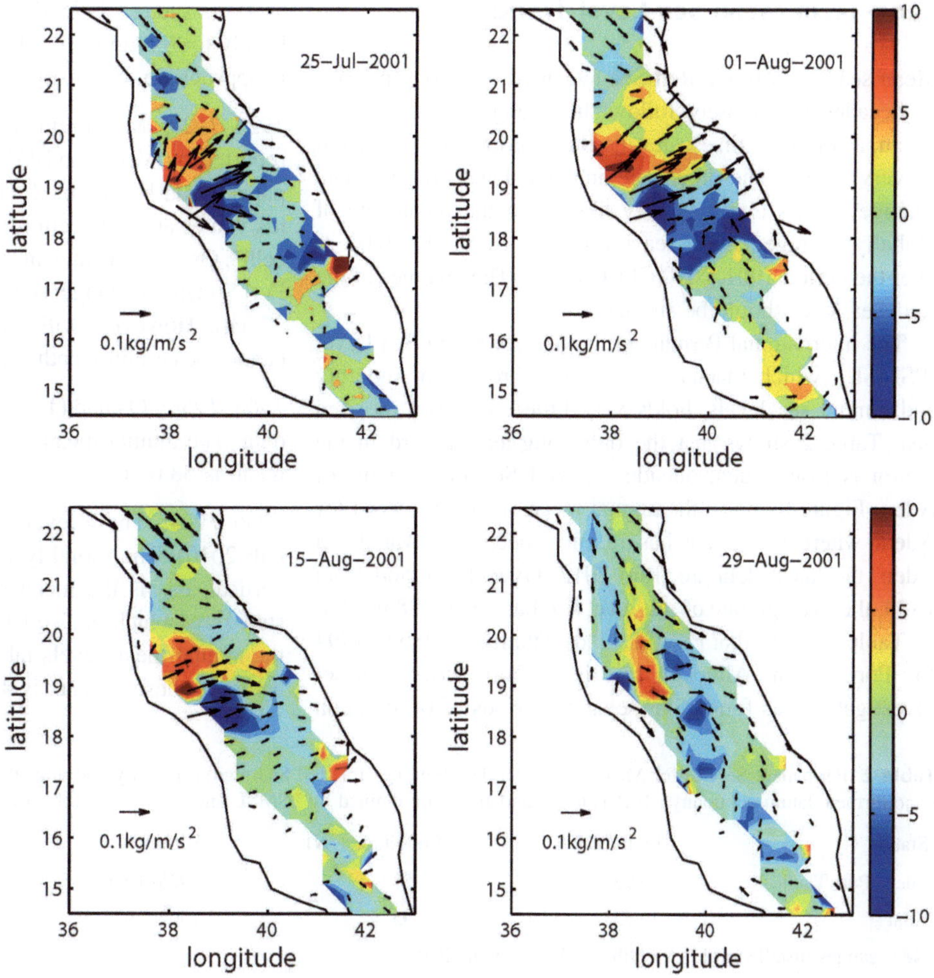

Fig. 8 The sequence development of the QuickSCAT wind stress and the wind stress curl (unit 10^{-7} N/m^3) to the same period in Fig. 4; cyclonic eddy develops with (−) SLA, and anticyclonic eddy develops with (+) SLA (Zhai and Bower 2013)

connection to the Gulf of Aden. This reversal of phase is also opposed to the seasonal cycle of air pressures and dominates over a normal ocean inverted barometer pressure effect on sea level (Osman 1984). Annual evaporation in the central Red Sea was estimated at around 2 m per annum (Morcos 1970; Ahmad and Sultan 1989), while there is very little precipitation or river inflow. Sofianos and Johns (2001) with altimeter data demonstrated that the annual cycle of sea level in the central and northern Red Sea can be explained as a balance between sea surface elevations and wind stress.

These sea-level variations are well reproduced in the data assimilation program SODA (http://www.atmos.umd.edu/~ocean; Carton and Giese 2008). SODA ocean modelling assimilates data from many sources including temperature/salinity profiles and altimetry data. Manasrah et al. (2009) have analysed the Red Sea SODA data output from 1958 to 2001. They verified the outputs by comparing them with the sea-level records for Port Sudan for 1986–1994. Sea-level data from three different parts (north, central, and south) for 43 years showed, compared to the other parts, that the sea level in the northern part of the Red Sea is permanently lower by about 7–8 cm. Overall, the sea levels in the three regions are significantly higher in the winter and spring compared with the summer and autumn. Furthermore, the mean sea level during the summer is significantly lower than that in the winter by about 10–12 cm. The increasing sea level from the north to the south and from summer to winter is related to the water mass exchanges with the Gulf of Aden during both seasons. Therefore, the coherent Red Sea response is due to the influx of water from the Gulf of Aden at the surface during winter with less salinity and warmer temperature which produces positive steric effect compared with the summer season when the Red Sea overflow water (RSOW) exits to the Gulf of Aden. A quasi 2.5–2.7-year oscillation was identified in these analyses, with varying amplitudes and without any identified forcing function. Mohamad (2012) has also analysed Red Sea levels as encapsulated in the SODA data. Some information on long-term changes in deep-water steric effects is included in Alraddadi (2013).

Long-Term Mean Sea-Level Trends

Mean sea levels in the Red Sea averaged over long periods of decades will closely follow the changes in the global ocean, as on these timescales the Bab-al-Mandab connection is very open. The Intergovernmental Panel on Climate Change reports that it is very likely that the mean rate of globally averaged mean sea levels rose by 0.19 ± 0.02 m over the period 1901–2010 (IPCC 2013). The changes of the Red Sea level should be similar.

The international Permanent Service for Mean Sea Level (PSMSL), which maintains the global archive of monthly and annual sea levels, holds very little data from the Red Sea. Table 2 shows that the only long-term record in the region is from Aden, outside the Red Sea in the Gulf of Aden. Figure 9 shows the annual mean sea-level record for Aden, where the gauge has recently been reactivated. At Aden, the latest data are from 2012. Over the period from 1880, the average rate of sea-level rise has been 2.88 mm/yr.

Table 2 shows that PSMSL holds data from 1986 to 1994 for Port Sudan. Although this is a short period, further investigation has found four separate periods of observation at Port Sudan: 1925–1928, 1962–1968, 1986–1994, and briefly for 6 months in 2013. In the context of this chapter, it is appropriate to consider these collectively in more detail.

1925–1928: Measurements of sea levels at Port Sudan are reported by Vercelli (1931). Unfortunately, no copy of this paper could be obtained. However, Osman (1984) says that his values are "21.8 cm higher than the value listed by Vercelli". Osman's mean value was 58.8 cm above the zero of the Port Sudan tide gauge, so that Vercelli's mean level would be 37 cm. However, without reading the earlier papers, we cannot be certain whether the same datum level applies.

1962–1968: Osman (1984) gives a detailed analysis of these data. The annual means are given in Table 3a. The overall mean is 58.6 cm.

1986–1994: These levels have been analysed in detail (Eltaib 2010). The monthly means are plotted in Fig. 3b. The vertical axis is the mean sea level, presumably to the tide staff zero, which is also Chart Datum. Table 3b summarizes the annual mean levels tabulated by the PSMSL. The average over these 7 years is 44.4 cm.

Table 2 Permanent Service for Mean Sea Level data holdings for Red Sea region monthly and annual mean sea levels. Metric quality means unconfirmed datum continuity. RLR is to a standard datum defined by PSMSL and has a more robust datum history

Station	Latitude (°E)	Longitude (°N)	Period	Completeness (%)	Quality
Suez (Port Taufiq)	29.93	32.57	1925–1986	32	Metric
Eilat	29.55	34.95	1962–1971	90	Metric
New gauges installed at Eilat North and Eilat South, 2010					RLR
Port Sudan	19.63	37.12	1986–1994	95	RLR
Aden	12.78	45.00	1879–2012	51	RLR
Djibouti	11.58	43.15	1970–2012	15	Metric

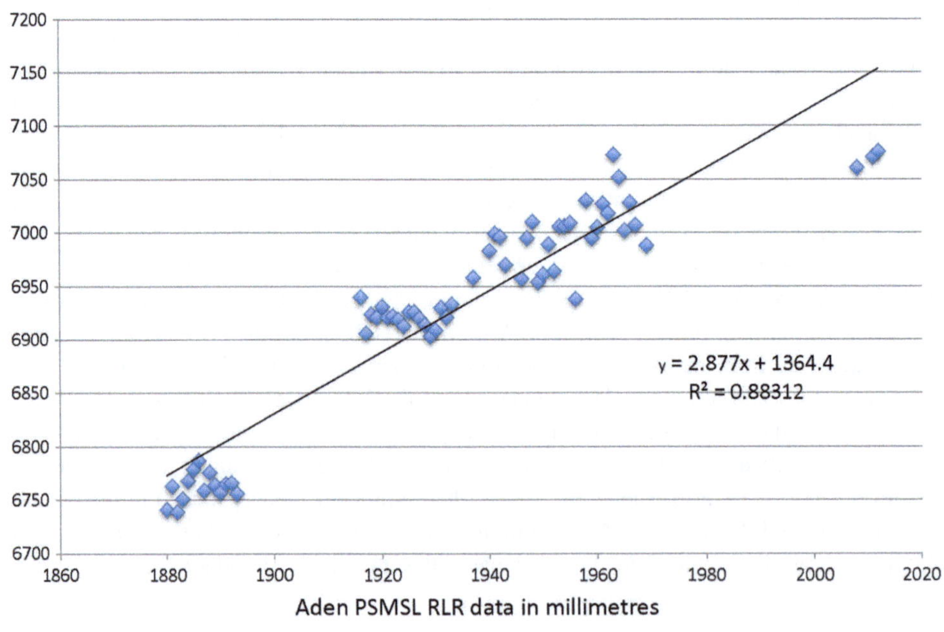

Fig. 9 The long-term sea-level record from Aden

Table 3 Annual (a) 1962–1968 and (b) 1987–1994 mean sea levels at Port Sudan

(a)	1962	1963	1964	1965	1966	1967	1968
Annual mean (cm)	60.3	62.7	56.0	57.8	59.6	54.9	58.8

(b)	1987	1988	1989	1990	1991	1992	1993	1994
Annual mean (cm)	45.2	45.7	44.9	42.4	45.7	38.2	45.8	50.2

Table 4 Summary of mean sea levels for all four measurement periods at Port Sudan

	1925–1928	1962–1968	1987–1994	2013 (part, adjusted)
Mean (cm)	37	58.8	44.4	48

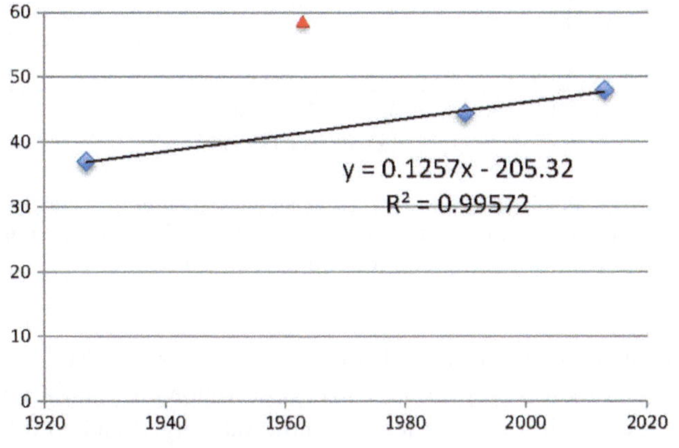

Fig. 10 Plot of sea-level annual means at Port Sudan, assuming the same datum for different epochs

2013: Measurements taken in 2013 were part of a study of amphidrome development in the vicinity of Port Sudan. The RBR-manufactured gauge measured sea bed pressures at a jetty adjacent to the permanent tide gauge location, from February to September. These seabed pressures are converted to readings from the permanent tide staff adjacent to the tidal building, using a series of staff readings taken at 0900 (L) every morning from 22 April to 31 July by Elfatih Bakry Ahmed Eltaib, the physical oceanographer on the staff of the adjacent Institute of Marine Research. Having only 6 months of 2013 data makes it impossible to draw conclusions about recent trends. Speculatively, we can adjust our half-year period over which the measurements were taken, to an average annual cycle of sea level, using the curve of average monthly mean sea levels at Port Sudan, in Osman (1984). On average, over these 6 months, the mean sea level is about 4 cm lower than the annual average level, so we might adjust our 43.71 cm upwards for an estimated annual mean of 48 cm. It would be valuable to continue these recent readings.

The mean sea levels of the four periods are summarized in Table 4 and plotted in Fig. 10.

In conclusion, the 2013 levels look reasonable in comparison with earlier data. However, the 1962–1968 levels seem 10–15 cm too high, beyond any statistically acceptable variability.

Tentatively ignoring the 1960s' data, and assuming the same datum (Chart Datum) for the other periods, gives a trend over the other measurements of 1.26 mm/yr from 1925 to 2013, equivalent to 12.6 cm per century, which agrees favourably with global trends over the same period. Local vertical land movement also affects measured trends.

The Geological Record

Mean sea-level changes are measured locally relative to some fixed land datums. Only with the advent of satellite altimetry has it been possible to measure sea levels in absolute geocentric terms. Of course, these sea-level measurements relative to the land include both the vertical sea surface and vertical land movements. Generally, over a few decades, sea levels change more rapidly than land levels, but over centuries and millennia, the land movements are increasingly important and eventually dominant. Apart from plate and crustal tectonic movements, by far the major influence on land levels is the adjustment to crustal loading that occurs during global glacial cycles, where polar ice mass becomes a globally distributed mass of sea water. The additional local water mass can depress local coastal areas substantially. Lambeck et al. (2011) have applied global ice load and crustal models to evaluate how the glacial cycles affect the Red Sea and land levels. They have looked in detail at relative sea levels in the Red Sea during the Holocene, during the last glacial maximum and during the last interglacial period (see also Bailey, this volume).

During the Holocene, that is, roughly the past 10,000 years, levels were perhaps a few metres lower than today, but not evenly so along the whole Red Sea, because of the different land loading effects. Figure 11a, b shows a section from Port Said to Socotra and the relative sea levels

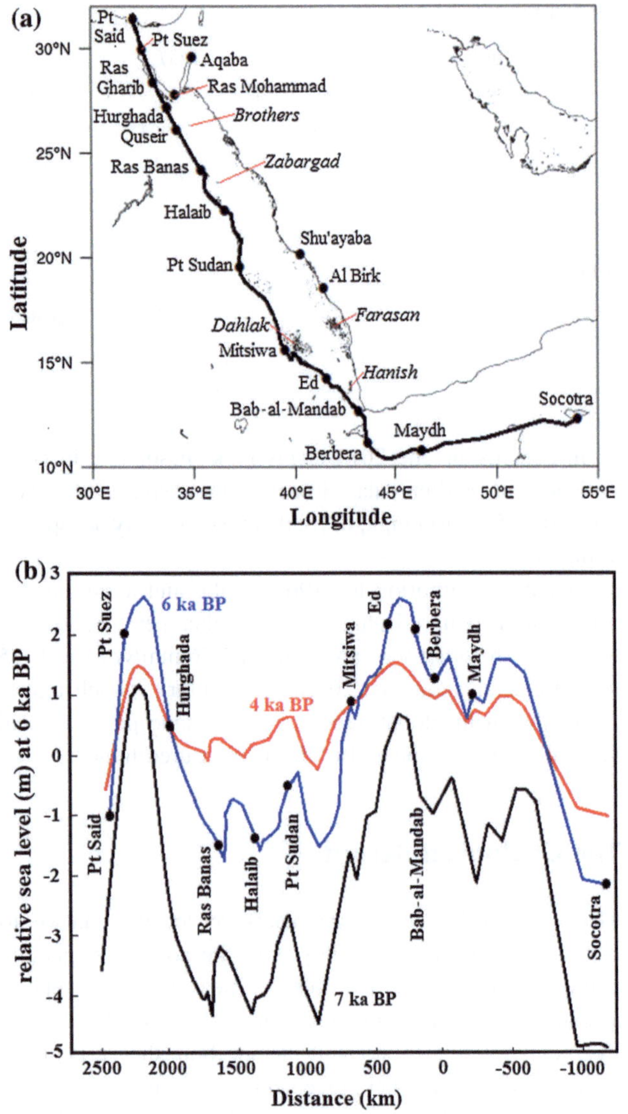

Fig. 11 **a** Location of selected sites and the section corresponding to the profiles in Fig. 11b (Reprinted from Lambeck et al. 2011, with permission from Elsevier) and **b** representative predictions for mid-Holocene sea-level change along the section in Fig. 11a (Reprinted from Lambeck et al. 2011, with permission from Elsevier)

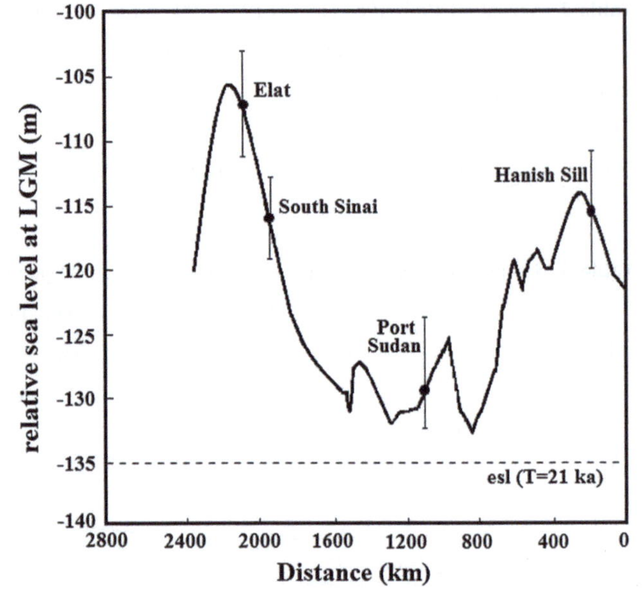

Fig. 12 Predicted LGM sea levels along the section in Fig. 11a. The *dashed horizontal line* is the model level at 21 ka BP (Reprinted from Lambeck et al. 2011, with permission from Elsevier)

along the section some 7,000 years ago (7 ka BP), 6,000 years ago and 4,000 years ago. There was a strong rise between 7,000 and 6,000 years ago and then a slower rise or fall, depending on location, to today's levels. There is observational evidence in the form of raised coral platforms to support this.

During the last glacial maximum, some 20,000 years ago, globally the water locked in the polar ice caps reduced sea levels by about 125 m. In the Red Sea, allowing for the regionally variable crustal response to reduced sea mass loading locally as isostatic loading adjustments, levels ranged from −110 m in south Sinai, to −128 m south of Port Sudan, relative to the present day. Figure 12 shows the relative curve, but note that the Gulf of Suez (shaded) was dry at these lower levels. Direct observational evidence for these much lower sea levels is now submerged and so difficult to find, though some submerged terraces have been tentatively identified, which may be interpreted as supplying evidence.

The last interglacial extreme maximum, some 125,000 years ago, when the polar ice caps were at their minimum extent, probably produced Red Sea levels some 5–7 m above today's mean sea levels, also with regional vertical land movement differences. This is confirmed by field evidence for reefs some 5–8 m above the present coast level, certainly in the northern and central Red Sea.

There is no evidence through the period of these studies that the Red Sea has been cut off from the Gulf of Aden. If this happened, given the very high local evaporation rates (about 2 m/year), and low precipitation, the Red Sea levels would fall relatively rapidly, with very high salinities as a result. However, the corresponding sediments have not been found. Nevertheless, levels at the southern entrance, the Hanish Sill, at the last glacial maximum some 20,000 years ago, were much reduced. Lambeck et al. (2011) estimate that the depth maximum at the sill may have been 25 ± 4 m and that the cross-sectional area of the much reduced connecting channel could have been only about 2 % of the cross-sectional area today.

As global sea levels have risen and fallen by many tens of metres over the geological Quaternary Period, the hydraulic connection between the Red Sea and the Gulf of Aden has increased and reduced accordingly. More limited water exchange at times when sea levels were lower, that is, at times of maximum glaciation, resulted in higher Red Sea salinities and changes in the oxygen isotope composition (Siddall et al. 2004). By relating the oxygen ^{16}O to ^{18}O isotope ratio in the foraminifera in Red Sea sediment cores, it has been possible to indirectly estimate the water depths at the Hanish Sill, which was the controlling factor for water exchange through this Gulf of Aden connection. These depths in turn may be considered to represent global sea levels. Hence, Red Sea sediments contain a valuable record of *global* sea-level changes through a succession of glacial cycles over the past 550,000 years (Siddall et al. 2003).

Tsunamis

The major tsunami disasters in the adjacent Indian Ocean make it a little surprising that very few tsunamis have been reported for the Red Sea. This is despite the region being one of the frequent seismic activities, as the central Red Sea axis is an opening mid-ocean ridge, with associated activity. The few records of actual tsunami generation in the region are summarized in Jordan (2008). They include the inundation at El Tor, Egypt, on the Sinai Peninsula in the Gulf of Suez on 11 July 1879. A landslide is a possible source for this tsunami. The second reported incident was at Mitsiwa, Eritrea, on 20 July 1884, when an earthquake occurred at sea offshore. Reportedly, sea waves built up in the Massawa harbour, mostly between the localities known as Taulud and Edaga Barai. The waves swept over a causeway, and ships in the harbour were seen swaying violently. Multiple flooding from the sea over land left dead fish onshore (Ambraseys et al. 1994). From this evidence, it appears that generally the magnitude and character of local earthquakes are not favourable for frequent tsunami generation.

Summary

Red Sea tides are relatively small but scientifically challenging. Weather effects are at least as important as tidal changes. Seasonal sea levels are out of phase with the general Northern Hemisphere pattern of maximum levels in late summer. Instead, seasonal changes are dominated by weather and water exchange processes. Despite an acute sparsity of sea-level observations, records from Port Sudan hint at increases over the past century compatible with global increases, but only if a continuous datum is assumed. Over geological periods, the effects of glacial cycles have raised levels by up to 7 m and lowered them by as much as 128 m, from today's levels.

Looking ahead, global mean sea level will continue to rise during the twenty-first century, and the Red Sea will follow this trend (IPCC 2013). The rate of rise will very likely exceed that observed globally during 1971–2010 (2.0 mm/yr), due to continuing global warming and loss of ice from glaciers and ice sheets. The impacts of these changes on local flooding risks are not easily determined because of the limited availability of sea-level data.

References

Abdelrahman SM (1997) Seasonal fluctuations of mean sea level at Gizan, Red Sea. Coastal Res 13:1166–1172

Ahmad F, Sultan SAR (1989) Surface heat fluxes and their comparison with the oceanic heat flow in the Red Sea. Oceanol Acta 12(1):33–36

Alraddadi TM (2013) Temporal changes in the Red Sea circulation and associated water masses. Ph.D. thesis, University of Southampton. http://eprints.soton.ac.uk/id/eprint/355542

Ambraseys NN, Melville CP, Adams RD (1994) The seismicity of Egypt, Arabia and the Red Sea. Cambridge University Press, New York, 181 pp

Carton JA, Giese B (2008) A reanalysis of ocean climate using simple ocean data assimilation (SODA). Mon Weather Rev 136:2999–3017

Chen C, Ruixiang L, Pratt L, Limeburner R, Beardsley R, Bower AS, Jiang H, Abualnaja Y, Liu X, Xu Q, Lin H, Lan J, Kim TW (2014) Process modeling studies of physical mechanisms of the formation of an anticyclonic eddy in the central Red Sea. J Geophys Res Oceans 119:1445–1464

Clifford M, Horton C, Schmitz J, Kantha LH (1997) An oceanographic nowcast/forecast system for the Red Sea. J Geophys Res 102:25101–25122

Cromwell D, Smeed DA (1998) Altimetric observations of sea level cycles near the Strait of Bab al Mandab. Int J Remote Sens 19:1561–1578

Defant A (1961) Physical oceanography, vol II. Pergamon Press, Oxford, 598 pp

Eltaib EBA (2010) Tides analysis in the Red Sea in port Sudan and Gizan. M.Sc., thesis, University of Bergen Geophysical Institute

Grace S (1930) The semi-diurnal tidal motion of the Red Sea. Monthly Notices of the Royal Astronomical Society, Supplement (March), 274 pp

IPCC (2013) Climate change. The physical science basis. http://www.ipcc.ch

Jarosz E, Murray SP, Inoue M (2005) Observations on the characteristics of tides in the Bab el Mandab Strait. J Geophys Res 110: C03015

Jordan BR (2008) Tsunamis of the Arabian peninsula: a guide of historical events. Sci Tsunami Hazards 27:31–46

Lambeck K, Purcell A, Flemming NC, Vita-Finzi C, Alsharekh AM, Bailey GR (2011) Sea level and shoreline reconstructions for the Red Sea: isostatic and tectonic considerations and implications for hominin migration out of Africa. Quatern Sci Rev 30:3542–3574

Limeburner R, Abualnaja Y, Beardsley R (2012) Bottom pressure variability in the Red Sea. EGU General Assembly 2012, held 22–27 Apr 2012 in Vienna, Austria:11907

Manasrah R, Hasanean HM, Al-Rousan S (2009) Spatial and seasonal variations of sea level in the Red Sea, 1958-2001. Ocean Sci J 44 (3):145–159

Mohamad KA (2012) Sea level variation in the Red Sea based on SODA data. M.Sc. thesis, University of Bergen Geophysical Institute

Monismith SG, Genin A (2004) Tides and sea level in the Gulf of Aqaba (Eilat). J Geophys Res 109:C04015

Morcos SA (1970) Physical and chemical oceanography of the Red Sea. Oceanogr Mar Biol Annu Rev 8:73–202

Osman MM (1984) Variations of at Port Sudan. International Hydrographic Review LXI, pp 137–144

Patzert WC (1974) Wind-induced reversal in Red Sea circulation. Deep Sea Res 21(2):109–121

Pedgley DE (1974) An outline of the weather and climate of the Red Sea. L'oceanographie physique de la Mer Rouge. CNEXO, Paris, pp 9–27

Proudman J (1953) Dynamical oceanography, Methuen. London, 410 pp

Pugh DT (1981) Tidal amphidrome movement and energy dissipation in the Irish Sea. Geophys J Roy Astron Soc 67:515–527

Pugh DT, Woodworth PL (2014) Sea-level science: understanding tides, surges, tsunamis and mean sea level changes. Cambridge University Press, New York, 395 pp

Pugh DT, Dixon DJ, Woodworth PL (2001) Marine impacts of potential climate change. J Fac Mar Sci, King Abdulaziz Univ 12:3–11

Siddall M, Rohling EJ, Almogi-Labin A, Hemleben C, Meischner D, Schmeizer I, Smeed DA (2003) Sea-level fluctuations during the last glacial cycle. Nature 423:853–858

Siddall M, Smeed DA, Hemleben C, Rohling EJ, Schmelzer I, Peltier WR (2004) Understanding the Red Sea response to sea level. Earth Planet Sci Lett 225:421–434

Sofianos SS, Johns WE (2001) Wind induced sea level variability in the Red Sea. Geophys Res Lett 28:3175–3178

Sofianos SS, Johns WE (2003) An Oceanic General Circulation Model (OGCM) investigation of the Red Sea circulation: 2. Three-dimensional circulation in the Red Sea. J Geophys Res 108:3066

Sultan SAR, Elghribi NM (2003) Sea level changes in the central part of the Red Sea. Indian J Mar Sci 32:114–122

Sultan SAR, Ahmad F, Elghribi NM (1995a) Sea level variability in the central Red Sea. Oceanological Acta 18(6):607–615

Sultan SAR, Ahmad F, El-Hassan A (1995b) Seasonal variations of the sea level in the central part of the Red Sea. Estuar Coast Shelf Sci 40(1):1–8

Sultan SAR, Ahmad F, Nassar D (1996) Relative contributions of external sources of mean sea-level variations at Port Sudan, Red Sea. Estuar Coast Shelf Sci 42(1):19–30

Vercelli F (1931) Nuovericerche sullen correnti marine nel Mar Rosso. AnnaliIdrografico 12:424–428

Zhai P, Bower A (2013) The response of the Red Sea to a strong wind jet near the Tokar Gap in summer. J Geophys Res 118(1):421–434

Air–Sea Interaction and Horizontal Circulation in the Red Sea

Amy S. Bower and J. Thomas Farrar

Abstract

This chapter discusses the horizontal circulation of the Red Sea and the surface meteorology that drives it, and recent satellite and in situ measurements from the region are used to illustrate properties of the Red Sea circulation and the atmospheric forcing. The surface winds over the Red Sea have rich spatial structure, with variations in speed and direction on both synoptic and seasonal timescales. Wintertime mountain-gap wind jets drive large heat losses and evaporation at some locations, with as much as 9 cm of evaporation in a week. The near-surface currents in the Red Sea exhibit similarly rich variability, with an energetic and complex flow field dominated by persistent, quasi-stationary eddies, and convoluted boundary currents. At least one quasi-stationary eddy pair is driven largely by winds blowing through a gap in the mountains (Tokar Gap), but numerical simulations suggest that much of the eddy field is driven by the interaction of the buoyancy-driven flow with topography. Recent measurements suggest that Gulf of Aden Intermediate Water (GAIW) penetrates further northward into the Red Sea than previously reported.

Air–Sea Interaction

Regional Meteorology

The water circulation of the Red Sea is driven by the exchange of momentum, heat, and freshwater with the overlying atmosphere. The continental landmasses surrounding the Red Sea exert a profound influence on the regional meteorology and air–sea fluxes and hence on the Red Sea circulation. The relatively narrow basin (maximum width 355 km) is located between Africa and the Arabian subcontinent, and the landmasses surrounding the Red Sea can be classified as arid land, desert, or semi-desert (Naval Oceanography Command Detachment 1993). The dry air flowing over the Red Sea from these arid landmasses helps to make it a site of intense evaporation, with the basin-averaged evaporation estimated to be about 2 m/yr (Sofianos et al. 2002).

A.S. Bower (✉) · J.T. Farrar
Department of Physical Oceanography, Woods Hole Oceanographic Institution, Woods Hole, MA, USA
e-mail: abower@whoi.edu

The coastal mountain chains bordering the Red Sea along both the African and Arabian coasts also exert significant influence on the regional meteorology, primarily by constraining the mean, large-scale surface winds to flow along the axis of the basin (Fig. 1; Patzert 1974). The seasonal winds blow toward the southeast all year in the northern part of the basin (north of 19–20°N), but in the southern part of the basin, the winds reverse from northwesterly in summer to southeasterly in winter under the influence of the two distinct phases of the Arabian monsoon (Fig. 1; Pedgley 1974; Patzert 1974; Clifford et al. 1997; Sofianos and Johns 2003; Jiang et al. 2009).

This seasonally reversing wind is associated with substantial spatial variations in mean wind speeds, with speed variations of about a factor of two between different parts of the basin (Fig. 1; using the 10-year climatology of QuikSCAT scatterometer winds of Risien and Chelton 2008). The monsoonal circulation and associated wind reversal in the southern part of the basin also contribute to distinct regional differences in the annual cycle of wind speed; the south-central part of the basin (16–20°N) experiences the strongest winds in summer, but the rest of the basin, especially the

N.M.A. Rasul and I.C.F. Stewart (eds.), *The Red Sea*,
Springer Earth System Sciences, DOI 10.1007/978-3-662-45201-1_19,

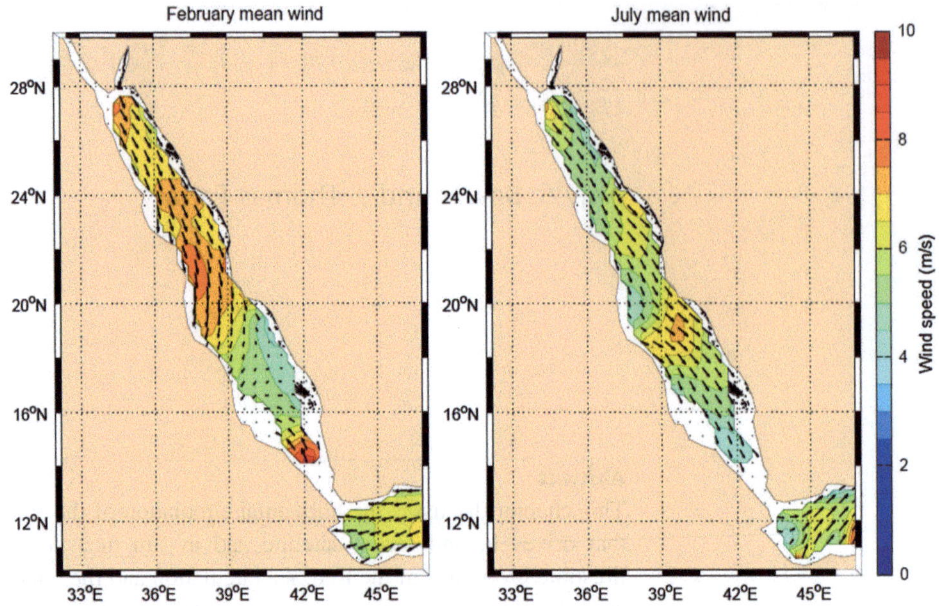

Fig. 1 Climatological monthly mean surface winds for February (*left*) and July (*right*), at the two extremes of the annual cycle of surface winds in the Red Sea. The *colored* contours indicate wind speed, and the vectors depict the direction and relative speed. The climatological means are based on data from the QuikSCAT satellite scatterometer from September 1999 to October 2009 (Risien and Chelton 2008)

northern and southern ends of the Red Sea, experiences the strongest winds in winter (Fig. 1). During winter, the surface winds converge in the south-central part of the basin (roughly 16–20°N), leading to relatively weak wind speeds (<4 m/s) in this region. Some of the converging air flows out of the basin through the Tokar Gap and other gaps in the coastal mountain ranges (e.g., Jiang et al. 2009). During summer, the prevailing winds are more spatially uniform, both in speed and direction.

Gaps in the coastal mountain chains also affect the surface wind field on smaller scales, with intense mountain-gap wind jets blowing across the Red Sea from Africa during summer (e.g., Hickey and Goudie 2007) and from the Arabian subcontinent during winter (Jiang et al. 2009). Figure 2, from Jiang et al. (2009), illustrates the structure of the surface wind field during the summertime and wintertime wind jet events in simulations carried out using the Weather Research and Forecasting (WRF) model. The effect of these strong transient wind events on the currents of the Red Sea is not well understood, but field measurements from one site in the central Red Sea suggest that the effect of these wind jets on the air–sea exchange of heat and freshwater is significant, as is discussed in the next subsection.

Heat and Freshwater Fluxes

The mean circulation of the Red Sea is believed to be driven primarily by surface buoyancy fluxes associated with the mean loss of heat and freshwater to the atmosphere, both of which act to make the surface waters of the Red Sea denser (Sofianos and Johns 2003). The freshwater budget of the Red Sea is dominated by evaporation, which tends to make the Red Sea saltier, and by the flow of water through the Strait of Bab-al-Mandab, where relatively freshwater flows in at the surface from the Gulf of Aden and salty Red Sea overflow water spills from the Red Sea to the Gulf of Aden at depth (e.g., Murray and Johns 1997).

The best available constraints on the areal-average surface fluxes of heat and freshwater in the Red Sea come from observations of currents and water properties at the Strait of Bab-al-Mandab, used together with the fact that the heat and freshwater transport through this strait must be balanced by the air–sea fluxes of heat and freshwater over the Red Sea (e.g., Tragou et al. 1999; Sofianos et al. 2002). Sofianos et al. (2002) estimated the areal-average heat and freshwater loss to the atmosphere to be 11 ± 5 W/m^2 and 2.06 ± 0.22 m/yr. More direct estimates of the basin-average surface heat and freshwater fluxes in the Red Sea from meteorological measurements and bulk flux formulas have proven problematic because of measurement and sampling errors in the meteorological observations available from voluntary observing ships and in the bulk formulas for solar and infrared radiative fluxes (e.g., Tragou et al. 1999).

Accurate and direct estimates of the air–sea fluxes of heat and freshwater at a site in the central Red Sea (22°9′N, 38°30′W; location indicated in Fig. 2) comes from detailed measurements of surface meteorology and solar and infrared radiation from a heavily instrumented air–sea interaction mooring deployed by the Woods Hole Oceanographic Institution (WHOI) in collaboration with the King Abdullah University of Science and Technology (KAUST). The surface buoy carried instruments to measure wind, humidity, temperature, precipitation, air pressure, sea surface temperature, solar (or “shortwave”) radiation, and infrared (or “longwave”) radiation (Farrar et al. 2009), which have been used together with the COARE bulk flux algorithm (Fairall

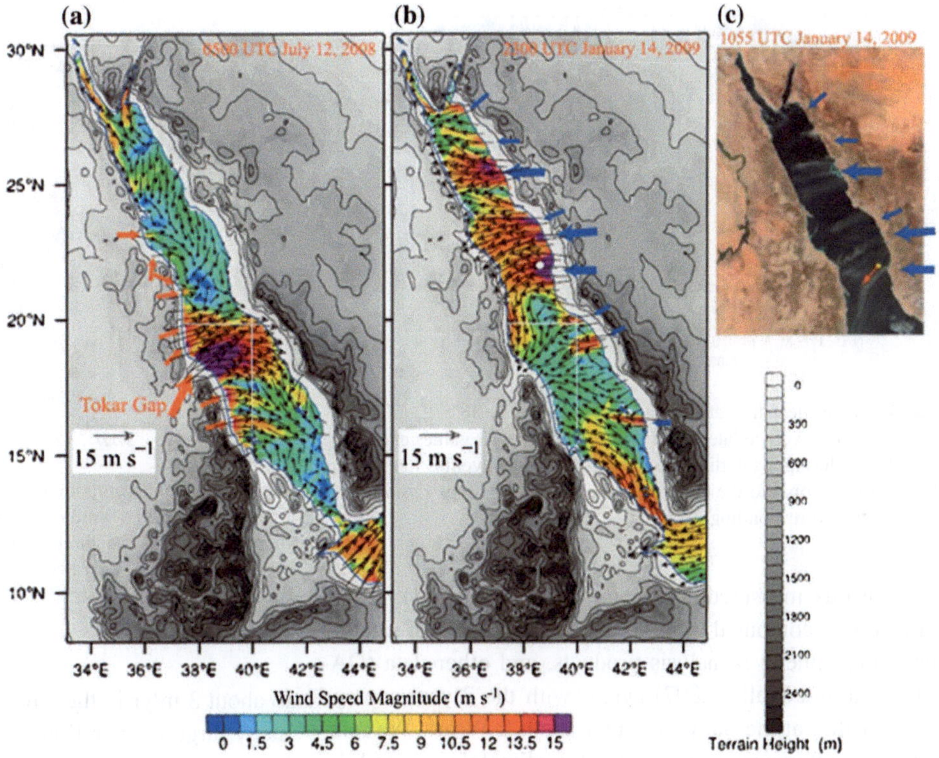

Fig. 2 Surface winds (10 m height) in high-resolution WRF model simulations for July 12, 2008 (**a**) and January 14, 2009 (**b**), (**c**) visible imagery taken by the MODIS instrument on NASA's Aqua satellite on January 14, 2009, showing dust plumes being carried over the Red Sea from the Arabian Peninsula. The location of the air–sea interaction buoy is shown by a *small circle* in panels (**b**) and (**c**). From Jiang et al. (2009)

et al. 2003) to estimate the air–sea heat fluxes of heat, momentum, and freshwater (Fig. 3). The mean evaporation rate at the site was about 1.98 m/yr, and the mean net heat flux was about −14 W/m^2.

These measurements at one location do not provide any direct constraints for the basin-average fluxes discussed in the previous paragraph, but they can provide some insight into the nature of the errors in available surface flux products. For example, the 2 years of buoy measurements are compared to two surface heat flux climatologies in Fig. 3. A flux climatology derived from the database of weather observations from ships (Josey et al. 1999) overestimates the net heat flux at the site by about 60 W/m^2, having a mean value that is about +44 W/m^2 (whereas the

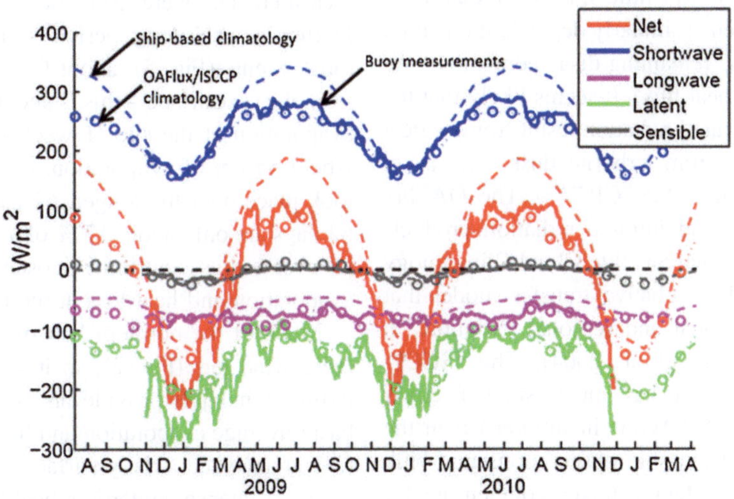

Fig. 3 Air–sea heat flux components (*solid lines*) estimated from measurements of surface radiation, and meteorology on an air–sea interaction buoy maintained in the central Red Sea by WHOI and KAUST (Farrar et al. 2009). The *dashed lines* show climatological values estimated from weather observations reported by volunteer observing ships (Josey et al. 1999), and the *dotted lines* with large circles show climatological values estimated from a combination of satellite data, atmospheric reanalysis products, and other data (Yu and Weller 2007) The estimates of shortwave (or solar) and longwave (or infrared) radiation from the ship-based climatology are based on bulk formulas rather than direct measurements, and these quantities account for most of the disagreement with the measured values

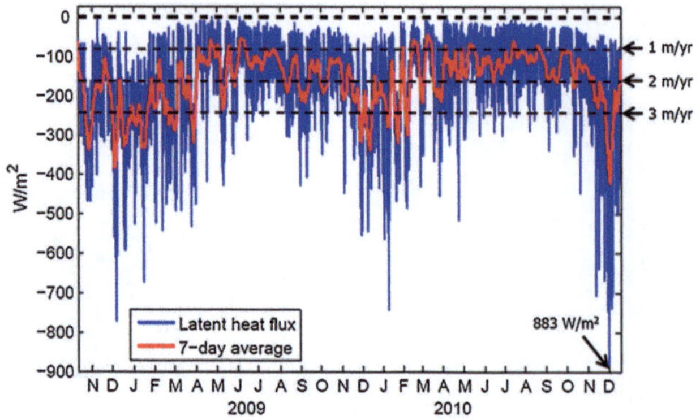

Fig. 4 Latent heat flux at the buoy site (22.15°N, 38.5°E). The *blue line* shows 1 h average latent heat fluxes, which reach values exceeding 500 W/m^2 during wintertime mountain-gap wind events. Latent heat flux is proportional to the evaporation rate, and the *black dashed lines* indicate the corresponding evaporation rates

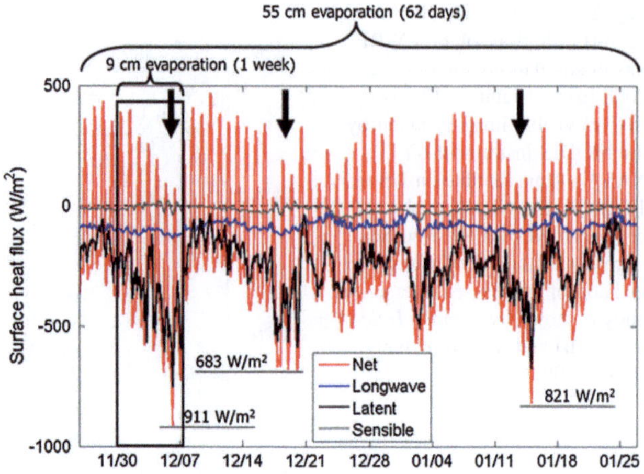

Fig. 5 Heat flux components at the buoy site during a period spanning three mountain-gap wind events in the winter of 2008–2009. The times of occurrence of the three wind events are indicated by *heavy black arrows*. During the first event in early December, there was about 9 cm of evaporation, about 4.5 % of the annual average evaporation at the site

net heat flux measured at the buoy was −14 W/m^2). A flux climatology computed from an objective analysis of satellite data, atmospheric reanalysis products, and other data (OA-Flux; Yu and Weller 2007) agrees with the 2-year-average net heat flux at the buoy to within about 2 W/m^2. We refer below to differences between the flux climatologies and the buoy records as "errors," but we note that some differences between the climatological surface fluxes and the fluxes over a particular 2-year time period should be expected. About half of the error (+30 W/m^2) in the ship-based climatology comes from the estimate of the solar radiation, which was estimated in the flux product from a bulk formula that depends on the fraction of the sky obscured by clouds. An additional +17 W/m^2 comes from the infrared radiation (where the plus sign means too little infrared radiation is coming from the sky), which is similarly dependent on cloud cover fraction. (Most of the remaining disagreement, 12 W/m^2, comes from the latent heat flux). It seems likely that the large errors in solar and infrared radiation result from neglect of regional aerosol effects from airborne dust (e.g., as in Fig. 2c). " can be changed to "ISCCP."‵‵ –>The OAFlux product uses surface solar and infrared radiation products estimated from the International Satellite Cloud Climatology Project (ISCCP) data and a radiative transfer model that explicitly accounts for regional and seasonal variations in atmospheric aerosols (Zhang et al. 2004). The ISCCP-derived radiative flux product has much smaller errors (−8.5 W/m^2 in solar and −5.3 W/m^2 in infrared) than the radiative fluxes estimated from ship observations and bulk formulas. These biases in radiative fluxes compensate for overestimation of the latent and sensible heat fluxes (by +15.2 and +1.0 W/m^2, respectively).

The 2-year record of latent heat flux shows a strong seasonal cycle in evaporation, with weekly average evaporation rates varying from about 1 m/yr in the summer months to about 3 m/yr in the winter months (Fig. 4). There are also many large evaporation events during the wintertime, with hourly average evaporative heat losses frequently exceeding 500 W/m^2 (Fig. 4). These evaporation events are associated with the occurrence of the wintertime mountain-gap wind events discussed by Jiang et al. (2009). The buoy data indicate that these events bring relatively cool (∼20 °C), dry (specific humidity of ∼7 g/kg) air from the Arabian subcontinent. For the three mountain-gap wind events occurring in December 2008 and January 2009 (Jiang et al. 2009), the peak hourly average heat losses estimated from the buoy measurements were 911, 683, and 821 W/m^2 (Fig. 5). During a week-long period surrounding the strongest of those events (Fig. 5), about 9 cm of water was evaporated from the sea surface—this is about 4.5 % of the annual mean evaporation at the site (1 week is about 1.9 % of a year). About 55 cm of evaporation, or 28 % of the annual mean, took place over the longer, 62 day period shown in Fig. 5; 62 days are only about 17 % of a year. The wintertime wind jet events thus contribute substantially to the annual mean evaporation and heat loss at the buoy site.

Much of the northern half of the Red Sea is affected by these wind jets (Fig. 2), so it seems likely that these wintertime wind jets provide an important contribution to the basin-average evaporation and heat loss. Papadopoulos et al. (2013) analyzed a daily surface flux product and found that the most intense wintertime heat loss in the northern part of the Red Sea (north of 25°N) tends to occur when the large-scale atmospheric conditions favor westward winds over the northern Red Sea. This occurs when there is a region of atmospheric high pressure that extends westward from the Siberian high to the Arabian subcontinent and the

Mediterranean Sea (Papadopoulos et al. 2013). These are the same conditions that are associated with the occurrence of the intense wintertime wind jet events (Jiang et al. 2009 and their Figure S1). The Papadopoulos et al. (2013) analysis identified a fairly close correspondence between the pattern of mean wintertime heat loss and the standard deviation of the heat flux, which suggests that atmospheric synoptic variability makes important contributions to the wintertime heat loss. However, the flux product analyzed by Papadopoulos et al. (2013) is too coarse to spatially resolve the individual wind jets, and individual examination of the most extreme heat loss events in the far northern part of the basin showed a range of synoptic conditions, including some that were conducive to westward wind jets and some that were not, so further study of the conditions driving the largest wintertime heat loss is still needed.

Horizontal Circulation

Our understanding of how wind stress and surface fluxes of heat and freshwater impact the three-dimensional structure of the Red Sea circulation and its variability is still rudimentary. This is due largely to the limited number of sustained in situ observations of currents and water properties. Satellite-borne sensors that have now measured sea surface temperature, color, and elevation for multiple years have been helpful in identifying some Red Sea circulation surface features and estimating average trends in sea surface temperature (e.g., Acker et al. 2008; Raitsos et al. 2011, 2013), but remote sensing measurements have one or more limitations, including low spatial resolution, no subsurface information, lack of coverage near coastlines, no direct current information, and lost data due to clouds and dust in the atmosphere.

The most comprehensive studies to date of the basin-scale Red Sea circulation and its variability have been mainly those based on analyses of output from ocean general circulation models (OGCMs), which are usually forced with relatively low-resolution (≥1°) global wind and buoyancy flux products from atmospheric re-analyses (Clifford et al. 1997; Eshel and Naik 1997; Sofianos and Johns 2003; Biton et al. 2008; Yao et al. 2014a, b). With the exception of Bab-al-Mandab, where an 18-month time series of the exchange flow between the Red Sea and the Gulf of Aden provides a benchmark for testing OGCM performance (Murray and Johns 1997; Sofianos et al. 2002), most data sets on currents and water properties are spatially and/or temporally compromised, allowing for only limited verification of model output (Clifford et al. 1997; Yao et al. 2014a, b). Furthermore, since the physical environment is of fundamental importance to other oceanographic processes, including primary production, biogeochemical cycles, coral reef ecosystem dynamics, and interaction between the coral reefs and the open Red Sea, the overall oceanographic knowledge base in the Red Sea is low compared to some other similarly sized ocean basins. Here, we review earlier observational and numerical model studies of the horizontal circulation in the Red Sea and present some new results collected during two major hydrographic expeditions to the Red Sea in 2010 and 2011.

Eddies and Gyres

Many studies of the basin-scale circulation in the Red Sea have focused on the two-dimensional (latitude depth) thermohaline-driven overturning circulation, which includes a two-way (winter) or three-way (summer) exchange flow through Bab-al-Mandab (e.g., Phillips 1966; Tragou and Garrett 1997; Murray and Johns 1997; see also Sofianos and Johns, this volume, for a complete discussion). The overturning is driven by a net evaporation on the order of 2 m/yr over the Red Sea (Ahmad and Albarakati, this volume). However, regional observations and numerical simulations have revealed that the overturning circulation, with its annual mean transport on the order of 0.4 Sv, is embedded in a complex, three-dimensional pattern of currents consisting of large, energetic, quasi-stationary eddies (also called gyres or sub-gyres by some authors), and narrow boundary currents, with transports that are an order of magnitude larger than the overturning circulation (e.g., Morcos 1970; Morcos and Soliman 1972; Maillard 1971; Sofianos and Johns 2003). Both clockwise (anticyclonic) and anti-clockwise (cyclonic) eddies appear in observations and model output, often with diameters similar to the width of the basin (about 100–200 km—much larger than the 40 km Rossby radius of deformation; Zhai and Bower 2013) and with the strongest currents in the upper several hundred meters.

In the first basin-scale study of these eddies, Quadfasel and Baudner (1993) used CTD and repeated XBT sections made along and across the central axis during 1983–1987 to show that the Red Sea is often filled with a north–south stack of eddies with surface geostrophic azimuthal speeds (relative to 400 dbar) of up to 0.6 m/s and circular transports of 1–3 Sv. This pattern of stacked eddies was recently confirmed with an along-axis transect of velocity measured using a vessel-mounted acoustic Doppler current profiler (ADCP) by Sofianos and Johns (2007). With their repeated transects, Quadfasel and Baudner (1993) found that anticyclonic eddies are more prevalent than cyclonic eddies and that the eddies appear to be quasi-stationary, trapped in four subregions of the Red Sea (in the latitude bands 17°–19°N, 19°–21.5°N, 21.5°–24°N and 24°–27°N). They argued that the eddies are generated when along-axis currents resulting from positive and negative wind stress curl on either side of the prevailing along-axis wind jet are deflected across the basin by coastal capes and promontories, forming closed circulation cells in the several sub-basins of the Red Sea.

On the other hand, Clifford et al. (1997) emphasized the importance of the *cross-basin* component of the wind field, arguing that this introduces larger wind stress curl and subsequently more eddies in the Red Sea. They ran two numerical experiments with a high-resolution (6-km grid spacing) primitive equation OGCM, one forced with low-resolution (40-km grid spacing) wind fields and another forced with higher-resolution (grid spacing about the same as the OGCM) wind fields that were corrected to more accurately follow the local orography. In both cases, more eddies appeared when there was a cross-basin component to the wind field. However, in the case with orographically steered winds, there was more wind stress curl and consequently more and stronger eddies in the model velocity fields. Jiang et al. (2009) recently confirmed the importance of mountain gaps on both sides of the Red Sea in funneling wind into intense cross-basin jets using the WRF model and a high-resolution regional atmospheric model.

The impact of one particular cross-basin wind jet, the Tokar Jet, on Red Sea eddy generation was recently studied by Zhai and Bower (2013). This eastward wind jet originates in the Tokar Gap on the Sudanese coast at about 19°N during most summers and can reach amplitudes of 20–25 m/s (Pedgley 1974; Jiang et al. 2009). Zhai and Bower used time series of scatterometer wind fields and altimeter-derived sea level anomalies (SLA) to identify a strong dipolar eddy that often appears in summer centered near 19°N with a horizontal scale of $\sim$140 km, shortly after the onset of the wind jet. They showed that the positive and negative wind stress curl pattern associated with the wind jet spins up a cyclonic eddy north of the wind jet axis and an anticyclonic eddy to the south and that the strength of the dipole is proportional to the strength of the wind jet. The wind jet exhibits significant interannual variability, which is reflected in the strength of the dipole. The dipole can persist for several months after the wind jet disappears and can have cross-basin surface currents up to 1 m/s. The dipole structure observed in SLA during August 2001 was confirmed with in situ current observations made with an ADCP, which also showed that the dipole currents extended to 100–200 m (see also Sofianos and Johns 2007). The Tokar Dipolar eddy is not typically evident in the various OGCM simulations because the wind fields used to force the models do not adequately resolve the narrow Tokar Wind Jet.

Unlike the Tokar Dipolar eddy, several of the other eddies identified by Quadfasel and Baudner (1993) and others do show up consistently in most OGCM simulations, although with varying intensity and spatial scale. Prime among these are the apparently persistent anticyclonic eddy centered near 23°N and the cyclonic circulation in the far northern Red Sea (Clifford et al. 1997; Sofianos and Johns 2003; Biton et al. 2008). Sofianos and Johns (2003) suggested that generation of these eddies does not require cross-basin winds, or any wind forcing for that matter, and that some eddies form when buoyancy forcing alone is applied. They employed the Miami Isopycnic Coordinate Ocean Model (MICOM), first forced with both climatological winds and thermohaline fluxes and then with the two forcing mechanisms separately. <Query ID="Q3" Text="Please check whether the usage of the terms "thermohaline forcing-only" and "thermohaline-only forcing" are OK." ->The model mean surface circulation in the thermohaline-only forcing case was in general most similar to that with the full forcing, suggesting that the wind forcing plays a minor role, at least for the climatological mean surface circulation. For example, the cyclonic eddy usually observed at the northern end of the Red Sea was successfully reproduced in the thermohaline-only forcing case but had the opposite polarity in the wind-only forcing case. On the other hand, the 23°N anticyclonic eddy, clearly evident in their full-forcing model run, does not appear as a closed circulation cell in either of the separate forcing experiments, nor apparently in their sum. This example highlights the possible nonlinear interaction of the wind and thermohaline forcing and the overall complexity of the dynamics of the eddies in the Red Sea.

More support for the idea that at least some eddies in the Red Sea are not directly wind driven was presented by Chen et al. (2014), who used an unstructured grid, Finite-Volume Community Ocean Model (FVCOM) configured for the Red Sea and Gulf of Aden to investigate the dynamics of the eddy centered near 23°N (recent observations of this eddy are described in Chen et al. and in Sect. "Boundary Currents"). Horizontal model resolution varied from 3–8 km, and there were 30 layers in the vertical. Both high-resolution (from WRF) and lower-resolution (from NCEP) wind stress and buoyancy fluxes were used in separate numerical experiments to force the model. Through a series of model runs with different combinations of wind and buoyancy forcing, they found that the 23°N anticyclonic eddy was well reproduced compared to observations with heat flux forcing only, leading them to conclude that the eddy forms as a result of basin-scale (not local) geostrophic adjustment to the seasonal cycle in the large-scale buoyancy flux over several months. A southward coastal current on the western boundary seemed to play an important role in setting the eddy's strength. Wind forcing played a minor, but not insignificant role, affecting the intensity and vertical structure of the eddy.

Boundary Currents

In addition to the eddies discussed above, there is some evidence for near-surface currents that extend over hundreds of kilometers along the eastern and western boundaries of the Red Sea. These boundary currents extend over much

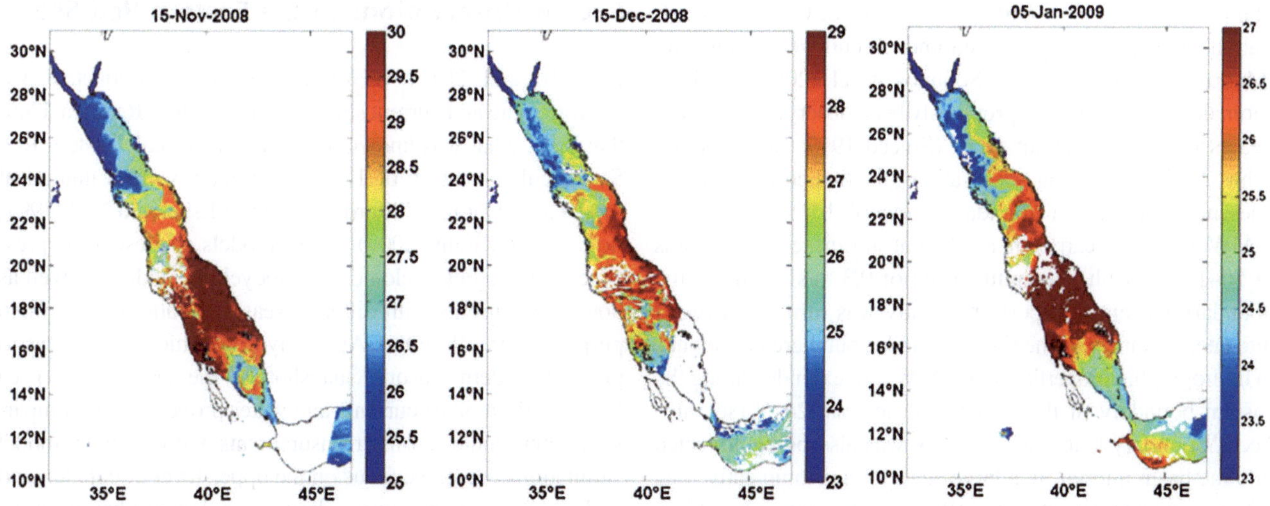

Fig. 6 Sea surface temperature imagery at three times during the winter of 2008–2009. The presence of a band of warm water along the eastern boundary of the northern half of the Red Sea suggests the presence of a northward-flowing boundary current carrying warm water from the south to the north. This boundary current has been seen in numerical simulations of the Red Sea circulation, but there has been only limited observational evidence for its existence

larger spatial scales than the eddies. Until recently, indications of boundary currents were confined almost entirely to results from numerical model studies. For example, the climatological mean winter circulation in the MICOM model configuration used by Sofianos and Johns (2003) has a northward western boundary current in the southern Red Sea that crosses to the eastern boundary at about 16°N and continues to the northern end of the basin as a narrow eastern boundary current. This model boundary current system transports relatively warm, low-salinity surface water entering the Red Sea through Bab-al-Mandab northward as it is cooled and made more saline by air–sea fluxes. In the model, the water returns southward in a subsurface layer that eventually exits the Red Sea as a high-salinity, dense overflow (Murray and Johns 1997). A similar pattern of surface boundary currents was found in the climatological winter model results of Biton et al. (2010) and Yao et al. (2014b).

Invoking some theoretical ideas of McCreary et al. (1986) on the dynamics of eastern boundary currents, Sofianos and Johns (2003) argued that north of some critical latitude, traditional western intensification by westward-propagating Rossby waves is prevented because the Rossby wave group velocity is weaker than the mean eastward geostrophic flow induced by the meridional gradient in surface density. This critical latitude hypothetically sets the crossover latitude of the boundary current system from the western to eastern boundary. In another numerical study using the MIT GCM, Eshel and Naik (1997) obtained a similar pattern of western and eastern boundary currents, although the crossover latitude of the western boundary current was much farther north (about 25°N) and was attributed to the collision of the northward western boundary current and the southward return flow of the eastern boundary current.

Sea surface temperature (SST) imagery from satellites supports the existence of an eastern boundary current in the northern half of the Red Sea resembling the one seen in the model simulations of Sofianos and Johns (2003). For example, Fig. 6 shows a sequence of infrared SST images taken from the Moderate Resolution Imaging Spectroradiometer (MODIS) instrument on NASA's Aqua satellite during the winter of 2008–2009. A narrow band of warm water is visible along the eastern boundary of the northern half of the Red Sea, suggesting the presence of a northward-flowing boundary current carrying warm water from the south to the north. Although the thin band of warm water is nearly continuous, it is very convoluted, with meanders that extend across the full width of the Red Sea (e.g., near 21°N on 15 November) and significant temporal variability.

Further confirmation of these and other boundary currents, and their seasonal variability, has been delayed by a lack of in situ observations near the Red Sea boundaries. One exception is the recent observation during a joint WHOI–KAUST hydrographic survey in September–October 2011 of a narrow, subsurface current in the southern Red Sea transporting relatively cold, low-salinity, high-nutrient Gulf of Aden Intermediate Water (GAIW) northward along the eastern boundary in the southern Red Sea (Churchill et al. 2015).

The jet originates in the subsurface inflow of GAIW through Bab-al-Mandab during the summer (Southwest) monsoon (Murray and Johns 1997; Sofianos et al. 2002), and its anomalous properties had previously been observed in the Red Sea as far north as about 19°N (Smeed 1997; Sofianos and Johns 2007). The recent observations, the first comprehensive ones near the eastern boundary north of 17°N, reveal that GAIW was being carried in a coherent subsurface jet at depths of 50–100 m with maximum speed of 0.3 m/s. Some of the boundary current transport apparently was turning eastward into deep channels on the Farasan Banks, but there is evidence in hydrographic properties that GAIW also extended along the eastern boundary of the Red Sea to at least 24°N (see also Sect. 'Boundary Currents"). GAIW was also being advected into the basin interior by a large-scale anticyclonic eddy. The subsurface jet transporting GAIW along the eastern boundary was recently successfully simulated in the modeling study of Red Sea summer circulation using the MIT GCM by Yao et al. (2014a). The pathways of GAIW through the Red Sea, which is the only oceanic supply of new nutrients (Souvermezoglou et al. 1989), likely have significant implications for the spatial and temporal distributions of primary production (Churchill et al. 2015).

Recent Observations in the Eastern Red Sea

In 2010 and 2011, WHOI and KAUST conducted two hydrographic and current surveys of the eastern Red Sea. Until that time, what was known about the basin-scale eastern Red Sea circulation north of 17°N came entirely from numerical model simulations (Clifford et al. 1997; Eshel and Naik 1997; Sofianos and Johns 2003). These models suggested the presence of numerous cyclonic and anticyclonic eddies, as well as boundary currents with different seasonal patterns. The main purpose of the WHOI–KAUST hydrographic surveys was to provide modern seasonal snapshots of the sea surface-to-sea floor distribution of currents and water properties for comparison with remote sensing measurements and numerical model simulations. Here, we focus on the upper-layer circulation and hydrography (temperature, salinity, and density) based on the in situ observations collected during these surveys.

The first survey was carried out from March 16 to 29, 2010, on the *R/V Aegaeo* and covered the eastern half of the Red Sea north of 22°N. A total of 111 full-depth conductivity–temperature–depth (CTD) profiles were obtained during the cruise, mainly along nine cross-slope transects, with an average station spacing of 10 km along the transects

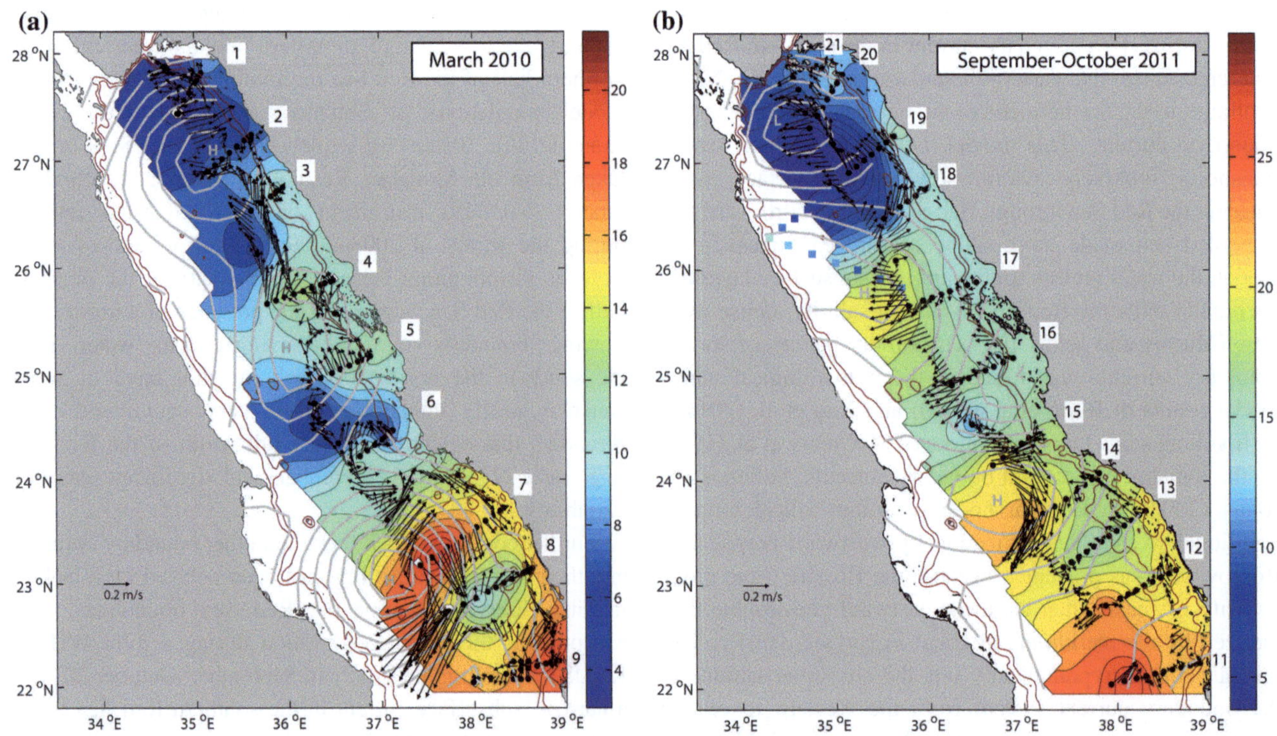

Fig. 7 **a** and **b** ADCP velocity vectors averaged horizontally in 5 km distance bins along the cruise track and vertically over the top 200 m during (**a**) March 2010 and (**b**) September–October 2011 survey cruises. *Color shading* indicates 10–200 dbar geopotential anomaly (×10 to display in dynamic cm). *Colored squares* in (**b**) also indicate dynamic height but for select CTD stations occupied during August 2001 (Sofianos and Johns 2007). *Gray* contours show average SLA for the 2-week period of the 2010 and 2011 in situ measurements, contoured every 2 cm. The 200 and 500 m bathymetric contours from ETOPO2 are also indicated with *magenta lines*

(Fig. 7a). The cruise began in Thuwal, Saudi Arabia, transited northward to transect 1, then worked the transects 1–9 southward and ended in Thuwal. Heavy winds during transect 1 interrupted the cruise while the vessel took shelter in the lee of an island near the entrance to the Gulf of Aqaba. These transects plus several more, numbered 11–22, were occupied again (from south to north) during the second half of the second survey, which took place from September 27, 2011 to October 10, 2011, on the same vessel (Fig. 7b). Transects 1–10 in 2011 were occupied in the southern Red Sea—here we focus only on the area common to both surveys, 22–28°N. A vessel-mounted ADCP provided continuous measurements of current speed and direction from 18 m down to about 600 m depth along the entire cruise track at a vertical resolution of 8 m during both cruises.

The average vertical profiles of current speed from the ADCP measurements during the two cruises (not shown) indicate a surface-intensified vertical structure with average surface values of 0.24–0.27 m/s and 600 m speeds of 0.06–0.07 m/s. On average, 75–95 % of the vertical shear occurred over the top 200 m of the water column. Therefore, to summarize the upper ocean density field and currents, Fig. 7 shows the geopotential anomaly at 10 m relative to 200 m at each CTD station (multiplied by 10 to express as dynamic height in dynamic cm) and ADCP velocity vectors averaged over the same vertical scale. The strongest circulation feature in the 2010 survey was an anticyclonic eddy centered near 23°N, with maximum 0–200 m average speed of 0.6 m/s. This is the same feature found in nearly all OGCM simulations and was identified by Quadfasel and Baudner (1993) as one of the more persistent eddies in the Red Sea. The circulation sense observed with the ADCP is consistent (assuming geostrophic balance to lowest order) with the slope of dynamic height on transects 7 and 8 (increasing offshore). The eddy also is clearly evident in the distribution of sea level anomaly, with a 10–12 cm sea level maximum elevation along transect 7 (Fig. 7a, gray contours). The 10 to 200 m dynamic height difference across the flank of the eddy was less, only 6–7 dynamic cm on CTD transects 7 and 8, but neither transect unambiguously reached the eddy center, where dynamic height would reach its maximum value. Furthermore, the eddy currents measured with the ADCP extended to nearly 400 m; the 10 to 400 m dynamic height (not shown) increased offshore by 12 dynamic cm along transect 7, quantitatively similar to the SLA.

Volume transport across transect 7 in the top 200 m estimated from ADCP velocities was −5.4 Sv (southward). This is an order of magnitude larger than the transport associated with the vertical overturning circulation, whose annual mean value is estimated from direct current observations in the strait to be about 0.4 Sv (Murray and Johns 1997; Sofianos et al. 2002). It is also larger than eddy transport estimates made by

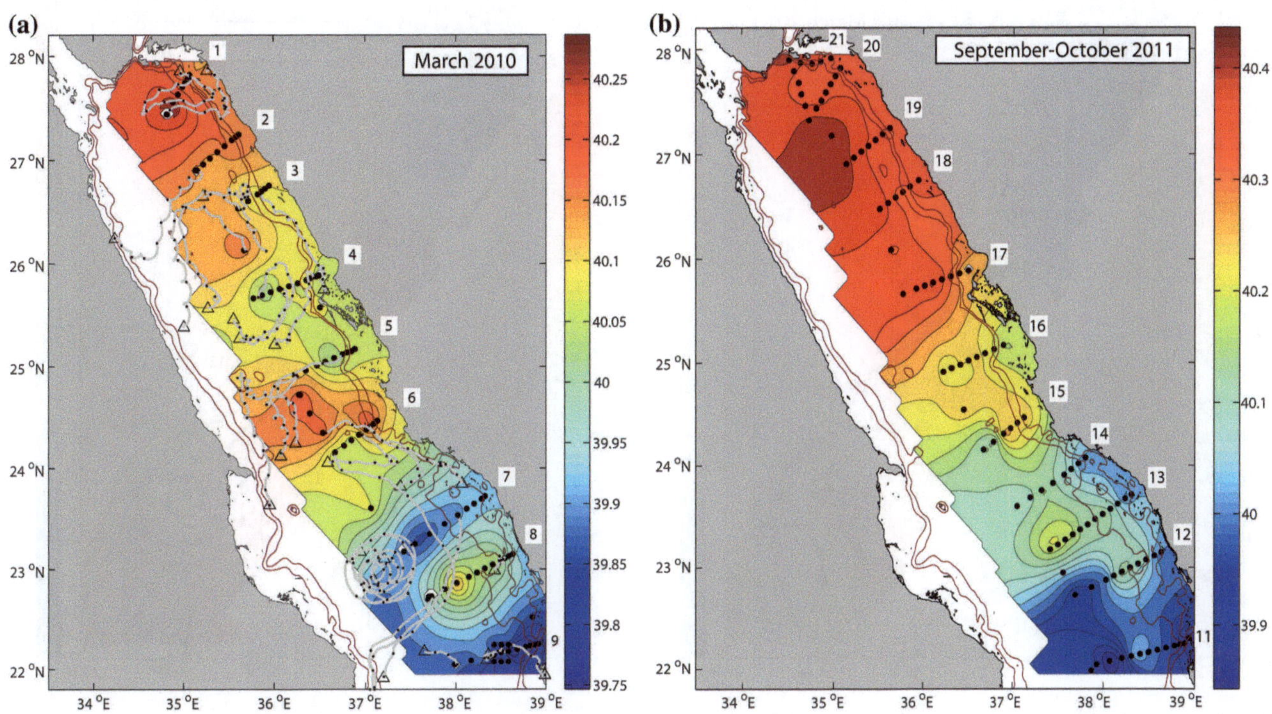

Fig. 8 **a** and **b** Vertically averaged (0–200 m) salinity (*color shading*) for the two survey cruises. Superimposed in (**a**) are the trajectories of 28 Davis-type surface drifters released during the 2010 cruise, only from the launch position to the end of March, with *small dots* shown once daily along the tracks

Quadfasel and Baudner (1993) using a 400 m level of no motion in their geostrophic velocity calculations. The tracks of three surface drifters released near the anticyclonic eddy center (Fig. 8a) confirm its closed circulation. One drifter circulated in the eddy for 138 days and had a mean looping radius and speed of 27 km and 0.34 m/s, respectively, and a mean looping period of 6.6 days.

There is no evidence of a continuous northward eastern boundary current during the March 2010 cruise, such as seen in the climatological mean winter simulation of Sofianos and Johns (2003) and inferred from the early winter SST images in Fig. 6. Rather a complex pattern of eddies and boundary current segments was observed. Dynamic height shows a slight increase (about 4–5 dynamic cm) toward the eastern boundary on transects 8 and 9 due to relatively fresher and warmer water in the upper 200 m within 30 km of the boundary (Figs. 8a and 9a), accompanied by a weak northward flow component. However, currents on transect 7 were southward right up to the eastern boundary, interrupting any possible continuous northward-flowing eastern boundary current. Such a boundary current, if it exists, may have been obscured by the more energetic eddy field in March 2010 (as suggested in model results from Yao et al. 2014b) or may be in the process of dying out after the peak in buoyancy forcing that occurs earlier in the winter (Fig. 8).

Farther north, ADCP velocities and dynamic height gradients are generally weaker and indicate a circulation pattern of eddies and/or meanders. On several transects, dynamic height was higher adjacent to the eastern boundary (and northward currents were measured with ADCP), resulting from warmer and fresher water near the boundary compared to offshore stations (e.g., transects 1, 3 and 5). Transports (10–200 m) across these transects were northward 1.8 Sv (5), 0.2 Sv (3), and 0.6 Sv (1). Drifter tracks and velocity vectors suggest a closed cyclonic circulation in the far northern Red Sea in the vicinity of transect 1. On transects 2 and 4, as on transect 7, lowest salinity and highest temperature were found offshore of rather than against the eastern boundary (Figs. 8a and 9a), and all observations indicate anticyclonic circulation with southward flow near the eastern boundary. In both cases, the SLA fields also indicate a local maximum (anticyclone), although the patterns of SLA and dynamic height/velocity do not line up as closely with each other as they did for the anticyclonic eddy near 23°N.

Of particular note is the southward drift of surface drifters west of the CTD transects. After low-pass filtering to remove inertial oscillations, typical southward speeds of these drifters were 0.2–0.3 m/s, with occasional higher values up to 0.5–0.6 m/s (see also Chen et al. 2014). The trajectories show little evidence of eddies and/or meanders in the

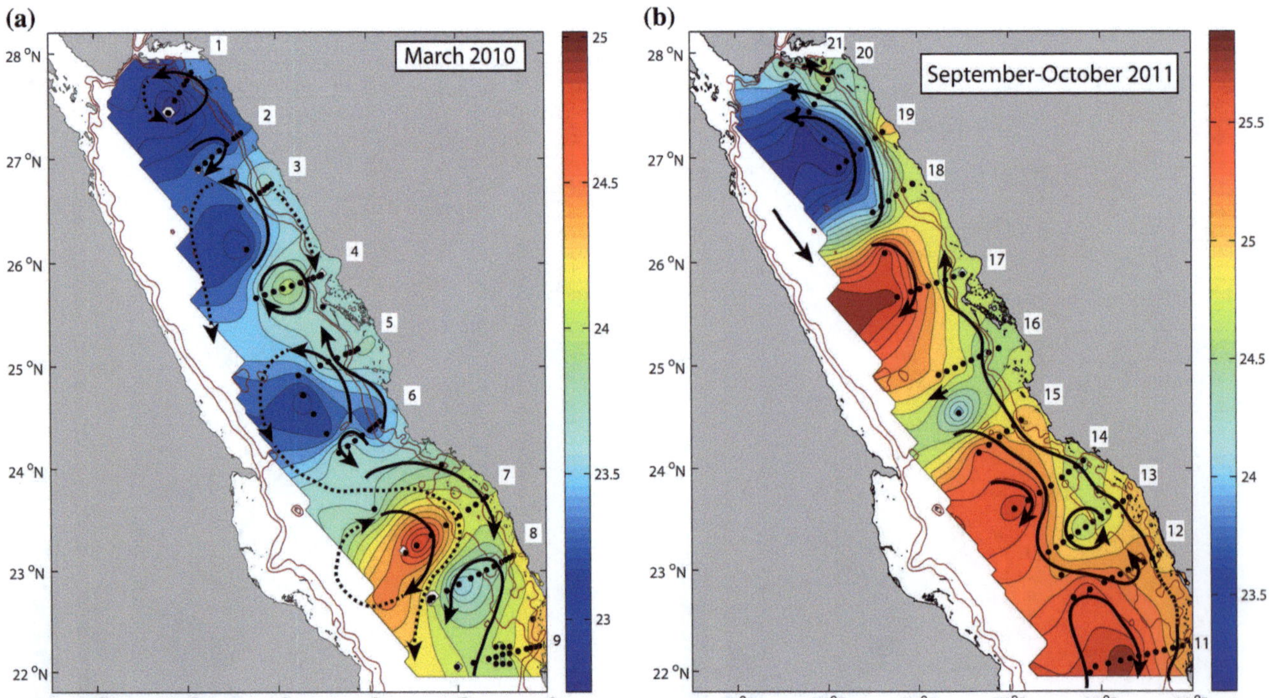

Fig. 9 **a** and **b** Schematic diagrams illustrating major circulation features during the two survey cruises, superimposed on the 0–200 m vertically averaged potential temperature (*color shading*). *Solid lines* are based on ADCP velocities and surface drifters (where the trajectories are consistent with ADCP vectors). *Dotted lines* indicate features only observed in drifters. Bathymetric contours are 200 and 500 m

western region until they reach the vicinity of the anticyclonic eddy centered near 23°N, when some of these southward-moving drifters were deflected eastward then southward around the perimeter of the anticyclonic eddy there. Chen et al. (2014) successfully captured this southward flow near the western boundary and its continuation around the 23°N eddy based on the trajectories of particles initialized in the northern region of their numerical simulations.

Figure 9a summarizes the circulation in the northeastern Red Sea during the March 2010 hydrographic survey. The strongest currents in the region were associated with several 100–200 km diameter anticyclonic eddies centered near 27°N, 25.5°N, and 23°N, with the 23°N anticyclonic eddy being the strongest. The eddies were separated by areas of cyclonic flow. Salinity generally increased northward and westward; minimum 0–200 m average salinity (39.75) was observed at the southern limit of the survey area, while the maximum (40.29) was observed in the interior northern region. Far from a simple monotonically increasing meridional structure, salinity was lower (and temperature was higher) in anticyclonic eddies compared to cyclonic ones. Although not fully resolved by these observations confined to the eastern Red Sea, the circulation pattern is suggestive of a meandering northward transport of relatively warm, low-salinity water near the eastern boundary, with offshore meanders forming eddies of similarly fresher water. On the other hand, the eddy at 23°N appears to have completely blocked (at least temporarily) any northward penetration of warmer, fresher water from the south, based on the wrapping of southward-moving surface drifters around its eastern flank. The eddy's location and strength may be seasonally modulated and not always pressed up against the eastern boundary (see below and Chen et al. 2014). Surface drifters provide strong evidence of persistent southward flow in the western part of the northern Red Sea. On the largest spatial scale, there is qualitative consistency between the in situ observations and the sea level anomaly field, with three anticyclonic features in the surface circulation, which roughly correspond to three regions of anticyclonic flow in the in situ observations. The best match in both location and amplitude is for the 23°N eddy. Not surprisingly, the in situ observations reveal smaller-scale currents that cannot be resolved by altimetry.

In the September–October 2011 survey, strong eddy-like circulation features were again present (Fig. 7b). Maximum 10–200 m average speeds were about 0.4 m/s, similar to those reported by Sofianos and Johns (2007) along an axial ADCP transect in August 2001. The strongest currents line up approximately with high and low SLA features, especially the cyclonic circulation centered near 27°N and the anticyclonic eddy near 24°N. Sea level and dynamic height anomalies are similar for these features: about 5 cm (24°N eddy) and 4 cm (27°N). Transports across transects 18, 19, 20, and 22 were 1.2, 1.8, 1.7, and 1.8 Sv, respectively, indicating a more vigorous and larger meridional scale cyclonic circulation in the northern region during this survey compared to the March 2010 survey, consistent with the model results of Sofianos and Johns (2003) and Biton et al. (2008).

Farther south, there was a slight increase in dynamic height (typically not more than a few dynamic cm) within about 30 km of the eastern boundary on transects 11–17. This is consistent with the directly measured velocities, which indicate a narrow, possibly continuous, band of northward flow. In some transects, the northward coastal flow is separated from offshore current features by a minimum in the middle of the transects (12–17). Salinity and temperature observations indicate that this coastal current transported relatively low-salinity, cold water (compared to offshore water) northward (Figs. 8b and 9b). This is likely indicative of a northward extension of GAIW, which enters the Red Sea as an intermediate layer (30–100 m) through Bab-al-Mandab during the summer months (Murray and Johns 1997; Smeed 1997; Sofianos et al. 2002; Sofianos and Johns 2007). Churchill et al. (2015) report on the GAIW distribution all along the eastern boundary of the Red Sea based on the full 2011 hydrographic survey data. Northward 0–200 m, along-slope transport in this boundary current (measured from the minimum transport in the middle of each transect) ranged from 1.0 Sv across transect 13 to 0.1 Sv across transect 15 and 0.2 Sv across transect 17. Vertical sections of velocity indicate that this northward boundary current has a subsurface maximum (see Churchill et al. 2015). From transect 18 northward, there was still northward along-slope flow across the transects, and the lowest salinities were found near the boundary, but the northward flow was not intensified against the eastern boundary as in transects farther south. The water near the boundary was warmer relative to the significantly colder water in the interior. The cold, high-salinity interior in the northern region produces a pool of minimum in dynamic height and relatively strong cross-basin velocities ($\sim$0.4 m/s) between transects 17 and 18. The western half of the cyclonic circulation was unfortunately not observed during the 2011 survey (no drifters). However, Sofianos and Johns (2007) observed southward flow in two transects near the western boundary in 2001.

Salinity averaged in the upper 200 m increased northward along the eastern boundary; the minimum was 39.83 on transect 11 and 40.37 on transect 20. Maximum salinity in the survey area was 40.44 in the cold, high-salinity pool centered near 27°N. Note that near-surface salinities were everywhere higher than in March 2010. This seems surprising at first given that the net freshwater flux to the atmosphere reaches an annual minimum during summer (Fig. 4; see also Sofianos and Johns 2002). Plots of salinity versus density for 2010 and 2011 in the northern Red Sea (not shown) reveal that salinity is higher in 2011 on all isopycnals down to at least 200 m. This suggests that the higher summer values are not solely the result of shoaling of isopycnals in the cyclonic interior of the northern Red Sea during summer. Rather, these results point to the importance of advection of fresher water into the region from the south during winter. Without more hydrographic observations like these, it is impossible to determine how interannual variability in the supply of fresher water to the northern Red Sea may impact salinity distributions. As summarized in Fig. 9b, the strongest currents in 2011 were associated with the anticyclonic circulation near 24°N and the cyclonic cell centered near 27°N. ADCP velocities and T-S measurements indicate a possibly continuous northward along-slope eastern boundary current transporting relatively cold, low-salinity water as far north as about 25.5°N. Low-salinity water was also found against the boundary north of that latitude, but a boundary-intensified current was not distinguishable from the generally northward flow associated with the offshore area of low dynamic height. While the strongest eddies correspond moderately well in location and amplitude with the SLA, the narrow eastern boundary current, with its dynamic height amplitude of at most a few dynamic cm over ~30 km, is not resolvable from altimetry.

Given the moderately good agreement between dynamic height and SLA for the basin-scale circulation features away from the boundaries, it is of interest here to use the nearly 20-year time series of SLA to investigate how the observed circulation features fit into the seasonal cycle of cyclonic and anticyclonic eddy circulations. This is illustrated in Fig. 10, which shows a Hovmoller diagram of the mean seasonal cycle of relative vorticity along the central axis of the northern Red Sea, calculated from the gradients in SLA. Positive (negative) relative vorticity corresponds to cyclonic (anticyclonic) circulation. The midpoints of the 2010 and 2011 surveys are indicated by vertical dashed lines. During March, the climatological mean vorticity shows an AC centered at about 23°N that is close to a climatological maximum amplitude at the time of year of the March 2010 survey. As pointed out earlier, this feature in the Hovmoller diagram corresponds well to the AC observed during the March 2010 cruise (Fig. 9a). Based on the climatology, this AC near 23°N apparently spins up quickly beginning in early December, reaches an initial maximum in March, translates northward to about 24°N, reaches a second peak in amplitude in mid-August, then spins down by the end of October.

The timing of the AC spin-up in early December is consistent with the arrival of relatively warm, low-salinity Gulf of Aden Surface Water, which begins to enter the Red Sea through Bab-al-Mandab in early September (Sofianos et al. 2002) (~1100 km to the south) assuming a mean northward advection rate of about 0.15 m/s. These results suggest, but do not confirm, that the 23°N AC eddy may spin up in this region every year as a result of the arrival of northward-flowing Gulf of Aden Surface Water interacting somehow with the local bathymetry and/or other eddies.

Immediately north of this AC, the vorticity climatology indicates near-zero vorticity at the time of year of the March 2010 survey, but just before the cruise period, a strong

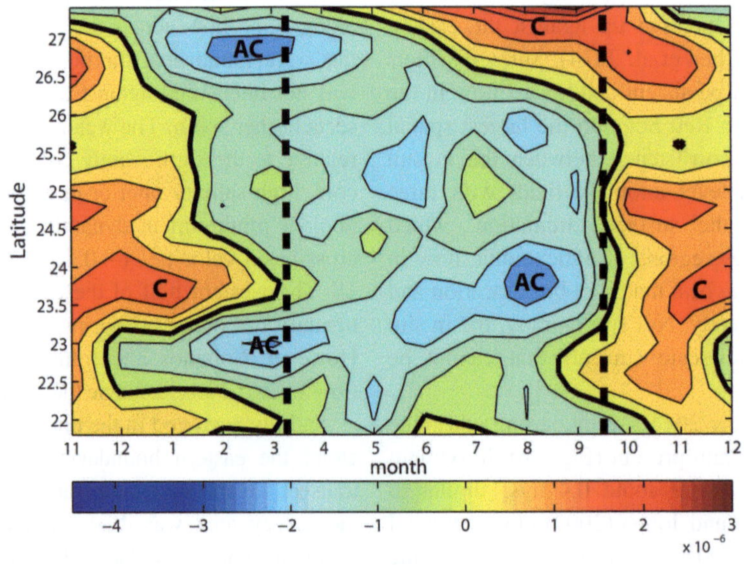

Fig. 10 Hovmoller diagram of annual mean relative vorticity along the Red Sea axis as a function of latitude and month of the year (marked at the midpoint of each month). Heavy *dashed lines* indicate the midpoint of each of the two hydrographic surveys. Negative relative vorticity (*blue*) indicates anticyclonic (*AC*) circulation, and positive (*red*) represents cyclonic (*C*) circulation. Relative vorticity was estimated from surface geostrophic velocity calculated from AVISO sea level anomaly maps

cyclonic eddy is evident in the climatology. The remnants of such a cyclonic eddy might be what are seen in the March 2010 ADCP velocity observations, centered near 24°N (Fig. 7a). This cyclonic eddy reaches its climatological maximum amplitude during November to January. The September–October 2011 survey caught several circulation features that can also be related to features in the SLA climatology, including a string of AC, C, and AC eddies across transects 11–15.

At 27°N, anticyclonic vorticity dominates in the climatology from mid-December to June, after which it appears to migrate southward and be replaced in the northern basin by cyclonic vorticity, which itself migrates southward in December. This switch in polarity of the circulation in the northern Red Sea between winter and summer is consistent with what was observed in the 2010 and 2011 current surveys, with anticyclonic circulation seen in March 2010 (transect 2, Fig. 7a) and cyclonic circulation dominating in October 2011 (transects 18, 19, 20, and 22; Fig. 7b).

Summary and Conclusions

Recent direct estimates of air–sea fluxes from a moored meteorological buoy in the east-central Red Sea have exposed a new phenomenon that may be responsible for much of the heat and freshwater flux out of the basin during winter, namely cold, dry mountain-gap wind jets blowing westward off the coast of Saudi Arabia. These measurements have also revealed shortcomings in the climatological air–sea flux products often used to force numerical models.

Ocean numerical model studies and observations over the past several decades have revealed that the horizontal circulation of the Red Sea is energetic and complex, dominated by boundary currents and quasi-stationary eddies, both anticyclonic and cyclonic. Transports associated with the eddies, typically 1–5 Sv, are an order of magnitude larger than the annual mean overturning transport ($\sim$0.4 Sv). The cross-basin component of the wind stress has been shown to be important in spinning up at least one eddy pair, while other eddies are reproduced in numerical model simulations forced only with buoyancy fluxes. The eddy formation mechanism in the latter case is still unclear. Satellite altimetry is reasonably successful in capturing the largest, most energetic eddies in the interior of the basin, but not the narrow (<50 km width) boundary currents. The nearly 20-year time series of altimetric measurements also reveals an annual cycle in the average location and strength of eddies, including the persistent 23°N eddy and the reversal in the northern circulation cell. The persistence of eddies at particular latitude bands is most likely due to the interaction of the buoyancy-driven circulation with capes and promontories in the bathymetry, although in at least one case eddy latitude is fixed by the location of an intense summer cross-basin wind jet. New observations near the eastern boundary north of 17°N point toward a greater northward extension of GAIW during late summer than previously reported, although it is significantly diluted by exchange with the interior and with coral reef channels along its path. While there is renewed interest for making new in situ observations of air–sea fluxes, currents, and hydrographic properties in the Red Sea, there remains a strong need for sustained, basin-scale measurements to thoroughly evaluate numerical models and to increase our understanding of the roles of wind and buoyancy forcing in driving Red Sea circulation and its variability on seasonal and longer time scales. Specifically, time series of currents and water properties near both boundaries, transects across the full width of the basin, and air–sea flux measurements at several latitudes are needed to validate numerical simulations and test ideas about the dynamics governing Red Sea circulation.

Acknowledgments Data collection during the WHOI-KAUST collaboration was made possible by Award Nos. USA00001, USA00002, and KSA00011 to the WHOI by the KAUST in the Kingdom of Saudi Arabia.

References

Acker J, Leptoukh G, Shen S, Zhu T, Kempler S (2008) Remotely-sensed chlorophyll a observations of the northern Red Sea indicate seasonal variability and influence of coastal reefs. J Mar Syst 69:191–204

Biton E, Gildor H, Peltier WR (2008) Red sea during the last glacial maximum: implications for sea level reconstruction. Paleoceanography 23(1). doi:10.1029/2007pa001431

Biton E, Gildor H, Trommer G, Siccha M, Kucera M, van der Meer MTJ, Schouten S (2010) Sensitivity of Red Sea circulation to monsoonal variability during the holocene: an integrated data and modeling study. Paleoceanography 25. doi:10.1029/2009pa001876

Chen C, Ruixiang L, Pratt L, Limeburner R, Beardsley R, Bower AS, Jiang H, Abualnaja Y, Liu X, Xu Q, Lin H, Lan J, Kim T-W (2014) Process modeling studies of physical mechanisms of the formation of an anticyclonic eddy in the central Red Sea. J Geophys Res Oceans 119. doi:10.1002/2013JC009351

Churchill JA, Bower A, McCorkle DC, Abualnaja Y (2015) The transport of nutrient-rich Indian Ocean water through the Red Sea and into coastal reef systems. J Mar Res (in press)

Clifford M, Horton C, Schmitz J, Kantha LH (1997) An oceanographic nowcast/forecast system for the Red Sea. J Geophys Res 102:25101–25122

Eshel G, Naik N (1997) Climatological coast jet collision, intermediate water formation, and the general circulation of the Red Sea. J Phys Oceanogr 27(7):1233–1257

Fairall CW, Bradley EF, Hare JE, Grachev AA, Edson JB (2003) Bulk parameterization of air-sea fluxes: Updates and verification for the COARE algorithm. J Clim 16:571–591

Farrar JT, Lentz S, Churchill J, Bouchard P, Smith J, Kemp J, Lord J, Allsup G, Hosom D (2009) King Abdullah University of Science and Technology (KAUST) mooring deployment cruise and fieldwork report, technical report. Woods Hole Oceanographic Institute, Woods Hole, Mass, 88 pp

Hickey B, Goudie AS (2007) The use of TOMS and MODIS to identify dust storm source areas: the Tokar delta (Sudan) and the Seistan basin (south west Asia). In: Goudie AS, Kalvoda J (eds) Geomorphological Variations. P3K, Prague, pp 37–57

Jiang H, Farrar JT, Beardsley RC, Chen R, Chen C (2009) Zonal surface wind jets across the Red Sea due to mountain gap forcing along both sides of the Red Sea. Geophys Res Lett 36, L19605

Josey S, Kent E, Taylor P (1999) New insights into the ocean heat budget closure problem from analysis of the SOC air–sea flux climatology. J Clim 12:2856–2880

Maillard C (1971) Etude hydrologique et dynamique de la Mer Rouge en hiver. Annales de l'Institut océanographique, Paris 498(2):113–140

McCreary JP, Shetye SR, Kundu PK (1986) Thermohaline forcing of eastern boundary currents: with application to the circulation off the west coast of Australia. J Mar Res 44:71–92

Morcos SA (1970) Physical and chemical oceanography of the Red Sea. Oceanogr Mar Biol Annu Rev 8:73–202

Morcos S, Soliman GF (1972) Circulation and deep water formation in the northern Red Sea in winter. UNESCO, L'Oceanographie Physique de la Mer Rouge, pp 91–103

Murray SP, Johns W (1997) Direct observations of seasonal exchange through the Bab el Mandeb Strait. Geophys Res Lett 24(21):2557–2560

Naval Oceanography Command Detachment (1993) U.S. Navy regional climatic study of the Red Sea and adjacent waters. NAVAIR 50-1C-562. National Oceanic and Atmospheric Administration, Asheville

Papadopoulos VP, Abualnaja Y, Josey SA, Bower A, Raitsos DE, Kontoyiannis H, Hoteit I (2013) Atmospheric forcing of the winter air-sea heat fluxes over the northern Red Sea. J Clim 26:1685–1701

Patzert WC (1974) Wind-induced reversal in Red Sea circulation. Deep Sea Res 21:109–121

Pedgley DE (1974) An outline of the weather and climate of the Red Sea. L'Oceanographie Physique de la Mer Rouge. United National Educational, Scientific, and Cultural Organization, Paris, pp 9–27

Phillips OM (1966) On turbulent convection currents and the circulation of the Red Sea. Deep Sea Res 13(6):1149–1160

Quadfasel D, Baudner H (1993) Gyre-scale circulation cells in the Red Sea. Oceanol Acta 16:221–229

Raitsos DE, Pradhan Y, Brewin RJW, Stenchikov G, Hoteit I (2013) Remote sensing the phytoplankton seasonal succession of the Red Sea. PLoS ONE 8(6):64909

Raitsos DE, Hoteit I, Prihartato PK, Chronis T, Triantafyllou G, Abualnaja Y (2011) Abrupt warming of the Red Sea. Geophys Res Lett 38(14), L14601. doi:10.1029/2011GL047984

Risien CM, Chelton DB (2008) A global climatology of surface wind and wind stress fields from eight years of QuikSCAT scatterometer data. J Phys Oceanogr 38:2379–2413

Smeed D (1997) Seasonal variation of the flow in the Strait of Bab al Mandab. Oceanol Acta 20(6):773–781

Sofianos SS, Johns WE (2002) An Oceanic General Circulation Model (OGCM) investigation of the Red Sea circulation: 1. Exchange between the Red Sea and the Indian Ocean. J Geophys Res 107 (C11):3196

Sofianos SS, Johns WE (2003) An Oceanic General Circulation Model (OGCM) investigation of the Red Sea circulation: 2. Three dimensional circulation in the Red Sea. J Geophys Res 108 (C3):3066

Sofianos SS, Johns WE (2007) Observations of the summer Red Sea circulation. J Geophys Res 112:C06025

Sofianos SS, Johns WE, Murray SP (2002) Heat and freshwater budgets in the Red Sea from direct observations at Bab el Mandeb. Deep Sea Res Part II 49:1323–1340

Souvermezoglou E, Metzl N, Poisson A (1989) Red Sea budgets of salinity, nutrients, and carbon calculated in the Strait of Bab el Mandab during the summer and winter season. J Mar Res 47:441–456

Tragou E, Garrett C (1997) The shallow thermohaline circulation of the Red Sea. Deep-Sea Res 44:1355–1376

Tragou E, Garrett C, Outerbridge R, Gilman G (1999) The heat and freshwater budgets of the Red Sea. J Phys Oceanogr 29:2504–2522

Yao F, Hoteit I, Pratt L, Bower A, Zhai P, Köhl A, Gopalakrishnan G, Rivas D (2014a) Seasonal overturning in the Red Sea: part 1. Model verification and summer circulation. J Geophys Res 119:2238–2262. doi:10.1002/2013JC009331

Yao F, Hoteit I, Pratt L, Bower A, Zhai P, Köhl A, Gopalakrishnan G, Rivas D (2014b) Seasonal overturning in the Red Sea: part 2. Winter circulation. J Geophys Res 119:2263–2289. doi:10.1002/2013JC009331

Yu L, Weller RA (2007) Objectively analyzed air-sea heat fluxes for the global ice-free oceans (1981–2005). Bull Am Meteorol Soc 88:527–539

Zhai P, Bower A (2013) The response of the Red Sea to a strong wind jet near the Tokar Gap in summer. J Geophys Res 118(1):421–434

Zhang Y, Rossow WB, Lacis AA, Oinas V, Mishchenko MI (2004) Calculation of radiative fluxes from the surface to top of atmosphere based on ISCCP and other global data sets: refinements of the radiative transfer model and the input data. J Geophys Res 109: D19105. doi:10.1029/2003JD004457

Water Mass Formation, Overturning Circulation, and the Exchange of the Red Sea with the Adjacent Basins

Sarantis Sofianos and William E. Johns

Abstract

The Red Sea experiences strong atmospheric forcing through both wind stress and air–sea buoyancy fluxes. Direct observations and modeling experiment show a robust response that consists of a strong and complicated three-dimensional circulation pattern with intense seasonal variability, involving water masses that are locally formed in the Red Sea or enter it from adjacent basins. Two thermohaline cells are identified related to intermediate (Red Sea Outflow Water—RSOW) and deepwater (Red Sea Deep Water—RSDW) formation processes. Results from numerical simulations indicate that the permanent cyclonic gyre in the northern end of the basin is the most probable location for the RSOW formation to take place, but further investigation with observations and numerical modeling techniques is needed to better understand the processes involved as well as the role of the Gulfs of Suez and Aqaba in the regional thermohaline circulation. The Red Sea circulation is closely linked to the flow pattern in the Strait of Bab-al-Mandab where the exchange of the Red Sea with the Indian Ocean takes place. The exchange is of an inverse estuarine type, which compensates for the strong heat and freshwater fluxes in the basin, but with very strong seasonal and synoptic variability related to remote and local forcing. Although important progress has been achieved during the last two decades in describing and understanding basic processes of the Red Sea dynamics, several features are not yet understood and explained. Further observational and modeling activities are required to improve our understanding of these processes and should be combined in an interdisciplinary approach to improve our monitoring and forecasting capabilities for environmental management and protection.

Introduction

The Red Sea is an elongated basin, oriented SE to NW, separating the African and Asian continents (Fig. 1). It extends from 12.5°N to 30°N with an average width of 220 km (Patzert 1974a). The mean depth of the basin is 524 m (Patzert 1974a), but with maximum recorded depths of almost 3,000 m. At its northern end, it bifurcates into two gulfs, the Gulf of Suez with average depth of 40 m and the Gulf of Aqaba with depths exceeding 1,800 m and a sill at the entrance of approximately 175 m (Neumann and McGill 1962).

Due to its arid setting, the Red Sea experiences one of the largest evaporation rates in the world's oceans. Combined with very small precipitation and negligible river runoff, it loses approximately 2 m/year of freshwater through its upper surface on an annual basis (Yegorov 1950; Neumann 1952; Privett 1959; Tragou et al. 1999; Bower and Farrar this volume). In order for the Red Sea's average salinity to remain in a steady state, this means that the Red Sea must import freshwater from adjacent basins. The transport through the Suez Canal, which connects the Mediterranean Sea with the Gulf of Suez and the Red Sea, is negligible, and therefore, the only significant connection between the Red

S. Sofianos (✉)
University of Athens, Athens, Greece
e-mail: sofianos@oc.phys.uoa.gr

W.E. Johns
Rosenstiel School of Marine and Atmospheric Science, University of Miami, Miami, USA

N.M.A. Rasul and I.C.F. Stewart (eds.), *The Red Sea*,
Springer Earth System Sciences, DOI 10.1007/978-3-662-45201-1_20,

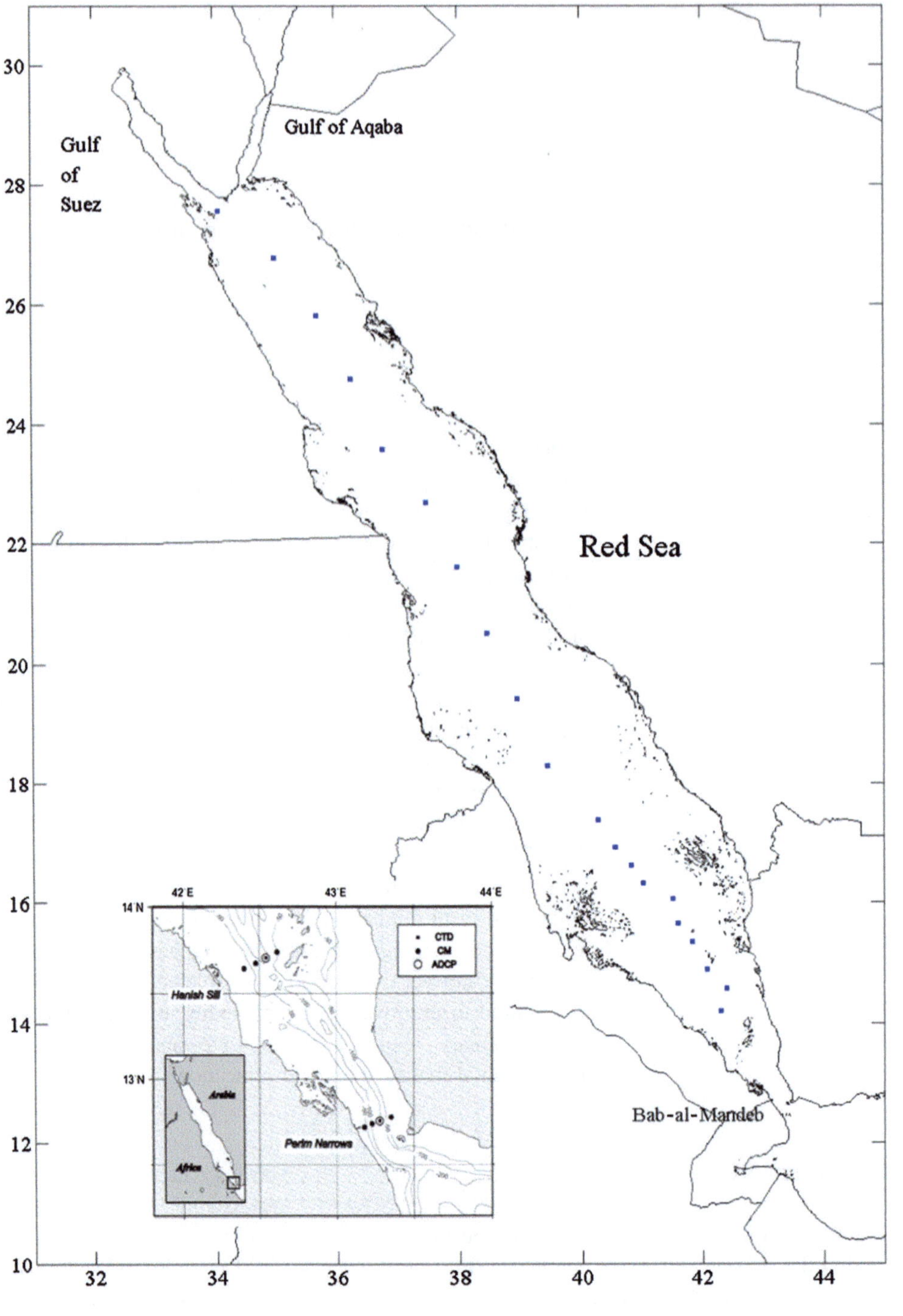

Fig. 1 Map of the Red. The Bab-al-Mandab Strait geography/topography is presented in the inset, including the locations of the current meter mooring arrays near the Hanish Sill and Perim Narrows deployed by Murray and Johns (1997). The *blue dots* are the stations occupied along the axis of the Red Sea in the 2001 cruise (Sofianos and Johns 2007)

Sea and the global ocean is the Strait of Bab-al-Mandab located at the southern end of the basin and connecting it to the Gulf of Aden and Indian Ocean. It is a shallow and narrow channel with a sill depth of 137 m (Werner and Lange 1975) and a minimum width of about 25 km.

Long-term observations of the flow and temperature/salinity characteristics at the Strait of Bab-al-Mandab (Murray and Johns 1997) indicate an equivalent annual freshwater flux into the Red Sea of 2.06 ± 0.22 m/year (Sofianos et al. 2002), which is consistent with the most recent evaporation rate estimates (Ahmad and Sultan 1987, 1989). The latent heat flux associated with the evaporation is very large (about 160 W/m^2); however, the net radiative heat gain nearly balances this loss, so that the annual mean heat loss from the surface of the Red Sea is small, O (10 W/m^2) (Patzert 1974b; Tragou et al. 1999; Sofianos et al. 2002; Ahmad and Albarakati this volume). The main buoyancy forcing that drives circulation in the Red Sea is therefore related to the high evaporation. Additionally, a seasonally changing wind stress pattern influences the Red Sea

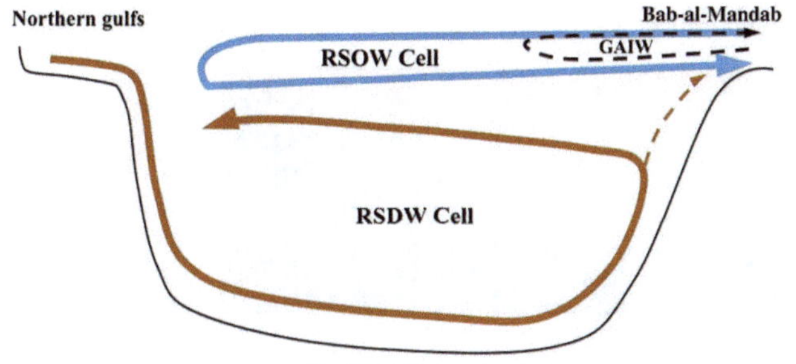

Fig. 2 Schematic representation of the main thermohaline cells in the Red Sea

circulation. The wind field is characterized by two regimes, one in the northern part of the Red Sea and one in the southern part. The former is in influenced by the eastern Mediterranean weather systems (Pedgley 1974) and the wind is from the northwest all around the year. In the south, the wind is controlled by the Indian monsoon system and reverses from southeast during winter (October–May) to northwest during summer (June–September).

Due to its limited connection with the world ocean and the excess of evaporation over precipitation and river runoff, the Red Sea acts as a typical concentration basin. An inverse estuarine circulation therefore occurs in the Strait of Bab-al-Mandab, with a relatively fresh surface inflow on the top of a deep outflow of higher salinity water (Fig. 2). Dense water mass formation and an associated overturning circulation within the basin are expected to close the inverse estuarine circulation cell driven by the thermohaline forcing. From the point of view of a physical oceanographer, the Red Sea is an advantageous experimental basin combining several interesting characteristics. It has a relatively simple shape, and its size is moderate. It experiences strong atmospheric forcing of both wind and thermohaline form, with strong seasonal and interannual variability. It is a semi-enclosed basin connected with the world ocean through a very shallow and narrow constriction, the Bab-al-Mandab Strait, whose exchange dynamics can affect processes over the whole basin.

Despite its scientific importance and its ecological and commercial significance, the Red Sea remains widely unexplored and several processes are still not well understood and are highly debated. During the last decades, a series of monitoring expeditions, processes-oriented studies and numerical modeling experiments have shed light on various processes and have allowed an improved understanding to be developed of the basin-wide circulation, water mass formation, thermohaline structure of the basin and its exchange with the Indian Ocean. Our current knowledge of these components is described in this chapter. We start from the identification of the water masses that participate in the Red Sea's general circulation. The next section is devoted to the identification of water mass formation processes and the pattern of overturning cells circulating the water masses in the basin. Finally, the exchange between the Red Sea and the open ocean, and its variability at various timescales, is presented in the last section.

The Water Masses of the Red Sea

Most of the available observations in the Red Sea have been sporadic in both time and space, making an attempt to construct a picture of the three-dimensional distribution of water mass characteristics in the Red Sea very difficult. The only exception is the area of the Strait of Bab-al-Mandab, which has attracted the interest of several observational efforts (Vercelli 1927; Siedler 1968; Maillard and Soliman 1986; Murray and Johns 1997). Until recent decades, north of the strait, inside the Red Sea basin, most of our knowledge of the distribution of oceanographic properties comes from a handful of expeditions covering very small parts of the basin (e.g., Vercelli 1931a, b; Morcos 1970; Morcos and Soliman 1974; Maillard 1974; Clifford et al. 1997). Some more recent expeditions covering a large part of the basin's hydrography (Quadfasel and Baunder 1993; Sofianos and Johns 2007) enable us to reconstruct a more comprehensive picture of the water mass structure in the Red Sea. Nevertheless, a more intensive and basin-wide observational strategy is still highly needed.

Much of the hydrographic data comes from sections taken along the axis of the Red Sea, such as the data shown in Fig. 3 from a summer 2001 survey (Sofianos and Johns 2007). A compilation of all historical stations collected in the Red Sea is shown in Fig. 4 in the form of temperature–salinity diagrams to identify the water masses. We can identify several water masses, either formed inside the basin, which have high salinities due to its concentration character, or imported from adjacent basins. Two water masses originating in the Gulf of Aden can be identified inside the Red Sea: the Gulf of Aden Surface Water (GASW), in the surface layer, and the Gulf of Aden Intermediate Water (GAIW), at intermediate depths (Fig. 3). They intrude into the basin during the winter and summer period, respectively, and are characterized by low salinities compared to the water masses formed in the Red Sea. The GASW forms the relatively fresh surface layer inflow in the Red Sea that corresponds to the upper part of the inverse estuarine exchange, associated with the fresh water loss from the surface of the sea. It is present only during the winter season and has temperature and salinity characteristics in the range of 25–30 °C and 36.0–36.5. On the other hand, the GAIW is present in the basin only during the summer season and forms an intrusion at intermediate depths, between the surface outflow and the deep outflow. It is identified in the southern half of the basin,

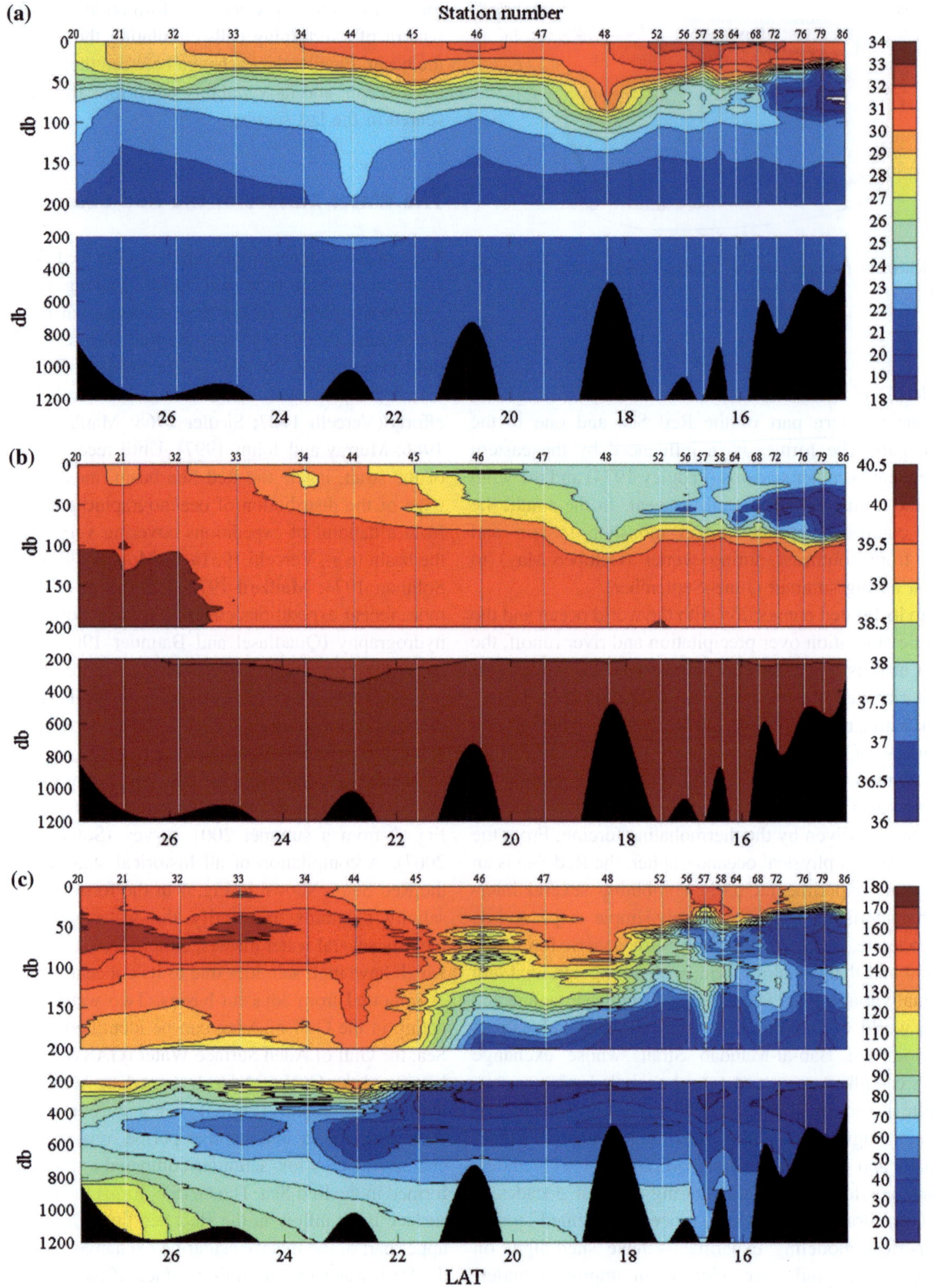

Fig. 3 **a** Potential temperature (°C), **b** salinity (psu), and **c** oxygen (μ mol/kg) concentration distribution along the axis of the Red Sea during summer 2001 (from Sofianos and Johns 2007). The stations correspond to Fig. 1 positions

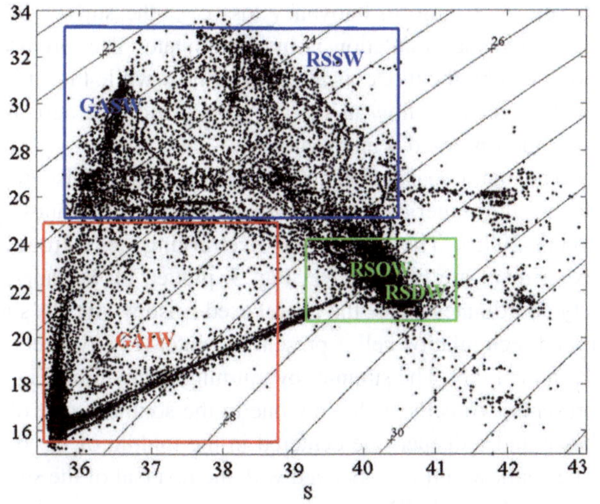

Fig. 4 Temperature–salinity (in practical salinity units) characteristics in the Red Sea derived from the NODC dataset with the addition of the 2001 cruise (the *red* area corresponds to GAIW, the *blue* area to the surface layers (RSSW and GASW) and the *green* area to the deep layers, namely the RSOW and the RSDW)

as a subsurface minimum in salinity and temperature which enters the basin with temperature and salinity values near 16.5 °C and 37.5, respectively. As it intrudes farther into the basin, it exhibits strong latitudinal gradients due to horizontal and vertical mixing, with potential temperature increasing to almost 30 °C and the salinity to over 38 (Fig. 3). At the same time, the depth of the salinity minimum gradually decreases, from around 75 m to very shallow depths (Sofianos and Johns 2007). The northern limit of the intrusion appears to be about 22°N. Although the strong circulation features may disperse lenses with GAIW to the northern part of the basin, no clear signal is detected north of 22°N. This latitude is close to latitude estimated from historical data by Smeed (1997) for the northernmost intrusion latitude at the end of the summer season (end of September). The mechanisms controlling the inflow of these water masses to the Red Sea will be discussed in Sect. "Exchange Between the Red Sea and Adjacent Basins".

The most important, locally formed water masses are the Red Sea Outflow Water (RSOW) and the Red Sea Deep Water (RSDW), the former occupying intermediate layers of the basin, while the latter fills the deep part of the basin. They are both characterized by high salinity values (over 40), being two of the most saline water masses of the world ocean. The RSDW is characterized by very homogeneous temperature and salinity values (Figs. 3 and 4) and fills the basin below the level of 150–250 m (this level deepens to the north and is modified by the strong circulation features present in the basin). On top of this water mass lies the RSOW flowing to the south to form the main core of the Red Sea deep outflow to the Gulf of Aden. The mechanisms involved in the formation of these two water masses are discussed in the next section. Finally, during summer, the surface waters outflowing from the basin through the Strait of Bab-al-Mandab acquire their characteristics in the Red Sea and are usually referred as the Red Sea Surface Water (RSSW).

The RSOW and RSDW have very similar temperature and salinity characteristics and for a long time they were identified as one water mass. It was only when basin-wide observations measuring other tracers such as oxygen, ^{14}C, and ^{3}He concentrations became available (e.g., Cember 1988; Quadfasel and Baunder 1993; Sofianos and Johns 2007) that it was possible to clearly distinguish between the two water masses. The RSDW is characterized by very low oxygen concentration, due to consumption while it spreads at bottom levels to the south and upwells and then recirculates to the north at higher levels (Fig. 2). The RSOW is a relatively younger water mass that contributes most of the high salinity waters to the outflow through the Bab-al-Mandab into the Indian Ocean. It is also characterized by the relatively higher oxygen concentration levels near 100–150 m in the southern Red Sea (Fig. 3c).

Water Mass Formation in the Red Sea and the Basin-wide Overturning Circulation

There is a wide agreement between investigators that the origin of the RSDW is located in the northern Gulfs of Suez and Aqaba. Most studies propose the Gulf of Suez as the main source of deep waters in the Red Sea basin (Woelk and Quadfasel 1996; Cember 1988; Maillard 1974; Wyrtki 1974). Studies in the Strait of Tiran (Hecht and Anati 1983; Murray et al. 1984; Manasrah et al. 2004), connecting the Gulf of Aqaba with the Red Sea basin, showed that there is also a contribution of waters from the Gulf of Aqaba to the deep layers of the Red Sea.

The deep waters of the Red Sea (deeper than 600 m) are renewed through sinking of dense waters formed in these northern gulfs. The time of this formation as well as the amount of RSDW produced is not well known. It is generally believed that it occurs during the peak of the winter season, when intense evaporation and surface cooling take place. Relatively small amounts of extremely dense water are produced in the gulfs and after mixing with ambient waters at the mouth of the gulfs sink to replenish the deep reservoir of the Red Sea. The RSDW formation rate at the Gulf of Suez, derived by direct observations (Maillard 1974), is on the order of magnitude of O (0.04 Sv) on annual basis (1 Sv = 10^6 m^3/s). Kuntz (1985), combining historical hydrography, Tritium and ^{3}He measurements in a two-dimensional advective–diffusive box model of the Red Sea, estimated the deep-water formation as low as 0.016 Sv. Cember (1988) estimated the formation rate at 0.05 Sv, based on ^{14}C measurements. Wyrtki (1974), based on a simple vertical advection diffusion model, estimated the deep-water formation at 0.06 Sv, while Woelk and Quadfasel (1996), using a one-dimensional plume model, estimated the Gulf of Suez outflow at 0.058 Sv, over a period of 7 months, but they also proposed that the deep-water formation process are presented only during winters with extreme forcing. The Gulf of Aqaba deep-water outflow is less investigated, although its contribution to the deep layers is thought to be of similar importance. Klinker et al. (1976) estimated the winter outflow from the gulf to the Red Sea at 0.069 Sv, Anati (1976) at 0.05 Sv, while Murray et al. (1984) measured the deep outflow over a period of 2 weeks and calculated a transport of 0.029 Sv. Biton and Gildor (2011), using a numerical model of the Gulf of Aqaba, estimated its contribution to be 0.0185 Sv. Overall, the total contribution from these two gulfs to the deep layers is estimated at the range of 0.05–0.1 Sv.

There are no direct observations available of RSOW formation. Sofianos and Johns (2003), using a numerical model approach, investigated the processes involved in the formation of this water mass. The results of the model simulations suggest that RSOW forms through open ocean convection that takes place in the northern Red Sea during the peak of the winter season. They proposed that it is associated with a cyclonic gyre that is centered at about 25°N. This feature is observed in several different studies (e.g., Clifford et al. 1997; Sofianos and Johns 2007) and seems to be a permanent pattern of the circulation in the northern basin. Inside this gyre, the mixed layer achieves high densities throughout the year, and at the peak of the winter season, the mixed layer has densities close to those of the RSOW layer.

Direct observations and modeling results agree that the waters formed inside the basin circulate in two basin-wide cells (Fig. 2): (i) an intermediate cell, carrying the RSOW toward the Strait of Bab-al-Mandab to form the main core of the deep Red Sea outflow and (ii) a deep cell recirculating the RSDW in deep layers of the basin, as can be seen from the oxygen distribution of the deep layers in Fig. 3. The interaction of the two cells and related mixing processes are not known. Another shallower cell as shown in Fig. 2 (black dashed line), generated during the summer season and related to the GAIW, is also hypothesized (Sofianos 2000).

The thermohaline forcing seems to be the most important mechanism driving the overall circulation in the Red Sea (Sofianos and Johns 2003). The pattern of the surface thermohaline cell is related to the buoyancy loss (mostly due to the very large evaporation rates), which increases the density to the north and produces a northward pressure gradient associated with the downward sloping of the sea surface, and a return flow at intermediate depths. The simplest theory for the thermohaline driven circulation of the Red Sea is the two-dimensional model by Phillips (1966), involving similarity solutions of a convectively driven model. The model is driven by a uniform buoyancy loss over the surface of the basin and the circulation is in steady state. The circulation produced by the model is a two-layer flow while the strength of the exchange flow at the sill is in qualitative agreement with the observed winter pattern.

The main seasonal thermohaline cells were also reproduced in numerical modeling simulations (Sofianos and Johns 2002). The meridional overturning cells from this experiment are presented in Fig. 5, plotted on latitude/density coordinates (since the model used is isopycnic). A single broad recirculation cell is present in winter (Fig. 5a) centered at about 17°N. The summer overturning circulation (Fig. 5b) presents two cells, a shallow one in the southern part of the basin and a deeper one centered in the middle of the basin. The shallow cell is associated with the reversal of the surface flow and the GAIW intrusion, in which water entering the basin in the GAIW layer upwells to feed the shallow outflow. The second, deeper, cell connects the intermediate inflow with the RSOW outflow, which is about a half of the strength in summer as it is in winter.

Taking into account the RSOW formation rate and the volume of the lower branch of the surface cell, the renewal time of the layer is on the order of a few years. The simulations showed that about two-third of the total amount of Red Sea deep outflow into the Gulf of Aden (about 0.25 Sv) is formed inside the cyclonic gyre at 25°N, while the remainder is formed north of 20°N latitude by mixing with the surrounding layers. Thus, it can be concluded that the dominant formation region for the RSOW is the northern Red Sea where the vertical structure and the water trapping related to the cyclonic gyre and the strong buoyancy forcing in winter provide a favorable environment for intermediate depth convection. These conclusions are also consistent with tracer distributions of ^{14}C and ^{3}He in the basin (Cember 1988).

The renewal timescale of the deep cell, involving RSDW, is, in contrast to its surface counterpart, on the order of decades. Based on ^{14}C and ^{3}He data from the GEOSECS Indian Ocean Expedition, Cember (1988) estimated a deepwater residence time of about 36 years. Woelk and Quadfasel (1996) used hydrographic data in a one-dimensional plume model and long-term atmospheric forcing to estimate a bulk residence time of the deep waters 73 between 40 and 90 years. They also proposed that the formation of deep waters is episodic and can take place only under extreme meteorological conditions ("convection winters"). These extreme years occur at a timescale of 4–7 years. The influence of this wintertime atmospheric forcing variability on the RSOW formation has not been fully investigated.

The RSDW layer circulation is relatively sluggish and it can be only traced by the oxygen depletion (Sofianos and Johns 2007) as it recirculates to the south in the lower branch and northward at mid-depths. There is little evidence for RSDW escaping through the bottom of the Strait of

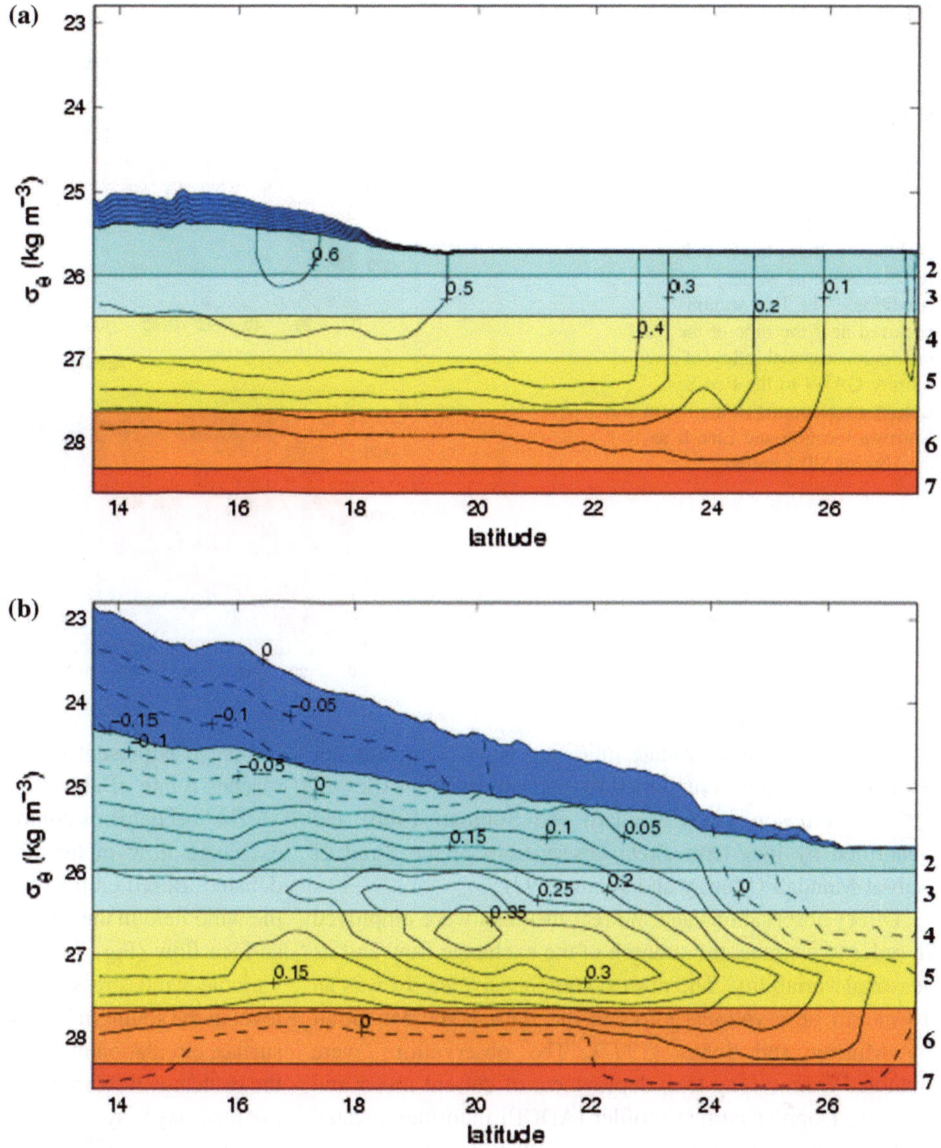

Fig. 5 Meridional overturning volume fluxes (in Sv) on latitude/density coordinates for **a** December and **b** June, from MICOM simulation (Sofianos and Johns 2002). *Colors* indicate different isopycnal model layers, and contours indicate the overturning cells (note that the two panels have different contour intervals). During the winter regime (**a**), RSOW is formed mainly in the northern Red Sea north of 23°N and flows southward toward the strait (on the *left* side of the figure) at densities (σ_θ) of between 26.5 and 27.8 kg m^{-3}. During the summer regime (**b**), inflow of GAIW at intermediate densities ($\sigma_\theta = 24.5 - 26.0$ kg m^{-3}) partly upwells into the southward-flowing RSSW south of about 20°N, and the remainder continues northward to supply sinking of RSOW between 22 and 26°N

Bab-al-Mandab, and thus, we can assume that this layer upwells into the base of the RSOW layer at a rate approximately equal to its formation rate. The contribution of this cell to the Red Sea outflow is minor compared to that of the RSOW formed by the intermediate depth convection, but it nevertheless plays an important role in the ventilation of the deep parts of the Red Sea.

Exchange Between the Red Sea and Adjacent Basins

The circulation pattern and water mass formation processes inside the Red Sea are reflected in its exchange with the adjacent basins. In winter, an inverse estuarine exchange flow dominates the circulation at the Strait of Bab-al-Mandab, with a relatively fresh surface inflow and a deep outflow of water with higher salinity. The diverse stratification in the strait during winter and summer seasons can be readily depicted in the salinity profiles in the strait. Two salinity sections corresponding to the winter and summer exchange patterns, from cruises in the Bab-al-Mandab (Murray and Johns 1997) are presented in Fig. 6. The winter pattern consists of two layers, with a relatively fresh layer on top of a hypersaline lower layer, while in summer, the stratification changes to a three-layer system with a very fresh layer at intermediate depths between two layers with much higher salinity. An Italian expedition on board the AMIRAGLIO MAGNAGHI carried out current measurements along the Strait of Bab-al-Mandab during summer (Vercelli 1931a, b) and for the first time demonstrated the presence of a three-layer exchange during this season, with surface and deep

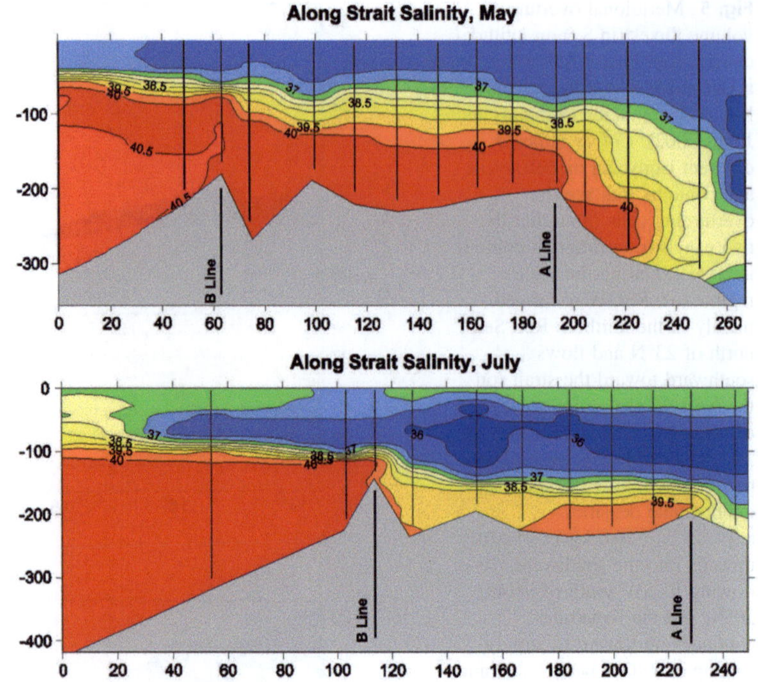

Fig. 6 Along strait salinity transects during May 1995 (*upper panel*) and July 1996 (*lower panel*), from along-strait hydrographic sections taken during the Murray and Johns (1997) Bab-al-Mandab study. The May section occurred during the late phase of the winter two-layer exchange pattern, but is still representative of winter conditions. The July section occurred near the time of the maximum seasonal inflow of low-salinity GAIW to the Red Sea. Line A corresponds to the Perim Narrows location and Line B to the Hanish Sill location

outflows and an intermediate inflowing layer. This summer exchange pattern was supported by later observations (Sewell 1934; Thompson 1939; Maillard and Soliman 1986) and quantified by long-term direct observations at the Strait of Bab-al-Mandab (Murray and Johns 1997).

Direct observations and indirect methods were employed in order to estimate the strength of the exchange flow and its seasonal variability. The most comprehensive study was an 18-month observational effort in the Bab-al-Mandab Strait by Murray and Johns (1997). The observations were obtained from a long-term deployment of current meter and acoustic Doppler current profiler (ADCP) moorings located at the Hanish Sill and Perim Narrows (Fig. 1). The mooring arrays were deployed in two 9-month settings, from June 1995 to March 1996 and from March to December 1996. The volume transport profile for the Perim Narrows derived from these observations, obtained by integrating the along-strait flow across the width of the channel, is presented in Fig. 7 (Johns and Sofianos 2012). It has been filtered with a 60-day low-pass filter to emphasize the seasonal cycle. The basic seasonal pattern of the exchange, two layers in winter (October–May) and three layers in summer (June–September), is clearly revealed. Based on this data, the monthly mean transport of the three layers was estimated (Sofianos et al. 2002) and is presented in Fig. 8, together with other historical estimates. The strength of the winter exchange achieves its maximum during February when the RSOW layer has an outflow rate of more than 0.6 Sv. This outflow becomes very weak in summer with a minimum of 0.05 Sv during August. During August, the strongest three-layer exchange is present in the strait, when the GAIW intrusion reaches a rate of about 0.3 Sv.

The mechanisms controlling the seasonal variability of the exchange flow in the Strait of Bab-al-Mandab are highly debated. Based on the obvious coincidence of the reversal of the wind field in the southern Red Sea with the reversal of the surface flow (Fig. 7), Patzert (1974a) hypothesized that the variable wind stress and the associated changes of the sea level in the southern Red Sea and the Gulf of Aden control the surface circulation in the region throughout the year as well as the exchange flow in the Strait of Bab-al-Mandab. To explain the three layer system and the associated GAIW intrusion at intermediate depths during summer, Patzert proposed that the northwest winds and a drop in the sea level at the north-western Gulf of Aden due to wind-induced upwelling forces a shallow surface layer outflow from the Red Sea. The same mechanism displaces the subsurface isopycnals upward, producing an opposite pressure gradient at intermediate depths that induces the intermediate intrusion of the GAIW. A different mechanism controlling the seasonal exchange flow between the Red Sea and the Gulf of Aden, based on the buoyancy fluxes over the Red Sea basin, was proposed by Sofianos (2000) using a modified version of the Phillips (1966) model. This model was also able to explain the summer season subsurface intrusion. Modeling experiments performed by Sofianos and Johns (2002) verified the relevance of the Patzert wind-driven mechanism. They also showed that the combination of the two forcing mechanisms, that is, wind stress and buoyancy flux, can reproduce the seasonal pattern and the amount of the inflowing water masses.

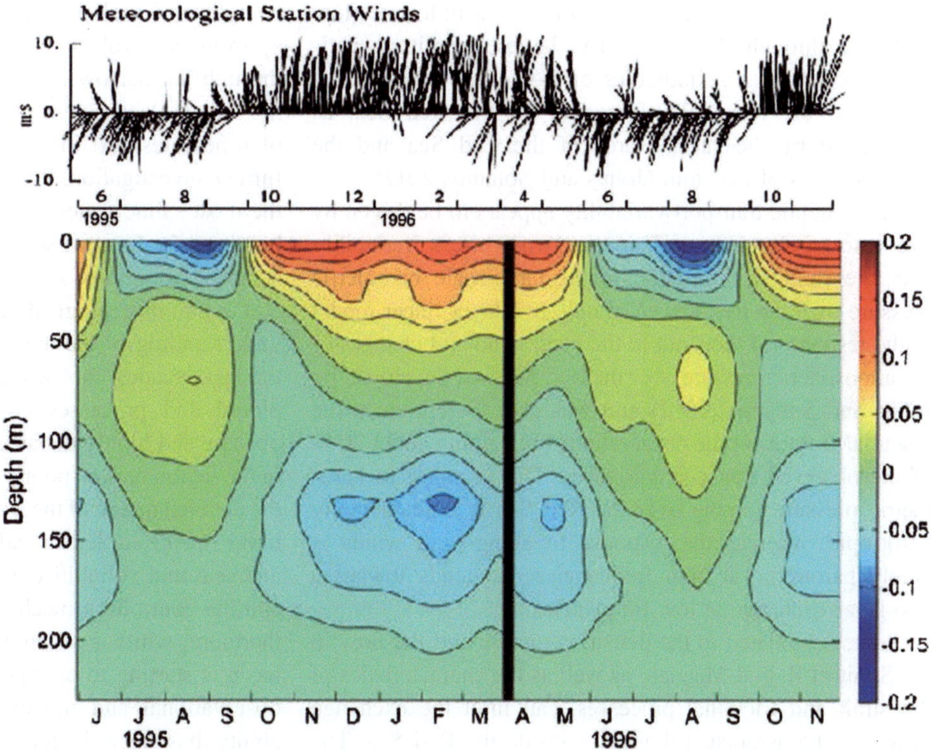

Fig. 7 The wind time series from a meteorological station located at the Strait of Bab-al-Mandab and the transport-per-unit-depth profile at the Perim Narrows section, after low-pass filtering with a 60 day filter to emphasize the seasonal exchange pattern through the strait (*units* 10^4 m^2 s^{-1}). Positive transports correspond to inflow to the Red Sea from the Gulf of Aden. The masked-out area (*black bar*) corresponds to the period when moorings were serviced midway through the experiment. (From Johns and Sofianos 2012)

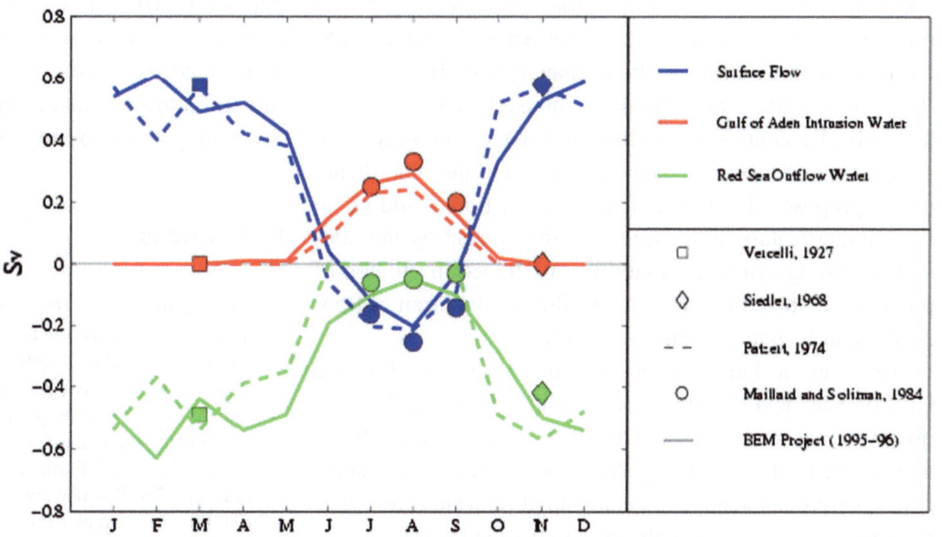

Fig. 8 Monthly mean transport (in Sv) for the three layers: surface (*blue*), GAIW (*red*), and RSOW (*green*). *Solid lines* represent the layer transport monthly mean values derived from Murray and Johns (1997), the *dashed lines* represent the seasonal layer fluxes estimated by Patzert (1974b), and other symbols represent various historical estimates

Averaging the volume transport of the three layers over the annual cycle, we can estimate the average strength of the exchange flow in Bab-al-Mandab. The annual mean RSOW outflow, which is a good measure of the overall strength of the exchange, was estimated at 0.36 Sv, and showed a large annual cycle ranging from over 0.6 Sv during the peak of the winter (February) to less than 0.1 Sv during the peak of the summer season (Fig. 8). The reversal of the surface flow and the intrusion of the GAIW were found to last for 4 months (June–September) when an average of 0.22 Sv of the GAIW enters the Red Sea basin.

In addition to the seasonal variability, the exchange experiences strong fluctuations at shorter timescales, including tidal (dominantly semidiurnal) and synoptic timescales ranging from a few days to several weeks. Transport variations through the strait on synoptic timescales (days to weeks) appear very strong, reaching values nearly twice as large as the mean rate of the exchange through the strait driven by annual evaporation over the Red Sea (Johns and Sofianos 2012). A very interesting feature is the large-amplitude variability of the transport through the strait that occurs at high frequencies that can reach amplitudes of up to

1 Sv (comparable in magnitude to the total inflow/outflow exchange through the strait). The largest amplitude variability occurs near periodicities of 5–6 days, which corresponds to the Helmholtz frequency of the Red Sea, as determined by the surface area of the Red Sea and the characteristics of the strait (Johns and Sofianos 2012).

The synoptic transport variability appears to be driven by two primary forcing mechanisms: wind stress variability over the strait and variation in the large-scale barometric pressure over the Red Sea. A simple two-layer linear model of the response of the strait to the local wind and changes of the atmospheric pressure over the Red Sea was developed by Johns and Sofianos (2012) and was able to reproduce the essential features of the observations (explaining about 70 % of the observed transport variance). The response to barometric pressure forcing over the Red Sea is fundamentally barotropic, whereas the response to along-strait winds is nearly barotropic at high frequencies and tends toward a two-layer exchange at low frequencies.

Processes related to the density structure and the flow in the Strait of Bab-al-Mandab, as well as the characteristics of the strait and frictional processes, can limit the exchange flow and the associated dynamics inside the Red Sea. The direct and detailed observations at the strait gave the opportunity to investigate these processes and evaluate the possibility of hydraulic control (Farmer and Armi 1988; Armi and Farmer 1988) in the exchange flow. It was found (Pratt et al. 1999, 2000) that the mean exchange flow is not hydraulically controlled in either the winter (two-layer) or summer (three-layer) exchange regimes. On the other hand, it was proposed that hydraulically critical flow could be an intermittent feature, influenced to a great extent by the tides and the strong synoptic variability. A detailed discussion of processes involved in the exchange flow in the Strait of Bab-al-Mandab is given by Smeed (2004).

There are no long-term observations at the mouth of the Gulf of Suez and the mouth of the Gulf of Aqaba. The few observational studies (Maillard 1974; Murray et al. 1984) were carried out for relatively short periods, and the exchange patterns were not studied in detail and at relevant timescales. Systematic monitoring of the two gulf mouths are of great importance for understanding the response of these regions to atmospheric forcing as well as for quantifying the contribution of the gulfs to the Red Sea thermohaline circulation.

Summary and Discussion

Due to its shape, size, and the intense atmospheric forcing of both wind and buoyancy flux, the Red Sea is a very advantageous experimental basin from a physical oceanographic viewpoint. It can be thought of as a miniature global ocean, with strong thermohaline and wind-driven circulation components, shelf and open ocean water mass formation and a strong and robust overturning circulation. Its exchange through the narrow and shallow Bab-al-Mandab strait with the Indian Ocean presents strong variability on a wide range of timescales and involves a variety of process that need further investigation. The outflow of the Red Sea is one of the most saline water masses of the world ocean, and it can be traced far from its source into the equatorial and southern subtropical regions of the Indian Ocean.

Despite its ecological importance and the advances in our understanding of the dynamics of the Red Sea made during the last decades, this sea nevertheless remains widely unexplored and processes controlling its circulation are still obscure and highly debated. Several aspects of the dynamics in the region remain poorly or only partly understood, such as the detailed nature of the water mass formation processes both in the open Red Sea and adjacent northern gulfs, the patterns and seasonal variability of the surface and deep circulation, and the controlling mechanisms of the strait exchange. Furthermore, while a coherent picture of the mean state of the basin is starting to emerge from the combination of various observational and modeling studies, the interannual variability has been largely neglected and only recently has attracted the attention of the oceanographic community (e.g., Alraddadi 2013). Long-term monitoring observations, complemented by numerical modeling techniques, are needed to understand the sensitivity of the circulation and water mass formation to wind and thermohaline forcing, and its connection with global- and regional-scale climate variability.

References

Ahmad F, Sultan SAR (1987) On the heat terms in the central region of the Red Sea. Deep Sea Res 34:1757–1760

Ahmad F, Sultan SAR (1989) Surface heat fluxes and their comparison with the oceanic heat flow in the Red Sea. Oceanol Acta 12:33–36

Alraddadi TM (2013) Temporal changes in the Red Sea circulation and associated water masses. University of Southampton, Doctoral thesis, Ocean and Earth Science, 198 pp

Anati DA (1976) Balance and transports in the Red Sea and the Gulf of Elat (Aqaba). Israel J Earth Sci 25:104–110

Armi L, Farmer DM (1988) The flow of Atlantic water through the Strait of Gibraltar. Prog Oceanogr 21:1–105

Biton E, Gildor H (2011) The general circulation of the Gulf of Aqaba (Gulf of Eliat) revisited: the interplay between the exchange flow through the Straits of Tiran and surface fluxes. J Geophys Res 116: C08020. doi:10.1029/2010JC006860

Cember RP (1988) On the sources, formation, and circulation of Red Sea Deep Water. J Geophys Res 93:8175–8191

Clifford M, Horton C, Schmitz J, Kantha LH (1997) An oceanographic nowcast/forecast system for the Red Sea. J Geophys Res 102:25101–25122

Farmer DM, Armi L (1988) The flow of Mediterranean water through the Strait of Gibraltar. Prog Oceanogr 21:1–105

Hecht A, Anati D (1983) A description of the Straits of Tiran in winter 1978. Israel J Earth Sci 32:149–164

Johns WE, Sofianos SS (2012) Atmospherically forced exchange through the Bab el Mandeb Strait. J Phys Oceanogr 42:1143–1157

Klinker J, Reiss Z, Kropach C, Levanon I, Harpaz H, Halicz E, Assaf G (1976) Observations on the circulation pattern in the Gulf of Elat (Aqaba), Red Sea. Israel J Earth Sci 25:85–103

Kuntz R (1985) Bestimmung der Tiefenwasserzirkulation des Roten Meeres anhand einer Boxmodellauswertung von Tritium-, ^{3}He- und Dalinitatsdaten. Ph.D. thesis, Ruprecht Karls University, Heidelberg, 76 pp

Maillard C (1974) Eaux intermediaires et formation d'eau profonde en Mer Rouge. L'oceanographie physique de la Mer Rouge. Cent. Natl. pour l'Exploitation des Oceans, Paris, pp 105–133

Maillard C, Soliman G (1986) Hydrography of the Red Sea and exchanges with the Indian Ocean in summer. Oceanol Acta 9:249–269

Manasrah RS, Badran M, Lass HU, Fennel W (2004) Circulation and the winter deep-water formation in the northern Red Sea. Oceanologia 46:5–23

Morcos SA (1970) Physical and chemical oceanography of the Red Sea. Oceanogr Mar Biol Ann Rev 8:73–202

Morcos S, Soliman GF (1974) Circulation and deep water formation in the northern Red Sea in winter (based on R/V Mabahiss sections, January-February, 1935). L'oceanographie physique de la Mer Rouge. Cent. Natl. pour l'Exploitation des Oceans, Paris, pp 91–103

Murray S, Hecht A, Babcock A (1984) On the mean flow in the Tiran Strait in winter. J Mar Res 42:265–284

Murray SP, Johns W (1997) Direct observations of seasonal exchange through the Bab el Mandeb Strait. Geophys Res Lett 24:2557–2560

Neumann J (1952) Evaporation from the Red Sea. Israel Explor J 2:153–162

Neumann AC, McGill DA (1962) Circulation of the Red Sea in early summer. Deep Sea Res 8:223–235

Patzert WC (1974a) Wind-induced reversal in Red Sea circulation. Deep Sea Res 21:109–121

Patzert WC (1974b) Volume and heat transports between the Red Sea and Gulf of Aden, notes on the Red Sea heat budget. L'oceanographie physique de la Mer Rouge. CNEXO, Paris, pp 191–201

Pedgley DE (1974) An outline of the weather and climate of the Red Sea. L'Oceanographie Physique de la Mer Rouge. Cent. Natl. pour l'Exploitation des Oceans, Paris, pp 9–27

Phillips OM (1966) On turbulent convection currents and the circulation of the Red Sea. Deep Sea Res 13:1149–1160

Pratt LJ, Johns W, Murray SM, Katsumata K (1999) Hydraulic interpretation of direct velocity measurements in the Bab al Mandab. J Phys Oceanogr 29:2769–2784

Pratt LJ, Deese HE, Murray SP, Johns W (2000) Continuous dynamical modes in straits having arbitrary cross sections, with applications to the Bab al Mandab. J Phys Oceanogr 30(10):2515–2534

Privett DW (1959) Monthly charts of evaporation from the Indian Ocean (including the Red Sea and the Persian Gulf). Q J Royal Meteorol Soc 85:424–428

Quadfasel D, Baunder H (1993) Gyre-scale circulation cells in the Red Sea. Oceanol Acta 16:221–229

Sewell RBS (1934) The John Murray expedition to the Arabian Sea. Nature 133:86–89

Siedler G (1968) Schichtungs- und Bewegungsverhaltnisse am Sudausgang des Roten Meeres. Meteor Forschungsergeb A 4:1–76

Smeed D (1997) Seasonal variation of the flow in the strait of Bab el Mandeb. Oceanol Acta 20:773–781

Smeed DA (2004) Exchange through the Bab el Mandab. Deep Sea Res Part II 51:455–474

Sofianos SS (2000) Circulation and water mass formation in the Red Sea, and the exchange with the Indian Ocean. Ph.D. thesis, University of Miami, Miami, Florida

Sofianos SS, Johns WE, Murray SP (2002) Heat and freshwater budgets in the Red Sea from direct observations at Bab el-Mandeb. Deep Sea Res Part II 49:1323–1340

Sofianos SS, Johns WE (2002) An Oceanic General Circulation Model (OGCM) investigation of the Red Sea circulation: 1. Exchange between the Red Sea and the Indian Ocean. J Geophys Res 107 (C11):3196. doi:10.1029/2001JC001184

Sofianos SS, Johns WE (2003) An Oceanic General Circulation Model (OGCM) investigation of the Red Sea circulation: 2. Three dimensional circulation in the Red Sea. J Geophys Res 108 (C3):3066. doi:10.1029/2001JC001185

Sofianos S, Johns WE (2007) Observations of the summer Red Sea circulation. J Geophys Res 112:C06025. doi:10.1029/2006JC003886

Thompson EF (1939) Chemical and physical investigations. The exchange of water between the Red Sea and the Gulf of Aden over the "Sill". John Murray expedition 1933–34, vol 2, Scientific Reports, 105 pp

Tragou E, Garrett C, Outerbridge R, Gilman G (1999) The heat and freshwater budgets of the Red Sea. J Phys Oceanogr 29:2504–2522

Vercelli E (1927) Richerche di oceanografia fisica eseguite della R. N. AMMIRAGLIO MAGNAGHI (1923–24), IV, La temperatura e la salinita. Ann Idrograf 11:1–66

Vercelli E (1931a) Nuove richerche sulli correnti marine nel Mar Rosso. Annali Idrografici 12:1–74

Vercelli E (1931b) Le esplorazioni talassografiche nel Mar Rosso. Atti del Primo Congresso di Studi Coloniali, Firenze, 23 pp

Werner F, Lange K (1975) A bathymetric survey of the sill area between the Red Sea and the Gulf of Aden. Geologisches Jahrbuch D 13:125–130

Woelk S, Quadfasel D (1996) Renewal of deep water in the Red Sea during 1982–1987. J Geophys Res 101:18155–18165

Wyrtki K (1974) On the deep circulation of the Red Sea. L'oceanographie physique de la Mer Rouge. Cent. Natl. pour l'Exploitation des Oceans, Paris, pp 135–163

Yegorov NI (1950) Calculation of the heat balance of the Red Sea. Meteorologiya i Gidrologiya 3:49–56

Heat Balance of the Red Sea

Fazal Ahmad and Alaa M.A. Albarakati

Abstract

The Red Sea is one of the few regions of the world where the balance of the surface heat fluxes at the air–sea interface can be compared favourably with the advective heat flux at Bab-al-Mandab. Various estimates of the surface heat fluxes and advective oceanic heat flux are discussed and their average values are derived. The annual average values for the absorbed solar radiation at the sea surface, net long-wave radiation, latent heat and sensible heat fluxes are, respectively, 220 ± 6, 70 ± 5, 165 ± 4 and −3 W m^{-2} for the Red Sea as a whole. This gives a heat loss of −12 W m^{-2} at the air–sea interface which is almost in agreement with the net advective flux of 13 ± 4 W m^{-2} to the Red Sea at Bab-al-Mandab. Because of its location in an arid area, the evaporation is high and precipitation is very low. The average precipitation along the eastern side of the Red Sea is about 0.05 m per year. The estimates of evaporation vary in time and space, and the average is about 2 m per year for the whole Red Sea. Based on the Comprehensive Ocean Atmospheric Data Set (COADS) from 2000 to 2010, the monthly averages of latent heat flux are calculated for the southern, central and northern regions of the Red Sea. The variability of the net long-wave radiation and sensible heat flux for these regions is also shown. The surface heat fluxes over the Red Sea show a wide temporal and spatial variation.

Introduction

In recent years, there has been growing interest in the meteorology and oceanography of the Red Sea and the adjacent waters. This interest has developed because of the recognition of the importance of air–sea interface processes. The atmosphere and the ocean form a coupled system, exchanging heat, momentum and water at the interface. A good understanding of the processes involved in the air–sea interaction is necessary for weather and climate prediction, and changes over days, weeks, years and centuries are influenced by this interaction. The transfer of heat from one compartment to the other is accomplished through the processes of conduction, evaporation and radiation.

Various methods of determining fluxes are summarized by Pond (1975) and Kraus (1972) that include the eddy correlation method, profile method, dissipation method and bulk aerodynamic method, and recently, comparative features of modern bulk flux algorithms to estimate surface fluxes have been discussed by Fairall et al. (2003) and Singh et al. (2005). Major uncertainties remain in our understanding of the fundamental processes of the air–sea interface particularly in heterogeneous and non-equilibrium conditions. More comprehensive parameterization of the surface processes is required as well as improvement in satellite retrievals and assimilations in numerical models. The Red Sea is one of the few regions where advective oceanic heat flux can be known to provide a useful comparison with the estimates of the net surface heat flux at the air–sea interface. The conditions in the Red Sea and the overlying atmosphere have adjusted themselves so that both the annual net exchange of heat with the atmosphere and the advective heat flux through Bab-al-Mandab must nearly

F. Ahmad (✉) · A.M.A. Albarakati
Faculty of Marine Sciences, King Abdulaziz University, Jeddah, Saudi Arabia
e-mail: faali@kau.edu.sa

N.M.A. Rasul and I.C.F. Stewart (eds.), *The Red Sea*,
Springer Earth System Sciences, DOI 10.1007/978-3-662-45201-1_21,

balance, when the minor terms such as conduction through the sea bed, condensation of water vapour, transformation of kinetic energy into heat and heating due to chemical processes are neglected.

The basic heat balance equation in its simplest form, for an area like the Red Sea over an annual cycle is the following:

$$Q_s - Q_b - Q_e - Q_h + Q_v = 0$$

Q_s	is the absorbed solar radiation at the sea surface
Q_b	is the net upward long-wave radiant energy from the sea
Q_e and Q_h	are latent and sensible heat fluxes from the sea surface
Q_v	is the net advective heat flux to the area

Absorbed Solar Radiation (Q_s)

The solar radiation incident on the sea surface depends on the time of day, the latitude and the atmospheric absorption and scattering. The direct and diffuse short wave radiation that reaches the sea surface is absorbed at the water surface and a small part is reflected from it. The amount of reflection depends on the solar zenith angle and sea surface roughness. The absorbed radiation is:

$$Q_s = Q_0(1 - \alpha)$$

where α is the reflection coefficient and Q_0 is the solar radiation incident on the sea surface.

Atmospheric scientists have developed methods for the estimation of incoming solar radiation, but the basic data over the oceans required as input to these methods are generally lacking. Hence oceanographers use various formulae needing input such as the solar altitude and cloud cover to estimate the insolation from the clear sky irradiance that reaches the sea surface (List 1958; Reed 1977; Sechel and Beaudry 1973; Rosati and Miyakoda 1988). From the clear sky values Q_0, Reed (1977) suggested a relation of the following form for the absorbed solar radiation Q_s:

$$Q_s = Q_0(1 - c_n n + 0.0019h)(1 - \alpha)$$

Here n is the cloud cover in tenths, h is the noon solar altitude in degrees, c_n is a cloudiness coefficient with value 0.62, and α is the reflection coefficient from the sea surface that varies with the time of day, being highest in the first and second hours after sunrise, and also in the last and second to last hours before sunset. It decreases towards noon (≈3 %). Ahmad and Sultan (1989) considered an average reflection of 6 % for the whole day over the Red Sea.

Under conditions of cloudiness of <0.3 a reduction of 5 % is suggested rather than use of the above equation (Reed 1977). Generally the monthly average cloud cover over the Red Sea is <0.3. da Silva et al. (1994) provided a tuning factor of 0.92 for the clear sky radiation and 1.04 for the cloudiness coefficient, giving:

$$Q_s = 0.92Q_0(1 - 1.04\,c_n n + 0.0019h)(1 - \alpha)$$

Applying the formula for clear sky insolation by Sechel and Beaudry (1973), called the Smithsonian formula, Al-Madani (2002) gave a value of 238 W m^{-2} for the insolation over the Red Sea after taking into account the cloud cover and the 11 % reduction due to aerosols as suggested by Tragou et al. (1999). Matsoukas et al. (2007) also reported an insolation of 238 W m^{-2} for the Red Sea. This amounts to absorbed radiation of about 224 W m^{-2}, after considering a reflection of 6 % at the sea surface for the Al-Madani (2002) and Matsoukas et al. (2007) values of 238 W m^{-2}.

The recorded solar radiation from Port Sudan (≈20°N, 41°E) between 1964 and 1968 gave a mean annual value of 225 W m^{-2} for the absorbed solar radiation after considering an albedo of 6 % at the sea surface. At Aden (13°N, 45°E) the recorded solar radiation from 1958 to 1964 was 246 W m^{-2} which gives absorbed radiation of 231 W m^{-2} (Bunker et al. 1982).

Other estimates for solar radiation by empirical formulae include 230 W m^{-2} by Hastenrath and Lamb (1979), which gives 216 W m^{-2} for the absorbed radiation. Bunker and Goldsmith (1979) and Bishop et al. (1997) gave values of 263 and 273 W m^{-2}, respectively. However, their calculations do not take into account the increased opacity over the Red Sea that is due to sand and dust. The appropriate values for the flux of solar radiation may then be lower than those calculated, by as much as by 10–20 W m^{-2} (Bunker et al. 1982). da Silva et al. (1994) reported a value of 230 W m^{-2} for insolation with aerosols corrections included. This amounts to absorbed solar radiation of 216 W m^{-2}. However, concentrations of natural and anthropogenic aerosols over the Red Sea vary both in space and time so that parameterization of aerosol effects as constants can lead to significant errors in estimating the insolation (Tragou et al. 1999).

Based on the measurements of solar radiation at inland stations along the eastern side of the Red Sea Ahmad and Sultan (1987, 1989) gave a value of 220 and 210 W m^{-2} for the absorbed solar radiation, respectively, for the central region and the Red Sea as a whole. Ahmad et al. (1989) reported an annual average of 214 ± 3 W m^{-2} near Jeddah and Abdulrahman and Ahmad (1994) gave a value of 216 W m^{-2} for the absorbed solar radiation at the sea surface for the whole Red Sea. From these studies, the annual average absorbed solar radiation at the sea surface for the Red Sea is 220 ± 6 W m^{-2}.

Net Upward Long-Wave Radiation (Q_b)

Besides absorbing and reflecting solar radiation the sea surface also emits radiation of a wavelength depending on its temperature and absorbs long-wave radiation transmitted downward from clouds and the atmosphere. The net upward long-wave radiation is at much larger wavelength than that of the solar radiation because the sea surface is far cooler than the sun's surface. Simpson and Paulson (1979) have discussed the considerable errors that can be made by using empirical formulae to estimate the net long-wave radiation. The net long-wave radiation at the sea surface depends on the vertical distribution of the temperature, humidity and cloud properties. It is common to derive Q_b with the aid of various empirical formulae or with a radiation transfer model (Zhang et al. 1995). The empirical formulae assume that the surface properties of temperature and humidity represent those of the atmospheric column and consider various parameterization schemes with constants derived from regression fitting to certain sets of observations. In general, only the cloud fraction is taken into consideration and the cloud type (high, medium or low) is ignored, thus increasing the uncertainty of calculations (Fung et al. 1984). However, empirical formulae are still widely used because they allow the computation of Q_b from routinely measured parameters such as sea surface temperature, air temperature, the water vapour pressure gradient over the sea surface and cloudiness.

The empirical formulae are of the type:

$$Q_{b_0} = \varepsilon\sigma T_0^4(a - b\sqrt{e_a})$$

Q_{b_0} is the net long-wave radiation under cloudless conditions
ε is the coefficient of emissivity average value of about 0.985 for the sea surface (Kraus 1972)
σ is the Stefan–Boltzmann constant
T_0 is the absolute sea surface temperature
a and b are constant and e_a is the water vapour pressure in the atmosphere at a standard height

At small absolute humidity, the dependence of net long-wave radiation on humidity should be used in a somewhat modified form. The net long-wave radiation is influenced considerably by the cloudiness and sea–air temperature difference. The effect of the cloudiness, latitude and sea–air temperature difference is usually accounted for by a relation of the type:

$$Q_b = Q_{b_0}(1 - cn^2) + 4\varepsilon\sigma T_0^3(T_0 - T_a)$$

where Q_b is the net long-wave radiation with cloudiness n (as a fraction of unity), the coefficient c depends on latitude and the sea–air temperature difference ($T_0 - T_a$) is in degree K.

The estimations of net long-wave radiation for the Red Sea vary considerably. The value given by Bunker (1976) is 48 W m^{-2} which is considered very low by Bunker et al. (1982). His own method for estimating the net long-wave flux produced a considerably increased value of 76 W m^{-2} for the Red Sea (Bunker and Goldsmith 1979). Hastenrath and Lamb (1979) calculated a value of 75 W m^{-2}. Other estimates include 57 W m^{-2} by Ahmad and Sultan (1987) for the central region of the Red Sea, 66 W m^{-2} by Ahmad and Sultan (1989) for the Red Sea as a whole and 66 ± 1 W m^{-2} by Ahmad et al. (1989) for the coastal waters near Jeddah. Abdulrahman and Ahmad (1994) gave a value of 63 W m^{-2}, Bignami et al. (1995) gave 76 W m^{-2}, Rossow and Zhang (1995) gave 104 W m^{-2}, while da Silva et al. (1994) gave 53 W m^{-2} for the Red Sea. However, the resolution of the Rossow and Zhang (1995) data set was 2.5° × 2.5° which is very coarse for the narrow Red Sea and the da Silva et al. (1994) value is derived from the Atlas of surface marine data. Josey et al. (1997) state that the Bignami et al. (1995) formula performs well in semi-enclosed seas. Al-Madani (2002) gave a value of 72 W m^{-2} for the Red Sea. Matsoukas et al. (2007) state that the net long-wave radiation does not exhibit large seasonal variations, having a value of around 70 W m^{-2} throughout the year.

Based on various estimates of net long-wave radiation for the Red Sea, the average is about 70 ± 5 W m^{-2}.

Latent and Sensible Heat Fluxes (Q_e and Q_h)

The heat abstracted by evaporation per unit area of the sea surface in unit time is as follows:

$$Q_e = LE$$

where L is the latent heat of vaporization and E is the rate of evaporation. The surface latent heat flux over the Red Sea is important for heat budget studies or as a basis for oceanic general circulation models. Conventionally, the latent heat flux has been estimated from bulk formulae that employ measurements of parameters such as sea surface temperature, saturation specific humidity, near surface humidity and wind speed.

The flux of sensible heat due to conduction is the heat transfer by turbulent convection between the sea surface and the overlying atmosphere. If the sea surface is warmer than the air, the air in contact with the water becomes more buoyant than the air above it, so it convects the heat upward into the atmosphere. When the sea surface is colder than the atmosphere, the air in contact with the sea cools becomes denser and hence more stable and more stratified. As such the conduction process becomes slower in carrying heat from the atmosphere into the sea. The transfer of heat between sea and atmosphere by conduction is more efficient when the sea is warmer than air.

Over the oceans, direct measurements of the surface fluxes are limited to a few specialized platforms, so extensive and large-scale measurements must rely on the parameterization of the surface fluxes that can use easily measured quantities obtained either in situ or remotely. In the bulk aerodynamic method, a bulk transfer coefficient c_s is defined so that the flux:

$$F_s = \rho c_s \bar{w} \overline{\Delta_s}$$

$\overline{\Delta_s} = \overline{S_0} - \overline{S_a}$	the mean difference of the property S between water and air
$\bar{w}$	is the mean wind speed
ρ	is the density of the medium

Usually, the transfer coefficients are related to a standard height or converted to that height. The complicated transfer processes are lumped together in the transfer coefficient c_s. Based on the bulk aerodynamic consideration, the equations for latent and sensible heat fluxes are as follows:

$$Q_e = \rho_a L c_e (q_0 - q_a) w$$
$$Q_h = \rho_a c_p c_h (T_0 - T_a) w$$

ρ_a	is air density
c_p	is the specific heat of air at constant pressure
$(T_0 - T_a)$	is the sea–air temperature difference and $(q_0 - q_a)$ is the difference between the saturated specific humidity at the sea surface temperature and the specific humidity at the air temperature, while w is the wind speed
c_e and c_h	are latent and sensible heat flux coefficients
L	is the latent heat of vaporization and is 2,500.8–2.3 t_0 kJ kg^{-1}, where t_0 is the sea surface temperature in °C

The coefficients depend on the sea–air temperature difference and the wind speed (Bunker 1976). Some researchers propose different values of c_h for stable and unstable conditions (stable $T_a > T_0$ and unstable $T_a < T_0$), but most researchers consider the same value for both conditions. Encompassed within the bulk method is more than 20 bulk transfer coefficient schemes that vary in complexity and sophistication.

The estimates of latent heat flux Q_e include 135 W m^{-2} by Bunker (1976), 109 W m^{-2} by Hastenrath and Lamb (1979) and 183 W m^{-2} by Bunker and Goldsmith (1979). Bunker et al. (1982) concluded that Hastenrath and Lamb value of 109 W m^{-2} is underestimated for the Red Sea and an average value of 159 W m^{-2} from Bunker (1976) and Bunker and Goldsmith (1979) can be considered for the Red Sea. Based on oceanographic and meteorological data for the Red Sea, other estimates are 165 W m^{-2} by Ahmad and Sultan (1987) for the central region of the Red Sea, 168 ± 8 W m^{-2} by Ahmad et al. (1989) for the coastal waters off Jeddah and Ahmad and Sultan (1989) calculated 169 W m^{-2} for the Red Sea as a whole based on their southern, central and northern regions. Abdulrahman and Ahmad (1994) estimated 163 W m^{-2} as an annual weighted average using the bulk aerodynamic method. Matsoukas et al. (2007) estimated an annual average latent heat flux of 164 W m^{-2}. Based on these studies, the annual average latent heat flux for the Red Sea is 165 ± 4 W m^{-2}.

Neumann (1952) gave a small negative value for the sensible heat flux for the Red Sea. His results show negative values for Bowen's ratio R from March to October with a minimum of −0.21 in May. The positive values of R in winter are smaller than 0.1, that is, smaller than the minimal values usually found in equatorial and tropical regions. Other estimates of the sensible heat flux are 5 W m^{-2} by Bunker (1976), 3 W m^{-2} by Bunker and Goldsmith (1979) and 3 W m^{-2} by Bunker et al. (1982). Hastenrath and Lamb (1979) derived a value of zero based on the sensible heat flux graphs of the southern and northern Red Sea. Ahmad and Sultan (1987) derived a value of −3 W m^{-2} for the central Red Sea, while Ahmad and Sultan (1989) calculated the sensible heat flux for the southern, central and northern Red Sea; their average value for the three regions is also −3 W m^{-2}. Abdulrahman and Ahmad (1994) also calculated a value of −3 W m^{-2} for the sensible heat flux. Ahmad et al. (1989) calculated a value of 5 ± 3 W m^{-2} for the coastal waters near Jeddah.

The sensible heat flux term Q_h, though small, shows great variability from 5 to −3 W m^{-2}. An annual average of −3 W m^{-2} can be considered for the whole Red Sea as the values of Bunker (1976), Bunker and Goldsmith (1979) and Bunker et al. (1982) are derived from the heat budget Atlas, whereas the values by Ahmad and Sultan (1987, 1989) and Abdulrahman and Ahmad (1994) are computed on the basis of real-time data. Therefore, using the following values:

Absorbed solar radiation (W m^{-2})	Net long-wave radiation (W m^{-2})	Evaporative heat flux (W m^{-2})	Sensible heat flux (W m^{-2})
220 ± 6	70 ± 5	165 ± 4	−3

the heat balance expression $(Q_s - Q_b - Q_e - Q_h)$ gives a deficit of −12 W m^{-2} at the air–sea interface. This deficit should be balanced by an advective heat flux to the Red Sea through Bab-al-Mandab. Based on the available comprehensive ocean atmosphere data set (COADS) and satellite observations including sea surface temperature, air temperature, dew point temperature and wind speed from 2000 to

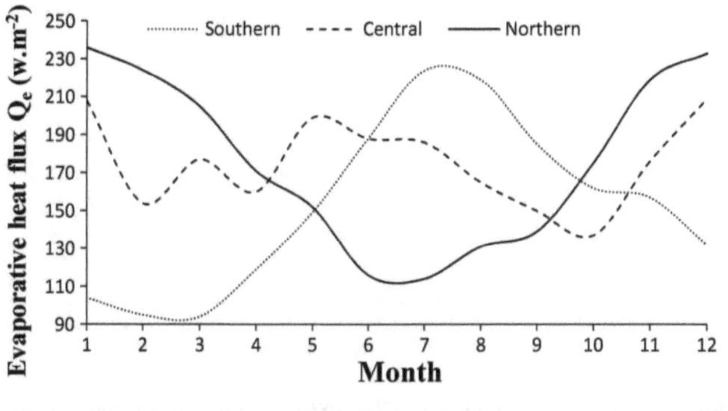

Fig. 1 Monthly variations of latent heat flux Q_e (W m^{-2}) for southern, central and northern regions of the Red Sea, based on meteorological data (2000–2010)

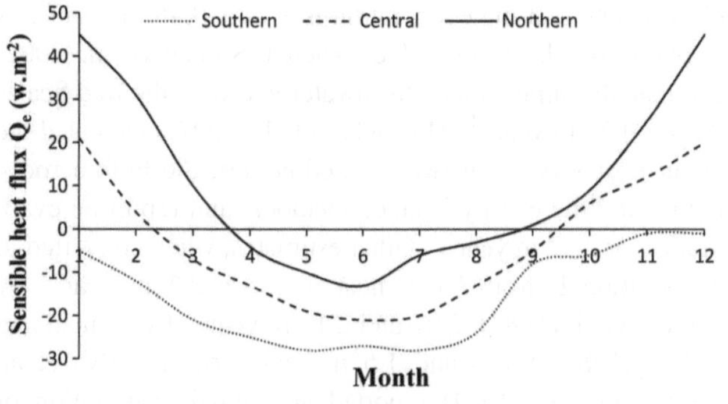

Fig. 2 Monthly variations of sensible heat flux Q_h (W m^{-2}) for the southern, central and northern regions of the Red Sea (after Ahmad and Sultan 1989)

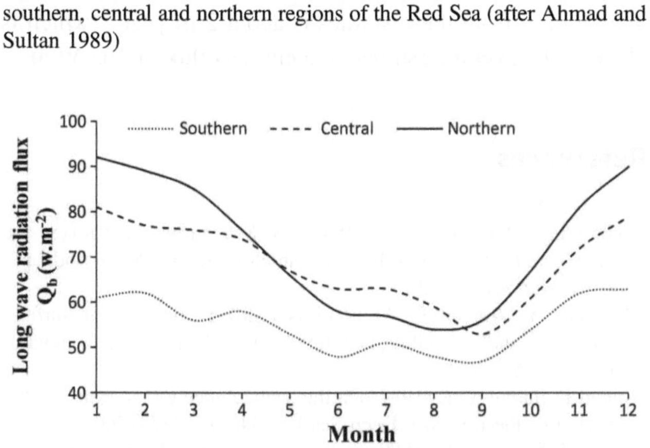

Fig. 3 Monthly variations of net long-wave radiation flux Q_b (W m^{-2}) for the southern, central and northern regions of the Red Sea (after Ahmad and Sultan 1989)

2010, the monthly variations of latent heat flux Q_e for the southern, central and northern regions of the Red Sea, based on the bulk aerodynamic method, are shown in Fig. 1. The southern region extends from Bab-al-Mandab to 18°N, the central region from 18° to 23°N and the northern region from 23°N to the Gulf of Suez. The monthly estimates of Q_h and Q_b for these regions are reproduced from Ahmad and Sultan (1989) and are shown in the Figs. 2 and 3. The fluxes show wide variability in time and space.

Advective Heat (Q_v)

The Red Sea is connected to the Gulf of Aden via Bab-al-Mandab. The shallowest section of Bab-al-Mandab is the Hanish Sill, which is located to the north of the narrowest passage near Perim Island. The greatest depth of the sill is about 137 m (Werner and Lange 1975). The sill depth maximum occurs within a deep passage that is only 6 km wide, outside of which, the depth is only of the order of 50 m (Murray and Johns 1997). The overall width of the Perim Narrows is about 18 km (Maillard and Soliman 1986).

The large-scale circulation within the Red Sea is governed by the thermohaline and wind forcing, each of which is extreme with respect to oceanic scales. The biannually reversing monsoon winds in the southern Red Sea give rise to a seasonally changing pattern of the upper layer circulation. These processes are reflected at Bab-al-Mandab. The winter exchange at Bab-al-Mandab consists of a surface inflow to the Red Sea compensated by the deep outflow; a typical case of basins where the evaporation exceeds precipitation. During summer, the two-layer flow of winter is replaced by a three-layer flow with surface layer to the south over the sill and a subsurface inflow and deep outflow. During the last decade, there have been a number of important developments in the understanding of the exchange flow in the Bab-al-Mandab Strait (Smeed 2000, 2004; Sofianos et al. 2002; Siddall et al. 2002). Sofianos et al. (2002) have given the monthly mean values of the inflow and outflow volumes at the Bab-al-Mandab Strait with average temperature and salinity of these volumes. From the inflow and outflow volumes, the advective heat flux can be calculated as:

$$Q_v = \frac{1}{A}(\rho_i v_i c_{p_i} t_i - \rho_o v_o c_{p_o} t_o)$$

Here, A is area of Red Sea, ρ is the density of water, v is the inflow or outflow volume, c_p is the specific heat of water and t is temperature. The subscripts i and o stand for inflow and outflow. During summer, the outflow consists of surface and deep outflows and the outflowing advective heat is based on the two layers.

Patzert (1974) used historical monthly temperature profiles from the National Oceanographic Data Center, combined with the monthly volume transports from KNMI (Koninklijk Nederlands Meteorological Institute), to give an advective heat flux of 7 W m^{-2} at Bab-al-Mandab. Tragou et al. (1999) gave a value of 8 ± 2 W m^{-2} by including a term based on the heat lost due to the volume of water that leaves the basin at the surface by evaporation. Ahmad and Sultan (1989) gave an annual average of 19 W m^{-2}. Sofianos et al. (2002) estimated a value of 11 ± 5 W m^{-2}. Based on the inflows and outflows from Sofianos et al. (2002), Abualnaja et al. (2011) calculated the advective heat flux of 35 W m^{-2} from the Gulf of Aden to the Red Sea. Their results when adjusted for the Gulf of Aden and Red Sea areas (Gulf of Aden area ≈220 × 10^3 km^2 and area of the Red Sea ≈450 × 10^3 km^2) amount to 17 W m^{-2}. Albarakati (2005), based on inflows and outflows from Sofianos et al. (2002), estimated a value of 13 W m^{-2}, considering E_{net} = 0.025SV, and 17 W m^{-2} with E_{net} = 0.03SV, where SV (Sverdrup) = 10^6 m^3 s^{-1} and E_{net} is excess of evaporation over precipitation for the Red Sea. These studies give an average value of 13 ± 4 W m^{-2} for the advective heat flux to the Red Sea which is in agreement with the balance of −12 W m^{-2} at the air–sea interface.

Evaporation and Precipitation

The Red Sea is a meridionally elongated basin and is surrounded by arid land and desert. Consequently, the evaporation is very high. Monthly average values of precipitation (1970–1986) in mm at five locations, namely Alwajh (26°14′N, 36°26′E), Yanbu (24°07′N, 38°03′E), Jeddah (21°30′N, 39°12′E), Almudailif (19°52′N, 41°03′E) and Jizan (16°52′N, 42°35′E), along the eastern side are given in Fig. 4 (data from Al-Jerash 1989) and show that precipitation is very low over the Red Sea and increases towards the south. The annual average along the eastern side of the Red Sea is about 0.05 m. Older estimates of evaporation from the Red Sea seem to be high. Maury (1855) gives 4.5 m year^{-1} and Vercilli (1927), based on on-board measurements, adopted the Port Sudan value of 3.5 m year^{-1} as the lowest possible value for the Red Sea.

Wust (1954) concluded that the actual evaporation at the sea surface amounts to about 55 % of evaporation measured on board. If Vercilli's value is reduced this way, it gives an evaporation of about 1.92 m year^{-1} which is close to recently calculated values. Using the heat balance equation and Bowen's ratio, Neumann (1952) estimated the annual evaporation in the Daedalus reef area as 2.15 m year^{-1}. Following the principle of conservation of volume and salt, Grasshoff (1969) estimated the annual evaporation as 2 m year^{-1}. Using the water balance, other estimates are 2.66 m year^{-1} (Bogdanova 1966) and 2 m year^{-1} (Siedler 1968). From the vertical water vapour flux in the lower atmosphere, Yegorov (1950) and Privett (1959) gave values of 2.3 and 1.83 m year^{-1}, respectively. Based on these studies, Morcos (1970) concluded that the average annual evaporation in the Red Sea is 2.10 m year^{-1}.

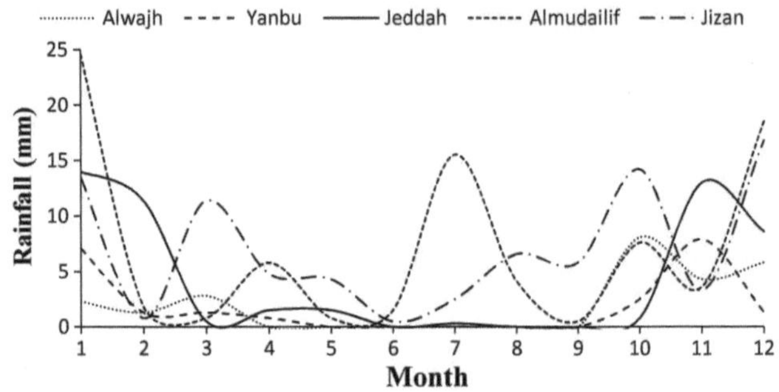

Fig. 4 Monthly averages (1970–1986) of rainfall in mm at five locations along the Red Sea (data from Al-Jerash 1989)

Behairy et al. (1981), using the aerodynamic method, estimated 2.05 m year^{-1} for the coastal region. Tragou et al. (1999) state that the evaporation is 1.74 ± 0.35 m year^{-1}, as implied from the water budget, whereas Sofianos et al. (2002) estimate the annual mean freshwater loss over the Red Sea as 2.06 ± 0.22 m year^{-1}. Matsoukas et al. (2007) calculated the latent heat flux using two methodologies, the bulk aerodynamic and the energy balance methods, and report an evaporation of 2.12 m year^{-1}. Other estimates, when converted to evaporation E from latent heat flux, are 2.3 m year^{-1} by Bunker et al. (1982), 2.07 and 2.13 m year^{-1} by Ahmad and Sultan (1987, 1989) and 1.5 m year^{-1} by da Silva et al. (1994). Osman (1984) reported an annual evaporation of 2.04 m from the coastal waters off Port Sudan. The average evaporation from these studies is about 2 m year^{-1} which is close to the average estimated latent heat flux of 165 W m^{-2}.

References

Abdulrahman SM, Ahmad F (1994) Red Sea surface heat fluxes and advective heat transport through Bab-el-Mandab. J King Abdulaziz Univ Mar Sci 6:3–13

Abualnaja Y, Ahmad F, Al-mtairi NA (2011) Balance of surface, advective and up-welling heat fluxes in the Gulf of Aden. Indian J Geo Mar Sci 40:42–47

Ahmad F, Sultan SAR (1987) On the heat balance terms in the central region of the Red Sea. Deep Sea Res 34(10):1757–1760

Ahmad F, Sultan SAR (1989) Sea surface fluxes and their comparison with the oceanic heat flow in the Red Sea. Oceanol Acta 12(1):33–36

Ahmad F, Sultan SAR, Moammar MO (1989) Monthly variations of net heat flux at the air-sea interface in coastal waters near Jeddah, Red Sea. Atmos Ocean 27(2):406–413

Albarakati AMA (2005) Advective heat transport to the Red Sea at the strait of Bab-el-Mandab. J King Abdulaziz Univ Mar Sci 16:133–140

Al-Jerash MA (1989) Data for climatic water balance in Saudi Arabia (1970–1986). Scientific Publishing Centre, King Abdulaziz University, Jeddah

Al-Madani SA (2002) Variation of heat fluxes at the air-sea interface in the Red Sea. M.Sc. thesis, King Abdulaziz University, Faculty of Marine Sciences, Jeddah, Saudi Arabia

Bignami FS, Marullo RS, Schiano ME (1995) Long wave radiation budget in the Mediterranean sea. J Geophys Res 100:2501–2514

Behairy AKA, Osman MM, Meshal AH (1981) Evaporation from the coastal waters in front of Jeddah. Jeddah J Mar Sci 1:35–45

Bishop JKB, Rossow WB, Datton EG (1997) Surface solar irradiance from the international satellite cloud climatology project 1983–1991. J Geophys Res 102(06):6883–6910

Bogdanova AK (1966) Oceanological Research. Soviet Geophysical Committee, Academy of Sciences, USSR 15:45–68

Bunker AF (1976) Computation of surface energy flux and annual air-sea interaction cycles of the North Atlantic Ocean. Mon Weather Rev 104:1122–1140

Bunker AF, Goldsmith RA (1979) Archived time series of Atlantic Ocean meteorological variables and surface fluxes. Woods Hole Oceanographic Institution Technical Report No. WHOI-79-3

Bunker AF, Charnock H, Goldsmith RA (1982) A note on the heat balance of the Mediterranean and Red Sea. J Ma Res Suppl 40:73–84

da Silva AM, Young CC, Levitus S (1994) Atlas of surface marine data 1994, NOAA Atlas, NESDIS 6-10. U.S. Government Printing Office, Washington DC (5 volumes)

Fairall CW, Bradley EF, Hare JE, Grachev AA, Edson JB (2003) Bulk parameterization of Air-Sea fluxes: updates and verification for the COARE algorithms. Am Meteorol Soc 16:571–591

Fung IY, Harrison DE, Lacis AA (1984) On the variability of net long wave radiation at the ocean surface. J Geophys Res 22(2):177–193

Grassohf K (1969) Red Sea: Gulf of Aden: chemical oceanography: hydrography: evaporation: water exchange: circulation: nutrients. RV METEOR, 1964–1965

Hastenrath S, Lamb PJ (1979) Climatic atlas of the Indian Ocean. University of Wisconsin Press, Madison

Josey SA, Oakley D, Pascal RW (1997) On estimating the atmospheric longwave flux at the ocean surface from ship meteorological reports. J Geophys Res 102(C13):27961–27972

Kraus EB (1972) Atmosphere-ocean interaction. Clarendon Press, Oxford, 275 pp

List RJ (1958) Smithsonian meteorological tables. Smithsonian Institution, Washington DC, 527 pp

Matsoukas C, Banks AC, Pavlakis KG, Hatzianastassiou N, Stackhouse Jr. PW, Vadavas I (2007) Seasonal heat budgets of Red and Black seas. J Geophys Res Oceans 112, C10017. doi:10.1029/2006JC003942

Maillard C, Soliman G (1986) Hydrography of the Red Sea and exchanges with the Indian Ocean in summer. Oceanol Acta 9 (3):249–269

Maury MF (1855) The physical oceanography of the sea and its meteorology. In: Leighly J (ed) Harvard Library, Cambridge, Mass

Morcos SA (1970) Physical and chemical oceanography of the Red Sea. Oceanogr Mar Biol Rev 8:73–202

Murray SP, Johns W (1997) Direct observations of seasonal exchange through the Bab-el-Mandab Strait. Geophys Res Lett 24(21):2557–2560

Neumann J (1952) Evaporation from the Red Sea. Israel Explor J 2:153–162

Osman MM (1984) Evaporation from coastal waters off Port Sudan. In proceedings of the symposium on coral reef environment of Red Sea, 14–18 Jan 1984

Patzert WC (1974) Wind-induced reversal in the Red Sea circulation. Deep Sea Res 21:109–121

Pond S (1975) The exchange of momentum heat and moisture at the ocean-atmosphere interface. In: Symposium proceedings of numerical models of ocean circulation. National Academy of Sciences, Washington DC, pp 26–28

Privett DW (1959) Monthly charts of evaporation from the North Indian Ocean, including the Red Sea and Persian Gulf. Q J Roy Meteorol Soc 85:424–428

Reed RK (1977) On estimating insolation over the ocean. J Phys Ocean 7:482–485

Rosati A, Miyakoda K (1988) A general circulation model for upper ocean simulation. J Phys Oceanogr 18:1601–1626

Rossow WB, Zhang YC (1995) Calculation of Surface and top of atmosphere radiative fluxes from physical quantities based on ISCCP data sets part II validation and first results. J Geophys Res 100:1167–1197

Sechel GR, Beaudry FH (1973) The radiations from sun and sky over the north Pacific Ocean (Abstract). Trans Am Geophys Union 54:1114

Siddall M, Smeed DA, Matthiesen S, Rohling EJ (2002) Modelling of the seasonal cycle of the exchange flow in Bab-el-Mandab, Red Sea. Deep Sea Res Part I 49:1551–1569

Siedler G (1968) Schichtungs- und Bewegungsverhaltnise am Sudausgang des Roten Meeres. Meteor Forschungsegeb 4:1–76

Simpson JJ, Paulson CA (1979) Mid ocean observations of atmospheric radiation. Q J Roy Meteorol Soc 105:487–502

Singh R, Joshi PC, Kishtawal CM (2005) A new technique for estimation of surface latent heat fluxes using satellite-based observations. Am Meteorol Soc 133(9):2692–2710

Smeed DA (2000) Hydraulic control in three-layer exchange flows: application the Bab-el-Mandab. Phys Oceanogr 30:2574–2588

Smeed DA (2004) Exchange through the Bab-el-Mandab. Deep Sea Res II 51(4–5):455–474

Sofianos SS, Johns WE, Murray SP (2002) Heat and freshwater budgets in the Red Sea from direct observations Bab-el-Mandab. Deep Sea Res Part II 49:1323–1340

Tragou E, Garrett C, Outerbridge R, Gilman C (1999) The heat and freshwater budget of the Red Sea. J Phys Oceanogr 29:2504–2522

Vercilli F (1927) The Hydrographic Survey of the R. N. Amrairaglio Magnaghi in the Red Sea. Annu Hydrogr 2:1–290

Werner F, Lange K (1975) A bathymetric survey of the sill area between the Red Sea and the Gulf of Aden. Geohgisches Jahrhiicli D13:125–130

Wust G (1954) Gestmaessige wechselbeziehungen zwischen ocean und atmosphere im der zonalen verteilung von oberflaechensalgehalt verdunstung und niederschlag. Archiv für Meteorologie, Geophysik und Bioklimatologie, Serie A 1:305–328

Yegorov NI (1950) Calculation of the heat balance of the Red Sea. Meteorologiia i Gidrologiia Moscow 3:49–56

Zhang YC, Rossow WB, Lacis AA (1995) Calculation of surface and top-of-atmosphere radiative fluxes from physical quantities based on ISCCP data sets, part I: method and sensitivity to input data uncertainties. J Geophys Res 100:1146–1165

Impacts of Climate Change on the Red Sea Region and its Watersheds, Saudi Arabia

Mohamed Sultan, Saleh Sefry, and Mazen AbuAbdallah

Abstract

This chapter evaluates the impacts of projected climate change (period: 2020–2100) on the hydrologic systems along the Eastern Red Sea coastal plain. Shuttle Radar Topography Mission (SRTM) data and Landsat Thematic Mapper (TM) scenes were used to delineate watersheds and their channel networks (155 watersheds) that collect precipitation over the Red Sea Hills in the Kingdom of Saudi Arabia (KSA) and drain toward the Red Sea coastal plain. Continuous rainfall-runoff models were constructed and calibrated for each of the 19 major watersheds (watershed area: 1,825–107,769 km^2; total area: 176,683 km^2). Regionalization techniques were applied to extrapolate catchment-specific parameters from proximal calibrated catchments (i.e., Wadi Girafi, which has an area of 5,057 km^2). The calibrated Soil Water and Assessment Tool (SWAT), and climatic inputs from the Climate Forecast System Reanalysis (CFSR) database, and from Tropical Rainfall Measuring Mission (TRMM) measurements were used to calculate the partitioning (runoff, evapotranspiration, and potential groundwater recharge) of precipitation for the time period from 1998 to 2010. TRMM data were validated by comparing them to rain gauge data (period: 1998–2012; number of rain gauge stations: >100) across the study area. Downscaled outputs of the Community Climate System Model (CCSM4) were used to simulate (for the period 2020–2100) the effects of climatic changes on the selected 19 Red Sea watersheds. Findings indicate that (1) the average annual precipitation is 130 mm and the total annual amount is 23×10^9 m^3; (2) with time, precipitation will increase over the northern and central watersheds and decrease over the southern watersheds; and (3) global warming is causing a rise in sea surface temperature, enhancing evaporation, intensifying northerly and northwesterly winds and precipitation over the northern and central watersheds, and possibly causing shifts in monsoonal fronts and modifying precipitation patterns over the southern watersheds.

Introduction

In arid and hyper-arid fragile deserts of the world, the hydrologic systems, the adjoining water bodies, and their ecosystems are affected significantly by climatic fluctuations. During wet climatic periods, fluvial systems and drainage networks develop, underlying aquifers recharge, rising groundwater tables discharge in lowlands and depressions, runoff and sediment load increase, and interactions between surface runoff and groundwater flow systems intensify. The opposite happens in dry periods. Runoff is reduced, surface drainage patterns dry up and may be buried under encroaching eolian deposits, aquifer recharge is reduced and localized, groundwater tables are lowered, and groundwater discharge decreases in lowlands.

Along the eastern and western coastal plains of the Red Sea, extensive watersheds collect precipitation from the adjoining Red Sea Hills and channel the collected runoff toward the Red Sea and its coastal plains as surface runoff

M. Sultan (✉)
Department of Geosciences, Western Michigan University, Kalamazoo, MI 49008, USA
e-mail: mohamed.sultan@wmich.edu

S. Sefry · M. AbuAbdallah
Saudi Geological Survey, Jeddah, Saudi Arabia

N.M.A. Rasul and I.C.F. Stewart (eds.), *The Red Sea*,
Springer Earth System Sciences, DOI 10.1007/978-3-662-45201-1_22,

and/or groundwater flow. As the runoff reaches the gently dipping coastal plain, it slows down and deposits its sediment load, in some cases forming deltas where the stream meets the sea shoreline. The alluvial aquifers flooring the channel networks are fed by infiltration from the runoff and by groundwater channeled along faults and shear zones where fractured basement rocks are found (Sultan et al. 2008). These hydrologic systems (surface and groundwater flow) deliver nutrients and/or sediments to the Red Sea coastal areas that support a wide range of ecosystems, but could also inhibit others. For example, the mangroves flourish in brackish water zones where fresh groundwater mixes with saline seawater. In contrast, coral reefs thrive in clear waters and are inhibited in coastal areas where sediments are introduced by runoff from nearby watersheds.

This chapter addresses the following questions: (1) How is precipitation partitioned over the Red Sea Hills in the investigated watersheds? (2) What climatic changes are expected to take place in this area by the end of this century? (3) How would these climatic changes affect the temporal and spatial distribution of precipitation over the Red Sea Hills, the partitioning of precipitation, and the amount of runoff and recharge?

Site Description

The Red Sea is an ocean in the making. It is more than 2,000 km long and separates the once-contiguous Arabian and Nubian Shields. Its width varies from 180 km in the north to 350 km in the south, before narrowing down to 28 km in the Strait of Bab-al-Mandab, where it joins the Indian Ocean (Fig. 1). On both the Arabian and Nubian sides, the Red Sea coastal plains give way to a long chain of mountains known as the Red Sea Hills.

The Red Sea Hills are composed largely of Neoproterozoic (550–900 Ma) volcano-sedimentary rock units of the Arabian–Nubian Shield (Sultan et al. 1990; Stern and Kroner 1993) that crop out along the Red Sea coastline in Egypt, Sudan, Ethiopia, and the Kingdom of Saudi Arabia (KSA). The opening of the Red Sea during the Oligocene was associated with extension and uplift; the latter exposed the basement complex that now crops out along the eastern (Fig. 1) and western margins of the Red Sea coastline.

In the KSA, the Red Sea Hills comprise two mountain ranges that are separated from one another in the vicinity of the city of Makkah. The elevation of the northern range seldom exceeds 2,100 m above mean sea level (a.m.s.l.) and decreases in the southern regions, reaching some 600 m a.m.s.l. (Fig. 1). The rugged mountain wall drops abruptly in the direction of the Red Sea, and the coastal plain is generally narrow (width: 2–50 km), especially in the northern sections (Fig. 1).

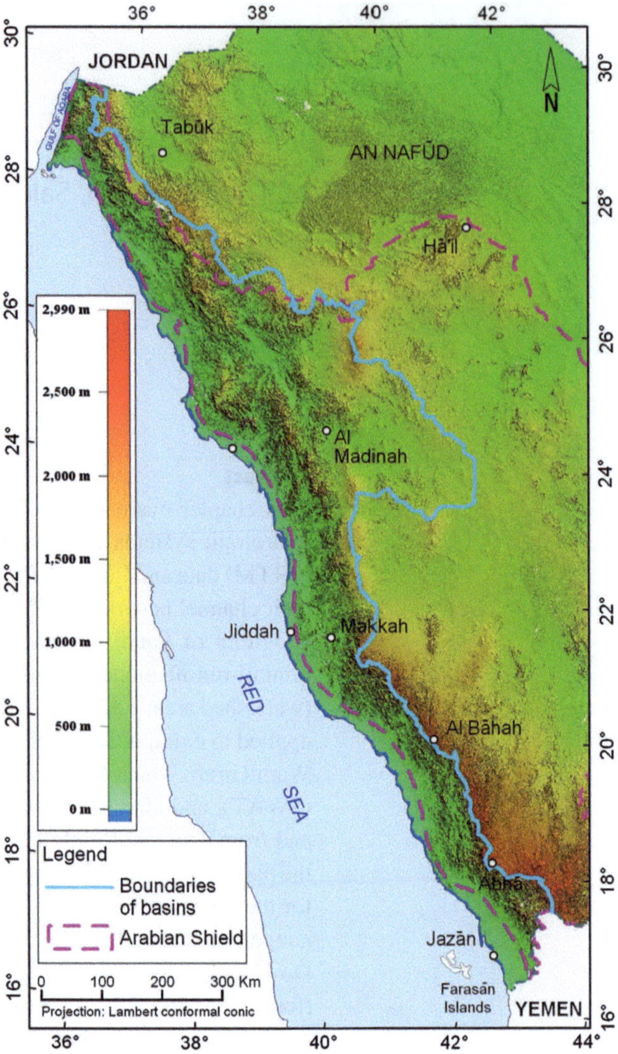

Fig. 1 Color-coded Digital Elevation Model (DEM) for the Arabian shield and the area occupied by the Red Sea coastal basins

The crystalline basement is unconformably overlain by thick sequences of sedimentary formations ranging in age from Cambrian to Recent. In the Arabian Peninsula, these stratigraphic sequences dip gently to the east and thicken in the same direction, reaching thicknesses of 6 km in the vicinity of the Arabian Gulf. The subject of this chapter is the watersheds that drain the Red Sea Hills in the KSA and discharge in the coastal plains of the Red Sea and the Gulf of Aqaba. The study area is bounded by the Red Sea (west), the drainage topographic divide (east), the Gulf of Aqaba (north), and the Saudi Arabian/Yemen political boundary (south).

Worldwide rainfall occurrence is closely linked with the general circulation of the atmosphere and with the distribution of mountain ranges and water bodies. Rainfall is prominent where air rises and is less abundant where it sinks; therefore, tropical regions receive high amounts of rainfall, whereas subtropical areas (e.g., the Arabian Peninsula) receive much less precipitation. The Red Sea coast of the KSA has a desert climate in general, but is affected by the moisture from the surrounding water bodies (Red Sea, Gulf of Aden) and by the high mountains (Red Sea Hills) in the east, which act as orographic cooling barriers.

Examination of the average annual (1998–2012) precipitation extracted from three-hourly Tropical Rainfall Measuring Mission (TRMM) data over the Arabian Shield (Fig. 2) shows that precipitation is largely concentrated over the mountain ranges and/or highlands, especially the southern parts of the Red Sea Hills, which receive an average rainfall of 500 mm/yr. In general, precipitation increases from north to south.

Two main wind direction systems bring precipitation to the investigated watersheds: the first system is the winter northerly to northwesterly winds, the "Shamal," that prevail over the northern and central parts of the investigated areas; the second is the summer monsoons that develop over the southern investigated watersheds (Alsharhan et al. 2001). Precipitation is collected by an extensive network of streams that channel runoff into the main valleys and eventually drain into the Red Sea coastal plain.

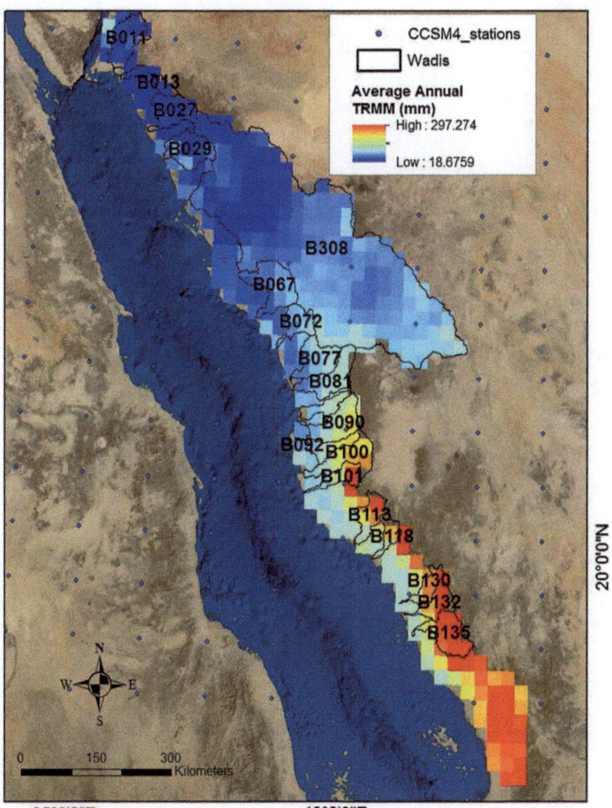

Fig. 2 Average annual precipitation (1998–2012) extracted from three-hourly TRMM data. Also shown are the locations of the 19 major (area: 1,825–107,769 km^2) watersheds and the Community Climate System Model version 4 (CCSM4) stations

The present study evaluates the projected climatic changes in the Red Sea region and its surroundings, searches for observations to validate these projections, and assesses the effects of these climatic changes on the distribution and amounts of precipitation over these areas and on the partitioned runoff toward the Red Sea coastal plain.

Methodology

The methodology adopted involved a six-step procedure: (1) the Shuttle Radar Topography Mission (SRTM) data and Landsat Thematic Mapper (TM) scenes were used to delineate most of the watersheds (155 watersheds) and their channel networks along the Red Sea coast within the KSA, and to extract their morphometric parameters; (2) TRMM-derived rainfall data were acquired over the study area and validated with available rainfall gauge data (from >100 stations) throughout the TRMM operational period (1998–2012); (3) continuous rainfall-runoff models were constructed for the 19 major (area: 107,769–1,825 km^2) watersheds; (4) regionalization techniques were applied to extrapolate catchment-specific parameters from a calibrated catchment that is proximal to the investigated watersheds; (5) calibrated SWAT models and climatic inputs from the Climate Forecast System Reanalysis (CFSR) database were employed to calculate the partitioning of precipitation into runoff, evapotranspiration, and potential groundwater recharge for the time period from 1998 to 2010; and (6) outputs of the Community Climate System Model (CCSM4) were used to simulate the partitioning of precipitation for the time period 2020–2100 following the application of a bias correction to the CCSM4 model outputs.

Defining Watersheds along the Red Sea Coast of Saudi Arabia

Watershed boundaries and channel networks were delineated from the SRTM Digital Elevation Model (DEM; spatial resolution: 90 m) data using the Watershed Modeling System (WMS) software package. In total, 155 watersheds (catchment area: >80 km^2) were delineated and their morphometric characteristics were identified using the ArcGIS and WMS software packages; all of these watersheds collect precipitation over the Red Sea Hills in the KSA and drain toward the Red Sea coastal plain. The distribution of the watersheds is shown in Fig. 3a through 3d, and their morphometric characteristics are given in Table 1; the watersheds were numbered from north to south (B001 through B155).

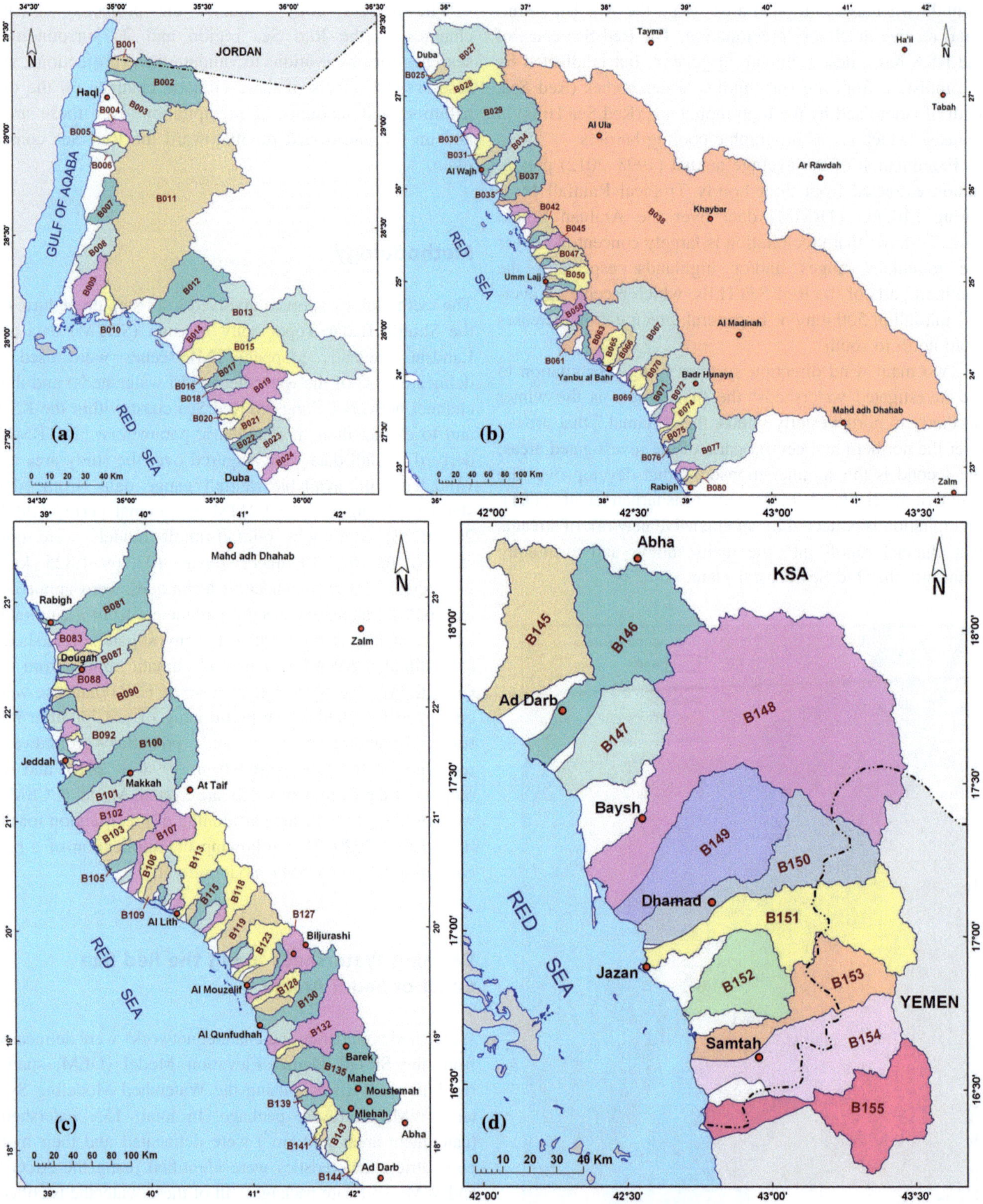

Fig. 3 Distribution of the Red Sea coastal watersheds. **a** watersheds B001–B024, **b** watersheds B025–B080, **c** watersheds B081–B144, and **d** watersheds B145–B155

Table 1 Morphometric characteristics of coastal watersheds

Basin code	Basin name	Area (km^2)	Basin length (km)	Basin slope (%)	Maximum elevation (m) a.m.s.l.
B001	Unknown	96	18	9.39	992
B002	Wadi Mabrak	847	67	20.98	1,662
B003	Wadi Umm Jurfayn	549	59	14.96	1,602
B004	Wadi Dabr	101	30	15.30	1,689
B005	Wadi Al-Hasha	90	24	23.40	1,454
B006	Wadi Kulayb	83	24	27.91	1,367
B007	Wadi Haqaf	264	42	27.09	1,455
B008	Wadi Al-Hamdah	330	42	10.88	635
B009	Wadi An Nakhalah	328	46	5.62	512
B010	Unknown	85	25	1.41	147
B011	Wadi Ifal	5,018	206	18.51	2,348
B012	Wadi Aynunah	825	67	19.93	1,368
B013	Wadi Shiqri	2,111	166	13.30	1,625
B014	Wadi Ratiyah	130	33	5.31	542
B015	Wadi Sadr	990	25	5.34	336
B016	Wadi Umm Siyayyil	85	36	15.60	2,086
B017	Wadi Zawji	221	30	11.24	957
B018	Wadi Qamirah	106	60	23.73	1,294
B019	Wadi Surr	579	12	8.61	365
B020	Wadi Al-Bayda	87	15	6.82	228
B021	Wadi Al-Maarrash	262	33	5.29	636
B022	Wadi Al-Ghal	131	45	12.14	697
B023	Wadi Zahkan	321	14	6.62	237
B024	Wadi Salma	197	20	7.07	273
B025	Wadi Al-Amud	222	36	6.14	744
B026	Wadi Al-Buharah	83	24	2.17	283
B027	Wadi Damah	4,428	158	17.05	1,607
B028	Wadi Al-Ayn	1,769	129	11.77	1,836
B029	Wadi Thalbah	4,985	151	10.69	1,344
B030	Wadi Antar	200	31	9.15	415
B031	Wadi Haramil	447	52	5.74	345
B032	Wadi Zaam	329	48	3.94	283
B033	Wadi Ash-Shijnah	309	42	3.61	255
B034	Wadi Al-Miyah	1,399	127	13.30	1,641
B035	Wadi Al-Arja	311	48	3.63	352
B036	Wadi Al-Marrah	252	39	3.44	373
B037	Wadi Aba Al-Ajaj	1,934	109	12.50	1,715
B038	Wadi Al-Hemdh	107,769	826	6.23	1,016
B039	Unknown	216	40	1.59	130
B040	Unknown	94	26	2.48	122
B041	Wadi Sabq	251	35	1.74	168
B042	Wadi Hayran	472	62	5.16	788
B043	Wadi Murran	157	36	2.20	216
B044	Unknown	144	35	2.48	254

(continued)

Table 1 (continued)

Basin code	Basin name	Area (km^2)	Basin length (km)	Basin slope (%)	Maximum elevation (m) a.m.s.l.
B045	Wadi Amq	402	63	9.10	712
B046	Wadi Khuff	249	43	3.38	478
B047	Wadi Marakh	907	88	10.71	1,165
B048	Unknown	90	19	3.13	247
B049	Wadi Al-Hail	280	58	4.95	743
B050	Wadi Simin	921	77	11.54	1,326
B051	Unknown	123	31	4.34	534
B052	Wadi Al-Ghumayr	238	47	10.36	1,053
B053	Wadi An-Nukhbar	162	38	6.13	516
B054	Wadi Quwaq	355	55	12.16	1,628
B055	Wadi Al-Makhamas	230	41	5.24	745
B056	Wadi Ad-Dahlah	162	26	4.99	286
B057	Wadi Samur	122	29	6.42	296
B058	Wadi Nabt	764	84	13.40	1,351
B059	Unknown	88	30	6.45	326
B060	Wadi Al-Buhayrah	199	46	6.51	630
B061	Wadi Al-Wadiyayn	369	69	5.10	510
B062	Wadi Ajil	509	77	11.63	1,144
B063	Wadi Khumal	898	72	12.11	1,616
B064	Wadi Al-Fari	148	24	1.55	113
B065	Wadi Aways	830	65	8.96	2,133
B066	Wadi Al-Makharr	712	87	14.56	2,224
B067	Wadi Yanbu An-Nakhl	6,569	211	22.63	1,662
B068	Wadi Shatb	112	36	1.65	247
B069	Wadi Ghurghur	402	38	1.97	155
B070	Wadi Al-Faqir	880	93	12.33	1,125
B071	Wadi Wasit	792	65	7.74	971
B072	Wadi Al-Safra	2,881	168	23.23	2,261
B073	Wadi As-Sarayr	384	49	3.76	490
B074	Wadi Al-Arayyish	988	108	15.74	1,735
B075	Wadi Al-Midan	601	70	8.72	1,412
B076	Unknown	491	72	13.87	1,312
B077	Wadi Masturah	4,755	193	15.66	996
B078	Wadi Al-Jaddah	224	51	2.70	198
B079	Wadi Rahab	108	27	3.28	176
B080	Wadi Al-Khamas	423	69	4.20	316
B081	Wadi Rabigh	5,241	217	9.24	1,359
B082	Unknown	132	28	0.96	39
B083	Wadi Al-Halq	740	65	4.79	311
B084	Wadi Thamrih	684	143	7.68	1,641
B085	Unknown	337	55	4.18	316
B086	Unknown	121	30	1.41	164
B087	Wadi Qidayd	2,280	142	14.97	1,619
B088	Wadi Al-Wisami	806	77	7.20	847

(continued)

Table 1 (continued)

Basin code	Basin name	Area (km^2)	Basin length (km)	Basin slope (%)	Maximum elevation (m) a.m.s.l.
B089	Wadi Thuwal	146	29	1.69	101
B090	Wadi Khulays	5,186	214	11.77	1,639
B091	Wadi Uwayja	354	38	2.19	154
B092	Wadi As Sughu	2,406	199	7.10	1,292
B093	Wadi Murayyakh	223	30	3.88	323
B094	Wadi Ghaya	92	27	9.47	454
B095	J_Airport	196	36	1.13	120
B096	Wadi Al-Asla	292	41	5.19	667
B097	Unknown	89	24	1.49	79
B098	Wadi Qaws	81	31	3.83	267
B099	Wadi Mathwab	110	27	4.63	216
B100	Wadi Fatimah	5,086	266	13.28	2,063
B101	Wadi Numan	2,404	165	16.06	2,559
B102	Wadi Malakan	2,039	132	12.90	2,408
B103	Wadi Al-Kharqah	943	87	4.23	478
B104	Wadi Dam	717	98	8.00	1,195
B105	Wadi Al-Misayrih	512	49	2.52	148
B106	Unknown	121	29	0.62	80
B107	Wadi Lamlam	1,881	143	23.57	2,317
B108	Wadi Sayah	727	85	12.82	1,551
B109	Wadi Markub	523	60	11.98	454
B110	Unknown	112	22	9.46	497
B111	Wadi Al-Ghallah	597	78	15.86	1,703
B112	Unknown	169	26	3.45	177
B113	Wadi Al-Lith	3,268	174	27.20	2,428
B114	Unknown	158	35	1.16	321
B115	Wadi Iyar	1,066	100	18.76	1,828
B116	Wadi Ar Rukhmi	335	45	2.02	489
B117	Wadi Silaq	231	51	4.52	1,200
B118	Wadi Halyah	1,825	143	24.92	2,208
B119	Wadi Iliyab	1,784	115	18.55	2,439
B120	Wadi Al-Mahanah	156	34	3.24	647
B121	Wadi Hifar	247	38	2.22	529
B122	Unknown	95	34	1.00	136
B123	Wadi Dawqah	1,430	110	21.49	2,358
B124	Wadi Qarma	419	72	8.93	920
B125	Unknown	395	91	15.48	2,137
B126	Wadi Ash Shaghz	167	40	1.06	152
B127	Wadi Al-Ahsabah	1,346	115	26.64	2,455
B128	Wadi Lumah	736	91	7.88	1,255
B129	Wadi Al-Hulayfah	462	54	1.37	291
B130	Wadi Qanunah	2,299	162	19.45	2,000
B131	Unknown	499	44	1.62	229
B132	Wadi Yabah	3,012	172	21.32	2,286

(continued)

Table 1 (continued)

Basin code	Basin name	Area (km^2)	Basin length (km)	Basin slope (%)	Maximum elevation (m) a.m.s.l.
B133	Wadi As-Sirr	150	36	1.44	256
B134	Wadi Ad-Dayqah	297	49	2.14	406
B135	Wadi Haly	5,160	216	22.90	2,699
B136	Wadi Al-Kuffayrah	475	61	4.47	540
B137	Wadi Shafqah	377	67	8.41	800
B138	Wadi Al-Halfah	89	21	2.41	285
B139	Wadi Amaq	350	49	4.88	739
B140	Wadi Dabsa	160	23	4.33	316
B141	Wadi Dhahban	510	66	10.67	873
B142	Wadi Najl	87	26	6.38	410
B143	Wadi Hamadah	946	75	11.87	1,198
B144	Wadi Nahab	569	69	6.45	1,020
B145	Wadi Rim	1,582	100	16.68	1,277
B146	Wadi Itwad	1,946	117	29.91	2,236
B147	Wadi Bayd	1,057	76	7.11	822
B148	Wadi Baysh	5,824	254	29.52	2,644
B149	Wadi Sabya	1,823	101	11.20	2,436
B150	Wadi Damad	1,389	127	26.92	2,709
B151	Wadi Jazan	1,809	146	19.84	2,818
B152	Wadi Mamlah	867	65	3.74	516
B153	Wadi Khulab	1,101	104	18.84	2,640
B154	Wadi Liyah	1,564	106	13.53	2,473
B155	Wadi Harad	1,495	123	18.68	1,905

Collecting and Analyzing the Rainfall Data from Rain Gauges

Accurate estimates of rainfall amounts and patterns are required to calculate the partitioning of precipitation into runoff, evaporation, transpiration, and recharge. Rainfall data were collected from 115 rain gauges that are distributed across the study area (Fig. 4). Collection of rain gauge data started as early as 1964 for a few of the investigated stations; unfortunately, the data are generally intermittent and covers varying time periods for the individual stations. A subset of the investigated rain gauges was selected; these gauges had near-continuous precipitation data during the TRMM operational period (1998–2012). Average annual precipitation amounts were extracted from these stations (Fig. 5) and were compared to those derived from TRMM data (Fig. 2). These comparisons revealed overall similarities in the patterns of precipitation over the investigated area. Given the differences in their spatial resolution (TRMM: 0.25° × 0.25°; rain gauge: <1 m), a one-to-one correspondence between these two datasets should not be expected. Moreover, the TRMM sensors acquire precipitation every 3 hours, so short-lived precipitation events that occur between TRMM acquisitions could be missed. This could explain why precipitation values from rain gauge data generally exceed those from TRMM data.

Construction, Calibration, and Regionalization of Continuous Rainfall-runoff Models

The construction and application of calibrated rainfall-runoff models on regional scales are often hindered by the general paucity of detailed field data on such scales. To minimize uncertainties related to data limitations, we utilized global remote sensing data that are readily available for the majority of the world's continents. Such applications are ideal for the arid and hyper-arid parts of the world, where remotely sensed observations are not hindered by atmospheric contributions from clouds and moisture as is the case over wetter parts of the world. The adopted methodologies address the uncertainties that arise from the scarcity of one or more datasets, including temporal and spatial rainfall depths, stream flow, and field data. A physically based, continuous, semi-distributed model, the Soil Water Assessment Tool (SWAT) was adopted; the model combines two scale levels:

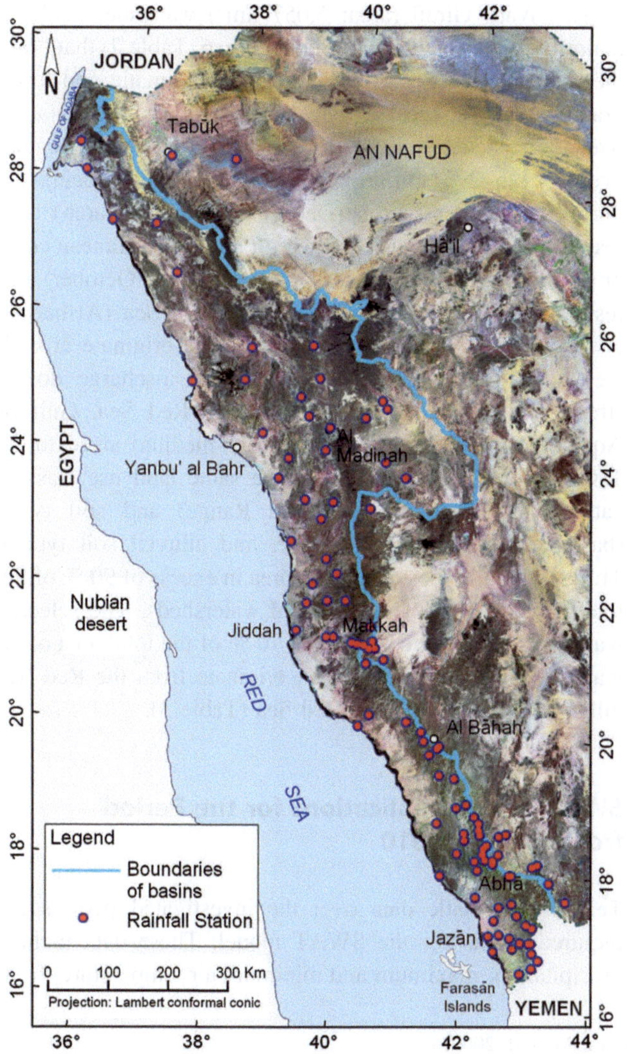

Fig. 4 Location map for the selected rain gauge stations

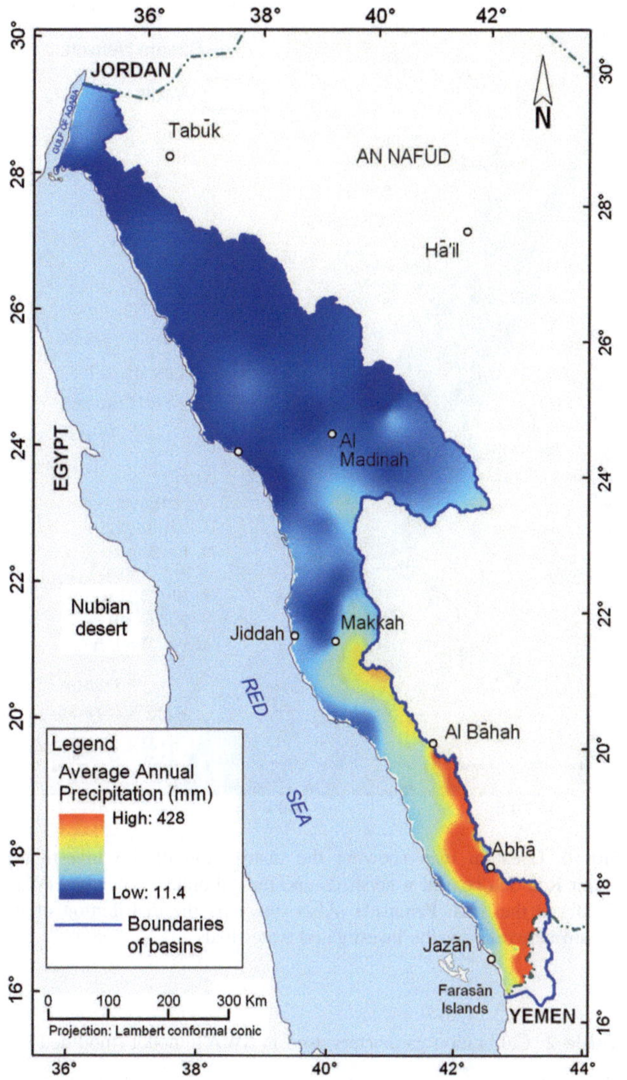

Fig. 5 Average annual precipitation (1998–2012) from selected rain gauges

(1) a fine scale that is represented by the uniform Hydrologic Response Units (HRUs), and (2) a coarse scale that is represented by the sub-basins; the latter aggregates HRUs that maximize the representation of lateral variations while maintaining reasonable computational efficiency.

Watersheds and channel networks were delineated from the SRTM data and verified by comparison with co-registered Landsat TM scenes and geologic maps. The Soil Conservation Service (SCS) curve number method (SCS 1985) was adopted to calculate upstream initial losses in the sub-basins. Runoff calculations were enabled by adopting the SCS unit hydrograph (SCS 1985) and by using the Riverside County lag-time equations for mountainous, foothill, and valley areas (Riverside County Flood Control 1978). Channel routing was conducted using the Muskingum routing method (McCarthy 1938), whereby the average Manning's coefficients for gravel-bed rivers were calculated using procedures that were developed by Jarrett (1984). Lane's method was adopted for estimating transmission losses (SCS 1985) in the stream networks draining the Red Sea Hills. The aerial extents of soil types and various geomorphic features (valleys, mountains, and foothills) were identified from co-registered Landsat TM, SRTM, and digital geologic mosaics.

Physical catchment descriptors such as geography, climate, catchment size, topography, geology, vegetation, land use, and stream network density were identified using regionalization techniques to investigate whether catchment-specific parameters extracted from calibrated catchments could be extrapolated to other ungauged proximal catchments. Catchment-specific parameters from one of the few gauged watersheds (Wadi Girafi; Fig. 6) in Sinai's Red Sea Hills were extrapolated to similar watersheds in the Arabian Red Sea Hills (see Table 2).

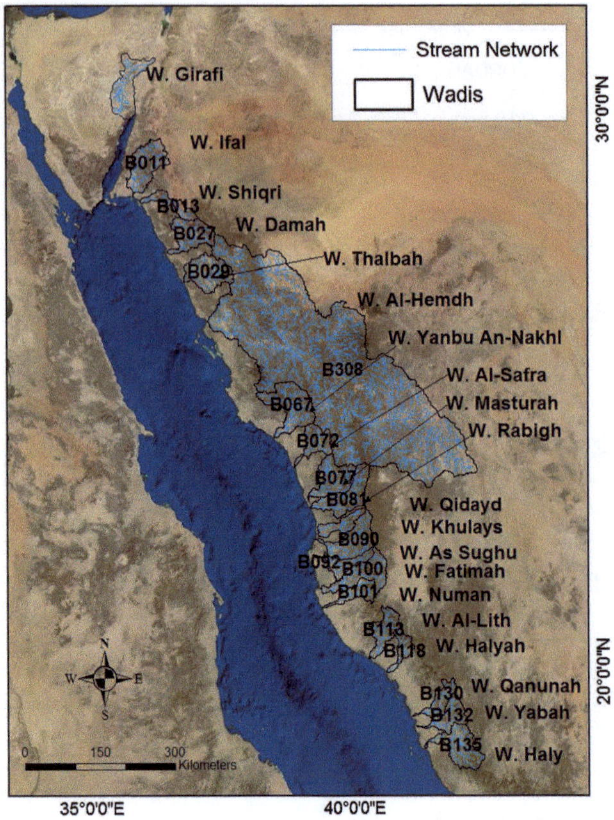

Fig. 6 Location map showing the distribution of the investigated major Red Sea coastal watersheds and the calibrated watershed (Wadi Girafi) in the Sinai Peninsula. Also shown is the distribution of the stream networks for the investigated watersheds

The Wadi Girafi (area: 5,057 km^2) watershed and the majority of the sizable watersheds (Fig. 6, Table 2) that drain the Arabian Red Sea Hills have similar climatic and topographic settings, areal extent, and soil types. Precipitation over the selected watersheds is sparse to moderate (annual averages 25–273 mm); precipitation generally occurs (1) during the winter season (December through March) as a result of southeast winds coming off the Mediterranean coast or (2) during the summer season (July through October) as a result of northwest winds off the Arabian Sea (Alsharhan et al. 2001). The investigated watersheds originate at high elevations (1,294–2,984 m a.m.s.l.) and discharge downstream into the adjacent water bodies (Red Sea, Gulf of Aqaba). The selected watersheds are medium sized (area: 1,825–107,769 km^2) and share the same land use classification (Southwestern U.S. Arid Range) and soil types (basement, limestone, sandstone, and alluvial soil types). These four soil types occupy an area in excess of 90 % of the total area of each of the selected watersheds. The selected watersheds constitute more than 70 % of the total area of all watersheds in the KSA, which originate from the Red Sea hills and drain toward the Red Sea (Table 3).

SWAT Model Applications for the Period from 1998 to 2010

Temporal climatic data over the investigated periods are required inputs for the SWAT model. These data include precipitation, maximum and minimum air temperature, solar

Table 2 Calibration parameters used in SWAT model (modified from Milewski et al. 2009)

Parameter	Value	Definition
CN (Alluvial)	63	SCS curve number
CN (Sandstone)	74	SCS curve number
CN (Limestone)	92	SCS curve number
CN (Precambrian)	94	SCS curve number
ALPHA_BF	1.0	Base flow alpha factor (days)
CH_K	150	Effective hydraulic conductivity in channel alluvium (mm/h)
SURLAG	15.0	Surface runoff lag coefficient (days)
SOL_AWC	Varies (0.01–0.50)	Available water capacity of the soil layer (mm/mm soil)
SOL_K	Varies (0.5–10)	Saturated hydraulic conductivity (mm/h)
CH_N1	Varies (0.035–0.05)	Manning's "*n*" value for the tributary channels
CH_N2	Varies (0.035–0.05)	Mannings's "*n*" value for the main channel
CANMX	0.0	Maximum canopy index
SOL_Z	Varies (1–1,000)	Soil depth (mm)
GW_DELAY	0.0	Groundwater delay time (days)
GWQMN	2,000	Threshold depth of water in the shallow aquifer required for return to occur (mm H_2O)

Table 3 Selected major (area > 1,825 km^2) watersheds

Basin code	Watershed name	Area (km^2)	Slope average (°)	Basin length (m)	Maximum elevation (m a.m. s.l.)	Soils (%)			
						Alluvium	Basement	Limestone	Sandstone
B038	Wadi Al-Hemdh	107,769	8.8	600,605	1,016	6	83	8	1
B067	Wadi Yanbu An-Nakhl	6,569	23.7	117,649	1,662	6	93	0	1
B081	Wadi Rabigh	5,241	7.79	143,757	1,359	11	87	0	1
B090	Wadi Khulays	5,186	12.3	131,761	1,639	9	89	0	1
B135	Wadi Haly	5,160	24.7	121,483	2,699	5	94	0	0
B100	Wadi Fatimah	5,086	14.5	148,768	2,063	15	83	0	1
B011	Wadi Ifal	5,018	19.9	124,498	2,348	12	60	11	15
B029	Wadi Thalbah	4,985	11.5	102,348	1,344	22	65	0	12
B077	Wadi Masturah	4,755	16.8	121,333	996	4	95	0	1
B027	Wadi Damah	4,428	18.3	106,134	1,607	11	68	20	1
B113	Wadi Al-Lith	3,268	31.3	109,488	2,428	8	90	0	1
B132	Wadi Yabah	3,012	22.4	105,929	2,286	0	100	0	0
B072	Wadi Al-Safra	2,881	33.1	100,774	2,261	10	90	0	0
B092	Wadi As Sughu	2,406	7.63	127,824	1,292	28	61	0	10
B101	Wadi Numan	2,404	20.1	123,106	2,559	24	75	0	1
B130	Wadi Qanunah	2,299	21.4	111,588	2,000	4	94	0	1
B087	Wadi Qidayd	2,280	43.2	109,393	1,619	29	69	0	1
B013	Wadi Shiqri	2,111	27.4	124,789	1,625	21	56	13	8
B118	Wadi Halyah	1,825	26.1	95,247	2,208	14	85	0	0

radiation, wind speed, and relative humidity. Unfortunately, the data from the existing meteorological stations across the study area are spatially and temporarily inadequate. The limited available climatic data from meteorologic stations were compared to data from the SWAT climatic database. With the exception of precipitation, a good correspondence was observed between the values of climatic parameters that were extracted from the CCSM4 and CFSR databases (Fig. 7). For this reason, precipitation was extracted from TRMM datasets (available from 1998 to the present) and the remaining climatic data were extracted from the CFSR database.

SWAT Model Applications for the Period from 2020 to 2100

Most of the continental or regional-scale hydrologic distributed grid models are land surface models (coupled with global climate models); these models provide vertically detailed representations of processes such as interception, canopy drip, snow accumulation and ablation, infiltration, surface and subsurface runoff, soil moisture, and the partitioning of evapotranspiration (ET) between canopy evaporation, transpiration, and soil evaporation. Lateral variations in soil types, vegetation, and terrain are accounted for in recently developed grid models by dividing grid cells into a mosaic of sub-grids. In this study, outputs from the newly released (1 April 2010, CCSM4.0, http://www.cesm.ucar.edu/models/ccsm4.0/) state-of-the-art CCSM4 model (grid size: 1.1° [longitude: ~110 km] by 0.9° [latitude: ~90 km]) were used. The most recent version, CCSM4, is made available to the community from the CCSM Website. The CCSM is a general circulation climate model with linked components (e.g., atmosphere, land, ocean, and sea ice) that exchange information and fluxes. It has been used to study paleoclimatic and historical climatic periods and was also employed to predict climate change over areas of interest.

Predicting future river runoff changes requires accurate meteorological input data for the hydrological models. We investigated the extent to which the CCSM4 outputs are consistent with observations throughout a reference period (1979–2005). A good correspondence was observed (in magnitude and phase) between the CCSM4 outputs and CFSR database in all but one parameter (precipitation). Precipitation is consistently high in the CCSM4 database, compared to values in the CFSR database and TRMM datasets (see Figs. 7 and 8).

To remove the bias in the CCSM4 precipitation data, we adjusted the CCSM4 daily precipitation to TRMM daily precipitation throughout the overlapping period between the two datasets (i.e., from 1998 to 2013) (Fig. 8). The extracted correction function for the reference period was then used to

Fig. 7 Comparison of 30-day running average (a) precipitation, (b) solar radiation, and (c) maximum temperature extracted from CCSM4 and CFSR databases for a selected location (longitude: 41.2500°E; latitude: 23.0890°N) within the study area

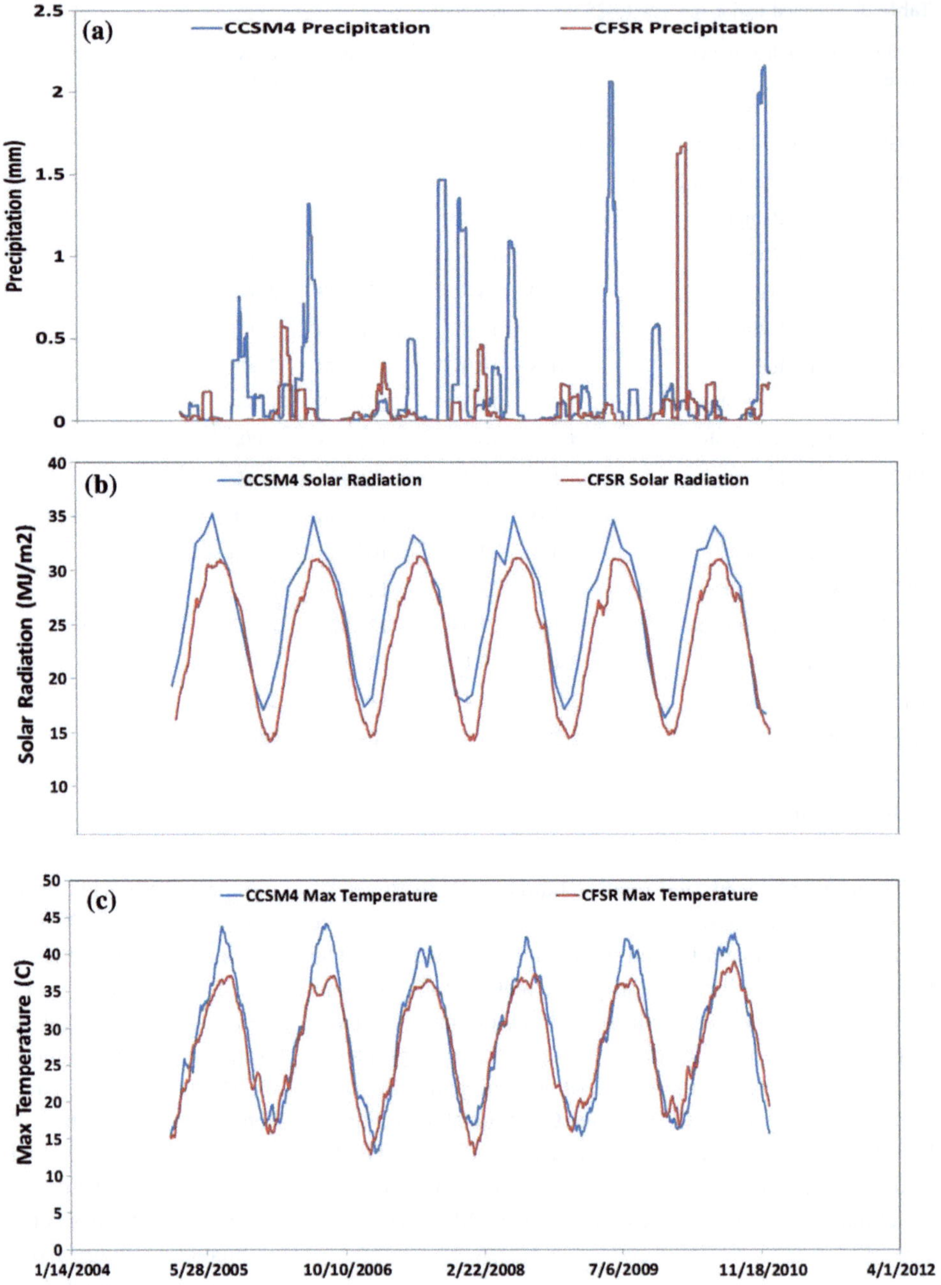

Fig. 8 Comparison of monthly (1998–2013) precipitation data extracted from TRMM and CCM4 databases for a selected location (longitude: 41.2500°E; latitude: 130.0890°N) within the study area. Also shown are the adjusted CCSM4 data that were normalized to the TRMM data from 1998 to 2013 for the same location

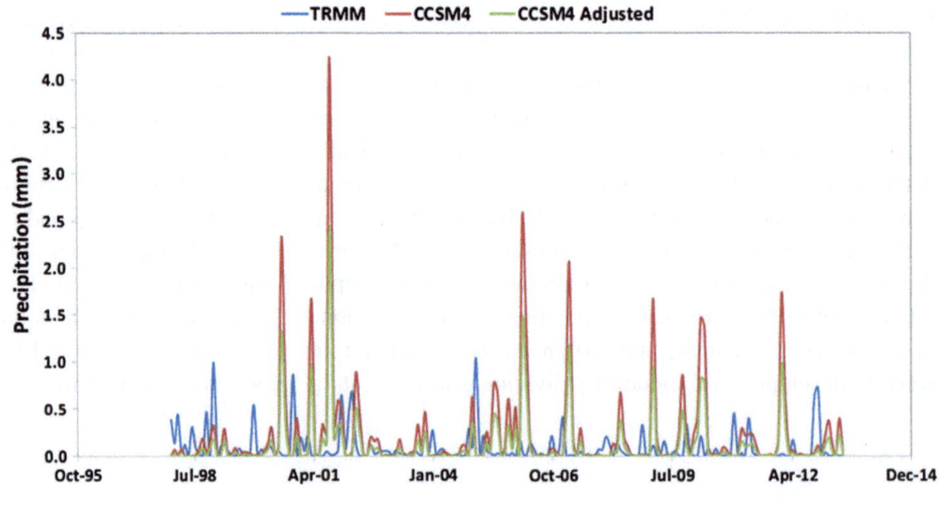

Table 4 Modeled average annual (1998–2010) values of hydrologic variables for the investigated watersheds

Basin Code	Watershed name	Area	Precipitation		Stream flow[a]		Potential recharge[b]		Initial losses[c]	
		(km^2)	(mm)	($\times 10^6$ m^3)	(mm)	(%)	(mm)	(%)	(mm)	(%)
B038	Wadi Al-Hemdh	107,769	58	6,197	19.1	33.1	7.19	12.5	31.3	54.37
B067	Wadi Yanbu An-Nakhl	6,569	58	379	0.86	1.49	3.85	6.67	53	91.84
B081	Wadi Rabigh	5,241	106	555	5.53	5.22	11.9	11.2	88.5	83.54
B090	Wadi Khulays	5,186	134	696	8.96	6.68	10.61	7.91	115	85.42
B135	Wadi Haly	5,160	273	1,407	33.8	12.2	157	57.7	82	30.07
B100	Wadi Fatimah	5,086	153	777	8.56	5.6	10.4	6.83	134	87.57
B011	Wadi Ifal	5,018	27	136	0.09	0.33	0.13	0.48	27	99.19
B029	Wadi Thalbah	4,985	54	270	0.6	1.11	3.06	5.65	50.5	93.25
B077	Wadi Masturah	4,755	85	405	4.77	5.6	7.67	9	72.8	85.4
B027	Wadi Damah	4,428	33	147	0.16	0.48	0.6	1.81	32.4	97.71
B113	Wadi Al-Lith	3,268	143	467	4.24	2.97	12.5	8.73	126	88.3
B132	Wadi Yabah	3,012	258	776	32.8	12.7	149	58	75.4	29.24
B072	Wadi Al-Safra	2,881	78	224	3.75	4.83	6.64	8.56	67.2	86.61
B092	Wadi As Sughu	2,406	150	361	3.29	2.19	8.82	5.88	137	91.93
B101	Wadi Numan	2,404	254	611	2.62	1.03	13.3	5.24	238	93.73
B130	Wadi Qanunah	2,299	226	520	44.3	19.6	107	47.3	74.9	33.11
B087	Wadi Qidayd	2,280	108	246	0.2	0.19	12.6	11.7	95	88.09
B013	Wadi Shiqri	2,111	25	53	0.29	1.16	0.43	1.71	24.4	97.13
B118	Wadi Halyah	1,825	243.4	444	6.04	2.48	16	6.57	221	90.94

[a] Stream flow = Surface runoff + Lateral flow + Return flow − Transmission losses
[b] Potential recharge = Percolation to shallow aquifer + Transmission losses − Revap from shallow aquifer − Return flow
[c] Initial losses = Precipitation − Potential recharge − Stream flow

Table 5 Modeled average annual (2020–2100) values of hydrologic variables for the investigated watersheds

Basin code	Watershed name	Area	Precipitation		Stream flow		Potential recharge		Initial losses	
		(km^2)	(mm)	($\times 10^6$ m^3)	(mm)	(%)	(mm)	(%)	(mm)	(%)
B038	Wadi Al-Hemdh	107,769	85	9,193	12	14	23.7	27.7	49.7	58.23
B067	Wadi Yanbu An-Nakhl	6,569	92	602	1.8	2	6.8	7.4	83	90.61
B081	Wadi Rabigh	5,241	134	702	1.9	1.4	12.3	9.2	120	89.43
B090	Wadi Khulays	5,186	108	558	0.6	0.6	5.2	4.9	102	94.56
B135	Wadi Haly	5,160	165	850	11	6.7	59.9	36.4	93.8	56.95
B100	Wadi Fatimah	5,086	144	731	1.5	1.1	15.9	11	126	87.91
B011	Wadi Ifal	5,018	42	212	0.3	0.6	0.9	2	41.1	97.35
B029	Wadi Thalbah	4,985	56	272	0.3	0.5	2.5	4.5	51.8	94.97
B077	Wadi Masturah	4,755	131	621	2.3	1.8	12.3	9.4	116	88.78
B027	Wadi Damah	4,428	57	252	0.6	1	2.9	5	53.6	93.98
B113	Wadi Al-Lith	3,268	175	572	7.3	4.1	21.7	12.4	146	83.44
B132	Wadi Yabah	3,012	184	554	5.9	3.2	95.6	52	82.4	44.82
B072	Wadi Al-Safra	2,881	96	276	1.5	1.5	7.6	7.9	86.9	90.6
B092	Wadi As Sughu	2,406	111	267	0.2	0.2	3.9	3.5	107	96.35
B101	Wadi Numan	2,404	171	410	5.9	3.5	21.5	12.6	143	83.91
B130	Wadi Qanunah	2,299	228	524	9.7	4.3	125.8	55.2	92.4	40.55
B087	Wadi Qidayd	2,280	111	254	1	0.9	6.4	5.8	104	93.33
B013	Wadi Shiqri	2,111	56	117	0.7	1.2	2.2	3.9	52.7	94.95
B118	Wadi Halyah	1,825	196	358	10.2	5.2	21.6	11	164	83.82

remove the bias in the CCSM4 precipitation data for the period 2020–2100. The bias correction was conducted using the following equation (Xu and Yang 2012):

$$\begin{aligned} P_{\text{bias-corrected}} = & \left(\text{Mean } P_{c\text{-future}}\right) \\ & + \left(\text{Mean } P_{\text{o}}\right) - \left(\text{Mean } P_{c\text{-past}}\right) \\ & + \left(\text{Mean } P'_{c\text{-future}}\right) \times \left(S_{\text{o}}/S_{c\text{-past}}\right), \end{aligned}$$

where Mean $P_{c\text{-future}}$ = a mean of future daily values over the future period (2020–2100); Mean P_{o} − Mean $P_{c\text{-past}}$ = a mean difference for the same past period (a mean daily value over the period of 1998–2013) between observed and computed precipitation; Mean $P'_{c\text{-future}}$ = a difference between a daily value and a mean daily value; S_{o} = a standard deviation of daily observed values for each day over the past period; and $S_{c\text{-past}}$ = a standard deviation of computed daily values for each day over the past period.

Discussion and Findings

Model results for the investigated period, 1998–2010, and the period 2020–2100, including the average annual amount of precipitation, stream flow, potential recharge, and initial losses are given in Tables 4 and 5, respectively. Inspection of these tables reveals a number of observations or findings. Increasing the proportion of areas occupied by basement and/or increasing precipitation amounts increases the proportion of stream flow. For example, Wadi Qanunah (watershed area occupied by basement: 94 %) and Yabah (100 % basement) receive 226 and 258 mm/yr of precipitation, of which 20 and 13 %, respectively, are partitioned as stream flow. In contrast, the watersheds that receive less precipitation and have a smaller proportion of their area occupied by basement outcrops have smaller proportions of stream flow. For example, Wadi Shiqri (56 % basement) and Wadi Ifal (60 % basement) receive 25.1 and 27.2 mm/yr of precipitation, respectively, of which only 1 % or less is partitioned as stream flow. The larger the amount of precipitation and the runoff, the greater the amount of transmission loss and potential recharge. For example, the precipitation and stream flow for Wadis Haly, Yabah, and Qanunah are high (precipitation exceeds 220 mm/yr, runoff exceeds 12 %), and so is the potential recharge (recharge exceeds 40 %) (Fig. 9).

Comparison of measured average annual precipitation for the periods 1998–2010 (Table 4) with the adjusted average annual CCSM precipitation for the period 2020–2100 (Table 5) shows a general increase in precipitation with time in the northern and central watersheds, and a decline in precipitation in the southern watersheds. For example, the average annual precipitation for Wadis Ifal, Shiqri, and Damah increased from 27 to 42 mm, 25 to 55 mm, and 33 to 57 mm, respectively, whereas the average annual precipitation for Wadis Haly, Yabah, Numan, and Fatimah decreased from 273 to 165 mm, 258 to 184 mm, 254 to 171 mm, and 153 to 144 mm, respectively.

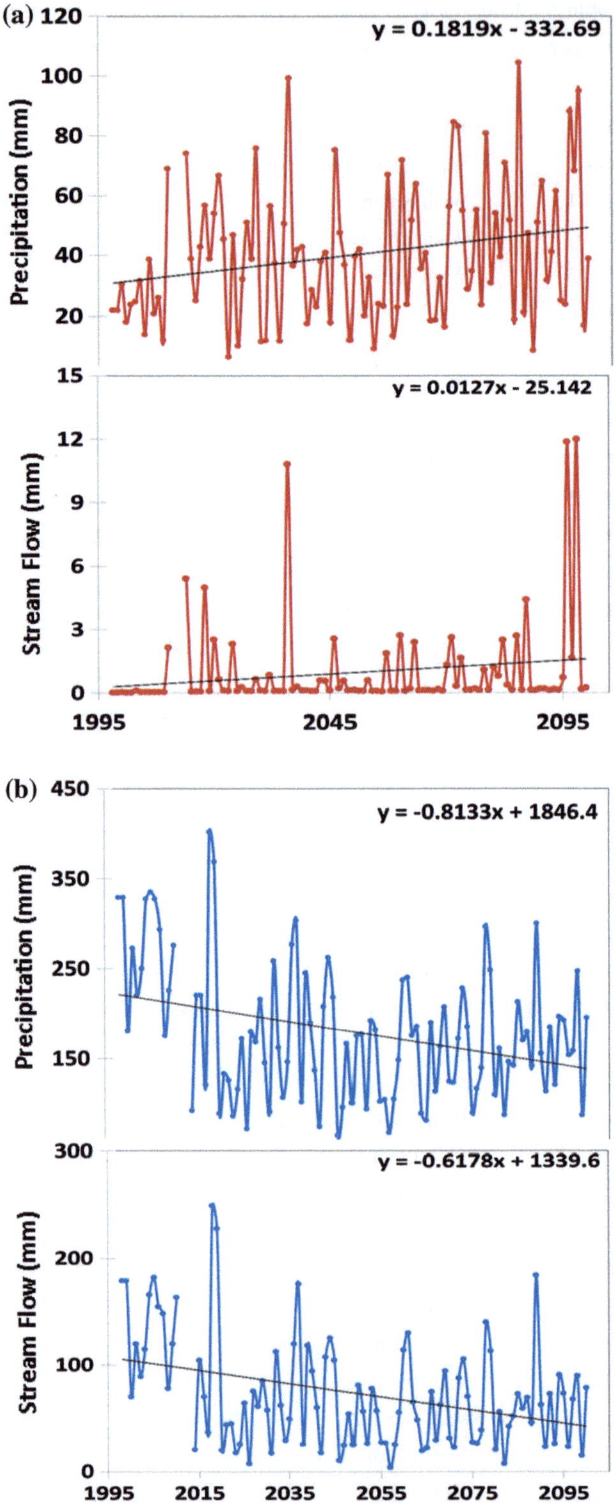

Fig. 9 Precipitation and runoff time series from 1998 to 2100 at **a** Wadi Ifal (northern Red Sea Hills), and **b** Wadi Haly (southern Red Sea Hills)

Observations discussed in this chapter support earlier suggestions that global warming could be causing a rise in sea surface temperature, enhancing evaporation, and intensifying northerly and northwesterly winds and precipitation over the northern and central watersheds. Global warming could cause shifts in monsoonal wind regimes that could modify precipitation patterns over the southern watersheds (Fouda and Gerges 1994).

Acknowledgments We wish to sincerely thank our colleagues Racha Elkadiri, Mohamed Ahmed, Talal Alharbi, and Mustafa Emil at the Department of Geosciences, Western Michigan University, Kalamazoo, and Eugene Yan of the Argonne National Laboratory, Argonne, Illinois, U.S., for their contributions toward making this chapter possible.

References

Alsharhan A, Rizk Z, Nairn A, Bakhit D, Alhajari S (2001) Hydrogeology of an arid region: the Arabian Gulf and adjoining areas. Elsevier Science, The Netherlands, 331 pp

Fouda M, Gerges M (1994) Implications of climate change in the Red Sea and Gulf of Aden region: an overview. UNEP Reg Seas Rep Stud 156:1–59

Jarrett R (1984) Hydraulics of high-gradient streams. J Hydraul Eng 110:1519–1539

McCarthy G (1938) The unit hydrograph and flood routing. Unpublished paper. U.S. Engineering Office, Providence RI. U.S. Army Corps of Engineers, New London, CT

Milewski A, Sultan M, Yan E, Becker B, Abdeldayem A, Soliman F, Gelil F (2009) A remote sensing solution for estimating runoff and recharge in arid environments. J Hydrol 373:1–14

Riverside County Flood Control (1978) Hydrology manual. Water conservation district. SCS (1985) Section 4: Hydrology. National engineering handbook. U.S. Department of Agriculture, Soil Conservation Service, Engineering Division, Washington, DC, USA

SCS (1985) Section 4: hydrology. National Engineering Handbook. U. S. Department of Agriculture, Soil Conservation Service, Engineering Division, Washington, DC, USA

Stern R, Kroner A (1993) Late Precambrian crustal evolution in NE Sudan: isotopic and geochronologic constraints. J Geol 101:555–574

Sultan M, Wagdy A, Manocha N, Sauck W, Abdel Gelil K, Youssef AF, Becker R, Milewski A, el Alfy Z, Jones C (2008) An integrated approach for identifying aquifers in transcurrent fault systems: the Najd Shear System of the Arabian Nubian Shield. J Hydrol 349(3–4):475–488. doi:10.1016/j.jhydrol.2007.11.029

Sultan M, Duncan IJ, Arvidson RE, Stern RJ, El Kaliouby B (1990) Extension of the Najd Shear System from Saudi Arabia to the Central eastern Desert of Egypt based on integrated field and Landsat observations. Tectonics 9:539–543

Xu Z, Yang Z (2012) An improved dynamical downscaling method with GCM bias corrections and its validation with 30 years of climate simulations. Am Meteorol Soc. doi:10.1175/JCLI-D-12-00005.1

Raised Coral Reefs and Sediments in the Coastal Area of the Red Sea

Abbas M. Mansour and Hashem A. Madkour

Abstract

Along the Red Sea, narrow coastal plains ascend directly into fault-bounded blocks within a few kilometers of the shoreline. Littoral areas on both sides of the Red Sea are characterized by mixed sedimentation relating to a complex system of fringing and barrier reefs and alluvial fans. These marine sediments are uplifted to altitudes rarely exceeding 50 m. However, although terraces are well developed on both sides of the basin there is no apparent correlation, the possible exception being the youngest level situated about 2 m above present sea level. Raised coral reefs result from either the corrosive action of waves or from local erosion by occasional torrents creating low cliffs and exposures just above high-tide level. Each reef unit exhibits in a short distance lateral facies changes, which begin at the shore with the beach facies, mainly composed of siliciclastics, and end at the reef crest zone with transition to the fore slope made up of carbonate sediments. A strong similarity can be noticed between sedimentary facies of ancient Pleistocene sediments and those now present in modern fringing reefs. Reefs with their siliciclastic associations occur in the form of repeated cycles reflecting tectonic effects and/or sea level changes. Reef sequences exhibit different degrees of diagenetic alteration, which are reflected by a gradual change of skeletal particles and the early-formed cement, from aragonite and high Mg-calcite to low Mg-calcite. Tectonism controls the areal distribution of the depositional systems and influences the number, thickness, extension, and elevations of the reef sequences. Each sequence in each area can be uniquely correlated to the overall (global) population of dated terraces. Coastal areas of the Red Sea are under stress from a variety of human activities and many have experienced widespread degradation, especially around Hurghada and Jeddah. Hotel, resort and other developments along the coast of Egypt are growing rapidly, destroying raised reefs and threatening valuable coral reef ecosystems. Some areas along the coast suffer from construction problems that are associated with coral reefs. These problems include ground settlement and low bearing capacity which are mainly due to the low shear resistance and high porosity of reef sediments. These problems greatly affect the safe and economic land utilization of the coasts. Prediction of the future changes along the Red Sea coast would give guide lines to what will happen due to the varying nature of the coast. Such predictions would have implications for future social and economic development along the coast. Effective and integrated coastal zone management programs are critical to sustaining the natural resources of the Red Sea.

A.M. Mansour (✉)
Geology Department, Faculty of Science, South Valley University, Qena 83523, Egypt
e-mail: aman@svu.edu.eg

H.A. Madkour
Red Sea Branch, National Institute of Oceanography and Fisheries, Hurghada 84511, Egypt

Introduction

Along the Red Sea, narrow coastal plains ascend directly into fault-bounded blocks within a few kilometers of the shoreline (Fig. 1). Along the shore there is a continuous band of

N.M.A. Rasul and I.C.F. Stewart (eds.), *The Red Sea*,
Springer Earth System Sciences, DOI 10.1007/978-3-662-45201-1_23,

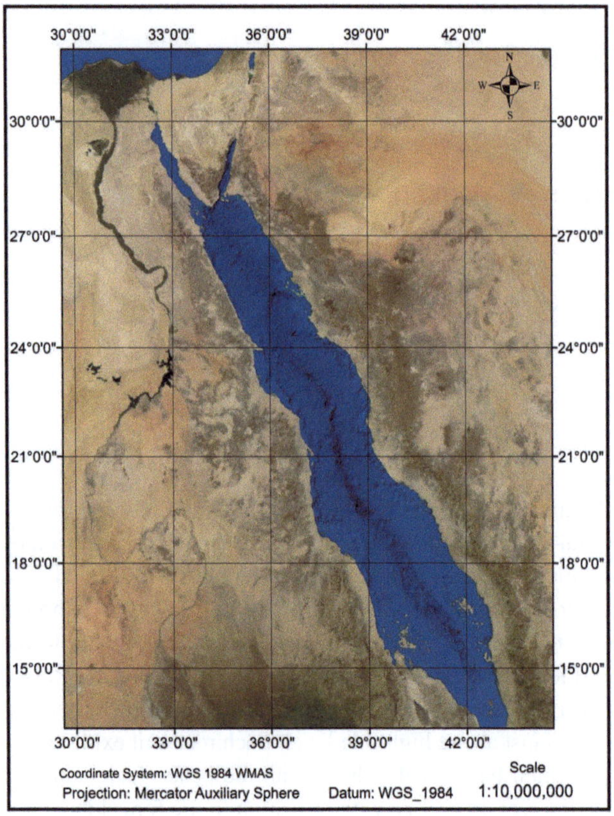

Fig. 1 Landsat image showing the narrow coastal plains that ascend directly into fault-bounded blocks within a few kilometers of the shoreline (Google map—2014 digitized by Arc map v.10)

emergent reef terraces and a sandy gravel surface which is inclined towards the sea, as described in 1888 by J. Walther, the founder of actuogeology. Littoral areas on both sides of the Red Sea are characterized by mixed sedimentation relating to a complex system of fringing and barrier reefs and alluvial fans. A series of raised fossil coral reefs higher than current sea level are one of the most striking features of the Red Sea coastline. There are at least two Pleistocene reef terraces; however, three are observed in several areas. Each exhibits a lateral facies development over a short distance, beginning at the shore with a beach of mainly siliciclastics, and ending at the reef zone with carbonate sediments (Mansour 2000a). Mansour (2000b) noted 6 terraces at Dishet El Dabaa (Fig. 2), south Hurghada. El-Asmar and Attia (1996) noted 4 terraces in the Sinai, and Guilcher (1988) refers to a series of 11 uplifted reefs on the Tiran Islands at the entrance of the Gulf of Aqaba extending vertically 320 m.

The Red Sea coastlines frequently extend out in the form of rocky headlands referred to, along the whole region, as Ras (e.g. Ras Banas, Ras Ghareb, Ras Hatiba). The coastlines and outlying fringing reefs are incised at irregular intervals by creeks known by the local people as Sharms (e.g. Sharm el-Sheikh). These creeks (or Sharms) are typically drowned discharge valleys. Recent coral reefs and elevated Pleistocene reefs terraces are also well developed along the coasts of the Red Sea.

The raised coral reefs and the sediments of the coastal area in the Red Sea are the main subjects of this chapter. Prediction of their future changes could be useful for the management and sustainable development of the coast. Several authors have discussed the ages and diagenesis of the Pleistocene reefs terraces along the Red Sea coast (Plaziat et al. 2008, with references therein; Manaa 2011).

Quaternary Coral Reefs

The Quaternary reefs appear strictly controlled by the Pleistocene cyclicity of ice ages which in turn modulate with the basin's hydrological conditions (Taviani 1998). It is proposed that the shallow-silled Red Sea basin was cyclically affected by more or less severe biotic turnovers as a consequence of periods of high environmental stress (temperature, salinity) during glaciations.

Raised coral reefs occur in the form of a ridge-furrow system, produced either by the corrosive action of waves or by local erosion by occasional torrents creating low cliffs and exposures just above high-tide level. Their thickness generally increases from north to south (Mansour 1993). Raised coral reefs are composed mainly of corals in their growth position and the preservation of corals is mostly good (Fig. 3).

Quaternary Coral Reefs on the Western Side of the Red Sea

The Quaternary sequences of the Egyptian Red Sea coast constitute a conspicuous belt flanking the Neogene succession in the NW part of the Red Sea rift. They include both alluvial fans and reefs. The former are the result of high basement relief and the presence of numerous scarps. Much of this detrital material is trapped in grabens formed by Quaternary and Neogene faulting (Purser et al. 1987). As on the Saudi Arabian coast, Pleistocene reefs may be uplifted to altitudes of about 50 m. The marine terraces along the Red Sea coast and the Gulf of Aqaba exhibit different elevations relative to the present sea level (Al-Sayari et al. 1984a, b; El-Asmar and Attia 1996).

The Pleistocene terraces are dissected by WSW-ENE trending dry wadis that act as conduits for coarse detritus to the sea only during flash floods. Reefs with their siliciclastic associations occur in the form of repeated cycles reflecting tectonic effects and/or sealevel changes. Pleistocene sequences comprise continental sediments and shallow-water carbonates and siliciclastics of different facies mostly similar to those of modern fringing reefs. However, this facies

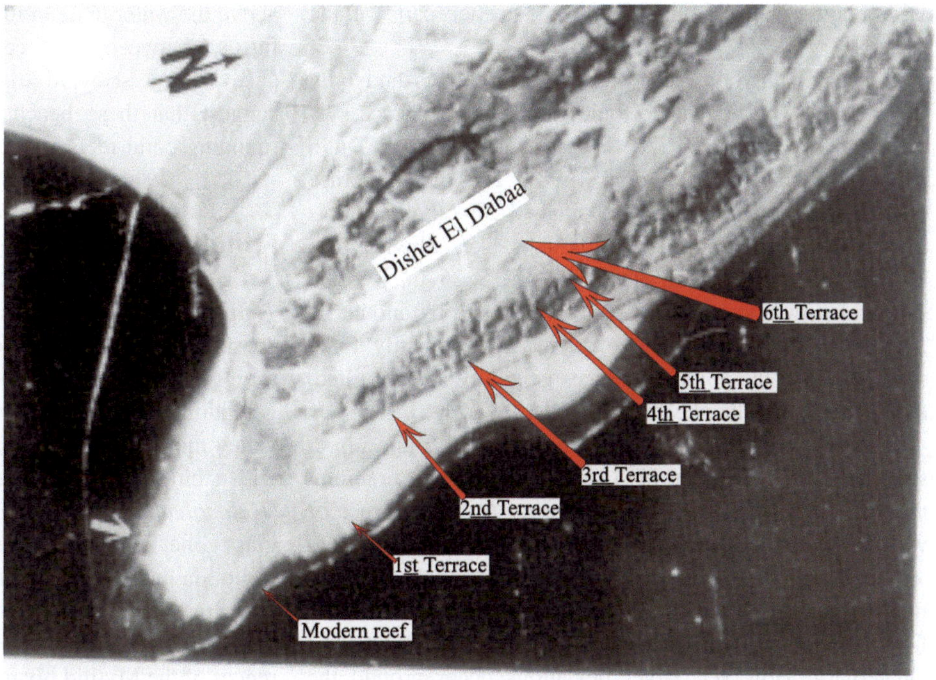

Fig. 2 Aerial photograph showing Quaternary reef terraces at Dishet El-Dabaa, 25 km south of Hurghada (Mansour 2000a)

Fig. 3 Raised coral reefs are composed mainly of corals in their growth position and the preservation of corals is mostly good

similarity probably reflects similar climatic conditions. All reef sequences reveal transgressive phases developed during the sea level rise, while the alluvial deposits are regressive sequences accumulated during the lowering of sea level.

Sequence boundaries of most reef units in most areas are depicted by an abrupt basinward shift of the coastal onlap. Other sequence boundaries are observed when alluvial channel deposits suddenly replaced coral reef (Mansour 2000a).

Fossil coral reefs, like the present fringing reefs, outcrop all along the Sudanese coast as well as everywhere around the Red Sea, except in the delta areas of the most important wadis, the Khor Baraka in the south and the Khor Dib in the north, where the silty alluvial deposits inhibit coral development. Generally narrow (a few hundred meters), the fossil coral reefs often form two or three parallel flat ridges, such as in the Marsas Fijja, Arus and lrayes areas.

Eastern Side of the Red Sea

Along the Red Sea coast of Saudi Arabia a sequence of marine and terrestrial terraces unconformably overlies Pliocene sediments. Alluvial fans relating to the mountainous hinterland and various fault scarps slope gently towards the sea. Partly reworked by wadi systems, they grade into coastal sabkhas developed as wide littoral plains or into reef assemblages along the narrower and steeper parts of the coast.

Caused mainly by eustatic sea level changes and only subordinately by vertical tectonics, three main terrace systems can be recognized (Al Sayari et al. 1984a, b; Dullo 1990; Dullo and Montaggioni 1998). The oldest terrace system whose reef platform is situated 30–50 m a.s.l. is associated with extreme coarse, boulder-like, alluvial fans. The second system, with elevations of 15–25 m, is often subdivided by a step situated at 15–18 m, and by another at 20–22 m. The terraces above 20 m are frequently tilted slightly, while the younger terraces generally are not tilted, though they may be affected by faults.

The third terrace occurs between 6 and 10 m and represents the last interglacial high sea level dated by uranium/thorium at about 95,000–120,000 years. A small terrace or wave cut notch located 2 m above modern sea level is related to the Holocene transgression. The terrace levels are relatively constant along the NW Arabian Red Sea coast confirming relatively minor vertical displacement during the Quaternary. However, larger deviations occur around the Midyan Peninsula, especially along the Gulf of Aqaba.

Facies Patterns of Recent Reefs

Beach

The beach is 3–8 m wide, a few cm to about 50 cm high and sloping 8–12°. The sediments are gravel and sand with abundant shell material swept up from the sea. Beach sediments at the mouths of valleys are mainly composed of quartz, feldspars and rock fragments. A few meters away from the mouths of the valleys, carbonate fragments are abundant in the beach sediments. Sometimes these materials are partly cemented by aragonite and/or Mg-calcite, transforming it into beach rock. The beach rocks exhibit clear color and coherence contrasts to beach sediments with a seaward dip. The beach sediments are flat with occasional heavy mineral concentrations, current shadows associated with bivalve shells, and various burrows of crustacea. At the valley mouths, beach sands extend and vary in width from a few tens of meters to a few hundred meters. The seaward margins of the sand facies grade into facies of the subtidal zone. Primary sedimentary structures of these sands include wave ripple cross-lamination. Crustacean burrows mostly destroy these structures and textures by mixing sediments.

Beach Rocks

Beach rocks are exposed in a narrow intertidal zone. They occur as conglomerate and/or sandstone layers at or just above the water table in the mouth of some wadis and in the furrows between coral reefs. They exhibit a typical appearance with a seaward dip of 10–20°. Field relationships suggest that these beach rocks are younger than the surrounding coral reefs.

Fringing Reef

Tidal flats are most widespread and conspicuous along the coast. In all cases they are rocky, built by coral limestone with a rugged surface. The contact to the subtidal zone is either gradational (e.g. at a valley mouth) or abrupt (e.g. away from valley mouths). The surface and its depressions are covered with a thin sand layer, sometimes with sparse seagrass growth (*Haloduleuninervis*, *Halophilastipulacea*) forming the sea grass zone. This zone is very rich with foraminifera, mainly Miliolida and Soritida which are a characteristic feature for this zone in the Pleistocene terraces. Ophiurids along with gastropods are abundant on hard substrates; massive occurrences of the Mytilid bivalve are also characteristic.

The Coral zone is a few meters wide and laterally increases on both sides of the valley mouths. It is characterized by a relatively even distribution of coral colonies. Corals grow on the margins of the channels just after a few tens of meters from the beach and become dense and variable seawards (Fig. 4). Leewards, the reef grades into a few meters wide intertidal zone. Seawards, the reef flat is sharply bordered by the reef slope, which is very steep (45–90°) leading to a complete disintegration of the reef and the development of a sandy bottom in the shallow parts grading into a coral carpet in the deeper (>15 m) waters. The reef flat is characterized by the common occurrence of the coral genera *Pocillopora*, *Stylophora* and other scleractinians such as *Favia*. The areas between the coral heads consist of a wavy, rocky bottom with a smooth surface covered by various calcareous (mainly coralline) and soft algae. At the reef slope, several species of the genera *Acropora* and *Porites* are dominant, and *Millepora* also exist. Some corals use dead coral heads as a substrate. Pools occur within the flat areas, which exhibit a dense cover of living scleractinians along their margins. The pool floor is covered by bioclasts from the reef slope during storms. A gentle inclined ramp and the abrupt onset of scleractinian growth characterize the reef crest. Within small fissures sea urchins thrive. Their thick spines are characteristic in the field and in thin sections for recognizing this zone in the Pleistocene terraces.

Facies of Pleistocene Reefs

The Pleistocene terraces occur in a narrow belt, with lateral facies changes, mostly parallel to the older Pliocene and the modern reef belt. The change of Pleistocene outline and topography in some places is due to the supply of clastic sediments during different climatic conditions. Based on field observations, sedimentological analysis and the distribution of fossils it can be stated that reef sequences comprise continental sediments and shallow-water carbonates and siliciclastics of different facies that mostly correspond to those of the modern environments.

Using this correspondence and the vertical and lateral succession in some areas along the coast (e.g. Wadi Guisses, Wadi Hamrawein) the distribution of rock units is clearly observed and each reef unit is represented by a facies association consisting of beach, back reef, reef crest and reef front (talus). This facies association is repeated in each terrace in the form of successive cycles, as was shown for the uplifted reefs on the Saudi Arabian coast by Dullo (1986, 1990).

Because of the commonly steep topography of the Red Sea coast, the sequence boundary of most reef units in most areas along the coast is depicted by an abrupt basin ward shift of the coastal on lap. This means that the sea level fall was extensive and rapid and falls below the physiographic sea margin and the entire coast may be exposed subaerially. Other reef boundaries unconformably resulted from the alluvial deposits. Sequence boundaries are clearly observed where alluvial channel deposits suddenly replace coral reef.

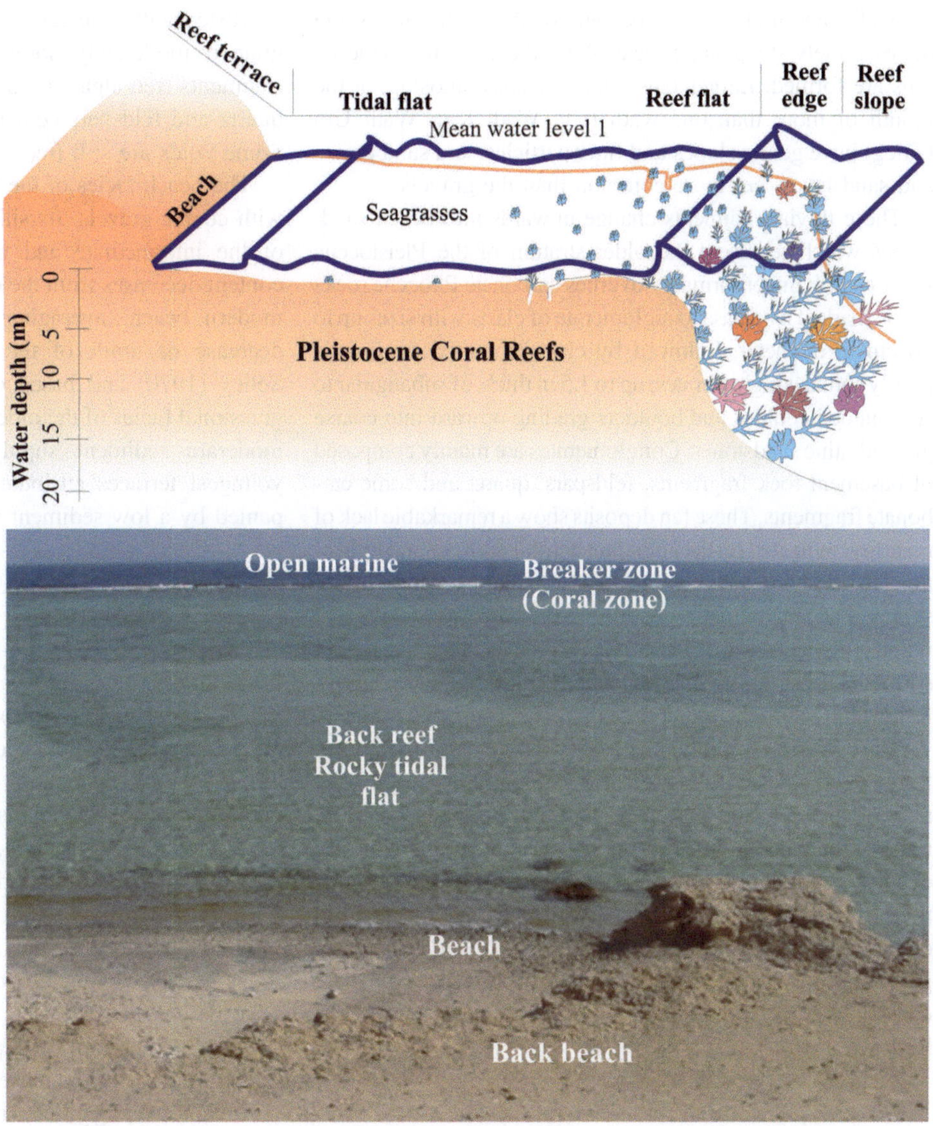

Fig. 4 Corals grow just after a few tens of meters from the beach and become dense and variable seawards (upper model, lower photo)

Genetically related sedimentary facies are attributed to various depositional systems. The various siliciclastics and carbonate depositional systems and facies are briefly outlined as follows.

Alluvial Fan

Impressive evidence of stream erosion and deposition exists in the areas of the Egyptian Red Sea and Gulfs of Suez and Aqaba (Issawi et al. 1971; Sellwood and Netherwood 1984). Infrequent rainfall and catchment areas of rocky desert highlands funnel flashfloods that build alluvial fans several square km in extent. Below the water surface submarine alluvial cones are formed, especially on the gentle slopes. On the extremely steep slope areas along the coast no subaerial fans are formed. Larger fans, alone or anastomosing, at the mouth of more than one wadi (e.g. Wadi Kid, Wadi Um Ghieg) have gentle slopes and finer particles, and sand grains and sand layers are more common than the gravels.

These fluvial sediments change upwards into beach sandstones which represent the older stratum of the Pleistocene sequences. It unconformably overlies carbonate Pliocene rocks and is marked by a basal conglomerate of clasts with sizes up to 30 cm in diameter, followed by clast supported polymictic poorly sorted conglomerates, up to 1.5 m thick, of subangular to subrounded cobbles and boulders grading upward into coarse grained lithic sandstones. Conglomerates are mainly composed of basement rock fragments, feldspars, quartz and some carbonate fragments. These fan deposits show a remarkable lack of maturity, indicated by the abundant detrital feldspar and little or no clay, implying a lack of chemical weathering. This demonstrates that arid climatic conditions prevailed at the time of initial fragmentation of the rocks. Moreover, their very poor sorting and lack of weathering suggests flashflood conditions, followed by drought, rather than permanent streams.

The Quaternary alluvial fan gravels are one of the preferred aggregates used for construction, buildings, highways, airport runways, and other structures. In different localities, most of them can be used as quarry sites due to their surface occurrence and proximity to roads. Deposits close to granitic rocks are very rich in feldspars, which can be used in ceramics. Alluvial fans are frequently excellent sources of ground water along the coast. The abundance of these coastal sediments can be useful in the development of the Red Sea area.

Beach

Successive fining upward deposits, 0.5–1.5 m thick, of a transgressive system dominate the beach. Its deposits change from conglomerate landwards to dominantly coarse to medium grained sandstones seawards. They show large scale unidirectional cross bedding dipping up to 5° seawards. The deposits are composed of quartz, feldspars, red-algal and foraminiferal packstones to grainstones and vary from intrasparite to biosparite rocks. However, in areas where beach deposits are in sharp contact with Pliocene sediments the dip angle of their lower contact is up to 15°. Upwards, they diminish progressively in dip which reaches 5° at their upper contact.

The beach facies of the intermediate reef (e.g. Wadi Hamrawein and Wadi Guisses) unconformably rests on alluvial channel deposits which form a siliciclastic foundation. This facies is 30–50 cm of conglomerates of cobbles and boulders of basement and carbonate rock fragments encrusted with calcareous red algae, and coarse to medium grained moderately sorted sandstones of carbonate rock fragments (red algae, foraminifera, echinoderm spines) with quartz and feldspars cemented by sparite and microsparite. Some voids are still free of cement.

The beach facies of the lower reef unit is mostly covered with coarse gravels. Its siliciclastic content is less than that of the intermediate and upper reef units. Generally, this content decreases from the older reef (averaging 57 %) to the modern beach (averaging 31 %), probably suggesting a decrease of uplift of the hinterland areas. According to Selley (1970) and based on the data available, the transgressional facies of the oldest terrace were accompanied by a moderate sediment supply, while the intermediate and youngest terraces attributed to transgression were accompanied by a low sediment supply.

Reefal Carbonates

Reef growth and its lateral seaward extension possibly reflect actual increments of subsidence or transgression of the sea. Lateral accretion continued initially without a significant break. The lateral facies variation leads to the recognition of three main parts. These are tidal flat or back reef, reef crest and reef front.

The back reef is subdivided into two main types

1. *Bioturbated-molluscan sandy limestone* is narrow (less than few meters) and is transitional from beach sandstones into dominantly carbonate reef deposits. It is 30 m wide and consists of a 1.5 m thick massive bed of pack stones composed of quartz, feldspars and biogenic clastics. Mollusks increase in amounts and size seaward. Field investigations, mineralogical composition, and petrographic study of samples collected along a profile of the older reef unit over a distance of about 30 m away from dominantly siliciclastic beach deposits show both the rapid decrease of siliciclastics (averaging 57–16 %) and the abundance of carbonate sediments and coral

Fig. 5 Reef front (talus) is clearly observed in the older reef in some areas (e.g. W. Hamrawein, W. Guisses and Ras Abu Soma). It is poorly bedded with unsorted reef fragments broken off by breaker erosion

heads. This rapid decrease of siliciclastics and the abundance of carbonate from the beach to coral zone are repeated in all reefs.

2. *Algal-molluscan limestone* is 15 m wide and is dominated by calcareous red algae with the appearance of a few small coral heads. Corals are locally encrusted by crustose red algae. The middle terrace still includes many calcareous algae and scleractinians, whereas the older terrace is poor in preserved fossils, many of which have been dissolved leaving molds filled with sparite cement, in which, of course, fossils have been present also prior to dissolution. Detrital fragments of carbonates, algal debris, echinoid spines, and anulids increase in size and amount toward the coral zone.

The reef crest (*coral zone*) is clearly observed in all reefs in most areas. It is a massive, wave-resistant zone about 10–15 m wide and about 3 m thick. It is dominated at the base by algal and coral debris and echinoid spines embedded in coarse bioclastics, followed upward by dominantly branched and domal corals in their growth position as well as calcareous red algae, mollusks and echinoids. Landwards of coral colonies, some corals use dead coral heads as a substrate. Seawards this coral zone interfingers with intervening talus reefal facies. Some geopetal fabrics of these coral colonies are filled with arkosic sand and skeletal debris.

The lower unit represents the biggest outcropping reef terrace with a higher diversity of coral species. It extends seawards and forms a steep wall bordering the present shoreline. Occasionally, the reef crest was eroded or tectonically subsided to the position of back reef. In some places it is bracketed by deep furrows, pits and pools occupied later by raised beach gravels and sabkha.

The reef front (*talus*) is clearly observed in the older reef in some areas (e.g. Wadi Hamrawein, Wadi Guisses and Ras Abu Soma). It is poorly bedded with unsorted reef fragments broken off by breaker erosion (Fig. 5). On the basis of clast sizes, a proximal and distal talus is present. The proximal part is dominated by angular coarse reef-derived detritus, with big reef blocks (50 cm) embedded in coarse bioclasts and algal debris, grading seawards into a distal part of finer coral-algal fragments (up to 3–5 cm), embedded in fine bioclasts and lime mudstone and siltstones. Rock types of biosparite range from wackstone at the base to grainstone, packstone and rudstone in the upper part. Terrigenous detritus may be mixed with slope reef facies.

Alluvial Channel

A progressive rapid fall of sea level terminated the deposition of the older reef terrace. This drop of sea level caused the incision of local channels into the pre-existing reefs by ephemeral fluvial streams. Detrital sediments filled local oval or semicircular channels about 10 m wide and 2 m high that unconformably overlie the distal part of the older reef (e.g. Wadi Hamrawein and Wadi Guisses). Channelized conglomerates of basement and carbonate detritus mark the lower contact of channels followed upward by fossil reef carbonate and coral fragments, embedded in pebbly siliciclastic conglomerates and algal bioclastics. This system can be interpreted as regression with a low sediment supply. High bioturbation is common in the upper part of this facies. Laterally on the margins of the channels, branched corals in growth position encrusted with calcareous red algae are common. However, floods of siliciclastics were enough to prevent coral growth along the axes of the channels.

Other Associations that Cover Coastal Pleistocene Reefs

During the last lowering of sea level erosional processes took place associated with tectonism. This resulted in the formation of a wide plain of gentle coastal morphology and occasional local depressions. Subsequently fluvial conditions prevailed in the area. This is demonstrated by large alluvial fans that funneled their loads of coarse clastics over pre-existing reefs. In areas where the paleo-shoreline is relatively far from the mountains there is a clear gradation seawards from dominantly alluvial fan deposits into beach conglomerates and sandstone sequences. In areas where the paleo-shoreline is near the mountains these coarse clastics were reworked by littoral hydrodynamics, as recorded by the beach gravels. Their gray color and coarse siliciclastics are a typical feature of these associations, which unconformably overlie and cut through the preceding Pleistocene reefs. Later on, parts of these coastal deposits were eroded forming lowland areas. These acted as pools filled by marine water, probably during flood episodes, which evaporated later under arid to semiarid conditions. This favoured the deposition of salt ponds and sabkhas.

1. *Alluvium* is represented by sporadic piles covering most parts of the reefs, and occasionally occurs on the low-lying Pliocene hills. The sediments consist of poorly sorted and immature large breccia and conglomerates (a few cm) intercalated with lenticular bodies of red sands and paleosol. It is believed that the sediments of this landward facies accumulated as proximal alluvial fans.
2. A *raised beach* is observed in some areas (e.g. Wadi Siatin, Wadi Hamrawein, Wadi Abu Hamra El Bahari and Wadi Guisses), and is absent in others (e.g. Ras Abu Soma and Dishet El Dabaa). It is a dark narrow belt and covers the upper part of the lower reef terrace. Seawards, this facies grades into conglomerates and sandstone layers. These deposits show a pronounced large scale cross bedding (8°E) suggesting an origin along gentle slopes. Closer to the shoreline, a prominent beach feldspathic sandstone is common.
3. *Evaporite facies or sabkha* occur on the upper part of the Pleistocene raised beach close to the sea, occupying small irregular depressions nearly parallel to the coast. The evaporites consist of about 0.5 m thick gypsum and anhydrite interbedded with calcareous fine muddy sands. The composition of the raised beach in some areas of less cemented gravel encouraged the formation of local erosional depressions during the subsequent lowering of sea level. These depressions were filled with evaporites as they become more restricted under an arid climate after a period of inundation. The restriction is probably related to the lowering of sea level or formation of beach barriers.

Composition of Pleistocene Reefs

Raised coral reefs are composed mainly of scleractinians in growth position and their preservation is mostly good (Fig. 3). The reef frame is usually massive and cavernous, with voids filled by marine cements and by internal sediment that is commonly perched on or within these cements. Skeletal remains of coralline algae, echinoderms, mollusks, and serpulids, with minor amounts of terrigenous materials constitute most of the sediments between corals. Raised coral reefs and beach rocks exposed along the coast of the Red Sea are generally composed of poorly sorted and well cemented framework grains. They consist almost exclusively of fragmented skeletal material derived primarily from corals, coralline red algae, echinoderms, mollusks and benthic foraminifera. Non-skeletal carbonate components such as ooids are very rare. Animal borings, mainly filled with brown sand to silt sized material also occur and are found to be continuous from the ground surface down to a few meters below the ground surface.

Patches of terrigenous materials with a lensoid shape within the raised coral reef horizon are also observed at two different levels at many places along the coast (Fig. 6). These patches are probably deposited in eroded areas created by wave action, and they are therefore younger than the surrounding coral reefs. These terrigenous bodies are few meters wide and few tens of centimeters thick and record the erosion history and the uplift of coral reefs at three different stages.

A comparison of the gross mineralogy of beach sands (Mansour 1991) and intertidal and marine bottom sediments (Piller and Mansour 1990) reveals that essentially the same constituents occur in both the beach rocks and raised coral reefs, but in significantly different proportions. The similarity of composition is explained by the derivation of most of these sediments from the adjacent coral reefs.

The mineralogy of the recent reef carbonates and Pleistocene reef sediments showed that the modern reef carbonates are composed of high Mg-calcite and aragonite with minor amounts of low Mg-calcite and traces of dolomite. On the other hand in the Pleistocene reef sediments low Mg-calcite, which is considered to have been formed by the meteoric digenesis of high Mg-calcite and aragonite, is the dominant mineral with varying proportions of aragonite (e.g. Dullo 1986, 1990). The carbonate content is more than 90 % and the non-carbonate material is dominantly quartz, minor

Fig. 6 Patches of terrigenous materials (*arrow*) within the coral reefs (Lower Terrace)

feldspar and heavy minerals in the sand fraction. Behairy (1980) recorded minor clay minerals and goethite in the clay fraction.

Diagenesis of Pleistocene Reefs

Detailed petrographic studies of thin sections of these reefs indicated that they are porous and slightly lithified skeletal limestone. Thin section investigations show similarities in mineralogical and petrographic characteristics as well as the distinct faunal assemblage of these sediments. Petrographic examination reveals the importance of cementation as a diagenetic process in the reef which contributes to the stability of reef structures. As mentioned before, in Pleistocene coral reef terraces along the Ras Mohamed-Sharm El-Sheikh coast (El-Asmar and Attia 1996) the petrographic examination reveals successive episodes of diagenesis from the younger to the older reef. Coral reefs of the lower terrace are cemented by carbonate of different textures. These are: (1) Calcareous mud matrix, (2) Cryptocrystalline high Mg-calcite (HMC) cement, and (3) Fibrous aragonite cement. The cement occurs mainly as micrite, and sometimes as microspar. Gypsum is present as cement, especially in the upper part of this reef. In the intermediate reef incipient transformation of Mg-calcite and aragonite into sparry calcite is recorded, whereas in the older reef sparry calcite is typically developed and all voids are filled with sparite cement.

Diagenetic cementation of sparry calcite is typically developed. The beach is characterized by well cemented calcareous sandstone with little porosity. Similar diagenetic effects were also observed in the upper parts of terrace three on the Saudi Arabian coast (Dullo 1986).

Calcareous mud matrix is the common cement in the lower reef, usually occurring as a groundmass in which the grains float and as cavity filling or micritic high Mg-calcite rims lining the interskeletal pore and the biogenic grains. Similar envelopes were observed in terrace "IV" of El-Asmar and Attia (1996), and in living corals from the Red Sea related to stage "II" of Gvirtzman and Friedman (1977). A cryptocrystalline submarine cement is precipitated within millimeters to centimeters of the surfaces of such high-energy deposits as reefs and beach rocks (Friedman 1985). Both intergranular and intragranular cryptocrystalline cement is common. Intergranular pores are seldom completely filled.

Diagenesis of beach facies records incipient transformation of Mg-calcite and aragonite into sparry calcite. Micrite rims are still present. The original high Mg-calcite of the micritic envelope becomes low Mg-calcite without a change in fabric. Any aragonite void filling that may have formed in the reef of the lower terrace is ultimately removed in this reef. Similar diagenetic observations were observed in terrace "III" of El-Asmar and Attia (1996) and the middle terrace (stage III) at south Sinai (Gvirtzman and Friedman 1977).

Another type of cement is the isopachous fringe of aragonite needle-like fibers nucleating the surface of the micritic envelope and/or the interskeletal pore forming palisade. In areas where cement fills a gap between parallel sides of two grains, aragonite laths grow on both sides parallel to each other and are closely interfingered. This fabric is presumably responsible for the hard and dense nature of the rocks (Khalaf 1988). In intergranular pores between more than three grains, aragonite cement is present as an isopachous fringe formed of interconnected laths growing from the surface of each grain towards the pore space. Similar diagenetic observations were observed in terrace "IV" (stage 2) of El-Asmar and Attia (1996) and the lower terrace (stage II) at south Sinai (Gvirtzman and Friedman 1977).

In the older reefs (upper ones) all voids are filled with sparite cement. Aragonite coral skeletons are transformed into a neomorphic sparry calcite mosaic. The neomorphic calcite crystals normally extend out of the original skeleton to fill partially the interskeletal pores, forming the cross-cutting calcite mosaic. This fabric is present either as a calcite cement crust of bladed dogteeth crystals characteristic of meteoric phreatic diagenesis, or as a micritic envelope of equant crystals of calcite growing on and lining the surface of the interskeletal pores. Similar diagenetic observations were observed in terrace "II" (stage 3) of El-Asmar and Attia (1996) and the upper terrace (stage IV) at south Sinai (Gvirtzman and Friedman 1977; Gvirtzman and Buchbinder 1978; Dullo 1984, 1986, 1990; M'Rabet et al. 1989; Strasser et al. 1992; Fathy 1994).

A thin hard layer of intraclastic beach rock observed only in the area north of Ras Gharib occurs at the surface of the intertidal zone. It consists of pelletoid sands, mainly faecal pellets, with quartz grains and smaller amounts of mollusks and foraminiferal shell fragments.

The occurrence of cemented fragments of pottery and glass artifacts in the beach rocks at sea level of the Red Sea suggests that lithification is currently active and rapid. Rapid lithification of beach rocks can be explained by the active growth of aragonite and high Mg-calcite crystals in the intertidal zone on the shore of the Red Sea (Mansour 1993).

Cryptocrystalline high Mg-calcite is the dominant cement, especially in the middle and upper horizons of coral reefs. The fibrous aragonite cement in the beach rocks and in the lower horizon of the raised coral reefs is common and concentrates in most areas of the rock. Aragonite cement in the raised coral reefs generally decreases upward and occurs only as small irregular aragonite patches in the upper horizon not concentrated in any particular part of the horizon. These details suggest that the dominance of high Mg-calcite in the upper horizon is due to the alteration of aragonite cement. The selective loss of aragonite over magnesium calcites has also occurred in Red Sea sediments (Friedman 1965).

Petrographic investigations do not show evidence for the alteration of aragonite to low Mg-calcite. Therefore, the relatively stable carbonates are believed to be of detrital origin, derived ultimately from ancient carbonates and calcareous rocks exposed along the Red Sea Coast. These detrital minerals are transported to the beach either by torrents or by winds.

Cements also play an important role in the mineral composition of beach rocks and raised coral reefs. The abundance of aragonite in the beach rocks and in the lower part of the raised coral reefs, although calcareous red algae of high Mg-calcite constitutes most of these rocks, is mainly due to the aragonite cement. There is a tendency for fibrous aragonite to occur more commonly in the beach rocks. According to Folk (1974), Milliman (1974), Bathurst (1975), Scholle (1978), Tucker and Wright (1990) and Montaggioni and Braithwaite (2009) the formation of such cement is favoured from pore water of marine composition. However, high Mg-calcite cement is more frequent in the coral reefs, especially above the high-tide level. The amount of cement generally increases in the coral reefs more than in the beach rocks.

Dating of Red Sea Raised Reefs

Much attention has been devoted to the precise dating of coral terraces in order to evaluate relative sea-level changes in regions considered tectonically "stable" or affected by a known and constant uplift rate. As mentioned before in Plaziat et al. (2008) the limited and episodic increase of rainfall and the relative tectonic stability of the shoreline suggest that the Egyptian reefs constitute an extremely favorable objective for a detailed study of the global climate and the instability of sea level during the late Quaternary highest stands (i.e. above Present sea level) for they were recorded by reef units referred to the Marine Isotopic Stages, MIS 7, MIS 5.5 and MIS 1 (= Mid Holocene Optimum) according to the $\delta^{18}O$ terminology (Martinson et al. 1987). Moreover, Plaziat et al. (2008) propose a nine stage diagrammatic reconstruction that illustrates the sedimentary results of around 200 ka of evolution of the Egyptian reefal shoreline (details in Plaziat et al. 1998a, b; Orszag-Sperber et al. 2001).

However, a general review of the dating of reefs on the coasts of the Red Sea, including those of Egypt, Jordan, Sudan, Eritrea, Saudi Arabia and Djibouti is given by Plaziat et al. (2008). A few Egyptian, Sudanese and Djibouti coral reefs were dated before 1980 (Butzer and Hansen 1968; Veeh and Giegengack 1970; Faure et al. 1980) while other dates have appeared in recent decades, many of them assigning an extremely wide range of ages to the lower reefs referred to as Late Pleistocene, from 150 to 50 ka. Because this lower reef unit is interpreted as being made up of "three onlapping reef cycles", Dullo (1990) suggested (his Fig. 21) that all the MIS (5e, 5c and 5a) high sea-level stands were appreciably above Present sea level. The lowermost Pleistocene "terrace", at +1.5 m, gave three ages (87.6, 86.6 and 57.6 ka) referred as a whole to the 5a substage (Plaziat et al. 2008). However, the last one is compared to a supposed stage 3 coral of the New Guinean record (Huon Peninsula, in Chappell and Shackleton 1986). Detailed studies of Egyptian reefs (Gvirtzman et al.1992; Gvirtzman 1994; El Moursi 1992; El Moursi et al. 1994) interpreted the younger dates as indicative of the probability that MIS 5c and MIS 5a reefs are a part of above sea-level outcrops, despite the absence of evidence of an adequate upheaval of the associated MIS 5e reef (Plaziat et al. 2008; cf. Reyss et al. 1993, Figure H2.35 in Plaziat et al. 1998a, b). On the other hand, an assumed tectonic activity during Holocene times (rift shoulder surrection or evaporate diapirism) in the pre-existing hemi-graben series induced Ibrahim et al. (1986) to mistake a late Pleistocene (5e) reef for a raised Holocene reef and to refer to gypsum residual tables of the same 5e substage (culminating more than 3 m above the Present littoral sabkhas) as Holocene salinas or sabkhas (see Orszag-Sperber et al. 2001). According to Manaa (2011) the reefs were probably formed during the major highstand of isotope 5e where the age of the upper reef is more likely to be 122.8 ka (MIS 5e), whereas the lower reef could be MIS 7 with no evidence of major tectonics in Rabigh area during the last 124 ka.

Reefs which formed during the interglacial period of isotope stage 7 (250,000–190,000 years B.P., Hays et al. 1976) have been identified by Gvirtzman and Buchbinder (1978) in Sinai, by Al-Rifaiy and Cherif (1988) in Jordan, and by Dullo (1990) on the Saudi Arabian side of the Gulf of Aqaba. According to Strasser et al. (1992), $^{230}Th/^{234}U$ dating in South Sinai places the older reef cycle between 350,000 and 270,000 years B.P., which corresponds to the interglacial period of isotope stage 9. The younger reef cycle has been dated between 140,000 and 60,000 years B. P. (isotope stage 5). At Zabargad Island, there are at least three systems of raised coral reefs (Hoang and Taviani 1991). The oldest terrace (>290,00 and 300,000 years B.P.) is found at +10 to +15 m. A 200,000 years B.P. high-sea stand is recorded by a terrace relict at +17 m on peridotite bedrock; the youngest system (125,000–138,000 years B.P.) is very well represented around the island, with terraces at about +6 to +8 m. Corals from Northern Brother yield ages of 132,000–135,000 and 204,000 years B.P. suggesting the existence of two systems of interglacial raised reefs. Both islands appear to have been tectonically quite stable since at least 125,000 years B.P. At Gebel el Zeit (southern Gulf of Suez rift basin), where terraces at elevations of +10 to 18 and +42 m have been radiometrically dated as 125 and 426 ka, respectively (Bosworth and Taviani 1996). However, as also mentioned before in Plaziat et al. (2008) most of the dates assigned to the raised reefs of the Red Sea may be considered much too imprecise, owing to the use of the α counting method and a loose selection of acceptable samples. The overall distribution of reef-growth ages may have been subject to the processes of rejuvenation that were invoked to interpret the broad deviations in the dates of Late Pleistocene 5e reefs on the Egyptian coast. A variable rejuvenation, related to local differences in the conditions that effected diagenesis, seems to be the key to the interpretation of the somewhat dispersed ages assigned to raised reefs.

Moreover, and as Dullo (pers. comm.) has added, most coral reef records have concerned sea-level high stands corresponding to the Last Interglacial period approximately 125 ka, and/or to the penultimate Interglacial (isotopic stage 7) from coral reef terraces exposed on the Huon Peninsula, Papua New Guinea (Chappell 1974, 2002; Bloom et al. 1974; Chappell and Veeh 1978; Stein et al. 1993; Chappell et al. 1996; Esat and Yokoyama 2006), Barbados (Mesolella et al. 1969; Edwards et al. 1987; Bard et al. 1990; Gallup et al. 1994; Potter et al. 2004), Sumba (Pirazzoli et al. 1993; Bard et al. 1996) and Mexico (Blanchon et al. 2009).

Impact of Civic Development on Coral Reefs

Coastal and marine resources are highly exploited in a non–sustainable way. The threats include habitat destruction, over-exploitation of living terrestrial and marine resources, environmental degradation from petroleum exploration and exploitation, significant risks from marine transportation, pollution from industrial activities, diverse environmental impacts from urban and tourism development and a series of emerging environmental issues associated with new types of economic development and uses of new technologies. Land reclamation and coastal road construction also affect shore zones and near shore waters. The impact of these activities can already be observed in areas subject to intensive use such as Dishet El-Dabaa (Fig. 7). Apart from areas of the sea that are lost, sediment loading of the water has increased and may affect coastal habitats in a similar manner to dredging (Aleem 1992; GEF 1997; UNEP 1997; Mansour 1999; Mansour et al. 2000a, b, 2005, 2011, 2013; PERSGA/GEF 2000, 2003; Madkour and Dar 2007; Madkour and Ali 2008; Madkour 2009, 2011; Madkour et al. 2012). Part of a highway to the north of Jeddah has been built on a reef flat, and any further building in this manner could cause serious losses to coastal habitats. Extensive coastline modifications have been carried out on an increasing scale in Hurghada and other areas.

Raised Coral Reefs and Construction Problems

The Red Sea coast is currently growing quickly due to flourishing diverse economic resources. Major harbours, cities and tourist villages in different countries along the Red Sea are built on these coral reefs (e.g. Hurghada, Safaga, Quseir, Port Sudan, Ashoaibah, Jeddah, and Rabigh harbours). The geological conditions and the type of foundations required are a key consideration in the structural and financial

Fig. 7 Coastal resources (e.g. raised coral reefs of Dishet El-Dabaa shown here) are highly exploited in a non–sustainable way

viability of a tall building. Within the coastal plain area the surface layer geology is primarily made up of coral reefs, sandy beach sediments, and other marine or alluvial deposits. Whilst these conditions do not prevent the construction of large hotels and tall buildings, they are not ideal. The hazards associated with the various ground conditions identified will influence the type and depth of foundations required (El-Shafie 2010). In the Marsa Ghaleb area along the Red Sea coast of Egypt some one floor buildings reveal many cracks (Fig. 8) and now whole buildings have been evacuated.

In recent years, however, the size of structures in some areas (e.g. Hurghada, Jeddah) has been increasing. Thus, evaluation of the engineering characteristics of the rock type has become an important issue for design and construction in this area. Construction problems associated with coral reefs include ground settlement and low bearing capacity, which are mainly due to low shear resistance and high porosity that are related to cavities in coral reefs and animal borings. Dewatering is a serious problem in these rocks due to their high porosity. These problems greatly affect the safe and economic land utilization of the coasts.

Ground improvement techniques such as compaction, deep compaction, using of stone columns, caissons, injection, and concrete piles are very important for construction in these areas. The selection of suitable methods depends on the economy and nature of the projects. However, adequate safety factors must be allowed for in the design of any construction founded on this rock type.

Predictions of Future Changes Along the Red Sea Coast

The population and economic growth in the coastal zone of the region has increased dramatically. A major contributor to growth in the coastal zone, with consequent impacts, is the rapidly expanding tourism industry. Prediction of the future changes along the Red Sea coast with its varying nature has implications for social and economic development along the coast. The associated land-use planning and mineral supply problems may be difficult to resolve. However, effective integrated coastal zone management programs are critical to sustaining the natural resources of the Red Sea.

The Red Sea is a very rich area from the resources point of view; it contains both living and non-living resources distributed over the terrestrial and marine sections. The data available on the resources of the area is scattered and mostly outdated. The scientific research on the natural resources and human impact is mostly at the national level and very few reports have been issued at the regional level. All types of habitat present in tropical coastal and marine areas are represented in the countries adjoining the Red Sea, with different degrees of distribution. All the threats to the natural resources arise from activities that are somehow related to the economic status of the countries.

Integration of planning, management and research in the coastal zone, on both land and sea is necessary to prevent pervasive degradation of the terrestrial and marine environments and to achieve ecologically sustainable use of coastal resources and conservation of these environments. All human activities if they are not done in a sustainable way can cause damage to the natural resources of the Red Sea area. The major factor that may reduce the damage from human activities is the increase in public awareness. The lack of laws and legal mechanism for implementation in most countries causes the environmental problems to build up and become disastrous. The lack of coordination between the countries may contribute in an indirect way to some of the problems. This demonstrates the importance of regional organizations such as Regional Organization for the Conservation of the Environment of the Red Sea and Gulf of Aden (PERSGA).

Fig. 8 One floor buildings on the lower terrace of Marsa Ghaleb area reveal many cracks

Because of the political and administrative differences between the countries in the area, the transfer of technology and experience between countries in the region is a way to ensure capacity building in the field of environmental protection. Cooperation between governments and owners of the land along the coast will solve many problems. The owners should be compensated for the damaged resources and the government should restrict property use to protect public health and for the safety and welfare of the people. Coastal and marine parks or even a zoning plan for controlling activities along the Red Sea are important and necessary in order to regulate the potentially damaging effects of development activities.

References

Aleem AA(1992) The occurrence of the sea-grass: Halophila stipulacea on the west coast of Egypt. Bull Fac Sci Univ Alexandria 4:79–84

Al-Rifaiy IA, Cherif OH (1988) The fossil coral reefs of Al-Aqaba, Jordan. Facies 18:2019–2230

Al-Sayari SS, Dullo WC, Hötzl H, Jado AR, Zötl JG (1984a) The Quaternary along the coast of the Gulf of Aqaba. In: Jado AR,Zötl JG (eds) Quaternary period of Saudi Arabia, vol 2, pp 32–47

Al-Sayari SS, Hötzl H, Moser H, Rauert W, Zötl JG (1984b) Quaternary from Dhuba to Al Wajh. In: Jado AR, Zötl JG (eds) Quaternary period of Saudi Arabia, vol 2, pp 66–82

Bard E, Hamelin B, Fairbanks RG (1990) U-Th ages obtained by mass spectrometry in corals from Barbados: sea-level during the past 130,000 years. Nature 346(6283):456–458

Bard E, Hamelin B, Arnold M, Montaggioni L, Cabioch G, Faure G, Rougerie F (1996) Deglacial sea-level record from Tahiti corals and the timing of global meltwater discharge. Nature 382:241–244

Bathurst RCG (1975) Development in sedimentology 12: carbonate Sediments and their diagenesis, 2nd edn. Elsevier Scientific Publishing Company, Amsterdam

Blanchon P, Eisenhauer A, Fietzke J (2009) Rapid sea-level rise and reef back-stepping at the close of the last interglacial highstand. Nature 458 (7240):881–884. doi:10.1038/nature07933

Behairy AKA (1980) Clay and carbonate mineralogy of the reef sediments, west coast of Saudi Arabia, vol 4. Bulletin of Faculty of Science, King Abdulaziz University, Jeddah, pp 265–279

Bloom AL, Broecker WS, Chappell JMA, Mattews RK, Mesolella KJ (1974) Quaternary sea level fluctuations on a tectonic coast: new $^{230}Th/^{234}U$ dates on the Huon Peninsula. N G Quatern Res 4(2):185–205

Bosworth W, Taviani M (1996) Late Quaternary reorientation of stress field and extension direction in the southern Gulf of Suez, Egypt: Evidence from uplifted coral terraces, mesoscopic fault arrays, and borehole breakouts. Tectonics 15(4):791–802

Butzer KW, Hansen CL (1968) The coastal plain of Mersa Alam. In: Butzer KW, Hansen CL (eds) Desert and river in Nubia:

geomorphology and prehistoric environments at the Aswan reservoir. University of Wisconsin Press, Madison, pp 395–432

Chappell J (1974) Geology of coral terraces, Huon Peninsula, New Guinea: a study of Quaternary tectonic movements and sea-level changes. Bull Geol Soc Am 85(4):553–570

Chappell J (2002) Sea level changes forced ice breakouts in the last glacial cycle: new results from coral terraces. Quatern Sci Rev 21 (10):1229–1240

Chappell J, Veeh HH (1978) $^{230}Th/^{234}U$ support of an interstadial sea level of −40 m at 30,000 yr BP. Nature 276:602–604

Chappell J, Shackleton NJ (1986) Oxygen isotopes and sea level. Nature 324(6093):137–140

Chappell J, Omura A, Esat T, McCulloch M, Pandolfi J, Ota Y, Pillans B (1996) Reconciliation of late quaternary sea levels derived from coral terraces at Huon Peninsula with deep sea oxygen isotope records. Earth Planet Sci Lett 141:227–236

Dullo WC (1984) Progressive diagenetic sequence of aragonite structures: Pleistocene coral reefs and their modern counterparts on the eastern Red Sea coast, Saudi Arabia. Paleontographica Am 54:254–260

Dullo WC (1986) Variation in diagenetic sequences: an example from Pleistocene coral reefs, Red Sea, Saudi Arabia. In: Schroeder JH, Purser BH (eds) Reef Diagenesis. Springer, Heidelberg, pp 77–90

Dullo WC (1990) Facies, fossil record, and age of Pleistocene reefs from the Red Sea (Saudi Arabia). Facies 22(1):1–45

Dullo WC, Montaggioni L (1998) Modern Red Sea coral reefs: a review of their morphologies and zonation. In: Purser BH, Bosence DWJ (eds) Sedimentation and tectonics in Rift Basins: Red Sea-Gulf of Aden. Springer, The Netherlands, pp 583–594

Edwards RL, Chen JH, Wasserburg GJ (1987) ^{238}U–^{234}U-^{230}Th-^{232}Th systematics and the precise measurement of time over the past 500,000 years. Earth Planet Sci Lett 81:175–192

El-Asmar HM, Attia GM (1996) Diagenetic trends in Quaternary coral reef terraces, Ras Mohammed-Sharm El Sheikh coast, Southern Sinai, Egypt. Sedimentology Egypt 4:19–31

El Moursi M (1992) Evolution quaternaire de la plainecotiere de la Mer Rouge entre Hurghada et Marsa Alam, Egypte. These nouveau doctorat, Universited'Aix-Marseille ii, no. 92AiX22087, 245 pp

El Moursi M, Hoang CT, El Fayoumy IE, Hegab O, Faure H (1994) Pleistocene evolution of the Red Sea coastal plain, Egypt: Evidence from uranium-series dating of emerged reef terraces. Quatern Sci Rev 4:345–359

El-Shafie M (2010) Sustainability versus mega urban development projects. Int J Civil Environ Eng Int J Eng Sci 10(4):1–7

Esat TM, Yokoyama Y (2006) Growth patterns of the last ice age coral terraces at Huon Peninsula. Global Planet Change 54:216–224

Fathy E (1994) Physiography and quaternary sedimentation of the coastal zone in the South Sinai. Egypt Acta Geologica Hungarica 37 (3–4):311–325

Faure H, Hoang CT, Lalou C (1980) Datations$^{230}Th/^{234}U$ des calcaires coralliens et mouvements verticauxà Djibouti. Bulletin de la Societe Geologique de France, Paris, series 7 22(6):959–962

Folk RL (1974) Petrology of sedimentary rocks. University of Texas, Hemphill Publication Company, Texas, 182 pp

Friedman GM (1965) Recent carbonate sediments of the Gulf of Aqaba, Red Sea, pp 67–68. Geological Society of America, Abstracts for 1964, Special paper 82, 400 pp

Friedman GM (1985) The problem of submarine cement in classifying reef rock: an experience in frustration. In: Schneidermann N, Harris PM (eds) Carbonate cements. Society for Sedimentary Geology Special Publication No. 36, Tulsa, pp 117–121

Gallup C, Edwards RL, Johnson RG (1994) The timing of high sea levels over the past 200,000 years. Science 263(5148):796–800

Global Environment Facility (1997) Report 2: baseline studies. Egyptian global environment facility Red Sea coastal and marine resource management project, Cairo, Egypt, 109 pp

Gvirtzman G (1994) Fluctuations of sea level during the past 400,000 years: the record of Sinai, Egypt (northern Red Sea). Coral Reefs 4:203–214

Gvirtzman G, Buchbinder B (1978) Recent and Pleistocene coral reefs and coastal sediments of the Gulf of Eilat. In: Guidebook 10th international sedimentological congress, Jerusalem, pp 162–191

Gvirtzman G, Friedman GM (1977) Sequence of progressive diagenesis in coral reefs: studies in geology. Bull Am Assoc Pet Geol 4:357–380

Gvirtzman G, Kronfeld J, Buchbinder B (1992) Dated coral reefs of southern Sinai (Red Sea) and their implication to late Quaternary sea levels. Mar Geol 1:29–37

Guilcher A (1988) Coral reef geomorphology. John Wiley, New York, 228 pp

Hays JD, Imnrie J, Shackleton NJ (1976) Variations in the Earth's orbit: pacemaker of the ice ages. Science 194:1121–1132

Hoang CT, Taviani M (1991) Stratigraphic and tectonic implications of Uranium-series-dated coral reefs from uplifted Red Sea islands. Quat Res 35:264–273

Ibrahim A, Rouchy JM, Maurin A, Guelorget O, Perthuisot JP (1986) Mouvements halocinétiques récents dans le golfe de Suez: L'exemple de la péninsule de Guemsah. Bulletin de la Société Géologique de France Paris ser 8 2(1):177–183

Issawi B, Francis M, El Hinnawi M, El Deftar T (1971) Geology of Safaga-Quseir coastal plain and of Mohamed Rahaba area. Annu Geol Surv Egypt 1:1–19

Khalaf FI (1988) Quaternary calcareous hard rocks and the associated sediments in the intertidal and offshore zones of Kuwait. Mar Geol 80:1–27

Madkour HA (2009) Impacts and mitigation of anthropogenic factors on terrestrial Red Sea environment. Review article, National Institute of Oceanography and Fisheries, Red Sea Branch, 72 pp

Madkour HA (2011) Impacts of human activities and natural inputs on heavy metal contents of many coral reef environments along the Egyptian Red Sea coast. Arab J Geosci 6:1739–1752. doi:10.1007/s12517-011-0482-5

Madkour HA, Ali MY (2008) Heavy metals in the benthic foraminifera from the coastal lagoons, Red Sea, Egypt: indicators of anthropogenic impact on environment (case study). Environ Geol 58:543–553. doi:10.1007/s00254-008-1529-0

Madkour HA, Dar MA (2007) The anthropogenic effluents of the human activities on the Red Sea coast at Hurghada Harbour (case study). Egyptian J Aquat Res 33(1):43–58

Madkour HA, El-Taher A, Ahmed NA, Mohamed A, El-Erin MT (2012) Contamination of coastal sediment in El-Hamrawein Harbour, Red Sea, Egyptian. J Environ Sci Technol, International Standard Serial Number 1994–7887, Asian Network for Scientific Information. doi:10.3923/jest.2012

Manaa A (2011) Late Pleistocene raised coral reefs in the Eastern Red Sea—Rabigh, Saudi Arabia. Master of science research thesis, School of Earth and Environmental Science, University of Wollongong, 190 pp, http://ro.uow.edu.au/theses/3501

Mansour AM (1991) Interplay of beach and coastal dune environments around Safaga Bay, Red Sea, Egypt. Bull Fac Sci Assiut Univ 20(2-f):127–147

Mansour AM (1993) Petrography of the raised coral reef and beach rocks of the Egyptian Red Sea coast. In: Bulletin of the Faculty of Science, Mansoura University, Egypt, Special Issue Symposium of the Quaternary and development in Egypt, pp 71–98

Mansour AM (1999) Changes of sediment nature by environmental impacts of Sharm Abu Makhadeg area, Red Sea, Egypt. Sedimentology Egypt 7:25–36

Mansour AM (2000a) Sedimentology of the Red Sea coast of Egypt. In: Soliman SM (ed) Book of the century, part 1, pp 15–44

Mansour AM (2000b) Quaternary reef terraces and their relation to tectonic/eustatics of the Red Sea region, Egypt. Sedimentology Egypt 8:19–33

Mansour AM, Nawar AH, Mohamed AW (2000a) Geochemistry of coastal marine sediments and their contaminant metals, Red Sea, Egypt: a legacy for the future and a tracer to modern sediment dynamics. Sedimentology Egypt 8:231–242

Mansour AM, Nawar AH, Mohamed AW (2000b) Shallow marine sediments, Red Sea Egypt: restricted distribution of coral debris. Sedimentology Egypt 8:63–74

Mansour AM, Nawar AH, Madkour HA (2005) Metals concentration of recent invertebrates along the Red Sea coast of Egypt: a tool for monitoring environmental hazards. Sedimentology Egypt 13:171–185

Mansour AM, Nawar AH, Madkour HA (2011) Metal pollution in marine sediments of selected harbours and industrial areas along the Red Sea coast of Egypt. Annalen des Naturhistorischen Museums in Wien Serie A 113:225–244

Mansour AM, Askalany MS, Madkour HA, Assran BS (2013) Assessment and comparison of heavy-metal concentrations in marine sediments in view of tourism activities in Hurghada area, northern Red Sea, Egypt. Egyptian J Aquat Res. doi: 10.1016/j.ejar.2013.07.004

Martinson DC, Pisias NG, Hays JD, Imbrie J, Moore TC, Shackleton NJ (1987) Age dating and the orbital theory of the ice ages: development of a high-resolution 0 to 300,000-year chronostratigraphy. Quatern Res 27(1):1–29

Mesolella KJ, Matthews RK, Broecker WS, Thurber DL (1969) The astronomical theory of climatic change: Barbados data. J Geol 77:250–274

Milliman JD (1974) Marine carbonates. Springer, Berlin, 375 pp

Montaggioni LF, Braithwaite CJR (2009) Quaternary coral reefs system: history, development processes and controlling factors. Elsevier, Oxford

M'Rabet A, Purser BH, Soliman M (1989) Diagenese compare de reef coralliensactuels et quaternaires de la cote egyptienne de la mer Rouge. Geol Mediterr 16:5–39

Orszag-Sperber F, Plaziat JC, Baltzer F, Purser BH (2001) Gypsum salina-coral reef relationships during the last interglacial (MIS 5e) on the Egyptian Red Sea coast: a Quaternary analogue for Neogene marginal evaporites. Sed Geol 140(1-2):65–85

PERSGA/GEF (2000) The status of coral reefs in Saudi Arabia. PERSGA, Jeddah

PERSGA/GEF (2003) Coral reefs in the Red Sea and Gulf of Aden. Surveys 1990 to 2000 summary and recommendations. PERSGA technical series no. 7, 171 pp

Piller WE, Mansour AM (1990) The northern Bay of Safaga (Red Sea, Egypt) An actuoupalaeontological approach. II sediment analysis and sedimentary facies. Beiträge zur Paläontologie Österreich Wien 16:1–102

Pirazzoli PA, Radtke U, Hantoro WS, Jouannic C, Hoang CT, Causse C, Borel Best M (1993) A one million-year-long sequence of marine terraces on Sumba Island, Indonesia. Mar Geol 109:221–236

Plaziat J-C, Baltzer F, Choukri A, Conchon O, Freytet P, Orszag-Sperber F, Raguideau A, Reyss J-L (1998a) Quaternary marine and continental sedimentation in the northern Red Sea and Gulf of Suez (Egyptian coast): influences of rift tectonics, climatic changes and sea-level fluctuations. In: Purser BH, Bosence D (eds) Sedimentary and tectonic evolution of rift basins: the Red Sea-Gulf of Aden. Chapman and Hall, London, pp 537–573

Plaziat J-C, Reyss J-L, Choukri A, Orszag-Sperber F, Baltzer F, Purser BH (1998b) Mise en evidence, sur la cote recifaled'Egypte, d'une regression inter-rompant le plus haut niveau du dernier interglaciaire (5e): un nouvelindice de variations glacio-eustatiques a haute frequence au Pleistocene ? Bulletin de la Societegeologique de France, Paris 169(1):115–125

Plaziat JC, Reyss JL, Choukri A, Cazala C (2008) Diagenetic rejuvenation of raised coral reefs and precision of dating. The contribution of the Red Sea reefs to the question of reliability of the Uranium-series datings of middle to late Pleistocene key reef-terraces of the world. Carnets de Geologie/Notebooks on Geology, Article 2008/04 (CG2008_A04)

Potter EK, Esat TM, Schellmann G, Radtke U, Lambeck K, McCulloch MT (2004) Suborbital-period sea-level oscillations during marine isotope substages 5a and 5c. Earth Planet Sci Lett 225:191–204

Purser BH, Soliman M, M'Rabet A (1987) Carbonate, evaporite, siliciclastic transition in Quaternary rift sediments of the northwest Red Sea. Sed Geol 53:247–267

Reyss JL, Choukri A, Plaziat JC, Purser BH (1993) Datationsradiochimiques des recifscoralliens de la rive occidentale du Nord de la Mer Rouge, premieres implications stratigraphiques et tectoniques. Comptes Rendus de l'Academie des Sciences, Paris, series 2 Sciences de la Terre 317(4):487–492

Scholle PA (1978) A color illustrated guide to carbonate rock constituents, textures, cements, and porosities. Am Assoc Pet Geol Mem 27:201

Selley RC (1970) Ancient sedimentary environments. Cornell University Press, Ithaca, New York, 237 pp

Sellwood BW, Netherwood RE (1984) Transformation of aragonite to calcite in marine gastropods. Can J Earth Sci 15:1861–1866

Stein M, Wasserburg GJ, Aharon P, Chen JH, Zhu ZR, Bloom AL, Chappell J (1993) Thermal ionisation mass spectrometry U-series dating and stable isotopes of the last interglacial event in Papua New Guinea. Geochim Cosmochim Acta 57(11):2541–2554

Strasser A, Strohmenger C, Davaud E, Bach A (1992) Sequential evolution and diagenesis of Pleistocene coral reefs, South Sinia, Egypt. Sed Geol 78:59–79

Taviani M (1998) Post-miocene reef faunas of the Red Sea: glacio-eustatic controls. In: Purser BH, Bosence DWJ (eds) Sedimentation and Tectonics in Rift Basins: Red Sea—Gulf of Aden. Springer, Netherlands, pp 574–582

Tucker ME, Wright VP (1990) Carbonate sedimentology. Wiley, Oxford

UNEP (1997) Assessment of land-based sources and activities affecting the marine environment in the Red Sea and Gulf of Aden. UNEP Regional Seas Reports and Studies, No. 166

Veeh HH, Giegengack R (1970) Uranium-series ages of corals from the Red Sea. Nature 226(5241):155–156

Geomorphology of Shallow Water Coral Reef Environments in the Red Sea

Gwilym Rowlands and Sam Purkis

Abstract

In this chapter, we examine the different processes that control the orientation and arrangement of shallow water coral reef environments in the Red Sea. Particular focus is paid to the diversity, distribution, and abundance of coral reefs in Saudi Arabia, where shallow water (<30 m) environments have been mapped into one of sixteen types of 'coral reef system'. Each of these types of coral reef system represents the terminal node in a decision tree that differentiates reef environments based on distinctive planar morphology, that is, viewed from above and audited from high-resolution satellite imagery. The wide variation in morphology of the Red Sea reefs is primarily governed by the width of the coastal shelf. Coral reef systems can be described as either 'shore-attached' or 'detached' on the basis of the depth of water separating them from a landmass. The morphology of attached systems ranges from the simple arrangement of fringing reefs and sediments to more complex forms that extend into the Red Sea basin and are incised by channels, large lagoons, and repetitive reef lineaments. There is also considerable diversity in morphology in detached coral reef systems, which are as abundant overall as those that are shore-attached. Several morphological types of coral reef system are restricted to narrow regions of latitude. Such distributional trends may be explained in process terms by the rift tectonics of the Red Sea basin, spatial variability in the presence of sub-seafloor evaporites, the input of siliciclastic (non-carbonate) detritus onto the coastal shelf via wadis, and eustatic sea-level control. These processes act in concert, but at different spatial and temporal scales, to deliver diverse coral reef morphologies throughout the Red Sea basin.

Introduction

Coral reefs are valued for their inherent beauty, high biological diversity, and productivity. Though both deep and cold-water coral communities exist, the term 'coral reef' is most synonymous with massive calcium carbonate structures found in photic tropical waters. Many human communities subsist on, or are tied economically to, coral reefs, which provide many goods and services. These include sources of nutrition, tourism appeal, fisheries, and shoreline protection. Ancient fossil reefs, meanwhile, when buried beneath impermeable stratigraphies, may develop into a valuable hydrocarbon reservoirs (Moberg and Folke 1999). An understanding of the diversity and distribution of coral reef morphology is therefore of interest to a large number of disciplines.

The study of coral reef morphology has a substantial scientific pedigree. Darwin's (1842) theory of atoll evolution, an outcome of his famous voyage on the HMS Beagle, was perhaps the first to capture the biological and geological essence of reef development. The theory recognized the importance of light, variations in relative sea-level, and longevity for reef development. Vertical growth by corals and other reef organisms can lead to the accretion of carbonate, which may be rapid enough to keep pace with the rising sea-level through an interglacial cycle, or falling sea-level as a result of subsidence.

G. Rowlands (✉) · S. Purkis
Oceanographic Center, Nova Southeastern University, Dania Beach, Fort Lauderdale, FL 33004, USA
e-mail: gwilym@rowlandsecology.com

N.M.A. Rasul and I.C.F. Stewart (eds.), *The Red Sea*,
Springer Earth System Sciences, DOI 10.1007/978-3-662-45201-1_24,

Three morphological terms, 'fringing reef', 'barrier reef', and 'atoll' were introduced by Darwin (1842) and remain much in use to this day. For example, in the Red Sea, the term barrier reef can be found describing the 'Little Barrier Reef' (Sheppard 1985) of Saudi Arabia and the term atoll in the name of *Sanganeb Atoll* (origin unknown) in Sudan. The Red Sea has been famed for having 'over 4,500 km of fringing reef' amounting to 'the longest continuous fringing reef' in the world (Longhurst and Pauly 1987, p. 20). Yet this statement by Longhust and Pauly (1987) may be considered an oversimplification, a product of limited geographic knowledge of morphological development and the basin-wide extrapolation of what was known from the most studied coral reefs in the gulfs of Aqaba and Suez (Sheppard et al. 1992). Describing Red Sea coral reef morphology under such generalities belies the complexity of reefs present both within its inshore and offshore environments (Guilcher 1988; Sheppard et al. 1992; Dullo and Montaggioni 1998; Rowlands et al. 2014). Guilcher (1988) and Sheppard et al. (1992) note how, in the context of the Red Sea rift, the use of terms such as 'atoll' and 'barrier' can also be confusing. Though many systems in the Red Sea bear an 'atoll-like' morphology, the mode of origin may be more aligned to tectonics of a rift basin and the mobilization of sub-seafloor evaporite (Dullo and Montaggioni 1998; Rowlands et al. 2014), as opposed to the progressive subsidence and upward reef growth around a volcano, as originally described by Darwin (1842). The limits of Darwin's three-pronged nomenclature are now realized, as noted by Toomey et al. (2013) who consider that 'few modern environments have the right combination of conditions to produce the Darwinian atoll progression' (p. 1). Work in many reef regions describes subtle morphological varieties and recognizes additional process roles for antecedent topography, as well as wind and wave and tidal climates (Fairbridge 1967; Maxwell 1970; Purdy 1974; Hopley 1982; Hopley et al. 1989; Purkis et al. 2002; Bosence 2005; Andréfouët et al. 2006; Purkis et al. 2010, 2012; Leon and Woodroffe 2013; Purkis et al. 2014; Rowlands et al. 2014).

In the context of a discussion of coral reef morphology, the term 'coral reef' is itself open to misinterpretation. In addition to coral, other calcareous organisms, both animal and plant, contribute to the volume of a reef (Hart and Kench 2007). Hereafter, we take the term '*coral reef system*' to describe '*the shallow water (*<30 m*) accumulation of both reef frameworks and associated carbonate sediments*'; an equivalent definition for what Rowlands et al. (2014) refer to as a 'carbonate system'.

Contemporary coral reef systems adopt morphologies that reflect past erosion and deposition. Coral reef systems are shaped by far more than island subsidence; coral growth and glacial sea-level cycles are just as important. Morphology is a temporally dynamic property of a coral reef, a fact recognized in Hopley's (1982) evolutionary progression of coral reef systems in the Australian Great Barrier Reef (GBR). Reef development proceeds through different stages from juvenile (unmodified antecedent platforms, submerged reefs, and irregular patch reefs) to mature (crescentic reefs and lagoonal reefs) and finally to senile (planar reefs) forms.

Despite the relatively small size of the Red Sea, a wide variety of morphological forms have been described (Sheppard et al. 1992; Dullo and Montaggioni 1998; Bosence 2005; Purkis et al. 2010, 2012; Rowlands et al. 2014). In the first part of this chapter, we focus on a case study of Saudi Arabia to illustrate the morphological diversity, distribution, and abundance of Red Sea coral reef systems. In our review, we do not concentrate on meter-scale morphologies such as spur and groove which may develop in high wave energy areas (cf. Friedman 1985; Sheppard et al. 1992), nor do we consider the vertical and planar distribution of facies or biotopes that form the thin biological veneer to a reef (cf. Montaggioni et al. 1986; Sheppard et al. 1992; Dullo and Montaggioni 1998; Bruckner et al. 2011; Rowlands et al. 2012; Klaus, this volume). Instead, the morphological diversity of whole coral reef systems (contiguous areas of reef frameworks and sediments) is considered. We examine how a typology, a logical scheme for the classification of coral reef systems based on their planform morphology, can be used to map coral reef systems within a geographic information system (GIS). Patterns of spatial distribution and abundance are then analyzed throughout the Saudi Arabian Red Sea to describe where, and in what quantity, different types of coral reef systems occur.

The last part of the chapter describes the underlying processes of morphological development. The orientation and arrangement of coral reef systems can be broadly explained by the input of siliciclastics as controlled by the hyper-arid Arabian climate, combined with the action of rift-related tectonics (Sheppard et al. 1992; Dullo and Montaggioni 1998; Bosence 2005; Purkis et al. 2012; Rowlands et al. 2014). Examples relating to the Saudi Arabian case study as well as from the western Red Sea are discussed.

The Morphological Diversity, Distribution, and Abundance of Coral Reefs in the Saudi Arabian Red Sea

Background

In recent years, efforts to characterize and map coral reef systems have been greatly assisted by the widespread availability of remote sensing data, including satellite- or airborne imagery, and ship-borne acoustic surveys (Hopley et al. 1989; Andréfouët et al. 2006; Leon and Woodroffe 2013; Rowlands et al. 2014). To examine the morphological diversity of coral reef systems in the Red Sea, we focus on Saudi Arabia whose coastline comprises more than two-thirds of the eastern margin of the basin. On a per country basis, Saudi Arabia also boasts the majority of coral reefs in the Red Sea (Spalding et al. 2001), while examination of Google Earth imagery suggests the reef morphologies identified in Saudi Arabia are also representative of those seen elsewhere in the basin.

A Typology for Describing the Morphology of Coral Reef Systems

Following Rowlands et al. (2014), a typology is described that partitions coral reef systems into one of sixteen distinct morphological end-members (Figs. 1 and 2). Within this typology, each end-member was identified based on the planform arrangement of both reef frameworks and sediments, viewed, and audited from above in satellite imagery (Fig. 1). The imagery used included Google Earth, as well as over 95,000 km^2 of meter resolution satellite imagery. This imagery was acquired under clear and calm sea conditions that allowed the seabed morphology to be appraised down to depths of $\sim$30 m. Because the entirety of the coastal shelf has been imaged from satellite, all of the coral reef systems in the Saudi Arabian Red Sea could be mapped out according to the rules of the typology.

The coral reef typology takes the form of a decision tree (Fig. 2). Criteria in the typology were guided by both statistical analyses of system properties as well as logical divisions and are designed to be both objective and reproducible (Rowlands et al. 2014). These criteria include differences in the spectral content and texture of satellite image pixels, analysis of the size and shape of reef systems as measured in a GIS, and water depth assessed from nautical charts (Rowlands et al. 2014). On the basis of this typology, maps describing the distribution and abundance of the different coral reef system morphologies in the Saudi Arabian Red Sea were then developed.

At the top of the typology are two 'forms' of coral reef system, attached and detached, which split into four 'groups' (narrow, expansive, tower, and platform). The terminal

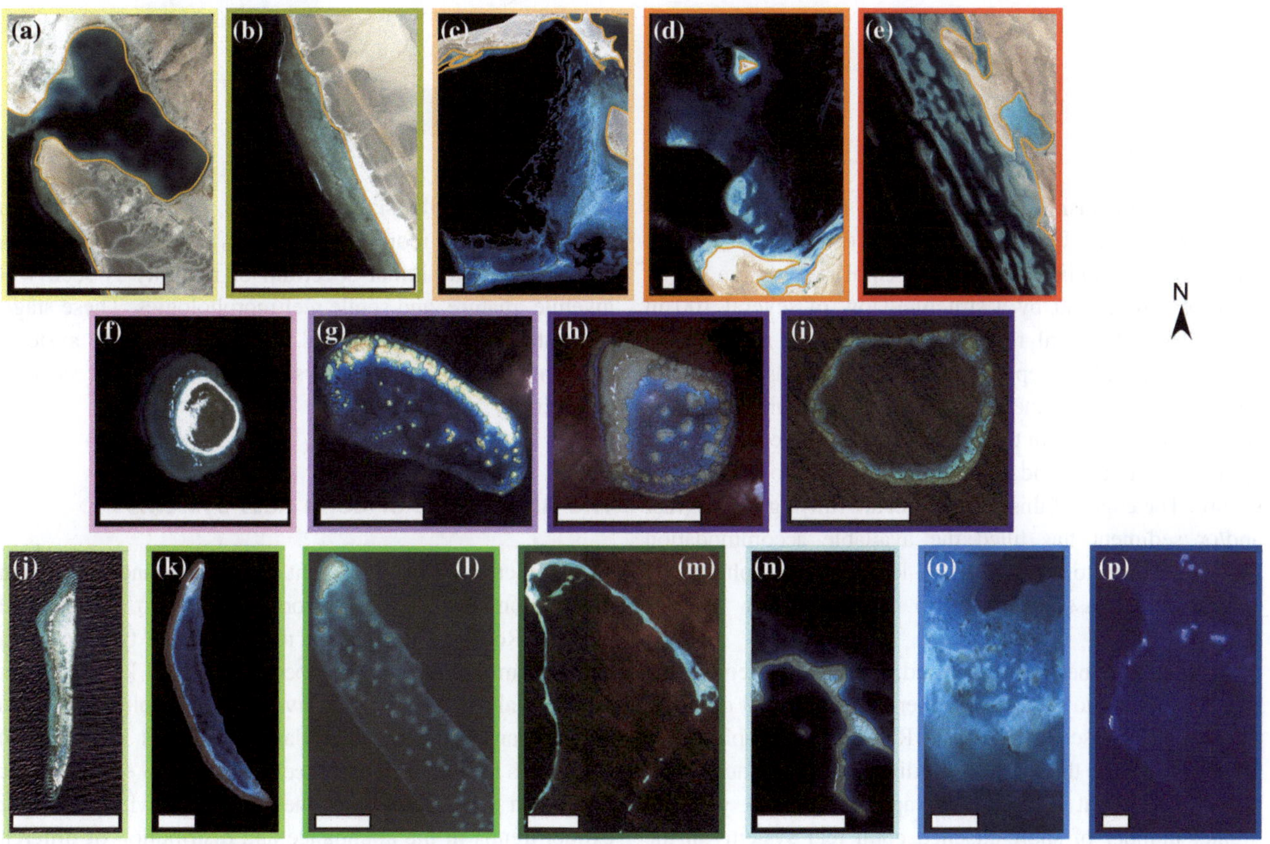

Fig. 1 Subsets of QuickBird satellite imagery illustrating the 16 morphological end-members of the coral reef typology. The *white horizontal scale bar* in each image represents 1 km. The first row (**a–e**) are attached coral reef systems, the second row (**f–i**) are the tower group of coral reef systems, and the third row (**j–p**) are the platform group of coral reef systems. See Fig. 2 for names of coral reef systems. Figure adapted from Rowlands et al. (2014)

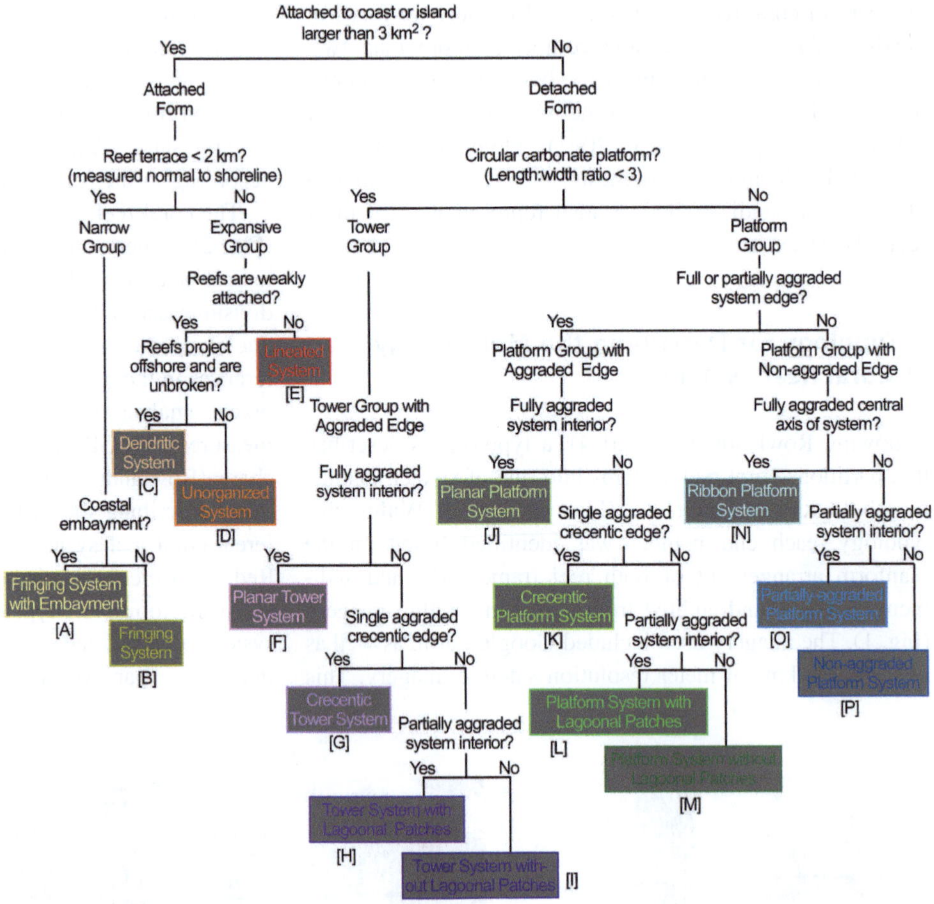

Fig. 2 Typology of Red Sea coral reefs. *Colored boxes* represent the 16 morphological end-members of the decision tree, while associated letters *[A–P]* correspond to image thumbnails in Fig. 1. Figure adapted from Rowlands et al. (2014)

nodes of the decision trees are the 16 morphological end-members (Fig. 2). Use of the typology can be illustrated by the following example. Consider a coral reef system that is separated from land by a structural low of several 100 m water depth. A coral reef system is defined as having a 'detached form' if separated from a landmass by water deeper than 50 m, the most appropriate contour that is consistently mapped in the nautical charts of the region. The system is circular and therefore, it falls into the 'tower group'. The edges of this reef system are fully aggraded (reef and/or sediment has filled the available accommodation space and built to, or above, sea-level), the morphological end-member classification is a 'planar tower system' (Fig. 1f).

With 16 end-members described, coral reef systems in the Red Sea appear to exhibit a greater morphological diversity than has been described in the GBR, Australia (Hopley et al. 1989). However, this increase in diversity can be accounted for by the slight difference in approach which yields a greater number of shore-attached coral reef systems in the Red Sea and recognition of the tower group of coral reef systems as distinct from the platform group. Where a similar typological scheme applied to the GBR by Hopley, a greater level of complexity would likely also be seen in that region. Nonetheless, for the size of the basin, coral reef systems in the Red Sea are morphologically diverse. Using Hopley's (1982) terms, a similar spectrum of maturity is seen with juvenile, mature, and senile reef morphologies. These stages of maturity are represented by non-aggraded systems, crescentic and lagoonal systems, and planar systems, respectively.

The Abundance of Coral Reef Systems

Mapping enables an assessment of the abundance and spatial distribution of each type of coral reef system. In the Saudi Arabian Red Sea, the sixteen morphological types of coral reef system (Fig. 2) were mapped onto a 1 × 1 km grid based on the analysis of a vast archive of high-resolution satellite imagery and nautical chart data (Rowlands et al. 2014). Under this technique, a coral reef measuring 6 × 4 km counts as 24 grid cells. Maps were then examined within a GIS to extract trends in the abundance and distribution of different types of coral reef system.

In Saudi Arabia, the area occupied by each morphological end-member is highly variable (Fig. 3). Coral reefs that are shore-attached are only slightly more abundant (10,554 km^2)

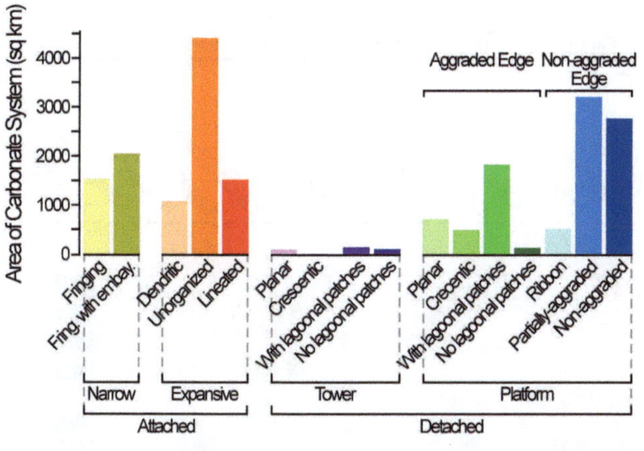

Fig. 3 Area occupied by each of the 16 morphological end-members in the Saudi Arabian Red Sea typology of coral reefs (*colored bars*). Attached forms of coral reef system occupy a similar area (10,554 km^2) to the detached forms (9,951 km^2). The tower group of coral reef systems occupies the least area (352 km^2), while the platform group occupies the greatest (9,599 km^2). The narrow group of coral reef systems (3,565 km^2) occupies around half the area of the expansive group (6,989 km^2). Figure adapted from Rowlands et al. (2014)

than those that are detached (9,951 km^2). Moreover, most of the attached reefs fall within the expansive group, where carbonate frameworks and sediments extend away from the coastline into the Red Sea basin, and are typified by complex morphologies incised by channels, large lagoonal areas, and repetitive reef lineaments (Fig. 1; Rowlands et al. 2014).

The overall abundance of shallow coral reef environments reflects the basin-scale geological processes of coastal shelf development (Fig. 4). On both the eastern and western sides of the Red Sea, the breadth of the continental shelf generally increases from north to south. The Saudi Arabian Red Sea coastal shelf, defined here as the 500-m bathymetric contour, can be partitioned therefore on the basis of its width into three distinct 'precincts', since there are marked increases in shelf width south of ~27.5 and ~20°N (Fig. 4a; Precinct A, B and C; Rowlands et al. 2014). Precinct A corresponds to coral reefs in the Gulf of Aqaba and has a mean shelf width of 4.4 km (SD: 1.0 km); precinct B extends from Tiran and Sanifir Islands to the coastline near the village of Al Lith and has a mean width of 27.9 km (SD: 12.8 km); while precinct C extends south of Al Lith through the Farasan Islands and has a mean width of 129.2 km (SD: 10.2 km).

The latitudinal abundance of Red Sea reefs is not an expression of the percentage of the shelf that they occupy (Fig. 4c), but rather echoes the pattern of shelf width (Fig. 4b). Where the continental shelf is narrow, coral reefs are themselves small and narrow. Where the shelf is wide, individual coral reef systems may be small or large, but a

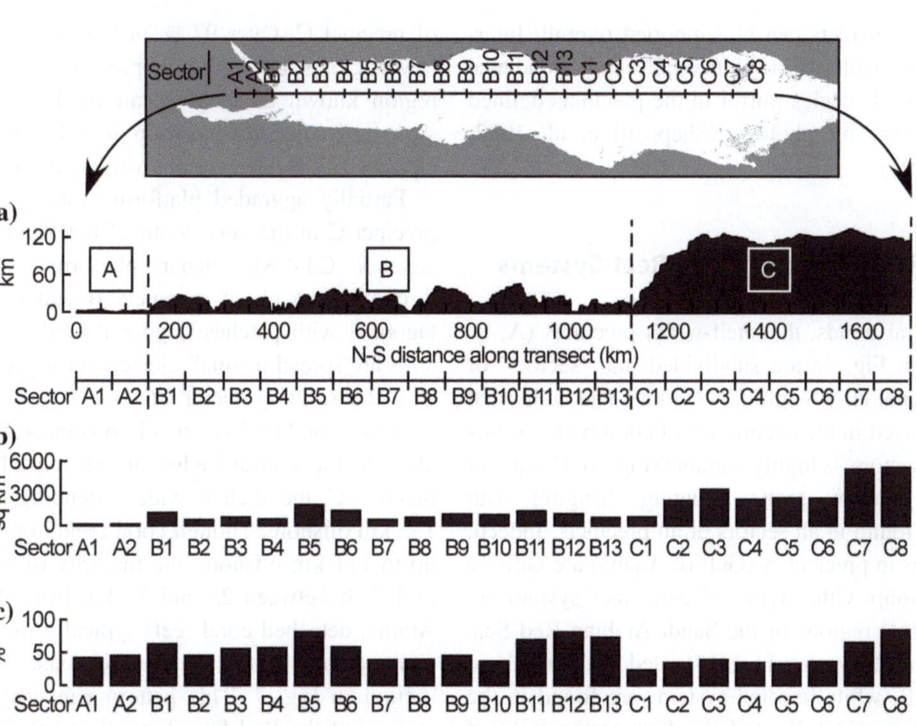

Fig. 4 **a** Minimum width of the Saudi Arabian Red Sea coastal shelf (500-m bathymetric contour) assessed normal to a ~1,700-km transect running north to south down the central axis of the Gulf of Aqaba and Red Sea (*upper image*). To assess the latitudinal and cross-shelf distribution of coral reef systems, the Saudi Arabian Red Sea is split into three precincts (*A*, *B* and *C*) on the basis of the width of the coastal shelf. Within these precincts are nested sectors of approximately equal N–S length (lower x-axis). **b** Area occupied by all types of coral reef system in each sector of the Saudi Arabian Red Sea. **c** Area occupied by coral reef systems in each sector expressed as a percentage of the total area of the coastal shelf in each sector. Figure adapted from Rowlands et al. (2014)

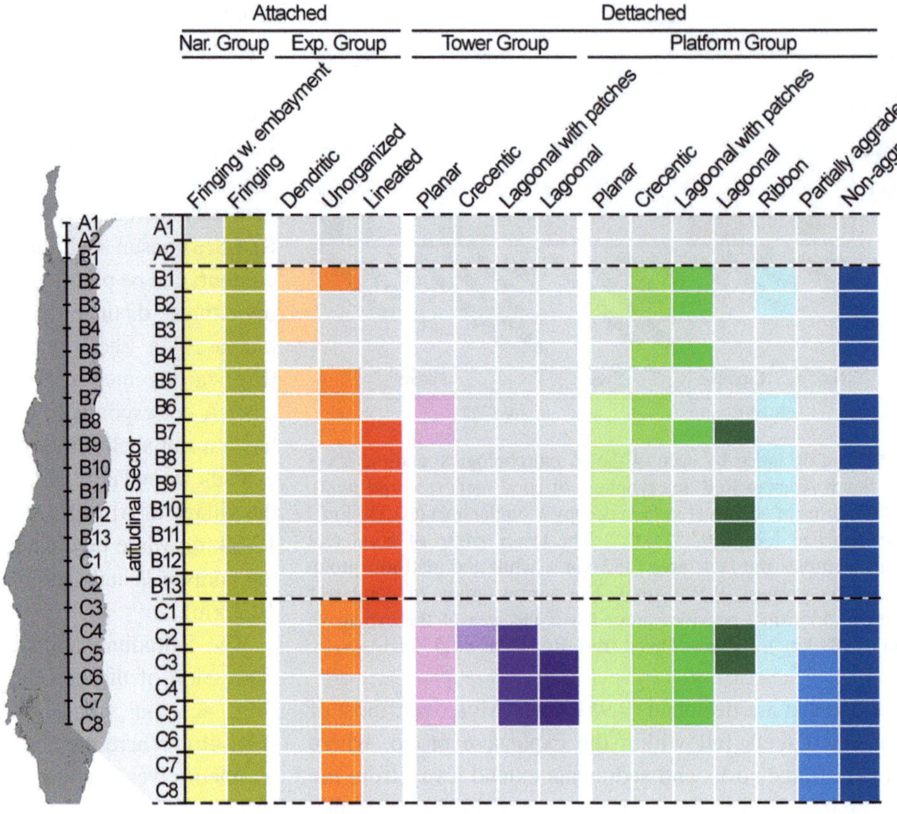

Fig. 5 The latitudinal distribution of coral reef systems in the Saudi Arabian Red Sea. The Red Sea basin is shown to the *left* of the plot. The presence (*colored boxes*) or absence (*grey boxes*) of coral reefs in the Saudi Arabian Red Sea is considered across three shelf-width precincts (*A*, *B*, and *C*) divided into sectors of approximate equal length (x-axis; Fig. 4). Figure adapted from Rowlands et al. (2014)

greater abundance of reefs can be supported overall. Interestingly, latitudinal patterns in the diversity and abundance of reef fish and coral species mirror in the precincts defined above by coral reef morphology (Sheppard et al. 1992; Roberts et al. 1992).

The Spatial Distribution of Coral Reef Systems

To examine spatial trends, the shelf-width precincts (A, B, and C) defined in Fig. 4a are subdivided into 'sectors' of approximately equal N–S length (75 ± 3 km).The latitudinal distribution, reflected in the occurrence of coral reef systems in each of these sectors, is highly variable (Fig. 5). Coral reef systems in the narrow group (fringing; fringing with embayment) are found in all sectors of all precincts. Indeed, coral reef systems in precinct A (Gulf of Aqaba) are entirely of the narrow group. Other types of coral reef system are restricted to distinct regions of the Saudi Arabian Red Sea. For instance, dendritic systems are restricted to the northern half of precinct B, while lineated systems are found in the southern half of precinct B, and the first sector (C1) of precinct C.

Unorganized systems are found in two clusters in the north of precinct B (sector B1, B5–B7) and throughout most of precinct C. Over 97 % of the coral reef systems in the tower group are found in precinct C (sectors C2–C5), a region known as the Farasan Banks. The few coral reef systems in the tower group found outside of precinct C (sectors B6, B7) are exclusively of the planar type.

Partially aggraded platform systems are only found in precinct C in the very south of the Saudi Arabian Red Sea (sectors C3–C8). Planar platform systems are found throughout much of precincts B and C, while crescentic, lagoonal with patches, lagoonal, and ribbon platform systems are spread in small clusters throughout precincts B and C, as far south as sector C6.

Many coral reef systems in precincts A and B are located close to the seaward edge of the coastal shelf (Fig. 6). In precinct C, the shelf is wide, extending to between 70 and 125 km offshore. Though coral reefs in precinct C are found up to 111 km offshore, the majority of systems are located mid-shelf between 25 and 75 km from the coast. In Saudi Arabia, detached coral reefs typically run in multiple lineaments parallel to the Red Sea rift axis, as is shown in cross section in Fig. 7. This pattern can also be seen in other regions of the Red Sea (Figs. 8 and 9; Purkis et al. 2012). The distance between reef systems measured across the Saudi Arabian coastal shelf ranges from 4 to 16 km, with a mean of 9.2 km (SD: 4.2 km) (Rowlands et al. 2014).

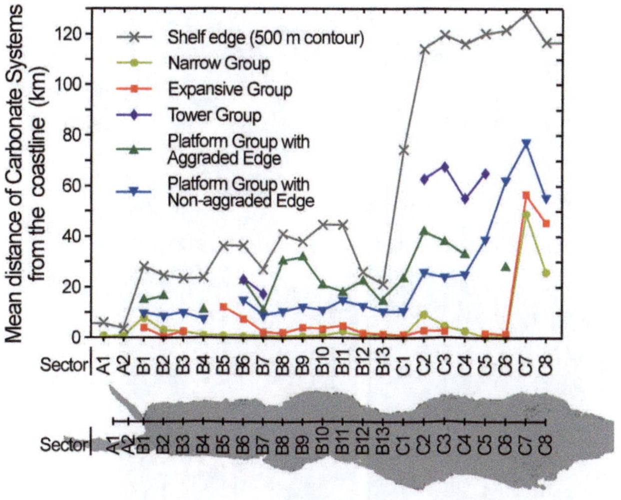

Fig. 6 Mean distances of coral reef systems from coastline, plotted in each latitudinal sector at the group level of the coral reef typology (Fig. 2); position of outer edge of the coastal shelf (500-m bathymetric contour) also plotted. Figure adapted from Rowlands et al. (2014)

The Morphological Development of Coral Reef Systems

The first-order control of the arrangement and orientation of shallow water coral reef environments in the Red Sea is explained by rift-related tectonics, while the input of siliciclastics as controlled by the hyper-arid Arabian climate exerts a second order of control (Bermert and Ormond 1981; Dullo and Montaggioni 1998; Bosence 2005; Purkis et al. 2012; Rowlands et al. 2014). Bosence (2005) identifies these controls as contributing to three seabed configurations which he viewed as important in the formation of shallow coral reef systems: (1) fault-block, (2) salt-diapir, and (3) delta-top.

Tectonics

Sedimentary processes on the margins of the Red Sea have been primarily controlled by tectonics since the opening of the rift (Purser and Hötzl 1988; Bosworth 1994; Bosence 2005; Bosworth, this volume). Both attached and detached reef systems (Figs. 1 and 2) orient along the grain of the main Red Sea rift axis (Purkis et al. 2012). By comparing the orientations of coral reef systems to the orientations of local fault trends, Purkis et al. (2012) describe the relationship of coral reef system orientation to fault control. Six regions are considered (Fig. 8); the Gubal Straits and Shalatan in Egypt, Trinkitat in Sudan, Dahlak and Halib in Eritrea, and the Farasan Banks in Saudi Arabia. In five of these six areas, a clear relationship is seen between the orientation of faults and the arrangement of coral reef systems (Figs. 9 and 10). The exception to this rule, Dahlak, is important since it emphasizes the morphological role of subsurface evaporite and salt-diapirism which is discussed in detail below.

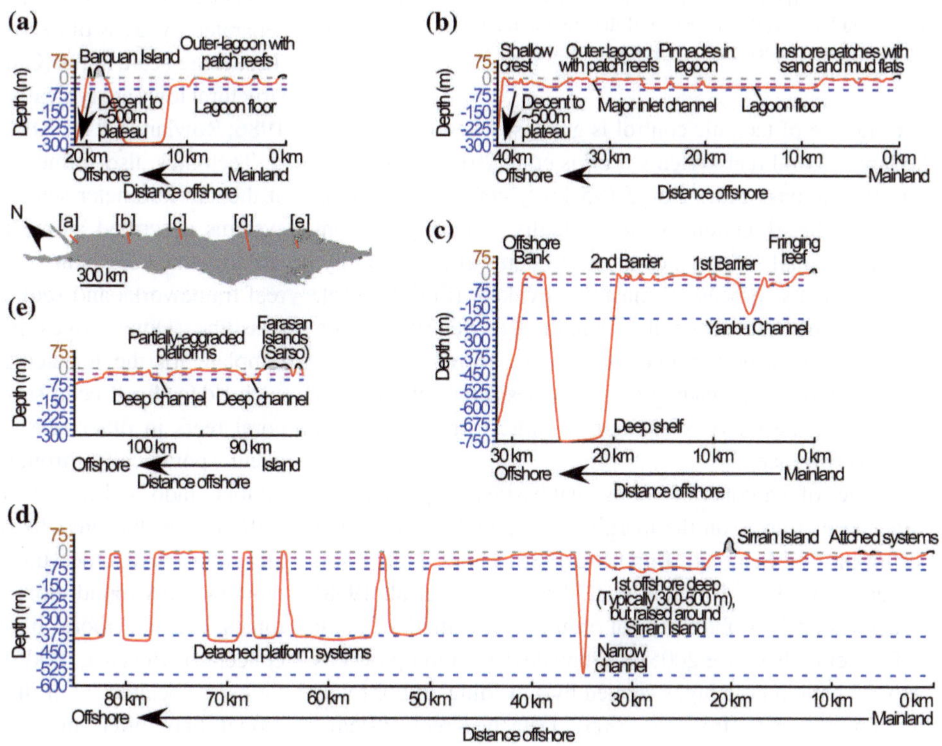

Fig. 7 Five cross-shelf profiles on the eastern Red Sea, illustrating the diversity of topography below sea-level. The coral reef systems identified in Fig. 1 occupy the upper ~25 m depths. Presence of terraces at ~25 and ~50 m below present day sea-level can be seen in many regions, and may correspond to Pleistocene still stands (Sheppard et al. 1992). Tower and platform reefs often rise rapidly from regions of the coastal shelf at hundreds of meters depth

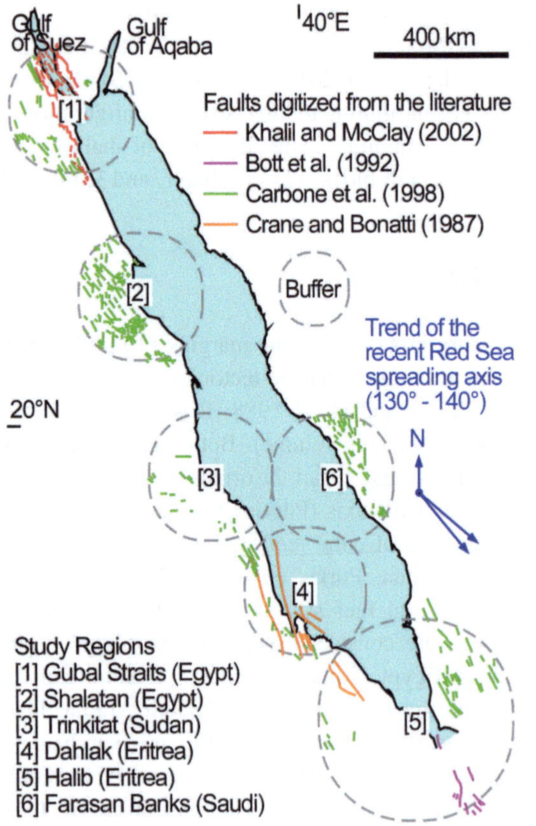

Fig. 8 Location of six focus areas distributed throughout the length of the Red Sea rift used to assess role of tectonics on the orientation of coral reef systems. The size and orientation of these coral reef systems are assessed across 1,600 km^2 in each area and compared to fault lineaments digitized from the literature (Crane and Bonatti 1987; Bott et al. 1992; Carbone et al. 1998; Khalil and McClay 2002) within a circular buffer of 150 km for five of the areas and a buffer of 250 km in Halib. Figure adapted from Purkis et al. (2012)

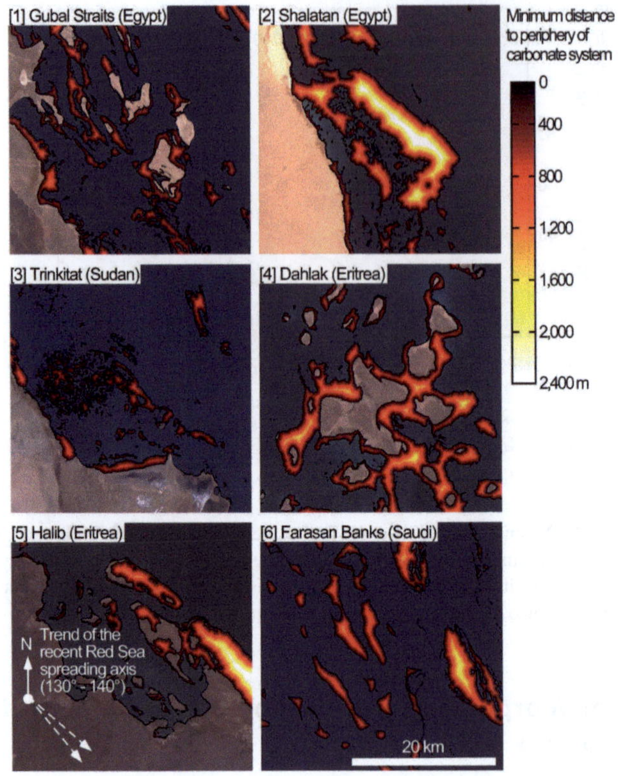

Fig. 9 Heat maps indicate the width of shallow coral reef systems for each of the six focus areas shown in Fig. 8; *brighter colors* indicate wider systems, and heat maps are overlaid on Landsat satellite image where *brown tones* correspond to land and *dark blue tones* correspond to deep (>30 m) water. Figure adapted from Purkis et al. (2012)

Evidence of tectonic control is greatest when considering the largest coral reef systems (Purkis et al. 2012; red dots on Fig. 10 compass plot). Large (>5 km^2) coral reef systems align to the direction of local faults, while the more numerous small (<5 km^2) coral reef systems align randomly, and secondary processes must be evoked (Purkis et al. 2012). If carbonate accumulation in the Red Sea rift is taken as an analog for now buried rift settings, knowledge of the orientation of the spreading axis provides information when considering the likely orientation of large carbonate bodies in the subsurface.

Studies of a number of extensional basins, but particularly Miocene platforms on the margins of the Gulf of Suez and northwest Red Sea (e.g., Burchette 1988; James et al. 1988; Purser et al. 1998; Bosence et al. 1998; Cross et al. 1998) reveal a model for fault control on the arrangement of coral reef systems (Bosence 2005). Carbonate deposition proceeds atop the bathymetric highs created by subsiding and/or rotating 'fault-blocks' (Bosence 2005). Detached reef systems, described in Figs. 1 and 2, nucleate atop such highs and are separated by areas of deep water (structural lows) defined by these large-scale faults (Guilcher 1988). In Saudi Arabia, such structural lows are typically 4–16 km wide (Montaggioni et al. 1986; Rowlands et al. 2014) and 100s of meters deep (Fig. 7).

Tectonics also influence coral reef system morphologies at the sub-kilometer scale. Consider the case of the lineated systems described above in Saudi Arabia (Fig. 1e), which are so named for being comprised of small repeating linear reef frameworks and sediment bodies aligned parallel to the coastline. Dune ridges have been offered as a potential template for the lineated systems (Rowlands et al. 2014); such ridges have been shown to be the nucleating point for coral reefs in other seas (Banks et al. 2007). Dune ridges occur sporadically throughout the length of western Saudi Arabia (Jado and Zötl 1984). However, as Rowlands et al. (2014) note, the crest-to-crest distances between reef units are almost five times those of the contemporary terrestrial dune systems found along the margin of the Red Sea. Tectonics are the more likely explanation. The fault-block concept of Bosence (2005) can be extended to smaller-scale faults associated with horst and graben structures. This model may offer an explanation for the morphology of

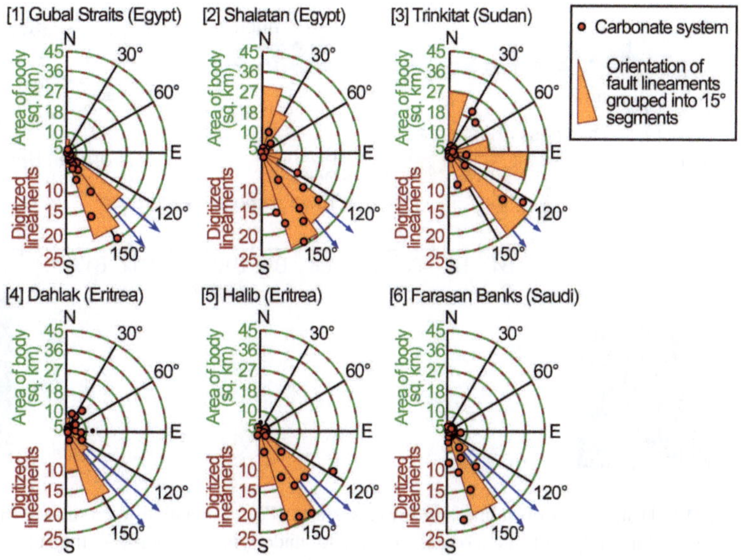

Fig. 10 Compass plots (polar coordinates) for the six focus areas indicating the orientation of fault lineaments (*orange segments*) and orientation of coral reef system (*red dots*). *Blue arrows* indicate the trend of the recent Red Sea spreading axis (Crane and Bonatti 1987). With the exception of Dahlak (Eritrea), coral reefs tend to orient according to fault lineaments. Figure adapted from Purkis et al. (2012)

lineated systems, as previously suggested by Sheppard et al. (1992) and Guilcher (1988).

The importance of tectonics in the formation of past and present coral reef systems can also be seen in the steep, almost vertical, slopes around many coral reef systems (Fig. 7) and in the exposed and elevated remains of former reef flats, coral heads, beach rock , reef terraces, and reef islands that exist along the coastline (Jado and Zotl 1984; Sheppard et al. 1992; Gvirtzman 1994; Dullo and Montaggioni 1998). As for submerged reef systems, many offshore islands orient according to the tectonic grain of the rift basin (Purkis et al. 2012). For example, the Saudi Arabian Farasan Islands run 'noticeably parallel' to the rift axis (Dabbagh et al. 1984, p. 213). The Farasan Islands are typically less than 20 m high and formed of tectonic uplift and shattering of a Plio–Pleistocene reef flat (Dabbagh et al. 1984); these islands, and their attendant shallow reefs and sediments, were molded secondarily by the creep of sub-seafloor evaporites. In the north of the Red Sea, Pleistocene terraces have been faulted, culminating in surfaces as high as 520 m above present sea-level at Tiran and Sanafir Islands (Brown 1970). Near Hurghada, Dullo and Montaggioni (1998) describe stepped horizontal Holocene terraces fringing the continental coastline and gently sloping surfaces around islands. Such morphological structures are interpreted as strong evidence for sharp, rapid, coastal uplift related to rifting, and tilting controlled by evaporite tectonics or block rotation (Bosworth 1994; Dullo and Montaggioni 1998).

Salt-Diapirs

The morphology of many coral reefs in the Red Sea is influenced by the presence of sub-seafloor evaporites (Bosence et al. 1998; Dullo and Montaggioni 1998; Bosence 2005; Purkis et al. 2012; Rowlands et al. 2014). Red Sea evaporites (Zeit and South Gharib formations) were laid down during the Mid- and Upper-Miocene (16.4–5.3 Ma) (Mitchell et al. 1992; Hovland, this volume). In this period, sea-level was lower and the Red Sea basin was restricted at its southern limit, the Bab-al-Mandab. Evaporites are sandwiched beneath Pliocene siliciclastics, above which sit the Pleistocene and Holocene carbonate stratigraphies (Bosence et al. 1998; Brown 1970; Mitchell et al. 1992). Because evaporites are less dense than the rock layers above them, they tend to rise, distorting, and/or piercing the overlying strata to form topographic highs. These highs, termed 'salt-diapirs', exhibit domed morphologies and in a marine setting may be topped by islands and coral reefs (e.g., Bosence 2005; Purkis and Riegl 2012).

Evaporite deposits that are both laterally expansive and vertically thick (exceeding several kilometers) have been measured on both sides of the Red Sea basin (Mitchell et al. 1992). Much of what is known of the distribution of evaporites in the Red Sea has been gleaned from seismic traverses and borings that extend into, and in many cases through, the Miocene evaporite layer (Fig. 11; Brown 1970; Mitchell et al. 1992; Bosence et al. 1998). The vertical thickness of evaporite is documented in borehole data (Mitchell et al. 1992). To inform on latitudinal variation in the distribution of sub-seafloor evaporite, Rowlands et al. (2014) spatially interpolate borehole data across the coastal shelf of the eastern side of the Red Sea (Fig. 11). Large salt deposits are described in the north of Saudi Arabia near Tiran and Sanafir Islands, as well as under the Farasan Banks, Farasan Islands, and offshore of Yemen.

The morphological control exerted by salt-diapirism can be sufficient to override control by rift tectonics (Purkis et al. 2012). Topographic highs associated with salt-diapirs

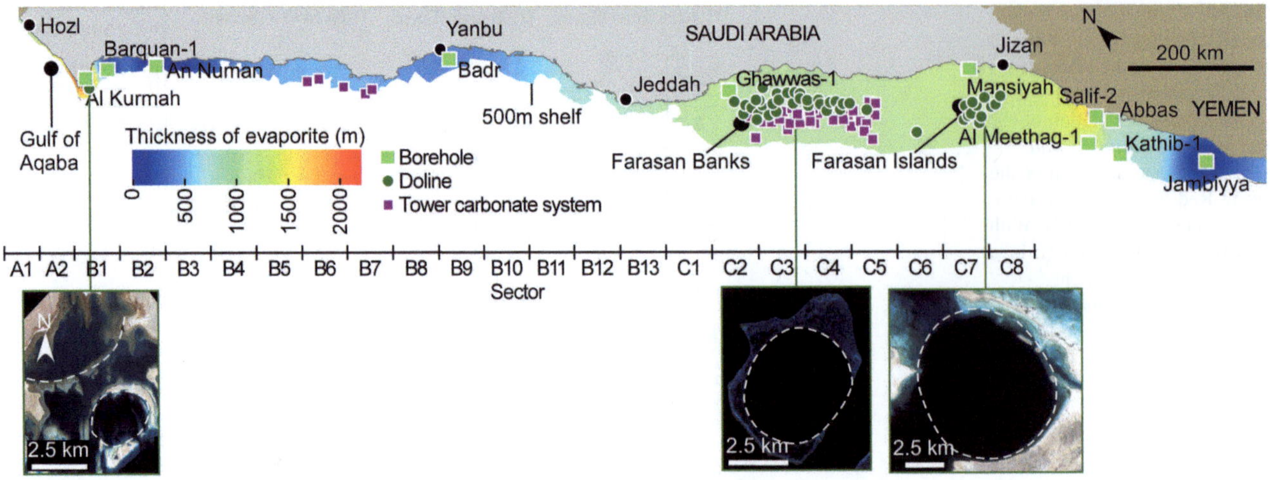

Fig. 11 Distribution of doline and tower systems (Fig. 1) relative to the shoreline, edge of the coastal shelf, and underlying evaporite formations. Dolines are circular depressions, as illustrated in QuickBird satellite image subsets at base of figure, while tower reefs are circular elevations as illustrated in Fig. 1. The base layer is the interpolated surface (inverse distance method) derived from the vertical thickness of the sub-seafloor evaporite layer based on borehole data (Mitchell et al. 1992; Bosence et al. 1998) in the eastern Red Sea. Horizontal axis shows the three shelf-width precincts (*A*, *B* and *C*) subdivided into sectors (see Fig. 4). Figure adapted from Rowlands et al. (2014)

develop into coral reef systems if they extend into the shallow photic waters conducive to carbonate production (Guilcher 1988; Sheppard et al. 1992; Bosence 2005). These salt-diapir platforms can develop into complex arrangements of shallow water separated by deep structural lows (Bosence et al. 1998). In cases where salt-diapirs have pierced the seabed and are then subaerially exposed, as is common during sea-level low-stands, meteoric dissolution of the evaporite rock may lead to the formation of a void through collapse. The resultant depressions, termed 'dolines', have characteristic circular or semicircular shapes (Bosence 2005; Fig. 11). Coral reef systems controlled by evaporite adopt circular, arcuate and amoeboid, and elongate shapes, punctuated with subcircular embayments, and deep reentrants and channels (Bosence 2005; Purkis et al. 2010; Rowlands et al. 2012). Salt-diapirism has not led, in the terms of Hopley (1982), to systems that are either more or less mature. In the Saudi Arabian Red Sea, salt-diapirism seems to have increased the morphological diversity of coral reefs in a given region, as can be seen in the Farasan Banks and may underlie the tower group morphology (precinct C, Figs. 5 and 11).

The cross-shelf patterns in evaporite thickness are better documented for the western margin of the Red Sea. Here, it is worthwhile considering the cross-shelf transect described by Mitchell et al. (1992) that runs from the coastline of Eritrea, beneath the Dahlak Islands (Fig. 8), and on toward the deep Red Sea axial trough. Along this transect, the evaporite layer steeply thickens away from the Eritrean coast, before passing into a 'salt zone' at the main axis of the evaporite basin (Mitchell et al. 1992). In this salt zone, vertical evaporite thickness ranges from 2,000 to 4,000 m. As the axial trough is approached, however, the evaporite stratigraphy 'thins and interfingers with volcanics in a transitional zone' (Mitchell et al. 1992). The Plio–Pleistocene Dahlak Islands are situated atop a vast salt dome (Bunter and Abdel Magid 1989; Mitchell et al. 1992; Carbone et al. 1998; Orzag-Sperber et al. 1998; Crossley and Cripps 1999), and adopt circular, semicircular and arcuate shapes (Purkis et al. 2012). The numerous large and closely spaced islands are the primary control over the orientation of shallow water carbonates (Purkis et al. 2012; Figs. 7 and 8).

Although there are no boreholes in the mid- and outer coastal shelf of the Farasan Islands and Farasan Banks in the eastern Red Sea (Fig. 5a; precinct C), the density of dolines in these areas suggests the presence of a significant salt basin —an 'eastern salt zone' (Fig. 11). Gravitational and magnetic fields in the Farasan Banks and Farasan Islands indicate the ascent of evaporite in the form of circular diapirs and elongate diapirs along tectonic lines parallel to the rift axis (Dabbagh et al. 1984; Guilcher 1988). The circular 'atoll-like' coral reef systems described above as falling into the tower group (Figs. 1 and 2) are located along the outer edge of the eastern salt zone (Fig. 6). These shallow water reefs appear to sit atop pedestals that rise hundreds of meters from the coastal shelf, buoyed upwards under the influence of evaporites (Fig. 7).

Tectonics and salt-diapirism elevate a significant portion of the coastal shelf in the southern Red Sea into the photic zone, as shown for Saudi Arabia above (Fig. 4). The Farasan Islands increase the length of coastline in sectors C7 and C8 by a factor of four. Narrow and expansive systems fringe these islands, almost doubling the area occupied by shore-attached coral reef systems in this part of the basin. Away from the epicenter of the archipelago, partially aggraded platform systems become more abundant (Fig. 5).

Siliciclastics

Where terrestrial siliciclastics (i.e., non-carbonate detritus) accumulate in the coastal zone, they may form topographic highs that extend into the photic zone. When sufficiently stable and undisturbed, these highs may host substantial carbonate accumulations. Bosence (2005) refers to this process as the 'delta-top' model of reef formation, which should not be confused with the '*deltaic*' morphology identified by Maxwell (1970) in the Pompey Complex of the GBR. The deltaic morphology is formed atop terrestrial debris organized in the marine environment by ebb and flood tidal deltas, while the delta-top model considers the formation of coral reefs atop alluvial fans. Rowlands et al. (2014) reflect on the dendritic systems (Fig. 12), named for the numerous finger-like projections of aggraded reef and sediment into the Red Sea basin. These projections are similar, in a manner, to 'bird's foot' river deltas (Coleman et al. 1998) and the alluvial fans of the Saudi coastal hinterland (Fig. 12a). Bird's foot deltas tend to be associated with large river channels and the continuous delivery of fine sediment (Blair and McPherson 1994). Persistent sediment loads, however, typically retard coral reef development (Fabricius 2005). Corals struggle to settle on mobile substrates and sediment in the water column reduces the depth to which sunlight is available for photosynthesis, while continuous high volumes of sediment can smother corals (e.g., Schlager and Purkis 2013). Alluvial flows into the Red Sea are, however, of a different type, with spatially and temporally variable delivery of coarse terrestrial siliciclastics.

The Red Sea rift basin is flanked by mountains. The climate is hyper-arid, though slightly wetter to the south in recent times. Storms, when they occur, can be severe, and the mountain topography channels large volumes of water into wadi beds (dry ephemeral rivers). Such storms have been observed to wash substantial infrastructure, such as concrete bridges, hundreds of meters downstream. Coarse siliciclastics are mobilized (Dullo and Montaggioni 1998) and transported seaward within these alluvial flows. The delivery of siliciclastics out of wadi mouths and into the Red Sea is highly episodic. The outflow and accumulation of siliciclastics creates bathymetric highs on the coastal shelf in the form of channel levees and alluvial debris deposits (Bosence 2005). Coral communities may be temporarily destroyed through sediment stress and scour at the time of such outflows. However, provided there are sufficient periods of stasis (10–100s years typical in the Red Sea; Hayward 1982) colonization and consolidation of debris by corals and calcareous organisms will occur (Dullo and Montaggioni 1998; Perry and Smithers 2009).

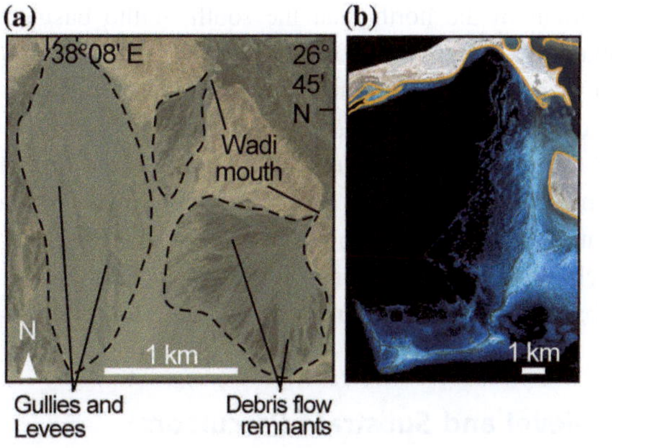

Fig. 12 **a** Examples of alluvial fans found just inland of the Saudi Arabian Red Sea coast. *Black dashed outline* depicts separate alluvial fan systems. The word 'wadi' is an Arabic term used to describe a valley or a dry ephemeral riverbed; **b** Image of a dendritic system. The arrangements of finger-like projections bear a striking morphological resemblance to the gullies, levees, and debris deposits of alluvial fans. Figure adapted from Rowlands et al. (2014)

Earlier in this chapter, the morphological similarity between dendritic systems and alluvial fans was identified (Fig. 12), suggesting the two may be linked in process terms. Rowlands et al. (2014) discuss the feasibility of such a link and consider four conditions that must be satisfied for it to occur. First, the morphology of the fan should be preserved in a submerged form atop the coastal shelf, such as would occur if deposition coincides with a sea-level low-stand; second, the drainage basin landward of the fan must be sufficiently large and steep to feed the volumes of sediment necessary to build a coherent system of channel levees; third, episodic heavy rainfall is required to mobilize quantities of sediment into the fan system; and fourth, the coastal shelf must be sufficiently wide for an alluvial fan to form.

The potential for sedimentary input has been estimated by Rowlands et al. (2014) across 2,120 drainage basins feeding into the Saudi Arabian Red Sea. According to their analysis, 80 % of the linear length of the mountain range abutting the Red Sea ultimately drains into only 10 % of the drainage basins. Slopes across much of these large (>1,000 km^2) drainage basins are steep (>25°). However, not all large drainage basins are associated with dendritic systems, which only occur in the northern half of precinct B (Fig. 5). It is worth noting that while the climate is hyper-arid today, historically weather patterns have been wetter. For example, during the Mid-Holocene, rainfall would have been greater in the northern Red Sea due to the southern migration of Mediterranean weather systems (Purkis et al. 2010), a period also coinciding with a lower sea-level. Historical patterns of rainfall may help explain why dendritic systems are more prevalent in the north than the south of the basin despite apparently wetter conditions to the south today. It is notable that dendritic systems are absent in the Gulf of Aqaba (precinct A, Fig. 5). Here, the coastal shelf is narrow (Fig. 4) and terrestrial sediment deposits are transported to hundreds of meters depth (Tibor et al. 2010). Growth of reefs is stunted at wadi mouths on both sides of the Red Sea down much of its length, which form small sandy sharms or embayments (Fig. 1a; Dullo and Montaggioni 1998).

Sea-level and Substrate Precursors

As has already been established for alluvial fans above, the process of sea-level oscillation is important to the morphology of shallow water coral reefs. Complicated by tectonic movements, Red Sea records of sea-level for the Late Holocene, the Last Glacial Maximum, the Last Interglacial, and for earlier interglacials are incomplete (Sheppard et al. 1992; Siddall 2004; Lambeck et al. 2011; Bailey, this volume). Nonetheless, many records in the Red Sea attribute reef terraces and wave notches and step formations to sea-level change (Dullo 1984; Gvirtzman 1994; Sheppard et al. 1992; Dullo and Montaggioni 1998). Reefs that fill the available accommodation during a high-stand provide a record of sea-level (e.g., Purkis et al. 2014), while during low-stands, the tops of coral reefs may be left exposed and subject to weathering, erosion, and a re-sorting of sediments.

Sea-level benchmarks have been established on both sides of the Red Sea at depths of 50–60 m below present day (Braithwaite 1982; Sheppard and Sheppard 1985; Bruckner et al. 2011; Rowlands et al. 2012). Arising from this base, many submerged patch reefs have an upper surface between 20 and 40 m below present day sea-level (Sheppard et al. 1992; Bruckner et al. 2011; Rowlands et al. 2012), while the floors of lagoons in shore-attached coral reef systems can be found at similar depths (Fig. 7). These terraces are, according to Sheppard et al. (1992), consistent with a period of the Pleistocene lasting some 100,000 years, during which time sea-level fluctuated between 20 and 60 m below present. Holocene coral reefs may adopt morphologies that reflect the antecedent template left by such erosional processes, for example, complex reticulated structures that arise as coral reef growth initiates on topographic highs, while sediments accumulate in topographic lows (Purkis et al. 2010; Schlager and Purkis 2013). Under a relatively stable sea-level scenario, and where carbonate budgets are positive, coral reef systems can be expected to fill available accommodation and transition toward more mature morphologies (Hopley 1982). Coral reef systems in the Red Sea do not tend to sort across the coastal shelf according to morphological maturity, as has been described, for example, in the Australian GBR (Hopley et al. 1989). That coral reef system morphology does not sort either latitudinally nor across the coastal shelf according to evolutionary maturity (see Fig. 6) is indicative of the tectonic dynamism of the Red Sea.

Research Perspectives

The Red Sea lags behind many other tropical marine regions in terms of the volume of coral reef research conducted (Berumen et al. 2013). Half of the work cited from the Red Sea is constrained geographically to coral reefs in the north of the Gulf of Aqaba which have a very simple morphology. Yet many areas of the Red Sea, such as the Farasan Banks, exhibit high morphological diversity in a relatively small geographic area. This situation lends itself to interesting avenues of research. For example, there is global interest in how coral reefs will respond to observed and predicted rates of sea-level rise (e.g., the keep-up, catch-up, give-up debate; Neumann and Macintyre 1985). With so many types of coral reef system in close proximity but at apparently different evolutionary stages, we can therefore ask why some coral reef systems are fully accommodated (planar systems that have filled to sea-level), while ostensibly similar systems are not. Work elsewhere indicates a range of models, and alternatives to the classic bucket-fill model must be considered (Adams and Hassler 2010; O'Leary and Perry 2010; Schlager and Purkis 2013). Work that explores antecedent topography, the sites, and rates of carbonate production, erosion, transport, and deposition is required. Toward a different application, coral reef morphology has been shown to influence the distribution of coral cover and exposure to thermal bleaching through the abundance of habitat, topography, and differences in wave exposure (Davis et al. 2011; Rowlands 2013). An understanding how different morphological processes operate in different areas of the Red Sea can assist those targeting research to identify and conserve healthy coral reefs.

Conclusion

A variety of processes have acted in the Red Sea across different spatial and temporal scales to deliver some of the most morphologically diverse coral reef systems in the world. While several reef morphologies exist at all latitudes of the Red Sea, others are confined to narrow ranges. The orientation and abundance of coral reef systems is primarily controlled by tectonics, and reefs align according to the grain of the rift axis. Other trends can be explained by the spatial arrangement of sub-seafloor evaporate stratigraphies, the delivery of siliciclastic sediment to the basin, and eustatic sea-level variation. Where seabed topography is controlled by sub-seafloor evaporite, reef morphologies may become circular, arcuate, amoeboid, or elongate, and are often punctuated by subcircular embayments, deep reentrants, and channels. Changes in sea-level influence the progression of morphological development for a coral reef and its attendant bodies of sediment. Many of today's shore-attached coral reefs sit atop alluvial fan systems deposited during sea-level low-stands. The present climate of the Red Sea region is hyper-arid, but has been punctuated by periods of high precipitation. Such pluvial events led to an increase in the delivery of siliciclastic sediments to the basin, which is spatially restricted by inland drainage patterns.

Acknowledgment Financial support was provided by the National Coral Reef Institute (NCRI) and the Khaled bin Sultan Living Oceans Foundation (KSLOF). Field support was provided by the crew of the M/Y Golden Shadow, through the generosity of HRH Prince Khaled bin Sultan. We thank Brett Thomassie and DigitalGlobe Inc. for assistance with satellite image acquisition. The use of Google Earth data, Esri ArcGIS base imagery, and the United States Geological Survey Landsat satellite imagery was critical to this work and is acknowledged.

References

Andréfouët S, Müller-Karger FE, Robinson JA, Kranenburg CJ, Torres-Pulliza D, Spraggins SA, Murch B (2006) Global assessment of modern coral reef extent and diversity for regional science and management applications: a view from space. In: Proceedings of 10th international coral reef symposium. Okinawa, Japan, pp 1732–1745

Adams EW, Hasler C-A (2010) The intrinsic effect of shape on the retrogradation motif and timing of drowning of carbonate patch reef systems (Lower Frasnian, Bugle Gap, Canning Basin, Western Australia). Sedimentology 57:956–984

Banks KW, Riegl BM, Shinn EA, Piller WE, Dodge RE (2007) Geomorphology of the Southeast Florida continental reef tract (Miami-Dade, Broward, and Palm Beach Counties, USA). Coral Reefs 26:617–633

Bermert G, Ormond R (1981) Red Sea coral reefs. Taylor and Francis, London, 192 pp

Berumen ML, Hoey AS, Bass WH, Bouwmeester J, Catania D, Cochran JEM, Khalil MT, Miyake S, Mughal MR, Spaet JLY, Saenz-Agudelo P (2013) The status of coral reef ecology research in the Red Sea. Coral Reefs 32:737–748

Blair TC, McPherson JG (1994) Alluvial fans and their natural distinction from rivers based on morphology, hydraulic processes, sedimentary processes, and facies assemblages. J Sediment Res 64:450–489

Bosence D (2005) A genetic classification of carbonate platforms based on their basinal and tectonic settings in the Cenozoic. Sediment Geol 175:49–72

Bosence DWJ, Al-Aawah MH, Davison L, Rosen BR, Vita-Finzi C, Whitaker E (1998) Salt domes and their control on basin margin sedimentation: a case study from the Tihamah Plain, Yemen. In: Purser BH, Bosence DWJ (eds) Sedimentation and tectonics of rift basins: Red Sea—Gulf of Aden. Chapman and Hall, London, pp 441–472

Bosworth W (1994) A model for the three-dimensional evolution of continental rift basins, north-east Africa. In: Schandelmeier H, Stern RJ (eds) Geology of Northeast Africa (part 2), Geologische Rundschau, vol 83(4), pp 671–688

Bott WF, Smith BA, Oakes G, Sikander AH, Ibraham AI (1992) The tectonic framework and regional hydrocarbon prospectivity of the Gulf of Aden. J Petrol Geol 15:211–243

Braithwaite CJR (1982) Patterns of accretion of reefs in the Sudanese Red Sea. Mar Geol 46:297–325

Brown GF (1970) Eastern margin of the Red Sea and the coastal structures in Saudi Arabia. Philos Trans R Soc Lond Ser A, Math Phys Sci 267:75–87

Bruckner A, Rowlands G, Riegl B, Purkis S, Williams A, Renaud P (2011) Khaled bin Sultan living oceans foundation atlas of Saudi Arabian Red Sea marine habitats. Panoramic Press, Phoenix, Arizona, 274 pp

Bunter MAG, Abdel Magid AEM (1989) The Sudanese Red Sea, 1: new developments in stratigraphy and petroleum geological evolution. J Petrol Geol 12:145–166

Burchette TP (1988) Tectonic control on carbonate platform facies distribution and sequence development: Miocene, Gulf of Suez. Sediment Geol 59:179–204

Carbone F, Matteucci R, Angelucci A (1998) Present-day sedimentation on the carbonate platforms of the Dahlak Islands Eritrea. In: Purser BH, Bosence DWJ (eds) Sedimentation and tectonics of rift basins: Red Sea—Gulf of Aden. Chapman and Hall, London, pp 523–536

Coleman JM, Roberts HH, Stone GW (1998) Mississippi river delta: an overview. J Coast Res 14:698–716

Crane K, Bonatti E (1987) The role of fracture zones during early Red Sea rifting: structural analysis using space shuttle radar and landsat imagery. J Geol Soc Lond 144:407–420

Cross NE, Purser BH, Bosence DWJ (1998) The tectono-sedimentary evolution of a rift margin carbonate platform: Abu Shaar, Gulf of Suez, Egypt. In: Purser BH, Bosence DWJ (eds) Sedimentation and tectonics of rift basins: Red Sea—Gulf of Aden. Chapman and Hall, London, pp 271–295

Crossley R, Cripps D (1999) Templates from mainland Africa and the Red Sea for interpreting the early evolution of the South Atlantic. In: Cameron NR, Bate RH, Clure VS (eds) The oil and gas habitats of the South Atlantic, Special Publication No. 153. The Geological Society of London, London, pp 85–96

Dabbagh A, Hötzl H, Schnier H (1984) Farasan Islands, in quaternary period in Saudi Arabia. In: Jado AR, Zottl JG (eds) Sedimentological, hydrological, hydrochemical, geomorphological, geochronological, and climatological investigations in Western Saudi Arabia. Quaternary period in Saudi Arabia, vol 2. Springer, Vienna, pp 212–220

Darwin CR (1842) The structure and distribution of coral reefs. Smith, Elder, London, 344 pp

Davis KA, Lentz SJ, Pineda J, Farrar JT, Starczak VR, Churchill JH (2011) Observations of the thermal environment on Red Sea platform reefs: a heat budget analysis. Coral Reefs 30:25–36

Dullo WC (1984) Progressive diagenetic sequence of aragonite structures: pleistocene coral reefs and their modern counterparts on the eastern Red Sea coast, Saudi Arabia. Palaeontogr Am 54:254–260

Dullo WC, Montaggioni L (1998) Modern Red Sea coral reefs: a review of their morphologies and zonation. In: Purser BH, Bosence DWJ (eds) Sedimentation and tectonics of rift basins: Red Sea—Gulf of Aden. Chapman and Hall, London, pp 583–594

Fabricius K (2005) Effects of terrestrial runoff on the ecology of corals and coral reefs: review and synthesis. Mar Pollut Bull 50:125–146

Fairbridge RW (1967) Coral reefs of the Australian region. In: Jennings JN, Mabbutt JA (eds) Landform studies from Australia and New Guinea. Australian National University, Canberra, pp 386–451

Friedman GM (1985) Geological and sedimentological framework. In: Friedman GM, Krumbein WE (eds) Hypersaline Ecosystems. Chap. 3, Gulf of Elat (Aqaba), the Gavish Sabkha. Springer, Berlin, pp 39–71

Guilcher A (1988) A heretofore neglected type of coral reef: the ridge reef, morphology and origin. In: Proceedings of the 6th international coral reef symposium. Townsville, pp 399–402

Gvirtzman G (1994) Fluctuations of sea-level during the past 400,000 years: the record of Sinai, Egypt (northern Red Sea). Coral Reefs 13:203–214

Hart DE, Kench PS (2007) Carbonate production of an emergent reef platform, Warraber Island, Torres Strait, Australia. Coral Reefs 26:53–68

Hayward AB (1982) Coral reefs in a clastic sedimentary environment: fossil (Miocene, SW Turkey) and modern (recent, Red Sea) analogues. Coral Reefs 1:109–114

Hopley D (1982) Geomorphology of the great barrier reef: quaternary development of coral reefs. Wiley, New York, 453 pp

Hopley D, Parnell KE, Isdale PJ (1989) The great barrier reef marine park: dimensions and regional patterns. Aust Geogr Stud 27:47–66

Jado AR, Zötl JG (1984) Sedimentological, hydrological, hydrochemical, geomorphological, geochronological, and climatological investigations in Western Saudi Arabia. quaternary period in Saudi Arabia, vol 2. Springer, Vienna

James NP, Coniglio M, Aissaoui DM, Purser BH (1988) Facies and geological history of an exposed miocene rift margin carbonate platform, Gulf of Suez, Egypt. Am Assoc Petrol Geol Bull 72:555–572

Khalil SM, McClay KR (2002) Extensional fault-related folding, northwest Red Sea, Egypt. J Struct Geol 24:743–762

Lambeck K, Purcell A, Flemming NC, Vita-Finzi C, Alsharekh AM, Bailey GN (2011) Sea-level and shoreline reconstructions for the Red Sea: isostatic and tectonic considerations and implications for hominin migration out of Africa. Quat Sci Rev 30:3542–3574

Leon JX, Woodroffe CD (2013) Morphological characterisation of reef types in Torres strait and an assessment of their carbonate production. Mar Geol 338:64–75

Longhurst AR, Pauly D (1987) Ecology of tropical oceans. Academic press, London 407 pp

Maxwell WGH (1970) Deltaic patterns in reefs. Deep-Sea Res 17:1005–1018

Mitchell DJW, Allen RB, Salama W, Abouzakm A (1992) Tectonostratigraphic framework and hydrocarbon potential of the Red Sea. J Petrol Geol 15:187–210

Moberg F, Folke C (1999) Ecological goods and services of coral reef ecosystems. Ecol Econ 29:215–233

Montaggioni LF, Behairy AKA, Sayed MKE, Yusuf N (1986) The modern reef complex, Jeddah area, Red Sea: a facies model for carbonate sedimentation on embryonic passive margins. Coral Reefs 5:127–150

Neumann AC, Macintyre IG (1985) Reef response of sea-level rise: keep-up, catch-up or give-up. In: Proceedings of the 5th international coral reef congress, vol 3. Tahiti, French Polynesia, pp 105–110

O'Leary MJ, Perry CT (2010) Holocene reef accretion on the Rodrigues carbonate platform: an alternative to the classic bucket-fill model. Geology 38:855–858

Orszag-Sperber F, Harwood G, Kendall A, Purser BH (1998) A review of the evaporites of the Red Sea-Gulf of Suez rift. In: Purser BH, Bosence DWJ (eds) Sedimentation and tectonics of rift basins: Red Sea—Gulf of Aden. Chapman and Hall, London, pp 409–425

Perry CT, Smithers SG (2009) Stabilisation of intertidal cobbles and gravels by Goniastreaaspera: an analogue for substrate colonisation during marine transgressions? Coral Reefs 28:805–806

Purdy EJ (1974) Reef configurations, cause and effect. In: Laporte LF (ed) Reefs in time and space: selected examples of the recent and ancient, Special Publication 18. Society of Economoic Paleontologists and Minerologists, pp 9–76

Purkis SJ, Kenter JAM, Oikonomou EK, Robinson IS (2002) High-resolution ground verification, cluster analysis and optical model of reef substrate coverage on Landsat TM imagery (Red Sea, Egypt). Int J Remote Sens 23:1677–1698

Purkis SJ, Rowlands GP, Riegl BM, Renaud PG (2010) The paradox of tropical karst morphology in the coral reefs of the arid middle east. Geology 38:227–230

Purkis SJ, Riegl BM (2012) Geomorphology and reef building in the SE Gulf. In: Coral reefs of the Gulf: adaptation to climatic extremes hardcover. Springer, Netherlands. ISBN 978-94-007-3007-6

Purkis SJ, Harris PM, Ellis J (2012) Patterns of sedimentation in the contemporary Red Sea as an analog for ancient carbonates in rift settings. J Sediment Res 82:859–870

Purkis SJ, Rowlands G, Kerr JM (2014) Unravelling the influence of water depth and wave energy on the facies diversity of shelf carbonates. Sedimentology. doi:10.1111/sed.12110

Purser BH, Hötzl H (1988) The sedimentary evolution of the Red Sea rift: a comparison of the northwest (Egyptian) and northeast (Saudi Arabian) margins. Tectonophysics 153:193–208

Purser BH, Barrier P, Montenat C, Orszag-Sperber F, Ottd'Estevou J, Plaziat JC, Philobbos E (1998) Carbonate and siliciclastic sedimentation in an active tectonic setting: miocene of the northwest Red Searift Egypt. In: Purser BH, Bosence DWJ (eds) Sedimentation and tectonics of rift basins: Red Sea—Gulf of Aden. Chapman and Hall, London, pp 239–270

Roberts CM, Shepherd ARD, Ormond RFG (1992) Large-scale variation in assemblage structure of Red Sea butterfly fishes and angel fishes. J Biogeogr 19:239–250

Rowlands G, Purkis S, Riegl B, Metsamaa L, Bruckner A, Renaud P (2012) Satellite imaging coral reef resilience at regional scale: a case-study from Saudi Arabia. Mar Pollut Bull 64:1222–1237

Rowlands G (2013) Remote sensing the diversity, distribution and resilience of coral reef environments. Ph.D. thesis, Nova Southeastern University, USA

Rowlands GP, Purkis SJ, Bruckner AW (2014) Diversity in the geomorphology of shallow-water carbonate depositional systems in the Saudi Arabian Red Sea. Geomorphology. doi:10.1016/j.geomorph.2014.03.014

Siddall M (2004) Understanding the Red Sea response to sea-level. Earth Planet Sci Lett 225:421–434

Sheppard CRC (1985) The unspoilt little barrier reef of Saudi Arabia. Sea Frontiers 31:94–103

Sheppard CRC, Sheppard ALS (1985) Reefs and coral assemblages of Saudi Arabia 1: the central part of the Red Sea at Madinat al Sinaiyah. Fauna Saudi Arabia 7:17–36

Sheppard C, Price A, Roberts C (1992) Marine ecology of the Arabian region: patterns and processes in extreme tropical environments. Academic Press, London

Schlager W, Purkis SJ (2013) Bucket structure in carbonate accumulations of the Maldive, Chagos and Laccadive archipelagos. Int J Earth Sci 102:2225–2238

Spalding M, Ravilious C, Green EP (2001) World atlas of coral reefs. University of California Press, Berkeley, 428 pp

Tibor G, Niemi TM, Ben-Avraham Z, Al-Zoubi A, Sade RA, Hall JK, Hartman G, Akawi E, Abueladas A, Al-Ruzouq R (2010) Active tectonic morphology and submarine deformation of the northern Gulf of Eilat/Aqaba from analyses of multibeam data. Geo-Mar Lett 30:561–573

Toomey M, Ashton AD, Perron JT (2013) Profiles of ocean island coral reefs controlled by sea-level history and carbonate accumulation rates. Geology 41:731–734

Coral Reefs and Communities of the Central and Southern Red Sea (Sudan, Eritrea, Djibouti, and Yemen)

Rebecca Klaus

Abstract

The Red Sea hosts a high diversity of modern coral reef morphologies and non-accreting coral communities which exist under a wide range of environmental regimes. The majority of research into the coral reefs in the Red Sea has focused on the steep fringing reefs' characteristic of the northern Red Sea and other reef types along the Arabian coastline. This chapter explores the less well-known coral reefs and other coral communities in the central region and southern region, along the coasts of Sudan, Eritrea, Djibouti, and Yemen. In the central region, the Red Sea widens and fringing reefs diminish in scale, as nearshore water depths shallow on account of the presence of broad shallow shelf platforms. Other more unusual reef types and coral communities occur, growing on carbonate, granitic, and volcanic lava flows. Environmental conditions become progressively more extreme towards the south and exhibit cross-shelf and pronounced seasonal gradients. Corals demonstrate a high degree of local acclimation and adaptation to these shifts in conditions. The central region and southern region are transitional zones between the optimal Red Sea reef conditions and the more marginal conditions characteristic of the Gulf of Aden and Arabian Sea.

Introduction

The Red Sea hosts a high diversity of scleractinian hard coral species, including endemics, and is recognised as a "centre of evolution" (Spalding et al. 2007). The Red Sea also supports a high diversity of both coral reef morphologies and non-accreting coral assemblages, some of which are globally unique (e.g. Guilcher 1988). Scleractinian species diversity, zonation and reef development patterns vary along the length of the Red Sea (Sheppard et al. 1992). The dramatic near-vertical fringing reefs, with drop-offs into clear deep water, for which the Red Sea is most famous, are mainly found in the northern and eastern Red Sea (e.g. Behairy et al. 1992). While fringing reefs continue to border much of the Arabian and African mainland coasts, the depth of the reefs is constrained by the presence of extensive shelf platforms (Fig. 1). These broad shelf platforms host a diversity of more unusual modern reef morphologies, and other mixed and monotypic non-reefal coral assemblages on non-biogenic and relict reef substrates occur. Given that there has been less research in the central and southern Red Sea regions, the following chapter aims to consolidate the current state of knowledge about the coral reefs and other coral communities along the coasts of Sudan, Eritrea, Djibouti and Yemen, adding new information where available.

Characteristics of the Central and Southern Red Sea

The Red Sea is about 2,000 km long, with maximum depths of 2,000 m along the deep central trench, and it is connected to the Indian Ocean through the narrow Bab-al-Mandab Strait, which extends from the shallow Hannish sill (160 m) to the island of Miyurn (Perim) (Fig. 1). The foundations for both past and modern reef growth originated from the rifting process which caused Arabia to split from Africa. Rifting was not

R. Klaus (✉)
Tropical Marine Ecosystems of the Future LOEWE Biodiversity and Climate Research Centre (BiK-F), Senckenberg Research Institute, Nature Museum Frankfurt, Marine Zoology, Senckenberganlage 25, 60325 Frankfurt, Germany
e-mail: rebecca.klaus@gmail.com

N.M.A. Rasul and I.C.F. Stewart (eds.), *The Red Sea*,
Springer Earth System Sciences, DOI 10.1007/978-3-662-45201-1_25,

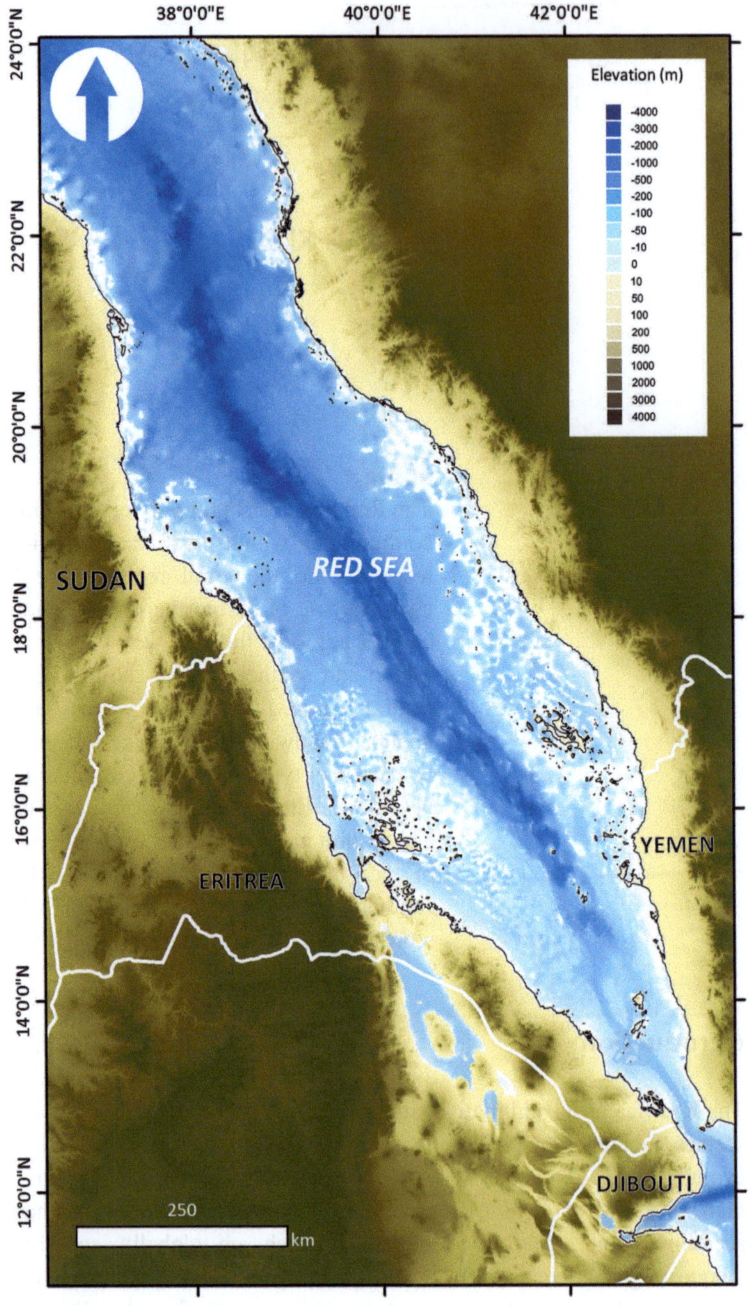

Fig. 1 Topography and bathymetry of the Red Sea, showing the wide coastal plains and broadening of the coastal shelf (*data source* SRTM30-plus arc-minute global topography model, Becker et al. 2009)

a continuous process and was interrupted by periods of volcanic activity, which created coastal formations in the north and far south and extruded the volcanic islands in the south, which remain active today, with the last eruption in the Zubair Islands occurring in 2011. Shorelines are mostly sedimentary and often formed from alluvial fans or uplifted fossil coral reef; although volcanic lava flows and Precambrian granitic rocky outcrops occur, they are restricted to smaller areas.

Factors that have contributed towards the variation in reef development and coral community composition along the length of the Red Sea have been discussed by various authors (Behairy et al. 1992; Braithwaite 1987; Head 1980; Scheer and Pillai 1983; Sheppard and Sheppard 1985, 1991; Sheppard et al. 1992; Medio et al. 2000). These factors include the influence of (i) geological and paleoclimatological processes and quaternary changes in sea level on the topography and availability of habitat suitable for reef growth, (ii) quaternary changes in sea level on speciation and evolutionary processes, (iii) present-day interactions between abiotic factors (e.g. sea water temperature, salinity, exposure to wind and wave-driven water movement, depth, light availability, particulates and nutrients) and biotic factors (e.g. recruitment, growth rates, competition, predation, bioerosion, symbiosis, nuisance species and disease) and (iv) anthropogenic impacts.

Many of the modern Red Sea reefs are considered to be growing on the antecedent topographic remains of Pliocene-Pleistocene carbonate platforms, the surfaces of which were shaped over time by tectonic forces, irregular patterns of uplift by the underlying Miocene evaporite deposits (salt domes), and periods of sediment deposition and subaerial dissolution during low sea stands (Angelucci et al. 1985; Sheppard et al. 1992; Bosence 2005). Reef structures began to form in the Red Sea during the Pliocene (2.6–5.3 Ma), when reef-forming organisms first entered from the Indian Ocean (Sheppard et al. 1992). Reef building continued during the Pleistocene (2.6–0.01 Ma). Sea levels then began to fall between 0.14 and 0.11 Ma and remained at 30–60 m below present levels for a sustained period of time between 0.11 and 0.03 Ma (Hopely 1982). During this time, there were at least eight periods when sea levels were stable, each of which lasted for approximately 2,000 years, which was sufficient time to allow for substantial reef growth (Hopley 1982; Pugh and Abualnaja, this volume; Bailey, this volume).

There were at least two periods of heavy rainfall during the Pleistocene that eroded the Red Sea mountains and resulted in sediments being deposited onshore where they formed large alluvial fans and offshore where they smothered existing reefs (Sheppard et al. 1992). The sediments that settled offshore were eventually consolidated, creating new stable structures that provided the basis for further reef accretion. Freshwater flows and subaerial dissolution of antecedent frameworks during low sea level stands cut through existing reefs, creating irregularities in the bathymetry that are still evident in coastal and reef formations.

Sea level changes resulted in alternate periods of flooding and evaporation in the Red Sea and the opening and closing of connections with the Mediterranean and Indian Ocean (Braithwaite 1987). Sea levels are thought to have fallen to a minimum of 120–150 m below present levels around 0.017 Ma before rising back to present levels around 0.007 Ma at start of the Holocene (see also Pugh and Abualnaja, this volume; Bailey, this volume). Whether or not the Red Sea was ever entirely closed off from the Indian Ocean when sea levels were at their lowest has been debated. The Red Sea most likely became hypersaline during this period and localised extinctions of marine flora and fauna most likely occurred, thereby influencing speciation and evolutionary processes. The theory then is that the marine ecosystems in the Red Sea were effectively "reset" when sea water temperatures and sea levels rose at start of the Holocene (Sheppard et al. 1992).

Present-day scleractinian species distribution and zonation patterns vary with environmental conditions (e.g. sea water temperature, exposure, light and nutrients), which shift with latitude within the narrow confines of the Red Sea. Early studies on the Arabian coast identified four distinct latitudinal subunits, which included the Gulf of Aqaba, and the northern, central and southern Red Sea regions (Ormond et al. 1984a, b, c; Sheppard 1985; Sheppard and Sheppard 1991; Sheppard et al. 1992; Roberts et al. 1992; Kemp 1998). A zone of vicariance or biogeographical discontinuity was identified at 20°N, separating species assemblages characteristic of the northern-central Red Sea from those characteristic of the southern Red Sea (Winterbottom 1985; Roberts et al. 1992). These studies identified 13 coral community types between the northern Red Sea and the Yemen border, associated with changes in depth, exposure and sedimentation (Sheppard 1985; Sheppard and Sheppard 1991), which were consolidated into four broad community types by DeVantier et al. (2000) for the Saudi Arabian central-northern Red Sea (Table 1). Later studies on the African coast found that the biogeographical barrier appeared further south, commencing at 16°N within Eritrean waters (Zekeria 2003). Changes were however already noticeable south of 18°N in Sudanese waters (Klaus et al. 2008a, b) associated with shifts in environmental conditions.

Environmental conditions in the central and southern Red Sea often exceed the limits considered tolerable for scleractinian coral survival and reef development (Coles 1988; Kleypas et al. 1999; Guinotte et al. 2003) (Fig. 2). The present-day climate throughout the Red Sea is hyper arid or arid (Table 2). Average precipitation on the coast is however highly variable, ranging from 36 mm per year at Halaib to >200 mm per year in Djibouti. Freshwater run-off rarely reaches the Red Sea, due to high evaporation over the coastal plain, although there are exceptions. The Tokar delta, for example, in southern part of the Red Sea State of Sudan, can receive substantial rainfall and run-off from seasonal streams originating in the Ethiopian and Eritrean highlands, which can transport up to 9 cm of new silt to the delta every year. The southern Tihama in Yemen and northern parts of Eritrea are also subject to heavy rains, resulting in seasonal flash flooding during the late summer monsoon and in spring. Sea water temperatures are warmer in the central and southern regions than in the north, but exhibit greater variability in the south due to the upwelling (Fig. 2). Sea water temperatures at depth remain consistently close to 21 °C throughout the year, and surface temperature in offshore areas typically exceeds 30 °C during the hot summer months. Sea water temperatures in nearshore areas reach 35–37 °C in Yemen (Turak et al. 2007) and 36–38 °C in Eritrea with a maximum of 45 °C inside sheltered lagoons (Ateweberhan and Prud'Homme Van Reine 2005). Average surface salinities in the Red Sea are 38 ‰ and range from 36.5 ‰ at Miyurn Island to >41 ‰ in the north, due to the limited influx of fresh water and high evaporation rates along the length of the Red Sea. Salinities are also higher in nearshore areas and at all latitudes during the summer for the same reasons.

Much of the variability in the environmental conditions along the length of the Red Sea can be explained by the seasonal wind patterns which result from the movement of the Intertropical Convergence Zone (Sheppard et al. 1992).

Table 1 Key features of the coral communities in the central and southern Red Sea (from Sheppard and Sheppard 1991; DeVantier et al. 2000; Turak et al. 2007)

No.	Location*	Red Sea (Sheppard and Sheppard 1991)	Saudi Arabia (DeVantier et al. 2000)	Yemen (Turak et al. 2007)
2	N–C	Shallow exposed forereef slope. Dominant coral species *Acropora hyacinthus, A. humilis*	Community A *Pachyseris–Leptoseris* community characteristic of lower slopes (cf. community 6 and 10 of Sheppard and Sheppard 1991) Community B *Acropora gemmifera–Leptoria* spp. community, characteristic of reef crests and upper reef slopes exposed to strong wave action (cf. community 2 and 9 of Sheppard and Sheppard 1991) Community C *Montipora spumosa–Pavona decussata* characteristic of turbid areas. (cf. community 4, 7, 11 and 13 of Sheppard and Sheppard 1991) Community D: massive faviid community characterising reef cusps. (cf. community 3, 5 and 8 of Sheppard and Sheppard 1991)	Community D2. Massive coral macroalgal community (massive *Porites* spp. and *Stylophra pistillata*) Community D1 exposed reefs with coralline red algae and monospecific coral stands (*Playgyra daedelea*–Faviidae) Community C: southern fringing reefs (*Montipora* –Faviidae) Community B2. Semi-protected island reefs in the north (*Goniastrea retiformis, Pavona cactus)* Community B1: south facing exposed island reefs (*G. retiformis, Echinopora gemmacea*) Community A: clear water communities facing open water (*Platygra Acropora humilis)*
4	N–C	Sheltered fringing reefs or backs of patch reefs. Dominant coral species *Porites lutea.*		
5	N–C	Shallow-mid depths in north, mid-deep in central area. All moderate exposure. Dominant coral species in subclusters included *Millepora* and *Goniopora.*		
6	All	Moderately exposed mid-depth (5–25 m) and 37 moderately turbid sites.		
7	C	Mid-depths patch reefs beside and among sand. Dominant coral species *Pocillopora damicornis, A. eurystoma, A. clathrata.*		
8	C	Patch and barrier reefs mid-depths. Dominant coral species *P. verrucosa, A. hemprichii*		
9	C	Barrier and very exposed fringing reefs. Shallow areas only. Subgroup A is most exposed, B intermediate, C. least exposed. Dominant coral species *Acropora* zones (with different species in each subgroup)		
10	C–S	Barrier and fringing reefs mid-depths. Dominant species *Porites* with *Montipora* or *Goniastrea pectinata.*		
11	S	Shallow reefs near mangroves. Dominant coral species *Porites.*		
12	S	Fringing reefs meeting sand at <5 m. Clear water. Dominant coral species staghorn *Acropora* spp. and *Porites.* spp.		
13	S	Red algal patch reef. No dominant species		

* N = Northern, C = Central, S = Southern

In the northern Red Sea above 20°N, winds blow fairly consistently throughout the year from the NNW and are funnelled south down the Red Sea by the high mountain ranges of both coasts, with occasional winds from the south during the winter months. In the southern Red Sea, conditions are influenced by the seasonally reversing Indian Ocean monsoon system. During the summer monsoon (June to September), hot dry winds blow across the African continent from the west and are funnelled south down the Red Sea in the same direction as winds in the northern Red Sea. During the winter monsoon (October to May), winds blow from the east along the southern coast of Arabia and Gulf of Aden and are funnelled north up the Red Sea in the opposite direction. An intermediate region develops between the NNW winds of the northern Red Sea and the SSE winds of the southern Red Sea, characterised by an area of low pressure, resulting in calmer conditions. The intermediate region varies in size and oscillates in position, before moving gradually southwards and dissipates as the prevailing wind changes from SSE to NNW. Winds in the southern Red Sea are strong throughout the year, particularly through the narrow Bab-al-Mandab Straits, but the winter SSE winds tend to be the strongest and have a significant influence on the local hydrodynamic conditions. Localised variations in wind patterns also occur, with occasional strong sea breezes during the summer, which hit the coast at oblique angles and can drive higher wave energy conditions and affect sea levels. Other specific wind patterns can carry dust storms from the Sahara across the Red Sea.

Winds have a strong influence on shallow marine communities in the central and southern Red Sea, due to their influence on shallow surface water circulation, exposure (hydrodynamic forces), mixing, sediment transport, sedimentation, water clarity, nutrient circulation and upwelling. While overall there is a net flow of water into the Red Sea from the Gulf of Aden through the Bab-al-Mandab, this

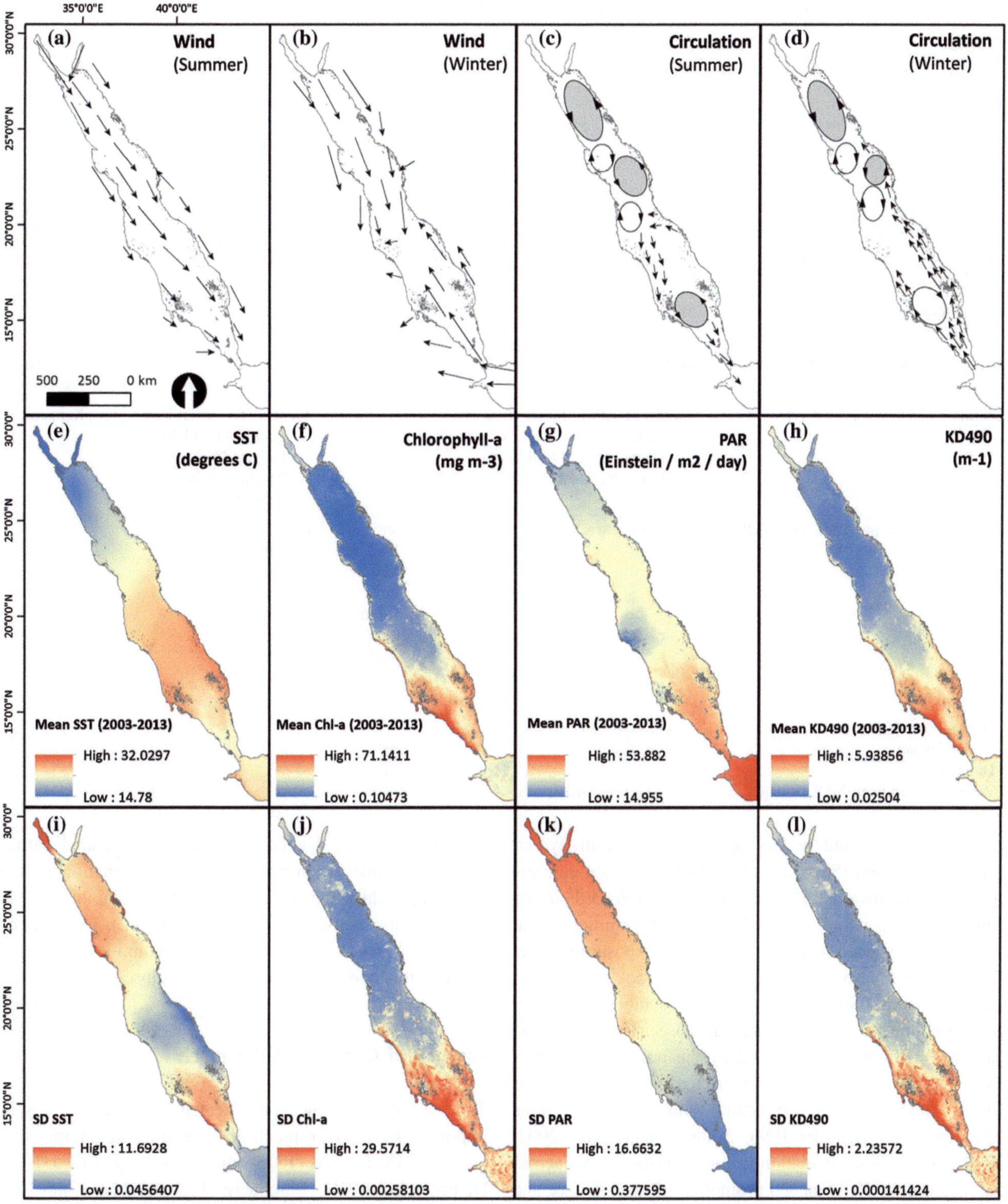

Fig. 2 Key environmental characteristics of the Red Sea environment showing **a** wind direction in summer and **b** winter; **c** Surface circulation in summer and **d** winter (Sofianos and Johns 2003); Aqua MODIS (2003–2013) **e** mean and **f** standard deviation sea surface temperature (°C); **g** mean and **h** standard deviation chlorophyll-a concentration (mg m^{-3}); **i** mean and **j** standard deviation photosynthetically active radiation (*PAR*) (Einstein/m^2/day) and; **k** mean and **l** standard deviation KD490 (diffuse attenuation at 490 nm m^{-1}). NASA level 3 standard mapped image (*SMI*) data, published by NASA Goddard Space Flight Center (*GSFC*) ocean colour group (Feldman and McClain 2013)

Table 2 Characteristics of the countries in the central and southern Red Sea

Country	Sudan	Eritrea	Djibouti	Yemen
Coastal length (km)	750	1,350	370	735
Exclusive economic zone (sq.km) (area outside Red Sea)	66,436	78,352	1,953 [5,034]	37,550 [507, 208]
Average annual rainfall (mm per annum)	252[a]	183.3[b]	163.5[c]	200[d]
Salinity (‰)	40−45[e]	36.5−40	36.5−39	36.5−38
Mean sea surface temperature (2003-2013)[f] (°C) (range)	26.54–28.95 [2.41]	26.17–29.98 [3.80]	27.76–28.38 [0.61]	27.28–30.17 [2.88]
Maximum sea surface temperature (2003–2013)[f] (°C) [range]	28.53–32.97 [4.45]	30.24–33.95 [3.69]	31.26–32.36 [1.11]	30.15–33.9 [3.76]
Mean chlorophyll-a (2003–2013)[f] (mg l^{-3})	0.13–5.02	0.23–35.60	0.60–1.7	0.49–9.12
Maximum chlorophyll-a[f] (2003–2013)[f] (mg l^{-3})	0.29–75.22	1.17–99.63	4.7–81.27	0.65–99.10
Marine-protected area (MPAs)	Sanganeb Atoll Marine National Park Dungonab and Mukkawar Island National Park		Moucha Territorial Park South Maskali Islands Integral Reserve Ras Siyyan-Isles des Sept Frères	Kamaran
Proposed marine-protected areas (MPAs)	Suakin Archipelago	Dissei-Madote, Museri, Ras Fatuma Sheikh Seid	Hamaras Art plage	

[a] Tokar
[b] Massawa
[c] Djibouti town
[d] al-Huddaydah
[e] Dungonab
[f] Values derived from MODIS Aqua satellite data (2003–2013)

varies seasonally and in fact operates as a multi-layered contraflow system (Sheppard et al. 1992). Surface currents are driven southwards in the summer and northwards in the winter (Fig. 2), which results in coral communities in the south being exposed to a seasonal shift in exposure regime. Winter winds also push nutrient and plankton-enriched waters from the Gulf of Aden into the southern Red Sea, generating the intense upwelling (Fig. 2). Surface circulation patterns are further complicated by the presence of a permanent cyclonic gyre in the northern Red Sea and a series of anti-cyclonic gyres in the central and southern regions (Quadfasel and Baunder 1993; Sofianos and Johns 2002, 2003, 2007). These circulation features transport water across the width of the Red Sea and promote mixing of surface and intermediate waters down to 300 m depth (Sofianos and Johns 2007), which likely have significant influence on connectivity and biogeography within the region. The other main factor that makes the environmental conditions in the central and southern Red Sea less favourable for coral growth and reef development is the shallow sedimentary nature of nearshore habitats, higher quantities of suspended particulates and sedimentation rates, which in combination with the upwelling reduce the water clarity and availability of photosynthetically active radiation (PAR) (Fig. 2). The influence of these various factors on the coral communities and reefs in the region is considered further in the subsequent sections.

History of Coral Reef Research

The southern Red Sea, and Yemen in particular, was an important location for much of the earliest taxonomic work on corals by Forskal (1775) and others (e.g. Edwards and Haime 1857–1860; Ehrenberg 1834; Esper 1788–1830; Forskal 1775; de Lamarck 1816; Klunzinger 1879). Description of coral collections from Sudan was published by Crossland (1913, 1952) and Gardiner (1909). Gravier (1910a, b, c, 1911) described collections from the Gulf of Tadjoura and the Strait of Bab-al-Mandab.

From the 1950s onwards, there were a number of important expeditions to the central and southern Red Sea including "Calypso" (1951/1952), "Formica" (1953), "Xarifa" (1957/1958), "University I" (1962, 1965), "Meteor" (1964/1965, 1987), "Sonne" (1977) and "Valdivia" (1979, 1981). Corals collected from the Farasan Bank during the first "Calypso" expedition were discussed by Nesteroff (1955) and Guilcher (1955). The coral reefs of the Dahlak Archipelago were described by Rossi (1954) after the "Formica" expedition and by Lewinsohn and Fishelson (1967); Oren (1962) and Wainwright (1965) after the "University I" expeditions. Berry et al. (1966) described the raised reefs in Sudan. Hans Hass produced a film and wrote a book about his diving experiences in Sudan (Hass 1952) and later led the "Xarifa" expedition (Hass 1961). In 1963, Captain Jacques-Yves Cousteau returned to Sudan with the "Calypso" where he filmed "The World without Sun" and constructed the famous underwater village on Sha'ab Rumi in Sudan (Cousteau and Diolé 1970; Diolé and Falco 1976), the remains of which are still present today. The majority of work published in the 1970s focussed on the northern Red Sea, although there were coral collections in the Dahlak Archipelago (Fishelson 1971), Wingate reef (Streicher 1980), Sanganeb atoll (Schroeder 1974), Towartit reef (Bemert and Ormond 1981), Massawa, Port Sudan and Al Hudaydah in Yemen (Kuehlman 1983) among others.

The first modern systematic account of reef-building corals in the Red Sea was published in the 1980s, which included descriptions of >180 species of coral from 54 genera collected during the "Xarifa" expedition among many others (Scheer and Pillai 1983). The next major account of coral taxonomy (Sheppard and Sheppard 1991) resulted from studies along the Saudi Arabian coast (Sheppard 1985, 1987; Sheppard and Sheppard 1985; Sheppard et al. 1992). Ten years later, corals collected on the Arabian side of the Red Sea (e.g. Saudi Arabia, DeVantier et al. 2000, and Yemen, Turak et al. 2007) were included in the most recent major revision of the taxonomy of scleractinian corals (Veron 2000, 2002). The studies completed by Turak and Brodie (1999) identified 221 scleractinian species in Yemen, including 71 species not previously reported from the southern Red Sea and 14 entirely new species distribution records, which demonstrates the lack of previous research in this southern region.

Research in the central and southern Red Sea reefs in recent years has largely been supported by non-governmental organisations or university led projects in Sudan (APF 2006; Bemert and Ormond 1981; Cousteau Society 2013; Head 1980; Kemp et al. 2002; Klaus et al. 2008a, b, Krupp et al. 1994; Mergner and Schuhmacher 1985; Reinicke et al. 2003; Schroder and Nasr 1983; Schumacher and Mergner 1985; Vine and Vine 1980), Yemen (Barratt et al. 1987a, 1987b; Klaus and Eisinger 2004; Turak and Brodie 1999; Turak et al. 2007), Eritrea (Hilman and Tsegay 1997; Rushdi et al. 1994; Tilot et al. 2008) and Djibouti (Barker et al. 2002; Barrat and Medley 1988; PERSGA 1998; Obura 1998; Pichon 1989). These projects have often been orientated towards building capacity for sustainable development and implemented in partnership with national governments, through the World Bank or United Nations Development Programme (UNDP) or the regional management authority PERSGA ("Regional Organisation for the Conservation of the Environment of the Red Sea and Gulf of Aden") with the support of international donors such as the European Commission or global environment facility (GEF). Although the results of some of these projects have been published, much of the information remains in the grey literature.

Character and Distribution of Reefs and Coral Communities

Sudan

The Sudanese Red Sea coastline extends over 750 km from the border with Egypt in the north (22°N) to the border with Eritrea in the south (18°N) and includes numerous uninhabited offshore islands. Sudan's Red Sea supports extensive coral growth and hosts both classical Red Sea fringing reefs and a broad diversity of more unusual reef formations (Fig. 3). Early reports suggested that Sudan's reefs were among the most diverse in the Red Sea (Ormond 1976) and by extension the entire Arabian and north-western Indian Ocean region.

Early studies on Sudanese reefs were mostly restricted to the vicinity of Port Sudan (Bemert and Ormond 1981; Edwards and Head 1987; Kuehlman 1983; Head 1980) and Sanganeb atoll (Head 1980; Krupp et al. 1994; Mergner and Schumacher 1985; Reinicke et al. 2003; Schumacher and Mergner 1985). As part of the PERSGA Marine Protected Areas Strategic Action Plan (MPA-SAP), surveys were completed in 1997 and 1999 (PERSGA/GEF 2003) and in 2002 in the vicinity of Sanganeb Marine National Park (SMNP) and Dungonab Bay (Kemp et al. 2002), which latter of which led to the designation of Dungonab Bay and Mukkawar Island National Park (DBMP) in 2004. In 2006, the Government of Sudan invited the African Parks Foundation (APF) to manage these MPAs, during which time permanently marked coral reef monitoring sites were established at 18 sites (APF 2006). In 2007, the Cousteau Society undertook a more geographically comprehensive assessment of the reefs between Dungonab and the Suakin Archipelago, during which 40 sites were surveyed, a further 4 of which were permanently marked as monitoring sites (Cousteau Society 2013; Klaus et al. 2008a, b). The following account

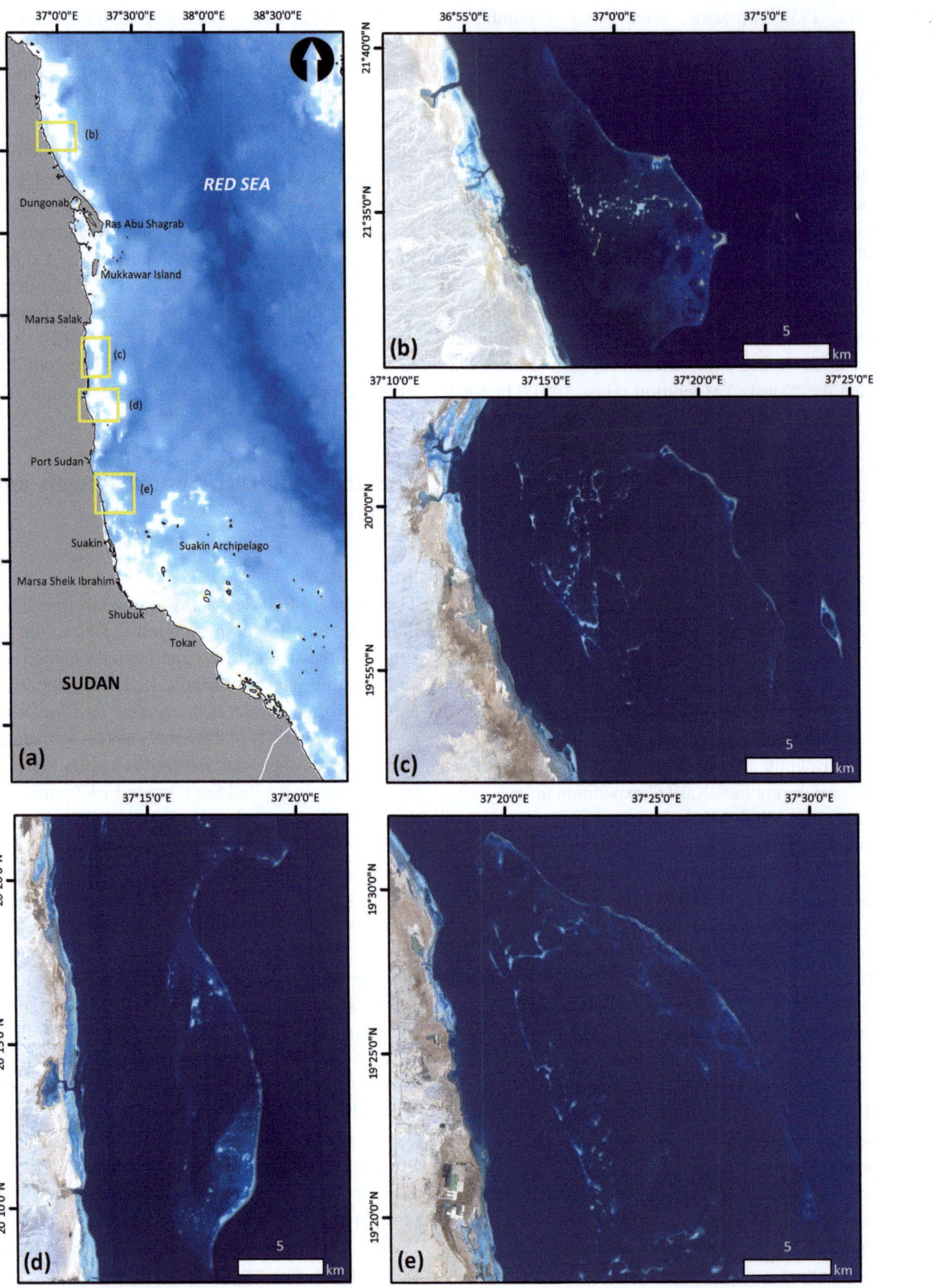

Fig. 3 Map illustrating **a** the coast of Sudan and the mainland fringing reefs and offshore barrier reef complexes of **b** Abu Fanadair, north of Dungonab, **c** the northern end of Wingate reef, also showing the location of Sha'ab Rumi, **d** Gorna, Sha'ab Sudi and blue belt reef complex, and **e** Towaritit Reef to the south of Port Sudan. Maps **b–d** derived from landsat-8 satellite images. Data available from the US Geological Survey (http://landsat.usgs.gov/landsat8.php)

summarises the observations from the 2002, 2006, and 2007 surveys (APF 2006; Cousteau Society 2013; Kemp et al. 2002; Klaus et al. 2008a, b).

Types of coral reef and coral assemblage
Sudan's reefs can be grouped into five broad types (Klaus et al. 2008a, b): (i) mainland fringing reefs; (ii) island fringing reefs; (iii) offshore "barrier-type" reef complexes, situated on a platform 1–14 km from the mainland, beyond a deeper coastal channel; (iv) a nearshore "barrier-type" reef, situated in the south near Shubuk; and (v) an assortment of other isolated offshore reef structures, including platform or "tower" reefs. According to the classification scheme of Rowlands and Purkis (this volume), these reefs would include various "attached" and "detached" reef systems, although not all of Sudan's reefs fit within the proposed scheme.

Mainland Fringing Reefs
A near-continuous fringing reef spans almost the entire length of the mainland coast of Sudan. The extent of reef development is variable (Fig. 4a–f), reflecting shifts in the antecedent topography, hydrodynamic (wind and wave) exposure, and freshwater inflows among other factors. The cross-reef profile from the shore and reef crest ranges from <200 to >2,500 m, increasing southwards. Shallow sediment filled lagoons of variable depth occur in embayments and provide a habitat for macroalgae, seagrass beds, corals and patch reefs within the deeper lagoons. The main discontinuities in the fringing reefs occur where there are passes associated with embayments or ancient river beds (variously called "marsa", "khors" and "wadis") or highly sedimentary coasts. The largest discontinuity occurs in front of the Tokar delta (18° 41′N to 18° 31′N) at the same latitude as the Suakin Archipelago, where the coast is formed from an alluvial fan and freshwater inflows are greatest.

On the northern Sudanese Red Sea, from the border with Egypt to Ras Abu Shagrab (at the southern tip of the Dungonab Peninsula), exposed steep classical fringing reef follows the contour of the coast and plunges from the reef crest to depths >50 m (Fig. 4a). The coast consists of uplifted undercut fossil reef rock, and lagoons have hard carbonate pavements with pockets of coarse sandy sediments and scattered coral colonies. Forereef slopes are typically rugose, with projecting overhangs, ledges and deep fissures, and colonised by massive and submassive *Porites* spp., *Galaxea fascicularis, Galaxea astreata,* faviids, occasional acroporids and soft corals (*Xenia, Heteroxenia, Sinularia, Lobophytum* and *Sarcophyton*) in sheltered locations. Breaks in the reef occur associated with steep-sided khors, carved into the fossil reef platform. Narrow shallow (<10 m) fringing reefs line the internal edges of these passes (Fig. 4a).

The reef systems around the southern tip of the Dungonab Peninsula consist of a complex network of patch and platform reefs among the islands, together with "dendritic" and "lineated" reef systems, according to the Rowlands and Purkis (this volume) classification scheme. The dendritic reefs extend perpendicular from the southern tip, while the lineated reefs orientate parallel to the southern shoreline. Coral cover in 2006 was low (15–21 %) and composed of small remnant colonies of *Porites* spp. massives, *Seriatopora, Pocillopora,* faviids, *Goniastrea, Echinopora* as well as *Mycedium, Merulina,* fungiids and acroporids.

Narrow sheltered fringing reefs line the internal edges the large deep-water embayment of Dungonab Bay and continue inside Khor Naiteb (Fig. 4b). These reefs extend to 3–8 m depth and withstand high water temperatures, high salinity (>40 ppt) and low-light conditions due to re-suspended sediments. Depth-related zonation patterns are compressed, such that shallow water species co-occur with species more characteristic of deep clear water communities, such as digitate, encrusting, branching and tabular acroporids, encrusting and massive poritids, faviids, mussids, Pectinids and oculinids, fungiids and soft corals. In 2006, these reefs were dominated by foliose, encrusting and branching forms of *Montipora* spp., as well as *Acropora* spp. and *Stylophora* spp., other common genera included *Porites, Pavona, Stylophora, Echinopora, Echinophyllia* and *Goniopora* (APF 2006). Coral and dense seagrass communities were present even at the furthest northern extent of the bay where water clarity was extremely low.

The coast south of Mohammed Qol to Port Sudan is partly sheltered by the shelf islands and offshore barrier reef-type complexes (Fig. 3c, d). The reefs become broader and shallower, and there are more embayments with fine soft sediment habitats, supporting seagrass beds and macroalgae. The reefs often have multiple fronts, with both inner and outer reef slopes, separated by a series of lagoonal pools of variable depths (Fig. 4c). Reef fronts are also often terraced descending in one or two steps, around 8–10 m or 15–20 m, before the slope ends on soft substrates in shallower water depths ($\sim$20–30 m) than to the north.

South of Port Sudan (Fig. 4d, e), the coast is partially protected by Towaritit reef and the Suakin shelf extension to the south. The distance from the shore to the reef edge exceeds 2,500 m, and the lagoons are deep enough to host a significant number of patch reefs. The reef is dissected by numerous reverted passes with both outer and inner slopes. Surveys at Suakin (Fig. 4e) and Marsa Sheik Ibrahim (Fig. 4f) in 2007 revealed similar reef structures and coral community composition. The outer forereef slopes were composed of dissected rock platform with terraces, at 8–10 and 15 m depth, and had a high per cent cover of tabular *Acropora*-dominated communities. Coral cover was highest at Marsa Sheik Ibrahim (55 %), where tiers of regularly sized tabular *Acropora clathrata* created a canopy shading a more diverse community of faviids, Pectinids and mussids beneath.

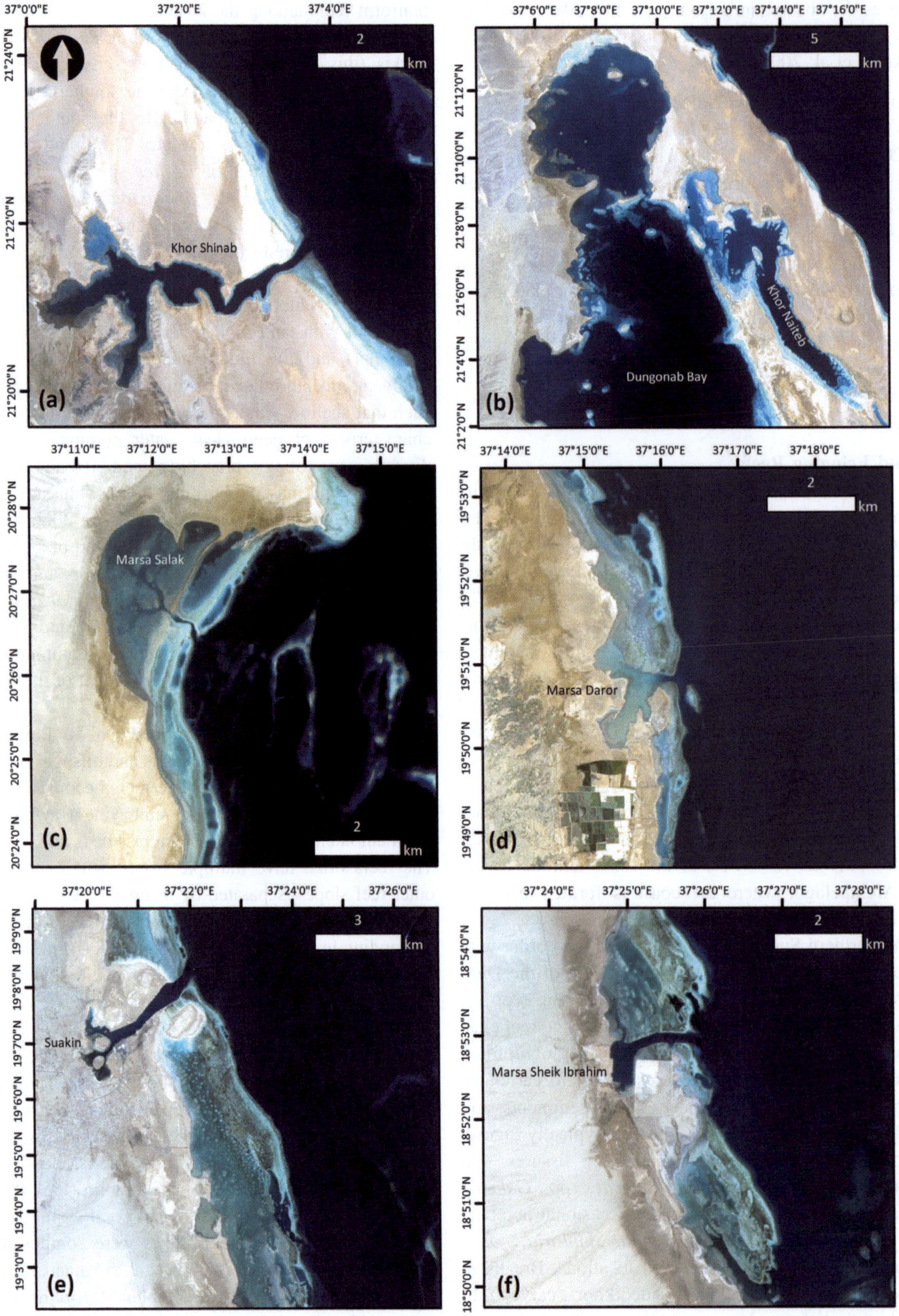

Fig. 4 Maps illustrating different morphologies of mainland fringing reefs within Sudanese waters types: **a** exposed fringing reef north of Dungonab Bay and the sheltered reefs lining the inner edges of Khor Shinab, and **b** sheltered reefs lining inside Dungonab Bay and Khor Naiteb. Moderately sheltered fringing reefs of **c** Marsa Salak and **d** Marsa Daror, and **e** Suakin and **f** Marsa Sheik Ibrahim. Maps **a**–**f** derived from landsat-8 satellite images. Data available from the US Geological Survey (http://landsat.usgs.gov/landsat8.php)

Island Fringing Reefs

There are three main groups of islands within Sudan's waters: (i) the northern shelf islands in the vicinity of Mohammed Qol (Mukkawar, Maytab Kebir, Maytab Sagir and Meshareif), (ii) the islets inside Dungonab Bay (Hysoit, Um-Tarda Island, Brasit Kebir and Um-ar Sheikh) and (iii) the offshore scattered islands of the Suakin Archipelago in the south. All the islands are uninhabited except for the occasional temporary fishing camp.

The islands near Mohammed Qol in the north are mostly low-lying (generally <1 m) with sand and halophytic vegetation overlying biogenic reef rock or slightly uplifted (1–2 m) undercut fossil reef. The two exceptions are Mukkawar Island and Mayteb Kebir, which are high rocky islands composed of uplifted sedimentary rocks and fossil reef. Mukkawar is the largest (12 km long by 3 km wide) and highest (90 m) island in Sudan. Mayteb Kebir has more modest proportions (1 km long by 0.4 km wide) and an elevation of 40 m. These islands sit on a shallow platform separated from the mainland by a deep-water channel and have fringing reefs and dendritic-like reef projections, most notably around Mukkawar Island. The islands that occur within Dungonab Bay itself are much smaller (<0.5 km^2) and surrounded by shallow fringing reefs typically dominated by reefs composed of massive *Porites* spp. (Fig. 4b).

The Suakin Archipelago in the south consists of a scattering of sand cays and coralline islands composed of fossil reef fringed by reefs, together with platform and tower reefs as described in the next section. These structures are situated in ~100–200 m water depth on an irregular shelf extension that most likely formed from the alluvial deposits from the Tokar delta when sea levels were lower. Fringing reefs around each islands had a similar profile and zonation pattern to the exposed classical fringing reefs, with moderately narrow reef flats and lagoons and steep terraced forereefs descending to >30 m. Various islands were surveyed in 2007 (Cousteau Society 2013; Klaus et al. 2008a, b), some of which are described below (Fig. 5a–d).

Seil Ada Kebir (Fig. 5a) is a small low-lying, teardrop-shaped coralline island encircled by a fringing reef, which widens northwards. The submerged reef flat shelves gradually onto a solid unevenly terraced forereef slope, with ledges at 8 and 15 m depth. Scleractinian species richness was among the highest recorded in 2007 (136 species) (Klaus et al. 2008a). The top of the reef to the south was colonised by *Stylophora* and *Pocillopora* colonies with dense turf algae, and large *Porites* bommies and *Millepora dichotoma* dominated the reef edge. Coral cover on the steep forereef slope was 46.6 % and included species characteristic of deeper water (low light) (e.g. *Leptastrea* spp. and *Pachyseris* spp.), together with *Acropora* tables, *Lobophyllia corymbosa* and faviids. The reef slope to the north was dominated by soft corals (26.2 %), particularly xeniids, interspersed by other encrusting organisms (sponges, *Palythoa*, ascidians). Hard coral cover was lower (24.8 %) and composed of *Porites, Pocillopora, Echinopora, Echinophyllia, Favites, Favia, Galaxea fascicularis* and *Acropora* (Klaus et al. 2008a).

Isle Harorayeat (Fig. 5b) is a pair of small, low-lying, vegetated islets, surrounded by fringing reef and separated by a narrow channel, with a lagoon ~100 m wide. In 2007, the reef slope to the south-east of the northern island descended onto a sand flat around 10 m and was colonised by tabular and staghorn *Acropora,* and scattered *Porites* bommies with a cover of 40.5 % (Klaus et al. 2008a). The reef slope in the channel between the islets was constructed from large (>3 m) massive *Porites* interspersed with thickets of staghorn *Acropora* beds and mixed with *Stylophora* and *Pocillopora,* with a cover of 50 % hard corals, ending on a sandy floor at 8–10 m (Klaus et al. 2008a).

Talla Talla Seghir (Fig. 5c) is a low-lying, oval island, encircled by a fringing reef that widens to the north-west, with a lagoon (30–200 m wide) composed of carbonate rock pavement and coarse gravely sand. On the south coast, the reef slope drops gently (30°) onto a plateau at 8–10 m. Coral cover was 46.4 % and dominated by acroporids (branching, staghorn, tabular), large (>5 m) massive *Porites* and *Diploastrea heliopora.* Similarly, on the northwest coast, the reef was dominated by large staghorn *Acropora* in the shallows and massive *Porites lobata* and *Porites lutea* ranging from moderate (1–2 m) to large (4–5 m) colonies on a plateau at 10 m depth. Coral cover was lower (21.1 %), and there was evidence of Crown-of-Thorns (CoTs) (*Acanthaster plancii*) impacts (Klaus et al. 2008a).

Talla Talla Kebir (Fig. 5d) is composed of three islets, enclosed within a broad fringing reef. In 2007, the reef flat was densely colonised by *Stylophora, Seriatopora, Pocillopora* and *Acropora* and the reef slope started around 3–4 m depth. The reef slope on the south-east coast was colonised by large patches of branching *Acropora* colonies dominated with smaller patches of *Pocillopora, Goniastrea* and faviids, and hard coral cover was 47.1 %. On the north eastern side of the island, the reef slope was composed of spur and groove formations between 4 and 12 m depth and ended on a sand and rubble floor at 20 m depth. Coral cover was 36.7 % and dominated by *Echinopora,* interspersed by *Acropora* and *Porites* colonies (Klaus et al. 2008a).

Nearshore "Barrier-type" Reef

In the vicinity of Shubuk (19°N) north of the Tokar delta, is a nearshore "barrier-type" reef system situated 15 km from the mainland (Fig. 5e). The most closely comparable reef formation is the "Little Barrier Reef" in Saudi Arabia (Sheppard 1985; DeVantier et al. 2000). The outer "barrier" at Shubuk runs parallel to the shoreline and adjoined a network of reticulated patch reefs separated by narrow channels, orientated perpendicular to the shore,

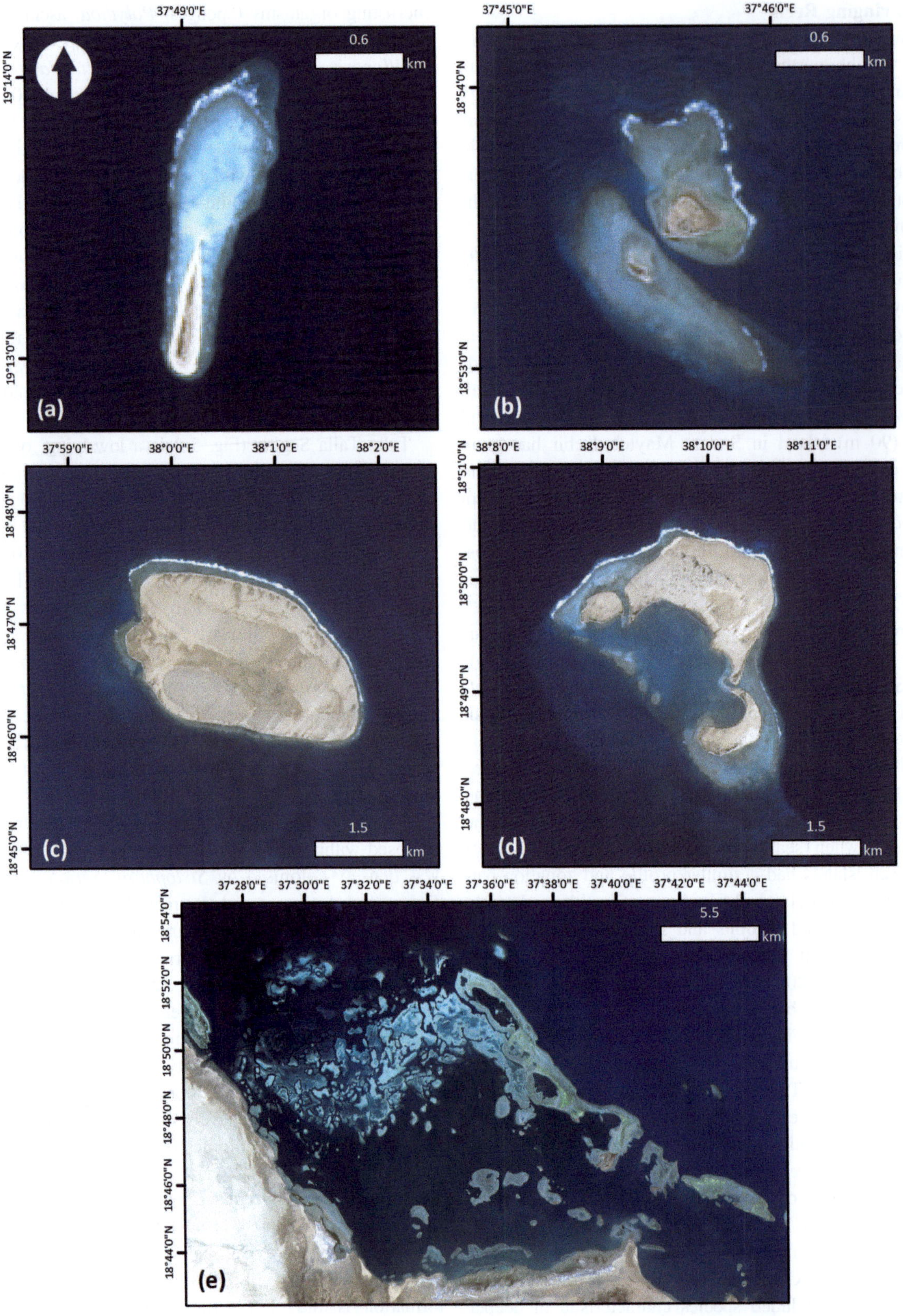

Fig. 5 Maps illustrating different morphologies of island fringing reefs within the Suakin Archipelago. **a** Seil Ada Kebir, **b** Isle Harorayeat, **c** Talla Talla Sagir, **d** Talla Talla Kebir and **e** the nearshore barrier reef at Shubuk, to the north of the Tokar delta. Maps **a–e** derived from landsat-8 satellite images. Data available from the US Geological Survey (http://landsat.usgs.gov/landsat8.php)

encompassing submerged patch reefs and islets with fringing reefs. The tops of these structures were colonised by *Sargassum* and *Turbinaria* macroalgae. Coral cover on the outer reef slope was 50.9 % and dominated by massive *Porites* and large (4 m) tabular *Acropora* colonies at the base of the reef at ~10–12 m (Klaus et al. 2008a). Water quality behind the "barrier" reef was highly turbid, and reef slopes typically ended at 5 m on silt. Coral cover on one of these reef slopes was the highest recorded for all sites in 2007 (64.8 %) (Klaus et al. 2008a). Coral community composition was dominated *Montipora* spp., massive and submassive poritidae, *and* other common species included the Red Sea endemic, *Mycedium umbra,* and other large polyped mussids, Pectinids and faviids (e.g. *Oxypora* spp., *Echinophyllia* spp., *Echinopora* spp. and *Goniopora* spp.) (Klaus et al. 2008a).

Offshore "Barrier-type" Reef Complexes

Offshore "barrier-type" reef complexes span the Sudanese coast from the border with Egypt in the north to south of Port Sudan and include (i) Sha'ab Elba, (ii) Sha'ab Halaka, (iii) Sha'ab Qumeira, (iv) Tailla reef complex, (v) Sha'ab Salak/Gorna reef complex, (vi) Sha'ab Sudi/Blue Belt, (vii) Wingate Reef and (viii) Towaritit Reef. These frameworks sit on top of a discontinuous shelf platform, which most likely formed as a fault block as a result of rifting during the Pleistocene, separated from the mainland by a deep-water channel with depths of between 70 and 400 m. Each complex is composed of an assemblage of irregular eroded pinnacle or columnar-shaped reefs, which coalesce at depth creating a larger framework. The seaward edge of each complex is formed by a series of these structures, which create the outer reef slope before the drop-off. Some of these complexes are composed of a type of reef that appears to be peculiar to the Red Sea, which Guilcher (1988) called "*ridge reefs*". Guilcher (1988) described these reefs from his observations around Port Sudan, Sinai (Egypt), the Farasan Islands and Jeddah (Saudi Arabia) and defined them as longitudinal ridges lying on the axis of the Red Sea, probably the result of a combination of normal faulting from the progressive opening of the Red Sea and upward movement of underlying salt domes (diapirs) along these faults (also see Rowlands and Purkis, this volume).

The coalesced exposed outer reef slopes of these complexes may be composed of large interlocking spurs ~5 m high and 6–7 m wide, separated by almost flat sand- and rubble-filled channels, while the inner reef slopes often have a more intricate topography. Zonation patterns on the outer slopes are similar to those found on exposed fringing reefs. At Sha'ab Suedi (Fig. 3c), live hard coral cover in 2007 (46 %) was dominated by *Acropora, Lobophyllia, Stylophora* and *Echinopora* and soft corals were also common (Klaus et al. 2008a). At Towaritit reef (Fig. 3e), the shallow parts of the spurs on the northern outer reef were composed of short staghorn acroporids and *Millepora dichotoma,* interspersed by carpets of xeniid soft corals. Live hard coral cover in 2007 was moderate (23–29 %), and there was evidence of a recent disturbance. Further north, within the Sha'ab Salak reef complex (Fig. 3d), although there were patches of healthy *Acropora,* live coral cover was lower (16–19 %) (Klaus et al. 2008a).

Platform, Tower reefs, "Atolls" and "Pseudo-Atolls"

Other isolated reef formations emerge from deep water within Sudan's Red Sea, and each reef has its own unique morphology. These include two of the most famous reef structures in the entire Red Sea, Sha'ab Rumi and Sanganeb Atoll, situated close to Port Sudan, and other structures occur in the vicinity of Dungonab and the Suakin Archipelago, some examples of which are described below.

Sanganeb (Fig. 6a) and Sha'ab Rumi (Fig. 6b) are semi-enclosed atoll-like structures situated in deep water to the north of Port Sudan. Sanganeb has been described by various authors (Mergner and Schumacher 1985; Schumacher and Mergner 1985; Reinicke et al. 2003). The use of the term "atoll" with reference to Sanganeb has remained controversial (e.g. Behairy et al. 1992), as similar reef structures are abundant along much of the Saudi Arabian Red Sea coast, though they are not usually referred to as atolls. Benthic cover in 2006 was dominated by hard and soft corals, as previously described (Reinicke et al. 2003), and live hard coral cover was highest on the sheltered outer reef to the east (47 %) (APF 2006).

Sh'ab Rumi (Fig. 6b) has a similar structure to Sanganeb. The outer reef slope on the west shelves steeply from ~1 m below the surface, onto a broad wide shelf at 8–12 m and at 30 m depth. The first of these terraces was the location of the Cousteau Conshelf project. The top part of the reef, from 1 m to 8 m, is steep heavily dissected and undercut, and the lower part is colonised by mixed hard and soft corals. Hard coral cover in 2007 was moderate (24–39.0 %) and matched by soft coral cover (25–28.5 %) (Klaus et al. 2008a).

Within the Suakin Archipelago, there are more isolated reef systems than islands, and these structures are found to a distance of 90 km offshore. Various reefs were surveyed in 2007 (Cousteau Society 2013; Klaus et al. 2008a, b) (Fig. 6c–f).

Sha'ab Anbar (Fig. 6c) is an 8.5-km-long elongated crescentic platform reef with a submerged reef flat surrounding a narrow deep central lagoon. The reef slope is terraced at 3–4 m depth followed by a near vertical wall. Live hard coral cover in 2007 was highest in the south (38 %) and dominated by *Acropora, Millepora, Platygyra, Porites, Echinopora, Goniastrea* and *Diploastrea,* accompanied by soft coral (13 %) (Klaus et al. 2008a).

Sha'ab Tawil (Fig. 6d) is a bean-shaped platform reef, with tail-like promontories that extend southwards. On the west, the reef shelves onto a sand platform covered with tabular *Acropora*. On the east, the reef slope was composed

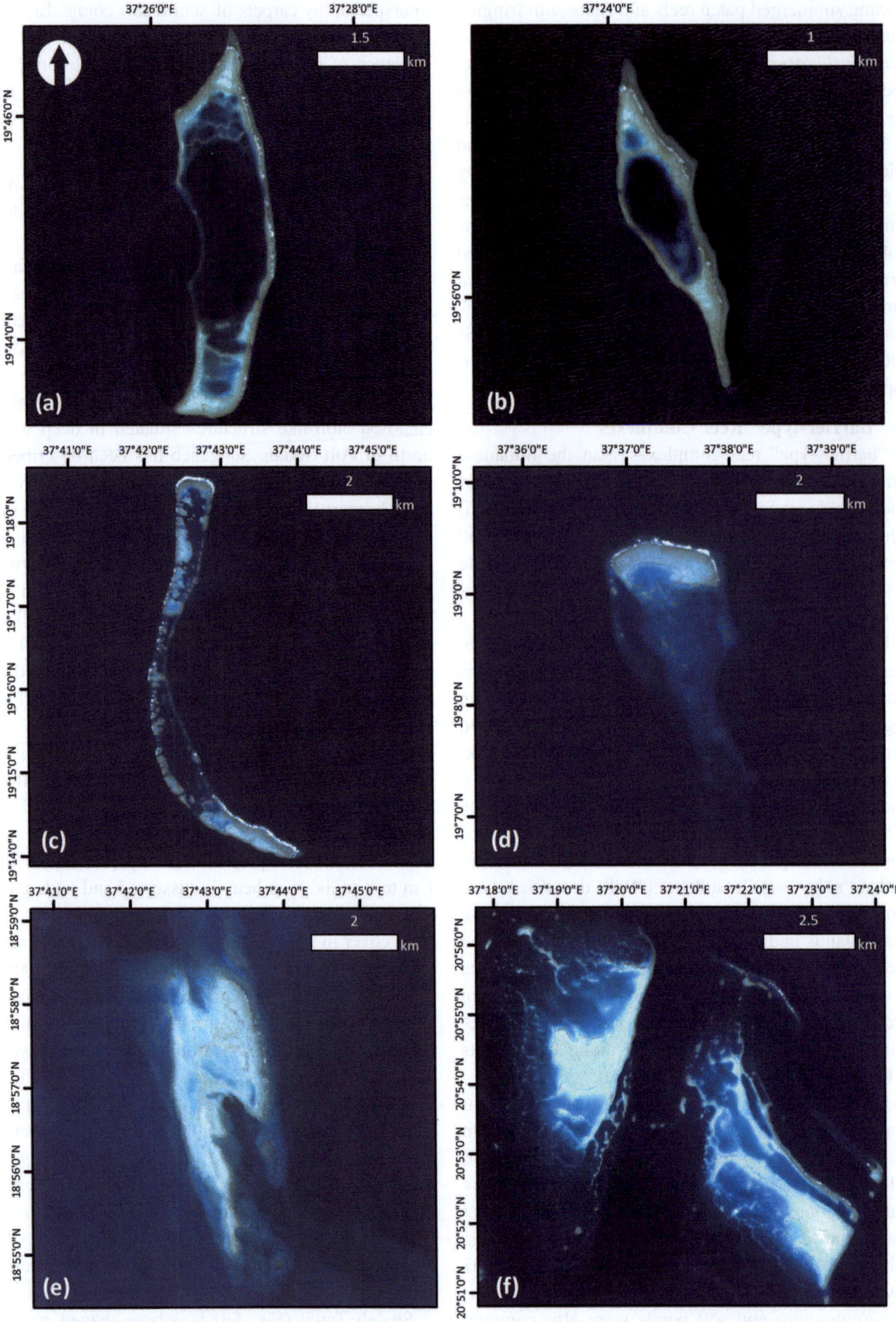

Fig. 6 Maps illustrating the diversity of isolated reef structures including the atoll-like structures **a** Sanganeb and **b** Sha'ab Rumi, and the platform reefs **c** Sha'ab Anbar, **d** Sha'ab Tawil, and **e** Dhanab Qirsh and **f** Shambaya in the north. Maps **a–f** derived from landsat-8 satellite images. Data available from the US Geological Survey (http://landsat.usgs.gov/landsat8.php)

of short spur and grooves between 6 and 10 m depth, with an interlocking series of patch reefs with variable topographic relief separated by sand channels (Klaus et al. 2008a).

Dhanab Qirsh (Fig. 6e) is an irregular oblong-shaped platform reef orientated north–south with a deep reticulated lagoon and two tail-like promontories to the south. The outer reef slope shelves onto a platform at around 7–10 m, and both the slope and platform are colonised by mixed acroporiid communities and massive *Porites*. On the eastern promontory, there were extensive terraces of tabular *Acropora* at ~9 m. The inner reef of the western promontory was gently sloped (10°) and covered in massive *Porites lobata* colonies, and dense beds of staghorn, tabular and corymbose *Acropora* spp. (Klaus et al. 2008a).

Status and Trends

Sudan is within what is considered to be the centre of biodiversity in the Red Sea, and scleractinian hard corals provide the dominant cover on most reefs (Fig. 7). Scleractinian species richness is high, with 265 species of stony coral recorded to date (in 2007), including 3 species of *Millepora* (and a further 23 unconfirmed records), from 56 genera, and 15 families (Klaus et al. 2008a, b), which is more than the number reported for the Saudi Arabian Red Sea (DeVantier et al. 2000) and Yemen (Turak et al. 2007) to date. Sites with the highest species richness often did not have the highest live coral cover, as these sites tended to be species poor communities in sheltered turbid areas.

Total hard coral cover on the reefs of Sudan ranged from 15 to > 65 % (35 %) although soft corals provide the dominant cover (~30 %) on some reefs (Klaus et al. 2008a, b), as observed elsewhere in the Red Sea (Reinicke et al. 2003). The fire coral, *Millepora,* although common, was often found in low abundances except on shallow exposed reefs (maximum cover 6 %). Fleshy macroalgae were not a significant component of benthic cover, although brown fucoid macroalgae *(Sargassum* sp. and *Turbinaria* sp.) were more common in the south.

Comparison of the survey results from 2002, 2006 and 2007 (Kemp et al. 2002; APF 2006; Cousteau Society 2013; Klaus et al. 2008a, b) revealed certain trends. In 2002, stony hard coral cover (including *Millepora* fire corals) was moderately high (38 % in 2002), but there was evidence of a recent (<5 years) severe disturbance event, with a high abundance of dead standing coral, and banks of rubble colonised by turf algae at some sites. The sites with the highest remaining coral cover were on exposed platform reefs (e.g. Shambaya, Sha'ab Qumeira and Merlot) and sheltered low-light sites (e.g. Dungonab Bay, Khor Shinab and Khor Naiteb).

The 2002 survey was the first broadscale systematic assessment of Sudanese reefs after the 1997/1998 widespread coral bleaching event that caused mass mortality of corals in the Indian Ocean and elsewhere due to anomalously high sea water temperatures and calm conditions over the summer months. Corals on fringing and offshore reefs in the Dungonab area were observed to bleach in 1998 (Andrea Bari of Dungonab Pearl Oyster Farm, pers. comm., February 2002). Indeed, the reefs were reported to be "*so bright white as to be visible from the shore*", while the weather conditions were "*calm and very hot*" (Kemp et al. 2002; Klaus et al. 2008b). Researchers from the Institute of Marine Science in Sudan and tourism operators also reported observing white corals (PERSGA/GEF 2003).

Coral cover at the sites surveyed around Dungonab remained stable between 2002 (ranging from 16.8 to 67.3 %, with an average 38.9 % for 18 sites) and 2006 (ranging from 15.4 to 61.5 %, averaging 32.4 % for 18 sites) (Kemp et al. 2002; APF 2006). Comparison of the 2006 and 2007 results revealed that stony coral cover was lower at sites north of Port Sudan (inshore 32.4 %, offshore 30.4 %) than in the south (inshore 49.7 %, offshore 39.3 %) (APF 2006; Klaus et al. 2008a, b). Many reefs south of Port Sudan supported diverse communities with a higher cover of stony hard corals, including bleaching susceptible genera that had been impacted further north. The results suggested that corals south of Port Sudan may not have been subject to severe bleaching conditions as occured in the north. Sites in the south with high coral cover also had higher abundances and diversity of corallivorous butterfly fishes (Cousteau Society 2013; Klaus et al. 2008a, b).

Other threats to the coral reefs of Sudan that have been recorded to date include corallivorous species, nuisance species (Fig. 8), fisheries-associated impacts, coastal development, maritime transportation, marine litter, and land-based sources of pollution (PERSGA/GEF 2003; APF 2006; Cousteau et al. 2013). The majority of the anthropogenic threats have only had localised impacts and are restricted to areas in proximity to Port Sudan, with the exception of fisheries-related impacts which were more widespread. Surveys using underwater visual census techniques in 2002, 2006 and 2007 found extremely low abundances or absences of many commercially important species of finfish, such as parrot fish, large snappers, groupers (*Epinephelus* spp. and *Plectropomus* spp.) and emperors (lethrinids), which are targeted by handline fisheries in Sudan (Kemp et al. 2002; APF 2006; Klaus et al. 2008a, b; Cousteau Society 2013). Low abundances or absences of commercially valuable gastropod molluscs and holothurians were also recorded, indicating widespread overfishing of these important macroinvertebrate groups.

Low abundances or complete absences of the common sea urchins (*Echinometra* spp. and *Diadema* spp.) were also recorded from many reefs, which was more unusual. Historical data on the distribution and density of sea urchins in the Red Sea, especially *Diadema setosum*, are limited. These black-spined urchins were abundant in sheltered locations,

Fig. 7 Photographs illustrating coral communities characteristic of the Sudanese Red Sea, including **a** shallow reef community dominated by *Acropora* sp. and **b** *Pocillopora* sp.; **c** Island fringing reef dominated by massive *Porites* (Dungonab Bay); **d** Fringing reef slope dominated by *Porites* sp.; **e** mixed *Acropora* spp., *Stylophora* spp., *Pocillopora* sp. community; **f** dominated by a high abundance of *Galaxea astreata;* and **g** shallow forereef under low-light (high-turbidity) conditions, inside Khor Naiteb. Photographs by J. Kemp

Fig. 8 Some of the commonly observed threats to corals observed in the central and southern Red Sea: **a** "Partial" coral bleaching on the upper surface of a *Goniastrea* coral colony, which are paler than the sides of the colony; **b** sediment collected at the centre of foliose coral colony (*Mycedium* sp.), **c** Red *Cliona* spp. sponge infestation; **d** grey (*Terpios* cf. *hoshinota*) sponge infestation on a colony of *Pocillopora* sp. **e** Crown-of-thorns starfish (*Acanthaster plancii*) feeding among a bed of staghorn *Acropora*, and the recent white feeding scars are clearly evident; **f** Coral colony damage caused by the corallivorous gastropod mollusc *Drupella* spp. Photographs by R. Klaus

such as the lagoon at Sha'ab Shubuk in 2007, but they were absent from many offshore reefs (APF 2006; Klaus et al. 2008a), as had previously been reported for Sanganeb (Schumacher et al. 1995). Reef tops without urchins were, uncharacteristically for the Red Sea, overgrown with small algae, a possible consequence of reduced grazing from both parrot fish and sea urchins. Whether or not the absence of urchins from the offshore atolls is a natural pattern due to a lack of connectivity with the nearshore systems requires further investigation. Minor CoTs outbreaks have been recorded in Sudan since the 1970s, and although they were recorded at low abundances during recent surveys, CoTs outbreaks were only recorded at two locations in the Suakin Archipelago (Klaus et al. 2008a, b). The corallivorous gastropod mollusc *Drupella* was also present at low densities and generally only observed at high densities in shallow areas of high density of *Stylophora* corals (Kemp et al. 2002; APF 2006; Cousteau Society 2013).

Sudan has declared two MPAs, Sanganeb atoll marine national park (SMNP) and Dungonab and DMNP (Table 2). SMNP was gazetted for nomination in 1994 and DMNP was gazetted in 2004. Both have been recommended as a potential World Heritage Site, although SMNP was first proposed in the 1980s (Vine and Vine 1980), it is still only on the tentative list for nomination. Neither MPA is effectively managed, and both require further support. Other areas that have been proposed for protection include the Suakin Archipelago and Abu Hashish. The Suakin Archipelago is one of several important island groups in the central and southern Red Sea, which warrants protection given the higher diversity of scleractinian corals and healthy condition of most reefs. The geographical span of the Archipelago substantially increases the range of environmental conditions to which reefs are exposed within the Sudanese Red Sea. The reefs apparently survived the previous 1997/1998 bleaching event, which increases their importance, and their location in proximity to deeper fresher well-mixed water may confer greater resilience under future climate change scenarios.

Eritrea

Eritrea's Red Sea coast extends over 1,350 km of the mainland from 18°N at the border with Sudan in the north to 12° 45′N at the border with Djibouti in the south (Fig. 9). Coastal waters are dominated by the broad shallow Dahlak platform, which is set off from the mainland by a deeper channel. The platform bulges to nearly 120 km offshore in the central region and narrows to ~20 km at the northern and southern extremes. The platform and central and southern coast are littered with 354 islands, which increase the total coastal length to 3,300 km (Hilman and Tsegay 1997; Jeudy de Grissac and Negussie 2007).

The mainland coast is mostly composed of tertiary or quaternary sedimentary deposits, overlaid in places with more recent basaltic lava flows. The nearshore and offshore islands are of carbonate, granitic and volcanic origins. The volcanic coastal formations and islands in the south are associated with the Red Sea Rift and Afar Rift, which are both parallel to the coast and have numerous fault lines. The carbonate islands are the eroded uppermost layer of the Plio-Pleistocene platform that was split by tectonics and then underwent irregular uplift caused by the underlying Miocene evaporite deposits (salt domes) (Angelucci 1981; Angelucci et al. 1985). The extent of uplifted fossil reefs within the Dahlak Archipelago is evidence that conditions were highly favourable for coral growth in the past (Angelucci 1981).

The coastal climate today is hot and arid. The coastal plains are crossed by more than 20 wadis, the most significant of which to the north of Massawa (Falcat, Karora, Hum, Abibado, Misho, Wakiro, Desset and Lebca) originate from the central and northern highlands and/or the eastern escarpment. Freshwater flows can reach this section of coast during the summer (July to September). To the south of the Buri Peninsula, the major river (Ramod-Regalle) drains into the Danakil depression. Small wadis originating from the top of the ridges separating this area from the Danakil depression feed into the coastal plain but have negligible impact on the coast.

Early studies concluded that conditions in Eritrean coastal waters were unsuitable for coral growth due to the shallow platform, highly sedimentary conditions, and seasonal upwelling, all of which were thought to favour algal growth over reef development (Oren 1962; Lewinsohn and Fishelson 1967). Wainwright (1965) commented on the "*scarcity of reef-building corals and the total lack of living, structural coral reefs in the Dahlak Islands*". More recent surveys found biogenic reefs along much of the mainland coast and around many of the islands, with reef development being more pronounced in proximity to deeper water, along the northern and southern mainland and around outer edges of the platform.

Eritrean coral reefs have been assessed during a number of projects including surveys of Dahlak Kebir and Norah Islands by Tel Aviv University, Asmara University and the Research Section of the Ministry of Fisheries (MoF) (1993), the University of Wales, Aberystwyth (UK) and MoF (1994); rapid assessment surveys conducted by the UNDP-GEF funded preparatory phase for the "*Eritrean Coastal Marine and Islands Biodiversity*" (ECMIB) project (1995–1997) implemented by the MoF; surveys around Massawa by the University of Newcastle (UK) and the MoF (1997); and most recently by the Eritrean Coastal Marine and Islands Biodiversity (ECMIB) project funded by GEF implemented by UNDP and by MoF (2004–2007). The following account is compiled from the reports from these projects, and in particular, the unpublished reports from the preparatory (1996–1997) (summarised in Hilman and Tsegay 1997) and implementation phase (2005–2007) of the UNDP-GEF funded ECMIB project, during which 96 coral reef stations

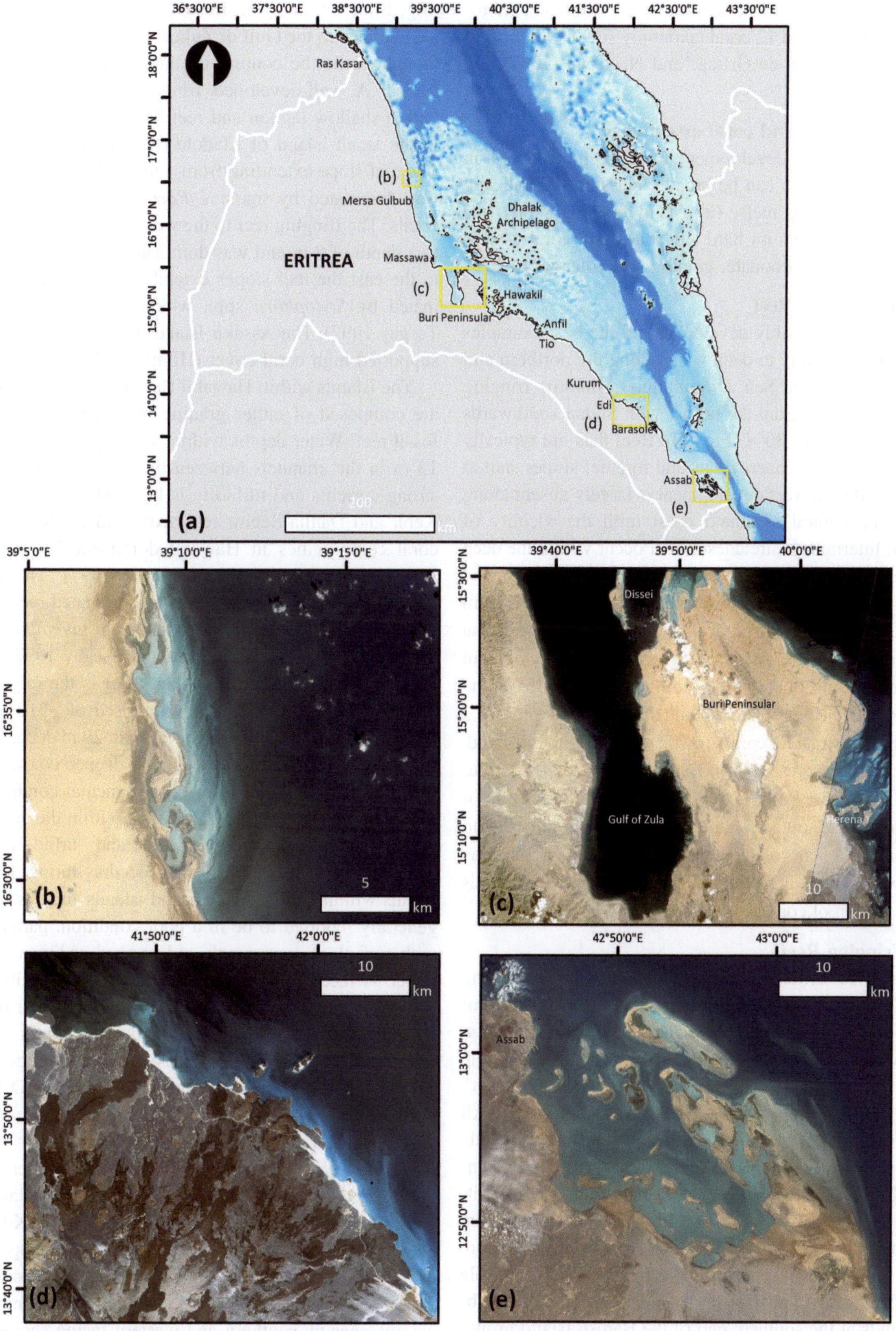

Fig. 9 Map illustrating **a** the coast of Eritrea and the mainland fringing reefs **b** north of Massawa, **c** Gulf of Zula and Buri Peninsula, **d** volcanic coastline between Edi and Barasole and **e** Assab. Maps **b–e** derived from landsat-8 satellite images. Data available from the US Geological Survey (http://landsat.usgs.gov/landsat8.php)

were surveyed (68 Reef Check surveys, 16 video and photo-quadrant transects, and 12 coral taxonomy sites), which were summarised by Jeudy de Grissac and Negussie (2007) and Tilot et al. (2008).

Types of coral reef and coral assemblage

Three types of reef development have been observed in Eritrean waters, which can be broadly classified as follows: (i) mainland fringing reefs, (ii) island fringing reefs, (iii) submerged patch reefs on hard substrates and coral assemblages growing on carbonate, granitic and volcanic rock.

Mainland Fringing Reef

Fringing reefs are weakly developed, and the best examples are found in proximity to deep water along the northern and southern Eritrean Red Sea. Discontinuous shallow fringing reefs edge the coast from the border with Sudan southwards to approximately 16° 30′N (Fig. 9b). Reef flats are typically dominated by fucoid macroalgae, and forereef slopes start at about 1.5 m depth. Reef development is largely absent along much of the central mainland coast until the vicinity of Massawa. Intermittent stretches of reef occur within the deep Gulf of Zula and around the Buri Peninsula (Fig. 9c).

South of the Buri Peninsula, from Tio (14° 36′N) to Kurum (14°N), the coast is sedimentary, backed by a broad alluvial plain, and both coral growth and fringing reef development are limited. From Edi (13° 54′N) to Barasole (13° 36′N), the coast is volcanic (Fig. 9d), and while the lava flows provide a more stable substrate for coral growth, reef build-up is limited. Fringing reefs are present from Barasole (13° 36′N) south to Assab (13°N) (Fig. 9e). The patchy discontinuous nature of the reef along the southern mainland coast is partially due to the presence of unconsolidated mobile sediments but also due to the exposure to upwelling. Further research along this southern stretch of coast is warranted.

Island Fringing Reefs

The islands of Eritrea can be divided into the following groups: (i) the nearshore deep-water islands around the tip of the Buri Peninsula and the entrance to the Gulf of Zula (e.g. Dissei, Madote, Shumma, Assakir Black and Assakir White islands) (Fig. 10a); (ii) the nearshore "shallow-water" islands, near Massawa (Sheik Sheid) and inside Hawakil Bay and Anfil Bay, on the eastern side of the Buri Peninsula (Hawakil, Baca, Dehlia, Aguiz, Delesen, Dergman) (Fig. 10b); (iii) the islands of the Dahlak Archipelago, situated on the shallow shelf platform (Fig. 10c); (iv) carbonate islands near the port of Assab (Fig. 10e); (v) the nearshore volcanic islands between Kurum (13° 30′N) and Barasole (14°N) (Fig. 10d); and (vi) the scattered volcanic Mohabbaka Islands at the southern end of the Hanish Island group.

The nearshore deep-water islands to the west and north of the Buri Peninsula are composed mostly of Precambrian granitic basement rock. The islands of Dissei and Madote (Fig. 10a), for example, situated west of the Buri Penisular at the entrance to the Gulf of Zula, are part of an undersea ridge thought to be the continental threshold of the rift valley in Africa. A well-developed fringing reef surrounds Dissei, with a shallow lagoon and reef flat that extends northwards to the small island of Madote (Hilman and Tsegay 1997). The reef slope extending from 4 to ~14 m with a 45° slope was dominated by massive *Porites* corals and xeniid soft corals. The fringing reef to the west around Madote extended to a depth of 8 m and was dominated by *Galaxea* sp., while to the east the reef slopes extended to 13 m and was dominated by *Stylophora* spp., which later died (Hilman and Tsegay 1997). The Assakir Islands (Fig. 10a) also reportedly supported high coral cover (Hilman and Tsegay 1997).

The islands within Hawakil Bay (Fig. 10b) and Afnil Bay are composed of either granitic basement rock or uplifted fossil reef. Water depths within both bays range from 0.5 to 13 m in the channels between the islands, where there are strong currents and turbidity is high. The islands of Dahlia Kebir and Dahlia Seghir reportedly had the best developed coral communities in Hawakil dominated by *Porites* and *Acropora* spp. (Hilman and Tsegay 1997). The island of Herena was encircled by a *Porites*-dominated reef, although *Acropora* had previously been abundant given the extent of branching coral rubble (Hilman and Tsegay 1997).

The Dahlak Archipelago (Fig. 10c) is the largest island group in the entire Red Sea, comprised of >210 low-lying islands, situated on a platform less than 20 m deep, separated from the mainland coast beyond a deeper coastal channel between 60 and 80 m deep. Environmental conditions tend to be less favourable to coral growth within the archipelago, where waters are shallowest, hot and turbid, and aerial exposure of the benthos often occurs during low tides. Corals within the Dahlaks around islands such as Norah are generally reported to be in a poor condition, partially dead, with turf algal overgrowth and trapped sediments on their upper surfaces (Hilman and Tsegay 1997). Healthier coral communities were found on patch reefs in deeper pools and channels between the islands and coral cover, and reef development increases towards the outer edges of the Dahlaks. For example, an island 170 m south of Dehil in the northern Dahlaks, which is situated in deeper water (~30 m) reportedly had a well-developed reef slope with high coral cover that increased with depth to 16 m and then ended at 20 m on bare sand (Hilman and Tsegay 1997). Islands in the far east of the Archipelago (e.g. Aucan, Dhu-Kurush and Mojeidi) and to the south-east (e.g. Museri) supported patchy coral growth on a sandy reef slope on north facing shorelines, with healthier, more extensive and diverse assemblages on southern and western facing shores.

The volcanic islands (e.g. Fig. 10c) further south did not exhibit significant reef development, but some of other islands in the far south near Assab had well-defined reefs

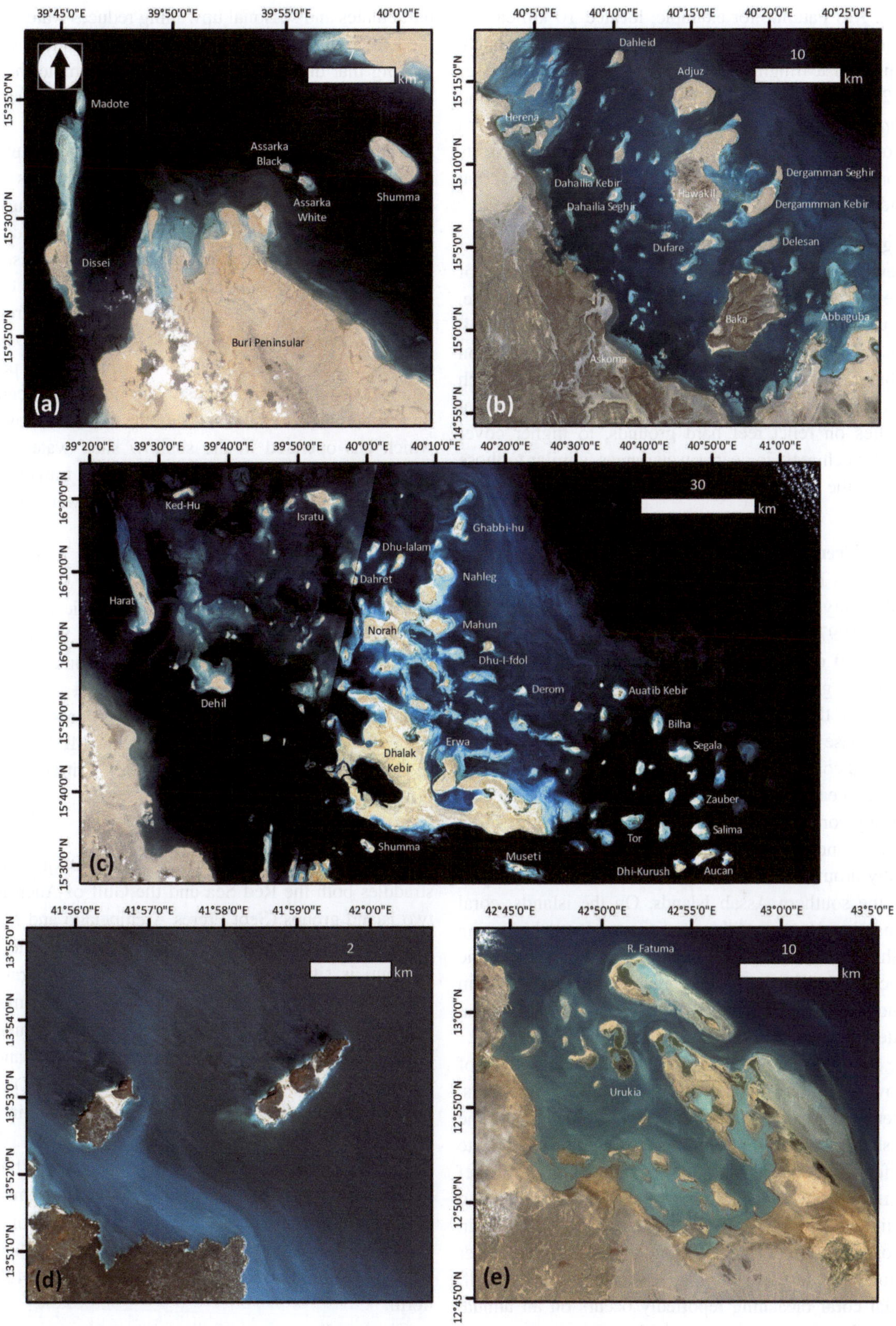

Fig. 10 Maps illustrating different morphologies of island fringing reefs in Eritrean Red Sea including **a** the islands of Dissei, Madote, Assarka Black and White and Shumma; **b** the Hawakil Archipelagoo; **c** the Dahlak Archipelago; **d** the volcanic islands near Edi and Barasole; and **e** Assab. Maps **a–e** derived from landsat-8 satellite images. Data available from the US Geological Survey (http://landsat.usgs.gov/landsat8.php)

(Fig. 10e). Ras Fatuma, for example, located 10 km east of Assab is the second largest island in the area next to Haleb. The island had a fringing reef with significant coral cover beneath 2 m depth, dominated by massive *Porites* spp. and *Stylohphora* spp. (Hilman and Tsegay 1997). The southern section of Eritrean coast and islands reportedly supports fewer coral species but often with moderate-to-high coral cover and health.

Submerged Patch Reefs and Coral Assemblages

While several unpublished reports mention the presence of patch reefs with coral assemblages growing on carbonate, granitic and volcanic rock substrates, there is limited detail available on the composition of the benthic communities on these submerged structures. Reports suggest that these patch reef communities range from low diversity, low cover coral communities on relict reef hard grounds, to higher cover monotypic patch reefs in deeper water areas, similar to those described in the shallow waters of the southern Farasans in Yemen.

Status and Trends

Recent surveys have increased knowledge about the distribution, diversity and status of the reefs of Eritrea (Hilman and Tsegay 1997; Zekeria 2003; Tilot et al. 2008) (Fig. 11). A total of 220 corals species have been recorded to date, from 39 coral genera (Tilot et al. 2008). Further surveys, particularly at the edges of the Dahlak shelf, will likely further increase the number of species recorded. Coral condition and coral cover ranged from 10 to 67.3 % (Zekeria 2003) up to nearly 100 % in some locations (Tilot et al. 2008). Stony coral cover and biogenic accretion both tend to be greater around the islands than on the mainland, but more specifically around the outer edges of the Dahlak Archipelago and the southern Assab Islands. On the islands, coral cover tended to be vary with orientation and was higher on the south and west than on the north facing coasts. The occurrence of fucoid macroalgae and seasonal variation in the abundance of macroalgae is a characteristic of Eritrean reefs (Ateweberhan et al. 2005a, b).

There are numerous constraints on the extent and depth of coral growth, and biogenic accretion within Eritrean waters. The extensive areas of unconsolidated sediments on the shallow shelf platform limit both the availability of substrate suitable for coral recruitment and the potential depth of reef development. Abrasion, smothering and episodic burial, and high sedimentation rates are all factors that can constrain coral growth and reduce coral condition. Both sea water temperatures and turbidity are higher over the shallow shelf, and partial coral bleaching reportedly occurs on an annual basis in shallow areas. Mass coral bleaching was the suspected cause of the mortality of many *Acropora* spp. and *Stylophora* spp. beds in recent years (Hillman and Tsegay 1997). Wind- and wave-driven re-suspension of fine particulates and seasonal upwelling reduces water clarity and the availability of PAR. The seasonal variations in nutrient loading that drive the upwelling also promote the seasonal increase in the abundance of macroalgae, which can compete with corals for space.

The fact that corals in Eritrean waters exist under such extremely stressful ambient conditions increases their susceptibility to other disturbances. Fortunately, much of Eritrea's extensive coast and numerous islands remain relatively pristine and unimpacted by human disturbances. The coastal population has remained low due to the hot inhospitable climate and limited access to fresh water (Hillman and Tsegay 1997). Tourist numbers have also remained low on account of the prolonged periods of political instability and unrest. The most significant impact on Eritrean reefs to date has been the industrial trawl fishery, which has operated in the shallow shelf waters for many years. Regulations were drafted to restrict trawling to water depths >30 m, but it is not known if these were implemented (Hillman and Tsegay 1997).

The Eritrean government identified four islands as potential MPAs (Dissei-Madote, Museri, Ras Fatuma and Sheikh Seid) on the basis of the ecological and potential tourism value. The plan was to develop Sheikh Seid and Dissei–Madote islands as Eritrea's two first MPAs, which may later be expanded into a larger area around the Buri peninsula (Tilot et al. 2008). Further progress towards realising the full implementation of these MPAs would help secure the rich and unusual habitats Eritrean coral reefs provide.

Djibouti

The Republic of Djibouti has a coast 370 km long, which straddles both the Red Sea and the Gulf of Aden and hosts two island groups (Sept Frères Archipelago and Musha and Maskali islands) (Fig. 12). The majority of the coast of Djibouti is situated within the Gulf of Tadjourah, a deep east-west oriented triangular trench with a maximum depth of 883 m. The mouth of the Gulf of Tadjourah starts near the border with Somalia in the south (11° 30′N) and extend 60 km north across to Ras Bir (11° 54′N). The capital, Djibouti town, is situated on the south coast within the Gulf, to the north of which are the Musha and Maskali islands. At the far western end of the Gulf connected to the Gulf of Tadjourah by a narrow channel is a shallow semi-enclosed basin, Ghourbet al-Kharab. The Red Sea section of coast, the focus of this account, extends for 160 km from Ras Bir (11° 54′N) in the south to the border with Eritrea (12° 42′N) in the north.

The Red Sea coast of Djibouti faces the narrow Straits of the Bab-al-Mandab (Fig. 12a, b). The most prominent feature along the otherwise near-straight coast is the rocky headland of Ras Siyyan (12° 27′N) (Fig. 12c). Ras Siyyan marks the

Fig. 11 Photographs illustrating coral communities characteristic of the Eritrean Red Sea, showing **a–b** a mixed *Stylophora* spp., *Porites* spp. and branching sponge community, with macroalgae; **c** faviid- and poritiid-dominated community with large colonies of *Echinopora fructilosa* sp. and massive *Lobophyllia* sp.; **d** *Montipora* spp.-dominated community; **e** *Porites*-dominated community, with evidence of previous impacts; and **f** a *Merulina*-dominated community, with branching sponges and fungiids on sand and rubble. Photographs by R. Klaus

narrowest point in the Strait of Bab-al-Mandab and is only 30 km from Ras Menheli on the Yemen coast. The Strait itself is composed of two channels, divided by the Yemeni island of Miyurn (Perim) (Fig. 12b). The western channel (Dact-el-Mayun) is 25 km wide and 310 m deep, and the eastern channel (Bab Iskender) is 3 km wide and 29 m deep. The narrow strait has a shallow sill, which is thought to have acted as a barrier during previous eustatic sea level changes. Climatic and oceanographic conditions within the Strait act as a semi-permeable barrier during the present day due to the influence of the restriction on the seasonal exchange of water masses between the Red Sea, Gulf of Aden and northern Indian Ocean waters. Situated west of Ras Siyyan is the Sept Frères Archipelago (Siyyan Himar, Hamra, Boeing, Ile du

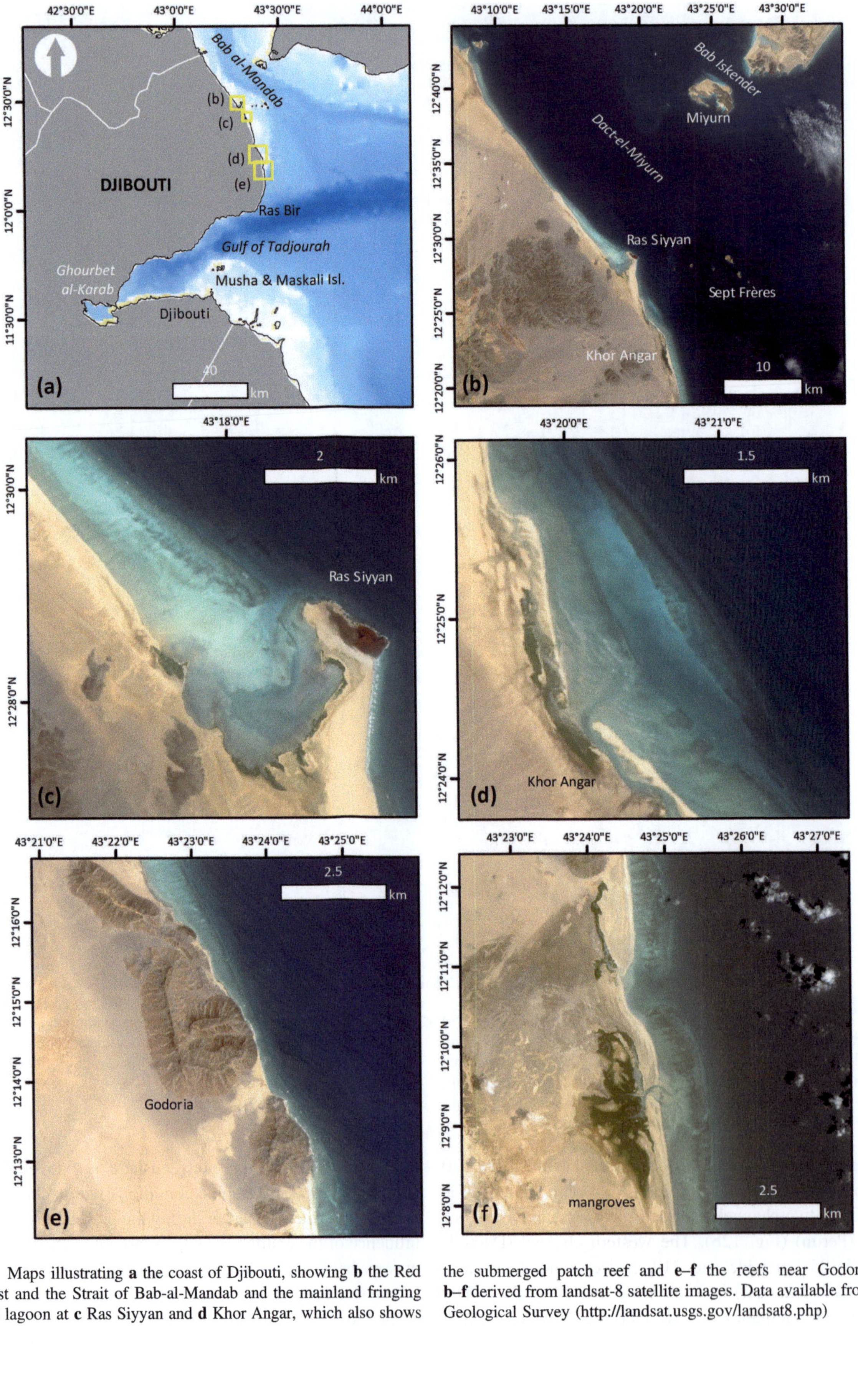

Fig. 12 Maps illustrating **a** the coast of Djibouti, showing **b** the Red Sea coast and the Strait of Bab-al-Mandab and the mainland fringing reef and lagoon at **c** Ras Siyyan and **d** Khor Angar, which also shows the submerged patch reef and **e–f** the reefs near Godoria. Maps **b–f** derived from landsat-8 satellite images. Data available from the US Geological Survey (http://landsat.usgs.gov/landsat8.php)

Bas, Tolka, Grande Ile, Ile de l'est and Ile du Sud). South of Ras Siyyan is Khor Angar, which empties fresh water and sediments into the Red Sea and the basaltic flows near Godoria (Fig. 12e).

Various reef assessment surveys in Djiboutian waters have been completed in the past decade (Obura 1998; PERSGA 1998; PERSGA/GEF 2003; Barker et al. 2002). Hunting Aquatic Resources completed surveys on behalf of PERSGA (PERSGA 1998; PERSGA/GEF 2003) at 185 locations at Moucha and Maskali islands, Khor Ambado, Djibouti, Sable Blanc, Godoria, Khor Angar, Sept Frères and Ras Siyyan. Obura (1998) completed surveys at 21 sites in the Gulf of Tadjourah, the Moucha and Maskali islands, and the Sept Frères Islands. Further surveys were also completed by the same author at Moucha and Maskali islands, Khor Ambado, Les Trois Plages, Sable Blanc, Ras Duan, Sept Frères, Recif d'Ambouli and a site off Tadjourah (PERSGA/GEF 2003). During the PERSGA MPA-SAP, 173 rapid assessment surveys and 6 quantitative surveys were completed in 2002 to characterise and map the coastal and marine habitats and associated communities at Sept Frères and the adjacent mainland coast (Barker et al. 2002).

Types of coral reef and coral assemblage

Shallow hard substrates along the Red Sea mainland and island coasts of Djibouti support coral communities. There are three main reef types, (i) mainland fringing reef, (ii) submerged patch reefs and (iii) island fringing reefs, and various different types of non-accreting coral "assemblages" on non-biogenic rock around the offshore islands.

Mainland Fringing Reefs

A long stretch of near-continuous fringing reef borders the Red Sea coast of Djibouti, extending over 35 km from Ras Doumera, at the border with Eritrea (Fig. 12a), to Ras Siyyan and south to Ras Bir. North of Ras Siyyan, the reef is backed by a broad reef flat and shallow lagoon, which widens southwards (Fig. 12a). The lagoon is predominantly shallow (1 m to 4 m deep) and sandy, providing habitat for scattered patches of macroalgae (*Padina* spp., *Caulerpa* spp., *Sargassum* spp. and *Turbinaria* spp.) and seagrass, and sparse (<5 %) but extensive *Pocillopora* sp. coral communities on patches of hard substrate (Barker et al. 2002).

Several discontinuities occur in the fringing reef. North of Ras Siyyan, a deep-water reef-edged channel leads into the southern end of the lagoon (Fig. 12c). The fringing reef continues as a narrow band around the northern and eastern edge of the headland, colonised by predominantly branching *Acropora* spp. and massive *Porites* spp. communities. Reef development weakens around southern edge of the headland, then increases and continues along the coast to Khor Angar, the location of the next major discontinuity (Fig. 12d). Coral cover along this section of reef ranged from 10 to 55 % with a mean value of 25 % (Barker et al. 2002). South of Khor Angar, the fringing reef continued south to the basaltic flows near Godoria (Fig. 12e). There is another major discontinuity by the large stand of mangroves to the south of Godoria (Fig. 12f), before the reef continues south to Ras Bir.

Submerged Patch Reefs

Situated approximately 1,500 m offshore from Khor Angar is a submerged elliptical-shaped patch reef, approximately 200 m wide and 3,500 m in length, orientated parallel to the mainland fringing reef, separated by a broad sand-filled channel (Fig. 12d). The top of the reef was between 2 and 3 m depth, and live stony coral cover was 38 %, dominated by *Acropora* spp. and *Pocillopora* spp. (Barker et al. 2002).

Island Fringing Reefs and Coral Assemblages on Volcanic Rock

Fringing reef development around the Sept Frères Islands is limited and interspersed by volcanic rock terraces and lava flows with non-accreting coral assemblages. Around each island, rocky substrates descend directly from the shoreline into the subtidal zone (Fig. 13a–d). Between 1 and 2 m depth, the substrate was typically covered with a thin layer of algal turf, while the deeper slope was colonised by *Porites* spp. and faviids. Between 3 and 15 m depth, branching and tabular *Acropora* spp. and *Pocillopora* spp. were more common. At the base of the slope, there was a broader platform with a sand-covered hard bottom with coral rubble. Outcrops and pinnacles on the platform provided additional hard substrate suitable for the growth of sparsely distributed faviids, *Acropora* spp. and *Porites* spp. The southern exposed coastlines were more directly affected by wave action, and although macroalgal cover was limited during the 2002 surveys, this may increase after the upwelling.

Hamra is a near-circular volcanic island, 380 m long by 260 m wide, situated 4.5 km from Ras Siyyan (Fig. 13a). In 2002, massive *Porties* sp.-dominated assemblages were found on the shallow hard substrates, with >60 % live hard coral cover (Barker et al. 2002). A shallow submerged projection to the south of the island was mapped as part of PERSGA MPA-SAP and requires further survey (Barker et al. 2002).

Boeing, a steep-sided oblong-shaped island, 400 m long by 250 m wide, situated 2 km east of Hamra (Fig. 13b), has a steep rocky shoreline that extends into the subtidal. The rocky slope shelves at an angle of 60° between 5 and 30 m and steeply to >50 m. Coral cover off the south-east coast in 2002 was dominated by massive *Porites spp.* and xeniid soft corals which provided >20 % of the total cover (Barker et al. 2002).

Tolka, a three-pronged island, 280 m long by 300 m, situated 3 km east of Boeing at the centre of the Sept Frères (Fig. 13b), has a shallow fringing reef and reef flat. In 2002, coral cover on the reef slope ranged between 32 and 66 %, the composition of which was dominated by branching

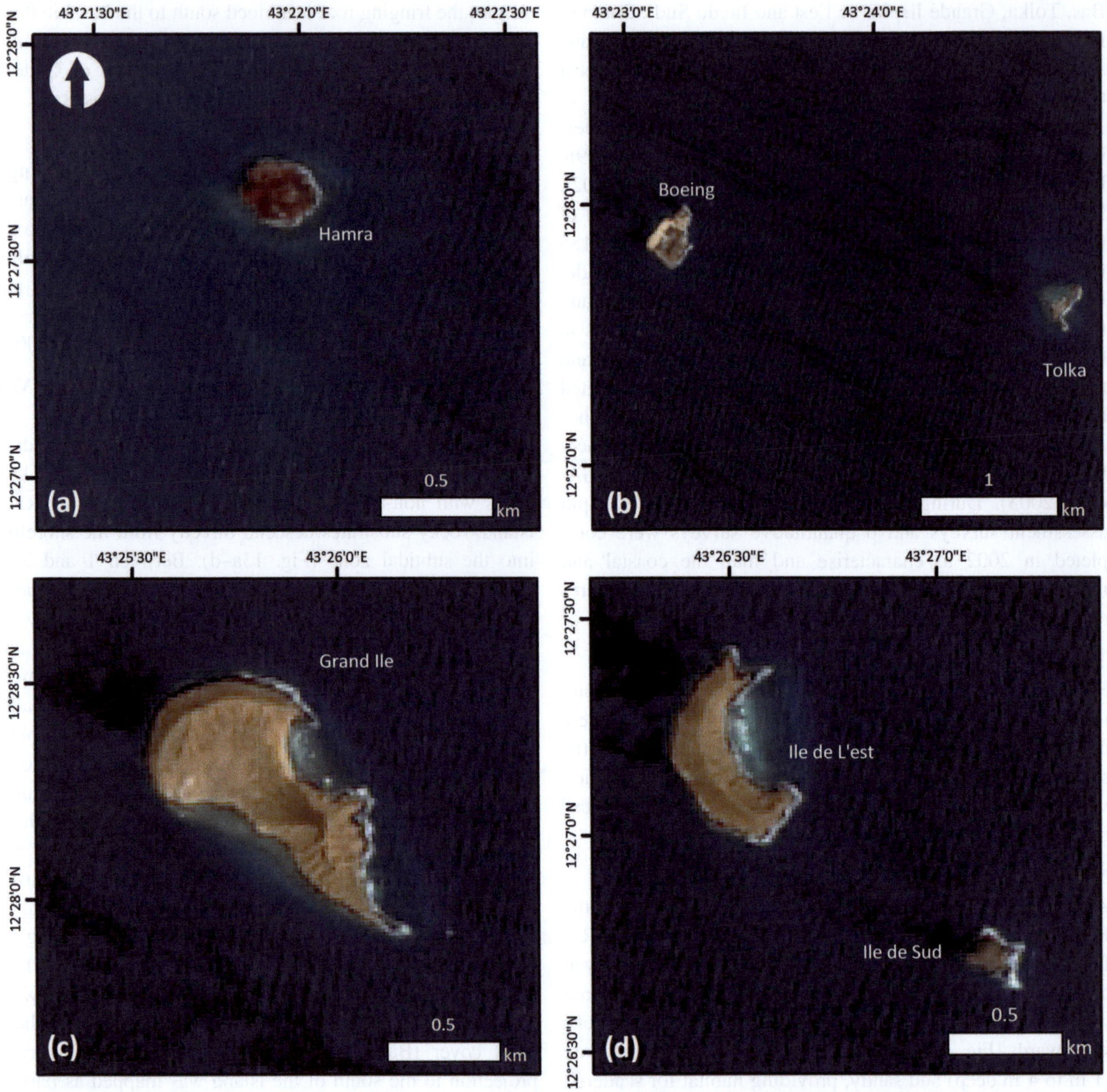

Fig. 13 Maps illustrating the islands of the Sept Frères Archipelago in the Djiboutian Red Sea including **a** Hamra, **b** Boeing and Tolka, **c** Grand Ile, **d** Ile de l'Est and Ile de Sud. Not shown is the seventh island in this group, Siyyan Himar, which is situated north of Hamra. Maps **a–d** derived from landsat-8 satellite images. Data available from the US Geological Survey (http://landsat.usgs.gov/landsat8.php)

(18 %) and tabular *Acropora* spp. (8.8 %), *Porites* spp. (11 %) and xeniid soft corals (12 %).

Grande Ile, the largest of all the islands in the Sept Frères, 1,300 m long by 700 m wide, situated 1.7 km north east of Tolka (Fig. 13c), has a rocky shoreline apart from two sandy beaches on the northeast and southwest coasts. Diverse coral communities colonised the shallow subtidal zone, dominated by massive *Porites* spp., branching and tabular *Acropora* spp., other *Stylophora* spp. and small massives. Hard coral cover in the south was 90 % in 1998 (PERSGA 1998) and 65 % in 2002 (Barker et al. 2002). Coral communities around Grand Ile showed evidence of recent disturbances, most likely due to anchor damage, but also CoTs, which were evident in 2002 (Barker et al. 2002).

Ile de l'Est, a steep-sided crescent-shaped island, situated 1.2 km south-east of Grande Ile (Fig. 13d), has a rocky shoreline apart from a narrow sandy beach on the east. The fringing reef encircling the island was dominated by a *Porites* reef that created a narrow discontinuous reef flat. The slope dropped gradually from 5 to 15 m forming an inclined plateau composed of sand and rubble and colonised by larger coral colonies. From 15 m the slope steepens and coral cover was dominated by *Porites* spp. and *Acropora* spp. Average hard coral cover was 48 %, and soft coral cover was dominated by xeniids (9 %) (Barker et al. 2002).

Ile de Sud the most southerly of the Sept Frères (Fig. 13d) only has a weakly developed reef. The slope inclined steeply at an angle of 60° towards a sandy bottom with coral patches. Cover by hard coral was dominated by massive *Porites* spp. and branching *Acropora* colonies and ranged between 13 and 64 %, with a mean cover of 47 %.

Status and Trends

Recent surveys have revealed the characteristics, composition and status of coral communities within Djibouti's Red Sea (Barker et al. 2002; Obura 1998; PERSGA 1998; Kotb et al. 2004, 2008). The wide fringing reef bordering the mainland coast is one of the best developed reef structures in the southern Red Sea, broken only by deep-water channels, areas of mobile sediment and freshwater input (Khor Angar and Godoria). A total of 167 coral species (including 3 species of black coral) have been recorded for the whole of Djibouti to date, with the highest diversity in the Red Sea section being found at Ile Grande (84 species) (Obura 1998). The diversity of corals found around the Sept Frères Islands may be due to their location in the midst of the Bab-al-Mandab, at the junction between two biogeographical provinces.

Coral cover in Sept Frères was >50 % at all sites in 1998 and 1999, ranging from 20 to 90 % at Hamra in Sept Frères (Obura 1998). Coral bleaching was not observed in Djibouti in 1998, but repeat surveys in late 1998 recorded 30 % bleaching-related coral mortality around Sept Frères, with no mortality observed in the Gulf of Tadjourah (Obura and Djama 2000, cited in PERSGA/GEF 2003). In 2002, dead hard coral was low at most sites within the Sept Frères, although there was evidence of previous bleaching-related impacts at Tolka and Boeing (Barker et al. 2002). Despite this, the surveys in 2002 found total hard coral cover ranging from 52 % to a maximum of 72 % at Tolka Island, and average live hard coral cover was 59.8 % (Barker et al. 2002). Coral cover around the Sept Frères was lower than reported by Sheppard et al. (1992) who observed a luxuriant acroporid-dominated community with close to 90 % cover at Tolka Island in Sept Frères. Coral cover was however lower (38 %) on the submerged barrier patch reef at Khor Angar on the mainland. Sheltered locations around Grande Ile and Khor Angar also showed evidence of mortality due to physical impacts, such as anchor damage (Barker et al. 2002).

Djibouti's reefs are under threat from domestic tourism, sewage discharges, shipping and associated spills and pollution. Anthropogenic pressures are particularly high around the capital city and lower in more remote locations, such as Sept Frères in the Red Sea (Obura 1998). Damage associated with tourism from boating, and the use of anchors, is increasing, although international tourism is still just developing. There is low-level subsistence fishing and limited exploitation of fish for live export, but aquarium fish collecting is increasing and there are other transboundary challenges due to illegal fishing by Yemeni fishers that target shark in Gulf of Aden and Red Sea. Shipping is an important commercial sector, and Djibouti is the major harbour for Ethiopia. A new port is currently under construction which will likely further increase the level of impacts.

The Government of the Republic of Djibouti has declared various MPAs. The Moucha Territorial Park ("Parc Territorial de Musha"), covering an area of 3 km^2, and South Maskali Islands Integral Reserve ("Reserve Integrale de Maskali Sud"), covering an area of 10 km^2, were designated in 1972 and 1980, respectively. In 2003, new legislation was prepared for the development of a national network of protected areas (Presidential Decree 2001-0098/PR/MHUEAT of 27 May 2001), which included the areas of Isles des Sept Frères–Ras Siyyan–Khor Angar and Godoria. A management plan was prepared for the Ras Siyyan–Isles des Sept Frères and Godoria complex during the PERSGA MPA-SAP (PERSGA 2004). The other two MPAs are both situated on the south coast and include Haramous to the east of Djibouti town and Art Plage to the west, but the present legal status of these areas is uncertain. The enforcement of all MPAs is at present limited, and reportedly, the areas are still used by artisanal fishers and subject to other threats such as anchor damage. Further support is now needed to help increase the effective implementation of these MPAs.

Yemen

The Republic of Yemen, located on the south-west corner of the Arabian peninsula, has a coastline of 2,200 km, 735 km of which borders the Red Sea, extending from the Saudi Arabian border (16° 23′N) to the Strait of Bab-al-Mandab (12° 30′N) (Fig. 14a). The mainland coast is mostly sedimentary, with a few rocky outcrops, and backed by the Tihama plain that extends from the high mountain ranges of western Yemen (elevation 3,800 m) to the shore. The Tihama is an important agricultural area, composed of large alluvial fans, coarse sediments and dunes, and soils with a high loam and silt content. Freshwater drains from the mountains through wadis but rarely reaches the coast except during extreme flood events, due to the high levels of evaporation across the plain.

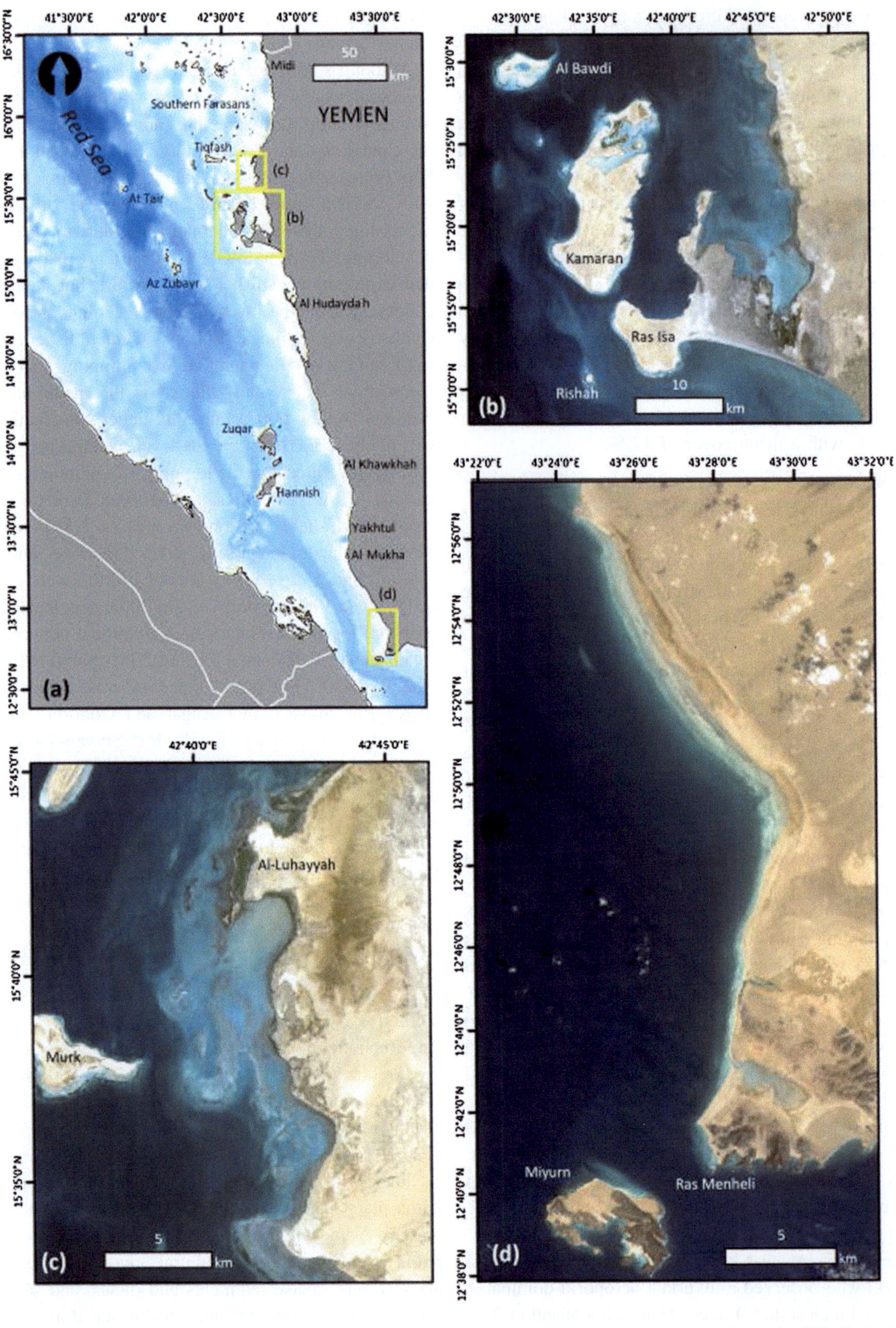

Fig. 14 Maps illustrating **a** the Red Sea coast of Yemen and mainland fringing reefs **b** around Ras Isa, **c** al-Luhayyah and **d** the southern coast to Ras Menheli at the Bab-al-Mandab. Maps **b–d** derived from landsat-8 satellite images. Data available from the US Geological Survey (http://landsat.usgs.gov/landsat8.php)

The northern mainland coast of Yemen is sheltered by a swarm of >100 islands, often referred to as the southern Farasan Islands, which are scattered across the shallow continental shelf extension between Ras Isa and Midi (Fig. 14a). The shelf itself extends to 80 km offshore near Hanish Island before narrowing to 20 km to the south of al-Mukha. South of Ras Isa, the coast is more exposed, except in the vicinity of the two large sand spits to the north and south of Al Hudaydah, which create more sheltered sedimentary habitats.

Early studies concluded that the broad shallow continental shelf was dominated by soft mobile sedimentary habitats, with poor coral diversity and reef development. The extent of contemporary coral growth and biogenic reefs throughout much of the Yemeni Red Sea was only revealed recently through survey work completed in 1986 (Barratt et al. 1987a), 1996–1998 (Turak and Brodie 1999; Turak et al. 2007), 2002 (Hassan et al. 2002), 2004 (Klaus and Eisinger 2004; Krupp et al. 2006) and 2005 (Zajonz et al. 2005).

Types of Coral Reef and Coral Assemblage

Three main types of reef development are found in the Yemeni Red Sea including (i) mainland fringing reefs, (ii) island fringing reefs and (iii) submerged patch reefs, and three types of non-accreting coral assemblage on (i) carbonate relict Pleistocene/Holocene reef deposits, (ii) volcanic rock, lava flow terraces and pinnacles and (iii) red coralline algal reefs (Turak and Brodie 1999; Turak et al. 2007). The distribution and benthic composition found on these structures are further described below.

Mainland Fringing Reefs

Mainland fringing reefs are typically constrained to a maximum depth of 3–4 m or 5–6 m by the presence of mobile soft sediments and the shallowness of the shelf. Major discontinuities in the reef occur where there are wadis, even though these only transport sediments and fresh water during flash floods. The most substantial areas of biogenic reef accretion occur in the north from al-Luhayyah to the headland at Ras Isa (15° 12′N) (Fig. 14b, c) and in the south between al-Khawkhah (13° 55′N) and Ras Menheli (12° 40′ N) at the Bab-al-Mandab (Fig. 14d). The southern reef is the longest continuous stretch of reef in the Yemeni Red Sea, mirroring the reef development on the Djiboutian coast, before continuing around Ras Menheli into the Gulf of Aden (Turak and Brodie 1999).

The cross-reef profile can vary in width from 100 to 1,000 m and may include a distinct lagoon and reef flat and reef crest. Lagoons where present are typically shallow (0.2–1.5 m depth), sand and rubble filled with small patches of seagrass (*Thallasodendron* sp., *Cymodecea rotundata* and *Halodule uninervis*), macroalgae (*Padina* sp., *Sargassum* sp., *Caulerpa* sp. and *Halimeda* sp.) and scattered coral colonies (*Porites* sp., *Stylophora pistillata*, *Montipora* spp. and faviids). Reef flats vary between 10 and 200 m in width and are typically composed of rubble and sand, although they may be composed of solid rock pavement and support coral growth, red coralline algae or macroalgae in more exposed locations. Reef crest development is limited, and more typically reef flats merge gradually into a more solid forereef slope, which then descends onto a flat sandy bottom between 3 and 6 m deep. Coral growth tends to be more vigorous in waters >1–5 m deep. In spring, there is often an increase in the growth of fleshy macroalgae such as *Sargassum*, *Padina* and *Dictyota* spp. on the top of the reef and in the lagoon which then dies back in the summer.

Turak and Brodie (1999) reported that the reefs near al-Luhayyah and Khawr Kathib showed little evidence of recent coral growth, and macroalgae provided the dominant cover with the exception of the Khawr Kathib. Surveys around Ras Isa in 2004 and 2005 revealed a healthy fringing reef with good coral growth, and a well-developed reef flat, reef crest and reef slope extending to 4–5 m depth (Klaus and Eisinger 2004; Zajonz et al. 2005). The Ras Isa reef is particularly well developed on the southern and south-western side of the headland and continues around the western tip of the peninsula. Fringing reef is also present around Ras Marsa to the south of Ras Isa (Fig. 14b).

The structural characteristics and community composition of the Ras Isa reef vary with exposure (Klaus and Eisinger 2004; Zajonz et al. 2005). The lagoon is of variable width and depth and composed mostly of limestone plateau with small patches of coral growth where the water is >1 m. Lagoon coral communities in the east were dominated by *Porites* colonies, often partially dead and overgrown by dense mats of zoanthids or the brown algae *Turbinaria* sp. Other areas are dominated by *Stylophora pistillata*, which may extend over the reef crest and reef slope to 2–3 m. The fore reef slopes are dominated by massive *Platygyra*, *Turbinaria* and *Goniopora* and the organ pipe octocoral *Tubipora musica*. The southern more exposed reef slope exhibited a low profile "spur and groove" orientated into the prevailing conditions interspersed by sand- and rubble-filled channels. The spurs are either overgrown with faviids, *Porites* spp. and xeniid soft corals, or with *Stylophora pistillata*, or more diverse coral communities with large colonies of *Montipora foliosa*, and acroporids. Further north coral growth is dominated by *Porites lobata* and *Acropora* cf. *clathrata* colonies, and *Stylophora* sp. Macroalgal growth was limited, and stony and soft corals were dominant or exhibited a mix of all the various taxa groups (Klaus and Eisinger 2004; Zajonz et al. 2005).

Turak and Brodie (1999) also described the reefs south of al-Khawkhah to the Bab-al-Mandab (Fig. 14d). The reef slope base was at 3–4 m deep, except near al-Mukha where there was a healthy patch of *Platygyra* colonies on a sandy

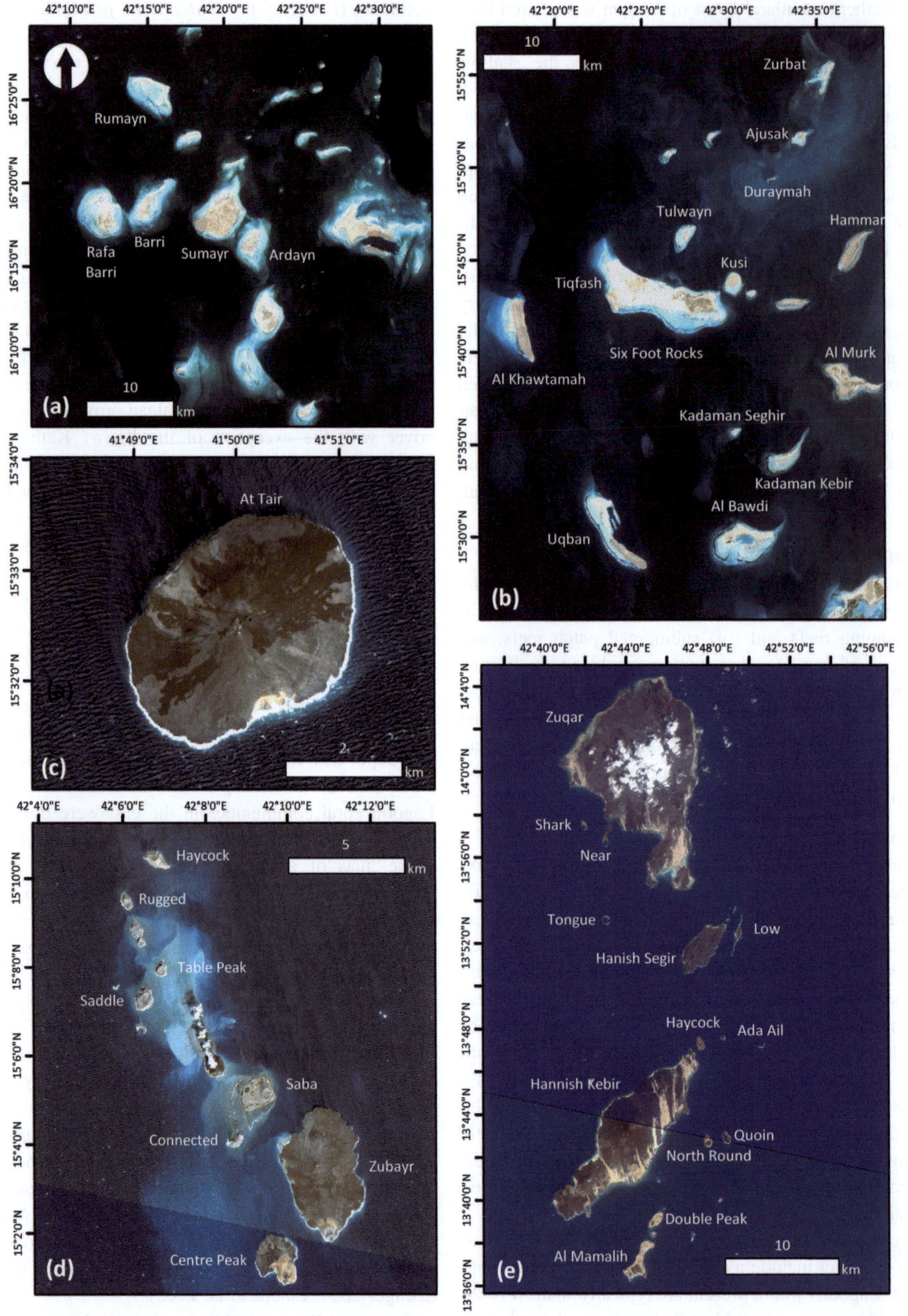

Fig. 15 Maps illustrating the island groups in the Yemeni Red Sea including **a–b** the inshore islands of the southern Farasans, including the Kamaran islands, and the offshore volcanic islands of **c** At Tair, **d** Az Zubayr Island group and **e** the Hanish and Zuqar Island group. Maps **a–e** derived from landsat-8 satellite images. Data available from the US Geological Survey (http://landsat.usgs.gov/landsat8.php)

bottom at 5–6 m depth, in relatively clear water. Massive *Porites* provided the dominant framework, although *Stylophora pistillata* was common and occurs together with stands of tabular and branching *Acropora* spp., subbranching *Montipora* and small massive and encrusting species. Other common species included *Platygyra daedalea, Acropora valida* and *A. humilis, Porites nodifera, Pavona cactus, Galaxea fascicularis, Acanthastrea echinata, Hydnophora* spp., *Favia favus, Goniastrea retiformis, Leptastrea purpurea* and *Echinopora gemmacea*. Barratt et al. (1987a) described a *G. fascicularis*-dominated coral community, which reportedly stretched for several kilometres south of al-Mukha. Subsequent surveys have yet to rediscover this reef, and it may have since died and been overgrown by algae (Turak and Brodie 1999).

Island Fringing Reefs

There are five main groups of islands in Yemen that can be broadly subdivided into the inshore shelf islands and offshore volcanic islands. The inshore shelf islands include (i) the islands of the southern Farasans, which includes the Kamaran Islands (Fig. 14a, b). The offshore volcanic islands are separated from the shelf by a deep basin (1,600 m) and include from north to south (ii) At Tair (Fig. 15c), (iii) the Az Zubayr Island group (Fig. 15d), (iv) the Hanish and Zuqar Island group (Fig. 15e) and (v) Miyurn (Fig. 14d). Biogenic reefs occur around most of these islands with the exception of At Tair. The extent of reef build-up varies, and where present, the reefs have the appearance of an extended reef flat with a gradual slope composed of topographically complex patches of reef substrate interspersed by sand and rubble areas.

The islands of the southern Farasans are situated in relatively shallow water ($\sim$40 m) on the shelf extension between Ras Isa northwards to the border with Saudi Arabia (Fig. 15a, b). The islands are composed of uplifted fossil reefs and range in size from <0.1 to 112 km^2. Kamaran (Fig. 14b), the largest island in Yemeni waters, is 22 km long (north to south) and 6 km wide (east to west), and the southern tip is <2 km away from the headland at Ras Isa. The highest inshore island of Tiqfash is 35 m above sea level (Fig. 15b), while other islands (e.g. al-Bawdi and Kadaman) are much lower with an elevation of 8–10 m.

The southern Farasans (Fig. 15a, b) are all at least partly edged by fringing reef, with the most extensive reef development being on the southern or southwest coast, although this may vary depending on the shape of the island and degree of exposure. Some of the islands close to shore are nearly fully encircled by fringing reef (e.g. al-Bawdi, Zurbat, al-Murk). Island reef profiles are similar to those found along the mainland. From the shore, there is a shallow lagoon with macroalgae or seagrass. The reef front may drop abruptly from <1 m to a sandy flat at 2–4 m deep or may have a gradual broken slope to the sand flat below.

In 2004, live hard coral cover ranged between 5 and 50 % around the Kamaran Islands (Klaus and Eisinger 2004). At Uqban Kebir, the fringing reef on the east was predominately covered with large tabular *Acropora clathrata*, branching acroporids (51.3 %) interspersed with carpets of the soft coral *Xenia* sp. and patches of bare substrate consisting of a mixture of sand and rubble (Klaus and Eisinger 2004). Other common coral species included the massive *Platygyra daedelae*, massive *Porites, Stylophora pistillata*, and foliose and encrusting *Montipora* spp. There were little algae or evidence of recent disturbance.

The offshore islands are high volcanic islands typically located in >500 m water depth, although the two largest islands, Zuqar (elevation 624 m) and Hanish al-Kebir (elevation 407 m), are situated in shallower water ($\sim$100 m) (Fig. 15e). The island of At Tair and the Az Zubayr Island group (Zubayr, Quoin, Haycock, Rugged, Table Peak, Saddle, Low, Saba, Connected, Centre Peak) are situated on the mid-Red Sea fault line (Fig. 15d). Fringing reefs are generally less well developed around these islands, but they are present around the Hanish and Zuqar Island groups and Miyurn to the south.

The Hanish and Zuqar Islands (Zuqar, Tongue, High, Shark Island, Pile and Quoin, Hanish al-Seghir, Hanish al-Kebir, Haycock, Ada Ail and Low) are approximately 50 km offshore from al-Khawkhah on the mainland (Fig. 15e). These islands were not surveyed during the first phase of the UNDP-GEF project (Turak and Brodie 1999), but were surveyed in 2004 (Klaus and Eisinger 2004). Hanish al-Kebir is approximately 18 km long and 6 km wide. The island is arid, almost entirely unvegetated, and there are no known sources of fresh water. One small patch of mangrove was found on an islet off the north-west coast, but there was no evidence of seagrass communities around this island group. A substantial portion of the northwest coast is edged by a narrow shallow discontinuous fringing reef broken by sandy bays. These reefs shelve directly from the shore, without the development of a reef flat, and continue on a gentle slope (20°–30°) to 8–15 m depth and end on sand. Coral communities typically consist of massive *Porites* in the shallows, merging into dense stands of branching acroporids between 5 and 8 m, with larger tabular acoporiids and more diverse massive and encrusting corals towards the base of the reef slope. In other locations, communities may be dominated by *Acropora* and *Montipora* corals, or *P. nodifera*.

Miyurn Island (Fig. 14d) is a volcanic island situated in the narrowest part of the Strait of Bab-al-Mandab is more exposed and subject to different oceanographic conditions than the islands further north, and is subject to periodic nutrient-rich cold-water upwelling from the Indian Ocean (PERSGA/GEF 2003). Reef accretion, where present, occurs on the basal volcanic rock and appears at least several metres thick on the more sheltered northern side, where there are extensive reef flats with diverse and healthy coral growth (Turak and Brodie, 1999). Monospecific *Stylophora* communities dominate on the eastern and western side, while there are *Porites*-dominated communities on the southern side, all of which survived the 1997/1998 coral bleaching event (Turak and Brodie 1999).

Submerged Patch Reefs
Turak and Brodie (1999) described large submerged patches west of Al Hudaydah, south of Ras Isa, and scattered between the southern Farasan Islands. The shallower parts of these patches were 6–8 m below the sea surface, although some patches grow to just below the sea surface and may be exposed during low tides. These patches were dominated by massive *Porites*, but other species of massive and encrusting growth forms, such as faviids and mussids, were also common. Turak and Brodie (1999) suggested that these patch reefs provide a substantial proportion of the total surface area of contemporary coral reef accretion in the Yemeni waters.

Coral Growth on Relict Reef Formations
The majority of coral growth in the Yemeni Red Sea occurs on relict Pleistocene or Holocene carbonate rock (Turak and Brodie 1999). These types of coral community were common in the southern Farasans, and community composition was dominated by massive *Porites* spp., *Stylophora pistillata* and *Goniastrea retiformis*. Other common corals included *Acropora hemprichi, A. clathrata, A. humilis* and *A. pharaonis, P. cactus, G. fascicularis, Lobophyllia* spp. and other encrusting, massive and foliaceous faviid species (Turak and Brodie 1999).

An example of this type of community was found on the south-west corner of Tiqfash Island (Fig. 15b), where coral growth started at 4 m depth and continued onto a plateau to 6–7 m depth (Klaus and Eisinger 2004). At 5–6 m depth, the rocky slope was colonised by densely coalescing coral colonies (62.7 %), interspersed by narrow sand-filled channels. Cover was dominated by massive *Porites* spp., foliose *Echinopora lamellosa, Sylophora pistillata,* tabular *Acropora clathrata*, branching acroporids, *P. cactus* and *Cyphastrea micropthalma,* and macroalgae and soft coral cover were low.

Other areas of non-reefal coral growth on relict reef rock were found in the sheltered or semi-sheltered conditions inside the embayment north of the Ras Isa headland and on the east of Kamaran Island (Zajonz et al. 2005). Distinct coral community types were found in close proximity, and these included mixed stony and soft coral communities, monotypic beds of *Porites* cf. *nodifera*, xeniids and *Pocillopora damicornis,* tabular *Acropora, Stylophora pistillata,* monospecific stands of *G. retiformis*, branching or foliose *Montipora*, *P. daedalea*, and *Stylophora pistillata* and *Psammocora contigua*.

Coral Communities on Volcanic Rock
Volcanic lava flows around the offshore islands and pinnacles typically supported higher diversity coral assemblages, irrespective of the degree reef build-up. These communities often consisted of small massive *Porites* spp. and faviids and mixed with encrusting *Montipora* and faviids *Leptastrea* and *Cyphastrea* spp., as well as pocilloporids, *Lobophyllia hemprichii*, *G. fascicularis* and *Hydnophora* spp. among others.

Turak and Brodie (1999) surveyed the rock pinnacles at Six Foot Rocks, south of Tiqfash Island (Fig. 15b), and Avocet Rock south of Al Hudaydah. Massive *Porites* accounted for the majority of the live coral cover and provided >80 % cover at one site. Nearly, all scleractinian coral families from the Red Sea were recorded, with faviidae being the most common and diverse. Klaus and Eisinger (2004) surveyed another location at Six Foot Rock in 2004 and found the rocky substrate was dominated by moderately sized ($\sim$1 m) colonies of *Stylophora pistillata,* and tabular *Acropora downingni,* although a number of different encrusting *Montipora* spp., *Echinopora* spp. and small colonies of faviids were also common.

In other locations, volcanic rock slopes may support monospecific *Stylophora* stands, mixed communities of *Millepora* and large tabular acroporids (*Acropora clathrata* and *A. cytherea*), *Pocillopora*, *Stylophora* and soft corals (xeniids and *Lithophyllum*) (Klaus and Eisinger 2004). Turak and Brodie (1999) also reported a *Siderastrea savignyana*-dominated community at a turbid site at At Tair Island (Fig. 15c), which hosted large colonies of *S. savignyana* (>1 m diameter), often partially buried in shifting sediments.

Status and Trends
Contemporary reef development is present along much of the mainland coast and islands of the Yemeni Red Sea (Turak and Brodie 1999; Turak et al. 2007; Klaus and Eisinger 2004; Zajonz et al. 2005). The extent of reef development, community composition and condition of corals is highly variable (Figs. 16 and 17). Reef development in Yemeni waters is constrained by the shallow sedimentary conditions on the shelf platform and similar environmental conditions as found in Eritrean waters. Island fringing reefs close to the shelf edge are generally better developed than inshore reefs. Yet all reef types and non-reefal coral assemblages, including monotypic beds, often co-occur in close proximity (Klaus and Eisinger 2004; Turak and Brodie 1999; Turak et al. 2007) (Fig. 18).

Fig. 16 Photographs illustrating coral communities characteristic of the Yemeni Red Sea including **a** mixed *Acropora* spp. and *Stylophora* sp.-dominated community (Ras Isa); **b** bed of *Acropora* sp.; **c** massive *Porites lutea* and tabular *Acropora* sp.; **d** soft coral-dominated community consisting of xeniids, and *Dendronepthea* sp.; **e** a coralline algal reef; and **f** a coral community with various faviids, poritids and *Acropora* spp. with the fucoid macroalgae *Sargassum*. Photographs by M. Esinger

A total of 223 scleractinian species have been reported (Turak and Brodie 1999; Klaus and Eisinger 2004; Zajonz et al. 2005; Turak et al. 2007; Benzoni et al. 2013). Scleractinian species diversity was highest around the offshore volcanic islands and on reefs at the seaward extent of the shelf (Turak et al. 2007), which are exposed to clearer and cooler waters than inshore areas. Low diversity communities, including high-cover monotypic communities, are common inshore and offshore. While the monotypic coral communities do not follow the same pattern of accretion as found on coral reefs, they do provide a highly complex and high coral cover >50 % habitat.

Surveys in the late 1990s by Turak and Brodie (1999) found live hard coral ranging between 3 and 29 % (average 11 %), compared to dead hard coral and macroalgae, which averaged 29 and 21 %, respectively. Although coral

Fig. 17 Photographs illustrating some of the coral communities characteristic of Yemen showing the following: **a** *Porites* sp., *Stylophora* sp., *Pavona* sp., with *Echinopora* spp., *Echinophyllia* spp. and branching *Acropora* sp. (Tiqfash Island); **b** a large colony of *Favia stelligera* with *Seriatopora hystrix*, *Porites* sp. and *Pavona* sp. (Tiqfash Island); **c** *Montipora* sp. and *Pavona* sp. and the soft coral *Sarcophyton* sp. (Uqban Seghir); **d** *Acropora* sp. and *Stylophora* sp. interspersed with macroalgae (Hannish Island); **e** Tabular *Acropora cytherea* and branching *Montipora* sp. with a carpet of *Sarcophyton* sp. (Addar Ail); and **f** Mixed *Stylophora* spp. hard corals and soft corals (Ras Isa). Photographs **a**, **b**, **f** by M. Esinger and **c**, **d**, **e** by R. Klaus

bleaching was not observed, the authors described how their results indicated the occurrence of two severe mass coral bleaching events in the Yemeni Red Sea. The first incident occurred during the summer of 1996 and affected the northern region, while the second occurred in 1998 and impacted the southern region (Turak and Brodie 1999; Turak et al. 2007). One site near al-Khawkhah, for example, had moderately high cover of *G. retiformis*, branching *Montipora* and massive *Porites* in 1996, but was mostly dead by 1997 (Turak and Brodie 1999). The same authors reported observing extensive areas of recently dead tabular *Acropora* communities and banks of rubble around the northern shelf

Fig. 18 A selection of coral species endemic to the Red Sea including **a** *Acropora maryae* (Veron 2000), **b** *Stylophora wellsi* (Scheer 1964), **c** *Echinopora fructilosa* (Ehrenberg 1834), **d** *Mycedium umbra* (Veron 2000), **e** *Symphyllia erythrea* (Klunzinger 1879) and **f** *Favia lacuna* (Veron, Turak and DeVantier, 2000). Photographs by R. Klaus

islands, and near-total mortality on reefs where healthy coral cover had been recorded ten years earlier (Barratt et al. 1987a). Similarly, at Ras Marsa, a formerly *Porites* cf. *nodifera*-dominated reef, surveyed in early 1998 (M. Eisinger, pers. obs.), was nearly completely overgrown by coralline red algae of the genera *Porolithon* and *Lithothamnion* in 2005 (Zajonz et al. 2005).

Both the 1996 and 1998 coral bleaching events in the Yemeni Red Sea were most likely caused by a sustained period of elevated sea water temperature in combination with high levels of irradiance, during calm weather conditions. Turak and Brodie (1999) reported a lack of recruitment following these events and suggested that macroalgae would proliferate on the bleaching impacted reefs limiting coral recovery. However, surveys completed 5 years later (Klaus and Eisinger 2004; Zajonz et al. 2005) found that although live hard coral cover was patchy, it was still extremely high in many places. Average hard coral cover was 46.8 % in 2004, ranging from 62.7 % at Tiqfash Island to 27.9 % at Addar Ail. Hard coral cover at other locations in the Kamarans remained close to 50 % (Uqban Kebir 51.1 % and Uqban Seghir 45.4 %). *Acropora* spp. provided 11.93 % of the cover on average and was highest at Uqban Kebir (24.2 %). These findings indicate that the impact of the bleaching events was variable.

Other threats to the coral communities in Yemen's Red Sea coral reefs include CoTs, *Drupella* and sponge and soft coral overgrowth, fisheries-related impacts as well as coastal development, catchment pollution and sedimentation, chemical pollution, marine litter, shipping and harbour activities, ballast water discharge and oil-related industries (PERSGA/GEF 2003; Klaus and Eisinger 2004). While CoTs outbreaks have affected offshore reefs, the densities reported to date have been relatively low (PERSGA/GEF 2003). The coral-feeding gastropod *Drupella* found throughout the Red Sea was commonly on compact branching forms of *Acropora*, tabular. There is limited information about the distribution and effects of coral diseases in Yemeni waters.

There has been a slow but gradual decline in catches of both pelagic and demersal fish stocks since late 1980s in Yemen (PERSGA/GEF 2003). Sharks (mostly Carcharinidae and Sphyrnidae) are also overfished. Coastal cities and towns have expanded, and unplanned settlements around cities such as Al Hodeidah increase the pressure and pollution of nearshore habitats. Waste waters from the coastal towns and cities and industries are discharged directly into the sea as very few coastal cities are served by public sewage networks. Tourism is limited given the political instability and poses a limited threat to reefs from either anchor or diver damage Power stations such as the one at Mokha discharge saline high-temperature water directly into the sea. Oil spill risks associated with shipping and off-loading are significant. A 409,000 tonnes deadweight, floating storage and off-loading vessel moored offshore from Ras Isa on the Al Salif Peninsula. This ship is supplied with crude oil by a pipeline from the shore, which is then transferred to crude oil tankers (PERSGA/GEF 2003).

The Government of Yemen declared MPAs in the Gulf of Aden (e.g. Socotra Archipelago), but to date, the only one protected has been declared for the Red Sea coast on Kamaran island. The Kamaran mangroves are one of the last "healthy" mangal ecosystems in Yemen and of critical importance within the wider Kamaran-As Salif ecosystem (Zajonz et al. 2010). Propositions to award the site as protected areas were approved by Prime Ministerial (Cabinet) Decree No. (310) of 2009. The decree classifies the area according to five zoning types, and a preliminary zoning map has been prepared, but there are no management plans or management activities (Zajonz et al. 2010). The Yemeni Island Promotion and Development Authority also prepared a master plan for Kamaran which was not in agreement with the one prepared by EPA (Zajonz et al. 2010). Given that there are plans to establish a new oil terminal and other developments, the protected area remains vulnerable.

Discussion

The central and southern Red Sea regions host a wide diversity of modern coral reef morphologies and other non-accreting coral communities, which range from classical fringing reefs to marginal coral communities characteristic of the Gulf of Aden and Socotra Archipelago (DeVantier et al. 2004; Klaus and Turner 2004). The wide range and diversity of coral reefs and other coral communities in this region reflects temporal and spatial variations in the geological and paleoclimatic history, and present-day abiotic and biotic factors. The following discussion is a non-exhaustive synopsis of some of the key factors that influence the distribution and development of reef types and other coral communities in the region.

Geological and Paleoclimatic Influences

Scleractinian corals are known to preferentially grow on carbonates or rough lava flows, while sedimentary rocks and unconsolidated sediments are less suitable (Cabioch et al. 1995). In the central–southern Red Sea regions, corals are found growing on karst or limestone, volcanic and granitic rock substrates. Coral communities are found growing within naturally highly sedimentary habitats, although these communities most likely settled on patches on prior to the accumulation of sediments.

The majority of modern reef frameworks in this region are formed upon the eroded fractured bathymetric highs of ancient carbonate structures that were formed during the Plio-Pleistocene. Reef frameworks are found throughout the region but are better developed in the central Red Sea. Monotypic coral community types on relict reef rock and nascent frameworks on volcanic or granitic rock substrates become increasingly common in the south. The mainland reef frameworks, and many of the offshore structures, follow the same orientation as the main Red Sea and Afar rift. More irregular distribution patterns occur as a result of local fault lines and differences in the magnitude of uplift due to salt diapirs and subsequent erosion and depositional processes (Bosence 2005; Purkis et al. 2012; Rowlands and Purkis this volume).

Variations in the magnitude of uplift are evident, particularly noticeable in the vicinity of Dungonab Bay in Sudan, the southern Farasans in Yemen and Dahlaks in Eritrea. The sculpting of the antecedent framework by the paleoclimate and most notably rainfall and sediment discharge created further complexities in the topography, and the basis for the reef formations found in the region today (Sheppard et al. 1992). Depositional processes also were important in the formation of shelf platforms. Outflow from the Tokar delta, for example, contributed towards the formation of the Suakin shelf platform in southern Sudan when sea levels were lower.

Abiotic and Biotic Interactions

Environmental conditions in the Red Sea vary with latitude and season. Seasonal variability is more pronounced in the southern Red Sea than in the north, on account of the reversing wind system and increased wind strength. Cross-shelf gradients also occur due to the broad extent of the shallow platforms. These factors create a wide diversity of environmental regimes that influence coral community composition and reef development as outlined below.

Sea Water Temperatures

Sea water temperatures in the central and southern Red Sea are warmer and more consistent than in the north, although they are more variable and cooler in the Straits of the Bab-al-Mandab, where cross-shelf gradients also occur. In sheltered nearshore habitats around the Farasans and Dahlaks, for example, water temperatures can be several degrees warmer than in the central basin due to insolation and reduced mixing across the broad shallow shelf platforms. The most significant variability in sea water temperatures in this region occurs along the coast of Eritrea, where coral communities are directly exposed to the seasonal wind-driven upwelling during the winter months.

Both hot and cold sea water temperatures can significantly influence the physiology of the coral holobiont. Heat stress causes an increase in metabolic activities, oxidative stress, membrane and protein damage, damage to host tissue, expulsion of symbiotic zooxanthellae and nutrient source. Cold stress slows metabolic activity, respiration and photosynthesis and can cause photoinhibition. Although there are clearly areas where Red Sea corals have adapted to cope with high sea water temperature, coral bleaching events due to excessively high sea water temperatures during the summer months have impacted the coral reefs in all countries in this region. Sustained exposure to anomalously high sublethal sea water temperatures has already reduced growth and net carbonate accretion rates elsewhere in the Red Sea (Cantin et al. 2010). The implications of this finding are more alarming where other factors are already constraining coral growth rates.

Light

Coral species zonation and the maximum depth of reef development are controlled by physical water depth (i.e. bathymetry) and factors that influence the suitability of the habitat for coral growth, including the depth of penetration of light through the water column (Kleypas et al. 1999). Light intensity is attenuated by 60–80 % within the top 10 m of clear waters (Kinzie 1973) and even further reduced where waters are turbid (Kirk 1977). A reduction in the availability of light, whether due to depth or turbidity, affects photosynthesis and respiration of autotrophic organisms (Rogers 1979, see Sawall and Al-Sofayni this volume for further discussion), coral distribution (Roy and Smith 1971; Titlyanov and Latypov 1991) and reef development (Kleypas et al. 1999). While certain corals are able to adapt to a wider range of light conditions, other species have more specific depth/light requirements (Fricke et al. 1987). Light availability is the primary factor controlling reef development and community composition in certain parts of the central–southern regions (see below).

Sedimentation

The extent of unconsolidated shallow sedimentary habitats is higher in the central and southern Red Sea than in the north due to the presence of the shelf platforms. Corals that exist within habitats in proximity to unconsolidated mobile sediments are vulnerable to abrasion, smothering and episodic burial. Sedimentation can also detrimentally impact coral larval recruitment (Harrigan 1972), survival and development (Gilmour 1999), and coral physiology and health and reef build-up (Johannes 1975; Rogers 1990). These factors all present a major constraint on the maximum depth of reef growth, as well as affecting the energetics of the coral holobiont. In Yemen, for example, the maximum depth of reef development on the shelf platform is 5–6 m, due to the high accumulation of sediments at the base of the reef.

Turbidity

High levels of turbidity as a result of suspended sediment loads restrict the amount of PAR reaching the seabed, influencing coral autotrophy, morphology, coral growth and other physiological processes (Glynn et al. 1986; Davies 1991; Anthony and Connolly 2004). Suspended sediment loads in this region originate from three main sources: the re-suspension of unconsolidated sediments due to wind and wave action and tidal current flows; episodic inflow of land-derived sediments during flash flooding; and influxes of wind-borne dusts. Wind speeds are stronger in the southern Red Sea, which increases sediment transport and the re-suspension of fine particulates. Seasonal changes in the wind direction in the southern Red Sea reverse the exposure patterns thereby creating a more dynamic mobile environment. Little is known about the volume of wind-blown African dusts settling in the Red Sea, and further work is needed to determine their relative contribution to local conditions.

Fresh water

Inputs of silts from terrestrial sources are a major potential source of turbidity and sedimentation near the mouths of khors, wadis and deltas. These ephemeral waterways accumulate fine particulates that are then flushed out following heavy rains. Flash flooding events occur throughout the Red Sea but have a more significant impact on shallow environments, as in the south, than where the subtidal topography is steeper. Fresh water is less dense than sea water, and fine suspended sediment particles can be transported kilometres away from the coast before they begin to settle. Suspended sediments carried in flood waters will cross over the tops of steep fringing reefs and settle offshore, while the broad shallow shelf platforms facilitate the deposition and accumulation of sediments.

Upwelling and Productivity

Productivity along the southern Red Sea coast of Eritrea far exceeds the rest of the Red Sea due to the seasonal upwelling. While studies along this coast have characterised the algal communities (Ateweberhan et al. 2005a, b), further work is needed to determine the influence of the upwelling and other factors, such as hydrodynamic exposure, on coral community composition and reef development along this coast. It is known that corals can tolerate a range of nutrient concentrations and that reef development is not necessarily restricted to oligotrophic waters (Kleypas et al. 1999; Szmant 2002). While elevated nutrients levels are generally considered to have a greater impact at the ecosystem level than at the organism level, a recent review concluded that nutrient enrichment beyond certain thresholds can negatively impact upon both the physiological performance of the coral organism and ecosystem functioning (D'Angelo and Wiedenmann 2014). In upwelling affected areas elsewhere in the southern Arabian region, elevated patches of hard substrates continue to support hard coral communities, often mixed with macroalgal communities, sponges or soft corals (Klaus and Turner 2004). The main factor limiting the occurrence of stony corals under these condition is the presence of mobile sediments and episodic disturbances.

Macroalgae

The seasonal occurrence of macroalgae, such as *Sargassum,* on the tops of reef frameworks is a characteristic feature of shallow marine communities in the southern Red Sea. Seasonal increase in algal biomass is a common feature in the southern Red Sea, Gulf of Aden (Ormond and Bainimoon 1994; Klaus and Turner 2004) and wider Arabian Region (e.g. Coles and Fadlallah 1991; Pichon 1971; Sheppard et al. 1992), typically driven by naturally elevated levels of nutrients caused by cold-water upwelling. Algal biomass typically increases following a lag period after the upwelling and occurs at the end of summer in the Gulf of Aden (Klaus and Turner 2004) and at the end of the winter in the southern Red Sea (Ateweberhan and Prud'Homme Van Reine 2005). These algae then naturally die back during the summer as sea water temperatures increase. In Eritrea, for example, algal biomass peaks during the winter season and decreases during the hot season (Ateweberhan et al. 2005a).

From Coral Reefs to Marginal Coral Communities

As the environmental regime becomes more stressful in the southern Red Sea (temperature variability, high turbidity, sedimentation, exposure, etc.), stony corals continue to provide the dominant cover on stable hard substrates. The coral communities that withstand these conditions often form dense high-cover non-accreting coral communities, without the three-dimensional framework of a "true" coral reef. These "marginal" communities are often dominated by one or more sediment-tolerant, fast-growing, aggressive, opportunistic or pioneering species, which may co-occur with soft corals, sponges and/or macroalgae as conditions become more extreme. More than one of these monotypic communities may occur in close proximity to one another, forming a complex patchwork. Species most commonly found to form these monotypic communities often include *Stylophora,* branching pocilloporiids, poritids, faviids such as *Goniastrea* spp., *Montipora* spp., *Platygyra* spp. and *Galaxea fasicularis* and *G. astreata, Siderastrea savignyana.* Interestingly, many of these species/genera are often among the early pioneers reportedly found on Pleistocene reefs in Kenya (Crame 1980, 1981).

Examples of these areas dominated by these types of marginal coral community are found throughout the Gulf of Aden (DeVantier et al. 2004) and the southern Arabian region (Wilson 2006) and indeed in many other locations around the world (Perry and Larcombe 2003). Key environmental factors that limit the development of reefs in the southern Arabian region include hydrodynamic exposure and episodic disturbance events. Reef build-up under these types of extreme conditions tends to be limited to sheltered environments. For example, other high stress environments further north in the central Red Sea, such as the high-turbidity high-temperature coral communities found within sheltered steep-sided khors and embayments, still tend to support narrow fringing reefs. These communities tend to host more species-rich coral communities, as sediment-tolerant species co-occur with species more commonly found in deeper water. Recruitment limitation and competition thus also appear to have a important role in controlling the composition of the communities. Coral larvae that successfully recruit onto a patch of hard substrate could potentially proliferate in the absence of competition from other species. The species most commonly found to form monospecific coral beds tend to be brooding species as opposed to spawners, which again suggests that recruitment may have a fundamental role.

Threats and Other Anthropogenic Impact

Coral reefs and other coral communities in the central and southern Red Sea are subject to various threats ranging from global scale climate change-related threats, such as increasing sea water temperatures and ocean acidification, to more localised impacts as a result of anthropogenic activities such as coastal development, shipping and harbour activities, oil- and gas-related activities, and fisheries. The present condition and species composition on the reefs in the central and southern Red Sea have been significantly influenced in recent years by the mass coral bleaching events that impacted these reefs in the late 1990s (e.g. DeVantier et al. 2005; Klaus et al. 2008b; PERSGA 2010). There were only a limited number of direct visual observations of coral bleaching during the mass coral bleaching event. These included reports of coral bleaching on the reefs around Dungonab in northern Sudan (Kemp et al. 2002) and at a similar latitude in Saudi Arabian reefs between August and September 1998 (DeVantier et al. 2005). Surveys revealed that corals in the Yemeni Red Sea were impacted by a severe bleaching in 1996 in advance of the main event in 1997 (Turak and Brodie 1999). The reefs and coral communities that were worse impacted in this region tended to be situated in shallow water (<10 m depth). Areas that appeared to withstand the extremely hot sea water temperatures and survived were situated in or close to the central deep-water channel or in more turbid environments, indicating that light was an important contributor to the severity of these events (Klaus et al. 2008b).

Corallivorous species (*Acanthaster plancii* and *Drupella* sp.) have been reported from all countries in the central and southern Red Sea. Some significant outbreaks of CoTs were reported from Sudan in the 1980s, and extensive coral mortality was reported on offshore reefs in Yemen during the late 1990s. Smaller more isolated outbreaks have since been observed in Sudan during surveys in 2002, 2006 and 2007. The lack of regular monitoring in the majority of these countries however makes it difficult to determine the relative importance of these outbreaks compared to other impacts such as the coral bleaching.

While the coastal environments of Sudan, Eritrea, Djibouti and Yemen are not entirely pristine, they are considerably less developed than many other parts of the Red Sea. The majority of anthropogenic impact is restricted to coastal port cities (e.g. Port Sudan, Massawa and al-Huddaydah), with the exception of fisheries-related impacts, including the use of certain destructive types of gear, such as trawling which are more widespread. Shipping and harbour activities present another suite of threats. The central and southern Red Sea are major shipping routes, and >20,000 ships pass through the Gulf of Aden and into the Straits of Bab-al-Mandab each year and calling at the ports in Djibouti, Yemen, Sudan and Eritrea. Existing port facilities in these countries are out of date, and many lack adequate reception facilities. Ships are known to dump waste and ballast into the Red Sea waters. Ballast waters can result in the formation of tar balls, and these have been found along the coastline in some parts of the Red Sea coast (PERSGA/GEF 2003). Solid waste from ship-based sources is often found washed up on beaches.

Despite the potential, coastal tourism in the central and southern Red Sea has been limited to date by political instability, civil war and security concerns. In Sudan, there is a liveaboard dive industry, with 12 boats permanently situated in Port Sudan and plans for new tourism developments. A new coastal tourism development is proposed for Dungonab and Mukkawar Island National Park in Sudan, which would centre on Mukkawar Island. This "Dubai" style development would involve the infilling of the shallow marine environments around the island resulting in the loss of coral reefs, turtle nesting beaches and habitat for associated marine fauna, including a globally significant aggregation of manta rays (*Manta alfredi* and *Manta birostris*), which were first reported by Hans Hass (Hass 1952) and have since remained resident in the area.

The expansion of oil and gas exploration and the construction of new refineries also pose a potentially significant threat. Poor maintenance of existing infrastructure has resulted in low level, but locally significant oil contamination from ballast water discharge, leaks from bunkering facilities, or spills while loading or unloading (PERSGA/GEF 2003). There is a plan to construct and operate a 60,000 barrel per day (bpd) refinery in Yemen, which would involve the construction of relevant infrastructure at Ras Isa. The existing terminal at Ras Isa handles the export of crude oil transported by pipeline from the Maarib oilfields and is situated directly adjacent to the MPA on Kamaran.

Conclusions

The central and southern Red Sea is a transitional zone, where shallow marine communities are subject to progressively more marginal conditions, similar to those found in the Gulf of Aden and Arabian Sea, which vary along both latitudinal and cross-shelf gradients. The spectacular coral reefs' characteristic of the northern Red Sea becomes shallower and gradually transition into non-accreting coral communities, accompanied by an increase in the seasonal abundance of macroalgae. The tipping point at which these shifts occur relates to synergistic meso-scale interactions between various abiotic factors (e.g. exposure, turbidity, sediments and nutrients) and their influence on biological controls (e.g. recruitment, growth, competition, predation and symbiosis) on local metapopulation dynamics. The high degree of local acclimatization and adaptation demonstrated by the corals in this region presents a unique research opportunity to improve understanding as to how reef systems elsewhere will react to increasingly marginal conditions that are anticipated to occur under future climate scenarios (Smith and Buddemeier 1992; Guinotte et al. 2003; Sheppard 2003). The extreme conditions that the unique coral communities in this region already withstand increase their susceptibility and vulnerability to other anthropogenic stressors. Although progress has been made towards developing appropriate strategies and actions (PERSGA 1998; PERSGA/GEF 2003), these have yet to be fully implemented and urgent action is needed to support and improve conservation management within the central and southern regions. The expansion and effective enforcement of a network of MPAs that encompass a wide diversity of the reef and coral community types spanning the different environmental regimes would help secure viable populations of reef taxa and enable them to withstand future anthropogenic and climate change-related disturbances.

References

African Parks Foundation (2006) Survey report: sudan marine parks expedition (2006). Dungonab Bay and Mukkawar Island National Park and Sanganab Atoll marine national park, vol 2. African Parks Foundation, Netherlands, 121 pp, 116 pp

Anthony KRN, Connolly SR (2004) Environmental limits to growth: physiological niche boundaries of corals along turbidity-light gradients. Oecologia 141:373–384

Angelucci (1981) Outline of geology and sedimentary environments of the Dhalaks, Southern Red Sea. Bollettino della Societa Geologica Italiana 99:405–419

Angelucci et al. (12 others) (1985) L'archipelago Delle Isole Dhalak Nel Mar Rosso Meridionale: Alcune Caratteristche Geologiche. Bollenttino della Societa Geographica Italiana, Series XI 2:233–262

Ateweberhan M, Bruggemann JH, Breeman AM (2005a) Seasonal dynamics of *Sargassum ilicifolium* (Phaeophyta) on a shallow reef flat in the southern Red Sea (Eritrea). Mar Ecol Prog Ser 292:159–171

Ateweberhan M, Bruggemann JH, Breeman AM (2005b) Seasonal patterns of biomass, growth and reproduction in *Dictyota cervicornis* and *Stoechospermum polypodioides* (Dictyotales, Phaeophyta) on a shallow reef flat in the southern Red Sea (Eritrea). Bot Mar 48:8–17

Ateweberhan M, Prud'homme van Reine WF (2005) A taxonomic survey of seaweeds from Eritrea. Blumea 50:65–111

Barker N, Galal N, Klaus R (2002) Survey of the proposed marine protected area at Iles des Sept Frères and Ras Siyyan, Djibouti. PERSGA marine protected area—strategic action programme (SAP). Unpublished report, 65 pp

Barrat L, Medley P (1988) Assessment of the aquarium reef fishery in Djibouti. FAO report (TCP/DJI/6755/A)

Barratt L, Dawson-Shepherd A, Ormond RFG, McDowall R (1987a) Former Yemen Arab Republic marine conservation survey. Vol. I. Distribution of habitats and species along the YAR coastline, vol 110. IUCN Red Sea and Gulf of Aden Environment Program/ TMRU York, UK, 2 pp

Barratt L, Dawson-Shepherd A, Ormond RFG, Mcdowall R (1987b) Former Yemen Arab Republic marine conservation survey. Vol. II. Preliminary coastal zone management recommendations for the YAR. IUCN Red Sea and Gulf of Aden Environment Program/ TMRU York, UK, 46 pp

Becker JJ, Sandwell DT, Smith WHF, Braud J, Binder B, Depner J, Fabre D, Factor J, Ingalls S, Kim S-H, Ladner R, Marks K, Nelson S, Pharaoh A, Trimmer R, Von Rosenberg J, Wallace G, Weatherall P (2009) Global bathymetry and elevation data at 30 arc seconds resolution: SRTM30_PLUS. Mar Geodesy 2(4):355–371. doi:10.1080/01490410903297766

Behairy AKA, Sheppard CRC, El-Sayed MK (1992) A review of the geology of coral reefs in the Red Sea. UNEP, Nairobi, Kenya, UNEP regional seas reports and studies, 36 pp

Bemert G, Ormond R (1981) Red Sea coral reefs. Kegan Paul International, London

Benzoni F, Pichon F, Dutrieux E, Chaîneau CH, Abdulaziz M, Al-Thary I (2013) The scleractinian fauna of Yemen: diversity and species distribution patterns. In: Proceedings of the 12th international coral reef symposium, Cairns, Australia, 15A Biodiversity and Systematics, 9–13 July 2012

Berry L, Whiteman AJ, Bell SV (1966) Some radiocarbon dates and their geomorphological significance, emerged reef complex of the sudan. Zeitschr Geomorphol NF 10:119–143

Bosence D (2005) A genetic classification of carbonate platforms based on their basinal and tectonic settings in the cenozoic. Sed Geol 175:49–72

Braithwaite CJR (1987) Geology and paleogeography of the Red Sea region. In: Edwards AJ, Head SM (eds) Key environments. Red Sea. Pergamon Press, Oxford, pp 22–44

Cabioch G, Montaggioni LF, Faure G (1995) Holocene initiation and development of New Caledonian fringing reefs. South-West Pacific. Coral Reefs 14:131–140

Cantin NE, Cohen AL, Karnauskas KB, Tarrant AM, and McCorkle, DC (2010) Ocean warming slows coral growth in the central Red Sea. Science 329. doi:10.1126/science.1190182

Coles S (1988) Limitations of reef coral development in the Arabian Gulf: temperature or algal competition? In: Proceedings 6 international coral reef symposium, vol 3, pp 211–216

Coles SL, Fadlallah YH (1991) Reef coral survival and mortality at low temperatures in the Arabian Gulf: new species specific lower temperature limits. Coral Reefs 9(4):231–237

Cousteau J-Y, Diolé P (1970) Life and death in a Coral Sea. Cassel, London

Cousteau Society (Chekchak T, Klaus R, eds) (2013) Toward a sustainable future for the Red Sea coast of sudan: part 1. Coastal and marine habitats Survey. Cousteau Society, Custodians of the Sea, New York

Crame JA (1980) Succession and diversity in the Pleistocene coral reefs of the Kenya coast. Palaeontology 23:1–37

Crame JA (1981) Ecological stratification in the Pleistocene coral reefs of the Kenya coast. Palaeontology 24:609–646

Crossland C (1913) Desert and water gardens of the Sudanese Red Sea. Cambridge, 158 pp

Crossland C (1952) Madreporaria, Hydrocollinae, Heliopora, and Tubipora. Sci Rep Great Barrier Reef Exped 1928–29(6):85–257

D'Angelo C, Wiedenmann J (2014) Impacts of nutrient enrichment on coral reefs: new perspectives and implications for coastal management and reef survival. Curr Opin Environ Sustain 7:82–93

Davies PS (1991) Effects of daylight variations on the energy budgets of shallow-water corals. Mar Biol 108:137–144

DeVantier LM, Turak E, Al-Shaikh KA (2005) Coral bleaching in the central-northern Saudi Arabian Red Sea, August–September 1998. In: Abuzinada A, Joubert E, Krupp F (eds) Proceedings of an international symposium on the extent and impact of coral reef bleaching in the arabian region. National Commission for Wildlife Conservation and Development, Riyadh, 5–9 Feb 2000, pp 75–90

DeVantier LM, De'ath G, Klaus R, Al-Moghrabi S, Abdulaziz M, Reinicke G, Cheung CPS (2004) Reef-building corals and coral communities of the Socotra Islands, Yemen: a zoogeographic 'crossroads' in the Arabian Sea. Fauna Arab 20:117–168

DeVantier LM, Turak E, De'Ath G, Al-Shaikh KA (2000) Coral communities of the centralnorthern Saudi Arabian Red Sea. Fauna Arab 18:23–65

Diolé P, Falco A (1976) Le memoires de Falco, chef plongeur de la Calypso. Paris, 294 pp

Edwards HM, Haime J (1857–1860) Histoire naturelle des Coralliaires. Paris, 1–3, 326, 632, 560 pp

Edwards AJ, Head SM (eds) (1987) Red Sea, key environments series, Oxford, UK, 441 pp

Ehrenberg CG (1834) Beiträge zur physiologischen Kenntnis der Corallenthiere im Allgemeinen und besonders im rothen Meernebst einem Versuch zur physiologischen Systematik deselben. Abhandlungen der Preußischen Akademie der Wissenschaften, Berlin 1832:225–380

Esper EJC (1788-1830) Die Pflanzenthiere 1-3, Nürnberg, Forsetzungen 1–2, pp 1–230

Feldman, GC, McClain, CR (2013) Ocean color web, aqua MODIS reprocessing, NASA Goddard Space Flight Center. In: Kuring N, Bailey SW (eds) 30th December 2013. http://oceancolor.gsfc.nasa.gov/

Fishelson L (1971) Ecology and distribution of the benthic fauna in the shallow waters of the Red Sea. Mar Biol 10:113–133

Forskal P (1775) Descriptiones Animalium, Avium, Amphiborium, Piscium, Insectorum, Vermium que in intinere orientali observavit Petrus Forskal. IV Corallia. Huaniae, pp 131–139

Fricke HW, Vareschi E, Schlichter D (1987) Photoecology of the coral *Leptoseris fragilis* in the Red Sea twilight zone (an experimental study by submersible). Oecologia 73:371–381

Gardiner C (1909) The madreporarian corals. I. Fungida pt. IV Tubiporidae. With a revision of its genera and species and an account of their geographical distribution. rans Linn Soc London 2nd ser Zool 12:257–290

Gilmour J (1999) Experimental investigation into the effects of suspended sediment on fertilisation, larval survival and settlement in a scleractinian coral. Mar Biol 135(3):451–462

Glynn PW, Howard LS, Corcoran E, Freay D (1986) Preliminary investigations into the occurrence and toxicity of commercial herbicide formulations in reef building corals. In: Jokiel PL, Richmond RH, Rogers RA (eds) Coral reef population biology

Gravier C (1910a) Sur les recifs coralliens de la Baie de Tadjourah et leurs Madreporaires: Gulf of Aden. Comptes Rendus Hebdomadaire des Seances de l'Academie des Sciences. Paris 151:650–652

Gravier C (1910b) Sur quelques formes nouvelles de Madreporaires de la Baie de Tabdjourah: Gulf of Aden. Bull Mus Nat Hist Paris 16:273–276

Gravier C (1910c) Sur quelques particularites biologique des recifs madreporiques de la Baie de Tadjourah: Gulf of Aden. C R Ass Avanc Sci 39:167–169

Gravier C (1911) Les recifs de coraux et les madreporaires de la Baie de Tadjourah: Gulf of Aden. Annales de l'Institut océanographique Paris 2:99

Guilcher A (1955) Géomorphologie de l'extrémité septentrionale du banc Farsan (Mer Rouge) Annales de l'Institut océanographique 30, pp 55–100

Guilcher A (1988) A heretofore neglected type of coral reef: the ridge reef. Morphology and origin. In: Proceedings of the 6th international coral reef symposium, Townsville, Australia, pp 399–402

Guinotte JR, Buddemeier RW, Kleypas J (2003) Future coral reef habitat marginality: temporal and spatial effects of climate change in the Pacific basin. Coral Reefs 22:551–558

Harrigan JF (1972) The planula larva of *Pocillopora damicornis*: lunar periodicity of swarming and substrate selection behavior. PhD thesis, University of Hawaii

Hass H (1952) Manta, Teufel in Roten Mer. Berlin, 180 pp

Hass H (1961) Expedition ins Unbekannte. Berlin/Frankfurt/Wien, 167 pp

Hassan M, Kotb MMA, Al-Sofyani AA (2002) Status of coral reefs in the Red Sea-Gulf Of Aden

Head SM (1980) The ecology of corals in the Sudanese Red Sea. PhD thesis, University of Cambridge

Hilman JC, Tsegay S (1997) Coral reef research and conservation in Eritrea: area profile. In: Western indian ocean regional workshop on coral reef research and conservation, international year of the reef 1997, Mombasa, Kenya, Feb 25–Mar 1 1997, 17 pp

Hopley D (1982) The geomorphology of the great barrier reef. Wiley Interscience, New York

Jeudy de Grissac A, Negussie H (2007) Eritrea's coastal marine and Island biodiversity conservation project. UNDP-GEF ECMIB project, Massawa 75 pp

Johannes RE (1975) Pollution and degradation of coral reef communities. In: Wood EJF, Johannes RE (eds) Tropical marine pollution. Elsevier Science Publications Company, Oxford, pp 1–181

Kemp J, Klaus R, Salem M, Awadalla Y, Saleh, B (2002) Survey of the proposed marine protected area at Dunganob Bay and Mukkawar Island, Sudan. PERSGA SAP Component 5A Regional Network of Marine Protected Areas, 164 pp

Kemp J (1998) Zoogeography of the coral reef fishes of the Socotra Archipelago. J Biogeogr 25:919–933

Kinzie RA (1973) The zonation of West Indian gorgonians. Bull Mar Sci 23:93–155

Kirk JTO (1977) Attenuation of light in natural waters. Aust J Mar Freshw Res 28:497–508

Klaus R, Eisinger M (2004) Coral reef monitoring surveys on the Red Sea coast of Yemen. In: Final draft report for the UNDP-GEF protection of marine ecosystems of the Red Sea coast of Yemen, 35 pp

Klaus R, Turner JR (2004) The marine biotopes of the Socotra Island group. Fauna Arab 20:45–115

Klaus R, Kemp JM, Samoilys M, Anlauf H, El Din S, Abdalla EO, Chekchak T (2008a) Biodiversity of the sudanese Red Sea. In:

Integrated coastal zone management chapter 2. Cousteau Society, unpublished report
Klaus R, Kemp JM, Samoilys M, Anlauf H, El Din S, Abdalla EO, Chekchak T (2008b) Ecological patterns and status of the reefs of Sudan. In: Proceedings of the 11th international coral reef symposium, Ft. Lauderdale, Florida, session number 18. 7–11 July 2008
Kleypas JA, McManus JW, Menez LAB (1999) Environmental limits to coral reef development: where do we draw the line? Am Zool 39:146–159
Klunzinger CB (1879) Eine zoolgische excursion auf ein Korrallenriff des Rothen Meeres. Verhandlungen der Zoologisch-Botanischen Gesellschaft in Wien 20:389–394
Kotb M, Abdulaziz M, Al-Agwan Z, Al-Shaikh K, Al-Yami H, Banajah A, DeVantier L, Eisinger M, Eltayeb M, Hassan M, Heiss G, Howe S, Kemp J, Klaus R, Krupp F, Mohamed N, Rouphael T, Turner J, Zajonz U (2004) Status of coral reefs in the Red Sea and Gulf of Aden in 2004. In: Wilkinson C (ed) Status of coral reefs of the world: 2004. Australian Institute of Marine Science, Townsville, Queensland, Australia, 301 pp
Kotb MMA, Hanafy MH, Rirache H, Matsumura S, Al-Sofyani AA, Ahmed AG, Bawazir G, Al-Horani FA (2008) Status of coral reefs in the Red Sea and Gulf of Aden Region
Krupp F, Türkay M, El Hag AGD, Nasr D (eds) (1994) Comparative ecological analysis of biota and habitats in littoral and shallow sublittoral waters of the Sudanese Red Sea. Forschungsinstitut Senckenberg, Frankfurt and Faculty of Marine Science and Fisheries, Port Sudan
Krupp F, Apel M, Hamoud A, Schneider W, Zajonz U (2006) Zoological survey in the Red Sea coastal zone of Yemen. Fauna Arab 21:11–32
Kuehlman DHH (1983) Composition and ecology of deep-water coral associations. Helgoländer Wiss. Meeresunters 26:183–204
de Lamarck JBPA (1816) Histoire naturelle des animaux sans vertèbres. Tome 2. Paris, 568 pp
Lewinsohn C, Fishelson L (1967) The second Israel South Red Sea expedition 1965. Report no. 3 (general report). Israel J Zool 16:59–68
Medio D, Sheppard, CRC, Gascoigne J (2000) The Red Sea. In TR McClanaha CRC Sheppard, DO Obura (eds) Coral reefs of the indian ocean their ecology and conservation. Oxford University Press, Oxford, pp 231–255
Mergner H, Schumacher H (1985) Quantitative Alalyse von Korallengemeinschaften des Sanganeb Atolls (mittleres Rotes Meer) I. Die Besiedslungs-Struktur hydrodynamisch unterschiedlich exponierter Auben-und Innenriffe. Helgolander Wisenschaftliche Meeresuntersuchungen 26:238–358
Nesteroff WD (1955) Les recif coralliens du Banc Farasan Nord (Mer Rouge). Ann Inst Oceangr 30, 46 pp
Obura D (1998) Marine and coastal assessment, Djibouti. Draft Report EARO/75545/389
Obura D, Djama (2000) Coral reef survey in Djibouti post bleaching. In: Tatwany H (ed). In: Proceedings of the international workshop on the extent and impact of coral bleaching in the Arabian Region. National Commission for Wildlife Conservation and Development, Riyadh
Oren OH (1962) The Israel South Red Sea expedition. Nature 194:1134–1137
Ormond RFG (1976) The Red Sea. In: Promotion of the establishment of marine parks and reserves in the northern indian ocean including the Red Sea and Persian Gulf. Papers and proceedings of the regional meeting held at Tehran, Iran, IUCN Publications, New Series 35, 6–10 Mar 1975, pp 115–123
Ormond RFG, Dawson Shepherd AR, Price ARG, Pitts JR (1984a) Report on the distribution of habitats and species in the Saudi Arabian Red Sea. No. 1 Kingdom of Saudi Arabia: IUCN/MEPA
Ormond RFG, Dawson Shepherd AR, Price ARG, Pitts JR (1984b) Report on the distribution of habitats and Species in the Saudi Arabian Red Sea. No. 2 Kingdom of Saudi Arabia: IUCN/MEPA
Ormond RFG, Dawson Shepherd AR, Price ARG, Pitts JR (1984c) Management of Red Sea coastal resources; recommendations for protected areas. Kingdom of Saudi Arabia: IUCN/MEPA
Ormond RFG, Bainimoon SA (1994) Ecology of intertidal macroalgal assemblages on the Hadramout coast of southern Yemen, an area of seasonal upwelling. Mar Ecol Prog Ser 105(1–2):105–120
Perry CT, Larcombe P (2003) Marginal and non-reef building coral environments. Coral Reefs 22:427–432
PERSGA (1998) Surveys of natural habitats and plans for their protection in Sudan. Hunting Aquatic Resources, London, Draft Final Report
PERSGA (2010) The status of coral reefs in the Red Sea and Gulf of Aden: 2009. PERSGA Technical Series No. 16, PERSGA, Jeddah
PERSGA/GEF (2003) Coral reefs in the Red Sea and Gulf of Aden. Surveys 1990 to 2000: summary and recommendation. The regional organization for the conservation of the environment of the Red Sea and Gulf of Aden, Technical Series No. 7, 136 pp
PERSGA/GEF (2004) Master plan for the Isles des Sept Frères/Ras Siyyan and Godoria marine protected area. Regional organization for the conservation of the environment of the Red Sea and Gulf of Aden
Pichon M (1971) Comparative study of the main features of some coral reefs of Madagascar, La Reunion and Mauritius. Symp Zool Soc Lond 28:185–216
Pichon M (1989) Rapport sur l'etat des recifs coralliens de Djibouti (effets de l'etoile de mer Acanthaster planci). Institut Superieur d'Etudes et de Recherches Scientifiques et Techniques (ISERST), Djibouti, Mission Francaise de Cooperation
Purkis SJ, Harris PM, Ellis J (2012) Patterns of sedimentation in the contemporary Red Sea as an analog for ancient carbonates in rift settings. J Sedim Res 82:859–870
Quadfasel D, Baunder H (1993) Gyre-scale circulation cells in the Red Sea. Oceanol Acta 16:221–229
Reinicke GB, Kroll SDK, Schumacher H (2003) Patterns and changes of reef-coral communities at the Sanganeb-Atoll (Sudan, Central Red Sea): 1980 to 1991. Facies 49:271–298
Roberts CM, Dawson Shepherd AR, Ormond RFG (1992) Large-scale variation in assemblage structure of Red Sea butterflyfishes and angelfishes. J Biogeogr 19:239–250
Rogers CS (1979) The effect of shading on coral reef structure and function. J Exp Mar Biol Ecol 41:269–288
Rogers CS (1990) Responses of coral reefs and reef organisms to sedimentation. Mar Ecol Prog Ser 62:185–202
Rossi L (1954) Madreporari, Stoloniferi e Milleporini. Rivista di Biologia Colon. 14:23–72
Roy KJ, Smith SV (1971) Sedimentation and coral reef development in turbid water: fanning Lagoon. Pac Sci 25:234–248
Rushdi AI, Abubakr MM, Hebba HMA (1994) Marine habitats of the Red Sea at Alurj-Alsalif and Dhubab-Yakhtul areas: their ecology, environment and management recommendations. UNDP/Yemen Republic Sanaa University, Faculty of Science, 117 pp
Scheer G (1964) Korallen von Abd-el-Kuri:Zoologische Jahrbücher. Atbteilung für Systematik 91(3):451–466
Schroeder JH (1974) The institute of oceanography in the democratic Republic of the Sudan. Report to the "National Council for Research", Manuscripy Berlin, pp 1–13
Scheer G, Pillai CSG (1983) Report on the stony corals from the Red Sea. Zoologica 45(133):19
Schroder JH, Nasr D (1983) The fringing reefs of Port Sudan: I. Morphology Sedimentology Zonation, Essener Georg
Schumacher H, Mergner H (1985) Quantitative Analyse von Korallengemeinschaften des Sanganeb-Atolls (mittleres Rotes Meer). II.

Vergleich mit einem Riffareal bei Aqaba (nordliches Rotes Meer) am Nordrande des indopazifischen Riffgurtels. Helgo Meeresunt 39:419–440

Schuhmacher H, Kiene W, Dulo W (1995) Factors controlling Holocene reef growth: an interdisciplinary approach. Facies 32:145–188

Sheppard CRC (1985) Reefs and coral assemblages of Saudi Arabia. 2. Fringing reefs in the southern region, Jeddah to Jizan. Fauna Saud Arab 7:37–58

Sheppard CRC (1987) Coral species of the indian ocean and adjacent seas: a synonymized compilation and some regional distributional patterns. Atoll Res Bull 307

Sheppard CRC (2003) Predicted recurrences of mass coral mortality in the Indian Ocean. Nature 425:294–297

Sheppard CRC, Sheppard ALS (1985) Reefs and coral assemblages of Saudi Arabia. 1. The central Red Sea at Yanbu al Sinaiyah. Fauna Saud Arab 7:17–36

Sheppard CRC, Sheppard ALS (1991) Corals and coral communities of Arabia. Fauna Saud Arab 12:3–170

Sheppard C, Price A, Roberts C (1992) Marine ecology of the arabian region: patterns and processes in extreme tropical environments. Academic Press, London, 359 pp

Smith SV, Buddemeier RW (1992) Global change and coral reef ecosystems. Ann Rev Ecol Syst 23:89–118. http://dx.doi.org/10.1146/annurev.es.23.110192.000513

Sofianos SS, Johns WE (2002) An oceanic general circulation model (OGCM) investigation of the Red Sea circulation: 1. Exchange between the Red Sea and the indian ocean. J Geophys Res 107 (C11):3196. doi:10.1029/2001JC001184

Sofianos SS, Johns WE (2003) An oceanic general circulation model (OGCM) investigation of the Red Sea circulation: 2. Three-dimensional circulation in the Red Sea. J Geophys Res 108 (C3):3066. doi:10.1029/2001JC001185

Sofianos SS, Johns WE (2007) Observations of the summer Red Sea circulation. J Geophys Res 112(C0):6025. doi:10.1029/2006JC003886

Spalding MD, Fox HE, Allen GR, Davidson N, Ferdana ZA, Finlayson M, Halpern BS, Jorge MA, Lombana AL, Lourie SA (2007) Marine ecoregions of the world: a bioregionalization of coastal and shelf areas. Bioscience 57:573–583

Streicher S (1980) 90 Tage im Korallenmeer. Rostock, 192 pp

Szmant AM (2002) Nutrient enrichment on coral reefs: is it a major cause of coral reef decline? Estuaries 25:743–766

Tilot B, Veron JEN, Jeudy de Grissac A (2008) The coral reefs of Eritrea: Little known gems. In: Kotb MMA, Hanafy MH, Rirache H, Matsumura S, Al-Sofyani AA, Ahmed AG, Bawazir G, Al-Horani FA (eds) Status of coral reefs in the Red Sea and Gulf of Aden Region

Titlyanov EA, Latypov YY (1991) Light dependence in scleractinian distribution in the sublittoral zone of South China Sea Islands. Coral Reefs 10:133–138

Turak E, Brodie J (1999) Coral and reef habitats. In: DouAbul A, Rouphael TS, Marchant R (eds) Protection of marine ecosystems of the Red Sea coast of Yemen. Hassell and Associates, AMSAT and UNOPS

Turak E, Brodie J, DeVantier L (2007) Reef-building corals and coral communities of the Yemen Red Sea. Fauna Arab 23:1–40

Veron JEN (2000) Corals of the World. In: Stafford-Smith M (ed) Australian Institute of Marine Science, vol 3

Veron JEN (2002) New species described in corals of the world. Australian Institute of Marine Science, Townsville, Monograph Series 11, 206 pp

Vine PJ, Vine MP (1980) Ecology of Sudanese coral reefs with particular reference to reef morphology and distribution of fishes. In: Proceedings of symposium on the coastal marine environment of the Red Sea, Gulf of Aden and tropical western Indian Ocean

Wainwright SA (1965) Reef community visited by the Israel South Red Sea expedition 1962. Report no. 9. Sea Fish Research Station. Haila Bull 38:40–53

Wilson SC (2006) Ecology of coral communities in a marginal environment: Southern Arabia. PhD thesis, Department of Biological Sciences, Warwick University, Warwick

Winterbottom R (1985) Revision of the congrogadid Haliophis (Pisces: Perciformes), with the description of a new species from Indonesia, and comments on the endemic fish fauna of the northern Red Sea. Can J Zool 63:209–217

Zajonz U, Eisinger M, Hamoud A (2005) Baseline biological and ecological survey of Ras Issa Peninsula and adjoining areas. Phase II environmental impact assessment. Proposed Ras Issa Refinery. DRAFT Support Document

Zajonz U, Deodatus F, Al-Harrani G (2010) Yemen strategic environmental assessment of coastal zone management. Ref. no. 100020474. Phase III Report (Progress Report Assessment and Analysis, Part 2, and Recommendations). March 2010 (revised June 2010), 221 pp

Zekeria AZ (2003) Butterflyfishes of the Southern Red Sea: ecology and population dynamics. University of Groningen, Department of Marine Biology, Haren, The Netherlands

Coral Reefs of the Red Sea with Special Reference to the Sudanese Coastal Area

Dirar Nasr

Abstract

Several features have favored the development of coral reefs in the Red Sea including the semi-enclosed nature, situation in an arid area with no permanent rivers or significant upwelling, its warm sea water and a reduced tidal range with moderate winds. The fact that the Red Sea coral reefs are the best developed in the western Indian Ocean is not surprising; more than 60 different genera of reef forming corals are found in the Red Sea alone with an exceedingly large recorded number of species. However, reef development varies from north to south in the Red Sea. North of 20°N reefs are well developed, occurring as narrow fringing reefs with steep slopes that drop into very deep water, particularly in the Gulf of Aqaba. South of 20°N the continental shelf widens and therefore reefs are less well developed vertically and often occur in more turbid water. Nevertheless, reef health is generally good throughout the Red Sea, with 30–50 % live coral cover at most locations and more than 50 % total cover on average. General threats to coral reefs and coral communities of the Red Sea include land filling and dredging for coastal expansion, destructive fishing methods, shipping and maritime activities, sewage and other pollution discharges, damage from the recreational scuba industry, global climate change, and insufficient implementation of legal instruments that affect reef conservation such as Marine Protected Areas (MPAs). The Sudanese reefs consist of three primary coral habitats along the Sudanese coastline: barrier reefs, fringing reefs and the Sanganeb atoll. They are considered to be in moderate to good health, with good fish fauna health. Raised fossil reefs that form coastal cliffs are characteristics of some sites such as Suakin and Dungonab Bay, while Sanganeb Marine National Park and Dungonab Bay–Mukawwar Island are the only MPAs in Sudan. Many of the present problems with coral reef conservation in Sudan are attributed to a lack of law enforcement, a lack of awareness, a weak legal framework, and the absence of surveillance. The crown-of-thorns starfish (COTS) *Acanthaster planci* was not recorded in plague numbers at any of the Sudan reefs. However, in 1999, bleached corals were estimated to cover 14 % of the substrate. In addition to the Jeddah Consolidated Convention, the Red Sea countries have become signatories to a number of international, regional, bilateral or multilateral agreements, and other legal instruments. Each country also possesses a relatively complete set of national laws and regulations. However, the implementation of these remains generally poor, and in some cases, there is no implementation or enforcement. The Regional Organization for the Conservation of the Environment of the Red Sea and Gulf of Aden (PERSGA) has made significant efforts to assist its member states, including the Sudan, to conserve their coastal and marine ecosystems and key species. Nevertheless, there is a need for further continued research on coral reefs and an information dissemination programme to enhance community participation and awareness.

D. Nasr (✉)
Red Sea University, Port Sudan, Sudan
e-mail: d_nasr47@hotmail.com

N.M.A. Rasul and I.C.F. Stewart (eds.), *The Red Sea*,
Springer Earth System Sciences, DOI 10.1007/978-3-662-45201-1_26,

Introduction

The unique setting of the Red Sea has attracted the attention of scientists and divers from all over the world. It is situated in an arid area and has been described in literature as being one of the world's most diverse and varied tropical marine habitats and communities (Chiffings 1995; PERSGA/GEF 1998, 2001). It is a semi-enclosed sea nearly isolated from the Indian Ocean with great biogeographic complexity and high levels of endemism (Chiffings 1995; PERSGA/GEF 2001), especially among reef fishes and invertebrates, the latter including a number of dinoflagellates and euphausiids (Roberts et al. 1992; Getahun 1998).

Extensive surveys on coral reefs have been carried out in the Red Sea as one of the most attractive ecosystems. They have been among the most photographed and most studied of any in the world (Head 1987). Yet, signs of deterioration have appeared as a result of human influences (PERSGA 2006) as well as the recent phenomenon of global warming. Accordingly, several conservation initiatives have been carried out on regional and national levels under the guidance of the Regional Organization for the Conservation of the Environment of the Red Sea and Gulf of Aden (PERSGA).

The Red Sea Coral Reefs

On a global scale, coral reefs are regarded as the most spectacular and diverse marine ecosystems on the planet, hosting hundreds of thousands of species, many of which are currently undescribed by science (Hoegh-Guldberg 1999). They are renowned for their extraordinary natural beauty, biological diversity, and high productivity. They represent crucial sources of income and resources through their role in tourism, fishing, building materials, coastal protection, and the discovery of new drugs and biochemicals (Carte 1996).

The coral reefs of the Red Sea are the best-developed reefs in the western Indian Ocean (PERSGA/GEF 2004a). Several factors are responsible for the development of such beautiful coral reefs in the Red Sea, for instance the absence of permanent rivers flowing into it bringing land sediments with their waters, and the absence of significant upwelling that brings the cold nutrient-rich bottom waters to the surface layers. Consequently, the amount of sediment is kept to a minimum and the Red Sea is regarded as one of the youngest and the deepest seas in the world despite its narrow shape (Vine 1985) (Fig. 1). The lower level of suspended sediments allows the penetration of sunlight further than in many other tropical seas (Vine 1985).

There is a clear north–south trend in the occurrence of the different community types and it appears that latitude, bathymetry, and coastal morphology are the underlying factors responsible for this pattern (Sheppard et al. 1992).

Fig. 1 Map of the Red Sea showing its *narrow shape* and surrounding countries (modified from Google Earth)

Additionally, the Red Sea is characterized by warm sea water, and a reduced tidal range with moderate winds indirectly related to the Monsoon. All these features have greatly favoured the development of coral reefs, particularly in the mid-portion of the Red Sea (Vine 1985). High evaporation and low precipitation also maintain the Red Sea as one of the most saline water masses of the world oceans. Due to the high evaporation and low precipitation, surface water salinities range from about 36.5 near Bab-al-Mandab to about 40-41 ‰ at the northern end of the main Red Sea basin (Alraddadi 2013). However, the salinity at the entrance of Bab-al-Mandab is 38 ‰ (Morcos 1970), the salinity at Port Sudan is 39 ‰ and at Dongonab Bay (180 km north of Port Sudan) it is 42 ‰ (records of the Institute of Marine Science of the Red Sea University, Port Sudan), and it reaches 42.5 ‰ at the Gulf of Aqaba. Therefore, the mean salinity of the Red Sea may be about 40.4 ‰.

The above description does not apply to the whole Red Sea; the Gulf of Suez, for instance, is shallow and sedimented and is not part of the main, deep Red Sea rift system (Sheppard et al. 1992). Reef development varies from north to south in the Red Sea. North of 20°N reefs are well developed, occurring as narrow fringing reefs with steep

Fig. 2 Narrow fringing reef north of Jeddah allowing sediments to be washed to the deep waters (photograph by M. Qurban)

slopes that drop into very deep water, particularly in the Gulf of Aqaba. In Egypt, however, the distribution and development of reef-building corals is mainly restricted by water temperature, sediment load, salinity and light intensity. These factors combined make the Gulf of Aqaba a more suitable reef habitat than the Gulf of Suez (PERSGA 2006). The fringing reefs in the central and northern Red Sea are mostly narrow allowing sediments to be washed away to the deep waters (Fig. 2), while in some areas, especially further south, reefs extend commonly 1 km to seaward with well-developed beaches (Sheppard et al. 1992).

In contrast, the coastal shelf is much wider in the southern Red Sea, so sediments are less readily lost to deeper water. The shallow bathymetry created by these sediment deposits in the southern Red Sea combined with their fine nature results in substrate instability and associated turbidity in exposed conditions which limits the opportunity for the development of coral reefs (MEPA/IUCN 1987). Even if conditions are suitable for the growth of corals, the continuous deposition of sediments would cover the growing corals leading to their smothering and death. With more deposition of sediments, the newly formed soft bottom creates a suitable condition for seagrass and algal growth. Thus, seasonal coverings of macroalgae are a feature of shallow coral reefs in the southern Red Sea (Fig. 3), and they often form the major coverage of hard substrates in areas too turbid for coral growth (Gladstone 2000; PERSGA/GEF 2001). An account of the increase in algal cover with the fall in coral cover in the southern Red Sea is given by Sheppard et al. (1992). However, reef diversity in the southern Red Sea is greatest offshore and includes platform and patch reefs, barrier reefs, coral cays, and extensive fringing reefs around island systems, especially the Farasan and Dahlak (Eritrea) groups of islands (PERSGA 2006).

The richness of hermatypic corals in the Red Sea was believed to include 180–200 species (Vine 1986; Sheppard and Sheppard 1991; Sheppard et al. 1992). However, a recent extensive study of the central-northern Red Sea coastline of Saudi Arabia extended this to probably 260 species, based on recently described species and range extensions (DeVantier et al. 2000a). Descriptions of the composition and status of corals, coral reefs, and coral communities of the Red Sea and Gulf of Aden are given by Sheppard and Sheppard (1991), Sheppard et al. (1992), DeVantier et al. (2000a, 2004), Kemp (1998) and Kemp and Benzoni (2000).

Status of Coral Reefs in the Red Sea

The status of corals and coral reefs in the Red Sea and Gulf of Aden has been described by several workers, for example Sheppard and Sheppard (1991), Sheppard et al. (1992), DeVantier et al. (2000a), Kemp (1998) and Kemp and Benzoni (2000). Some of these works have been compiled by PERSGA (2006), based primarily on PERSGA/GEF (2003a, b).

Fig. 3 Macroalgae forming the major coverage of hard substratum north of Jazan (photograph by M. Qurban)

Reef health is generally good throughout the Red Sea and Gulf of Aden, with 30–50 % live coral cover at most locations and more than 50 % total cover on average (PERSGA 2006). These values are within the range of other reefs surveyed globally as part of the Reef Check programme. Coral diversity and reef-associated fauna were considered among the highest in the Indian Ocean region (PERSGA 2006). Between 1998 and 1999, in the central-northern Red Sea living coral cover ranged from less than 10 % to more than 75 % and averaged 35 % (PERSGA 2006). Most of these reefs were in good condition (DeVantier et al. 2000a) with little evidence of human impact apart from on reefs near urban areas.

The greatest development of corals in the Egyptian coast occurs at the tip of the Sinai Peninsula at Ras Mohammed and between Hurghada and Safaga (PERSGA 2006). However, live coral cover varies from 55 % in exposed areas to 85 % in sheltered areas (PERSGA/GEF 2003b). The diversity of reef-building corals, however, is distributed in the following pattern: Gulf of Aqaba (47 genera, 120 species), Gulf of Suez (25 genera, 47 species), northern Red Sea (45 genera, 128 species), central portion of the Egyptian Red Sea (49 genera, 143 species) and southern Egyptian Red Sea (31 genera, 74 species) (PERSGA/GEF 2003b). Fifty-three species of reef-building corals were found at reefs off Hurghada (Ammar and Amin 2000). The fringing reefs of the Jordanian coast border up to 50 % of the Jordanian coastline supporting a high diversity of coral and associated fauna. They contain at least 158 coral species from 51 genera (PERSGA/GEF 2003b).

Compared to other PERSGA countries, a lot of information is available with regard to the species composition and assemblage diversity of Saudi Arabian Red Sea coral reefs. Saudi Arabian reefs comprise of at least 260 species of scleractinian corals from 68 genera and 16 families, in addition to at least 30 taxa of soft corals, fire corals, zoanthids and gorgonians (PERSGA 2006). Coral reefs fringe much of the Saudi Arabian Red Sea coastline and offshore islands. The best-developed barrier reef system in the region occurs along the seaward margin of the Al-Wajh Bank. South of Jeddah reefs become less well developed along the coast (PERSGA 2006). Generally, the coral reefs of the Saudi Arabian Red Sea are in good condition, with the exception of those near Jeddah and Yanbu (PERSGA/GEF 2003b).

With regard to Yemen's Red Sea only about 25 % of the coastline supports coral reefs. The most highly developed reefs occur offshore, in the vicinity of the many islands that characterize the southern Red Sea (PERSGA/GEF 2003b). A total of 176 species of stony corals have been recorded from the Red Sea coast of Yemen, with richness at individual sites ranging from 1 to 76 species. At least 19 new records have been identified for the southern Red Sea. Reefs in the northern Yemeni Red Sea showed low live coral cover (average 17 %), high dead coral (DC) cover (average 34 %) and a high percentage cover of macroalgae (20 %). The northern and central Yemeni coast and near shore islands had very low live coral cover (3 %) and high DC cover (average 34 %) (PERSGA/GEF 2003b).

In Djibouti, most coral reefs are in average to good condition (PERSGA/ALECSO 2003) in relation to the mean abundance of fish and invertebrate indicator species and average of impact levels and substrate cover. Living hard coral cover across reefs averaged 39 % with a maximum of 80 % (PERSGA 2006). One hundred and sixty-seven species of corals have been recorded from Djibouti, including three species of black coral (PERSGA 2006). The highest diversity of corals was recorded from Arta Plage, Gulf of Tadjoura (93 species) followed by Iles des Sept Frères (84 species) and Trois Plages, Gulf of Tadjoura (75 species) (PERSGA/GEF 2003b). The corals of Djibouti are a unique assemblage of species due to the confluence of several biogeographic zones including the tropical warm water biota of the Indian Ocean and Red Sea and species common to cold water upwelling habitats of the Somali and Arabian Sea regions (PERSGA 2006). The southern coast, however, has poorly developed reefs due to cold water upwelling from the Indian Ocean.

Status of Coral Reefs in Sudan

The coastline of the Sudanese Red Sea is characterized by a large number of inlets, *mersas* and bays. In addition to others, it encompasses three primary coral habitats: barrier reefs, fringing reefs and the Sanganeb atoll (Fig. 4). The Sudanese coral reefs (Fig. 5) have attracted the attention of many national and international scientists. For instance, species lists for Sanganeb (Schroeder and Scheer 1981) and Wingate (Scheer and Pillai 1983) are available, and the reefs in the vicinity of Port Sudan have been extensively described (Schroeder and Nasr 1983). The Wingate and Towartit reef complex has been studied extensively by Vine and Vine (1980). Comparative ecological analysis of biota and habitats in littoral and shallow sub-littoral waters has been carried out in the Sudanese Red Sea with an emphasis on the Sanganeb Atoll (Krupp et al. 1994b).

Most of the coast is bordered by fringing reefs 1–3 km wide that are separated by deep channels from a barrier reef 1–14 km offshore. The outer barrier drops steeply to several hundred metres depth. Previous studies along these reefs suggested they are among the most diverse and spectacular in the Red Sea (Head 1980; IUCN/UNEP 1985; Krupp et al. 1994b; Ormond and Edwards 1987; Schroeder 1981; Schroeder et al. 1980; Vine and Vine 1980; Vine 1985, 1986). One of the most unique reef structures in the Sudanese Red Sea is the Sanganeb atoll (Krupp 1990), whose steep slopes rise from a sea floor more than 800 m deep (Fig. 6).

Some of these authors referred to these reefs, in the past, as being among the most diverse in the Red Sea (UNEP/IUCN 1988). However, the reality is more complex due to the interplay of many physical and environmental factors (Vine and Vine 1980). Others considered these reefs to be in moderate to good condition (depending on the abundance of fish and invertebrate indicator species, the average of impact levels and substrate cover), despite recent reports of extensive coverage of algae over a high proportion of the fringing reefs. The condition of Sudan's coral reefs was assessed in 1997 (PERSGA/

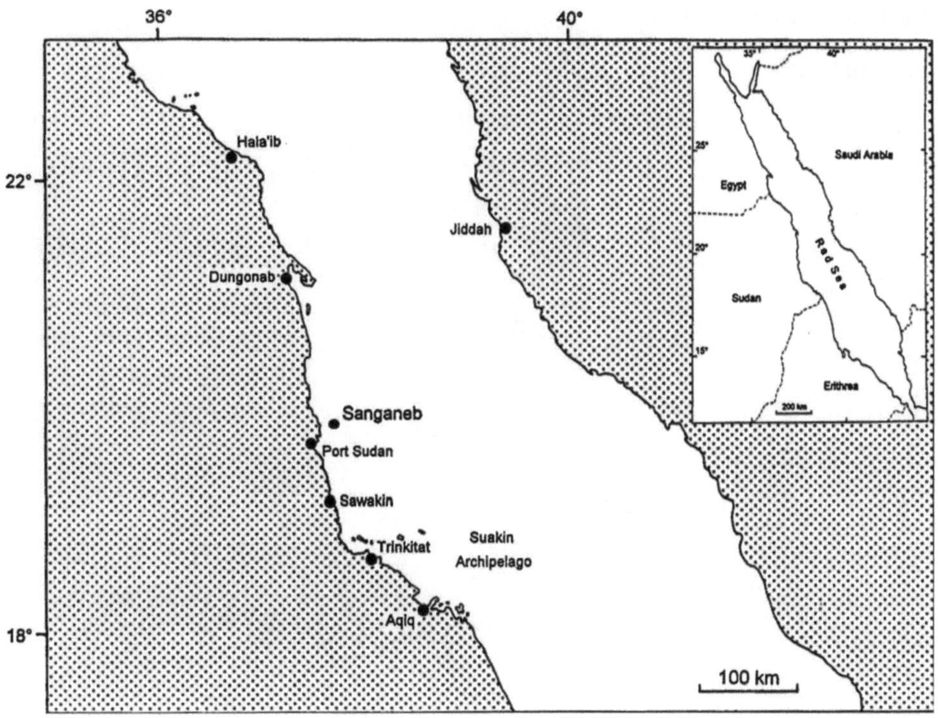

Fig. 4 The Sudanese coast with sites mentioned in the text (after Pilcher and Nasr 2000)

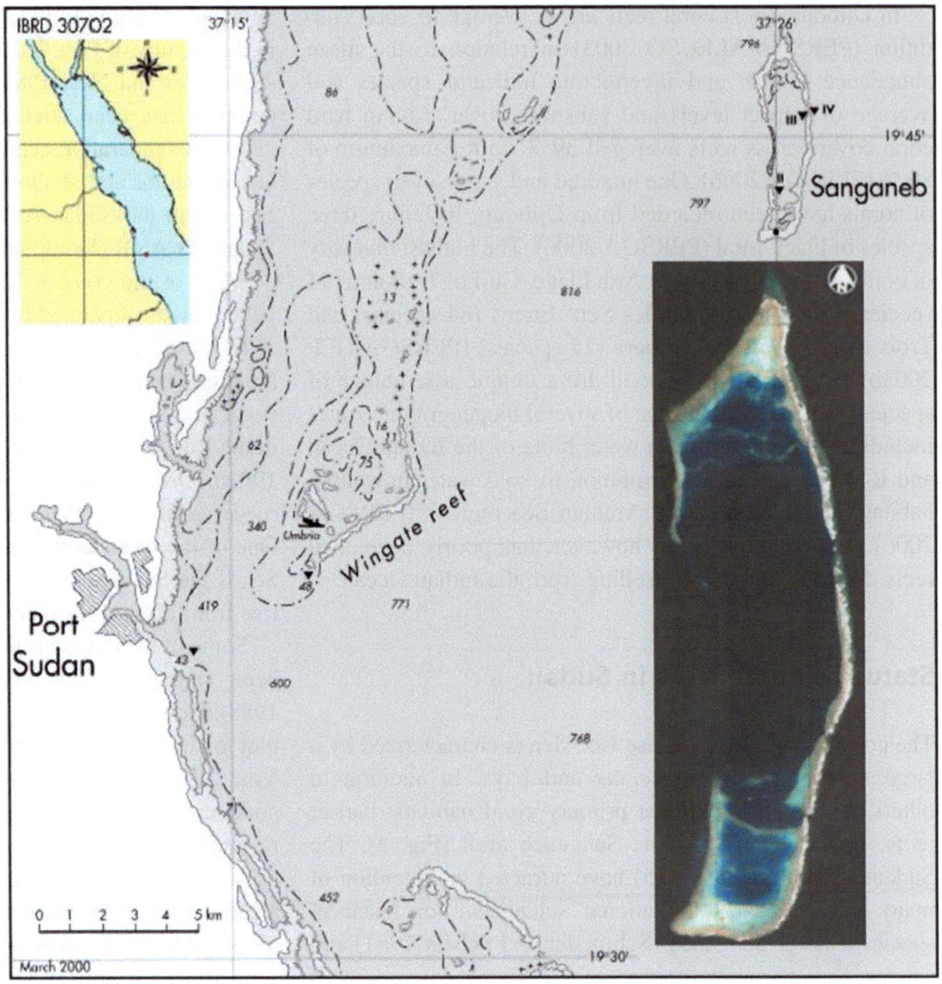

Fig. 5 Reefs in the vicinity of Port Sudan (modified from PERSGA/GEF 2001)

Fig. 6 Sanganeb atoll, a marine national park (photograph by Hans and Nasr)

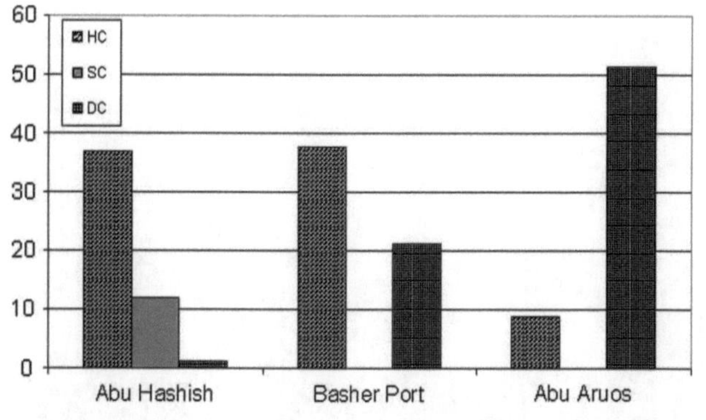

Fig. 7 Average percentage of coral cover in the three selected sites on the Sudanese Red Sea coast (after Nasr and Al-Sheikh 2000)

ALECSO 2007), 1999 (Nasr and Al-Sheikh 2000, PERSGA/GEF 2003b) and in 2008 (PERSGA 2010). Surveys by Nasr and Al-Sheikh (2000) at Abu Hashish Jetty (at Port Sudan) found that the percentage of hard live coral (HC) ranged from 23.5 % at 10 m depth and 50 % at 5 m, while DC ranged from 2.5 % at 10 m depth and 0 % at 5 m depth suggesting that the area was comparatively healthy (Fig. 7). However, key indicator species were abundant and diversity appeared high relative to other Red Sea sites (PERSGA/GEF 2003b).

The offshore reefs of Suakin have also been described as being of high diversity offshore with decreasing diversity as one proceeds towards the coastal fringing reef (Head 1980). Recent observations during June 2012 by the author showed that the reefs at the Sanganeb atoll, Shaab Rumi and the wreck of the Umbria, in the vicinity of Port Sudan city, are still in good condition (Fig. 8a–c). The wreck of the Umbria has been completely covered by various colonies of soft and hard corals, with cabins of the wreck providing perfect shelter and nursery grounds for various coral fish and other marine organisms.

The famous Conshelf 2 underwater living experiment conducted at Shaab Rumi reef by the Cousteau team of divers in 1964 has provided a unique opportunity to study growth rates of various corals and to investigate the factors affecting their settlement and development on what has turned out to be an almost perfectly designed artificial reef (Fig. 9, Vine 1985).

The coastline of some parts of both Dungonab Bay and Suakin harbour is characterized by raised fossil reefs forming coastal cliffs (Fig. 10). A combination of sea-level change and tectonic movements that have elevated certain sections of previously flourishing reefs above the present day sea level have resulted in such fossilized coral reefs (Vine 1985). In both areas, these fossils form coastal cliffs that are under continual erosion from the effects of waves, rainfall, wind and sun (Vine 1985). Although fragments of molluscs and branching corals cover the surfaces of these cliffs, massive corals such as *Porites* or the brain coral *Platygyra* are quite distinguishable as they effectively resist breakdown and remain as evidence of once flourishing prehistoric coral reefs (Vine 1985).

Threats Facing Coral Reefs

Coral reefs are being endangered by a diverse range of human-related threats (Hoegh-Guldberg 1999). The continued deterioration of coral reefs as a result of human influences has been observed by several writers (Bruno et al. 2007; Hughes et al. 2003). Such influences include eutrophication, increased sedimentation, over-exploitation of marine species and physical destruction by reef users (Sebens 1994), in addition to overfishing and land development causing widespread changes in reef ecosystems over the past two centuries (Hughes 1994). Such threats are compounded by the more recent, superimposed impacts of global climate change (Hughes et al. 2003). Many reef ecologists suspect that anomalously high ocean temperatures contribute to the increased incidence and severity of disease outbreaks (Bruno et al. 2007) causing global decline of reef-building corals; an estimated 30 % are already severely damaged, and close to 60 % may be lost by 2030 (Wilkinson 2002).

Moreover, increasing atmospheric carbon dioxide is expected to alter the alkalinity of the world's oceans over the next century, creating challenging conditions for growth and survival of corals and other carbonate secreting organisms (PERSGA/GEF 2003b). Coral bleaching and disease have already increased greatly in frequency and magnitude over the past 30 years (Hoegh-Guldberg 1999, Hughes et al. 2003). Present predictions are that calcification rates may slow by as much as two-third over the next 50 years, with the potential for catastrophic effects on reef growth and marine biodiversity in general (Kleypas et al. 1999).

Although large parts of the Red Sea are still in a pristine state (PERSGA 2006), coral reefs of the region are exposed to a variety of threats including land filling and dredging for coastal expansion, destructive fishing methods, shipping and maritime activities, sewage and other pollutants discharges, damage from the recreational scuba diving industry, lack of public awareness and insufficient implementation of legal instruments that affect reef conservation (PERSGA/GEF 2003b). The depth distribution of coral damage suggested that most of the damage in shallow sections of reefs was caused by anchors (PERSGA 2006). These rapidly increasing environmental threats require immediate action to protect the region's coastal and marine environment.

The Report on the State of Marine Environment in the Red Sea and Gulf of Aden (PERSGA 2006) has focused on the particular threats that affect coral reefs in coastal areas of PERSGA member states. These include pollution, port activities, intense coastal development (Jordan), high levels

Fig. 8 **a** The wreck of the Umbria completely covered by various colonies of soft and hard corals (photograph by Hans and Nasr). **b** A "Micro-atoll" formed by *Porites* at Shaab Rumi (photograph by Hans and Nasr). **c** The steep reef slope at Sanganeb Marine National Park showing healthy corals and coral community (photograph by Hans and Nasr)

of coral usage, bleaching, crown-of-thorns starfish (COTS), or sedimentation on approximately 10 % of reefs (Egypt), fishing practices that damage coral reefs and the discharge of high temperature brine from power stations (Yemen) in addition to tourism and sewage discharges (Djibouti). Threats to Saudi Arabia's coral reefs originate primarily through industrial development and maritime transport, including oil spills, land filling, pollutant discharges and effluents from desalination activities (PERSGA 2006). The most acute damage to reefs is localized around major urban areas. With regard to Sudanese coastal areas, due to the limited industrial development along the Sudanese coastline threats are localized at the few urban centres, such as dredging and land filling for the extension of the main Port and the port at Suakin, as well as land-based pollution from the petroleum industry, although the tourism sector contributes to damage of the reefs by anchors and flipper damage.

Coral Die-off

Anthropogenic pressures on the Sudanese reefs are low, with those most affected being the fringing reefs along the coast in the vicinity of Port Sudan and Suakin (Pilcher and Nasr 2000). However, over 80 % of the shallow coastal fringing coral reef sites surveyed in 1997 had a high percentage of thin algal film cover, averaging 28.8 %. Such algal film

Fig. 9 The "garage" of Cousteau's Conshelf 2 underwater living experiment conducted at Shaab Rumi reef by the Cousteau team of divers in 1964 (photograph by Hans and Nasr)

Fig. 10 Raised fossil reefs forming coastal cliffs characterize parts of the coasts of Dungonab Bay and old Suakin, with close-up in inset image (photographs by Hans and Nasr)

cover was not found to affect those in depths greater than 10 m where reefs contain healthy colonies of framework corals (Pilcher and Nasr 2000). Similar observations were recorded by Vine and Vine (1980) who extensively studied the Towartit reef complex and reported poor hard coral growth and high algal covering on the fringing reef south of Port Sudan. The origin of the high algal film cover was attributed to a thermal event, possibly through runoff of high temperature waters from the lagoon (PERSGA/ALECSO 2007). The die-off event is consistent with reports of similar events in Saudi Arabia and Eritrea at the same time (PERSGA/GEF 1998). The algal cover did not affect larger colonies of *Porites* and *Acropora*, and it was suggested that the reef may recover from this within a span of decades, rather than centuries. The coral die-off has been observed even at the Sanganeb reefs which are far from the impacts of coastal activities (about 35 km NE of Port Sudan city), and also at Shaab Rumi situated north of Sanganeb atoll (Fig. 11). This observation suggests that anthropogenic coastal activities are not the only cause of coral die-off and that it is a natural phenomenon that requires further investigations.

Fig. 11 Coral die-off seen at various parts of the Sudanese reefs (photograph by Hans and Nasr)

Although coral diseases have never been investigated in the Sudanese waters, it is probable that competition between coral communities for light and space has resulted in weakening neighbouring colonies making them susceptible to diseases. Corals compete directly by damaging the tissue of neighbouring colonies with tentacles and digestive filaments (Lang and Chornesky 1990). These encounters usually cause lesions and local tissue necrosis (Lang and Chornesky 1990) that could facilitate pathogen transmission and colony infection (Bruno et al. 2007). The results of investigations in the Great Barrier Reef (GBR) showed that major outbreaks of the coral disease white syndrome only occurred on reefs with high coral cover after especially warm years (Bruno et al. 2007). Competitive interactions among corals on the GBR are relatively rare when cover is below 50 % (Connell et al. 2004). However, little progress has been made in identifying the causative agents for marine diseases or in applying standard epidemiological methods to assess impact or mode of transmission (Harvell et al. 1999).

Coral Bleaching

Coral bleaching is the disruption of symbiosis between coral hosts and photosynthetic microalgal endosymbionts, referred to as zooxanthellae (Brown 1997). The bleaching process is the loss of the zooxanthellae and/or their photosynthetic pigments causing the coral to lose colour (Rosenberg and Ben-Haim 2002). Bleaching may occur at local scales (Egana and DiSalvo 1982; Goreau 1964) or as a mass bleaching (Hoegh-Guldberg and Salvat 1995; Brown 1997). Six major episodes of coral bleaching have occurred since 1979 with massive mortalities of coral affecting reefs in every part of the world (Hoegh-Guldberg 1999).

The coral bleaching of 1998 was the most geographically extensive and severe in recorded history (Wilkinson et al. 1999), causing significant mortality worldwide (Baird and Marshall 1998). An estimated 16 % of the world's reef-building corals died (Wilkinson 2000). The impact of thermal stress on reefs can be dramatic, with the almost total removal of corals in some instances (Hoegh-Guldberg 1999; Brown 1997; Mumby 2001). This is why it is considered by most reef scientists to be a serious challenge to the health of the world's coral reefs (Hoegh-Guldberg 1999). In studies on coral diseases, some scientists describe coral bleaching as a disease—a process resulting in tissue damage or alteration of function, producing visible physiological or microscopic colour symptoms (Rosenberg and Ben-Haim 2002). On the other hand, coral biologists do not agree with this statement (Peters 1984; Hayes and Goreau 1998; Richardson 1998). However, the subject is debatable and no definite conclusions have been reached.

In the Red Sea and Gulf of Aden region, coral bleaching caused extensive coral mortality, including near total mortality on several reefs in 1998 (Pilcher and Alsuhaibany 2000; Spalding et al. 2001), with the exception of Jordan, possibly as a result of Jordan's more northerly latitude (PERSGA/GEF 2003b). Bleaching caused mass coral mortality in the central-northern Saudi Arabian Red Sea in August–September 1998 when sea-surface temperatures exceeded 31 °C, which was more than 2 °C above the mean monthly average (DeVantier et al. 2000b). Bleaching was patchily distributed and highly variable in intensity. The most intense bleaching occurred near Rabigh, where more than 65 % of total coral cover was bleached or had died

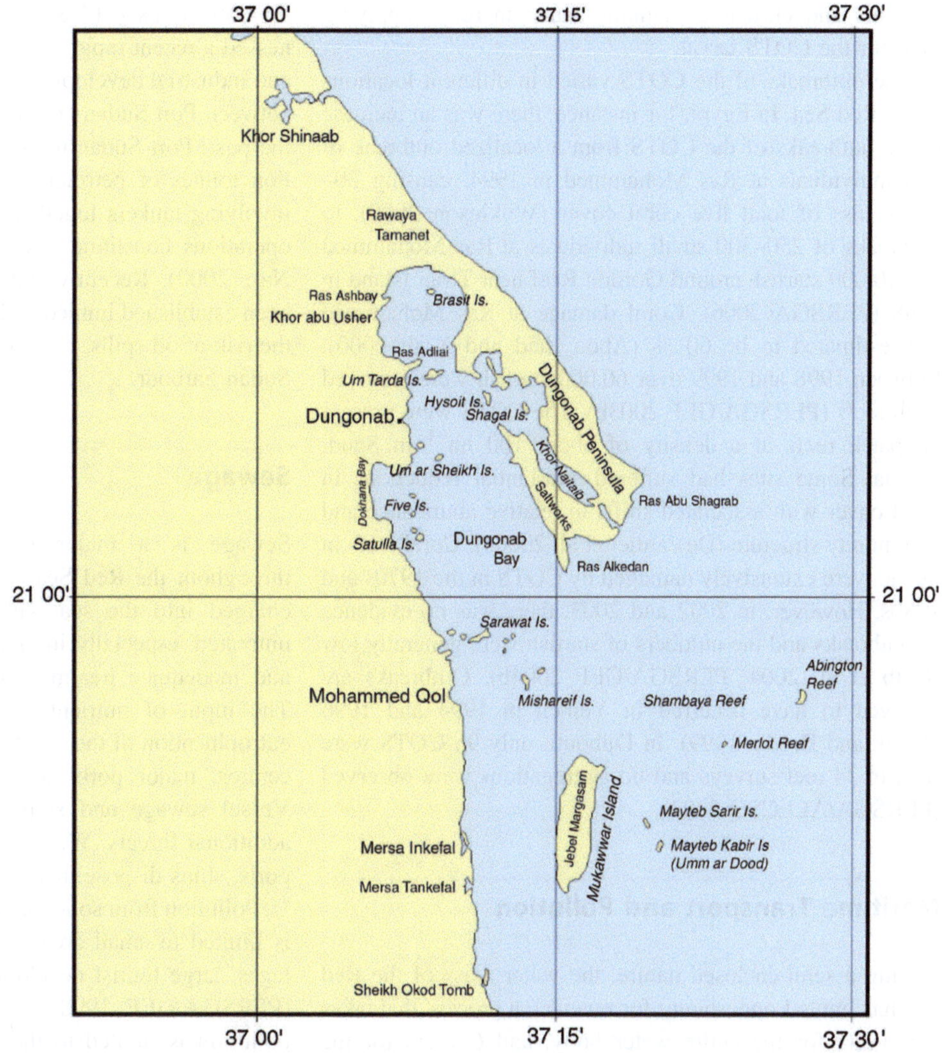

Fig. 12 Dungonab Bay/ Mukawar Island MPA (after PERSGA/GEF 2004b)

recently (PERSGA 2006). Significant levels of coral mortality were observed along the southern Red Sea, where at some sites (e.g., Abalat Islands) live coral cover declined from 80 % in 1993 to about 10 % in 1999 (PERSGA 2006).

Coral reefs in the Dungonab Bay–Mukawwar Island MPA that lies at 180 km north of Port Sudan (Fig. 12) were assessed in 2002 (PERSGA/GEF 2004b). Major differences in the health of coral communities were present between parts of the MPA. The coverage of living coral was generally greatest within Dungonab Bay. The greater coverage of DC outside Dungonab Bay was attributed to the effects of the 1998 coral bleaching event. Corals within Dungonab Bay may be pre-adapted to greater ranges in sea-surface temperature because of the semi-enclosed nature of the bay. Additionally, it is possible that Dungonab Bay may not have experienced the elevated sea-surface temperatures because it is somewhat isolated from the main body of the Red Sea, unlike coastal areas of the MPA (PERSGA/GEF 2004b). Thus, Dungonab Bay may be regarded as a refuge for corals from future bleaching events.

The Crown-of-thorns Starfish (COTS)

The COTS (*Acanthaster planci*) feeds on coral tissue also resulting in coral bleaching. This starfish has been observed in low densities in several areas in the Sudanese waters without causing any problem to coral reefs. However, densities of COTS greater than 30 mature individuals per hectare consume living coral tissue at a rate greater than the rate of replenishment and the coverage of DC increases (PERSGA 2006). Outbreaks have been reported since the 1970s from Australia, Japan, Palau and Fiji (PERSGA 2006). Such outbreaks have been reported for many years in the Red Sea and some authors believe these to be associated with the removal of predators (by fishing) such as puffer fish and triggerfish (Ormond and Campbell 1971, 1974; Moore 1990) while others (Vine 1970, 1973) attribute such fluctuations to natural causes. Natural causes responsible for the outbreak of the COTS are described in some papers (e.g., Birkeland 1982), such as rain water run-off bringing nutrients and enriching plankton fauna thereby supplying food for the COTS larvae.

The outbreaks of the COTS varied in different locations in the Red Sea. In Egypt, for instance, there was an increase in the outbreaks of the COTS from a localized outbreak of 200 individuals at Ras Mohammed in 1994, causing 20–30 % loss of total live coral cover (Wilkinson 2000), to outbreaks of 250–300 small individuals at Ras Mohammed and 10,000 starfish around Gordon Reef near Tiran Island in 1998 (PERSGA 2006). Coral damage at Ras Mohammad was estimated to be 60 % (Abou Zaid and Kotb 2000). Between 1998 and 1999 over 60,000 starfish were removed from reefs (PERSGA/GEF 2003b). The COTS were present on patch reefs at a density of about 100 ha^{-1} in Saudi Arabia. Some sites had suffered substantial reductions in coral cover with associated shifts in relative abundance and community structure (DeVantier et al. 2000b). Coral reefs in Sudan were extensively damaged by COTS in the 1970s and 1980s. However, in 2002 and 2003, there was no evidence of outbreaks and the numbers of starfish were generally low (Kotb et al. 2004; PERSGA/GEF 2004b). Outbreaks are believed to have occurred in Yemen in 1994 and 1996 (Turak and Brodie 1999). In Djibouti, only 96 COTS were seen in 34 reef surveys and no aggregations were observed (PERSGA/ALECSO 2003).

Maritime Transport and Pollution

Having a semi-enclosed nature, the water mass of the Red Sea has limited opportunity for renewal, a process that takes 200 years for the entire water body, and 6 years for the surface layer (Sheppard et al. 1992). Hence, the Red Sea is vulnerable to all sorts of pollution especially by oil and other pollutants from land-based activities. The Red Sea is regarded as one of the busiest shipping routes in the world where about 25,000–30,000 ship transits occur annually, mostly involving the transport of petrochemical products (Gladstone et al. 1999), including more than 100 million tonnes of oil (UNEP 2002). Hence, the potential for major oil spills and disasters at sea is high (PERSGA/GEF 1998). The coral reef systems also pose several problems to navigation taking into account insufficient tanker safety specifications and poor navigation aids (Pilcher and Nasr 2000). The absence of adequate reception facilities at most ports in the Red Sea presents another risk of pollution by ballast water and bunker oil discharge. Chronic oil pollution has already been observed in the immediate vicinity of some major Red Sea ports as a result of operations at oil terminals or discharges from power plants (Gerges 2002).

The coastal and marine biodiversity on the Sudanese coast is in moderate to good condition; the mean percentage of living substrate cover is 32.66 for hard corals and 13.52 for soft corals (PERSGA 2010). Nevertheless, it has witnessed a recent rapid growth of maritime transport activities and industrial development, particularly in the coastal stretch between Port Sudan and Suakin (Nasr and Eltayeb 2012). In the past, Port Sudan used to handle approximately 1.5 million tonnes of petroleum products annually and accidents involving tankers together with discharges from unloading operations constitute a serious pollution risk (Pilcher and Nasr 2000). Recently, three oil exporting terminals have been established immediately south of Port Sudan increasing the risk of oil spills, but decreasing oil contamination in Port Sudan harbour.

Sewage

Sewage is a major source of coastal contamination throughout the Red Sea (UNEP/PERSGA 1997). It is discharged into the sea with poor treatment or sometimes untreated, especially in areas with rapid population growth and inadequate treatment facilities (PERSGA/GEF 1998). The input of nutrient-rich sewage water also results in eutrophication of the coastal waters around some population centres, major ports and tourist facilities (Gerges 2002). Vessel sewage and ship discharges of solid waste pose additional threats. Without waste reception facilities at the ports, ships dispose their waste directly into the sea.

Pollution from solid waste is a major problem, although it is limited to small areas around urban centres, coastal villages, large tourist developments and major shipping lanes (PERSGA/GEF 1998; Gladstone et al. 1999). Chemical pollution is limited to the vicinity of industrial zones and facilities (PERSGA/GEF 1998), which usually discharge their effluents directly into the sea. These industries include phosphate mines, desalination plants, chemical industrial installations and oil production and transportation facilities.

Coastal Development

Growing human populations, coastal urbanisation and tourist development as well as increasing oil and gas exploitation and transport in the region are expected to place increasing pressures on the health of the Red Sea (PERSGA/GEF 1998; Gladstone et al. 1999). Most of the environmental threats and impacts can be prevented by proper environmental planning and management, use of environmental assessments and through the enforcement of appropriate regulations, most of which are already in place.

Habitat destruction as a result of coastal development is localized. The extension of Port Sudan and the port of Suakin, which involved dredging and land filling, resulted in severe sedimentation pressure on coral reefs. In Suakin, parts of the coastal fringing reef have been removed for the extension of the port. A new port has been constructed at O'Seif and a fourth one is planned at Agig, and further reef damage is expected at these locations (Pilcher and Nasr 2000). The establishment of a proposed Economic Free Zone (EFZ), which will cover 600 km^2 between Port Sudan and Suakin, may also impact diverse coral reefs at Towartit, which are located immediately in front of the planned EFZ. Heavy industries, petrochemical industries, fish processing factories, slaughter houses with a capacity of 3,000 head per day, tanneries, and warehouses will be established in the area (Pilcher and Nasr 2000).

Coral Reefs Conservation Initiatives

For coral reef conservation to improve and to be effective in the region, there is a need for increased public awareness, increased implementation and enforcement of national and international legal instruments, and the execution of coastal management plans that integrate coastal development, pollution control and tourism with the maintenance of environmental quality in marine habitats (Pilcher and Nasr 2000). The Regional Organization for the Conservation of the Environment of the Red Sea and Gulf of Aden (PERSGA) has played a key role in most initiatives formulated to conserve the marine resources of the Red Sea and Gulf of Aden including coral reefs. Concentrating on its long-term objective to safeguard the coastal and marine environments and ensure sustainable use of its resources, PERSGA has been active in promoting regional cooperation and providing support to national environmental plans through meetings, surveys, environmental assessments, legal developments and training workshops.

Legal Aspects

One of the most significant achievements of PERSGA was the development of international laws protecting the environment of the region. The Regional Convention for the Conservation of the Environment of the Red Sea and Gulf of Aden (the 1982 Jeddah Convention), the Protocol Concerning Regional Cooperation in Combating Pollution by Oil and other Harmful Substances in Cases of Emergency (1982), and the Action Plan for the Conservation of the Environment of the Red Sea and Gulf of Aden were signed and ratified by PERSGA member states; they provide the legal framework for cooperation in marine environmental issues, focusing on the prevention, reduction and fight against pollution. Major functions of PERSGA include the implementation of the Jeddah Convention, the Action Plan and the Protocol (PERSGA 2005).

New Legal Instruments were developed by PERSGA following a regional review of all marine environment-related legislation. In this context, three additional important protocols have been developed and signed by member states, namely the Protocol on Biodiversity and the Establishment of Regional Network of MPAs (2005), the Protocol on the Protection of the Marine Environment from Land-Based Sources (2005) and the Regional Protocol Concerning Technical Cooperation to borrow and transfer Experts, Technicians, Equipment and Materials in Cases of Emergency (2009).

The Red Sea countries have become signatories to a number of international, regional, bilateral or multilateral agreements and other legal instruments. Each country also possesses a relatively complete set of national laws and regulations. However, the implementation of these remains generally poor, and in some cases there is no implementation or enforcement (PERSGA/GEF 2003b).

Action and Contingency Plans

PERSGA has succeeded in the preparation of several action plans on regional and national levels in cooperation and coordination with member states and relevant international organizations. The objectives of these action plans were the conservation and sustainable use of key habitats and species in the Red Sea and Gulf of Aden region. These action plans were preceded by the preparation of a set of standard survey methods for key species and habitats (PERSGA/GEF 2004a) and collection of critical baseline data on key habitats and species through surveys and a series of training workshops.

The action plans relevant to coral reefs included the Regional Action Plan for the Conservation of Coral Reefs in the Red Sea and Gulf of Aden (PERSGA/GEF 2003a) and the Regional Programme of Action for the Protection of the Marine Environment from Land-based Activities (UNEP/PERSGA 1997). As an implementation of this regional programme of action, PERSGA supported the preparation of national programmes of action for the protection of the marine environment from land-based activities in member states. The countries of the region also prepared their national oil spill contingency plans and ICZM plans. For Sudan, PERSGA prepared the National Oil Spill Contingency Plan (PERSGA/GEF 2003a), a national ICZM plan, as well as surveys of habitats and plans for their protection in Sudan (PERSGA/ALECSO 2007).

Marine Protected Areas (MPAs)

MPAs are currently the best management tool for conserving coral reefs and many other marine systems. MPAs range from ineffective "paper parks", to multiple-use areas with varying degrees of protection, to marine reserves or no-take areas (Hughes et al. 2003). The Red Sea contains a number of MPAs (PERSGA 2006), and the formation of a regional network of MPAs has been proposed by PERSGA (PERSGA/GEF 2001).

During the implementation of the GEF supported project "Strategic Action Programme (SAP) for the Red Sea and Gulf of Aden", PERSGA succeeded in the establishment of a network of MPAs in the region, and the development of a Regional Master Plan which formed a regionally agreed framework for the planning and management of each MPA in the network (PERSGA/GEF 2002). With GEF support and PERSGA execution, existing MPAs have been enhanced, and new ones established, accelerated through a range of ecological and socioeconomic workshops, training and surveys. Further, PERSGA prepared a set of site-specific master plans with management guidelines for each of the new proposed MPAs based on the results of the field surveys.

Given Sudan's relatively small coastline, two important protected areas exist, with Sanganeb being a world-renowned marine protectorate. Sanganeb has yielded a wealth of information on Sudan's marine habitat and is the centre of much of the country's research into coral reef ecosystems (Pilcher and Nasr 2000). Dungonab Bay and Mukkawar Island National Park include within their boundaries, a highly diverse complex of coral reefs, mangroves, seagrass beds, beaches, and intertidal areas.

Sanganeb and Dungonab Bay/Mukawwar Island MPAs were declared by the Government of Sudan in 1990 and 2004, respectively. Both MPAs are not designated as no-take areas (NTAs) although NTAs provide the most effective protection of corals and coral communities from destructive activities such as uncontrolled coastal development; they also serve as a spatial refuge for a portion of the stock from which larvae and adults can disperse to neighbouring deteriorated areas. Dungonab Bay/Mukawwar Island MPA, at present, is regarded as a multiple-use area being used mostly by local fishermen, while Sanganeb MPA, with varying degrees of protection, is mostly visited by tourists.

Monitoring and Management Capacity to Conserve Coral Reefs

Capacity building. Regional capacity in the marine sciences has been improved through numerous workshops and training courses provided by PERSGA through an Annual Training Programme aiming at improving the performance of all stakeholders in the region in managing the marine and coastal environments as laid down in the Jeddah Convention (1982) and its attached subsequent Regional Action Plans and Regional Protocols. Such courses included, for instance, sewage management, fish stock assessment, environmental inspection, ballast water management, monitoring of invasive species, persistent organic pollutants, contingency planning in case of pollution, and economic valuation of marine resources. These training courses have been organized in collaboration with various international and regional organizations. However, routine follow-up monitoring from selected training should be performed immediately following the training programme to examine how the skills learned are being used within Sudan.

The on-the-ground projects. Under the framework of on-the-ground projects in the Member States, PERSGA implemented a number of projects at national levels which focused, for instance, on education for sustainable development (Jordan), eco-tourism (Egypt), coral reef conservation in Farasan Islands (Saudi Arabia), environmental awareness (Djibouti), and the development of national systems for contingency planning and marine pollution control (Sudan, Yemen).

The Regional Environmental Monitoring Programme (REMP). The Regional Environmental Monitoring Programme aims to provide information on the status of sea water to be delivered to the PERSGA's Information Centre in order to discuss patterns of changes in reporting the current status of the marine environment in the region and take the necessary decisions accordingly at the national and regional levels. The programme continuously and periodically follows physical and chemical changes of sea water in the Red Sea and Gulf of Aden region, where reports of monitoring data are collected from Member States as well as supporting countries to raise their technical and human resources to carry out collecting operations and sample analyses.

PERSGA's strategy to cope with the impacts of climate change. The Strategy is a regional initiative that aims to develop integrated programmes implemented through national strategies and includes a package of measures and procedures for the coastal and marine environments; these measures have been included within the main PERSGA programme components to facilitate the application process. These actions, when implemented as proposed, are regarded as an integrated plan by which to assess the viability and impact of climate change on environments, natural resources and economic development in coastal areas, in addition to enhanced capacities required for the implementation of plans to cope with these changes.

Main Conclusions

1. The Red Sea is renowned for its extraordinary natural beauty, biological diversity and high productivity especially with regard to coral reef ecosystems.

2. Reef health is generally good throughout the Red Sea with coral diversity and reef-associated fauna being considered among the highest in the Indian Ocean region.
3. General threats to coral reefs of the Red Sea include land filling and dredging for coastal expansion, shipping and maritime activities, sewage and other pollutants discharges, lack of public awareness and insufficient implementation of legal instruments.
4. Coral bleaching in the Red Sea caused extensive coral mortality, including near total mortality on several reefs in 1998.
5. PERSGA has played a key role in most initiatives formulated to conserve the marine resources of the Red Sea and Gulf of Aden including coral reefs.
6. Although the Red Sea countries have become signatories to a number of international, regional, bilateral or multilateral agreements, the implementation of these remains generally poor, and in some cases, there is no implementation or enforcement (PERSGA/GEF 2003b).
7. Much of the infrastructure needed for regular monitoring and effective management of coral reef resources in the Red Sea countries is available. Nevertheless, many of the present problems can be attributed to a widespread lack of law enforcement, a lack of awareness among law enforcement authorities and the absence of surveillance.
8. Although climate change is by definition a global issue, coral bleaching, die-off, diseases and resilience need further investigations in the Red Sea. A major problem in conservation of reef resources in most countries of the Red Sea is funding for research and management efforts.
9. In spite of PERSGA's efforts in training, authorities responsible for the management of MPAs in the Red Sea still lack experience in doing so in most countries.
10. The International Convention for the Prevention of Pollution from Ships (MARPOL) has not yet been ratified by all countries, because of a lack of port reception facilities. Member countries should assist PERSGA in its efforts to declare the Red Sea Special Area by ratifying MARPOL in addition to joining other related international conventions such as the International Convention on Oil Pollution Preparedness, Response and Co-operation (1990), the International Convention Relating to Intervention on the High Seas in Cases of Oil Pollution Casualties (1969), the Civil Liability Convention (1969), the International Oil Pollution Compensation Fund, and the Convention on the Prevention of Marine Pollution by Dumping of Wastes and Other Matter (London Convention).
11. An environmental awareness and education programme for various target groups are urgently needed to enhance public participation in environmental initiatives.

References

Abou Zaid MM, Kotb MMA (2000) Human-induced and natural impacts to the Egyptian Red Sea coral reefs. In: Abuzinada AH, Joubert E, Krupp F (eds) Proceedings of an international symposium on the extent of coral bleaching in the Arabian region, Riyadh, Kingdom of Saudi Arabia, PERSGA, Jeddah; ROPME, Kuwait; NCWCD, Riyadh, 5–9 Feb 2000, pp 58–64

Alraddadi TM (2013) Temporal changes in the Red Sea circulation and associated water masses, University of Southampton, Ocean and Earth Science, PhD Thesis, 198 pp

Ammar MSA, Amin EM (2000) Community structure and species diversity of reef-building corals at the marine biological station, Hurghada, Red Sea, Egypt. In: Abuzinada AH, Joubert E, Krupp F (eds) Proceedings of an international symposium on the extent of coral bleaching in the Arabian region, Riyadh, Kingdom of Saudi Arabia (5–9 Feb 2000), pp 43–57

Baird AH, Marshall PA (1998) Mass bleaching of corals on the Great Barrier Reef. Coral Reefs 17:376

Birkeland C (1982) Terrestrial runoff as a cause of outbreaks of *Acanthaster planci* (Echinodermata, Asteroidea). Mar Biol 69:175–185

Brown BE (1997) Coral bleaching: causes and consequences. Coral Reefs 16:129–138

Bruno JF, Selig ER, Casey KS, Page CA, Willis BL, Harvell CD, Sweatman H, Melendy AM (2007) Thermal stress and coral cover as drivers of coral disease outbreaks. PLoS Biol 5(6):e124. doi:10.1371/journal.pbio.0050124

Carte BK (1996) Biomedical potential of marine natural products. Bioscience 46:271–286

Chiffings AW (1995) Marine region 11 Arabian Seas. In: Kelleher G, Bleakley C, Wells S (eds) A Global representative system of marine protected areas. Volume III, Central Indian Ocean, Arabian Seas, East Africa and East Asian Seas. Great Barrier Reef Marine Park Authority; World Bank; World Conservation Union (IUCN), pp 39–70

Connell JH, Hughes TE, Wallace CC, Tanner JE, Harms KE (2004) A long-term study of competition and diversity of corals. Ecol Monogr 74:179–210

DeVantier LM, Turak E, Al-Shaikh KA, De'ath G (2000a) Coral communities of the central-northern Saudi Arabian Red Sea. Fauna Arabia 18:23–65

DeVantier LM, Turak E, Al-Shaikh KA (2000b) Coral bleaching in the central-northern Saudi Arabian Red Sea—September 1998. In: Abuzinada AH, Joubert E, Krupp F (eds) Proceedings of an international symposium on: the extent of coral bleaching in the Arabian region, Riyadh, Kingdom of Saudi Arabia, 5–9 Feb 2000, pp 75–90

DeVantier L, De'ath G, Klaus R, Al-Moghrabi S, Abdulaziz M, Reinicke GB, Cheung C (2004) Reef-building corals and coral communities of the Socotra Archipelago, a zoogeographic 'crossroads' in the Arabian Sea. Fauna Arabia 20:117–168

Egana AC, DiSalvo LH (1982) Mass expulsion of zooxanthellae by Easter Island corals. Pac Sci 36:61–63

Gerges MA (2002) The Red Sea and Gulf of Aden action plan—facing the challenges of an ocean gateway. Ocean Coast Manage 45:885–903

Getahun A (1998) The Red Sea as an extension of the Indian Ocean. In: Sherman K, Okemwa EN, Ntiba MJ (eds) Large marine ecosystems of the Indian Ocean—assessment, sustainability and management. Blackwell Science, Cambridge, US, pp 277–283

Gladstone W, Tawfiq N, Nasr D, Andersen I, Cheung C, Drammeh H, Krupp F, Lintner S (1999) Sustainable use of renewable resources and conservation in the Red Sea and Gulf of Aden: issues, needs and strategic actions. Ocean Coast Manage 42:671–697

Gladstone W (2000) Ecological and social basis for management of a Red Sea marine protected area. Ocean Coast Manage 43:1015–1032

Goreau TF (1964) Mass expulsion of zooxanthellae from Jamaican reef communities after hurricane Flora. Science 145:383–386

Harvell CD, Kim K, Burkholder JM, Colwell RR, Epstein PR, Grimes DJ, Hofmann EE, Lipp EK, Osterhaus ADME, Overstreet RM, Porter JW, Smith GW, Vasta GR (1999) Emerging marine diseases—climate links and anthropogenic factors. Science 285:1505–1510

Hayes RL, Goreau NI (1998) The significance of emerging diseases in the tropical coral reef eco system. Revista de Biología Tropical 46:173–185

Head SM (1987) Corals and coral reefs of the Red Sea. In: Edwards AJ, Head SM (eds) Red Sea. Pergamon Press, Oxford, pp 128–151

Head SM (1980) Ecology of corals in the Sudanese Red Sea. PhD thesis, University of Cambridge, UK

Hoegh-Guldberg O, Salvat B (1995) Periodic mass bleaching of reef corals along the outer reef slope in Moorea, French Polynesia. Mar Ecol Prog Ser 121:181–190

Hoegh-Guldberg O (1999) Climate change, coral bleaching and the future of the world's coral reefs. Mar Freshw Res 50:839–866

Hughes TP (1994) Catastrophes, phase shifts, and large-scale degradation of a Caribbean coral reef. Science 265:1547–1551

Hughes TP, Baird AH, Bellwood DR, Card M, Connolly SR, Folke C, Grosberg R, Hoegh-Guldberg O, Jackson JBC, Kleypas J, Lough JM, Marshall P, Nyström M, Palumbi SR, Pandolfi JM, Rosen B, Roughgarden J (2003) Climate change, human impacts, and the resilience of coral reefs. Science 301:929–933

IUCN/UNEP (1985) Management and conservation of renewable marine resources in the Red Sea and Gulf of Aden Region. UNEP regional seas reports and studies no. 64, UNEP, Nairobi

Kemp JM (1998) Zoogeography of the coral reef fishes of the Socotra Island Group. J Biogeogr 25(5):919–933

Kemp JM, Benzoni F (2000) A preliminary study of coral communities in the northern Gulf of Aden. Fauna Arab 18:67–86

Kleypas J, Buddemeier RW, Archer D, Gattuso JP, Langdon C, Opdyke BN (1999) Geochemical consequences of increased atmospheric carbon dioxide on coral reefs. Science 284:118–120

Kotb M, Abdulaziz M, Al-Agwan Z, Al-Shaikh K, Al-Yami H, Banajah A, DeVantier L, Eisinger M, Eltayeb M, Hassan M, Heiss G, Howe S, Kemp J, Klaus R, Krupp F, Mohamed N, Rouphael T, Turner J, Zajonz U (2004) Status of coral reefs in the Red Sea and Gulf of Aden in 2004. In: Wilkinson C (ed) Status of coral reefs of the World: 2004, vol 1. Australian Institute of Marine Science, Townsville

Krupp F, Türkay M, El Hag AG, Nasr DH (eds) (1994b) Comparative ecological analysis of biota and habitats in littoral and shallow sublittoral waters of the Sudanese Red Sea. Forschungsinstitut Senckenberg, Frankfurt and Faculty of Marine Science and Fisheries, Port Sudan, 89 pp

Krupp F (1990) Sanganeb—ein Unterwasser Nationalpaerkim Roten Meer. Nat Mus 120:405–409

Lang JC, Chornesky EA (1990) Competition between scleractinian reef corals—a review of mechanisms and effects. In: Dubinsky Z (ed) Ecosystems of the world, vol 25, Coral reefs Elsevier, Amsterdam, pp 209–252

MEPA/IUCN (1987) Saudi Arabia: assessment of coastal zone management requirements, MEPA, vol 7, Jeddah

Moore R (1990) Persistent and transient populations of the crown-of-thorns starfish. In: Bradbury RH (ed) Acanthasterplanci, Acanthaster and the coral reef: a theoretical perspective. Springer, Berlin, pp 236–277

Morcos SA (1970) Physical and chemical oceanography of the Red Sea. Oceanogr Mar Biol Rev 8:73–202

Mumby PJ (2001) Unprecedented bleaching-induced mortality in Porites spp. at Rangiroa Atoll, French Polynesia. Mar Biol 139(1):183–189

Nasr D, Al-Sheikh K (2000) Assessment of coral reefs in the Sudanese Red Sea in the context of coral bleaching. In: Tatwany H (ed) Proceedings of the international workshop on the extent and impact of coral bleaching in the Arabian Region. National Commission for Wildlife Conservation and Development, Riyadh

Nasr DH, Eltayeb M (2012) Vulnerability and adaptations to climate change in the Sudanese Red Sea coast. Report submitted to the Higher Council for Environment and Natural Resources, Ministry of Environment, Forestry and Physical Development, Khartoum, Sudan

Ormond RFG, Edwards A (1987) Red Sea Fishes. In: Edwards AJ, Head SM (eds) Red Sea. Pergamon Press, Oxford, pp 252–287

Ormond RFG, Campbell AC (1971) Observations on *Acanthaster planci* and other coral reef echinoderms in the Sudanese Red Sea. Symp Zool Soc Lond 28:433–454

Ormond RFG, Campbell AC (1974) Formation and breakdown of *Acanthaster planci* aggregations in the Red Sea. In: Proceedings of 2nd International Coral Reef Symposium 1:569–619

PERSGA/GEF (1998) Strategic Action Program (SAP) for the Red Sea and Gulf of Aden, vol 1, main report. World Bank, Washington, DC, 90 pp

PERSGA/GEF (2001) Strategic Action Program for the Red Sea and Gulf of Aden, vol 2, Country reports. PERSGA, Jeddah and the World Bank, Washington DC, 205 pp

PERSGA/GEF (2002) The Red Sea and Gulf of Aden regional network of marine protected areas. Regional Master Plan. PERSGA technical series no 1, PERSGA, Jeddah

PERSGA (2005) Environmental monitoring program for the Red Sea and Gulf of Aden. PERSGA, Jeddah

PERSGA (2006) State of the marine environment. Report for the Red Sea and Gulf of Aden. PERSGA, Jeddah

PERSGA (2010) The status of coral reefs in the Red Sea and Gulf of Aden: 2009. PERSGA technical series no 16, PERSGA, Jeddah

PERSGA/ALECSO (2003) Survey of habitats in Djibouti and plans for their protection. PERSGA technical series no 5 (English), PERSGA, Jeddah

PERSGA/ALECSO (2007) Survey of habitats in Sudan and plans for their protection. PERSGA technical series no 13, PERSGA, Jeddah

PERSGA/GEF (2003a) Regional action plan for the conservation of coral reefs in the Red Sea and Gulf of Aden. PERSGA technical series no 3, PERSGA, Jeddah

PERSGA/GEF (2003b) Coral reefs in the Red Sea and Gulf of Aden. Surveys 1990–2000 summary and recommendations. PERSGA technical series no 7, PERSGA, Jeddah

PERSGA/GEF (2004a) Standard survey methods for key habitats and key species in the Red Sea and Gulf of Aden. PERSGA technical series no 10. PERSGA, Jeddah, 310 pp

PERSGA/GEF (2004b) Survey of the proposed marine protected area at Dungonab Bay and Mukawwar Island, Sudan. Report for PERSGA, PERSGA, Jeddah

Peters EC (1984) A survey of cellular reactions to environmental stress and disease in Caribbean scleractinian corals. Hegol Wiss Meeresunters 37:113–137

PilcherN Alsuhaibany A (2000) Regional status of coral reefs in the Red Sea and the Gulf of Aden. In: Wilkinson C (ed) Status of coral reefs of the world 2000. Australian Institute of Marine Science, Townsville, pp 35–54

Pilcher N, Nasr D (2000) The status of coral reefs in Sudan-2000. A report submitted to PERSGA for the preparation of the regional status of coral reefs in the Red Sea and Gulf of Aden. In: Wilkinson C (ed) Status of coral reefs of the world 2000. Australian Institute of Marine Science, Australia, pp 35–54

Richardson LL (1998) Coral diseases: what is really known? Trends Ecol Conserv 13:438–443
Roberts CM, Shepherd ARD, Ormond RFG (1992) Large-scale variation in assemblage structure of Red Sea butterfly fishes and angel fishes. J Biogeogr 19:239–250
Rosenberg E, Ben-Haim Y (2002) Environ Microbiol 4(6):318–326
Scheer G, Pillai CSG (1983) Report on the stony corals from the Red Sea. Zoologica 130:1–198
Schroeder JH (1981) Man versus reefs in the Sudan: threats, destruction, protection. In: Proceedings of Fourth International Coral Reef Symposium Manila 1:252–253
Schroeder JH, Scheer G (1981) Corals of Sanganeb Reef, collected by Scroeder JH, identified by Scheer G. Institute of Oceanography, Port Sudan. Typescript, 6 pp
Schroeder JH, Nasr DH (1983) The fringing reefs of Port Sudan, Sudan. 1. Morphology, sedimentology and zonation. Geogr Arb 6:29–44
Schroeder JH, Nasr DH, Idris FH (1980) Coral reef conservation in the Sudanese Red Sea. In: Proceedings of the symposium on the coastal and marine environment of the Red Sea, Gulf of Aden and Tropical Western Indian Ocean, Khartoum, pp 163–178
Sebens KP (1994) Biodiversity of coral reefs: what are we losing and why? Am Zool 34:115–133
Sheppard CRC, Sheppard ALS (1991) Corals and coral communities of Arabia. Fauna Saudi Arab 12:3–170
Sheppard C, Price A, Roberts C (1992) Marine ecology of the Arabian region: patterns and processes in extreme tropical environments. Academic Press, London
Sofianos SS, Johns W, Murray SP (2002) Heat and freshwater budgets in the Red Sea from direct observations at Bab al Mandab. Deep-Sea Res II 49(7–8):1323–1340
Spalding MD, Ravilious C, Green EP (2001) World atlas of coral reefs. World Conservation Monitoring Centre, University of California Press, Berkeley
Turak E, Brodie J (1999) Coral and reef habitats. In: DouAbul A, Rouphael TS, Marchant R (eds) Ecosystems of the Red Sea coast of Yemen. Protection of marine ecosystems of the Red Sea coast of Yemen. Hassell and Associates, AMSAT and UNOPS
UNEP/IUCN (1988) Coral reefs of the world, vol 2: Indian Ocean, Red Sea and Gulf. UNEP regional seas directories and bibliographies. IUCN, Gland, Switzerland and Cambridge, UK/UNEP, Nairobi
UNEP (2002) Coastal and marine areas—Africa. In: Global environment outlook 3. United Nations Environment Programme, Nairobi, pp 188–190
UNEP/PERSGA (1997) Assessment of land-based sources and activities affecting the marine environment in the Red Sea and Gulf of Aden. Regional organisation for the conservation of the environment of the Red Sea and Gulf of Aden, UNEP Regional Seas Reports and Studies 166
Vine PJ (1970) Densities of *Acanthaster planci* in the Pacific Ocean. Nature 228(5269):341–342
Vine PJ (1973) Crown of Thorns (*Acanthaster planci*) plagues: the natural causes theory. Smithsonian Institute, Atoll Research Bulletin 166, Washington DC
Vine PJ, Vine MP (1980) Ecology of the Sudanese coral reefs with particular reference to reef morphology and distribution of fishes. In: Proceedings of the symposium on the coastal and marine environment of the Red Sea, vol 1. Gulf of Aden, and Tropical Western Indian Ocean, Khartoum, pp 87–140
Vine PJ (1985) The Red Sea. Immel Publishing, London and Jeddah 128 pp
Vine PJ (1986) Red sea invertebrates. Immel Publishing, London 224 pp
Wilkinson CR (ed) (2000) Status of coral reefs of the world. Global coral reef monitoring network. Australian Institute of Marine Science, Townsville
Wilkinson C (ed) (2002) Status of coral reefs of the world. Australian Institute of Marine Science, Townsville
Wilkinson CR, Linden O, Cesar H, Hodgson G, Rubens J, Strong AE (1999) Ecological and socio economic impacts of 1998 coral mortality in the Indian Ocean: An ENSO impact and a warning of future change? Ambio 28:188–196

The Status, Threats, and Resilience of Reef-Building Corals of the Saudi Arabian Red Sea

Andrew W. Bruckner and Alexandra C. Dempsey

Abstract

The Saudi Arabian Red Sea (SARS) contains diverse shallow water coral reef systems that include attached (fringing and dendritic reefs) and detached (platform, patch, tower, ribbon, and barrier reefs) reef systems extending up to 90 km offshore. To better understand the current status of coral reefs in SARS, the Living Oceans Foundation conducted assessments of representative reef environments in the Farasan Islands (2006), Ras Al-Qasabah (2007), Al Wajh and Yanbu (2008), and the Farasan Banks (2009). A combination of belt transects and quadrats was used to assess the diversity, size structure, partial mortality, condition, and recruitment of the dominant reef-building corals. Most sites had high structural complexity, with up to 52 genera of scleractinian corals recorded from a single region. Living corals varied in abundance and cover by region, habitat, and depth, with the highest species richness documented in the south (Farasan Banks), followed by Al Wajh and Yanbu and lowest at Ras Al-Qasabah. On most reefs, a single species was dominant. The reef architecture was constructed by massive and columnar *Porites*, with unusually large (1–4 m diameter) colonies in shallow water (up to 80 % live cover in 2–10 m depth) and a deeper reef *Porites* framework that was mostly dead. *Porites lutea* was the single most abundant coral throughout SARS, and the dominant species on leeward reef crests and slopes, while reef slopes and deeper coral carpets were predominantly *Porites columnaris* and *P. rus*. Faviids (*Goniastrea* and *Echinopora*) were the next most abundant corals, especially in areas that had experienced a disturbance, although these were small (most <15 cm diameter) and made up a small fraction of the total live coral cover. Multi-specific *Acropora* assemblages often formed large thickets, but these were restricted in distribution. *Pocillopora* was the dominant taxon in Yanbu, widespread in Al Wajh, and much less common in northern and southern sites. Coral cover throughout the region averaged about 20 %, with higher cover (often >50 %) in shallow water and rapid decline with increasing depth. In each region, many reefs (15–36 %) showed signs of damage and had less than 5 % live coral cover. These degraded sites were characterized by extensive dead skeletons in growth position, substrates colonized by thick mats of turf algae and soft corals (*Xenia*), and surviving massive and plating corals that were subdivided by partial mortality into numerous small (<10 cm) ramets. Mortality was attributed to bleaching events, disease, and outbreaks of corallivores occurring over the last 10–15 years. Several sites also exhibited signs of recent mortality from crown of thorns sea stars (*Acanthaster*), coral-eating snails, and coral disease. In many cases, the *Porites* framework had been recolonized by

A.W. Bruckner (✉) · A.C. Dempsey
Khaled bin Sultan Living Oceans Foundation, 8181 Professional Place, Landover, MD 20785, USA
e-mail: bruckner@livingoceansfoundation.org

N.M.A. Rasul and I.C.F. Stewart (eds.), *The Red Sea*,
Springer Earth System Sciences, DOI 10.1007/978-3-662-45201-1_27,

faviids, acroporids, and other corals and these had subsequently died. Most degraded areas appeared to be rebounding, as substrates had high cover of crustose coralline algae (CCA), little macroalgae, and high numbers of coral recruits and juvenile corals.

Introduction

The Red Sea contains unique, biologically diverse coral reefs that exhibit striking contrasts in their biology, environmental attributes, and degree of human influence. The reefs form part of the greater Indo-Pacific region and are the most biologically diverse reef communities outside of the southeast Asia coral triangle. Coral species overlap with Indian Ocean fauna and also contain about 10 % endemic species, with over 260 reported coral species in 68 genera and 16 families (Sheppard et al. 1992). The small geographic range of several species suggests long periods of isolation, its extreme environment, and the limited exchange that occurs through the Bab-al-Mandab Strait (Persga 2006). Red Sea coral reefs are some of the most northerly reef communities and are subjected to harsh climatic conditions including temperature and salinity extremes and high solar radiation (DeVantier and Pilcher 2000). Inflowing surface currents and winds vary seasonally, affecting water temperature, salinity, upwelling, and wave action and exchange with the Indian Ocean (Behairy et al. 1992). The Red Sea is also unique for having unusually warm temperatures that extend throughout the water column.

Most of the Red Sea is characterized by fringing coral reefs that extend nearly 2,000 km from the Gulfs of Suez and Aqaba in the north to Bab-al-Mandab and Gulf of Aden in the south (Persga 2006). Fringing reefs form a narrow border a few kilometers wide along the shoreline and surrounding islands, dropping steeply into the depths of the Red Sea. Fringing reefs are 5,000–7,000 years old (Behairy et al. 1992) and largely formed by a framework of *Acropora* and *Porites* corals.

Off much of Saudi Arabia, the coastal shelf is wider and it supports extensive sea grass, coral reef, and mangrove habitats, with shallow reef communities found from the shoreline to distances of up to 50–100 km offshore (Bruckner et al. 2011a). These areas support a wide range of reef morphologies, including barrier reefs, platform reefs, patch reefs, and atoll-like tower reefs. The offshore reef formations are unusual in that they defy classic (i.e., Darwinian) coral reef classification schemes. Their development is attributed to (1) the high levels of tectonic activity that characterize the area, (2) salt diapirs, (3) hyper-arid climate punctuated by extreme rainfall events during low sea-level stands, and (4) sediment delivery through wadis and other drainage basins (Bruckner et al. 2011a; Rowlands et al. 2014). The wide range of environmental conditions, latitudinal gradient, and the underlying geologic structure all contribute to the development and persistence of these highly variable coral habitats (Bosence 2005). Their future survival may be integrally linked to the degree of anthropogenic influences.

Human pressures associated with two centuries of rapid industrialization and precipitous economic growth increasingly threaten the fragile coral reef environments of the Red Sea. According to the 2011, *Status of Coral Reefs of the World*, the Red Sea is extremely vulnerable to environmental threats posed by large coastal populations, overfishing, pollution, and coastal development (Burke et al. 2011). Nearly 60 % of the reefs in the Red Sea are at risk from landfilling and dredging, port activities, sewage and other pollution, and tourism (Burke et al. 2011). Global stressors associated with climate change are also predicted to have major compounding impacts on Red Sea coral reefs. Mass coral mortality occurred due to bleaching in the central to northern Saudi Arabian Red Sea in late 1998 (DeVantier and Pilcher 2000), but most showed promising signs of recovery a decade later (Wilkinson 2008). In 2005, coral reefs were reported to be generally in good condition, with 30–50 % live coral cover at most locations and more than 50 % total cover on average at 5 m depth (Persga 2006). While no bleaching was noted between 2006 and 2009, a mass bleaching event occurred in central Saudi Arabia in 2010 (Riegl et al. 2013). Thermal stress and ocean acidification are projected to increase threat levels up to nearly 90 % by 2030; by 2050, these climate change impacts, combined with current local impacts, will push all reefs in the Middle East to threatened status, with 65 % at high, very high, or critical risk (Burke et al. 2011).

This chapter provides an overview of the current status of reef-building corals on different reef systems throughout the Saudi Arabian Red Sea, with emphasis on five regions with prominent offshore reef systems. The distribution, abundance, cover, population structure, and condition of corals are compared within and among these five regions. A detailed description of the major biotic threats affecting the coral communities is also presented. The data included in this report represent a collaborative effort of multiple scientists that participated in a four-year research mission conducted by the Khaled bin Sultan Living Oceans Foundation.

Methods

Coral community structure was examined on 129 reefs in five regions (Fig. 1) including Ras Al-Qasabah ($n = 19$), Al Wajh ($n = 21$), Yanbu ($n = 11$), the Farasan Banks ($n = 56$),

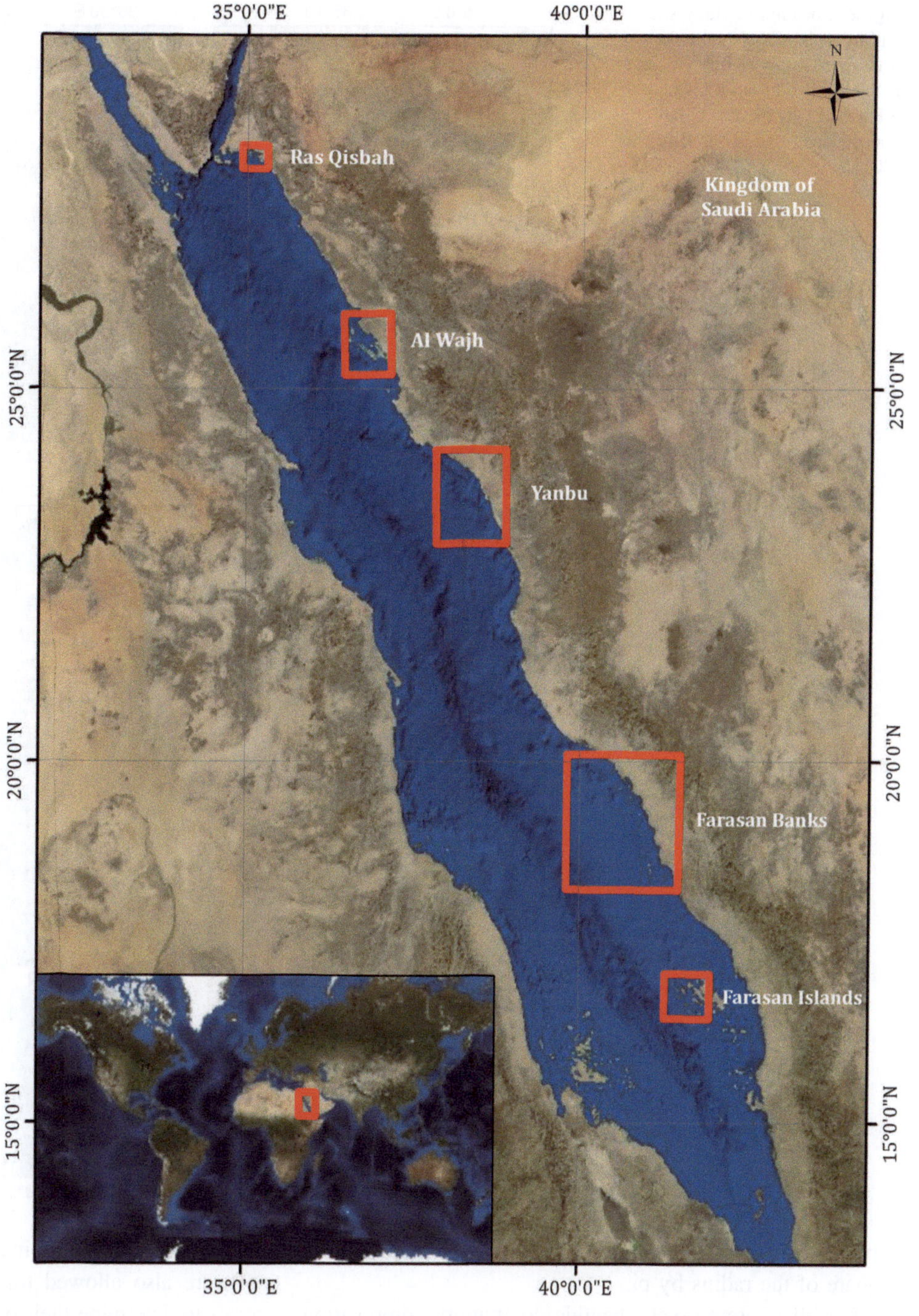

Fig. 1 Five major areas examined by the Khaled bin Sultan Living Oceans Foundation during the four-year research mission. Base image provided by Google

and the Farasan Islands (n = 22) (Fig. 2). On each reef, depth-stratified surveys (3–5, 10–12, and 15–20 m) were conducted using photographic transects, belt transects, and quadrats. Four measures were recorded for corals: (1) coral cover; (2) coral taxa (to genus) and growth form; (3) coral size class distributions; and (4) coral condition (extent of mortality and cause of mortality) (Table 1). The abundance, size, and amount of partial mortality (subdivided into recent and old mortality) of all corals 4 cm or larger in size (recorded to genus) were assessed along either 10 m by 1 m or 25 m by 1 m belt transects using a 1-m bar marked in 1-cm intervals for scale. Sampling for corals smaller than 4 cm was done using a minimum of six 25 by 25 cm quadrats per transect, with each quadrat located at predetermined intervals (e.g., at 0, 5, 10, 15, 20, and 25 m). Measurements varied slightly among sites, depending on the observer. Coral size was measured to the nearest cm on reefs examined in Ras Al-Qasabah, Farasan Islands, and Farasan Banks, while corals assessed in Al Wajh and Yanbu were lumped into six size classes. Coral abundance is presented for each taxon as a percent of all corals recorded within transects and the total planar surface area occupied by that

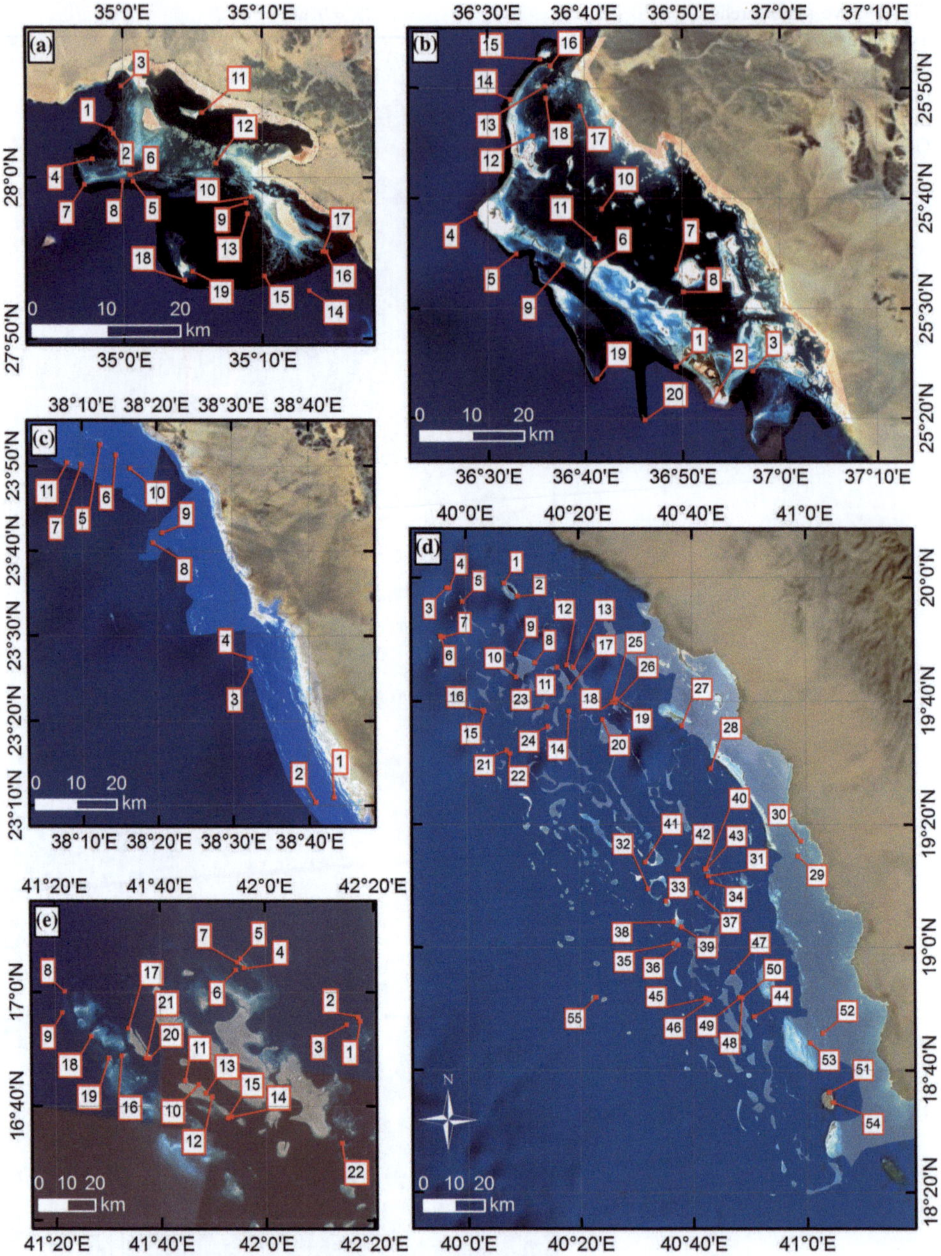

Fig. 2 Location of study sites within each of the five regions. **a** Ras Al-Qasabah; **b** Al Wajh; **c** Yanbu; **d** Farasan Banks; **e** Farasan Islands. For GPS coordinates of sites, see Bruckner (2011b)

taxa. Planar surface area was calculated by multiplying the square of the radius by pi.

For all phototransects, benthic community composition, including cover and size (planar surface area) of corals, cover of other animals (subdivided into soft corals, sponges, and other invertebrates), cover of algae subdivided by functional group (turf, macroalgae, crustose coralline algae, erect coralline algae, and cyanobacteria), and substrate type (rubble, sand, and uncolonized pavement) were assessed using coral point count (CPCE) software developed by the National Coral Reef Institute (NCRI). Cover was determined by recording the benthic attribute located directly below a random point, with 30–50 points per photograph (depending on the size of the image; for example, 30 points for a 0.5-m^2 photograph and 50 points for a 1-m^2 photograph). This software also allowed tracing of the outline of individual corals to determine their planar surface area and size of live versus dead portions of the colony.

Results

Coral Community Structure

A total of 55 genera were recorded in the 5 regions. The Farasan Banks exhibited the highest diversity (54 genera), followed by Al Wajh (52), Yanbu (49), Farasan Islands (46), and Ras Al-Qasabah (44). The spatial distribution of the

Table 1 Resilience assessments for stony corals included in SARS surveys

Measure	Description
Stony coral	Genus-level identification of the composition, abundance, cover, size structure, condition, and level of recruitment within representative habitats
Taxa and growth form	The diversity and structural complexity of a site. Communities with higher numbers of functional groups (e.g., acroporids, faviids, pocilloporids, and poritids) and growth forms (e.g., branching, plating, massive corals), and redundancy of these groups may support more associated species and be more resilient
Cover	Measure of amount of living coral. Small changes in cover are difficult to document, but abrupt changes may reflect a major disturbance
Size structure	Maturity and ecological state of the taxa. Dominance of large corals may be a sign of stable environmental conditions and long-term persistence of the community; a dominance by small corals suggests frequent disturbance or recovery from a recent disturbance
Recruitment	Abundance of recruits reflects the reef's potential for growth and recovery after major disturbances and the influx of genetic diversity
Fragmentation	Level of physical disturbance and potential for asexual propagation. Locations of fragments (accumulations in sand channels or on reef) and fragment condition (no tissue loss; fusion to the substrate; presence of new growth) reflects the potential survival and contribution of fragments to recovery
Dead standing coral	Amount of dead standing coral can be used to hindcast past disturbance events up to a decade or more
Old mortality	Presence of dead areas on corals that are colonized by other biotic agents (e.g., skeleton is not white). Species-specific differences reflect the life history strategies, population dynamics and susceptibility to various physical and biological factors. Old mortality may increase with colony size
Recent mortality	Extent of recent mortality (white skeleton) reflects the severity, timing, and duration of a stressor. Large white areas indicate rapid ongoing tissue loss, while a thin band of white skeleton and a wide band of green, algal-colonized skeleton suggests the event is near its end
Bleaching	Reef-wide bleaching may be associated with recent or ongoing temperature anomalies; extent of recent mortality indicates the duration and severity of the temperature stress
Disease	Spatial patterns of disease reflect potential for spread and level of contagion. Distinction between background mortality (chronic stressors) and of disease outbreaks (acute mortality)
Corallivores	Abundance of coral predators (*Drupella* and *Coralliophila* snails, *Acanthaster* sea stars) indicative of the amount of recent mortality and possible changes to coral community structure and species diversity due to chronic high-level infestations or outbreaks
Competition and overgrowth	Negative factors inhibiting the growth and recovery of corals and causes of chronic mortality (e.g., algal competition) and bioerosion (e.g., sponges). Extent and composition of competitor may reflect status of fish communities (herbivores), nutrient loading, and sedimentation
Coral associates	Obligate corallivores (butterflyfish) provide an indication of the health of the coral community or a particular taxa
	Abundance and diversity of fishes and invertebrates within coral branches are indicative of the topographic complexity and health of the site
	An abundance of territorial damselfish within branches and large algal lawns indicate possible overfishing of piscivores
	High densities of sea urchins may be associated with extensive bioerosion of reef substrates and low recruitment and survival of newly settled corals; macroalgae may increase in with the absence of sea urchins, especially sites with few herbivorous fishes

dominant coral species varied by region, site, and habitat/depth. The dominant genera throughout all sites were *Porites* and *Acropora*, with *P. lutea* being the single most common species (Fig. 3). On offshore habitats in the central Red Sea, the three consistently dominant coral genera over all pooled sites were *Acropora* (average cover over all transects 3 %), *Porites* (2.75 %), and *Pocillopora* (1.3 %). Large mono-specific assemblages of *Pocillopora* were found on shallow fore-reefs in Yanbu; this genus was present but not among the five most dominant species in Al Wajh and Farasan Banks and was rare in Ras Al-Qasabah and Farasan Islands. In Al Wajh, *Montipora* was the second most abundant genera, primarily on offshore reefs, while branching and table acroporids were most common on inshore and turbid reefs. Acroporids were only rarely seen on offshore sites of Al Wajh, with the exception of small digitate acroporids and tightly branched, low-lying colonies in reef crests and shallow fore-reefs (Fig. 3b).

Tabular acroporids formed dense thickets in shallow water (<8 m) on offshore sites at the northern and northwestern end of the bank system in the Farasan Islands, while dense *Acropora* communities were found throughout nearshore and

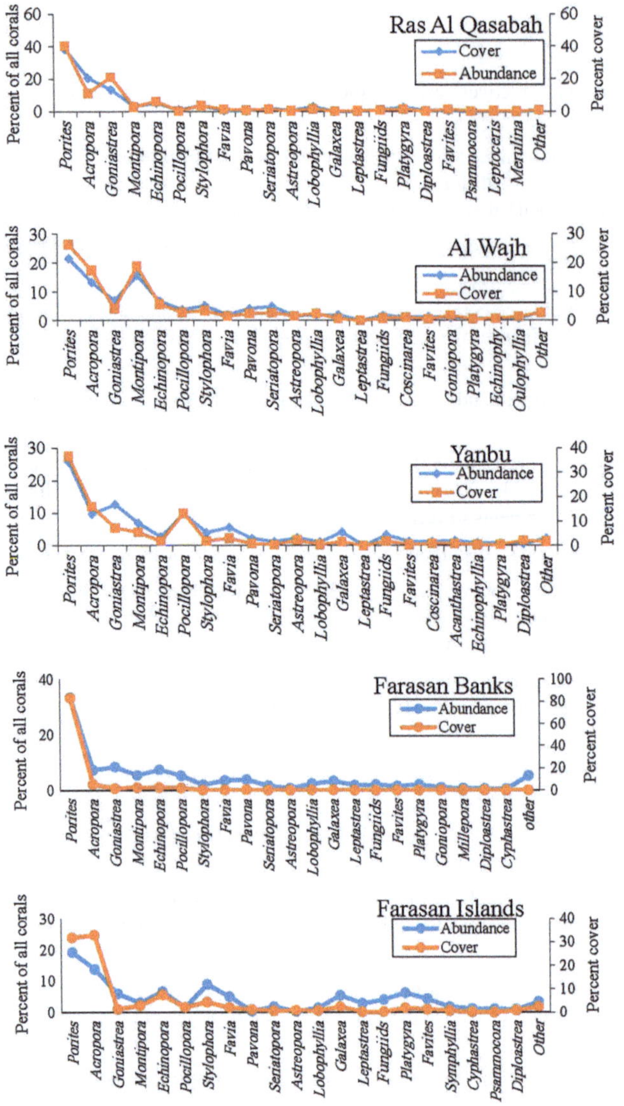

Fig. 3 Relative abundance and area occupied by all of the coral genera identified within belt transects. Each genus is illustrated as a percent of all genera recorded and the total estimated planar surface area of all corals in that genera. **a** Ras Al-Qasabah; **b** Al Wajh; **c** Yanbu; **d** Farasan Banks; **e** Farasan Islands

midshelf locations in the Farasan Banks; *Acropora* was rare overall in Ras Al-Qasabah. Coral assemblages on offshore vertical walls either formed encrusting, foliaceous, and plate-like colonies of *Montipora*, *Echinopora*, *Pavona*, *Turbinaria*, *Leptoseris*, and *Pachyseris*, or small submassive colonies, such as *Stylophora marmorata*; 22 % of offshore locations had large dead stands of foliaceous corals still in growth position. Small *Goniastrea* colonies (recruits and remnants) were the most abundant coral in lagoonal areas of Al Wajh and throughout Ras Al-Qasabah, especially where most of the *Porites* colonies had died.

Coral Cover

The benthic environment in most locations was dominated by scleractinian corals, which were highest at the Farasan Banks (mean cover = 29.4 % ± 13.4 SE; range across sites = 4–60 %) and was lowest at Ras Al-Qasabah (mean cover = 20.4 % ± 10.4 SE; range = 2–40.1 %) (Fig. 4). Dead standing corals also comprised a significant portion of the total cover (mean cover = 10.4 % across all sites). Between 15 and 36 % of sites in each region showed signs of damage and had significantly less living coral cover (1–5 %) and higher cover of dead standing skeletons (14–20 %). This included all of Ras Al-Qasabah, most lagoonal reefs in Al Wajh, and several remote midshelf and offshore locations in Yanbu, the Farasan Banks, and the Farasan Islands.

In all regions except Ras Al-Qasabah, coral cover was usually the highest in the shallow reef crest and fore-reef on outer fore-reef locations (1–5 m depth; 30–70 %). Coral cover declined with depth, with 20–30 % cover at middepths (7–12 m depth), and <5–30 % at the base of the reef slope (15–30 m), especially on midshelf and outer sites. Throughout the Saudi Arabian Red Sea (SARS), *Porites* made up 25 % or more of the living coral cover and over 80 % of the coral cover in the Farasan Banks. The cover of *Acropora* exceeded *Porites* only in the Farasan Islands, mainly due to the dominance of large colonies on offshore locations (Fig. 7b). In all locations, the substrate and dead skeletons were carpeted in soft corals (predominantly *Xenia*;14–20 % cover), encrusting sponges, and other non-coral invertebrates, especially in degraded areas and on deeper reefs (>20 m depth), and these were often competing with hard corals for space and overgrowing surviving remnants. Deep (>20 m) offshore reefs in the Farasan Banks also had low cover of scleractinian corals and large assemblages (30–50 % cover) of organ pipe coral (*Tubipora musica*). Shallow locations in the Farasan Islands, nearshore reef flat communities in Farasan Banks, and areas in Ras Al-Qasabah had high cover of macroalgae and moderate cover of turf algae and crustose coralline algae, while macroalgae were rare elsewhere (Fig. 4).

Coral Size Structure

Corals in all locations were mostly small in size (over 90 % <20 cm diameter), with medium-sized corals (20–40 cm) making up most of the cover (Fig. 5). These differences are due to the non-linear relationship between coral size and surface area as a result of a very high numerical abundance of recruits, juvenile corals, and tissue remnants. Recruits and juvenile corals (<5 cm diameter) were more

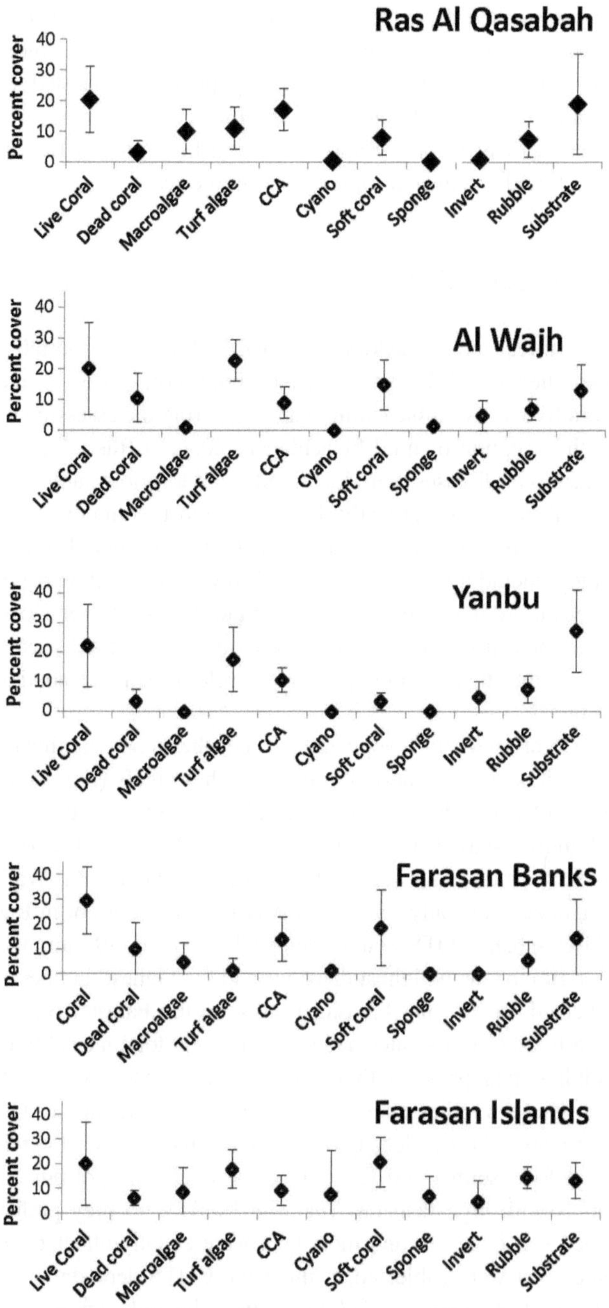

Fig. 4 Benthic cover subdivided into live coral cover, dead coral, algal functional groups (macroalgae, turf algae, crustose coralline algae, and cyanobacteria), benthic invertebrates (soft coral, sponges, and all other invertebrates, pooled), and rubble and uncolonized substrate (sand and pavement). **a** Ras Al-Qasabah; **b** Al Wajh; **c** Yanbu; **d** Farasan Banks; **e** Farasan Islands

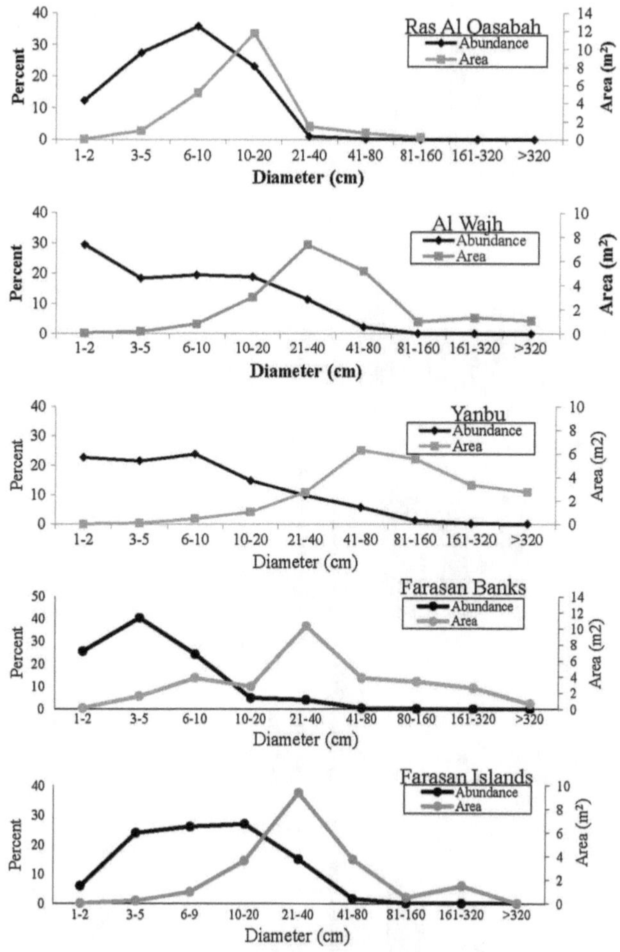

Fig. 5 The total abundance of corals per size class and the total colony area for all scleractinian corals (pooled species). **a** Ras Al-Qasabah; **b** Al Wajh; **c** Yanbu; **d** Farasan Banks; **e** Farasan Islands

abundant than all other size classes at Al Wajh (48 % of all corals; Fig. 5b), Yanbu (44 %; Fig. 5c), and Farasan Banks (38 %; Fig. 5d). Although small corals dominated numerically in these locations, most of the cover consisted of 30–40 cm diameter colonies at the Farasan Banks and 21–80 cm at Al Wajh, while much larger corals (40–160 cm) made up most of the cover at Yanbu. Ras Al-Qasabah (Fig. 5a) and the Farasan Islands (Fig. 5e) had much lower numbers of recruits and juveniles. Even though Ras Al-Qasabah was dominated by small corals (most corals were 6–9 cm diameter with 10–20 cm colonies making up most of the cover), these tended to be surviving tissue remnants on much larger skeletons. While 3–20 cm colonies were most numerous in the Farasan Islands, the bulk of the cover consisted of colonies that were 30–75 cm. The mean colony size also varied greatly between taxa. The largest corals overall were *Porites* and *Acropora*, with low numbers of very large (>1 m) *Diploastrea*, *Galaxea*, and *Montipora* colonies seen in some locations, while faviids (*Favia*, *Favites*, *Platygyra*, *Leptastrea, and Cyphastrea*) tended to be small (<10 cm). The size structures for the four most abundant genera (*Porites*, *Acropora*, *Goniastrea*, and *Echinopora*) are shown in Fig. 6.

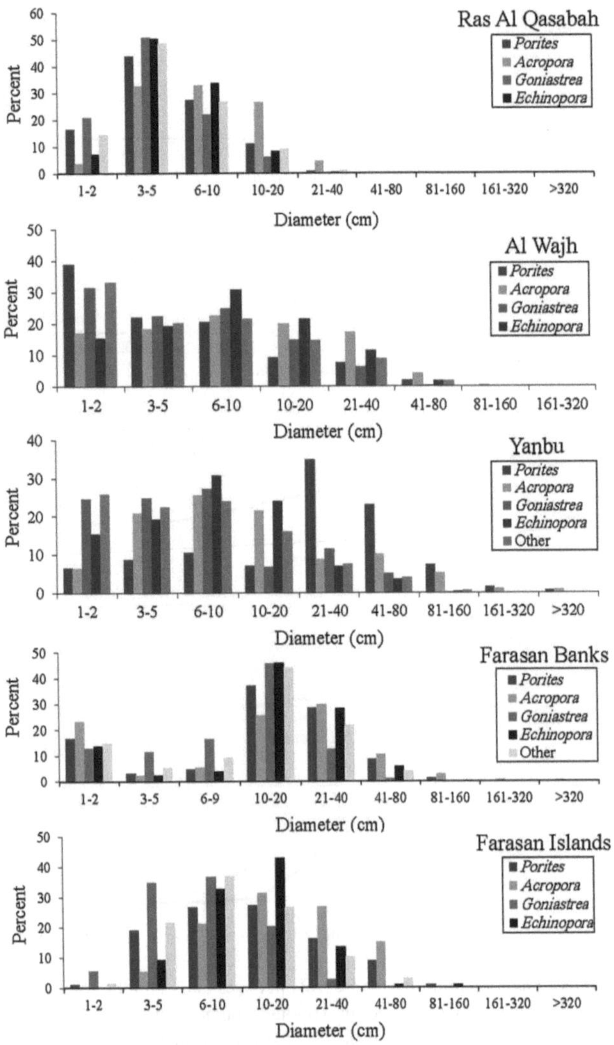

Fig. 6 Size frequency distribution of the four dominant genera of corals and all other species pooled. **a** Ras Al-Qasabah; **b** Al Wajh; **c** Yanbu; **d** Farasan Banks; **e** Farasan Islands

Coral Recruitment

Coral recruitment was lowest at Ras Al-Qasabah and the Farasan Islands, with moderate levels of recruitment documented at fore-reef sites in Yanbu and both barrier reef and lagoonal sites at Al Wajh. The Farasan Banks had the highest recruitment overall, especially in degraded sites; an average density of 3.3 ± 1.7 juvenile corals (<4 cm diameter) per quadrat and ~13 juvenile corals (>4 cm diameter) per m^2 was documented. Poritids were the most common recruiters (26 %), followed by the faviids (20 %), *Pavona* (12 %), and the acroporids (11 %). Seven genera made up three quarters of all recruits (74.7 %). The distribution of recruitment frequency reflected the dominance of large corals: *Porites* recruited more commonly than *Acropora* and also covered more space on the reefs. The commonly recruiting faviids, which have comparable growth rates to *Porites*, nevertheless covered less space on the reef. Almost the entire encountered *Leptastrea* population consisted of small colonies. Although *Montipora* recruited more frequently than *Acropora* and also has a high growth rate, it was found to cover far less space on the reefs.

Coral Condition

The three major factors responsible for coral mortality identified in SARS are coral bleaching, coral disease, and corallivory. Coral bleaching was rare during the assessments with exception of mild bleaching in 2006, at Ras Al-Qasabah. Coral diseases were identified in all locations, although prevalence was generally low. The most common coral diseases reported for the Indo-Pacific were observed in this study, including white syndrome, brown band, yellow band, skeletal eroding band, and black band disease. Several colonies had sustained extensive partial mortality from diseases, suggesting that diseases play a key role in reef degradation (Fig. 7).

Coral predation appeared to be the most significant chronic source of mortality affecting these reefs (Fig. 8). In particular, localized outbreaks of *Drupella* snails and crown of thorns starfish (COTS) were noted. Active COTS outbreaks were observed in the Farasan Islands, Al Wajh (lagoonal sites only), and the Farasan Banks, but not in Ras Al-Qasabah. COTS counts ranged from 10 to 498 animals per 100 m^2, with highest densities and the highest number of affected reefs at the Farasan Banks. In the Farasan Banks, nearly 20 % of the sites had sustained near total mortality of adult corals prior to these surveys, and extensive recent mortality at other sites was attributed to ongoing COTS outbreaks. At low density, COTS preferred acroporid corals as a food source and fed only occasionally on poritids. In areas with high densities (Farasan Banks), no prey preference was observed and most taxa were equally affected. At sites that were subjected to intensive COTS depredation in the past, *Diploastrea heliopora* was the only large coral (diameter ~1 m) that had survived, with high numbers of small (10–20 cm diameter) massive colonies of *Echinopora* and isolated tissue remnants of *Goniastrea.* Several sites that had been damaged prior to these surveys also exhibited extensive mortality on the deeper parts of the fore-reef (>5 m) while healthy coral communities remained in shallow water (2–5 m), suggesting that damage was not due to past bleaching events. Lesions from coral-eating snails (*Drupella*) were observed frequently on *Pocillopora*, followed by *Acropora*, while *Porites* and faviids were mostly affected by *Coralliophila*. In general, *Drupella* caused large patches of tissue loss that expanded across the corals in a band-like manner while *Coralliophila* typically inhabited depressions

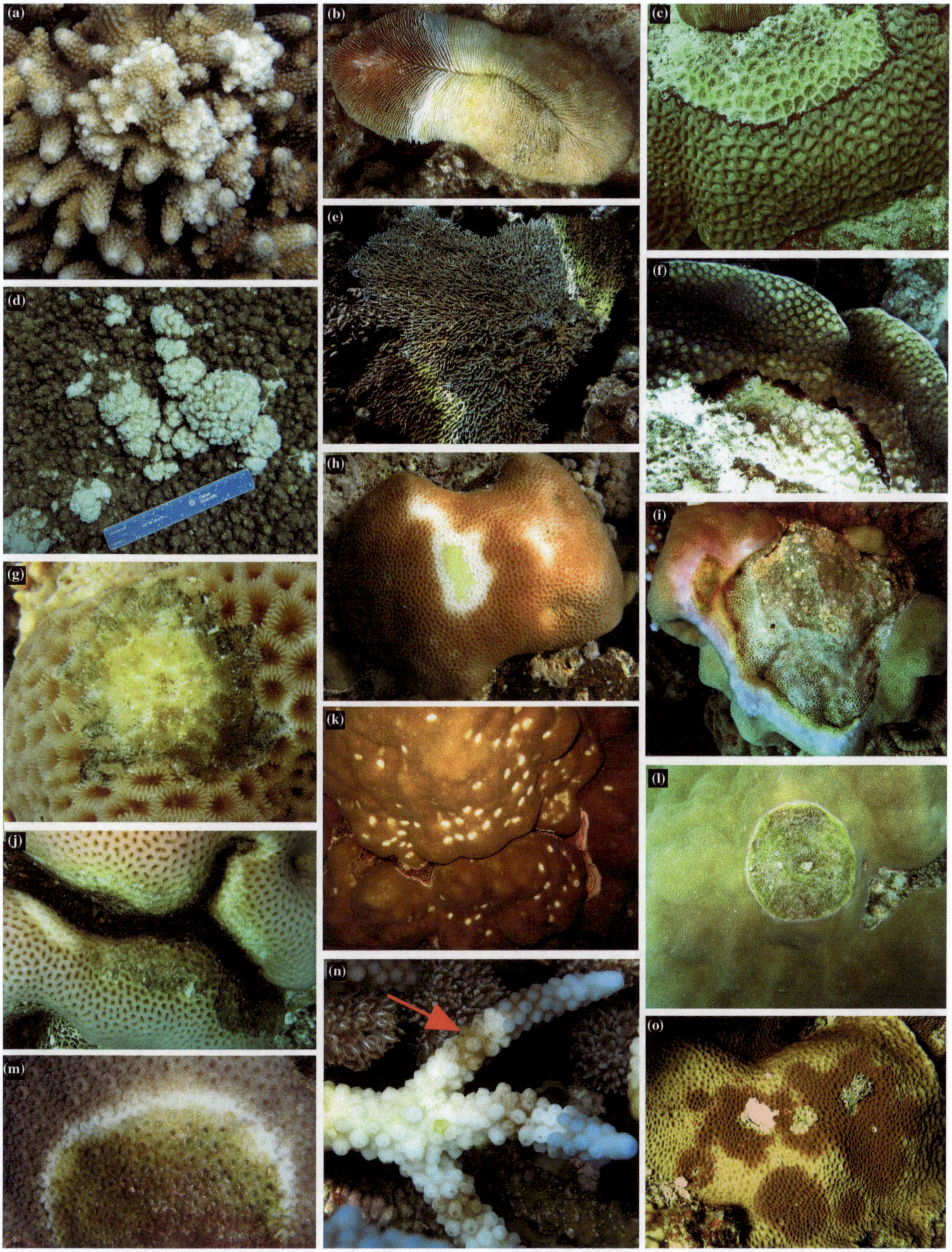

Fig. 7 Coral diseases observed in the Red Sea. **a** Growth anomaly (hyperplasia) on *Acropora gemmifera*. **b** White syndrome on *Ctenactis echinata*. **c** Black band disease on *Favites* spp. **d** Tumors (neoplasia) on a table *Acropora*. **e** White syndrome on *Acropora hyacinthus*. **f** Black band disease on *Echinopora*. **g** Skeletal eroding band on *Goniastrea*. **h** Undescribed multifocal lesion on *Goniastrea*. **i** Pink line disease on *Porites lutea*. **j** Skeletal eroding band on *Goniastrea*. **k** Ulcerative white spots (PUWS) on *Porites*. **l** Pink line disease on *Porites*. **m** White syndrome on *Favia*. **n** Brown band disease on *Acropora*. **o** Dark spots disease on *Goniastrea*

Fig. 8 Characteristic signs of predation observed in the Red Sea. **a** *Coralliophila* gastropods feeding on *Porites*. **b** *Drupella* gastropods feeding on *Pocillopora verrucosa*. **c** *Stylophora* with gastropod predation. **d** Close-up of *Drupella* snails on *Acropora*. **e** *Acanthaster* (crown of thorns starfish) feeding on *Echinopora*. **f** COTS feeding on *Porites*. **g**, **h**, and **j** Damselfish algal lawns on *Porites* and *Platygyra*. **i** Spot biting by parrotfish on *Porites*. **k** *Pocillopora* with branch ends bitten off by parrotfish. **l**, **m**, **n**, and **o** Focused biting by parrotfish on *Leptoria*, *Galaxea*, and *Porites*

Fig. 9 Nuisance species. **a** Fan sponge (possibly *Phyllospongia*) that commonly overgrows corals in the Farasan Islands and Farasan Banks, shown here covering a *Porites lutea*. **b** *Xenia*, a soft coral that monopolizes substrate, overgrows corals, and inhibits recruitment. *Xenia* was a dominant constituent of deeper reef communities throughout the SARS and is seen here overgrowing a *Porites lutea* in the Farasan Islands. **c** Unidentified encrusting sponge that covered large areas of reef and overgrew corals in deeper locations in the Farasan Banks. **d** Mat of *Palythoa* (colonial zoanthids) overgrowing *Porites lutea*. **e** Cyanobacterial mat on a colony of *Goniastrea*. **f** Encrusting/bioeroding sponge killing a *Porites lobata* colony. **g** Brown macroalgae covered large portions of reef flat communities in nearshore and southern localities. **h** *Millepora* hydrozoan coral overgrowing *Galaxea*

in a colony, especially between lobes, and only rarely caused visible tissue loss. *Drupella* appeared to control the upper size limit of *Pocillopora* colonies in Yanbu, as most corals over 10 cm were infested with snails and few intact colonies over 20 cm were found.

A number of nuisance species (e.g., *Xenia*, clionid sponges, encrusting sponges, leather coral, *Millepora*, zoanthids, and *Peyssonnelia* brown algae) competed and overgrew corals and monopolized substrates. Throughout SARS, the deeper *Porites* framework was largely dead and most of the skeletons were carpeted in mats of *Xenia*. In the Farasan Banks and Farasan Islands, encrusting sponges often formed large mats on open substrates and they were frequently observed overgrowing and bioeroding corals. Macroalgae were generally uncommon, but in parts of Ras Al-Qasabah, shallow nearshore sites in the Farasan Banks, and throughout the Farasan Islands, a substantial amount of substrate was colonized by macroalgae. Damselfish algal lawns were especially common in areas with few predatory fishes (Fig. 9).

Discussion

The assessments conducted in the Saudi Arabian Red Sea, including measures of coral diversity, size structure, condition, causes of mortality, and other benthic attributes present a first indication of the status, vulnerability, and resilience of these reefs. Knowledge of the demographic structure of a coral population can provide insight into the effects of disturbance and resilience of reefs. Coral populations are normally highly positively skewed, with dominance by the smallest size classes and an exponential decrease with increasing colony size (Bak and Meesters 1998, 1999). Larger colonies typically become rarer in a population, but they have a decreased probability of total colony mortality relative to smaller colonies (Hughes and Jackson 1985; Soong and Lang 1992). One consequence of this is that smaller colonies suffer total mortality more frequently than larger colonies, while larger colonies sustain higher levels of partial mortality over multiple disturbances, but have a greater chance of survival (Babcock and Mundy 1996; Hall and Hughes 1996). Thus, size frequency distributions of coral populations are key indicators for categorizing changes in coral communities in response to acute and chronic disturbances and may help identify differences between populations exposed to differing degrees of environmental stressors.

In addition to differences in survival, colony size is an indicator of the reproductive potential of the population. Corals show a positive relationship between colony size and fecundity, with smaller colonies allocating more energy to growth and maintenance and focusing on reproduction only after achieving a critical threshold size (Szmant-Froelich 1985; Harrison and Wallace 1990). Reproductively mature colonies may also regress in size below that minimum threshold, becoming non-reproductive (Szmant 1991). Competitive abilities and regenerative potential also increase with size (Meesters et al. 1994). By combining size structure with estimates of mortality, it is possible to predict whether a population is more likely to continue to contribute to recruitment. A reef consisting of healthy, larger, older colonies with low levels of partial mortality is likely to produce higher numbers of gametes than a reef dominated by juveniles or older colonies with high levels of partial mortality. A population dominated by small size classes may indicate high mortality in intermediate and larger corals (Nyström et al. 2008). Large colonies that have been reduced into tissue isolates that are below the minimum size needed for reproduction may expend more energy on growth and maintenance and less on reproduction (Bruckner 2012).

In SARS, knowledge of the extent of partial and whole colony mortality provided an indication of the severity of past disturbances, resistance of the corals, how much recovery has occurred, and the likelihood that the reef will survive future disturbances. For instance, several offshore reefs in Al Wajh were dominated by large, massive *Porites* colonies with few juveniles; these areas likely have not experienced a recent large-scale disturbance. This suggests that, despite what appears to be a fairly stable community, these reefs may lack resilience to future disturbances, as new corals are not settling or replacing older colonies. Most reefs in Ras Al-Qasabah and lagoonal reefs in Al Wajh had few large corals, several canopy layers of dead corals in growth position, and a high abundance of small tissue remnants, suggesting these reefs have been perturbed very frequently. In contrast, a large number of midshelf and offshore reefs in the Farasan Banks had few large corals, high numbers of skeletons in growth position, and an unusually high number of recruits, indicating a mass die-off from a previous disturbance, with signs of rapid recovery.

Ras Al-Qasabah

Reefs throughout the region showed evidence of a past (10–15 years) mortality event that affected 50–75 % of the corals and was responsible for the loss of most of the columnar *Porites* colonies. The *Porites* framework was subsequently colonized by acroporids, faviids, and *Lobophyllia* in shallow water, and foliose and plating *Echinopora* and *Turbinaria* colonies in deeper water but most of these corals had 3–9 years before these surveys, and skeletons remained in growth position. Living coral cover is now lower than in all other regions; surviving corals include highly fragmented remnants of large *Porites* colonies, small massive faviid corals (*Goniastrea* and *Echinopora*) and encrusting *Montipora* and branching corals (primarily *Stylophora* and *Seriatopora*). There were also high cover of *Millepora*, *Xenia*, and *Peyssonnelia*, large areas of uncolonized reef framework, and considerable amounts of rubble.

Coral bleaching events appear to be the major stress responsible for extensive losses of coral at Ras Al-Qasabah. First, extensive mortality was apparent in all sites and depths including shallow water, unlike other regions examined in this study which had prolific coral growth in shallow water (1–5 m) and mortality restricted to middepth and deep sites. Furthermore, during the four-year mission, bleaching was only documented in Ras Al-Qasabah. One contributing factor is that the region is exposed to temperature extremes and it lacks the protective buffering afforded by the upwelling of cooler waters because the eastern boundary current that flows northwards up the Saudi coastline may suppress wind-driven upwelling of deep water. Mortality will therefore be higher than expected during periods of unusual thermal anomalies and more severe than the relatively more healthy neighboring Tiran area (Ras Qisbah) off the Sinai Peninsula. Sites at the edge of the platform along the extensive bank and reticulate reef systems were also affected by pooling of water over shallow reefs, and high outflows of sediment and super-heated water may prevent recovery of these sites and exacerbate the impacts of future bleaching events. The richest coral cover and diversity remain in areas near channels, as tide-driven flow through restricted channels promotes flushing

Al Wajh and Yanbu

The coral community structure at Yanbu and Al Wajh was dominated by small tissue remnants of *Porites columnaris* and high numbers of recruits and juveniles of *Acropora*, *Montipora*, *Goniastrea*, *Favia*, and *Pocillopora*. Although the sites show evidence of a past mortality event, they appear to have stabilized and are now recovering. While many of these reefs are vulnerable to future perturbations, they also show high resilience. Al Wajh has an extensive bank and barrier reef system, and an unusually large lagoonal system with a high number of shallow and differentiated habitats (e.g., grassbed, mangrove, patch reef, and fringing reef). The highest coral cover occurred on the barrier reef and on reefs located near channels, while most lagoonal patch reefs exhibited low cover. The lagoonal reefs appear to have experienced a mass mortality event 6–9 years before these surveys were completed, with coral communities dominated by small, fragmented tissue remnants along with a moderate number of faviid, poritid, and acroporid colonies. These disturbances appear to have affected many of the important framework corals, as evidenced by the extensive amount of dead coral and bioerosion of these colonies. Lagoonal habitats are likely to experience more highly fluctuating temperatures in the shallows, higher sediment transport, lower visibility, stronger/more variable currents, a greater degree of ponding of waters, and moderate to high levels of turbidity. The reefs at Yanbu include extensive shallow coastal marine habitats with large stands of mangroves, submerged grass beds, and a complex offshore barrier reef system consisting of a string of linear reef complexes parallel to the coast and separated by a major shipping channel. Water temperatures are likely to be cooler on offshore sites due to the steeper linear slopes, and thus, bleaching impacts will be less severe. Shading by the reef structure was pronounced; the linear reef system runs north/south, with a high degree of shading on vertical faces and slopes, particularly for the steeper west-facing slopes in the morning. Yanbu's reefs as well as the outer barrier reef sites of Al Wajh are both likely to be exposed to high wave energy and have few sheltered locations.

Farasan Banks

The Farasan Banks contain the most extensive and diverse reef environments in SARS, with fringing reefs along the mainland coast, nearshore algal reefs, circular or elongated patch and platform reefs, barrier reefs, coral pinnacles, and emergent or submerged atoll-like structures. Shallow and deeper reef habitats were dominated by a high cover of living *Porites*. These taxa was both numerically dominant and often the largest corals in the population, although nearly one-third of the sites examined contained large areas of dead *Porites* framework as well as signs of recent mortality from disease and predation. Nevertheless, most mortality was patchy and restricted to middepth and deeper locations, while shallow sites often had 50–70 % living coral cover, suggesting bleaching played a minor role in disturbances. The site experiences large latitudinal and seasonal extremes in temperature, and above-average salinities, two factors that would typically contribute to high mortality of corals. Still, the presence of flourishing coral communities in many of the sites suggests local populations are well adapted to temperature-induced stress, with local environmental extremes likely to be close to their physiological tolerance maxima. In this case, local coral populations may be particularly vulnerable to climate change.

On a positive note, the midshelf and offshore reefs in most sites were surrounded by deep water and strong currents (seasonally changing direction) and exhibited numerous physical attributes (such as steep vertical walls and overhangs) that are likely to moderate temperature extremes. Furthermore, turbid conditions on nearshore reefs reduce UV penetration, also mitigating the potential for coral bleaching. Although many sites appeared degraded, with a near total loss of adult colonies, the region exhibited numerous additional factors that confer resilience. First, the number of coral recruits was 2–10 times higher than that observed elsewhere in the Red Sea and up to 100 times higher than that reported from adjacent areas in the Indian Ocean. Secondly, substrate quality was high and conducive to coral settlement, with low amounts of unstable rubble, very low macroalgal cover, relatively low cover of soft corals, and higher amounts of crustose coralline algae than observed in other areas along the Saudi Arabian Red Sea coastline.

Farasan Islands

Coral communities in the Farasan Islands were largely restricted to windward and leeward coral crests, located seaward of the major islands, while nearshore habitats were largely algae-covered pavements, sand flat communities, seagrass meadows, mangroves, patch reefs, and macroalgal-dominated fringing reefs. Live corals typically formed small patches in shallow water (<8 m) surrounded by rubble fields and sediment covered algal pavements, with low cover of living corals in deeper water, except for offshore locations. Most of the sites with the highest living coral cover were located to the west and northwest of Farasan Kabir, especially those sites on Al Baghlah Bank, Massad Island, Shuma, and Lajhan. Located far offshore, they have much less siltation than nearshore habitats. In contrast, reefs off the main islands, including Zufaf Island and Dushuk Island, were much more degraded. These sites had higher abundances of soft corals, encrusting and bioeroding sponges, macroalgae, and cyanobacteria. This community shift may be the result of a number of stressors, namely land-based nutrient loading and/or reduced herbivory. While many sites showed signs of previous disturbance events evidenced by extensive rubble, dead coral, and colonization by early recruiting soft corals, the high proportion of small colonies throughout these disturbed sites is indicative of the success of recent successful recruitment events and progressive regeneration.

Conclusion

Like elsewhere, SARS coral reefs have been subjected to serious disturbances over the past few decades including the 1998 bleaching event (DeVantier and Pilcher 2000). Sites examined during these surveys had been impacted by ongoing COTS outbreaks and had clearly suffered from past bleaching. Causes of coral mortality could be clearly identified in most cases; in other cases, low coral cover and high numbers of dead and mostly dead colonies in growth position bore testimony of disturbances that had occurred in the recent past (10–15 years). Impacts to these sites occurred in all habitats and reef zones although the severity of damage was highly variable. Many reefs appeared to have been disturbed, undergone recovery, and then damaged again, while others showed few signs of degradation. The variability between sites and within individual reefs, coupled with the number of degraded reefs, suggests that local disturbances acting on individual reefs (e.g., COTs and disease outbreaks) have been more common than region-wide (e.g., mass bleaching) events.

While diversity remained fairly high, the occurrence of completely denuded reefs and low remaining species diversity on these sites suggests that some of the rarer coral species identified from the region during earlier studies are likely to disappear (Sheppard 1985; Sheppard and Sheppard 1985, 1991; Devantier et al. 2000). Although coral community differentiation with regards to zone-forming genera was still apparent throughout SARS, there appears to be a loss of the largest colonies, especially some of the slower growing taxa that were reported in previous studies (Kleemann 1992; Riegl and Piller 1997, 1999; Devantier et al. 2000). While numerous unusually large skeletons were encountered, in degraded areas, surviving colonies consisted largely of small tissue remnants. Further, there has been a general shift from *Acropora* and *Porites* in shallow water to faster growing *Stylophora* and *Pocillopora*, and from large *Porites* to smaller faviid colonies in deep water (Riegl et al. 2013). These factors highlight an increase in community homogenization and decline in average size of coral colonies, as reported for other parts of the Red Sea (Riegl et al. 2012).

Although these reefs have undergone extensive degradation over the last several decades, most reefs show a promising trajectory of recovery. The dominant reef frame builders are massive and submassive colonies of *Porites.* This taxon is more tolerant to thermal stress than most other genera and also less susceptible to mortality from disease. Coral diseases were present, but relatively uncommon, with exception of a disease affecting massive *Porites* colonies. Several outbreaks of *Acanthaster planci* (crown of thorns) were noted in midshelf reefs, and these sea stars were negatively impacting sites with the highest live cover of branching acroporid corals.

Nevertheless, their target species are among the fastest growing corals and they exhibited very high levels of recruitment. A high diversity of reef-building corals exists, including multiple growth forms and resistant taxa, many functional groups, and a wide size range extending from the small (juvenile and recruits) to large-sized colonies. In areas with widespread mortality of adult corals, the dead colony structure remained largely intact, and substrate quality was high, with high cover of crustose coralline algae. Further, the presence of unusually high numbers of juvenile corals and recruitment levels that often exceeded those reported from other locations in the Red Sea and Indian Ocean suggests these reefs are undergoing rapid recovery. Reefs also exhibited a number of environmental and physical factors that confer resilience. In general, reefs in central and southern SARS had relatively high structural complexity, several canopy layers (which helps shade corals), low amounts of loose rubble (which makes the substrate unstable and reduces the settlement of corals), and little macroalgae (which overgrows and smothers slower growing corals).

Although most damaged areas were showing promising signs of recovery during these surveys, many areas are at a critical threshold and could fail to recover from future disturbances. Firstly, the region is under increasing pressure from artisanal and subsistence fisheries (Bruckner 2011a; Bruckner et al. 2011b) and most of the fishing occurs within the lagoon and nearshore sites, where corals are under the greatest stress from temperature perturbations and land-based stresses. As populations continue to expand in coastal areas, the consequences of human threats are likely to worsen. The presence of cyanobacterial mats, a rarity of detritivores (e.g., sea cucumbers), a scarcity of large herbivores, coupled with high cover of *Xenia* (carpeting much of the *Porites* framework), increases in other nuisance species (encrusting and bioeroding sponges, damselfish algal lawns) provide strong evidence that substrate quality may be declining. The five regions examined in this study contain the most unique and extensive coral reef ecosystems found in the Red Sea, and they support extensive associated nursery habitats (e.g., mangroves and seagrass beds). As climate-related stressors in the Red Sea continue to increase (Cantin et al. 2010), it is critical that human impacts to these areas are reduced and more consideration is given for the inclusion of these areas in a network of marine protected areas.

Acknowledgments The four-year research mission to the Saudi Arabian Red Sea could never have been completed without the leadership, vision, and generosity of His Royal Highness Prince Khaled bin Sultan. We are deeply appreciative of his financial support and for the generous use of his research vessel, M/Y Golden Shadow. His vision of Science Without Borders® was materialized in the research mission to the Red Sea through the partnerships and involvement by the Saudi Wildlife Authority, Nova Southeastern University, the International Union for Conservation of Nature (IUCN), Coastal Ocean Research and Development in the Indian Ocean (CORDIO), and Blue Ventures. Data presented here were summarized from benthic transects conducted by Bernhard Riegl, David Obura, Alastair Harris, and Andrew Bruckner. All phototransects were conducted by Annelise Hagan and Philip Renaud. The Saudi Wildlife Authority (formerly the National Commission for Wildlife Conservation and Development) under the leadership of the Secretary General, His Excellency Prince Bander bin Saud, provided assistance to KSLOF throughout this research mission. All permits for the work were obtained through the assistance of Mr. Omar Khushaim. None of this research would have been possible without dedicated involvement by Captain Nick Gilbert, Captain Mike Hitch and the devoted officers, and crew of the M/Y Golden Shadow. I am grateful for the detailed comments provided by four anonymous reviewers.

References

Babcock R, Mundy C (1996) Coral recruitment: consequences of settlement choice for early growth and survivorship in two scleractinians. J Exp Mar Biol Ecol 206:179–201

Bak RPM, Meesters EH (1998) Coral population structure: the hidden information of colony size-frequency distributions. Mar Ecol Prog Ser 162:301–306

Bak RPM, Meesters EH (1999) Population structure as a response of coral communities to global change. Am Zool 39:56–65

Behairy AKA, Sheppard CRC, El-Sayed MK (1992) A review of the geology of coral reefs in the Red Sea. UNEP regional seas reports and studies no. 152, Nairobi, 39 pp

Bosence D (2005) A genetic classification of carbonate platforms based on their basinal and tectonic settings in the Cenozoic. Sediment Geol 175:49–72

Bruckner A (2011a) Khaled bin sultan living oceans foundation habitat mapping and characterization of coral reefs of the Saudi Arabian Red Sea: 2006–2009. Panoramic Press, Arizona, 140 pp

Bruckner A (2011b) Khaled bin Sultan living oceans foundation habitat mapping and characterization of coral reefs of the Saudi Arabian Red Sea: 2006–2009. Final report part II, Ras Qisbah, Al Wajh, Yanbu, Farasan Banks and Farasan Islands. Panoramic Press, Arizona, 248 pp

Bruckner AW (2012) Static measures of resilience of Caribbean coral populations. Rev Biol Trop 60:39–57

Bruckner A, Riegl B, Purkis S, Rowlands G, Williams A, Renaud P (2011a) Khaled bin sultan living oceans foundation atlas of Saudi Arabian Red Sea marine habitats. Panoramic Press, Arizona, 262 pp

Bruckner AW, Faisal M, Alnazary H (2011b) Paradigm shift for fisheries management to enhance recovery, resilience, and sustainability of coral reef ecosystems in the Red Sea. Sustainable fisheries: multi-level approaches to a global problem. American Fisheries Society, Maryland, pp 85–111

Burke L, Reytar K, Spalding M, Perry A (2011) Reefs at risk revisited. World Resources Institute, Washington, D.C

Cantin NE, Cohen AL, Karnauskas KB, Tarrant AM, McCorkle DC (2010) Ocean warming slows coral growth in the central Red Sea. Science 329(5989):322–325

De Vantier L, Pilcher N (2000a) The status of coral reefs in Saudi Arabia. Glob Coral Reef Monit Netw (GCRMN)

De Vantier L, Tourak E, Al-Shaikh K, Death G (2000) Coral communities of the central-northern Saudi Arabian Red Sea. Fauna Arab 18:23–66

Hall VR, Hughes TP (1996) Reproductive strategies of modular organisms: comparative studies of reef building corals. Ecology 77:950–963

Harrison PL, Wallace CC (1990) Reproduction, dispersal and recruitment of scleractinian corals. In: Dubinsky Z (ed) Ecosystems of the world: coral reefs. Elsevier, Amsterdam, pp 133–207

Hughes TP, Jackson JBC (1985) Population dynamics and life histories of foliaceous coral. Ecol Monogr 55:141–166

Kleemann K (1992) Coral communities and coral-bivalve associations in the northern Red Sea at Safaga. Egypt Facies 26:1–10

Meesters EH, Noordeloos M, Bak RPM (1994) Damage and regeneration: links to growth in the reef-building coral *Montastrea annularis*. Mar Ecol Prog Ser 112:119–128

Nyström M, Graham NAJ, Lokrantz J, Norström A (2008) Capturing the cornerstones of coral reef resilience: linking theory to practice. Coral Reefs 27:795–809

PERSGA (2006) State of the marine environment, report for the Red Sea and Gulf of Aden. Pergsa, Jeddah

Riegl B, Piller WE (1997) Distribution and environmental control of coral associations in northern Safaga Bay, Red Sea. Egypt Facies 36:141–162

Riegl B, Piller WE (1999) Coral frameworks revisited: reefs and coral carpets in the northern Red Sea. Coral Reefs 18(3):241–254

Riegl B, Bruckner A, Rowlands G, Purkis S, Renaud P (2012) Red Sea coral reef trajectories over 2 decades suggest increasing community homogenization and decline in coral size. PLoSONE 7(5):e38396

Riegl B, Berumen M, Bruckner A (2013) Coral population trajectories, increased disturbance and management intervention: a sensitivity analysis. Ecol Evol. doi:10.1002/ece3.519

Rowlands G, Purkis SJ, Bruckner A (2014) Diversity in the geomorphology of shallow-water carbonate depositional systems in the Saudi Arabian Red Sea. Geomorphology. doi:10.1016/j.geomorph.2014.03.014

Sheppard CRC (1985) Reefs and coral assemblages of Saudi Arabia, 2: fringing reefs of the southern region, Jeddah to Jizan. Fauna Saudi Arab 7:37–58

Sheppard CRC, Sheppard ALS (1985) Reefs and coral assemblages of Saudi Arabia 1: the central Red Sea at Yanbu al Sinaiyah. Fauna Saudi Arab 7:17–36

Sheppard CRC, Sheppard ALS (1991) Corals and coral communities of Arabia. Fauna Saudi Arab 12:3–170

Sheppard C, Price A, Roberts C (1992) Marine ecology of the Arabian region: patterns and processes in extreme tropical environments. Academic Press, London

Soong K, Lang JC (1992) Reproductive integration in reef corals. Biol Bull 183:418–431

Szmant-Froelich AM (1985) The effect of colony size on the reproductive ability of the Caribbean coral *Montastrea annularis* (Ellis and Solander). In: Proceedings of 5th International Coral Reef Congress, vol 4, pp 295–300

Szmant AM (1991) Sexual reproduction by the Caribbean reef corals *Montastrea annularis* and *M. cavernosa*. Mar Ecol Prog Ser 74:13–25

Wilkinson C (2008) Status of coral reefs of the world: 2008. Global Coral Reef Monitoring Network and Reef and Rainforest Research Center, Townsville

Biology of Red Sea Corals: Metabolism, Reproduction, Acclimatization, and Adaptation

Yvonne Sawall and Abdulmohsin Al-Sofyani

Abstract

Coral reefs are the most abundant shallow water ecosystems in the Red Sea, harboring a high species diversity and habitat complexity over large environmental gradients. At the same time the semi-enclosed ocean basin and its partly extreme environmental conditions may promote species evolution being distinct from Indo-Pacific coral reefs. Extreme conditions are found in the southern Red Sea, where temperatures reach up to 33 °C in summer and where nutrient input is high. Mechanisms of organism adjustment to these conditions are of particular interest in the light of climate change research. Towards the north, conditions become more 'coral-promoting' finally reaching temperatures between 21–27 °C (winter-summer) and clear waters at the northern end of the Red Sea (Gulf of Aqaba). In this chapter, we summarize the current knowledge about the biology of shallow water, symbiotic, reef-building corals of the Red Sea. We start with an overview on the environmental conditions of the Red Sea, the history of coral reef research in this region and a general introduction into coral biology, before we describe the ecophysiology of Red Sea corals. Coral ecophysiology is presented in the context of varying environmental conditions over depth (e.g., light), between seasons, and over latitudes (e.g., light, temperature, nutrients). Mechanisms and patterns of coral reproduction are discussed in the context of seasonal and latitudinal environmental changes. Finally, we briefly describe anthropogenic influences on Red Sea coral reefs. Acclimatization mechanisms of corals to changing conditions over a depth gradient (mainly light reduction) have been well studied in the Gulf of Aqaba and include the following metabolic adjustments with depth: (i) an upregulation of light-harvesting pigments (chlorophyll a) and a downregulation of photo-protective pigments (xanthophyll), (ii) an increase of heterotrophy, and (iii) a decrease of metabolic activity (e.g., calcification and growth). In addition, a change in the symbiont composition (*Symbiodinium* clade and/or type) over depth was observed in some coral species. Seasonal environmental changes (mainly light availability, temperature, nutrients) lead to various metabolic responses of the corals, including (i) changes in zooxanthellae pigmentation and density and (ii) changes in the metabolic activity. In particular, changes in calcification and growth rates can be observed with lowest rates during low temperatures in winter. Interestingly, however, this reverses in the southern Red Sea, where calcification rates are higher in winter than in summer. This kind of latitudinal shift is also evident in the timing of reproduction, which occurs earlier in the year (January–March) in the south compared to the north (March–August). This indicates that growth and reproduction are strongly linked to

Y. Sawall (✉)
Helmholtz Center of Ocean Research (GEOMAR), Kiel, Germany
e-mail: ysawall@geomar.de

A. Al-Sofyani
King Abdulaziz University (KAU), Jeddah, Saudi Arabia

N.M.A. Rasul and I.C.F. Stewart (eds.), *The Red Sea*,
Springer Earth System Sciences, DOI 10.1007/978-3-662-45201-1_28,

temperature, following a single temperature optimum, which occurs at different times throughout the year from north to south. Furthermore, this hints towards a high phenotypic plasticity (acclimatization) rather than local genetic adaptation of the investigated coral species. A clear shift in the genetic population structure from north to south in another coral species, however, indicates local adaption. Adjusting mechanisms need to be further understood in order to provide indication for predicted climate change effects.

Introduction

Corals (class: Anthozoa) belong to the phylum Cnidaria in the animal kingdom, as do Hydrozoa, Cubozoa (box jellyfish), and Scyphozoa (true jellyfish). Cnidarians are characterized by the existence of generally two cell layers (epithelia), the ectoderm and the endoderm, which are held together by the gelatinous extracellular matrix, the mesoglea. Furthermore, they possess stinging cells for the purpose of defense and/or to capture prey, which are called nematocycsts or cnides. These specialized cells are phylum specific (therefore 'Cnidaria'). Corals are mostly sessile animals after larval settlement and metamorphosis consisting of polyps which are either solitary or colony forming. Most coral species are colonial consisting often of 100 to 1,000s of polyps with a polyp size ranging from a millimeter (e.g., some *Acropora* and *Porites* species) to several centimeters (e.g., species of the family Meandrinidae). Corals with the largest polyps are found among the mainly solitary living coral family Fungiidae. Anthozoans are further divided into the subclasses Octocorallia and Hexacorallia, according to the number of septa within a polyp. Octocorallia include inter alia the groups of soft corals, leather corals, blue corals, and gorgonians. Hexacorallia include the well-known reef-building corals, the scleractinians, as well as sea anemones and other less prominent groups (e.g., Centantharia [tube-dwelling anemones] and Zoantharia). Scleractinians are found throughout large geographic regions, in cold and warm, as well as in deep and shallow waters. They are best known, however, from tropical shallow waters, where most species live in symbiosis with unicellular algae and function as the most important bioengineers of extensive, structurally complex, and diverse coral reefs. These reefs are found along most parts of the Red Sea coast (Rowlands and Purkis, this volume) and hence form the most important ecosystem of the Red Sea. Therefore, we will focus on the biology of shallow water tropical corals in this chapter and will use the term 'coral' as an equivalent to scleractinian shallow water tropical corals, if not stated differently.

The preferred environment for corals is a clear, light-penetrated, and warm environment, which explains the occurrence of extensive coral reefs in clear shallow water (0–50 m) between the latitudes 30°N and 30°S of the equator (Kleypas et al. 1999). The typical temperature range for corals lies between 18 and 30 °C (minimum and maximum of weekly mean temperature), and the preferred nutrient regime features low nutrient concentrations (=low particle concentration), in order to allow high light penetration (Bell 1992; Kleypas et al. 1999). These conditions are met in the Gulf of Aqaba and the northern Red Sea, but are exceeded in the central and southern Red Sea. In the Gulf of Aqaba, sea surface temperatures vary between 20 and 27 °C, while they fluctuate between 28 and 33 °C in the southern Red Sea (Fig. 1). Chlorophyll *a* concentrations are up to 4 $\mu g\ l^{-1}$ in the southern Red Sea due to high nutrient input, which is almost an order of magnitude higher than what is usually considered to be a typical reef environment (Bell 1992). Nutrients are gradually depleted toward the central Red Sea (indicated by chlorophyll *a* in Fig. 1). Despite the strong latitudinal environmental gradients in the Red Sea, including rather extreme environmental conditions in the central and southern Red Sea (high temperature, high-nutrient input), corals are growing in high abundance, diversity, and structural complexity throughout the Red Sea (Sheppard and Sheppard 1991; Sawall et al. 2014b; Bruckner and Dempsey, this volume; Rowlands and Purkis, this volume). This implies that Red Sea corals feature a broad physiological flexibility and/or Red Sea-specific genetic adaptations in order to be able to inhabit even the 'coral-challenging' environments. Since coral reefs worldwide are threatened by anthropogenically caused shifts in sea surface temperature, eutrophication, and ocean acidity (ICPP 2007; Rabalais et al. 2009), the Red Sea offers an ideal place to investigate potential acclimatization or adaptation mechanisms in corals to some of the predicted scenarios. If corals of the Red Sea do feature special mechanisms to resist, for example, high temperatures, they may serve as a 'reservoir' of particularly resistant coral populations during ocean warming (Fine et al. 2013).

In this chapter, we provide an overview about the physiology of Red Sea corals, including their metabolism and reproduction. We look at mechanisms of acclimatization (phenotypic adjustments) and/or adaptation (genetic differentiation) to shifting environmental conditions, most importantly light, temperature, and nutrients. Most published work is available from the Gulf of Aqaba so far; however, a few studies from the main Red Sea allow us to provide a rough overview of potential adjusting mechanisms of corals

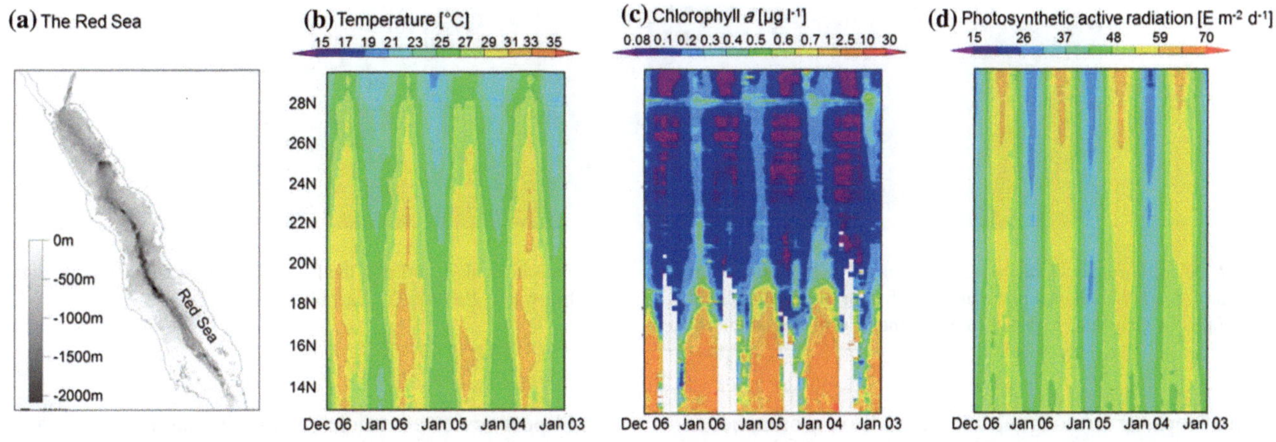

Fig. 1 Environmental conditions of the Red Sea (**a**) from north (28°N) to south (14°N). Spatial–temporal variation of the following: **b** sea surface temperature, **c** chlorophyll *a*, and **d** photosynthetic active radiation from north to south (*y*-axis), and from January 2003 to December 2006 (*x*-axis). *Data* derived from satellite images of NASA, Giovanni online data system, Ocean Color Radiometry, monthly averaged data set MODIS-Aqua 4 km

to the latitudinal environmental gradient. In addition, we will briefly address potential threats to corals and coral reefs in the Red Sea.

Progress of Coral Reef Research in the Red Sea

For more than 200 years, coral reefs of the Red Sea have been investigated (cf. Mergner 1984). There have been frequent expeditions by researchers collecting and describing various reef animals, as well as reef communities, habitats, and structures (see references in Mergner 1984; Bruckner and Dempsey, this volume). However, initial studies on the behavior and physiology of scleractinian corals were mainly conducted in other geographic regions, such as the Great Barrier Reef (GBR) and the Caribbean. There, researchers discovered the symbiotic relationship between the coral host animal and the unicellular algae (zooxanthellae) living inside the host tissue (Vaughan 1919; Yonge 1940). Later, it was recognized that there is a vital nutrient exchange between the host and symbionts (Muscatine and Porter 1977) and that zooxanthellae contribute substantially to the corals' energy needs by the translocation of photosynthetically produced energy-rich C compounds to the host (Muscatine and Cernichiari 1969). Studies on coral biology in the Red Sea started in the 1970s, where researchers working in the Gulf of Aqaba significantly enhanced the knowledge about coral reproduction (Rinkevich and Loya 1979a, b; Shlesinger and Loya 1985; Rinkevich 1989; Shlesinger et al. 1998) and photo-acclimation processes of the zooxanthellae (Falkowski and Dubinsky 1981; McCloskey and Muscatine 1984; Porter et al. 1984; Dubinsky et al. 1990). The highly abundant r-strategic Red Sea coral *Stylophora pistillata* (Loya 1976) became the preferred model species for various field and laboratory studies up to the present time (see Box 1).

While there is extensive knowledge on coral biology, including reproduction, metabolism, and acclimatization mechanisms from the Gulf of Aqaba, the main Red Sea is still strongly underinvestigated due to low accessibility during recent decades. The few studies that are published from the main Red Sea include those on coral growth and reproduction at the Egyptian coast (Kotb 2001; Mohamed et al. 2007; Hanafy et al. 2010) and studies on reproduction (Bouwmeester et al. 2011a, b) and physiology from the central Red Sea, Jeddah, Saudi Arabia (Al-Sofyani and Davies 1993; Harland et al. 1993; Al-Sofyani and Floos 2013; Bayer et al. 2013; Furby et al. 2013; Jessen et al. 2013; Roder et al. 2013; Liew et al. 2014). Comparative studies along the north–south gradient in the Red Sea are still very rare (Al-Sofyani 1987) and have only recently been intensified by the authors of this chapter. New facilities, such as the King Abdullah University of Science and Technology (KAUST) 100 km north of Jeddah, and new international collaboration of Saudi Arabian universities and institutes allow the urgently needed coral research in the Red Sea to be extended.

Box 1: Model species *Stylophora pistillata*
The best studied coral in the Red Sea is the common coral *Stylophora pistillata* of the family Pocilloporidae (example specimens from different regions of the Red Sea are shown in Fig. 2). It is a cosmopolitan species and has an exceptionally wide range of habitats in the Red Sea, including shallow and deep, protected and exposed, and clear and turbid waters, and can be dominant at shallow exposed reef slopes (Sheppard and Sheppard 1991). Depending on its habitat conditions (light, exposure, turbidity), its growth form varies; for example, it grows (i) short, thick, light brown branches in high-light shallow exposed reef slopes or

Fig. 2 *Stylophora pistillata* in 3–5 m depth. **a** In the Gulf of Aqaba (Maqna) on a rather shaded place in the reef, **b** in the northern Red Sea (Al-Wajh) in clear and high-light environment, **c** in the southern Red Sea (Farasan Banks), and **d** in the far southern Red Sea (Farasan Island) in turbid water. *Pictures* Y. Sawall

reef tops, or (ii) long, slender, dark brown branches in shaded protected understories or cavities (Fig. 2). This implies a very high acclimatization potential, which was confirmed by studies revealing a high photo-acclimation and trophic plasticity not only in Red Sea specimens (Dubinsky et al. 1990; Mass et al. 2007), but also in *S. pistillata* of other regions (e.g., Japan—Titlyanov et al. 2001a). The ability to grow in temperatures above 32 °C seems to be restricted to Arabian populations, since *S. pistillata* of the GBR were found to be very sensitive to temperatures above 30 °C, causing rapid coral bleaching (Hoegh-Guldberg and Smith 1989). This hypothesis is supported by the fact that Arabian populations are genetically distinct from SE Asian and GBR populations, as well as from the eastern coast of Africa (Keshavmurthy et al. 2013). *S. pistillata* is a brooder with gonads found almost throughout the year and a larvae release during several months within the first half of the year (Shlesinger and Loya 1985).

S. pistillata from the Red Sea is still until now the preferred model species for in situ investigation as well as for laboratory-based experiments addressing various physiological aspects of hermatypic corals. *S. pistillata*'s career as the 'laboratory rat' of corals started during the late 1970s, when Loya described its characteristics as an r-strategist in a paper published in 'Nature' (Loya 1976). Those characteristics include for example a large distribution range, fast growth, and a high population turnover, which are ideal conditions for culturing. Furthermore, it is a robust and easy-to-satisfy species under culture conditions. Numerous mini-colonies (nubbins) have been created by cutting the apical parts of branches, which heal within weeks (C. Ferrier-Pagès, pers. comm.), and used for experiments to answer fundamental physiological and eco-physiological questions. *S. pistillata* from the Red Sea has appeared in about 100 publications at the Monaco Scientific Center alone. A Web of Science search (October 2013) found the species name in the title of 114 peer-reviewed articles, while the majority of them stem from countries working on Red Sea specimens (Fig. 3). The genome of the Monaco Scientific Center clone is currently being sequenced, assembled, and annotated by the laboratory of Christian R. Voolstra at the King Abdullah University of Science and Technology (KAUST), Saudi Arabia.

The Symbiotic Relationships of Corals

In order to understand the effects of different environmental conditions on corals, it is important to know that most corals live in symbiosis with unicellular algae, commonly referred to as zooxanthellae, as well as with coral-associated microorganisms mainly situated in/on the coral mucus layer covering the outside of the coral polyp. The entity of organisms involved in symbiosis is called the coral holobiont (Rosenberg et al. 2007). All components of the coral holobiont, basically, have the ability to acclimatize or adapt to changing environmental conditions (Gates and Edmunds 1999; Rosenberg et al. 2007; Howells et al. 2012). In particular, the relationship between the coral host and the zooxanthellae is believed to be the reason for their great success throughout the last 200 million years (Stanley 2003) and their mutual relationship has been studied extensively within the last several decades. More recent work recognized the significance of bacterial communities associated with corals (Reshef et al. 2006; Rosenberg et al. 2007; Zilber-Rosenberg and Rosenberg 2008; Mouchka et al. 2010).

Most corals form a colony consisting of numerous polyps, which cover the coral skeleton. The skeleton is permanently created at the base of the polyp, where aragonite ($CaCO_3$) is formed in the extracellular calcifying fluid between skeleton and tissue (Tambutté et al. 2011). Zooxanthellae are photosynthetic dinoflagellates of the genus *Symbiodinium* and are located in the endoderm of the coral polyp (Fig. 4) in numbers of up to more than 1 million cells per cm^2 of coral surface. The symbiosis allows the coral to derive energy and

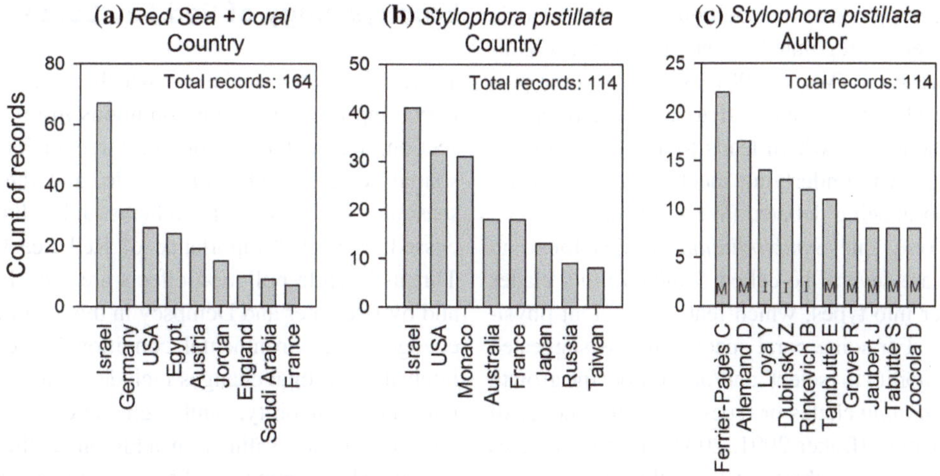

Fig. 3 Publication statistics (peer-reviewed articles) based on Web of Science search results in October 2013. Search criteria: words found in the title of publications: **a** Red Sea and coral, **b** and **c** *Stylophora pistillata*. Publications sorted by **a** and **b** country, and by **c** author. All results with more than 5 records (**a** and **b**) and more than 7 records (**c**) are presented. *Capital letters within the bars* of figure **c** indicate the country of the author with M = Monaco and I = Israel

nutrients in two ways, (i) heterotrophically by the coral polyp, which captures prey with its tentacles, and (ii) autotrophically by the zooxanthellae (Fig. 4). Furthermore, it guarantees the efficient recycling of nutrients within the coral polyp, hence avoiding the loss of nutrients in a clear, low-nutrient environment (Muscatine and Porter 1977). In shallow, high-light environments, corals cover most of their energy needs by photosynthesis, where CO_2 is reduced to energy-rich organic C compounds (Muscatine et al. 1984; Davies 1991). While there are still large gaps in knowledge concerning the translocation rates of nutrients and compounds between the host and zooxanthellae, as well as about the actual nature of compounds being translocated (Davy et al. 2012; Lesser 2013), general acceptance is found for the following processes: From the coral host, inorganic nutrients such as ammonium as well as respired CO_2 are transferred to the zooxanthellae, which they use to build up organic compounds such as amino acids, glycerol, glucose, and lipids. Those are partly used for zooxanthellae growth and partly transferred back to the coral host. Most likely, not only inorganic nutrients, but also organic compounds (e.g., amino acids, glycoproteins, and lipids) are translocated from the host to the symbiont (Lesser 2013). The number of symbionts is controlled by the coral host mainly by zooxanthellae expulsion, degradation, and regulation of nutrient supply (Falkowski et al. 1993; Titlyanov et al. 1996; Lesser 2013). The latter process may be disrupted in the case of uncontrolled nutrient supply from the environment (i.e., eutrophication), leading to an uncontrolled increase of zooxanthellae, which is potentially harmful for the coral host (Falkowski et al. 1993). The energy in the coral host is allocated into various metabolic processes, including calcification, biomass growth, reproduction, stress prevention, and mucus release (Fig. 4; Edmunds and Davies 1989; Davies 1991; Leuzinger

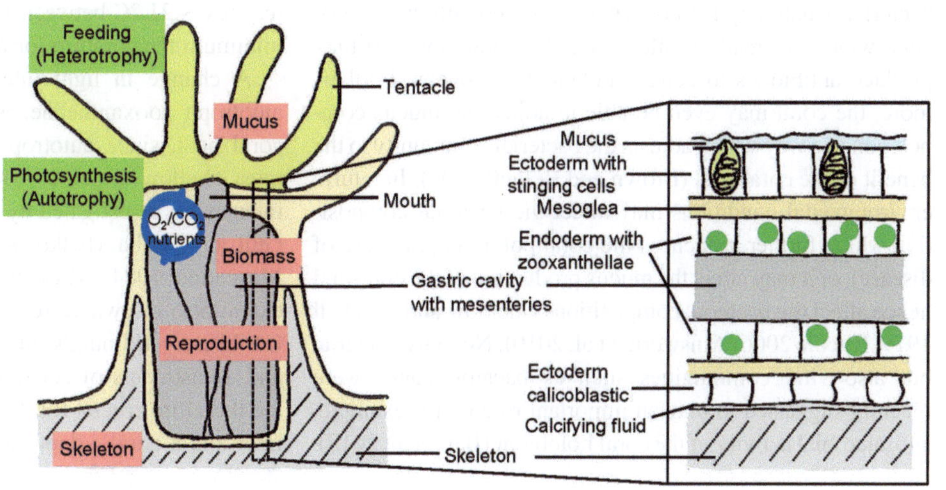

Fig. 4 Structure of a coral polyp (cross section) including the main energy acquisition (*green boxes*) and energy storing/consuming (*red boxes*) processes. Intense recycling of O_2, CO_2, and nutrients is indicated by the *blue cyclic arrow*. The *right* picture presents a schematic drawing of the cell layers (cross section). Drawn by Y. Sawall

et al. 2012). These processes are delimited by the genetic constitution, as well as regulated by environmental conditions. Temperature stress (Weis 2008), as well as some pollutants (Jones et al. 2003), can lead to an impairment of zooxanthellae functioning, which leads to a loss of zooxanthellae (bleaching). The underlying mechanism of zooxanthellae decimation is being debated (Weis 2008).

There is not only one *Symbiodinium* species found in corals, but there are several so-called *Symbiodinium* clades subdivided further into types, which feature different physiological strengths. For some coral species, it was suggested that they have the ability to 'select' and/or promote the growth of certain *Symbiodinium* clades or types as a consequence of environmental changes (Baker 2001, 2003). This guarantees the maintenance of the symbiotic relationship by selecting more suitable types for the prevailing conditions. For example, within a given coral species and reef, corals were found to harbor *Symbiodinium* species from different clades in different depths which allows optimizing photosynthesis in the different light environments (Frade et al. 2008a). Another example is provided by the change of *Symbiodinium* clades after a bleaching event, where some coral species were found to be able to switch their *Symbiodinium* population to rather heat-resistant clades (Baker 2001; Jones et al. 2008). Other coral species revealed a high species-specific selection for a *Symbiodinium* type or a large-scale geographic pattern in the association with *Symbiodinium* types (Baker et al. 2004; Goulet 2006; LaJeunesse et al. 2010).

Bacteria, associated with the coral mucus, living inside the coral tissue and the coral skeleton, and their potential mutual relationship with the coral host have received growing attention within the last 2 decades, when investigations on coral diseases were intensified (e.g., Reshef et al. 2006; Rosenberg et al. 2007; Bayer et al. 2013; Roder et al. 2014). The advantages of this symbiosis are suggested to be manifold (Ritchie 2006; Ainsworth et al. 2010): (i) The coral host provides substrate for bacteria growth in the form of organic material released by the coral (mucus), (ii) the coral cultures bacteria, which they feed on, (iii) the bacteria produce nutrients, which are made available for the coral, and (iv) they produce antibiotics to control potential pathogens. Furthermore, the coral may even be able to adjust the mucus composition in a way which adjusts the bacterial community to the benefit of the coral host (Brown and Bythell 2005). In return, environmental conditions may affect the bacterial composition which further influences the coral polyp (e.g., in case of disease), or it may affect the mucus production of the coral and hence affect the bacterial composition (Ducklow and Mitchell 1979; Ritchie 2006; Ainsworth et al. 2010). Not only bacteria, but also viral communities such as bacteriophages were recently suggested to play an important role in, for example, antimicrobial activity of the coral holobiont (Barr et al. 2013).

Ecophysiology of Red Sea Corals

There are several scales on which corals need to adjust to prevailing environmental conditions in the Red Sea. In this section, we will focus on the three major scales: the two spatial scales, depth and latitude, and the temporal scale, seasonality. Cross-shelf gradients and pollution will be discussed in Sect. "Importance of Red Sea Coral Reefs and Threats", while pollution effects are also described by Nasr and by Bruckner and Dempsey in this volume. Depth-related changes in environmental conditions concern mainly light intensity, seasonal changes include in particular temperature, nutrient availability, and light intensity, and latitudinal changes include salinity in addition to the aforementioned factors. The response of the coral to these parameters, however, may differ depending on whether the change occurs on a seasonal or on a latitudinal basis. Seasonal/temporal changes in the coral metabolism can be ascribed to phenotypic acclimatization mechanisms, while latitudinal/spatial variation may be ascribed to both, phenotypic acclimatization and/or genetic adaptation. The latter may then further account for latitudinal differences in seasonal responses.

Depth

The most important factor that changes with depth is light. The northern half of the Red Sea harbors one of the clearest coral reef waters. Here, during winter, 1 % of surface radiation (photosynthetic active radiation—PAR) penetrates down to about 95 m depth (Stambler 2005) and even down to 150 m depth in summer (estimated from $k_{d\lambda 490}$ and PAR values derived from satellite images of NASA August, September 2011, Giovanni online data system, Ocean Color Radiometry, MODIS-Aqua). In contrast, in the central Red Sea, 1 % of PAR penetrates down to about 70 m depth and in the south to only 20 m depth in summer (NASA derived). Temperature also gradually decreases with depth until it reaches ~21 °C between 150 and 200 m depth, which is the minimum temperature for the entire Red Sea (Siedler 1969).

A change in light intensity has a direct effect on the autotroph zooxanthellae, which in consequence affects the coral host, since autotrophy is the main energy source in most shallow water corals. This relationship was initially intensively investigated by researchers in the Red Sea, in the Gulf of Aqaba (Falkowski et al. 1984; McCloskey and Muscatine 1984; Muscatine et al. 1984), and to a lower extent also elsewhere (e.g., Hawaii—Muscatine et al. 1981, Caribbean—Edmunds and Davies 1989). Light availability and adjustments of zooxanthellae to different light regimes are therefore crucial for the coral holobiont and to determine inter alia a coral's depth distribution.

Light (PAR) is captured by the photo-pigments of the photosystems within the chloroplasts of the symbionts, and the energy is transferred to metabolic energy carrier (ATP, NADPH/H$^+$) along the electron transport chain. The energy is used to build up energy-rich carbohydrates by the uptake and reduction of CO_2 in the Calvin Cycle (Davy et al. 2012). The efficiency of the photosynthetic apparatus depends largely on the abundance of photo-accessory pigments, such as chlorophyll *a* and *c* and peridinin, found in dinoflagellates (zooxanthellae), as well as on the enzymatic activity of the proteins involved in photosynthesis. Protection of zooxanthellae from too much light is provided by photo-protective pigments, the xanthophylls. Hence, light changes can trigger an up- or downregulation of one or the other pigment, a process which is called photo-acclimation (Titlyanov et al. 2001b; Dubinsky and Stambler 2009; Mass et al. 2010).

A number of comparative studies on light and shade-acclimated corals (Falkowski and Dubinsky 1981; Dubinsky et al. 1984; Porter et al. 1984) and studies along a depth gradient (McCloskey and Muscatine 1984; Mass et al. 2007) in the Gulf of Aqaba demonstrated photo-acclimation processes of the zooxanthellae within the coral *S. pistillata*. Further studies in the central Red Sea showed a similar photo-acclimation behavior for other coral species (e.g., *Pocillopora verrucosa*—Ziegler et al. 2014, *Seriatopora hystrix* and *Lobophyllia corymbosa*—Floos 2012). These studies showed, for example, that the chlorophyll *a* content per zooxanthella doubled with a light reduction of 50–60 % in 5 versus 20 m depth (Mass et al. 2007; Ziegler et al. 2014), that the relative abundance of xanthophyll was reduced by about 20 % (Ziegler et al. 2014), and that the photosynthetic efficiency (α) was increased by 20–30 % (Mass et al. 2007; Ziegler et al. 2014), thereby maintaining high photosynthetic rates. Indeed, Stambler et al. (2008) showed that photo-acclimation processes over a depth range from 5 to 75 m can result in similar productivities of corals.

Not only photo-acclimation, but also adaptive processes (=change of genetic constitution) of the coral can counterbalance decreased light intensities by accommodating different zooxanthellae clades or types with different physiological strengths (Iglesias-Prieto et al. 2004; Frade et al. 2008b). Thus, *S. pistillata* was found to harbor exclusively *Symbiodinium* types of clade A in 5 m depth, but *Symbiodinium* types of clades A and/or C in 17 m depth (Lampert-Karako et al. 2008) and a change in C types deeper than 50 m depth in the Gulf of Aqaba (Daniel 2006 in Mass et al. 2007). This is in line with findings in the Caribbean, where the coral species *Montastrea annularis* and *M. faveolata* harbored predominantly clades A and B in 0–6 m depth and clade C in 6–14 m depth (Rowan and Knowlton 1995). This implies that *Symbiodinium* types of clade A have a greater capability to handle high light and possibly higher temperature fluctuations than types of clade C, while types of clade C features better light-harvesting capabilities needed in deeper waters. Interestingly, however, in the central Red Sea, almost exclusively *Symbiodinium microadriaticum*, a type of clade A, was found in *Pocillopora verrucosa* in shallow (5 m) as well as deep (20 m) waters (Ziegler et al. 2014) and type C15 was found almost exclusively in *Porites* species between 1 and 60 m depth (Ziegler et al. 2015). This demonstrates that the distribution of *Symbiodinium* types is not only determined by the environmental conditions but also by the coral species.

When photosynthetic input cannot meet the corals' energy demand in case of limited acclimatization/adaption ability to low-light conditions, heterotrophy becomes more important (Muscatine et al. 1989; Houlbrèque and Ferrier-Pagès 2008). The ability to switch between auto- and heterotrophy (called trophic plasticity) strongly depends on the coral species, which restricts their depth distribution and ecological niche (Anthony and Connolly 2004; Houlbrèque and Ferrier-Pagès 2008). For example, *S. pistillata* is found as deep as 65 m in the Gulf of Aqaba, inferring a large trophic plasticity, where additional heterotrophic input was found to be indispensable in waters deeper than 20 m (Mass et al. 2007). However, even though a high trophic plasticity allows existence in deeper waters, metabolic rates (e.g., respiration—Mass et al. 2007) and coral growth rates are reduced in deeper waters (Table 1) (with exceptions, cf. Stambler et al. 2008). Reasons for this may be found in a faster and more efficient uptake of autotrophic energy versus heterotrophic energy, a limitation of heterotrophy through a limitation of food availability in clear waters and a decrease in temperature which lowers metabolic activity. Coral species which inhabit a specific depth zone, however, often feature limited trophic plasticity, as well as different susceptibilities to wave action and to fluctuating environments in the shallow reef (e.g., tides, high light stress; Loya 1972; Sheppard 1982).

Hermatypic zooxanthellae-containing corals, which grow under extremely low light intensities (mesophotic zone, irradiance ~0.5–5 % of surface irradiance), were investigated in the Gulf of Aqaba, and recently also in the central Red Sea. In the Gulf of Aqaba, some coral species were found to feature a particularly high distribution range, between ~5 and 100 m (e.g., *Echinophyllia aspera*, *Oxypora lacera*, *Leptoseris explanata*), while others were found only deeper than 50 m (e.g., some species of the genus *Leptoseris* and *Alveopora*; Fricke and Schuhmacher 1983). Adjusting mechanisms to low-light conditions include a flattened growth to maximize light harvesting (Fricke and Schuhmacher 1983) and the production of pigment granules by the coral host (*Leptoseris fragilis*), which transforms and reflects the light with a short to a long wavelength suitable for photosynthesis (Schlichter et al. 1986). Corals in the twilight zone generally feature a very low metabolism

Table 1 Coral skeletal growth and calcification of selected coral species along depth profiles in the Red Sea

Coral species	Region	Depth						Data source
		1–5 m	10 m	15 m	30 m	50 m	65 m	
Growth: linear extension in mm year^{-1}								
Stylophora pistillata	NNC	9.24		7.48	6.51			Kotb (2001)
Pocillopora. damicornis	NNC	7.39		6.60	Absent			Kotb (2001)
Acropora granulosa	NNC	6.34		9.24	5.90			Kotb (2001)
Growth: mg skeleton d^{-1}								
Stylophora pistillata	CE	39.0	11.5	(Corals transplanted from 1 to 10 m depth)				Al-Sofyani (1991)
Echinopora gemmacea	CE	13.5	11.3					Al-Sofyani (1991)
Calcification µmol $CaCO_3$ cm^{-2} coral surface 24 h^{-1} (averaged over summer and winter)								
Stylophora pistillata	Gu-A	~4.8	~7.0			~2.4	~0.7	Mass et al. (2007)

Regions North–north central/southern tip of Sinai (*NNC*), central-east/Jeddah (*CE*), Gulf of Aqaba/Eilat (*Gu-A*)

(e.g., photosynthetic rate, respiration, growth), which allows them to cope with the low energy input (Fricke et al. 1987).

Hermatypic corals (but without zooxanthellae) occurring in depths between 200 and 700 m and temperatures of ~21 °C were recently described from the central Red Sea (Roder et al. 2013). The main differences found between the Red Sea and other deep-sea habitats for hermatypic corals (e.g., Norway, New Zealand) are high temperatures and low oxygen concentrations in the Red Sea. Corals living under these conditions seem to be adapted by featuring extremely low metabolic rates and possessing very little biomass compared to relatives in other deep-sea habitats (Roder et al. 2013).

Seasons

The most important environmental changes between seasons pertain light, temperature, and nutrient availability. Winter and summer incident light intensities differ more than 2.5-fold (~25–63 E m^{-2} day^{-1}) in the northern Red Sea and about 1.8-fold (~30–55 E m^{-2} day^{-1}) in the southern Red Sea (Fig. 1d). Winter and summer temperatures differ between 20 and 27 °C in the north (Gulf of Aqaba) and between 28 and 33 °C in the south (Fig. 1b). Water chlorophyll *a* data indicate higher nutrient availability in winter than in summer, varying between 0.05 and 0.3 µg chlorophyll *a* l^{-1} in the north and between 0.1 and 0.5 in the central Red Sea (Fig. 1c). In the south, summer values are unavailable; however, some snap sampling during an expedition in September 2011 and March 2012 at the Farasan Islands revealed that in contrast to the majority of the Red Sea, higher chlorophyll *a* concentrations occurred in September (3.0 µg chlorophyll *a* l^{-1}) compared to March (0.8 µg chlorophyll *a* l^{-1}, unpublished data, Y. Sawall). This is in line with a higher light intensity in winter compared to summer (1.3-fold in 5 m depth) at the Farasan Islands (SE Farasan, Table 2).

These changes affect various components of the coral as demonstrated by studies in the Gulf of Aqaba (Mass et al. 2007; Naumann et al. 2010), in the northern Red Sea (Kotb 2001; Mohamed et al. 2007), in the central Red Sea (Al-Sofyani and Davies 1993; Floos 2007, 2012), and along the latitudinal environmental gradient (Sawall et al. 2014a, Sawall et al. accepted). The most investigated parameter is coral growth and/or calcification, which is almost consistently lower in winter than in summer (Table 3). This can mainly be ascribed to lower temperatures in winter, based on previous findings, where a strong positive correlation between calcification and temperature was found (Lough and Barnes 2000; Marshall and Clode 2004). A reason for this may be found in higher metabolic/enzyme activity at higher temperatures (see e.g., Ip et al. 1991). An exception, however, is evident in the south, where calcification was higher in winter, most likely due to rather unfavorable conditions during summer, as discussed in the next paragraph (Sect. "Latitudes").

The photosynthetic rate (standardized to coral surface area) is mainly the product of pigment concentration per zooxanthellae, photosynthetic efficiency, density of zooxanthellae, instantaneous light intensity, and temperature. These features are influenced by seasonally varying environmental factors in different ways: Pigment concentration and photosynthetic efficiency are mainly adjusted to light availability, whereas zooxanthellae density is mainly regulated by nutrient availability, provided either in the form of inorganic nutrients from the surrounding water or in the form of transferred inorganic and/or organic nutrients from the coral host (Stambler et al. 1991; Falkowski et al. 1993). While temperature generally increases the metabolic activity (including photosynthesis), it can affect the concentration of xanthophyll, responsible for photo-protection, as well as the photosynthetic efficiency (Warner et al. 1996; Sawall et al. 2014a). However, in case the temperature tolerance is exceeded, an increase in zooxanthellae damage and consequently a loss of zooxanthellae is likely to occur (Jones et al.

Table 2 Seasonal variation in metabolic parameters of zooxanthellae in the Red Sea

Coral species	Region (depth)	Light intensity	Temperature (°C)	Metabolic parameters				Data source
				Zooxanthellae density	Chlorophyll *a* per cell	Photo synth. efficiency (α)	Photosynthetic rate per cm^2	
Stylophora pistillata	Gu-A (5 m)	×0.5	20–27	×2.5	×4	×0.8	×0.7	Mass et al. (2007)
Stylophora pistillata	CE (1 m)	×0.5	25.5–31			×1	×0.6	Al-Sofyani and Davies (1993)
Echinopora gemmacea	CE (3 m)	×0.5	25.5–31			×1.3	×1.2	Al-Sofyam and Davies (1993)
Seriatopora hystrix	CE (5 m)	×0.6	27–32	×15		×3	×2	Floos (2012)
Lobophyllia corymbosa	CE (5 m)	×0.6	27–32	×7		×0.6	×0.4	Floos (2012)
Pocillopora verrucosa	Gu-A (5 m)	×0.8	21.5–27	×3.1	×0.7	×1.2	×1.2	Sawall et al. (2014a)
Pocillopora verrucosa	NE (5 m)	×0.8	23–29	×2	×0.7	×0.7	×1	Sawall et al. (2014a)
Pocillopora verrucosa	CE (5 m)	×1	25.5–30	×1.5	×0.6	×1.1	×0.9	Sawall et al. (2014a)
Pocillopora verrucosa	SE (5 m)	×1.3	28–32	×0.7	×0.8	×0.9	×1.4	Sawall et al. (2014a)

All parameters are presented as relative changes from summer to winter (× = fold change/multiplied), except temperature. Light intensity and temperatures are monthly averaged data measured at corresponding time and depth. *Regions* Gulf of Aqaba/Eilat or Maqna (*Gu-A*), central-east/Jeddah (*CE*), northeast/Al-Wajh (*NE*), southeast/Farasan Islands (*SE*)

Table 3 Seasonal variation in coral skeletal growth and calcification of selected coral species in the Red Sea

Temporal patterns (seasons)						
Coral species	Region (depth)	Season				Data source
		Spring	Summer	Autumn	Winter	
Growth: linear extension in mm 3 months^{-1}						
Stylophora pistillata	NNC (5 m)	2.16	3.15	2.43	1.71	Kotb (2001)
Pocillopora damicornis	NNC (5 m)	1.89	1.89	2.25	1.53	Kotb (2001)
Acropora granulosa	NNC (5 m)	2.07	1.44	2.25	0.72	Kotb (2001)
Acropora humilis	NNW (5 m)	1.97	2.04	1.59	1.44	Mohamed et al. (2007)
Stylophora pistillata	NNW (5 m)	1.65	1.34	1.47	1.38	Mohamed et al. (2007)
Growth: weight gain in mg skeleton d^{-1} nubbin^{-1}						
Stylophora pistillata	CE (1 m)		56.0		21.6	Al-Sofyani and Davies (1993)
Echinopora gemmacea	CE (3 m)		13.5		13.6	Al-Sofyani and Davies (1993)
Seriatopora hystrix	CE (5 m)		1.6		1.5	Floos (2012)
Lobophyllia corymbosa	CE (5 m)		7.3		5.7	Floos (2012)
Calcification: μmol $CaCO_3$ cm^{-2} coral surface d^{-1}						
Stylophora pistillata	Gu-A (5 m)		~4.0		~0.8	Mass et al. (2007)
Pocillopora verrucosa	Gu-A (5 m)		9.66 (Sep)		1.20 (Mar)	Sawall et al. (accepted)
Pocillopora verrucosa	NE (5 m)		4.12 (Sep)		0.89 (Mar)	Sawall et al. (accepted)
Pocillopora verrucosa	CE (5 m)		3.29 (Sep)		2.51 (Mar)	Sawall et al. (accepted)
Pocillopora verrucosa	SE (5 m)		3.23 (Sep)		4.62 (Mar)	Sawall et al. (accepted)

Regions North–north central/southern tip of Sinai (*NNC*), north–northwest/Hurghada (*NNW*), central-east/Jeddah (*CE*), Gulf of Aqaba/Eilat or Maqna (*Gu-A*), northeast/Al-Wajh (*NE*), southeast/Farasan Islands (*SE*)

1998; Weis 2008). This process is further promoted under high light stress. Photo-acclimation processes were shown to be completed within 4–6 weeks, as cross-transplantation experiments between different light environments demonstrated (Falkowski and Dubinsky 1981; Ziegler et al. 2014).

Table 2 summarizes observed changes in zooxanthellae characteristics of corals from in situ experiments in the Red Sea. It includes the relative change in light intensity and the range of temperature for the corresponding season/month and depth. Reliable nutrient-related data from the corresponding sampling sites is largely missing and was therefore not included in the table, but the general trend (as described above) showed fourfold–fivefold higher water chlorophyll *a* concentrations in winter, but a reverse picture far south. This explains the higher zooxanthellae densities in winter, found from the Gulf of Aqaba to the central Red Sea, and the higher densities in summer at the southernmost region SE Farasan (Table 2). The chlorophyll *a* concentration per zooxanthellae strongly increased in *S. pistillata* from summer to winter in the Gulf of Aqaba, which is expected as a consequence of reduced light availability in winter (Mass et al. 2007). However, the reverse was observed for *P. verrucosa* in various regions of the Red Sea independent of an increase or decrease in light availability (Table 3). It needs to be mentioned that the study of Sawall et al. (2014a) took place in September and March, during the ends of the summer and winter seasons, which flattens the light effects (light intensity was similar in both months). Therefore, the change in cell-specific chlorophyll *a* concentration may rather be attributed to changes in nutrient availability than to changes in light. It was suggested that nutrient depletion in September leads to a reduction in zooxanthellae number, but an increase in cell-specific photo-pigments in order to maintain high photosynthetic performance (Sawall et al. 2014a). In contrast to the rather clear seasonal pattern in zooxanthellae densities and pigmentation, the photosynthetic efficiencies and photosynthetic rates showed a very inconsistent pattern over species and regions. This does not allow a general conclusion, but rather demonstrates the complex multifactorial responses and adjustments to various interacting environmental factors. In addition, different types of *Symbiodinium* from different clades, which are either species specific or region specific, may explain different acclimatization mechanisms. For example, in the Gulf of Aqaba, *S. pistillata* and *P. verrucosa* experienced similar environmental conditions during the seasonal comparison, but responses varied strongly. *S. pistillata* revealed a low photosynthetic efficiency and photosynthetic rate in winter despite high pigmentation and zooxanthellae density (Mass et al. 2007), which may be ascribed to low metabolic activity as a consequence of low temperatures (20 °C; Coles and Jokiel 1977). A reversed pattern was evident in *P. verrucosa* in the Gulf of Aqaba (Sawall et al. 2014a). This may partly be explained by differences in *Symbiodinium* composition: *S. pistillata* in the Gulf of Aqaba in 5 m depth harbors predominantly *Symbiodinium* types of clade A (Lampert-Karako et al. 2008), while *P. verrucosa* harbors mainly types of clade C (Sawall et al. 2014a).

Also coral mucus release rates and composition vary between different environmental conditions, as well as with changes in the coral metabolism such as photosynthetic rates (Brown and Bythell 2005; Bythell and Wild 2011). Coral mucus has various functions. Among others, it serves as a protective shield against settling particles, harmful substances, and potential pathogens, and it is a mechanism to 'excrete' excess photosynthetically fixed carbon (Brown and Bythell 2005; Bythell and Wild 2011). It can either be released as particulate matter (strings, which are retained on a 0.7-μm-pore size filter) or as dissolved matter (smaller than 0.7 μm), while the preference for one or the other was suggested to be species specific (Naumann et al. 2010). We are aware of two studies from the Red Sea that assessed the effect of seasonally changing environmental conditions on coral mucus release. The study of Naumann et al. (2010) was conducted in Aqaba, Jordan, and included 6 dominant coral genera. The authors found that most corals feature highest mucus release rates in summer associated with high C/N ratios. Increased mucus release was confirmed by the study of Sawall et al. (accepted) for the species *P. verrucosa* from the Gulf of Aqaba and various regions along the Saudi Arabian Red Sea coast. Additionally, C/N ratios were found to be almost doubled in summer (between 13 and 18) compared to winter (between 8 and 10, unpublished data, Y. Sawall). The concomitant increase in mucus release and C/N ratios in summer means that more photosynthetically fixed carbon is translocated to mucus production. This relationship was also found along a depth gradient (Crossland 1987) and over a day–night cycle (Naumann et al. 2010). However, not only light availability/photosynthetic rates, but also temperature may be related to seasonal changes in mucus release. On the one hand, increasing temperature enhances metabolic processes, potentially including mucus production as well (Naumann et al. 2010). On the other hand, increased temperature is known to increase microbial activity (e.g., White et al. 1991) including potentially the virulence of pathogens (e.g., Rosenberg and Ben-Haim 2002; Case et al. 2011), which may in turn trigger higher mucus release in order to protect the coral from potential pathogens. Since photosynthetic rates were shown to be variable between seasons, in contrast to a rather consistently higher mucus release and higher C/N ratio in summer, it was suggested that temperature plays a dominant role in seasonal variations of mucus release in Red Sea corals (Sawall et al. accepted).

Latitudes

Strong temperature (north: 20–27 °C; south: 28–33 °C) and nutrient gradients (indicated by ~15-fold increase of annual averaged water chlorophyll *a* from north to south) are found from the northern to the southern Red Sea. The temperature gradient is stronger from the northern to the central Red Sea, while the nutrient gradient occurs mainly from the central to southern Red Sea (Fig. 1b and c). Light intensities also show latitudinal variations (Fig. 1d), which is related firstly to differences in the altitude of the sun, and secondly to a change in turbidity. Turbidity increases toward the south due to increased nutrient concentrations, which leads up to threefold lower light intensities in a given depth around the Farasan Islands compared to the northern Red Sea (estimated from $k_{d\lambda490}$ and PAR values derived from satellite images of NASA, Giovanni online data system, Ocean Color Radiometry, MODIS-Aqua). Furthermore, salinity increases from the south (37 ‰) to the north (41 ‰).

In most geographic regions worldwide, where complex and diverse tropical shallow water coral communities occur, weekly maximum temperatures in summer are usually below 30 °C. In case temperatures exceed those normally experienced temperatures for an extended period of time (1–3 °C for several weeks or months), many corals become stressed, lose their zooxanthellae (bleaching), possibly develop signs of disease, and finally die (Brown 1997; Fitt et al. 2001). In the southern Red Sea, naturally occurring temperatures are exceptionally high with 31–33 °C over several months in spring and summer each year, without visually affecting corals. Concomitantly, nutrient input is high. In general, high nutrient supply, in contrast to high temperatures, is harmful on an ecosystem level rather than on an organism level. This is because nutrients support the growth of fast-growing algae or soft-bodied filter feeders, which potentially outcompete hermatypic corals (Bellwood et al. 2004). Algae and soft-bodied filter feeder are particularly successful when corals are already weakened and less competitive, for example, through temperature stress (Szmant 2002). On an organism level, however, the effect of high nutrients was shown to vary; it can either be harmful, neutral, or beneficial (Koop et al. 2001; Fabricius 2005; Sawall et al. 2011; Jessen et al. 2013).

Only recently, systematic investigations on coral population structures and performance along the strong latitudinal environmental gradients have been started, within the 'Jeddah Transect Project' (since 2010). Here, colleagues in Saudi Arabia (King Abdulaziz University [KAU], King Abdullah University of Science and Technology [KAUST]) and Germany (Helmholtz-Institute for Ocean Research [GEOMAR]) work together in order to explore the underlying mechanisms of adaption and/or acclimatization to these gradients and to extreme conditions in the southern Red Sea. A first approach to answer the question whether corals show genetic structuring may be provided by the assessment of population structures. This is currently done with two coral species, which belong to the same coral family (Pocilloporidae), but feature different reproduction modes. *S. pistillata* is a brooding species (release of developed larvae), and *P. verrucosa* is a broadcast spawning species (larval development in the water) (see below Sect. "Reproduction modes"), which infer a wider larval dispersal for the broadcast spawner and therefore a lower differentiation in population structure. First results, based on the haplotype identification with the mitochondrial ORF marker (i.e., 'open reading frame with unknown function'; Flot et al. 2008), indicate that morph types commonly associated with *Stylophora pistillata* in the Red Sea belong to different well-established molecular lineages and suggest that local adaptation and geographical speciation had played a major role in the structuring of populations as well as in the intra- and interspecific differences found within the current *S. pistillata* morph types along the Red Sea (Banguera-Hinestroza et al. 2012). In contrast, analyses of *P. verrucosa* populations using the same mitochondrial region, as well as microsatellite data, show no obvious latitudinal trend, (Table 4; Banguera-Hinestroza et al. 2012; Robitzch et al. accepted; Sawall et al. accepted). The pattern of associated dominant *Symbiodinium* clades/types found in 5 m depth seems similar to population structures of the coral hosts. *Symbiodinium* clades of *S. pistillata* are bound to the two different linages, meaning that *Symbiodinium* ITS2 types of clade C occurred predominantly in association with the northern linage, and types of clade A (mainly A1, *S. microadriaticum*) mainly with the southern linage (Banguera-Hinestroza et al. 2012). In *P. verrucosa,* clade A1 is dominant and rather homogenously distributed throughout the Red Sea, however, with

Table 4 Latitudinal and seasonal variation in metabolic parameters of the coral *Pocillopora verrucosa* in 5 m depth

	Coral host characteristics					**Zooxanthellae characteristics**				
	Biomass per cm²	Calcification per cm² and day	Mucus release per cm² and day	Tissue growth per cm² and day	Coral haplotypes	Zooxanthellae density per cm²	Chlorophyll *a* per cell	Xantho/chlorophyll *a* ratio	Photosynthesis per cm² and day	Zooxanthellae clades
N–S					H1-H7					C1 (A1); A1 (C1, D1); A21 (A1)

Top to bottom (north/Gulf of Aqaba to south/Farasan Islands), dark gray (summer), light gray (winter), increase/decrease (triangle), variable along the latitudes/no clear trend (snake line), homogeneous along latitudes (box). Data from Sawall et al. (2014a), Sawall et al. (accepted) and Sawall (unpublished data)

Fig. 5 Incubation chambers used to measure metabolic rates of Red Sea corals. The more complex setup (**a** *left*, and **b** → metabolic chambers) is used to measure photosynthetic rates, respiration (if covered with a black cloth), and calcification rates; the less complex setup (**a** *right* → mucus chambers) is used to measure mucus release rates. **Metabolic chambers**: Each of the 4 cylindrical acrylic chambers contains a battery-run stirrer, a water inlet, and outlet with one-way valves in front and behind the chamber, respectively, a fragment holder and an oxygen sensor (optode). All 4 oxygen sensors are connected to a battery-run data logger. An UW-PAR sensor is installed next to the chambers at the same height and connected to the same data logger. Each water inlet is connected to a programmable battery-run pump via tubing, which pumps the surrounding water through the chambers at a given interval. **Mucus chambers**: Same acrylic chambers and stirrers as the metabolic chambers, however, with an exchangeable Teflon membrane in the lids which allows exchange of gases, but not of dissolved or particulate matter and with a water outlet only. *Pictures* C. Lieberum and Y. Sawall

exceptions at the most northern region where mainly different clade C types occurred, and at the most southern region where a new type of clade A (A21) dominated (Table 4; Sawall et al. 2014a). *Symbiodinium* types from clade A were previously found to be an ideal shallow water clade, able to tolerate high temperature fluctuations and particularly able to adjust its photo-protective pathways (Caribbean—Reynolds et al. 2008).

Looking into how corals perform along the latitudinal environmental gradient, in situ investigations along the 2,000-km-long Saudi Arabian Red Sea coast revealed a rather inconsistent response of the widely distributed shallow water coral *P. verrucosa* to environmental conditions (Fig. 5 shows the in situ incubation setup developed at GEOMAR and used to measure metabolic rates). While calcification rates and mucus release revealed clear latitudinal trends (in summer and winter), zooxanthellae characteristics showed a latitudinal trend only within a certain season or not at all (Table 4). Calcification decreased from north to south in summer and showed the opposite trend in winter (Table 4), thereby revealing a strong temperature dependency, with highest calcification rates at around 28 °C (Sawall et al. accepted). This result indicates that *P. verrucosa* may not be adapted to local prevailing conditions within the Red Sea (as supported by the homogenous population structure, as well), since it suggests that specimens from all regions of the Red Sea calcify best at the same temperature. Lower calcification rates at temperatures above 28 °C during summer in the central and southern regions were found to co-occur with a slight decrease in photosynthesis in summer, as well as with a particularly high mucus release (all rates in carbon cm^{-2} day^{-1}). This may indicate that calcification at high temperatures is reduced on the one hand by a limited energy gain through photosynthesis and on the other hand by a high energy/carbon translocation into mucus production. In contrast, reduced calcification at lower temperatures can be linked to a generally lower metabolic activity at lower temperatures (Lough and Barnes 2000; Marshall and Clode 2004). Of additional note, however, are the particularly high calcification rates found in the coldest region (twofold higher, Gulf of Aqaba), if compared to corals from the main Red Sea at similar temperatures (~27/28 °C), which may indicate regional adaptation to rather low temperatures (Sawall et al. accepted). This adaptation may be found in the differing *Symbiodinium* composition.

Regarding annual growth rates, a study by Lieberum (2011) on the common massive coral species *Porites lutea* revealed no obvious pattern along the latitudinal environmental gradient. Coral cores showed growth rates between 9 and 15 mm $year^{-1}$ from 2007 to 2010 and a slight tendency toward highest growth rates around the central Red Sea. This might demonstrate that each region experiences optimum (=maximum performance) and suboptimum (=lowered performance) temperatures at different times of the year, leading, however, to similar annual growth rates after all. In the central Red Sea, optimum temperatures or close-to-optimum temperatures may persist over the longest period, which may explain slightly higher growth rates in this region. In this context, however, it needs to be mentioned that a decrease in coral growth over the last 2 decades has been observed in the central Red Sea due to warming (Cantin et al. 2010). Whether this is the case in other regions of the Red Sea as well remains to be investigated.

The strong increase in mucus release from north to south, particularly during summer (>5-fold, Table 4), is most likely the response to both increasing nutrients and increasing temperature (Sawall et al. accepted; see discussion above in Section "Seasons"). Increased nutrient input causes an increase of particle loads in the water, which settle on the coral surface and thereby trigger increased mucus release (Telesnicki and Goldberg 1995). Furthermore, increased nutrient supply increases microbial biomass including potential pathogens in the water (Dinsdale et al. 2008), which may pressure their settlement on the coral surface. Increased temperature is known to increase microbial activity (White et al. 1991; Shiah and Ducklow 1994), which may also apply to microbes on the coral surface. In order to increase efficiency in the removal of unwanted microbes, corals may increase their mucus release. A substantial amount of carbon/energy is therefore allocated into mucus production, which reduces energy availability, for example, for coral growth. Heterotrophy could potentially compensate for the energy loss through mucus release; however, trophic plasticity seems to be limited in *P. verrucosa*, as indicated by the lack of a latitudinal trend in biomass and tissue growth (Table 4) and by a rather narrow depth distribution (between 1 and 20 m in the central Red Sea; Ziegler et al. 2014).

Looking at the zooxanthellae characteristics (Sawall et al. 2014a), it was found that the zooxanthellae densities increased from north to south in summer (~fourfold), however not in winter where zooxanthellae densities were generally higher regardless of site. In contrast to that, photopigments showed a latitudinal trend only in winter, where cellular chlorophyll *a* increased and xanthophyll/chlorophyll ratios decreased from north to south, while they were generally higher and rather homogenous in summer (Table 4). The latitudinal trend in zooxanthellae density (positively) and in xanthophyll/chlorophyll *a* ratio (negatively) correlated both with temperature, while the cell-specific chlorophyll *a* concentration could not be related to any environmental factor, neither temperature, nor nutrients or light (Sawall et al. 2014a). A different behavior was found predominantly at the most southern region, the Farasan Islands (e.g., chlorophyll *a* concentrations per cell were always highest and equal in both seasons). The low and inconclusive latitudinal pattern in zooxanthellae physiology, but the strong and mostly conclusive seasonal patterns (see above Section "Seasons") is in contrast to the environmental conditions (particularly nutrient availability), which varied more strongly spatially than temporally. This can have the following reasons: (i) a shift in zooxanthellae type composition as found in the most northern and most southern region (Table 4), (ii) adaptation of a specific species/clade to regional conditions in most parts of the main Red Sea, and/or (iii) counterbalancing mechanisms of the coral host, thereby creating a 'zooxanthellae-friendly' environment within the coral independent from ambient conditions (Leggat et al. 2011).

The potential effect of different salinities on Red Sea corals is so far largely unknown. One laboratory study resulted in significantly higher performances of *S. pistillata* incubated at 38 ‰ in contrast to corals incubated at 40 ‰ over 3 weeks, although the corals originated from the Gulf of Aqaba (41 ‰; Ferrier-Pagés et al. 1999). The fact, however, that the corals in this study have been acclimatized to 38 ‰ under laboratory conditions over several years made them most likely sensitive to a higher salinity at least over a short-term exposure of 3 weeks. We assume that, in contrast to the corals cultured over many years in the laboratory, corals growing in the Red Sea are well acclimatized (or adapted) to higher salinities in the northern regions. This is based on the few available studies on the effect of hypersalinity, which indicate a rather low effect of salinities relevant for the Red Sea (37–41 ‰; Coles and Jokiel 1978; Muthiga and Szmant 1987). This hypothesis, however, remains to be proven.

The Red Sea with its multiple latitudinal environmental gradients offers an ideal natural laboratory to explore adaptation versus acclimatization mechanisms of corals to different environmental conditions. It also serves to explore potential future global change scenarios on coral reefs, since predicted conditions partly already occur in the central (high temperature) and southern regions (high temperature, high nutrients). Therefore, further studies on population structures of Red Sea corals, as well as the conduction of synchronic experiments (reciprocal transplantation, common garden manipulation experiments) are strongly encouraged in order to assess adaptation versus acclimatization mechanisms of corals to different environmental regimes and to assess potential mechanisms of corals to deal with future global change (Sanford and Kelly 2011). For this, improved technologies, for example, genetic tools (next-generation sequencing, improved molecular maker in population genetics), as well as the integration of all components of the coral holobiont (host, zooxanthellae, microbes), are required to advance our understanding in the mechanisms underlying adaptation, acclimatization, and tolerance limits.

Coral Reproduction

Reproduction Modes (Sexual)

Coral life starts with a planktonic planula larva, which becomes demersal and starts searching for suitable settlement space after several days (Harrison and Wallace 1990). The preferred settlement substrate is preconditioned (e.g., biofilm, crustose red algae), solid, and provides cryptic microhabitats. Once a suitable place is found, the larva attaches itself to the substrate and starts to metamorphose into a polyp. It almost immediately starts the secretion of a skeleton, as well. Corals that form a colony grow by a process called budding, where the polyps divide and thereby create new polyps. Depending on the life history traits of a coral species (e.g., slow vs. fast growing), the coral reaches the age of fecundity after several years and starts to reproduce mostly once per year, but possibly also biannually or even monthly (Richmond and Hunter 1990).

Planula larvae are either released by the mother coral (brooder, internal fertilization) or they develop in the water column (broadcast spawner, external fertilization). The gametes can be produced by gonochoric (separate sexes) or by hermaphrodite corals (male and female gametes within one colony). In the Red Sea, most investigated coral species were found to be hermaphrodite broadcast spawners (18 out of 24 in the Gulf of Aqaba) and only a few were hermaphrodite brooders (3 of 24) or gonochoric broadcast spawners (3 of 24; Table 5; Shlesinger and Loya 1985; Shlesinger et al. 1998). This is in accordance with the overall Indo-Pacific pattern (except Eastern Pacific; Baird et al. 2009), although intraspecific variability of reproduction modes was found, for example, in the coral *P. verrucosa,* which is a broadcast spawner in the Red Sea (Shlesinger and Loya 1985), but a brooder in the central Pacific (Richmond and Hunter 1990). Within the Red Sea, however, the reproduction mode of a given species seems to be consistent from north to south (Table 5), although sexuality of some species may change between years. This was found during a long-term study on the hermaphrodite brooder *S. pistillata,* where some colonies were hermaphrodite in one year and male or even sterile in another year (Rinkevich and Loya 1987). Inconsistencies in reproduction modes might be due to energy limitations and/or stress conditions in some years (Rinkevich and Loya 1987), since reproduction is a major energy expenditure. A substantial amount of energy for reproduction is derived from photosynthesis (Rinkevich 1989; Leuzinger et al. 2012).

Gametogenesis and Spawning

Most coral species feature an annual reproduction cycle, but also biannual and monthly reproduction was found in some coral species and/or geographic locations (Richmond and Hunter 1990). Usually, the development of the oocytes and ripening of the eggs (oogenesis) starts first, followed by the onset of testis development for sperm production (spermatogenesis) 1–3 months later (Fig. 6; Richmond and Hunter 1990; Shlesinger et al. 1998). The location of oocyte and testis development inside the coral polyp, as well as the gonad morphology, the number of gonads and the number of oocytes per gonad was found to be rather consistent within each coral family across geographic regions in the Indo-Pacific including the Red Sea (Shlesinger et al. 1998). For most coral families, gonad development takes place in the mesenteries of the polyp, only corals of the family Pocilloporidae develop gonads on stacks reaching from the mesenteries into the gastric cavity (Al-Sofyani 1987; Shlesinger et al. 1998; Floos 2007). In broadcast spawning corals, the number of oocytes produced per polyp is positively related to the polyp size resulting in 20–75 eggs per polyp in Pocilloporidae, Acroporidae, and Poritidae, 100–700 eggs per polyp in Faviidae, and up to 3,500 eggs per polyp in Mussidae and Oculinidae (Shlesinger et al. 1998). The number of produced eggs in the brooder *S. pistillata* is exceptionally low with 1–5 eggs per polyp (Shlesinger et al. 1998; Mohamed et al. 2007).

While these features seem to be rather conservative, there are pronounced differences in the timing and length of gametogenesis and spawning of a given species between different geographic regions. These differences were related to various extrinsic environmental factors, such as temperature, wind intensity (calm vs. stormy season; van Woesik 2010), rainfall, insolation, tides, and day length, which vary seasonally with different intensities in different regions (reviewed by Baird et al. 2009). The fine-tuning of gamete release, which occurs in synchrony in order to guarantee fertilization success, is better understood than the triggers for the onset of gametogenesis. Thus, gamete release was found to be triggered by a certain combination of lunar phase, tide, and diurnal rhythm (Baird et al. 2009), and a potential underlying mechanism was found to be related to blue-light-responsive receptors (cryptochromes) able to 'sense' moon light intensity (Levy et al. 2007). Sensitivity to these cues makes mass spawning events possible, as they occur, for example, in the GBR and Western Australia, where the majority of coral species spawn within only a few nights per year (Babcock et al. 1986; Oliver et al. 1988; Babcock et al. 1994). Interestingly, however, different environmental triggers were suggested for the GBR (temperature—Harrison et al. 1984) and for Western Australia (day length—Babcock et al. 1994). Inter-regional differences in spawning behavior, however, may not only be related to external, but may also be related to intrinsic factors, meaning differences in genotypes due to regional adaptation. Therefore, a complex interplay between extrinsic and intrinsic factors most likely explains the timing of gametogenesis and spawning in Red Sea corals, which lack multispecies synchronized spawning aside from *Acropora* species in the main Red Sea (Table 5). Thus, so far, only *Acropora* species were found to spawn simultaneously in the northern (end April/ beginning of May; Hanafy et al. 2010) and central Red Sea (mid-April; Bouwmeester et al. 2011b), although not in the Gulf of Aqaba (throughout May, June, and July; Table 5).

Although the data for the main Red Sea are still very limited, there are clear signs of shifts in the timing and length of the gametogenesis process for a given species along the latitudinal gradient. For the brooding species *S. pistillata*, as well as for the broadcast spawners *Goniastrea retiformis*,

Table 5 Timing of coral spawning either directly observed or derived from gamete presence/absence in histological sections

Region	Species	Moon phase 1=new, 5=full	J	F	M	A	M	J	J	A	S	O	N	D	Data source
Gulf of Aqaba	*Stylophora pistillata* (B)	5-6 (1-8)	▒	▒	■	■	■	■						▒	Shlesinger & Loya 1985
[29°30′N, 35°55′E]															Zakai et al. 2006
	Seriatopora caliendrum (B)	8-1							■	■	■	■	■		Shlesinger & Loya 1985
	Alveopora daedaiae (B)	6-8										■	■	■	Shlesinger & Loya 1985
	Pocillopora verrucosa	1							■	■					Shlesinger & Loya 1985
	Galaxea fascicularis	6-7							■	■	■				Shlesinger & Loya 1985
	Goniastrea retiformis	7								■					Shlesinger & Loya 1985
	Platygyra lamellina	1-2							■						Shlesinger & Loya 1985
	Favia favus	6-7							■						Shlesinger & Loya 1985
	Astreopora myriophthalma	5							■	■					Shlesinger & Loya 1985
	Acropora hyacinthus	3							■						Shlesinger & Loya 1985
	Acropora scandens	5						■							Shlesinger & Loya 1985
	Acropora humilis	7						■							Shlesinger & Loya 1985
	Acropora eurystoma	5					■								Shlesinger & Loya 1985
	Acropora variabilis								■						Shlesinger et al. 1998
	Acropora hemprichi								■						Shlesinger et al. 1998
	Montipora erythraea									■					Shlesinger et al. 1998
	Porites lutea									■					Shlesinger et al. 1998
	Goniopora savignyi									■					Shlesinger et al. 1998
	Lobophyllia hemprichi							■							Shlesinger et al. 1998
	Acanthastrea echinata								■						Shlesinger et al. 1998
	Cyphystrea microphthalma									■					Shlesinger et al. 1998
	Pavona varians										■				Shlesinger et al. 1998
	Cynarina lacrymalis (deep)						■								Shlesinger et al. 1998
	Favites pentagona (deep)						■								Shlesinger et al. 1998
	Fungia granulosa (deep)	1-8								■	■				Kramasky-Winter et al. 1998
	Fungia scutaria	5							■	■	■	■			Kramasky-Winter et al. 1998
NNW (Hurghada)	17 *Acropora spp.* (**synchr.**)	6/7 ('08), 1/2 ('09)				■									Hanafy et al. 2010
[27°10′N, 33°49′E]	*S. pistillata* (B)				■	■	■	■							Mohamed et al. 2007
	Acropora humilis					■									Mohamed et al. 2007
NW (Marsa Alam)	12 *Acropora spp.* (**synchr.**)	6/7 ('08)				■									Hanafy et al. 2010
[24°41′N, 35°05′E]															
NE (Al-Wajh)	*Stylophora pistillata* (B)		▒ ?	▒ ?	▒ ?	▒ ?	▒ ?	▒ ?	■						Al-Sofyani 1987
[25°54′N, 36°45′E]	*Porites lutea*									■ ?					Al-Sofyani 1987
	Goniastrea retiformis								■						Al-Sofyani 1987
	Platygyra lamellina					■									Al-Sofyani 1987
CE (Yanbu)	*Pocillopora verrucosa*	1					■								Fadlallah 1985
[23°55′N, 38°15′E]															
CE (Thuwal)	*Pocillopora verrucosa*	1					■								Bouwmeester et al. 2011a
[22°13′N, 38°58′E]	10 *Acropora spp.* (**synchr.**)	4				■									Bouwmeester et al. 2011b
CE (Jeddah)	*Stylophora pistillata* (B)		■	■	■	■	▒							■	Al-Sofyani 1987, 1991
[21°50′N, 38°58′E]	*Seriatopora hystrix* (B)						■								Floos 2012
[21°42′N, 39°05′E]	*Pocillopora verrucosa*						■								Floos 2007
	Pocillopora damicornis						■								Floos 2007
	Porites lutea									■ ?					Al-Sofyani 1987
	Goniastrea retiformis					■									Al-Sofyani 1987
	Platygyra lamellina				■										Al-Sofyani 1987
	Lobophyllia corymbosa							■							Floos 2012
	Echinopora gemmacea											■			Al-Sofyani 1991
SE (Farasan)	*Stylophora pistillata* (B)		▒ ?	▒ ?	▒ ?	▒ ?	▒ ?	▒ ?	■						Al-Sofyani 1987
[16°48′N, 42°26′E]	*Porites lutea*											■ ?			Al-Sofyani 1987
	Goniastrea retiformis				■										Al-Sofyani 1987
	Platygyra lamellina			■											Al-Sofyani 1987

Listed after region from north to south for brooding species (B) and broadcast spawning species. Species growing below 15 m depth (deep). Synchronized spawning of multiple species (synchr.). Moon phases are presented as numbers 1–8 with 1 = new moon and 5 = full moon and the remaining numbers in between. Month: January (J) to December (D). No entry = no information available. *Stylophora pistillata*: peak of larvae release (black), occasional larvae release (gray). Uncertainties in timing of spawning are indicated by '?'

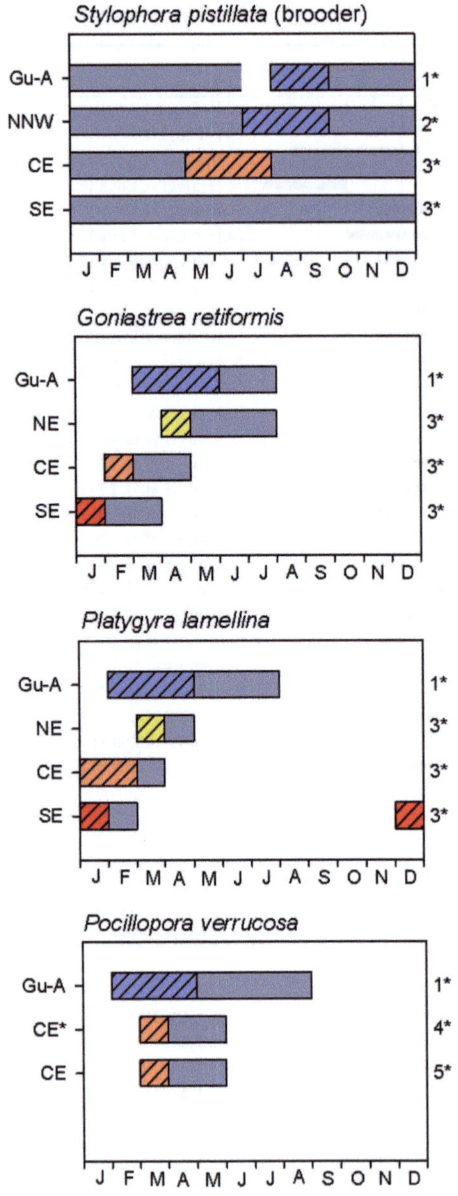

Fig. 6 Length and timing of gametogenesis in selected species. X-axis = month from January (J) to December (D). y-axis = region (*Gu-A* Gulf of Aqaba, *NNW* north–northwest [Hurghada], *NE* northeast [Al-Wajh], *CE* central-east [Jeddah], *CE** central-east [Yanbu], *SE* southeast [Farasan]). *Striped bar* oogenesis only, *gray bar* oogenesis and spermatogenesis concomitantly. *Data source 1** Shlesinger and Loya (1985), *2** Mohamed et al. (2007), *3** Al-Sofyani (1991), *4** Al-Sofyani (1987), *5** Fadlallah (1985), *6** Floos (2007)

Platygyra lamellina, and *P. verrucosa,* an earlier onset of gametogenesis from north to south was evident (Fig. 6). This is in contrast to the findings from the GBR, where a shift in spawning was not observed between the latitudes 14.5 and 23.5°S, despite pronounced latitudinal differences in the temperature regimes (Oliver et al. 1988). Here, gametogenesis occurred during increasing temperatures (spring) ending in a multispecies spawning event at the end of spring (November) to which most coral species contributed. The difference between the GBR and the Red Sea might be explained by the overall lower temperature regime in the GBR (<30.5 °C; Oliver et al. 1988) compared to the Red Sea (up to 33 °C), which in the GBR does not cause a shift in spawning seasonality. In the Red Sea, however, temperature seems to explain the shifts in reproduction seasonality over the latitudes very well. Indeed, all broadcast spawning coral species investigated over the latitudes revealed fidelity of spawning to a specific temperature (Fig. 7). This was most pronounced in the species *Platygyra lamellina*, spawning at 26.5 ± 0.5 °C, followed by *Goniastrea retiformis* spawning at 27.6 ± 1.0 °C and *Porites lutea* spawning at 29.4 ± 1.1 °C (Fig. 7). The *Acropora* species *A. humilis* revealed a lower temperature fidelity, which might indicate that other spawning cues are more important to trigger gamete release, which guarantee simultaneous spawning of diverse *Acropora* species. Therefore, in the Red Sea, optimum temperatures for spawning are found within the upper temperature regimes in the north (in summer) and during lower temperature regimes in the south (winter/spring) entailing a shift in the seasonality of spawning from north to south (Figs. 6 and 7). The lower temperature limits for gametogenesis and spawning may be driven by generally lower metabolic activity at low temperatures, while the upper limits may be driven by potentially stressful warm water conditions (Jokiel and Guinther 1978; Jokiel and Coles 1990; Edmunds et al. 2011), occurring with increasing extent during the summer in the south. Furthermore, productivity (photosynthesis) is elevated in summer compared to winter in the Gulf of Aqaba (e.g., doubled in the coral *S. pistillata* in July compared to February; Mass et al. 2007), thereby increasing energy available for reproduction (Rinkevich 1989). Toward the south, conditions become rather stressful during the summer months, requiring substantial energy/carbon translocation toward mucus production (Sawall et al. accepted) and potentially other stress-related processes (e.g., production of heat stress proteins), which forces timing of reproductive efforts into winter and spring months. In summary, corals seem to wait for the right moment, where increased photosynthesis but still non-stressful temperatures occur, ensuring a maximum of energy available for reproduction. This may also account for a decrease in the length of gametogenesis from north to south as observed in *G. retiformis* and *P. verrucosa* (Fig. 6). The brooder *S. pistillata* releases planulae over several months per year (peak between March and July; Table 5; Zakai et al. 2006), lacking obvious temperature preferences (Fig. 7). This is in line with findings from other geographic regions, where brooders of the family Pocilloporidae revealed reproduction seasons over several months or even all year round (Richmond and Hunter 1990). The reasons for this are unknown, but it might be speculated that the fully developed, rather large larvae released by the mother colony make the larvae

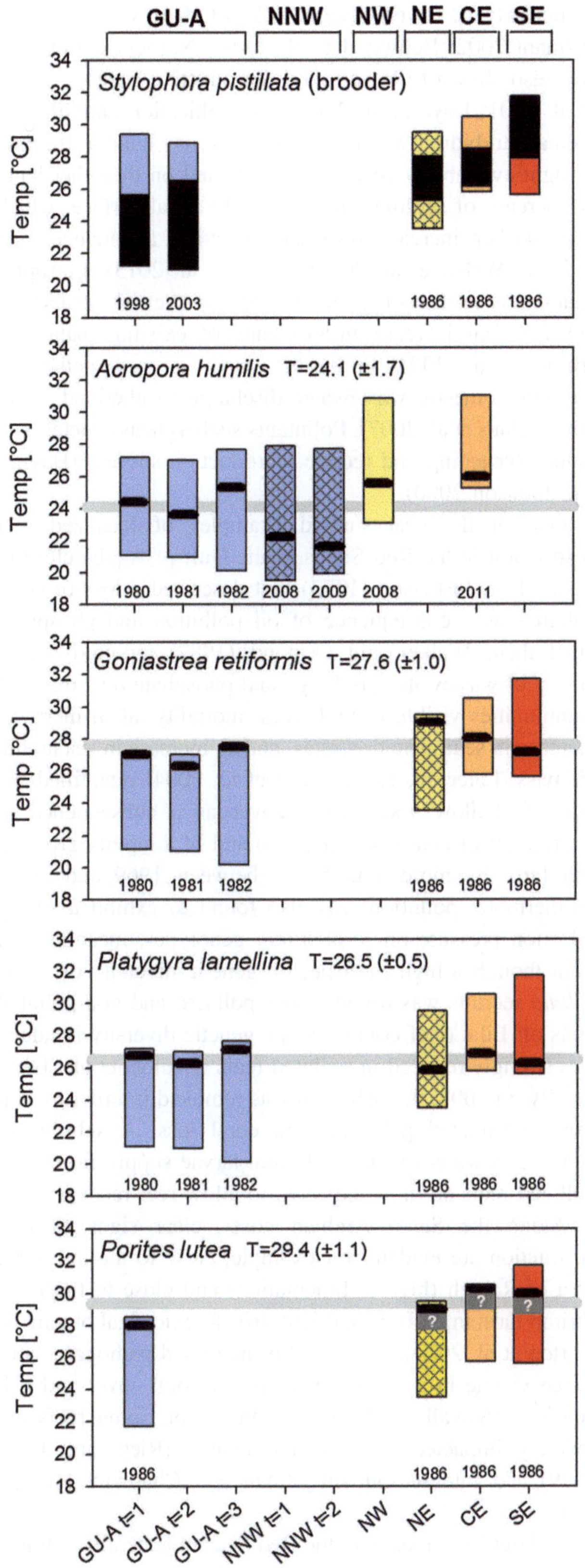

Fig. 7 Temperature at spawning (*black bar*) of selected species. Minimum and maximum temperature (lower and upper limit of bar) of the corresponding year. X-axis = region: Gulf of Aqaba (GU-A); north–northwest/Hurghada (NNW); northwest/Marsa Alam (NW); northeast/Al-Wajh (NE); central-east/Jeddah, Thuwal (CE); southeast/Farasan (SE); different years (t = 1–3). Same colors but different patterns indicate different regions but similar temperature regimes over the year, average temperature at spawning (T = mean (±SDV), *gray line*). Data source for coral spawning: Shlesinger and Loya (1985), Shlesinger et al. (1998), Hanafy et al. (2010), Al-Sofyani (1987), Bouwmeester et al. (2011b). Data source for temperature regimes: Integrated Global Ocean Service System (IGOSS), GOSTA gisst22 sst and Reyn SmithOlv2 monthly sst

less sensitive to temperature variability. Another reason might be found in the particularly low number of larvae per polyp and reproduction cycle, where a release of larvae at times when competition with recruits from other coral species is low enhance the recruitment success of pocilloporid brooders.

Whether enhanced nutrient concentrations in the south have a direct negative effect on the reproduction behavior of corals in the south (not due to increased mucus production) remains an open question up to now. It may only be speculated that particularly high nutrient concentrations far south during summer (Farasan Islands) may additionally favor reproduction during winter/spring, avoiding potentially negative nutrient effects on gamete production (Loya et al. 2004) and fertilization success (Harrison and Ward 2001).

Importance of Red Sea Coral Reefs and Threats

Coral reefs are among the most diverse ecosystems on earth (Connell 1978). The most important bioengineers of coral reefs are the hermatypic (=reef building) corals, which form extensive three-dimensional structures by their permanent buildup of calcareous skeletons. These complex structures offer a habitat for various reef organisms (Connell 1978), including species which are of great importance for the livelihood and economy for human coastal populations (Costanza et al. 1997; Brander et al. 2007). In the Red Sea, coral reefs fringe large parts of the coastline and islands, occur as barrier reefs, and exist as patch reefs (Rowlands and Purkis, this volume). Thus, they form the most important ecosystem in the Red Sea in terms of productivity and diversity, which further supports the economy of bordering countries by providing food and recreational services. Furthermore, they provide protection to the shorelines from erosion.

Although coral reefs in the Red Sea are still in a comparatively good condition, they are threatened by climate change as well as localized impacts (Kotb et al. 2004; PERSGA 2006; Nasr, this volume; Bruckner and Dempsey, this volume), likewise coral reefs in many other regions of the world (Hughes et al. 2003; Wilkinson 2004; Hoegh-Guldberg et al. 2007). Coral bleaching as a consequence of heat-stress-driven zooxanthellae loss has been recorded from the central Red Sea in 1998 at the Saudi Arabian coast (DeVantier et al. 2000) and Sudanese coast (Kemp et al. 2002) and in 2010 again at the Saudi Arabian coast (Furby et al. 2013). The northern Red Sea and the Gulf of Aqaba, in contrast, were less affected by the bleaching events (at least in 1998), due to generally cooler waters and coastal upwelling (DeVantier et al. 2000; Tilot de Grissac et al. 2000). The most susceptible coral species were found within the genera *Millepora* (hydrozoan), *Acropora,* and some rare species (DeVantier et al. 2000) as well as within the families Oculinidae and Agaricidae (Furby et al. 2013). Least affected and therefore rather tolerant species belong to the families Pocilloporidae, Poritidae, and Mussidae (DeVantier et al. 2000; Furby et al. 2013). Bleaching in the southern Red Sea in 1998 was described as severe, although quantitative data are missing (Kotb et al. 2004). However, the high bleaching susceptibility of southern corals might indicate that the corals in the south thrive in a marginal environment. This means that although corals are able (adapted) to cope with exceptionally high temperatures in the south, they may be hypersensitive to further temperature increase as it occurred during the bleaching event in 1998. Further support for this hypothesis is found in the decreased calcification rates and increased stress response (mucus release) of the coral *P. verrucosa* in the southern Red Sea during summer (Sawall et al. accepted). In the north, however, particularly in the Gulf of Aqaba, corals were found to feature a high bleaching threshold (Fine et al. 2013). It was suggested that this region could serve as a refuge for corals during ocean warming (Fine et al. 2013). Ocean warming, which proceeded dramatically over the last 15 years in the Red Sea (up to 1 °C; Raitsos et al. 2011), did not only lead to short-term bleaching events in particularly 'hot' years, but also led to a significant reduction in coral growth ($\sim$30 %) over the last two decades, as evidenced in the central Red Sea (Cantin et al. 2010). It was even predicted that corals would stop growing in that region by the year 2070, if warming continues at the same pace (Cantin et al. 2010).

Localized reef destruction is mostly visible where larger cities and industries cause eutrophication, pollution, and siltation (Loya and Kramarsky-Winter 2003; Kotb et al. 2004; PERSGA 2006; Bruckner and Dempsey, this volume). These cities include, for example, Yanbu, Jeddah, and Jizan on the Saudi Arabian coast, Aqaba and Eilat in the Gulf of Aqaba, some tourist cities along the Egyptian coast, and the Port of Sudan on the Sudanese coast. The effects on corals can be manifold, with the most prominent effects described as follows. Eutrophication enhances the growth of potential space and light competitors, thereby repressing coral growth, particularly if overfishing (lack of herbivores) occurs (Szmant 2002; Bellwood et al. 2004). Nutrient enrichment was also shown to lower coral reproductive success (Koop et al. 2001; Loya et al. 2004). Eutrophication and siltation increase turbidity, which causes on the one hand a decrease of light availability (Fabricius 2005) and on the other hand an increase of settling particles on the coral surface, which may further increase bacterial diversity, abundance, and activity (Weber et al. 2006; Jessen et al. 2013). Eutrophication was further suggested to increase the risk of disease, either by an increase in virulence of existing pathogens (Bruno et al. 2003) or by the import of new pathogens through domestic wastewater discharge (Sutherland et al. 2004; Klaus et al. 2007). Pollutants such as heavy metals can induce bleaching and reduce reproduction success (Haynes and Johnson 2000).

One of the best studied examples of localized reef destruction in the Red Sea is in the Gulf of Aqaba close to Eilat. Here, Fishelson (1973) first described a loss of coral colonies as a consequence of oil pollution and phosphate dust; then, Walker and Ormond (1982) reported severe effects of wastewater discharge and phosphate dust on coral communities visible in high coral mortality, algal increase, increase in sedimentation rates, and an increase in sea urchin biomass. Later on, Loya (Kotb et al. 2004) quantified the loss of shallow water coral cover as a consequence of wastewater discharge (until 1995) and of a rapidly growing fish farm by more than 50 % between 1969 and 2000. Furthermore, pollution was also found to exhibit a strong selection pressure on *S. pistillata* genotypes, meaning that even though a high intraspecific genetic diversity of *S. pistillata* recruits was found across polluted and non-polluted reefs off Eilat (reef connectivity), genetic diversity in adults was strongly reduced in polluted reefs (Zvuloni et al. 2008). Finally, in 2008, the fish farm was removed in order to stop fish farm-related pollution and coral loss. A subsequent increase in water quality and coral larvae supply from non-polluted reefs are now expected to allow reef recovery.

Along the Saudi Arabian coast, clear signs of reef destruction are evident, for example, close to the industrial area of Rabigh (high sedimentation) and close to the large shrimp farm in Al-Lith (nutrient-driven toxic algal blooms—Kürten et al. 2014), and possibly increased pathogen abundance visible in a strong reduction of coral cover and fish densities (Sawall et al. 2014b). In Egypt, some reefs are severely impacted by dynamite fishing (Riegl and Luke 1999) and intense tourism (Zakai and Chadwick-Furman 2002).

Although coral reefs in the Red Sea are generally still in a rather good condition compared to other geographic regions (e.g., Caribbean), increasing anthropogenic pressure may soon lead to an increase of reef destruction and loss. Climate change effects are expected to be stronger in the southern and central reefs, since these corals already thrive under marginal conditions, as well as in reefs which are already weakened by local stressors. Therefore, increasing coastal development, increasing tourism, and possibly also increasing fishing pressure need to be well managed in order to at least reduce irresistible climate change-driven warming and ocean acidification effects on coral reefs. On the other hand, corals from the southern Red Sea may hold the key for thermo-tolerance, which demonstrates that corals are able to adapt to extreme environments to a certain degree, although the question of pace remains. We therefore encourage particular protection efforts on the southern reefs, which have the potential to serve as a reservoir of thermo-tolerant populations for central and northern reefs during the next few decades.

References

Ainsworth TD, Thurber RV, Gates RD (2010) The future of coral reefs: a microbial perspective. Trends Ecol Evol 25:233–240

Al-Sofyani A (1987) Studies on growth and reproduction of Red Sea scleractinian. M.Sc. thesis, King Abdulaziz University, 107 pp

Al-Sofyani A (1991) Physiology and ecology of *Stylophora pistillata* and *Echinopora gemmacea* from the Red Sea. PhD thesis, University of Glasgow, 164 pp

Al-Sofyani A, Davies PS (1993) Seasonal variation in production and respiration of Red Sea corals. In: Richmond RH (ed) 7th international coral reef symposium. University of Guam Press, Mangilao, Guam, pp 351–357

Al-Sofyani A, Floos YAM (2013) Effect of temperature on two reef-building corals *Pocillopora damicornis* and *P. verrucosa* in the Red Sea. Oceanologia 55:917–935

Anthony KRN, Connolly SR (2004) Environmental limits to growth: physiological niche boundaries of corals along turbidity-light gradients. Oecologia 141:373–384

Babcock RC, Wills BL, Simpson CJ (1994) Mass spawning of corals on a high latitude coral reef. Coral Reefs 13:161–169

Babcock RC, Bull GD, Harrison PL, Heyward AJ, Oliver JK, Wallace CC, Willis BL (1986) Synchronous spawnings of 105 scleractinian coral species on the great barrier reef. Mar Biol 90:379–394

Baird AH, Guest JR, Willis BL (2009) Systematic and biogeographical patterns in the reproductive biology of Scleractinian corals. Annu Rev Ecol Evol Syst 40:551–571

Baker AC (2001) Ecosystems: reef corals bleach to survive change. Nature 411:765–766

Baker AC (2003) Flexibility and specificity in coral-algal symbiosis: diversity, ecology, and biogeography of symbiodinium. Annu Rev Ecol Evol Syst 34:661–689

Baker AC, Starger CJ, McClanahan TR, Glynn PW (2004) Coral reefs: corals' adaptive response to climate change. Nature 430:741

Banguera-Hinestroza E, Sawall Y, Al-Sofyani A, Wham D, Schnetzer J, Roder C, LaJeunesse T, Voolstra C (2012) The genetic diversity of the coral-dinoflagelate diversity in the Red Sea. Abstract. In: 12th international coral reef symposium, Cairns, Australia

Barr JJ, Auro R, Furlan M, Whiteson KL, Erb ML, Pogliano J, Stotland A, Wolkowicz R, Cutting AS, Doran KS, Salamon P, Youle M, Rohwer F (2013) Bacteriophage adhering to mucus provide a non–host-derived immunity. Proc Natl Acad Sci 110:10771–10776

Bayer T, Neave MJ, Alsheikh-Hussain A, Aranda M, Yum LK, Mincer T, Hughen K, Apprill A, Voolstra CR (2013) The microbiome of the Red Sea coral Stylophora pistillata is dominated by tissue-associated Endozoicomonas bacteria. Appl Environ Microbiol 79:4759–4762

Bell PRF (1992) Eutrophication and coral reefs—some examples in the Great Barrier Reef lagoon. Water Res 26:553–568

Bellwood DR, Hughes TP, Folke C, Nystrom M (2004) Confronting the coral reef crisis. Nature 429:827–833

Bouwmeester J, Berumen ML, Baird AH (2011a) Daytime broadcast spawning of *Pocillopora verrucosa* on coral reefs of the central Red Sea. Galaxea J Coral Reef Stud 13:23–24

Bouwmeester J, Khalil MT, La Torre P, Berumen ML (2011b) Synchronous spawning of *Acropora* in the Red Sea. Coral Reefs 30:1011

Brander LM, Van Beukering P, Cesar HSJ (2007) The recreational value of coral reefs: a meta-analysis. Ecol Econ 63:209–218

Brown BE (1997) Coral bleaching: causes and consequences. Coral Reefs 16:S129–S138

Brown BE, Bythell JC (2005) Perspectives on mucus secretion in reef corals. Mar Ecol Prog Ser 296:291–309

Bruno JF, Petes LE, Drew Harvell C, Hettinger A (2003) Nutrient enrichment can increase the severity of coral diseases. Ecol Lett 6:1056–1061

Bythell JC, Wild C (2011) Biology and ecology of coral mucus release. J Exp Mar Biol Ecol 408:88–93

Cantin NE, Cohen AL, Karnauskas KB, Tarrant AM, McCorkle DC (2010) Ocean warming slows coral growth in the central Red Sea. Science 329:322–325

Case RJ, Longford SR, Campbell AH, Low A, Tujula N, Steinberg PD, Kjelleberg S (2011) Temperature induced bacterial virulence and bleaching disease in a chemically defended marine macroalga. Environ Microbiol 13:529–537

Coles SL, Jokiel PL (1977) Effects of temperature on photosynthesis and respiration in hermatypic corals. Mar Biol 43:209–216

Coles SL, Jokiel PL (1978) Synergistic effects of temperature, salinity and light on the hermatypic coral *Montipora verrucosa*. Mar Biol 49:187–195

Connell JH (1978) Diversity in tropical rain forests and coral reefs. Science 199:1302–1310

Costanza R, D'Arge R, Groot RD, Farber S, Grasso M, Hannon B, Limburg K, Naeem S, O'Neill RV, Paruelo J, Raskin RG, Sutton P, Belt MVD (1997) The value of the world's ecosystem services and natural capital. Nature 387:253–260

Crossland CJ (1987) In situ release of mucus and DOC-lipid from the corals *Acropora variabilis* and *Stylophora pistillata* in different light regimes. Coral Reefs 6:35–42

Davies PS (1991) Effect of daylight variations on the energy budgets of shallow-water corals. Mar Biol 108:137–144

Davy SK, Allemand D, Weis VM (2012) Cell biology of Cnidarian-Dinoflagellate symbiosis. Microbiol Mol Biol Rev 76:229–261

DeVantier L, Turak E, Al-Shaikh K (2000) Coral bleaching in the central-northern Saudi Arabian Red Sea, August–September 1998. In: International workshop on the extent and impact of coral bleaching in the Arabian Region

Dinsdale EA, Pantos O, Smriga S, Edwards RA, Angly F, Wegley L, Hatay M, Hall D, Brown E, Haynes M, Krause L, Sala E, Sandin SA, Thurber RV, Willis BL, Azam F, Knowlton N, Rohwer F (2008) Microbial ecology of four coral atolls in the Northern Line Islands. PLoS ONE 3:e1584

Dubinsky Z, Stambler N (2009) Photoacclimation processes in phytoplankton: mechanisms, consequences, and applications. Aquat Microb Ecol 56:163–176

Dubinsky Z, Falkowski PG, Porter JW, Muscatine L (1984) Absorption and utilization of radiant energy by light-adapted and shade-adapted colonies of the hermatypic coral *Stylophora-Pistillata*. Proc Roy Soc London B Biol Sci 222:203

Dubinsky Z, Stambler N, Ben-Zion M, McCloskey LR, Muscatine L, Falkowski PG (1990) The effect of external nutrient resources on the optical properties and photosynthetic efficiency of *Stylophora pistillata*. Proc Roy Soc London B Biol Sci 239:231–246

Ducklow HW, Mitchell R (1979) Bacterial populations and adaptations in the mucus layers on living corals. Limnol Oceanogr 24:715–725

Edmunds PJ, Davies PS (1989) An energy budget for *Porites porites* (Scleractinia), growing in a stressed environment. Coral Reefs 8:37–43

Edmunds PJ, Cumbo V, Fan TY (2011) Effects of temperature on the respiration of brooded larvae from tropical reef corals. J Exp Biol 214:2783–2790

Fabricius KE (2005) Effects of terrestrial runoff on the ecology of corals and coral reefs: review and synthesis. Mar Pollut Bull 50:125–146

Fadlallah YH (1985) Reproduction i the coral *Pocillopora verrucosa* on the reefs adjacent to the industrial city of Yanbu (Red Sea, Saudi Arabia). In: Proceedings of the 5th international coral reef symposium, vol 4. Tahiti, pp 313–318

Falkowski PG, Dubinsky Z (1981) Light-shade adaptation of *Stylophora pistillata*, a hermatypic coral from the Gulf of Eilat. Nature 289:172–174

Falkowski PG, Dubinsky Z, Muscatine L, Porter JW (1984) Light and the bioenergetics of a symbiotic coral. Bioscience 34:705–709

Falkowski PG, Dubinsky Z, Muscatine L, McCloskey L (1993) Population control in symbiotic corals. Bioscience 43:606–611

Ferrier-Pagés C, Gattuso J-P, Jaubert J (1999) Effect of small variations in salinity on the rates of photosynthesis and respiration of the zooxanthellate coral *Stylophora pistillata*. Mar Ecol Prog Ser 181:309–314

Fine M, Gildor H, Genin A (2013) A coral reef refuge in the Red Sea. Glob Change Biol 19:3640–3647

Fishelson L (1973) Ecology of coral reefs in Gulf of Aqaba (Red Sea) influenced by pollution. Oecologia 12:55–67

Fitt W, Brown B, Warner M, Dunne R (2001) Coral bleaching: interpretation of thermal tolerance limits and thermal thresholds in tropical corals. Coral Reefs 20:51–65

Floos YAM (2007) Physiological and ecological studies of Pocillopora damicornis and Pocillopora verrucosa in Sharm Ubhur, Jeddah, Saudi Arabia. M.Sc. thesis, King Abdulaziz University, Jeddah, 101 pp

Floos YAM (2012) Physiological and ecological studies of the two reef building coral species *Seriatopora hystrix* and *Lobophyllia corymbosa* in the Red Sea. Ph.D. thesis, King Abdulaziz University, Jeddah, 165 pp

Flot J-F, Magalon H, Cruaud C, Couloux A, Tillier S (2008) Patterns of genetic structure among Hawaiian corals of the genus *Pocillopora* yield clusters of individuals that are compatible with morphology. CR Biol 331:239–247

Frade P, Englebert N, Faria J, Visser P, Bak R (2008a) Distribution and photobiology of *Symbiodinium* types in different light environments for three colour morphs of the coral *Madracis pharensis*: is there more to it than total irradiance? Coral Reefs 27:913–925

Frade PR, De Jongh F, Vermeulen F, Van Bleijswijk J, Bak RPM (2008b) Variation in symbiont distribution between closely related coral species over large depth ranges. Mol Ecol 17:691–703

Fricke HW, Schuhmacher H (1983) The depth limits of Red Sea stony corals: an ecophysiological problem (a deep diving survey by submersible). Mar Ecol 4:163–194

Fricke HW, Vareschi E, Schlichter D (1987) Photoecology of the coral *Leptoseris fragilis* in the Red Sea twilight zone (an experimental study by submersible). Oecologia 73:371–381

Furby KA, Bouwmeester J, Berumen ML (2013) Susceptibility of central Red Sea corals during a major bleaching event. Coral Reefs 32:505–513

Gates RD, Edmunds PJ (1999) The physiological mechanisms of acclimatization in tropical reef corals. Am Zool 39:30–43

Goulet TL (2006) Most corals may not change their symbionts. Mar Ecol Prog Ser 321:1–7

Hanafy MH, Aamer MA, Habib M, Rouphael AB, Baird AH (2010) Synchronous reproduction of corals in the Red Sea. Coral Reefs 29:119–124

Harland AD, Navarro JC, Spencer Davies P, Fixter LM (1993) Lipids of some Caribbean and Red Sea corals: total lipid, wax esters, triglycerides and fatty acids. Mar Biol 117:113–117

Harrison PL, Ward S (2001) Elevated levels of nitrogen and phosphorus reduce fertilisation success of gametes from scleractinian reef corals. Mar Biol 139:1057–1068

Harrison PL, Wallace CC (1990) Reproduction, dispersal and recruitment of scleractinian corals. In: Dubinsky Z (ed) Coral Reefs. Elsevier Science, Amsterdam, pp 133–207

Harrison PL, Babcock RC, Bull GD, Oliver JK, Wallace CC, Willis BL (1984) Mass spawning in tropical reef corals. Science 223:1186–1189

Haynes D, Johnson JE (2000) Organochlorine, heavy metal and polyaromatic hydrocarbon pollutant concentrations in the Great Barrier Reef (Australia) environment: a review. Mar Pollut Bull 41:267–278

Hoegh-Guldberg O, Smith GJ (1989) The effect of sudden changes in temperature, light and salinity on the population density and export of zooxanthellae from the reef corals *Stylophora pistillata* Esper and *Seriatopora hystrix* Dana. J Exp Mar Biol Ecol 129:279–303

Hoegh-Guldberg O, Mumby PJ, Hooten AJ, Steneck RS, Greenfield P, Gomez E, Harvell CD, Sale PF, Edwards AJ, Caldeira K, Knowlton N, Eakin CM, Iglesias-Prieto R, Muthiga N, Bradbury RH, Dubi A, Hatziolos ME (2007) Coral reefs under rapid climate change and ocean acidification. Science 318:1737–1742

Houlbrèque F, Ferrier-Pagès C (2008) Heterotrophy in tropical Scleractinian corals. Biol Rev 84:1–17

Howells EJ, Beltran VH, Larsen NW, Bay LK, Willis BL, van Oppen MJH (2012) Coral thermal tolerance shaped by local adaptation of photosymbionts. Nat Clim Change 2:116–120

Hughes TP, Baird AH, Bellwood DR, Card M, Connolly SR, Folke C, Grosberg R, Hoegh-Guldberg O, Jackson JBC, Kleypas J, Lough JM, Marshall P, Nystrom M, Palumbi SR, Pandolfi JM, Rosen B, Roughgarden J (2003) Climate change, human impacts, and the resilience of coral reefs. Science 301:929–933

ICPP (2007) In: Bernstein L, Bosch P, Canziani O, Chen Z, Christ R et al. (eds) Climate change 2007: synthesis report. An assessment of the intergovernmental panel on climate change. Cambridge University Press, Cambridge, UK, 73 pp

Iglesias-Prieto R, Beltrán VH, LaJeunesse TC, Reyes-Bonilla H, Thomé PE (2004) Different algal symbionts explain the vertical distribution of dominant reef corals in the eastern Pacific. Proc Roy Soc B Biol Sci 271:1757–1763

Ip YK, Lim ALL, Lim RWL (1991) Some properties of calcium-activated adenosine triphosphatase from the hermatypic coral *Galaxea fascicularis*. Mar Biol 111:191–197

Jessen C, Villa L, Javier F, Bayer T, Roder C, Aranda M, Wild C, Voolstra CR (2013) In-situ effects of eutrophication and overfishing on physiology and bacterial diversity of the Red Sea Coral *Acropora hemprichii*. PLoS One 8:e62091

Jokiel PL, Guinther EB (1978) Effects of temperature on reproduction in the hermatypic coral *Pocillopora damicornis*. Bull Mar Sci 28:786–789

Jokiel PL, Coles SL (1990) Response of Hawaiian and other Indo-Pacific reef corals to elevated temperature. Coral Reefs 8:155–162

Jones AM, Berkelmans R, van Oppen MJH, Mieog JC, Sinclair W (2008) A community change in the algal endosymbionts of a scleractinian coral following a natural bleaching event: field evidence of acclimatization. Proc Roy Soc B Biol Sci 275:1359–1365

Jones RJ, Hoegh-Guldberg O, Larkum AWD, Schreiber U (1998) Temperature-induced bleaching of corals begins with impairment of the CO2 fixation mechanism in zooxanthellae. Plant Cell Environ 21:1219–1230

Jones RJ, Muller J, Haynes D, Schreiber U (2003) Effects of herbicides diuron and atrazine on corals of the Great Barrier Reef, Australia. Mar Ecol Prog Ser 251:153–167

Kemp J, Klaus R, Salem M, Awadalla Y, Saleh B (2002) Survey of the proposed marine protected area at Dunganob Bay and Mukkawar Island, Sudan. PERSGA, Jeddah, SAP Component 5 A

Keshavmurthy S, Yang S-Y, Alamaru A, Chuang Y-Y, Pichon M, Obura D, Fontana S, De Palmas S, Stefani F, Benzoni F, MacDonald A, Noreen AME, Chen C, Wallace CC, Pillay RM, Denis V, Amri AY, Reimer JD, Mezaki T, Sheppard C, Loya Y, Abelson A, Mohammed MS, Baker AC, Mostafavi PG, Suharsono BA, Chen CA (2013) DNA barcoding reveals the coral "laboratory-rat". Stylophora pistillata encompasses multiple identities, Sci Rep 3

Klaus JS, Janse I, Heikoop JM, Sanford RA, Fouke BW (2007) Coral microbial communities, zooxanthellae and mucus along gradients of seawater depth and coastal pollution. Environ Microbiol 9:1291–1305

Kleypas JA, McManus JW, Menez LAB (1999) Environmental limits to coral reef development: where do we draw the line? Am Zool 39:146–159

Koop K, Booth D, Broadbent A, Brodie J, Bucher D, Capone D, Coll J, Dennison W, Erdmann M, Harrison P, Hoegh-Guldberg O, Hutchings P, Jones GB, Larkum AWD, O'Neil J, Steven A, Tentori E, Ward S, Williamson J, Yellowlees D (2001) ENCORE: the effect of nutrient enrichment on coral reefs. Synthesis of results and conclusions. Mar Pollut Bull 42:91–120

Kotb M (2001) Growth rates of three reef-building corals species in the Northern Red Sea, Egypt. Egypt J Aquat Biol Fish 5:165–185

Kotb M, Abdulaziz M, Al-Agwan Z, Al-Shaikh K, Al-Yami H, Banajah A, DeVantier L, Eisinger M, Eltayeb M, Hassan M, Heiss G, Howe S, Kemp J, Klaus R, Krupp F, Mohamed N, Rouphael T, Turner J, Zajonz U (2004) Status of coral reefs in the Red Sea and Gulf of Aden in 2004. In: Wilkinson C (ed) Status of coral reefs of the world: 2004. Australian Institute of Marine Science, Townsville, 301 pp

Kürten B, Al-Aidaroos AM, Struck U, Khomayis HS, Gharbawi WY, Sommer U (2014) Influence of environmental gradients on C and N stable isotope ratios in coral reef biota of the Red Sea, Saudi Arabia. J Sea Res 85:379–394

LaJeunesse TC, Pettay DT, Sampayo EM, Phongsuwan N, Brown B, Obura DO, Hoegh-Guldberg O, Fitt WK (2010) Long-standing environmental conditions, geographic isolation and host-symbiont specificity influence the relative ecological dominance and genetic diversification of coral endosymbionts in the genus *Symbiodinium*. J Biogeogr 37:785–800

Lampert-Karako S, Stambler N, Katcoff DJ, Achituv Y, Dubinsky Z, Simon-Blecher N (2008) Effects of depth and eutrophication on the zooxanthella clades of *Stylophora pistillata* from the Gulf of Eilat (Red Sea). Aquat Conserv 18:1039–1045

Leggat W, Seneca F, Wasmund K, Ukani L, Yellowlees D, Ainsworth TD (2011) Differential responses of the coral host and their algal symbiont to thermal stress. PLoS ONE 6:e26687

Lesser MP (2013) Using energetic budgets to assess the effects of environmental stress on corals: are we measuring the right things? Coral Reefs 32:25–33

Leuzinger S, Willis B, Anthony KN (2012) Energy allocation in a reef coral under varying resource availability. Mar Biol 159:177–186

Levy O, Appelbaum L, Leggat W, Gothlif Y, Hayward DC, Miller DJ, Hoegh-Guldberg O (2007) Light-responsive cryptochromes from a simple multicellular animal, the coral *Acropora millepora*. Science 318:467–470

Lieberum C (2011) Ist eine Rekonstruktion rezenter Umweltbedingungen im Roten Meer anhand von Korallenbohrkernen möglich? M. Sc. thesis, IFM-GEOMAR und Chrsitian-Albrechts-Universität zu Kiel, 99 pp

Liew YJ, Aranda M, Carr A, Baumgarten S, Zoccola D, Tambutté S, Allemand D, Micklem G, Voolstra CR (2014) Identification of MicroRNAs in the coral *Stylophora pistillata*. PLoS ONE 9:e91101

Lough JM, Barnes DJ (2000) Environmental controls on growth of the massive coral Porites. J Exp Mar Biol Ecol 245:225–243

Loya Y (1972) Community structure and species diversity of hermatypic corals at Eilat, Red Sea. Mar Biol 13:100–123

Loya Y (1976) The Red Sea coral *Stylophora pistillata* is an r strategist. Nature 259:478–480

Loya Y, Kramarsky-Winter E (2003) In situ eutrophication caused by fish farms in the northern Gulf of Eilat (Aqaba) is beneficial for its coral reefs: a critique. Mar Ecol Prog Ser 261:299–303

Loya Y, Lubinevsky H, Rosenfeld M, Kramarsky-Winter E (2004) Nutrient enrichment caused by in situ fish farms at Eilat, Red Sea is detrimental to coral reproduction. Mar Pollut Bull 49:344–353

Marshall AT, Clode P (2004) Calcification rate and the effect of temperature in a zooxanthellate and an azooxanthellate scleractinian reef coral. Coral Reefs 23:218–224

Mass T, Einbinder S, Brokovich E, Shashar N, Vago R, Erez J, Dubinsky Z (2007) Photoacclimation of *Stylophora pistillata* to light extremes: metabolism and calcification. Mar Ecol Prog Ser 334:93–102

Mass T, Kline DI, Roopin M, Veal CJ, Cohen S, Iluz D, Levy O (2010) The spectral quality of light is a key driver of photosynthesis and photoadaptation in *Stylophora pistillata* colonies from different depths in the Red Sea. J Exp Biol 213:4084–4091

McCloskey LR, Muscatine L (1984) Production and respiration in the Red Sea coral *Stylophora pistillata* as a function of depth. Proc Roy Soc London Series B Biol Sci 222:215–230

Mergner H (1984) The ecological research on coral reefs of the Red Sea. Deep Sea Res Part A Oceanogr Res Pap 31:855–884

Mohamed TAA, Kotb M, Ghobashy AFA, Deek MS (2007) Reproduction and growth rate of two scleractinian coral species in the Northern Red Sea, Egypt. Egypt J Aquat Res 33:70–86

Mouchka ME, Hewson I, Harvell CD (2010) Coral-associated bacterial assemblages: current knowledge and the potential for climate-driven impacts. Integr Comp Biol 50:662–674

Muscatine L, Cernichiari E (1969) Assimilation of photosynthetic products of zooxanthellae by a reef coral. Biol Bull 137:506–523

Muscatine L, Porter JW (1977) Reef corals: mutualistic symbioses adapted to nutrient-poor environments. Bioscience 27:454–460

Muscatine L, McCloskey LR, Marian RE (1981) Estimating the daily contribution of carbon from zooxanthellae to coral animal respiration. Limnol Oceanogr 26:601–611

Muscatine L, Porter JW, Kaplan IR (1989) Resource partitioning by reef corals as determined from stable isotope composition. Mar Biol 100:185–193

Muscatine L, Falkowski PG, Porter JW, Dubinsky Z (1984) Fate of photosynthetic fixed carbon in light- and shade-adapted colonies of the symbiotic coral *Stylophora pistillata*. Proc Roy Soc London Series B Biol Sci 222:181–202

Muthiga NA, Szmant AM (1987) The effect of salinity stress on the rates of aerobic respiration and photosynthesis in the hermatypic coral *Siderastrea siderea*. Biol Bull 173:539–551

Naumann M, Haas A, Struck U, Mayr C, el-Zibdah M, Wild C (2010) Organic matter release by dominant hermatypic corals of the Northern Red Sea. Coral Reefs 29:649–659

Oliver JK, Babcock RC, Harrison PL, Willis BL (1988) Geographic extent of mass coral spawning: clues to ultimate causal factors. In: Proceedings of the 6th international coral reef symposium, vol 2. Australia, pp 803–810

PERSGA (2006) State of the marine environment: report for the Red Sea and Gulf of Aden. Regional Organization for the Conservation of the Environment of the Red Sea and Gulf of Aden (PERSGA), Jeddah, Saudi Arabia, 260 pp

Porter JW, Muscatine L, Dubinsky Z, Falkowski PG (1984) Primary production and photoadaptation in light- and shade-adapted colonies of the symbiotic coral, *Stylophora pistillata*. Proc Roy Soc London Series B Biol Sci 222:161–180

Rabalais NN, Turner RE, Díaz RJ, Justić D (2009) Global change and eutrophication of coastal waters. ICES J Mar Sci: Journal du Conseil 66:1528–1537

Raitsos DE, Hoteit I, Prihartato PK, Chronis T, Triantafyllou G, Abualnaja Y (2011) Abrupt warming of the Red Sea. Geophys Res Lett 38

Reshef L, Koren O, Loya Y, Zilber-Rosenberg I, Rosenberg E (2006) The coral probiotic hypothesis. Environ Microbiol 8:2068–2073

Reynolds JM, Bruns BU, Fitt WK, Schmidt GW (2008) Enhanced photoprotection pathways in symbiotic dinoflagellates of shallow-water corals and other cnidarians. Proc Natl Acad Sci 105:13674–13678

Richmond RH, Hunter CL (1990) Reproduction and recruitment of corals: comparison among the Caribbean, the Tropical Pacific and the Red Sea. Mar Ecol Prog Ser 60:185–203

Riegl B, Luke KE (1999) Ecological parameters of dynamited reefs in the northern Red Sea and their relevance to reef rehabilitation. Mar Pollut Bull 37:488–498

Rinkevich B (1989) The contribution of photosynthetic products to coral reproduction. Mar Biol 101:259–263

Rinkevich B, Loya Y (1979a) Reproduction of the Red-Sea coral *Stylophora-Pistillata*. 1. Gonads and planulae. Mar Ecol Prog Ser 1:133–144

Rinkevich B, Loya Y (1979b) Reproduction of the Red-Sea coral *Stylophora-Pistillata*. 2. Synchronization in breeding and seasonality of planulae shedding. Mar Ecol Prog Ser 1:145–152

Rinkevich B, Loya Y (1987) Variability in the pattern of sexual reproduction of the coral *Stylophora pistillata* at Eilat, Red Sea: a long-term study. Biol Bull 173:335–344

Ritchie KB (2006) Regulation of microbial populations by coral surface mucus and mucus-associated bacteria. Mar Ecol Prog Ser 322:1–14

Robitzch V, Banguera E, Sawall Y, Al-Sofyani A, Voolstra CR (accepted) Absence of genetic differentiation in the coral Pocillopora verrucosa along environmental gradients of the Saudi Arabian Red Sea. Frontiers in Marine Science

Roder C, Berumen ML, Bouwmeester J, Papathanassiou E, Al-Suwailem A, Voolstra CR (2013) First biological measurements of deep-sea corals from the Red Sea. Sci Rep 3

Roder C, Arif C, Bayer T, Aranda M, Daniels C, Shibl A, Chavanich S, Voolstra CR (2014) Bacterial profiling of white plague disease in a comparative coral species framework. ISME J 8:31–39

Rosenberg E, Ben-Haim Y (2002) Microbial diseases of corals and global warming. Environ Microbiol 4:318–326

Rosenberg E, Koren O, Reshef L, Efrony R, Zilber-Rosenberg I (2007) The role of microorganisms in coral health, disease and evolution. Nat Rev Microbiol 5:355–362

Rowan R, Knowlton N (1995) Intraspecific diversity and ecological zonation in coral-algal symbiosis. Proc Natl Acad Sci 92:2850–2853

Sanford E, Kelly MW (2011) Local adaptation in marine invertebrates. Annu Rev Mar Sci 3:509–535

Sawall Y, Teichberg MC, Seemann J, Litaay M, Jompa J, Richter C (2011) Nutritional status and metabolism of the coral *Stylophora subseriata* along a eutrophication gradient in Spermonde Archipelago (Indonesia). Coral Reefs 30:841–853

Sawall Y, Al-Sofyani A, Banguera-Hinestroza E, Voolstra C (2014a) Spatio-temporal analyses of zooxanthellae physiology of the coral *Pocillopora verrucosa* along large-scale nutrient and temperature gradients in the Red Sea. PLoS ONE, doi:10.1371/journal.pone.0103179

Sawall Y, Al-Sofyani A, Kürten B, Al-Aidaroos AM, Hoang BX, Marimuthu N, Khomayis HS, Sommer U, Gharbawi WY, Wahl M (2014b) Coral communities, in contrast to fish communities, maintain a high assembly similarity along the large latitudinal gradient along the Saudi Red Sea coast. Journal of Ecosystem and Ecography S4: 003, doi:10.4172/2157-7625.S4-003

Sawall Y, Al-Sofyani A, Hohn S, Banguera-Hinestroza E, Voolstra C, Wahl M (accepted) Extensive phenotypic plasticity of a Red Sea coral over a strong latitudinal temperature gradient suggests limited acclimatization potential to warming. Scientific Reports

Schlichter D, Fricke HW, Weber W (1986) Light harvesting by wavelength transformation in a symbiotic coral of the Red Sea twilight zone. Mar Biol 91:403–407

Sheppard CRC (1982) Coral populations on reef slopes and their major controls. Mar Ecol Prog Ser 7:83–115

Sheppard CRC, Sheppard ALS (1991) Corals and coral communities of Arabia. Fauna Saudi Arabia 12:3–170

Shiah F-K, Ducklow HW (1994) Temperature and substrate regulation of bacterial abundance, production and specific growth rate in Chesapeake Bay, USA. Mar Ecol Prog Ser 103:297–308

Shlesinger Y, Loya Y (1985) Coral community reproductive patterns—Red-Sea versus the Great Barrier Reef. Science 228:1333–1335

Shlesinger Y, Goulet TL, Loya Y (1998) Reproductive patterns of scleractinian corals in the northern Red Sea. Mar Biol 132:691–701

Siedler G (1969) General circulation of water masses in the Red Sea. In: Degens ET, Ross DA (eds) Hot brines and recent heavy metal deposits in the Red Sea. Springer, New York, pp 131–137

Stambler N (2005) Bio-optical properties of the northern Red Sea and the Gulf of Eilat (Aqaba) during winter 1999. J Sea Res 54:186–203

Stambler N, Levy O, Vaki L (2008) Photosynthesis and respiration of hermatypic zooxanthellate Red Sea corals from 5 to 75 m depth. Isr J Plant Sci 56:45–53

Stambler N, Popper N, Dubinsky Z, Stimson J (1991) Effects of nutrient enrichment and water motion on the coral *Pocillopora damicornis*. Pac Sci 45:299–307

Stanley GD (2003) The evolution of modern corals and their early history. Earth Sci Rev 60:195–225

Sutherland KP, Porter JW, Torres C (2004) Disease and immunity in Caribbean and Indo-Pacific zooxanthellate corals. Mar Ecol Prog Ser 266:273–302

Szmant A (2002) Nutrient enrichment on coral reefs: Is it a major cause of coral reef decline? Estuaries Coasts 25:743–766

Tambutté S, Holcomb M, Ferrier-Pagès C, Reynaud S, Tambutté É, Zoccola D, Allemand D (2011) Coral biomineralization: from the gene to the environment. J Exp Mar Biol Ecol 408:58–78

Telesnicki GJ, Goldberg WM (1995) Effects of turbidiy on the photosynthesis and respiration of two south Florida reef coral species. Bull Mar Sci 57:527–539

Tilot de Grissac V, Saadalla E, Saleh B, Afifi A, Audalla Y (2000) Exploratory reef assessment of the offshore islands of the Egyptian Red Sea. In: Proceedings of the 9th international coral reef symposium vol 2, pp 867–872

Titlyanov E, Titlyanova T, Leletkin V, Tsukahara J, van Woesik R, Yamazato K (1996) Degradation of zooxanthellae and regulation of their density in hermatypic corals. Mar Ecol Prog Ser 139:167–178

Titlyanov EA, Titlyanova TV, Yamazato K, van Woesik R (2001a) Photo-acclimation of the hermatypic coral Stylophora pistillata while subjected to either starvation or food provisioning. J Exp Mar Biol Ecol 257:163–181

Titlyanov EA, Titlyanova TV, Yamazato K, van Woesik R (2001b) Photo-acclimation dynamics of the coral Stylophora pistillata to low and extremely low light. J Exp Mar Biol Ecol 263:211–225

van Woesik R (2010) Calm before the spawn: global coral spawning patterns are explained by regional wind fields. Proc Roy Soc B Biol Sci 277:715–722

Vaughan TW (1919) Corals and the formation of coral reefs. Annu Rep Smithson Inst 1917:189–276

Walker DI, Ormond RFG (1982) Coral death from sewage and phosphate pollution at Aqaba, Red Sea. Mar Pollut Bull 13:21–25

Warner ME, Fitt WK, Schmidt GW (1996) The effects of elevated temperature on the photosynthetic efficiency of zooxanthellae in hospite from four different species of reef coral: a novel approach. Plant Cell Environ 19:291–299

Weber M, Lott C, Fabricius KE (2006) Sedimentation stress in a scleractinian coral exposed to terrestrial and marine sediments with contrasting physical, organic and geochemical properties. J Exp Mar Biol Ecol 336:18–32

Weis VM (2008) Cellular mechanisms of Cnidarian bleaching: stress causes the collapse of symbiosis. J Exp Biol 211:3059–3066

White P, Kalff J, Rasmussen J, Gasol J (1991) The effect of temperature and algal biomass on bacterial production and specific growth rate in freshwater and marine habitats. Microb Ecol 21:99–118

Wilkinson C (2004) Coral reefs of the world: 2004. Australian Institute for Marine Science, Townsville

Yonge CM (1940) The biology of reef-building corals. British Museum (Natural History), London

Zakai D, Chadwick-Furman NE (2002) Impacts of intensive recreational diving on reef corals at Eilat, northern Red Sea. Biol Conserv 105:179–187

Zakai D, Dubinsky Z, Avishai A, Caaras T, Chadwick NE (2006) Lunar periodicity of planula release in the reef-building coral *Stylophora pistillata*. Mar Ecol Prog Ser 311:93–102

Ziegler M, Roder C, Büchel C, Voolstra CR (2014) Limits to physiological plasticity of the coral *Pocillopora verrucosa* from the central Red Sea. Coral Reefs 33:1115–1129

Ziegler M, Roder C, Büchel C, Voolstra CR (2015) Mesophotic coral depth acclimatization is a function of host-specific symbiont physiology. Frontiers in Marine Science doi:10.3389/fmars.2015.00004

Zilber-Rosenberg I, Rosenberg E (2008) Role of microorganisms in the evolution of animals and plants: the hologenome theory of evolution. FEMS Microbiol Rev 32:723–735

Zvuloni A, Mokady O, Al-Zibdah M, Bernardi G, Gaines SD, Abelson A (2008) Local scale genetic structure in coral populations: a signature of selection. Mar Pollut Bull 56:430–438

Taxonomic, Ecological and Historical Considerations on the Deep-Water Benthic Mollusc Fauna of the Red Sea

Ronald Janssen and Marco Taviani

Abstract

The semi-enclosed and narrow Red Sea basin is characterized by bathyal zones in its axial sectors. It is determined by extreme hydrological parameters regarding its deep-water salinity and temperature which are a serious challenge to be coped with by deep-water benthos. Besides, it is separated from the adjacent Indian Ocean (Gulf of Aden) by a remarkably shallow sill that not only opposes easy transit for deep-water benthos but also exerts a strong control on the basin's hydrology budget during sea-level fluctuations, likely causing pulsing basin-wide extinctions at times of low stands. Among the relevant macrobenthic groups inhabiting the deep Red Sea, Mollusca stand out as the more diverse phylum. Although the full taxonomic appreciation of the Red Sea deep-water molluscs is still unresolved, as many as 262 species are recorded to date from depths below 400 m (163 Gastropoda, 94 Bivalvia, 4 Scaphopoda and 1 Polyplacophora). Part of this fauna is represented by eurybathic species with a wide bathymetric range. A substantial aliquot is equipped with a larval strategy (planktotrophy) in principle enabling the crossing of the shallow sill from the Gulf of Aden. Various taxa occur also in the Indo-West Pacific, and only a few are putatively considered as Red Sea endemics.

Introduction

The astonishing advancement of deep-sea exploration in the last 40 years has refined and sometimes revolutionized (one being the discovery of chemosynthetic ecosystems) our appreciation of the ecology, diversity and evolution of benthos living at great depth (Rowe 1983; Wilson and Hessler 1987; Gage and Tyler 1991; Tunnicliffe 1991; Young and Eckelbarger 1994; Van Dover 2000; Rex et al. 2005a; Judd and Hovland 2007; Koslow 2007; Foucher et al. 2009; McClain et al. 2009; Glover et al. 2010; Taviani 2011). It is quite clear that the deep-sea realm is far from being a slowly evolving low-diversity biome mainly, if not solely, relying upon surface-produced organic fall as hypothesized by earlier scholars (see reviews by Koslow 2007 and Glover et al. 2010). On the contrary, the current view is that the deep-sea benthos is highly complex, diverse and governed by the interaction of multiple factors (Hessler and Sanders 1967; Dayton and Hessler 1972; Rex 1983; Tyler 1988; Gage and Tyler 1991; Grassle 1991; Ruhl and Smith 2004; Rex et al. 2005b; Levin and Dayton 2009; Glover et al. 2010; Rex and Etter 2010). On a global scale, a major ecological physiognomy of the deep-sea benthos is to inhabit a rather homogeneous environment, nutrient poor, and with a remarkable stability of salinity and temperature ranges (e.g. Brunn 1957; Sanders and Hessler 1969; Thiel 1975, 1989; Gage 1978). Only a few large deep-water basins of the present Earth divert from these basically stable hydrologic conditions, regarding either temperature or salinity or both, that is, the peri-Antarctic seas, the Mediterranean Sea and the Red Sea (Ekman 1953).

R. Janssen (✉)
Sektion Malakologie, Senckenberg Forschungsinstitut und Naturmuseum, Senckenberganlage 25, 60325 Frankfurt am Main, Germany
e-mail: ronald.janssen@senckenberg.de

M. Taviani
ISMAR-CNR, UOS Bologna, Via Gobetti 101, 40129 Bologna, Italy
e-mail: marco.taviani@bo.ismar.cnr.it

M. Taviani
Biology Department, Woods Hole Oceanographic Institution, 266 Woods Hole Road, Woods Hole, MA 02543, USA

N.M.A. Rasul and I.C.F. Stewart (eds.), *The Red Sea*,
Springer Earth System Sciences, DOI 10.1007/978-3-662-45201-1_29,

The semi-enclosed Red Sea basin attains maximum depths well in excess of 2,500 m in its axial trough, and vast parts of its main trough are below 200 m, thus being suitable for the settlement of deep-water benthic life (Taviani et al. 2007). In comparison with the average situation in the ocean, the salinity of the Red Sea from the surface down to its deepest stretches is anomalous. By reaching values as high as 41 ‰ below a pycnocline at about 100–150 m (Morcos 1970; Rützen-Kositzkau 1999; Quadfasel 2001), the Red Sea deep water is, in fact, the saltiest marine water on Earth, besides the extreme values observed in some undersea brine pools or lakes, some of which are also present in the Red Sea axial trough (Antunes et al. 2011). Moreover, a temperature of 21.5 °C is observed basin-wide throughout the water column below the thermocline which is homothermic even at depths exceeding 2,000 m (Morcos 1970). Such values are aberrant when compared to temperatures met in the adjacent Gulf of Aden where temperature decreases with depth from about 13 °C at 500 m to about 3 °C below 2,000 m (Türkay 1996; Thiel et al. 1987; Al Saafani and Shenoi 2007).

To this, we should also add the fact that the Red Sea is separated from the Gulf of Aden, the north-western extension of the Indian Ocean, by a remarkably narrow and shallow sill (Hanish Sill, 137 m; Werner and Lange 1975) which is an efficient hydrological barrier, and very sensitive to sea-level fluctuations (Siddall et al. 2003; Lambeck et al. 2011).

Not to be neglected is also the potential role played by cold upwelling waters around the Indian Ocean side of the Arabian Peninsula (Abdel-Rahman 2006; Morcos and Abdallah 2012). They could in fact represent a further barrier to the spreading from the Gulf of Aden of benthic molluscs and/or their meroplanktic larval stages into the Red Sea.

Combined all together, it is intuitive that these factors offer per se a serious constraint for the quality of the deep-sea stenoecious benthos and render this basin a unique and formidable laboratory in which to test, for instance, the relevance of abiotic stressors in shaping the deep-sea benthos diversity and evolution (Türkay 1996; Taviani et al. 2007). Lamentably, not enough is known about the Red Sea deep benthos due to various unfavourable geo-politic circumstances which over the decades in one way or another hampered the systemic research on this region. Most information available at present still stems mainly from old expeditions in the late nineteenth and twentieth centuries, beginning with the pioneer expedition of the Austrian ship *Pola* in 1895–1898, followed much later by German surveys devoted to assessing potential environmental impacts related to metalliferous mud mining (Thiel 1979, 1980; Türkay 1996), and to a minor extent by Italian, plus other expeditions in more recent times. Yet, published data on deep-sea benthic life is still scarce and often patchy. We have some incomplete information about taxonomic groups such as ostracods (Bonaduce et al. 1983), foraminifera (Edelman-Furstenberg et al. 2001), cumaceans (Mühlenhardt-Siegel 2008) and a few others. Regarding megafauna and macrofauna, two taxa have attracted more attention, namely deep-water corals (Marenzeller 1907; Fricke and Hottinger 1983; Fricke and Schuhmacher 1983; Fricke and Knauer 1986; Taviani et al. 2007; Roder et al. 2013; Qurban et al. 2014) and molluscs (Sturany 1900a, b, 1901, 1904; Janssen 1989; Rützen-Kositzkau 1999). Holoplanktic and meroplanktic larval mollusc shells are common components of Red Sea bottom sediments contributing substantially to thanatocoenoses at bathyal depths. The holoplanktic mollusc fauna has been recently revised by Janssen (2007).

The Red Sea deep-sea benthos is clearly facing at the same time challenges and opportunities coping with a highly peculiar if not extreme environment. It has been hypothesized that the stenoecious marine biota of the Red Sea has been annihilated in this basin at the time of the last glaciation, in response to intolerable high-salinity conditions at the times of sea level lows (Taviani 1994, 1998a, b, with references therein). The present shallow and deep benthic fauna might, therefore, represent the post-glacial re-colonization (Gvirtzman et al. 1977; Taviani 1998a, c; Taviani et al. 2007). Firstly, the Red Sea is separated from the Indian Ocean by a very shallow sill (about 137 m) that imposes a strong geometric filter to potential propagules of the deep-sea Indian Ocean species trying to enter the basin (Janssen and Taviani 1985). Thus, deep-sea benthic organisms should rely upon special reproductive and dispersal mechanisms to cross the sill. Secondly, present oceanographic conditions in the Red Sea itself are very challenging for the survival of larvae and later establishment of viable populations in the bathyal zone, because of the unusual higher salinity and temperature of the deep-water than in any other marine basin on the planet.

The scope of this paper was to present a first critical account of the mollusc fauna inhabiting at present the Red Sea at depths exceeding 400 m, discussing taxonomic, ecological and biogeographical traits. The choice of such a bathymetric threshold is merely dictated by the fact that no information is virtually available regarding deep-sea molluscs distributed in the 200–400 m range.

The Mollusc Fauna of the Deep Red Sea: Composition, Distribution and Relationships

Much could be learned about the colonization of this unique environment by using molluscs as tracers. In fact, the phylum Mollusca is the most diversified group of macrobenthic organisms inhabiting the deep Red Sea, overwhelming

Fig. 1 Location of deep-water stations (>400 m) of the Austrian *Pola*-expeditions (modified after Türkay 1986)

corals, bryozoans, crustaceans, brachiopods and other groups in terms of diversity. Furthermore, their shells often retain memory records of their early developmental stages, thus informing us about larval strategies functional to understanding dispersal mechanisms (Sabelli and Taviani 2014).

First information on deep-water molluscs was gained by the pioneer expedition of the Austrian ship *Pola* in 1895–1898, whose molluscs were studied by Sturany (1900a, b, 1901, 1904: Figs. 1 and 2). Later, several German oceanographic expeditions surveyed and sampled Red Sea deep-water habitats, that is, the missions of research vessels *Sonne*

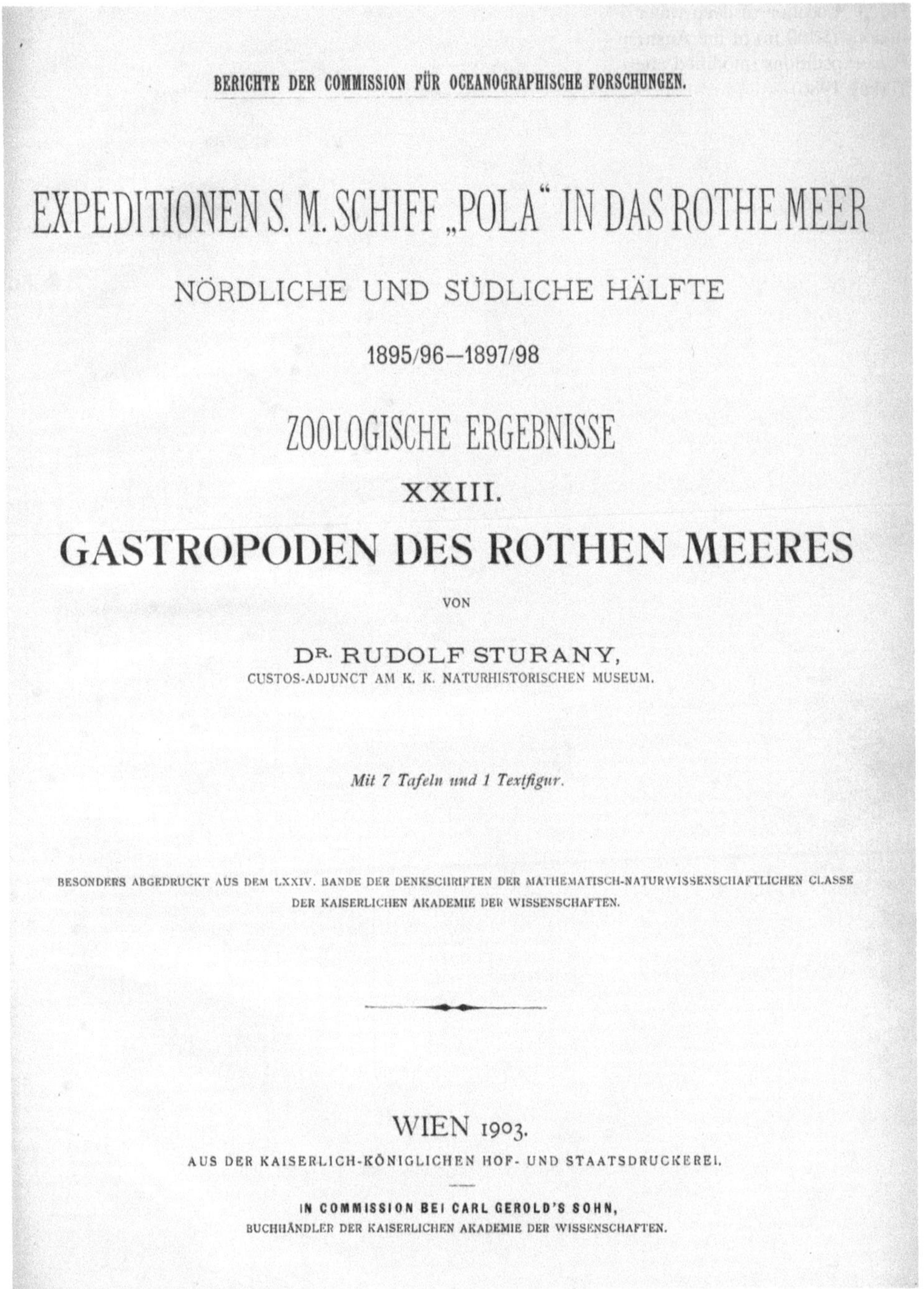
BERICHTE DER COMMISSION FÜR OCEANOGRAPHISCHE FORSCHUNGEN.

EXPEDITIONEN S. M. SCHIFF „POLA" IN DAS ROTHE MEER

NÖRDLICHE UND SÜDLICHE HÄLFTE

1895/96–1897/98

ZOOLOGISCHE ERGEBNISSE

XXIII.

GASTROPODEN DES ROTHEN MEERES

VON

DR. RUDOLF STURANY,

CUSTOS-ADJUNCT AM K. K. NATURHISTORISCHEN MUSEUM.

Mit 7 Tafeln und 1 Textfigur.

BESONDERS ABGEDRUCKT AUS DEM LXXIV. BANDE DER DENKSCHRIFTEN DER MATHEMATISCH-NATURWISSENSCHAFTLICHEN CLASSE DER KAISERLICHEN AKADEMIE DER WISSENSCHAFTEN.

WIEN 1903.

AUS DER KAISERLICH-KÖNIGLICHEN HOF- UND STAATSDRUCKEREI.

IN COMMISSION BEI CARL GEROLD'S SOHN,

BUCHHÄNDLER DER KAISERLICHEN AKADEMIE DER WISSENSCHAFTEN.

Fig. 2 Frontispiece of the publication of Sturany devoted to the Gastropoda of the *Pola* expedition. NB: There is controversy about the correct publication dates of the works of Sturany. Separate prints bear the publication dates 1899 for the bivalve part and 1903 for the gastropod part. However, actually, there is no unquestionable evidence that such dates are correct and reprints had really been published earlier than the respective volumes of the "Denkschriften". At present, we prefer therefore to refer to the dates of publication in the "Denkschriften" in agreement with the prevailing custom in the literature

(1977), *Valdivia* (1979, 1981) and *Meteor* (1987, 1995) (see Thiel 1980; Türkay 1996). The material from German collections forms the basis of the present article (Fig. 3). Molluscs were also obtained by the Italian expeditions MR79 (Cableship *Salernum*, 1979) and MR83 (RV *Urania*, 1983). Although this latter material is still unpublished, it was the source of the putative deep-sea endemic architectonicid gastropod *Pseudotorinia yaroni* described by Bieler (1993).

It should be noted that there is a considerable lack of comprehensive studies on the composition of the deep Red Sea mollusc fauna proper in comparison with that of the adjacent Gulf of Aden. An exception in this respect is the study of Rützen-Kositzkau (1999), mostly focused on taphonomic and ecological aspects.

Therefore, most observations discussed herein are based on our own observations and the evaluation of the rich material collected by the expeditions mentioned before, above all the German ones between 1977 and 1995. Many data, although sometimes of preliminary character, are presented here for the first time. Taxonomic work is in progress but, for many species identifications, further comparative work is necessary. Because of taxonomic problems, there is

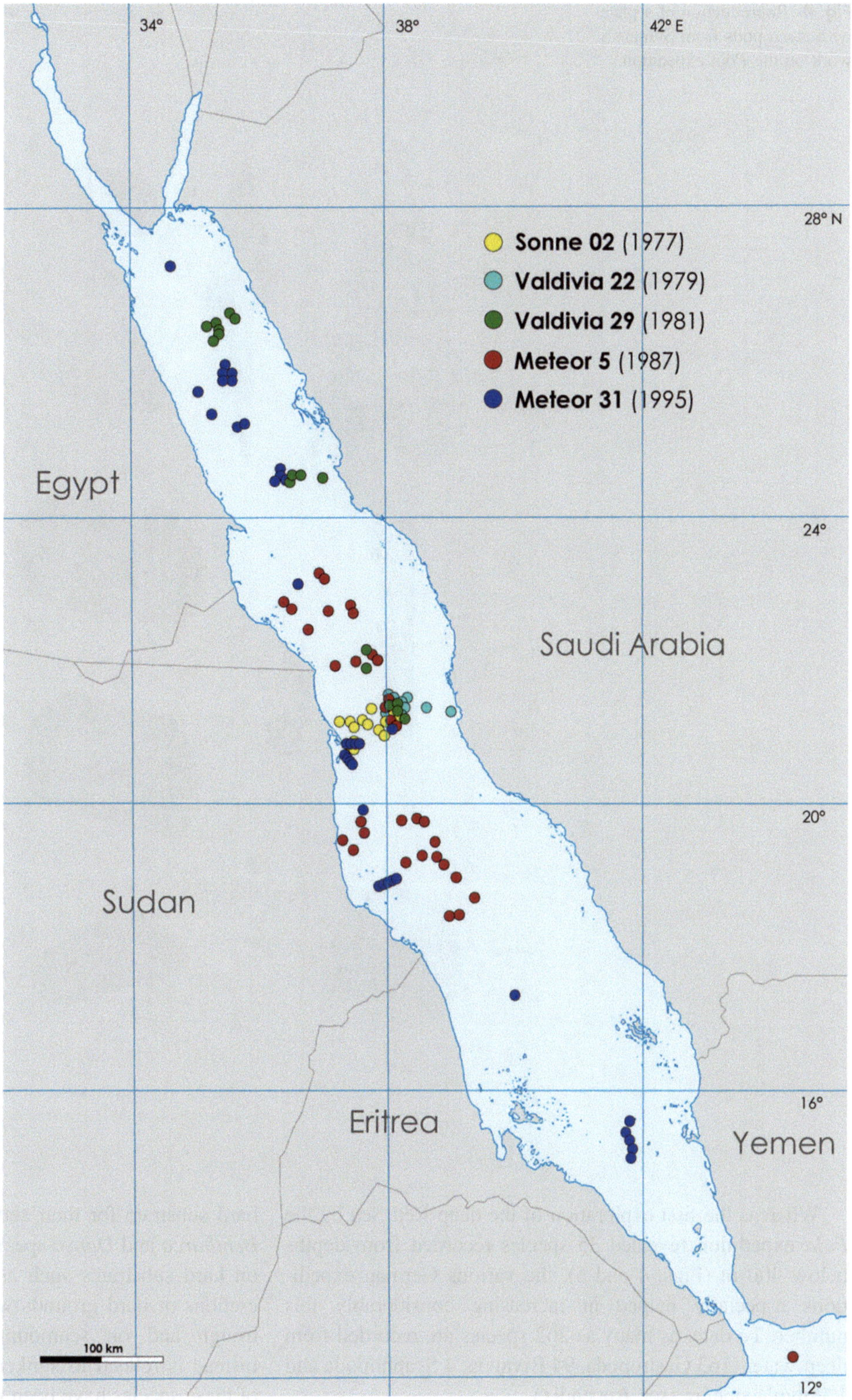

Fig. 3 Location of deep-water stations from the German *Sonne*, *Valdivia* and *Meteor* expeditions

still much uncertainty about the distribution and relationships of certain species, in particular those so far considered endemic to the Red Sea. Altogether, the samples collected by these expeditions provide basic observations on taxonomy, diversity and bathymetric range useful to categorize the Red Sea deep-water mollusc fauna.

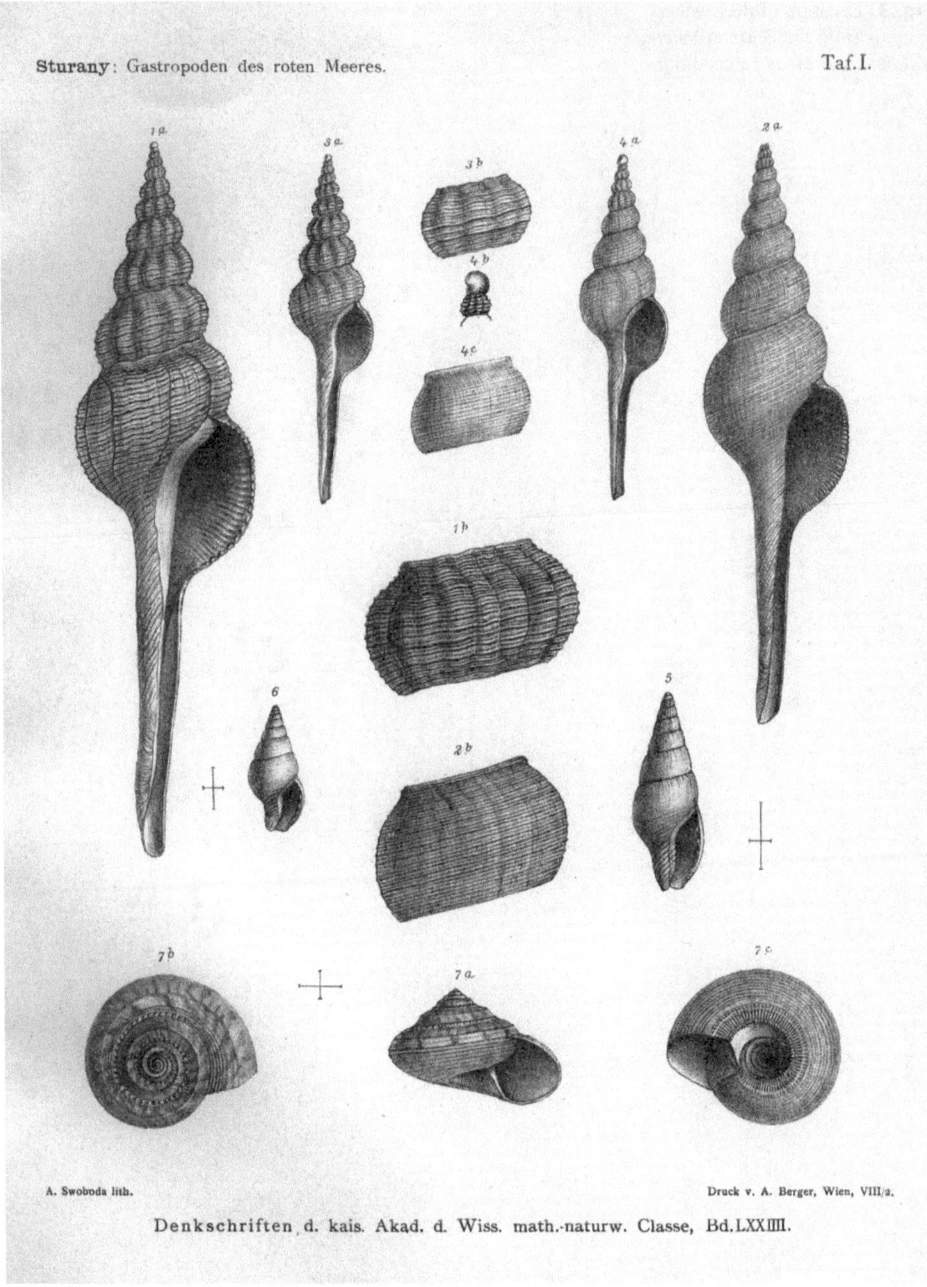

Fig. 4 Reproduction of a plate with gastropods from Sturany's work on the *Pola* expedition

Whereas the first exploration of the deep Red Sea by the *Pola* expedition revealed 35 species recorded from depths below 400 m (Figs. 4 and 5), the various German expeditions especially helped in increasing considerably this number. To date, as many as 262 species are recorded from deep-water (163 Gastropoda, 94 Bivalvia, 4 Scaphopoda and 1 Polyplacophora: see Appendix).

A substantial share of the known Red Sea deep-water mollusc fauna inhabits fine-grained bottoms, living either infaunally (the majority of bivalves and scaphopods) or as vagrants (the majority of gastropods). An exception is provided by three species [*Bentharca asperula* (Dall 1881), *Dimya* sp. and *Xylophaga* sp. 1], whose life habit requires hard substrate for their settlement instead of soft sediment. *Bentharca* and *Dimya* species are byssally fixed or cemented on hard substrates such as small stones or carbonate concretions or hard grounds which are widespread in the axial trough and on seamounts (Taviani 1998a). *Xylophaga* instead is related to sunken wood. Accordingly, specimens of these species have been found only at a smaller number of stations: *Bentharca asperula* at 48, *Dimya* sp. at 22 and *Xylophaga* sp. 1 at only 8 stations compared to about 50–70 stations for the other species.

So far, only 35 species have been collected alive, a mere 13.36 % of the entire deep-sea mollusc fauna. Table 1 shows the abundance and ranking of species, arranged as the total

Fig. 5 Reproduction of a plate with bivalves from Sturany's work on the *Pola* expedition

number of specimens and identifying shells versus live-recorded individuals. The very low fraction of living individuals among the total amount of specimens of each species is striking. Eight species contribute more than 51 % of all recorded specimens, while 76 % of the total material is represented by only 20 species, that is, 7.6 % of the species hitherto known from depths exceeding 400 m. The remaining 242 species (accounting for 92 % of the fauna) are found in much smaller numbers and in a limited number of stations. Nearly 60 % of the species, namely 158 species, are found at very few, often only one or two stations, and the possibility of downbasin transport of empty shells for some of them cannot be a priori excluded. A similar pattern is shown by the steadiness of distribution (Table 2); the most abundant species do also occur at most stations, not showing any distinct patchiness. These recurrent taxa seem to represent, therefore, the core of the deep-water mollusc fauna of the Red Sea. Overall, our data indicate a remarkably low population density. Such low biomass in the Red Sea benthic standing stocks, in spite of a primary production being far from negligible, is an established fact (Thiel et al. 1987; Pfannkuche 1993).

Table 1 Most common deep-sea molluscs in the Red Sea benthos

Rank (dead)	Rank (living)	Species	No. of specimens (total)	Living specimens	% of total specimens	Cumulative (%)
1	10	*Limopsis elachista*	2,025	7	11.40	11.40
2	8	*Nassarius lathraius*	1,614	11	9.10	20.50
3	10	*Cardiomya alcocki*	1,243	7	7.00	27.50
4	6	*Cuspidaria steindachneri*	1,202	19	6.77	34.27
5		*Jupiteria* sp.	983	4	5.54	39.81
6		*Clathurella pertabulata*	722	0	4.07	43.88
7	5	*Cuspidaria brachyrhynchus*	705	25	3.97	47.85
8	4	*Cetoconcha intracta*	629	59	3.54	51.39
9	8	*Parvamussium siebenrocki*	611	11	3.44	54.83
10	2	*Bentharca asperula*	568	256	3.20	58.03
11	9	*Pseudoneaera thaumasia*	499	8	2.81	60.84
12		*Bathyarca* sp.	443	6	2.49	63.33
13	1	*Dimya* sp.	409	312	2.30	65.63
14		*Costellaria casta*	400	0	2.25	67.88
15		*Gemmula* aff. *monilifera*	320	4	1.80	69.68
16	3	*Xylophaga* sp. 1	270	200	1.52	71.20
17		*Corbula* sp.	228	0	1.28	72.48
18		*Leiocithara* sp.	217	0	1.22	73.70
19		*Cuspidaria dissociata*	201	3	1.13	74.83
20		*Rhinoclama* sp.	197	2	1.11	75.94
21		*Haliris* sp.	175	0	0.98	76.92
22		*Gadila* sp.	168	0	0.94	77.86
23	7	*Fusinus bifrons*	164	15	0.92	78.78
24		*Lucidestea* ? sp.	156	0	0.88	79.66
25		*Cardiomya* sp.	155	0	0.87	80.53

Bathymetric Trends

The bathymetric distribution of the most common species shows that nearly all present a wide depth range, extending from waters shallower than 400 m even to depths exceeding 1,900–2,000 m (Table 3). However, due to the fact that the great majority of stations are located in the deep-water, the living records are mostly confined to a depth between about 400 and 1,400 m, only very few species having been observed living in depths exceeding 1,900 m. Although the more common species belong to bivalve groups which usually have their main depth range in deep-water (Fig. 6), such as Limopsidae, *Bathyarca*, Propeamussiidae and septibranchs (Cuspidariidae, Poromyidae, Verticordiidae), the depth ranges show that none of them is confined to true bathyal or abyssal depths. Some of them are even recorded from much shallower depths outside the Red Sea. Most of them also occur in depths above 400 m. The same pattern is shown by the commoner gastropods (Fig. 7) belonging to the families Nassariidae, Costellariidae, Fasciolariidae and "Turridae" s.l., which do not even belong to families with an exclusive bathyal distribution.

This pattern suggests that we are not really dealing with "deep-sea species" sensu stricto but rather with eurybathic taxa. In conclusion, judging from the mollusc component, it appears that the Red Sea seems deprived of a "true" deep-sea fauna, that is, species exclusively known from depths well in excess of 500 m. In this respect, it is noteworthy that there are practically no species shared between the deep Red Sea and the cold deep-water of the Gulf of Aden (see below).

The Red Sea deep-water mollusc fauna can be suitably clustered into three distinct groups of bathymetric significance (Janssen 1989). A first group of species appears more characteristic of a depth interval between about 400 and 600 m. Typical species are *Propeamussium steindachneri* (Sturany 1901), *Ilanga illustris* (Sturany 1900), *Nassarius steindachneri* (Sturany 1900) and *Conus grangeri* (Sowerby

Table 2 The most evenly distributed species

Rank	Rank (living)	Rank (abundance)	Species	No. of stations	% of stations	No. of stations with living specimens
1		1	*Limopsis elachista*	74	77	4
2	3	7	*Cuspidaria brachyrhynchus*	72	75	9
3		2	*Nassarius lathraius*	70	73	3
	5	3	*Cardiomya alcocki*			5
4	5	9	*Parvamussium siebenrocki*	67	70	5
5	4	4	*Cuspidaria steindachneri*	64	67	7
6		6	*Clathurella pertabulata*	62	64	0
7	1	8	*Cetoconcha intracta*	60	62	16
8	5	11	*Pseudoneaera thaumasia*	59	61	5
	4	12	*Bathyarca* sp.			7
9		19	*Cuspidaria dissociata*	56	58	2
10		5	*Jupiteria* sp.	55	57	2
11		21	*Haliris* sp.	54	56	0
12		14	*Costellaria casta*	53	55	0
13	2	10	*Bentharca asperula*	48	50	15
		20	*Rhinoclama* sp.			2
14		25	*Cardiomya* sp.	46	48	1

1900) (synonym: *Conus batheon* Sturany 1904). A second group occurs mainly between about 600 and 1,500 m, characterized by *Cuspidaria dissociata* Sturany 1901, *Visayaseguenzia compsa* (Melvill 1904), *Costellaria casta* (H. Adams 1872) and *Gadila* sp. A third group of species displays a wide bathymetric range, with taxa distributed between 400 and 2,000 m and deeper. The majority of species, including also the most abundant, belongs to this category. To this stock pertain *Jupiteria* sp., *Bathyarca anaclima* (Melvill and Standen 1907), *Limopsis elachista* Sturany 1901, *Parvamussium siebenrocki* (Sturany 1901), *Cetoconcha intracta* (Sturany 1901), *Pseudoneaera thaumasia* (Sturany 1901), *Cardiomya alcocki* (E.A. Smith 1894), *Cuspidaria brachyrhynchus* Sturany 1901, *Cuspidaria steindachneri* Sturany 1901, *Murex forskoehlii* (Röding 1798), *Nassarius lathraius* (Sturany 1900), *Fusinus bifrons* (Sturany 1900) and *Gemmula* aff. *monilifera* (Pease 1860). Only 10 species (e.g. *Parvamussium siebenrocki*, *Cardiomya alcocki*, *Cuspidaria steindachneri*, *Cuspidaria brachyrhynchus* and *Fusinus bifrons*) are recorded from a depth below 2,000 m, the deepest record being 2,120 m. However, it must be emphasized that these species groups do not represent well-defined associations, but only show a certain consistency in their abundance and distribution.

Based upon species abundance, Rützen-Kositzkau (1999) differentiated five species associations in the deeper water of the Red Sea:

1. *Jupiteria* sp. 1—*Limopsis elachista* association between 1,351 and 1,898 m
2. *Limopsis elachista*—*Nassarius lathraius* association between 343 and 646 m
3. *Ilanga illustris*—*Limopsis elachista* association between 237 and 567 m
4. *Nassarius lathraius*—*Cardiomya potti* association between 557 and 685 m
5. *Nassarius lathraius*—*Limopsis elachista* association between 308 and 1,410 m

However, these show a broad overlap of their depth ranges, plus a considerable number of species in common. Therefore, it has been already argued by Rützen-Kositzkau (1999: 88) that these presumed associations are better interpreted as reflecting faunal responses to some local sedimentary parameters (such as differing mineral contents and granulometry) rather than a bathymetric control. Our own observations on the occurrence of species show that nearly all the more common species occur together at the same stations over their whole depth range. At present, it can only be speculated whether a structured pattern perhaps might be detected by some sophisticated statistical analyses.

Any kind of bathymetric zonation or species associations necessarily would be based on differences in ecological parameters, such as bottom type, temperature and food supply. The rather uniform structure and conditions of the deep Red Sea environment, however, seem to speak against

Table 3 Distribution in depth

Species	Depth range in the Red Sea
Limopsis elachista	428–**490–1975**–2120
Nassarius lathraius	332–**507–1032**–1977
Cardiomya alcocki	428–**490–1975**–2120
Cuspidaria steindachneri	**363–778**–2120
Jupiteria sp.	428–**740–517**–1977
Clathurella pertabulata	428–2120
Cuspidaria brachyrhynchus	**363–1032**–2160
Cetoconcha intracta	**363–2120**
Parvamussium siebenrocki	56–**490–847**–2120
Bentharca asperula	**537–1135**–1528
Pseudoneaera thaumasia	**363–1175**–1972
Bathyarca sp.	**428–1424**–1977
Dimya sp.	740–**753–1135**–1377
Costellaria casta	58–**490–507**–1672
Gemmula aff. *monilifera*	212–**363–383**–1977
Xylophaga sp. 1	363–**519–600**–1975
Corbula sp.	487–1480
Leiocithara sp.	487–1977
Cuspidaria dissociata	428–**772–779**–1558
Rhinoclama sp.	578–**1972–1975**–1977
Haliris sp.	487–1977
Gadila sp.	487–1135
Fusinus bifrons	487–**696–1972**–2120
Lucidestes ? sp.	428–1128
Cardiomya sp.	363–**428–459**–2120

Records of living specimens in bold, based only on material collected by the above-mentioned German cruises

the possibility of forming a structured pattern of the mollusc fauna. In summary, there are no strong arguments thus far to substantiate a convincing bathymetric zonation of the deep Red Sea benthic mollusc fauna.

Chemosynthetic Habitats

The Red Sea is characterized by multiple situations which are suitable to sustain deep-sea chemosynthetic habitats and, therefore, associated chemosymbiotic fauna. For instance, this basin contains active submarine volcanoes, hydrocarbons, brine lakes and pools (e.g. Bonatti et al. 1984; Hovland and Judd 1988; Michaelis et al. 1990; Antunes et al. 2011; Ligi et al. 2011; Ligi et al., this volume; Schmidt et al., this volume). All such environments are potential sites for chemosynthesis and therefore specialized metazoan life. For example, the similarly semi-enclosed Mediterranean hosts a variety of parallel situations well exploited by chemosymbiotic molluscs (Taviani 2014). Admittedly, the shallower sill and stressing hydrologic conditions may interfere with the development of well-structured seep/vent communities in the Red Sea. Recently, Batang et al. (2012) reported on the very first discovery of active cold seeps (Tuwal Seeps) with associated macrofauna between 840 and 850 m off the central Red Sea continental margin. An unusual density of *Corbula* clams was noticed as well as another still undetermined bivalve. What is most remarkable is that salinity in the brine of the Tuwal Seeps attains 74 ‰. Furthermore, the occurrence of certain species such as the vent gastropod *Provanna* sp. and a rather large *Myrtea* sp. identified at very few stations (e.g. VA29, stations 740b, 741 at 1,480 and 1,465 m respectively) provides indications for the existence of widespread communities in the Red Sea. In fact, provannids are typical elements of hydrothermal habitats elsewhere in the oceans and lucinids, such as *Myrtea*, are known for their chemosymbiosis with sulphide oxidizing bacteria. These species are apparently associated with the "usual" non-vent common Red Sea deep-water species at stn VA 29-741, but this fact may only reflect contrasting mosaic-like adjacent microhabitats or pulsing times of fluid seepage.

Biogeographic Aspects

Due to the persisting situation of yet unresolved taxonomic questions, it is difficult to be conclusive on geographical relationships of the Red Sea deep-water species. However, among the most common species inhabiting this basin, there are several taxa which occur outside the Red Sea. Both *Limopsis elachista* and *Bathyarca anaclima* are recorded from 73 to 285 m in the Gulf of Oman (Melvill and Standen 1907), *Cardiomya potti* as originally described from the Red Sea by Sturany (1901) is synonymized with *C. alcocki* E.A. Smith (Poutiers and Bernard 1995) which is recorded as being widespread in the Indo-West Pacific in depths of 60–1,150 m. *Cuspidaria steindachneri*, too, is recorded with an Indo-West Pacific distribution and a depth range of 106–1,308 m. *P. thaumasia* is known also from the Indian Ocean (Poutiers and Bernard 1995). Not less than 10 of 16 pectinoidean species of the deep Red Sea occur even outside of that region (Dijkstra and Janssen 2013). The gastropod species *Gemmula* aff. *monilifera* and *Costellaria casta* are equally widely distributed in the Indo-West Pacific.

Propeamussium steindachneri, *Parvamussium siebenrocki*, *Cetoconcha intracta*, *Cuspidaria brachyrhynchus*, *Fusinus bifrons* and *Nassarius steindachneri* have for long been considered endemic to the Red Sea. They have been found in the Gulf of Aden (Rützen-Kositzkau 1999, own obs.), but never deeper than 500 m. Most of these species have been found at stn. 287 (in 472–479 m) in the inner Gulf of Aden under the hydrological influx of the Red Sea outflow (see below). This observation is important since,

Fig. 6 Characteristic deep-sea bivalves from the Red Sea. **a** *Propeamussium steindachneri* (383 m; *left valve*, h = 12.6 mm), **b** *Propeamussium steindachneri* (479 m, Gulf of Aden, from outflow of Red Sea water; *right valve*, h = 10.6 mm), **c** *Parvamussium siebenrocki* (1,377 m; *left valve*, h = 6. 4 mm, *right valve*, h = 4.9 mm), **d** *Cuspidaria steindachneri* (1,228 m; *left valve*, l = 22.8 mm, *right valve*, l = 23.5 mm), **e** *Cuspidaria brachyrhynchus* (1,995 m; *left valve*, l = 12.2 mm, *right valve*, l = 13.5 mm), **f** *Cardiomya alcocki* (487 m; *left valve*, l = 6.5 mm, *right valve*, l = 6.6 mm). (Photographs by S. Hof, Senckenberg Research Institute, Frankfurt a. M)

differently from the Red Sea, the Gulf of Aden cold deep-sea realm hosts a deep-water mollusc fauna sensu stricto (e.g. *Limopsis pelagica* E.A. Smith 1885, *Propeamussium alcocki* (E.A. Smith 1894), *P. caducum* (E.A. Smith 1885), *P. meridionale* (E.A. Smith 1885), *Amygdalum politum* (Verrill and Smith 1880), *Myonera caduca* (E.A. Smith 1894), *Stellaria testigera digitata* (Martens 1878)) and many deep-water turrid gastropods (Knudsen 1967; Sysoev 1996). Regarding diversity versus depth, there is a decrease in species numbers with increasing depth (Table 4). While numbers of species are comparable for both regions, the Gulf of Aden shows a slightly higher diversity. This is even more important if the very different numbers of stations is accounted for. Compared to the neighbouring Gulf of Aden as a part of the Indian Ocean, it is obvious that the semi-enclosed deep Red Sea only has a rather impoverished and not very diverse mollusc fauna. Although only 13 stations in the Gulf of Aden have been sampled (*Meteor* 5 expedition) in contrast to 97 stations in the Red Sea, these stations yielded a total of 148 species compared to 194 species in the Red Sea. On average, every station in the Gulf yielded 11.3 species in contrast to only 2.0 in the Red Sea (own obs., based on *Sonne, Valdivia* and *Meteor*-5 expeditions).

As said, there are, therefore, several species formerly deemed to be endemic to the deep Red Sea, e.g. *Fusinus bifrons, Nassarius steindachneri, Cuspidaria brachyrhynchus* and *Propeamussium steindachneri*. However, as has been shown above, all these species are now known to have a wider geographic distribution (such as *P. steindachneri*) at least in the Indian Ocean or have been found just outside the Red Sea proper in the neighbouring Gulf of Aden. This occurrence could be uniquely linked to the outflow of warm and salty Red Sea water into the Gulf of Aden rather than being regarded as geographically endemic (see also Türkay 1996). However, *Fusinus bifrons* in the meantime has been recorded from localities in the Indian Ocean (Snyder 2002) and is no longer Red Sea endemic. Another presumed endemic is the gastropod *Pseudotorinia yaroni* (Bieler 1993), but its true geographic range could be much wider than supposed since shells tentatively attributed to this elusive taxon are occasionally reported from the Philippines (unpublished data).

Fig. 7 Emblematic deep-sea gastropods from the Red Sea. **a** *Fusinus bifrons* (863 m; h = 79.3 mm), **b** *Nassarius steindachneri* (459 m; h = 22.9 mm), **c** *Nassarius lathraius* (487 m; h = 9.0 mm). (Photographs by S. Hof, Senckenberg Research Institute, Frankfurt a. M)

Results regarding deep-water crustacean decapods (Türkay 1986) agree fairly well with the pattern shown by Mollusca. Two main groups of species can be distinguished, that is, species common to both the Red Sea and the Indian Ocean (accounting for about one-third of the total), most of them being eurybathic and reaching greater depths in the Red Sea than in the Indian Ocean, and species only known at present from the Red Sea. Of these, however, for about half of the species it is not certain that they were true endemics. For the remaining approximately 30 %, very probable true endemism can be assumed. Concerning the ichthyofauna, Klausewitz (1986) assumed that a different degree of specific differentiation could be related to a different age of immigration to the Red Sea during the Pleistocene.

Türkay (1996) indicated a degree of endemism attaining at least 30 % of the total fauna. As for the mollusc fauna, due to the still unresolved taxonomy and systematic relationships of the greater part of the observed species at present, this conclusion cannot be confirmed. Of 142 taxa hitherto identified to species level, only 20 species, that is 14 % of the entire fauna, are considered endemics as far as is known. It is equally important to consider that no genomic studies have been made on Red Sea endemics to control their links (and co-specificity in some cases) with extra-basinal potential counterparts.

In general, the data presented here demonstrate the obvious distinctiveness of the mollusc fauna of the deep Red Sea which is in sharp contrast to the neighbouring Gulf of Aden regarding species diversity and composition. The character of this fauna must be explained as resulting from the peculiar physical factors constraining the benthic fauna of the Red Sea.

The Colonization of the Red Sea Deep-Sea

A major question regarding the Red Sea deep-water mollusc fauna (but it could be extrapolated to other groups as well) is the following. Is this fauna as a whole represented by viable (reproductive) Red Sea populations, or is it at least partly dependent upon its "mother" Indian Ocean? It is, in fact, possible that part at least of the present deep-water Red Sea mollusc fauna is not represented by stable and self-reproducing populations, but results from meroplanktic larval influx from the adjacent Gulf of Aden via shallow water inflow (Janssen and Taviani 1985; Taviani 1985). Therefore, it cannot be excluded that we are at times facing the existence of pseudopopulations, rather than viable permanent populations, a case similar to what has been postulated for the Mediterranean Sea by Bouchet and Taviani (1992). The type of larval development exhibited by the deep-water molluscs of the Red Sea has not yet been examined systematically and is unknown, at least for most of the bivalves. However, some data have been provided by Rützen-

Table 4 Diversity arranged by bathymetric ranges

Depth (m)	Red Sea: no. of species (living)	No. of stations	Gulf of Aden: no. of species (living)	No. of stations
400–800	27 (6)	34	30 (14)	2
800–1,200	18 (3)	28	20 (7)	3
1,200–1,600	8 (3)	21	16 (9)	3
1,600–2,000	12 (4)	12	6 (1)	2
>2,000	8 (1)	2	12 (8)	3

Comparison Red Sea–Gulf of Aden (for species numbers, only stations sampled with comparable gear are considered; number of species recorded living in brackets)

Kositzkau (1999) that documents that there is a rather high percentage of species with planktotrophic development among gastropods. This is in accordance with observations on deep-water faunas in other regions (Bouchet and Warén 1979; Bouchet 1976; Jablonski and Lutz 1980, 1983). Among the 25 commonest species of the gastropods found (Table 1), only *Fusinus bifrons* has a non-planktotrophic development. Obviously, this accounts for a limited distribution of that species in the Red Sea and the neighbouring parts of the Gulf of Aden and Indian Ocean. Among bivalves, Rützen-Kositzkau (1999) stated dominance of non-planktotrophic ontogeny which is in agreement with observations made by Knudsen (1967, 1970) and others. Since a consistent part of the non-planktotrophic species seems apparently distributed also in rather shallow environments (therefore in principle capable of crossing a shallow sill at the metamorphosed stage), we cannot exclude a later migration from shallow settings to the warm deeper habitats in the Red Sea.

However, among the 25 commonest species, at least *Limopsis elachista* and *Bathyarca anaclima* follow a planktotrophic development strategy, whereas species such as *Jupiteria* sp., *Bentharca asperula*, *Cardiomya alcocki* and *Pseudoneaera thaumasia* are non-planktotrophic. Nevertheless, at least *Bentharca asperula* and *Cardiomya alcocki* are widespread in the Indo-West Pacific region. A larval supply from the Indian Ocean can be admitted as a likely hypothesis for planktotrophic species. This could be an easy mechanism to facilitate re-colonization of the Red Sea basin by deep-water elements after glacial crises thought responsible for basin-wide faunal annihilation (Taviani 1998c, with references). This mechanism admittedly would not fully explain the occurrence of the putative endemics in the Red Sea fauna. Various, not mutually exclusive, scenarios can be envisaged. The first would basically question the existence of Red Sea endemics as a whole. Endemism would only be apparent, simply reflecting our incomplete knowledge of the deep Indian Ocean fauna due to a bias in sampling. Another option is that the peculiar stress conditions of the Red Sea are forcing adaptive speciation on aliquots of the Indian Ocean faunal stocks at a short temporal rate computable over the few thousand years of the Holocene (cf. the analogy with the Gulf of Gabès in the Mediterranean Sea: Sabelli and Taviani 2014). Thirdly, the level of hyperhaline conditions imprinting basin wide on the Red Sea water masses during glacial times has been overestimated by paleo-oceanographers and did not reach up to lethal conditions for the stenoecious benthos.

None of these hypotheses is supported by any conclusive evidence, requiring further comparative biological exploration of the deep fauna inside and outside the Red Sea as a fundamental prerequisite to unravel the dispersal patterns and evolutionary trends.

Final Remarks

As reported in this overview on deep-water molluscs, it is documented that the Red Sea hosts a relatively diverse, somewhat peculiar, deep-water fauna that shares some traits with its counterparts in the Indo-West Pacific.

Many questions still remain regarding the Red Sea deep-water benthos, as interpreted through its mollusc component:

Firstly, the taxonomic appreciation of many species is still unresolved since many taxa are known so far by their shells only and no biomolecular studies have been undertaken to check the Red Sea taxa versus their putative extra-basin occurrences to assess their specific validity.

Secondly, no experimental work has been done or is yet planned to test the planktotrophic hypothesis by intercepting meroplankton on both sides of Bab-al-Mandab and along the length of the Red Sea and checking the postulated presence of such trans-basinal deep-sea elements.

Thirdly, the fact that the Red Sea hosts some chemosymbiotic elements in its deep-water mollusc fauna warrants a dedicated exploration of its hydrothermal and potential cold seepage situations.

Moreover, in general, there is basic lack of information, and therefore of understanding, on the faunal metabolic requirements to cope with demands set by the extreme hydrologic conditions of the basin deep waters (cf. Seibel and Drazen 2007; Naumann et al. 2013). In such a perspective, the deep Red Sea is really a formidable laboratory where many important biological issues could be analysed and tested under natural conditions.

Acknowledgments We are grateful to masters, crew and colleagues onboard German and Italian research ships during the expeditions responsible for collecting the great majority of mollusc samples considered in this study. Thanks are due to Najeeb M.A. Rasul for the kind invitation to present our data at the workshop held in Jeddah in 2013 and for editorial handling. Sigrid Hof is thanked for the preparation of the figures. We acknowledge Henk Dekker, Bruno Sabelli and Hjalmar Thiel for their constructive criticism that greatly helped improving the quality of the paper. This article contributes to PRIN2012 Programme (Project 20125JKANY_002, P.I. M.Ligi) and is Ismar-CNR, Bologna scientific contribution n. 1824.

Appendix

Preliminary list of benthic mollusc species from the deep Red Sea (> 400 m) (compiled by the senior author)

Enumeration of species is based exclusively on reliable autochthonous records. Species recorded by fragments only or by single, often badly preserved specimens which obviously have been transported from shallow water environments are not taken into account.

Polyplacophora
Ischnochitonidae
Ischnochiton sp.

Bivalvia
Nuculidae
Nucula consentanea Melvill and Standen 1907
Manzanellidae
Nucinella sp.
Nuculanidae
Jupiteria sp.
Nuculana (Lembulus) sculpta (Issel 1869)
Crenellidae
Dacrydium sp.
Solamen vaillanti (Issel 1869)
Mytilidae
Amygdalum aff. *watsoni* (Smith 1885)
Brachidontes pharaonis (Fischer 1870)
Idas sp.
Modiolus sp.
Septiferidae
Septifer forskali (Dunker 1855)
Arcidae
Acar plicata (Dillwyn 1817)
Arca sp.
Barbatia sulcata (Lamarck 1819)
Bathyarca anaclima (Melvill and Standen 1907)
Bathyarca sp.
Bentharca asperula (Dall 1881)
Bentharca sp.
Ribriarca polycymoides (Thiele & Jaeckel 1931)
Limopsidae
Limopsis elachista Sturany 1901
Gryphaeidae
Parahyotissa numisma (Lamarck 1819)
Ostreidae
Ostrea deformis Lamarck 1819
Pteriidae
Pteria sp.
Malleidae
Malvufundus regulus (Forsskal 1775)
Pectinidae
Cryptopecten nux (Reeve 1853)
Delectopecten alcocki (E.A. Smith 1904)
Delectopecten musorstomi Poutiers 1981
Spondylidae
Spondylus proneri Lamprell & Healy 2001
Anomiidae
Pododesmus caelata (Reeve 1859)
Anomia achaeus Gray 1850
Dimyidae
Dimya sp.
Limidae
Ctenoides annulata (Lamarck 1819)
Limea pectinata H. Adams 1870
Entoliidae
Pectinella aequoris Dijkstra 1991
Propeamussiidae
Cyclopecten erythraeensis Dijkstra and Janssen 2013
Cyclopecten meteorae Dijkstra and Janssen 2013
Parvamussium formosum (Melvill and Standen 1907)
Parvamussium scitulum (E.A. Smith 1885)
Parvamussium siebenrocki (Sturany 1901)
Parvamussium thyrideum (Melvill and Standen 1907)
Propeamussium steindachneri (Sturany 1901)
Similipecten eous (Melvill and Standen 1907)
Carditidae
Cardites variegata Bruguière 1792
Lucinidae
Anodontia aff. *edentula* (Linnaeus 1758)
Codakia sp.
Ctena divergens (Philippi 1850)
Lamellolucina dentifera (Jonas 1846)
Lucinidae sp. 1
Lucinidae sp. 2
Myrtea ? sp.
Myrtea sp. 1
Myrtea sp. 2
Parvilucina sp.
Thyasiridae
Thyasira sp.
Cardiidae
Fulvia australis (Sowerby 1834)
Fragum nivale (Reeve 1845)
Papillocardium papillosum (Poli 1795)
Tellinidae
Angulus flaccus (Römer 1871)
Arcopella isseli (H. Adams 1871)
Moerella lactea (H. Adams 1871)
Psammotreta praerupta (Salisbury 1934)
Semelidae
Ervilia scaliola (Issel 1869)
Leptomyaria etesiaca (Hedley 1909)
Semele fragillima (Issel 1869)
Syndosmya cistula Melvill and Standen 1907
Solecurtidae
Azorinus coarctatus (Gmelin 1791)

Kelliellidae
Kelliella sp.
Ungulinidae
Diplodonta subrotundata Issel 1869
Veneridae
Paphia undulata (Born 1778)
Tapes cf. *deshayesi* (Hanley 1844)
Basterotiidae
Basterotia borbonica (Deshayes 1863)
Galeommatidae
Amphilepida ? sp.
Kellia ? sp.
Pholadidae
Xylophaga sp. 1
Xylophaga sp. 2
Teredinidae
Teredinidae sp. 1
Teredinidae sp. 2
Corbulidae
Corbula sp.
Varicorbula sp.
Cetoconchidae
Cetoconcha intracta (Sturany 1901)
Cuspidariidae
Cardiomya alcocki (Smith 1894)
Cardiomya sp.
Cuspidaria brachyrhynchus Sturany 1901
Cuspidaria corrugata Prashad 1932
Cuspidaria dissociata Sturany 1901
Cuspidaria hindsiana (A. Adams 1864)
Cuspidaria steindachneri Sturany 1901
Pseudoneaera minor Thiele & Jaeckel 1931
Pseudoneaera thaumasia Sturany 1901
Rhinoclama sp.
Parilimyidae
Panacca sp.
Verticordiidae
Haliris sp.
Euciroa sp.
Lyonsiidae
Lyonsia sp.

Scaphopoda

Dentaliidae
Graptacme sp.
Fustiariidae
Pseudantalis sp.
Gadilidae
Gadila sp.
Polyschides sp.

Gastropoda

Fissurellidae
Emarginula sp. 1
Emarginula sp. 2
Rimula cumingii A. Adams 1853
Zeidora (*Nesta*) *nesta* (Pilsbry 1891)
Zeidora calceolina A. Adams 1860
Lepetellidae
Lepetella simplicior (Melvill 1912)
Lepetella ? sp.
Anatomidae
Anatoma agulhasensis (Thiele 1925)
Anatoma japonica (A. Adams 1862)
Scissurellidae
Scissurella reticulata Philippi 1853
Scissurella rota Yaron 1983
Scissurella sp.
Sinezona singeri Geiger 2006
Sukashitrochus dorbignii (Audouin 1826)
Seguenziidae
Visayaseguenzia compsa (Melvill 1904)
Chilodontidae
Danilia sp.
Herpetopoma xeniolum (Melvill 1918)
Perrinia stellata (A. Adams 1864)
Solariellidae
Ilanga illustris (Sturany 1900)
Zetela mutabilis (Schepman 1908)
Trochidae
Fossarina mariei (Fischer 1890)
Pagodatrochus variabilis (H. Adams 1873)
Pseudostomatella papyracea (Gmelin 1791)
Areneidae
Arene echinacantha (Melvill & Standen 1903)
Turbinidae
Bolma sp.
Skeneidae
Cirsonella sp. 1
Cirsonella sp. 2
Leucorhynchia crossei Tryon 1888
Skeneidae sp. 1
Skeneidae sp. 2
Cocculinidae
Cocculinidae ? sp. 1
Cocculinidae ? sp. 2
Provannidae
Provanna ? sp.
Cerithiidae
Cerithium gloriosum Houbrick 1992
Turritellidae
Turritella sp.
Naticidae
Natica sp.
Iravadiidae
Ceratia sp.
Rissoidae

Lucidestea ? sp.
Rissoina sp.
Tornidae
Circulus octoliratus (Carpenter 1856)
Cochliolepis sp.
Teinostoma sp.
Bursidae
Bufonaria gnorima (Melvill 1918)
Bufonaria albivaricosa (Reeve 1844) ? (fide Sturany) = *gnorima* ?
Cassidae
Semicassis saburon (Bruguière 1792) (fide Sturany) = *faurotis* (Jousseaume 1888) ?
Tonnidae
Tonna galea (Linnaeus 1758)
Hipponicidae
Cheilea cicatricosa (Reeve 1858)
Xenophoridae
Stellaria solaris (Linnaeus 1764)
Epitoniidae
Epitonium cf. *scalare* (Linnaeus 1758)
Epitonium cultellicostatum (Boury 1913)
Epitonium deflersi (Jousseaume 1912) ?
Epitonium umbilicatum (Pease 1869)
Opalia bicarinata (Sowerby 1844)
Epitoniidae ? sp.
Eulimidae
Eulima (s.l.) sp. 1
Eulima (s.l.) sp. 2
Melanella sp. 1
Melanella sp. 2
Niso sp.
Triphoridae
Euthymella concors (Hinds 1843)
Inella sp.
Metaxia sp.
Triphora adamsi (Deshayes 1863)
Viriola tricincta (Dunker 1860)
Cerithiopsiidae
Cerithiopsis sp.
Newtoniellidae
Cerithiella ? sp.
Buccinidae
Pollia rubiginosa (Reeve 1846) (fide Sturany)
Colubrariidae
Colubraria cf. *tenera* (Gray 1839)
Columbellidae
Euplica festiva (Deshayes in Laborde 1834)
Mitrella erythraeensis Sturany 1900
Mitrella nomanensis (Sturany 1900)
Fasciolariidae
Fusinus bifrons (Sturany 1900)
Nassariidae
Nassarius lathraius (Sturany 1900)
Nassarius steindachneri (Sturany 1900)
Note There is controversy about the taxonomy and nomenclature of the *Nassarius* species described by Sturany. The view adopted here is that of the senior author, based on examination of type material. Further study is necessary to solve the problems.
Muricidae
Murex forskoehlii Röding 1798
Costellariidae
Costellaria casta (H. Adams 1872)
Cystiscidae
Granulina sp.
Mitridae
Mitra gonatophora Sturany 1903
Olividae
Ancilla eburnea (Deshayes 1830)
Ancilla lineolata (A. Adams 1853)
Ancilla cinnamomea Lamarck 1801 ? (fide Sturany)
Borsoniidae
Microdrillia circumvertens (Melvill and Standen 1901)
Tomopleura reevei (C.B. Adams 1850)
Clathurellidae
Clathurella pertabulata (Sturany 1903)
Cochlespiridae
Thatcheriasyrinx orientis (Melvill 1904)
Conidae
Conus grangeri G.B. Sowerby III 1900
Drilliidae
Clavus inchoatus (Sturany 1903)
Clavus siebenrocki (Sturany 1903)
Clavus sp. 1
Clavus sp. 2
Splendrillia sp. 1
Splendrillia sp. 2
Horaiclavidae
Paradrillia nannodes (Sturany 1903)
Mangeliidae
Antiguraleus sp.
Leiocithara sp.
Pseudoraphitoma ? sp.
Pseudoraphitoma kilburni Morassi & Bonfitto 2001
Pseudoraphitoma sp. 1
Pseudoraphitoma sp. 2
Pseudomelatomidae
Compsodrillia ? sp. 1
Compsodrillia ? sp. 2
Inquisitor ? sp.
Ptychobela cf. *flavidula* (Lamarck 1822)
Ptychobela sp.
Raphitomidae

Aliceia sp.
Daphnella sp.
Favriella sp.
Mioawateria sp.
Pseudodaphnella ? sp.
Rimosodaphnella ? sp.
Taranidaphne amphitrites (Melvill & Standen 1903)
Taranidaphne dufresnei Morassi & Bonfitto 2001
Teretia sp.
Veprecula vepratica (Hedley 1903)
Turridae
Gemmula aff. *monilifera* (Pease 1860)
Gemmula sp. 1
Gemmula sp. 2
Unedogemmula indica (Röding 1798)
Xylodisculidae
Xylodiscula sp.
Acteonidae
„*Acteon*“ sp. 1
„*Acteon*“ sp. 2
„*Acteon*“ sp. 3
Architectonicidae
Pseudotorinia yaroni Bieler 1993
Solatisonax acutecarinata (Thiele 1925)
Pyramidellidae
Chrysallida comacum (Melvill 1910)
Chrysallida pupula A. Adams 1861
Chrysallida sp.
Miralda gemma (A. Adams 1861)
Odostomella chorea (Hedley 1909)
Odostomia anabathmis Melvill 1910
Odostomia eutropia Melvill 1899
Odostomia sp.
Ondina? sp.
Oscilla appeliusi (Hornung & Mermod 1925)
Pyramidella pulchella (A. Adams 1854)
Pyramidellidae gen. sp. indet.
Pyrgulina comacea Melvill 1910
Pyrgulina tenerrima (Melvill 1906)
Syrnola massauensis (Hornung & Mermod 1924)
Syrnola sp. 1
Syrnola sp. 2
Syrnola sp. 3
Turbonilla sp. 1
Turbonilla sp. 2
Turbonilla sp. 3
Turbonilla sp. 4
Turbonilla sp. 5
Turbonilla sp. 6
Turbonilla sp. 7
Turbonilla sp. 8
Turbonilla sp. 9
Turbonilla sp. 10
Turbonilla sp. 11
Turbonilla sp. 12
Ringiculidae
Ringicula sp.
Haminoeidae
Haminoea sp.
Diniatys dentifer (A. Adams 1850)
Cylichnidae
Mnestia sp.
Roxania lithensis (Sturany 1903)
Philinidae
Phanerophthalmus smaragdinus (Rüppell & Leuckart 1830)
Philine sp.
Retusidae
Retusa sp.
Volvulella sp.

References

Abdel-Rahman AI (2006) A dual effect of upwelling and easterly jet stream on desert formation in southern and eastern parts of Yemen. In: Proceedings of the 2nd International Conference on Water Resources and Arid Environment, pp 1–22

Al Saafani MA, Shenoi SSC (2007) Water masses in the Gulf of Aden. J Oceanogr 63:1–14

Antunes A, Ngugi DK, Ulrich SU (2011) Microbiology of the Red Sea (and other) deep-sea anoxic brine lakes. Environ Microbiol Rep 3:416–433

Batang ZB, Papathanassiou E, Al-Suwailem A, Smith C, Salomidi M, Pethakis G, Alikunhi NM, Smith L, Mallon F, Yapici T, Fayad N (2012) First discovery of a cold seep on the continental margin of the central Red Sea. J Mar Syst 94:247–253

Bieler R (1993) Architectonicidae of the Indo-Pacific (Mollusca, Gastropoda). Abh Naturwiss Ver Hamburg (NF) 30:1–376

Bonaduce G, Ciliberto B, Minichelli G, Masoli M, Pugliese N (1983) The Red Sea benthic ostracodes and their geographical distribution. In: Maddocks RF (ed) Proceedings of the 8th International Symposium on Applications of Ostracoda, 26–29 July 1982, Department of Geosciences, University of Houston, pp 472–491

Bonatti E, Colantoni P, Della Vedova B, Taviani M (1984) Geology of the Red Sea transitional region (22°N–25°N). Oceanol Acta 7:385–398

Bouchet P (1976) Mise en évidence des stades larvaires planctoniques chez des gastéropodes prosobranches des étages bathyal et abyssal. Bull Mus Natl Hist Nat Paris 277:947–972

Bouchet P, Taviani M (1992) The Mediterranean deep-sea fauna: pseudopopulations of Atlantic species? Deep Sea Res 39:169–184

Bouchet P, Warén A (1979) Planktotrophic larval development in deep-water gastropods. Sarsia 64:37–40

Brunn A (1957) Deep sea and abyssal depths. Geol Soc Am Mem 67:641–672

Dayton PK, Hessler RR (1972) Role of biological disturbance in maintaining diversity in the deep-sea. Deep Sea Res 19:199–208

Dijkstra HH, Janssen R (2013) Bathyal and abyssal Pectinoidea from the Red Sea and Gulf of Aden (Bivalvia: Propeamussiidae, Entoliidae, Pectinidae). Arch Molluskenkunde 142:181–214

Edelman-Furstenberg Y, Scherbacher M, Hemleben C, Almogi-Labin A (2001) Deep-sea benthic foraminifera from the central Red Sea. J Foramin Res 31:48–59

Ekman S (1953) Zoogeography of the Sea. Sidgwick and Jackson, London, 440 pp

Foucher J-P, Westbrook GK, Boetius A, Ceramicola S, Dupré S, Mascle J, Mienert J, Pfannkuche O, Pierre C, Praeg D (2009) Structures and drivers of cold seep ecosystems. Oceanography 22:58–74

Fricke HW, Hottinger L (1983) Coral bioherms below the euphotic zone in the Red Sea. Mar Ecol Prog Ser 11:113–117

Fricke HW, Knauer B (1986) Diversity and spatial pattern of coral communities in the Red Sea upper twilight zone. Oecologia 71:29–37

Fricke HW, Schuhmacher H (1983) The depth limits of Red Sea stony corals: an ecophysiological problem (a deep diving survey by submersible). Mar Ecol 4:163–194

Gage JD (1978) Animals in deep-sea sediments. Proc R Soc Edinb 76B:77–93

Gage JD, Tyler PA (eds) (1991) Deep-Sea Biology: a natural history of organisms at the deep-sea floor. Cambridge University Press, Cambridge, 504 pp

Glover AG, Gooday AJ, Bailey DM, Billett DSM, Chevaldonné P, Colaço A, Copley J, Cuvelier D, Desbruyères D, Kalogeropoulou V, Klages M, Lampadariou N, Lejeusne C, Mestre NC, Paterson GLJ, Perez T, Ruhl H, Sarrazin J, Soltwedel T, Soto EH, Thatje S, Tselepides A, Van Gaever S, Vanreusel A (2010) Temporal change in deep-sea benthic ecosystems: a review of the evidence from recent time-series studies. Adv Mar Biol 58:1–95

Grassle JF (1991) Deep-sea benthic biodiversity. Bioscience 41:464–469

Gvirtzman G, Buchbinder B, Sneh A, Nir Y, Friedman GM (1977) Morphology of the Red Sea fringing reefs: a result of erosional pattern of the last-glacial low-stand sea level and the following Holocene recolonization. Mém Bureau Rech Géol Minières 89:480–491

Hessler RR, Sanders HL (1967) Faunal diversity in the deep-sea. Deep-Sea Res 14:65–78

Hovland M, Judd AG (1988) Seabed pockmarks and seepages: impact on geology, biology and the marine environment. Graham and Trotter, London, 239 pp

Jablonski D, Lutz RA (1980) Molluscan larval shell morphology: ecological and paleontological applications. Top Geobiol 1:323–377

Jablonski D, Lutz RA (1983) Larval ecology of marine benthic invertebrates: paleobiological implications. Biol Rev 58:21–89

Janssen AW (2007) Holoplanktonic Mollusca (Gastropoda) from the Gulf of Aqaba, Red Sea and Gulf of Aden (Late Holocene-Recent). The Veliger 49:140–195

Janssen R (1989) Preliminary report on the deep-water mollusk fauna of the Red Sea. In: Abstracts 10th international malacological congress, Tübingen, 115 pp

Janssen R, Taviani M (1985) Factors constraining the composition of the Red Sea mollusc fauna. In: Abstracts 4th deep-sea biology symposium, Hamburg, 23–29 June 1985

Judd AG, Hovland M (2007) Seabed fluid flow; impact on geology, biology, and the marine environment. Cambridge University Press, Cambridge, 492 pp

Klausewitz W (1986) Zoogeographic analysis of the vertical distribution of the deep Red Sea ichthyofauna, with a new record. Senckenb Marit 17(4/6):279–292

Knudsen J (1967) The deep-sea Bivalvia. The John Murray Expedition 1933–34 Scientific Reports 11:237–343, pls 1–3

Knudsen J (1970) The systematics and biology of abyssal and hadal Bivalvia. Galathea Report 11:241, pls 20

Koslow JA (2007) The silent deep: the discovery, ecology and conservation of the deep-sea. University of Chicago Press, Chicago, 288 pp

Lambeck K, Purcell A, Flemming NC, Vita-Finzi C, Alsharekh AM, Bailey GN (2011) Sea level and shoreline reconstructions for the Red Sea: isostatic and tectonic considerations and implications for hominin migration out of Africa. Quatern Sci Rev 30:3542–3574

Levin LA, Dayton PK (2009) Ecological theory and continental margins: where shallow meets deep. Trends Ecol Evol 24:606–617

Ligi M, Bonatti E, Taviani M (2011) L'Oceano del Faraone. Darwin 95:90–95

Ligi M, Bonatti E, Rasul N (this volume) Seafloor spreading initiation: geophysical and geochemical constraints from the Thetis and Nereus Deeps, central Red Sea

Marenzeller E von (1907) Expedition S.M. Schiff "Pola" in das Rote Meer, nördliche und südliche Hälfte 1895/96–1897/98 - Zoologische Ergebnisse 25; Tiefseekorallen. Denkschr mathem-naturwiss Cl k Akad Wissensch Wien 80:13–25

McClain CR, Rex MA, Etter RJ (2009) Patterns in deep-sea macroecology. In: Witman JD, Roy K (eds) Marine macroecology. University of Chicago Press, Chicago, pp 65–100

Melvill JC, Standen R (1907) The Mollusca of the Persian Gulf, Gulf of Oman and Arabian Seas as evidenced mainly through the collections of Mr. F.W. Townsend, 1893–1906, with descriptions of new species, Part 2. Pelecypoda. In: Proc zool Soc London (1906), 54, pp 783–848

Michaelis W, Jenisch A, Richnow HH (1990) Hydrothermal petroleum generation in the Red Sea sediments from the Kebrit and Shaban deeps. Appl Geochem 5:103–105

Morcos SA (1970) Physical and chemical oceanography of the Red Sea. Oceanogr Mar Biol—Annu Rev 8:73–202

Morcos SA, Abdallah AM (2012) Oceanography of the Gulf of Aden John Murray-Mabahiss expedition 1933–1934 revisited. Egypt J Aquat Res 38:77–91

Mühlenhardt-Siegel U (2008) The Cumacea of the Red Sea and Gulf of Aden, with the description of four new species and one genus. Zootaxa 1828:1–17

Naumann MS, Orejas C, Ferrier-Pages C (2013) High thermal tolerance of two Mediterranean cold-water coral species maintained in aquaria. Coral Reefs 32:749–754

Pfannkuche O (1993) Benthic standing stock and metabolic activity in the bathyal Red Sea from 17°N to 27°N. Mar Ecol 14:67–79

Poutiers J-M, Bernard FR (1995) Carnivorous bivalve molluscs (Anomalodesmata) from the tropical western Pacific Ocean, with a proposed classification and a catalogue of Recent species. In: Bouchet P (ed) Résultats des Campagnes Musorstom, vol 14. Mém Mus natl Hist nat Paris, 167, pp 107–187

Quadfasel D (2001) Red Sea circulation. In: Steele JH, Thorpe SA, Turekian KK (eds) Encyclopedia of ocean sciences. Academic Press, San Diego, pp 2366–2376

Qurban MA, Krishnakumar PK, Joydas TV, Manikandan KP, Ashraf TTM, Quadri SI, Wafar M, Qasem A, Cairns SD (2014) *In-situ* observation of deep water corals in the northern Red Sea waters of Saudi Arabia. Deep-Sea Research Part I, doi:10.1016/j.dsr.2014.04.002

Rex MA (1983) Geographic patterns of species diversity in the deep-sea benthos. In: Rowe GT (ed) The Sea 8. Wiley, New York, pp 453–472

Rex MA, Etter RJ (2010) Deep-Sea biodiversity: pattern and scale. Harvard University Press, Cambridge, 354 pp

Rex MA, Crame JA, Stuart CT, Clarke A (2005a) Large-scale biogeographic patterns in marine mollusks: a confluence of history and productivity? Ecology 86:2288–2297

Rex MA, McClain CR, Johnson NA, Etter RJ, Allen JA, Bouchet P, Waren A (2005b) A source-sink hypothesis for abyssal biodiversity. Am Nat 165:163–178

Roder C, Berumen ML, Bouwmeester J, Papathanassiou E, Al-Suwailem A, Voolstra CR (2013) First biological measurements

of deep-sea corals from the Red Sea. Scientific Reports 3, Article number 2802
Rowe GT (ed) (1983) Deep-sea biology. Wiley, New York, 560 pp
Rützen-Kositzkau B von (1999) Taphonomie und Biogeographie des hartteiltragenden Makrobenthos im Tiefwasser des Roten Meeres. Beringeria 24:1–150
Ruhl HA, Smith KL Jr (2004) Shifts in deep-sea community structure linked to climate and food supply. Science 305:17006–17011
Sabelli B, Taviani M (2014) The making of the Mediterranean molluscan biodiversity. In: Goffredo S, Dubinsky Z (eds) The mediterranean sea: its history and present challenges. Springer, Dordrecht, pp 285–306
Sanders HL, Hessler RR (1969) Ecology of the deep-sea benthos. Science 163:1419–1424
Schmidt M, Al-Farawati R, Botz R (this volume) Geochemical classification of brine-filled Red Sea deeps
Seibel BA, Drazen JC (2007) The rate of metabolism in marine animals: environmental constraints, ecological demands and energetic opportunities. Philos Trans R Soc B 362:2061–2078
Siddall M, Rohling EJ, Almogi-Labin A, Hemleben C, Meischner D, Schmelzer I, Smeed DA (2003) Sea-level fluctuations during the last glacial cycle. Nature 423:853–858
Snyder AA (2002) Fusinus dovpeledi, a new species (Gastropoda: Fasciolariidae) from the Red Sea, and range extension for two other species. Nautilus 116:56–58
Sturany R (1900a) Diagnosen neuer Gastropoden aus dem Rothen Meere, als Vorläufer einer Bearbeitung der gesammten von S.M. Schiff "Pola" gefundenen Gastropoden. Anz k Akad Wissensch Wien, mathem-naturwiss Classe 37:197–201
Sturany R (1900b) Diagnosen neuer Gastropoden aus dem Rothen Meere, als Vorläufer einer Bearbeitung der gesammten von S.M. Schiff "Pola" gefundenen Gastropoden. (Fortsetzung.). Anz k Akad Wissensch Wien, mathem-naturwiss Classe 37:208–212
Sturany R (1901) Expedition S.M. Schiff "Pola" in das Rothe Meer, nördliche und südliche Hälfte. 1895/96 und 1897/98. Zoologische Ergebnisse XIV. Lamellibranchiaten des Rothen Meeres. Denkschr mathem-naturwiss Cl k Akad Wissensch Wien 69:255–295, pls 1–7
Sturany R (1904) Expedition S.M. Schiff "Pola" in das Rothe Meer, nördliche und südliche Hälfte. 1895/96 und 1897/98. Zoologische Ergebnisse XXIII. Gastropoden des Rothen Meeres. Denkschr mathem-naturwiss Cl k Akad Wissensch Wien 74:209–283, pls 1–7
Sysoev AV (1996) Deep-sea conoidean gastropods collected by the John Murray Expedition, 1933–34. Bull Nat Hist Mus Lond (Zool) 62(1):1–30
Taviani M (1985). The planktotrophic larval development: A strategy enabling deep-sea benthos to colonize epicontinental seas separated by shallow sills. In: Abstracts 4th Deep Sea Biology Symposium, Hamburg, 23–29 June 1985
Taviani M (1994) The ever changing climate: late Quaternary paleoclimatic modifications of the Red Sea region as deduced from coastal and deep-sea geological data. In: Proceedings of the Egyptian-Italian seminar "on geosciences and archeology in the Mediterranean countries", Geol Surv Egypt Spec Publ 70:193–200
Taviani M (1998a) Axial sedimentation of the Red Sea Transitional Region (22°–25°N): pelagic, gravity flow and sapropel deposition during the late Quaternary. In: Purser BH, Bosence DWJ (eds) Sedimentation and tectonics of rift basins: Red Sea-Gulf of Aden. Chapman and Hall, London, pp 467–478
Taviani M (1998b) Post-Miocene reef faunas of the Red Sea: glacio-eustatic controls. In: Purser BH, Bosence DWJ (eds) Sedimentation and tectonics of rift basins: Red Sea-Gulf of Aden. Chapman and Hall, London, pp 574–582
Taviani M (1998c) Stable tropics not so stable: climatically-driven extinctions of reef-associated molluscan assemblages (Red Sea and Western Indian Ocean; Last Interglaciation to Present). In: Camoin GF, Davis G (eds) Reefs and Carbonate platforms in the Pacific and Indian oceans, vol 25. Special Publication International Association Sedimentologists, pp 69–76
Taviani M (2011) The deep-sea chemoautotroph microbial world as experienced by the Mediterranean metazoans through time. In: Reitner J, Qûeric N-V, Arp G (eds) Advances in Stromatolite Geobiology, vol 2131. Lecture notes in earth sciences, Springer, Heidelberg, pp 277–295
Taviani M (2014) Marine chemosynthesis in the Mediterranean Sea. In: Goffredo S, Dubinsky Z (eds) The Mediterranean Sea: Its history and present challenges. Springer, Dordrecht, pp 69–83
Taviani M, López Correa M, Zibrowius H, Montagna P, McCulloch M, Ligi M (2007) Last Glacial deep-water scleractinian corals from the Red Sea. Bull Mar Sci 81:361–370
Thiel H (1975) The size structure of the deep-sea benthos. Internationale Rev ges Hydrobiol Hydrographie 60:575–606
Thiel H (1979) First quantitative data on Red Sea deep benthos. Mar Ecol Prog Ser 1:347–350
Thiel H (1980) Benthic investigations of the deep Red Sea. Cruise Reports: R.V. "Sonne"– Meseda I (1977), R.V. "Valdivia" – Meseda II (1979). Cour Forschungsinst Senckenberg 40:1–35
Thiel H (1989) Structural aspects of the deep-sea benthos. Ambio Spec Rep 6:25–31
Thiel H, Pfannkuche O, Theeg R, Schriever G (1987) Benthic metabolism and standing stock in the Central and Northern deep Red Sea. Mar Ecol 8:1–20
Tunnicliffe V (1991) The Biology of hydrothermal vents: ecology and evolution. Oceanogr Mar Biol 29:319–407
Türkay M (1986) Crustacea Decapoda Reptantia der tiefsee des Roten Meeres. Senckenb Marit 18:123–185
Türkay M (1996) Composition of the deep Red Sea macro- and megabenthic invertebrate fauna. Zoogeographic and ecological implications. In: Uiblein F, Ott J, Stachowitsch M (eds) Deep-sea and extreme shallow-water habitats: affinities and adaptations, Biosystematics and Ecology series, 11. pp 43–59
Tyler PA (1988) Seasonality in the deep-sea. Oceanogr Mar Biol Annu Rev 26:227–258
Van Dover CL (2000) The Ecology of deep-sea hydrothermal vents. Princeton University Press, New Jersey, pp 1–424
Werner F, Lange K (1975) A bathymetric survey of the sill area between the Red Sea and the Gulf of Aden. Geol Jahrb D13:125–130
Wilson GDF, Hessler RR (1987) Speciation in the deep sea. Annu Rev Ecol Syst 18:185–207
Young CM, Eckelbarger KJ (1994) Reproduction, larval biology, and recruitment of the deep-sea benthos. Columbia University Press, New York, pp 1–336

Sea Slugs: Unexpected Biodiversity and Distribution

Nathalie Yonow

Abstract
Sea slugs have been making a comeback in recent years, with international research producing papers on their taxonomy, biochemistry, biology, and biogeography. Several large expeditions have been made to the western Pacific, resulting in numerous papers and books. Sadly, the Red Sea has not benefitted from all this funding, and a Google Scholar search reveals only the older papers, all of which have already been covered in my book (Yonow 2008a) and last paper (Yonow 2008b). The book generated much interest and to date, a further 73 species have been reliably recorded from the Red Sea. The original lists of species with their zoogeographical distributions are updated with these records and corrections, and are provided in this chapter. Some previously unnamed species now have names which, in many cases, reflect known distributions; others have no names, but photographic records have been confirmed from elsewhere, and more research needs to be done to see whether any of the older names apply or whether they are indeed new species, but at least they are well recorded. Finally, many really are species new to science, which are known either to be endemic to the Red Sea or to have wider distributions, while a small group are so complex that the literature and photographic database are insufficient to determine their distributions. This chapter will present some of the species "discovered" since 2007, with comments on their probable identifications and distribution implications. A discussion of changes in distribution patterns and endemism over the last seven years concludes the chapter.

Introduction

Sea slugs are marine gastropod molluscs which, through evolution, have lost their shells in the more advanced lineages. This loss of shell has been driven by an increase in active methods of defence—biological, mechanical, chemical, as well as physical. The latest works covering the Red Sea opisthobranch fauna were published in tandem, each with tables and figures which can be used together, with summaries of almost every aspect currently known about these molluscs (Yonow 2008a, b). Opisthobranchs have been making a comeback in the last years, with funded international research producing papers on their taxonomy, biochemistry, biological products, biogeography, and molecular genetics. Sadly, the Red Sea has not benefitted from this new wave of study, and a literature search specifically for this chapter reveals only the older papers, all of which have already been covered in the two references cited above. The newer papers covering natural products of single species are not reviewed here.

Much recent research in the western Pacific has focused on higher systematics using molecular analyses. As a result, the name Opisthobranchia has become obsolete (Schrödl et al. 2011) and we will be dependent upon their common name, sea slugs, to define this particular group of marine animals. The validity of Opisthobranchia as a monophyletic taxon has been under discussion for the last few decades, suggesting that both the Opisthobranchia and the Pulmonata are artificial assemblages of marine slugs or snails with or

N. Yonow (✉)
Department of Biosciences, Swansea University, Singleton Park, Swansea, Wales SA2 8PP, UK
e-mail: n.yonow@swansea.ac.uk

N.M.A. Rasul and I.C.F. Stewart (eds.), *The Red Sea*,
Springer Earth System Sciences, DOI 10.1007/978-3-662-45201-1_30,

without shells, which show tendencies of shell reduction and internalisation, and have a detorted, externally bilaterally symmetrical body, usually with at least one pair of head tentacles or a head shield. Certainly, there has been a long-standing discussion around the placement of several groups, for example the pyramidellids (parasitic species), the primitive shelled *Acteon* (with both a heavy shell and an operculum) and some related Acteonoidea (now completely removed from the sea slugs), and the tiny sand-dwelling Acochlidia (more closely related to the Pulmonata than previously thought). Indeed, the genus *Akera* was alternately placed in the Anaspidea (sea hares) and the Cephalaspidea (head-shield slugs), while the Pteropoda are now thought to be more closely related to the Anaspidea than the Acochlidia or Pulmonata. A review of these theories and how they have changed over the last three centuries was published recently (Wägele et al. 2013).

One may think it difficult to write a chapter on recent research on the sea slug fauna of the Red Sea when it is lacking; however, the book "Sea Slugs of the Red Sea" (Yonow 2008a) generated much interest, and several underwater photographers have since reviewed their slide collections and scanned all opisthobranchs from this region, while other photographers have been actively looking for sea slugs. One particular diver/collector has so engaged with these animals that he has systematically photographed and collected many species that were not identified in the book. Identification of these photographic records has resulted in a further 73 species being newly and reliably recorded from the Red Sea, as well as providing a number of name changes and updates in zoogeographical distributions (see Table 1). Some previously unnamed species were indeed new to science and have now been assigned names that also reflect known distributions; others are still without a name, but photographic records have been confirmed from elsewhere, and more research needs to be completed to see whether older names apply or whether they are new species, but at least some of these are recognised and can be assigned distributions. Finally, many really do appear to be species that are new to science, either known to have wider distributions or known to be endemic to the Red Sea, while a small group are so complex that the literature and photographic database are insufficient to determine their nomenclature or their distributions at this present time. Only the species which are new records or have name or distribution changes from the lists provided in Yonow (2008a, b) are summarised in Table 1.

Only a selection of species will be discussed in detail here, with a graphic representation of all the preliminary data. Of all the newly recorded species, only five species are non-nudibranchs, one belonging to the Pleurobranchida and four to the Sacoglossa. The remaining 68 new records all belong to the Nudibranchia, one of the more morphologically advanced orders of sea slugs. The graph (Fig. 1) shows that the general distribution of zoogeographical affinities has changed somewhat in the last five years; fewer species in the Red Sea are considered as endemic (26 %), followed by a reduction in numbers to 24 % of species with distributions limited to the Indo-West Pacific. One category has shown a dramatic increase since the previous analyses, that of species occurring in the wider Indo-Pacific. This large increase in Indo-Pacific records is mainly due to recent reliable records provided from Hawaii (http://seaslugsofhawaii.com/general/species-list.html), and not because so many new species have been recorded. Many Red Sea, western Indian Ocean, and Indo-West Pacific species have now been recorded from Hawaii, which have initiated small changes in the ratios of these other regions in the chart. One conclusion that can be drawn from these preliminary results is that once species have been flagged, either identified in scientific works or published on reliable internet sites simply as *Genus* sp., their distributions are proving to be wider than previously thought.

Of the Mediterranean species, the small changes are due to only one species, *Pyrunculus fourierrii* (Audouin, 1826), being reassigned from its previously unknown distribution, and a new species record, *Kaloplocamus ramosus* (Cantraine, 1835), which is newly recorded from photographs only (Fig. 2). These are two of only four Mediterranean species that have apparently migrated through the Suez Canal (i.e., an anti-Lessepsian migration), in the opposite direction of the majority of migrating species (Lessepsian migration; currently, 18 sea slugs have travelled from the Red Sea to the Mediterranean Sea (http://www.ciesm.org/atlas/appendix3.html). It is thought that the spread of this Mediterranean species of *Kaloplocamus* throughout the Indo-Pacific is due to shipping, presumably since its prey is a species of fouling bryozoan. It is, of course, equally or more probable that this species has entered the Red Sea from the Indo-Pacific and not the Mediterranean; however, it is unlikely that the occurrence of this species is due simply to natural range extension through the Suez Canal.

Pleurobranchida

Berthella martensi (Pilsbry, 1896) is a widely distributed Indo-Pacific species which is present in almost every general geographic sea slug collection; incredibly, it had not been recorded previously from the Red Sea. *Berthella* belongs to a non-nudibranch group which always has its single gill on the right side under the mantle, rolled rhinophores on the head, a trapezoidal oral veil, and sometimes a small internal shell; *B. martensi* is known to feed on sponges, but there is no detailed work on its feeding habits. The species can vary tremendously in colour in the Indo-Pacific, from cream with

Table 1 List of Red Sea species which have been renamed, have revised distributions, or are new records since 2008

The version published here includes only the changes to the lists of published records of all species recorded in the Red Sea (Yonow 2008a, b). This list includes the newly recorded species as well as those species on the original lists which have any changes (in identification, distribution, nomenclature, etc.). The list has been colour coded to show endemic species and geographical distributions of each species. The genera in blue are the genera newly recorded for the Red Sea. The species are listed in the same generic order as in the full lists to enable easy comparison.

KEY : **Endemic species.** **Western Indian Ocean.** **Indian Ocean.** **Indo-West Pacific.** **Indo-Pacific.** **West Pacific.** **Circumtropical.** **Mediterranean.**

CEPHALASPIDEA	
Tornatina inconspicua **(H. Adams, 1852)**	
Tornatina simplex **(A. Adams, 1850)**	
Pyrunculus fourierrii (Audouin, 1826)	**Mediterranean**
Chelidonura cf. *inornata* Baba, 1949	**W Pacific Ocean**
Philinopsis ctenophaga Gosliner, 2011	**W Pacific Ocean**
Philinopsis speciosa (Pease, 1860)	**Indo-West Pacific**
Philinopsis falciphallus Gosliner, 2011	**W Pacific Ocean**
Odontoglaja mosaica Gosliner, 2011	**W Indian Ocean**
Haminoea pemphis **(Philippi, 1847)**	
Bulla arabica Malaquias & Reid, 2008	**W Indian Ocean**
ANASPIDEA	
Aplysia argus Rüppell & Leuckart, 1828	**Indo-West Pacific**
PLEUROBRANCHIDA	
Berthella martensi (Pilsbry, 1896)	**Indo-Pacific**
SACOGLOSSA	
Lobiger *viridis* Pease, 1863	**Mediterranean (+ Indo-Pacific)**
Elysia mercieri Pruvot-Fol, 1930	**W Pacific Ocean**
Costasiella *kuroshimae* Ichikawa, 1993	**Indo-West Pacific**
Placida *cremoniana* (Trinchese, 1892)	**Circumtropical**
NUDIBRANCHIA	
Doridina - Phanerobranchs	
Trapania euryeia Gosliner & Fahey, 2008	**Indo-Pacific**
Trapania palmula Gosliner & Fahey, 2008	**W Pacific Ocean**
Okenia rhinorma Rudman, 2007	**Indo-West Pacific**
Gymnodoris okinawae Baba, 1936	**Indo-Pacific**
Gymnodoris **sp. 1**	
Gymnodoris **sp. 2**	
Polycera aff. *melanosticta* Miller, 1996	**Indo-West Pacific**
Kaloplocamus *ramosus* (Cantraine, 1835)	**Mediterranean (+ Indo-Pacific)**
Aegires **sp. nov.**	
Notodoris *citrina* Pruvot-Fol, 1930	**Indo-West Pacific**
Doridina - Cryptobranchs	
Alliodoris sp.	unknown
Hallaxa / Geitodoris sp.	unknown
Discodoris caerulescens (Bergh, 1888)	**Indo-West Pacific**
Discodoris cebuensis Bergh, 1877	**Indo-West Pacific**
Discodoris sp.	unknown
Paradoris **sp. 1**	
Paradoris **sp. 2**	
Peltodoris murrea (Bergh, 1905)	**Indo-West Pacific**
Peltodoris rubra Eliot, 1904	**Indian Ocean**
Aldisa sp. nov.	**W Indian Ocean**
Atagema **sp.**	
Carminodoris estrelyado Gosliner & Behrens, 1998	**Indo-West Pacific**
Doris sp. 1	unknown
Doris sp. 2	unknown
Doris sp. 3	unknown

(continued)

Table 1 (continued)

Sclerodoris sp. 1	unknown
Sclerodoris sp. 2	unknown
Asteronotus mimeticus Gosliner & Valdes, 2002	**W Pacific Ocean**
Asteronotus spongicolus Gosliner & Valdes, 2002	**Indo-West Pacific**
Jorunna labialis Eliot, 1908	**W Indian Ocean**
Jorunna parva (Baba, 1938)	**Indo-West Pacific**
Jorunna sp.	unknown
Jorunna / Rostanga sp.	unknown
Rostanga *lutescens* (Bergh, 1905)	**Indo-Pacific**
Thordisa *tahala* Chan & Gosliner, 2007	**Indo-West Pacific**
Platydoris pulchra Eliot, 1904	**Indian Ocean**
Platydoris inframaculata (Abraham, 1877)	**Indo-West Pacific**
Diversidoris *aurantionodulosa* Rudman, 1987	**Indo-West Pacific**
***Glossodoris* sp. nov.**	
Doridina - Porostomes	
Dendrodoris albopurpura Burn, 1957	**Indo-Pacific**
Dendrodoris denisoni (Angas, 1864)	**Indo-Pacific**
***Phyllidiopsis* sp.**	
Phyllidiella cf. *zeylanica* (Kelaart, 1859)	**Indian Ocean**
Dendronotina	
***Tritonia* sp.**	
Tritoniopsis sp.	**Indo-West Pacific**
Tritoniid gen. nov. & sp. nov.	
Bornella cf. *johnsonorum* Pola et al., 2009	**Indo-Pacific**
Lomanotus sp. nov. 1	**W Pacific Ocean**
Lomanotus sp. nov. 2	**W Pacific Ocean**
***Lomanotus* sp. nov. 3**	
Crosslandia viridis Eliot, 1902	**Indo-West Pacific**
Melibe viridis Kelaart, 1858	**Indo-West Pacific**
Arminina	
Dermatobranchus cf. *caeruleomaculatus* Gosliner & Fahey, 2011	**Indo-West Pacific**
Dermatobranchus cf. *leoni* Gosliner & Fahey, 2011	**W Pacific Ocean**
Janolus mirabilis Baba & Abe, 1970	**Indo-Pacific**
Janolus savinkini Martynov & Korshunova, 2012	**Indo-West Pacific**
Janolus sp. nov.	unknown
Madrella gloriosa Baba, 1949	**Indo-West Pacific**
Aeolidina	
***Facelina* sp. nov. 1**	
***Facelina* sp. nov. 2**	
***Facelina* sp. nov. 3**	
Aeolidiopisis cf. *ransoni* Pruvot-Fol, 1956	**W Pacific Ocean**
Limenandra barnosii Carmona et al., 2013	**W Pacific Ocean**
Limenandra confusa Carmona et al., 2013	**Indo-Pacific**
Limenandra fusiformis (Baba, 1949)	**W Pacific Ocean + Oman**
Limenandra rosanae Carmona et al., 2013	**Pacific Ocean**
Protoaeolidiella *atra* Baba, 1955	**W Pacific Ocean**
Pleurolidia *juliae* Burn, 1966	**Indo-West Pacific**
***Phidiana* sp.**	
Favorinus mirabilis Baba, 1955	**Indo-West Pacific**
Phyllodesmium cf. *koehleri* Burghardt et al., 2008	**W Indian Ocean**
Cuthona sp. nov. 1	**Indo-West Pacific**
***Cuthona* sp. nov. 2**	
Trinchesia *yamasui* (Hamatani, 1993)	**Indo-West Pacific**
Cratena aff. *affinis* (Baba, 1949)	**Indo-West Pacific**
Herviella *albida* Baba, 1966	**W Pacific Ocean**
***Eubranchus* sp.**	

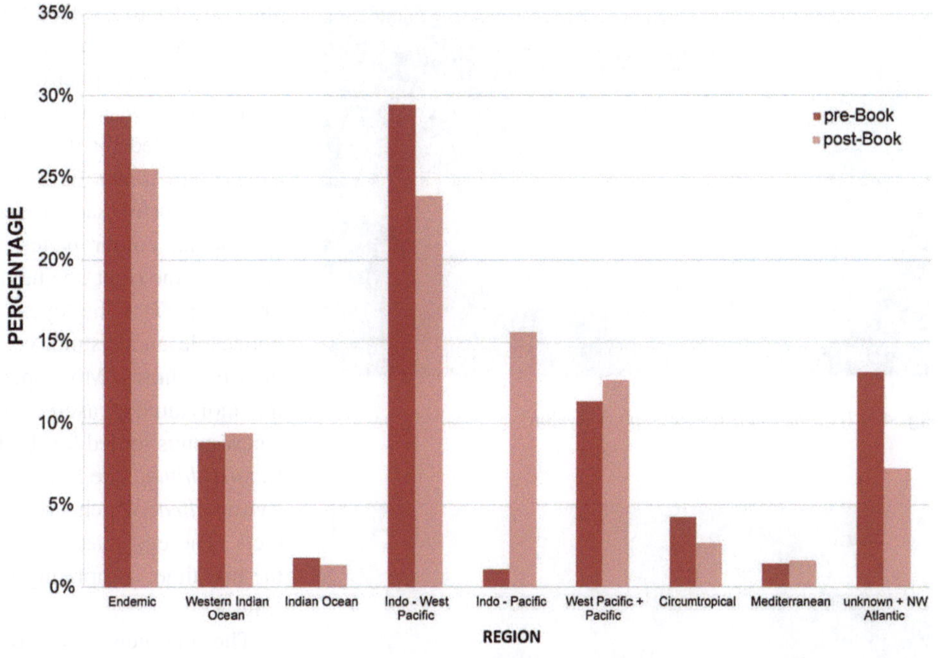

Fig. 1 Comparison of the geographical distributions of all species before and after 2008

small brown spots to vivid orange with large brown patches or pustules. The three separate sections of the mantle (which can be seen clearly in this photograph) can be individually autotomised in self-defence (Fig. 3).

Sacoglossa

Elysia mercieri Pruvot-Fol, 1930 is a small sacoglossan rarely recorded even from its native region, the western Pacific (Fig. 4). The Sacoglossa are a group of highly evolved sea slugs which only feed on algae, their radula having been modified to contain a single row of dagger-like teeth with which they stab algal cells; they then suck out the contents with a strong muscular pharynx. Morphologically, they are extremely varied, and some species retain a shell.

Lobiger viridis Pease, 1863 (Fig. 5) feeds on the alga *Caulerpa* throughout the Indo-Pacific and unfurls its four parapodial tentacles to ward off predators: on their inner surfaces, they are brightly coloured and, if persistently disturbed, are cast off and continue to wriggle, distracting the predator from the animal itself. This is the first time a species of *Lobiger* has been recorded from the Red Sea.

Costasiella kuroshimae Ishikawa, 1993, originally described from Okinawa, southern Japan and subsequently recorded throughout the Indo-West Pacific, is now recorded from the Red Sea (Fig. 6). However, it is probably not an

Fig. 2 *Kaloplocamus ramosus* (photograph E Bojanowski)

Fig. 3 *Berthella martensi* (photograph S Kahlbrock)

Fig. 4 *Elysia mercieri* (photograph S Ayalon)

Fig. 5 *Lobiger viridis* (photograph S Ayalon)

alien species to the Red Sea, but is so small that it has most likely been overlooked: its algal turf habitats are not a priority for photographer/divers. Indeed, this individual and the three other new sacoglossan records were found in clumps of algae collected for the local aquarium in Eilat, Israel. *Costasiella kuroshimae* is a striking if small species, one of the more colourful sacoglossans; the head detail shows the spectacular colour pattern and the distinctive black-tipped tentacles, the right one having been damaged and grown two tips (Fig. 7). The eyes of *Costasiella* are immediately noticeable and distinctive, set very close together between the rhinophores. Most species are less than 5 mm long, but this individual measured just more than 6 mm. *Costasiella* is a new genus record for the Red Sea. Many species, including *C. kuroshimae*, are limited to one food source, the green alga *Avrainvillea*, which grows in sandy patches between coral reefs. The egg mass of *C. kuroshimae* is recorded here for the first time, comprising four flattened whorls of white eggs (Fig. 8).

The final new sacoglossan record is *Placida cremoniana* (Trinchese, 1892), currently recorded as circumtropical, but originally described from the Mediterranean Sea (Fig. 9). This genus is newly recorded in the Red Sea; however, it is not yet known whether there is one single species which has been distributed around the world by shipping a very long time ago, or whether it comprises different species with more limited distributions. The current advances in molecular phylogenetics are breaking down circumglobal species into geographically distinct species (Alexander and Valdés 2013; Churchill et al. 2014). Certainly, some species will be circumtropical in distribution, as their food sources have been shown to have been recorded around the world.

Fig. 6 *Costasiella kuroshimae* (photograph S Ayalon)

Fig. 7 *Costasiella kuroshimae* (photograph S Ayalon)

Fig. 8 *Costasiella kuroshimae* (photograph S Ayalon)

Fig. 9 *Placida cremoniana* (photograph S Ayalon)

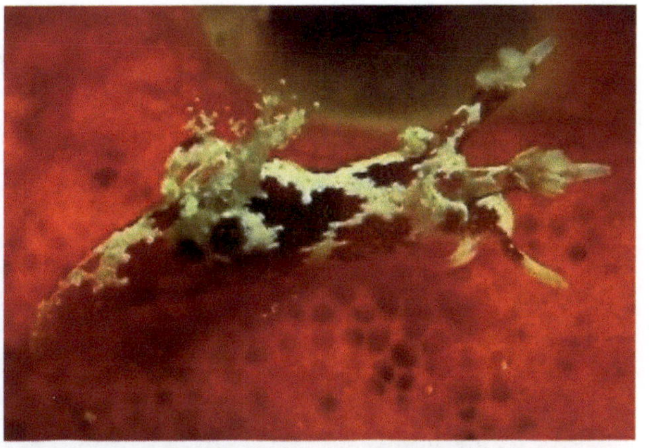

Fig. 10 *Trapania euryeia* (photograph S Kahlbrock)

Fig. 11 *Aegires* sp. nov. (photograph S Kahlbrock)

Nudibranchia—Doridina

The genera *Trapania*, *Aegires*, *Notodoris*, and *Gymnodoris* all belong to the group of dorids called phanerobranchs. These are usually elongated nudibranchs in which the gills and rhinophores do not retract into dorsal pockets, amongst other characters, and are sometimes protected by extensions of the mantle, as in the first three species. *Trapania euryeia* Gosliner and Fahey, 2008 was originally described from both the Indian and Pacific oceans and is now proving widespread in its distribution (Fig. 10). All species of *Trapania* are found in sponge complexes upon which their microscopic prey lives, the tiny filter-feeding animals belonging to the phylum Kamptozoa (Entoprocta). *Aegires* sp. nov. has elevated processes in the form of tubercles, protecting the vulnerable rhinophores and gills (Fig. 11). Species of this genus feed on calcareous sponges and are variously coloured, often matching the colour of their sponge prey. There are several records in the northern Red Sea of this small species, which is well camouflaged with its disruptive purple and green colouration. It appears to be endemic to the Red Sea—several other unnamed species from the Pacific Ocean are uniformly violet. *Notodoris citrina* Pruvot-Fol, 1930 is a well-known and widely distributed Indo-West Pacific species, belonging to a group which all have yellow pigment as the ground colour (Fig. 12). These species are usually invisible on their prey, commonly reported to feed on bright yellow calcareous sponges belonging to the genus *Leucetta*.

The species of *Gymnodoris* are difficult to identify and are in need of a critical taxonomic revision. Many species from the Red Sea, and indeed from the Indo-Pacific, are unidentified at present, but one of the three new Red Sea records is known: *Gymnodoris okinawae* Baba, 1936 is recorded throughout the Indo-Pacific and is known to feed on other opisthobranchs (Fig. 13). It grows to approximately 40 mm, and the body is covered in orange spots or lines, and small white tubercles. Its spawn is a vivid orange-yellow coil of 1.5 coils in the western Indian Ocean, where it spawns

Fig. 12 *Notodoris citrina* (photograph c/o D Kipling)

Fig. 13 *Gymnodoris okinawae* (photograph S Kahlbrock)

during the summer months (http://seaslugs.free.fr/nudibranche/a_intro.htm).

The highest number of new records, 23 species to be precise, belongs to the group of cryptobranch dorids. These are dorso-ventrally flattened sea slugs with the gills and rhinophores retractable into pockets below the mantle. They may be smooth or tuberculate, soft or firm, and some are well camouflaged, while others are vibrantly coloured. They feed on a variety of sessile animals, and the radula is usually broad with many hooked teeth per row. The genus *Asteronotus* has a surprising story in the Red Sea: two cryptic species were formally described from the Indian Ocean and western Pacific in 2002; however, both species had been photographed from the Jeddah area of the Red Sea long before that, in the late 1980s and early 1990s, albeit only once. *Asteronotus mimeticus* Gosliner and Valdés, 2002 is a medium-sized dorid (approximately 40 mm) which is very

Fig. 14 *Asteronotus mimeticus* (photograph W Pridgen)

well camouflaged on its sponge prey, in this case *Phyllospongia* sp. (Fig. 14). *Asteronotus spongicolus* Gosliner and Valdés, 2002 is a smaller species which feeds on the same sponge prey in the Red Sea (Fig. 15). There are only two photographic records of these two species: is it overlooked because it is so cryptic? Is it simply a very rare species (although its food is not)? It will be interesting to see what records of these species the future will bring, but as they live on the underside of sponges (http://www.seaslugforum.net/), it is unlikely that they will ever be frequently recorded.

Aldisa sp. nov. is another of the many new records of dorids (Fig. 16). It has been found along the western Indian Ocean coastline to South Africa for a number of years, but still has not received a specific name. Its external morphology is distinctive and different from other species of the genus, with a bright orange-red and yellow colour pattern and sculpturing with large pits along the midline resembling sponge oscula (the large openings through which water flows in and out of the sponge). The ridges and crests further help camouflage this small species—the recently collected specimen in this photograph measures only 15 mm in

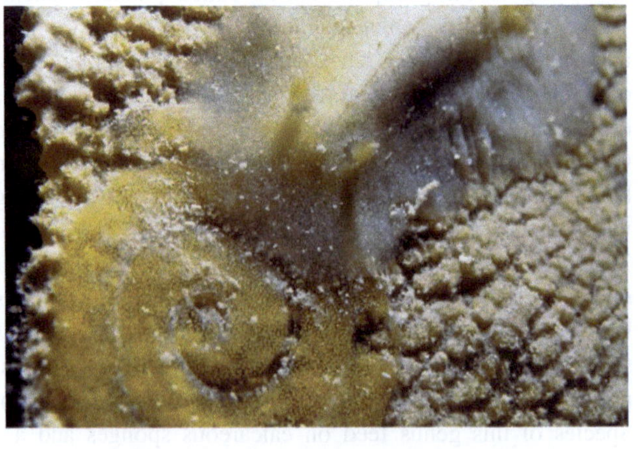

Fig. 15 *Asteronotus spongicolus* (photograph W Pridgen)

Fig. 16 *Aldisa* sp. nov. (photograph S Kahlbrock)

Fig. 17 *Carminodoris estrelyado* (photograph S Kahlbrock)

preservative, and retains the spectacular sculpture of the living specimen.

Carminodoris estrelyado Gosliner and Behrens, 1998 has been frequently recorded in the Indo-West Pacific since its formal description, and it has been recently recorded in the Red Sea (Fig. 17). *Platydoris inframaculata* (Abraham, 1877) is another large dorid, attaining 180 mm in Indo-West Pacific oceans, with only two reliable photographic records from the Red Sea (Fig. 18). While the latter is a very old species, another more recently described Indo-West Pacific species is a new record from the Red Sea: *Thordisa tahala* Chan and Gosliner, 2007 (Fig. 19). It is quite possible that these species have always been in the Red Sea, merely not recorded, although one cannot imagine how a nudibranch nearly 200 mm in length could be overlooked!

Many of the colourful cryptobranch dorids belong to the family Chromodorididae: they display a vast array of bright colours and striking patterns, and all feed on siliceous sponges. *Diversidoris aurantionodulosa* Rudman, 1987 was a surprising discovery among the new records, as

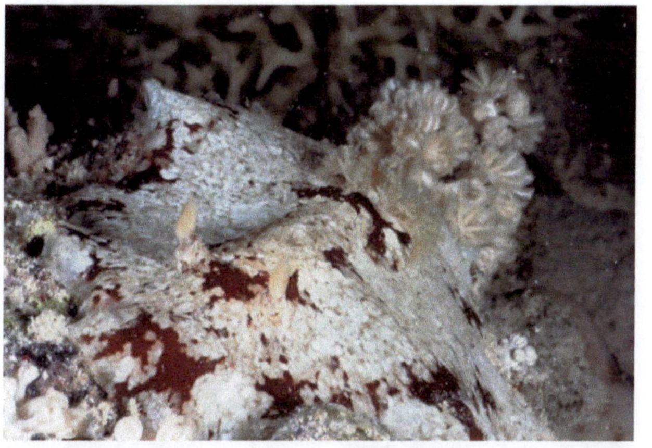

Fig. 18 *Platydoris inframaculata* (photograph E Koehler)

Fig. 19 *Thordisa tahala* (photograph D van Beeck)

chromodorids are easily spotted and fairly well documented (Fig. 20). This is a small species, 20 mm in length, and is found throughout the Indo-West Pacific region. The genus is additionally a new record for the Red Sea and comprises

Fig. 20 *Diversidoris aurantionodulosa* (photograph E Bojanowski)

Fig. 21 *Glossodoris* sp. nov. (photograph S Kahlbrock)

Fig. 22 *Dendrodoris denisoni* (photograph S Kahlbrock)

only this species, distinguished by the tiny tubercles on the dorsum and a wide undulating mantle margin. The gills are small, but the pocket that contains them when they are retracted is very large compared to that of the other chromodorid genera.

The most spectacular chromodorid is without a name at present, as no specimens were collected until very recently, although it has been photographed several times in the Red Sea (Fig. 21). It is relatively large by chromodorid standards, and certainly, if it is present, it cannot be overlooked. It is currently assigned to the genus *Glossosdoris* due to the permanent primary undulations on the mantle margin and less permanent secondary undulations. Without doubt, this species is new to science, and it is also known to be endemic to the Red Sea: there are no similar Indo-Pacific species. How this species escaped detection is unknown; certainly, in the period from 1750 to the 1950s, scientific collections were not made by divers and were therefore unselective, consisting of indiscriminate trawl and grab samples. However, after this period, diving was becoming a widespread activity and yet no one appears to have photographs dating earlier than approximately 2004–2005. Curiously, two other species of chromodorid were described in the early 1900s and never seen again till a few years ago, and now they are photographed regularly: *Chromodoris charlottae* (named by Schrödl, 1999, but recognised by Eliot 1911) and *Hypselodoris dollfusi* (Pruvot-Fol, 1933). What is the explanation for their sudden reappearance in such numbers?

One group of dorids, the porostomes, have lost the radula during the course of their evolution, and instead have suctorial mouthparts which they use to insert into the sponge and digest it, and then suck up the resulting "soup." *Dendrodoris denisoni* (Angas, 1864) is one such species (Fig. 22), and another "surprise" record for the Red Sea: it is commonly found in the western Pacific, with few records from the western Indian Ocean (http://seaslugs.free.fr/nudibranche/a_intro.htm) and very recently the Gulf of Oman (G Smith pers. comm.), but also extending its distribution eastward to Hawaii. This is clearly a recent introduction, as it is hardly cryptic or small, and truly cannot be overlooked if it is present on the reef. Coincidentally, the first records for Oman and the Persian Gulf are from the same period as the first Red Sea photograph. Another large dendrodorid newly recorded, also with a wide Indo-Pacific distribution, is *Dendrodoris albopurpura* Burn, 1957 (Fig. 23). This species has a confused taxonomic history, but is now clearly defined and recognisable. The dark patches are characteristic, as are the soft tubercles. The body is very elongated, and the gills are located very far posteriorly. It is capable of great stretching and moves in a leech-like fashion.

The other family of suctorial sponge feeders with a large presence in the Red Sea is the Phyllidiidae. To date, there are twelve described and well-documented species, but three additional species belonging to different genera have been recorded in the last five years. One of these is *Phyllidiopsis*

Fig. 23 *Dendrodoris albopurpura* (photograph S Kahlbrock)

Fig. 24 *Phyllidiopsis* sp. (photograph E Bojanowski)

sp. illustrated in Fig. 24. As no specimens have been collected, its precise identity cannot be confirmed. It is most similar to *P. sinaiensis* (Yonow, 1998) described from the Red Sea and a sister species to *P. krempfi* Pruvot-Fol, 1957 from the Indo-West Pacific, but the exceptionally compound tubercles and extensive black patterning do not conform to the documented variability of these species; both species are large and have bicoloured rhinophores characteristic of the genus and clearly visible in this photograph.

A smaller species belonging to the genus *Phyllidiella* has been photographed (Fig. 25), resembling a known Indian Ocean species, but without specimens it is impossible to ascertain its identity. The same difficulty applies to a photograph of a species of *Doriopsilla* (Fig. 26): this genus has been recorded only once before in the Red Sea, but the identity of the species is unconfirmed (White 1951 as *Doriopsilla* sp.). This genus is so difficult and understudied that it is not even possible to say whether this new photograph matches the recorded species.

Fig. 26 *Doriopsilla* sp. (photograph D van Beeck)

Dendronotina

The Dendronotina are proving a very interesting suborder of nudibranchs. The nine new records comprise one new genus, five new species, and two pelagic species widely recorded throughout the Indo-West Pacific. The small cryptic species could safely be assumed to have always been present—improvements in diving techniques as well as in digital photographic equipment are probably the reason for their late discoveries. However, the remaining larger species are more of an enigma, as is the new genus. The species belonging to the new genus has been photographed several times (Fig. 27), and finally, a specimen has been collected, so in time its identification and relationships to the other

Fig. 25 *Phyllidiella* cf. *zeylanica* (photograph S Kahlbrock)

Fig. 27 Tritoniidae gen. nov. and sp. nov. (photograph S Kahlbrock)

Fig. 28 *Tritonia* sp. (photograph S Kahlbrock)

Fig. 29 *Bornella* cf. *johnsonorum* (photograph S Kahlbrock)

members of the dendronotid genera will be established. It lives on the large soft corals *Sarcophyton* and *Lobophyllia*, both of which are common in the Red Sea and typical food items of many dendronotids. Its shape is similar to that of *Tritoniopsis*, but the rhinophoral sheath is much more complicated, and the gills appear very much reduced. The body is covered in pustules, as are the rhinophores and gills, presumably to protect them from predators.

Another more typical tritoniid is recorded as new for the Red Sea, but the genus *Tritonia* is not well known (Fig. 28), with many similar species remaining undescribed throughout the Indo-Pacific (e.g. Gosliner et al. 2008). This particular group of species are small, with white bodies and distinct small branching gills in a single series along each side of the dorsum; the distal parts of the gills often appear curled over. The rhinophores have a simple smooth sheath with a decorated margin. Surprisingly, this appears to be the first record of a species belonging to the genus *Tritonia* in the Red Sea, and as so many tritoniid species are endemic, it is here presumed that this small cryptic species of *Tritonia* is also endemic.

Bornella cf. *johnsonorum* Pola et al., 2009 is the second species of the genus *Bornella* to be recorded from the Red Sea (Fig. 29). It appears similar to a species originally described from the Marshall Islands in the Pacific several years ago and subsequently recorded in La Réunion in the western Indian Ocean (http://seaslugs.free.fr/nudibranche/a_intro.htm); both are translucent with opaque white spots. However, this Red Sea specimen is orange in colour with an overlay of dense white spots, and the rhinophoral sheath appears to lack some smaller processes; otherwise, the shapes of the gills and their processes are the same.

The Lomanotidae is a small group of dendronotid species with distinctive characters—an elongated and much reduced body, and instead of the branching gills typical of the dendronotids, they have undulating lobes or a row of small appendages along each side. Only one species, the aptly named *Lomanotus vermiformis* Eliot, 1908 is presently known from the Indo-Pacific. It was originally described from the Red Sea, but is now known to have a very wide distribution (Yonow 2008a). There is a second species recorded from the Red Sea, which does not match any of the photographic records of the dozens of species illustrated in NudiPixel (http://www.nudipixel.net/species/lomanotus_sp/). Without specimens, it is impossible to attempt to establish whether these first two photographs (Figs. 30 and 31) are of the same species, with similar (undescribed) forms occurring in the western Pacific. The hydroid is a plumulariid, and hopefully, a specimen will be collected in the not-too-distant future. A third species, with appendages instead of lobes, was photographed in the Egyptian Red Sea and is once again without a name (Fig. 32).

Fig. 30 *Lomanotus* sp. (photograph S Kahlbrock)

Fig. 31 *Lomanotus* sp. (photograph E Bojanowski)

Fig. 32 *Lomanotus* sp. (photograph E Bojanowski)

The final dendronotid included in this review of new records to the Red Sea since the book (Yonow 2008a) is *Melibe viridis* Kelaart, 1858 (Fig. 33). This species was described many years ago, but it has not been previously recorded in the Red Sea, although it has been frequently recorded throughout the Indo-Pacific and the Mediterranean Sea. Incredibly, it is thought to have entered the Mediterranean through the Red Sea even though it has never been recorded there (until now). It grows to an enormous size—this one is approximately 30 cm, but individuals have been recorded up to 120 cm in length! Several photographic records now exist, and in 2012, a juvenile was photographed and collected, which could mean that *M. viridis* may be establishing itself, at least in the northern part of the Egyptian Red Sea. *Melibe* is an interesting genus because its constituent species have a much enlarged oral hood with which they sweep and collect small crustaceans on which they feed, either from sea grasses, the seabed, or directly from the plankton. Similarly to the porostomes, tethyds have lost their radular teeth in a parallel course of evolution.

Fig. 33 *Melibe viridis* (photograph E Bojanowski)

Arminina

This suborder is now recognised to be an artificial grouping of all the species which do not fit into any other suborder. It is not well studied, but with the increased numbers of divers and photographers, more species are coming to light. Members of this suborder do not occur in great numbers, perhaps less than one-tenth the abundance of the other nudibranchs, and so it will take time to accumulate the amount of material required for identification and revisionary work. However, individual records do serve to increase awareness, and their deposition in museums means that eventually enough material will accumulate for comprehensive studies. All species are benthic, sometimes rather small and drably coloured, but they can also be large and brightly coloured. *Dermatobranchus caeruleomaculatus* Gosliner and Fahey, 2011 is one of the more vivid species and has recently been recorded in Oman (G Smith pers. comm.); however, the species recorded from the Red Sea is different, with a continuous purple band around the margin instead of bright blue spots (Fig. 34). Whether this is a significant difference remains to be seen.

A second distinctive species of *Dermatobranchus* was photographed in June 2013, representing yet another new record from the Red Sea (Fig. 35). Once again, it is similar to two species recorded in the western Pacific. *Dermatobranchus pustulosus* (van Hasselt, 1824) differs in lacking the distinctive black and orange markings on the oral veil and in having ochre rhinophores with a different shape. *Dermatobranchus leoni* Gosliner and Fahey, 2011 differs in having a red central dorsum with longitudinal black lines laterally. The rhinophores are similar, but the foot markings differ. Both named species are relatively large, which correlates with the image of this Red Sea species.

Fig. 34 *Dermatobranchus* cf. *caeruleomaculatus* (photograph S Kahlbrock)

Fig. 35 *Dermatobranchus* cf. *leoni* (photograph S Kahlbrock)

Janolus savinkini Martynov and Korshunova, 2012 on the other hand, is brightly coloured and covered in cerata (Fig. 36). One of the distinguishing features of the genus, which tells it apart physically from the aeolids, is that the cerata are present around the front on the head, in front of the often spectacularly lamellated rhinophores. These rhinophores bear a pleated fan-shaped sensory organ between them, visible in Fig. 37, whose shape is often helpful in identification. This species, which has been photographed throughout the Indo-West Pacific, is very common in Indonesia and the Philippines and is usually found on an erect calcified arborescent bryozoan, possibly *Tricellaria* sp., on which it presumably feeds. There are currently only two photographic records from the Red Sea.

Two more species of *Janolus* are newly recorded for the Red Sea. One is *Janolus mirabilis* Baba and Abe, 1970 a rather strange species with exceptionally large cerata that is

Fig. 36 *Janolus savinkini* (photograph S Kahlbrock)

widely recorded throughout the Indo-Pacific. The second appears to be a new species; it resembles no described species but is very small and may be a juvenile (Fig. 38). The long diagonally lamellated rhinophores with cerata arranged in front of them place the species in *Janolus*; however, there is only one photographic reference.

A third genus belonging to the Arminina which also has cerata around the front of the head is *Madrella*, which can be distinguished from *Janolus* by the long smooth stalk of the rhinophore with a small lamellated bulb perched on the top. Species of *Madrella* have the ability to shed their cerata when threatened or disturbed, which grow back rapidly. *Madrella gloriosa* Baba, 1949 is recorded for the first time from the Red Sea, having been photographed in 2007

Fig. 37 *Janolus savinkini* (photograph E Bojanowski)

Fig. 38 *Janolus* sp. nov. (photograph S Kahlbrock)

Fig. 39 *Madrella gloriosa* (photograph D van Beeck)

(Fig. 39). Soon after, there was a further record from Kenya (http://www.nudipixel.net/species/madrella_gloriosa/) but nothing thereafter. The species appears to be very rarely recorded, even in the western Pacific.

Aeolidina

The Aeolidina comprises the second largest group of new records of sea slugs in the Red Sea since 2008, with 17 species being newly recorded in the last five years. This is hardly surprising since many feed on fouling organisms, especially hydroids, and these are easily transferred from port to port by shipping. The species are evenly divided between well-known Pacific and Indo-Pacific species and new or unidentified species. They belong to twelve genera, of which four are new records for the Red Sea. The three species of *Facelina* illustrated here are not named; not only do they appear to be new species but they also seem to be endemic to the Red Sea (Figs. 40, 41 and 42). Two of them are particularly distinctive and have been photographed several times in the Egyptian Red Sea.

Fig. 40 *Facelina* sp. nov. (photograph S Kahlbrock)

Fig. 41 *Facelina* sp. nov. (photograph O Lederman)

Fig. 42 *Facelina* sp. nov. (photograph E Bojanowski)

Fig. 43 *Limenandra barnosii* (photograph D van Beeck)

A recent paper on the genus *Limenandra* has clarified some of the problems and provided names for previously known but unnamed species; two species are already known from the Red Sea (Yonow 2000), and now, a further three species have been recorded, sadly by photographs only. However, *Limenandra barnosii* Carmona et al., 2013 (Fig. 43) and *L. confusa* Carmona et al., 2013 (Fig. 44) are unmistakeable and it will be interesting to see how rapidly further records are accumulated. The species belonging to this genus are characterised by a long thin body bearing pigmented circles, and the cerata arranged in clusters. The rhinophores have irregular papillae on their posterior surfaces.

Protoaeolidiella atra Baba, 1955 (Fig. 45) and *Pleurolidia juliae* Burn, 1966 (Fig. 46) are very unusual aeolids: both are extremely elongated with two rows of short cerata, somewhat resembling the dendronotid genus *Lomanotus* (see Fig. 32 above). Both are obligate predators of athecate hydroids belonging to the genus *Solanderia*, and when aligned along a branch of the colony, they are extremely well camouflaged from potential predators. The first species is only known from the western Pacific, and the second from the Indo-West Pacific, so these discoveries represent considerable range extensions, and both the species and the genera are new records for the Red Sea. The hydroid *Solanderia secunda* form *crosslandi* (Thornely, 1908) is the only species which has been recorded in the Red Sea (Vervoort 1993).

Fig. 44 *Limenandra confusa* (photograph S Pikarski)

Fig. 45 *Protoaeolidiella* (photograph S Kahlbrock)

Phidiana sp. is known only from a single sighting in northern Egypt (Fig. 47). It is distinctive with its translucent body, orange cerata dusted with white, and the white oval marking on the head. The long recurved propodial tentacles and the longer oral tentacles are marked with white and pale ochre, as are the smooth rhinophores. It was photographed in a clump of thecate hydroids; if this hydroid is the diet of

Fig. 46 *Protaeolidia juliae* (photograph B Picton)

Fig. 47 *Phidiana* sp. (photograph E Bojanowski)

Phidiana sp. and an invasive species, we might expect a spread of its nudibranch predator. The photographs are not clear enough to identify the hydroid, and the aeolid does not appear to match any published descriptions or illustrations; it is considered endemic until further information is available.

Favorinus mirabilis Baba, 1955 joins the other two species of *Favorinus* recorded in the Red Sea (Fig. 48). This species is distinctive, very small (up to 10 mm), and only known from a series of photographs of three individuals in the north of the Gulf of Eilat. Colour is somewhat variable in the species throughout its distribution, but the pigmentation is fairly consistent—white speckling on the cerata, rhinophores, body, and oral tentacles, and dark red spots on the ends of the cerata. The orange lines on the head are characteristic but can sometimes be masked by white pigment (Fig. 49). Both Eilat and Aqaba are large ports, and probably this species has been transported, as larvae, from the Pacific on large ships; it has been reported recently from the western Indian Ocean (http://seaslugs.free.fr/nudibranche/a_intro.htm). The species feeds on the eggs of other sea slugs so would have no difficulties in establishing itself in new locations.

Fig. 48 *Favorinus mirabilis* (photograph B & S Koretz)

Fig. 49 *Favorinus mirabilis* (photograph B & S Koretz)

Phyllodesmium cf. *koehleri* Burghardt et al., 2008 is very interesting because it is also found in the western Indian Ocean, and these records both differ from the Pacific records and photographs (see http://seaslugs.free.fr/nudibranche/a_intro.htm for the Indian Ocean records and http://www.seaslugforum.net/showall/phylkoeh for the Pacific records). The main differences (tuberculate vs. wrinkled rhinophores, less vs. "spikier" cerata) between this and the nominal species may be a function of diet, but not enough information is available at present. In the Pacific, *P. koehleri* has been recorded as feeding on the Nephtheidae, while the photographs from the Indian Ocean and Red Sea (Fig. 50) show it on sediments and other mixed substrates.

Many species of *Cuthona* are now being recorded in the Red Sea. They are generally very small aeolids, approximately 10 mm, often very colourful, and almost always found on hydroids, normally a single species which comprises their specific food item. The two species illustrated here are good examples; while one has been photographed in the Indo-West Pacific before (Fig. 51), the other seems to be endemic to the Red Sea, and both appear to be unnamed species. The lovely orange endemic species (Fig. 52) has been photographed numerous times, always on the same hydroid, *Sertularella* sp., one of a group of three-dimensional erect thecate hydroids.

Trinchesia yamasui (Hamatani, 1993) is a widespread Indo-West Pacific species newly recorded in the northern Red Sea (Fig. 53). It is variable in colour pattern, but the Red Sea individuals are clearly identified by the dark

Fig. 50 *Phyllodesmium* cf. *koehleri* (photograph D van Beeck)

Fig. 51 *Cuthona* sp. nov. (photograph S Kahlbrock)

pigmentation on the head and the colourful banding on the cerata. There is a broad translucent foot and slightly recurved oral tentacles. *Trinchesia yamasui* is thought to feed exclusively on stinging hydroids of the genus *Aglaophenia*, of which at least five species are recorded from the Red Sea (Vervoort 1993).

While many juvenile species of nudibranchs differ from their adult form, and indeed often feed on different prey, a number seem to be miniature versions of the adults (Fig. 54). One such example is this species of *Cratena,* which belongs to a group of species all related to *C. affinis* (Baba, 1949); the nominal species is only known from Japan (http://www.seaslugforum.net/showall/crataffi), characterised by two plate-like rings on the rhinophores, white speckling all over the body and tentacles, white lines on the cerata, and bright

Fig. 52 *Cuthona* sp. nov. (photograph E Bojanowski)

Fig. 53 *Trinchesia yamasui* (photograph E Bojanowski)

Fig. 54 *Cratena* cf. *affinis* (photograph S Ayalon)

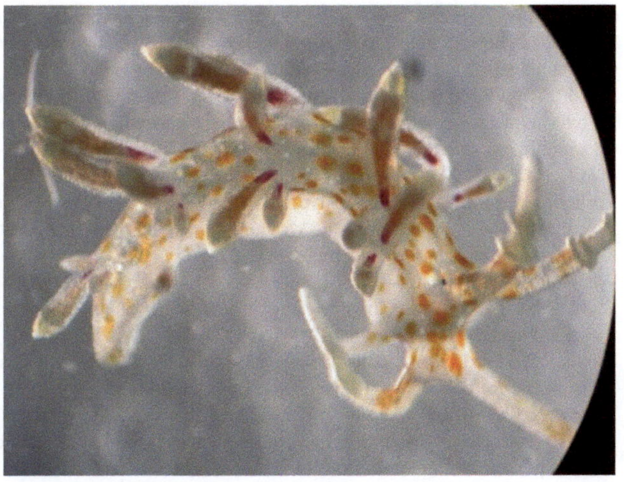

Fig. 55 *Cratena* cf. *affinis* (photograph S Ayalon)

pink spots at the base of the cerata (Fig. 55). Although only tiny individuals, they clearly have some of the markings and colours of the adult. These Red Sea individuals differ from the nominal *C. affinis* in pigmentation (they lack the white lines on the cerata) and from the recorded *C.* aff. *affinis* (found in Japan and South Africa, http://www.seaslugforum.net/showall/cratcfaffi) in pigmentation and having circular ridges on the rhinophores instead of two ridges of papillae.

Discussion

The ramifications of this preliminary analysis of new species recorded in the Red Sea since 2008 are twofold, taxonomic and biogeographical. In the first instance, the selection of species described above forms part of ongoing taxonomic work on the species composition of sea slugs in the Red Sea. It is hoped that publications such as this will highlight the significance of these species and that specimens will be collected for examination in due course. Establishing a faunal list for one of the historically best studied seas in the world is certainly a difficult task, and one that is introduced here with some of the more interesting and colourful examples.

The second aim is to discuss the distribution of species recorded in the Red Sea. While it is difficult to form conclusions on endemism and distribution, it is safe to assume that while some species have always been in the Red Sea (native), others are limited to and only found here (endemic), while still others are certainly new, either by natural range expansions or as alien introductions. Much more work is necessary (e.g. collections, dissections, food samples, and genetic analyses), to establish identities and origins. Much work has been published on Mediterranean aliens, which is a current ongoing environmental issue, analysing different methods of dispersal (e.g. Zenetos et al. 2012). Incidental research into Pacific aliens is provided in several papers, and some interesting points are worth mentioning here; shipping activity began in the 1800s, and with it the transport of species on hulls and in ballast water (Carlton 1987), although with new IMO shipping regulations and improved antifouling paints, these methods should be reducing the impact. Both summaries indicate that shipping, aquaculture, and aquarium escapees are the major sources of introduced species.

The assigned distributions of the species without names are to some extent arbitrary; those which have been recorded or photographed elsewhere are easily assigned as such (see Table 1 and Fig. 56). Those which are similar to (cf.) or related to (aff.) a species have been assigned the distribution of that nominal species despite the differences. All those which seem to have no similar records elsewhere have been categorised as endemic. In some cases, this is quite clear and accurate, whereas in others, such as for several aeolids, this is probably unrealistic; the usual prey of many aeolids is hydroids, and these and their larvae are easily transported by ships. On the other hand, the photographic databases available on the Internet are now very extensive, and comparisons are relatively straightforward if time-consuming. The small

Fig. 56 The geographical distributions of the new records of sea slugs found in the Red Sea since 2008

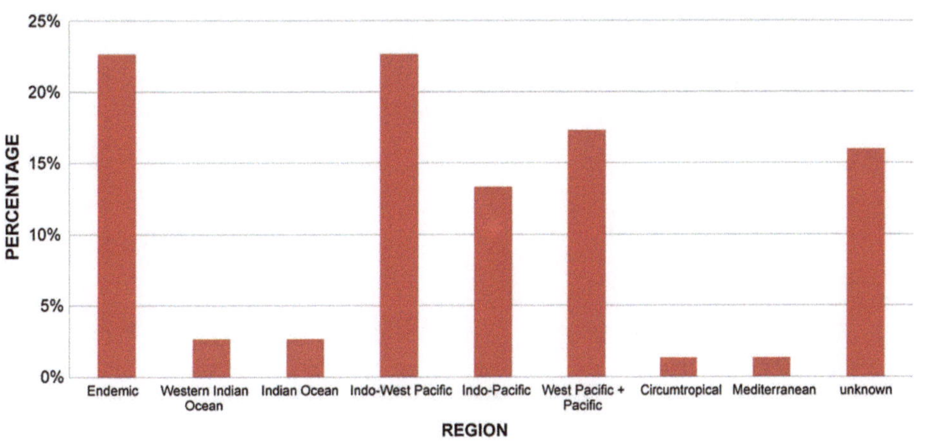

cryptic dorids have been assigned as of "unknown" distribution; despite some major works on dorids, we are a long way from having a complete picture. Most flattened dorids feed on, under, and in sponges and are necessarily very well camouflaged and not particularly photogenic. It is probable that many have a much wider distribution than currently available. Of the (total) 60 non-chromodorid cryptobranch dorids recorded in the Red Sea, only 15 species are endemic; the majority of the named species have a wider distribution, and one would expect the unnamed species to have similar distributions.

Acknowledgments I am grateful to all the photographers who have been combing through their collections and dive logbooks to provide me with as much information on sea slugs from the Red Sea as possible. I am most grateful to Sven Kahlbrock who has carefully and diligently collected and preserved many specimens over the last five years. I would like to thank Dr. Richard Willan for his thoughts on identifications of a number of the species which are new records—his experience and willingness to help are gratefully acknowledged. Professor John Ryland and Dr. Bernard Picton are duly thanked for assistance in the identification of the hydroids and bryozoans, so critical to any study of nudibranchs feeding on them. The three referees provided useful comments to this chapter.

References

Alexander J, Valdés A (2013) The ring doesn't mean a thing: molecular data suggest a new taxonomy for two Pacific species of sea hares (Mollusca: Opisthobranchia, Aplysiidae). Pac Sci 67(2):282–294

Carlton JT (1987) Patterns of transoceanic marine biological invasions in the Pacific Ocean. Bull Mar Sci 41(2):452–465

Churchill CKC, Valdés A, Foighil DO (2014) Afro-Eurasia and the Americas present barriers to gen flow for the cosmopolitan neustonic nudibranch *Glaucus atlanticus*. Mar Biol. doi:10.1007/s00227-014-2389-7

Eliot CNE (1911) Chromodorids from the Red Sea, collected and figured by Mr. Cyril Crossland. Proc Zool Soc, London 1068-1072

Gosliner TW, Behrens DW, Valdés A (2008) Indo-Pacific nudibranchs and sea slugs: a field guide to the world's most diverse fauna. Sea Challengers Natural History Books, Washington, 425 pp

Schrödl M, Jörger KM, Klussmann-Kolb A, Wilson NG (2011) Bye bye "Opisthobranchia"! A review on the contribution of mesopsammic sea slugs to euthyneuran systematics. Thalassas 27(2):101–112

Vervoort W (1993) Report on hydroids (Hydrozoa, Cnidaria) in the collection of the Zoological Museum, University of Tel-Aviv, Israel. Zoologische Mededelingen 67:537–565

Wägele H, Klussmann-Kolb A, Verbeek A, Schrödl M (2013) Flashback and foreshadowing—a review of the taxon Opisthobranchia. Org Divers Evol. doi:10.1007/s13127-013-0151-5

White KM (1951) On a collection of molluscs, mainly nudibranchs, from the Red Sea. Proc Malacol Soc Lond 28:241–253

Yonow N (2000) Red Sea Opisthobranchia 4: the orders Cephalaspidea, Anaspidea, Notaspidea, and Nudibranchia: Dendronotacea and Aeolidacea. Fauna Arab 18:87–131

Yonow N (2008a) Sea slugs of the Red Sea. Pensoft Publishers, Sofia, 304 pp

Yonow N (2008b) Opisthobranchs of the Gulf of Eilat and the Red Sea: an account of similarities and differences. In: Por F (ed) Aqaba—Eilat, the improbable Gulf. Environment, biodiversity and preservation. The Hebrew University Magnes Press, Jerusalem, pp 177–196

Zenetos A, Gofas S, Morri C, Rosso A, Violanti D, García Raso JE, Çinar ME, Almogi-Labin A, Ates AS, Azzurro E, Ballesteros E, Bianchi CN, Bilecenoglu M, Gambi MC, Giangrande A, Gravili C, Hyams-Kaphzan O, Karachle PK, Katsanevakis S, Lipej L, Mastrototaro F, Mineur F, Pancucci-Papadopoulou MA, Ramos Esplá A, Salas C, San Martín G, Sfriso A, Streftaris N, Verlaque M (2012) Alien species in the Mediterranean Sea by 2012. A contribution to the application of European Union's Marine Strategy Framework Directive (MSFD). Part 2. Introduction trends and pathways. Mediterr Mar Sci 13(2):328–352

http://seaslugs.free.fr/nudibranche/a_intro.htm

http://seaslugsofhawaii.com/general/species-list.html

http://www.ciesm.org/atlas/appendix3.html

http://www.nudipixel.net/genus/

http://www.seaslugforum.net/

Marine Turtles of the Red Sea

Agnese Mancini, Islam Elsadek, and Magdy A.N. El-Alwany

Abstract

Marine turtles are long-lived reptiles that appeared on Earth in the late Triassic. There are seven extant species worldwide, five of which can be found in the Red Sea: the green turtle, the hawksbill turtle, the loggerhead turtle, the olive ridley turtle and the leatherback turtle. Marine turtles—as all top predators—have a prominent role in maintaining balanced and healthy ecosystems, in particular seagrass beds and coral reefs, but also in transporting nutrients towards naturally nutrient-poor ecosystems (the nesting beaches), and providing food and transportation for other marine species (e.g., barnacles and commensal crabs). Marine turtles also play an important role in the economy of the tourism industry. In fact, because they can usually be observed in coastal areas frequented by people, marine turtles are the primary attraction for numbers of snorkellers and divers, and contrary to other pelagic species, they are easily accessible even from the shore. In the Red Sea, very few data are available on marine turtles: green and hawksbill turtles are known to feed and nest in the area, although quantitative data are not precise. It is estimated that approximately 450–550 and 450–650 females of green and hawksbill turtles, respectively, nest every year along the Red Sea coast. Loggerhead turtles are known to nest outside the Gulf of Aden on the Socotra Islands (Yemen) in great numbers but no nesting activity has been reported within the Red Sea. Only one nest from an olive ridley turtle has been reported in the region, from 2006 in Eritrea. The leatherback turtle is only sporadically seen feeding, usually following jellyfish blooms, but no nesting activity has been recorded. Data for turtles in their feeding grounds and in-water habitats are greatly lacking for most countries. In Egypt it was found that there are at least 453 green turtles using 13 shallow bays as feeding and developmental areas. No regional assessment has been undertaken to quantify the impact of human-induced mortality on marine

A. Mancini (✉)
Hurghada Environmental Protection and Conservation Association (HEPCA), Hurghada, Egypt
e-mail: agnese@hepca.com; amancini1979@gmail.com

I. Elsadek · M.A.N. El-Alwany
Department of Marine Science, Faculty of Science, Suez Canal University, Ismailia, Egypt

I. Elsadek
Nature Conservation Sector, Marsa Alam Office, Marsa Alam, Egypt

A. Mancini
Boomerang for Earth Conservation, Antony, France

N.M.A. Rasul and I.C.F. Stewart (eds.), *The Red Sea*,
Springer Earth System Sciences, DOI 10.1007/978-3-662-45201-1_31,

turtles in the region; however, major threats have been identified as bycatch, habitat destruction, directed harvest and pollution. Marine turtles are legally protected in all countries bordering the Red Sea both through national laws and international agreements; however, enforcement at sea and on nesting beaches is substantially lacking.

Introduction

Marine turtles are long-lived reptiles that appeared in the late Triassic ca. 200 million years ago (Pritchard 1997). At present, only seven species survive worldwide (Marquez 1990). All marine turtles share similar life cycles (although there are differences between species) that include a long pelagic maturation period and late sexual maturity, short- to long-distance migrations from feeding to breeding areas, 1–5 year intervals between reproductive seasons, large clutch sizes with high mortality rates during the first life stages (Van Buskirk and Crowder 1994). Over-exploited for centuries, marine turtle populations have declined sharply worldwide (Groombridge and Luxmoore 1989) which has resulted in an inability to fulfil their ecological role (Bjorndal and Jackson 2003). Currently, all seven species of marine turtles feature in the IUCN Red Lists of endangered fauna and flora (IUCN 2013) and are included in Appendix 1 of the Convention on International trade of Endangered Species (CITES 2013).

Ecological Importance of Marine Turtles

Many marine turtle populations have declined drastically in the past centuries due mainly to direct harvesting and in more recent years to incidental captures in fisheries (Jackson 1997; Jackson et al. 2001; Bjorndal and Jackson 2003). As a consequence of this decline that started centuries ago and because we are lacking in most cases historical data on abundance, we are unable at present to fully understand the ecological role of marine turtles (Bjorndal and Jackson 2003). However, we know that even at decreased levels, marine turtle populations still play fundamental roles that can be summarized as follows: (1) maintaining balanced ecosystems such as seagrass beds and coral reefs, (2) transporting nutrients to naturally poor ecosystems (like the nesting beaches), (3) providing habitats for certain species and (4) contributing substantially to the marine food webs (Wilson et al. 2010). As an example, green turtles increase the productivity and nutrient contents of seagrass beds by grazing and cropping the longer leaves (Bjorndal 1980; Thayer et al. 1984). Hawksbill turtles, on the other hand, feed mainly on sponges that usually tend to compete with reef-builder corals for space (Meylan 1988; Leon and Bjorndal 2002), thus regulating the colonization of reefs by other species (Wilson et al. 2010). At nesting beaches, marine turtles are known to be an important transporter of nutrients to and from spatially separated habitats (Bjorndal and Jackson 2003; Vander Zanden et al. 2012), enriching the nesting beaches through unhatched eggs, eggshells and unborn hatchlings that can be used by predators, detritivores and decomposers but also plants (Bouchard and Bjorndal 2000). Furthermore, marine turtles are usually associated with other species (epibionts) that live attached to the turtle carapace (sometimes in a parasitic relationship, sometimes in a commensalism relationship), thus obtaining transportation, food and protection (Dellinger et al. 1997; Pfaller et al. 2006). Marine turtles also function as top predators regulating, for example, the abundance of jellyfish populations (Houghton et al. 2006), they provide for food to a variety of cleaning fish species (Losey et al. 1994) and are themselves an important source of food for fish such as sharks, wrasses and groupers but also seabirds feeding on hatchling in the nesting beaches (Gulko and Eckart 2004).

Economic and Cultural Importance of Marine Turtles

Marine turtles have been a valuable resource for coastal communities in the Red Sea and worldwide not only as a major source of food and oil, but also as a cultural symbol (Frazier 2003). In Egypt, at the time of the Pharaohs, some ports along the Red Sea flourished thanks to an intense trade in hawksbill turtle shells that were exported to other countries (Parsons 1972). Frazier (2003) reported that some of the oldest artefacts representing marine turtles were found in the Middle East and Arabian Peninsula, implying that these animals had an important role in the local culture. More recently, turtle shells were used to build roofs by some of the local tribes (M. Gad, pers. comm.).

While at present the cultural importance of these animals might have decreased, marine turtles still play an important role in the economy of many countries around the Red Sea, as they act as flagship species (see Frazier (2005) for some examples). While there is no absolute definition of flagship species, most conservationists agreed in defining them as charismatic animals having a social, economical and/or cultural importance that can be used to enhance conservation efforts (Frazier 2005). Marine turtles, with other charismatic species such as marine mammals, corals and associated coral reef fish, usually attract hundreds of thousands of people every year to the Red Sea to dive and snorkell (Gladstone et al. 2012). Contrary to other pelagic species, marine turtles in their feeding grounds are easily accessible to most divers and snorkellers, even less experienced ones (Rayan and Mancini, unpublished). On the other hand, marine turtles in their nesting grounds may attract thousands of visitors every year, with the potential for producing high incomes through eco-tourism (Tisdell and Wilson 2002). This kind of tourism is not yet exploited and/or regulated in the Red Sea; however, there is good potential for it to develop (A. Mancini, personal observation).

Review of Marine Turtles in the Red Sea

Five of the seven sea turtle species existing in the world can be found in the Red Sea: the green turtle (*Chelonia mydas*), the hawksbill turtle (*Eretmochelys imbricata*), the leatherback turtle (*Dermochelys coriacea*), the loggerhead turtle (*Caretta caretta*) and the olive ridley turtle (*Lepidochelys olivacea*) (Ross 1985; Ross and Barwani 1982; Frazier and Salas 1984). Green and hawksbill turtles are commonly seen, nesting and feeding grounds and migrations in the region (Frazier and Salas 1984). The remaining three species have been recognized by fishermen (Frazier and Salas 1984) or reported by recreational and professional divers (Mancini, pers. obs.).

Following is a synopsis of marine turtles of the Red Sea per species.

Green Turtles (*Chelonia mydas*)

Green turtles are widely distributed in the tropical and subtropical waters, usually preferring near-shore bays and continental waters (Marquez 1990). They are known to forage on shallow seagrass beds and undertake long migrations from rookeries to feeding areas (Marquez 1990; Bjorndal 1997). Usually, these migrations can be performed along the coastline or across oceanic regions with some exceptions [*e.g.,* Ascension Island population (Marquez 1990)]. Green turtles are considered as one of the two most abundant species in the Red Sea and are known to nest and feed in the region (PERSGA 2004; see Fig. 1).

Nesting Activities

No nesting activity was recorded in Jordan (PERSGA 2004). In Saudi Arabia, green turtles nest from August to November almost exclusively at Ras Baridi, a 6-km stretch of the shoreline, and on the Farasan Islands. The total annual nesting population was estimated to be around 150 individuals (Miller 1989; Al-Merghani et al. 2000; Pilcher and Al-Merghani 2000; Pilcher 1999). In Egypt, the nesting season is thought to extend from May to September with a peak at the end of July and beginning of August (Hanafy and Sallam 2003). Three main nesting sites have been identified along the Egyptian coast: Tiran Island, Zabargad Island and Wadi Gemal-Hamata National Park (Hanafy 2012). However, nesting activity has also been described as widespread and low-density (Miller 2004). The annual nesting female populations are estimated to be around 200 nesting females for Zabargad Island (ElSadek et al. 2013) and around 250 for the whole Egyptian Red Sea Coast (HEPCA 2012). Recent surveys reported declining trends from the Wadi Gemal-Hamata National Park area (HEPCA 2012). In Sudan, a colony of approximately 50 nesting females was found on Seil Ada Kebir Island that is part of the Suakin Archipelago (PERSGA 2004). In Eritrea, green turtles are known to nest on the Dahlac Islands, but no quantitative data are available (Teclemariam et al. 2009). In Djibouti, recent reports indicate that green turtles nest at Maskali and Moucha Islands from January to April with a peak in February and March (Al-Mansi et al. 2003a). Along the Yemeni Red Sea coast, green turtles are known to nest sporadically on most sandy beaches, although higher concentrations can be found on the Gulf of Aden coast (PERSGA 2004).

Foraging Grounds and Movements

While regionally important nesting areas of this species have already been identified, information on the abundance and trends in the feeding grounds is substantially lacking (PERSGA 2004). In Jordan, green turtles are considered as rare probably due to the fact that seagrass meadows are scarce in the Gulf of Aqaba (Al Zibdah 2007). Recent work in Egypt estimated that there are approximately 453 green turtles in 13 shallow coastal areas (HEPCA 2012) although this is probably an underestimate of the real population size as only a few bays have been surveyed. Four green turtles were tagged on Zabargad Island in 2009, Egypt, two of which moved southward towards Sudan and Eritrea, while one stayed around the nesting area and one moved north towards south Sinai (Attum et al. 2014). No quantitative data are available for Saudi Arabia; however, it is very likely that green turtles are using the extensive seagrass beds located all along the coast, in particular around Ras Baridi, where also the dugong is frequently observed (Marsh et al. 2002). In Djibouti, two green turtles with tags from Sri Lanka were found by fishermen (Al-Mansi et al. 2003b). In Yemen, green turtles are known to use the seagrass beds around

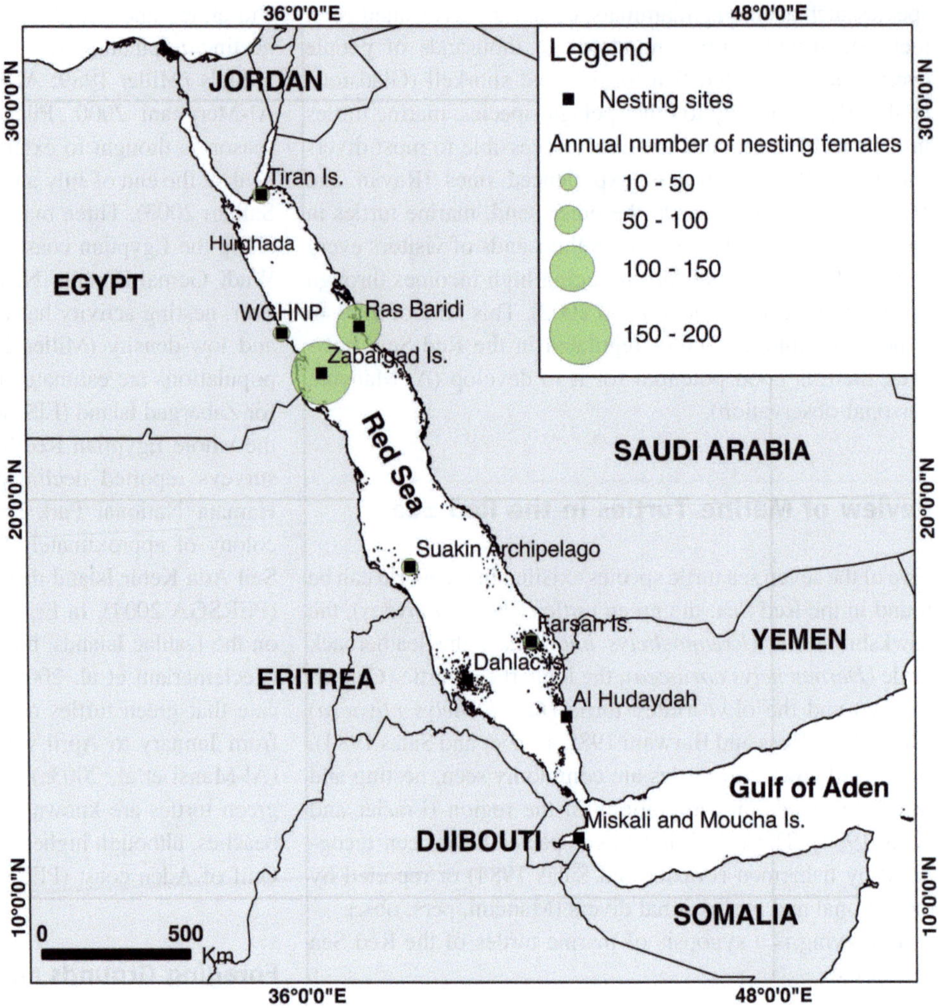

Fig. 1 Known nesting sites for green turtles in the Red Sea. When available, the annual number of nesting females is included in the legend

Birim Island which is also considered as an important waterway for green turtles entering the Red Sea (Al-Mansi et al. 2003a). Two green turtles have been tracked into the Red Sea after nesting on Masirah Island in Oman (Rees et al. 2012). One turtle established a foraging area in Eritrea while the other turtle was still migrating north along the Saudi coast when the tag stopped transmitting.

Hawksbill Turtles (*Eretmochelys imbricata*)

Hawksbill turtles have a circum-global distribution, inhabiting tropical and to a lesser extent subtropical waters of the central Atlantic and Indo-Pacific regions (Marquez 1990; Mortimer and Donnelly 2008). Hawksbill turtles are known to live in clear waters and feed on a number of invertebrates including corals and sponges (Marquez 1990; Meylan 1988). In the Red Sea, hawksbill turtles have been observed nesting and feeding (PERSGA 2004) (Fig. 2).

Nesting Activities

Mortimer and Donnelly (2008) estimated the population of hawksbill turtles in the Red Sea to be around 450–650 females per year. No nesting activity has been recorded in Jordan (PERSGA 2004). In Saudi Arabia, aerial surveys revealed low-density nesting activity by hawksbill turtles on the Farasan Islands (Miller 1989). The nesting season is reported to run from February to May, and an annual nesting female population of less than 50 individuals has been estimated (Miller 1989; PERSGA 2004). In Egypt, the hawksbill turtle nesting season is thought to run from May to September with two peaks, one in June and one in August (Hanafy and Sallem 2003). Most important nesting grounds for this species have been found on in-shore islands with Giftun Kebir being the most important site. On this island, an annual nesting population of 200 females was reported (Hanafy and Sallam 2003); however, recent surveys estimated that the population is now only 20–30 females per

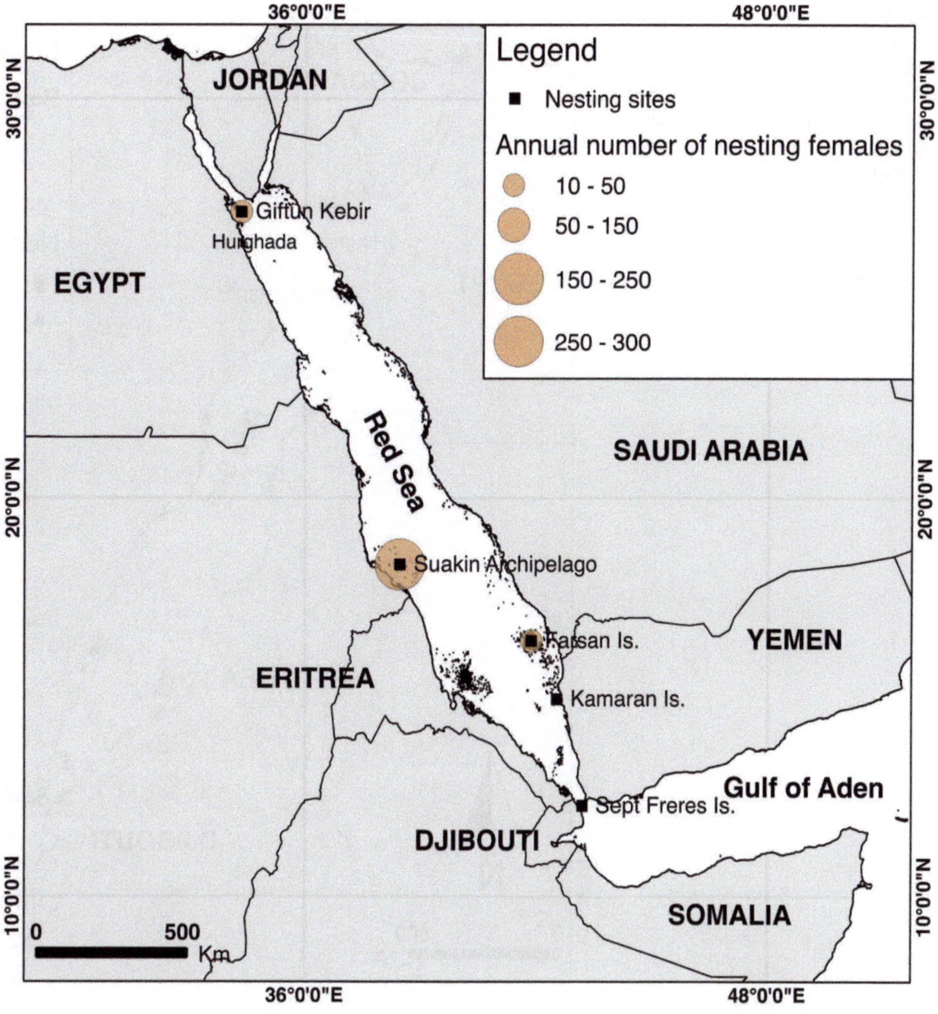

Fig. 2 Known nesting sites for hawksbill turtles in the Red Sea. When available, the annual number of nesting females is included in the legend

year (HEPCA 2012). In Sudan, hawksbill turtle nesting activity was recorded on the Suakin Archipelago from March to July (PERSGA 2004). Moore and Balzarotti (1977) estimated that 300–350 hawksbill turtles nested annually in Sudan. In Eritrea, hawksbill turtles are known to nest regularly but mostly on the Dahlac Islands (Hillman and Gebremariam 1996; Temcelmariam 2013). In Djibouti, hawksbill turtles are thought to nest from March to June on Ras Siyyan and Sept Frères Island (PERSGA 2004). Along the Yemeni Red Sea coast, nesting activities have been reported on the Kamaran Islands, Makran, and Perim Island (Green 1996; Ross and Barwani 1982).

Foraging Grounds and Movements

Foraging grounds for hawksbill turtles are usually located on fringing reefs. Little information is available at present on the population abundance. In the northern Egyptian Red Sea, including the south Sinai and Hurghada/Gouna areas, a citizen science-based project revealed that hawksbill turtles are the species most commonly observed by divers and they can be observed all-year round with a peak during summer months (from June to August); however, the increased number of observations might be due to an increased number of dives (Mancini, unpublished data). In the southern Egyptian Red Sea, the population of hawksbill turtles was estimated as being around 10–15 individuals; however, this figure is considered to be extremely conservative as the surveys took place mostly in seagrass areas (HEPCA 2012). In the Egyptian Red Sea, all observed hawksbill turtles were sub-adult and adult individuals, with a straight carapace length range of 60–80 cm (HEPCA 2012). In Jordan, a study revealed similar patterns with the hawksbill turtle being the only species observed at 13 dive sites (Al Zibdah 2007). All turtles in this study were classified as sub-adults or adults, similar to the results obtained in the Egyptian Red Sea (Al Zibdah 2007). In other areas, including Sudan, Eritrea and Saudi Arabia, hawksbill turtles have been observed in their feeding grounds; however, no quantitative information is available (PERSGA 2004). The population trend is not known, but it is believed that the

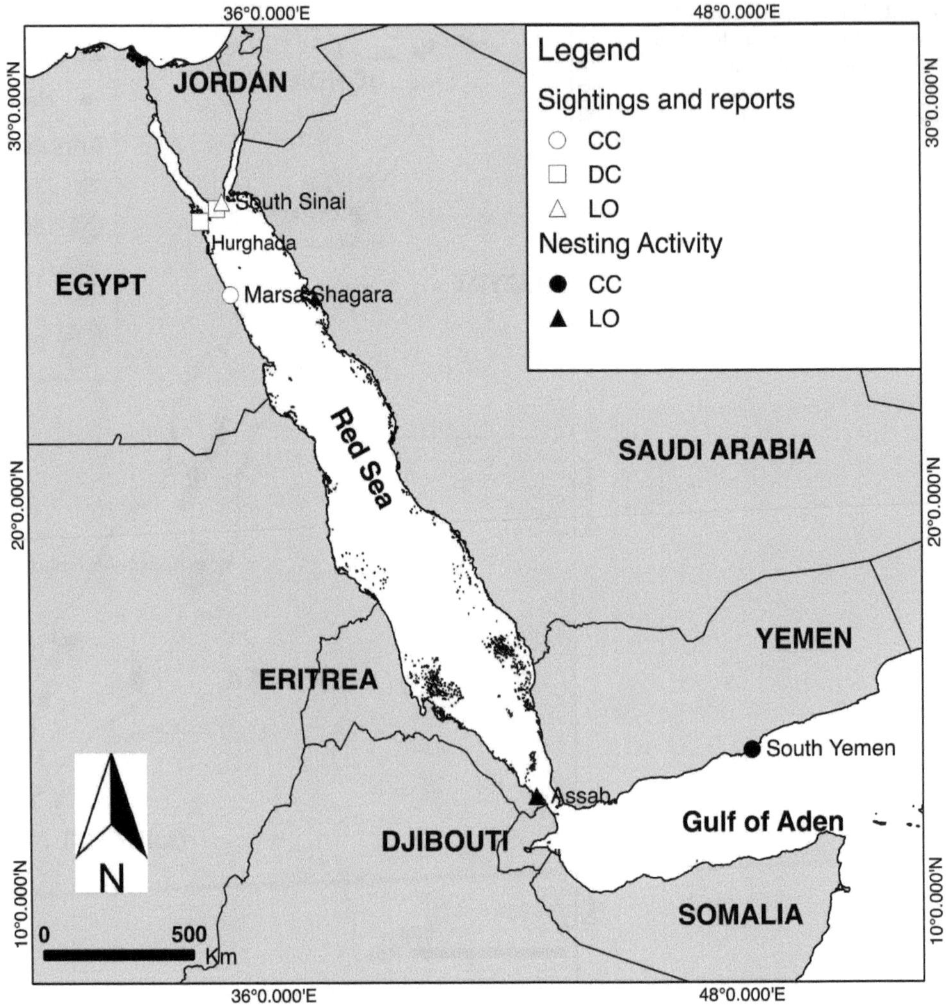

Fig. 3 Distribution of reported nesting activities and sightings of loggerhead (CC), leatherback (DC) and olive ridley (LO) turtles in the Red Sea. Only site-specific sightings and nesting activities are shown on the map

actual population is smaller than it used to be, mainly due to direct intake for the shell trade (Parsons 1972; Groombridge and Luxmoore 1989).

Loggerhead Turtles (*Caretta caretta*)

The loggerhead turtle is widely distributed in the tropical and sub-tropical waters around the world. They are carnivorous and feed predominantly on invertebrates, particularly small crustaceans (Marquez 1990). While this species is known to nest on Socotra Island (Yemen) and Masirah Island (Oman, Arabian Sea), and to use feeding grounds within the Gulf of Aden (Baldwin et al. 2003), sightings in the Red Sea are rare (Fig. 3).

Nesting Activities

No nesting activity for this species has been reported within the Red Sea. The closest known nesting ground is located on Socotra Island (Yemen) (Baldwin et al. 2003) and sporadically on the southern coast of Yemen, in the Gulf of Aden (Saad 2002).

Foraging Grounds and Movements

Reports of loggerhead turtles come mainly from Yemen (Pilcher and Saad 2006); however, they have also been observed in Sudan (Al-Mansi et al. 2003a) and Djibouti (Al-Mansi et al. 2003b). A recent video taken from a professional diver in Marsa Shagara (southern Egyptian Red Sea) featured a loggerhead turtle swimming northward (Mancini, pers. obs.), thus confirming the previous assertion (Frazier and Salas 1984) of the species' presence also in the northern part of the Red Sea.

Olive Ridley Turtles (*Lepidochelys olivacea*)

The olive ridley turtle (*Lepidochelys olivacea*) is a pan-tropical species usually seen in continental coastal waters although not close to the shore (Marquez 1990). This species is well known for its mixed reproductive strategy that includes the famous *arribadas* (when thousands of olive ridley turtles leave the sea together to nest on a short stretch of beach) (Abreu-Grobois and Plotkin 2008). This species is considered to be the most abundant worldwide, but it is only sporadically seen in the Red Sea region (Marquez 1990; Abreu-Grobois and Plotkin 2008; see Fig. 3).

Nesting Activities

Only one record of nesting activity of an olive ridley turtle in the Red Sea has been reported from Assab, Eritrea in 2005 (Pilcher et al. 2006a).

Foraging Grounds and Movements

The olive ridley turtles are known to feed on a variety of prey (mostly invertebrates and small fish). Very few reports are available for this species in the Red Sea waters. Frazier and Salas (1984) cited five records of this species for the Egyptian Red Sea. More recently, three olive ridley turtles were reported by divers in South Sinai (Mancini, unpublished data) although no picture was available.

Leatherback Turtles (*Dermochelys coriacea*)

Leatherback turtles (*Dermochelys coriacea*) are the most widely distributed species, known to occupy tropical to sub-polar waters (Marquez 1990). While their nesting sites are located in tropical areas (see Sarti-Martinez (2000) for a review), feeding grounds may be located thousands of miles away (Hughes et al. 1998). Leatherback turtles are known to feed mainly on soft-body prey such as jellyfish and salps, and they have been observed aggregating where swarms of these species are formed (James et al. 2005). Although jellyfish aggregations are known to happen seasonally (Frazier and Salas 1984), leatherback turtles (*Dermochelys coriacea*) are only sporadically seen in the Red Sea (Fig. 3).

Nesting Activities

No nesting activity for this species has been recorded within the countries bordering the Red Sea (Frazier and Salas 1984; Ross 1985; Miller 1989; Gasperetti et al. 1993; PERSGA 2004; Pilcher et al. 2006b).

Foraging Grounds and Movements

Leatherback turtles have been observed swimming and feeding in the Red Sea on a few occasions in the Egyptian waters (Frazier and Salas 1984; Mancini, pers. obs.), in Saudi Arabia (Gasperetti et al. 1993; Ross 1985), in Sudan (Al-Mansi et al. 2003a), in Djibouti (Al-Mansi et al. 2003b) and in Yemen (Saad 2002) although the species is thought to be extremely rare.

Main Threats Faced by Marine Turtles in the Red Sea

Worldwide, marine turtles face numerous threats during all their life stages. Natural mortality is usually high during the first life stages: eggs are easy prey for foxes, raccoons and crabs, etc., while full nests are susceptible to the hazards of the weather conditions and tidal levels (e.g., Boulon 1999; Engeman et al. 2003) and hatchlings are easy prey for birds, crabs and large fish (Frith 1975; Stancyk 1982; Witherington and Salmon 1992). It is thought that only one out of 1,000 hatchlings will survive to adult age (Dobbs 2001). Once marine turtles reach maturity and therefore a certain size, the number of natural threats and predators is drastically reduced and usually limited to certain species of sharks and marine mammals (e.g., Fergusson et al. 2000; Heithaus et al. 2008).

However, in recent years, human-induced threats have dramatically increased the mortality rates of marine turtles, in all life stages. Various authors indicate direct harvest of eggs and meat for consumption (e.g., Fleming 2001; Tomillo et al. 2008; Mancini and Koch 2009) and incidental fishery (Lewison et al. 2004; Koch et al. 2006; Peckham et al. 2008; Wallace et al. 2010) as major causes of mortality worldwide. Other major threats come from the destruction of nesting and feeding grounds (e.g., Crain et al. 1995, 1996), pollution (including oil spills) and ingestion of marine debris (e.g., Plotkin and Amos 1990; Bugoni et al. 2001; Derraik 2002; Tomas et al. 2002), and minor impacts from collisions with recreational boats (Davenport and Davenport 2006; Hazel and Gyuris 2006; Hazel et al. 2007; Denkinger et al. 2013).

In its Action Plan for Marine Turtles of the Red Sea and Gulf of Aden, PERSGA (2004) addressed consumption and bycatch as the main threats but also insisted on the necessity of protecting marine turtles' nesting and feeding habitats and regulating vehicles and vessel speed for the same habitats. While general guidelines have been formulated already (PERSGA 2004), no regional or national studies have been undertaken to quantify the impact of different threats on marine turtle populations in the Red Sea.

The following is a synopsis of available threat data in the area, classified according to the type of threat.

Review of Impacts

Directed Harvest

Directed harvest for meat and egg consumption is known to have happened for centuries, especially in the southern areas of the Red Sea (Frazier and Salas 1984; Pilcher et al. 2006a). Currently, this threat is considered to have a very low impact in the northern part of the Red Sea (PERSGA 2004), but it is thought to be a major threat in the southern Red Sea (Howe et al. 2004). Sporadic reports of green turtles killed for consumption have been recorded in south Sinai where 4 carapaces were found in a dive site (Mancini, pers. obs.). It appears, however, that direct exploitation of marine turtles might be more widespread: in summer 2012, a boat was stopped in the southern Egyptian Red Sea transporting 20 carapaces of hawksbill turtles to be sold to tourists and/or to the Asian market where the powder obtained from the bones is used for medical purposes (ElSadek and Mancini, pers. obs.). There are also unverified reports from tourist cities in the northern Red Sea of salesmen receiving hundreds of carapaces per month (ElSadek, pers. comm.). In Saudi Arabia, consumption of eggs and meat (considered to be an aphrodisiac) has been reported (Miller 1989). In Sudan, direct harvest of turtles for their meat is thought to be common (Hariri et al. 2000). In Eritrea, during a survey in fishing camps, 67 carapaces of turtles killed for consumption were found, 40 of them being identified as green turtles, 10 as hawksbill turtles and 17 unidentified (Howe et al. 2004). In Djibouti, direct harvest of turtles for consumption of eggs and meat is considered to be widespread (Hariri et al. 2000). In Yemen, egg consumption is common in the nesting grounds on the Gulf of Aden coast; however, marine turtle are also exploited heavily for their fat, used to produce oil that is thought to cure asthma (Hariri et al. 2000; Moqbil 2007).

Bycatch

Few data are available on incidental fishery of marine turtles in the Red Sea by industrial fleets. Teclemariam et al. (2009) reported that 21 turtles were caught in four Egyptian trawlers in the period between December 2004 and February 2005; only one was found dead while the others were released. The same study reported that 3,342 marine turtles were caught accidentally between 1994 and 2004 in Eritrean waters by industrial trawlers (a large majority of them recorded as green turtles) with 78 % of the turtles captured alive (Teclemariam et al. 2009). In Sudan, marine turtles have been described as a typical bycatch product in large-scale shark fisheries by foreign vessels (Hamann et al. 2000). In Yemen, trawlers and long-liners have reported high bycatch rates of marine turtles (Pilcher and Saad 2006).

Data on bycatch in artisanal fishery are scarce. Rouphael et al. (2013) carried out social surveys in the Gabel Elba protected area (southern Egyptian Red Sea) and reported that marine turtles are frequently caught in gillnets (80 % of the interviewed fishermen) and sometimes retained for consumption. In Sudan and Djibouti, fisheries seem to be a limited threat (Al-Mansi et al. 2003a, b). Moqbil (2007) reported marine turtle bycatch in the sardine shore nets in Yemen.

Destruction of Habitats

The consequences of habitat destruction are particularly evident at the turtles' nesting and shallow coastal sites. Coastal development and increased light pollution are considered the primary cause of the sharp decline in the number of nests along the southern Egyptian Red Sea coast (Hanafy 2012; HEPCA 2012). Increased light pollution has been recorded in most major tourist sites: Ras Mohammed National Park (Egypt), Sanganeb and Sept Freres Islands (Sudan) and Moucha and Maskali (Djibouti) (PERSGA 2004). Dredging and filling have been reported as a common cause for the destruction of shallow marine habitats used as foraging grounds by green turtles in particular, but no quantitative data are currently available (PERSGA 2004).

Pollution and Marine Debris

The main sources of pollution in the Red Sea are oil spills, discharges from boats at sea and within harbours, sewage and litter. Oil spills are known to be very frequent in the Red Sea and have been reported to have serious lethal effects on coral reefs (Rinkevich and Loya 1979); no studies are, however, available on the mortality of marine turtles in the Red Sea due to oil spills. The only available data from the region on marine turtle mortality due to pollution are from Saudi Arabia where cement dust pollution in Ras Baridi was found to reduce hatchling emergence success to 40 % (Pilcher 1999). Heavy metal concentrations have been found to be unexpectedly high, especially in areas close to harbours (Abu-Hilal and Badran 1990). Heavy metals can be lethal for marine turtles if concentrations reach toxic levels (Godley et al. 1999); however, there are no data from the Red Sea. Marine debris is also known to have an impact on the mortality rate of marine turtles (e.g., Tomas et al. 2002) and litter is a major problem in the Red Sea (Abu-Hilal and

Al-Najjar 2004); however, there are no studies addressing the impact of such threats on marine turtles.

Management and Conservation

Due to their life cycle characterized by slow growth, late sexual maturity and high mortality rates during their first developmental stages, marine turtles are considered to be extremely vulnerable to exploitation (Bjorndal and Zug 1995) and it was demonstrated that population recoveries may take decades (Musick 1999). Currently, all the five species of marine turtles in the Red Sea are listed in the IUCN Red List either as critically endangered [leatherback and hawksbill turtles (Sarti-Martinez 2000; Mortimer and Donnelly 2008)], endangered [loggerhead and green turtles (MTSG 1996; Seminoff 2004)] or vulnerable [olive ridley turtle (Abreu-Grobois et al. 2007)]. Due to their vulnerability, marine turtles are protected in most countries bordering the Red Sea by national and international laws.

International Conventions

During their life cycle, marine turtles occupy a variety of habitats, from terrestrial areas to high seas and coastal shallow waters, and usually migrate between countries. For this reason, international cooperation is critical for their survival (Tiwari 2002). During the past decades, the number of international instruments addressing environmental protection in general and marine turtles in particular increased consistently (Frazier 2002; Hykle 2002). The role and value of these instruments reside in promoting harmonization of legislation among countries and the setting of minimum standards. In countries where environmental legislations are not fully developed or can be considered ambiguous, the acceptance of these international instruments usually forces the local governments to adopt more stringent rules.

Regional and international treaties can be divided into three categories: (1) those based on the species, (2) those targeting the protection of the general environment or biodiversity, or more specifically, the conservation of certain habitats and finally (3) there are fisheries-related instruments that usually provide guidelines for the protection of marine species (FAO 2004). Even if their effectiveness for marine turtle conservation has been debatable (Frazier 2002; Hykle 2002; Tiwari 2002), many of these instruments are binding for the parties and need to be implemented within national systems through the adoption of adequate legislation (FAO 2004). Hereafter, we discuss the most relevant of these instruments within the Red Sea region (see Table 1 for a summary of Red Sea countries that are signatories of international treaties).

Species-based treaties:

- African Convention on the Conservation of Nature and Natural Resources (Algiers 1968) about the preservation of wildlife in its natural environment lists in annex A (protected species) all marine turtles.
- Convention on International Trade in Endangered Species of wild animals (CITES, Washington, 1973): this convention aims at regulating international trade of different species of threatened animals. Animals are listed into three appendixes and are granted different degrees of protection. All species of marine turtles are listed in appendix I which forbids trade of these species in all signatory countries except in exceptional circumstances. In order to be legally binding, each signatory country must adopt national legislation under the framework established by the CITES.
- Convention on the Conservation of Migratory Species of Wild Animals (CMS, Bonn 1979): this convention aims to conserve terrestrial, aquatic and avian migratory

Table 1 Main international treaties on conservation of marine turtles and role of countries

	Egypt	Saudi Arabia	Jordan	Djibouti	Sudan	Eritrea	Yemen
Algiers	X (R)			X (R)	X (R)		
CITES	X (ACS)	X (ACS)	X (ACS)	X (ACS)	X (ACS)	X (ACS)	X (R)
CMS	X (P)	X (P)	X (P)	X (P)	X (AS)	X (P)	X (P)
CBD	X	X (ACS)	X	X (ACS)	X (ACS)	X (ACS)	X (ACS)
Ramsar	X		X	X	X		X
MARPOL	X			X			
Jeddah	X	X	X	X	X		X
UNCLOS	X (R)	X (R)	X (R)	X (R)	X (R)		X (R)
PERSGA	X	X	X	X	X		X

ACS accession,
R ratified,
P Party

species throughout their range. All species of marine turtles are listed in Appendix 1 (listing migratory species threatened with extinction) and Appendix 2 (migratory species for which conservation status would benefit from international cooperation). This convention is an intergovernmental treaty that becomes legally binding when agreements are signed and included into national legislation. Memoranda of Understanding (MoU) are less formal instruments (and non-legally binding ones). Some countries bordering the Red Sea are part of the MoU on Marine Turtles and their Habitats of the Indian Ocean and South-East Asia.

- Convention on Biological Diversity (CBD, Rio 1992): this is an internationally binding treaty aiming at conserving biodiversity in signatory countries, promoting sustainable use of resources and fair sharing of benefits from genetic resources. The convention should be implemented in signatory countries by national committees that have to prepare national action plans and ensure their implementation.

Habitat-based treaties (this is a non-exhaustive list):

- Convention on Wetlands of International Importance especially as Waterfowl Habitat (Ramsar 1971) on the importance of keeping the ecological character of the Wetlands of International Importance and planning for the sustainable use of all of the wetlands in their territories.
- International Convention for the Prevention of Pollution from Ships (MARPOL, London 1973) aimed at preventing and/or minimizing pollution from ships produced by accident or during their normal routine operations.
- Regional Convention for the Conservation of the Marine Environment of the Red Sea and the Gulf of Aden Environment (Jeddah 1982) that highlight the need for regional collaboration to protect the fragile Red Sea and Gulf of Aden ecosystems.
- United Nations Convention on the Law of the Sea (UNCLOS, Montego Bay 1972) aiming at establishing guidelines for the use of marine resources.

Legal Framework for Marine Turtle Management and Conservation by Country

Egypt

In Egypt, marine turtles and their most commonly used habitats are protected by many national laws and decrees:

- Law 53/1966 on Agriculture: This aims at organizing agricultural production, protecting agriculture and animal abundance, without minimizing the agricultural area while conserving its fertility. It protects mostly birds and tortoises but refers also to endangered animals (including marine turtles).
- Law 102/1983 on Protected areas: It sets the legal framework for the creation and management of protected areas. Every human activity in protected areas is to be strictly controlled. In particular, it forbids every activity "which will lead to the destruction or deterioration of the natural environment or harm the biota (terrestrial, marine or freshwater), or which will detract from the aesthetic (beauty) standards within protected areas", therefore protecting marine turtle habitats such as coral reefs and seagrass beds.
- Law 124/1983 on Fisheries: It regulates different aspects of the fishing activities setting standards for fishing techniques, banning destructive fishing methods, introducing a licence system for fishermen, regulating catches in freshwater bodies and at sea. It also includes a set of regulations for aquaculture.
- Law 4/1994 on the Environment: This is the most important law for the protection of marine turtles in Egypt. It states clearly in article 28 that "It is forbidden to hunt, kill, or catch the species of wild birds and animals determined in the executive regulations of this Law or to possess, transport, circulate with, sell or offer to sell such birds and animals either dead or alive". The law includes marine turtles and other species mentioned in international conventions for which Egypt is a signatory country.
- Prime Ministerial decree 264/1994: It regulates construction along the shoreline. No construction or development is allowed without permission from Egyptian Environmental Affairs Agency (EEAA).
- Prime Ministerial Decree 1068/1983, amended by Decree 2035/1996: It declares Ras Mohammed Reserve, Sanafir and Tiran islands as protected areas because they include important nesting and feeding areas for marine turtle populations.
- Prime Ministerial decree 134/2003: It establishes Wadi Gemal-Hamata National Park, which includes major inshore nesting sites.
- Decree Ministry of Agriculture 1403/1990: It forbids catching or killing any kind of reptiles in any way including the green turtle.

Jordan

In Jordan, the national legislation does not directly protect marine turtles but there are a number of laws that directly or indirectly protect coral reefs and other marine habitats. The most important are the Environment Protection Law No. 52 of 2006 (that replaced the Environment Protections Law No. 12 of 1995), and the Environment Protection Law No. 1 of 2003 that includes the establishment of the General Administration for the Environmental Protection setting its main aims and competencies. It also includes fines and penalties for those who break the environmental legislation.

Saudi Arabia

There are no legal instruments in Saudi Arabia targeting specifically marine turtles; however, as Saudi Arabia is a signatory party of many international conventions, these instruments define the legal framework for marine turtle protection. Furthermore, some existing national laws broadly address issues related to marine conservation, including protection of habitats used by marine turtles:

- Environmental Protection Standards Document No. 1401-01 (1402 H);
- The Council of Ministers Decision No. 271, 23/11/1404: It obliges "the use of best available technology to reduce pollutant emissions (such as cement dust)".
- The Royal Decree No. 7/505 M 28/3/1406 on fisheries regulations: This decree not only sets regulations for limiting fisheries but also divides responsibilities and management among different agencies. The Ministry of Agriculture has jurisdiction for fishery activities and issues permits for extensive filling of submerged lands that have been granted in the Eastern Province. The Meteorology and Environmental Protection Agency (MEPA) is responsible for setting standards for the environmental protection and for carrying out a programme of environmental impact assessment. The MEPA became the Presidency of Meteorology and Environment (PME), and its duties and responsibilities are detailed in the Royal Decree No.71 M/8903.
- Royal Decree No. 33, 27/7/137: It established the Saudi Arabian Coast Guard as the body having jurisdiction on the area between the border of the territorial seas (12 miles offshore) and 10 km inland.
- The Royal Decree No. M/22, dated 12/9/1406, established the National Commission for Wildlife Conservation and Development (NCWCD) as being responsible for the management of protected areas.

Djibouti

In Djibouti, national laws and decrees protect both marine turtles and their habitats:

- Decree 80–62/PR/MCTT of 25 May 1980: It provides the legal framework for the protection of the seabed and the marine fauna. The capture of marine mammals and turtles is illegal, as well as the trade with or export of these animals.
- Fishery laws (as drafted before the country's independence) as in articles 148, 149 and 220–225 (enacted by law 212/AN/82), and Articles 16–19 of Law No. 52/AN/78 (1979). These articles regulate the fishery and particularly forbid the use of certain destructive practices like the use of explosives and poisons. Fishery is regulated according to fishing zones and seasonal closures. This set of articles also establishes a fine of 50,000 Djibouti francs and 6 months jail for whoever is caught fishing for a marine turtle.

Eritrea

In Eritrea, habitats used by marine turtles are protected by the Marine and Coastal Integrated Management Proclamation (1995) that includes regulations about the management and protection of the coastal zone, the environmental effects of projects, pollution, and control of human activities and also establish sets of offences and penalties. Since 1998, according to the Eritrean Fisheries Proclamation 104/1998, direct harvest and domestic trade of endangered and protected species (including all species of marine turtles, as stated in the Protected Species Legal Notice 39/1998) are prohibited. The fisheries proclamation also sets rules for fishing activities, for trawlers in particular, trawling is possible only 4 miles from the islands and 8 miles from the coastline, at a minimum depth of 30 m. Furthermore, all trawlers must be equipped with a Turtle Excluder Device (TED); however, this regulation has not yet been implemented. The use of destructive fishing techniques (chemicals and explosives) is also strictly forbidden.

Sudan

The national legislation of the Republic of Sudan does not address marine turtles specifically; however, there are environmental laws that support control measures for pollution, navigation and protection of habitats like coral reefs:

- Marine Fisheries Ordinance (1937) is being updated; this set of regulations limits fishing activities and use of destructive fishing methods. No specific mention of marine turtles is made.
- The Environmental Health Act (1975) does not refer to marine turtles specifically but sets out regulations about water pollution and disposal of waste and sewage.
- Wildlife and National Parks Protection Act (1986) aims at defining and protecting national parks, and promoting rational use of wildlife. It also implements the Convention

on International Trade in Endangered Species of Wild Fauna and Flora (CITES) agreement of 1973.

- The Environmental Protection Act (2001) includes regulations for the protection of the environment and "to conserve the purity, natural components and equilibrium of the environment". The act also includes articles about the correct use of resources and affirms the connection between natural resources and development. It also establishes that environmental protection is competence of the High Council for Environment and Natural Resources.

Yemen

In Yemen, protection and conservation of marine turtles and their habitats is granted through different laws and regulations:

- Prime Ministerial Decree 94/1990: It established the creation of the Environmental Protection Council (EPC). This is the official government agency in charge of developing and implementing the national environmental policy, and therefore, it is the prime agency responsible for marine turtle conservation.
- The Presidential Resolution on Fishing, Exploitation and Protection of Living Aquatic Resources (Law No. 42) (1991): It represents the main set of regulations for all fishery activities. It states clearly that killing of marine turtles is strictly forbidden. The law also details which destructive fishing practices are illegal in Yemeni waters.
- The Ministerial Decree for Specifications of Fishing Vessels and Gear (No. 101) (1995) addresses indirectly marine turtles by regulating fishing gears and techniques.
- Prime Ministerial Decree No. 4 (1996) established Socotra as a protected area and developed a High Committee for Development of Socotra headed by the Deputy Prime Minister and Minister of Planning and Development.

Regional Organizations

The signatory countries of the Jeddah Convention (1982) united into a regional organization for the Conservation of the Environment of the Red Sea and Gulf of Aden (PERSGA). All bordering countries of the Red Sea excluding Eritrea are part of this organization that aims at promoting a rational use of living and non-living resources in the Red Sea. In 2004, PERSGA produced a regional action plan for marine turtles (PERSGA/GEF 2004) setting priorities for the research and conservation of these species. While the action plan clearly underlines the necessity of a prompt implementation of the procedures listed in the document, to date, no country has signed a memorandum of understanding to implement such activities.

References

Abreu-Grobois A, Plotkin P (IUCN SSC Marine Turtle Specialist Group) (2008) *Lepidochelys olivacea*. In: IUCN 2012. IUCN red list of threatened species, version 2012.2. www.iucnredlist.org. Downloaded on 3 June 2013

Abu-Hilal AH, Badran MM (1990) Effect of pollution sources on metal concentration in sediment cores from the Gulf of Aqaba (Red Sea). Mar Pollut Bull 21(4):190–197

Abu-Hilal AH, Al-Najjar T (2004) Litter pollution on the Jordanian shores of the Gulf of Aqaba (Red Sea). Mar Environ Res 58(1):39–63

Al Zibdah MK (2007) Population status and conservation of marine turtles at Jordan's Gulf of Aqaba, Red Sea. Testudo 6:58–66

Al-Mansi A, Nasser NA, Aden A (2003b) The marine turtles in the Republic of Djibouti: their biology and conservation. PERSGA Technical Report, Jeddah, 26 pp

Al-Mansi AM, Bilal SA, Abdullah EO, Elamin SM (2003a) The marine turtles in the Republic of Sudan: their biology and conservation. PERSGA Technical Report, Jeddah, 26 pp

Al-Merghani M, Miller JD, Pilcher NJ, Al-Mansi A (2000) The green and hawksbill turtles in the Kingdom of Saudi Arabia: synopsis of nesting studies 1986–1997. Fauna Arab 18:369–384

Attum O, Kramer A, Mahmoud T, Fouda M (2014) Post-nesting migrations patterns of Green Turtles (Chelonia mydas) from the Egyptian Red Sea Zoology in the Middle East, 60(4)

Baldwin R, Hughes GR, Prince RIT (2003) Loggerhead turtles in the Indian Ocean. In: Bolten AB, Witherington BE (eds) Loggerhead Sea turtles. Smithsonian Institution Press, Washington, pp 218–232

Bjorndal KA (1980) Nutrition and grazing behavior of the green turtle *Chelonia mydas*. Mar Biol 56:147–154

Bjorndal KA (1997) Foraging ecology and nutrition of marine turtles. In: Lutz PL, Musick JA (eds) The biology of marine turtles. CRC Press, Boca Raton, pp 199–231

Bjorndal KA, Jackson JBC (2003) Roles of marine turtles in marine ecosystems: reconstructing the past. In: Lutz PL, Musick JA, Wyneken J (eds) The biology of marine turtles, vol II. CRC Press, Boca Raton, pp 259–273

Bjorndal KA, Zug GR (1995) Growth and age of marine turtles. In: Bjorndal KA (ed) Biology and conservation of marine turtles. Smithsonian Institute Press, Washington, pp 599–600

Bouchard SS, Bjorndal KA (2000) Marine turtles as biological transporters of nutrients and energy from marine to terrestrial ecosystems. Ecology 81:2305–2313

Boulon RH (1999) Reducing threats to eggs and hatchlings: In situ protection. In: Eckert KL, Bjorndal KA, Abreu-Grobois FA and Donnelly M (eds) Research and management techniques for the conservation of marine turtles. IUCN/SSC Marine Turtle Specialist Group Publication No. 4

Bugoni L, Krause L, Petry VM (2001) Marine debris and human impacts on marine turtles in southern Brazil. Mar Pollut Bull 42 (12):1330–1334

CITES (2013) Convention on International Trade of Endangered Species. Available online: www.cites.org. Consulted on 12 Feb 2013

Crain DA, Bolten AB, Bjorndal KA (1995) Effects of beach nourishment on marine turtles: review and research initiatives. Restor Ecol 3:95–104

Davenport J, Davenport JL (2006) The impact of tourism and personal leisure transport on coastal environments: a review. Estuar Coast Shelf Sci 67(1–2):280–292

Dellinger T, Davenport J, Wirtz P (1997) Comparisons of social structure of Columbus crabs living on loggerhead marine turtles and inanimate flotsam. J Mar Biol Assoc UK 77(1):185–194

Denkinger J, Parra M, Muñoz JP, Carrasco C, Murillo JC, Espinosa E, Rubianes F, Koch V (2013) Are boat strikes a threat to marine turtles in the Galapagos marine reserve? Ocean Coast Manag 80:29–35

Derraik JG (2002) The pollution of the marine environment by plastic debris: a review. Mar Pollut Bull 44(9):842–852

Dobbs K (2001) Marine turtles in the great barrier reef world heritage area: a compendium of information and basis for the development of policies and strategies for the conservation of marine turtles. Great Barrier Reef Marine Park Authority, Townsville, Queensland, Australia

ElSadek I, Mancini A, Hanafy MH, Girondot M (2013) Green turtle nesting activities at a major rookery in the Southern Egyptian Red Sea. In: Tucker T, Belsiks L, Panagopoulou A, Rees A, Frick M, Williams K, LeRoux R, and Stewart K (eds) Proceedings on the 33rd annual symposium on sea turtle conservation and biology. NOAA technical memorandum NOAA NMFS-SEFSC-645

Engeman RM, Martin RE, Constantin B, Noel R, Woolard J (2003) Monitoring predators to optimize their management for marine turtle nest protection. Staff Publications, USDA National Wildlife Research Center. Paper 215. http://digitalcommons.unl.edu/icwdm_usdanwrc/215

Fergusson IK, Compagno LJ, Marks MA (2000) Predation by white sharks *Carcharodon carcharias* (Chondrichthyes: Lamnidae) upon chelonians, with new records from the Mediterranean Sea and a first record of the ocean sunfish *Mola mola* (Osteichthyes: Molidae) as stomach contents. Environ Biol Fishes 58(4):447–453

Fleming EH (2001) Swimming against the tide: Recent surveys of exploitation, trade, and management of marine turtles in the northern Caribbean. TRAFFIC North America, Washington DC, 185 pp

Food and Agriculture Organization (FAO) (2004) Report of the technical consultation on sea turtles conservation and fisheries. FAO Fisheries Report. No. 765. Bangkok, Thailand, 29 Nov–2 Dec 2004. Rome, FAO. 2005, 31 pp

Frazier J (2002) Marine turtles and international instruments: the agony and the ecstasy. J Int Wildl Law Policy 5(1–2):1–10

Frazier J (2003) Prehistoric and ancient historic interactions between humans and marine turtles. In: Lutz PL, Musick JA and Wyneken J (eds) The Biology of marine turtles, vol. 2. CRC Press, Boca Raton, Florida, pp 1–38

Frazier J (2005) Marine turtles: the role of flagship species in interactions between people and the sea. MAST 3(2) and 4(1):5–38

Frazier J, Salas S (1984) The status of marine turtles in the Egyptian Red Sea. Biol Conserv 30:41–67

Frith CB (1975) Predation upon hatchlings and eggs of the green turtle *Chelonia mydas*, on Aldabra Atoll, Indian Ocean. Atoll Res Bull 185:11–12

Gasperetti J, Stimson AF, Miller J, Ross JD, Gasperetti PR (1993) Turtles of Arabia. Fauna Saudi Arab 13:170–367

Gladstone W, Curley B, Shokri MR (2012) Environmental impacts of tourism in the Gulf and the Red Sea. Mar Pollut Bull Available online: http://dx.doi.org/10.1016/j.marpolbul.2012.09.017, 15 Oct 2012

Godley BJ, Thompson DR, Furness RW (1999) Do heavy metal concentrations pose a threat to marine turtles from the Mediterranean sea? Mar Pollut Bull 38(6):497–502

Green D (1996) Marine turtles of North Yemen. In: Keinath JA, Barnard DE, Musick JA and Bell BA (eds) In: Proceedings of the 15th annual workshop on sea turtle biology and conservation. NOAA Tech Memo NMFS-SEFSC-387:116

Groombridge B, Luxmoore R (1989) The green turtle and hawksbill (Reptilia: Cheloniidae): world status exploitation and trade. CITES Secretariat, Lausanne, 601 pp

Gulko DA, Eckert KL (2004) Marine turtles: an ecological guide. Mutual Publishing, Honolulu, HI, 128 pp

Hamann M, Limpus C, Hughes G, Mortimer L, Pilcher NJ (eds) (2000) Assessment of the conservation status of the leatherback turtle in the Indian Ocean and South-East Asia. IOSEA species assessment vol 1

Hanafy HM (2012) Nesting of marine turtles on the Egyptian beaches of the Red Sea. Egypt J Aquat Biol Fish 16(2):59–71

Hanafy MH, Sallam A (2003) Status of marine turtles nesting on the Egyptian beaches of the Red Sea. National Report to PERSGA, 45 pp

Hariri KI, Nichols P, Krupp F, Mishrigi S, Barrania A, Ali AF, Kedidi SM (2000) Status of the living marine resources in the Red Sea and Gulf of Aden region and their management. Regional Organisation for the Conservation of the Environment of the Red Sea and Gulf of Aden (PERSGA), Strategic Action Programme for the Red Sea and Gulf of Aden, Jeddah, Saudi Arabia

Hazel J, Gyuris E (2006) Vessel-related mortality of marine turtles in Queensland, Australia. Wildl Res 33:149–154

Hazel J, Lawler IR, Marsh H, Robson S (2007) Vessel speed increases collision risk for the green turtle *Chelonia mydas*. Endang Species Res 3(2):105–113

Heithaus MR, Wirsing AJ, Thomson JA, Burkholder DA (2008) A review of lethal and non-lethal effects of predators on adult marine turtles. J Exp Mar Biol Ecol 356(1):43–51

HEPCA (2012) An assessment of marine turtles in the Red Sea. Unpublished report, HEPCA, Hurghada, Egypt

Hillman JC, Gebremariam T (1996) The status of sea turtle conservation in Eritrea. In: Humphrey SL and Salm RV (eds) Status of sea turtle conservation in the western Indian Ocean. UNEP regional seas reports and studies No. 16, IUCN/ UNEP, Nairobi, pp 41–56

Houghton JDR, Doyle TK, Wilson MW, Davenport J, Hays GC (2006) Jellyfish aggregations and leatherback turtle foraging patterns in a temperate coastal environment. Ecology 87(8):1967–1972

Howe SA, Asfaha B, Kemp JM (2004) Turtle strandings in the southern Eritrean Red Sea. Mar Turt Newsl 103:4–7

Hughes GR, Luschi P, Mencacci R, Papi F (1998) The 7000-km oceanic journey of a leatherback turtle tracked by satellite. J Exp Mar Biol Ecol 229(2):209–217

Hykle D (2002) The convention on migratory species and other international instruments relevant to marine turtle conservation: pros and cons. J Int Wildl Law Policy 5:105–120

IUCN (2013) IUCN red list of threatened species. Available online: http://www.redlist.org. Consulted on 12 Feb 2013

Jackson JBC (1997) Reefs since Columbus. Coral Reefs 16:S23–S32

Jackson JBC, Kirby MX, Berger WH, Bjorndal KA, Botsford LW (2001) Historical overfishing and the recent collapse of coastal ecosystems. Science 293:629–638

James MC, Myers RA, Ottensmeyer CA (2005) Behaviour of leatherback marine turtles, *Dermochelys coriacea*, during the migratory cycle. Proc Royal Soc B Biol Sci 272(1572):1547–1555

Koch V, Nichols WJ, Peckham H, de la Toba V (2006) Estimates of sea turtle mortality from poaching and bycatch in Bahia Magdalena, Baja California Sur, Mexico. Biol Conserv 128(3):327–334

Leon YM, Bjorndal KA (2002) Selective feeding in the hawksbill turtle, an important predator in coral reef ecosystems. Mar Ecol Prog Ser 245:249–258

Lewison R, Freeman SA, Crowder LB (2004) Quantifying the effects of fisheries on threatened species: the impact of pelagic longlines on loggerhead and leatherback marine turtles. Ecol Lett 7:221–231

Losey GS, Balazs GH, Privitera LA (1994) Cleaning symbiosis between the wrasse, *Thalassoma duperry*, and the green turtle, *Chelonia mydas*. Copeia 5:684–690

Mancini A, Koch V (2009) Sea turtle consumption and black market trade in Baja California Sur, Mexico. Endang Species Res 7(1):1–10

Marine Turtle Specialist Group (MTSG) (1996) *Caretta caretta*. In: IUCN 2012. IUCN red list of threatened species, version 2012.2. www.iucnredlist.org. Downloaded on 8 June 2013

Marquez R (1990) Marine turtles of the world. An annotated y illustrated catalogue of sea turtle species known to date. FAO species catalogue, FAO fisheries synopsis, vol 11(125), 81 pp

Marsh H, Penrose H, Eros C, Hugues J (2002) The Dugong (*Dugong dugon*) status reports and action plans for countries and territories in its Range. Final Report, United Nations Environment Programme, Nairobi, Kenya

Meylan A (1988) Spongivory in hawksbill turtles: a diet of glass. Science 239(4838):393–395

Miller JD (1989) Marine turtles, volume 1: an assessment of the conservation status of marine turtles in the Kingdom of Saudi Arabia. Report No. 9, MEPA, Jeddah, Saudi Arabia, 289 pp

Miller JD (2004) Report on marine turtles of Wadi El Gemal-Hamata Park, southern Red Sea coast, Egypt. Report to the Red Sea Authority, Hurghada, Egypt

Moore RJ, Balzarotti MA (1977) Report of 1976 expedition to Suakin Archipelago (Sudanese Red Sea). Results of marine turtle survey and notes on marine and bird life. Unpublished, 27 pp

Moqbil AA (2007) Marine turtle activities in the Republic of Yemen. IOSEA profile of the month. Available: http://www.ioseaturtles.org/pom_detail.php?id=54. Consulted on 13 February 2013

Mortimer JA, Donnelly M (2008) *Eretmochelys imbricata*. In: IUCN red list of threatened species, version 2010.1. Available at www.iucnredlist.org

Musick JA (1999) Ecology and conservation of long-lived marine animals. Am Fish Soc Symp 23:1–10

Parsons J (1972) The hawksbill turtle and the tortoiseshell trade. Etudes de géographie tropicale offertes à Pierre Gourou, Mouton. Paris, La Haye

Peckham SH, Maldonado-Diaz D, Koch V, Mancini A, Gaos A, Tinker MT, Nichols WJ (2008) High mortality of loggerhead turtles due to bycatch, human consumption and standings at Baja California Sur, Mexico, 2003–2007. Endang Species Res 5(2–3), pp 171–183

PERSGA (1982) Regional Convention for the conservation of the red sea and Gulf of Aden Environment. Jeddah, 14 February 1982

PERSGA (2004) Regional action plan for the conservation of marine turtles and their habitats in the Red Sea and Gulf of Aden. PERSGA, Jeddah

Pfaller JB, Bjorndal KA, Reich KJ, Williams KL, Frick MG (2006) Distribution patterns of epibionts on the carapace of loggerhead turtles, *Caretta caretta*. Marine Biodivers Rec 1, e36. doi:10.1017/S1755267206003812

Pilcher NJ (1999) Cement dust as a cause of sea turtle hatchling mortality at Ras Baridi, Saudi Arabia. Mar Pollut Bull 38:966–969

Pilcher NJ, Al-Merghani M (2000) Reproductive biology of the green turtle *Chelonia mydas* at Ras Baridi, Saudi Arabia. Herpetolog Rev 32:142–149

Pilcher NJ, Saad M (2006) Status of leatherback turtles in Yemen. In: Hamann M, Limpus C, Hughes G, Mortimer L and Pilcher NJ (eds) Assessment of the conservation status of the leatherback turtle in the Indian Ocean and South-East Asia. IOSEA Species Assessment, vol 1

Pilcher NJ, Mahmud S, Howe S, Tecklemariam Y, Weldeyohannes S (2006) An update on Eritrea's marine turtle programme and first record of olive ridley turtle nesting in the Red Sea. Mar Turt Newsl 111:16

Pilcher, NJ, Mahmud S, Tecklemariam J (2006b). Status of leatherback turtles in Eritrea. In: Hamann M, Limpus C, Hughes G, Mortimer L and Pilcher NJ (eds) Assessment of the conservation status of the leatherback turtle in the Indian Ocean and South-East Asia. IOSEA Species Assessment, vol 1, pp 40–42

Plotkin P, Amos AF (1990) Effects of anthropogenic debris on marine turtles in the northwestern Gulf of Mexico. In: Shomura RS and Yoshida HO (eds) Workshop on the fate and impact of marine debris, Honolulu, pp 736–743

Pritchard PCH (1997) Evolution, phylogeny, and current status. In: Lutz PL, Musick JA (eds) The biology of marine turtles. CRC Press, Boca Raton, pp 115–123

Rees AF, Al-Kiyumi A, Broderick AC, Papathanasopoulou N, Godley BJ (2012) Each to their own: inter-specific differences in migrations of Masirah Island turtles. Chelonian Conserv Biol 11:243–248

Rinkevich B, Loya Y (1979) A Laboratory experiments on the effects of crude oil on the Red Sea coral *Stylophora pistillata*. Mar Pollut Bull 10(11):328–330

Ross JP (1985) Identification of marine turtles in the Red Sea. J Saudi Arab Nat Hist Soc 2:12–21

Ross JP, Barwani MA (1982) Review of marine turtles in the Arabian Area. In: Bjorndal KA (ed) Biology and conservation of marine turtles. Smithsonian Institution Press, Washington, DC, pp 373–382

Rouphael T, Abdulla A, Attum O, Marshall N, Ghazali U (2013) Do marine protected areas in the Red Sea afford protection to *dugongs* and marine turtles? J Biodivers Endang Species 1:102. doi:10.4172/jbes.1000102

Saad MA (2002) Status of marine turtles in Yemen: Survey report. PERSGA/SAP Technical Document 1981. In: Sanders MJ and Kedidi SM (eds) Summary review of the Red Sea commercial fisheries, catches and stock assessments. FAO/UNDP RAB/77/008/19, Jeddah, 12 pp

Sarti Martinez AL (Marine Turtle Specialist Group) (2000) *Dermochelys coriacea*. In: IUCN 2012. IUCN Red list of threatened species, version 2012.2. www.iucnredlist.org. Downloaded on 3 June 2013

Seminoff JA (2004) *Chelonia mydas*. In: IUCN 2012. IUCN Red List of Threatened Species, version 2012.2. www.iucnredlist.org. Downloaded on 3 June 2013

Stancyk SE (1982) Non-human predators of marine turtles and their control. In: Bjorndal KA (ed) Biology and conservation of marine turtles. Smithsonian Institution, pp 139–152

Teclemariam Y, Giotom M, Mengstu T, Abraha H, Mahmud S (2009) An update on marine turtles in Eritrea, Red Sea. Indian Ocean Turtle Newsl 9:6–10

Temcelmariam Y (2013) Marine turtle update from the Eritrean Red Sea. Indian Ocean Turtle Newsl 18:2–5

Thayer GW, Bjorndal KA, Ogden JC, Williams SL, Zieman JC (1984) Role of larger herbivores in sea grass communities: Functional ecology of sea grass ecosystems: a perspective on plant-animal interactions. Estuaries 7(4):351–376

Tisdell C, Wilson C (2002) Ecotourism for the survival of marine turtles and other wildlife. Biodivers Conserv 11:1521–1538

Tiwari M (2002) An evaluation of the perceived effectiveness of international instruments for sea turtle conservation. J Int Wildl Law Policy 5(1–2):145–156

Tomas J, Guitart R, Mateo R, Raga JA (2002) Marine debris ingestion in loggerhead marine turtles, *Caretta caretta*, from the Western Mediterranean. Mar Pollut Bull 44:211–216

Tomillo PS, Saba VS, Piedra R, Paladino FV, Spotila JR (2008) Effects of illegal harvest of eggs on the population decline of leatherback turtles in Las Baulas Marine national park. Costa Rica. Conserv Biol 22(5):1216–1224

United Nation (UN) (1992) Report on the United Nations Conference on Environment and Development. Rio de Janeiro, 3–14 June 1992

UNESCO (1971) Convention on wetlands of International importance, especially waterflow habitats. Ramsar, 2 February 1971

Van Buskirk J, Crowder LB (1994) Life-history variation in marine turtles. Copeia, pp 66–81

Vander Zanden HB, Bjorndal KA, Inglett PW, Bolten AB (2012) Marine-derived nutrients from green turtle nests subsidize terrestrial beach ecosystems. Biotropica 44:294–301

Wallace BP, Lewison RL, McDonald SL, McDonald RK, Kot CY, Kelez S, Bjorkland RK, Finkbeiner EM, Crowder LB (2010) Global patterns of marine turtle bycatch. Conserv Lett (3) 3:131–142

Wilson EG, Miller KL, Allison D, Magliocca M (2010) Why healthy oceans need marine turtles: the importance of marine turtles to marine ecosystems. Oceana Report, July 2010. Available at http://oceana.org/en/news-media/publications/reports/why-healthy-oceans-need-sea-turtles-the-importance-of-sea-turtles-to-marine-ecosystems

Witherington BE, Martin RE (1996) Understanding, assessing, and resolving light-pollution problems on sea turtle nesting beaches. FMRI technical report TR-2. Florida Marine Research Institute, St. Petersburg, Florida, 73 pp

Witherington BE, Salmon M (1992) Predation on loggerhead turtle hatchlings after entering the sea. J Herpetolog 26:226–228

Phytoplankton of the Red Sea

Amany A. Ismael

Abstract

Although the phytoplankton of the Red Sea has been studied since 1900, our information is still inadequate and information is scattered; the last review of the plankton was done by Halim (1969). The primary goal of this chapter is to give an overview of the phytoplankton of the entire Red Sea, not only of its composition and distribution, but also of the primary production and biomass. Earlier sources of information relating to both expeditions and individual works on this subject are considered in this review. The Red Sea is an oligotrophic basin characterized by special features due to its enclosed position. The phytoplankton of the entire Red Sea currently comprises 389 species and varieties, an increase of 181 species since Halim's review. Both the Gulf of Suez and Gulf of Aqaba are less diverse than the main Red Sea. There is a gradual decrease in phytoplankton richness from the southern Red Sea to the Gulf of Suez. Primary production and chlorophyll biomass increases from north to south, consistent with the distribution of nutrients.

Introduction

The Red Sea is a narrow basin of water located between 12.5°N and 30°N with a total area of approximately 4.51×10^5 km^2. The maximum depth in this sea is more than 2,500 m (Smeed 2004). The Red Sea is an oligotrophic water body characterized by special features due to its semi-closed position. The Red Sea connects the Arabian Sea and Mediterranean Sea through the Suez Canal in the north and Bab-al-Mandab in the south. There are many stories about the name of the Red Sea, but the one favored by phytoplankton scientists is that "the Red Sea owes its name to the occurrence of intense phytoplankton blooms, especially of the colony-forming filamentous cyanobacteria *Trichodesmium*, discoloring its surface waters" (Belogoroskaya 1970). There is no river inflow to the Red Sea, and evaporation is high at more than 210 cm/yr (Shaikh et al. 1986). Salinity increases from south to north (Edwards 1987). The water column of the Red Sea is highly stratified; the thermocline extends from 50 to 250–300 m (Edwards 1987). Nutrient concentrations generally increase from north to south, although the Red Sea waters are considered highly deficient in major nutrients (Weikert 1987). The most nutrient-poor waters are in the northern Red Sea, while the southern part receives nutrients from the Gulf of Aden (Acker et al. 2008). As a result, seasonal inflow of the surface water of the Gulf of Aden appears to play an important role in the maintenance of pelagic organisms and their diffusion within the Red Sea basin.

The direction of flow is determined by the monsoons. In summer, during the SW monsoon, winds prevail along the whole of the Red Sea, causing a surface flow of Red Sea water into the Gulf of Aden. In winter, the NE monsoon generates a current in the opposite direction, so during the period from November to March, there is a flow from the Gulf of Aden into the southern part of the Red Sea (Halim 1984; Kimor 1973).

Although there are phytoplankton records from the Red Sea from before 1900 (Ehrenberg 1838), the information available is scarce and scattered, and much more attention has been given to the hydrography, benthic fauna and fish.

A.A. Ismael (✉)
Oceanography Department, Faculty of Science,
Alexandria University, Alexandria, Egypt
e-mail: amany_3@yahoo.com

N.M.A. Rasul and I.C.F. Stewart (eds.), *The Red Sea*,
Springer Earth System Sciences, DOI 10.1007/978-3-662-45201-1_32,

Qualitative studies of phytoplankton based on a few samples were carried out by Cleve (1900, 1903), Ostenfeld and Schmidt (1901), and Schröder (1906). The German Deep-Sea Expedition (Karsten 1907) studied the distribution of phytoplankton from six stations located in the northern and southern Red Sea up to 100 m depth. Matzenauer (1933) studied the dinoflagellate (excluding *Ceratium*) from 11 samples in the Red Sea. The first determinations of phytoplankton productivity of the Red Sea were carried out during ATLANTIS Cruise 242, in 1958 (Yentsch and Wood 1961), while during ATLANTIS Cruises 8 (1963) and 15 (1965), chlorophyll concentrations were determined in the western Indian Ocean and Red Sea (Yentsch 1965; McGill and Lawson 1966). The first review article of the plankton of the Red Sea was made by Halim (1969). The primary production of the Red Sea was examined by Koblentz-Mishke et al. (1970). The distribution and ecology of the genus *Ceratium* in the central and northern Red Sea were studied by Dowidar (1983a). The distribution and production of phytoplankton in the Red Sea off Jeddah were studied by Shaikh et al. (1986). Lenz et al. (1988) examined the diatom assemblage in the open waters of Sudan. Kimor (1990) reported the primary production and species composition of microplankton in the northern Red Sea, Gulf of Suez, and Levantine Basin of the Mediterranean. The seasonal fluctuations of biomass and productivity of phytoplankton of the Somali Current, Gulf of Aden, and southern Red Sea were determined by Baars et al. (1998). In addition, the seasonal changes of phytoplankton and zooplankton in relation to nutrient inflow in the Red Sea were studied by Getahun (1998). The phytoplankton distribution and the grazing effect on phytoplankton in the Gulf of Aqaba and the northern Red Sea were studied by Sommer (2000) and Sommer et al. (2002). The biological dynamics of the northern Red Sea using chlorophyll-a data generated by SeaWiFS was investigated by Acker et al. (2008). Most recently, the phytoplankton composition in relation to environmental parameters was studied by Touliabah et al. (2010), and the interannual variability of chlorophyll by Elawad (2012).

This chapter reviews the present knowledge of phytoplankton distribution, productivity, and biomass in the whole Red Sea, since the last review by Halim (1969).

Red Sea Proper

The phytoplankton of the Red Sea is entirely tropical and consists of 377 species and varieties (Table 1). It is considerably richer in dinoflagellates, with 168 species (Figs. 1 and 2, Table 1) than in diatoms (137 species). The Indo-Pacific species, namely *Neoceratium carriense* (*Ceratium carriense*), *Neoceratium massiliense* (*C. massiliense*), and *Neoceratium trichoceros* (*C. trichoceros*), are the major components within the phytoplankton community for the Red Sea proper and the Gulf of Suez (Halim 1969; Ismael 2005). The dinoflagellate is present at all times except in February. It appears that the Red Sea dinoflagellates constitute an indigenous population which does not depend on the winter monsoon and inflow from the Gulf of Aden. Chrysophyceae are scarce in the Red Sea, represented by *Dictyocha calida*, *Dictyocha fibula*, and *Dictyocha octonaria* (Touliabah et al. 2010). Coccolithophores are represented by *Rhabdosphaera claviger* (Ostenfeld and Schmidt 1901) and *Syracosphaera cordiformis* which dominate the Gulf of Suez (Ismael 2005). Sukhanova (1969) established important differences in the diversity of species in the different parts of the Red Sea. The central part of the Red Sea is the poorest in species with a more diverse population in the northern and southern parts. Generally, phytoplankton abundance decreases from the southern Red Sea to the Gulf of Suez (Halim 1969), but both the Gulf of Suez and Gulf of Aqaba exhibit less diversity than the main Red Sea (Halim 1984). Dense blooms of the dinoflagellate *Noctiluca scintillans* (*Noctiluca miliaris*) often occur in the Red Sea, particularly in the warm season (Halim 1969). A bloom of *Noctiluca scintillans* was reported also off the coast of Saudi Arabia during winter (Mohamed and Mesaad 2007). However, several studies show that winter is the most productive season throughout the Red Sea and suggest a close correlation between phytoplankton productivity and conditions resulting from the winter monsoon. This winter peak is clearly dominated by diatoms, genera *Bacteriastrum*, *Chaetoceros*, *Nitzschia*, and *Rhizosolenia*. *Chaetoceros coarctatus* with its *Vorticella microstoma* is characteristic of the plankton of the entire Red Sea. Although Kimor (1973) reported the almost complete absence of diatom blooms in the Red Sea, massive blooms of *Hemidiscus cuneiformis*, *Rhizosolenia calcar-avis*, *Rh. imbricata* (*Rh. Shrubsolei*), and *Guinardia flaccida* have been recorded during winter (Halim 1969; Shaikh et al. 1986). The northern Red Sea shows a different trend. It is characterized by the dominance of autotrophic picoplankton, while larger cells are scarce in this region. Four different groups of phytoplankton are important in the annual cycle of the northern Red Sea. These are as follows: diatoms which are present most of the year and are the predominant group in winter, blue-green algae (*Trichodesmium* spp.) and nanoplankton (*Prochlorococcus* spp. and *Synechococcus* spp.) which dominate in late spring through summer, reaching up to 10^8 cells l^{-1} and contribute more than 95 % to the total chlorophyll (Chl-a) (Stambler 2005), and finally dinoflagellates in the fall (Seeberg-Elverfeldta et al. 2004). Phytoplankton abundance off Saudi Arabian coasts ranged from 0.12×10^6 to 2.6×10^6 m^{-3} (Touliabah et al. 2010). The water column throughout the Red Sea is highly stratified, with a thermocline extending

Table 1 Phytoplankton population of the Red Sea and its distribution through the Red Sea Proper (R.S.P.), Gulf of Suez (G.S.), and Gulf of Aqaba (G.A.)

Species	R. S. P	G. S.	G. A.
Diatoms			
Actinocyclus curvatulus Janisch in A. Schmidt	+		
Actinocyclus exiguus Fryxell and Semina	+		
Actinocyclus octonarius Ehrenberg	+		
Actinocyclus subtilis (Gregory) Ralfs	+		
Actinoptychus senarius[a] (Ehrenberg) Ehrenberg	+		
Actinoptychus vulgaris Schuman	+		
Amphora laevis Gregory	+		
Asterolampra marylandica[a] Ehrenberg	+	+	
Asteromphalus arachne (Brébisson) Ralfs	+		
Asteromphalus cleveanus Grunow	+		
Asteromphalus flabellatus[a] (Brébisson) Greville	+		
Asteromphalus heptactis[a] (Brébisson) Ralfs	+		
Azpeitia Africana (Janisch ex A.Schmidt) G. Fryxell and T.P. Watkins	+		
Azpeitia barronii G Fryxell and T.P. Watkins	+		
Azpeitia neocrenulata G Fryxell and T.P. Watkins	+		
Azpeitia nodulifera (A. Schmidt) G. Fryxell and P.A. Sims	+		
Bacillaria Paxillifera[a] (Müller) Hendey		+	
Bacteriastrum delicatulum[a] Cleve	+	+	
Bacteriastrum elongatum Cleve	+		
Bacteriastrum furcatum[a] Shadbolt	+		
Bacteriastrum hyalinum Lauder	+	+	
Biddulphia tuomeyi (Bailey) Roper		+	
Campylodiscus decorus Brébisson		+	
Cerataulina bicornis[a] (Ehrenberg) Hasle	+	+	
Cerataulina Pelagicai[a] (Cleve) Hendey	+		
Chaetoceros atlanticus Cleve	+		
Chaetoceros affinis[a] Lauder	+	+	
Chaetoceros anastomosans[a] Grunow	+		
Chaetoceros coarctatus[a] Lauder	+	+	+
Chaetoceros compactus[a] Schütt	+		
Chaetoceros compressus[a] Lauder	+	+	+
Chaetoceros contortus[a] Pavillard	+		
Chaetoceros curvisetus Cleve	+	+	+
Chaetoceros dadayi Pavillard			+
Chaetoceros decipiens[a] Cleve	+	+	+
Chaetoceros denticulatum[a] Lauder	+		
Chaetoceros diadema[a] (Ehrenberg) Gran	+		
Chaetoceros didymus[a] Ehrenberg	+	+	
Chaetoceros diversus[a] Cleve	+	+	
Chaetoceros exiguous[a] Cleve??			
Chaetoceros laciniosus[a] Schütt	+	+	+
Chaetoceros lorenzianus[a] Grunow	+	+	+

(continued)

Table 1 (continued)

Species	R. S. P	G. S.	G. A.
Chaetoceros messanense[a] Castracane	+		
Chaetoceros neapolitanum[a] Schroeder	+		
Chaetoceros peruvianus[a] Brightwell	+	+	
Chaetoceros pseudocurvisetum Mangin	+		
Chaetoceros rostratus[a] Lauder	+		
Chaetoceros schmidtii[a] Ostenfeld	+		
Chaetoceros seychellarus[a] Karsten	+		
Chaetoceros socialis Lauder		+	
Chaetoceros sumatranum[a] Karsten	+		
Chaetoceros tetrastichon[a] Cleve	+		
Climacodium biconcavum[a] Cleve	+		
Climacodium frauenfeldianum[a] Grunow	+	+	+
Cocconeis lyra[b] Schmidt		+	
Cocconeis pediculus Ehrenberg	+		
Cocconeis placentula Ehrenberg	+		
Cocconeis scutellum Ehrenberg	+	+	
Cocconeis stauroneiformes (Rabenhorst) Okuno	+		
Corethron cryophilum[a] Castracane	+		
Coscinodiscus centralis Ehrenberg	+	+	+
Coscinodiscus gigas[a] Ehrenberg	+	+	
Coscinodiscus granii Gough		+	
Coscinodiscus janischii Schmidt	+		
Coscinodiscus oculus-iridis Ehrenberg			+
Coscinodiscus perforatus Ehrenberg	+	+	+
Coscinodiscus radiatus[a] Ehrenberg	+	+	+
Coscinodiscus thorii Duda	+		
Cyclotella litoralis Lange and Syvertsen	+		
Cyclotella meneghiniana Kützing	+	+	
Cylindrotheca closterium (Ehrenberg) Lewin and Reimann		+	+
Cymatonitzschia marina Simonsen	+		
Dactyliosolen antarcticus[a] Castracane		+	
Dactyliosolen blavyanus[b] (Peragallo) Hasle		+	
Dactyliosolen hyalinus[a] Cleve	+		
Dactyliosolen mediterraneus[a] Peragallo			+
Detonula pumila[a] (Castracane) Gran	+		
Diploneis constricta (Grunow) Cleve		+	
Diploneis crabro (Ehrenberg) Ehrenberg	+	+	+
Eucampia cornuta[a] (Cleve) Grunow	+		
Eucampia hemiauloides[a] Ostenfeld	+		
Fragilariopsis doliolus (Wallich) Medlin and Sims	+		
Grammatophora oceanica Ehrenberg		+	
Guinardia flaccida[a] Peragallo	+	+	+
Guinardia striata[a] (Stolterfoth) Hasle		+	
Gyrosigma balticum (Ehrenberg) Rabenherst		+	

(continued)

Table 1 (continued)

Species	R. S. P	G. S.	G. A.
Helicotheca tamesis[a] (Schrubsole) Richard	+		
Hemiaulus hauckii Grunow in van Heurck	+	+	+
Hemiaulus heibergi[a] Cleve??	+		
Hemiaulus sinensis Greville	+	+	+
Hemidiscus cuneiformis[a] Wallich	+	+	
Lauderia annulata[a] Cleve	+	+	+
Leptocylindrus danicus[a] Cleve	+	+	
Leptocylindrus mediterraneus (Peragallo) Hasle	+		
Lyrella lyroides (Hendey) Mann		+	
Mastogloia rostrata (Wallich) Hustedt	+		
Melosira granulata (Ehrenberg) Ralfs	+	+	
Meuniera membranacea[a] (Cleve) Silva	+	+	+
Navicula directa (Smith) Ralfs	+		
Navicula florinae Mørller		+	
Navicula perminuta Grunow	+		
Neostreptotheca subindica[a] Von Stosch	+		
Nitzschia bicapitata Cleve	+		
Nitzschia bilobata[b] Smith	+		
Nitzschia braarudii Hasle	+		
Nitzschia longissima[a] (Brébisson) Ralfs	+	+	+
Nitzschia sicula (Castracane) Hustedt	+		
Nitzschia sigma (Kützing) Smith	+		
Odontella sinensis[a] (Greville) Grunow		+	
Planktoniella sol[a] (Wallich) Schütt	+		+
Pleurosigma angulatum (Queckett) Smith	+		+
Pleurosigma directum Grunow	+		
Pleurosigma formosum[b] Smith		+	
Pleurosigma normanii Ralfs			+
Pleurosigma salinarum (Grunow) Grunow	+		
Proboscia alata[a] (Brightwell) Sundström	+	+	+
Proboscia truncate[a] (Karsten) Nöthig and Ligowski	+	+	
Pseudo-nitzschia australis Fringuelli	+		
Pseudo-nitzschia delicatissima (Cleve) Heiden	+		+
Pseudo-nitzschia fraudulenta (Cleve) Hasle	+		
Pseudo-nitzschia granii Hasle	+		
Pseudo-nitzschia prolongatoides Hasle	+		
Pseudo-nitzschia pungens[a] (Grunow ex Cleve) Hasle	+	+	
Pseudonitzschia seriata[a] (Cleve) Peragallo	+	+	
Pseudo-nitzschia subcurvata (Hasle) Fryxell	+		
Pseudotriceratium punctatum (Wallich) Simonsen	+		
Rhizosolenia acicularis Sundström	+		
Rhizosolenia bergonii[a] Peragallo	+	+	
Rhizosolenia calcar-avis[a] Schultze	+	+	+
Rhizosolenia clevei Ostenfeld		+	

(continued)

Table 1 (continued)

Species	R. S. P	G. S.	G. A.
Rhizosolenia fragillima[a] Gran??	+		
Rhizosolenia hebetata Bailey	+	+	+
Rhizosolenia hyalina[a] Ostenfeld	+	+	
Rhizosolenia imbricate[a] Brightwell	+	+	+
Rhizosolenia ostenfeldii Sundström	+		+
Rhizosolenia pungens Cleve-Euler	+		
Rhizosolenia quadrijuncta[a] Peragallo??	+		
Rhizosolenia robusta[a] Norman	+		
Rhizosolenia semispina[a] Hensen	+		
Rhizosolenia setigera[a] Brightwell	+		+
Rhizosolenia styliformis[a] Brightwell	+	+	+
Rhizosolenia temperei Peragallo	+		
Roperia tesselata (Roper) Grunow ex Pelletan	+		
Shionodiscus oestrupii[a] (Ostenfeld) Alverson, Kang and Theriot	+		
Stellarima stellaris (Roper) Hasle and Sims		+	
Stephanopyxis turris[a] (Greville) Ralfs ex Pritchard	+		
Streptotheca indica[a] Karsten	+		
Striatella delicatula[a] (Kützing) Grunow ex Van Heurck	+	+	+
Surirella fastuosa[b] Ehrenberg		+	
Surirella gemma[b] Ehrenberg	+	+	+
Synedra crystalline[a] (C. Agardh) Kützing	+		
Synedra ulna Ehrenberg		+	
Thalassionema bacillare (Heiden) Kolbe	+		
Thalassionema frauenfeldii[a] (Grunow) Tempère and Peragallo	+		
Thalassionema nitzschioides[a] (Grunow) Grunow ex Hustedt	+	+	+
Thalassiosira aestivalis Gran and Angst	+		
Thalassiosira anguste-lineata[a] (Schmidt) Fryxell and Hasle	+		
Thalassiosira decipiens (Grunow) E.G. Jørgensen	+		+
Thalassiosira delicatula Ostenfeld in Borgert	+		
Thalassiosira eccentrica[a] (Ehrenberg) Cleve	+	+	+
Thalassiosira endoseriata Hasle and G. Fryxe	+		
Thalassiosira ferelineata Hasle and G.Fryxell	+		
Thalassiosira hyalina (Grunow) Gran	+		
Thalassiosira leptopus (Grunow) Hasle and G.Fryxell	+		
Thalassiosira lineate[a] (Schmidt) Fryxell and Hasle	+	+	
Thalassiosira mala Takano	+		
Thalassiosira minima Gaarder	+		
Thalassiosira monile[a] Cleve??	+		
Thalassiosira oceanica Hasle	+		
Thalassiosira oestrupii (Ostenfeld) Hasle	+		
Thalassiosira pacifica Gran and Angst	+		
Thalassiosira poro-irregulata Hasle and Heimdal	+		
Thalassiosira sacketii Fryxel	+		
Thalassiosira subtilis (Ostenfeld) Gran	+	+	

(continued)

Table 1 (continued)

Species	R. S. P	G. S.	G. A.
Thalassiosira symmetrica Fryxell and Hasle	+		
Thalassiothrix longissima[a] Cleve and Grunow	+	+	
Triceratium antediluvianum[b] (Ehrenberg) Grunow		+	
Dinoflagellates			
Akashiwo sanguineum (Hirasaka) Hansen and Møestrup	+	+	
Amphidinium turbo[a] Kofoid and Swezy	+		
Amphisolenia bidentata[a] Schröder	+	+	+
Amphisolenia globifera[a] Stein	+		
Amphisolenia palmata[a] Stein	+		
Blepharocysta splendor-maris[a] (Ehrenberg) Ehrenberg	+		
Ceratium arietinum var. *gracilentum* (Jörgensen) Sournia		+	+
Ceratium horridum var. *buceros*[a] (Zacharias) Sournia	+	+	+
Ceratocorys armata (Schütt) Kofoid	+	+	+
Ceratocorys bipes[a] (Cleve) Kofoid	+	+	
Ceratocorys gourretii[a] Paulsen	+	+	+
Ceratocorys horrida[a] Stein	+	+	+
Cithuristes regius Stein		+	
Cladopyxis brachiolata[a] Stein	+	+	
Dinophysis caudata[a] Saville- Kent	+	+	+
Dinophysis hastate[a] Stein	+		
Dinophysis michaelis[a] Ehrenberg	+		
Dinophysis miles[a] Cleve	+		
Dinophysis monacantha Kofoid and Skogsberg		+	
Dinophysis ovum Schütt	+	+	
Dinophysis parvula (Schütt) Balech	+		
Dinophysis punctata[a] Jörgensen	+		
Dinophysis schuettii Murray and Whitting	+		
Dinophysis similis[a] Kofoid et Skogsberg	+		
Dinophysis sphaerica[a] Stein	+		
Dinophysis tripos[a] Gourret	+	+	
Dinophysis uracantha[a] Stein	+	+	
Diplpopsalis lenticula[a] Bergh	+		
Diplpopsalis saecularis[a] Murray and Whitting??	+		
Dissodium asymmetricum[a] (Mangin) Loeblich III	+		
Gonidoma polyedricum[a] (Pouchet) Jörgensen	+	+	+
Gonidoma sphaericum[a] Murray and Whitting	+	+	+
Gonyaulax digensis Kofoid	+		
Gonyaulax glyptorhynchus Murray and Whitting			+
Gonyaulax grindleyi Reinecke	+		
Gonyaulax hyaline[a] Ostenfeld and Schmidt	+		+
Gonyaulax kofoidii Pavillard			+
Gonyaulax minima[a] Matzenauer	+		+
Gonyaulax monocantha Pavillard	+		
Gonyaulax orientalis[a] Lindemann	+		

(continued)

Table 1 (continued)

Species	R. S. P	G. S.	G. A.
Gonyaulax ovalis Schiller	+		
Gonyaulax ovata[a] Matzenauer	+		+
Gonyaulax pacifica Kofoid	+		
Gonyaulax polygramma[a] Stein	+	+	+
Gonyaulax scrippsea Kofoid		+	+
Gonyaulax spinifera[a] (Claparède and Lachmann) Diesing	+	+	
Gonyaulax turbynei[a] Murray and Whitting	+		
Gonyaulax verior Sournia	+		
Gymnodinium galaeformis[a] Matzenauer	+		+
Gyrodinium nasutum (wulff) schiller		+	
Histoneis crateriformis[a] (Stein) Kofoid et Skogsberg	+		
Histoneis depressa Schiller			+
Neoceratium arietinum[a] (Cleve) Gomez, Moreira and Lopez-Garcia	+	+	+
Neoceratium azoricum[a] (Cleve) Gomez, Moreira and Lopez-Garcia	+	+	
Neoceratium belone[a] (Cleve) Gomez, Moreira and Lopez-Garcia	+	+	+
Neoceratium bigelowii (Kofoid) Gomez, Moreira and Lopez-Garcia	+		
Neoceratium breve[a] (Ostenfeld and Schmidt) Gomez, Moreira and Lopez-Garcia	+	+	
Neoceratium candelabrum[a] (Ehrenberg) Gomez, Moreira and Lopez-Garcia	+	+	+
Neoceratium carnegiei[a] (Graham and Bronikowsky) Gomez, Moreira and Lopez-Garcia	+	+	+
Neoceratium concilians Jörgensen	+	+	
Neoceratium contortum[a] (Gourret) Gomez, Moreira and Lopez-Garcia	+	+	+
Neoceratium contrarium[a] (Gourret) Gomez, Moreira and Lopez-Garcia	+	+	+
Neoceratium declinatum (Karsten) Gomez, Moreira and Lopez-Garcia	+	+	+
Neoceratium deflexum (Kofoid) Gomez, Moreira and Lopez-Garcia	+	+	+
Neoceratium dens[a] (Ostenfeld et Schmidt) Gomez, Moreira and Lopez-Garcia	+		
Neoceratium egyptiacum[a] (Halim) Gomez, Moreira and Lopez-Garcia	+	+	
Neoceratium euarcuatum (Jörgensen) Gomez, Moreira and Lopez-Garcia	+		
Neoceratium extensum[a] (Gourret) Gomez, Moreira and Lopez-Garcia	+	+	+
Neoceratium falcatum (Kofoid) Gomez, Moreira and Lopez-Garcia		+	
Neoceratium furca[a] (Ehrenberg) Gomez, Moreira and Lopez-Garcia	+	+	+
Neoceratium fusus[a] (Ehrenberg) Gomez, Moreira and Lopez-Garcia	+	+	+
Neoceratium gibberum (Gourret) Gomez, Moreira and Lopez-Garcia	+	+	+
Neoceratium gravidum[a] (Gourret) Gomez, Moreira and Lopez-Garcia	+		+
Neoceratium hexacanthum[a] (Gourret) Gomez, Moreira and Lopez-Garcia	+	+	+
Neoceratium horridum[a] (Gran) Gomez, Moreira and Lopez-Garcia	+	+	+
Neoceratium humile (Jörgensen) Gomez, Moreira and Lopez-Garcia	+	+	+
Neoceratium incisum[a] (Karsten) Gomez, Moreira and Lopez-Garcia	+		+
Neoceratium inflatum (Kofoid) Gomez, Moreira and Lopez-Garcia	+	+	+
Neoceratium karstenii[a] (Pavillard) Gomez, Moreira and Lopez-Garcia	+	+	+
Neoceratium kofoidii[a] (Jörgensen) Gomez, Moreira and Lopez-Garcia	+	+	+
Neoceratium limulus[a] (Gourret) Gomez, Moreira and Lopez-Garcia	+		+
Neoceratium lineatum[a] (Ehrenberg) Gomez, Moreira and Lopez-Garcia	+	+	
Neoceratium longipes (Bailey) Gomez, Moreira and Lopez-Garcia		+	

(continued)

Table 1 (continued)

Species	R. S. P	G. S.	G. A.
Neoceratium longirostrum (Gourret) Gomez, Moreira and Lopez-Garcia	+	+	
Neoceratium longissimum (Schröder) Gomez, Moreira and Lopez-Garcia	+		
Neoceratium lunula (Schimper ex Karsten) Gomez, Moreira and Lopez-Garcia	+	+	+
Neoceratium macroceros[a] (Ehrenberg) Gomez, Moreira and Lopez-Garcia	+	+	+
Neoceratium massiliense (Gourret) Gomez, Moreira and Lopez-Garcia	+	+	+
Neoceratium minutum (Jörgensen) Gomez, Moreira and Lopez-Garcia	+	+	+
Neoceratium pavillardi (Jörgensen) Gomez, Moreira and Lopez-Garcia	+	+	
Neoceratium pentagonum[a] (Gourret) Gomez, Moreira and Lopez-Garcia	+	+	
Neoceratium platycorne (Daday) Gomez, Moreira and Lopez-Garcia	+		+
Neoceratium praelongum (Lemmermann) Gomez, Moreira and Lopez-Garcia	+		+
Neoceratium pulchellum (Schröder) Gomez, Moreira and Lopez-Garcia	+	+	+
Neoceratium ranipes (Cleve) Gomez, Moreira and Lopez-Garcia	+		+
Neoceratium reflexum[a] (Cleve) Gomez, Moreira and Lopez-Garcia	+	+	+
Neoceratium symmetricum (Pavillard) Gomez, Moreira and Lopez-Garcia	+	+	
Neoceratium tenue[a] (Ostenfeld and Schmidt) Gomez, Moreira and Lopez-Garcia	+		
Neoceratium teres (Kofoid) Gomez, Moreira and Lopez-Garcia	+	+	+
Neoceratium trichoceros[a] (Ehrenberg) Gomez, Moreira and Lopez-Garcia	+	+	+
Neoceratium tripos[a] (Müller) Gomez, Moreira and Lopez-Garcia	+	+	+
Neoceratium uncinus (Sournia) Gomez, Moreira and Lopez-Garcia	+	+	+
Noctiluca scintillans[a] (Macartney) Kofoid and Swezy	+	+	
Ornithocercus magnificus[a] Stein	+		+
Ornithocercus quadratus[a] Schütt	+	+	+
Ornithocercus splendidus[a] Schütt	+		
Ornithocercus steinii[a] Schütt	+		
Oxytoxum constrictum[a] (Stein) Bütschli	+		
Oxytoxum gladiolus[a] Stein	+		+
Oxytoxum longiceps Schiller		+	
Oxytoxum milneri[a] Murray and Whitting	+		+
Oxytoxum scolopax[a] Stein	+	+	+
Oxytoxum sphaeroideum[a] Stein	+		+
Oxytoxum tesselatum[a] (Stein) Schütt	+		
Phalacroma argus[a] Stein	+		+
Phalacroma cuneus[a] Schütt	+		+
Phalacroma doryphorum[a] Stein	+	+	+
Phalacroma favus Kofoid and Michener		+	
Phalacroma operculatum[a] Stein	+		+
Phalacroma prodictyum[a] Stein	+		
Phalacroma rapa[a] Jörgensen	+		
Podolampas bipes[a] Stein	+	+	
Podolampas palimpes[a] Stein	+	+	+
Prorocentrum compressum[a] (Bailey) Abé ex Dodge	+	+	+
Prorocentrum gibbosum[a] (Schiller) Schiller	+		
Prorocentrum gracile[a] Schütt	+	+	+

(continued)

Table 1 (continued)

Species	R. S. P	G. S.	G. A.
Prorocentrum micans[a] Ehrenberg	+	+	+
Prorocentrum oblongum[a] (Schiller) Abé	+	+	
Prorocentrum obtuse Pavillard	+	+	+
Prorocentrum scutellum Schröder		+	
Protoceratium areolatum kofoid	+		
Protoceratium reticulatum[a] (Claperede and Lachmann) Bütschli	+		
Protoceratium spinulosum (Murray et Whitting) Schiller	+		
Protoperidinium abei[a] (Paulsen) Balech	+		
Protoperidinium africanoides[a] (Dangeard) Balech	+	+	
Protoperidinium asymmetricum[a] Balech	+		
Protoperidinium breve Paulsen	+	+	
Protoperidinium brochi (Kofoid and Swezy) Balech	+	+	+
Protoperidinium cerasus (Paulsen) Balech	+	+	+
Protoperidinium conicum[a] (Gran) Balech	+	+	+
Protoperidinium crassipes[a] (Kofoid) Balech	+	+	+
Protoperidinium depressum[a] (Bailey) Balech	+	+	+
Protoperidinium diabolus[a] (Cleve) Balech	+	+	+
Protoperidinium divergens[a] (Ehrenberg) Balech	+	+	+
Protoperidinium elegans[a] (Cleve) Balech	+		
Protoperidinium ellipticum[a] Halim	+		
Protoperidinium globulus[a] (Stein) Balech	+	+	+
Protoperidinium gracile[a] Gran and Braarud	+		
Protoperidinium grande[a] (Kofoid) Balech	+	+	+
Protoperidinium inclinatum (Balech) Balech	+		
Protoperidinium inflatum[a] (Okamura) Balech	+		
Protoperidinium leonis[a] (Pavillard) Balech	+	+	+
Protoperidinium matzenaueri[a] (Böhm) Balech	+		
Protoperidinium nipponicum[a] (Abé) Balech	+	+	
Protoperidinium obtusum[a] (Karsten) Park and Dodge	+		
Protoperidinium oceanicum[a] (VanHöffen) Balech	+	+	+
Protoperidinium orientale[a] (Matzenauer) Balech	+		
Protoperidinium ovatum[a] Pouchet	+		+
Protoperidinium pallidum[a] (Ostenfeld) Balech	+	+	+
Protoperidinium pedunculatum (Schütt) Balech	+		
Protoperidinium pellucidum (Bergh ex Loeblich and Loeblich III	+		+
Protoperidinium puctulatum (Paulsen) Balech		+	
Protoperidinium quarnerense (Schröder) Balech		+	
Protoperidinium schutti[a] Halim	+		
Protoperidinium sinaicum[a] (Matzenauer) Balech	+	+	+
Protoperidinium solidicorne[a] (Mangin) Balech	+	+	
Protoperidinium sphaericum[a] (Murray et Whilling) Balech	+	+	+
Protoperidinium sphaeroids[a] (Dangeard) Balech	+		
Protoperidinium steinii[a] (Jörgensen) Balech	+	+	+

(continued)

Table 1 (continued)

Species	R. S. P	G. S.	G. A.
Protoperidinium subpyriform[a] (Dangeard) Balech	+		
Protoperidinium tristylum[a] (Stein) Balech	+		
Protoperidinium venustum[a] (Matzenauer) Balech	+		
Pyrocystis elegans Pavillard	+		
Pyrocystis ellipsoids[a] (Haeckel) Lemmermann	+		
Pyrocystis fusiformis[a] Wyville-Thomson ex Murray	+	+	+
Pyrocystis hamulus[a] Cleve	+	+	+
Pyrocystis lunula[a] (Schütt) Schütt	+		
Pyrocystis noctiluca Murray ex Schütt	+		
Pyrocystis rhomboids Matzenauer			+
Pyrocystis robusta Kofoid			+
Pyrodinium bahamensis var. compressum[a] (Böhm) Steidinger, Tester and Taylor	+	+	
Pyrophacus horologicum[a] Stein	+	+	+
Spiraulax jollifei[a] (Murray et Whitting) Kofoid	+		
Spiraulax kofoidii Graham	+		
Triposolenia bicornis Kofoid	+		
Coccolithophores			
Rhabdosphaera claviger[a] Murray and Blackman	+		
Syracosphaera cordiformis Schiller		+	+
Cyanobacteria			
Lyngbya majuscule Harvey		+	
Merismopedia punctata Smith		+	
Oscillatoria sp.[a]		+	
Prochlorococcus sp.	+		+
Richelia intracellularis[a] Schmidt	+	+	+
Synechococcus sp.	+		+
Trichodesmium clevei (Schmidt) Anagnostidis et Komárek	+		
Trichodesmium erythraeum[a] (Ehrenberg) Gomont	+	+	+
Trichodesmium radians (Wille) Golubic	+		
Trichodesmium thiebaulti[a] Gomont ex Gomont	+		
Chrysophyceae			
Dictyocha anuspulchra (Schiller) Ling and Takahashi	+		
Dictyocha calida Poelchau	+		
Dictyocha fibula Ehrenberg	+	+	
Dictyocha messanensis Haeckel	+		
Dictyocha octonaria Ehrenberg	+		
Chlorophyceae			
Ankistrodesmus falcatus (Corda) Ralfs		+	
Chlorella vulgaris Beijerinck		+	
Halosphaera viridis[a] Schmitz	+		
Monactinus simplex (Meyen) Corda		+	
Pediastrum clathratum (Schröder) Lemmermann		+	
Scenedesmus quadricauda Chodat		+	

(continued)

Table 1 (continued)

Species	R. S. P	G. S.	G. A.
Scenedesmus bijuga (Chodat) Leite		+	
Scenedesmus dimorphus (Turpin) Kützing		+	
Staurastrum paradoxum Meyen ex Ralfs		+	
Euglenophyceae			
Euglena sp.		+	

[a] Species recorded by Halim (1969)
[b] Immigrant Species through Suez Canal
?? Uncertain species

from 50 to 250–300 m. Nutrient renewal to the upper layers is therefore suppressed, especially in the northern part because of the persistent thermocline. As a result, primary production of the surface waters is dependent on the nutrient concentrations in the upper 200–300 m of the water column. This is why the upper water column in the northern part of the Red Sea is highly oligotrophic (Acker et al. 2008). Primary production and plankton abundance decrease from the southern Red Sea (0.41 g m^{-2} d^{-1}) to the northern part (0.16 g m^{-2} d^{-1}). Average annual phytoplankton production north of 17°N ranges from 250–500 mg m^{-2} d^{-1} (Yentsch and Wood 1961; Koblentz-Mishke et al. 1970), while the productivity in the central Red Sea is about 170 mg m^{-2} d^{-1} (Dowidar 1983b). Vertically, the surface chlorophyll is generally less than in the deep water (Stambler 2005). The phytoplankton biomass in the Red Sea shows a layer of maximal chlorophyll-a (0.36 μg l^{-1}) at around 50–60 m, followed by a decrease in its concentration to less than 0.03 μg l^{-1} toward the surface. The major seasonal peak of chlorophyll (Chl-a) starts in December at 75 m water depth and then moves upwards in the water column during January and February. In the northernmost part of the Red Sea (24°−27.4°N), the remotely sensed chlorophyll-a data (from SeaWiFS) show that, despite the normal low chlorophyll-a concentration (0.1–0.2 mg m^{-3}) in these oligotrophic waters, there is a characteristic seasonal bloom in March–April ranging from 0.6 to 0.9 mg m^{-3} (Acker et al. 2008).

Gulf of Aqaba

The Gulf of Aqaba is about 170 km long, 14–26 km wide and 1830 m deep (Morcos 1970). It is located between the Sinai Desert and the western Arabian Desert and is separated from the northern Red Sea by a shallow still (240 m) at the Strait of Tiran (Post et al. 2002). Salinity is very high, about 40.2 to > 41 ‰ (Klinker et al. 1978). The Gulf is poor in nutrients. Phytoplankton abundance in the Gulf is dependent mainly upon nutrient supply from the Red Sea at the upper 150 m (Klinker et al. 1978). The Gulf alternates from being deeply mixed in winter to being stably stratified in summer. Deep winter mixing brings nutrients into the nutrient-depleted eutrophic zone causing a phytoplankton bloom in spring. However, during recent decades, eutrophication from anthropogenic sources such as sewage and fish farms located in the northern part of the Gulf has contributed significantly to phytoplankton abundance (Stambler 2005). Phytoplankton distribution patterns with regard to the mixing regime are as follows: (i) taxa exhibiting consistent positive responses to winter mixing (Cryptophyceae, Chlorophyceae), (ii) taxa with short-term positive responses to deep mixing (diatoms), (iii) taxa showing negative responses to winter mixing and positive responses to summer stratification (Prochlorophyceae), and (iv) taxa (Dinophyceae, Prymnesiophyceae, and Chrysophyceae) without or with unclear seasonality (Li et al. 1983; Zohary and Robarts 1998). The number of phytoplankton species decreases with depth and the peak of the diatoms occurs at the end of February–March (Kimor and Golandsky 1977). The diatoms consist mostly of *Chaetoceros* and *Leptocylindrus* species found throughout the photic layer down to 100 m depth. *Proboscia alata* (*Rhizosolenia alata*) dominate the community in the upper 10 m, with densities of 30,000 cells m^{-3} (Post et al. 2002). Dinoflagellates belong mostly to the Gonyaulacales (*Neoceratium* and *Protoperidinium*) and Dinophysiales (*Dinophysis, Phalacroma,* and *Ornithocercus*). The genus *Neoceratium* is best represented by *Neoceratium fusus* and *Neoceratium trichoceros* which are the most abundant. Cyanobacteria are represented by various species of *Trichodesmium* (Zohary and Robart 1998; Post et al. 2002). Eukaryotes dominate in winter, *Synechoccus* in spring and fall, and *Prochlorococus* in summer (Lindell and Post 1995). There is a strong spring bloom and a weaker fall bloom in the Gulf, both of which exhibit considerable interannual variability with a maximum concentration of 4×10^7 cells l^{-1} (Stambler 2005).

Primary production measurements are extremely scarce in the Gulf of Aqaba. Productivity ranges from 19.7 to 111 mg C m^{-2} d^{-1} with a maximum in winter (Levanon-Spanier et al. 1979). Chlorophyll distribution throughout the Gulf shows the same pattern of standing crop, except for the

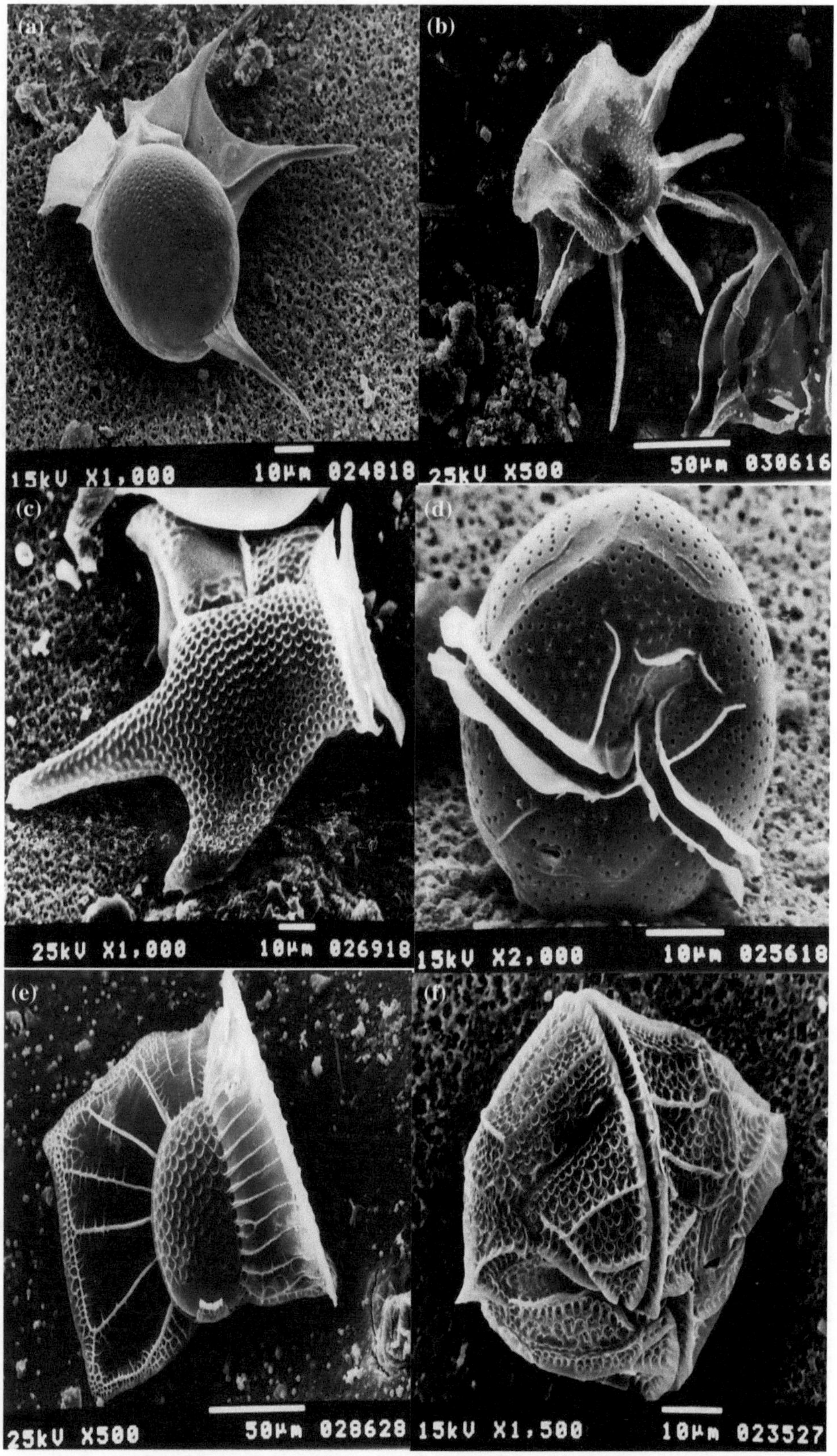

Fig. 1 Examples of dinoflagellate species recorded from the Red Sea. **a** *Dinophysis shutii*, **b** *D. tripos*, **c** *Ornithocercus quadratus*, **d** *Protoperidinium crassipes*, **e** *P. divergens*, **f** *Podolampas bipes*, **g** *Ceratocorys horrida*, **h** *Gonidoma sphaericum*, **i** *Gonyaulax polygramma*, **j** *G. turbynei*, **k** *Spiraulax kofoidi*, and **l** *Neoceratium egyptiacum*

Fig. 1 (continued)

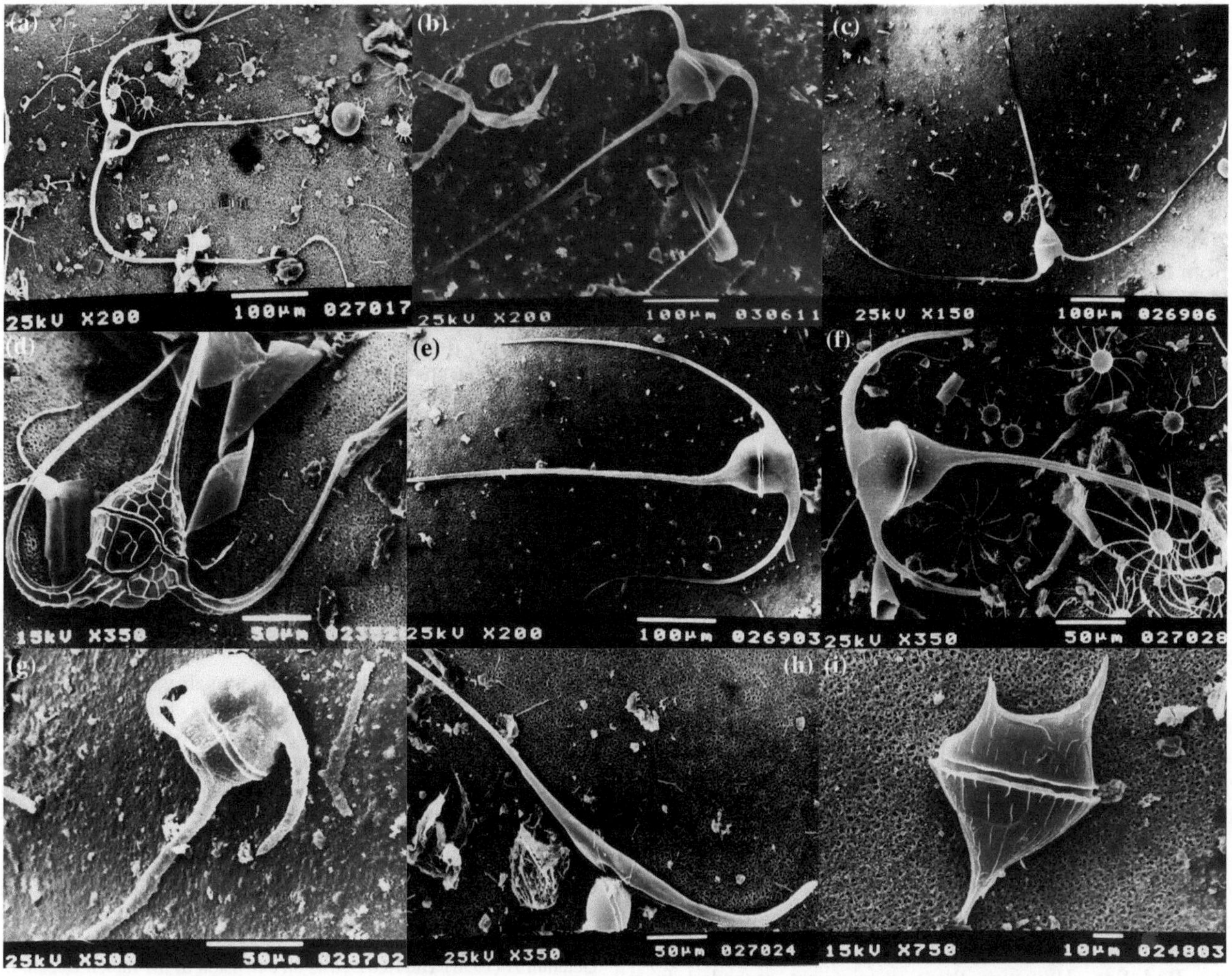

Fig. 2 Examples of leading species recorded from the Red Sea. **a** *Neoceratium trichoceros*, **b** *Neoceratium massiliense*, **c** *Neoceratium carriense*, **d** *Neoceratium hexacanthum*, **e** *Neoceratium karsteni*, **f** *Neoceratium tripos*, **g** *Neoceratium concilians*, **h** *Neoceratium falcatum*, and **i** *Neoceratium pentagonum*

upwelling side of the Gulf, which is higher, with an average of > 0.5 mg chl-a m^{-3}, than the rest of the Gulf (Labiosa et al. 2003). Biomass is generally low, rarely exceeding 1 mg Chl-a m^{-3}. In summer, surface chlorophyll is low and fluctuates between 0.02 and 0.04 μg l^{-1}, and at the end of winter, chlorophyll increases to 1.2 μg l^{-1}, with a deep chlorophyll maximum at 80 m (Klinker et al. 1978; Genin et al. 1995). During the spring bloom, chlorophyll reaches its maximum of 3 mg chl-a m^{-3} (Badran and Foster 1998; Rasheed et al. 2002; Labiosa et al. 2003). The bulk of the biomass is comprised of ultraplankton (<8 μm cell diameter) belonging to three major taxonomic groups, cyanobacteria, prochlorophytes, and eucaryotic algae which contribute > 90 % of the chlorophyll biomass (Lindell and Post 1995; Al-Najjar et al. 2007).

Gulf of Suez

The Gulf of Suez lies at the head of the Red Sea, extends for 300 km to the northwest and is 32 km wide. It ranges in depth from 55 to 90 m and drops suddenly to over 500 m as it joins the Red Sea (Wild 1989). The Gulf supports a number of oil and gas industries. In addition, it is one of the most important fishing grounds within the Red Sea (Wild 1989). The Suez Canal, which connects the Gulf to the Mediterranean, forms an international shipping route from the Indian Ocean to the Mediterranean. As a result, gulf waters along the navigation channel are affected by turbidity caused by powerful ship propellers and the occasional release of waste and ballast water from ships (Ismael 2005).

Salinity in the Gulf of Suez is comparatively very high, 42.4 ± 0.9 ‰ (Wild 1989). Despite the shallowness of the Gulf, studies of the Gulf of Suez phytoplankton are few (Ostenfeld and Schmidt 1901; Cleve 1903; Nassar and Hamed 2003; Ismael 2005; Shams El-Din et al. 2005). The Gulf of Suez is oligotrophic, with a low standing crop with maximum values of 7×10^3–9.9×10^3 cells l^{-1} (Ismael 2005; Madkour et al. 2010). Chlorophyll-a is low, ranging from 0.05 to 6.91 μg l^{-1} along the coastal waters of the Gulf (Fahmy et al. 2005).

The phytoplankton population consists of 187 species. The distribution of phytoplankton species in both the northern and intermediate parts of the Gulf is similar, but the southern part is different. The abundance of dinoflagellate species increases from north to south in the Gulf. *Trichodesmium erythraeum* was recorded only in the intermediate and southern parts of the Gulf. Its contribution to the total community increases southwards toward the Red Sea proper (Ismael 2005). Although neritic forms dominate the Gulf population, a number of species characteristic of the open Red Sea and Indian Ocean also appear, namely *Neoceratium humile* (*C. humile*), *Protoperidinium sinaicum,* and *Pyrodinium bahamensis* var. *compressum*. In addition, the Gulf of Suez is characterized by the presence of an endemic species, *Neoceratium egyptiacum* (*Ceratium egyptiacum* Halim), which has not been recorded elsewhere. Seven diatom species recorded from the gulf, *Dactyliosolen blavyanus*, *Triceratium antediluvianum*, *Cocconeis lyra*, *Pleurosigma formosum*, *Nitzschia bilobata*, *Surirella fastuosa,* and *S. gemma* have also not been reported from the Indian Ocean or the Red Sea. However, these species are recorded from different parts of the Mediterranean Sea and therefore may have transgressed south to the Gulf through the Suez Canal (Sournia et al. 1979; Halim 1969). Some studies (Nassar and Hamed 2003; Shams El-Din et al. 2005) show that the coastal waters of the northern Gulf are characterized by the presence of brackish water species though in small amounts, *Ankistrodesmus falcatus*, *Chlorella vulgaris Pediastrum clathratum, Monactinus simplex*, *Scenedesmus quadricauda*, *S. bijuga*, *S. dimorphus*, and *Staurastrum paradoxum*. The appearance of these species may be due either to freshwater discharge in this area resulting from urbanization or from ships at anchor, waiting to transit the Suez Canal.

References

Acker J, Leptoukh G, Shen S, Zhu T, Kempler S (2008) Remotely-sensed chlorophyll a observations of the northern Red Sea indicate seasonal variability and influence of coastal reefs. J Mar Syst 69:191–204

Al-Najjar T, Badran MI, Richter C, Meyerhoefer M, Sommer U (2007) Seasonal dynamics of phytoplankton in the Gulf of Aqaba, Red Sea. Hydrobiologia 579:69–83

Baars MA, Schalk PH, Veldhuis MJW (1998) Seasonal fluctuations in plankton biomass and productivity in the ecosystems of the Somali Current, Gulf of Aden, and Southern Red Sea. In: Sherman et al. (eds) Large marine ecosystems of the Indian Ocean: assessment, sustainability, and management. Blackwell Science, Oxford, pp 143–174

Badran MI, Foster P (1998) Environmental quality of the Jordanian coastal waters of the Gulf of Aqaba, Red Sea. Aquat Ecosyst Health Manage 1:75–89

Belogorskaya EV (1970) Qualitative and quantitative distribution of phytoplankton in the Red Sea and Gulf of Aden in October–November 1963. Biol Morya (Vladivost)/Mar Biol 21:133–152

Cleve PT (1900) Plankton from the Red Sea. Öfvers K VetenskAkad Förh 57 (9):1025–1038

Cleve PT (1903) Report on plankton collected by Mr. Thorild Wulff during a voyage to and from Bombay. Arkiv för Zoology Stockholm 1:329–381

Dowidar NM (1983a) The genus Ceratium from the Red Sea. J Fac Mar Sci 3:514–545

Dowidar NM (1983b) Primary production in the central Red Sea off Jiddah. In: Latif AF, Bajoumi AR, Thompson MF (eds) Proceedings international conference on marine science in the Red Sea, Bulletin of Institute of Oceanography and Fisheries, A.R.E., Egypt 9, pp 160–170

Edwards FJ (1987) Climate and oceanography. In: Edwards AJ, Head SM (eds) Key environments: Red Sea. Pergamon Press, Oxford, pp 45–70

Ehrenberg CG (1838) Die Infusionsthierchen als vollkommene Organismen. Ein Blick in das tiefere organische Leben de Natur. Verlag von Leopold Voss, Leipzig, pp i–xvii, 1–548, pls 1–64

Elawad AES (2012) Study of inter-annual variability of chlorophyll in the Red Sea. MSc thesis, University of Bergen, Geophysical Institute, 66 pp

Fahmy MA, Sheriadah MA, Aboul Soeud A, Abdel Rahman SM, Shindy M (2005) Hydrography and chemical characteristics of the coastal water along the Gulf of Suez. Egypt J Aquat Res 31:1–14

Genin A, Lazar B, Brenner S (1995) Vertical mixing and coral death in the Red Sea following the eruption of Mount Pinatubo. Nature 377:507–510

Getahun A (1998) The Red Sea as an extension of the Indian Ocean. In: Sherman et al. (eds) Large marine ecosystems of the Indian Ocean: assessment, sustainability, and management. Blackwell Science, Oxford, pp 277–283

Halim Y (1969) Plankton of the Red Sea. Oceanogr Mar Biol Ann Rev 7:231–275

Halim Y (1984) Plankton of the Red Sea and the Arabian Gulf. Deep-Sea Res 31:969–982

Ismael AA (2005) Phytoplankton of the Gulf of Suez and the effect of ship traffic. J Egypt Acad Soc Environ Develop (D- Environmental studies) 6(1):75–92

Karsten G (1907) Das indishe Phytoplankton. Wissenschaftliche Ergebnisse Valdivia 2:221–548

Kimor B (1973) Plankton relations of the Red Sea, Persian Gulf and Arabian Sea. In: Zeitzschel B (ed) The biology of the Indian Ocean, vol 3. Springer, New York, pp 221–232

Kimor B (1990) Microplankton of the Red Sea, Gulf of Suez and the Levantine Basin of the Mediterranean. Bulletin de l'Institut Océanographique Monaco 7:29–38

Kimor B, Golandsky B (1977) Microplankton of the Gulf of Elat: aspects of seasonal and bathymetric distribution. Mar Biol 42:55–67

Klinker J, Reiss Z, Kropach C, Levanon I, Harpaz H, Shapiro Y (1978) Nutrients and biomass distribution in the Gulf of Aqaba (Eilat), Red Sea. Mar Biol 45:53–64

Koblentz-Mishke OJ, Volkovinsky VV, Kabanova JG (1970) Plankton primary production of the world ocean. In: Wooster WS (ed) Scientific Exploration of the South Pacific. National Academy of Sciences, Washington, DC, pp 183–193

Labiosa RG, Arrigo KR, Genin A, Monismith SG, van Dijken G (2003) The interplay between upwelling and deep convective mixing in determining the seasonal phytoplankton dynamics in the Gulf of Aqaba: evidence from SeaWiFS and MODIS. Limnol Oceanogr 48 (6):2355–2368

Lenz J, Schneider G, El Hag AGD, Gradinger R, Fritsche P, Moigis A, Pillen T, Rolke M, Weisse T (1988) Planktological data from the Central Red Sea and the Gulf of Aden; R.V. Meteor, cruise no. 5/2, Jan–Mar 1987. Berichte aus dem Institut für Meereskunde an der Christian-Albrechts-Universität Kiel 180, 200 pp

Levanon-Spanier I, Padan E, Reiss Z (1979) Primary production in a desert enclosed sea, the Gulf of Eilat (Aqaba), Red Sea. Deep-Sea Res 26:673–685

Li WKW, Subba Rao DV, Harrison WG, Smith LC, Cullen JJ, Irwin B, Platt T (1983) Ultraphytoplankton in the eastern Mediterranean Sea: towards deriving phytoplankton biomass from flow cytometric measurements of abundance, fluorescence and high scatter. Science 219:292–295

Lindell D, Post AF (1995) Ultraphytoplankton succession is triggered by deep winter mixing in the Gulf of Aqaba (Eilat). Red Sea Limnol Oceanogr 40(6):1130–1141

Madkour FF, El-Sherbiny MM, Aamer MA (2010) Phytoplankton population along certain Egyptian coastal regions of the Red Sea. Egypt J Aquat Biol Fish 14(2):95–109

Matzenauer L (1933) Die Dinoflagellaten des Indischen Ozeans. Bot Arch 35:437–509

McGill DA, Lawson TJ (1966) The distribution of chlorophyll in the western Indian Ocean during the Northeast monsoon period. Woods Hole Oceanogr Inst 66–12:69

Mohamed ZA, Mesaad I (2007) First report on *Noctiluca scintillans* blooms in the Red Sea off the coasts of Saudi Arabia: consequences of eutrophication. Oceanologia 49(3):337–351

Morcos SA (1970) Physical and chemical oceanography of the Red Sea. Oceanogr Mar Biol Ann Rev 8:73–202

Nassar MZ, Hamed MA (2003) Phytoplankton standing crop and species diversity in relation to some water characteristics of Suez Bay (Red Sea), Egypt. Egypt J Aquat Biol Fish 7(3):25–48

Ostenfeld CH, Schmidt J (1901) Plankton fra det Röde Hav Aden bugten. Videnskabelige Meddelelser Dansk Naturhistorisk Forening Kjobenhaven, pp 141–182

Post AF, Dedej Z, Gottlieb R, Li H, Thomas DN, El-Absawi M, El-Naggar A, El-Gharabawi M, Sommer U (2002) Spatial and temporal distribution of Trichodesmium spp in the stratified Gulf of Aqaba, Red Sea. Mar Ecol Prog Ser 239:241–250

Rasheed M, Badran MI, Richter C, Huette M (2002) Effects of reef framework and bottom sediment on nutrient enrichment in a coral reef of the Gulf of Aqaba, Red Sea. Mar Ecol Prog Ser 239:277–285

Schröder B (1906) Beitrage zur Kenntnis der Phytoplanktons warmer Meeres Vierteljahrsschrift der Naturforschenden Gesellschaft in Zürich 51, pp 319–377

Seeberg-Elverfeldta IA, Carina B, Lange CB, Arzc HW, Paötzolda J, Pike J (2004) Preservation of siliceous microplankton in surface sediments of the northern Red Sea. Mar Micropaleontol 51:193–211

Shaikh EA, Roff JC, Dowidar NM (1986) Phytoplankton ecology and production in the Red Sea off Jiddah, Saudi Arabia. Mar Biol 92:405–416

Shams El-Din NG, Nassar MZ, Abd El-Rahman NS (2005) Surveillence studies on plankton in the northern part of the Red Sea during winter and summer, 2002. J Egypt Ger Soc Zool 48D:49–77

Smeed DA (2004) Exchange through the Bab el Mandeb. Deep Sea Res Part II 51(4):455–474

Sommer U (2000) Scarcity of medium-sized phytoplankton in the northern Red Sea explained by strong bottom-up and weak topdown control. Mar Ecol Prog Ser 197:19–25

Sommer U, Berninger UG, Böttger-Schnack R, Cornils A, Hagen W, Hansen T, Al-Najjar T, Post AF, Schnack- Schiel SB, Stibor H, Stübing D, Wickham S (2002) Grazing during early spring in the Gulf of Aqaba and the northern Red Sea. Mar Ecol Prog Ser 239:251–261

Sournia A, Grall JR, Jacques G (1979) Diatomées et diniflagellés planctoniques d'une coupe meridienne dans le sud de l'Océan Indien (campagne « antiprod I » du Marion-Dufresne, mars 1977). Bot Mar 22:183–198

Stambler N (2005) Bio-optical properties of the Northern Red Sea and the Gulf of Eilat (Aqaba) during winter 1999. J Sea Res 54(3):186–203

Sukhanova IN (1969) Some data on the phytoplankton of the Red Sea and the western Gulf of Aden. Oceanology 9:243–247

Touliabah HE, Abu El-Kheir WS, Kuchari MG, Abdulwass NIH (2010) Phytoplankton composition at Jeddah coast—Red Sea, Saudi Arabia in relation to some ecological factors. J King Abdulaziz Univ Sci 22 (1):115–131

Weikert H (1987) Plankton and the pelagic environment. In: Edwards AJ, Head SM (eds) Key Environments: Red Sea. Pergamon Press, Oxford, pp 90–111

Wild K (1989) Oil pollution in the Gulf of Suez, Egypt and a related study on Coral reef communities and algal symbiosis. University College of North Wales, Bangor, Marine Biology Project, 143 pp

Yentsch CS (1965) Distribution of chlorophyll and phaeophytin in the open ocean. Deep Sea Res 12:653–666

Yentsch CS, Wood L (1961) Measurements of primary productivity in the Red Sea, Gulf of Aden and Indian Ocean. Woods Hole Oceanographic Institution, Ref. 61-6, Appendix 8:6

Zohary T, Robarts RD (1998) Experimental study of microbial P-limitation in the eastern Mediterranean. Limnol Oceanogr 43:387–395

Mangroves of the Red Sea

Ahmed S.M. Khalil

Abstract

Mangroves of the Red Sea have particular biogeographic and ecological significance, as they constitute the boundary for mangrove distribution in the Indo-Pacific region. *Avicennia marina* is the most abundant mangrove species in the Red Sea region. *Rhizhophora mucronata* stands coexist in a few areas. Both species show various growth performances in the region, depending on local environmental conditions. Previous studies reported the earlier existence of *Ceriops tagal* and *Bruguiera gymnorizha* in several areas of the region; however, recent reports indicate that the existence of these species is currently confined to Djibouti and Eritrea. The vast majority of the mangrove stands in the region are mono-specific, consisting of *A. marina,* which typically forms narrow forests along the shoreline and on nearshore and offshore islands, or fringing tidal creeks and channels. Although they are mostly narrow, such forests vary considerably in extent from a few tens of meters to several kilometers along the shore. Many mangrove areas in the region were reported to be degraded at various rates. Unless intervention through restoration and conservation takes place, degradation of mangroves will possibly have adverse impacts on fisheries, coastal stability, and on adjacent habitats connected to mangroves such as coral reefs and sea-grass beds. The main obstacles to conservation efforts in the region include the lack of institutional and human capacities, inadequacy in legislation and weak compliance, gaps in information and data, poor awareness, and rapid coastal development. This chapter presents a brief overview of the available information on mangroves of the Red Sea with emphasis on their distribution, characteristics, significance, and status. The chapter also describes and discusses mangrove conservation efforts and challenges facing them in the region.

Background

The Red Sea mangroves have special historical value being the substance for the earliest clues and accounts describing mangroves by ancient scholars. The mangrove trees growing on the Red Sea coast were described by Theopratus in his "Historia Plantarum" (350 BC) and Pliny in his "Historia-Naturalis" (AD 77). In 1230 AD, the Arabian botanist Abu-al-Abbas al-Nabaty gave an account of a journey through Arabia, in which he described *Rhizophora* mangrove and named it as "*kendela,*" and *Avicennia* mangrove, which he named "quorm." Both names persist in Arabic and Swahili to the present day (MacNae 1968). From ecological and biogeographic perspectives, the Red Sea mangroves have a significant value because they represent the most northerly location of their distribution range in the Indo-Pacific (PERSGA/GEF 2004). Mangroves might represent a valuable resource for the region, given the general aridity and scarcity of perennial woody vegetation in its coastal areas (Abu Ghararah and Khalil 2012).

A.S.M. Khalil (✉)
Faculty of Science, University of Khartoum, Khartoum, Sudan
e-mail: ahmed.khalil@persga.org

A.S.M. Khalil
Living Marine Resources and Climate Change, Regional Organization for the Conservation of the Environment of the Red Sea and Gulf of Aden (PERSGA), Jeddah, kingdom of Saudi Arabia

N.M.A. Rasul and I.C.F. Stewart (eds.), *The Red Sea*,
Springer Earth System Sciences, DOI 10.1007/978-3-662-45201-1_33,

Mangroves represent one of the major ecosystems of the biosphere, dominating vegetation in tropical coasts (Spalding et al. 1997), and providing a wide range of ecological services and goods (Saenger 2002a). Mangroves help stabilize shorelines and reduce the impacts of natural disasters such as tsunamis and hurricanes. They serve as nursing grounds for marine biota and provide food, medicine, fuel, and building materials for coastal communities. They also sequester and store significant amount of carbon which help mitigation of climate change and enrich the ocean by providing organic inputs (Dittmar 1999; Jennerjahn and Ittekot 2002; Murray et al. 2012).

Mangroves have a wide range of traditional and commercial uses, for example, as timber and domestic firewood, which require tree cutting and felling. In many tropical areas, removal of mangroves takes place in favor of aquaculture, land reclamation, and other purposes (Saenger et al. 1996; Spalding et al. 1997). The growing evidence of mangrove degradation around the world has led to an increasing recognition that the ecosystem is threatened (e.g., Mastaller 1996, 1997; FAO 2007). This has stimulated efforts by scientists in the last decades of the twentieth century to investigate various aspects of the mangrove ecosystem, which has inspired conservation activities by relevant international, regional, and national agencies.

Worldwide, numerous studies have focused on mangroves covering a range of their ecological and biological aspects, such as their distribution, ecosystem dynamics, and associated communities (e.g., reviews by Por and Dor 1984; Robertson and Alongi 1992; Schwamborn and Saint-Paul 1996; Saenger 2002a), along with assessment studies reporting on the status of mangroves, impacts of their degradation on the marine environment, and restoration issues (e.g., Ruwa 1988; Fondo and Martens 1998; Sasekumar and Chong 1998; Saenger et al. 1996). Compared to other parts of the world, mangroves of the Red Sea region have received less attention by research and assessment studies. Limited knowledge has been available on the mangroves, and their status in this region from surveys and studies carried out over the last few decades (e.g., PERSGA/GEF 2004). This chapter provides a brief overview of the Red Sea mangroves based on the relatively meager studies and surveys carried out hitherto. The compiled information in this overview focuses on mangrove distribution, extent, status, associated communities, and their value to the region. It also describes and discusses their conservation efforts and challenges and updates information provided by previous reviews of mangroves in the region, considering recent studies and related emerging topics, such as the potential role of the region's mangroves in the implementation of the ecosystem-based management and adaptations to the impacts of climate change on coastal areas of the region.

Distribution and Extent

The distribution of mangroves on the Red Sea coasts is greatly influenced by the geomorphology of the coastal area. Mangroves are often associated with the seaward terminations of valleys, locally known as *marsas*. Rabaa (1980) reported that there was a close relationship between the formation of the *marsas* on the Red Sea coast and the paleodrainage system of land, which conforms with Whiteman (1971) that the *marsas* are erosional features. Today, perennial rivers are lacking in the Red Sea. Instead, numerous valleys originate in the adjacent mountains, which collect rainwater during occasional floods, bringing it across the coastal plains to the seashore together with alluvium. These create many estuary-like formations suitable for mangrove development. The mangroves typically form narrow forests (occasionally wide in some areas) along the shoreline and on nearshore and offshore islands, or fringing creeks and channels associated with such formations on the coast. Consequently, the distribution of mangrove forests along the Red Sea coast is sporadic (Mohamed 1984; Khalil 2001; PERSGA/GEF 2004).

Although they are mostly narrow, mangrove forests in the Red Sea vary in extent considerably from some tens of meters to well over several kilometers along the shore and inlets (Mohamed 1984; Price et al. 1987; PERSGA/GEF 2004). Generally, more dense stands occur in southern parts of the region, where conditions become more favorable for mangrove growth, such as relatively higher amounts of surface runoff, muddier substratum, more nutrient supply, gentle slope of the shore, and wider intertidal areas (Mohamed 1984; Price et al. 1987; Khalil 2001; PERSGA/GEF 2004). Sato et al. (2005) argued that the influence of seasonal rainwater inputs via valleys on mangrove growth in such arid coasts of the Red Sea is largely due to their role in furnishing nutrients from land. Sheppard et al. (1992) suggested that the increase in density of mangroves in southern parts of the Red Sea could be attributed to the availability of more nutrients in the water.

Information on mangrove cover in the Red Sea has relatively improved with the conduct of more comprehensive surveys and the application of remote sensing techniques in recent studies (e.g., PERSGA/GEF 2004; Kumar et al. 2010a, b; PERSGA/SWC 2010; Nagi et al. 2012; Fatoyinbo and Simard 2012; Elsebaie et al. 2013). However, some discrepancies still exist in the estimates for mangrove cover (Table 1). Most consistent estimates for the total mangrove cover in the Red Sea appear to range between 13,500 and 14,900 ha, although high estimates reach 35,500 ha.

The northerly boundary for mangrove distribution along the eastern coast of the Red Sea is Sharm Zubeir (Lat. 27° 25′N) in Saudi Arabia. The mangrove distribution extends

Table 1 Area estimates for mangrove cover on the Red Sea coasts as reported in some literature (in hectares rounded to next hundred)

Country	Low estimate	High estimate	References
Djibouti	1,000	1,700	FAO (2007), Spalding et al. (2010), Fatoyinbo and Simard (2012)
Egypt	500	500	FAO (2007), Spalding et al. (2010), Chevallier (2013)
Eritrea	6,400	10,000	FAO (2007), Spalding et al. (2010), MLWE (2012)
Saudi Arabia	3,500–5,000	20,000	PERSGA/SWC (2010), Spalding et al. (2010), Kumar et al. 2010a, b) FAO (2007)
Sudan	500	1,000	Wilkei (1995), PERSGA/GEF (2004), Spalding et al. (2010)
Yemen	1,500	2,300	Kumar et al. (2011), Nagi et al. (2012)
Total	13,400–14,900	35,500	

Source modified from PERSGA/UNEP (2013)

southward along the Saudi coast to beyond Jizan into Yemen. The Red Sea coast in Saudi Arabia supports many sites with mangrove stands. Price et al. (1987) recorded mangrove growths at more than 40 localities along the Red Sea coast and offshore islands. In a satellite-based survey, PERSGA and SWC (2010) reported more than 100 mangrove stands along the Saudi Red Sea coast with an estimated total area of around 3,500 ha. Other estimates by Kumar et al. reach 5,000 ha. Mangrove development increases toward the south, coinciding with the gradual disappearance of stony corals and increased availability of muddier substrate and rainwater (Price et al. 1987). The southern coastal strip extending from Al-Laith to the Yemeni border supports the most dense mangrove stands in Saudi Arabia, where they fringe a considerable part of the shoreline (e.g., at KhorAmq, Shuqaiq). Mangrove stands are also found on some inshore islets and the Farasan Archipelago (Al-Wetaid and Khalil 2003; PERSGA/GEF 2004; Elsebaieet al. 2013).

The Red Sea coast in Yemen supports relatively well-developed mangroves. Different estimates were reported for the total mangrove cover in the country. Satellite-based studies by Kumar et al. (2011) and Nagi et al. (2012) provide estimates of 1,500 and 2,255 ha, respectively. The northern strip of the coast extending from Midi to Hodeida supports most of the mangrove wealth and largest stands in the country, particularly in the Maidi, Al-Luhaia, Habl, Al-Buhais, and Ibn Abbas areas. Stands growing in the southern coastal strip are relatively sparse and narrow, occurring mainly in the Yakhtul, Dhubab, and Ghuraira areas to the north of Bab-al-Mandab. The Yemeni Red Sea coast supports most of the mangrove stands, comprising more than 90 % of the mangrove cover of the country. The remaining few mangrove stands growing on the Gulf of Aden occur at two locations on the mainland coast and a third location at the southern coast of Socotra Island (PERSGA 2002b; PERSGA/GEF 2004; Nagi et al. 2012).

On the western coast of the Red Sea, mangrove distribution extends a bit further north than on the eastern coast, reaching up to Nabq at Sinai Peninsula in Egypt (Lat. 28°12′ N). Mangroves in Egypt have received more studies compared to other parts of the region, so relatively more robust information on mangrove distribution and their associated communities are available for the country. Mangrove stands were reported at 23 sites along the Egyptian Red Sea coast. Most of them are small with total mangrove cover of around 500 ha (Saenger 2002b; Galal 1999, studies on the coastal ecology and management of Nabq Protected Area, South Sinai, Egypt. Unpublished PhD thesis, University of York; PERSGA/GEF 2004; FAO 2007).

A total of 14 mangrove areas have been reported along the Red Sea coast in Sudan. The stands comprising these areas are generally small, covering altogether around 500–600 ha. They form narrow fringes along bays, lagoons and sheltered coastlines, and on nearshore islets. It is estimated that an area at least equivalent in size has been lost and could be rehabilitated (Wilkei 1995; PERSGA/GEF 2004). The southern coastal strip extending south of Suakin to the borders with Eritrea, which represents estuaries of the two large Goub and Ashat valleys supports the most dense mangrove stands in the country (Khalil 1994; Wilkei 1995; PERSGA/GEF 2004).

The Red Sea coast of Eritrea supports well-developed mangroves. Estimates for total mangrove cover of the country vary from 6,400 (FAO 2007) to 10,000 ha reported by the Ministry of Land, Water, and Environment of the State of Eritrea (MLWE 2012). Mangrove stands occur at various sites of the coast, but the largest are found mainly around Assab and the nearby islands, and between Tio and Massawa (De Grissac and Negussie 2007).

The coast of Djibouti, which lies at the junction of the Red Sea and the Gulf of Aden, supports the most well-grown and diverse mangrove stands in the Red Sea region. The number of mangrove stands in Djibouti is limited, but each of them is relatively large, particularly those found at Khor Angar, Godoria on the northern part of the coast, and Moucha Island offshore from Djibouti city. Estimates of 800–1,000 ha were reported for the total mangrove cover in the country (FAO 2007; MUHAT 2013), although some satellite-based estimates reach 1,700 ha (Fatoyinbo and Simard 2012). For Djibouti, mangroves is considered a valuable resource comprising around 17 % of the limited forest cover, which represents only 0.2 % of the land area in the country (World Bank 2014).

Species and Growth Performance

Avicennia marina, which is more adapted to the harsh conditions of high salinity and aridity of the Red Sea coastal areas, is the dominant mangrove species in the region. The vast majority of the mangrove stands in the region are monospecific, consisting of this mangrove species. Significant *Rhizhophora mucronata* stands coexist in a few areas of the region, particularly in Djibouti, Eritrea, Yemen, and Saudi Arabia. The largest mangrove forests of the region are found in Djibouti, specifically at Mucha Island, Khor Angar, and Godoria.

In Saudi Arabia, *R. mucronata* was recorded by Price et al. (1987) in several mangrove areas comprising Wajh Bank, Umm al Qandal Island, Gharaniq Island, Zifaf Island, Shi'b Abu Al-Liqa, and the Farasan Islands. At all these sites *A. marina* also coexists; however, *R. mucronata* occurs in separate zones located at the seaward site of the forests. In more recent surveys, Al-Wetaid and Khalil (2003), PERSGA/GEF (2004) extended this list to include eleven areas with *R. mucronata* stands in the country. The most significant *R. mucronata* stands are those found on the southern parts of the coast, Farasan Islands, and Al-Wajh Bank (PERSGA/GEF 2004). In Yemen, among 29 mangrove sites surveyed in the country, *R. mucronata* was reported at two sites including Kamaran Island and a small islet near Hodeida. The former supports the largest *R. mucronata* stands in Yemen (PRSGA/GEF 2004; Nagi et al. 2012).

Previous surveys reported the earlier existence of two other mangrove species, *Ceriopstagal* and *Bruguiera gymnorhiza* in several Red Sea countries (Andrews 1950). However, recent reports indicate that the current existence of *C. tagal* is confined to Eritrea and Djibouti, while *B. gymnorhiza* exists in Djibouti only (PERSGA 2004; FAO 2007).

Both *A. marina* and *R. mucronata* species show various degrees of growth performance in the region depending on local conditions (see Plates 1 and 2). *A. marina* growing in the Red Sea region has variable heights ranging between 0.5 and 6 m and girth at breast height (GBH) in the range of 10–100 cm. The tallest stands such as those found at Al-Urj (Yemen), Maskali Island (Djibouti), Jizan (Saudi Arabia), and Arakiyai (Sudan) reach up to 8–12 m in height and around 200 cm GBH (PERSGA 2002a, b, c; Khalil 2010; PERSGA/GEF 2004).

The best developed *R. mucronata* stands in the region, which are found at Khor Angar, Godoria, and Mucha Island in Djibouti support trees reaching heights of 9–13 m. Moderate *R. mucronata* stands found in Yemen (Kamaran Island) and Saudi Arabia (Jizan coast) support trees that reach heights of 7–8 m, while smaller trees reaching 4–5 m are found at a few other places that include less developed stands of the region. PERSGA (2002a, b, c), PERSGA/GEF (2004) provides detailed assessments of mangrove sites on the Red Sea coast, including a description of tree density, height, and diameter.

Ecological Importance and Economic Value to the Region

In general, mangroves and other linked coastal ecosystems, such as tidal marshes sea grasses and coral reefs provide a range of ecosystem goods and services. As an ecosystem, the mangrove has the advantage of offering almost all of the four categories of environmental services indicated by the Millennium Assessment Report (2006), comprising regulating, provisioning, cultural, and supporting services. As to regulating services, mangroves provide shoreline protection, atmospheric and climate regulation, and flood and erosion control. Regarding provisioning services (goods and products), mangroves provide wood and timber for cooking fuel, fish processing, salt production, charcoal, construction, and thatching. As to cultural services (non-material benefits), mangroves have a significant esthetic value and represent a rich resource for recreation/tourism, ointments, and traditional medicines. The supporting services (natural processes that maintain other ecosystem services) of mangroves are significant, such as their roles in nutrient cycling, provision of fish nursery habitats, sediment trapping, water filtration, and waste treatment (Chevallier 2013).

Currently, particular emphasis is put on the advantages of mangroves for adaptation to the impacts of climate change, as well as its mitigation. Mangroves are known for their enormous adaptation potential such as mitigating storm and wave damage; for example, they could largely absorb tsunami waves, withstand annual sea level fluctuations, and greatly reduce erosion. They have a documented high ability to sequester and store atmospheric carbon, while the mangrove soils have a particularly high carbon content. Conversely, degradation of mangroves can lead to the release of long-term stored carbon in their soil and further carbon emissions (Murray et al. 2012; World Bank 2010).

Limited productivity data are available for the mangroves of the Red Sea. However, Galal (1999, Studies on the coastal ecology and management of Nabq Protected Area, South Sinai, Egypt. Unpublished PhD thesis, University of York)

Plate 1 *Avicennia marina* from different locations in the Red Sea and Gulf of Aden coasts. Note variation in the species growth performance among different sites depending mainly on availability of favorable conditions (after PERSGA 2011). **a** Al-Luhayiah (Yemen), **b** Ras Siyyan (Djibouti), **c** Khor Amq (Saudi Arabia), **d** Midi (Yemen), **e** Al-Urj (Yemen), and **f** Near Jizan (Saudi Arabia)

has investigated the aboveground biomass (AGB) and annual litter fall (ALF) for the mangroves of Nabq on the Sinai Peninsula and reported values of 5.4 t ha^{-1} for short ($\sim$1 m) *Avicennia marina* communities and up to 74 t ha^{-1} for tall ($\sim$4 m) communities. These AGB estimates compare readily with other *Avicennia marina* communities of similar heights and suggest that despite the climatic constraints, mangrove standing stock in the Red Sea is similar to other regions at similar latitudes. The ALF for short and tall *Avicennia marina* communities is 0.78 t ha^{-1} yr^{-1} and 4.82 t ha^{-1} yr^{-1}, respectively (Galal, 1999, Studies on the coastal ecology and management of Nabq Protected Area, South Sinai, Egypt. Unpublished PhD thesis, University of York). Again, these data compare readily with ALF of *Avicennia marina* communities in other regions. For the Red Sea, the only other data on ALF available are that from *Avicennia marina* communities just north of Jeddah, Saudi Arabia (Saifullah et al. 1989). They reported ALF values

Plate 2 *Rhizophora mucronata* from different locations in the Red Sea and Gulf of Aden coasts showing variation in the species growth performance depending on availability of its favorable conditions (after PERSGA 2011). **a** Doghm Sabq (Saudi Arabia), **b** Farasan Islands (Saudi Arabia), **c** Kamaran Island (Yemen), **d** Biash, South Suqaiq (Saudi Arabia), **e** Godoria (Djbouti), and **f** Muchi Island (Djibouti)

ranging from 4.41 to 6.53 t ha^{-1} $year^{-1}$, estimates not significantly different to those from Nabq. Given a total mangrove area of ~14,000 ha in the Red Sea, the annual contribution of mangroves to inshore organic matter is around 39,000 tonnes, and although not large, this input appears to be significant for inshore productivity. In another study, Abohassan et al. (2012) estimated the aboveground and belowground biomass production of two *A. marina* stands at Shuaiba and Yanbu areas on the Red Sea coast of Saudi Arabia. Overall, the AGB of the two sites was 14.77 ha^{-1}, while the belowground biomass was 67.8 ha^{-1}, which indicated a high biomass allocation to the roots.

Mangroves of the Red Sea, although they are relatively less luxuriant than other parts of the Indo-Pacific, play an important role as nurseries for several commercial fish species in the region (Khalil 1994; Khalil and Krupp 1994). They also protect coral reefs by trapping sediment loads of the seasonal rainwater influx (Wilkei 1995) and enhance stability of shoreline sediments, particularly in the southern Red Sea (Sheppard et al. 1992).

Mangroves are found on the coastline throughout the Red Sea, while assessments revealed that there is a great potential for extending mangrove cover in the region through conservation and restoration. Therefore, mangroves can play an important role in the region's adaptation to climate change impacts on coastal areas (Abu Ghararah and Khalil 2012). They also contain and sequester considerable amounts of carbon, especially when compared to the poor terrestrial ecosystems in the region. A recent study conducted by PERSGA for assessing blue carbon in the region estimated that carbon stock for mangroves of the Red Sea could reach 1,550 t CO_2/ha^{-1}, while the carbon sequestration rate could reach 5.0 t $CO_2/ha^{-1}/year^{-1}$ (PERSGA/UNEP 2013). Referring to Table 1 above, this suggests that the Red Sea mangroves could have a carbon stock equivalent of 22 million t CO_2 at low estimates and 55 million t CO_2 at high estimates of the mangrove cover of the region. Also, their total carbon sequestration rate could reach 70,000 t $CO_2/ha^{-1}/year^{-1}$ for low estimates and 177,500 t $CO_2/ha^{-1}/year^{-1}$ for high estimates of the mangrove cover.

An important ecological value of mangroves to the region is their support of biodiversity in coastal areas, either directly through hosting mangrove-associated biota, or indirectly through protection of coral reefs, the most crucial cornerstone for biodiversity and living resources in the Red Sea. Mangroves also protect the extensive shallow flats seawards from erosion, excessive sedimentation, and turbidity. These provide favorable habitats for algal growths like *Digenea* sp. and *Cystoseira* sp., supporting rich populations of commercially important gastropods of *Strombus* sp. and *Lambis* sp., which are collected for their meat and *Dufra* (local perfumes extracted from the operculum of the animal). Sea cucumbers are also collected from shallow waters around mangrove areas (PERSGA/GEF 2004). The role of mangroves for fishery and consolidation of adjacent habitats in the region is tentatively recognized. It needs to be investigated through long-term studies and monitoring.

Mangroves in the Red Sea support rich associated communities that are typically found in such habitats, while richer floral and faunal assemblages are found in the relatively more extensive and dense mangroves of the southern parts of the coast (Price et al. 1987; PERSGA/GEF 2004). The mangrove-associated biota in the Red Sea has been the subject of several studies (e.g., Fishelson 1971; Por et al. 1977; Por and Dor 1984; Dor and Levy 1984; Price et al. 1987; Khalil 1994, 2001; Abdel-Wahab 2005). Summarizing the findings of Fishelson (1971), Por et al. (1977), Dor (1984), Price et al. (1987), Sheppard et al. (1992) listed more than 257 species from Red Sea mangroves belonging to 10 major groups, which included mangroves, algae, foraminifera, sponges, coelenterates, polychaetes, crustaceans, mollusks, echinoderms, and fishes. However, most of these studies were based on mangrove stands in the northern Red Sea, particularly in Sinai, which are apparently less dense and diverse than those found on the southern Red Sea.

In more recent studies, some other mangrove-associated biota were investigated. Khalil (2001) listed 40 species of free-living marine nematodes from mangroves areas in Sudan, dominated by deposit feeders, mainly *Terschillingia* sp. and *Daptonema* sp. The same study reported more than 16 meiofaunal groups from mangrove soils, which were dominated by the nematodes, followed by foraminifera and copepods, respectively. Meiofaunal densities reported from Red Sea mangroves are comparable to those reported from mangrove habitats elsewhere in the world. Abdel-Wahab (2005) identified thirty-nine fungal species on decayed wood of *A. marina* in Egypt. The most frequent fungus was *Swampomyce sarmeniacus*, while other common species included *Hypoxylon* sp., *Lineolata rhizophorae*, *Kallichromatethys*, *Swampomyces aegyptiacus,* and *Lulworthia grandispora*. Common fungi in the northern sites were different from those recorded from southern sites. The diversity of fungi recorded in Red Sea mangroves is comparable to that recorded from subtropical mangroves and lower than that recorded from tropical mangroves.

Price et al. (1987) concluded that the mangroves of the Red Sea represent composite habitats containing both hard and soft bottom substrates and inhabited by fauna typical of each. Some mangrove characteristic species occur on dense mangroves of the southern Red Sea. For example, the mudskippers, *Periopthalmus koelreuteri,* were not recorded by Price et al. (1987) north of latitude 16° 44′ 25″ on the western coast of the Red Sea.

The mangrove vegetation supports a quantitatively rich algal, and mobile and sessile macrofaunal species. Among the macrofauna, the most conspicuous is the grapsid crab *Metopograpsus* sp. Other common species are the medusa *Cassiopeia* sp. and some water spiders. The trunk and pneumatophores provide habitats for intertidal fauna like *Littorina* sp. and barnacles. The outer mud and sand flats landwards, where only dwarfed *Avicennia* shrubs may be found, become entirely exposed and dry at low tide and partially submerged during high tide. The most conspicuous macrofauna are *Uca* sp., *Dotilla* sp., *Ocyopode* sp., and the terrestrial hermit crab *Coenobita* sp. These species are not specific to mangrove areas and can be found elsewhere along the Red Sea shore. As this zone is situated landwards, it is often subjected to physical stress by grazing camels and pollution by domestic solid wastes, especially at the mangrove sites occurring near population settlements. Such impacts are widely reported in surveys of Red Sea mangroves (Sheppard et al. 1992; PERSG/GEF 2004).

Some studies reported poorer animal associations at the degraded mangroves disturbed by heavy camel grazing and woodcutting, compared to the relatively undisturbed mangrove areas. The abundance of the grapsid crab *Metopograpsus* sp. appears to be largely influenced by the density and condition of the mangrove. Well-developed and relatively undisturbed mangroves typically support dense populations of *Metopograpsus* sp. Lower numbers occur at the mangrove areas suffering from degradation and disturbance (PERSGA/GEF 2004).

The mangrove lagoon and associated channels are inhabited by a number of fish species, most of which are not true residents and utilize the mangroves as nursery grounds. Studies in the Sudanese Red Sea (Khalil 1994; Khalil and Krupp 1994) have reported that several commercially important fish species, which are caught as adults in areas adjacent to the mangroves, are dependent upon the mangroves as nurseries and refuges. Among the mangrove areas the diversity and abundance of fish species were reported to be significantly higher at the more extensive and dense stands. Khalil (1994) described the relations of fish species to mangroves in the Red Sea and identified three main categories, true residents, which spend their entire life cycle in the mangroves such as *Aphanias dispar, Gerres oyena,* and some gobiids; closely associated species, comprising species found in the mangroves as juveniles and/or subadults that apparently utilize the mangroves as nurseries or feeding grounds, such as *Acanthopagrus berda, Chanos chanos, Crenidens crenidens, Hypoatherina temminckii, Leiognathus equulus, Terapon jarbua, Pomadasys commersonni,* and some Mugilid species; and loosely associated species, which occur in the mangroves as occasional visitors in search for food or refuge, such as *Sillago sihama, Thryssabaelama,* and many others.

A variety of bird fauna is associated with the mangroves and the surrounding wetlands. These habitats serve as important staging and wintering areas of migratory waterfowl, including shorebirds and seabirds, as well as breeding and nesting sites for some species"Kindly note that Plate 3 has been cited in the text but it is not provided. Please check and confirm." –>. *Pelecanus rufescens*, which is very common in the region, and *Ardea goliath* probably breed in the mangrove (Evans 1994; PERSGA/GEF 2004).

Studies assessing the economic value of mangrove in the region are lacking, except for a single assessment conducted by FAO on Egypt's mangroves (FAO 2002). The total economic value (TEV) of the mangroves could reach US $91,000/ha/yr for Ras Mohammed and US$24,000/ha/yr for Nabq Protected Area. Most other mangroves were estimated to have a value in the order of US$13,000/ha/yr for fisheries and perhaps in the order of US$13,000/ha/yr for nonuse values. However, the study assumed several uncertainties due to limited information available, while its estimates of TEV seemed to be rather higher than most of those obtained in other similar studies (FAO 2002). In general, putting many assumptions is common in economic analysis, where it is often difficult to obtain cost and price information, particularly in the case of non-market evaluation.

Impacts and Threats

As they grow in a very hostile environment, the Red Sea mangroves are very sensitive to over-exploitation (Price et al. 1987). Surveys carried out in previous studies, however, had often reported mangrove degradation in the area, mainly from excessive browsing by camels and overcutting (e.g., Mohamed 1984; Wilkei 1995; Khalil 2001; PERSGA 2002a, b, c). Other impacts coming from pollution, coastal reclamation, and coastal development associated with modification of hydrological regimes in the coastal areas were also reported (Mandura 1997; PERSGA/GEF 2004; PERSGA 2006). Detailed characteristics and status of mangroves in the Red Sea were provided by PERSGA/GEF (2004) and PERSGA (2006). A summary of the impacts on mangroves of the Red Sea is given in Table 2.

Camel grazing was reported in most mangroves in the region as a major problem causing mangrove degradation, excluding only a few stands growing in protected islets. The intensity of camel browsing varied from low at some areas to moderate at most of the mangrove stands and severe at some others. This depends on the size of the camel herds present in the area, accessibility of the site, and the availability and condition of other pastoral resource in the coastal area (PERSGA/GEF 2004).

In areas where heavy grazing occurs, severe impacts on the mangroves include a considerable reduction in the green parts, stunted growth of mangrove trees, and damage to the seedlings and pneumatophores. Cutting was reported to have more severe impacts due to the generally limited size of the stands. In some areas, cutting has lead to a reduction in the number of trees, denuded patches within mangrove forests, abundance of multi-stemmed trees, barren depression in which the hydrological regimes were modified, and compaction of the surface soil at the impacted sites. In addition, cutting provided passages to the inner dense parts of the stands for camels, making these parts more accessible for grazing. Thus, the combined effects of grazing and cutting have accelerated degradation of several mangrove areas near major population settlements in the region (PERSGA/GEF 2004).

Mass mortality of mangrove trees appears to be a serious problem in the mangrove areas along the southern coasts of Yemen and Sudan, and the Gulf of Aden coast in Djibouti, where mortalities include considerable parts of some mangrove stands. In some others, localized pockets of dead

Table 2 Summary of the impacts on mangroves of the Red Sea and Gulf of Aden

Category of impact	Effect
Direct impacts on mangroves	
Removal for constructing roads, houses, hotels, and other infrastructure	Destruction
Cutting (e.g., for wood and charcoal)	Degradation
Browsing by animals	Degradation
Human trampling, disturbance, and souvenir collection	Degradation
Indirect impacts on mangroves	
Over fishing (e.g., on adjacent reef flats)	Loss of use value
Hydrological changes (e.g., through coastal roads and jetties)	Degradation/Dry-up
Accumulation of solid waste	Degradation
Pollution (e.g., oil spills, tar balls phosphates)	Degradation
Impacts on associated flora and fauna	
Disturbance	Change in species composition and diversity
Alteration of habitats	Change in community structure and functional properties
	Overall decline in ability to provide ecosystem goods and services

Source Khalil 2001; PERSGA/GEF 2004; PERSGA/UNEP 2013

standing trees were reported due to various kinds of known and unknown stresses. Top dying of uppermost and outermost sun branches is common among trees in many severely disturbed mangrove areas in the region.

The major cause of the mass mortality and top dying of the mangrove trees appeared to be localized modifications in the topography of the coastal area, leading to diverting tidal water away from mangroves, decreasing water levels in the mangrove channels/lagoons, and reduction in the mangrove area inundated by tidal water in the upper reaches of the stands. This can mostly be attributed to construction activities involving dredging and sediment dumping in the shore areas, diverting tidal water to feed salt pans through constructed canals, and excessive sedimentation of sand dunes leading to burial of tidal channels, through which seawater overflows to the mangroves. In many areas, degradation of the mangrove and terrestrial vegetation has led to reduced trapping capacity and excessive movement of sand dunes from land (PERSGA/GEF 2004).

On the other hand, the mangrove distribution along the Red Sea coasts is greatly influenced by the amounts of surface runoff and alluvium deposited at the mouths of the valleys by seasonal floods. Therefore, damming of freshwater and excessive groundwater pumping is a potential threat that may lead to degradation of the mangroves at the seaward termination. This can be attributed to increased intrusion of saltwater, decline of the alluvium in sediment load depositing at the mouths, enhancing sand infilling, and reduction of the amounts of seasonal supply of nutrients from the land (PERSGA/GEF 2004).

Pollution by domestic solid wastes (e.g., polyethylene sacs and bottles, plastic, and metal cans) was also reported as a problem in many mangrove areas of the Red Sea. It is particularly encountered where disposal in large quantities occurs near population centers of the major coastal towns and villages. Solid wastes may have physical impacts by enfolding the young seedlings and pneumatophores, blocking tidal channels, and causing disturbance to the mangrove-associated fauna.

Some mangrove stands in the region suffer from pollution by sewage water (PERSGA/GEF 2004). Such stands are dominated by stunted, multi-stemmed trees, with high incidence of dead pneumatophores (Mandura 1997). Other potential pollution threats to mangroves in the Red Sea include oil spills and outputs from aquaculture, particularly shrimp farms. Mangroves are particularly vulnerable to oil, both through direct toxicity from the lighter hydrocarbon fractions and by interfering with the root ventilation system, whereby oxygen is delivered to their roots. The Red Sea region forms one of the major thoroughfares for international maritime traffic between Asia Pacific and Europe. It is also the world's largest producer and exporter of oil, most of which is transported by sea. These factors place the region's mangroves at high risk (Saenger and Khalil 2007).

The growing shrimp farming industry in the region may pose serious impacts on the mangroves. Intensive shrimp farms have been in general reported as problematic and unsustainable, mainly because of rapid degradation of water quality caused by effluents from the shrimp farms themselves and consequent spread of diseases. Given the growing market demand and the need to raise income from export, the shrimp farming industry may grow in the region in the future. Therefore, sound environmental policies including reliable environmental impact assessment studies, as well as monitoring of compliance and possible impacts on mangroves, are required (PERSGA/GEF 2004; PERSGA 2006).

Table 3 Average dissimilarities and groups responsible for difference between non-cleared, partially cleared, and cleared mangrove sites in the Red Sea during summer and winter.

	Non-cleared site		Partially cleared site	
Partially cleared site	*Summer*	*Winter*		
	Av. diss. = 35.81 %	Av. diss. = 35.19 %		
	Copepoda ↑	Nematoda ↓		
	Nematoda ↓	Copepoda ↑		
	Foraminifera ↓	Polych. larvae ↑		
	Acari ↓	Acari ↓		
	Ostracoda ↑	Turbellaria ↑		
Cleared site	*Summer*	*Winter*	*Summer*	*Winter*
	Av. dis. = 33.75 %	Av. diss. = 39.5 %	Av. diss. = 29.64 %	Av. diss. = 36.96 %
	Foraminifera ↓	Copepoda ↑	Nematoda ↓	Copepoda ↑
	Acari ↓	Nematoda ↓	Ostracoda ↓	Nematoda ↓
	Ostracoda ↓		Copepoda ↓	
	Copepoda ↑			
	Nematoda ↓		Foraminifera ↑	

Av. diss. Average dissimilarity. Arrows indicate increased (↑) or decreased (↓) abundance (*source* Khalil 2001)
Groups are ranked in order of importance to their contribution to the dissimilarity. These are results of similarity percentage (SIMPER) analysis based on square root transformed data

Studies assessing the impacts of mangrove degradation on the ecosystem and associated communities in the Red Sea are generally scarce. Khalil (2001) investigated the response of meiofauna to mangrove deforestation in the Red Sea coast of Sudan. At deforested mangrove sites, sediment sorting, mean grain size, water, and organic contents significantly decreased, while variations in sediment characteristics between shoreward and seaward parts of the intertidal zone were intensified. Correlated shifts in the structure of meiofauna occurred, which were attributed to modification of the sediment depositional and reworking processes, obviously due to the loss of mangrove cover (Table 3).

Khalil (2001) found that the variability in community structure was generally increased within mangrove impacted sites. Schatzberger and Warwick (1998) provided experimental evidence supporting the argument of Warwick and Clarke (1993) that variability is indicative of community stress. This empirical evidence supports the findings by Khalil (2001) that the mangrove deforestation has subjected the meiofauna to environmental stress, indicated by the increased variability.

Both of the measured changes in the community structure and habitat modification indicated a gradual shifting from the typical mangrove meiofaunal composition. Regarding nematodes as an example, the main change was a decrease in the abundance of selective deposit feeders, which reflects the decline in their food readily available in the microbial rich mangrove sediments (Table 4). This indicates that the mangrove areas under deforestation stress lose their characteristics of high mud, good sorted sediment, high organic, detritus/microbial, and water contents. Consequently, the efficiency of the ecosystem as a nursery for marine organisms deteriorates. This conclusion resides on the assumption that the species composition is representative of community function (Austen et al. 1998) and partly reflects the whole ecosystem function. Seasonal variability was also exaggerated at deforested sites, which indicates the positive role of the mangrove in ameliorating the impact of the extreme seasonal changes in such arid coastal environments (Khalil 2001).

Conservation and Management

Regional cooperation for the protection of the marine environment in the Red Sea has been established since the early 1970s. In 1982, a regional convention (Jeddah Convention) was adopted, which has led to the establishment of the Regional Organization for the Conservation of the Environment of the Red Sea and Gulf of Aden (PERSGA) in 1996. PERSGA has seven member states (Djibouti, Egypt, Jordan, Saudi Arabia, Somalia, Sudan, and Yemen). It executes multi-linked program activities addressing sustainable management and conservation of marine resources in the region, of which mangrove is one of the key components (PERSGA 2011). The main previous and current conservation efforts by PERSGA addressing mangroves are summarized in box 1.

Table 4 ANOVA results for variation in the abundance of nematode feeding types (%) between sites and between stations, in summer and winter samples collected from three mangrove areas in the Red Sea

Nematode feeding type	ANOVA results			
	Summer		Winter	
	Between sites	Between stations	Between sites	Between stations
Selective deposit feeders	Sig.*	Sig.*	Sig.*	Sig.*
	NC, PC > CL	PC3, PC4, NC2, NC3, NC4 > all other stations.	NC > PC > CL	PC4, NC1, NC3, NC4 > all others except NC2
				NC2 > CL1, CL2, CL4, PC1, PC2, PC3
Non-selective deposit feeders	Sig.*	Sig.*	Sig.*	Sig.*
	CL > PC, NC	CL1, CL2, CL3 > all other stations	CL > PC > NC	CL2, CL3, CL4, PC1 > all others except PC2
				PC2 > PC4
Epigrowth feeders	Sig.*	Sig.*	Sig.*	Sig.*
	PC > NC > CL	PC2, NC1 > all other stations	CL > NC	CL1 > PC4 > all other stations
Predator/omnivores	Ns.	Sig.*	Sig.*	PC3 > all others except PC2
		CL4 > all other stations	PC > CL, NC	PC2 > all rest except NC2

CL1, CL2, CL3, CL4 are stations at Site CL (cleared mangrove site) representing high intertidal, mid-intertidal, low intertidal and subtidal zones, respectively. Similarly, PC1–PC4 are stations at Site PC (partially cleared mangrove site) and NC1-NC4 are stations at Site NC (non-cleared mangrove site) in the Red Sea. *Sig.** significant at $p < 0.01$, *Ns.* not significant. Critical value for the pairwise test is $p = 0.05$ (*source* Khalil 2001)

The major objective of the "Regional Action Plan for Conservation of Mangroves in the Red Sea and Gulf of Aden" (Saenger and Khalil 2007) is to protect mangroves from degradation and to ensure restoration of the degraded areas and sustainable use of the resource. The framework of the plan includes some key components: integrated coastal zone management planning for the conservation of mangroves, education and awareness, marine protected areas, ecologically sustainable mangrove utilization, impact of shipping and marine pollution, research, monitoring, and economic valuation.

Box 1. Main previous and current efforts by PERSGA addressing mangroves of the Red Sea and Gulf of Aden

- Developing regional standard survey methods (SSMs) for Inter-tidal and Mangrove Habitats in the Red Sea and Gulf of Aden, and training of specialists on conducting mangrove monitoring (2001).
- Conducting assessments and developing country reports and a regional report on the status of mangroves during 2002–2003, and developing regional and country status reports.
- Developing the Regional Action Plan (RAP) for Conservation of Mangroves in the Red Sea and Gulf of Aden, and National Action Plans (NAPs) for member states during 2004–2005.
- Establish a Regional Network for Protected Areas in the Red Sea and Gulf of Aden Region, most of them include important mangrove sites.
- Develop Regional Guidelines for Mangrove Restoration in the Red Sea and Gulf of Aden in 2010.
- Conducting a preliminary assessment of blue carbon in the Red Sea and Gulf of Aden in 2013.
- Organizing regular capacity building workshops on approaches and tools for mangrove management and restoration.
- Implementing demonstration projects addressing mangrove restoration in some member states.

Most Red Sea countries have attempted mangrove restoration projects during the last two decade, achieving different levels of success. Yet most of these projects remain limited in scale, compared to available potential resources and the area suitable for mangrove restoration in the region (PERSGA 2011). It is required to establish sustainable programs that consider building capacities and knowledge of mangrove restoration techniques, research, sustainable financial support, and participation of local communities in mangrove restoration activities.

Although several steps in action planning for mangrove conservation were taken at both regional and national levels, effective implementation is hindered by major shortcomings in capacity development and financial resources. Other complexities may include lack of awareness, insufficient public involvement, the rapid pace of development and changing state of the region, large relatively unstudied areas and gaps in information, difficulty in monitoring and assessment, a poor legal framework, and inadequate compliance and enforcement (PERSGA/UNEP 2014). Sound interventions are needed to overcome such difficulties and facilitate effective implementation of the regional and national plans for conservation of mangroves in the Red Sea region.

References

Abdel-Wahab M (2005) Diversity of marine fungi from Egyptian Red Sea mangroves. Bot Mar 48(5):348–355

Abohassan R, Okia C, Agea J, Kimondo J, McDonald M (2012) Perennial biomass production in arid mangrove systems on the Red Sea coast of Saudi Arabia. Environ Res J 6:22–31

Abu Ghararah ZH, Khalil ASM (2012) Ecosystem based adaptations in the Red Sea and Gulf of Aden. Climate Action Publication Edit. 2012, Climate Action Program, UK UNEP

Al-Wetaid A, Khalil ASM (2003) Status of Red Sea mangrove in Kingdom of Saudi Arabia. PERSGA, Jeddah

Andrews FW (1950) The flowering plants of Anglo-Egyptian Sudan (I). T. Buncle and Company Limited, Arbroath

Austen MC, Widdicombe S, Villano-Pitacco N (1998) Effects of biological disturbance on diversity and structure of meiobenthic nematode communities. Mar Ecol Prog Ser 174:233–246

Chevallier R (2013) Balancing development and coastal conservation: mangroves in Mozambique. Research Report 14, The Governance of Africa's Resources Program (GARP) of the South African Institute of International Affairs (SAIIA), Johannesburg

De Grissac AJ, Negussie K (2007) Eritrea's coastal marine and island biodiversity conservation project. Ministry of Fisheries of the State of Eritrea and United Nations Development Programme (UNDP) Office, Eritrea

Dittmar T (1999) Outwelling of organic matter and nutrients from a mangrove in North Brazil: evidence from organic tracers and flux measurements (PhD thesis). Zentrumfür Marine Tropenökologie (ZMT) Contributions 5. Center for Tropical Marine Ecology, Bremen

Dor I (1984) Epiphytic blue-green algae (cyanobacteria) of the Sinai mangal: consideration on vertical zonation and morphological adaptations. In: Por FD, Dor L (eds) Hydrobiology of the mangal. Dr. W. Junk Publishers, The Hague, pp 35–54

Dor I, Levy I (1984) Primary productivity of the benthic algae in the hard-bottom mangal of Sinai. In: Por FD, Dor L (eds) Hydrobiology of the mangal. Dr. W. Junk Publishers, The Hague, pp 179–191

ElsebaieIH Aguib ASH, Al Garni D (2013) The role of remote sensing and GIS for locating suitable mangrove plantation sites along the southern Saudi Arabian Red Sea coast. Int J Geosci 4:471–479

Evans MI (1994) Important Bird Areas in the Middle East. Birdlife Conservation Series No. 2. Birdlife International, Cambridge

FAO (2002) Socio-economic assessment and economic valuation of Egypt's mangroves. Consultancy Report TCP/EGY/0168 (A), Rome

FAO (2007) The World's mangroves 1980–2005. FAO Forestry Paper 153, Rome

Fatoyinbo T, Simard M (2012) Mapping of Africa's mangrove forest extent, height and biomass with ICE sat/GLAS and SRTM data fusion. Int J Remote Sens 34:668–681

Fishelson I (1971) Ecology and distribution of the benthic fauna in the shallow waters of the Red Sea. Mar Biol 10:113–133

Fondo EN, Martens EE (1998) Effects of mangrove deforestation on macrofaunal densities, Gazi Bay, Kenya. Mangroves Salt Marshes 2:75–83

Jennerjahn TC, Ittekot V (2002) Relevance of mangroves for the production and deposition of organic matter along tropical continental margins. Naturwissenschaften 89:23–30

Khalil ASM (1994) An Ecological study on fishes of the mangrove ecosystems of the Sudanese Red Sea. M.Sc. thesis, Department of Zoology, University of Khartoum, Sudan

Khalil ASM (2001) Response of meiofauna to mangrove deforestation in arid coastal habitats of the Red Sea (Sudan) with emphasis on free-living marine nematodes. ZMT Contributions, Number 13. PhD thesis, Zentrum für Marine Tropenökologie, Bremen, 82 pp

Khalil ASM (2010) Mangroves of the Red Sea and Gulf of Aden. Presentation at regional workshop on mangrove management and restoration, FRIM, Kuala Lumpur in Malaysia, 26–29 July 2010

Khalil ASM, Krupp F (1994) Fishes of the mangrove ecosystem. In: Comparative ecological analysis of biota and habitats in littoral and shallow sublittoral waters of the Sudanese Red Sea. Report for the period of Apr 1991–Dec 1993, Faculty of Marine Science and Fisheries, Port Sudan, and ForschungsinstitutSenckenberg, Frankfurt

Kumar A, Khan M, Muqtadir A (2010a) Distribution of mangroves along the Red Sea coast of the Arabian Peninsula: part1: the northern coast of western Saudi Arabia. Earth Sci India 3(I):28–42

Kumar A, Khan M, Muqtadir A (2010b) Distribution of mangroves along the Red Sea coast of the Arabian Peninsula: part 2: the southern coast of western Saudi Arabia. Earth Sci India 3(III):154–162

Kumar A, Khan M, Muqtadir A (2011) Distribution of mangroves along the Red Sea coast of the Arabian Peninsula: part 3: Coast of Yemen. Earth Sci India 4(II):29–38

MacNae W (1968) A general account of the fauna and flora of the mangrove swamps and forests in the Indo-West-Pacific Region. Adv Mar Biol 6:73–270

Mandura AS (1997) A mangrove stand under sewage pollution stress. Mangroves Salt Marshes 1:255–262

Mastaller M (1996) Destruction of mangrove wetlands—causes and consequences. Nat Res Dev 43(44):37–57

Mastaller M (1997) Mangroves: the forgotten forest between land and sea. Tropical Press Sdn. Bhd, Kuala Lumpur

MLWE (2012) Eritrea's five years action plan (2011–2015) for the great green wall initiative (GGWI). Ministry of Land, Water and Environment (MLWE), Asmara

Mohamed BF (1984) Ecological observations on the mangroves of the Red Sea shores of the Sudan. Hydrobiologia 110:109–111

MUHAT (2013) Seconde Communication Nationale de la République de Djibouti à la Convention cadre des Nations Unies sur les Changements Climatique. Ministère de l'Habitat, de l'Urbanisme et de l'Environnement Direction de l'Aménagement du Territoire et de l'Environnement, République de Djibouti, Déc 2013

Murray B, Watt C, Cooley D, Pendleton L (2012) Coastal blue carbon and the United Nations framework convention on climate change: current status and future directions. Nicholas Institute for Environmental Policy Solutions, Duke University, Durham

Nagi HM, Khanbari KM, Al-Sameh A (2012) Estimating total area of mangrove habitats in the Republic of Yemen. Using remote sensing and GIS. Faculty of Science Bulletin 24, Sana'a University, Yemen, pp 75–84

PERSGA (2002a) Status of mangroves in the Republic of Djibouti. Country report PERSGA, Jeddah. (Prepared by Khalil ASM)

PERSGA (2002b) Status of mangroves in the Republic of Yemen. Country report PERSGA, Jeddah. (Prepared by Khalil ASM)

PERSGA (2002c) Status of mangroves in the Republic of Sudan. Country report PERSGA, Jeddah. (Prepared by Khalil ASM)

PERSGA (2006) Status of Marine Environment in the Red Sea and Gulf of Aden. PERSGA, Jeddah. Prepared by Gladstone W, Facy R, Harriri K

PERSGA (2011) Regional guidelines for mangrove restoration in the Red Sea and Gulf of Aden. PERSGA Guidelines No. GD-0010, PERSGA, Jeddah, pp 1–34. (Prepared by Saenger P, Khalil ASM)

PERSGA/GEF (2004) Status of Mangroves in the Red Sea and Gulf of Aden. Technical Series Number 11, PERSGA, Jeddah, 67 pp. (Prepared by Khalil ASM)

PERSGA/SWC (2010) Mangrove distribution map along the Red Sea Coast of Saudi Arabia. PERSGA, Jeddah

PERSGA/UNEP (2013) Assessment of blue carbon policies in the Red Sea and Gulf of Aden (Draft Report). PERSGA, Jeddah. Prepared by Michaelowa A

PERSGA/UNEP (2014) A framework for ecosystem-based management (EBM) and a regional strategy to establishing long-term processes to support EBM implementation in the Red Sea and Gulf of Aden (Draft Report). PERSGA, Jeddah. Prepared by Abdallah A

Por FD, Dor I, Amir A (1977) The mangal of Sinai: limits of an ecosystem. Helgolander wissenschaftliche Meeresuntersuchungen 30:295–314

Por FD, Dor I (eds) (1984) Hydrobiology of the Mangal. Dr. W. Junk Publishers, The Hague

Price ARG, Medley PAH, McDowall RJ, Dawson-Shlephard AR, Hogarth PJ, Ormond RFG (1987) Aspects of mangal ecology along the Red sea coast of Saudi Arabia. J Nat Hist 21:449–464

Robertson AI, Alongi DM (eds) (1992) Tropical mangrove systems coastal and Estuarine studies 41. American Geophysical Union, Washington

Rabaa SMA (1980) Geomorphological characteristics of the Red Sea coast with special emphasis on formation of "Marsas" in the Sudan. In: Proceedings of symposium on the coastal and marine environment of the Red Sea, Gulf of Aden and Western Indian Ocean, Khartoum 9–14 Jan 1980. University of Khartoum, Sudan, pp 53–72

Ruwa RK (1988) Changes in patterns of macrofaunal distribution in mangrove ecosystems at the Kenyan coast due to natural and unnatural causes. In: Kenyan Belgium project, 4th quarterly report, pp 98–116

Saenger P (2002a) Mangrove ecology, silviculture and conservation. Kluwers Academic Publishers, Dordrecht

Saenger P (2002b) Ecological assessment of mangroves in Egypt. FAO Report TCP/EGY/0168(A), Cairo, pp 1–28

Saenger P, Khalil ASM (2007) Regional action plan for the conservation of mangroves. In: PERGA/GEF Regional action plan for the conservation of marine turtles, seabirds and mangroves in the Red Sea and Gulf of Aden. Technical Series 12, PERSGA, Jeddah, pp 101–129

Saenger P, Sankare Y, Perry T (1996) Effects of pollution and overcutting on mangroves. GEF large marine ecosystem project for the Gulf of Guinea. UNDP report, Abidjan

Saifullah SM, Khafaji AK, Mandura AS (1989) Litter production in a mangrove stand of the Saudi Arabian Red Sea coast. Aquat Bot 36:79–86

Sasekumar A, Chong VC (1998) Faunal diversity in Malaysian mangroves. Glob Ecol Biogeogr Lett 7:57–60

Sato G, Fisseha A, Gebrekiros S, Abdul Karim H, Negassi S, Fischer M, Yemane E, Teclemariam J, Riley R (2005) A novel approach to growing mangroves on the coastal mud flats of Eritrea with the potential for relieving regional poverty and hunger. Wetlands 25 (3):776–779

Schratzberger M, Warwick RM (1998) Effects of physical disturbance on nematode communities in sand and mud: a microcosm experiment. Mar Biol 130:643–650

Schwamborn R, Saint-Paul U (1996) Mangroves—forgotten forests? Nat Resour Dev 43(44):13–36

Sheppard CRC, Price ARG, Roberts CM (1992) Marine Ecology of the Arabian Region: patterns and processes in extreme tropical environments. Academic Press, London

Spalding M, Blasco F, Field SD (1997) World Mangrove Atlas. The International Society for Mangrove Ecosystems, Okinawa

Spalding M, Kainuma M, Collins L (2010) World atlas of mangroves. A collaborative project of ITTO, ISME, FAO, UNEP-WCMC, UNESCO-MAB, UNU-INWEH and TNC. Earthscan, London, 319 pp

Warwick RM, Clarke KR (1993) Increased variability as a symptom of stress in marine communities. J Exp Mar Biol Ecol 172:215–226

Wilkei ML (1995) Mangrove conservation and management in the Sudan. FAO Report. Ministry of Environmesnt and Tourism, Khartoum & FAO

Whiteman AJ (1971) The geology of the Sudan Republic. Clarendon Press, Oxford

World Bank (2010) Convenient solutions to an inconvenient truth: ecosystem-based approaches to climate change. The International Bank for Reconstruction and Development/The World Bank, Washington

World Bank (2014) http://data.worldbank.org/indicator/AG.LND.FRST.ZS. Accessed on 12 April

The Evolution of the Red Sea as a Human Habitat During the Quaternary Period

Geoff Bailey

Abstract

This chapter summarises current knowledge about the deep history of human occupation in the Arabian Peninsula and more specifically examines the likely role of the Red Sea escarpment and coastal region both as a major zone of human occupation in early prehistory and as a key pathway for the movements of people and the transmission of cultural ideas between Africa and Eurasia. This is a highly topical issue in the international literature at present both because of new archaeological investigations that are providing new dates for early Stone Age settlements in various parts of the Arabian Peninsula and because of genetic studies that highlight the southern Red Sea and southern Arabian Peninsula as a major 'corridor' of early human settlement and connection between Africa and Asia. The time range of these processes covers at least the past 150,000 years and could extend to 1 million years or more and therefore places a high premium on new understandings about the impact of climate change, sea-level change and other geological processes on the suitability of different areas of the Arabian landscape for human settlement and dispersal. This chapter discusses the archaeological and climatic evidence for Quaternary occupation, the effect of sea-level changes on the possibility of sea crossings of the southern Red Sea, the evidence for coastal archaeological settlements demonstrating early human interest in the exploitation of marine resources and seafaring, and new investigations in the Farasan Islands region that are searching for traces of submerged landscapes and archaeological sites formed at periods of lower sea level.

Introduction

The Red Sea coastal region and western escarpment of the Arabian Peninsula form the gateway between the African continent and the rest of the world. As such, they are of pivotal importance in understanding the great narrative of human dispersal from an original homeland in Africa. All the evidence currently available in the form of human fossils, stone tools, geochronology and genetic evidence shows that all human populations distributed throughout the world today had an origin that is deeply rooted in Africa, and that during the Quaternary period[1] there have been at least two major episodes of human expansion out of Africa (Oppenheimer 2003; Bellwood 2013). The first, associated with the genus *Homo* (*Homo ergaster* or *Homo erectus*), took place after about 2 million years ago, resulting in the expansion of these archaic human populations across southern Europe and Asia, extending from southern Britain in the west to Northeast China, and southwards as far as Indonesia (Grine et al. 2009). A second major dispersal occurred with the development in Africa of anatomically modern humans (AMH), *Homo sapiens sapiens* (White et al. 2003), who are thought to have evolved from earlier archaic populations in Africa and then dispersed across the Old World, replacing

G. Bailey (✉)
Department of Archaeology, University of York, King's Manor, York YO1 7EP, UK
e-mail: geoff.bailey@york.ac.uk

[1] The Quaternary period is defined as beginning at 2.58 million years and comprises two epochs, the Pleistocene, lasting until 11,500 cal BP (calibrated radiocarbon years before present), marking the conventional end of the Last Ice Age, and the Holocene, marking the period up to the present day (Gibbard et al. 2010).

N.M.A. Rasul and I.C.F. Stewart (eds.), *The Red Sea*,
Springer Earth System Sciences, DOI 10.1007/978-3-662-45201-1_34,

the pre-existing populations there, such as the Neanderthals, possibly with some admixture (Green et al. 2010), and expanding further than before to colonise the cold high latitudes of northern Eurasia, the Americas and the southern continent of Sahul (Australia, New Guinea and Tasmania).

Yet the Arabian Peninsula has been persistently discounted as an attraction to very early human settlement, except for a narrow corridor through the Sinai Peninsula in the north, on the grounds that the Red Sea would have acted as a barrier further south and that the extensive desert and semi-arid conditions of the interior would have been a further disincentive to human occupation. However, it has long been known that wetter conditions with lakes and extensive grasslands periodically and repeatedly spread across the desert interior during the past 2 million years and more (Edgell 2006; Vincent 2008), and the possibility of a 'southern corridor' across the southern end of the Red Sea, perhaps at a time when a land crossing was possible during periods of low sea level, has also been raised, particularly for AMH dispersal (Lahr and Foley 1994).

AMH dispersal must have occurred after about 200,000 years ago (the earliest date for currently known AMH remains in Africa), and before about 50,000 years ago, which is when modern humans first reached Australia. Moreover, the entry into Sahul involved sea crossings over distances of at least 60 km, which implies the use of simple boats or rafts and some skill in seafaring clearly linked to familiarity with marine resources such as fish and shellfish (O'Connell et al. 2010; O'Connor 2010; O'Connor et al. 2011). This has given rise to the hypothesis that the earliest populations to depart from Africa also had these skills and used them to facilitate dispersal along coastlines, across other sea barriers such as the southern end of the Red Sea, and around the rim of the Indian Ocean into the Indian subcontinent and Southeast Asia, an hypothesis of dispersal frequently linked, on the basis of indirect inference from mapping of genetic relationships among present-day populations, to a date of 60,000 years ago (Stringer 2000; Walter et al. 2000; Macaulay et al. 2005; Mellars 2006; Mellars et al. 2013). However, the basis for this hypothesis has been strongly contested on both archaeological and genetic grounds (Bailey 2009; Petraglia 2012; Boivin et al. 2013). The pathway of AMH dispersal into the Arabian Peninsula, whether via the Sinai Peninsula in the north or across the southern end of the Red Sea, the date of its occurrence, whether at 60,000 or earlier, and the mode of dispersal, whether or not involving some degree of dependence on marine resources and seafaring, are all matters of opinion in need of further investigations and observations.

The language used in describing these patterns of human expansion sometimes implies that they represented 'events', involving purposeful migrations of people trekking (or sailing) over long distances with some distant goal in view, involving a time span of decades or centuries, rather on the analogy of the earliest Europeans to enter North America in the C17th AD. However, this is likely to be a misconception in most cases. The expansion of the human habitat and the colonisation of new territory visible in the archaeological record are better viewed as a long-term process involving incremental range extensions of small social groups and their offspring into adjacent areas to fill the available habitat and involving many human generations and perhaps many hundreds or even thousands of years to accomplish the colonisation of new territory on a continental scale. The actual rate of expansion would, of course, have been determined in the first instance by the natural rate of population increase, but even very slow rates of population growth would have populated large areas quite rapidly on Quaternary timescales given suitable environmental conditions. The term 'dispersal' is more appropriate for such a process.

Another issue is the question of whether human populations would only have expanded beyond some pre-existing limit when driven to do so by a deterioration of environmental conditions within their existing habitat (e.g. Carto et al. 2009). Climate change during the Quaternary is clearly a major variable that is likely to have influenced patterns of early human evolution and dispersal (Maslin and Christensen 2007; DeMenocal 2011). However, populations are more likely to have expanded when environmental and climatic conditions improved, making larger areas available for human settlement, and to have contracted when conditions worsened, although examples of the latter could also occur and have been invoked, for example, in the first colonisation of the island of Cyprus by seafarers seeking new offshore resources during the climatic downturn of the Younger Dryas some 11,000 years ago (Ammerman et al. 2011). More potent facilitating or delaying factors would have been the availability of favourable environments with familiar resources in adjacent terrain, the adaptability of the pre-existing populations and their capacity to cope with new resources and environmental conditions, and the presence of physical, climatic and environmental barriers to further expansion. These barriers, in their turn, are likely to have changed significantly on Quaternary timescales because of changes in climate and changes of topography and palaeogeography brought about by changes in sea level, volcanism and tectonic activity (Petraglia and Rose 2009; Petraglia et al. 2011; Bailey and King 2011; Lambeck et al. 2011; Bosworth, this volume). All of these variable factors—of climate, sea level and tectonics—would have been powerfully active in the Red Sea and Arabian context.

The aim of this chapter is to summarise current knowledge about the deep history of human occupation in the Arabian Peninsula, the nature of changes in climate and sea level and their likely impact on the accessibility and suitability of the region for human occupation and the likely role

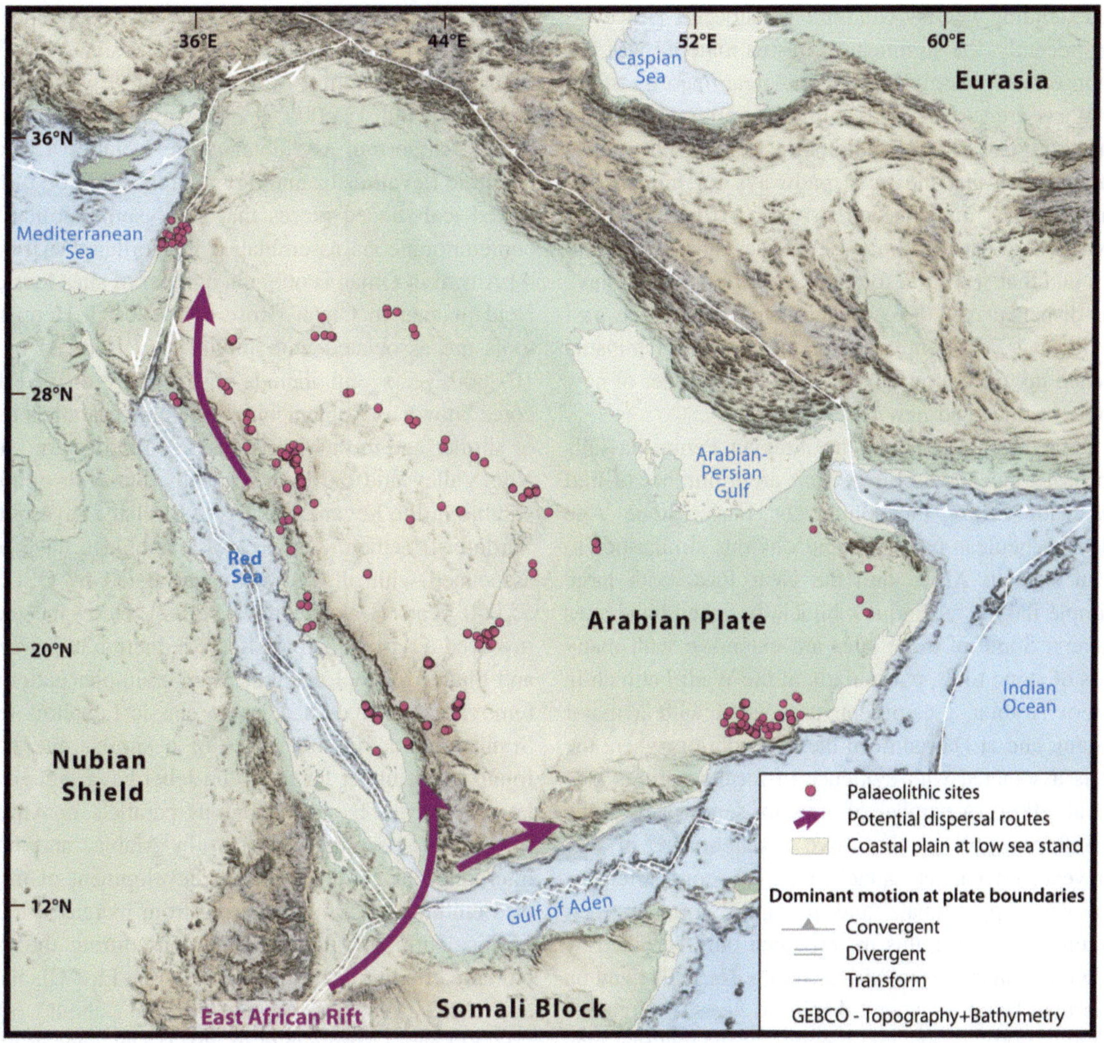

Fig. 1 Map of the Arabian Peninsula showing major geological and geographical features and distribution of Stone Age find spots (courtesy of Maud Devès)

of the Red Sea escarpment and coastal region as a major zone of human occupation, as a source of terrestrial and marine food resources and as a key pathway for the movements of people and the transmission of cultural ideas between Africa and Eurasia. Particular attention will focus on sea-level change, since this affects the likelihood of sea crossings at the southern end of the Red Sea, the visibility of coastal archaeological evidence, and also the varying productivity of the Red Sea as a source of marine resources for human consumption. It also affects the evidence for coastal archaeological settlements demonstrating early human interest in the exploitation of marine resources and seafaring. Also discussed here are new investigations in the Farasan Islands region that are searching for traces of submerged landscapes and archaeological sites formed at lower periods of sea level.

Quaternary Archaeology

Geographically speaking, the Arabian Peninsula forms a key stepping stone between Africa and Eurasia (Fig. 1), and one would expect an early history of Stone Age occupation, with human populations entering the Peninsula from the north via the Sinai Peninsula, or from the south across the southern end of the Red Sea. In fact, early Stone Age material has long been known. The Comprehensive Archaeological Survey Program of Saudi Arabia (CASP), which took place during the late 1970s and 1980s (Zarins et al. 1979, 1980, 1981), and the work of American and Russian archaeologists working at about the same time (Whalen et al. 1983, 1984, 1986, 1988; Amirkanov 1991), demonstrated the presence of many finds of Stone Age artefacts, indicating a human presence extending well back into the Pleistocene. However, until very recently, the significance of this material was not widely appreciated, mainly because the great majority of the stone artefacts are surface finds without stratigraphic integrity or accurate and reliable radiometric dates, and because of the assumption that the main pathways for human dispersal and cultural transmission between Africa and Asia were always through the Sinai–Levantine land corridor to the North and that semi-arid to desert conditions would have inhibited dispersal into the Arabian Peninsula. In the past decade, however, different areas of the Arabian Peninsula have opened up to exploration, resulting in a number of new and important finds and new dates (Fig. 1).

The Stone Age material in the Arabian Peninsula falls into two broad categories. First, there are a number of find locations with Lower Palaeolithic or Early Stone Age material of Acheulean type showing obvious similarities to Acheulean sites in Africa and the Near East, with large cores, simple flaking techniques, bifacially flaked hand axes and cleavers. Some of these sites are extensive with many thousands of stone tools, particularly at the Wadi Fatimah in the western Arabian escarpment, associated with a major river system, and at Dawadmi in the arid interior, where the material is associated with evidence of former springs (Petraglia et al. 2009). A number of uranium-series dates at the site of Saffaqah at Dawadmi give a date range of 60–204,000 years on calcrete attached to the stone artefacts, giving a very approximate and minimum age for their manufacture. Otherwise, this material can only be dated by analogy with similar sites in Africa and the Near East and on that basis may be as early as 1.4 million years, though Petraglia et al. (2009) suggest that the bulk of the material may date from 800,000 years onwards. This material is clearly associated with an early episode or episodes of human expansion out of Africa, though whether dispersal occurred via the North or the South cannot be determined from the available archaeological evidence.

A second category of finds comprises stone tool assemblages of Middle Stone Age or Middle Palaeolithic type, characterised by more complex and efficient patterns of core preparation and core reduction to produce regular-shaped flakes and blades. These sites include Jubbah in the north of Saudi Arabia, where Stone Age material has been found in stratigraphic context alongside a now-dry lake in the Nefud Desert, with optically stimulated luminescence (OSL) dates of the associated sand sediments of 75,000 years (Petraglia et al. 2011). The stone tools include material that is typical of Middle Palaeolithic or Middle Stone Age technology found widely across the Near East and North Africa, including Levallois cores, a distinctive technology for removing successive flakes of predetermined shape from a specially prepared nodule, some of which are retouched along the edge to form notched and denticulated pieces. At Jebel Faya in the United Arab Emirates, a sequence of stone tool industries has been found in deposits originally accumulated in the mouth of a rock shelter, with earliest OSL dates of about 125,000 years (Armitage et al. 2011). The earliest stone tool assemblage here includes a Middle Palaeolithic Levallois technology with the addition of bifacially flaked leaf-shaped pieces that show some similarities with contemporaneous assemblages in Northeast Africa. Aybut Al Auwal in Oman is one of a number of sites located on the Nejd plateau in Oman (Rose et al. 2011). Here, the stone tools are associated with fluvial gravels dated by OSL to 107,000 years and include distinctively shaped Levallois cores known as Nubian cores, with close parallels to material of similar age in the desert oases of the Eastern Sahara, the Nile Valley and the Red Sea Hills, but with nothing comparable in the Levant. Finally, at Shi'bat Dihya in the Wadi Surdud of Yemen, a stone tool assemblage has been found associated with alluvial sediments dated by OSL at about 55,000 years (Delagnes et al. 2012). Here, the technology involved a type of prepared-core reduction to produce flakes and blades for tool use, but using a simpler pattern of flake removals than that used in the classic Levallois technique. Similar stone tool industries of a similar age have been found in the upper levels of the Jebel Faya sequence. These later industries have no obvious parallels in Africa or the Levant, or elsewhere, as yet, in Arabia, and have been interpreted as evidence for the development of more localised cultural traditions resulting from increased localisation and isolation of human populations during the more arid climatic conditions at the height of the Last Glacial (MIS 3 and MIS 2), approximately from about 60,000 to 20,000 years (Armitage et al. 2011; Delagnes et al. 2012).

Quaternary Climate Change

A critical variable affecting the likelihood of early human occupation is climate change, particularly increases in precipitation. Many of these early Stone Age sites are found in the desert interior in regions that have too little surface water today to support human life without modern technological aids. It is clear from this fact, and from the presence of fossil tufas indicating former spring activity, dried up lake beds and wadis filled with alluvial sediments indicating much greater stream competence than today, that conditions must have been periodically wetter in the past, allowing the spread of grasslands, grazing mammals and human hunters into the desert interior and multiplying the pathways for movements and contact across the Peninsula from the Red Sea escarpment to the Gulf Coast. Mapping of palaeochannels, dating of lake sediments and analyses of proxy climate indicators in speleothem sequences are beginning to create a clearer picture of the temporal and spatial pattern of climate variability,

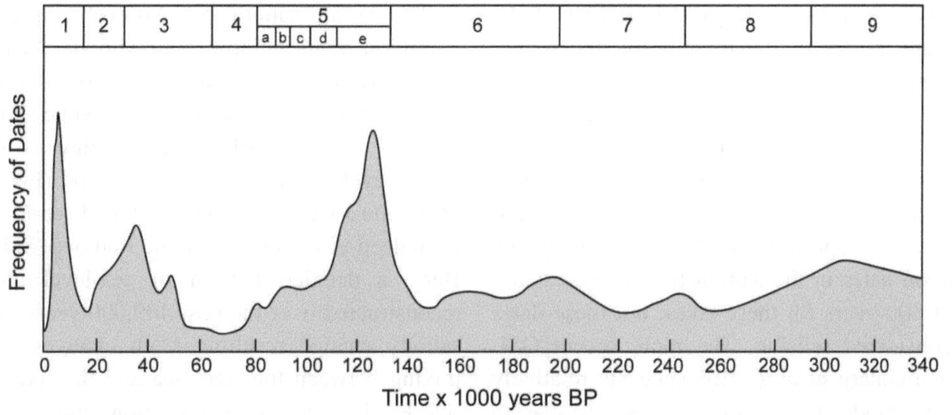

Fig. 2 Variations in climate in the Arabian Peninsula over the past 350,000 years (after Parker 2009). The *curve* represents the probability of wetter climate intervals based on the number of dated indicators for a given period. *Peaks* in the *curve* represent well-dated wet intervals rather than unusually wet intervals. *Numbered intervals* indicate marine isotope stages. For further details, see Parker (2009)

at least for the last 350,000 years (McClure 1976; Schultz and Whitney 1986; Sanlaville 1992; Parker et al. 2006; Fleitmann et al. 2007; Parker and Rose 2008; Parker 2009; Crassard et al. 2013; Rosenberg et al. 2013; Fig. 2).

The principal driver of climate change in the Arabian Peninsula during the Quaternary was shifts in the inter-tropical convergence zone (ITCZ) associated with the glacial–interglacial cycle, which led periodically to northward incursion of the Indian Ocean monsoon (IOM) weather system. Today, rainfall associated with the IOM falls mainly on the southern edge of the Arabian Peninsula, especially in the coastal regions of Oman, and extends as far north as Jizan in Saudi Arabia, with high rainfall over the Asir and Yemeni highlands in the southwest. This northward shift during the Pleistocene brought increased rainfall to present-day desert regions throughout Arabia, activating networks of stream channels and creating numerous shallow lakes of greater or lesser extent across desert regions such as the Rub al-Khali, the Mundafan Basin and the Nafud in the north. An additional influence may have been the extension southwards of the Mediterranean cyclonic system, bringing winter rainfall into areas further south than is the case today.

The onset, timing and duration of these wetter episodes are matters of ongoing investigation. As a very general rule, there is a broad correlation between the marine isotope curve and increased precipitation, with the wettest conditions in interglacials (odd-numbered MIS) and maximum aridity in glacial periods (even-numbered MIS). But when sufficient dates are available, the pattern in detail appears more complex. Over the past 130,000 years, there is a clear and strong correlation between wetter periods and the early part of the interglacial cycle, in MIS 5e at 130–120 ka, and again in the early Holocene (the early part of MIS 1) from about 11.5–6 ka (Fig. 2). The latter part of the Last Interglacial, MIS 5a, 82–74 ka was also wet. MIS 4 witnessed the onset of increased aridity but was punctuated by at least one wetter interval at 61–58 ka (Parton et al. 2013). During MIS 3, associated with progressive onset of Northern Hemisphere glaciation, the position is less clear. The predominant climatic signal is one of increased aridity, reaching its peak at the Last Glacial Maximum at 22 ka, but a number of wet intervals have been claimed between 40 ka and 20 ka and a short-lived wetter phase at about 15–13 ka on the basis of radiocarbon-dated alluvial and lacustrine sediments (see, in particular, Parker 2009 for details). However, more recent OSL dating of lacustrine sediments does not show any evidence of these wetter intervals (Crassard et al. 2013; Rosenberg et al. 2013), and it appears that the radiocarbon dates are unreliable because of contamination by younger carbon. Going further back in time, both the MIS 7 and MIS 9 interglacials were associated with wetter conditions indicated by OSL dating of lacustrine sediments and U/Th dating of speleothems, centred at about 193 ka and 319 ka, respectively. MIS 6 was another arid period, but one punctuated by brief episodes of increased precipitation at 147, 152, 160, 170 and 174 ka according to Parker (2009). Earlier still, in the Middle Pleistocene, arid indicators are present between 700 and 560 ka, and wet indicators between about 560 and 319 ka, but the broad duration of these climatic zones may be misleading because of scarcity of dates and cannot be resolved into a more detailed record to compare with the later part of the Quaternary record. Earlier again, carbon isotope evidence ($\delta^{13}C$) in Early Pleistocene faunas associated with palaeolake deposits in the western Nafud desert indicates an open savannah landscape with lakes, buffalo and hippopotamus between about 3.5 and 1.2 Ma (Thomas et al. 1998).

There are difficulties in generalising with confidence and in detail from this evidence, except perhaps for the very end of the Quaternary sequence, given the irregular and generally sparse spatial and chronological sampling of the climatic record, and the increasing margins of uncertainty associated with dating methods as one goes further back in time. It remains unclear to what extent the spot dates for wet indicators such as lake and river sediments, for example in MIS 6, represent short-lived wet intervals separated by more arid conditions, or irregular sampling of the same prolonged and continuous climatic episode. McClure (1976), on the basis of radiocarbon dates in the Rub al-Khali, suggested a duration of about 800 years for these lakes, but these dates are no longer considered reliable. The more recent OSL dating work of Rosenberg et al. (2013) suggests relatively short durations for periods of lake formation, but the margin of error in the dating method does not allow further refinement. The best chronological indicator comes from annual laminations of lake sediment in Nafud, which indicate a duration of about 1,400 years for lake formation during MIS 9. An additional complexity is the possibility that the spread of wetter conditions induced by the IOM was time transgressive from south to north. Continuous records from speleothems and the deep-sea marine record are subject to their own uncertainties, and discrepancies between them and the discontinuous terrestrial record may reflect either problems of dating, or divergence of local climate conditions in particular landscape settings from a generalised climate curve derived from marine sediments or speleothems. Allowing for these uncertainties, the broad pattern seems to be one of persistent and periodic spread of wetter conditions throughout the Quaternary, with the wettest and longest intervals occurring during the early part of interglacials, and perhaps periodic bursts of increased precipitation but of shorter duration at other times.

Sea-Level Change and Sea Crossings

A factor of potential importance in determining the likely attractiveness and importance of coastal regions is the possibility afforded by marine and oceanographic conditions in the Red Sea for the development of simple methods of seafaring and the availability of easily accessible marine resources such as inshore fish and intertidal shellfish. Here, however, we face a fundamental problem and that is the fact that for long periods of the Pleistocene epoch under review here, sea levels were much lower than present. This has serious implications for reconstructing the position and configuration of past coastlines and island archipelagos and hence the need for sea crossings, the nature of varying oceanographic conditions and their impact on currents and marine productivity within the Red Sea Basin, and the visibility of archaeological evidence for human activity on these ancient but now-submerged shorelines.

The broad pattern of sea-level change over the past 2 million years is now well known. Analysis of oxygen isotope variations in foraminifera stratified in deep-sea sediments has demonstrated that the isotope signal closely maps changes in ocean volume in response to the glacial–interglacial cycle of the Quaternary period (Fig. 3). Indeed, one of the most recent and most detailed sea-level curves comes from deep-sea sediments in the Red Sea (Siddall et al. 2003). Here, a detailed pattern of sea-level change has been reconstructed over the past 400,000 years, using a model of salinity change resulting from changes in the degree of mixing between the Red Sea and the Indian Ocean. When sea level was low, ocean flow into the Red Sea was restricted, and the salinity of the Red Sea was elevated above the ocean average because of higher rates of evaporation. One important consequence of this analysis is to demonstrate

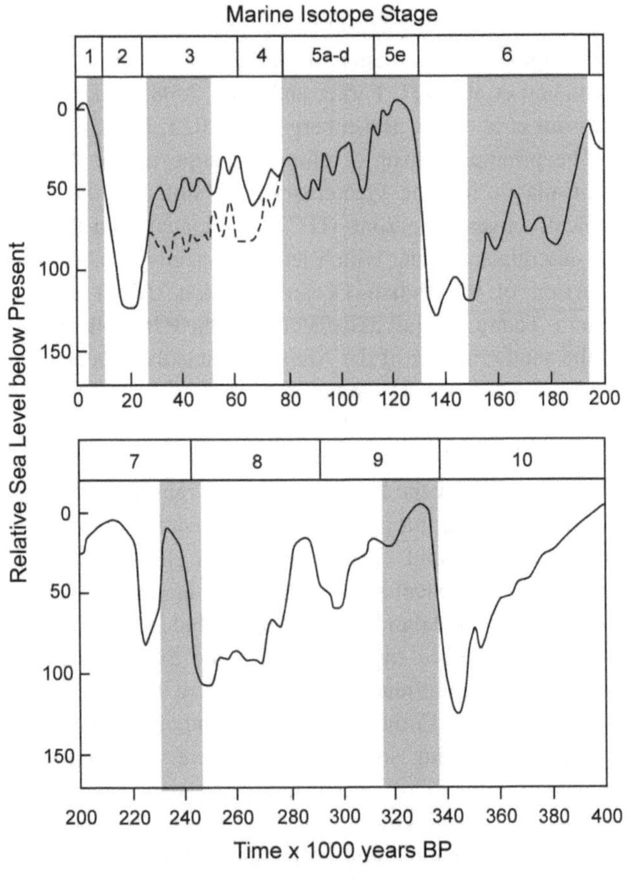

Fig. 3 Variations in eustatic sea level (ocean volume) according to the deep-sea isotope record over the past 350,000 years. *Sources* Chappell and Shackleton (1986), Shackleton (1987), Imbrie et al. (1984), Lambeck and Chappell (2001), Waelbroek et al. (2002), Sidall et al. (2003), Lambeck et al. (2011). The *dotted part* of the *curve* shows the original isotope readings and the *solid line* above it the sea-level curve after removal of temperature effects from the isotope readings. *Shaded columns* give an approximate indication of the relationship between sea-level position and climate change as derived from Fig. 2. See also Fig. 6 and the text for further discussion of climate data

that salinity in the Red Sea over this time range never reached very high values that would imply evaporation within a closed basin. In other words, at no time in the past 400,000 years was it possible for people to cross the southern end of the Red Sea between Northeast Africa and the Arabian Peninsula across dry land.

In order to pursue these implications, it is necessary to convert the sea-level curve into maps of shoreline position at different stages of the glacial–interglacial cycle. This is a complex procedure and cannot be done simply by mapping sea-level positions against modern seabed bathymetry, because the coastal crust has been warped and flexed by hydro-isostatic loading and unloading of water masses on the continental shelf, by large-scale propagation of the Red Sea rift and tectonic uplift of the rift flanks, by localised processes of salt doming and withdrawal, by variable accumulations of sediment on the sea bed and by localised volcanic activity (Bosworth, this volume; Ligi et al., this volume; Hovland et al., this volume; Pugh and Abualnaja, this volume). Modelling of these processes in the southern Red Sea, incorporating all available dates for raised or submerged beaches throughout the Red Sea region, has been carried out by Kurt Lambeck working with the Saudi–UK archaeological team working in the region (Lambeck et al. 2011).

The reconstruction of coastal palaeogeography in the southern Red Sea, and the data and theoretical models on which it is based, are set out in detail in Lambeck et al. (2011) and are only briefly summarised here (see also Pugh and Abualnaja, this volume). The key sources of required information are the following: (1) bathymetry of the shallowest part of the southern Red Sea, namely the Hanish Sill; (2) variations in ocean volume with time (variations in eustatic sea level); (3) variations in the isostatic loading of the Red Sea region over time and in different parts of the Red Sea region; (4) earth model parameters on the thickness and viscosity of the underlying mantle in the Red Sea region, which determine the way in which the Earth's crust responds locally to isostatic loading effects; and (5) rates of tectonic uplift. Variable (1) is known approximately from actual measurements taken by Werner and Lange (1975) and the minimum depth is 137 m, although the possibility of deeper channels cannot be excluded (Lambeck et al. 2011, pp. 3570–3571). Variable (2) is known from information on raised shorelines in far-field locations such as Western Australia where isostatic and tectonic effects are absent or independently known and can be further corroborated against the marine isotopic record of changes in ice volume. Isostatic deformation, variable (3), is usually associated with regions close to the continental ice sheets of the Northern Hemisphere, where crustal depression by ice sheets and rebound following deglaciation (glacio-isostatic) can elevate shorelines by hundreds of metres. In more distant locations such as the Red Sea, the glacio-isostatic effect is small though not entirely absent and is supplemented by additional effects resulting from the loading and unloading of water masses on the continental shelf (hydro-isostatic effects). These effects can be modelled, taking into account glacio-isostatic effects and the best available data on hydro-isostatic loading in adjacent sea basins, and can have a significant impact. Variable (4) is known but with some degree of uncertainty that can be assessed by applying different assumptions within a range of likely possibilities. Variable (5) is the unknown variable in this equation and is estimated by comparing the position of the shoreline at selected time intervals (Early Holocene, Last Glacial Maximum, Last Interglacial), as predicted by isostatic modelling, with the actual elevation of dated shoreline features. The latter are represented by a large and somewhat scattered sample of data points available for this purpose from the Red Sea, including dates from uplifted coral terraces that are especially prominent in the northern Red Sea, coral reefs or other shoreline features that are close to the present shoreline elevation and some that are now submerged. Any resulting differences between predicted and observed shoreline elevations can be used to estimate the effect of tectonic uplift associated with long-term rifting, and this amounts to at least 0.1 mm year^{-1} (or 1 m per ten thousand years) in the northern Red Sea. Once these adjustments have been made, they can be applied to the reconstruction of shoreline configurations at selected time intervals and in different regions.

The shortest sea crossing at modern sea level today is about 29 km across the Bab-al-Mandab Strait, a crossing that is almost impossible except with seaworthy boats and navigational skills, something that is generally assumed by archaeologists to have been feasible only during recent millennia. However, palaeogeographic modelling shows that a narrow channel would have extended for about 100 km from the Bab-al-Mandab Strait in the south to the Hanish Islands in the north and that the shallowest part of the channel is over the Hanish Sill (Fig. 4). Moreover, the shape of the channel in the Hanish region is such that for more than half the period of time encompassed by a single glacial cycle, that is to say whenever sea level was lower than 50 m below present, for a period of about 60,000 years in every 100,000 year cycle, sea crossings of about 4 km or less were possible via small islands in the Hanish region (Fig. 5). During these periods of lower sea level, crossings via island hopping across the Hanish Sill could easily have been accomplished with little risk by swimming or simple rafting. This greatly increases the likelihood of human crossings, whether by accident or intention, not only over an extensive proportion of the past 250,000 years, but much further back into human prehistory.

Further research is needed on the precise bathymetry of the channel region of the Hanish Sill to test in more detail these palaeogeographical reconstructions and the models on

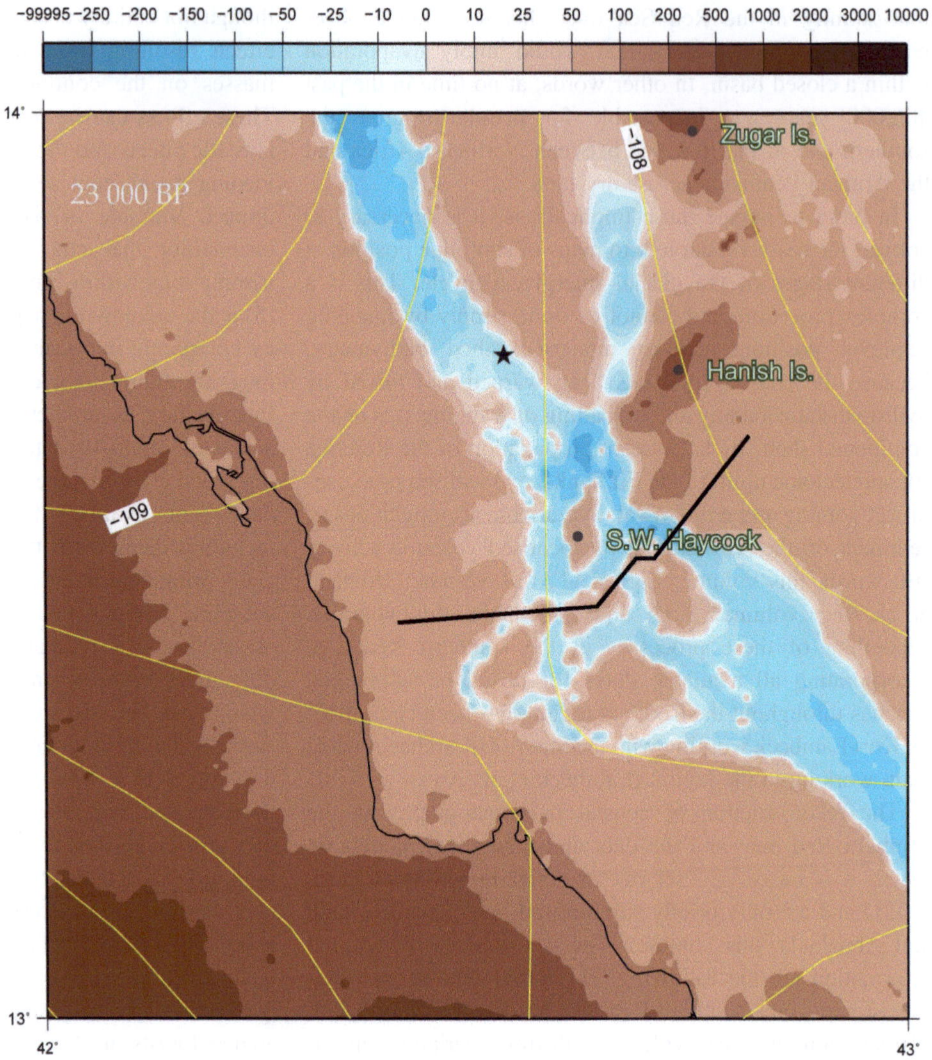

Fig. 4 Map showing the position of palaeoshorelines at the southern end of the Red Sea during the maximum sea-level low stand at the Last Glacial Maximum. The shortest sea crossings at this period would have been in the region of the Hanish Sill, via the Haycock Islands, as shown by the black line (after Lambeck et al. 2011)

which they are based. More well-dated palaeoshoreline features, whether from submerged shorelines that are now under water, or from raised coral reefs such as the impressive sequence visible at Umm Lajj in the north (Vincent 2008, Fig. 2.3), are also needed to refine and test the models. Another question is the effect that the formation of a long and constricted channel at lowered sea level between the Bab-al-Mandab and the Hanish Islands would have had on tidal currents and the extent to which these would have been faster and more hazardous for human sea crossings. Answers to this question must remain speculative for the moment, but new research on palaeotidal modelling is now under way that may help to constrain the likely pattern of sea currents during these periods of lower sea level (Lambeck, pers. comm., September 2013).

It is of interest to compare the periods when sea crossings would have been most easily feasible with minimal technology, as identified by the above reconstructions, and those periods of increased rainfall when the territory available for human occupation would have been most attractive and most extensive (Fig. 6). The comparisons can only be approximate because of uncertainties in the dating and duration of some of the climatic intervals and smoothing effects in the drawing of the sea-level curve. Of course, the absence of easy sea crossings in the south during periods of favourable climate would not have deterred a human presence in the Arabian Peninsula, since entry into the Peninsula and dispersal within it would always have been possible from the north via the Sinai route. The point is that a conjunction of favourable climates and easy sea crossings would have afforded the maximum opportunities for dispersal and interpopulation contact between Africa and Arabia, whereas the absence of one or the other, and even more so the absence of both, would have tended to reinforce bottlenecks in dispersal and regional isolation. On these grounds, the most striking feature of Fig. 6 is that periods of wetter climate and periods of easy sea crossing are almost largely mutually exclusive. A favourable conjunction of

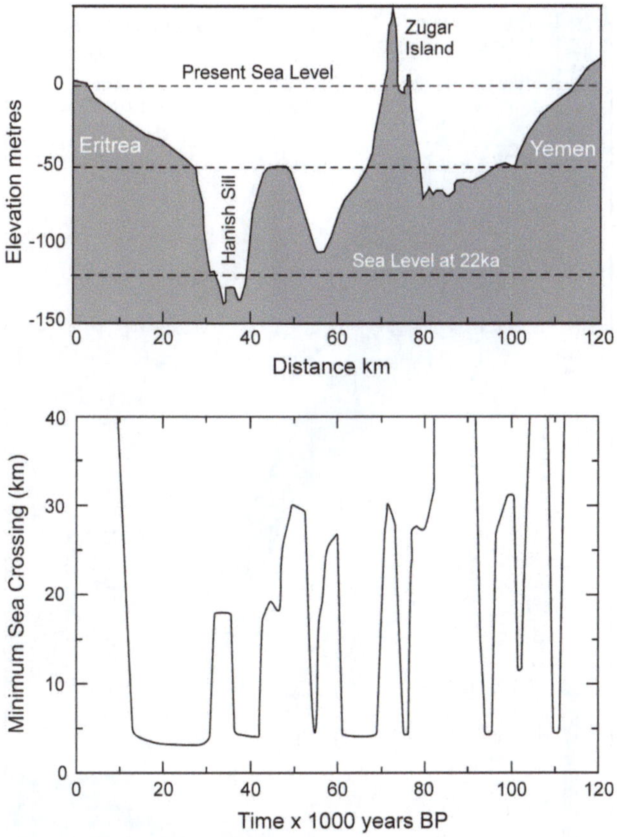

Fig. 5 Cross section of the Hanish Sill region. The cross section is marked by the *black line* in Fig. 4. The *upper* diagram shows the effect of channel geometry on sea crossing distances at different sea-level positions and the *lower* diagram the periods when short sea crossings would have been possible during the past 120,000 years (after Lambeck et al. 2011)

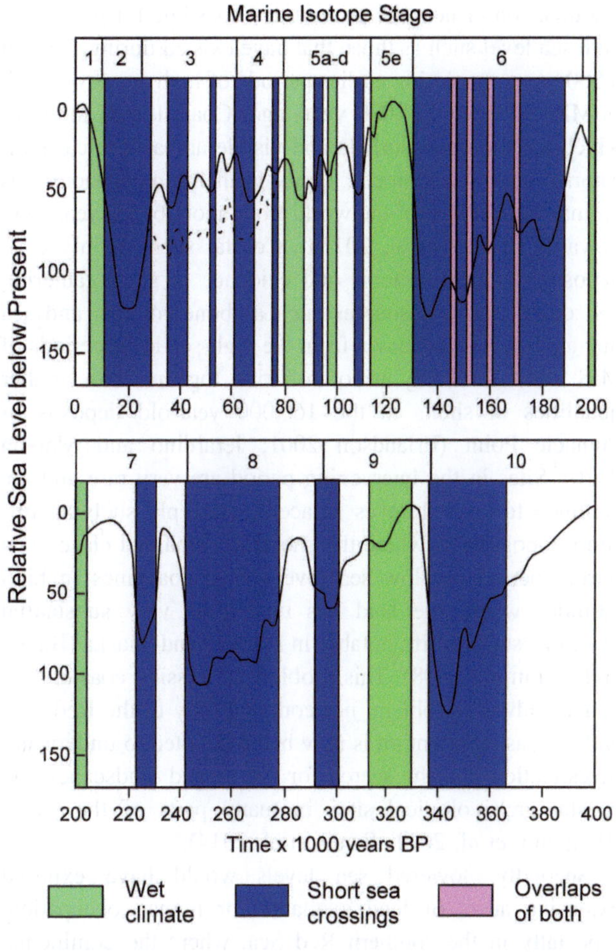

Fig. 6 A comparison of wet periods when climatic conditions would have been most favourable and sea level stands sufficiently lowered to facilitate easy sea crossings at the southern end of the Red Sea over the past 350,000 years. *Blue* short sea crossings; *Green* wetter climates; *Pink* overlap of periods with short sea crossings and wetter climates. Climate data are taken from Fig. 2, modified according to the more recent work of Rosenberg et al. (2013) and data on sea crossings from Lambeck et al. (2011)

circumstances may have occurred in the wetter intervals of MIS 5 but only for very short intervals. The longest period of overlap occurred in MIS 6, between about 140 and 180 ka, if Parker's (2009) dates are followed, but these wet intervals are not recorded in Rosenberg et al.'s (2013) data. Interestingly, the most favourable climatic periods during the interglacials of MIS 1, 5e, 7 and 9 and the late glacial episodes in MIS 2 appear to be periods when sea crossings would have been most difficult or would have required skilful seafaring abilities. Least favourable periods, combining maximum aridity with high sea levels, appear to be rare, the most notable example being the latter part of MIS 1 (the Late Holocene after about 6 ka), and perhaps part of MIS 7, but the climate record is incomplete in this period to say nothing of earlier MIS stages. Other periods of combined aridity and high sea level may have existed, but they are either disguised by poor resolution in the dating, or else they were of short duration.

Coastal Archaeological Sites and Offshore Landscapes

Sea-level change is not only of importance in narrowing the crossing distances between opposite shorelines. It also has two additional archaeological consequences. First, any evidence for the use of marine resources and a maritime way of life must be invisible except during periods of high sea level like the present. Palaeoshorelines formed when sea level was lower than the present are now deeply submerged and some distance offshore, and we can only expect to see archaeological evidence of coastal settlements and use of marine resources on or near the present-day coastline for periods of high sea level such as those that have existed during the past 6,000 years, or during earlier periods of high sea level such as MIS 5e about 130,000 years ago. Coastal shell mounds, which are the most durable and visible archaeological indicator of coastal settlement, appear in their tens of thousands on many coastlines of the world from about 6,000 years ago onwards (Bailey et al. 2013a). Coastal sites, mainly cave deposits, with stone tools and evidence of shell gathering and other marine resources such as bones of fish and sea mammals, are also known from the high-sea-level periods of MIS 5 and 4, notably in South Africa, together with smaller quantities of shells in the 160,000 year-old deposits of Pinnacle Point (Erlandson 2001; Jerardino and Marean 2010). Sites in the intervening period are very rare and are confined to coastal caves adjacent to steeply shelving offshore topography, where the shoreline remained close even during periods of low sea level, or to coastlines in high latitudes where the land has undergone very substantial glacio-isostatic uplift, notably in Norway and Alaska (Bailey and Flemming 2008). This problem of missing coastal sites is a worldwide problem, not confined only to the Red Sea, and increasing attention is now being devoted to underwater investigation and the search for submerged landscapes and coastal archaeological sites in many parts of the world (Benjamin et al. 2011; Evans et al. 2014).

Secondly, lowered sea levels would have exposed extensive areas of land available for human occupation, especially in the southern Red Sea, where the continental shelf is quite shallow (Fig. 7). Here, an additional increment of land some 100 km wide extended seaward of the present coastline, and there was a similar extension of land on the African side offshore of Eritrea. Both shelves host a concentration of islands—the Dahlak Islands on the Eritrean side, and the Farasan Islands on the Saudi Arabian side, formed by salt tectonics, which would presumably have projected as a clump of low hills in this extensive coastal terrain when sea level was lower. Moreover, both theoretical considerations (Faure et al. 2002) and inspection of the available bathymetry suggest that this now-submerged landscape may have been quite attractive for plant and animal life, and hence for humans, because of a complex topography with fault-bounded basins, extensive spring lines and accumulation of groundwater in solution hollows, providing attractive conditions in these coastal regions even during periods of relatively arid climate.

We are now exploring these ideas through new investigations in the region of the Farasan Islands. These islands were connected to the mainland when sea level was about 50 m below present or lower. Their archaeological interest stems from the fact that there are more than 3,000 coastal archaeological sites, most of them shell middens, but covering a wide range of site types from surface finds with few or a limited scatter of shells to large mounds up to 5 m high, some of which contain almost nothing but discarded mollusc shells, while others include remains of fish bones and land mammals in addition to the mollusc shells (Bailey et al. 2013b; Fig. 8). These are not to be confused with the Farasan Banks, which are further to the north, or the 'shell banks' of the geological literature (Dabbagh et al. 1984). They are not natural deposits of shells thrown up by storms but the remains of settlements of prehistoric coastal peoples who collected the marine molluscs for food and perhaps also for fish bait. These sites began forming at or soon after about 6,000 years ago and clearly relate to the establishment of modern sea level. As such, they are recent in date, but at the same time, they give a useful insight into the visible archaeological features that are associated with a fully maritime way of life involving seafaring, fishing and shell gathering among other activities. They are a useful benchmark for the sorts of evidence that we would expect to see in relation to hypotheses for the existence much further back in the prehistoric period of maritime peoples around the coastlines of the Red Sea and the Arabian Peninsula, given

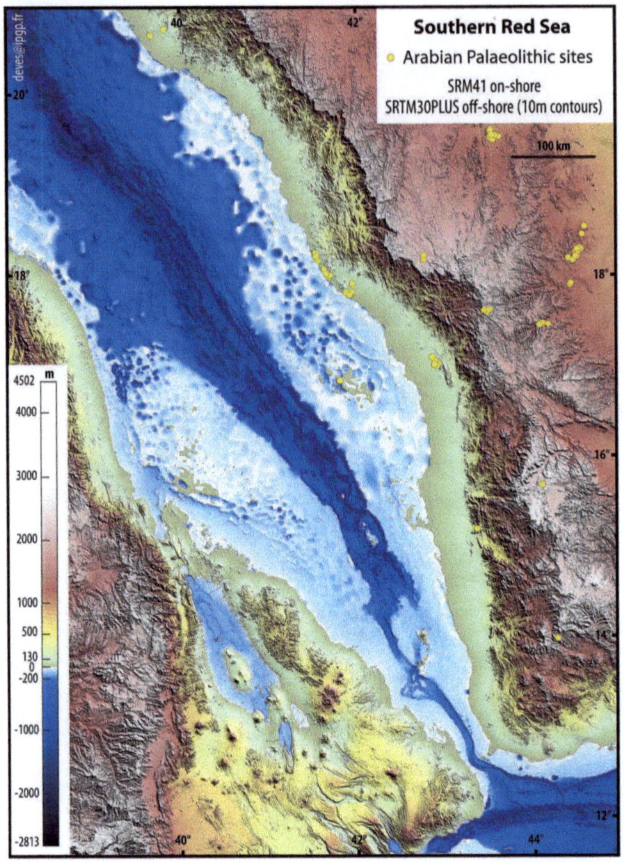

Fig. 7 Enhanced satellite imagery of the southern Red Sea showing the position of the Farasan Islands, and the extent of the submerged landscape at maximum sea-level regression during glacial periods, and the general nature of the seabed topography. ASTER GDEM is a product of METI and NASA (courtesy of Maud Devès)

Fig. 8 Shell mound in Janaba Bay on the main island of Farasan, showing the position of the shell mound on a modern shoreline undercut by marine erosion (photograph by Hans Sjoeholm 2006)

that consumption of molluscs was a universal accompaniment to shoreline settlement or activity in the Arabian context and that in any case, the shells of the molluscs are the primary factor giving visibility to shoreline activity of whatever type.

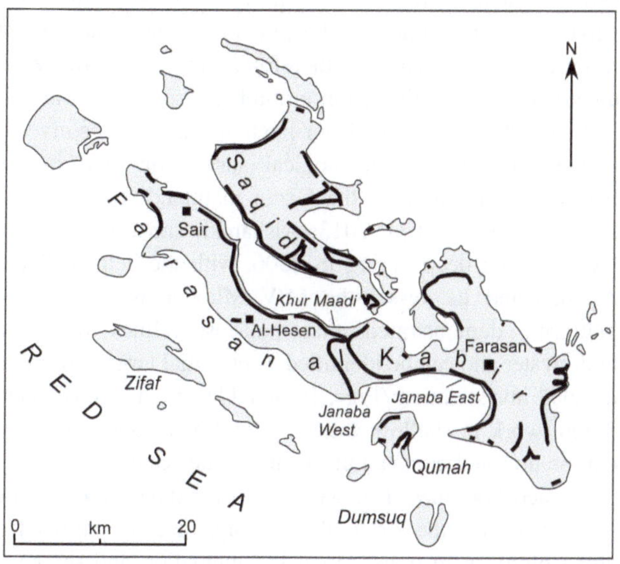

Fig. 9 Distribution of shell mounds on the Farasan Islands. The number and density of sites is such that they cannot be shown individually on a map of this scale. Instead, a more or less continuous line is used to provide a representation of the distribution. Sites that appear to be inland of the present-day shoreline are located on palaeoshorelines that are now located inland because of tectonic uplift and sediment infilling of shallow bays and channels

The majority of the sites are on the large islands of Farasan Kabir and Saqid, but there are also shell mounds on the island of Qumah and on some of the smaller islets to the north of Farasan (Fig. 9). The largest mounds are up to 4–5 m high and the largest extend for hundreds of metres along the shoreline. The bulk of these shell deposits appear to have been formed between about 6,000 and 5,000 years ago. Later deposits also exist, extending up to the Islamic era, but these are thinner shell deposits or shell scatters, often associated with remains of buildings made from naturally cemented beach rock or coral. The reasons for this difference in volume of shell deposits remain unclear, but part of the explanation may have to do with the existence of an ecological window of opportunity for extensive beds of marine molluscs in shallow embayments that came into being for only a short period and were then subsequently filled in with sand.

Excavations show that the shell mounds comprise a wide range of shallow water and intertidal molluscan species variously associated with sandy substrates or coral reefs. The dominant species is a small gastropod, *Strombus fasciatus*, but other common species are the large gastropods, *Pleuroploca trapezium* and *Chicoreus ramosus*, and bivalve species such as *Chama reflexa*, *Spondylus marisrubri* and *Pinctada* sp. Extensive ash layers are interleaved with the shell deposits, representing the remains of fireplaces, and fish bones and the bones of gazelle have also been recovered from some shell layers, along with very rare and isolated finds of stone tools and potsherds.

The huge quantity of mollusc shells implied by the number and volume of shell deposits does not necessarily mean that the mounds were created by very large numbers of people, or by people who subsisted mainly on shell food. The amount of food represented by the shells is actually relatively small, once one takes into account the high ratio of shell to meat in most shell species, and the time over which the mounds have accumulated. Detailed measurements show that the amount of food represented by these large shell deposits may be as little as 5–10 % of total food intake and that the impressive appearance of the resulting shell mounds is mainly the result of the large amounts of debris created by shell gathering and the durability and resistance to decay of the dead shells compared with most other food remains or by-products of human activity.

Since the Farasan shell mounds are clearly late in date, the question arises as to whether similar evidence can be found or might be expected in association with earlier periods of high sea level. An obvious candidate for inspection is the high-sea-level period of MIS 5e. Walter et al. (2000) have reported evidence of marine mollusc shells collected as food at the 130,000 year-old site of Abdur on the Eritrean coast in association with stone tools and mammal bones. Full details have not been published so that the quantity of material and the status of the mollusc shells as food items have yet to be evaluated. Similar sites occur on the Arabian side of the Red Sea in the vicinity of the extensive lava fields found in the coastal region of Al Birk. Here, stone tools of Middle Stone Age and Early Stone Age type have been found on elevated coral terraces that refer to earlier periods of high sea level, most probably MIS 5 (Bailey et al. 2007a; Bailey 2009). However, the artefacts so far recovered are surface finds and cannot be dated with confidence or associated with remains of shell food or other subsistence activity. Ecological conditions that allow the accumulation of substantial shell mounds like those of the Farasan Islands are only patchily distributed along any given coastline. Moreover, even shells are vulnerable to fragmentation and dispersal if exposed on the surface for many millennia. Hence, the absence of thick or extensive shell deposits on coastlines where stone artefacts have been found is not decisive refutation of an interest in marine resources by the stone tool makers. What is needed is archaeological material in stratified context, where there are good chances of obtaining geochronological control and contextual information on environment and subsistence, including marine shells discarded as food, or other indicators of marine exploitation. Sites with these features of MIS 5 age or older are rare, and usually only preserved and recovered in cave deposits, such as those recorded in South Africa and on parts of the Mediterranean coastline. Abdur is an exception, showing that relevant material can be preserved in open-air locations. The search for similar material in Saudi Arabia is ongoing (Devès et al. 2013).

An even greater challenge is the question of whether coastal archaeological material was deposited on palaeo-shorelines that are now submerged. The fact that the Farasan shell mounds appear almost exactly at the moment when sea level stopped rising and shorelines became visible on the present-day land surface is a strong indicator that earlier coastal sites could have existed and have escaped discovery so far for no better reason than that they are now submerged, and no one has yet gone underwater to look for them. That possibility, in its turn, demands underwater investigation. One possible objection to pursuing this hypothesis is that during periods of lower sea level, the reduced inflow of water from the Indian Ocean and the resulting increase in salinity would have inhibited marine productivity and reduced the availability of marine resources. Sidall et al. (2003) have identified 'aplanktonic' levels in the deep-sea sediment sequence associated with low-sea-level episodes. The more general significance of such evidence is unclear. Many marine species can tolerate conditions of high salinity, as is clearly demonstrated by the presence of an endemic marine fauna in the Red Sea that has persisted through many cycles of sea-level change. That people might have been deterred from exploiting the resources of the sea during the periods of low sea level, such as those of early MIS 1, and MIS 2–4, because marine resources were absent is highly speculative and can only be tested by searching under water for relevant evidence. In any case, whether or not people present in the region during periods of lowered sea level took an interest in marine resources or ignored them, it is certain that an extensive terrestrial landscape would have existed with potential for the pursuit of plant and animal resources on land. There is, therefore, a strong imperative for underwater exploration from a number of points of view.

Since 2006, we have been pursuing underwater investigations alongside archaeological survey on land in the Farasan Islands and on the Gizan mainland (Bailey et al. 2007a, b; Devès et al. 2013; Alsharekh and Bailey 2014). Experimental deep diving in 2006, with the help of Saudi Aramco and its vessel the M/V Midyan as an offshore platform, demonstrated that palaeoshorelines do exist underwater and can be identified as potential targets for more detailed investigation (Figs. 10 and 11). We also conducted diving work in shallow water in 2008 and 2009, including exploration and excavation of underwater caves and undercut palaeoshorelines. This work is slow and painstaking. The way in which the original landscape and its associated archaeology were transformed by inundation and sea-level rise is not yet well understood. Many sites may be washed away, degraded or buried beneath marine sediments. Others may be well preserved because of a complex topography that provided protection from the full force of wave action during inundation or by burial beneath marine sediment followed by subsequent partial erosion and exposure. We know from

Fig. 10 Palaeoshoreline showing characteristic features at a depth of about 20 m (photograph by Hans Sjoeholm 2006)

extensive underwater investigations elsewhere that underwater features can be preserved, often with excellent conditions of organic preservation, and can survive even in high-energy coastal environments and through several cycles of sea-level change (Benjamin et al. 2011; Evans et al. 2014).

There is every reason to suppose that similar features are preserved in the Red Sea region. The next essential step in underwater exploration is to examine the seabed more extensively in the Farasan region for traces of palaeoshorelines and other topographic features, using the full range of modern technological aids including seismic and acoustic

Fig. 11 A diver recording features and collecting samples from a submerged palaeoshoreline during a deep dive to a depth of over 60 m using mixed gas (trimix) technology (photograph by Trevor Jenkins 2006)

recording, remotely operated vehicles, and sediment coring. Suitable equipment has now been sourced for this phase of investigation, and a preliminary survey took place in June 2013 using the facilities of the R/V Aegaeo of the Hellenic Centre for Marine Research. The survey carried out a series of transects extending from the edge of the continental shelf to shallower channels between the Farasan Islands and the Gizan mainland. Surveys included multibeam bathymetry to characterise seabed topography, sub-bottom profiling to identify geological and geomorphological structures such as faulting and sediment thickness, a side-scan survey to highlight surface anomalies, the use of a remotely operated vehicle with cameras for visual inspection and the collection of a large sample of sediment cores. It is too early to report in detail on that survey, but preliminary results indicate that the landscape that was exposed when sea level was low was a landscape of considerable topographic complexity, with fault-bounded basins that trapped water and sediment, deep solution hollows that could also have collected freshwater and an extensive network of drainage channels.

It is axiomatic that when climate change resulted in the extension of wetter and greener conditions into the desert interior, those improved conditions would have applied also to coastal regions. Conversely, when the interior became uninhabitable because of increased aridity, coastal regions would have remained relatively well-watered because of capture of orographic rainfall on the Red Sea escarpment and the presence of high water tables and springs. In short, coastal regions would always have been attractive relative to the desert interior, regardless of climatic conditions, and especially during more arid climatic episodes, providing the stable, core regions of human settlement from which populations would have periodically expanded into the interior or retreated again according to the pattern of climatic change as summarised in Fig. 6. As is clear from this diagram, many periods of arid climate coincided with low sea levels, and so did some of the wetter episodes, including the Early Holocene wet period, when sea levels were rising at the end of the last glaciation but were still as much as 50 m below the present. If this hypothesis of relative coastal attractiveness is correct, then we should expect archaeological sites to occur in greater number in coastal regions than in the interior, regardless of whether marine resources were part of the coastal subsistence economy or not, and with greater persistence, both when sea level was high and when it was low. During periods of low sea level, the coastline of the southern Red Sea shifted by as much as 100 km seaward of its present position, exposing an extensive coastal region with many potential attractions for human settlement including relative abundance of water supplies and terrestrial plant and animal resources. This is likely to have shifted the geographical focus of human population to the now-submerged landscapes of the region. This in its turn reinforces the need for underwater investigation, without which we will remain in ignorance about the nature of these submerged landscapes, the records they contain of changes in palaeoenvironment, palaeoclimate and sea level, and of course their archaeological significance in illuminating a long and crucial period of human history in the Arabian Peninsula.

Conclusion

During the past decade, there has been a steady increase in field surveys and investigations of the Stone Age prehistory of the Arabian Peninsula including the Red Sea region and of palaeoenvironmental and palaeoclimatic investigations directed by archaeological questions or relevant to them. These investigations are highlighting the central geographical importance of the Arabian Peninsula in the earliest stages of human prehistory and its role in creating the foundations for later developments in more recent millennia. They indicate that fluctuations in climate and sea level have periodically expanded the opportunities for human settlement and dispersal throughout the Arabian Peninsula and then contracted them, providing a highly dynamic context for the early history of human settlement. As the primary gateway of human contact between Africa and Eurasia, the Red Sea region is of particular interest and importance. New research includes surveys for Stone Age archaeological sites dating far back into the Quaternary, studies of palaeoclimate from onshore and offshore sediment sequences and speleothems and studies of sea-level change and its palaeogeographical and marine–oceanographic consequences, excavation of coastal shell mounds, underwater exploration of submerged landscapes and palaeoshorelines, and improved geochronological control. This intersection of many different scientific interests provides a potentially fertile ground for interdisciplinary collaboration and a need for such collaboration if new hypotheses and lines of future enquiry are to be pursued effectively.

Acknowledgments Field observations on the archaeology and submerged landscapes of the Gizan region and the Farasan Islands have been made possible through the support and collaboration of the Saudi Commission for Tourism and Antiquities (SCTA), in particular its President, HRH Prince Sultan bin Salman bin AbdulAziz al Saud, theVice President, Professor Ali Al-Ghabban, and the Director General, Jamal al Omar. Underwater work has been supported by permits from the Department of General Survey of the Ministry of Defense and further facilitated through the assistance of Saudi Aramco, Shell Companies Overseas, the Saudi British Bank and the Hellenic Centre for Marine Research (HCMR). The work has been supported by grants from the European Research Council (Advanced Grant 269586 DISPERSE under the 'Ideas' Specific Programme of FP7), the Natural Environment Research Council (NERC), UK, under its EFCHED (Environmental Factors in Human Evolution and Dispersal) Programme, the British Academy and the National Geographic Society. This field research is a joint Saudi–UK project in collaboration with Professor Abdullah Alsharekh of King Saud University and also draws on the specialist expertise and assistance of a number of colleagues, whose assistance is gratefully acknowledged: Saud Al-Ghamdi and Abdarrazzack Al Maamary (King Saud University), Maud Devès and Geoffrey King (Institut de Physique du Globe, Paris), Nicholas Flemming and Garry Momber (National Oceanography Centre, and Marine Archaeology Trust, Southampton), Kurt Lambeck (Australian National University), Eva Laurie, Robyn Inglis and Matthew Meredith-Williams (University of York), Najeeb M.A. Rasul (Saudi Geological Survey), Dimitris Sakellariou (HCMR, Greece) and Claudio Vita-Finzi (Natural History Museum, London). Thanks are also due to Simon Holdaway and Michael Petraglia for additional information and comments that have improved the text. This is DISPERSE contribution no. 13.

References

Alsharekh AM, Bailey GN (eds) (2014) Coastal Archaeology in Southwest Saudi Arabia and the Farasan Islands: 2004–2009 investigations. Saudi Commission for Tourism and Antiquities, Riyadh

Ammerman AJ, Howitt Marshall D, Benjamin J, Turnbull T (2011) Underwater investigations at the early sites of Apros and Nissi Beach on Cyprus. In: Benjamin J, Bonsall C, Pickard K, Fischer A (eds) Submerged prehistory. Oxbow Books, Oxford, pp 263–271

Amirkhanov KM (1991) The palaeolithic of Southern Arabia. Nauka, Moscow (In Russian)

Armitage SJ, Jasim SA, Marks AE, Parker AG, Usik VI, Uerpmann HP (2011) The southern route "Out of Africa": evidence for an early expansion of modern humans into Arabia. Science 331:453–456

Bailey GN (2009) The Red Sea, coastal landscapes and hominin dispersals. In: Petraglia MD, Rose JI (eds) The evolution of human populations in Arabia. Springer, Dordrecht, pp 15–37

Bailey GN, Flemming NC (2008) Archaeology of the continental shelf: marine resources, submerged landscapes and underwater archaeology. Quatern Sci Rev 27:2153–2165

Bailey GN, King GCP (2011) Dynamic landscapes and human dispersal patterns: tectonics, coastlines and the reconstruction of human habitats. Quatern Sci Rev 30:1533–1553

Bailey GN, King GCP, Flemming NC, Lambeck K, Momber G, Moran LJ, Al-Sharekh AM, Vita-Finzi C (2007a) Coastlines, submerged landscapes and human evolution: the Red Sea Basin and the Farasan Islands. J Isl Coast Archaeol 2:127–160

Bailey GN, Al-Sharekh AM, Flemming NC, Lambeck K, Momber G, Sinclair AGM, Vita-Finzi C (2007b) Coastal prehistory in the southern Red Sea Basin, underwater archaeology and the Farasan islands. Proc Semin Arabian Stud 37:1–16

Bailey GN, Hardy K, Camara A (eds) (2013a) Shell energy: mollusc shells as coastal resources. OxBow, Oxford

Bailey GN, Meredith-Williams MGM, Alsharekh A (2013b) Shell mounds of the Farasan Islands, Saudi Arabia. In: Bailey GN, Hardy K, Camara A (eds) Shell energy: mollusc shells as coastal resources. OxBow, Oxford, pp 241–254

Bellwood P (2013) First migrants: ancient migration in global perspective. Wiley-Blackwell, Chichester

Benjamin J, Bonsall C, Pickard K, Fischer A (eds) (2011) Submerged prehistory. Oxbow, Oxford

Boivin N, Fuller DQ, Dennell R, Allaby R, Petraglia MD (2013) Human dispersal across diverse environments of Asia during the Upper Pleistocene. Quatern Int 300:32–47

Carto SL, Weaver AJ, Hetherington R, Lam Y, Wiebe EC (2009) Out of Africa and into an ice age: on the role of global climate change in the late Pleistocene migration of early modern humans out of Africa. J Hum Evol 56:139–151

Chappell J, Shackleton NJ (1986) Oxygen isotopes and sea level. Nature 32:137–140

Crassard R, Petraglia MD, Drake NA, Breeze P, Gratuze B, Alsharekh A, Arbach M, Groucutt HS, Khalidi L, Michelsen N, Robin CJ, Schiettecatte J (2013) Middle Palaeolithic and Neolithic occupations around Mundafan palaeolake, Saudi Arabia: implications for climate change and human dispersals. PLoS ONE 8(7):1–23 e69665

Dabbagh A, Hotzl H, Schnier H (1984) Farasan Islands. General considerations and geological structure. In: Jado AR, Zötl JG (eds) Quaternary period in Saudi Arabia, vol 2. Springer, New York, pp 212–220

Delagnes A, Tribolo C, Bertran P, Brenet M, Crassard R, Jaubert J, Khalidi L, Mercier N, Nomade S, Peigné S, Sitzia L, Tournepiche J-F, Al-Halibi M, Al-Mosabi A, Macchiarelli R (2012) Inland human settlement in southern Arabia 55,000 years ago. New evidence from the Wadi Surdud Middle Paleolithic site complex, western Yemen. J Hum Evol 63:452–474

DeMenocal PB (2011) Climate and human evolution. Science 311:540–541

Devès M, Inglis R, Meredith-Williams M, Al Ghamdi S, Alsharekh AM, Bailey GN (2013) Palaeolithic survey in Southwest Saudi Arabia: methodology and preliminary results. Adumatu 27:7–30

Edgell HS (2006) Arabian deserts. Nature, origin and evolution. Springer, Dordrecht

Erlandson JM (2001) The archaeology of aquatic adaptations: paradigms for a new millennium. J Archaeol Res 9:287–350

Evans A, Flemming NC, Flatman J (eds) (2014) Prehistoric archaeology of the continental shelf: a global review. Springer, New York

Faure H, Walter RC, Grant DR (2002) The coastal oasis: ice age springs on emerged continental shelves. Global Planet Change 33:47–56

Fleitmann D, Burns SJ, Mangini A, Mudelsee M, Kramers J, Villa I, Villa I, Neff U, Al-Subbary AA, Buettner A, Hippler D, Matter A (2007) Holocene ICTZ and Indian monsoon dynamics recorded in stalagmites from Oman and Yemen (Socotra). Quatern Sci Rev 26:170–188

Gibbard PL, Head MJ, Walker MJC, the Subcommission on Quaternary Stratigraphy (2010) Formal ratification of the Quaternary System/Period and the Pleistocene Series/Epoch with a base at 2.58 Ma. J Quat Sci 25:96–102

Green RE, Krause J, Briggs AW, and 53 other co-authors (2010) A draft sequence of the Neandertal genome. Science 328(5979):710–722

Grine FE, Fleagle JG, Leakey RE (eds) (2009) The first humans—origin and early evolution of the genus Homo. Springer, Dordrecht

Imbrie J, Hays JD, Martinson DG, McIntyre A, Mix AC, Morley JJ, Pisias NG, Prell WL, Shackleton NJ (1984) The orbital theory of Pleistocene climate: support from a revised chronology of the marine $\delta^{18}O$ record. In: Berger AL, Imbrie J, Kukla G, Saltzmann B (eds) Milankovitch and climate, Part 1. Reidel, Dordrecht, pp 269–305

Jerardino A, Marean CW (2010) Shellfish gathering, marine paleoecology and modern human behaviour: perspectives from cave 13PPB, Pinnacle Point, South Africa. J Hum Evol 59:412–424

Lahr M, Foley R (1994) Multiple dispersals and modern human origins. Evol Anthropol 3(2):48–60

Lambeck K, Chappell J (2001) Sea level change through the last glacial cycle. Science 292:679–686

Lambeck K, Purcell A, Flemming N, Vita-Finzi C, Alsharekh A, Bailey GN (2011) Sea level and shoreline reconstructions for the Red Sea: isostatic and tectonic considerations and implications for hominin migration out of Africa. Quatern Sci Rev 30(25–26):3542–3574

McClure H (1976) Radiocarbon chronology of late Quaternary lakes in the Arabian Desert. Nature 263:755–756

Macaulay V, Hill C, Achilli A et al (2005) Single, rapid coastal settlement of Asia revealed by analysis of complete mitochondrial genomes. Science 308:1034–1036

Maslin MA, Christensen B (2007) Tectonics, orbital forcing, global climate change, and human evolution in Africa: introduction to the African paleoclimate special volume. J Hum Evol 53:443–464

Mellars PA (2006) Going east: new genetic and archaeological perspectives on the modern human colonization of Eurasia. Science 313:796–800

Mellars PA, Gori KC, Carr M, Soares PA, Richards MB (2013) Genetic and archaeological perspectives on the initial modern human colonization of southern Asia. In: Proceedings of the National Academy of Science 110:10699–10704

O'Connell JF, Allen J, Hawkes K (2010) Pleistocene Sahul and the origins of seafaring. In: Anderson A, Barrett JH, Boyle K (eds) The global origins and development of seafaring. McDonald Institute for Archaeological Research, Cambridge, pp 57–80

O'Connor S (2010) Pleistocene migration and colonization in the Indo-Pacific region. In: Anderson A, Barrett JH, Boyle K (eds) The global origins and development of seafaring. McDonald Institute for Archaeological Research, Cambridge, pp 41–55

O'Connor S, Ono R, Clarkson C (2011) Pelagic fishing at 42,000 years before the present and the maritime skills of modern humans. Science 334:1117–1121

Oppenheimer S (2003) Out of Eden: the peopling of the world. Constable, London

Parker AG (2009) Pleistocene climate change in Arabia: developing a framework for hominin dispersal over the last 350 ka. In: Petraglia MD, Rose JI (eds) The evolution of human populations in Arabia. Springer, Dordrecht, pp 39–49

Parker AG, Rose JI (2008) Climate change and human origins in southern Arabia. Proc Semin Arabian Stud 38:25–42

Parker AG, Goudie AS, Stokes S, White K, Hodson MJ, Manning M, Kennet D (2006) A record of Holocene climate change from lake geochemical analyses in southeastern Arabia. Quatern Res 66 (3):465–476

Parton A, Farrant AR, Leng MJ, Schwenninger J-L, Rose JI, Uerpmann H-P, Parker AG (2013) An early MIS 3 pluvial phase in Southeast Arabia: climatic and archaeological implications. Quatern Int 300:62–74

Petraglia MD (2012) Toba volcanic super-eruption of 74,000 years ago: climate change, environments and evolving humans. Quatern Int 258:1–4

Petraglia MD, Rose JI (eds) (2009) The evolution of human populations in Arabia. Springer, Dordrecht

Petraglia MD, Drake NA, Alsharekh AM (2009) Acheulean landscapes and large cutting tools assemblages in the Arabian Peninsula. In: Petraglia MD, Rose JI (eds) The evolution of human populations in Arabia. Springer, Dordrecht, pp 103–116

Petraglia MD, Alsharekh AM, Crassard E, Drake NA, Groucutt H, Parker AG, Roberts RG (2011) Middle Paleolithic occupation on a marine isotope stage 5 lakeshore in the Nefud Desert, Saudi Arabia. Quat Sci Rev 30(13–14):1555–1559

Rose J, Usik V, Marks A, Hilbert Y, Galletti C, Parton A, Geiling JM, Cerný V, Morley M, Roberts RG (2011) The Nubian complex of Dhofar, Oman: an African Middle Stone Age industry in southern Arabia. PLoS ONE 6(11):e28239

Rosenberg TM, Preusser F, Risberg J, Plikk A, Kadi KA, Matter A, Fleitmann D (2013) Middle and Late Pleistocene humid periods recorded in palaeolake deposits of the Nafud desert, Saudi Arabia. Quatern Sci Rev 70:109–123

Sanlaville P (1992) Changements climatiques dans la Péninsule Arabique durant le Pléistocène Supérieur et l'Holocène. Paléorient 18(1):5–26

Schultz E, Whitney JW (1986) Upper Pleistocene and Holocene lakes in the An Nafud, Saudi Arabia. Hydrobiologia 143:175–190

Shackleton NJ (1987) Oxygen isotopes, ice volume and sea level. Quatern Sci Rev 6:183–190

Siddall M, Rohling EJ, Almogi-Labin A, Hemleben C, Meischner D, Schmelzer I, Smeed DA (2003) Sea-level fluctuations during the last glacial cycle. Nature 423:853–858

Stringer C (2000) Coasting out of Africa. Nature 405:24–27

Thomas H, Geraads D, Janjou D, Vaslet S, Memesh A, Billiou D, Bocherens H, Dobigny G, Eisenmann V, Gayet M, Lapparent de Broin F, Petter G, Halawani M (1998) First Pleistocene fauna from the Arabian Peninsula: An Nafud desert, Saudi Arabia. Comptes rendus Academie des Sciences, Paris, Sciences de la Terre et des Planèts 326:145–152

Vincent P (2008) Saudi Arabia: an environmental overview. Taylor and Francis, London

Waelbroeck C, Labeyrie L, Michel E, Duplessy JC, Lambeck K, McManus JF, Balbon E, Labracherie M (2002) Sea-level and deep water temperature changes derived from benthic foraminifera isotopic records. Quatern Sci Rev 21:295–305

Walter RC, Buffler RT, Bruggemann JJ, Guillaume MMM, Berhe SM, Negassi B, Libsekal Y, Cheng H, Edwards RL, von Gosel R, Neraudeau D, Gagnon M (2000) Early human occupation of the Red Sea coast of Eritrea during the Last Interglacial. Nature 405:65–69

Werner G, Lange K (1975) A bathymetric survey of the sill area between the Red Sea and the Gulf of Aden. Geologisches Jarbuch D 13:125–130

Whalen NM, Sindi H, Wahidah G, Siraj-Ali JS (1983) Excavation of Acheulean sites near Saffaqah in al-Dawadmi (1402/1982). Atlal, J Saudi Arabian Archaeol 7:9–21

Whalen NM, Siraj-Ali JS, Davis W (1984) Excavation of Acheulean sites near Saffaqah, Saudi Arabia, 1403 AH 1983. Atlal, J Saudi Arabian Archaeol 8:9–24

Whalen NM, Siraj-Ali JS, Sindi H, Pease DW (1986) A lower Pleistocene site near Shuwayhitiyah in northern Saudi Arabia. Atlal, J Saudi Arabian Archaeol 10:94–101

Whalen NM, Siraj-Ali J, Sindi HO, Pease DW, Badein MA (1988) A complex of sites in the Jeddah-Wadi Fatimah area. Atlal, J Saudi Arabian Archaeol 11:77–85

White T, Asfaw B, DeGusta D, Gilbert H, Richards DG, Suwa G et al (2003) Pleistocene *Homo sapiens* from Middle Awash, Ethiopia. Nature 423:742–747

Zarins J, Ibrahim M, Potts D, Edens C (1979) Saudi Arabian archaeological reconnaissance 1978. The preliminary report on the third phase of the Comprehensive Archaeological Survey Program —the coastal province. Atlal, J Saudi Arabian Archaeol 3:9–42

Zarins J, Whalen N, Ibrahim M, Abd al Jawad M, Khan M (1980) The Comprehensive Archaeological Survey Program. Preliminary report on the central and southwestern provinces. Atlal, J Saudi Arabian Archaeol 4:9–36

Zarins J, Al-Jawad Murad A, Al-Yish KS (1981) The Comprehensive Archaeological Survey Program. The second preliminary report on the southwestern province. Atlal, J Saudi Arabian Archaeol 5:9–42

Authors' Biographies

Najeeb M.A. Rasul
Technical Advisor
Center for Marine Geology, Saudi Geological Survey, Jeddah, Saudi Arabia
rasul.nm@sgs.org.sa

Dr. Najeeb M.A. Rasul holds a Master's degree in Marine Geology, and has a Ph.D. in Geological Oceanography from the University of Wales, UK. He is presently a Technical Advisor at the Saudi Geological Survey (SGS) in Jeddah. For the past 15 years he has been associated with several organizations and has held key research positions at Geological and Geophysical Research Systems in Canada, the Challenger Division for Seafloor Processes at the National Oceanography Centre, Southampton, UK., Environment Canada in Burlington, Canada, and the Center for Marine Geology at the SGS, where he was the Head of the center from 2003 to 2010. He has conducted research cruises with European Union Research centres, including the Institute of Marine Sciences (ISMAR), Italy and the IOC-UNESCO (TTR programme) and has been a EUROMARGINS and Saudi Scientific Coordinator. He has been involved in several research projects and cruises as chief scientist, and scientific investigator in the Pacific, Indian and Atlantic Oceans, including the Mid-Atlantic Ridge, Strait of Gibraltar, Gulf of Cadiz, Mediterranean Sea, Tyrrhenian Sea, Alboran Sea, Henderson Lake, Lake Erie, Arabian Gulf, Strait of Hormuz, Arabian Sea, and currently the Red Sea for the last 10 years. His research interests have concentrated on the fields of both shallow and deep water sedimentary processes, depositional mechanisms and environmental aspects. These include research on modern underwater sedimentary features and processes involved in their formation, deep sea turbidite systems, sediment dispersal and pathways, sediment plumes, fluxes and deposition, contaminants and chemical pathways for pollution and their environmental impact in both time and space, as well as marine mineral resources. He has also worked on the formation and exploration of oil and gas in deltas and offshore regions.

Ian C.F. Stewart
Consultant
Stewart Geophysical Consultants Pty. Ltd., Adelaide, South Australia
stewgeop@senet.com.au

Dr. Ian Stewart received a Ph.D. in seismology from the University of Adelaide, South Australia. From 1974 until

N.M.A. Rasul and I.C.F. Stewart (eds.), *The Red Sea*,
Springer Earth System Sciences, DOI 10.1007/978-3-662-45201-1,

1982 he was with the Department of Earth Sciences at Memorial University of Newfoundland, Canada, where he taught geophysics. He then carried out research and worked on potential fields for Saudi Aramco in the Exploration Department in Dhahran, Saudi Arabia until 1996, and later for Normandy Exploration in Australia for 2 years in gold exploration. He was the Geophysical Technical Adviser to the Saudi Geological Survey, Jeddah, between 2003 and 2013. He is an independent geophysical consultant based in Adelaide, applying proprietary methods for processing gravity and magnetic data for numerous mining and oil companies in Australia, the Middle East, Africa and North and South America.

Zohair A. Nawab
President
Saudi Geological Survey,
Jeddah, Saudi Arabia
nawab.za@sgs.org.sa

Dr. Zohair Nawab has been the President of the Saudi Geological Survey, Kingdom of Saudi Arabia since 2006. He has a Ph.D. from the University of Western Ontario in London, Ontario, Canada as well as a Diploma in Applied Geology from the South Dakota School of Mines and Technology in Rapid City, South Dakota, USA. From 1978 to 1985 he was the Assistant Secretary General for the Saudi-Sudanese Red Sea Commission, and was a member of the Saudi Arabian Delegation to the UN Law of the Sea Conference and to the UN Sea-Bed Authority for many years. In 2001 he became the Deputy Minister for Mineral Resources, and became a Senior Advisor in the Ministry of Petroleum and Mineral Resources in 2003. He has also served as a board member of the Saudi Company for Precious Metals and the Saudi Arabian Mining Company (Ma'aden), the Saudi Geological Survey, as well as the Steering Committee for Railway Expansion Projects. He is the co-author of three books and has published many papers.

Enrico Bonatti
Professor
Lamont Doherty Earth Observatory (LDEO),
Columbia University, USA and
Istituto Di Scienze Marine-CNR (ISMAR), Italy
enrico.bonatti@bo.ismar.cnr.it

Dr. Enrico Bonatti holds degrees in Geology from the University of Pisa, Italy. In 1960 he was at Yale University with a Fullbright Fellowship. He then moved to Scripps Institution of Oceanography at the University of California as an Assistant Research Geologist. In 1965 he moved to the University of Miami School of Marine Science where in 1971 he became a Full Professor of Marine Geology. In 1975 he moved to Lamont Doherty Earth Observatory of Columbia University, U.S., as Doherty Senior Scientist. In 2002 he became Professor of Geodynamics at the University of Rome La Sapienza, and Director of the Institute of Marine Sciences of the Italian Research Council (CNR). He is presently an Associate at the Institute of Marine Sciences of the CNR and a Special Scientist at Columbia University Lamont Doherty Earth Observatory. He is a member of the Russian Academy of Science, of the European Academy of Science and a Fellow of the American Geophysical Union. He was awarded the Patterson Medal in Oceanography by the Royal Swedish Academy of Science and the Feltrinelli Prize by the Italian Accademia dei Lincei. He has led over 30 oceanographic expeditions in the Atlantic, Pacific, Indian and Antarctic Oceans, in the Red Sea and the Mediterranean Sea. He has participated in DSDP-IODP drilling legs. He dived several times to the floor of the Atlantic with the Alvin and Nautile submersibles. He has done field work in the Afar Rift (Ethiopia), Polar Urals (Russia), Tunguska (Siberia), Zabargad Island (Red Sea) and elsewhere. He has done research on various aspects of the geology of the oceans and has written about 200 scientific papers, including many in journals such as Nature and Science. He has also published many more popular articles in journals such as Scientific American.

Anna Cipriani
Assistant Professor
University of Modena and Reggio Emilia, Italy and Adjunct Associate Research Scientist at Lamont Doherty Earth Observatory, Columbia University, USA
anka@ldeo.columbia.edu

Dr. Anna Cipriani holds B.A., M.A. and Ph.D. degrees in Geology and Earth and Environmental Sciences from the University of Padua (Italy) and Columbia University (USA). Currently, she is an Assistant Professor of Geochemistry at the University of Modena and Reggio Emilia (Italy) and is an Adjunct Associate Research Scientist at the Lamont-Doherty Earth Observatory, Columbia University. She has participated in a number of major expeditions in the central Atlantic, in the Peri-Antarctic Ocean and in the Red Sea, including an ODP leg. Her research focuses on mantle geochemistry and more specifically on what the isotope geochemistry of the oceanic lithosphere can tell us on mantle composition and melting below mid ocean ridges, and on processes of generation of the oceanic crust.

Luca A. Lupi
Research Scientist
University of Pisa, Pisa, Italy
lupive@tiscalinet.it

Luca Lipi is an explorer, photographer and writer. After studying geology and volcanology at the University of Pisa he founded Vulcano Esplorazioni, and through this society he has organized many scientific expeditions and explored many volcanoes in the world, especially in East Africa, Iceland and America. He has written scientific articles, numerous publications and books on volcanoes that have been translated into many languages and has produced popular science documentaries. He is the author of the international project "Encyclopaedia Aethiopica", managed by the University of Hamburg. He is considered to be one of the leading world experts on the geology, geography and history of the Afar region, which he has extensively studied and explored (http://www.dancalia.it). He is undertaking the first complete and detailed geological and route map of this area (1:950,000–1:400,000 scales) as well as the first geological raised relief map of this area (http://www.lac-cartografia.it). He has lectured extensively in Italy and elsewhere on the Afar region.

William P. Bosworth
Senior Geological Advisor
Apache Egypt Companies, Egypt
bill.bosworth@apachecorp.com

William (Bill) Bosworth is a Senior Geological Advisor for Apache Egypt Companies based in Cairo, Egypt. He has over 30 years petroleum exploration experience, mostly in North and East Africa and the Near and Middle East. His education included Bachelors and Masters of Science degrees at Rensselaer Polytechnic Institute and a Ph.D. in structural geology at the State University of New York at Albany, USA. Prior to joining the oil industry, he taught at Colgate University and worked on research projects in the northeastern United States and Canada. Later research interests focused on continental rift systems, intra-plate stress fields, and the geodynamics of North Africa. He is an Associate Editor of the Journal of African Earth Sciences and of Marine and Petroleum Geology. Bill has been an active member of the American Association of Petroleum Geologists and is presently president-elect for its Africa Region. In 2014 he received the AAPG Distinguished Service Award for his work in support of Egyptian university students. He is a Board Member of the Egyptian Petroleum Exploration Society and frequently leads field trips to the Red Sea and Gulf of Suez.

Marco Ligi
Senior Research Scientist
Istituto Di Scienze Marine-CNR (ISMAR), Italy
marco.ligi@bo.ismar.cnr.it

Dr. Marco Ligi is presently a Senior Research Scientist at the Institute of Marine Sciences, National Research Council (CNR) in Bologna. His scientific career has included La Spezia at ENEA as a consultant in numerical cartography and database management systems, at Rice University, Houston, as a one-year Associate Researcher on multichannel seismic processing, and more than 20 years at the Institute of Marine Sciences. His research interests started with numerical cartography, but soon shifted to geophysics and structural aspects of the origin and evolution of ocean basins, ranging from geophysical modelling to the tectonics of oceanic transform plate boundaries, to the transition from continental to oceanic rifting in the Red Sea. In particular, his present scientific activity is focused on the geophysics of basic processes at accretionary plate boundaries such as mantle flow and melting modelling, multichannel seismic data acquisition, processing and interpretation, multibeam acquisition and processing, and inversion and modelling of gravity data along mid ocean ridges. He has led several oceanographic expeditions in the Mediterranean, Atlantic and Indian Oceans and has written over 80 scientific papers of which 43 are in ISI journals and 4 are in journals such as Nature and Science. He has supervised several graduate students from the Universities of Bologna and Rome, Italy.

Axel Ehrhardt
Senior Research Scientist
Bundesanstalt für Geowissenschaften und Rohstoffe (BGR),
Federal Institute for Geosciences and
Natural Resources, Hannover, Germany
axel.ehrhardt@bgr.de

Dr. Axel Ehrhardt has been a research scientist at the Federal Institute of Geosciences and Natural Resources (BGR), Germany, since 2004, where he is responsible for reflection seismic data acquisition, processing and management. He attended the University of Münster and then undertook a Diploma Thesis at the Alfred Wegener Institute for Polar and Marine Research in Bremerhaven. He received his Ph.D. in 2004 from the Institute of Geophysics at the University of Hamburg, where his research involved the acquisition, processing and interpretation of high resolution seismic data from the northern Red Sea and Gulf of Aqaba in order to map fault systems and sedimentary patterns associated with the tectonic development of the region. He has also acquired and mapped seismic data from several other parts of the world, including the Arctic, South Atlantic, Indian Ocean, South China Sea and the Mediterranean.

Christian P. Hübscher
Professor
CEN - Center for Earth System Research and Sustainability
Institute of Geophysics, University of Hamburg
Bundesstrasse 55, 20146 Hamburg, Germany
Christian.Huebscher@zmaw.de

Dr. Christian Hübscher has been the head of the marine seismic and hydroacoustic working group at the University of Hamburg since 1998. His first experiences in geophysical field work were in 1990 when he participated in an Antarctic expedition as a diploma student from the University of Münster. Marine geophysics became his passion as a Ph.D. student at the Alfred Wegener Institute for Polar and Marine Research in Bremerhaven. After several years as an Assistant Professor at the University of Bremen he moved to Hamburg where he concentrated on morphogenetic geological processes in the marine environment. During more than 40 marine research cruises and field campaigns he investigated active plate and salt tectonics, volcanism, fluid escape structures and current controlled sedimentation in various settings worldwide.

Neil C. Mitchell
Reader
School of Earth, Atmospheric and Planetary Sciences, University of Manchester, UK.
neil.mitchell@manchester.ac.uk

Dr. Neil Mitchell is a marine earth scientist who holds a readership at the University of Manchester. He gained his D.Phil. from the University of Oxford in 1989 for his work on the Indian Ocean Triple Junction and has since broadened his interests in mid-ocean ridges, faulting, sedimentation, erosion and growth of volcanic structures, supported by research fellowships held at Lamont-Doherty Earth Observatory, and the Universities of New Brunswick, Durham, Oxford and Cardiff. In Manchester, he is pursuing a range of research on seabed morphology in a variety of settings including continental margins and volcanic islands.

Robert E. Reilinger
Principal Research Scientist
Massachusetts Institute of Technology (MIT),
Cambridge, Massachusetts, USA
reilinge@erl.mit.edu

Dr. Reilinger is a Principal Research Scientist at the Department of Earth, Atmospheric, and Planetary Sciences, Massachusetts Institute of Technology. His research has focused on using geodesy, primarily the Global Positioning System (GPS) and Interferometric Synthetic Aperature Radar (InSAR), to observe crustal motions and deformations. For the last 25 years his research has been focused on the Arabia-Africa-Eurasia plate system and "greater" Mediterranean region. Results of this research are relevant to understanding the dynamics of plate motions, lithospheric rheology, and earthquake/tsunami and volcanic hazards. Reilinger has published more than 70 papers in the refereed literature since 1991.

Simon C. McClusky
Senior Fellow
Australian National University (ANU),
Research School of Earth Sciences (RSES)
Canberra, ACT, Australia
simon.mcclusky@anu.edu.au

Dr. Simon McClusky received his Bachelor of Surveying (B. Surv.) with first class honours and his Ph.D. from the University of New South Wales (UNSW) School of Surveying. McClusky then worked as a postdoctoral researcher in the Geodesy and Geodynamics group at the Department of Earth, Atmospheric and Planetary Sciences, Massachusetts Institute of Technology (MIT) before taking a position as Research Scientist within the same group, working there for 17 years. In 2010 he returned to Australia, joining the Geodesy and Geodynamics group at the Research School of Earth Sciences, Australian National University (ANU), as a Senior Fellow. His early research focused on improving the precision, accuracy and usability of geodetic techniques, primarily the Global Positioning System (GPS) for Earth science research applications. Subsequently his research has primarily focused on exploiting geodetic techniques to improve our understanding of the Earth system. Of particular interest has been precisely quantifying the kinematics of crustal deformation in the Arabia-Africa-Eurasia zone of tectonic plate interaction. Results of this research are necessary constraints for models that describe the dynamics of plate tectonics, the rheology of the Earth's lithosphere and mantle, and earthquake/tsunami and volcanic hazards. More recently he has been investigating using a combination of space geodetic techniques to explore the interaction of climate and global mass balance, specifically looking at the spatial and temporal relationships between ice sheet mass changes, continental hydrology variability, glacial isostatic adjustment (GIA) and sea level changes.

Abdullah ArRajehi
Research Scientist
King Abdulaziz City for Science & Technology, (KACST) Riyadh, Saudi Arabia, Saudi Arabia
arrajehi@kacst.edu.sa

Dr. ArRajehi is a Research Scientist at the National Center of Petroleum and Gas, King Abdulaziz City for Science & Technology. His Ph.D. is in Civil Engineering/Space Geodesy. His thesis is on innovative strategies of processing and analyzing regional networks consisting of stations using the Global Positioning System (GPS). At KACST, He has been involved in giving lectures at the local universities as well as undertaking research. His research has been focused on the application space geodesy and other remote sensing means in the study of geophysics and the local environment. Since 1997, his research has been focused on the Red sea expansion and the resulting crustal motions and deformations. Working closely with colleagues within KSA, neighboring countries, and beyond, such as the USA and Turkey, a number of papers were published on the Arabia-Africa-Eurasia plate system and "greater" Mediterranean region. Results of this research are relevant to understanding the dynamics of plate motions, lithospheric rheology, and earthquake/tsunami and volcanic hazards.

Salah El Hadidy Youssef
Technical Advisor
National Center for Earthquakes and Volcanoes, Saudi Geological Survey, Jeddah, Saudi Arabia
el_hadidy.sy@sgs.org.sa

Dr. Salah El Hadidy is a Technical Advisor in seismology at the National Center for Earthquakes and Volcanoes, Saudi Geological Survey, Jeddah, Saudi Arabia. He received a M.Sc. in seismology from Cairo University and a Ph.D. in seismology (seismology and seismotectonics) from Ain Shams University, Cairo, Egypt. He was an Associate Professor at the National Research Institute for Astronomy and Geophysics (NRIAG), Helwan, Egypt and the Director of the Earthquake Disaster Reduction Center at NRIAG. He has been a reviewer for national and international seismological journals and has supervised many Ph.D. and M.Sc., theses in seismology. Among his other research activities he was responsible for establishing the Egyptian National Seismic Network.

Ghebrebrhan M. Ogubazghi
Professor
Department of Earth Sciences,
Eritrea Institute of Technology
P.O. Box 12767, Mai Nefhi, Asmara, Eritrea
ogubazghi_ghebrebrhan@yahoo.com

Dr. Ghebrebrhan Ogubazghi is currently professor and head of the geophysics research team at the Eritrea Institute of Technology. He received the B.Sc., M.Sc., and the Ph.D. degrees, respectively, from the universities of Addis Ababa (Ethiopia), Nice (France), and Toulon (France). He was with the Geophysical Observatory of the Addis Ababa University from 1977 to 1992, and did research in optimizing the signal processing system of the new generation of clear air meteorological radars. In 1992, he joined the University of Asmara in Eritrea where his research focussed on seismology and seismic hazard assessment. From 2004 up to present he has been with the Eritrea Institute of Technology. He was an Associate of the International Centre of Theoretical Physics in Trieste, Italy, from 1991 to 1998, and was visiting scholar at the universities of Toulon, Kyoto (Japan), Bergen (Norway), and Botswana. He has held academic leadership positions at the University of Asmara and the Eritrea Institute of Technology.

Berhe Goitom
Seismologist
Eritrea Institute of Technology, Eritrea
berhegoitom14@gmail.com

Berhe Goitom is a lecturer in geophysics at the Eritrea Institute of Technology, Eritrea. He graduated from the University of Asmara in 1985 with a B.Sc., majoring in Physics and mathematics. In 1990 he received an M.Sc., in Exploration Geophysics from Indian Institute of Technology, Kharagpur, India. From 1985 to 1993 he worked as a geophysicist in the Ethiopian Institute of Geological Surveys where he conducted different geophysical surveys for mineral exploration. In 1993 he joined the Eritrean Geological Surveys as a senior Geophysicist and head of the Geophysics Unit involved in leading Geophysical Projects dealing with mineral exploration, groundwater, geothermal and engineering applications. In July 2011 he moved to the Eritrea Institute of Technology, Eritrea and is working as a Lecturer in Geophysics and is involved in seismic stations deployment, maintenance and seismic data analysis. Currently he is doing his Ph.D. on seismology and seismic hazard analysis at the University of Bristol, UK. His current research interests include seismic data analysis, seismicity of volcanic areas and seismic hazard assessment.

Sigurjón Jónsson
Associate Professor
King Abdullah University of Science and Technology (KAUST), Thuwal, Saudi Arabia
sigurjon.jonsson@kaust.edu.sa

Dr. Sigurjón Jónsson received the B.S. and M.S. degrees in geophysics from the University of Iceland, Reykjavík, Iceland, in 1994 and 1996 respectively, and a M.S. in electrical engineering and a Ph.D. in geophysics from Stanford University in 2002. His thesis work focused on using Interferometric Synthetic Aperture Radar (InSAR) measurements to constrain deformation models of volcanic and seismic processes. He was awarded the Reginald A. Daly Postdoctoral Fellowship at the Department of Earth and Planetary Sciences, Harvard University, in 2002, where he continued applying InSAR to various geophysical problems. From 2004 to 2009 he was a Senior Researcher and Lecturer with the Institute of Geophysics, ETH Zurich, Switzerland. Since 2009 he has been an Associate Professor in geophysics at the King Abdullah University of Science and Technology (KAUST), Thuwal, Saudi Arabia.

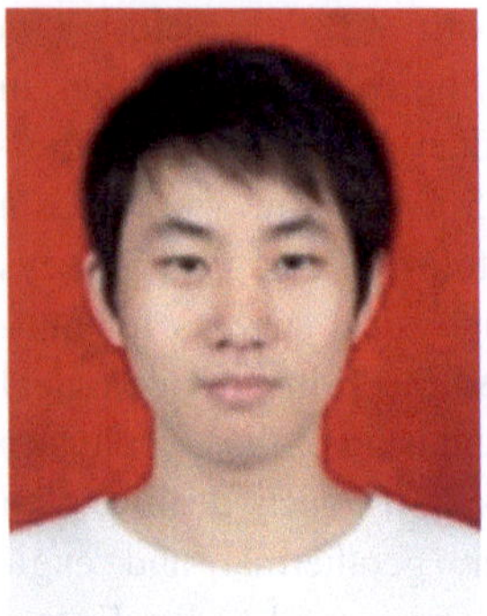

Wenbin Xu
Research Associate
Earth Science and Engineering
King Abdullah University of Science and Technology (KAUST), Thuwal 23955–6900, Saudi Arabia
wenbin.xu@kaust.edu.sa

Wenbin Xu received his B.Sc., degree in surveying engineering and his M.Sc., degree in photogrammetry and remote sensing from the Central South University, Changsha, China, in 2008 and 2011, respectively. Since 2011 he has been working towards his Ph.D. degree in Earth Science and Engineering at King Abdullah University of Science and Technology (KAUST), Thuwal, Saudi Arabia. His main research interests are on the use of space geodesy (including InSAR and GPS) for monitoring crustal deformation, InSAR atmospheric correction, and InSAR time series analysis.

Martin Hovland
Senior Research Scientist
Tech Team Solutions, Stavanger, Norway
Martin.Hovland@techteamsolutions.no

Dr. Philos Martin Hovland M.Sc., Ph.D. FGS, has been working for the Norwegian energy company Statoil ASA from 1976 to 2012 as senior engineer and marine geology specialist. He was simultaneously associated with the Universities of Tromsø and Bergen (Norway), as lecturer and adjunct professor, from 2010 (at the Centre for Geobiology, University of Bergen). Since 1987 he has also been member of the IODP (Integrated Ocean Drilling Program) expert panel for safety and the environment (EPSP). Through his research on features and processes in the seafloor, he has published four books and more than 100 peer reviewed papers. They range in topics from Deep-Sea Geohazards, including pockmarks, seepage, gas hydrates, slides, and mud volcanoes—to the enigmatic Norwegian deep-water coral reefs (*Lophelia* reefs). He also continues to work on natural high-temperature/high pressure (HTHP) systems (hydrothermal systems), including the formation of salt (NaCl) on Earth and Mars, serpentinization, and abiotic hydrocarbon formation. He is currently a senior consultant/advisor with the Norwegian company Tech Team Solutions, Stavanger. His working title is Professor Emeritus.

Håkon G. Rueslåtten
Senior Research Scientist
Numerical Rocks, N-7000 Trondheim, Norway
hg.rues@gmail.com

Dr. Håkon Rueslåtten graduated from the University of Oslo in 1976 with a Ph.D. in Geology and continued as a Researcher in the university's Department of Geology until 1978. In 1978–1979 he worked as a consultant in hydrogeology in an engineering company, NOTEBY A/S. Between 1979–1984 he held an Associate Professor position in Engineering Geology, University of Trondheim. In 1984 he joined the Department of Petroleum Technology in SINTEF, Trondheim, to establish a multi-disciplinary project team working on reservoir characterization, and in the period 1986–2005 he was a Research Advisor at Statoil's R&D facility in Trondheim. Joining Numerical Rocks in 2005, Håkon now holds the position of Chief Geoscientist. His special field of competence is within reservoir characterization, mineralogy and geochemistry.

In 2002, he was a member of a small group of Statoil scientists who looked into aspects of hydrothermal processes. During this work, the group studied the unique properties of supercritical water (SCRIW); a fluid that is capable of dissolving organic matter but cannot dissolve common sea salts at specific PT-conditions. The results from this study form the basis for new research into the formation of hydrothermal salts in various sub-surface geological environments.

Hans K. Johnsen
R&D Manager
Det norske oljeselskap ASA, Norway
Hans.Konrad.Johnsen@detnor.no

Dr. Hans Johnsen has been the R&D manager for Det norske oljeselskap ASA since 2008. He obtained a M.Sc., in 1976 and a Ph.D. in 1982 from the Norwegian Institute of Technology in Trondheim. From 1983 to 1992 he was the managing director of petroleum consultant PETRECO As. He was the manager of the Centre for Applied Gas Technology in Statoil from 1992 to 1995, and later the project manager for SOFC fuel cell development in Statoil from 1995 to1996. From 1996 to 1999 he was a member of the R&D Strategic Advisory Board in Statoil, and then the R&D manager, New Ideas, Radical Research Program for Statoil until 2008.

Peter Feldens
Postdoc
Institute of Geosciences, Kiel University, Otto-Hahn Platz 1, 24118 Kiel, Germany
pfeldens@geophysik.uni-kiel.de

Dr. Peter Feldens is a marine geologist, who received his doctorate from Kiel University in 2011 for work dealing with the impact of the 2004 Indian Ocean Tsunami on the inner continental shelf off Thailand, and for research on the evolution of the south-western Baltic Sea during the Holocene. Currently, he works in the Marine Geophysics and Hydroacoustics department of the Institute of Geophysics at Kiel University, using a combination of geological and geophysical methods to investigate different aspects of continental shelf and slope areas.

Mark Schmidt
Senior Research Scientist
GEOMAR Helmholtz Centre for Ocean Research Kiel, Wischhofstr. 1-3, 24148, Kiel, Germany
mschmidt@geomar.de

Dr. Mark Schmidt is a member of the Marine Biogeochemistry Division of the German research institute GEOMAR. He started his scientific career in the 1990s as a physical chemist (Ph.D. in atmospheric chemistry), and then changed to earth sciences (habilitation in geosciences in 2006). In recent years his research interests have covered the use of stable isotopes to reconstruct biogeochemical processes in the earth sciences. He is a specialist on mineral formation at cold and hot marine vent areas. In addition, he has investigated trace gas-formation, -degradation, and -transport processes in most of the oceans worldwide. He has participated in six international research cruises into the Red Sea and has become an expert on brines and hydrocarbon formation for this area.

Radwan Al-Farawati
Associate Professor
Marine Chemistry Department, Faculty of Marine Sciences, King Abdulaziz University (KAU),
Jeddah, Saudi Arabia
rfarawati@kau.edu.sa

Dr. Radwan Al-Farawati earned a B.Sc., in Marine Chemistry in 1990 from the Faculty of Marine Sciences, King Abdulaziz University, Jeddah, Saudi Arabia. He received his Ph.D. in oceanography in 1998 from the Oceanography Laboratories, Liverpool University, UK. He subsequently joined the Department of Marine Chemistry, Faculty of Marine Sciences, King Abdulaziz University, Saudi Arabia. Since 1999 he has been the Chairman of the Marine Chemistry Department. His research interests cover a wide area related to the marine environment. However, his main research interest is on the distribution of nutrients and trace metals in the coastal waters of the eastern Red Sea, and the speciation of dissolved trace metals in seawater. He has participated in research carried out jointly by KAU and IFM-Geomar to study the geochemistry of methane gas in the hot brines and shallow areas of the Red Sea.

Reiner Botz
Senior Research Scientist
Institute of Geosciences University of Kiel,
Ludewig-Meyn-Straße 10, 24118 Kiel, Germany
rb@gpi.uni-kiel.de

Dr. Reiner Botz is with the Institute of Geosciences, University of Kiel, Germany. He has been working for more than 30 years in geochemistry and is a specialist in isotope geochemistry. His research covers a wide range of scientific approaches that clarify the processes in mineral formation, hydrocarbon formation, and related secondary (diagenetic) processes. He has published more than 70 papers in international journals on multidisciplinary marine and terrestrial research that combines geochemical, mineralogical, microbiological, and geophysical studies. His contributions on organic geochemistry of the Red Sea brine sediments and authigenic carbonates in the Red Sea cover the most recent findings in that field of research.

Pierre M. Anschutz
Professor
EPOC: Environnements et Paleoenvironnements Oceaniques et Continentaux, Universite Bordeaux,
CNRS, Talence, France
p.anschutz@epoc.u-bordeaux1.fr

Dr. Pierre Anschutz has been at the University of Bordeaux, France, since 1996, where he is currently a Professor. He received his Ph.D. in 1993 in geochemistry from the University of Strasbourg, followed by lecturing at that university, then a post-doctoral position at INRS, Rimouski, Canada until 1996. His research is mainly on redox processes in natural marine environments, at the interface between geochemistry and biology, in order to understand early diagenesis, sedimentary geochemical records, and biogeochemistry in coastal environments. This has included studies of early diagenesis, benthic biogeochemistry at the sediment-water interface, coastal and tidal ecosystems, the genesis and recording of paleo-oceanographic proxies, the geochemical dynamics of inorganic contaminants, and oceanic hydrothermalism in the Red Sea. He has published about 80 papers in the refereed literature since 1993.

Hjalmar Thiel
Professor
University of Hamburg, Institute for Hydrobiology and Fishery, and Alfred Wegener Institute for Polar and Marine Research, Bremerhaven, Germany
hjalmar.thiel@hamburg.de

Dr. Hjalmar Thiel is an Emeritus Professor at the University of Hamburg, Germany. He attended the Universities of Hamburg, Bonn and Kiel, studying zoology, marine science and botany. He received his Ph.D. from the University of Kiel in 1962. From 1962 to 1995 he held various positions at the Institute for Hydrobiology and Fishery Science at the University of Hamburg, where he taught courses in marine hydrobiology, especially that of the deep sea, as well as the environmental impacts of deep-sea usage. Then from 1995 to 1998 he was associated with the Alfred Wegener Institute for Polar and Marine Research in Bremerhaven. His research has included quantitative studies of the distribution of deep-sea benthos in various biogeographical regions under different ecological conditions. He has evaluated the environmental impact of the deep-sea disposal of atomic waste products, the mining of metalliferous sediments from the deep Red Sea, and the mining of polymetallic nodules in the Pacific Ocean. He has also studied man-made coastal lagoon ecology related to nature conservation. From 1999 to 2005 he was involved in the preparation for the establishment of Marine Protected Areas and Science Priority Areas on the High Seas.

Ludwig Karbe
Senior Research Scientist and Lecturer
University of Hamburg,
Zoological Institute and the Institute for Hydrobiology and Fishery Science, Germany

Dr. Ludwig Karbe studied at the University of Hamburg, Germany, specialising in zoology, botany, and aquatic and fishery science. He obtained his Ph.D. there in 1960, and from 1959 to 1998 held various positions at the University of Hamburg at the Zoological Institute and the Institute for Hydrobiology and Fishery Science. This included teaching courses in zoology, aquatic biology, and water and aquatic resources management. His research focussed on sympatric speciation in fish populations and phenomena associated with changing ecological conditions in freshwater and marine environments. He has also evaluated anthropogenic impacts and carried out risk assessments in freshwater and marine environments, including the Red Sea, with special reference to the eco-toxicological and human-health related effects of contaminants. He was the German delegate to the OECD Water Management Policy Group (Group of Experts in Biological Testing).

Horst Weikert
Senior Research Scientist and Lecturer
University of Hamburg, Institute for Hydrobiology and Fishery Science, Germany

Dr. Horst Weikert attended the Universities of Münster and Kiel, Germany, specialising in the fields of oceanography, zoology, botany and limnology, and he obtained his Ph.D. from the University of Kiel in 1970. From 1969 to 1971 he was associated with the Institute for Marine Science in Bremerhaven, and from 1971 to 2007 was at the Institute for Hydrobiology and Fishery Science at the University of Hamburg. While at Hamburg he taught courses in marine hydrobiology, particularly zooplankton ecology, and structures and trophic interactions in open ocean and deep water communities. His research activities included the quantitative analysis of the composition and distribution of shelf and deep sea zooplankton communities in temperate and tropical seas, focussing on key species, and biogeochemical fluxes in response to climatic change. He has also been involved in the evaluation of the potential impact induced in pelagic communities by deep-sea mining of metalliferous sediments in the Red Sea and polymetallic nodules in the Pacific Ocean.

Ahmed Rushdi
Research Director
Chair of Green Energy Research, College of Food and Agriculture Sciences, King Saud University (KSU), Riyadh, Saudi Arabia
arushdi@ksu.edu.sa

Dr. Ahmed Rushdi has been an Associate Professor at King Saud University in Riyadh since 2008. After graduating from Sana'a University in Yemen, he obtained a Ph.D. in 1989 from the College of Oceanography at Oregon State University, USA. From 1989 to 1997 he was with Sana'a University, where he became Head of the Department of Oceanography in 1991. During this time he was involved with several projects to establish research in oceanography and studied the marine environment in Yemen. From 1995 he also held research positions at the University of Michigan and Scripps Institution of Oceanography in San Diego. Since 1998 he has also been an Associate Professor (Senior Research and courtesy positions) at Oregon State University. He has received several awards, including USAID (1983–1989), UNDP (1994–1995) and Fullbright (1995–1996) Scholarships, and has published extensively in international scientific journals.

David Pugh
Marine Science Consultant
National Oceanography Centre
Joseph Proudman Building 6 Brownlow Street
Liverpool L3 5DA, UK
d.pugh@me.com

Dr. David Pugh is a marine science consultant based in Chester, UK. Following a Ph.D. in geodesy and geophysics from the University of Cambridge in 1968 he worked at the Proudman Oceanographic Laboratory, Merseyside. In 1984 he became Head of Oceanography, Hydrology and Meteorology, Science Division, NERC. He has also served as Head of Information and Scientific Services in the former Institute of Oceanographic Sciences (1987–1992) and as Secretary to the United Kingdom Government Committee on Marine Science and Technology (1992–2003).

Elected President of the Intergovernmental Oceanographic Commission of UNESCO (2003–2007), he had previously been the Founding Chairman of the IOC Global Sea Level network, GLOSS. In addition to leading and serving many international organisations within the UN and ICSU, he has maintained an active program of research associated with Liverpool University and the National Oceanography Centre (NOC), Liverpool where he is a visiting Professor. His interests include tides, surges, mean sea level, coastal management and climate change, the economics of marine activities related to GDP, and the history of sea level.

Awards include Fellowships of the Institute of Physics, the International Association of Geodesy, and the Institute of Marine Engineering, Science and Technology, the 2003 Presidents Award of the Society for Underwater technology, and a 2009 Shackleton Memorial Scholarship. He has been a Trustee and Council member of the IMarEST and the UK Marine Conservation Society. In the 2003 he was awarded an OBE for services to marine sciences. He is the author of Tides, Surges and Mean Sea Level (1996) and Changing Sea Levels (2004). His recent work includes "Socio-economic indicators of marine related activities in the UK Economy" published by The Crown Estates in June 2008; and "Troubled Waters: Ocean science and governance", co-edited with Geoff Holland, in 2010. Recently, he has worked (UK NOC) and published on sea levels in the Falklands, and tides in Loch Ness, Scotland, with Philip Woodworth. Together they are preparing a new book *on Tides, surges, tsunami and mean sea level* to be published by Cambridge University Press in 2014.

Yasser Abualnaja
Senior Scientist
Red Sea Science & Engineering Research Center
P.O. Box 2134, 4700 King Abdullah University of Science and Technology (KAUST),
Thuwal, Saudi Arabia
yasser.abualnaja@kaust.edu.sa

Dr. Yasser Abualnaja has been a senior scientist in the Red Sea Research Center (RSRC) at King Abdullah University of Science and Technology (KAUST) since September 2009. He obtained a Ph.D. in Physical Oceanography field from Florida Tech. in the US in 2005, as well as two M.Sc. degrees in Physical Oceanography and Coastal Zone Management, after which he joined the Physical Oceanography Department in King Abdulaziz University in Jeddah. He has participated in joint projects between KAUST and several international organisations, such as the Woods Hole Oceanographic Institution, including taking part in research cruises. His oceanography research interests have focussed on the air-sea interaction, estimating air-sea heat fluxes and error analyses, water mass, salt and heat exchange, the air-sea boundary layer response to atmospheric forcing, observation and analysis of meso-scale ocean circulation, and integrated coastal zone management.

Amy S. Bower
Senior Scientist
Woods Hole Oceanographic Institution (WHOI), USA
abower@whoi.edu

Dr. Amy S. Bower is a Senior Scientist in the Department of Physical Oceanography at the Woods Hole Oceanographic Institution. She earned her Ph.D. in Oceanography from the University of Rhode Island Graduate School of Oceanography in 1988 and began her career at WHOI as a postdoctoral scholar. She was awarded tenure at WHOI in 1999. Her research is focused on boundary currents and eddies and their role in maintaining the distributions of water properties in the ocean. She uses many observational tools in her research, including subsurface acoustically tracked floats, surface drifters, moored current meters, hull-mounted and lowered acoustic doppler current profilers as well as hydrographic profilers. Over the past several years, she has been leading the first large-scale hydrographic and current surveys of the eastern Red Sea.

John T. Farrar
Research Scientist
Woods Hole Oceanographic Institution (WHOI), USA
jfarrar@whoi.edu

Dr. John T. Farrar is an Associate Scientist in the Department of Physical Oceanography at the Woods Hole Oceanographic Institution. He obtained his Ph.D. in physical oceanography from the Massachusetts Institute of Technology and the Woods Hole Oceanographic Institution in 2007, where he was the recipient of the MIT Presidential Fellowship. He took a postdoctoral research position at the Woods Hole Oceanographic Institution and joined the faculty in 2008. He had previously earned the S.M. degree from the Massachusetts Institute of Technology and B.S. (physics) and B.A. (philosophy) degrees from the University of Oklahoma. He has led three oceanographic research cruises in the Red Sea and has participated in eight other oceanographic research cruises. His research interests include air-sea interaction and exchange, dynamics of the upper ocean, dynamics of the tropical oceans and atmosphere, oceanic internal waves and eddies, and ocean observing and instrumentation. He has authored or co-authored 14 peer-reviewed scientific journal articles on these topics and has presented more than 20 papers at international scientific conferences.

Sarantis Sofianos
Assistant Professor
National and Kapodistrian University of Athens, Greece
sofianos@oc.phys.uoa.gr

Prof. Sarantis Sofianos is an assistant professor at the University of Athens, Faculty of Physics, and coordinator of the Ocean Physics and Modeling group. After graduating from the University of Athens he obtained his Ph.D. in Meteorology and Physical Oceanography from the University of Miami in 2000. His research is focused on the dynamics of semi-enclosed seas, coastal processes and climate variability using seagoing and numerical modeling techniques, including operational oceanography. He participates in the board of the Mediterranean Oceanography Network for the Global Ocean Observing System.

William E. Johns
Professor
Division of Meteorology and Physical Oceanography,
Rosenstiel School of Marine and Atmospheric Science,
University of Miami, 4600 Rickenbacker Causeway,
Miami, FL 33149, USA
bjohns@rsmas.miami.edu

Dr. William E. Johns is a seagoing oceanographer specializing in the use of long-term moored instrumentation to study ocean circulation. His research involves studies of the large-scale wind-driven circulation, with emphasis on the dynamics of western boundary currents, and on deep flows related to the global thermohaline circulation and climate variability. Dr. Johns also studies the dynamics of straits and the exchange processes between the oceans and marginal seas, including marginal sea overflows. Research programs are currently underway in several regions of the Atlantic Ocean. Previous active research areas include the Red Sea, the Arabian Gulf, and the Gulf of Aden in the western Indian Ocean.

Fazal Ahmad
Professor
Faculty of Marine Sciences,
King Abdulaziz University (KAU),
Jeddah, Saudi Arabia
faali@kau.edu.sa

Dr. Fazal Ahmad is a graduate of the University of Karachi, and received his Ph.D. from the University of Liverpool, UK, for his thesis on the circulation and mixing in the Irish Sea. He then joined the University of Karachi, where he was a member of several academic bodies. In 1983 he joined the Department of Marine Physics, Faculty of Marine Science at King Abdulaziz University in Jeddah, Saudi Arabia. He has actively participated in scientific cruises to the Irish Sea, North Sea, Arabian Gulf and the Red Sea. His publications span a wide range of topics including tidal analysis, residual currents, air-sea interaction processes, formation of the thermocline and fronts in coastal areas. Many of his publications relate to the Irish Sea, Red Sea, Arabian Gulf and adjacent waters.

Alaa Albarakati
Professor
Faculty of Marine Sciences,
King Abdulaziz University (KAU),
Jeddah, Saudi Arabia
aalbarakati@kau.edu.sa

Dr. Alaa Albarakati is a graduate of the Faculty of Marine Sciences, King Abdulaziz University, Jeddah (KAU), and in 2000 received his Ph.D. from the University of Manchester Institute of Science and Technology in the UK for his thesis on "Modeling of the Red Sea circulation". His present research interests include the heat balance at the air-sea interface, hydrography, currents in coastal areas, the formation of the seasonal thermocline, and small-scale circulation models related to coastal areas of the Red Sea and Arabian Gulf. He participated in a KAUST–Woods Hole research program during 2008 and 2010 Red Sea cruises. He developed a 3-D hydrodynamic model for the Red Sea, which was published in 2001. He also played an important role in the Jeddah Transect project, which is joint scientific research between KAU and IFM-GEOMAR to investigate the microstructure of temperature variations in the Atlantis II Deep. He is a member of several scientific boards, including the editorial board of the KAU Journal, the scientific board for Maritime Disaster Management and is a board member of the Saudi General Commission for Surveys. In 2011 he received an award for High Achieving Academics in KAU.

Mohamed I. Sultan
Professor
Department of Geosciences
Western Michigan University
Kalamazoo, Michigan, USA
mohamed.sultan@wmich.edu

Dr. Mohamed I. Sultan is a Professor and Chair of the Department of Geosciences at Western Michigan University. He graduated from Ain Shams University in Cairo with a B.Sc., in geology in 1974 and a M.Sc., in stratigraphy in 1978. He obtained his Ph.D. in geochemistry from Washington University in St. Louis in 1984, after which he was a research associate there until 1988. From 1988 until 1996 he was a senior research scientist in the NASA earth and Planetary Remote Sensing Facility in the McDonnell Center for the Space Sciences at Washington University. He was the project manager for International Programs in the Environmental Research Division at the Argonne National Laboratory in Illinois from 1996 to 2002. From 2002 to 2004 he was a professor in the Department of Geology in the University at Buffalo (SUNY) in Amherst, NY, and since 2004 he has been at Western Michigan University. He is also an adjunct professor at several other universities and is an associate editor of the Geological Society of America Bulletin. He has used remote sensing, GIS, geochemistry, geochronology, hydrology, surface runoff and groundwater flow modeling, and field geology to study a wide range of geological and environmental problems. Recent projects include the influences of natural processes, global change, and regional human activities on the water cycle, as well as monitoring and modeling the impacts of changes in landscape and land cover associated with major engineering projects in the Tigris–Euphrates watershed and the migration of dunes and wetland formation at Cape Cod National Sea Shore due to environmental disturbance. Other projects include assessments of renewable groundwater resources in arid and semi-arid areas (Egypt, Saudi Arabia, Pakistan), and modeling storage variability in artificial lakes (Lake Nasser).

Saleh A. Al-Sefry
Hydrogeologist
Manager of Applied Geology Department
Saudi Geological Survey.
Jeddah, Saudi Arabia
sefry.sa@sgs.org.sa

Engr. Saleh Al-Sefry has been the manager of the Applied Geology Department in the Saudi Geological Survey, Jeddah, since 2003. He received a M.Sc., in Applied Geology (Hydrogeology) from King Abdulaziz University, Jeddah, in 2006. He was the hydrogeologist of the Makkah Al-Mukarramah Technical Projects Division, in the Ministry of Petroleum and Mineral Resources from 1993 until 2000. In 2001 he joined the Saudi Geological Survey, where he has worked on strategic ground water storage in Wadi Fatimah, aquifer water quality in the Umm er Radhuma formation, and flood hazard studies within Saudi Arabia.

Mazen M. Abu Abdullah
Technical Advisor
Applied Geology Department, Saudi Geological Survey.
Jeddah, Saudi Arabia
Abuabdullah.mm@sgs.org.sa

Mazen Abu Abdullah has been the GIS and Databases Adviser to the Saudi Geological Survey, Jeddah, since 2005. After receiving a B.Sc., in Mathematical Science (IT Section) from the University of Damascus, Syria, in 1993, he joined the Arab Center for the Studies of Arid Zones and Dry Lands (ACSAD), where he worked in the water resources division on the development of software for analyzing pumping tests and to link GIS with mathematical models of groundwater and surface water, and the development of water resources database management systems. As well as contributing to water resource studies in various parts of the Arab world as well as training technicians on the use of GIS and mathematical models, he then worked with other divisions in ACSAD to develop systems to document and analyze environmental data such as vegetation and to estimate the efficiency of the use of saline water in irrigation for crops in arid and semi-arid zones. In 2011 he obtained a diploma in Integrated Water Resources Management from the UN University in cooperation with Arabian Gulf University, Manama, Bahrain. Since joining the SGS he has participated in hydrological and hydrogeological studies in addition to engineering geology projects.

Abbas Mansour
President
South Valley University, Qena, Egypt
aman@svu.edu.eg

Dr. Abbas Mansour is the President of South Valley University in Egypt, a position that he has held since 2006. After graduating in 1980 with a degree in geology from Assiut University, he obtained a M.Sc., in 1985, and then a Ph.D. in 1989, also from Assiut University, where he studied marine sediments in the Bay of Safaga in the Red Sea. From 1980 he has held faculty positions at Assiut University and South Valley University, and became Head of the Geology Department at the latter institution in 2000, where he has held several administrative posts. He has worked with several researchers in Austria, the U.S. and elsewhere and has been an environmental consultant for numerous projects within Egypt, including the Red Sea area. He also has been a peer reviewer for quality assurance and a reviewer for several national and international journals of sedimentary geology, environmental geology and marine geology.

Hashem Madkour
Senior Research Scientist
National Institute of Oceanography and Fisheries,
Red Sea Branch, Hurghada, Egypt
madkour_hashem@yahoo.com

Dr. Hashem Madkour has been with the Red Sea branch of the National Institute of Oceanography and Fisheries at Hurghada, Egypt since 1994, where he became an Associate Professor in 2010. After graduating in geology in 1988 from Assiut University, he received a M.Sc., in Marine Geology in 2000 from South Valley University, and then a Ph.D. from South Vally University in 2004 for work on geochemical and environmental studies of marine sediments and invertebrates of the Red Sea coast in Egypt. He has participated in and published on numerous studies of the Red Sea coast and its environment.

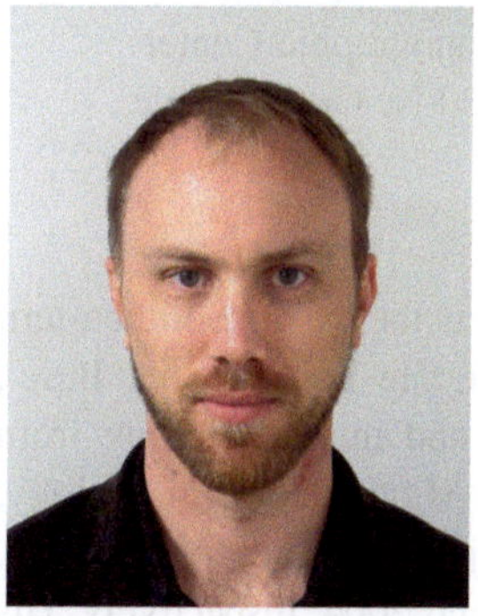

Gwilym P. Rowlands
Director
Rowlands Ecology Ltd.,
38 Fairlie Crescent, Birmingham B38 8DY, UK
gwilym@rowlandsecology.com

Dr. Gwilym Rowlands is an expert in the spatial ecology of tropical reef environments. In 2003 Gwilym received his B.Sc., (Hons.) in biology from the University of Nottingham, UK, in 2006 an M.Res. in Ecology and Environmental Management from the University of York, UK, and in 2013 his Ph.D. in Oceanography and Marine Biology from Nova Southeastern University, Florida. Gwilym is well versed in seafloor mapping, particularly in hyperspectral, multispectral, and acoustic mapping, along with techniques for advanced spatial analysis. As a research fellow of the Khaled bin Sultan Living Oceans Foundation, Gwilym's recent research has focused on the diversity, distribution and genesis of contemporary coral reef environments in the Saudi Arabian Red Sea, and spans both ecological and geological disciplines. Gwilym has written and contributed to several academic papers and books on this subject, including a recently published atlas of Red Sea marine habitats.

Samuel J. Purkis
Professor
National Coral Reef Institute Oceanographic Center
Nova Southeastern University 8000 N. Ocean Drive,
Dania Beach FL 33004, USA
purkis@nova.edu

Dr. Sam Purkis is a Professor at the Nova Southeastern University Oceanographic Center. He received his B.Sc., (Hons.) degree in oceanography and marine biology from Southampton University, UK, the M.Sc., degree in geology, and the Ph.D. degree from the Vrije Universiteit Amsterdam, The Netherlands, in 1998, 2001, and 2004, respectively. His work stretches from field monitoring to software development and mathematical simulation. Sam has authored more than seventy scientific publications including several books. Well versed in state-of-the-art technological solutions for seabed mapping, his expertise is relevant to a broad spectrum of marine applications.

Rebecca A. Klaus
Research Scientist
Tropical Marine Ecosystems of the Future LOEWE Biodiversity and Climate Research Centre (BiK-F), Senckenberg Research Institute and Nature Museum Frankfurt, Marine Zoology Senckenberganlage 25, 60325 Frankfurt a.M., Germany
rebecca.klaus@gmail.com

Dr. Rebecca Klaus is a coral reef ecologist who has been working on conservation management related projects in the Western Indian Ocean and Arabian Region (Red Sea, Gulf of Aden and Arabian Gulf) for the past 15 years. Her research interests cover coral reef ecology, coral bleaching and the effects of climate variability and change on ecosystem function, with a particular interest in marginal coral communities as a window into the future. Her consultancy work crosses spatial scales from baseline assessments of biodiversity and monitoring, through to biotope mapping of shallow tropical marine habitats using remote sensing technologies, and marine protected area and integrated coastal zone management design and planning. Her first involvement in the Red Sea was on the PERGSA-GEF funded "Marine Protected Areas-Strategic Action Plan (MPA-SAP)" (Sudan, Yemen, Djibouti and Somalia). She has since had the opportunity to continue to work on the coral reefs in the Red Sea of Sudan, Eritrea and Yemen.

Dirar Nasr
Associate Professor
Red Sea University, Port Sudan, Sudan
d_nasr47@hotmail.com

Dr. Dirar Nasr, a zoology graduate from the University of Khartoum, wrote his Ph.D. thesis in 1979 on the 'Biology of the pearl oyster Pinctada margaritifera on the Sudanese Red Sea coast'. Pursuing an academic career at the Institute of Oceanography (Sudan), he was rapidly promoted to Senior Researcher, Associate Professor and then, after one year, to Director. He attended several courses on various aspects of marine biology supported by PERSGA, UNESCO and the British Council. He gained a wide range of marine experience and has been a supervisor for several higher studies at the Faculty of Graduate College of the University of Khartoum and has been an External Examiner for Masters Degrees at the University of Khartoum and the Red Sea University. He participated in four scientific cruises focusing particularly on the environmental aspects of the exploitation of the metalliferrous sediments from the Atlantis II Deep in addition to his participation in several regional and international conferences. His association with PERSGA can be traced back to 1976 when he was a member of the Sudanese delegation at the Jeddah II Conference. He also took the important role of rapporteur for both the 1981 Legal and Environmental Experts Meeting and for the 1982 Conference of Plenipotentiaries when the Jeddah Convention (1982) was signed. He has been the PERSGA Coordinator (1995–2005). He is, at present, a lecturer at the Red Sea University (Port Sudan, Sudan) and an Environmental Consultant. His fields of interest are coral reefs, mangrove communities and human impact on the sea.

Andrew W. Bruckner
Chief Scientist
Khaled bin Sultan Living Oceans Foundation,
Science Without Borders®
8181 Professional Pl, Suite 215, Landover,
MD 20785, USA
bruckner@livingoceansfoundation.org

Dr. Andrew Bruckner is the Chief Scientist with the Living Oceans Foundation currently implementing a 5 year Global Reef Expedition to map, characterize and assess remote coral reefs in the Caribbean, Pacific and Indian Ocean and Red Sea. He received his M.S. in marine biology from Northeastern University, Boston, M.A. in 1988, and his Ph.D. from the University of Puerto Rico in 1999. Prior to joining the Foundation, he worked for the NOAA Coral Reef Conservation Program as a coral reef ecologist. He is the co-lead of NOAA's Coral Disease and Health Consortium. and has worked closely with resource managers and government agencies in the U.S. and internationally in developing conservation, management and restoration actions for coral reefs through legislation, international (CITES) regulations, development of sustainable management guidelines, and on-the-ground monitoring, research and restoration activities. During his tenure at NOAA, Dr. Bruckner received a Presidential Early Career Award from the U.S. government, and three bronze medals for his work that lead to the development of a National Action Plan to Conserve Coral Reefs. His research and policy efforts contributed to the listing of elkhorn coral (*Acropora palmata*) and staghorn coral (*Acropora cervicornis*) on the U.S. Endangered Species Act, based on his research and conservation efforts during the 2005 Caribbean coral bleaching event. He serves as a Councilor of the International Society for Reef Studies, and is a Science Advisor to SECORE, a consortium of over 50 public aquaria and zoos in the United States and Europe and to CORAL, a popular journal for aquarium and coral reef enthusiasts.

Alexandra C. Dempsey
Coral Reef Ecologist
Khaled bin Sultan Living Oceans Foundation
Science Without Borders®
8181 Professional Pl, Suite 215, Landover,
MD 20785, USA
adempsey@nova.edu

Alexandra C. Dempsey is currently the Coral Reef Ecologist for the Khaled bin Sultan Living Oceans Foundation. She completed her M.Sc., in marine environmental biology at Nova Southeastern University Oceanographic Center (2013). She was a coral biologist research assistant with the National Coral Reef Institute. She began employment with the GIS/ Remote Sensing lab in spring 2009. Alexandra has worked on a number of different projects including: the geostatistical comparison of modern and ancient carbonate reef systems, ecological, geological and bathymetric mapping from satellites, analysis of acoustic sub-seabed profiles, as well as sediment collection and analysis. She has worked with a host of different project sponsors including Shell, ExxonMobil, and The Nature Conservancy. Alexandra has been working closely with the Khaled bin Sultan Living Oceans Foundation on their Global Reef Expedition throughout the Atlantic and Pacific Oceans. She holds two Bachelors of Art and Science degrees in Environmental Science and Philosophy from the College of the Holy Cross, Worcester, Massachusetts. She graduated in 2008 with Phi Sigma Tau academic honors. She has completed research internships with Mote Marine Laboratory's Landscape Ecology Division, Center for Marine Resource Management Turks and Caicos, and Cincinnati Zoo's Aquatic Horticulture Department.

Yvonne Sawall
Research Scientist
Benthic Ecology - Research Division Marine Ecology,
Helmholtz Centre for Ocean Research Kiel (GEOMAR),
Düsternbrookerweg 20, 24105 Kiel, Germany
ysawall@geomar.de

Dr. Yvonne Sawall is a coral reef ecologist with particular interest in the acclimatization mechanisms of reef building corals to different prevailing and changing environmental conditions. She received her Ph.D. in 2010 from the University of Bremen (Germany), where she studied coral resistance to eutrophication and pollution, coral recruitment patterns and coral recovery after destruction (by a tsunami and by blast fishing activities) in Thailand and Indonesia. Since 2011, she holds a post-doctoral position at the Helmholtz Centre for Ocean Research (GEOMAR) in Kiel (Germany), where she investigates acclimatization/adaptation mechanisms of Red Sea corals in collaboration with the King Abdulaziz University in Jeddah (Saudi Arabia) and King Abdullah University for Science and Technology (KAUST, Saudi Arabia). For this, she conducted 3 large expeditions along the entire Saudi Arabian Red Sea coast which is characterized by a strong latitudinal temperature and nutrient gradient. Currently, she also investigates ecophysiological aspects of temperate calcifying organisms.

Abdulmohsin A. Al-Sofyani
Associate Professor
Head of the Department of Marine Biology,
Faculty of Marine Sciences, King Abdulaziz University (KAU), Jeddah, Saudi Arabia
asofyani@kau.edu.sa

Dr. Abdullah Mohsin Al-Sofyani is a coral reef ecologist studying coral physiology and coral reef community structures. He received his master degree in coral biology from the King Abdulaziz University (KAU, Saudi Arabia) in 1988 and his Ph.D. working on coral physiology in 1991 from the University of Glasgow (Great Britain). He joined the scientific panel investigating the effect of pollution in the Arabian Gulf during the Gulf crisis in 1990 and 1991. Since 1992, he is employed by the KAU, working as a lecturer and scientist, and currently holds an associate professor position. He participated in the Jeddah Transect project (2010–2013), a joint project between KAU and GEOMAR, leading 3 large expeditions along the Saudi Arabian Red Sea coast. Within this project the effects of nutrients and temperature variations (latitudinal and near-shore versus off-shore) on corals and coral reef communities were investigated. During the RV PELAGIA cruise 64PE351 in April 2012, he studied the corals of the mesophotic zone in the Red Sea. Furthermore, Dr. Al-Sofyani is chairman of the research co-ordination committee of the faculty, which organizes various symposiums on the Sewage Pollution Profile of Jeddah and on the Coastal Zone Management.

Ronald Janssen
Research Scientist
Senckenberg Forschungsinstitut und Naturmuseum Frankfurt a. M., Germany
Ronald.Janssen@senckenberg.de

Dr. Ronald Janssen studied geology, palaeontology and zoology from 1970 to 1974 at the University of Marburg/Lahn and received his Ph.D. with a taxonomic revision of a fossil bivalve fauna. He has held the position of the curator and head of the molluscan section at the Senckenberg Research Institute and Natural History Museum in Frankfurt, Germany, since 1977. His main research interests are taxonomy, systematics and (paleo-)biogeography of European Tertiary molluscs, Recent molluscs of the Arabian seas and bivalves of hydrothermal vent communities. He has participated in various research expeditions to the South China Sea, the Arabian Gulf and to Socotra Island. He is the scientific editor of the malacological journal Archiv für Molluskenkunde.

Marco Taviani
Senior Scientist
ISMAR-CNR
(Institute of Marine Sciences - National Research Council (ISMAR-CNR), Bologna, ISMAR-CNR), Via Gobetti 101, 40129 Bologna, Italy
marco.taviani@bo.ismar.cnr.it

Dr. Marco Taviani is Research Director at the Italian National Research Council Institute of Marine Sciences (ISMAR-CNR) in Bologna since 2001. After graduating from the University of Bologna in 1976 he held various research positions and fellowships until 1989 in the Laboratory of Marine Geology at Bologna, at Lamont-Doherty Geological Observatory, Columbia University, New York, and also at Louisiana State University and the University of Houston. Research lecturer at various USA, China and European Universities. He became a tenured researcher at the CNR in Italy in 1982 and received a Ph.D. in palaeontology from the University of Roma in 1987. His recent research has focussed upon marine ecosystems (especially deep water corals), Cenozoic-Recent marine paleontology and paleoecology (Antarctica, Mediterranean, Red Sea,

Western Indian Ocean), extreme environments (polar, cold seeps), bio-sedimentology (biogenic carbonate factories), paleoclimatology and paleoceanography based upon sediment cores (Mediterranean, Red Sea), drillholes (Antarctica, Mediterranean), corals (Mediterranean, Red Sea, Atlantic Ocean, Antarctica) and other natural archives, including multiple geochemical approaches. He has participated in over 50 oceanographic missions in the Mediterranean, Red Sea, Atlantic Ocean and Antarctica), often as chief scientist. He was the National Coordinator in 1991–1993 for the E.U. RED SED (The Red Sea and Gulf of Aden Sedimentology), and in 1994–1997 for the E.U. TESTREEF programme (Temporal and SpaTial variability of western Indian Ocean REEFs). He has been the coordinator of other national and international projects (such as the Euromargins Moundforce: Forcing of Carbonate Mounds and Deep Water Coral Reefs along the NW European Continental Margin and WP of the E.U. HERMES, HERMIONE, COCONET programmes). He has published over 300 scientific papers, as well as book chapters and various popular science articles in Italian and international magazines. He is in the UNEP list of coral experts and has been member of the CITES Scientific Commission (Italy).

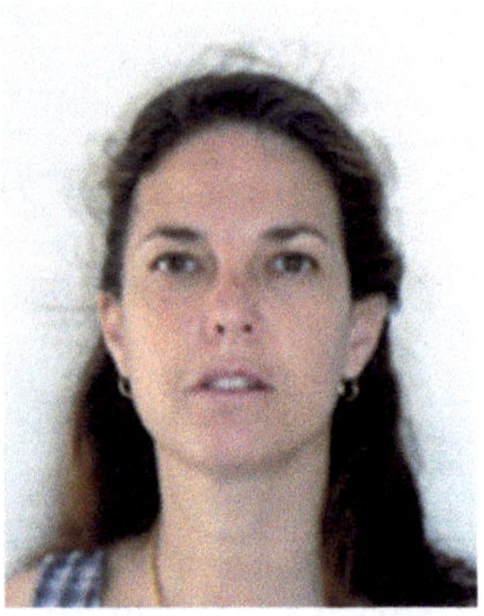

Nathalie Yonow
Research Scientist
School of Biological Sciences, University of Wales,
Singleton Park, Swansea SA2 8PP, Wales, UK
n.yonow@swansea.ac.uk

Dr. Nathalie Yonow is a marine biologist who has specialized in the taxonomy and biogeography of tropical marine opisthobranchs for more than 25 years, following an obsession with marine biology since a child born in Iran, growing up in Saudi Arabia and Greece, and studying in Italy and Wales. She earned both her B.Sc., and Ph.D. in Swansea University (Wales, UK) where she now teaches. She has been on expeditions collecting opisthobranchs from the Red Sea, Mauritius, and the Maldives, as well as studying collections from Ambon, Socotra, Chagos, Réunion, and the Persian Gulf that form the substance of many publications, and is on-going research. These attractive molluscs are important in studies on species diversity as well as zoogeographical associations of marine organism assemblages, and more significantly for their chemical components as many species acquire, concentrate, or synthesize organic chemicals for defence. Her specialist regions are the Red Sea (monograph "Sea Slugs of the Red Sea" 2008) and the Indian Ocean, covered in numerous publications. She has also written the opisthobranch chapters for several books, including "The Marine Fauna of the British Isles and North-West Europe" (1990) and "A Guide to the Seashores of Eastern Africa" (first ed. 1997, third ed. 2011).

Agnese Mancini
Research Scientist
Hurghada Environmental Protection and Conservation Association, New Marina, El Sakkala - Building B2
Hurghada- Red Sea- Egypt
P.O. Box 144 Hurghada- Red Sea- Egypt
agnese@hepca.com

Dr. Agnese Mancini is a marine biologist and currently works as a consultant for the IUCN in the Indian Ocean. She received a Ph.D. in Marine and Coastal Science from the Autonomous University of Baja California Sur in Mexico in 2009. Since 2004 she has been studying marine turtles in the Indian Ocean, in the Eastern Pacific Ocean and finally in the Red Sea. From 2010 to 2013 she worked as marine turtle specialist for the Hurghada Environmental Protection and Conservation Association (HEPCA) in Egypt. Her research has focussed on different aspects of marine turtle ecology and biology, including the abundance and distribution in the feeding grounds, mortality patterns, bycatch reduction, short-term movement and habitat use, and abundance and trends in nesting grounds. Current research interests include biotelemetry, genetic analysis and social implications of marine turtle fishery.

Islam Elsadek
Environmental Researcher
Department of Marine Science, Faculty of Science,
Suez Canal University, Ismailia, Egypt
islamelsadek2003@yahoo.com

Islam Elsadek is an Environmental Researcher at Gebal Elba Protected Areas working for the Egyptian Environmental Affairs Agency (EEAA). As Environmental Researcher at Gebal Elba Protected Area, Islam is in charge of monitoring of natural resources in the protected area including coral reefs, mangroves, wetlands and megafauna species like marine mammals, birds, sharks and marine turtles. Islam is also currently enrolled as a Master student at the University of Suez Canal, Ismailia, Egypt. He is the first Egyptian student actively studying marine turtles in their feeding and nesting grounds in the Egyptian Red Sea. His master thesis includes some ground-breaking information about the distribution, abundance, behaviour and molecular biology of local populations of marine turtles, green and hawksbill turtles in particular.

Magdy A.N. El-Alwany
Associate Professor
Department of Marine Science, Faculty of Science,
Suez Canal University, Ismailia, Egypt
magdy.elalwany@yahoo.com

Dr. Magdy El-Alwany is the Director of the Marine Environmental Center in Sharm El-Sheikh (a Research Unit in Suez Canal University). He is also an Associate Professor in the Department of Marine Science, Suez Canal University, Ismailia, Egypt. He was awarded a scholarship from ÖAD to study for his Ph.D. from 2000 to 2003 at the Institute of Zoology and Limnology, University of Innsbruck, Austria. He held a Post-Doctoral position from 2003 to 2006 in the Department of Marine Biology, University of Vienna. He is a long time activist in the conservation of marine birds, turtles and mammals in the Red Sea. He wrote a text book on marine turtles in the Red Sea that was published in Saudi Arabia in 2008.

Amany A.H.M. Ismael
Professor
Oceanography Department, Faculty of Science,
Alexandria University, Egypt
amany_3@yahoo.com

Dr. Amany Abdel Hamid Ismael is at present the Chair of Oceanography Department, Faculty of Science, Alexandria University, Egypt. Throughout her career she has held various academic and managerial positions, including Co-Chair of the IOC-HANA network (Harmful Algae in North Africa) from 2010 to 2012, and has been a member of the Crisis and Disasters Unit of the Faculty of Science, Alexandria University, since 2010. She has been a Vice Manager of the Quality Assurance Unit of Faculty of Science since 2011. Dr. Ismael has been the Egyptian Coordinator for Harmful Algae in the IOC-HANA Network since 2004. She received the Alexandria University Award for Scientific encouragement in 2008. She has participated in many research projects, such as RAF/7/012 "Using Nuclear Analytical Techniques to Support Harmful Algal Bloom Management in the Context of Climate and Environmental Change". She has published over 25 scientific publications in international journals, periodicals and reports and has participated in many international and regional conferences and meetings, and has received awards in recognition of her contributions.

Ahmed S. M. Khalil
Regional Programme Coordinator
Living Marine Resources and Climate Change
The Regional Organization for the Conservation of the Environment of the Red Sea and Gulf of Aden (PERSGA)
P.O. Box 53662, Jeddah 21583, Saudi Arabia
ahmed.khalil@persga.org

Dr. Ahmed S.M. Khalil holds a Ph.D. in marine ecology from the Institute of Tropical Marine Ecology, University of Bremen, Germany (2001), M.Sc., (1995) and B.Sc., (1990) in zoology from the University of Khartoum, Sudan. He has worked as Assistant Professor of marine ecology at the University of Khartoum and Director of Suakin Marine Laboratory since 2002. He had been lead consultant to the Regional Organization for the Conservation of the Environment of the Red Sea and Gulf of Aden (PERSGA) for conducting regional mangrove assessments and development of Regional and National Action Plans for Conservation of Mangroves in the Red Sea and Gulf of Aden region during 2002–2004. He has joined PERSGA as the Regional Program Coordinator for Living Marine Resources and Climate Change since 2007. During the last 12 years, Dr. Khalil worked extensively in conducting research, assessments and developing strategies and action plans in a range of disciplines including conservation of marine biodiversity, fisheries management, marine pollution, vulnerability assessment and adaptations to the impacts of Climate Change, often as part of larger projects and programs in collaboration with several national, regional and international partners in the Red Sea and Gulf of Aden region. He also published several papers, book chapters, articles, reviews, technical reports and guidelines in these fields.

Geoff N. Bailey
Anniversary Professor
Department of Archaeology, The King's Manor
University of York YO1 7EP, UK
geoff.bailey@york.ac.uk

Geoff Bailey took his BA in Archaeology in the University of Cambridge in 1970 and his Ph.D. in 1976, and stayed in Cambridge as British Academy Postdoctoral Research Fellow, University Lecturer, and Senior Tutor of Clare Hall. In 1996 he was appointed to the Chair of Archaeology at the University of Newcastle upon Tyne, and moved to his present position in 2004. His research interests are in coastal prehistory, shell middens, submerged landscapes, Quaternary-scale landscape evolution, and the human impact of sea-level change and active tectonics. He has field experience in many parts of the world, across Africa, the Americas, Australia, Europe, the Far East and SW Asia, and has led major multi-national and multi-disciplinary projects in Australia, Greece, Saudi Arabia and the UK in collaboration with marine and earth scientists. He is Chairman of the EU Trans-domain COST network SPLASHCOS (Submerged Prehistoric Archaeology and Landscapes of the Continental Shelf) and Principal Investigator of the DISPERSE Project (Dynamic Landscapes, Coastal Environments and Human Dispersals) funded by an Advanced Grant from the European Research Council. He has published 16 books and over 150 papers, and is a Fellow of the Society of Antiquaries, Corresponding Member of the German Archaeological Institute, and an elected member of the Academia Europaea.

MIX
Papier aus verantwortungsvollen Quellen
Paper from responsible sources
FSC® C105338

If you have any concerns about our products,
you can contact us on
ProductSafety@springernature.com

In case Publisher is established outside the EU,
the EU authorized representative is:
Springer Nature Customer Service Center GmbH
Europaplatz 3, 69115 Heidelberg, Germany

Printed by Libri Plureos GmbH
in Hamburg, Germany